Amyloid Proteins

Volume 1

edited by Jean D. Sipe

Further Titles of Interest

Johannes Buchner, Thomas Kiefhaber (eds.)

Protein Folding Handbook

(5 Volumes)
2005
ISBN: 3-527-30784-2

Knut H. Nierhaus, Daniel N. Wilson (eds.)

Protein Synthesis and Ribosome Structure
Translating the Genome

2004
ISBN: 3-527-30638-2

Giovanni Cesareni, Mario Gimona, Marius Sudol, Michael Yaffe (eds.)

Modular Protein Domains

2004
ISBN: 3-527-30813-X

Jean-Charles Sanchez, Garry L. Corthals, Denis F. Hochstrasser (eds.)

Biomedical Applications of Proteomics

2004
ISBN : 3-527-30807-5

R. John Mayer, Aaron J. Ciechanover, Martin Rechsteiner (eds.)

Protein Degradation
Vol. 1: Ubiquitin and the Chemistry of Life

2005
ISBN: 3-527-30837-7

Amyloid Proteins

The Beta Sheet Conformation and Disease

edited by Jean D. Sipe

Volume 1

WILEY-VCH Verlag GmbH & Co. KGaA

Editor

Jean D. Sipe
8406 North Brook Lane
Bethesda, MD 20814-2615
USA

Cover illustration includes images kindly provided by David Teplow, Department of Neurology, University of California Los Angeles, Los Angeles, CA, USA

Library of Congress Card No.: applied for

British Library Cataloguing-in-Publication Data: A catalogue record for this book is available from the British Library.

Die Deutsche Bibliothek – CIP Cataloguing-in-Publication Data: A catalogue record for this publication is available from Die Deutsche Bibliothek

Printed on acid-free paper

Composition K+V Fotosatz GmbH, Beerfelden
Printing Strauss GmbH, Mörlenbach
Bookbinding Litges & Dopf Buchbinderei GmbH, Heppenheim
Cover Design Gunther Schulz, Fußgönheim

Printed in the Federal Republic of Germany

ISBN-13: 978-3-527-31072-2
ISBN-10: 3-527-31072-X

Contents

Amyloid Proteins. The Beta Sheet Conformation and Disease. J. D. Sipe

ISBN: 3-527-31072-X

Preface

The folding of proteins into unique three dimensional structures is integral to their specific biological functions within the body. Of the tens of thousands of proteins encoded within the human genome, fewer than thirty are known to share the feature of being susceptible to increased folding of the polypeptide backbone into the beta sheet conformation and assembly into amyloid fibrils. In each case, the increased beta sheet folding is associated with a clinically distinct disease or disorder, one of the amyloidoses. The pathological consequences of amyloid fibril formation are implicated in a wide range of both common and rare diseases, including Alzheimer's disease and other brain disorders, adult onset (type II) diabetes mellitus, plasma B-cell dsycrasias, long term hemodialysis, hereditary polyneuropathies and hereditary periodic fever syndromes.

During the latter half of the twentieth century, chemical and physical studies of amyloid fibrils that had been isolated and purified from amyloid laden tissues and organs led to the recognition that there is a unique association between the chemical identity of the fibril forming protein and the pattern of localized or systemic clinical symptoms. However, despite their biochemical and clinical differences, each of the amyloidoses shares the common pathophysiologic features of an amyloidogenic protein precursor, permissive host genetic background, abnormalities in proteolysis that permit accumulation of precursor protein and fibril intermediates, and alteration in the chemistry of the extracellular matrix. Each of the more than 20 chemically distinct types of amyloid deposits contains a common set of extracellular matrix constituents, glycosaminoglycans and non-fibril forming proteins, serum amyloid P component and apolipoprotein E.

Until now, to some extent, clinical studies and basic studies have proceeded in parallel. Now, we are poised to integrate and utilize our knowledge of protein structure, physiology and pathology to forestall or ameliorate the clinical consequences of the tendency of amyloid fibril forming proteins to undergo increased folding into the beta sheet conformation. Outside the body, using conditions that alter protein folding, amyloid fibrils have been created from many more than 30 proteins; this is an indication of the key role of the local tissue environment in triggering amyloid fibril formation within the body.

Amyloid Proteins. The Beta Sheet Conformation and Disease. J. D. Sipe

ISBN: 3-527-31072-X

These volumes bring together preeminent amyloid clinicians and basic scientists to consider our present knowledge in terms of those structural and thermodynamic features which, over time, lead to amyloid fibril formation, deposition and disease. The authors present an overview of amyloidosis and amyloid proteins today, including the history of amyloid investigation, the internationally accepted nomenclature and the anatomic and clinical clues as to why amyloid fibrils form within the body. Protein folding, unfolding and refolding are considered in terms of thermodynamics, posttranslational modification and lipid association, and the influence of the extracellular matrix, serum amyloid P component and apolipoprotein E. Pathways to amyloid fibril formation are considered in terms of folding of natively unfolded proteins and unfolding of natively folded, globular proteins. The use of computational approaches to derive potential structures of amyloid fibril intermediates is presented. How the process of amyloid fibril formation causes damage to organs and tissues of the body is considered in terms of oligomeric fibril forming intermediates and cellular toxicity and brain dysfunction. Most of the amyloid proteins are considered individually in terms of current knowledge of structure, function and metabolism. Some of the more recently identified forms of amyloid, including medin, lactoferrin, apoA-IV or keratoepithelin (Table 1.1) have not been considered in detail here. It is anticipated they will be the subject of greater study in the future and that additional chemical forms of amyloid, particularly localized, will be identified in future studies. There is also within this volume the call for development of molecular diagnostics and targeted therapeutics in the amyloidoses. It is to that end that this volume is dedicated, to the acceleration of progress in understanding the contribution of the beta sheet conformation to the etiology and pathophysiology of disease, in order to enable prevention and better informed treatment of the amyloidoses.

I would like to express my gratitude to the contributing authors, to Frank Weinreich, Wiley-VCH, for his invitation to edit this volume and for his continuous support and encouragement, to Waltraud Wüst, Wiley-VCH for her able and amiable management of the publication process and to Byron Caughey and Margaret Johns for helpful discussions about the content of this volume.

Bethesda, Maryland
April, 2005

Jean D. Sipe

List of Contributors

Prof. Dr. Ilia V. Baskakov
Chapter 4
University of Maryland
Biotechnology Institute
Medical Biotechnology Center
725 W. Lombard St.
Baltimore, MD 21201
USA

Prof. Dr. Vittorio Bellotti
Chapter 24
Dipartimento di Biochimica
Universita degli Studi di Pavia
Via Taramelli 3b
I-27100 Pavia
Italy

Prof. Dr. Wilfredo Colon
Chapter 12
Department of Chemistry
& Chemical Biology
Rensselaer Polytechnic Institute
Cogswell Building, Room 126
110 Eighth Street
Troy, NY 12180
USA

Prof. Dr. Ana Margarida Damas
Chapter 21
IBMC-Molecular Structure Unit
Universidade do Porto
Rua do Campo Alegre, 823
4150-180 Porto
Portugal

Dr. Philippe Derreumaux
Chapter 18
Laboratoire de Biochimie Theorique –
UPR 9080 CNRS
Institut de Biologie Physico-Chimique
11 rue Pierre et Marie Curie
75005 Paris
France

Prof. Dr. Anthony L. Fink
Chapter 11
Department of Chemistry &
Biochemistry
University of California, Santa Cruz
1156 High Street
Santa Cruz, CA 95064-1099
USA

Prof. Dr. Jorge A. Ghiso
Chapter 19
Departments Pathology & Psychiatry
New York University School
of Medicine
550 First Avenue, TH-432
New York, NY 10016-6402
USA

Prof. Dr. Gilles Grateau
Chapter 25
Pierre et Marie Curie University
Service de medicine interne
Hôpital Hotel-Dieu
1, place du parvis Notre Dame
75181 Paris cedex 04
France

Amyloid Proteins. The Beta Sheet Conformation and Disease. J. D. Sipe

ISBN: 3-527-31072-X

Prof. Dr. Geoffrey Howlett
Chapter 6
Department of Biochemistry &
Molecular Biology
University of Melbourne, Parkville
Melbourne, Victoria 3010
Australia

Prof. Dr. Mariusz Jaskolski
Chapter 27
Department of Crystallography
Faculty of Chemistry
Adam Mickiewicz University
Grunaldzka 6
60-780 Poznan
Poland

Prof. Dr. Bruce Kagan
Chapter 14
Department of Psychiatry
Neuropsychiatric Institute
University of California, Los Angeles
760 Westwood Plaza
Los Angeles, CA 90024
USA

Prof. Dr. David C. Kilpatrick
Chapter 8
Scottish National Blood Transfusion
Service
National Science Laboratory
Ellen's Glen Road
Edinburgh, EH17 7QT
Scotland, UK

Prof. Dr. Robert Kisilevsky
Chapter 2
Department of Pathology
& Molecular Medicine
Queens University
Kingston, ON K7L 3N6
Canada

Prof. Dr. Barbara Kluve-Beckerman
Chapter 22
Department of Pathology &
Laboratory of Medicine
Indiana University
School of Medicine
635 Barnhill Drive, MS-128
Indianapolis, IN 46202
USA

Prof. Dr. Mary Jo LaDu
Chapter 10
ENH Research Institute
Northwestern University
1801 Maple Avenue, Suite 6240
Evanston, IL 60201
USA

Prof. Dr. Peter J. Neame
Chapter 10
Department of Biochemistry
& Molecular Biology
University of South Florida
3500 East Fletcher Avenue, Suite 302
Tampa, FL 33613
USA

Prof. Dr. Melanie R. Nilsson
Chapter 5
Department of Chemistry
McDaniel College
Eaton Hall, 2 College Hill
Westminster, MD 21157
USA

Dr. Ruth Nussinov
Chapters 13, 23
National Cancer Institute
Laboratory of Experimental &
Computational Biology
SAIC Frederick Inc.
P.O. Box B, Bldg. 469, Rm 149
Frederick, MD 21702-1201
USA

Prof. Dr. Sheena E. Radford
Chapter 26
School of Biochemistry &
Molecular Biology
University of Leeds
Mt. Preston St.
Leeds, LS2 9JT
UK

Prof. Dr. David C. Seldin
Chapter 15
Department of Medicine
Boston University Medical Center
EBRC420
650 Albany Street
Boston, MA 02118
USA

Prof. Dr. Jean D. Sipe
Chapter 2
Department of Biochemistry
Boston University
School of Medicine
8406 North Brook Lane
Bethesda, MD 20814-2615
USA

Dr. Fred J. Stevens
Chapter 30
Biosciences Division
Argonne National Laboratory
9700 Cass Avenue
Argonne, IL 60439
USA

Prof. Dr. Fabrizio Tagliavini
Chapter 16
Istituto Nazionale Neurologico
Carlo Besta
Via Celoria 11
20133 Milano
Italy

Prof. Dr. David Teplow
Chapter 17
Department of Neurology
University of California, Los Angeles
710 Westwood Plaza
(Reed CII 1A)
Los Angeles, CA 90095
USA

Prof. Dr. Gunilla T. Westermark
Chapter 28
Department of Cell Biology
University Hospital
Linköping University
58185 Linköping
Sweden

Prof. Dr. Per Westermark
Chapter 1
Departments of Genetics & Pathology
University Hospital
Uppsala University
75185 Uppsala
Sweden

Prof. Dr. Steve P. Wood
Chapter 9
Division of Biochemistry
and Molecular Biology
School of Biological Sciences
University of Southampton
Bassett Crescent East
Southampton SO16 7PX
UK

Part I
Overview of Amyloidosis and Amyloid Proteins

Amyloid Proteins. The Beta Sheet Conformation and Disease. J. D. Sipe

ISBN: 3-527-31072-X

1
Amyloidosis and Amyloid Proteins: Brief History and Definitions

Per Westermark

Es gibt fast kein Problem in der allgemeinen und speziellen Pathologie, das sich über Jahrhunderte in einer so sphinxhaften Weise verhalten hat wie die Amyloidose. [There is almost no problem in general and systemic pathology that over the centuries has behaved in such a sphinx-like way as amyloidosis.] Letterer (1966) [1]

1.1
Early History

History is not an absolute science. It is rather a somewhat subjective interpretation of available data and, when it comes to more recent history, of ones own memories, all set on a background of the spirit of the age. This should be remembered when reading this short history.

1.1.1
Initial Studies

The early history of amyloid and amyloidosis is fascinating, and the definition and nature of the amyloid-related alteration was the subject of intense debate during the 19th century. The interested reader is referred to several comprehensive and well-written reviews (e.g. [2–5]). This chapter addresses, primarily, the recent history of the amyloid proteins. However, it is hardly possible, or particularly fruitful, to distinguish between the history of amyloid proteins and that of amyloid itself. Anyone who wants an insight into the modern history of amyloid and amyloidosis will find an invaluable source in the volumes of the proceedings of the 10 international symposia on amyloid and amyloidosis, starting with that covering the First International Symposium on Amyloidosis, Groningen, The Netherlands (1967). The most recent volume covers the 10th International Symposium on Amyloidosis, Tours, France (2004). In addition, the excellent vol-

Amyloid Proteins. The Beta Sheet Conformation and Disease. J. D. Sipe

ISBN: 3-527-31072-X

ume resulting from the International Course on Amyloidosis, Groningen, The Netherlands (1986) contains a wealth of historical information [6].

When we talk about the history of amyloidosis, we often start with a reference to Rudolf Virchow (Fig. 1.1), who first used the term "amyloid" for a structural body in human tissues [7, 8]. "Amyloid", however, had been coined earlier and had been used in botany, and the disease amyloidosis, without this name, was well known among post-mortem tissue dissectors. It is apparent from Virchow's papers, which reference his own studies and those of pathologists such as von Rokitansky and Meckel, that systemic amyloidosis was well known at that time; conditions such as "lardaceous or cholesterin disease" (Speck- oder Cholesterin-krankheit) [9] and "wax-spleen" (Wachs-milz) were mentioned [7]. Malpighi recognized "sago spleen" and other investigators the "lardaceous spleen", referring to the distinct macroscopic appearances of two variant patterns of amyloid deposition.

Fig. 1.1 Rudolf Virchow.

Virchow used a water solution of iodine in combination with hydrated sulfuric acid as a stain for cellulose in the human body [10, 11]. A cellulose-like substance had earlier been described in lower animals. Virchow found that corpora amylacea in ependyme and choroid plexus showed a typical cellulose reaction with iodine, and stated in his first report that "no doubt regarding the cellulose nature is possible" [10]. Afterwards, Virchow and also Meckel [7] tested tissues corresponding to what we today call systemic amyloidosis and found a similar reaction to iodine as had been observed with corpora amylacea. Thus, Virchow found that the wax-like deposits and degeneration of spleen, liver and kidneys, in cases of what must have been amyloid A (AA) amyloidosis due to chronic infectious diseases, showed a starch-like reaction with iodine. This initiated a debate as to whether the iodine reaction depended on cellulose or on cholesterin.

The hypothesis of the cellulose nature of amyloid did not stand for long. Friedreich and Kekulé dissected out amyloid-rich segments from the spleen of a patient with amyloidosis, probably AA in nature [12]. In contrast to Virchow, they performed quite elegant direct chemical analyses of material extracted in different ways and came to the definitive conclusion that the main substance was protein in nature. This was confirmed by Hanssen [13], who showed that amyloid is digestible with pepsin. Ironically, 100 years later it was found that corpora amylacea in reality contain little protein and are essentially polysaccharide in nature [14].

The nature of the amyloid protein or proteins was a puzzle for a long time. Furthermore, it was debated whether the amyloid substance developed locally from underlying cells. It was also suggested that the protein originated from blood and that the specific protein was precipitated in organs by abnormal amounts of sulfuric acid present locally [15]. This is not too far from today's theory that circulating proteins interact, not with sulfuric acid, but with glycosaminoglycans [16, 17].

1.1.2
Different Chemical Forms of Amyloid: Early Studies

Systemic amyloidosis was initially regarded as a complication of chronic infectious diseases such as tuberculosis, syphilis and osteomyelitis. Later, it became clear that amyloidosis could also occur after the onset of non-infectious chronic inflammatory disorders, e.g. rheumatoid arthritis. However, quite early, single cases of amyloidosis without any obvious predisposing disease were described [18, 19]. In one case, Wild noted that, in addition to the absence of any additional disease in his patient, the distribution of amyloid was quite remarkable [19], best comparable with what is today known as AL amyloidosis. Interestingly, Soyka described cardiac amyloidosis without predisposing diseases, particularly at a high age [18]. Certainly, these cases must have been examples of transthyretin-derived senile systemic amyloidosis. Remarkable variation in the clinical manifestation of primary amyloidosis was noted quite early [20, 21]. The first investigator to identify "primary amyloidosis" as a distinctive group was Lubarsch

[20]. We now know that this group originally included not only AL amyloidosis, but also different familial forms of amyloidosis, derived from several proteins, and the transthyretin-derived senile systemic amyloidosis.

Although, after Virchow's initial studies, amyloidosis was typically found to be generalized, there are early descriptions of characteristic localized AL amyloidosis, particularly from the conjunctiva [22]. For example, Vossius described two cases of localized, tumor-like amyloidosis of the conjunctiva [23]. Very large inter-individual variation is also evident in the localized forms of AL amyloidosis and this is the obvious reason for the large number of case reports of AL amyloidosis appearing in the medical literature over the years.

One popular theory of the origin of amyloid was that amyloid represents an antigen–antibody precipitate [24], perhaps depending on an autoantigen [25]. It was possible to identify immunoglobulin and complement proteins in amyloid deposits by immunohistochemistry [26]. However, in extracts of amyloid, it was not possible to demonstrate γ-globulin (immunoglobulin) in reasonable amounts to explain the nature of amyloid [27, 28]. For a long time it was believed that the composition of all amyloids is one and the same, and, in the older literature, other possibilities are not discussed. However, Gellerstedt, in 1938, noted that amyloid in the islets of Langerhans differed in tinctorial properties as compared with vascular secondary amyloid in cases where both alterations were observed [29]. He obviously understood that the two types of deposits were different, although he did not explicitly state that they must contain different proteins. It was not until the first direct protein sequence analyses were performed (Section 1.2.1) that the complex chemical nature of the amyloid deposits started to become apparent.

1.1.3
Amyloid Staining Methodology

The initial method used to identify amyloid was that of iodine staining, introduced by Virchow. This method was soon replaced by metachromatic stains like crystal violet. The use of the most important histological staining marker for amyloid, Congo red, was introduced by Bennhold [30]. This direct dye for cotton had been used in the textile industry since 1884. Congo red was observed to have a strong affinity for amyloid deposits and there was the observation that its clearance from plasma could be used as a diagnostic method for amyloidosis. However, the value of Congo red in histology turned out to be much higher. An important discovery was made by Divry and Florkin in 1927 who noticed the enhanced birefringence of amyloid deposits after staining with Congo red [31]. It was suggested that this property of amyloid depends on an ordered arrangement of the elongated Congo red molecules in the amyloid, indicating that, in fact, the substance is not amorphous, as earlier described, but has an organized substructure [31–34]. A standardized Congo red staining method was introduced by Puchtler et al. [35] and is still used. Additional staining methods have been and are still used. The most important of these is probably Thioflavin T or S.

There were once divided opinions as to which of the two stainings Congo red or Thioflavin S is more specific for amyloid, but each method now has its own role today in the study of amyloidosis [36].

1.2 Amyloid Proteins – Modern History

The modern history of amyloidosis can be said to have started with the discovery by Cohen and Calkins, using electron microscopy, that amyloid, which is hyaline and structureless under the light microscopy, has a characteristic fine fibrillar ultrastructure [37]. This finding was confirmed in several other studies [38–42] and pointed to a specific structural organization of the constituent molecules – a concept that at that time was completely unknown.

The amyloid was found to contain "rigid", unbranched fibrils, around 10 nm in diameter and of undetermined length. The fibrils were usually without orientation, but when close to cells appeared in parallel bundles, sometimes perpendicular and close to cell membranes [43, 44]. This fibrillar organization was taken by some researchers as an indication that the fibrils were made by these cells [45], but others were hesitant about this. This situation gave rise to the famous argument raised by Bywaters that on pictures of San Sebastian transfixed with arrows "he looked a bit sick too, but nobody had suggested he was secreting them" [46]. Careful electron microscopic studies by Shirahama and Cohen showed that the amyloid fibrils, irrespective of origin, were composed of even thinner subelements, designated protofibrils [47–49].

1.2.1 The Amyloid Proteins

The first amyloid component to be identified, although not based on amino acid sequence data, was a soluble protein, which could be extracted from amyloid-laden tissues. When rabbits were immunized with extracted amyloid fibrils, antibodies to this protein were detected [50] that also recognized an immunoreactive component in plasma. The amyloid protein in tissues was consequently designated amyloid plasma component or P component (later AP). The circulating plasma counterpart, later shown to be identical, was called "serum amyloid P component" (SAP). SAP was shown to be identical to the AP particle that had been identified in tissue amyloid deposits by electron microscopy and described as a pentagonal structure [51]. SAP has been shown to be a ubiquitous part of all of the chemical types of amyloid deposits. The SAP protein binds non-covalently and calcium dependently to ligands [52] such as β-pleated sheet fibrils. SAP is a normal constituent of the glomerular basement membrane [53] and elastic microfibrils [54].

The nature of the amyloid fibril was long an enigma. Methods to concentrate amyloid fibrils and to make good, representative fibrillar extracts of systemic amyloid deposits were developed by Cohen et al. [48] and Pras et al. [55]. Further advancement and analysis of constitutive fibril subunits was difficult, however. Benditt et al. extracted secondary amyloid materials directly from tissues, without prior purification of fibrils [27]. They identified a protein component that was soluble in 6 M urea and exhibited an unusual amino acid composition, in that both cysteine and threonine were lacking. At the time, it was not possible to characterize this protein further and the investigators discussed many different possibilities, including a virus protein.

Amyloid was long believed to be a singular substance, perhaps of some unspecific degenerative origin. The relationship to other structureless deposits, particularly "hyaline", was often discussed in the literature and a process of transformation from hyaline to amyloid was suggested. Even when it had been shown by electron microscopy that the light microscopically amorphous amyloid consists of fine fibrils of a characteristic appearance, amyloid deposits were usually regarded as one unique kind of substance. The demonstration of a consistent cross-β-pleated sheet structure in amyloid fibrils was an important step in our understanding how amyloid is formed [56, 57]. The β-pleated sheet structured fibril seems to be the basis of the unusual resistance of all kinds of amyloid to degradation and, therefore, the progressive deposition of the material. It had been shown already that insulin could be converted into a fibrous form under conditions that are denaturing to the secondary structure of proteins [58]. It was shown that insulin fibrils [59] as well as synthetic fibrils made from other small proteins [60] all had the properties of amyloid fibrils, including affinity for Congo red and green birefringence when viewed under polarizing light microscopy. Glenner incorporated the characteristic X-ray diffraction pattern of amyloid fibrils into the definition of amyloid and proposed that the term "amyloid" be used primarily as a generic adjectival term to indicate the presence of non-branching, 80- to 100-Å fibrillar (linear or concentric) proteinaceous deposits demonstrated to have either Congo red birefringence or a β-pleated sheet X-ray diffraction pattern, together with the chemical nature of the fibril (if known) and the site or tissue of origin or deposition noted [60].

The first definitive proof of a chemically specific protein constituent of amyloid fibrils came from the studies of Glenner et al. They showed that, in cases of primary and myeloma-associated systemic amyloidosis, the fibril protein originated from homogeneous immunoglobulin light chains [61]. Glenner et al.'s work had an enormous immediate impact and, since, at the time, amyloid was widely believed to be a single substance, all of the clinical forms of amyloid were initially regarded to be of immunoglobulin origin. When reading the literature from the early 1970s, it is evident that there was a very high degree of international competition between four or five prominent amyloid research groups and the citations do not always seem to be fully correct. Later, Benditt wrote [62] that he had difficulties with the publication of an important paper in which he presented evidence for the existence of multiple chemical classes of amyloid substance [63]. Benditt's group showed that a unique protein, which had a characteristic amino acid composition and a uniform electrophoretic mobility, was present in all cases of typical "secondary" amyloidosis, but absent in cases with other kinds of amyloid. They called this component protein A and the protein(s) extracted from other kinds of amyloidosis, such as what is now known as AL, protein B [63]. Protein A (later designated as amyloid A or AA) was characterized by amino acid sequence in 1971 (see below). Based on varying amino acid compositions and electrophoretic mobilities, Benditt et al. also suggested that, in fact, there would likely be several different protein B forms [63]. Subsequent developments in amyloid research have shown that they were absolutely right.

The pioneering work of the laboratories of Benditt and Glenner showed that the most fruitful method for elucidation of the nature of the amyloidoses should be recognition of putative clinically specific amyloid diseases, and isolation from tissues and purification of the corresponding amyloid fibril proteins, followed by their chemical identification by amino acid sequence analysis. This approach was immediately initiated. A third amyloid fibril protein was soon identified by amino acid sequence analysis and was shown to be derived from calcitonin (or possibly procalcitonin), occurring in the amyloid of medullary thyroid carcinoma [64]. This observation was clear evidence of the diversity in the nature of the amyloid fibril.

1.2.2 Specific Amyloid Fibril Proteins

1.2.2.1 Protein AA and its Precursor, Serum AA

Benditt et al.'s pioneering papers from the pre-amyloid protein sequencing era clearly showed that the major protein associated with amyloidosis secondary to inflammatory diseases has specific properties, including a constant electrophoretic mobility and unusual amino acid composition [27, 63, 65, 66]. Most remarkable was the lack of threonine. The definitive proof of a unique nature came with the purification of the major protein and the demonstration of a unique N-terminal amino acid sequence of the protein designated protein A [67]. This was soon verified by several other groups [68–70]. The full amino acid sequence of amyloid protein A from several patients was soon published, indicating a 76-amino-acid protein [70, 71]. However, it has subsequently become clear that protein A of different lengths, but with the same N-terminus, exists [68, 72–74].

The nomenclature for this unique protein varied in publications from different groups. Since a *Staphylococcus* protein already was called "protein A", alternative names were proposed, such as protein AS [69], ASF [70] or AUO [68]. This confusion was ultimately resolved at the Second International Symposium on Amyloidosis, Helsinki (1975), where the foundation of a modern amyloid nomenclature based on biochemistry was created [75]. At this time, protein A became protein AA.

Protein AA was soon found to have a circulating counterpart in plasma [76, 77] now called serum AA (SAA). SAA, today known to be a protein family with several members of which two are circulating acute-phase reactants, was thus discovered through the ability of structurally related protein(s) to aggregate into amyloid fibrils. In plasma fractionated by gel filtration under physiological conditions, SAA appeared as a large protein, around 180–200 kDa [78], but a low-molecular-weight component, around 12 kDa, could be isolated by gel filtration under denaturing conditions [79, 80]. The explanation for this difference in protein mass came when it was found that SAA is an apolipoprotein, mainly associated with high-density lipoprotein, in humans [81], mice [82] and rabbits [83]. The blood plasma protein SAA was shown to be about 40% larger than the 76-amino-acid residue AA protein that was first described. This larger size apparently depended on a C-terminal extension in the SAA molecule [84]. Subsequently, it was established that human SAA is a 104-amino-acid molecule of which amyloid protein A corresponds to the major N-terminal part [85]. The acute-phase SAA is produced by the liver, but extrahepatic expression of SAA was demonstrated. Thus, it was not immediately accepted that the circulating SAA is the precursor of the amyloid protein. Eventually, however, direct animal experimental work showed unequivocally that circulating SAA is converted into amyloid fibrils [86, 87].

A systemic form of amyloidosis, resembling human secondary amyloidosis, had long been known to occur in many different mammals (for reviews, see [88, 89]), the most well known being that seen in mice. Secondary amyloidosis was also a common outcome in horses which had been immunized for antiserum production. At an early date, it was found to be possible to induce secondary amyloidosis in laboratory animals, including rabbits, hens [90] and mice [91, 92]. Protein AA was shown to be the major amyloid fibril component also in these species ([93]; for review, see [89]). Most important for future experimental studies was the finding that the experimentally inducible amyloid in mice is associated with protein AA [94]. As with human, a plasma component, antigenically identical to the tissue-derived protein AA, could be extracted in several animal species after induction of inflammation [95]. Also, as in humans, the plasma component appeared in a high-molecular-weight form in its native state, but could be extracted as a 12-kDa protein after denaturation [96]. Sipe et al. [96] seem to be the first to name SAA as an acute-phase protein. Today we know SAA as one of the most sensitive acute-phase reactants.

1.2.2.2 Immunoglobulin-derived Amyloid (AL and AH)

Cases with simultaneous occurrence of multiple myeloma and amyloidosis were described at an early date. The chemical nature of this amyloid was the subject of many studies and immunoglobulin was a natural candidate. Many attempts to extract significant amounts of immunoglobulin from the corresponding amyloidotic tissues were unsuccessful. Added to the difficulties was the still widespread belief that, chemically, amyloid was either one specific substance or a non-specific degeneration product. To make the situation even more confusing, the amyloid associated with myelomatosis was sometimes called "secondary".

One of the most important advances in amyloid history was the purification of the fibrillar protein from tissues of a patient with primary amyloidosis and the subsequent demonstration by Edman degradation that the N-terminal amino acid sequence corresponded to a monoclonal immunoglobulin light chain [97]. Several N-terminal sequences obtained from other individuals with the same technique confirmed the initial report [61, 98]. The findings fitted nicely with the demonstration that immunoglobulin light chains contain two sets of β-sheets, of which one comprises most of the variable region [99]. Some principles rapidly became clear. The amyloid in primary and myeloma-associated amyloidosis is biochemically identical, and consists of an N-terminal fragment of a monoclonal immunoglobulin chain. The fragment varies in length and, in rare occasions, whole light chains constitute the major fibril protein. A novel immunoglobulin light chain subtype was discovered by its unusual preponderance to form amyloid fibrils [100–102]. Later, it was shown that, in rare instances, monoclonal immunoglobulin heavy chains may make up amyloid [103, 104] and that also the constant region of light chains is amyloidogenic [105–107].

1.2.2.3 **Transthyretin**

Familial amyloidosis with varying clinical manifestations was described from many parts of the world long before biochemical characterization of amyloid was possible [108]. The first description of a familial amyloidosis was probably that of Ostertag [109]. Many familial amyloid forms from Portugal, Japan, Sweden, the USA and other countries had a progressive polyneuropathy as a major indication. In 1978, Costa et al. [110] showed that the fibrils in the Portuguese type were associated with prealbumin, which was the earlier name for transthyretin. The name prealbumin refers to the electrophoretic mobility of the protein. The primary structure of human transthyretin had been determined in 1974 [111] and its crystal structure was published in 1978 [112]. Transthyretin was found to have a high degree of β-structure. Soon, many reports verified the transthyretin nature of the amyloid fibril in many, but not all, of the hereditary amyloid forms [113–115]. Further analyses showed that in all of the cases of ATTR there was a mutation creating an amino acid substitution. The most common was found to be V30M [116–120]. A continuous stream of publications on new transthyretin variants, amyloidogenic or even protective against amyloidosis, has appeared until the present [121]. It has become increasingly clear that familial transthyretin amyloidosis is spread all over the world.

In addition to the different familial forms, including the common V122I mutation [120, 122], transthyretin was found to be the fibril protein identified in senile systemic amyloidosis [123]. Senile systemic amyloidosis is probably the most common of all the systemic amyloidoses and is of great theoretical interest since it is so obviously connected to aging. In contrast to the familial forms, transthyretin in this senile systemic amyloid form is of the wild-type [124].

1.2.2.4 Other Biochemical Forms of Familial Amyloidosis

The identification throughout the world of more families with amyloid syndromes and the development of more efficient, sensitive methods in protein analyses, combined with molecular biologic methods, led to the identification not only of new transthyretin variants associated with amyloidosis, but also new and unexpected amyloid proteins. The strategy was generally the same: identification of the family, purification of the major amyloid fibril protein and amino acid sequence analysis followed by sequencing of the specific gene. In this way it has become relatively easy to rapidly determine the specific genetic cause of many familial amyloidoses. These include amyloidosis derived from cystatin C [125, 126], apolipoprotein A-I [127], apolipoprotein A-II [128], fibrinogen [129], gelsolin [130, 131], lysozyme [132], ABri [133] and ADan [134].

1.2.2.5 β_2-Microglobulin (β_2M)

β_2M shares strong structural similarities with immunoglobulin light and heavy chain constant regions [135], and is part of the HLA class I complex. The protein has a high degree of β-sheet conformation and is a small molecule, thus fitting well as an amyloid fibril protein precursor. Indeed, in 1985, Gejyo et al. [136] and directly afterwards Gorevic et al. [137] showed that β_2M is the fibril protein in amyloidosis occurring as a complication to long-term hemodialysis, a disease that was described almost simultaneously [138–140]. Both full-length β_2M and fragments thereof were found in the amyloid deposits [141]. The amyloid disease had a peculiar systemic distribution, with destructive arthropathy as a major manifestation. An important cause of the disease is increased plasma concentration of β_2M in individuals on dialysis. Fortunately, this is an amyloid that is disappearing due to better treatment of kidney failure.

1.2.2.6 Specific Amyloid Forms in the Central Nervous System

Alzheimer and Amyloid The plaques in the cerebral cortex, described by Alzheimer [142], and the cerebral amyloid angiopathy, associated with Alzheimer's disease and aging [143], were entities known for long time, but not initially the subject of any intense interest. Amyloid was rarely, if ever, mentioned as important in the pathogenesis of the disease. The turning point came with the biochemical characterization of the amyloid fibril protein, initially from the vascular amyloid [144, 145]. Purification of cerebral plaque amyloid was more difficult, but, by application of wool technology with solubilization in formic acid, Masters et al. [146] succeeded in characterizing the fibril protein, which turned out to be the same as Glenner had found in angiopathy. Glenner called the protein Aβ, while the name used by Masters was A_4. After some confusion, the name of the protein has become Aβ. Aβ was found to be an internal fragment of a much larger protein, the Aβ protein precursor (AβPP). The further development of the field has put Aβ protein at the center of the pathogenesis of Alzheimer's disease [147, 148].

Spongiform Encephalopathies Amyloid itself is probably more of an epiphenomenon in the different spongiform encephalopathies (kuru, Creutzfeldt-Jakob disease, Gerstmann-Sträussler-Scheinker disease), but is characteristic of some types [149]. The history of the causative agent in these and in related animal diseases is very fascinating, and contains some of the most beautiful achievements in medicine. It started with Gajdusek's field studies in New Guinea where the disease kuru was identified in a small isolated Papuan population [150]. Gajdusek found evidence that the disease was transmitted by ritualistic cannibalism and he was also able to transmit the disease from human to chimpanzees [151], thereby proving its contagious properties. Later, Prusiner found that the transmissible agent, prion, is not a virus but a protein [152]. The prion protein aggregates and forms amyloid-like fibrils *in vitro*, and is the major component of spongiform plaques amyloid [153].

1.2.2.7 Polypeptide Hormone-derived ("Endocrine") Amyloid

It has been known for a long time that amyloid may be deposited in some hormone-producing tissues. The first described example was amyloid in the islets of Langerhans, although this was initially called hyaline [154, 155] and the amyloid nature was accepted much later [29, 156]. Later, amyloid was described in other endocrine tissues and in polypeptide hormone-producing tumors, such as medullary carcinoma of the thyroid. This "endocrine" amyloid was suggested by Pearse et al. [157] to be derived from non-functional parts of pro-hormones. A third class of amyloid fibril proteins was therefore suggested [157, 158]. Pearse et al. based this assumption on histochemical and microspectrofluorometric studies of amyloid in an insulinoma and in three medullary carcinomas, which indicated lack of both tyrosine and tryptophan. At that time the structure of pro-insulin had been determined [159] and the C-peptide shown to lack aromatic amino acid residues [160], but the more complicated precursor of calcitonin was unknown. However, further studies on amyloid from a medullary carcinoma clearly showed the presence of tyrosine [161] and amino acid sequence analysis of the amyloid protein from the same case showed identity with calcitonin [64]. A larger size of the fibril protein and an amino acid composition diverging from that of calcitonin was interpreted as a sign of the presence of pro-calcitonin in the amyloid, but no further amino acid sequence was obtained.

1.2.2.8 Islet Amyloid Polypeptide

The initial studies with medullary carcinoma showed that polypeptide hormones may give rise to amyloid fibrils *in vivo*. A close contact occurred between bundles of amyloid fibrils and β cells, resembling that seen between suspected fibril-forming cells and amyloid in experimental AA amyloidosis [162], and therefore islet amyloid was believed to be derived from proinsulin. The strong association between localized amyloid in the islets of Langerhans and Type 2 diabetes [156, 163, 164] made analysis of this kind of amyloid highly warranted.

It took a long time and hard work to purify the amyloid protein, and this was first done from an insulin-producing tumor [165]. Surprisingly, the major amyloid fibril protein was a previously unknown polypeptide with partial identity with calcitonin gene-related peptide (CGRP), initially called islet amyloid peptide and later islet amyloid polypeptide (IAPP). Further analyses showed that IAPP consists of 37 amino acid residues, and that it is the major protein also in amyloid of human and feline islets of Langerhans [166, 167]. The findings were verified by another group, which called the protein "diabetes-associated peptide" [168]. A later name has been "amylin". IAPP was found to be a normal product of islet β cells, and is stored and released together with insulin [169, 170]. Several structural features of IAPP indicated a hormonal nature including C-terminal amidation [171, 172] and it is now accepted as a β-cell hormone, the first discovered since insulin [173]. The identification of IAPP started a new branch in diabetes mellitus research. The interested can go to several reviews [174–176] and to Chapter 28 in this book. IAPP, together with Aβ, became popular model molecules for amyloid fibril formation.

1.3 Classification of Amyloid Diseases

Until affinity for Congo red and green birefringence after this staining, combined with a characteristic fine fibrillar ultrastructure, were generally accepted to be diagnostic for amyloidosis, there was a discussion about what should be included in the group of amyloidoses. The designation "amyloid" was usually reserved for systemic amyloidosis and for tumoral localized amyloid (today known to be of immunoglobulin origin). For many years there was a discussion of the nature of hyaline and its relationship to amyloid. It was even suggested that hyaline (which we know today is usually composed of collagen) could become transformed into amyloid. Scattered, small hyaline alterations were shown to occur in certain organs and some, but not all, of them are today included in the amyloid group. A good example is the amyloid of the islets of Langerhans, initially described as hyalinization [154]. The similarity of the histological appearance of the islet alteration and deposits in systemic amyloidosis was realized early, and the designation "para-amyloidosis" was coined for these alterations [29]. The following are the previously most commonly used categorizations of amyloid deposits.

1.3.1 Reimann's Classification

An early classification, that is partly used even today, is that of Reimann et al. [177] who divided the amyloid types into four categories:

(1) Primary amyloidosis
(2) Secondary amyloidosis

(3) Tumor-forming amyloidosis
(4) Amyloidosis associated with multiple myeloma.

1.3.2 King's Classification

Another classification, which is less commonly referred to, is that of King [178], who divided the amyloidoses into two groups, one with "typical amyloidosis", i.e. with an organ distribution seen in what we today know as AA amyloidosis, and "atypical amyloidosis", which is a group including all other cases. The designation "atypical amyloidosis", discriminating an amyloid from the deposition pattern of systemic amyloidosis in conjunction with a chronic inflammatory disease, was already used earlier (e.g. [25]) and is still seen occasionally. With modern knowledge, these designations should be avoided.

1.3.3 Classification of Missmahl et al.

A third classification is that of Missmahl et al. This was also suggested before the biochemical era of amyloidosis and was based on polarization findings with Congo red. These researchers noted that amyloid infiltration occurs either in association with collagen fibrils or with reticulin fibrils and based a differentiation of types on this difference [179, 180]. The classification into the pericollagen form, now known to include AL amyloidosis, and the perireticulin form, particularly including AA amyloidosis, did not survive for long.

1.3.4 Modern Classification

The nomenclature based on the chemical nature of amyloid proteins should in principle replace all earlier systems. However, the original classification with primary, secondary and familial systemic amyloidoses and localized amyloidosis is surprisingly difficult to eradicate. Unfortunately, this classification, although simple, often leads to confusion and misunderstanding. An example is the use of "secondary" amyloidosis for amyloid associated with multiple myeloma, which is chemically identical with that in primary amyloidosis, i.e. AL amyloidosis. It can also be noted that none of the older classifications include familial amyloidosis as one separate group. We know today that the group of familial amyloidosis is a highly heterogeneous one, containing many biochemically different amyloids.

1.3.4.1 The Present Classification of Amyloid Fibril Proteins

The modern classification had its beginnings at the Second International Symposium on Amyloidosis, Helsinki (1974) [75]. Here, it was decided that the designation of all amyloid forms should be based on their chemical composition.

Table 1.1 Amyloid fibril proteins and their precursors in human (from [182], slightly modified)

Amyloid	Precursor protein	Systemic (S) or localized (L)	Syndrome or involved tissues	Reference
AL	immunoglobulin light chain	S, L	primary myeloma associated	61
AH	immunoglobulin heavy chain	S, L	primary myeloma associated	103
ATTR	transthyretin	S	familial senile systemic	110
		L?	tenosynovium	
$A\beta_2M$	β_2-microglobulin	S	hemodialysis	136
		L?	joints	
AA	(apo)serum AA	S	secondary, reactive	67
AApoAI	apolipoprotein AI	S	familial	127
		L	aortic	
AApoAII	apolipoprotein AII	S	familial	128
AGel	gelsolin	S	familial	130
ALys	lysozyme	S	familial	132
AFib	fibrinogen α chain	S	familial	129
ACys	cystatin C	S	familial	126
ABri[a]	ABriPP	S	familial dementia, British	133
		L?		
AApoAIV[c)]	apolipoprotein AIV	S	senile	184
$A\beta$	$A\beta$ protein precursor ($A\beta PP$)	L	Alzheimer's disease, aging	144
APrP	prion protein	L	spongioform enceph-alopathies	152
ACal	(pro)calcitonin	L	C-cell thyroid tu-mors	64
AIAPP	islet amyloid poly-peptide	L	islets of Langerhans insulinomas	165
AANF	atrial natriuretic factor	L	cardiac atria	185
APro	prolactin	L	aging pituitary prolactinomas	186
AIns	insulin	L	iatrogenic	187
AMed	lactadherin	L	senile aortic, media	188
AKer	kerato-epithelin	L	cornea; familial	189
ALac	lactoferrin	L	cornea; familial	190
A(tbn)[b, c)]	tbn	L	Pindborg tumors	191

a) ADan comes from the same gene as ABri and has an identical N-terminal sequence. ADan is therefore not included in the nomenclature as a separate protein.
b) To be named.
c) Proteins that are preliminary.

The principle was created that all amyloid fibril proteins should be named "protein A" with a suffix identifying the specific protein molecule. The amyloid type and disease should then be named from the protein. Thus the term AA amyloidosis should replace secondary amyloidosis, and AL amyloidosis should replace the previously used names primary and myeloma-associated amyloid. At the time of the foundation of this classification, only the two chemical types of amyloidosis, AA and AL, were known with certainty, although it was suspected that the composition of amyloid would not be as uniform as earlier often believed. However, probably no one had then imagined the enormous heterogeneity of the human amyloid substances that later has been found to be the case.

The first real amyloid Nomenclature Committee was founded at the Third International Symposium on Amyloidosis, Povoa de Varzim, Portugal (1979). When this meeting was held, two more amyloid fibril proteins had been described, transthyretin and (pro)calcitonin, and it was now more definitely understood that there were more to be discovered. In addition to protein AA and AL, preliminary terms were decided for familial amyloid proteins (AF), amyloid in endocrine tissues (AE) and amyloid associated with aging (AS; S for senile) [181]. These designations have been dropped since many of the amyloid fibril proteins are now known. An amyloid Nomenclature Committee has been working since the meeting in Povoa de Varzim and is now formally elected by the Board of the newly formed International Society of Amyloidosis. To be accepted as an amyloid fibril protein, the protein must be definitely shown to be the major component of a distinctive amyloid deposit and the nature of the protein identified by amino acid sequence. The data should also have been published in a major scientific journal. Table 1.1 lists the amyloid fibril proteins so far identified [182].

1.4
What is Amyloid?

We now return to the starting point: how to define amyloid? Amyloid was originally described as an *in vivo* phenomenon and amyloidosis as a disease characterized by deposition of this material. With increasing knowledge of the nature of amyloid, new problems have arisen. It is possible to make Congophilic β-pleated sheet fibrils from synthetic peptides corresponding to known amyloid proteins or segments thereof. Are these fibrils amyloid? It is even possible to make similar fibrils from normally occurring peptides never found in amyloid or from completely laboratory-designed peptides. With increasing frequency, all of these kinds of fibrils are called amyloid in the scientific literature. It should be remembered that the amyloid, deposited in tissues, does not only contain the fibrils made from a single, pure protein species, but also additional proteins such as SAP, and glycosaminoglycans and proteoglycans. How these components are associated with the fibrils and what importance they may have in amyloidogenesis and in the persistence of the amyloid are questions that are insufficiently answered. The Nomenclature Committee of the International Society of Amyloidosis has discussed this problem and suggests that the designation "amyloid" should only be used for the abnormal, *in vivo* deposited material. Also, by definition, amyloid is mainly extracellular, which means that cellular inclusions, e.g. in Parkinson's disease, are not amyloid. *In vitro* produced fibrils should be called "amyloid-like" [182] (or amylog as suggested by Buxbaum [183]).

Acknowledgments

Supported by the Swedish Research Council.

References

1 Letterer, E. Zur Ätiologie und formalen Pathogenese der Amyloidose. *Nova Acta Leopoldina* **1966**, *31*, 11–22.
2 Kyle, R.A. and E.D. Bayrd. Amyloidosis: review of 236 cases. *Medicine* **1975**, *54*, 271–299.
3 Glenner, G.G. Amyloid deposits and amyloidosis. The β-fibrilloses. *N Engl J Med* **1980**, *302*, 1283–1292 and 1333–1343.
4 Cohen, A.S. General introduction and a brief history of amyloidosis. In *Amyloidosis*, J. Marrink and M.H. van Rijswijk (eds). Martinus Nijhoff, Dordrecht, **1986**, pp. 3–19.
5 Sipe, J.D. and A.S. Cohen. Review: history of the amyloid fibril. *J Struct Biol* **2000**, *130*, 88–98.
6 Marrink, J. and M.H. van Rijswijk (eds). *Amyloidosis*. Martinus Nijhoff, Dordrecht, **1986**.
7 Virchow, R. Zur Cellulose-Frage. *Virchows Arch Pathol Anat* **1854**, *6*, 416–426.
8 Virchow, R. Ueber den Gang der amyloiden Degeneration. *Virchows Arch Pathol Anat* **1855**, *8*, 364–368.
9 Meckel, H. Die Speck- oder Cholesterinkrankheit. *Ann Charité Krankenhaus* **1853**, *4*, 264–320.
10 Virchow, R. Ueber eine im Gehirn und Rueckenmark des Menschen aufgefundene Substanz mit der chemischen Reaction der Cellulose. *Virchows Arch Pathol Anat* **1854**, *6*, 135–138.
11 Virchow, R. Weitere Mittheilungen über das Vorkommen der pflanzlichen Cellulose beim Menschen. *Virchows Arch Pathol Anat* **1854**, *6*, 268–271.
12 Friedreich, N. and A. Kekulé. Zur Amyloidfrage. *Virchows Arch Pathol Anat* **1859**, *16*, 50–65.
13 Hanssen, O. Ein Beitrag zur Chemie der amyloiden Entartung. *Biochem Z* **1908**, *13*, 185–198.
14 Sakay, M., J. Austin, F. Witmer and L. Trueb. Studies of corpora amylacea. I. Isolation and preliminary characterization by chemical and histochemical techniques. *Arch Neurol* **1969**, *21*, 526–544.
15 Fahr, T. Pathologische Anatomie des Morbus Brightii. In *Handbuch der speziellen pathologischen Anatomie und Histologie*, F. Henke and O. Lubarsch (eds). Julius Springer, Berlin, **1925**, pp. 156–472.
16 Leupold, E. Cited by Fahr [15].
17 Kisilevsky, R. Review: amyloidogenesis – unquestioned answers and unanswered questions. *J Struct Biol* **2000**, *130*, 99–108.
18 Soyka, J. Ueber die amyloide Degeneration. *Prag Med Wschr* **1876**, *1*, 165–171.
19 Wild, C. Beitrag zur Kenntnis der amyloiden und der hyalinen Degeneration des Bindegewebes. *Beitr Pathol Anat Physiol* **1886**, *1*, 175–200.

20 Lubarsch, O. Zur Kenntnis ungewöhnlicher Amyloidablagerungen. *Virchows Arch Pathol Anat* **1929**, *271*, 867–889.

21 Symmers, W. S. C. Amyloidosis. Five cases of primary generalized amyloidosis and some other unusual cases. *J Clin Pathol* **1956**, *9*, 212–228.

22 Raehlmann, E. Ueber hyaline und amyloide Degeneration der Conjunctiva des Auges. *Virchows Arch Pathol Anat*, **1882**, *87*, 325–335.

23 Vossius, A. Ueber amyloide Degeneration der Conjunctiva. *Beitr Pathol Anat* **1889**, *4*, 335–360.

24 Löschke, H. Vorstellung über das Wesen von Hyalin und Amyloid auf Grund von serologischen Versuchen. *Beitr Pathol Anat* **1927**, *77*, 231–239.

25 Letterer, E. Neue Untersuchungen über die Entstehung des Amyloids. *Virchows Arch Pathol Anat* **1934**, *293*, 34–72.

26 Vogt, A. Immunhistochemische Befunde am Amyloid. *Nova Acta Leopoldina* **1966**, *31*, 37–41.

27 Benditt, E. P., D. Lagunoff, N. Eriksen and O. A. Iseri. Amyloid. Extraction and preliminary characterization of some proteins. *Arch Pathol* **1962**, *74*, 323–330.

28 Calkins, E., A. S. Cohen and D. Gitlin. Immunochemical determinations of gamma globulin content of amyloid. *Fedn Proc* **1958**, *183*, 1202–1203.

29 Gellerstedt, N. Die elektive, insuläre (Para-)Amyloidose der Bauchspeicheldrüse. Zugleich ein Beitrag zur Kenntnis der „senilen Amyloidose". *Beitr Pathol Anat* **1938**, *101*, 1–13.

30 Bennhold, H. Eine specifische Amyloidfärbung mit Kongorot. *Münch Med Wochenschr* **1922**, *69*, 1537–1538.

31 Divry, P. and M. Florkin. Sur les propriétées optiques de l'amyloide. *CR Soc Biol (Paris)* **1927**, *97*, 1808–1810.

32 Romhányi, G. Über die submikroskopische Struktur des Amyloids. *Zentralbl Allg Pathol* **1943**, *80*, 411.

33 Romhányi, G. Über die submikroskopische Struktur des Amyloid. *Schweiz Z Pathol (Pathol Microbiol)* **1949**, *12*, 253–262.

34 Missmahl, H. P. and M. Hartwig. Polarisationsoptische Untersuchungen an der Amyloid-substanz. *Virchows Arch Pathol Anat* **1953**, *324*, 489–508.

35 Puchtler, H., F. Sweat and M. Levine. On the binding of Congo red by amyloid. *J Histochem Cytochem* **1962**, *10*, 355–364.

36 Mandema, E., L. Ruinen, J. H. Scholten and A. S. Cohen (eds). *Amyloidosis*, Excerpta Medica, Amsterdam, **1968**.

37 Cohen, A. S. and E. Calkins. Electron microscopic observations on a fibrous component in amyloid of diverse origins. *Nature* **1959**, *183*, 1202–1203.

38 Frühling, L., J. Kempf and A. Porte. Structure et formation de la substance amyloide dans l'amylose expérimentale de la Souris. Étude au microscope électronique. *C R Acad Sci (Paris)* **1960**, *250*, 1385–1386.

39 Caesar, R. Die Feinstruktur von Milz und Leber bei experimenteller Amyloidose. *Z Zellforsch* **1960**, *52*, 653–673.

40 Battaglia, S. Electronenoptische Untersuchungen am Leberamyloid der Maus. *Beitr Pathol Anat* **1962**, *126*, 301–320.

41 Cohen, A. S., A. Frensdorff, S. Lamprecht and E. Calkins. A study of the fine structure of the amyloid associated with familial Mediterranean fever. *Am J Pathol* **1962**, *41*, 567–578.

42 Heefner, W. A. and G. D. Sorenson. Experimental amyloidosis. I. Light and electron microscopic observations of spleen and lymph nodes. *Lab Invest* **1962**, *11*, 585–593.

43 Cohen, A. S. and E. S. Cathcart. Casein-duced experimental amyloidosis. In *Methods and Achievements in Experimental Pathology*, E. Bajusz and G. Jasmin (eds). Karger, Basel, **1972**, pp. 207–242.

44 Shirahama, T. and A. S. Cohen. Intralysosomal formation of amyloid fibrils. *Am J Pathol* **1975**, *81*, 101–116.

45 Gueft, B. and J. J. Ghidoni. The site of formation and ultrastructure of amyloid. *Am J Pathol* **1963**, *43*, 837–854.

46 Bywaters, E. G. L. Discussion. In *Amyloidosis*, E. Mandema, L. Ruinen, H. Scholten and A. S. Cohen (eds). Excerpta Medica, Amsterdam, **1968**, p. 80.

47 Shirahama, T. and A. S. Cohen. High-resolution electron microscopic analysis of the amyloid fibril. *J Cell Biol* **1967**, *33*, 679–708.

48 Shirahama, T. and A. S. Cohen. Reconstitution of amyloid fibrils from alkaline extracts. *J Cell Biol* **1967**, *35*, 459–464.

49 Shirahama, T. and A. S. Cohen. A brief review of the ultrastructure of amyloid. In *Amyloidosis*, J. Marrink and M. H. van Rijswijk (eds). Martinus Nijhoff, Dordrecht, **1976**, pp. 51–57.

50 Cathcart, E. S., F. R. Comerford and A. S. Cohen. Immunologic studies on a protein extracted from human secondary amyloid. *N Engl J Med* **1965**, *273*, 143–146.

51 Bladen, H. A., M. U. Nylen and G. G. Glenner. The ultrastructure of human amyloid as revealed by the negative staining technique. *J Ultrastruct Res* **1966**, *14*, 449–459.

52 Pepys, M. B., A. C. Dash, E. A. Munn, A. Feinstein, M. Skinner, A. S. Cohen, H. Gerwurz, A. P. Osmand and R. H. Painter. Isolation of amyloid P component (protein AP) from normal serum as a calcium-dependent binding protein. *Lancet* **1977**, *i*, 1029–1031.

53 Dyck, R. F., C. M. Lockwood, M. Kershaw, N. McHugh, M. L. Baltz and M. B. Pepys. Amyloid P-component is a constituent of normal human glomerular basement membrane. *J Exp Med* **1980**, *152*, 1162–1174.

54 Breathnach, S. M., S. M. Melrose, B. Bhogal, F. C. de Beer, R. F. Dyck, G. Tennent, M. M. Black and M. B. Pepys. Amyloid P component is located on elastic microfibrils in normal human tissue. *Nature* **1981**, *293*, 652–654.

55 Pras, M., M. Schubert, D. Zucker-Franklin, A. Rimon and E. C. Franklin. The characterization of soluble amyloid prepared in water. *J Clin Invest* **1968**, *47*, 924–933.

56 Eanes, E. D. and G. G. Glenner. X-ray diffraction studies on amyloid filaments. *J Histochem Cytochem* **1968**, *16*, 673–677.

57 Bonar, L., A. S. Cohen and M. Skinner. Characterization of the amyloid fibril as a cross-β protein. *Proc Soc Exp Biol Med* **1969**, *131*, 1373–1375.

58 Waugh, D. F. A fibrous modification of insulin. I. The heat precipitate of insulin. *J Am Chem Soc* **1946**, *68*, 247–250.

59 Westermark, P. On the nature of the amyloid in human islets of Langerhans. *Histochemistry* **1974**, *38*, 27–33.

60 Glenner, G. G., E. D. Eanes, H. A. Bladen, R. P. Linke and J. D. Termine. β-pleated sheet fibrils. A comparison of native amyloid with synthetic protein fibrils. *J Histochem Cytochem* **1974**, *22*, 1141–1158.

61 Glenner, G. G., W. Terry, M. Harada, C. Isersky and D. Page. Amyloid fibril proteins: proof of homology with immunoglobulin light chains by sequence analysis. *Science* **1971**, *172*, 1150–1151.

62 Benditt, E. P. Amyloid protein AA and its precursor, the acute phase protein(s) apoSAA: a perspective. In *Amyloidosis*, J. Marrink and M. H. van Rijswijk (eds). Martinus Nijhoff, Dordrecht, **1986**, pp. 101–106.

63 Benditt, E. P. and N. Eriksen. Chemical classes of amyloid substance. *Am J Pathol* **1971**, *65*, 231–252.

64 Sletten, K., P. Westermark and J. B. Natvig. Characterization of amyloid fibril proteins from medullary carcinoma of the thyroid. *J Exp Med* **1976**, *143*, 993–998.

65 Benditt, E. P. and N. Eriksen. Chemical similarity among amyloid substances associated with long standing inflammation. *Lab Invest* **1962**, *26*, 615–625.

66 Benditt, E. P. and N. Eriksen. Amyloid, III. A protein related to the subunit structure of human amyloid fibrils. *Proc Natl Acad Sci USA* **1966**, *55*, 308–316.

67 Benditt, E. P., N. Eriksen, M. A. Hermodson and L. H. Ericsson. The major proteins of human and monkey amyloid substance: common properties including unusual N-terminal amino acid sequences. *FEBS Lett* **1971**, *19*, 169–173.

68 Ein, D., S. Kimura, W. D. Terry, J. Magnotta and G. G. Glenner. Amino acid sequence of an amyloid fibril protein of unknown origin. *J Biol Chem* **1972**, *247*, 5653–5655.

69 Husby, G., K. Sletten, T. E. Michaelsen and J. B. Natvig. Alternative non-immunoglobulin origin of amyloid fibrils. *Nature (New Biol)* **1972**, *238*, 187.

70 Levin, M., E.C. Franklin, B. Frangione and M. Pras. The amino acid sequence of a major nonimmunoglobulin component of some amyloid fibrils. *J Clin Invest* **1972**, *51*, 2773–2776.

71 Sletten, K. and G. Husby. The complete amino-acid sequence of non-immunoglobulin amyloid fibril protein AS in rheumatoid arthritis. *Eur J Biochem* **1974**, *41*, 117–125.

72 Møyner, K., K. Sletten, G. Husby and J.B. Natvig. An unusually large (83 amino acid residues) amyloid fibril protein AA from a patient with Waldenström's macroglobulinaemia and amyloidosis. *Scand J Immunol* **1980**, *11*, 549–554.

73 Sletten, K., G. Husby and J.B. Natvig. The complete amino acid sequence of an amyloid fibril protein AA of unusual size (64 residues). *Biochem Biophys Res Commun* **1976**, *69*, 19–25.

74 Westermark, G.T., P. Westermark and K. Sletten. Amyloid fibril protein AA. Characterization of uncommon subspecies from a patient with rheumatoid arthritis. *Lab Invest* **1987**, *57*, 57–64.

75 Cohen, A.S., E.C. Franklin, G.G. Glenner, J.B. Natvig, E.F. Osserman and O. Wegelius. Nomenclature. In *Amyloidosis*, O. Wegelius and A. Pasternack (eds). Academic Press, London, **1976**, p. ix.

76 Levin, M., M. Pras and E.C. Franklin. Immunologic studies of the major nonimmunoglobulin protein of amyloid. I. Identification and partial characterization of a related serum component. *J Exp Med* **1973**, *138*, 373–380.

77 Husby, G. and J.B. Natvig. A serum component related to nonimmunoglobulin amyloid protein AS, a possible precursor of the fibrils. *J Clin Invest* **1974**, *53*, 1054–1061.

78 Sipe, J.D., T.F. Ignaczak, P.S. Pollock and G.G. Glenner. Amyloid fibril protein AA: purification and properties of the antigenically related serum component as determined by solid phase radioimmunoassay. *J Immunol* **1976**, *116*, 1151–1156.

79 Linke, R.P., J.D. Sipe, P.S. Pollock, T.F. Ignaczak and G.G. Glenner. Isolation of a low-molecular-weight serum component antigenically related to an amyloid fibril protein of unknown origin. *Proc Natl Acad Sci USA* **1975**, *72*, 1473–1476.

80 Rosenthal, C.J., E.C. Franklin, B. Frangione and J. Greenspan. Isolation and partial characterization of SAA – an amyloid-related protein from human serum. *J Immunol* **1976**, *116*, 1415–1418.

81 Benditt, E.P. and N. Eriksen. Amyloid protein SAA is associated with high density lipoprotein from human serum. *Proc Natl Acad Sci USA* **1977**, *74*, 4025–4028.

82 Benditt, E.P. and N. Eriksen. Amyloid protein SAA is an apoprotein of mouse plasma high density lipoprotein. *Proc Natl Acad Sci USA* **1979**, *76*, 4092–4096.

83 Skogen, B., A.L. Borresen, J.B. Natvig, K. Berg and T.E. Michaelsen. High-density lipoprotein as carrier for amyloid-related protein SAA in rabbit serum. *Scand J Immunol* **1979**, *10*, 39–45.

84 Bausserman, L.L., P.N. Herbert and K.P.W. J. McAdam. Heterogeneity of human amyloid A proteins. *J Exp Med* **1980**, *152*, 641–656.

85 Sletten, K., G. Marhaug and G. Husby. The covalent structure of amyloid-related serum protein SAA from two patients with inflammatory disease. *Hoppe Seylers Zeitschr Physiol Chem* **1983**, *364*, 1039–1046.

86 Husebekk, A., B. Skogen, G. Husby and G. Marhaug. Transformation of amyloid precursor SAA to protein AA and incorporation in amyloid fibrils *in vivo*. *Scand J Immunol* **1985**, *21*, 283–287.

87 Tape, C., R. Tan, M. Nesheim and R. Kisilevsky. Direct evidence for circulating apoSAA as the precursor of tissue AA amyloid deposits. *Scand J Immunol* **1988**, *28*, 317–324.

88 Zschiesche, W. and W. Jakob. Pathology of animal amyloidoses. *Pharmacol Ther* **1989**, *41*, 49–83.

89 Johnson, K.H., P. Westermark, K. Sletten and T.D. O'Brien. Amyloid proteins and amyloidosis in domestic animals. *Amyloid: Int J Exp Clin Invest* **1996**, *3*, 270–289.

90 Maximow, A. Ueber die experimentell hervorgerufene Amyloid-Entartung der

Leber. *Virchows Arch Pathol Anat* **1898**, *153*, 353–401.

91 Domagk, G. Untersuchungen über die Bedeutung des retikuloendothelialen Systems für die Vernichtung von Infektionserregern und für die Entstehung des Amyloids. *Virchows Arch Pathol Anat* **1924**, *253*, 594–638.

92 Morgenstern, Z. Zur Frage über Amyloidose und Resorption. *Virchows Arch Pathol Anat* **1926**, *259*, 698–725.

93 Benditt, E. P. and N. Eriksen. The chemical classes of amyloid substance in men, monkeys and ducks. *Protides Biol Fluids* **1973**, *20*, 81–85.

94 Eriksen, N., L. H. Ericsson, N. Pearsall, D. Lagunoff and E. P. Benditt. Mouse amyloid protein AA: homology with nonimmunoglobulin protein of human and monkey amyloid substance. *Proc Natl Acad Sci USA* **1976**, *73*, 964–967.

95 Anders, R. F., J. B. Natvig, K. Sletten, G. Husby and K. Nordstoga. Amyloid-related serum protein SAA from three animal species: comparison with human SAA. *J Immunol* **1977**, *118*, 229–234.

96 Sipe, J. D., K. P. W. J. McAdam, B. F. Torain and P. S. Pollock. Isolation and structural properties of murine SAA – the acute phase serum precursor of amyloid AA. *Immunol Commun* **1977**, *6*, 1–12.

97 Glenner, G. G., J. Harbaugh, J. I. Ohms, M. Harada and P. Cuatrecasas. An amyloid protein: the amino-terminal variable fragment of an immunoglobulin light chain. *Biochem Biophys Res Commun* **1970**, *41*, 1287–1289.

98 Lian, J. B., M. Skinner, M. D. Benson and A. S. Cohen. Fractionation of primary amyloid fibrils. Characterization and chemical interaction of the subunits. *Biochim Biophys Acta* **1977**, *491*, 167–176.

99 Schiffer, M., R. L. Girling, K. R. Ely and A. B. Edmundson. Structure of a lambda-type Bence-Jones protein at 3.5-Å resolution. *Biochemistry* **1973**, *12*, 4620–4631.

100 Husby, G., J. B. Natvig and K. Sletten. New, third class of amyloid fibril protein. *J Exp Med* **1974**, *139*, 773–778.

101 Sletten, K., J. B. Natvig, G. Husby and J. Juul. The complete amino acid sequence of a prototype immunoglobulin-λ light-chain-type amyloid-fibril protein AR. *Biochem J* **1981**, *195*, 561–572.

102 Solomon, A., B. Frangione and E. C. Franklin. Bence Jones proteins and light chains of immunoglobulins. Preferential association of the $V_{\lambda VI}$ subgroup of human light chains with amyloidosis AL(λ). *J Clin Invest* **1982**, *70*, 453–460.

103 Eulitz, M., D. T. Weiss and A. Solomon. Immunoglobulin heavy-chain-associated amyloidosis. *Proc Natl Acad Sci USA* **1990**, *87*, 6542–6546.

104 Tan, S. Y., I. E. Murdoch, T. J. Sullivan, J. E. Wright, O. Truong, J. J. Hsuan, P. N. Hawkins and M. B. Pepys. Primary localized orbital amyloidosis composed of the immunoglobulin gamma heavy chain CH3 domain. *Clin Sci* **1994**, *87*, 487–491.

105 Engvig, J. P., K. E. Olsen, R. E. Gislefoss, K. Sletten, O. Wahlström and P. Westermark. Constant region of a κ III immunoglobulin light chain as a major AL-amyloid protein. *Scand J Immunol* **1998**, *48*, 92–98.

106 Solomon, A., D. T. Weiss, C. L. Murphy, R. Hrncic, J. S. Wall and M. Schell. Light chain-associated amyloid deposits comprised of a novel kappa constant domain. *Proc Natl Acad Sci USA* **1998**, *95*, 9547–9551.

107 Wally, J., G. Kica, Y. Zhang, T. Ericsson, L. H. Connors, M. D. Benson, J. J. Liepnieks, J. Murray, M. Skinner and R. L. Comenzo. Identification of a novel substitution in the constant region of a gene coding for an amyloidogenic kappa1 light chain. *Biochim Biophys Acta* **1999**, *1454*, 49–56.

108 Andrade, C., S. Araki, W. D. Block, A. S. Cohen, C. E. Jackson, Y. Kuroiwa, J. Nissim, E. Sohar, V. A. McKusick and M. W. Van Allen. Hereditary amyloidosis. *Arthritis Rheum* **1970**, *13*, 902–915.

109 Ostertag, B. Demonstration einer eigenartigen familiären „Paramyloidose". *Zentralbl Pathol* **1933**, *56*, 253–254.

110 Costa, P. P., A. S. Figueira and F. R. Bravo. Amyloid fibril protein related to prealbumin in familial amyloidotic polyneuropathy. *Proc Natl Acad Sci USA* **1978**, *75*, 4499–4503.

111 Kanda, Y., D. S. Goodman, R. E. Canfield and F. J. Morgan. The amino acid sequence of human plasma prealbumin. *J Biol Chem* **1974**, *249*, 6796–6805.

112 Blake, C. C. F., M. J. Geisow, S. J. Oatley, B. Rérat and C. Rérat. Structure of prealbumin: secondary, tertiary and quaternary interactions determined by Fourier refinement at 1.8 Å. *J Mol Biol* **1978**, *121*, 339–356.

113 Benson, M. D. Partial amino acid sequence homology between a heredofamilial amyloid protein and human plasma prealbumin. *J Clin Invest* **1981**, *67*, 1035–1041.

114 Skinner, M. and A. S. Cohen. The prealbumin nature of the amyloid protein in familial amyloidotic polyneuropathy (FAP) – Swedish variety. *Biochem Biophys Res Commun* **1981**, *99*, 1326–1332.

115 Tawara, S., S. Araki, K. Toshimori, H. K. Nakagawa and S. Ohtaki. Amyloid fibril protein in type I familial amyloidotic polyneuropathy in Japanese. *J Lab Clin Med* **1981**, *98*, 811–822.

116 Tawara, S., M. Nakazato, K. Kangawa, H. Matsuo and S. Araki. Identification of amyloid prealbumin variant in familial amyloidotic polyneuropathy (Japanese type). *Biochem Biophys Res Commun* **1983**, *116*, 880–888.

117 Saraiva, M. J. M., S. Birken, P. P. Costa and D. S. Goodman. Amyloid fibril protein in familial amyloidotic polyneuropathy, Portuguese type. Definition of molecular abnormality in transthyretin (prealbumin). *J Clin Invest* **1984**, *74*, 104–119.

118 Dwulet, F. E. and M. D. Benson. Characterization of a transthyretin (prealbumin) variant associated with familial amyloidotic polyneuropathy type II (Indiana/Swiss). *J Clin Invest* **1986**, *78*, 880–886.

119 Westermark, P., K. Sletten and B.-O. Olofsson. Prealbumin variants in the amyloid fibrils of Swedish familial amyloidotic polyneuropathy. *Clin Exp Immunol* **1987**, *69*, 695–701.

120 Gorevic, P. D., F. C. Prelli, J. Wright, M. Pras and B. Frangione. Systemic senile amyloidosis. Identification of a new prealbumin (transthyretin) variant in cardiac tissue: immunologic and biochemical similarity to one form of familial amyloidotic polyneuropathy. *J Clin Invest* **1989**, *83*, 836–843.

121 Connors, L. H., A. Lim, T. Prokaeva, V. A. Roskens and C. E. Costello. Tabulation of human transthyretin (TTR) variants, 2003. *Amyloid: J Protein Folding Disorder* **2003**, *10*, 160–184.

122 Jacobson, D. R., P. Gorevic and J. N. Buxbaum. A homozygous transthyretin variant associated with senile systemic amyloidosis: evidence for a late-onset disease of genetic etiology. *Am J Hum Genet* **1990**, *47*, 127–136.

123 Westermark, P., J. B. Natvig and B. Johansson. Characterization of an amyloid fibril protein from senile cardiac amyloid. *J Exp Med* **1977**, *146*, 631–636.

124 Westermark, P., K. Sletten, B. Johansson and G. G. Cornwell III. Fibril in senile systemic amyloidosis is derived from normal transthyretin. *Proc Natl Acad Sci USA* **1990**, *87*, 2843–2845.

125 Grubb, A. and H. Löfberg. Human γ-trace, a basic microprotein: amino acid sequence and presence in the adenohypophysis. *Proc Natl Acad Sci USA* **1982**, *79*, 3024–3027.

126 Cohen, D. H., H. Feiner, O. Jensson and B. Frangione. Amyloid fibril in hereditary cerebral hemorrhage with amyloidosis (HCHWA) is related to the gastroentero-pancreatic neuroendocrine protein, gamma trace. *J Exp Med* **1983**, *158*, 623–628.

127 Nichols, W. C., F. E. Dwulet, J. Liepnieks and M. D. Benson. Variant apolipoprotein AI as a major constituent of a human hereditary amyloid. *Biochem Biophys Res Commun* **1988**, *156*, 762–768.

128 Benson, M. D., J. Liepnieks, M. Yazaki, T. Yamashita, K. Hamidi Asl, B. Guenther and B. Kluve-Beckerman. A new human hereditary amyloidosis: the result of a stop-codon mutation in the apolipoprotein AII gene. *Genomics* **2001**, *72*, 272–277.

129 Benson, M.D., J. Liepnieks, T. Uemichi, G. Wheeler and R. Correa. Hereditary renal amyloidosis associated with a mutant fibrinogen alpha-chain. *Nature Genetics* **1993**, *3*, 252–256.

130 Maury, C.P. and M. Baumann. Isolation and characterization of cardiac amyloid in familial amyloid polyneuropathy type IV (Finnish): relation of the amyloid protein to variant gelsolin. *Biochim Biophys Acta* **1990**, *1096*, 84–86.

131 Levy, E., M. Haltia, I. Fernandez-Madrid, O. Koivunen, J. Ghiso, F. Prelli and B. Frangione. Mutation in the gelsolin gene in Finnish hereditary amyloidosis. *J Exp Med* **1990**, *172*, 1865–1867.

132 Pepys, M.B., P.N. Hawkins, D.R. Booth, D.M. Vigushin, G.A. Tennet, A.K. Soutar, N. Totty, O. Nguyen, C.C. F. Blake, C.J. Terry, T.G. Feest, A.M. Zalin and J.J. Hsuan. Human lysozyme gene mutations cause hereditary systemic amyloidosis. *Nature* **1993**, *362*, 553–557.

133 Vidal, R., B. Frangione, A. Rostagno, S. Mead, T. Revesz, G. Plant and J. Ghiso. A stop-codon mutation in the BRI gene associated with familial British dementia. *Nature* **1999**, *399*, 776–781.

134 Vidal, R., T. Revesz, A. Rostagno, E. Kim, J.L. Holton, T. Bek, M. Bojsen-Moller, H. Braendgaard, G. Plant, J. Ghiso and B. Frangione. A decamer duplication in the 3′ region of the BRI gene originates an amyloid peptide that is associated with dementia in a Danish kindred. *Proc Natl Acad Sci USA* **2000**, *97*, 4920–4925.

135 Becker, J.W. and G.N.J. Reeke. Three-dimensional structure of β_2-microglobulin. *Proc Natl Acad Sci USA* **1985**, *82*, 4225–4229.

136 Gejyo, F., T. Yamada, S. Odani, Y. Nakagawa, M. Arakawa, T. Kunitomo, H. Kataoka, M. Suzuki, Y. Hirasawa, T. Shirahama, A.S. Cohen and K. Schmid. A new form of amyloid protein associated with chronic hemodialysis was identified as β_2-microglobulin. *Biochem Biophys Res Commun* **1985**, *129*, 701–706.

137 Gorevic, P.D., T.T. Casey, W.J. Stone, C.R. DiRaimondo, F. Prelli and B. Frangione. Beta-2 microglobulin is an amyloidogenic protein in man. *J Clin Invest* **1985**, *76*, 2425–2429.

138 Bardin, T., D. Kuntz, J. Zingraff, M.-C. Voisin, A. Zelmar and J. Lansaman. Synovial amyloidosis in patients undergoing long-term hemodialysis. *Arthritis Rheum* **1985**, *28*, 1052–1058.

139 Morita, T., M. Suzuki, A. Kamimura and Y. Hirasawa. Amyloidosis of a possible new type in patients receiving long-term hemodialysis. *Arch Pathol Lab Med* **1985**, *109*, 1029–1032.

140 Munoz-Gómez, J., E. Bergadá-Barado, R. Gómez-Pérez, E. Llopart-Buisán, E. Subías-Sobrevía, J. Rotés-Querol and M. Solé-Arqués. Amyloid arthropathy in patients undergoing periodical haemodialysis for chronic renal failure: a new complication. *Ann Rheum Dis* **1985**, *44*, 729–733.

141 Linke, R.P., H. Hampl, H. Lobeck, E. Ritz, J. Bommer, R. Waldherr and M. Eulitz. Lysine-specific cleavage of β_2-microglobulin in amyloid deposits associated with hemodialysis. *Kidney Int* **1989**, *36*, 675–681.

142 Alzheimer, A. Ueber eine eigenartige Erkrankung der Hirnrinde. *Centrabl Nervenheilk Psychiatr* **1907**, *30*, 177–179.

143 Pantelakis, S. Un type particulier d'angiopathie sénile du système nerveux central: L'angiopathi congophile. Topographie et fréquence. *Monatsschr Psychiatr Neurol* **1954**, *128*, 219–256.

144 Glenner, G.G. and C.W. Wong. Alzheimer's disease: initial report of the purification and characterization of a novel cerebrovascular amyloid protein. *Biochem Biophys Res Commun* **1984**, *120*, 885–890.

145 Glenner, G.G. and C.W. Wong. Alzheimer's disease and Down's syndrome: sharing of a unique cerebrovascular amyloid fibril protein. *Biochem Biophys Res Commun* **1984**, *122*, 1131–1135.

146 Masters, C., G. Simms, N.A. Weinman, G. Multhaup, B.L. McDonald and K. Beyreuther. Amyloid plaque core protein in Alzheimer disease and

Down syndrome. *Proc Natl Acad Sci USA* **1985**, *82*, 4245–4249.

147 Selkoe, D. J. Alzheimer's disease: a central role for amyloid. *J Neuropath Exp Neurol* **1994**, *53*, 438–447.

148 Selkoe, D. J. The origins of Alzheimer disease: A is for amyloid. *J Am Med Ass* **2000**, *283*, 1615–1617.

149 Masters, C. L., D. C. Gajdusek and C. J. Gibbs Jr. Creutzfeldt-Jakob disease virus isolations from the Gerstmann-Sträussler syndrome. With an analysis of the various forms of amyloid plaque deposition in the virus-induced spongiform encephalopathies. *Brain* **1981**, *104*, 559–588.

150 Gajdusek, D. C. and V. Zigas. Degenerative disease of the central nervous system in New Guinea; the endemic occurrence of kuru in the native population. *N Engl J Med* **1957**, *257*, 974–978.

151 Gajdusek, D. C., C. J. Gibbs Jr and M. Alpers. Experimental transmission of a Kuru-like syndrome to chimpanzees. *Nature* **1966**, *209*, 794–796.

152 Bolton, D. C., M. P. McKinley and S. B. Prusiner. Identification of a protein that purifies with the scrapie prion. *Science* **1982**, *218*, 1309–1311.

153 Prusiner, S. B., M. P. McKinley, K. A. Bowman, D. C. Bolton, P. E. Bendheim, D. F. Groth and G. G. Glenner. Scrapie prions aggregate to form amyloid-like birefringent rods. *Cell* **1983**, *35*, 349–358.

154 Opie, E. L. On relation of chronic interstitial pancreatitis to the islands of Langerhans and to diabetes mellitus. *J Exp Med* **1901**, *5*, 397–428.

155 Weichselbaum, A. and E. Stangl. Zur Kenntnis der feineren Veränderungen des Pankreas bei Diabetes mellitus. *Wien Klin Wochenschr* **1901**, *14*, 968–972.

156 Ehrlich, J. C. and I. M. Ratner. Amyloidosis of the islets of Langerhans. A restudy of islet hyalin in diabetic and nondiabetic individuals. *Am J Pathol* **1961**, *38*, 49–59.

157 Pearse, A. G. E., S. W. B. Ewen and J. M. Polak. The genesis of apudamyloid in endocrine polypeptide tumours: histochemical distinction from immunamyloid. *Virchows Arch B Zellpathol* **1972**, *10*, 93–107.

158 Westermark, P. Amyloid of human islets of Langerhans. I. Isolation and some characteristics. *Acta Pathol Microbiol Scand C* **1975**, *83*, 439–446.

159 Steiner, D. F. and P. E. Oyer. The biosynthesis of insulin and a probable precursor of insulin by a human islet adenoma. *Proc Natl Acad Sci USA* **1967**, *57*, 473–480.

160 Ko, A. S. C., D. G. Smyth, J. Markussen and F. Sundby. The amino acid sequence of the C-peptide of human proinsulin. *Eur J Biochem* **1971**, *20*, 190–199.

161 Westermark, P. Amyloid of medullary carcinoma of the thyroid: partial characterization. *Upsala J Med Sci* **1975**, *80*, 88–92.

162 Westermark, P. Fine structure of islets of Langerhans in insular amyloidosis. *Virchows Arch A* **1973**, *359*, 1–18.

163 Bell, E. T. Hyalinization of the islets of Langerhans in diabetes mellitus. *Diabetes* **1952**, *1*, 341–344.

164 Westermark, P. Quantitative studies of amyloid in the islets of Langerhans. *Upsala J Med Sci* **1972**, *77*, 91–94.

165 Westermark, P., C. Wernstedt, E. Wilander and K. Sletten. A novel peptide in the calcitonin gene related peptide family as an amyloid fibril protein in the endocrine pancreas. *Biochem Biophys Res Commun* **1986**, *140*, 827–831.

166 Westermark, P., C. Wernstedt, E. Wilander, D. W. Hayden, T. D. O'Brien and K. H. Johnson. Amyloid fibrils in human insulinoma and islets of Langerhans of the diabetic cat are derived from a neuropeptide-like protein also present in normal islet cells. *Proc Natl Acad Sci USA* **1987**, *84*, 3881–3885.

167 Westermark, P., C. Wernstedt, T. D. O'Brien, D. W. Hayden and K. H. Johnson. Islet amyloid in type 2 human diabetes mellitus and adult diabetic cats contains a novel putative polypeptide hormone. *Am J Pathol* **1987**, *127*, 414–417.

168 Cooper, G. J., A. C. Willis, A. Clark, R. C. Turner, R. B. Sim and K. B. M.

Reid. Purification and characterization of a peptide from amyloid-rich pancreases of type 2 diabetic patients. *Proc Natl Acad Sci USA* **1987**, *84*, 8628–8632.

169 Lukinius, A., E. Wilander, G.T. Westermark, U. Engström and P. Westermark. Co-localization of islet amyloid polypeptide and insulin in the B cell secretory granules of the human pancreatic islets. *Diabetologia* **1989**, *32*, 240–244.

170 Johnson, K.H., T.D. O'Brien, D.W. Hayden, K. Jordan, H.K.G. Ghobrial, W.C. Mahoney and P. Westermark. Immunolocalization of islet amyloid polypeptide (IAPP) in pancreatic Beta cells by means of peroxidase-antiperoxidase (PAP) and protein A-gold techniques. *Am J Pathol* **1988**, *130*, 1–8.

171 Betsholtz, C., V. Svensson, F. Rorsman, U. Engström, G.T. Westermark, E. Wilander, K.H. Johnson and P. Westermark. Islet amyloid polypeptide (IAPP): cDNA cloning and identification of an amyloidogenic region associated with species-specific occurrence of age-related diabetes mellitus. *Exp Cell Res* **1989**, *183*, 484–493.

172 Nishi, M., T. Sanke, S. Seino, R.L. Eddy, Y.S. Fan, M.G. Byers, T.B. Shows, G.I. Bell and D.F. Steiner. Human islet amyloid polypeptide gene: complete nucleotide sequence, chromosomal localization, and evolutionary history. *Mol Endocrinol* **1989**, *3*, 1775–1781.

173 Banting, F.G., C.H. Best, J.B. Collip and J.J.R. Macleod. The effect of pancreatic extract (insulin) on normal rabbits. *Am J Physiol* **1922**, *62*, 162–176.

174 Westermark, P., K.H. Johnson, T.D. O'Brien and C. Betsholtz. Islet amyloid polypeptide – a novel controversy in diabetes research. *Diabetologia* **1992**, *35*, 297–303.

175 Kahn, S.E., S. Andrikopoulos and C.B. Verchere. Islet amyloid. A long-recognized but underappreciated pathological feature of type 2 diabetes. *Diabetes* **1999**, *48*, 241–253.

176 Höppener, J.W., B. Ahrén and C.J. Lips. Islet amyloid and type 2 diabetes mellitus. *N Engl J Med* **2000**, *343*, 411–419.

177 Reimann, H.A., R.F. Koucky and C.M. Eklund. Primary amyloidosis limited to the tissue of mesodermal origin. *Am J Pathol* **1935**, *11*, 977–988.

178 King, L.S. Atypical amyloid disease; with observations on a new silver stain for amyloid. *Am J Pathol* **1948**, *24*, 1095–1115.

179 Heller, H., H.-P. Missmahl, E. Sohar and J. Gafni. Amyloidosis: its differentiation into peri-reticulin and peri-collagen types. *J Pathol Bacteriol* **1964**, *88*, 15–34.

180 Missmahl, H.-P. Reticulin and collagen as important factors for the localization of amyloid and the use of polarization microscopy as a tool in the detection of the composition of amyloid. In *Amyloidosis*, E. Mandema, L. Ruinen, J.H. Scholten and A.S. Cohen (eds). Excerpta Medica, Amsterdam, **1968**, pp. 22–29.

181 Benditt, E.P., A.S. Cohen, P.P. Costa, E.C. Franklin, G.G. Glenner, G. Husby, E. Mandema, J.B. Natvig, E.F. Osserman, E. Sohar, O. Wegelius and P. Westermark. Guidelines for nomenclature. In *Amyloid and Amyloidosis*, G.G. Glenner, P.P. Costa and A.F. Freitas (eds). Excerpta Medica, Amsterdam, **1980**, pp. xi–xii.

182 Westermark, P., M.D. Benson, J.N. Buxbaum, A.S. Cohen, B. Frangione, S. Ikeda, C.L. Masters, G. Merlini, M.J. Saraiva and J.D. Sipe. Amyloid fibril protein nomenclature – 2002. *Amyloid: J Protein Fold Dis* **2002**, *9*, 197–200.

183 Buxbaum, J.N. Diseases of protein conformation: what do *in vitro* experiments tell us about *in vivo* diseases? *Trends Biochem Sci* **2003**, *28*, 585–592.

184 Bergström, J., C.L. Murphy, D.T. Weiss, A. Solomon, K. Sletten, U. Hellman and P. Westermark. Two different types of amyloid deposits – apolipoprotein A-IV and transthyretin – in a patient with systemic amyloidosis. *Lab Invest* **2004**, *84*, 981–988.

185 Johansson, B., C. Wernstedt and P. Westermark. Atrial natriuretic peptide

deposited as atrial amyloid fibrils. *Biochem Biophys Res Commun* **1987**, *148*, 1087–1092.

186 Westermark, P., L. Eriksson, U. Engström, S. Eneström and K. Sletten. Prolactin-derived amyloid in the aging pituitary gland. *Am J Pathol* **1997**, *150*, 67–73.

187 Dische, F. E., C. Wernstedt, G. T. Westermark, P. Westermark, M. B. Pepys, J. A. Rennie, S. G. Gilbey and P. J. Watkins. Insulin as an amyloid-fibril protein at sites of repeated insulin injections in a diabetic patient. *Diabetologia* **1988**, *31*, 158–161.

188 Häggqvist, B., J. Näslund, K. Sletten, G. T. Westermark, G. Mucchiano, L. O. Tjernberg, C. Nordstedt, U. Engström and P. Westermark. Medin: an integral fragment of aortic smooth muscle cell-produced lactadherin forms the most common human amyloid. *Proc Natl Acad Sci USA* **1999**, *96*, 8669–8674.

189 Korvatska, E., H. Henry, Y. Mashima, M. Yamada, C. Bachmann, F. L. Munier and D. F. Schorderet. Amyloid and non-amyloid forms of 5q31-linked corneal dystrophy resulting from keratoepithelin mutations at Arg-124 are associated with abnormal turnover of the protein. *J Biol Chem* **2000**, *275*, 11465–11469.

190 Ando, Y., M. Nakamura, H. Kai, S. Katsuragi, H. Terazaki, T. Nozawa, T. Okuda, S. Misumi, N. Matsunaga, K. Hata, T. Tajiri, S. Shoji, T. Yamashita, K. Haraoka, K. Obayashi, K. Matsumoto, M. Ando and M. Uchino. A novel localized amyloidosis associated with lactoferrin in the cornea. *Lab Invest* **2002**, *82*, 757–766.

191 Solomon, A., C. L. Murphy, K. Weaver, D. T. Weiss, R. Hrncic, M. Eulitz, R. L. Donnell, K. Sletten, G. Westermark and P. Westermark. Calcifying epithelial odontogenic (Pindborg) tumor-associated amyloid consists of a novel human protein. *J Lab Clin Med* **2003**, *142*, 348–355.

2
Anatomic and Clinical Clues to *In Vivo* Mechanisms of Amyloidogenesis

Vittorio Bellotti, Laura Obici, Robert Kisilevsky and Giampaolo Merlini

2.1
Introduction

Within the last 20 years it has been recognized that altered protein conformation can be at the basis of some of the most important disorders that affect humans. By far the most common example is amyloid formation, a process that likely plays a central pathogenetic role in diseases as diverse as Alzheimer's disease, adult-onset (Type 2) diabetes mellitus, joint destruction as a consequence of long-standing hemodialysis, systemic sequelae of persistent acute inflammation and immunoglobulin light chain deposits secondary to their production in B cell dyscrasias, to cite but a few examples. Although the proteins involved in each of these disorders vary greatly in structure, function and conformation, present evidence suggests that the conformation of each of these proteins in its amyloid state, i.e. fibrils, is remarkably similar. This similarity in conformation is thought to be responsible for the unique structural and staining properties that characterize an amyloid deposit.

In vitro studies with purified amyloidogenic proteins/peptides have shed considerable light on the common thermodynamic and kinetic parameters that govern the process of amyloid fibril formation, a process that includes conformational instability, the appearance of abnormally high concentrations of metastable folding intermediates and the nidus formation that precipitates fibril growth.

Since many of the amyloidogenic proteins/peptides circulate in plasma, based on the *in vitro* data assembled over the last 10–15 years, one might expect that amyloid deposits, regardless of the disease or protein concerned, would have a similar anatomic distribution in each of these disorders. Nevertheless, it is abundantly clear that the amyloid in a particular disease is deposited with anatomic specificity. It is this anatomic localization and the, as yet, unidentified structural/biochemical features of the protein/peptide oligomers that correlate with the clinical and pathophysiological manifestations of the disease concerned. At present, however, little is known about the molecular mechanisms that play a role in the *in vivo* formation and deposition of amyloid fibrils in specific target tissues, leading to organ dysfunction.

Amyloid Proteins. The Beta Sheet Conformation and Disease. J. D. Sipe

ISBN: 3-527-31072-X

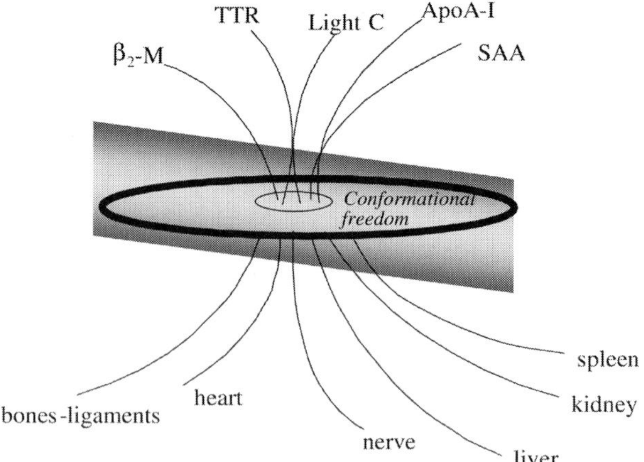

Fig. 2.1 Various globular proteins in the upper part of the scheme (some prototypic examples are cited) can rearrange their tertiary structure and converge into a similar restricted conformation, i.e. amyloid. However, notwithstanding this common conformation, the amyloid fibrils accumulate in many different target tissues.

It is therefore necessary to consider a scenario (Fig. 2.1) in which heterogeneous proteins cause substantially different clinical syndromes by each affecting different tissues, but they do so through common mechanisms which result in similar but unique histopathological products – the various forms of amyloid.

It is our purpose in this chapter to consider what the clinical and pathological features observed in selected examples of localized and systemic amyloidosis may indicate about mechanistic events in amyloidogenesis, how these events relate to the evidence obtained *in vitro*, and how they may account for the anatomic specificity and altered local pathophysiology that is responsible for the clinical features that characterize the different forms of amyloid.

2.2 AA Amyloidogenesis

Amyloid that follows on medical disorders such as rheumatoid arthritis, tuberculosis or cystic fibrosis, in which there is persistent acute (in a cellular, not temporal, sense) inflammation, has been termed AA amyloidosis. The frequency of this long-term complication of inflammation has decreased considerably in the last 100 years as more effective methods became available to treat the antecedent conditions. Although now relatively rare, AA amyloidosis, and the animal models that were developed for its study, has nevertheless provided a wealth of information about amyloidogenesis generally.

The polypeptide characterizing AA amyloid was among the first such proteins identified [1], hence the AA (amyloid A) designation. Using immunological methodology, it was shown that the AA peptide is related to a plasma protein, serum amyloid A (SAA) [2], and that the fibril-forming polypeptide in AA amyloid deposits is derived from this circulating component [3, 4]. The precise mechanism(s) by which circulating SAA is converted into the fibrillar tissue deposit has not yet been discovered. However, analyses of the *in vitro* kinetics of amyloid fibril formation generally, the kinetics of *in vivo* AA deposition, the AA fibril-associated tissue components, the organs and anatomically specific sites within such organs where AA amyloid is deposited, and identification of inhibitors of *in vivo* AA amyloidogenesis provide sufficient data to construct several working mechanistic hypotheses.

AA amyloidosis is not usually seen in the absence of a substantial increase in the plasma concentration of SAA [5]. Clearly an adequate precursor pool is a necessary factor. It is the dramatic and rapid rise of SAA concentration as part of the systemic response to local tissue injury that provides this circulating pool of AA precursor for amyloidogenesis. If this were the only factor required for this process, since acute inflammation is so common during the life of every person, AA amyloidogenesis would be a much more common disorder than it actually is. Furthermore, since SAA is a circulating plasma protein, its concentration is likely to be similar in all organs of the body. Yet, AA amyloid deposition is now relatively rare and, as shown by clinical as well as experimental studies, initial deposits occur first in the spleen, followed by the liver and subsequently in other organs [5, 6]. Local factors in such organs must therefore play a critical role in determining whether or not SAA will undergo the conformational changes required for amyloid deposition. *In vivo* murine experimental studies using standard (on the order of weeks) and rapid (on the order of days) AA induction protocols have provided substantial information on the nature of some of these local factors.

Induction of murine AA amyloid using standard protocols has shown that a fibrillogenic nucleating entity, termed amyloid enhancing factor (AEF), appears in tissues 24–48 h before histological demonstrable amyloid [7]. In the absence of an inflammatory reaction (i.e. in the absence of increased SAA levels), AEF administered to mice fails to provoke AA amyloid. AEF administered together with an inflammatory stimulus induces rapid (36–48 h) AA deposits in the spleen and liver [6]. The kinetics of AA amyloid accumulation in spleen (and also liver) is identical when using standard protocols or AEF-primed deposition, except for the marked shortening of the lag period when using AEF [6]. The kinetics of *in vivo* amyloid deposition is consistent with data from *in vitro* studies indicating nucleation-dependent fibrillogenesis.

It may be argued that such a nucleation event determines, in part, the organ and anatomic specificity with which AA amyloid appears in tissues. Experimentally, AEF is usually injected intravenously and is, therefore, available systemically. However, AEF is a particulate and is efficiently cleared by perifollicular cells in the spleen, Kupffer cells in the liver and other organs with significant populations of macrophages [8]. This may explain the organ and anatomic specificity when using exogenous AEF, but it does not explain the same organ and anatomic distribution of AA amyloid when using non-AEF induction protocols.

An additional factor determining organ and anatomic distribution may relate to a putative physiological function of SAA. SAA is found predominantly in the high-density lipoprotein (HDL) fraction of plasma [9]. HDL/SAA has a higher affinity for macrophages than HDL alone [10] and activated macrophages have more binding sites for HDL/SAA than non-activated macrophages [10]. Several potential macrophage SAA receptors have been identified [11–14]. These quantitative and affinity factors relate to a potential role of SAA in cholesterol mobilization from macrophages [15, 16], and, in so doing, may effect a very high local SAA concentration at sites in the spleen and liver where such phagocytic cells exist. Such a high local SAA concentration may make these sites particularly prone to AA amyloid deposition.

The molecular events responsible for assembly of the putative localized SAA into fibrils may be determined by components of the extracellular matrix. Among these is a set of structural constituents of basement membranes, serum amyloid P (SAP), collagen IV, laminin and heparan sulfate proteoglycan (HSPG). These constituents, as well as apolipoprotein E (ApoE), have been identified in many, if not all, forms of amyloid [17]. Furthermore, the earliest site of deposition of several forms of amyloid is in association with basement membranes [18–25]. Several of these components, i.e. the HSPG perlecan, collagen IV and laminin, have high affinities for SAA [26–28], but, among these, only HSPG has received considerable experimental attention.

Basement membranes (BM) are found throughout the body and it is not immediately clear why an amyloid precursor such as SAA should have an affinity for BM structural components in one anatomic location rather than another. One possible explanation relates to SAA's binding affinity for the individual BM components rather than to components that have already started to assemble themselves into nascent BM membrane intermediates. For example, SAA binds with high affinity to laminin, but fails to do so when laminin is already complexed with entactin [28], a BM protein that links several of the structural constituents. One can envision at least two scenarios where the individual BM components become available for SAA binding. In the first case, existing BM components may begin to disassemble, providing "free" entities to associate with SAA. Alternatively, there may be an increase in *de novo* synthesis of the BM components and these may be available to other ligands until such time as they find their natural partners. Based on the levels of splenic mRNA for these BM components during the initial phase of amyloidogenesis the second alternative appears more likely. Splenic perlecan, collagen IV, and laminin β- and γ-chain mRNAs are significantly elevated at the earliest stages of amyloidogenesis, and may precede histological demonstrable AA amyloid deposits [29, 30].

In contrast to amyloid fibrils generated *in vitro*, HSPG is an integral structural component of AA fibrils *in vivo* [31, 32]. *In vitro* the heparan sulfate glycosaminoglycan (HS) moiety of HSPG exerts a conformational influence on the amyloidogenic isoform of SAA, provoking an increase in β-sheet structure, the characteristic conformation of proteins in amyloid fibrils [33]. HS may exert this influence by at least three possible mechanisms, although not all of these have been explored in detail. Given the high sulfate content of HS, the pH of the microenvironment adjacent to HS may be acidic. Such changes in pH may affect the stability of the native structure of the amyloidogenic SAA isoform during its interaction with HS. There may be direct binding of HS to SAA that may affect its structure [26, 33]. These interactions may play a role in generating the nucleating event for fibrillogenesis and/or they may alter the thermodynamic barrier between the SAA in its native conformation and those intermediate conformations that are amyloidogenic. In support of the critical role of HS in AA amyloidogenesis, agents that inhibit the binding of HS and SAA are effective inhibitors of AA amyloidogenesis in cell culture and *in vivo* [34], as are agents that truncate the biosynthesis of HS [35]. Furthermore, transgenic mice that overexpress heparanase show resistance to AA amyloidogenesis but only in those tissues where this enzyme is expressed (Łi, J. P., Escobar-Galvis, M. L., Gong, F., Zhang, X., Zcharia, E., Metzger, S., Vlodavsky, I., Kisilevsky, R. and Lindahl, U., manuscript submitted).

2.3
β_2-Microglobulin (β_2M) and the Amyloid Deposition in Hemodialysis

Amyloidosis caused by β_2M represents a further example of a systemic form of the disease in which specific interactions between the amyloid precursor protein and some putative microenvironmental factors might play a unique role in determining the anatomic distribution of the fibrillar deposits. β_2M amyloidosis occurs in patients undergoing chronic hemodialytic treatment for end-stage renal failure, representing the only iatrogenic form of systemic amyloidosis. The well-recognized causative role of hemodialysis through the progressive increase of the β_2M plasma concentration demonstrates clearly the importance of an adequate precursor pool in the amyloidogenic process in a way that is more definite than for any other acquired amyloidosis. Moreover, the tissue localization in this amyloidosis is extremely peculiar, as amyloid involves almost exclusively the musculoskeletal apparatus, with deposits being localized in close proximity to collagen fibrils. These features make β_2M amyloidosis an invaluable model for studying the mechanisms involved in *in vivo* amyloidogenesis.

2.3.1
The Post-translation Modifications of β_2M in Naturally Occurring Amyloid Fibrils

Several post-translational modifications have been proposed as possible promoters of β_2M fibrillogenesis and, possibly, its tissue deposition, but only a few have been chemically characterized in the natural protein. Extensive oxidation of β_2M can certainly favor the process of aggregation [36], but, to date, only the oxidation of Met99 has been demonstrated in naturally derived fibrils. Furthermore, it cannot be excluded that this oxidation is caused during the Aβ_2M purification procedure and, thus, represents an artifact. β_2M glycation has been emphasized particularly because it was demonstrated that such glycated species are able to trigger localized inflammation [37] and their presence could be responsible for the infiltration of inflammatory cells that surround this particular type of amyloid deposit. However, the amount of fibrillar β_2M that is glycated is minimal and the glycation does not apparently enhance the propensity of β_2M to make fibrils [38]. Deamidation of Asn17 was discovered in β_2M purified from the fluid exchanged in the course of the hemodialytic procedure. This modification was considered to be responsible for the appearance of the β_2M double electrophoretic band seen under native conditions [39]. Amino acid sequencing of fibrillar β_2M has not confirmed the presence of deamidation; however, it is noteworthy that, *in vitro*, the replacement of this Asn with Asp accelerates the aggregation kinetics of β_2M [40]. Proteolytic re-modeling of β_2M was proposed by Linke on the basis of amino acid sequence data of natural fibrils extracted from different patients [41]. We have obtained results consistent with those reported by Linke, and the combination of amino acid sequencing and mass spectrometry made clear that the truncation of the sixth peptide bond is present in approximately 30% of natural fibrillar β_2M [42]. The possible pathogenic role of the proteolytic event as well as the effect of destabilization of strand A in β_2M is discussed further in Chapter 26.

2.3.2 The Interaction of β_2M with Collagen and Other Matrix Components

The peculiar tissue specificity of the amyloidosis associated with chronic hemodialysis was evident from the initial clinical description of this syndrome [43] and long before the chemical characterization of the amyloidogenic protein [44]. This type of amyloidosis is associated with a peculiar osteoarthropathy that includes the carpal tunnel syndrome, trigger finger, arthritis, fractures and bone cyst. The amyloid material is localized in very close proximity to collagen fibers both in the deposits, in tendons, as well as in the connective tissue where the amyloid follows the parallel orientation of the collagen fibrils. A possible direct role of cells constituting part of the synovial membrane has been proposed and intracellular deposit of β_2M fibrils is not an unusual finding in macrophage-like synoviocytes.

2.3.3 The Molecular Target of β_2M Amyloid Fibrils

The β_2M amyloid fibrils are largely insoluble in physiological buffers and their structure is remarkably similar to all the systemic amyloidoses. For this reason the Aβ_2M tissue specificity, and the search for a molecular target at such sites, should probably be restricted to the tissue target of the immediate fibril precursor or the fully folded native protein and its natural isoforms. This insight was initially achieved by Homma et al. [45] who determined a measurable affinity of native β_2M for collagen-type molecules. These authors showed that the triple helix of type I, or type II, collagen was capable of binding β_2M, but the binding parameters were not measured. Using surface plasmon resonance technology we measured the affinity of β_2M for triple-helical collagen and obtained evidence for an affinity in range of 100 μM for wild-type β_2M and a micromolar affinity for the β_2M species naturally truncated at the N-terminal end, ΔN6 β_2M. The binding parameter analysis included the k_{on} and k_{off}. This allowed us to establish that the increased affinity of the truncated species for collagen was mainly caused by the reduction of the k_{off}. We have also demonstrated that the affinity of the truncated species for collagen is favored by the reduction of pH from 7.5 to 6.5. Fluctuation of the binding properties is a function of pH, which opens the possibility that sudden pH modifications, in a range that might be expected in physiopathological conditions, could cause rapid and significant changes of local β_2M concentration in the solvent surrounding the collagen fiber. However, on the basis of the strength of interaction with collagen, a significant gradient of β_2M concentration in the solvent surrounding the collagen should be particularly considered in the case of ΔN6. A hypothetical exponential decay of protein concentration that becomes negligible a few nanometers from the collagen surface might be postulated and drawn as presented in Fig. 2.2.

The dependence of the kinetics of protein aggregation as a function of the concentration of protein precursor and time of incubation has been well demonstrated in *in vitro* studies [46] and biologically proven at the clinical level [47]. Likewise, we can state that the concentration of β_2M around the collagen is influenced by the binding itself and environmental modifications able to influence the affinity. Other elements, present in connective tissue in substantial quantity, could also represent a target of β_2M. In particular Ohashi et al. [48] have discovered that heparan sulfate binds β_2M with an affinity similar to that measured in the case of collagen. The role of glycosaminoglycans is probably of considerable importance in the mechanism of β_2M oligomerization. It has been shown that the addition of heparin can stabilize β_2M fibrils *in vitro* and protect them from depolymerization [49]. The presence of heparin can prime fibrillogenesis of wild type β_2M *in vitro* even at neutral pH [50]. Heparin interacts with collagen with a high affinity through a cluster of basic residues localized at the N-terminal end of the triple helix [51], and the interaction is sustained mainly by the electrostatic interaction between the negatively charged heparin and positively charged collagen. This type of interaction should theoretically generate a gradient of counter-ions (heparin) that is ideally represented in Fig. 2.2. We could speculate that the combination of β_2M concentration and heparin around the collagen surface can significantly increase the risk of oligomerization. Moreover, β_2M has a natural propensity to make oligomers. We have recently deduced from nuclear magnetic resonance studies [52] and subsequently substantiated by dynamic laser light scattering that β_2M forms oligomers of heterogeneous size at neutral pH and physiological temperature when the protein concentration reaches levels

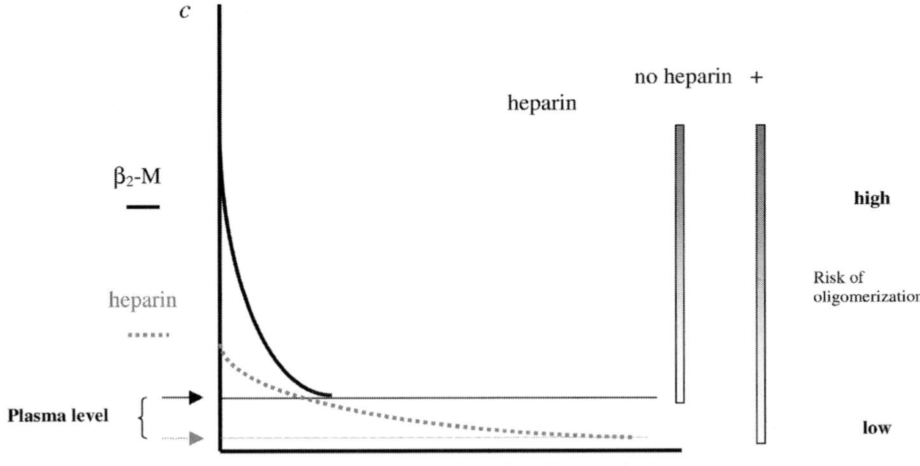

Fig. 2.2 Putative gradient of protein concentration surrounding the surface of a collagen fibril assuming a dissociation constant that would generate a higher protein concentration on the collagen surface than in plasma. The β_2M adhering to the collagen surface would be equivalent to β_2M free + β_2M bound. The total β_2M declines as a function of the distance from the surface (*r*) and the electrostatic potential attracting the two molecules. The theoretical curve is derived by assuming a prominent electrostatic interaction. Heparans that bind collagen mainly through electrostatic forces would also be distributed along a gradient of concentration that declines from the collagen surface.

usually detected in patients undergoing hemodialysis. The critical concentration for the formation of these oligomers is highly dependent on pH, and there is a linear correlation between the tendency to make oligomers and the reduction of pH in the range of 6–7 which can be found in various types of articular inflammations. Oligomerization also depends on the temperature and factors such as metal ions; in particular, the presence of copper ions *in vitro* facilitates β_2M self-assembly [53].

2.4
Other Amyloid Proteins Display Unique Tissue Specificity

The examples of Aβ_2M and AA are, in certain respects, unique because, for these proteins, we have data on putative molecular targets, the collagen types I and II, and heparan sulfate respectively. Unfortunately, analogous information for other types of amyloidosis is generally lacking. In the amyloidoses caused by transthyretin (TTR), apolipoprotein (Apo) AI and immunoglobulin light chains, in which the site of protein synthesis and the site of amyloid deposition is known, a biologic mechanism that favors the amyloid deposition in target organs should be ascertainable. In several of these cases, the tissue specificity is dramatic. Examples include mutant TTRs and nerve involvement [54], and others with heart or meningeal involvement. This also holds true in the case of immunoglobulin light chain and ApoAI variants responsible for hereditary amyloidosis.

ApoAI, the main protein constituent of HDL, causes both systemic and localized forms of amyloidosis in humans. The amyloidogenic properties of this protein and the molecular mechanisms involved in its *in vivo* deposition as fibrils have been investigated to a lesser extent than other forms of amyloid deposition. At present, most available information on this disease is based on the characterization of the biochemical composition of amyloid fibrils obtained *ex vivo*, and on clinical and experimental observations on lipid and lipoprotein metabolism associated with mutant ApoAI in patients. While awaiting further insights on amyloidogenic properties of recombinant ApoAI variants, and from cellular and animal models, there are some clinical and pathological features of ApoAI amyloidosis that exhibit interesting peculiarities and allow one to propose several mechanisms as to the pathogenesis and anatomic localization of this form of the disease.

Systemic AApoAI results from the occurrence of mutations in the apolipoprotein AI gene. Twelve variants have been identified to date. Amyloidogenic mutations are distributed along the first 90 N-terminal residues (eight variants) or clustered in the sequence 173–175 (four variants). Characterization of *ex vivo* fibrils has shown that they are formed almost exclusively by N-terminal fragments of the mature peptide (1–83 or 1–94). These fibril-forming peptides can, therefore, include in their sequence the mutation or a peptide representing the wild-type sequence. In this respect, the genetics of ApoAI amyloidosis is similar to hereditary Alzheimer's disease and cerebral amyloid angiopathy. These latter cases are associated with AβPP mutations occurring either inside or outside the amyloidogenic Aβ peptide. Moreover, as in Alzheimer's disease, ApoAI amyloidogenesis requires a proteolytic event. It is likely that the proteolysis at this site imparts a high susceptibility for amyloid deposition. In Fig. 2.3 the bidimensional electrophoresis of protein extracted from a heart biopsy of a patient affected by ApoAI amyloidosis is compared to his corresponding plasma proteins after immunostaining with an anti-ApoAI antibody.

This highly sensitive technique raises the possibility that the ApoAI fragments that are extremely abundant in the affected tissue are generated in the tissue-sites containing the amyloid deposits. However, since peptide fragments in general are cleared from plasma very rapidly, it is too early to exclude the possibility that such fragments may also be formed in plasma but are present at a concentration too low to be detected with existing immunological techniques. The putative proteases and the conformational events that make ApoAI susceptible to proteolysis are unknown. The possibility that a proteolytic fragment might also be formed physiologically during ApoAI metabolism is suggested by

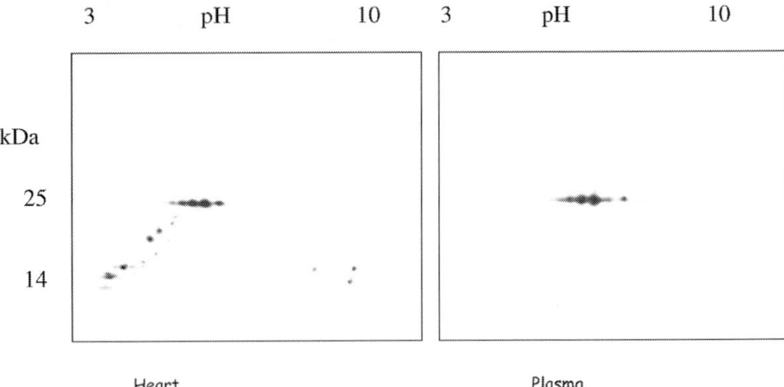

Fig. 2.3 Immunoblots using anti human ApoAI and bidimensional electrophoresis (combination of isoelectric focusing and SDS-PAGE) of proteins from plasma and heart biopsy from patient PIC affected by ApoAI amyloidosis caused by the mutation Leu174Ser.

the presence of amyloid fibrils derived from the N-terminal domain of wild-type ApoAI [55] in atherosclerotic plaques of elderly subjects. The pathogenetic mechanism involved in ApoAI amyloid deposition is, in part, suggested by knowledge of the metabolism of ApoAI and unique pathological features of its deposition.

Although virtually any organ can be the site of ApoAI deposits, this form of amyloid-related disease is most frequently observed in the kidneys, liver, adrenal glands and testis; the latter is an unusual site when compared to other types of amyloidoses. Insufficiency of hormonal synthesis in the testis and adrenal glands is part of the clinical history in a number of such patients and these features appear to develop as a consequence of amyloid deposition. A correlation exists between steroidogenic tissues and the formation of amyloid derived from ApoAI. This protein is functionally involved in the cholesterol trafficking of these tissues; however, the insufficiency of hormone synthesis seen in such tissues should not be attributed to a loss of ApoAI function caused by the presence of mutations, but rather to the tissue damage caused by the process of amyloid deposition. The transport of cholesterol via ApoAI toward and from the steroidogenic cell is very active. It has been demonstrated that ApoAI has a prominent role in removing excess cholesterol from the Sertoli cells through the ABC1 receptor [56], but ApoAI can also supply cholesterol to the cells through the SRBI receptor [57]. This metabolic trafficking is likely to sustain a local high concentration of the amyloidogenic precursor.

The reason why ApoAI is defined as an exchangeable lipoprotein revolves about its ability to influence either cholesterol removal or donation. This process of lipid exchange is associated with conformational modification of ApoAI, and it has been predicted that, in the lipid-free state, ApoAI should loose part of its secondary and tertiary structure. The N-terminus of the protein appears particularly disordered and flexible. It has been necessary to remove the first 40 residues to get successful crystallization [58] and we have shown that the polypeptide 1–93 displays the behavior of the natively unfolded protein [59]. The mutations described to date, apart from Gly26Arg, are predicted to all have the capacity to reduce its α-helical propensity. This could have the effect of shifting the equilibrium of folded–unfolded states toward the partially unfolded state. It is in just those tissues where the very active metabolic function of ApoAI is required that these types of conformational modification in the mutant protein can be easily cleaved, or might display a high propensity to self-aggregate.

The localization of ApoAI fibrils to the testis offers another unique aspect to our discussion. Testicular localization of ApoAI amyloid affects testosterone synthesis and testosterone is known to be a negative modulator of ApoAI levels [60]. In consequence, one can imagine that the lack of testosterone will increase the production of the mutant amyloidogenic precursor with possible negative consequences on the amyloidogenic process.

In the case of immunoglobulin light chains, the first molecular basis of organ selectivity was probably put forward by Isobe and Osserman [61] who proposed a possible antibody recognition mechanism by certain immunoglobulin light chains toward unknown molecular elements of the target tissue. However, such a specific antibody activity has been recognized in only a few cases. Of remarkable interest is the study reported by Harris et al. [62], who described the affinity of the light chain Mcg for nascent collagen fibrils. More recently, a possible interaction between the amyloidogenic immunoglobulin and the cubilin/megalin receptors has been described. This interaction, observed by Herrera's group, does not imply a classical antibody–antigen interaction [63]. These authors propose that the formation of amyloid in the kidney might be guided by the receptor that allows the internalization of light chains into mesangial cell. The target cells, in this case, apparently process the potentially amyloidogenic proteins and in so doing acquire the conformation suitable for local aggregation and fibril formation.

In the case of transthyretin and ApoAI, the possible role of a specific receptor should be considered, particularly because of the transport functions that these proteins perform. A high-affinity receptor for TTR was discovered in the liver where it most likely is involved in a catabolic function [64]. However, the clinical/pathologic evidence, i.e. a lack of amyloid deposits in the liver of patients with TTR variants, indicates the need of further studies to clarify the role of this receptor.

2.5 Local Production of Amyloidogenic Protein can Dictate the Occurrence of Localized Amyloidosis

Among the amyloidoses are many examples in which the amyloid fibrils are deposited in the same tissue where the amyloidogenic proteins are produced. This is likely the case in several neurological disorders such as Alzheimer's disease, the spongioform encephalopathies and other neurodegenerative syndromes. Some extracerebral amyloidoses are also localized to the sites of the amyloidogenic protein production. A particularly close association exists between the site of synthesis and site of deposition of certain hormones such as the calcitonin gene-related peptides (CGRP) (pro)calcitonin and amylin. Calcitonin can cause localized amyloid deposition in the thyroid in calcitonin secreting medullary carcinoma. Amylin is co-secreted with insulin from the β cells of pancreatic islet and the related amyloid fibrils, when present, are deposited in pancreatic islets adjacent to the islet cell. The *in vitro* mechanism of structural transition has been well investigated with both of these peptides, and, in both cases, the role of oligomeric nuclei in priming the fibril formation is well established. In the case of calcitonin, the time and concentration dependence of fibrillization can be described mathematically by a single exponential fitting and the minimal concentration reported in the *in vitro* experiments is 1 mg/ml [65]. If we consider that calcitonin in plasma does not exceed 0.1 ng/ml, we can easily imagine why the critical concentration for the fibril formation could be reached only in very close proximity to the site of calcitonin production where the concentration should be very high. A gradient of amyloidogenic protein concentration can be envisaged also, in this case, from the cell surface toward the extracellular space. It would be quite illuminating and interesting to discover the calcitonin concentration in closest proximity to the cells of a medullary carcinoma. Another interesting model is represented by localized AL amyloidosis. Local production of amyloidogenic light chains and their deposition as amyloid fibrils can occur along the respiratory tract, and in the bladder, urethra, head, neck and skin. The localized plasma cell clone produces amyloidogenic light chains, which deposits in the immediate proximity of the synthesizing cells. This may be caused by two, possibly concurring factors, the extreme instability of the light chains and/or their interactions with tissue components/receptors. One challenging hypothesis is that these components/receptors might be involved in both the homing of plasma cells and in the light chain deposition. In systemic AL one would expect that the bone marrow, the site of amyloidogenic light chain production, would be involved by massive amyloid deposition. Actually, this is rarely observed. It is possible that the sinusoidal circulating system of the bone marrow allows the rapid clearance of the newly synthesized amyloidogenic light chains and that the bone marrow environment does not express relevant receptors.

In the category of localized amyloidosis, one must certainly include Alzheimer's disease, in which clinically significant amyloid deposits are limited to the brain. In this case too, the aggregation kinetics are time and concentration dependent, and the primary structure of the amyloidogenic peptide can have enormous influence on the determination of the critical concentration required for the fibril assembly. There is clear evidence that the Aβ peptide 1–42, in comparison to 1–40, displays more rapid kinetics of protofibril formation [66]. The exact critical concentration required *in vivo* for the formation of fibrils is unknown and there is a remarkable discrepancy between the peptide concentration used in *in vitro* experiments where Aβ1–40 or Aβ1–42 are used at a concentration above 10 μM, whereas the concentration of the Aβ peptides in the cerebrospinal fluid is in the nanomolar range with an approximate ratio Aβ1–42/Aβ1–40 of 1/5. The concentration of the Aβ peptide in the brain at the site of fibril formation is unknown, but it may be speculated that it is significantly higher than in the CSF. The formation of Aβ oligomeric nuclei may be primed by peptides that possess the highest propensity to aggregate and, possibly, the lowest critical concentration threshold for oligomerization. Proteomic studies of post-mortem brain specimens suggest that, *in vivo*, the earliest fibrillar aggregates are constituted by the full length Aβ1–42 and Aβ1–42 truncated of its first two or three residues at N-terminal. The information about the Aβ composition of the earliest amyloid plaques was inferred from brain specimens of subjects who died in the absence of clinical signs of Alzheimer's disease, but who possessed, histologically, classical amyloid plaques [67]. The Aβ1–42 nuclei may function as a template and facilitate fibril growth by the assembly of both Aβ1–40 and Aβ1–42 monomers. Afterwards, the assembly of Aβ peptides, onto protofibril nuclei, configures a first-order reaction linearly dependent on peptide concentration, and the kinetics is similar for Aβ1–40 and Aβ1–42 [68].

2.6 Conclusions

The basis of the various clinical syndromes and the heterogeneous tissue targeting observed in amyloid diseases is presently limited by a paucity of robust *in vivo* molecular data. However, the ability to explain such clinical heterogeneity is actually a key issue for a greater comprehension of the mechanisms involved in human amyloidogenesis. In our view, a continuous effort to critically combine the growing amount of *in vitro* data with lessons obtained from clinical and detailed histopathological studies will provide further insights into the biologic basis of these diseases.

In this chapter, we have discussed the clinical and pathological features of some systemic and localized forms of amyloidosis that suggest the existence of unique *in vivo* factors underlying the tissue specificity observed in these diseases, and related the role of these putative factors to evidence obtained *in vitro*.

In vitro models have established that the process of fibrillogenesis requires a critical concentration of the precursor protein and/or the generation of protein conformers that are able to polymerize into amyloid fibrils when reaching a threshold concentration, the level of which still remains largely unknown for each of the amyloidogenic proteins.

We propose that the anatomical sites affected in these diseases may actively promote the process of fibril deposition by generating a high local concentration of amyloidogenic conformers through complex, but specific, molecular interactions mediated by local factors, such as matrix components and cellular receptors. Ionized atoms and molecules in these tissues can also promote conformational modifications, whereas other local factors may induce covalent modifications, such as proteolytic cleavages. Taken together, these factors/mechanisms are expected to contribute to the formation of highly amyloidogenic species and fibril seeds that are able to accelerate the local kinetics of amyloid deposition.

Although this topic is still based on great uncertainty, the effort to combine *in vitro* and *in vivo* evidence will eventually contribute to the discovery of more detailed links between the structural basis of protein fibrillogenesis and the role of the biologic environment in catalyzing a hazardous and complex biochemical/biophysical reaction.

Acknowledgments

This work was supported by Fondazione Cariplo and Ministero della Salute (grant 8920301), by MIUR (FIRB and COFIN), by the Canadian Institutes for Health Research (MOP-3153), by a Detweiler Travelling Fellowship from the Royal College of Physicians and Surgeons of Canada (to R. K.), and by a Haust Travel award from the Department of Pathology and Molecular Medicine, Queen's University. We thank Professor Patrizia Mangione for providing the proteomic analysis of amyloidogenic ApoAI, and Professors Amos Maritan, Flavio Seno and Antonio Trovato of the University of Padova for the helpful discussion.

References

1 Benditt, E. P., Eriksen, N., Hermodson, M. A. and Ericsson, L. H. The major proteins of human and monkey amyloid substance: common properties including unusual N-terminal amino acid sequences. *FEBS Lett* **1971**, *19*, 169–173.

2 Isersky, C., Page, D. L., Cuatrecasas, P., DeLillis, R. A. and Glenner, G. G. Murine amyloidosis: immunologic characterization of amyloid fibril protein. *J Immunol* **1971**, *107*, 1690–1698.

3 Husebekk, A., Skogen, B., Husby, G. and Marhaug, G. Transformation of amyloid precursor SAA to protein AA and incorporation in amyloid fibrils *in vivo*. *Scand J Immunol* **1985**, *21*, 283–287.

4 Tape, C., Tan, R., Nesheim, M. and Kisilevsky, R. Direct evidence for circulating apoSAA as the precursor of tissue AA amyloid deposits. *Scand J Immunol* **1988**, *28*, 317–324.

5 Hawkins, P. N. Serum amyloid P component scintigraphy for diagnosis and monitoring amyloidosis. *Curr Opin Nephrol Hypertens* **2002**, *11*, 649–655.

6 Kisilevsky, R. and Boudreau, L. The kinetics of amyloid deposition: I. The effect of amyloid enhancing factor and splenectomy. *Lab Invest* **1983**, *48*, 53–59.

7 Axelrad, M. A., Kisilevsky, R. and Beswetherick, S. Acceleration of amyloidosis by syngeneic spleen cells from normal donors. *Am J Pathol* **1975**, *78*, 277–284.

8 Axelrad, M. A., Kisilevsky, R., Willmer, J., Chen, S. J. and Skinner, M. Further characterization of amyloid enhancing factor. *Lab Invest* **1982**, *47*, 139–146.

9 Benditt, E. P. and Eriksen, N. Amyloid protein SAA is associated with high density lipoprotein from human serum. *Proc Natl Acad Sci USA* **1977**, *74*, 4025–4028.

10 Kisilevsky, R. and Subrahmanyan, L. Serum amyloid A changes high density lipoprotein's cellular affinity: a clue to serum amyloid A's principal function. *Lab Invest* **1992**, *66*, 778–785.

11 Su, S.B., Gong, W.H., Gao, J.L., Shen, W.P., Murphy, P.M., Oppenheim, J.J. and Wang, J.M. A seven-transmembrane, G protein-coupled receptor, FPRL1, mediates the chemotactic activity of serum amyloid A for human phagocytic cells. *J Exp Med* **1999**, *189*, 395–402.

12 Artl, A., Marsche, G., Lestavel, S., Sattler, W. and Malle, E. Role of serum amyloid A during metabolism of acute-phase HDL by macrophages. *Arterioscler Thromb Vasc Biol* **2000**, *20*, 763–772.

13 Le, Y.Y., Gong, W.H., Li, B.Q., Dunlop, N.M., Shen, W.P., Su, S.B., Ye, R.D. and Wang, J.M. Utilization of two seven-transmembrane, G protein-coupled receptors, formyl peptide receptor-like 1 and formyl peptide receptor, by the synthetic hexapeptide WKYMVm for human phagocyte activation. *J Immunol* **2000**, *163*, 6777–6784.

14 Liang, T.S., Wang, J.M., Murphy, P.M. and Gao, J.L. Serum amyloid A is a chemotactic agonist at FPR2, a low-affinity *N*-formylpeptide receptor on mouse neutrophils. *Biochem Biophys Res Commun* **2000**, *270*, 331–335.

15 Tam, S.P., Flexman, A., Hulme, J. and Kisilevsky, R. Promoting export of macrophage cholesterol: the physiological role of a major acute-phase protein, serum amyloid A 2.1. *J Lipid Res* **2002**, *43*, 1410–1420.

16 Kisilevsky, R. and Tam, S.P. Macrophage cholesterol efflux and the active domains of serum amyloid A2.1. *J Lipid Res* **2003**, *44*, 2257–2269.

17 Kisilevsky, R. and Fraser, P.E. Aβ amyloidogenesis: unique or variation on a systemic theme? *Crit Rev Biochem Mol Biol* **1997**, *32*, 361–404.

18 de Koning, E.J.P., Hoppener, J.W.M., Verbeek, J.S., Oosterwijk, C., Vanhulst, K.L., Baker, C.A., Lips, C.J.M., Morris, J.F. and Clark, A. Human islet amyloid polypeptide accumulates at similar sites in islets of transgenic mice and humans. *Diabetes* **1994**, *43*, 640–644.

19 Delgado, W.A. and Arana Chavez, V.E. Amyloid deposits in labial salivary glands identified by electron microscopy. *J Oral Pathol Med* **1997**, *26*, 51–52.

20 Gallo, C.D., Mauri, J.M., Poveda, R., Castelao, A.M., Gonzalez, M.T., Carreras, M. and Alsina, J. Renal amyloidosis associated with antiglomerular basement membrane disease. *Nephron* **1990**, *56*, 335–336.

21 Horiguchi, Y., Fine, J.D., Leigh, I.M., Yoshiki, T., Ueda, M. and Imamura, S. Lamina densa malformation involved in histogenesis of primary localized cutaneous amyloidosis. *J Invest Dermatol* **1992**, *99*, 12–18.

22 Kawai, M., Kalaria, R.N., Harik, S.I. and Perry, G. The relationship of amyloid plaques to cerebral capillaries in Alzheimer's disease. *Am J Pathol* **1990**, *137*, 1435–1446.

23 Lyon, A.W., Narindrasorasak, S., Young, I.D., Anastassiades, T., Couchman, J.R., McCarthy, K. and Kisilevsky, R. Co-deposition of basement membrane components during the induction of murine splenic AA amyloid. *Lab Invest* **1991**, *64*, 785–790.

24 Natte, R., Yamaguchi, H., Maat-Schieman, M.L.C., Prins, F.A., Neeskens, P., Roos, R.A.C. and van Duinen, S.G. Ultrastructural evidence of early non-fibrillar Aβ 42 in the capillary basement membrane of patients with hereditary cerebral hemorrhage with amyloidosis, Dutch type. *Acta Neuropathol* **1999**, *98*, 577–582.

25 Perlmutter, L.S. and Chui, H.C. Microangiopathy, the vascular basement membrane and Alzheimer's disease – a review. *Brain Res Bull* **1990**, *24*, 677–686.

26 Ancsin, J.B. and Kisilevsky, R. The heparin/heparan sulfate-binding site on apo-serum amyloid A: implications for the therapeutic intervention of amyloidosis. *J Biol Chem* **1999**, *274*, 7172–7181.

27 Ancsin, J.B. and Kisilevsky, R. Laminin interactions with the apoproteins of acute-phase HDL: preliminary mapping of the laminin binding site on serum amyloid A. *Amyloid* **1999**, *6*, 37–47.

28 Ancsin, J.B. and Kisilevsky, R. Characterization of high affinity binding between laminin and the acute-phase protein, serum amyloid A. *J Biol Chem* **1997**, *272*, 406–413.

29 Ailles, L., Kisilevsky, R. and Young, I.D. Induction of perlecan gene expression

precedes amyloid formation during experimental murine AA amyloidogenesis. *Lab Invest* **1993**, *69*, 443–448.

30 Woodrow, S. I., Stewart, R. J., Kisilevsky, R., Gore, J. and Young, I. D. Experimental AA amyloidogenesis is associated with differential expression of extracellular matrix genes. *Amyloid* **1999**, *6*, 22–30.

31 Snow, A. D. and Kisilevsky, R. A close ultrastructural relationship between sulphated proteoglycans and AA amyloid fibrils. *Lab Invest* **1988**, *57*, 687–698.

32 Snow, A. D., Bramson, R., Mar, H., Wight, T. N. and Kisilevsky, R. A temporal and ultrastructural relationship between heparan sulfate proteoglycans and AA amyloid in experimental amyloidosis. *J Histochem Cytochem* **1991**, *39*, 1321–1330.

33 McCubbin, W. D., Kay, C. M., Narindrasorasak, S. and Kisilevsky, R. Circular dichroism and fluorescence studies on two murine serum amyloid A proteins. *Biochem J* **1988**, *256*, 775–783.

34 Kisilevsky, R., Lemieux, L. J., Fraser, P. E., Kong, X. Q., Hultin, P. G. and Szarek, W. A. Arresting amyloidosis *in vivo* using small-molecule anionic sulphonates or sulphates: implications for Alzheimer's disease. *Nat Med* **1995**, *1*, 143–148.

35 Kisilevsky, R., Szarek, W. A., Ancsin, J. B., Elimova, E., Marone, S., Bhat, S. and Berkin, A. Inhibition of amyloid A amyloidogenesis *in vivo* and in tissue culture by 4-deoxy analogues of peracetylated 2-acetamido-2-deoxy-α and -β-D-glucose: implications for the treatment of various amyloidoses. *Am J Pathol* **2004**, *164*, 2127–2137.

36 Capeillere-Blandin, C., Delaveau, T. and Descamps-Latscha, B. Structural modifications of human beta 2 microglobulin treated with oxygen-derived radicals. *Biochem J* **1991**, *277*, 175–182.

37 Miyata, T., Inagi, R., Iida, Y., Sato, M., Yamada, N., Oda, O., Maeda, K. and Seo, H. Involvement of beta 2-microglobulin modified with advanced glycation end products in the pathogenesis of hemodialysis-associated amyloidosis. Induction of human monocyte chemotaxis and macrophage secretion of tumor necrosis factor-alpha and interleukin-1. *J Clin Invest* **1994**, *93*, 521–528.

38 Hashimoto, N., Naiki, H. and Gejyo, F. Modification of beta 2-microglobulin with D-glucose or 3-deoxyglucosone inhibits A beta 2M amyloid fibril extension *in vitro*. *Amyloid* **1999**, *4*, 256–264.

39 Odani, H., Oyama, R., Titani, K., Ogawa, H. and Saito, A. Purification and complete amino acid sequence of novel beta 2-microglobulin. *Biochem Biophys Res Commun* **1990**, *168*, 1223–1229.

40 Kad, N. M., Thomson, N. H., Smith, D. P., Smith, D. A. and Radford, S. E. Beta$_2$-microglobulin and its deamidated variant, N17D form amyloid fibrils with a range of morphologies *in vitro*. *J Mol Biol* **2001**, *313*, 559–571.

41 Linke, R. P., Hampl, H., Lobeck, H., Ritz, E., Bommer, J., Waldherr, R., Eulitz, M. Lysine-specific cleavage of beta 2-microglobulin in amyloid deposits associated with hemodialysis. *Kidney Int* **1989**, *36*, 675–681.

42 Bellotti, V., Stoppini, M., Mangione, P., Sunde, M., Robinson, C., Asti, L., Brancaccio, D. and Ferri, G. Beta2-microglobulin can be refolded into a native state from *ex vivo* amyloid fibrils. *Eur J Biochem* **1998**, *258*, 61–67.

43 Jain, V. K., Cestero, R. V. and Baum, J. Carpal tunnel syndrome in patients undergoing maintenance hemodialysis. *J Am Med Ass* **1979**, *242*, 2868–2869.

44 Gejyo, F., Yamada, T., Odani, S., Nakagawa, Y., Arakawa, M., Kunitomo, T., Kataoka, H., Suzuki, M., Hirasawa, Y., Shirahama, T. A new form of amyloid protein associated with chronic hemodialysis was identified as beta 2-microglobulin. *Biochem Biophys Res Commun* **1985**, *129*, 701–706.

45 Homma, N., Gejyo, F., Isemura, M. and Arakawa, M. Collagen-binding affinity of beta-2-microglobulin, a preprotein of hemodialysis-associated amyloidosis. *Nephron* **1989**, *53*, 37–40.

46 Podlisny, M. B., Walsh, D. M., Amarante, P., Ostaszewski, B. L., Stimson, E. R., Maggio, J. E., Teplow, D. B. and Selkoe, D. J. Oligomerization of endogenous and synthetic amyloid beta-protein at nanomolar levels in cell culture and stabiliza-

tion of monomer by Congo red. *Biochemistry* **1998**, *37*, 3602–3611.

47 Merlini, G. and Bellotti, V. Molecular mechanisms of amyloidosis. *N Engl J Med* **2003**, *349*, 583–596.

48 Ohashi, K., Kisilevsky, R. and Yanagishita, M. Affinity binding of glycosaminoglycans with beta$_2$-microglobulin. *Nephron* **2002**, *90*, 158–168.

49 Yamamoto, S., Yamaguchi, I., Hasegawa, K., Tsutsumi, S., Goto, Y., Gejyo F. and Naiki H. Glycosaminoglycans enhance the trifluoroethanol-induced extension of beta 2-microglobulin-related amyloid fibrils at a neutral pH. *J Am Soc Nephrol* **2004**, *1*, 126–133.

50 Yamaguchi, I., Suda, H., Tsuzuike, N., Seto, K., Seki, M., Yamaguchi, Y., Hasegawa, K., Takahashi, N., Yamamoto, S., Gejyo, F. and Naiki H. Glycosaminoglycan and proteoglycan inhibit the depolymerization of beta2-microglobulin amyloid fibrils *in vitro*. *Kidney Int* **2003**, *64*, 1080–1088.

51 Sweeney, S. M., Guy, C. A., Fields, G. B. and San Antonio, J. D. Defining the domains of type I collagen involved in heparin-binding and endothelial tube formation. *Proc Natl Acad Sci USA* **1998**, *95*, 7275–7280.

52 Corazza, A., Pettirossi, F., Viglino, P., Verdone, G., Garcia, J., Dumy, P., Giorgetti, S., Mangione, P., Raimondi, S., Stoppini, M., Bellotti, V. and Esposito, G. Properties of some variants of human beta2-microglobulin and amyloidogenesis. *J Biol Chem* **2004**, *279*, 9176–9189.

53 Eakin, C. M., Attenello, F. J., Morgan, C. J. and Miranker, A. D. Oligomeric assembly of native-like precursors precedes amyloid formation by beta-2 microglobulin. *Biochemistry* **2004**, *43*, 7808–7815.

54 Saraiva, M. J. Transthyretin mutations in hyperthyroxinemia and amyloid diseases. *Hum Mutat* **1995**, *5*, 191–196.

55 Westermark, P., Mucchiano, G., Marthin, T., Johnson, K. H. and Sletten, K. Apolipoprotein A1-derived amyloid in human aortic atherosclerotic plaques. *Am J Pathol* **1995**, *147*, 1186–1192.

56 Selva, D. M., Hirsch-Reinshagen, V., Burgess, B., Zhou, S., Chan, J., McIsaac, S., Hayden, M. R., Hammond, G. L., Vogl, A. W. and Wellington, C. L. The ATP-binding cassette transporter 1 mediates lipid efflux from Sertoli cells and influences male fertility. *J Lipid Res* **2004**, *45*, 1040–1050.

57 Trigatti, B. L., Krieger, M. and Rigotti, A. Influence of the HDL receptor SR-BI on lipoprotein metabolism and atherosclerosis. *Arterioscler Thromb Vasc Biol* **2003**, *23*, 1732–1738.

58 Borhani, D. W., Rogers, D. P., Engler, J. A. and Brouillette, C. G. Crystal structure of truncated human apolipoprotein A-I suggests a lipid-bound conformation. *Proc Natl Acad Sci USA* **1997**, *94*, 1291–1296.

59 Andreola, A., Bellotti, V., Giorgetti, S., Mangione, P., Obici, L., Stoppini, M., Torres, J., Monzani, E., Merlini, G. and Sunde, M. Conformational switching and fibrillogenesis in the amyloidogenic fragment of apolipoprotein a-I. *J Biol Chem* **2003**, *278*, 2444–2451.

60 Tall, A. R. Plasma high density lipoproteins. Metabolism and relationship to atherogenesis. *J Clin Invest* **1990**, 379–384.

61 Isobe, T. and Osserman, E. F. Patterns of amyloidosis and their association with plasma-cell dyscrasia, monoclonal immunoglobulins and Bence-Jones proteins. *N Engl J Med* **1974**, *290*, 473–477.

62 Harris, D. L., King, E., Ramsland, P. A. and Edmundson, A. B. Binding of nascent collagen by amyloidogenic light chains and amyloid fibrillogenesis in monolayers of human fibrocytes. *J Mol Recognit* **2000**, *4*, 198–212.

63 Teng, J., Russell, W. J., Gu, X., Cardelli, J., Jones, M. L. and Herrera, G. A. Different types of glomerulopathic light chains interact with mesangial cells using a common receptor but exhibit different intracellular trafficking patterns. *Lab Invest* **2004**, *84*, 440–451.

64 Sousa, M. M. and Saraiva, M. J. Internalization of transthyretin. Evidence of a novel yet unidentified receptor-associated protein (RAP)-sensitive receptor. *J Biol Chem* **2001**, *276*, 14420–14425.

65 Arvinte, T., Cudd, A. and Drake, A. F. The structure and mechanism of forma-

tion of human calcitonin fibrils. *J Biol Chem* **1993**, *268*, 6415–6422.

66 Jarrett, J.T., Berger, E.P. and Lansbury, P.T., Jr. The carboxy terminus of the beta amyloid protein is critical for the seeding of amyloid formation: implications for the pathogenesis of Alzheimer's disease. *Biochemistry* **1993**, *32*, 4693–4697.

67 Sergeant, N., Bombois, S., Ghestem, A., Drobecq, H., Kostanjevecki, V., Missiaen, C., Wattez, A., David, J.P., Vanmechelen, E., Sergheraert, C. and Delacourte, A. Truncated beta-amyloid peptide species in pre-clinical Alzheimer's disease as new targets for the vaccination approach. *J Neurochem* **2003**, *85*, 1581–1591.

68 Naiki, H., Hasegawa, K., Yamaguchi, I., Nakamura, H., Gejyo, F. and Nakakuki, K. Apolipoprotein E and antioxidants have different mechanisms of inhibiting Alzheimer's beta-amyloid fibril formation *in vitro*. *Biochemistry* **1998**, *37*, 17882–17889.

Part II
Protein Structure and the Beta Pleated Sheet Conformation

Amyloid Proteins. The Beta Sheet Conformation and Disease. J. D. Sipe

ISBN: 3-527-31072-X

3
The β-pleated Sheet Conformation and Protein Folding: A Brief History

Jean D. Sipe

3.1
Introduction

The amyloid proteins (see Table 1.1 in Chapter 1) can undergo extensive refolding, usually later in life, in one or more tissues and organs of the body, such that the polypeptide backbone assumes a greater degree of β-pleated sheet conformation. If unchecked, the increasingly abundant β-pleated sheet polypeptide strands subsequently self-assemble into amyloid fibrils, 8–10 nm in diameter with the polypeptide backbone perpendicular to the fibril axis (reviewed in [1, 2]). Studies of the amyloid proteins in relationship to the amyloidoses, a group of biochemically and clinically heterogeneous disorders, have led to the identification of previously unknown peptides and proteins. These include the cerebral amyloid β protein, Aβ, and its precursor protein AβPP; the amyloid A protein and its precursor, the cytokine-regulated serum amyloid A (SAA); the previously unrecognized diabetes associated islet amyloid polypeptide (IAPP); prion proteins that cause scrapie and other neurodegenerative diseases (PrP); and more than 80 transthyretin (TTR) variants, nearly all of which are associated with ATTR.

The naturally occurring amyloid proteins mainly have been identified in humans; most, but not all, are associated with a clinical level of disease [1, 2]. However, amyloidosis is widespread throughout the animal kingdom; one of the most well-known forms involves the prion protein that is at the root of the transmissible spongiform encephalopathies, which are known in humans as Creutzfeldt–Jacob disease, in cows as bovine spongiform encephalopathies and in sheep as scrapie. Reactive or systemic secondary amyloidosis involving the amyloid A protein has been reported in a large number of vertebrate species [3].

In addition to the amyloidoses, there are other diseases that involve refolding of the polypeptide backbones of proteins, such as α-synuclein in Parkinson's disease and huntingtin in Huntington's disease, in which amorphous aggregates or inclusion bodies are formed rather than the highly ordered amyloid fibrils. In the body, these non-fibrillar aggregates are not predominantly in the β-pleated sheet conformation, although the isolated or corresponding synthetic

Amyloid Proteins. The Beta Sheet Conformation and Disease. J. D. Sipe

ISBN: 3-527-31072-X

proteins often have been observed to form β-sheet amyloid-like proteins *in vitro*. A particularly unique type of linear aggregation via a β-pleated sheet linkage is associated with the serpinopathies. Mutations in the serine protease inhibitor (serpin) α_1-antitrypsin in liver are associated with hepatic cirrhosis and emphysema, and mutation in neuroserpin in brain results in a familial dementia [4, 5]. These non-fibrillar aggregates and linear polymers are distinct from the highly ordered amyloid fibrils, and the type of conformational change that leads to these aggregates will not be considered in this volume, except to consider differences from the amyloid proteins.

3.2
The β-pleated Sheet Structure of the Amyloid Fibril

Amyloid fibril structure has been studied since at least 1639, when the gross morphology of tissues was described as lardaceous liver, waxy liver, and spongy and "white stone" containing spleens (see Chapter 1; reviewed in [6, 7]). It was not until 1854 that Rudolph Virchow, using the best medical knowledge and scientific technology available at the time, introduced and popularized the term amyloid to denote the positive iodine staining reaction of a subset of macroscopically abnormal tissues. Subsequently, light microscopic studies with polarizing optics demonstrated the inherent birefringence of amyloid deposits. This finding enabled investigators such as Bennhold, Divry, Missmahl and Romhanyi to utilize the intense increase in birefringence after staining with Congo red dye in the diagnosis and clinical classification of the amyloid diseases (reviewed in [1, 2, 6, 7]). More recently, Congo red fluorescence has been reported to be preferable to Congo red birefringence as a tool for early, sensitive and specific diagnosis of amyloidosis [8].

In 1959, electron microscopic examination by Cohen et al. of amyloidotic tissues revealed the presence of fibrils, indeterminate in length and, invariably, 8–10 nm in width [9]. The two properties of Congophilia and fibrillar morphology served as enabling criteria with which the biochemical era of amyloid investigations was

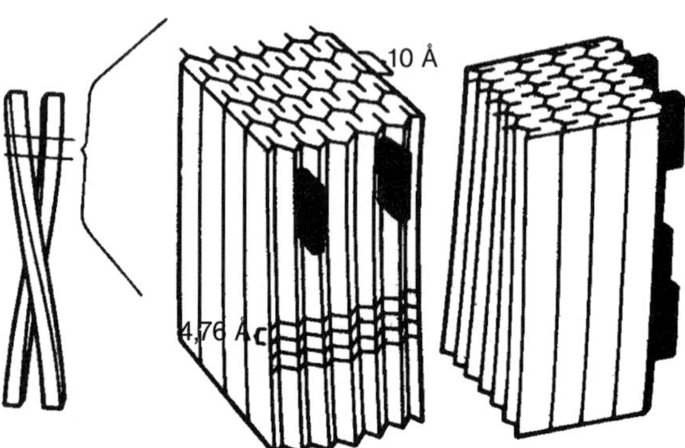

Fig. 3.1 Cooper's model [23] of the twisted β-pleated sheet of paired filaments of the amyloid fibril as adapted by Glenner [10]. The sites of binding of the planar dye Congo red are indicated by the presence of stippled blocks. (Figure 1 from [10], reprinted with permission.)

launched [10]. X-ray diffraction analysis (reviewed in [11] by Glenner et al.) has revealed that, *ex vivo* and *in vitro*, amyloid fibrils share the common feature of being ordered in the β-pleated sheet conformation with the direction of the polypeptide backbone perpendicular to the fibril axis (cross-β structure) (Fig. 3.1). Atomic-resolution structures are not available for full-length amyloid fibrils because of the lack of diffraction quality three-dimensional crystals, due at least in part to the insolubility of the intact amyloid fibrils. Recently, the three-dimensional structure (Fig. 3.2) (see also Chapter 7) of amyloid-like fibrils formed *in vitro* from the synthetic peptide WTIAALLSPYS corresponding to residues 105–115 of TTR fibrils was determined using magic angle spinning (MAS) nuclear magnetic resonance

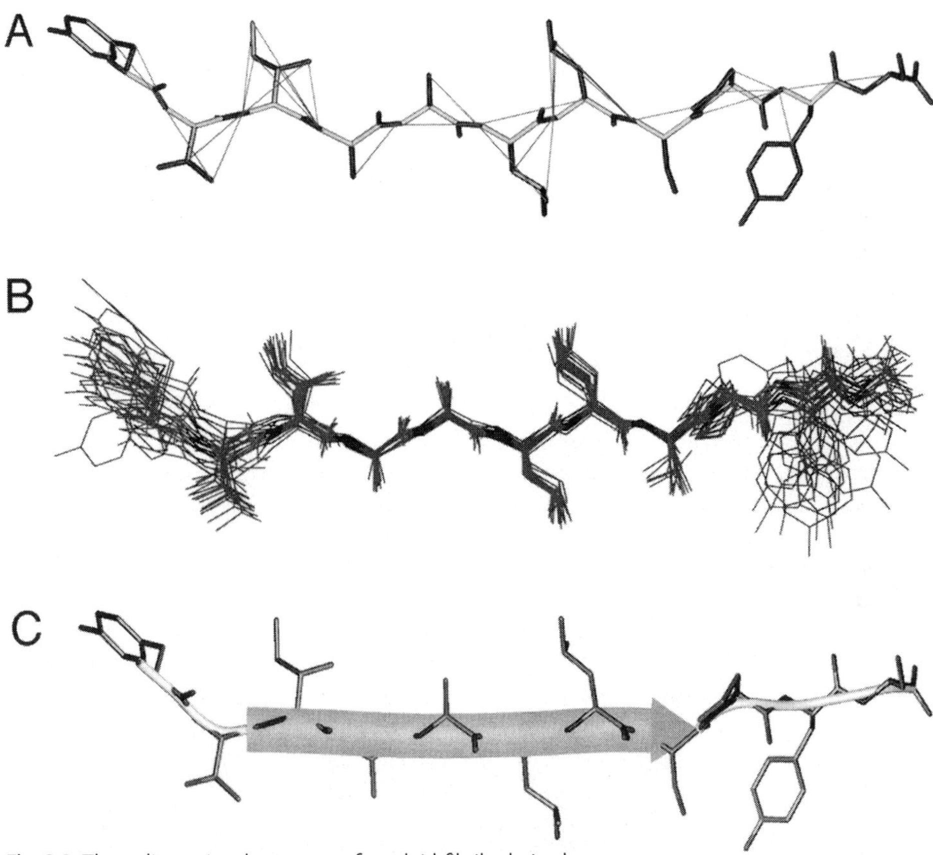

Fig. 3.2 Three-dimensional structure of amyloid fibrils derived from a synthetic peptide corresponding to residues 105–115 of TTR as determined by MAS NMR. (A) Structure of TTR(105–115). (B) Ensemble of 20 superimposed NMR structures of TTR(105–115). (C) Ribbon representation of the structure of TTR(105–115) in the amyloid fibril with side-chains shown as stick models. (Fig. 3 from [12], reprinted with permission. Copyright 2004 National Academy of Sciences, USA)

(NMR) [12]. The synthetic fibrils were comprised of fully extended β-pleated sheets, highly ordered, crystalline in regularity, with four sheets per protofilament. TTR (105–115) adopts an extended β-strand conformation in the amyloid fibrils, such that both the main- and side-chain torsion angles are close to their optimal values. It was reported that the structure of the peptide in the fibrillar form has a degree of long-range order that is generally associated only with crystalline materials and Jaroniec et al. suggested that the atomic level findings provide an explanation of the unusual stability of this form of polypeptide assembly. Others have described amyloid fibrils as one-dimensional crystals [13].

3.3
Polypeptide Backbone Folding: Steric Considerations

Proteins are linear polymers of L-α-amino acids linked by amide bonds, –NHCO–, which are called peptide bonds. Each of the products of the estimated 31,780 protein coding genes in the human genome [14] has the same polypeptide backbone, spanning from the usually unbonded primary amine group (N-terminus) to the unbonded carboxylic acid group (C-terminus) (Fig. 3.3). It is worth noting that a very small fraction, of the order of 30, of the protein-coding gene products has been found to form amyloid fibrils in nature.

More than 50 years of scientific study using model building, microscopy, fluorescence, spectroscopy, X-ray diffraction and NMR indicate that the α-helix and the β-pleated sheet conformations (Fig. 3.4A and B) are two major types of periodic folding of the polypeptide backbone that are compatible with the steric constraints posed by the planar nature of the peptide bond backbone of the polypeptide chain and the asymmetric nature of the tetrahedral carbon atom. If the polypeptide backbone of a protein exhibits no periodicity, the structure is known as a random coil or random chain (Fig. 3.4C) [15].

Pauling et al. constructed models of polypeptide conformation that would permit hydrogen bonding of each of the peptide bond carbonyl and imino groups, except for proline, to participate in hydrogen bonding [16, 17]. They incorporated into the models interatomic distances and bond angles corresponding to those measured for amino acids and peptides. Taking into consideration constraints posed by the tetrahedral configuration of the carbon atom and the asymmetry of the α-carbon atom for all amino acids except for glycine, they predicted two folding options for the polypeptide backbone of proteins: the α-helix and the β-pleated sheet conformation (Fig. 3.4A and B). The pitch of the α-helix is such that there is a rise of 3.6 amino acid residues for each turn of the helix. The number of peptide bond N–H and C=O hydrogen bonds is maximal with bonding to the fourth amino acid residue down the chain. There is intrastrand hydrogen bonding that is parallel to the axis of the helix; the distance between adjacent amino acids along the axis of the polypeptide backbone is 1.5 Å. The net effect of the α-helix is to superimpose a higher degree of proximity of the amino acid side-chains (R-groups, Fig. 3.3).

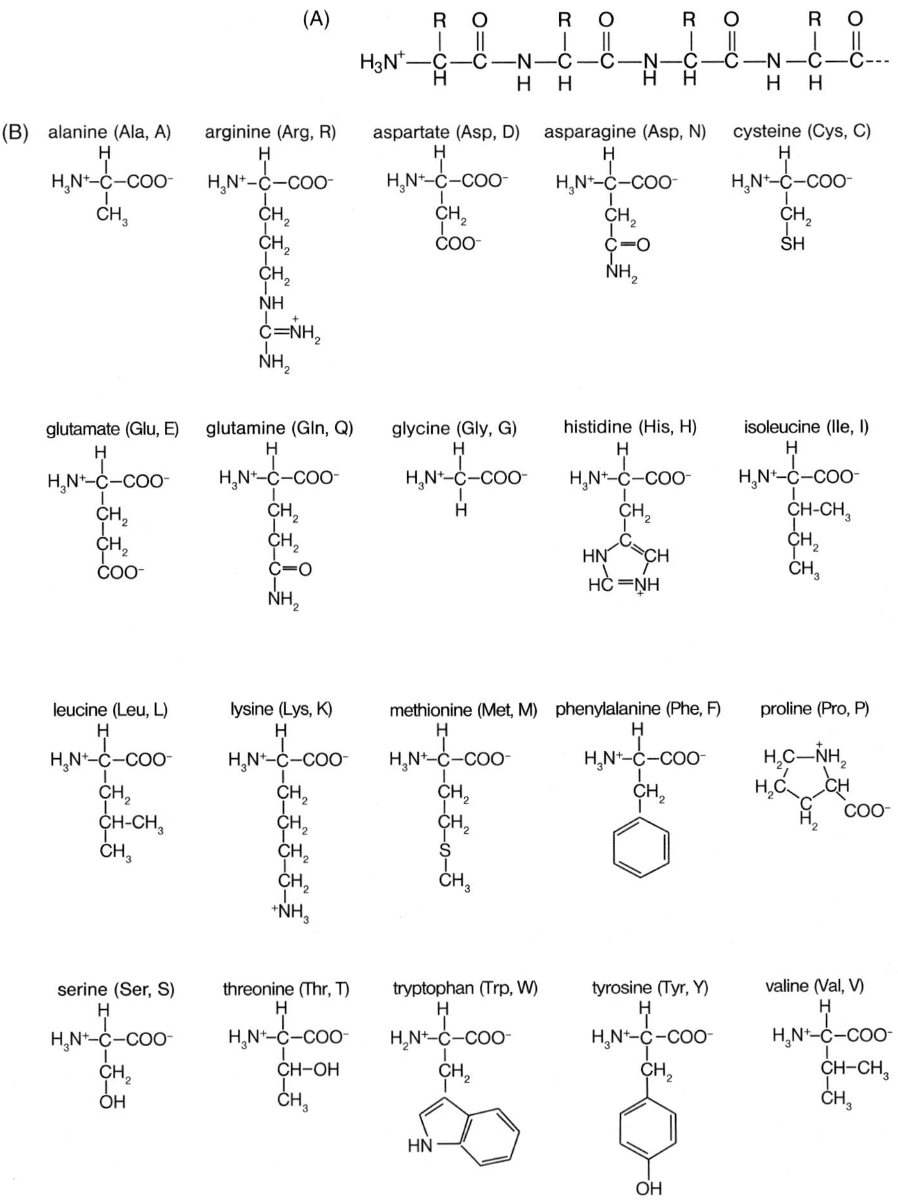

(A)

(B)

alanine (Ala, A)
arginine (Arg, R)
aspartate (Asp, D)
asparagine (Asp, N)
cysteine (Cys, C)
glutamate (Glu, E)
glutamine (Gln, Q)
glycine (Gly, G)
histidine (His, H)
isoleucine (Ile, I)
leucine (Leu, L)
lysine (Lys, K)
methionine (Met, M)
phenylalanine (Phe, F)
proline (Pro, P)
serine (Ser, S)
threonine (Thr, T)
tryptophan (Trp, W)
tyrosine (Tyr, Y)
valine (Val, V)

Fig. 3.3 (A) Chemical representation of the polypeptide structure using standard chemical abbreviations. (B) Chemical structures of R groups.

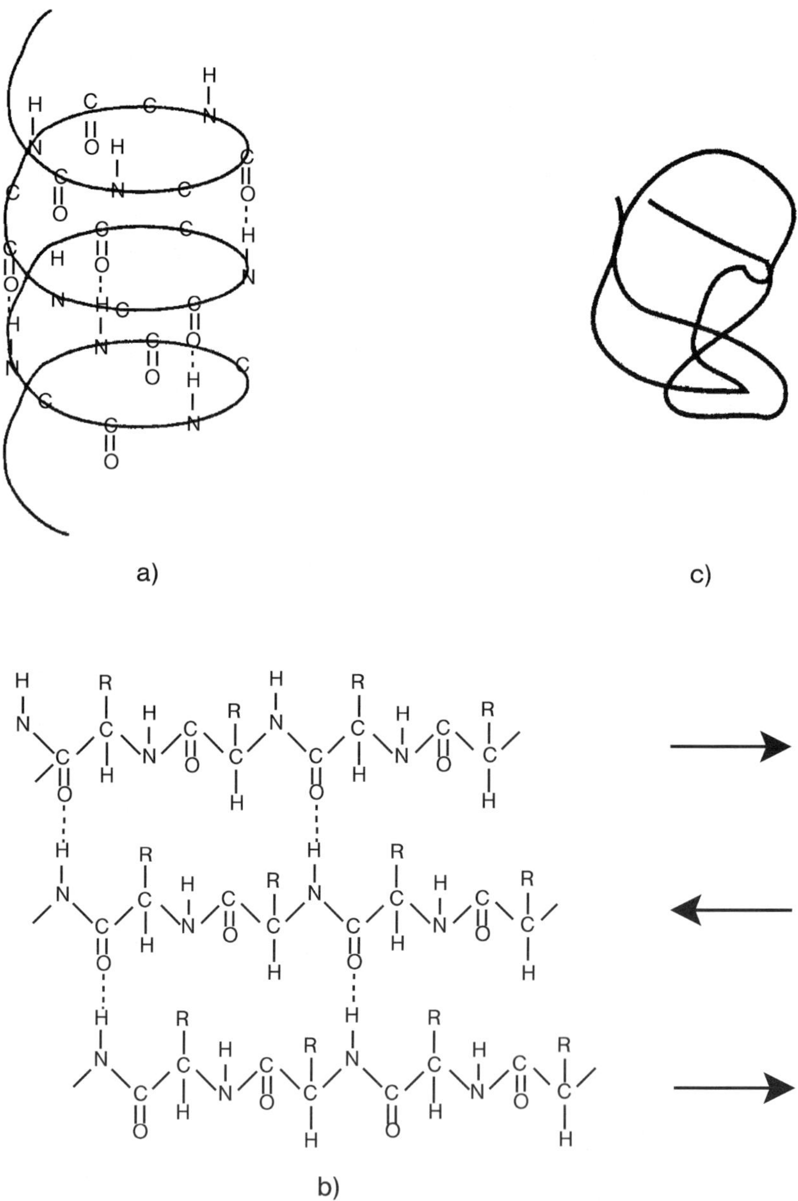

Fig. 3.4 Secondary structure of proteins: (A) α-helix, (B) β-pleated sheet and (C) random coil. (Reprinted from [15] with permission).

In contrast to the α-helix, a polypeptide in the β-pleated sheet conformation is almost fully extended; the axial distance between adjacent amino acids is 3.5 Å. Unlike the intrastrand hydrogen bonding of the α-helix, the β-pleated sheet conformation is stabilized by hydrogen bonds between N–H and C=O groups in different polypeptide strands. The orientation of adjacent chains in the β-pleated sheet conformation can be either parallel with adjacent chains running in the same direction or antiparallel with adjacent chains in the opposite direction. The antiparallel β-pleated sheet structure is favored when R groups are sufficiently large such that assumption of the α-helix conformation is sterically hindered. The side-chain groups will then be located above and below the pleated sheet with hydrogen bonding occurring within the sheet.

If side-chain structure permits, polypeptides in the β-pleated sheet conformation can self assemble into "one-dimensional crystals" involving a very large number of sheets. The capacity for unlimited interchain hydrogen bonding in the absence of structural restraints is thought to be what drives assembly of susceptible proteins into amyloid fibrils. The structure of amyloid fibrils reflects aggregation of strands of β-pleated sheet polypeptides into a long cross-β assembly with the orientation of the strands perpendicular to the fibril axis. Recent analyses of naturally occurring structural motifs (β-barrel, β-bulge, β-helix, β-propeller, single β-sheets, β-sandwiches) involving the β-pleated sheet fold have identified edge strands with a number of structural features that interrupt the self-assembly of β-pleated sheet strands. Structural features that result in edge strands include the presence of a charged side-chain within the hydrophobic region of the edge strand or the β-barrel and β-bulge that serve to limit interaction with other β-pleated sheet edges and thus to limit the number of β-sheets in a stack [18]. It is thought that edge strands have evolved to guard against uncontrolled propagation of the β-pleated sheet conformation that would interfere with the physiological function that is maintained by specific, controlled protein folding.

The peptide bond is planar, with the hydrogen of the substituted amine *trans* or opposite to the carbonyl carbon. The planar nature of the peptide bond constrains rotation of the carbon and nitrogen atoms; this limits folding options for the polypeptide backbone [15] (Fig. 3.5). In polypeptides (polyamino acids), the peptide bond has partial double-bond character with bond length of 1.32 Å as compared with the 1.49 Å length of a carbon to nitrogen single bond (C–N) and the 1.27 Å length of a carbon to nitrogen double bond (C=N). There is a high barrier of energy to rotation about the peptide bond and the freedom of movement of adjacent amino acid residues in polypeptides is limited. Ramachandran et al. [19] defined two angles about which the main polypeptide chain can rotate on either side of rigid peptide units as (1) ϕ as the degree of rotation at the bond between the nitrogen and carbon atoms of the main chain and (2) ψ as the degree of rotation between α-carbon and carbonyl carbon atoms (Fig. 3.5). The conformation of the polypeptide backbone is specified by all of the ϕ and ψ values for each amino acid residue in each amino acid residue in the polypeptide, and, thus, steric hindrance limits the values of ψ and ϕ that each individual amino acid residue in a polypeptide can assume. The Ramachandran

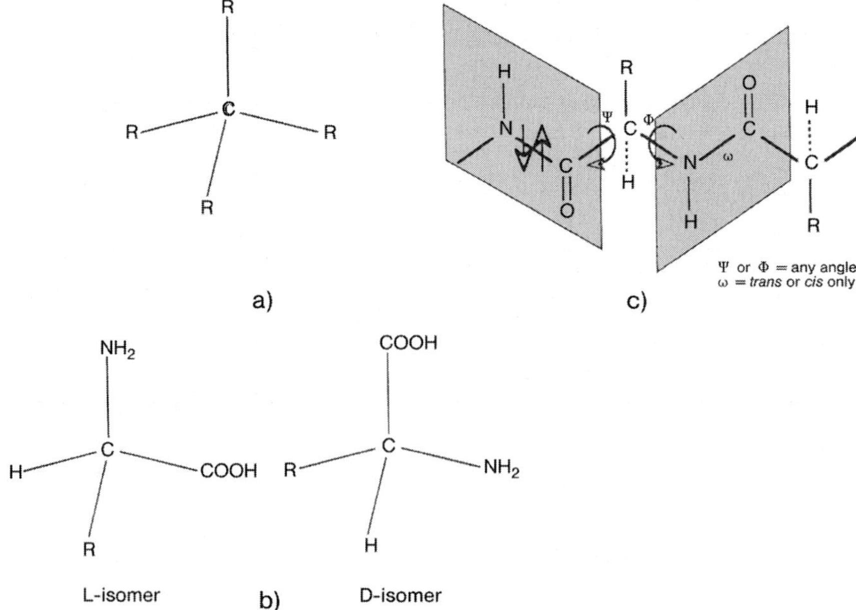

Fig. 3.5 Steric constraints on folding of the polypeptide backbone are posed by (A) tetrahedral bonding of carbon atoms, (B) asymmetry of the carbon atom in all amino acid residues except glycine and (C) the planar nature of the peptide bond. (From [15], reprinted with permission.).

approach extended the earlier studies of Pauling et al. and provides additional evidence that folding of the polypeptide backbone is significantly dependent on steric considerations.

Ramachandran plots of ψ versus ϕ for polymers of all of the amino acids except glycine and proline show a limited number of allowed conformations (Fig. 3.6). In addition to steric constraints posed by the covalent bonding of the polypeptide backbone and the charge and size of the side-chains, folding of the polypeptide backbone is also influenced by other factors. These include: (1) the chemical structure of the amino acid side-chains; (2) the local chemical environment, particularly the structure of water, in view of the extensive hydrogen bonding interactions between the polypeptide backbone and side-chains and water molecules; and (3) electrical fields that result from charged molecules in cells, tissues and organs, and from multiphase systems in which an electrically charged lipid membrane separates the aqueous intracellular and extracellular compartments.

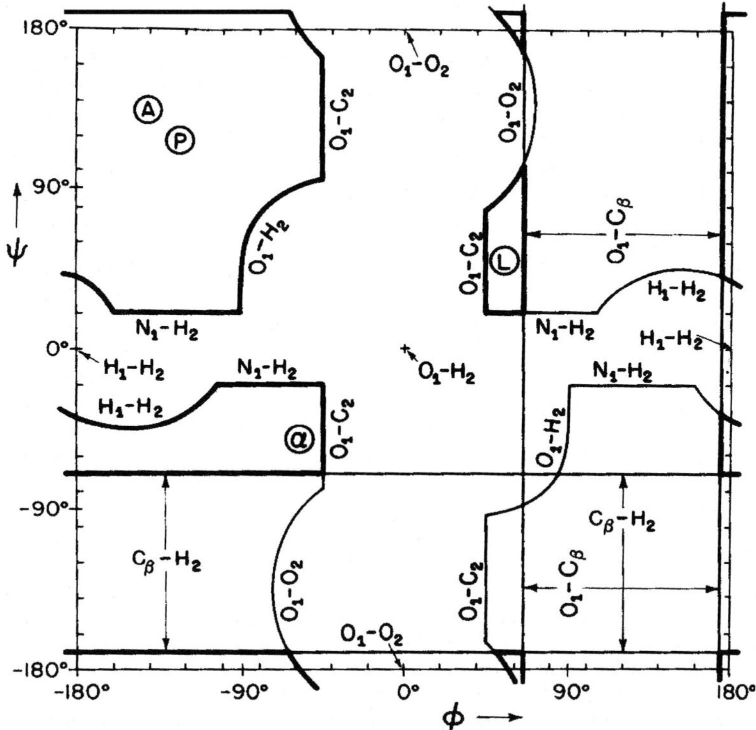

Fig. 3.6 Diagram illustrating the steric effects producing the Ramachandran plot as adapted from [25] and printed as figure 6.3 in [20] (reprinted with permission). The plot of the rotation angles ψ versus ϕ (Fig. 3.5) depicts the boundaries between allowed and forbidden regions obtained from molecular models of the type Pauling et al. used [16, 17]. The allowed region for parallel and antiparallel β-pleated sheets is in the upper left-hand corner; the right-handed α-helix is in the lower left-hand corner and the left-handed α-helix is in the upper right-hand corner. (Copyright 1977 Journal of Biological Chemistry.)

3.4 Polypeptide Backbone Folding: The Environment

Unlike the primary structure of a protein (polyamino acid), which has been generated by covalent bond formation, the secondary structure or folding of the polypeptide backbone is controlled by non-covalent interactions: ionic attractions, hydrogen bonds, hydrophobic interactions and van der Waals forces (Table 3.1). The covalent peptide linkage imparts a structure to the polypeptide chain, and forces the non-polar and polar side-chains into a restricted number of orientations. The inter- and intrapolypeptide strand molecular interactions are electronic in nature, and, while lower in energy than covalent bonds, are significant in number.

Table 3.1 Energy of chemical bonds.

Bond	**Energy (kcal/mol)**
Covalent	
single	50–110
double	150
triple	200
Hydrogen	5
Ionic	4–7
van der Waals	1

The thermodynamic forces that influence polypeptide folding in aqueous solution serve to minimize the disruption of hydrogen bonding of bulk water. The folding of a polymer is critically dependent on the effect of its solvent on non-covalent interactions. There is a wide range in the chemistry of amino acid side-chains – aliphatic, hydrophilic, charged and rigid cyclic, a single hydrogen (Fig. 3.3). In an aqueous environment, the exterior of a folded polypeptide is surrounded by a sheath of water that is disrupted from participating in the normal hydrogen bonding pattern of bulk water as shown in Fig. 3.7. The structure of water adjacent to the α-helix is clathrate pentagonal, forming the hydrotactoid [15]. The hydrotactoid is the ordered water structure surrounding the entire polypeptide and is very different in structure from water surrounding small non-polar solutes. The intrastrand hydrogen bonding of the α-helix is less disruptive of the hydrogen bonding of water than is the hydrogen bonding of the extended β-pleated sheet. It is the interaction of side-chains (Fig. 3.8) with water involving the above non-covalent interactions to generate a thermodynamically stable structure that results in a unique protein structure.

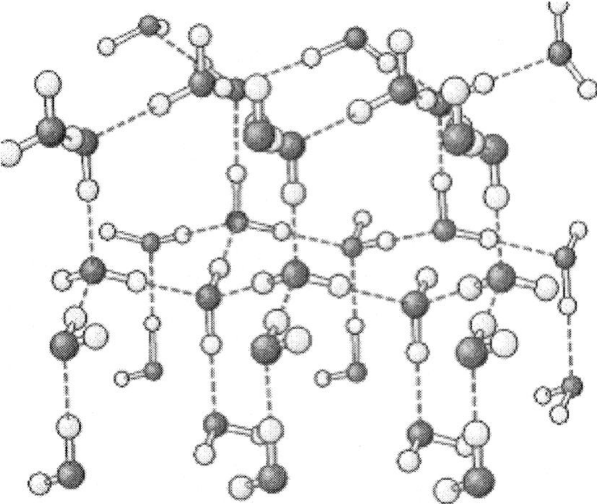

Fig. 3.7 Hydrogen bonds (dotted lines) are formed between water molecules. (Fig. 1.11 from [24], reprinted with permission.)

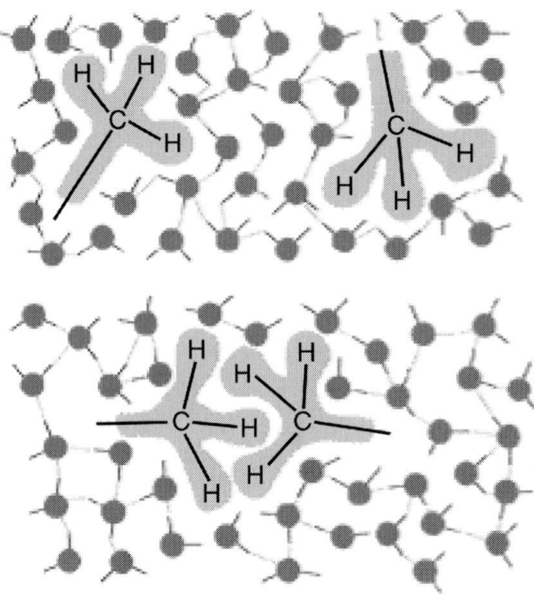

Fig. 3.8 Disruption of hydrogen bonding between water molecules by amino acid side-chains.

Thermodynamic studies have shown that the polypeptide side-chain structure in the context of solvent favors folding; otherwise the polypeptide would be fully extended as random coils, i.e. unfolded to permit maximum interactions of amino acid monomers with solvent, rather than seeking interactions within itself [20]. The fully extended, random coil polypeptide backbone structure is observed only *in vitro*.

In general, it is a feature of protein structure that the surface of a folded protein is charged or polar and the interior is hydrophobic. As discussed above, the three-dimensional structure of proteins is influenced by the side-chains of the 20 amino acid residues, within the constraints of the polypeptide backbone. Protein folding is aided by enzymes and chaperones, and in some cases is context dependent [21, 22]. Minor and Kim [22] designed an 11-amino-acid sequence that folded as an α-helix when inserted into one position within the IgG-binding domain of protein G and as a β-pleated sheet when inserted into another region of the molecule. This was a direct demonstration that non-local interactions can determine the secondary structure of peptide sequences that are similar in length to peptide segments that correspond to the amyloidogenic or the fibril-forming sequence of many of the amyloid proteins.

3.5
Conclusion

In nature, amyloid fibril formation is relatively rare, given the thousands of proteins the body produces. The folding options of the polypeptide backbone are limited by the constraints imposed by covalent bonding within polypeptides. The side-chains are key determinants of folding into the β-pleated sheet conformation and, by means of edge strands, of blocking uncontrolled self-assembly of β-sheets into amyloid fibrils. A greater understanding of the context within which proteins refold into amyloid fibrils will be valuable to the design of prevention and treatment strategies.

References

1 Sipe, J. D. Amyloidosis. *Annu Rev Biochem* **1992**, *61*, 947–975.
2 Sipe, J. D. Amyloidosis. *Crit Rev Clin Lab Sci* **1994**, *31*, 325–354.
3 Husby, G., G. Marhaug, B. Dowton, K. Sletten and J. D. Sipe. Serum amyloid A (SAA): biochemistry, genetics and the pathogenesis of AA amyloidosis. *Amyloid* **1994**, *1*, 119–137.
4 Chow, M. K., D. A. Lomas and S. P. Bottomley. Promiscuous β-strand interactions and the conformational diseases. *Curr Med Chem.* **2004**, *11*, 491–499.
5 Miranda E., K. Romisch and D. A. Lomas. Mutants of neuroserpin that cause dementia accumulate as polymers within the endoplasmic reticulum. *J Biol Chem* **2004**, *279*, 28283–28291.
6 Cohen, A. S. General introduction and a brief history of amyloidosis. In *Amyloidosis*, J. Marrink and M. H. van Rijswijk (eds). Martinus Nijhoff, Dordrecht, **1986**, pp. 3–19.
7 Sipe, J. D. and A. S. Cohen. History of the amyloid fibril. *J Struct Biol* **2000**, *130*, 88–98.
8 Sipe, J. D. Meeting Report: Amyloid and Amyloidosis, The 2nd Romhanyi Memorial Symposium, April 24, 2004, Pecs, Hungary. *Amyloid*, **2004**, *11*, 213–215.
9 Cohen, A. S. and E. Calkins. Electron microscopic observations on a fibrous component in amyloid of diverse origins. *Nature* **1959**, *183*, 1202–1203.
10 Glenner, G. G. Amyloid deposits and amyloidosis. The fibrilloses. *New Engl J Med* **1980**, *302*, 1283–1292 and 1333–1343.
11 Glenner, G. G., E. D. Eanes, H. A. Bladen, R. P. Linke and J. D. Termine. β-Pleated sheet fibrils. A comparison of native amyloid with synthetic protein fibrils. *J Histochem Cytochem* **1974**, *22*, 1141–1158.
12 Jaroniec, C. P., C. E. MacPhee, V. S. Bajaj, M. T. McMahon, C. M. Dobson, and R. G. Griffin. High-resolution molecular structure of a peptide in an amyloid fibril determined by magic angle spinning NMR spectroscopy. *Proc Natl Acad Sci USA* **2004**, *101*, 711–716.
13 Jarrett, J. T. and P. T. Lansbury, Jr. Seeding "one-dimensional crystallization" of amyloid: a pathogenic mechanism in Alzheimer's disease and scrapie? *Cell* **1993**, *73*, 1055–1058.
14 Gee, H. A journey into the genome: what's there. *Nature* **2001**, *Science Update*, 14 Feb.
15 Bergethon, P. R. and E. R. Simons. *Physical Biochemistry*. Springer, New York, **1990**.
16 Pauling, L., R. B. Corey and H. R. Branson. The structure of proteins: two hydrogen-bonded helical configurations of the polypeptide chain. *Proc Natl Acad Sci USA* **1951**, *37*, 205–211.
17 Pauling, L. and R. B. Corey. Configurations of polypeptide chains with favored

orientation around single bonds: two new pleated sheets. *Proc Natl Acad Sci USA* **1951**, *37*, 729–740.

18 Richardson, J. S. and D. C. Richardson. Natural β-sheet proteins use negative design to avoid edge-to-edge aggregation. *Proc Natl Acad Sci USA* **2002**, *99*, 2754–2759.

19 Ramachandran, G. N., C. Ramakrisnan and V. Sasisekharan. Stereochemistry of polypeptide chain configurations. *J Mol Biol* **1963**, *7*, 95–99.

20 J. Kyte. *Structure in Protein Chemistry.* Garland, New York, **1995**.

21 Trombetta, E. S. and A. J. Parodi. Quality control and protein folding in the secretory pathway. *Annu Rev Cell Dev Biol* **2003**, *19*, 649–676.

22 Minor, D. L. and P. S. Kim. Context-dependent secondary structure formation of a designed protein sequence. *Nature* **1996**, *380*, 730–734.

23 Cooper, J. H. Selective amyloid staining as a function of amyloid composition and structure: histochemical analysis of the alkaline Congo red, standardized toluidine blue, and iodine methods. *Lab Invest* **1974**, *31*, 232–238.

24 Berg, J. M., J. L. Tymoczko and L. Stryer (eds). *Biochemistry*, 5th edn. Freeman, San Francisco, CA, **2003**.

25 Mandel, N., G. Mandel, B. L. Trus, J. Rosenberg, G. Carlson and R. E. Dickerson. Tuna Cytochrome cat 20 Å resolution. *J Biol Chem* **1977**, *252*, 4619–4636

Part III
Protein Folding, Unfolding and Refolding

Amyloid Proteins. The Beta Sheet Conformation and Disease. J. D. Sipe

ISBN: 3-527-31072-X

4
Thermodynamics and Protein Folding

Ilia V Baskakov

4.1
Introduction

More than 20 systemic and neurodegenerative maladies have been linked to the formation of ordered protein aggregates [1, 2]. When the deposited polymeric form is sufficiently ordered to bind Congo red and Thioflavin T, the term amyloid is used to define these types of aggregation [3]. A common feature of amyloid aggregates is formation of β-sheet-rich polymeric forms organized into highly ordered fibrils or plaques [4]. Recent studies have demonstrated that a broad variety of proteins unrelated to any known conformational disease can adopt β-sheet-rich amyloid forms *in vitro* and *in vivo* [5–8]. Medin, a proteolytic fragment of lactadherin, is an example of polypeptides that form amyloid deposits *in vivo* (amyloid deposits of medin were found in aorta of virtually all individuals studied older than 60 years), but has not been linked to any pathological processes [9]. Amyloidogenic proteins are now found in a variety of organisms including prokaryotes, plants, insect and mammals [10–13]. No consensus sequences that would predetermine the ability to form amyloid have been identified in any of these classes of proteins. Even though the amyloidogenic proteins show no obvious sequence similarity, they share a similar conformational property when converted into the amyloid fibrils, i.e. thermodynamically stable cross-β-pleated sheets [14, 15]. This finding has led to the hypothesis that the ability to fold into amyloid forms is not a unique property of certain proteins associated with degenerative maladies; rather, it may well be a feature of polypeptides in general [16].

4.2
Thermodynamic versus Kinetic Control of Protein Folding

In his 1972 Nobel Prize lecture Christian Anfinsen described the "thermodynamic hypothesis" of protein folding [17]. This hypothesis states that "the three-dimensional structure of a native protein in its normal physiological milieu is

Amyloid Proteins. The Beta Sheet Conformation and Disease. J. D. Sipe

ISBN: 3-527-31072-X

the one in which the Gibbs free energy of the whole system is lowest; that is, that the native conformation is determined by the totality of interatomic interactions and hence by the amino acid sequence, in a given environment". Although the thermodynamic hypothesis has now been widely established and protein folding is commonly thought to be controlled by thermodynamic preferences, it has been understood by many, including Anfinsen and others, that kinetic issues can alter the folding landscape [18]. Whereas most small globular proteins will refold spontaneously *in vitro* to a native conformation, *in vivo* folding often exploits auxiliary molecules and defined subcellular compartments to avoid the deposit of ordered misfolded aggregates. What drives conversion of natively folded proteins into alternative misfolded conformations? Why do some proteins assemble into amyloid fibrils at some point within a protein's lifetime, while others do not?

Direct comparison of the thermodynamic stability of the native state with that of the β-sheet-rich amyloid state is impossible due to the insolubility, heterogeneity and high degree of polymerization of the amyloid fibril aggregates. On the other hand, the high resistance of the amyloid fibril aggregates to denaturation by detergents and to thermal and solvent-induced denaturation serves as a clever illustration of the extremely high thermodynamic stability and remarkable physical properties that the β-pleated sheet conformation imparts to amyloid fibrils. It is noteworthy that nature has learned to exploit these unusual properties of amyloid structures for a variety of physiological functions. For example, the major structural component of the shells of many insects and fish is of amyloid [11, 13]. To protect the developing embryo from temperature variation, mechanical pressure, proteases, bacteria, viruses and dehydration, the ability to construct complex amyloid structures evolved in these organisms through natural evolution and selection [11, 13]. Other examples of naturally occurring amyloid structures include extracellular curly fibrils expressed by *Escherichia coli* and *Salmonella*. These fibrils are involved in the colonization of bacteria on surfaces and in biofilm formation [19]. Furthermore, mammalian melanocytes produce a glycoprotein that polymerizes into amyloid-like fibrils, on which melanins are sequestered and concentrated during the multistage process of melanosome biogenesis [20]. These observations illustrate that amyloid formation is an evolutionary preserved biological pathway used to produce natural biomaterials with important physiological functions.

While the direct thermodynamic analysis of insoluble amyloid structures is quite complicated, numerous studies have demonstrated that, in addition to insoluble amyloid fibrillar forms, many amyloidogenic proteins also adopt soluble oligomeric β-sheet-rich states [21–25]. Some of these β-sheet-rich oligomeric forms lie on the kinetic pathways to the amyloid fibrils, while others are off the kinetic pathways [24–26]. A direct comparison of the thermodynamic stability of the native and misfolded β-sheet-rich oligomeric isoform illustrates that the native state is not the lowest energy state [27, 28]. It is shown in Fig. 4.1 that, when the unfolded state is used as a reference in the free energy diagram, the β-sheet-rich state is thermodynamically more stable than the native state. Even

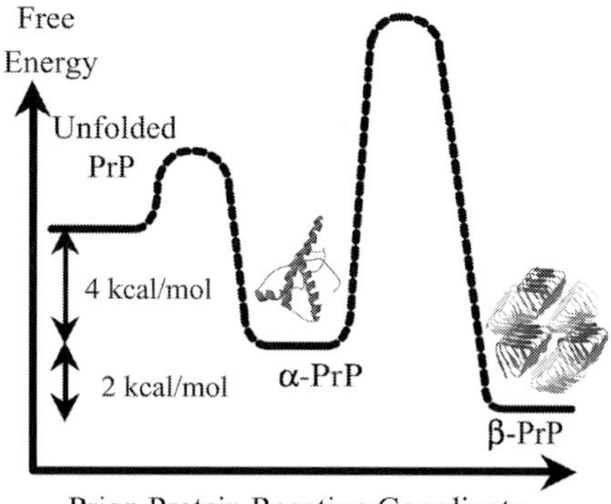

Fig. 4.1 Schematic free energy diagram of the conformational transition of the prion protein [27]. The α-helical monomeric isofrom is not the lowest energy state. However, the conformational transition form the α-helical isoform to a β-sheet-rich oligomeric form is controlled by a large energetic barrier and, therefore, is prevented within the protein lifetime.

though the β-oligomeric states exhibit a thermodynamic stability higher than that of the native states, they might not be the true global energy minimum states, because the β-oligomeric species may undergo an additional time-dependent transition to highly polymeric amyloid forms [24, 25].

Why is the thermodynamically more stable β-sheet-rich isoform not accessible during folding under native conditions? It has been demonstrated that the rate of folding to the β-sheet-rich oligomeric isoform is slower by several orders of magnitude than the rate of folding to the native conformation [27]. To prevent the conformational conversion, the native state has to be separated by a large energetic barrier from the alternative β-sheet-rich state (Fig. 4.1). Although the free energy diagram does not provide a view of the actual kinetic pathways for the conformational transition, several important observations can be made regarding the origin of the energetic barrier. First, the native form has to unfold substantially on the route to the β-sheet-rich oligomeric isoform. Indeed, several studies have demonstrated that a substantial portion of the energetic barrier is linked to the partial unfolding of the native conformations [27, 29, 30]. The conversion to β-sheet-rich forms can be accelerated by shifting the native–unfolded equilibrium toward the unfolded state [6, 30–32]. Furthermore, the connection between the structural complexity of the pre-transition state and the energetic barrier is demonstrated by numerous observations that conversion of polypeptides with low structural complexity into β-sheet-rich isoforms occurs spontaneously and does not require partially denaturing conditions [21, 33]. A significant contribution to the energetic barrier seems to be associated with the process of assembly. Several studies demonstrated that the accumulation of a β-rich conformation is coupled with oligomerization. Analyses of the kinetic traces indicate that the process of folding to the β-oligomeric isoforms represents a transition with an apparent reaction order of 3 or higher [27, 34]. This high order of reaction suggests that the conformational transition will depend dramatically upon the concentration of the transition state.

4.3
What Thermodynamic Forces are Responsible for the Exceptional Stability of Amyloid Aggregates?

It has been more than 40 years since Kauzmann described the thermodynamic forces responsible for the folding of proteins to the native conformations [35]. Such forces include hydrogen bonding, electrostatic, van der Waals, conformational entropy and hydrophobic interactions. Upon folding, groups accessible to solvent in the denatured state became newly buried in the native state. On transfer of the denatured state from a denaturing solution environment to physiological conditions, collapse of the denatured state occurs, causing removal of hydrophobic groups from water. This is believed to be an important driving force for folding of proteins into unique native conformations. For hydrophobic forces to play the major role in protein folding, the fraction of the hydrophobic groups destined for burial must be significant in the denatured states of proteins. Fig. 4.2 presents a analysis of the classes of groups that are buried upon protein folding in terms of their relative proportion as a function of protein molecular mass. Hydrophobic groups account for more than 20% of all buried groups upon protein folding and their relative proportion is consistently maintained, regardless of protein molecular mass. This result strongly supports the importance of the role that hydrophobic effects play in protein folding.

Remarkably, it is evident from Fig. 4.2 that it is the peptide backbone that comprises the largest numerical fraction of groups newly buried on protein folding (nu-

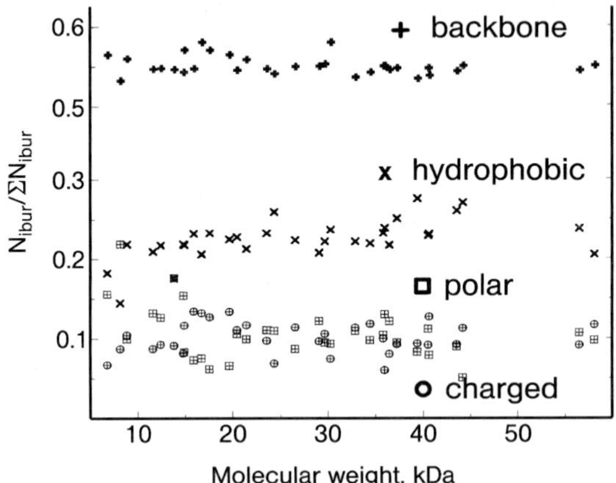

Fig. 4.2 Number fraction of backbone and side-chain classes that are newly buried upon folding proteins of different molecular masses to their native states. In addition to the peptide backbone class, all side-chains are grouped into three classes: hydrophobic, polar and charged side-chains. Crystallographic coordinates data were used from the PDB databank in evaluating the number of each group exposed in the native states of the proteins used in the calculations [38].

merical fraction = 55%). Whether the removal of the peptide backbone units from contact with water molecules is thermodynamically favorable or not has been widely debated [36]. The essential feature of the peptide backbone is its ability to form inter- and intramolecular hydrogen bonds. By forming hydrogen bonds between NH and CO groups in the folded state, polypeptides minimize the unfavorable effects of removing polar groups from water. The requirement that NH and CO groups of the peptide backbone must either form hydrogen bonds with water or with each other provides a strong structural constraint on the pathways toward possible folded states. This holds, especially, for ordered β-sheet-rich aggregates, where the role of hydrophobic interactions seems to be unsubstantial. Analyses of microcrystals of the amyloidogenic peptides demonstrated that amyloid fibril structures are highly hydrogen bonded, nearly anhydrous and densely packed β-sheets [37]. One can speculate that amyloid structures are formed in a way that optimizes formation of hydrogen bonds between strands of newly buried polypeptide backbone.

The ability of many proteins to adopt alternative β-sheet-rich polymeric folding *in vitro* and *in vivo* argues that the mechanism involved and the thermodynamic forces that stabilize amyloid conformations have to be generic in nature. The fact that the numerical fraction of the newly buried peptide backbone is large and constant with respect to molecular mass illustrates that, under favorable conditions, burial of the peptide backbone could counteract the unfavorable exposure of hydrophobic groups. Considering that the number of newly buried peptide backbone groups predominates over that of newly buried side-chains lends support to the concept that it is hydrogen bonding of the peptide backbone that provides generic force for the stabilization of amyloid fibril forms. As the amyloid fibril aggregates grow and, correspondingly, the ratio of surface accessible to solvent versus the volume occupied by protein fabric decreases, the stabilizing effect increases. In the way that the hydrophobic force is considered one of the major determinants of the unique native conformation, it seems that the chemical nature of the polypeptide backbone, with its capacity for forming inter- and intramolecular hydrogen bonds is the central determinant of amyloid fibril formation [38].

4.4 Single Polypeptide Chain–Multiple β-Sheet-rich Abnormal Isoforms

Numerous biophysical studies have revealed that amyloidogenic proteins can adopt conformationally distinct non-native β-sheet-rich isoforms *in vitro* [26, 39]. For example, amyloid fibril formation by the prion protein occurs through a pathway different from the one that leads to formation of the β-oligomeric species [26, 40]. Regardless of which abnormal β-isoforms are biologically relevant, the ability of amyloidogenic proteins to form distinct abnormal conformers reflects the complexity of the energetic landscape of folding, as well as high conformational plasticity. How do different conformers arise from the same amino acid sequence in the absence of cellular cofactors or templates?

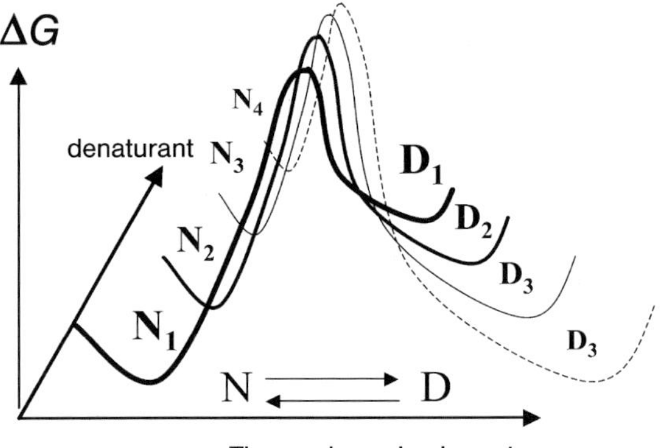

Fig. 4.3 Free energy diagram illustrating the thermodynamically variable characters of native and denatured states. Not only do the relative populations of the native (N) and the denatured (D) states change as a function denaturant concentration, but the physical properties of both states are variable depending of solvent conditions. [N_1, N_2, N_3, N_4] and [D_1, D_2, D_3, D_4] illustrate the thermodynamically variable character of native and denatures states, respectively.

Recent biophysical studies of model proteins demonstrated that physical properties of the native and denatured states of some proteins may change gradually with environmental conditions [41]. The gradual change within the native and denatured species is consistent with the thermodynamically variable model, which postulates that the thermodynamic character of the native and/or denatured ensemble changes continuously as a function of the solvent environment [42, 43] (Fig. 4.3). The gradual change of the thermodynamic character is not always accompanied by conformational rearrangements of secondary structure. Thus, many proteins that display a thermodynamically variable behavior of the native ensembles did not show any changes as monitored by circular dichroism (CD) [42, 44]. However, such changes can be observed by other techniques such as nuclear magnetic resonance (NMR) [45], proton uptake/release [42], size-exclusion chromatography [41, 44] and dynamic light scattering [34]. One may speculate that the diversity of potential misfolding pathways of amyloidogenic proteins is linked to the variable thermodynamic behavior. Thus, a gradual environment-dependent change in the physical properties of the native and/or denatured ensembles may bias the adoption of particular pathway of misfolding under different pathological conditions.

4.5
Does the Process of Prion Propagation Differ from Formation of Ordered Amyloid Aggregates?

Among many amyloid-related maladies one subclass of the conformational disorders, prion diseases, seems to be distinguished by certain peculiar features:

(1) The most unorthodox feature of prion disease is the existence of an *infectious isoform* of the prion protein, PrP^{Sc} [46, 47]. PrP^{Sc} propagates its abnormal conformation in an autocatalytic manner using the *normal isoform* (PrP^{C}) of the same protein as a substrate. In addition to the transmission of prion diseases in mammals, the phenomenon of self-propagating conformational transition has been described for prion proteins in yeast and in fungi [48, 49]. In all cases, the abnormal protein conformation acts either as the transmissible agent of disease or as a heritable determinant of phenotype. Reconstitution of mammalian prion infectivity *in vitro* has been difficult to accomplish for many years. This problem raised growing skepticism over the sufficiency of PrP alone to form an infectious agent. Recent work, however, demonstrated that the amyloid form of recombinant PrP induced a transmissible form of prion diseases in transgenic mice, providing the first compelling evidence for the "protein-only hypothesis" of prion propagation in mammals [50].

(2) Efficient self-propagation of prions requires identity or high homology between the amino acid sequences of PrP^{C} and PrP^{Sc}, implying there is a high species specificity associated with their interaction.

(3) Another prominent feature of prion propagation is the "strain" phenomenon. When PrP^{C} is converted into pathogenic isoforms, a single unique amino acid sequence is capable of adopting conformationally distinct states, which are known as "strains" of PrP^{Sc}.

Interestingly, despite differences in the primary and tertiary structures of yeast and mammalian prions, the yeast prions display all of the characteristic features of mammalian prion proteins [51, 52]. Remarkably, recent studies illustrated that amyloid formation by non-prion proteins can also exhibit some properties that are thought to be peculiar for prion propagation. Thus, the systemic amyloidosis caused by the amyloid deposition of the serum amyloid A protein can be transmitted from animal to animal by a prion-like mechanism [53]. A species barrier, previously thought to be a distinctive feature of prion propagation, was also observed for the non-prion protein α-synuclein [54]. The growth of amyloid fibrils by $A\beta$ peptides displayed striking specificity to cross-seeding by heterologous fibrils [55]. Since the process of amyloid formation is a general property of the polypeptide backbone, it would be of great interest to know how many amyloidogenic proteins are capable of self-propagating conformational transition in a prion-like manner (Fig. 4.4).

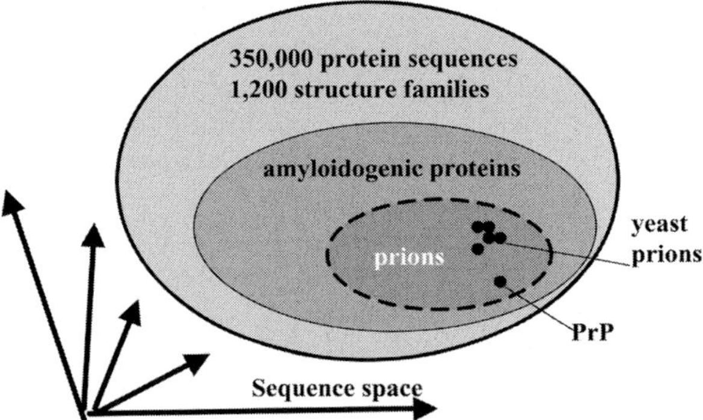

Fig. 4.4 The multidimensional sequence space defines all possible amino acids sequences. Among all species, only approximately 350,000 polypeptide sequences appear to have evolved through natural evolution and selection (http://protomap.cornell.edu/). Because the ability to adopt an amyloid conformation seems to be encoded in the physical/chemical nature of the polypeptide backbone, a substantial fraction of naturally occurring proteins would be expected to be amyloidogenic. The number of proteins that are capable of propagating abnormal conformations in a prion-like manner remains to be determined.

4.6 Prion Propagation is an Autocatalytic Process

Two competing models have been proposed to explain the autocatalytic conversion of prion folding: (1) the nucleation–polymerization model (NPM) [56] and (2) the template assisted model (TAM) [57, 58] (Fig. 4.5). Using different terminology, "heterogeneous nucleation" versus "templating", both models employ an autocatalytic mechanism to explain the process of prion transmission and replication. Phenomenologically, the presence of a catalyst (nucleus or template) accelerates the process of conversion, either by avoiding a rate-limiting step in NPM or by lowering of the energy barrier in TAM. Both models predict low occurrences of sporadic form of prion disease, and, correspondingly, a low rate of spontaneous conversion, since spontaneous or non-seeded formation of the first catalytic center is inaccessible either thermodynamically (according to NPM) or kinetically (as follows from TAM). Exogenous administration of catalyst (PrP^{Sc} in infectious form of disease) shortens the incubation time to such extent that disease develops within the human or animal lifetime. However, neither model explicitly discusses the mechanism of multiplication of the catalytic centers in the time course of the disease.

Time-dependent multiplication of catalytic centers is a key feature of any autocatalytic reaction and may have a dramatic effect on the final outcome of these processes. In a simplified mathematical equation, the reaction velocity of

The nucleation-polymerization model

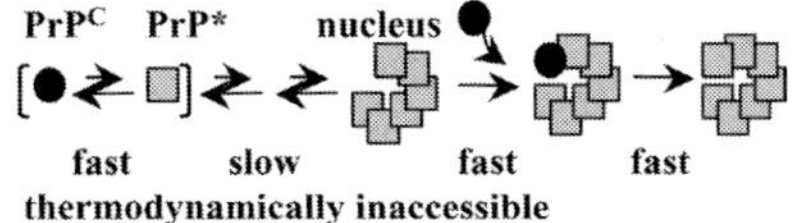

The template assisted model

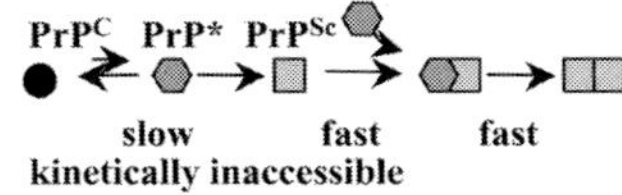

Fig. 4.5 Schematic diagram showing the nucleation-dependent and the template-assisted mechanisms of the conversion of PrP^C to PrP^{Sc}. Briefly, NPM postulates that the rate-limiting step is the formation of a nucleus: an oligomeric aggregate of PrP. The nucleus is thermodynamically unstable and this makes the spontaneous process a very rare event. However, as the nucleus is formed, further polymerization is facile. Exogenous addition of PrP^{Sc} bypasses the formation of the nucleus. TAM, on the other hand, argues that PrP^C is separated from PrP^{Sc} by a substantial energy barrier. The high barrier precludes the formation of PrP^{Sc} under normal conditions. However, the process of conversion is facilitated by an exogenous administration of PrP^{Sc}, which acts as a catalyst and lowers the energy barrier.

an autocatalytic process is determined by the multiplication coefficient (r), which is proportional to the rate of generation of new catalytic centers divided by the rate of their loss or clearance in the time course of the process. When $r>1$ the reaction rate increases exponentially, while the reaction decays when $r<1$ (Table 4.1). Such mechanism postulates a threshold effect when r is very close to 1. In this case slight changes in experimental parameters may switch the reaction from a decay mode to an autoacceleration mode and *vice versa*. This mechanism predicts that the autocatalytic reaction can be induced even at sub-

Table 4.1 The autocatalytic mechanism predicts three possible outcomes with respect to prion disease, depending on the multiplication coefficient (r).

	$r>1$	$r\sim 1$	$r<1$
Essential kinetic features	exponential growth of the reaction rate	steady-state level	decay of the reaction
Kinetics follows formal mechanism of:	branched chain reactions	enzyme catalyses	apparent first-order kinetics
Clinical form of disease	development of prion disease	subclinical form of disease	clearance of prions

threshold conditions ($r < 1$) by exogenous addition of catalytic centers. The proposed model is in agreement with the hypothesis that the progression of prion disease and the final outcome are controlled by fine dynamic balance between the rate of self-propagation and the rate of clearance of PrP^{Sc}. It would be of great interest to know what mechanisms account for multiplication of catalytic centers *in vivo*. The ability to control the rate of multiplication of catalytic isoforms of PrP would be a powerful strategy to treat prion diseases.

4.7 Conformational Diversity of Self-propagating Prion Aggregates

The "protein-only hypothesis" postulates that the given PrP amino acid sequence must be capable of adopting separate states when it is converted into pathological isoforms called "strains" of PrP^{Sc}. The existence of different prion "strains" has been recognized for a long time in the history of prion diseases [59, 60]. Each prion "strain" is associated with distinct neuropathologic features, such as the length of incubation time and the distribution of neuronal vacuolation in the brain [61]. These features are stable during serial transmission of PrP^{Sc} in a given host species. In the absence of another molecule that assists conversion, the formation of different strains by the same protein can only be accomplished either by covalent modification of PrP or through the adoption of multiple conformations.

During the past several years, considerable evidence has been amassed indicating that the properties of prion "strains" are enciphered in the conformation of PrP^{Sc} [62–66]. According to the template assisted model the abnormal PrP^{Sc} isoform provides a conformational template and guides conversion of PrP^{C} into a conformer identical to PrP^{Sc}. Therefore, the conformational properties of each individual "strain" are inherited by nascent PrP^{Sc}. Although supported by numerous experiments carried out *in vivo*, the principle of formation of conformationally distinct self-propagating aggregates by the prion protein within the same primary sequence has yet to be demonstrated *in vitro*. The questions of fundamental importance are: (1) what factors determine *the conformational diversity of self-propagating protein aggregates*, (2) what is the role *of the primary structure of protein versus template and environmental factors in this process*, and (3) *how do new conformers ("strains") arise?*

4.8 High Species Specificity of Prion Propagation

Numerous experiments on transmission of prions between different species demonstrated that prion replication is highly specific with respect to the primary structure of interacting isoforms, PrP^{C} and PrP^{Sc} [67]. When the amino acid sequence of the PrP^{C} of a recipient animal is identical to the sequence of

PrP^{Sc} of a donor animal, the disease has the shortest incubation time. In contrast, if prions are transmitted from one species to another, the incubation time is longer in the first passage and the newly infected animals develop atypical clinical signs and unusual histopathology [68]. Once the initial passage has been accomplished, the incubation period shortens in subsequent passages and becomes fixed, as does histopathology. This phenomenon is called a "species barrier" [69]. From studies in transgenic animals, two factors have been identified that contribute to the species barrier: (1) the difference in PrP sequences between the donor species and the recipient species, and (2) the "strains" of PrP^{Sc}. It has been noticed that, when transmitted to a different species, some "strains" overcome the species barrier more easily than others [70].

To explain the species barrier, Caughey et al. proposed that when both homologous and heterologous PrP^C are present, the heterologous PrP^C (PrP^C with a sequence different from PrP^{Sc}) could bind to PrP^{Sc} without conversion. By binding to PrP^{Sc}, the heterologous PrP^C inhibits binding and conversion of homologous PrP^C [71]. In this model, the process of conversion, rather than binding, requires amino acid compatibility between PrP^C and PrP^{Sc}. Although this simple model rationalizes a substantial body of experimental data on the species barrier, it fails to recognize the strain specificity of the species barrier. The ques-

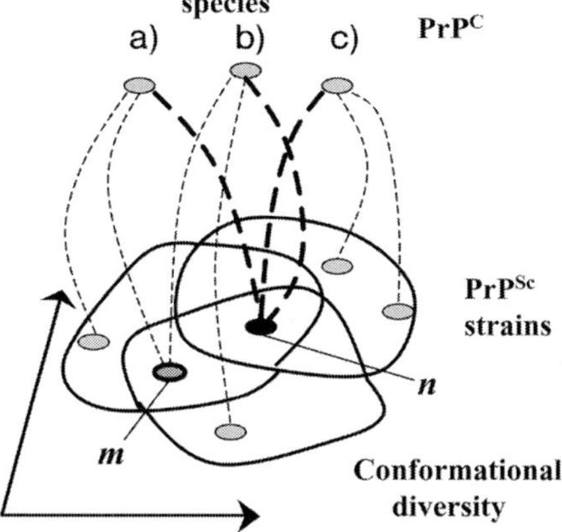

Fig. 4.6 The new conceptual framework postulates that (1) primary structure of PrP of each individual species (A, B and C) determines conformational diversity of "strains". (2) However, a template specifies particular conformation within given conformational space. (3) Subsets of conformers ("strains") formed by PrP molecules from different species may overlap suggesting that some PrP^{Sc} "strains" are more universal and can be shared by two or more species, whereas other "strains" are more species-specific. For instance, "strain" n is shared by species A, B, and C, while "strain" m is shared by species A and B. Only those "strains" that occupy overlapping areas can be propagated by more than one species.

tions are: (1) why can some *"strains"* from donor species be transmitted to recipient species, while other "strains" cannot, and (2) why some, but not all, recipient species propagate certain *"strains"* from donor species? It is clear now that the strain and the species barrier phenomena are closely connected. Therefore, the experimental data related to both phenomena have to be discussed in the context of a common model. Fig. 4.6 represents a new conceptual framework that explains the strain specificity of the species barrier. This framework may provide general guidance for research conducted on this topic.

4.9 Conclusions

The observations that many proteins are able to adopt alternative amyloid-like folding require us to revisit many basic issues of protein folding, such as the role of kinetic traps in the folding pathway, the complexity of the energy landscape and the position of the native state in this landscape. An alternative view of protein folding presented in Fig. 4.7 proposes a complex energy surface with multiple energy minima. Most polypeptides are trapped in a native state that corresponds to a local free energy minimum for the entire protein lifetime, whereas global energy minima are occupied by β-sheet-rich states, which are kinetically inaccessible under physiological conditions and at indefinite protein dilution. However, despite high kinetic barriers, some proteins, including PrP, α-synuclein, parkin, tau and other, find a route to β-sheet-rich multimeric structures with unfortunate consequences. If a β-sheet-rich amyloid structure is an intrinsic preference, particularly at high protein concentration, then compart-

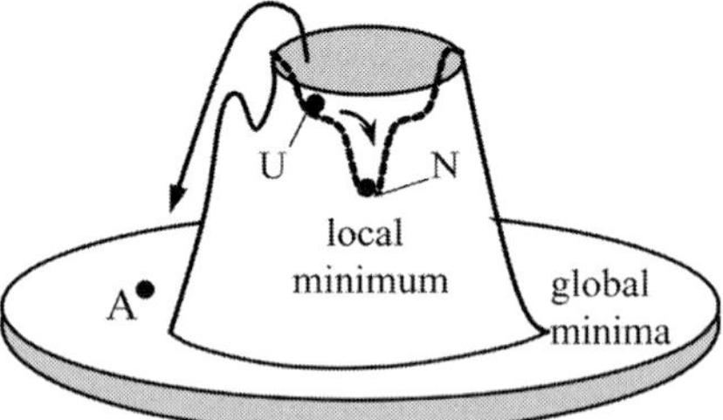

Fig. 4.7 An idealized "volcano energy landscape" of protein folding, to illustrate how a protein could fold fast to the native state, which might not the lowest energy state. Most proteins remain "trapped" in their native states for the entire protein lifetime, because alternative β-sheet-rich states are separated by a high energetic barrier. Opposite to the native form, the β-sheet-rich state of each individual protein may adopt more than one conformation within given conformational space. N = native state, U = unfolded state; A = amyloid state.

mentalization of proteins during folding and clearance of misfolded proteins should play critical roles in cellular health. In addition, amyloidogenic side-chain patterns, such as alternating polar and non-polar amino acid residues, will be avoided through natural selection [72].

With many fundamental rules that govern protein folding having been discovered, new questions have to be addressed in the nearest future. How general is the phenomenon of self-propagating conformational transition? Can the prion hypothesis be expanded to non-prion proteins? Are individual non-prion proteins capable of assembly into conformationally distinct self-propagating aggregates ("strains")? To what extent is sequence identity (or high homology) between native and self-propagating abnormal isoforms necessary for efficient self-propagation? What is the role of the template in conformational diversity of self-propagating β-sheet-rich states?

References

1 Carrell, R.W. and Lomas, D.A. Conformational disease. *Lancet* **1997**, *350*, 134–138.

2 Prusiner, S.B. Shattuck Lecture – neurodegenerative diseases and prions. *N Engl J Med* **2001**, *344*, 1516–1526.

3 Sipe, J.D. and Cohen, A.S. Review: history of the amyloid fibril. *J Struct Biol* **2000**, *130*, 88–98.

4 Kelly, J.W. The alternative conformations of amyloidogenic proteins and their multi-step assembly pathways. *Curr Opin Struct Biol* **1998**, *8*, 101–106.

5 Chiti, F., Webster, P., Taddei, N., Clark, A., Stefani, M., Ramponi, G. and Dobson, C.M. Designing conditions for *in vitro* formation of amyloid protofilaments and fibrils. *Proc Natl Acad Sci USA* **1999**, *96*, 3590–3594.

6 Ramirez-Alvarado, M., Merkel, J.S. and Regan, L. A Systematic exploration of the influence of the protein stability on amyloid fibril formation *in vitro*. *Proc Natl Acad Sci* **2000**, *97*, 8979–8984.

7 Yutani, K., Takayama, G., Goda, S., Yamagata, Y., Maki, S., Namba, K., Tsunasawa, S. and Ogasahara, K. The process of amyloid-like fibril formation by methionine aminopeptidase from a hyperthermophile, *Pyrococcus furiosus*. *Biochemistry* **2000**, *39*, 2769–2777.

8 Guijarro, J.I., Sunde, M., Jones, J.A., Campbell, I.D. and Dobson, C.M. Amyloid fibril formation by an SH3 domain. *Proc Natl Acad Sci USA* **1998**, *95*, 4224–4228.

9 Haggqvist, B., Naslund, J., Sletten, K., Westermark, G., Mucchiano, G., Tjernberg, L.O., Nordstedt, C., Engstrom, U. and Westermark, P. Medin: an integral fragment of aortic smooth muscle cell-produced lactadherin forms the most common human amyloid. *Proc Acad Natl Sci USA* **1999**, *96*, 8669–8674.

10 Konno, T. Multistep nucleus formation and a separate subunit contribution of the amyloidogenesis of heat-denatured monellin. *Protein Sci* **2001**, *10*, 2093–2101.

11 Iconomidou, V.A., Vriend, G. and Hamodrakas, K. Amyloid protects the silk moth oocyte and embryo. *FEBS Lett* **2000**, *479*, 141–145.

12 Chapman, M.R., Robinson, L.S., Pinkner, J.S., Roth, R., Heuser, J., Hammar, M., Normark, S. and Hultgren, S.J. Role of *Escherichia coli* curli operons in directing amyloid fiber formation. *Science* **2002**, *295*, 851–855.

13 Podrabsky, J.E., Carpenter, J.E. and Hand, S.C. Survival of water stress in annual fish embryos: dehydration avoidance and egg envelope amyloid fibers. *Am J Physiol* **2001**, *280*, R123–R131.

14 Jimenez, J.L., Nettleton, E.J., Bouchard, M., Robinson, C.V., Dobson, C.M. and Saibil, H. The protofilament structure of

insulin amyloid fibrils. *Proc Acad Natl Sci USA* **2002**, *99*, 9196–9201.

15 Jimenez, J. L., Guijarro, J. I., Orlova, E., Zurdo, J., Dobson, C. M., Sunde, M. and Saibil, H. Cryo-electron microscopy structure of an SH3 amyloid fibril and model of the molecular packing. *EMBO J* **1999**, *18*, 815–821.

16 Dobson, C. M. Protein misfolding, evolution and disease. *Trends Biochem Sci* **2002**, *24*, 329–332.

17 Anfinsen, C. B. Principles that govern the folding of protein chains. *Science* **1973**, *181*, 223–230.

18 Baker, D. and Agard, D. A. Kinetics versus thermodynamics in protein folding. *Biochemistry* **1994**, *33*, 7505–7509.

19 Chapman, M. R., Robinson, L. S., Pinkner, J. S., Roth, R., Heuser, J., Hammar, M., Normark, S. and Hultgren, S. J. Role of *Escherichia coli* curli operones in directing amyloid fiber formation. *Science* **2002**, *295*, 851–855.

20 Berson, J. F., Theos, A. C., Harper, D. C., Tenza, D., Raposo, G. and Marks, M. S. Proprotein convertase cleavage liberates a fibrillogenic fragment of a resident glycoprotein to initiate melanosome biogenesis. *J Cell Biol* **2003**, *161*, 521–533.

21 Baskakov, I. V., Aagaard, C., Mehlhorn, I., Wille, H., Groth, D., Baldwin, M. A., Prusiner, S. B. and Cohen, F. E. Self-assembly of recombinant prion protein of 106 residues. *Biochemistry* **2000**, *39*, 2792–2804.

22 Nettleton, E. J., Tito, P., Sunde, M., Bouchard, M., Dobson, C. M. and Robinson, C. V. Characterization of the oligomeric states of insulin in self-assembly and amyloid fibril formation by mass spectrometry. *Biophys J* **2000**, *79*, 1053–1065.

23 Huang, T. H. J., Yang, D.-S., Plaskos, N. P., Go, S., Yip, C. M., Fraser, P. E. and Chakrabartty, A. Structural studies of soluble oligomers of the Alzheimer beta-amyloid peptide. *J Mol Biol* **2000**, *297*, 73–87.

24 Walsh, D. M., Hartley, D. M., Kusumoto, Y., Fezoui, Y., Condron, M. M., Lomakin, A., Benedek, G. B., Selkoe, D. J. and Teplow, D. B. Amyloid beta-protein fibrillogenesis: structure and biological activity of protofibrillar intermediates. *J Biol Chem* **1999**, *274*, 25945–25952.

25 Serag, A. A., Altenbach, C., Gingery, M., Hubbell, W. L. and Yeates, T. O. Identification of a subunit interface in transthyretin amyloid fibrils: evidence for self-assembly from oligomeric building blocks. *Biochemistry* **2001**, *40*, 9089–9096.

26 Baskakov, I. V., Legname, G., Baldwin, M. A., Prusiner, S. B. and Cohen, F. E. Pathway complexity of prion protein assembly into amyloid. *J Biol Chem* **2002**, *277*, 21140–21148.

27 Baskakov, I. V., Legname, G., Prusiner, S. B. and Cohen, F. E. Folding of prion protein to its native α-helical conformation is under kinetic control. *J Biol Chem* **2001**, *276*, 19687–19690.

28 Hammarstrom, P., Wiseman, R. L., Powers, E. T. and Kelly, J. W. Prevention of transthyretin amyloid disease by changing protein misfolding energetics. *Science* **2003**, *299*, 713–716.

29 Khurana, R., Gillespie, J. R., Talapatra, A., Minert, L. J., Ionescu-Zanetti, C., Millett, I. and Fink, A. L. Partially folded intermediates as critical precursors of light chain amyloid fibrils and amorphous aggregates. *Biochemistry* **2001**, *40*, 3525–3535.

30 Chiti, F., Taddei, N., Bucciantini, M., White, P., Ramponi, G. and Dobson, C. M. Mutational analysis of the propensity for amyloid formation by a globular protein. *EMBO J* **2000**, *19*, 1441–1449.

31 Hamada, D. and Dobson, C. M. A kinetic of beta-lactoglobulin amyloid fibril formation promoted by urea. *Protein Sci* **2002**, *11*, 2417–2426.

32 Nielsen, L., Khurana, R., Coats, A., Frokjaer, S., Brange, J., Vyas, S., Uversky, V. N. and Fink, A. L. Effect of environmental factors on the kinetics of insulin fibril formation: elucidation of the molecular mechanism. *Biochemistry* **2001**, *40*, 6036–6046.

33 Kundu, B., Maiti, N. R., Jones, E. M., Surewicz, K. A., Vanik, D. L. and Surewicz, W. K. Nucleation-dependent conformational conversion of the Y145Stop variant of human prion protein: structural clues for prion propagation. *Proc Acad Natl Sci USA* **2003**, *100*, 12069–12074.

34 Sokolowski, F., Modler, A. J., Masuch, R., Zirwer, D., Baier, M., Lutsch, G., M. D. A., Gast, K. and Naumann, D. Formation of critical oligomers is a key event during conformational transition of recombinant Syrian hamster prion protein. *J Biol Chem* **2003**, *278*, 40481–40492.

35 Kauzmann, W. Some factors in the interpretation of protein denaturation. *Adv Protein Chem* **1959**, *14*, 1–63.

36 Honig, B. and Cohen, F. E. Adding backbone to protein folding: why proteins are polypeptides. *Fold Des* **1996**, *1*, R17–R20.

37 Balbirnie, M., Grothe, R. and Eisenberg, D. S. An amyloid-forming peptide from the yeast prion Sup35 reveals a dehydrated beta-sheet structure for amyloid. *Proc Acad Natl Sci USA* **2001**, *98*, 2375–2380.

38 Bolen, D. W. and Baskakov, I. V. The osmophobic effect: natural selection of a thermodynamic force in protein folding. *J Mol Biol* **2001**, *310*, 955–963.

39 Lu, B. Y. and Chang, J. Y. Isolation of isoforms of mouse prion protein with PrP^{Sc}-like structural properties. *Biochemistry* **2001**, *40*, 13390–13396.

40 Baskakov, I. V. Autocatalytic conversion of recombinant prion proteins displays a species barrier. *J Biol Chem* **2004**, *279*, 586–595.

41 Baskakov, I. V. and Bolen, D. W. Monitoring the sizes of denatured ensembles of staphylococcal nuclease proteins: implications regarding *m* values, intermediates, and thermodynamics. *Biochemistry* **1998**, *37*, 18010–18017.

42 Bolen, D. W. and Yang, M. Effects of guanidine hydrochloride on the proton inventory of proteins: implications on interpretations of protein stability. *Biochemistry* **2000**, *39*, 15208–15216.

43 Yang, M., Liu, D. and Bolen, D. W. The peculiar nature of the guanidine hydrochloride-induced two-state denaturation of staphylococcal nuclease: a calorimetric study. *Biochemistry* **1999**, *38*, 11216–11222.

44 Baskakov, I. V., Legname, G., Gryczynski, Z. and Prusiner, S. B. The peculiar nature of unfolding of human prion protein. *Protein Sci* **2004**, *13*, 586–595.

45 Kuwata, K., Li, H., Yamada, H., Legname, G., Prusiner, S. B., Akasaka, K. and James, T. L. Locally disordered conformer of the hamster prion protein: a crucial intermediate of PrP^{Sc}? *Biochemistry* **2002**, *41*, 12277–12283.

46 Gajdusek, D. C. Spontaneous generation of infectious nucleating amyloids in the transmissible and nontransmissible cerebral amyloidoses. *Mol Neurobiol* **1994**, *8*, 1–13.

47 Prusiner, S. B. Prion diseases and the BSE crisis. *Science* **1997**, *278*, 245–251.

48 Wickner, R. B. [*URE3*] as an altered *URE2* protein: evidence for a prion analog in *Saccharomyces cerevisiae*. *Science* **1994**, *264*, 566–569.

49 Maddelein, M. L., Dos Reis, S., Duvezin-Caubet, S., Coulary-Salin, B. and Saupe, S. J. Amyloid aggregates of the HET-S prion protein are infectious. *Proc Acad Natl Sci USA* **2002**, *99*, 7402–7407.

50 Legname, G., Baskakov, I. V., Nguyen, H.-O. B., Riesner, D., Cohen, F. E., Dearmond, S. J. and Prusiner, S. B. Synthetic wild-type mammalian prions. Paper presented at the *First International Conference of the European Network of Excellence Neuroprion*, Paris, May 25, **2004**.

51 Chien, P. and Weissman, J. S. Conformational diversity in a yeast prion dictates its seeding specificity. *Nature* **2001**, *410*, 223–227.

52 Tanaka, M., Chien, P., Naber, N., Cooke, R. and Weissman, J. S. Conformational variations in an infectious protein determine prion strain differences. *Nature* **2004**, *428*, 323–328.

53 Lundmark, K., Westermark, G. T., Nystrom, S., Murphy, C. L., Solomon, A. and Westermark, P. Transmissibility of systemic amyloidosis by a prion-like mechanism. *Proc Acad Natl Sci USA* **2002**, *99*, 6979–6984.

54 Rochet, J. C., Conway, K. A. and Lansbury, P. T., Jr. Inhibition of fibrillization and accumulation of prefibrillar oligomers in mixtures of human and mouse alpha-synuclein. *Biochemistry* **2000**, *39*, 10619–10626.

55 O'Nuallain, B., Williams, A. D., Westermark, P. and Wetzel, R. Seeding specificity in amyloid growth induced by hetero-

logous fibrils. *J Biol Chem* **2004**, *279*, 17490–17494.

56 Harper, J. D. and Lansbury, P. T., Jr. Models of amyloid seeding in Alzheimer's disease and scrapie: mechanistic truths and physiological consequences of the time-dependent solubility of amyloid proteins. *Annu Rev Biochem* **1997**, *66*, 385–407.

57 Cohen, F. E., Pan, K.-M., Huang, Z., Baldwin, M., Fletterick, R. J. and Prusiner, S. B. Structural clues to prion replication. *Science* **1994**, *264*, 530–531.

58 Cohen, F. E. and Prusiner, S. B. Pathologic conformations of prion proteins. *Annu Rev Biochem* **1998**, *67*, 793–819.

59 Fraser, H. and Dickinson, A. G. Scrapie in mice. Agent–strain differences in the distribution and intensity of grey matter vacuolation. *J Comp Pathol* **1973**, *83*, 29–40.

60 Dickinson, A. G. and Fraser, H. Modification of the pathogenesis of scrapie in mice by treatment of the agent. *Nature* **1969**, *222*, 892–893.

61 Hecker, R., Taraboulos, A., Scott, M., Pan, K.-M., Torchia, M., Jendroska, K., Dearmond, S. J. and Prusiner, S. B. Replication of distinct scrapie prion isolates is region specific in brains of transgenic mice and hamsters. *Genes Dev* **1992**, *6*, 1213–1228.

62 Bessen, R. A. and Marsh, R. F. Distinct Prp properties suggest the molecular basis of strain variation in transmissible mink encephalopathy. *J Virol* **1994**, *68*, 7859–7868.

63 Telling, G. C., Parchi, P., Dearmond, S. J., Cortelli, P., Montagna, P., Gabizon, R., Mastrianni, J., Lugaresi, E., Gambetti, P. and Prusiner, S. B. Evidence for the conformation of the pathologic isoform of the prion protein enciphering and propagating prion diversity. *Science* **1996**, *274*, 2079–2082.

64 Safar, J., Wille, H., Itri, V., Groth, D., Serban, H., Torchia, M., Cohen, F. E. and Prusiner, S. B. Eight prion strains have PrP^{Sc} molecules with different conformations. *Nat Med* **1998**, *4*, 1157–1165.

65 Caughey, B., Raymond, G. J. and Bessen, R. A. Strain-dependent differences in β-sheet conformations of abnormal prion protein. *J Biol Chem* **1998**, *273*, 32230–32235.

66 Peretz, D., Scott, M., Groth, D., Williamson, A., Burton, D., Cohen, F. E. and Prusiner, S. B. Strain-specified relative conformational stability of the scrapie prion protein. *Protein Sci* **2001**, *10*, 854–863.

67 Scott, M., Ridley, R. M., Baker, H. F., Dearmond, S. J. and Prusiner, S. B. Transgenetic investigations of the species barrier and prion strains. In *Prion Biology and Diseases*, Prusiner, S. B. (ed.). Cold Spring Harbor Laboratory Press, Cold Spring Harbor, NY, **1999**, pp. 307–347.

68 Pattison, I. H. Experiments with scrapie with special reference to the nature of the agent and the pathology of the disease. In *Slow, Latent and Temperate Virus Infections, NINDB Monograph 2*, Gajdusek, D. C., Gibbs, C. J., Jr. and Alpers, M. P. (eds). US Government Printing Office, Washington, DC, **1965**, pp. 249–257.

69 Pattison, I. H. and Jones, K. M. Modification of a strain of mouse-adapted scrapie by passage through rats. *Res Vet Sci* **1968**, *9*, 408–410.

70 Scott, M. R., Groth, D., Tatzelt, J., Torchia, M., Tremblay, P., Dearmond, S. J. and Prusiner, S. B. Propagation of prion strains through specific conformers of the prion protein. *J Virol* **1997**, *71*, 9032–9044.

71 Horiuchi, M., Priola, S. A., Chabry, J. and Caughey, B. Interactions between heterologous forms of prion protein: binding, inhibition of conversion, and species barriers. *Proc Natl Acad Sci USA* **2000**, *97*, 5836–5841.

72 Broome, B. M. and Hecht, M. H. Nature disfavors sequences of alternating polar and non-polar amino acids: implications for amyloidogenesis. *J Mol Biol* **2000**, *296*, 961–968.

5
Role of Post-translational Chemical Modifications in Amyloid Fibril Formation

Melanie R. Nilsson

5.1
Introduction

From a purely naïve point of view, it would seem that an obvious connection between normal protein degradation reactions (i.e. normal protein aging) and diseases associated with aging (e.g. amyloidosis) may exist. This connection, however, has received little attention. The lack of research in this area stems, in part, from the difficulties associated with detecting these subtle modifications and, in part, from the proposition that these modifications are not relevant to disease. The "conformational hypothesis" (the predominant view) suggests that amyloid fibril formation results from destabilization of the native protein as a result of changes in the local solution environment (e.g. pH, ionic strength, accessory proteins) [1–3]. An alternative view, which I will call the "chemical modification hypothesis", suggests the protein conformational change is facilitated by a change in the primary structure (i.e. chemical modification) of the protein. The local solution environment can dramatically affect the types of chemical modifications that occur, so the solution environment is still a critical factor in fibril formation. Both hypotheses are consistent with Anfinsen's view (cited in [1–3]) that the primary structure and aqueous environment govern the three-dimensional fold of a protein. In the conformational change hypothesis, the alteration in three-dimensional structure arises from changes in the aqueous environment, whereas in the chemical modification hypothesis, the altered conformation is a result of an altered primary sequence. It is quite possible that both hypotheses may be true, but applicable to different proteins. However, it is important to resolve which mechanism occurs in each disease because unraveling the mechanism of aggregation will have a significant impact on evaluating appropriate therapeutic strategies. The purpose of this chapter is to present the evidence that supports the view in which post-translational chemical modification of the primary sequence may be an important mechanism of amyloid fibril formation.

The primary structure of a protein refers to the covalent arrangement of atoms in a protein. This includes not only the sequence of amino acids, but also

Amyloid Proteins. The Beta Sheet Conformation and Disease. J. D. Sipe

ISBN: 3-527-31072-X

post-translational processing (e.g. formation of disulfides, proteolysis, glycosylation, lipid modification, phosphorylation) in addition to non-enzymatically regulated chemical modifications such as deamidation, racemization and isomerization. Table 5.1 lists some chemical modifications that are known to occur in proteins [4].

It is important to approach any relationship between primary structure and protein conformation with caution, since these relationships can be extremely difficult to predict. For example, it is not surprising that myoglobin and lysozyme have very different folds (myoglobin is all α-helical and lysozyme is a mixed α/β fold) since the primary structures are not similar. However, there are proteins with significantly different primary structures that all have the same basic fold (e.g. several non-homologous proteins fold into an α/β (TIM) barrel [5]). Another insightful example is the case of mutant sequences of transthyretin (TTR). Some mutations of TTR destabilize the protein conformation and facilitate amyloid deposition. However, there are also mutations that stabilize the

Table 5.1 Modification of amino acids in proteins (reprinted from [4], with permission)

Amino Acid	Type of modification
Alanine	GPI-anchoring
Arginine	*N*-methylation, ADP-ribosylation, deimination
Asparagine	glycosylation, GPI-anchoring, deamidation
Aspartic acid	phosphorylation, methylation, isomerization, racemization, GPI-anchoring
Cysteine	cystine formation, selenocysteine formation, heme linkage, myristoylation, palmitoylation, *S*-nitrosylation, oxidation to C-α-formylglycine, deamination, GPI-anchoring
Glutamic acid	γ-carboxyglutamate formation, methylation, polyglycylation
Glutamine	deamidation, crosslinking, pyroglutamate formation, ADP-ribosylation, *N*-methylation
Glycine	GPI-anchoring, myristoylation
Histidine	methylation, diphthamide formation, phosphorylation, deamination
Lysine	*N*-acetylation, *N*-methylation oxidation, hydroxylation, crosslinking, palmitoylation, biotinylation, deamination
Phenylalanine	hydroxylation
Proline	hydroxylation, glycosylation
Serine	phosphorylation, glycosylation, acetylation, GPI-anchoring, oxidation to C-α-formylglycine, dehydration and formation of a thioether link
Threonine	phosphorylation, glycosylation, dehydration and formation of a thioether link
Tyrosine	iodination, phosphorylation, sulfation, flavin linkage, nucleotide linkage, *o*-tyrosine, chloro-, nitrotyrosine and dityrosine formation
N-terminus	formylation, acetylation, pyroglutamate formation
C-terminus	GPI-anchoring, amidation, polyglycylation

native conformation and suppress amyloid formation, and polymorphisms that have no significant impact on amyloid formation. (For a review on ATTR, see [6].) These examples emphasize that it is not easy to predict the impact a given chemical modification will have on amyloid fibril formation and, furthermore, it cannot be assumed *a priori* that chemical modifications will play a primary role in all amyloid diseases.

Post-translational modifications encompass both enzymatic and non-enzymatic modifications. Enzymatically regulated modifications are a normal part of protein processing and typically are required for protein folding, stability or function. Non-enzymatic chemical modifications, in contrast, are generally not part of normal protein processing and result simply from the favorability of these reactions in the chemical environment of the body. Both types of modifications result in a change in the primary structure of the protein and, therefore, can potentially contribute to amyloid fibril formation.

The role of enzymatically regulated modifications in amyloid fibril formation is unclear. There are two possible scenarios by which these modifications could affect fibril formation – one in which the correct processing events do not occur and the other in which there is modification at the wrong place or at the wrong time. The absence of the modifications necessary for normal processing could play a role in amyloid formation by either destabilizing the native fold or by the presence of an unprocessed region that may initiate deposition. For example, incompletely processed islet amyloid polypeptide (IAPP), in which the pro-hormone is not completely proteolytically cleaved, has been identified in amyloid deposits [7–9]. The incorrectly processed peptide contains IAPP with an additional N-terminal peptide. The presence of the N-terminal segment could contribute to amyloid formation by destabilizing IAPP (possibly by preventing the formation of the proper storage form [10]) or the N-terminal peptide may bind to components of the extracellular matrix and act as an initiation site for fibril formation [11]. Similarly, evidence for the precursors of calcitonin [12] and atrial natriuretic peptide [13] have been identified in amyloid deposits. Enzymatically regulated modifications may also contribute to fibril formation by occurring at the wrong time or in the wrong place. For example, tau, the protein component of neurofibrillary tangles in Alzheimer's disease, is hyperphosphorylated [14]. Phosphorylation is an enzymatically regulated process but, in this case, occurs in excess.

The role of non-enzymatic chemical modifications in amyloid fibril formation is also unclear. Proteins spontaneously degrade *in vivo* and *in vitro* by a variety of mechanisms. Some of the more common reactions are racemization, oxidation of methionine, and deamidation of asparagine and glutamine. *In vitro* these modifications are frequently ignored and considered of minor significance. One notable exception, however, is the stability of peptide drugs, which has received considerable attention in an attempt to prolong their shelf life. *In vivo* the body normally protects itself against these modifications by either preventing the modification, repairing the modified sequences or clearing the damaged proteins. It has been proposed, for example, that the deamidation of asparagine to aspartic/isoaspartic acid may serve as a type of molecular clock to estimate the amount of time a protein has been in the cell [15]. Presumably, after a specified period of time, a signal will be sent to degrade the protein.

Alterations due to the labile nature of amino acids can lead to additional functionality, but there is also a danger that these reactions will result in negative consequences. This is reminiscent of the mutability of DNA that allows for evolution but also can lead to disease. Like random mutations in DNA, there is no way to easily predict the outcome of a chemical modification, but the probability of a damaging modification that is not repaired will statistically increase with age (for an overview of chemical damage associated with aging, see [16]). This is a very attractive proposition to explain the familial versus sporadic variability of amyloid disease progression. Familial cases of amyloid diseases occur when a genetic mutation leads to a protein sequence that is more prone to aggregation than the wild-type sequence. There are also, however, late-onset variations in which there are no mutations and the mechanism of destabilization of the native protein is unclear. One possible explanation is that chemical modifications which result from normal age-related protein degradation reactions may destabilize the proteins in a similar fashion as inherited point mutations.

5.2 Common Modifications that May Play a Significant Role *In Vivo*

5.2.1 Cleavage by Proteases or Non-enzymatic Hydrolysis

The cleavage of a peptide or protein can occur as a result of protease activity or non-enzymatic hydrolysis. Four major classes of proteases (aspartic acid, serine, cysteine and metalloproteases) have been identified and named according to the catalytic portion of the active site. Proteases are commonly stored in lysosomes and, therefore, protein cleavage associated with lysosomes has been implicated for many years in amyloidogenesis. For example, Glenner et al. in 1971 proposed a primary role for proteolysis in AL amyloid formation: "The fact that 'amyloid' fibrils can be created from some Bence–Jones proteins at a physiologic temperature in the presence of a proteolytic enzyme having an acidic pH optimum suggests that one possible pathogenetic mechanism for amyloid formation may be by means of intralysosomal catheptic digestion of light polypeptide chains of immunoglobulins" [17].

Cleavage of the amide backbone of a peptide or protein, however, does not need to be mediated by proteases. Acidic conditions can be used to facilitate cleavage of a protein into constituent amino acids in preparation for amino acid analysis. Although the conditions employed in this technique are not found in the body, hydrolysis of many amide bonds occurs under physiologically relevant conditions. Amide hydrolysis can be catalyzed by acid or base, but the mechanisms and labile sites are substantially different. In several cases in which the amyloid fibril component is a fragment of a larger precursor, specific proteases have not been identified in the generation of the amyloidogenic fragment, suggesting that some proteins may be cleaved by non-enzymatic reactions.

A survey of the amyloid literature results in a multitude of examples in which cleavage of the protein backbone is implicated in amyloid fibril formation. For example, in the most common form of amyloid deposition (which occurs in the aortic smooth muscle), the major protein component is medin, a small fragment of lactadherin [18]. More recently, unusual fragments of kerato-epithelin have been identified in corneal amyloid deposits [19]. Additional examples in which cleavage by proteolysis or hydrolysis may play a significant role in amyloid fibril formation are highlighted below.

Non-pathological proteolytically induced amyloid fibril formation has been identified in melanosome biogenesis [20]. Pmel17 is a protein expressed in melanocytes and, as part of normal processing, is cleaved into two unequal pieces (the 80-kDa Mα fragment and the 28-kDa Mβ fragment) by a furin-like convertase. The Mα fragment subsequently forms insoluble fibrils which act as a scaffold for melanin pigments. The full-length protein is unable to form fibrils and assembly only occurs upon cleavage. It is speculated that this tight control mechanism prevents aberrant fibril formation. Other physiologically relevant amyloid processes, such as curli formation in *Escherichia coli*, are also carefully regulated events [21].

Lithostathine (also known as pancreatic stone protein, S2) is a 144-residue secretory protein that generates a 133-residue amyloidogenic fragment (S1, pancreatic thread protein) upon spontaneous autocleavage or cleavage with trypsin [22]. S1 has been identified in the brain of patients with Alzheimer's disease [23] and in pancreatic stones [24]. Cleavage of lithostathine to S1 is a pre-fibrillogenic event and is an interesting example of fibril formation that can be induced by proteolysis or by an autocatalytic mechanism.

Gelsolin is an actin-binding protein [25]. In familial amyloidosis of the Finnish type (FAF), there is a mutation at residue 187 in which an aspartic acid is converted to an asparagine or tyrosine [26, 27]. The mutations have mild effects on the thermodynamic stability of the protein [28] but lead to a susceptibility to proteolytic cleavage [29]. Cleavage results in the production of an amyloidogenic fragment that encompasses residues 173–223 or 173–225 [26]. In contrast, the wild-type protein is not cleaved and does not form amyloid fibrils. The amyloidogenic fragment has been observed in the cerebrospinal fluid (CSF) [30] and identified as the major circulating form of patients with FAF [31, 32], suggesting that cleavage is the driving force in FAF amyloid deposition.

Cystatin C (also known as γ-trace basic protein) is an inhibitor of cysteine proteases. In hereditary cystatin C amyloid angiopathy (HCCAA or HCHWA-I), there is a mutation at position 68 in which a leucine is replaced with a glutamine [33]. Amyloid deposits of this variant protein contain cystatin C lacking the first 10 residues [33]. The soluble truncated form has not been observed in detectable quantities in the CSF [34], but the loss of the 10-residue fragment is consistent with the proposed domain-swapped dimer mechanism of cystatin fibril formation [35]. The dimer interface occurs between strands 2 and 3 (residues 42–59) [35], and the loss of the first 10 residues would eliminate the edge strand (strand 1) and potentially facilitate deposition. In sporadic cerebral amyloid angiopathy with cystatin C deposition (SCCAA), there is no mutation present and full-length cystatin C is found in amyloid deposits. However, in SCCAA it is speculated that cystatin C co-precipitates with the Aβ peptide [36].

N-terminal fragments of different apolipoproteins are associated with a wide variety of amyloid deposits. Fragments of apolipoprotein AIV (ApoAIV) have been co-localized with TTR in cardiac amyloid and the authors have proposed that this fragment may "serve as a local nidus for fibrillogenesis" [37]. An N-terminal fragment of apolipoprotein AI (ApoAI) is the major fibril protein in atherosclerotic plaques and a fragment of variant ApoAI is the major amyloid constituent in FAP Type III (Iowa) [38–40]. Similarly, the N-terminal portion of the apolipoprotein serum amyloid A (ApoSAA) has been isolated from amyloid deposits and this fragment has been generated *in vitro* upon digestion with cathepsin B, suggesting that proteolysis is a pre-fibrillogenic event [41, 42]. Hatters and Howlett propose a mechanism by which apolipoproteins may be deposited. They suggest that "oxidative processes may promote protein truncation or modification of apolipoproteins in a manner that perturbs their lipid binding properties and thus by default promote the amyloidogenic folding options" [43].

Familial British dementia (FBD) is associated with amyloid fibrils composed of C-terminal fragments of α- and β-tubulin [44] and fragments of a protein of unknown function named BRI [45, 46]. In FBD, a mutation in the *BRI* gene results in the production of a protein (BRI-L) that is extended by 11 amino acids at the C-terminus. Cleavage of this protein near the C-terminus results in the release of a 34-amino-acid peptide that readily forms amyloid fibrils [45]. Cleavage of BRI-L is speculated to be mediated by furin [47, 48]. A similar proteolytic event is involved in the formation of a different amyloidogenic peptide derived from BRI that occurs in familial Danish dementia [48].

β_2-Microglobulin (β_2M) amyloidosis occurs in patients undergoing long-term dialysis treatment. N-terminally truncated forms of β_2M are common in *ex vivo* amyloid. Cleavage occurs between K6 and I7 (around 25%), and to a lesser extent between residues 10–11, 17–18, 19–20, 86–87 and 98–99 [49, 50]. Truncated β_2M, which corresponds to a loss of the first six residues, has a high tendency to aggregate *in vitro* [49, 51]. Therefore, it is possible that truncation of β_2M destabilizes the native fold and leads to amyloid fibril formation.

TTR amyloid formation occurs in familial amyloid polyneuropathy (FAP) and senile systemic amyloidosis (SSA). FAP is associated with mutations in TTR [52], while SSA is associated with a fragment of wild-type TTR that results from cleavage between residues 46 and 52 [53]. It has been proposed that the mutations observed in FAP destabilize the native TTR fold and facilitate deposition. The lack of mutations observed in the A and H β-strands of TTR and the demonstration that a synthetic peptide of the A-strand can form amyloid fibrils *in vitro* [54, 55] led to the hypothesis that the A-strand (and H-strand) may be critical for fibril formation [56]. However, if the A-strand plays a critical role in the fibril structure, removal of this strand would be detrimental to fibril formation. The observation that around 80% of wild-type TTR in *ex vivo* amyloid fibrils lacks the N-terminal region (residues 1–46, 49, 52 or 53) [53] which contains the A-strand (residues 10–20) is inconsistent with this view. Therefore, wild-type TTR and mutant TTR may form fibrils via different pathways. Mutations of TTR could lead to global destabilization of the native fold that facilitates amyloid fibril formation, resulting in the relatively early age of onset observed in FAP. In contrast, wild-type TTR may be cleaved as a result of age-related chemical modification or aberrant proteolysis, which would be consistent with the older age of onset of SSA.

α-Synuclein (also known as NACP) has been identified in inclusions associated with Parkinson's disease, dementia with Lewy bodies (DLB), Alzheimer's disease, multiple system atrophy (MSA) and amyotrophic lateral sclerosis (ALS) (reviewed in [57]). Lewy bodies contain α-synuclein and *in vitro* studies reveal α-synuclein fibrils have all the hallmarks of amyloid [58]. Truncated forms of α-synuclein have been reported in Parkinson's disease and DLB [59, 60]. Fragments of α-synuclein (around 2–4 kDa shorter than the full-length protein) have been identified in Lewy bodies, but not in a normal control [60]. Fragments that are around 4 and 10 kDa shorter than full-length α-synuclein have been observed in soluble fractions of Parkinson's disease, DLB and normal control extracts [59], suggesting these fragments may be normal breakdown products of α-synuclein. However, the smaller fragment was also identified in poorly soluble fractions suggesting this fragment may play a role in fibril formation [59]. Immunochemical analysis indicates that the 4- and 10-kDa truncated fragments are derived from the middle region of α-synuclein [59], which is consistent with the previously identified NAC peptide isolated from the brains of patients with Alzheimer's disease [61].

Significant Aβ N-terminal and C-terminal heterogeneity has been observed in Alzheimer's disease amyloid deposits. Aβ is generated by cleavage of a larger precursor, the amyloid β precursor protein (AβPP). In addition to Aβ, cleavage of AβPP can generate two other peptides, C100 and p3 [62]. All three peptides (Aβ, p3 and C100) have been implicated in amyloid formation. Intact C100 is competent to form amyloid fibrils which, upon non-specific proteolysis, results in fibrils composed of Aβ [63]. The peptide p3 is the major component of "preamyloid", indicating an important role of p3 in amyloidogenesis [64]. Ragged N- and C-termini of Aβ have been found in plaques. The C-termini reported include V39, V40, A42 and T43 [65]. Aβ which ends in A42 is highly amyloidogenic and represents the earliest C-terminus observed in deposits, suggesting that an increase in the ratio of A42 to other peptides may facilitate fibril formation [66, 67]. Aβ peptides have been identified with N-termini beginning with –I6, –V3, D1 or isoD1, A2, pE3, F4, S8, G9, pE11 and L17. The earliest N-termini in deposits have been speculated to be L17 (p3) followed by deposition of pE3 and D1 [64, 66].

The full-length human prion protein (PrP^C) consists of residues 23–231 after the signal sequence is removed from the primary translation product and the GPI anchor attached. A protein truncated at approximately residue 80–90 is significantly elevated in Creutzfeldt-Jakob disease [68]. It has been proposed that heterogeneity at the N-terminus could contribute to the strain variation observed in prion diseases [69].

Neuronal inclusions observed in Huntington's disease are fibrillar [70] and Congo red birefringent [71, 72]. The inclusions contain huntingtin and a 40-kDa N-terminal fragment of huntingtin [70]. *In vitro* experiments on N-terminal fragments of huntingtin have led to the suggestion that "the proteolytic cleavage event that yields this 'toxic' fragment would also provide a rate-limiting step in Huntington's disease onset" [73]. Proteolysis has also been proposed as an important factor in other polyglutamine disorders [74]. The neurotoxicity observed in Huntington's disease and other polyglutamine diseases may be mediated by the binding of polyglutamine proteins (e.g. TATA-binding protein) to the huntingtin deposit which could deplete the necessary supply of these proteins in the brain [72].

Aberrant processing of viral proteins may lead to amyloidogenic fragments that can nucleate amyloid deposition [75]. This has been championed in the case of glycoprotein B (gB) from herpes simplex virus 1 (HSV1). HSV1 DNA has previously been identified in the same regions of the brain that are affected by Alzheimer's disease [76, 77] and more recently a fragment of gB has been demonstrated to accelerate Aβ fibril formation *in vitro* [75].

5.2.2
Deamidation, Isomerization, Racemization and Protein L-Isoaspartyl Methyltransferase (PIMT)

Deamidation of asparagine and glutamine is common under physiological conditions (for reviews, see [78–81]). Both the reaction rate and mechanism are influenced by a host of factors, including solution conditions (e.g. pH, ionic strength, buffers, temperature), primary sequence, secondary structure and tertiary structure. The inherent chemical and thermal lability of these residues is highlighted in the observed degradation products of peptide drugs upon aging (e.g. calcitonin [82], pramlintide [83], insulin [84]) and in the low number of Q/N repeats in thermophiles compared to mesophiles [85]. Deamidation reactions can lead to fragmentation of the amide backbone or result in subtle backbone or side-chain modifications. For example, asparagine deamidation can lead to the formation of L- or D-aspartic acid, L- or D-isoaspartic acid or fragmentation of the protein chain (Fig. 5.1) [86, 87]. Deamidation of glutamine can yield glutamic acid or, if the residue is at the N-terminus, pyroglutamic acid.

The conversion of an asparagine to an aspartic/isoaspartic acid or glutamine to glutamic/pyroglutamic acid results in a protein that has the equivalent of a post-translational point mutation. From a protein folding point of view, the chemical modification could affect the thermodynamic stability of the native fold, the kinetics of folding or unfolding, or the overall flexibility of the molecule. Deamidation of ribonuclease A (N67isoAsp), for example, significantly decreases the folding rate [88] while deamidation of lysozyme (N103D or N106D)

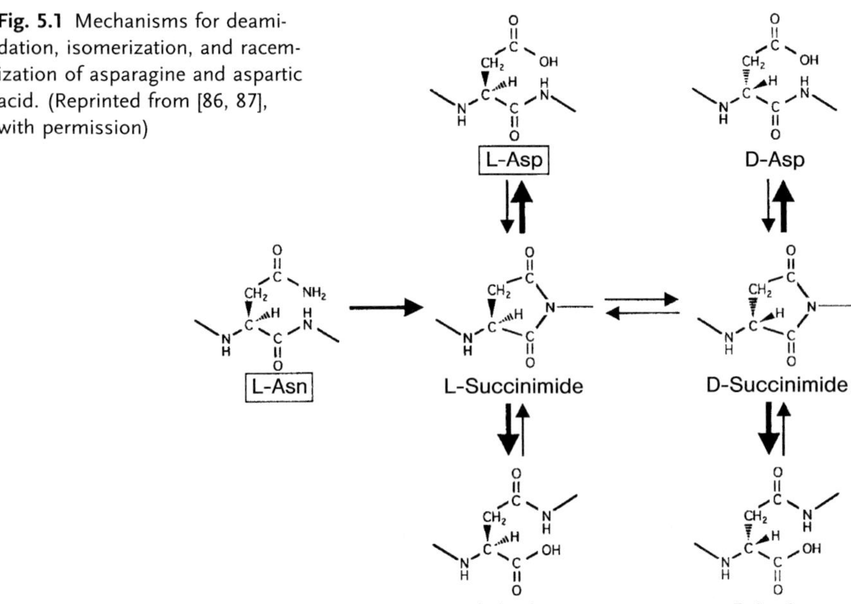

Fig. 5.1 Mechanisms for deamidation, isomerization, and racemization of asparagine and aspartic acid. (Reprinted from [86, 87], with permission)

increases the susceptibility to proteolysis [89]. The destabilization of the native state or enhanced flexibility could result in proteins that are more prone to aggregation.

Deamidation leads to a change in net charge of the protein. Under physiological conditions (pH $\sim$ 7.4), asparagine and glutamine are neutral, but aspartic acid and glutamic acid are negatively charged. Model studies suggest that aggregation is favored when the net charge of the protein is closer to zero [90], so deamidation may facilitate amyloid fibril formation of basic proteins by altering the net charge. Several other post-translational chemical modifications can alter the net charge of a protein, including C-terminal amidation, N-terminal acetylation, advanced glycation (or glycosylation) end-product (AGE) formation and arginylation [91].

Polyglutamine repeat diseases [e.g. Huntington's disease, spinalbulbar muscular atrophy (Kennedy's disease), dentatorubral-pallidoluysian atrophy, and spinocerebellar ataxias 1, 2, 3 (Machado-Joseph disease), 6, 7 and 17] are characterized by the aggregation of glutamine-rich peptides/proteins [92, 93]. Huntington's disease is the most common glutamine-repeat disease, and *in vitro* analysis of glutamine-rich peptides demonstrates that these deposits are fibrillar, β-sheet rich and bind Thioflavin T [94]. Polyglutamine peptides are likely candidates for deamidation, but the effect of deamidation on fibril formation has not yet been investigated. Polyglutamine peptides are, however, susceptible to transglutaminases which can catalyze deamidation of glutamine or crosslinking between glutamine and other amines (e.g. lysine) [95]. These chemical modifications could potentially play a role in fibril formation by promoting nucleation or by stabilizing the resulting fibril.

Glutamine- and asparagine-rich domains are common in yeast prions, and it is speculated that these repeats may provide interaction specificity [85]. These proteins, however, will also likely be susceptible to deamidation. Deamidation of PrP^C has been shown to occur spontaneously [96] and results in a protein that readily forms protease K-resistant aggregates upon exposure to Cu^{2+} [97]. Aggregation of other prions may be mediated by a similar pathway.

β_2M extracted from amyloid deposits has been reported to be deamidated at positions 17 and 42 (Asn to Asp/isoAsp) [98, 99]. *In vitro* experiments suggest that Asn 17 to Asp is unlikely to accelerate amyloid fibril formation [100], but the other deamidated species have not been investigated. Similarly, deamidation and isomerization of tau has been identified in paired helical filaments, although no specific role in the aggregation process has been identified for these modifications [101].

In familial cataract, amyloid fibrils composed of crystallins are thought to precede cataract formation *in vivo* [102] and this process is supported by *in vitro* studies [103]. The crystallins are a group of proteins that reside in the human lens. These proteins do not turn over as the lens ages, so chemical modifications of crystallins are common in both normal and cataractous lenses. Investigations to elucidate differences between normal and cataractous lenses have revealed elevated deamidation of γS-crystallin in cataractous lenses [104] at asparagine 143 [105].

Significant levels of glutamine deamidation have been reported in *ex vivo* Aβ amyloid deposits. The glutamine at position 3 of Aβ is significantly deamidated (10–50%) to yield an N-terminal pyroglutamic acid (pE3 Aβ) [65, 106–108]. Pyroglutamic acid at position 11 (pE11 Aβ) has also been reported [65]. The highly variable amounts reported for pE3 Aβ in the brain may, in part, be due to the different localization of this peptide [65, 108]. pE3 Aβ is common in diffuse amyloid plaques (proposed to be an early event in amyloid deposition) and the ratio of Aβ to pE3 Aβ directly correlates with age (younger subjects have higher levels of pE3 Aβ in plaques) [65, 109]. pE3 Aβ has also been shown to aggregate more readily than Aβ *in vitro* [107, 110]. Taken together, this data suggests that the formation of pE3 Aβ is a pre-fibrillogenic event which can induce amyloid fibril formation of pE3 Aβ followed by the subsequent deposition of Aβ onto these deposits [109].

Isomerization of aspartic acid and asparagine (concurrent with deamidation) is one of the most common non-enzymatic chemical modifications of proteins *in vivo*. This reaction occurs most readily in regions of high flexibility and with preference to sequences that contain an aspartic acid or asparagine followed by a glycine. Isomerization results in a change in conformation of the peptide/protein backbone [111] which could facilitate fibril formation. Aβ has three aspartic acid residues (D1, D7 and D23) and isomerization of each residue has been reported in *ex vivo* amyloid deposits [108, 112, 113]. If isomerization occurs as a secondary phenomenon to exposed Asp residues after Aβ fibril formation, a relatively equal distribution of chemically modified Aβ would be anticipated. Interestingly, the distribution of modified forms of Aβ differ between neuritic and vascular amyloid [108, 114]. This variable localization has, therefore, led to the hypothesis that Asp isomerization is a pre-fibrillogenic event [114]. A direct role in nucleation of fibril formation is supported by the observation that peptides with isoAsp at D1 and D7 have a higher β-sheet content [115], and isomerized D23 forms fibrils more readily than unmodified Aβ [113]. Furthermore, the presence of isoAsp (or D-Asp) can prevent protein turnover [112], thereby increasing the local concentration. Alternative explanations for the role of Asp isomerization in Alzheimer's disease have been proposed in which a central role is attributed to the succinamide intermediate that is transiently populated during the conversion of Asp to isoAsp. The succinamide intermediate, unlike Asp and isoAsp, is neutral and the loss of charge has been postulated to facilitate aggregation [116]. Furthermore, the succinamide residue adopts a geometry similar to a type II′ β-turn and may result in the formation of cytotoxic Aβ β-hairpins [117].

Racemization occurs when the stereochemistry at the α carbon is altered from the L configuration to the D configuration. This process is quite common [118] and can be catalyzed by racemases or occur non-enzymatically. Non-enzymatic racemization of amino acids in proteins has been explored as a possible dating technique to estimate the age of shells and other biomineralized materials [119]. Bacteria use racemases to generate cell wall peptides that contain both L- and D-amino acids which are relatively resistant to proteases. Some racemase inhibitors (e.g. cycloserine) produce neurological side-effects which may be a result of interactions with the recently discovered human serine racemase in the brain [120]. The presence of racemized amino acids can affect the protein conformation, have an impact on the folding/unfolding rates, or alter the thermodynamic stability. Although D-amino acids can destabilize β-sheets, peptides with D-amino acids have been shown to readily form amyloid fibrils [121].

Racemization of aspartic acid and serine residues has been observed in Aβ isolated from *ex vivo* amyloid fibrils [122]. Aspartic acid is very susceptible to spontaneous racemization and D-Asp Aβ has been identified *in vitro* [123]. D-Asp Aβ aggregates faster than unmodified Aβ, suggesting a possible role for racemization in amyloid fibril nucleation [123, 124]. D-Ser at position 8 likewise facilitates aggregation, but D-Ser at position 26 does not [124]. However, D-Ser 26 Aβ can be proteolyzed into a smaller, highly toxic fragment that may play a role in Alzheimer's disease [125].

PIMT is an intracellular repair enzyme that facilitates the conversion of D-Asp or L-isoAsp into L-Asp [126]. A lower amount of PIMT activity has been correlated with an earlier age of death [127], suggesting that there may be a relationship between the accumulation of chemically modified proteins and mortality. PIMT knock-out mice exhibit a 9-fold increase in L-isoAsp and a phenotype of fatal epilepsy [128]. Likewise, the examination of human hippocampal tissue reveals dramatically reduced levels (50% lower) of PIMT activity and a 2-fold increase in damaged tubulin in epileptic versus control samples [129]. Cortical samples show similar levels of PIMT activity compared to controls [129], which is consistent with an earlier Alzheimer's disease study [127]. Thus, local depletion of PIMT may be an interesting area of investigation to determine if excess deamidation or isomerization plays an active role in amyloid fibril formation.

5.2.3 Oxidative Damage

The reactants and products of oxidation reactions are highly variable, resulting in a staggering array of possible protein modifications that may impact amyloid deposition. Proteins can be oxidized by reacting with metals, lipids and other small molecules [16]. Oxidative damage has been proposed to play a role in a variety of late-onset diseases including Alzheimer's disease [130]. For example, lipid peroxides can react with cysteine, histidine, and lysine residues to generate advanced lipoxidation end-products (ALEs) which can affect protein structure, charge, hydrophobicity and crosslinking [16]. Baynes and Thorpe "propose that atherosclerosis is an age-related disease characterized by accelerated lipid peroxidation and lipoxidative aging of proteins in the vascular wall" [16]. Abnormal metabolites (e.g. ketoaldehydes) have also been proposed to similarly react with proteins and affect aggregation [131].

Cysteine side-chains are frequently oxidized to form disulfide bonds as part of normal protein processing, but aberrant disulfide bonds may facilitate amyloid formation [132]. For example, normal lenses contain αA-crystallin in which Cys131 and Cys142 are partially reduced, but cataractous lenses contain only oxidized Cys131 and Cys142 which may impact lens transparency [133]. The release of monomeric TTR from *ex vivo* amyloid upon reduction suggests that there may be some disulfide bonds formed between TTR monomers in the fibrils [53]. (Note: Cys10 is the only cysteine in wild-type TTR.) The identification of a Cys10Arg TTR mutation that results in FAP [134] and a Cys10Ala mutant that forms amyloid fibrils *in vitro* [135], however, indicates that an intermolecular disulfide bridge is not necessary for the initiation of amyloid fibril formation. The Cys10 crosslinks may instead play a role in the thermodynamic stabilization of the fibrils.

Cysteine residues on proteins can also react with free cysteine, cysteinylglycine, glutathione or sulfite *in vivo*. Disturbances in cysteine metabolism have been identified in several neurological disorders, including Alzheimer's disease, Parkinson's disease, motor neuron disease [136] and Hallervorden-Spatz disease [137]. Although no direct correlation to amyloid fibril formation has been elucidated, it has been proposed that altered thiol metabolism may contribute to the overall oxidative damage observed in these diseases. A more direct relationship between free thiols and amyloid fibril formation has been identified in AL amyloidosis and FAP. Protein extraction from *ex vivo* amyloid fibrils has revealed cysteinylation of Cys214 in κ1 light chain in AL amyloidosis [138], and conjugation of Cys10 of TTR to sulfite, cysteine, cysteinylglycine and glutathione in FAP [139]. Modification of Cys10 with cysteine, cysteinylglycine or glutathione results in destabilization of TTR and enhanced fibril formation [140]. Significantly, these modifications result in TTR molecules that readily form fibrils at pH 4.8, a condition under which wild-type TTR does not aggregate [140].

Methionine oxidation has been reported for several amyloid fibril proteins including lysozyme [141], PrP^{Sc} [142], β_2M [98] and Aβ [143, 144]. While earlier studies suggested that Met35 oxidation of Aβ increased the rate of aggregation [145], more recent work indicates that Met35 oxidation decreases the rate of Aβ fibril formation [146–148]. Oxidation of methionine residues in α-synuclein results in a decrease in aggregation, but this inhibition is eliminated in the presence of metals [149–151]. In view of the present evidence, the role of methionine oxidation in amyloidosis remains unclear. It is possible that, in some cases, the oxidation may be protective by slowing fibril formation.

Tyrosine is susceptible to a variety of oxidation reactions that can lead to the formation of chlorotyrosine, nitrotyrosine, *o*-tyrosine and dityrosine [16]. Tyrosine nitration of α-synuclein in Parkinson's disease and tau in Alzheimer's disease has been reported and postulated to contribute to disease progression by affecting fibril formation or stability [152, 153]. Chlorination of tyrosine occurs upon exposure to HOCl, a strong oxidizing agent that is produced by neutrophils and monocytes [154]. Tyrosyl chlorination can promote protein aggregation [155] and has been implicated in formation of atherosclerotic plaques [156]. Dityrosine formation can form covalent crosslinks between peptides and it has been proposed that dityrosine formation of Aβ may stabilize amyloid fibrils in Alzheimer's disease [157].

5.2.4
AGEs

Carbohydrates can react with protein amino groups (e.g. N-terminus, lysine, arginine) to form a Schiff base that can rearrange to an Amadori product and, finally, an AGE. The time required to reach equilibrium is on the order of hours for Schiff base formation, days for Amadori products and weeks to months for AGEs. For an excellent review of AGE chemistry, see Bucala and Cerami [158]. AGEs, therefore, are predominantly identified in proteins with a low turnover rate *in vivo*. The formation of an AGE may result in the attachment of an adduct to the protein, protein crosslinking (e.g. an arginine–lysine crosslink or dilysine crosslink) [16] or fragmentation of the backbone [159].

The presence of AGE-modified proteins has been observed in cataracts [158], FAP [160], AA amyloidosis [161], Type 2 diabetes mellitus [162], Alzheimer's disease [163] and patients undergoing long-term dialysis treatment [164]. AGE modification may promote amyloid fibril formation by decreasing the solubility as a result of the alteration in the net charge of the protein or AGE crosslinks could stabilize the fibrils. AGE modifications of both Aβ [163] and tau [165] have been identified in Alzheimer's disease. AGE modification likely occurs early in the disease process [166] and may nucleate amyloid deposition as a result of accelerating Aβ aggregation [167]. In contrast, AGE-modified IAPP does not dramatically alter fibril formation kinetics, but does lead to fibrils that are more cytotoxic [168].

5.2.5
Phosphorylation

Alzheimer's disease is characterized by the presence of Aβ amyloid deposits and neurofibrillary tangles (NFTs) in the brain. Similarly, inclusion body myositis (IBM) is characterized by the deposition of Aβ amyloid and tau filaments in muscle. Ultrastructurally, the tau filaments are very similar in Alzheimer's disease and IBM [169]. In Alzheimer's disease, these filaments are called paired helical filaments (PHFs) and are the principal structural component of NFTs. In IBM, these filaments accumulate in muscle fibers and are called twisted tubulofilaments (TTFs). PHFs and TTFs are both composed of hyperphosphorylated tau [14, 169]. Phosphorylation of tau is not required for the formation of PHFs *in vitro*, but fibrils composed of phosphorylated tau resemble authentic PHFs more closely than do fibrils generated from non-phosphorylated tau [170]. This data suggests that phosphorylation may be an important pre-fibrillogenic event. AGE modification of tau, however, has been proposed to play a greater role in PHF insolubility than phosphorylation [171]. Therefore, it is possible that phosphorylation may nucleate fibril formation and AGE modification may stabilize the resultant fibrils against degradation.

5.3
Proposed Mechanisms by which Chemical Modifications may Affect Amyloid Deposition

Virtually all amyloid deposits contain some protein that is chemically modified. However, the identification of chemically modified protein in *ex vivo* amyloid does not *a priori* indicate that the modifications play a causative role. Therefore, it is necessary to consider the possible ways in which chemical modifications may affect amyloid deposition. There are three fundamentally different ways in which chemical modifications may contribute to amyloidosis. A chemical modification may affect the rate at which fibrils are formed (kinetic effect), the relative stability of the fibrils (thermodynamic effect) or lead to an increase in fibril toxicity (cytotoxic effect).

In order to have a kinetic effect on amyloid fibril formation, the chemical modification must occur prior to fibrillogenesis and accelerate the slow step of fibril formation. Most of the mechanistic models of amyloid fibril formation require two basic steps, nucleation and growth (for a review of current models, see [172]). Nucleation involves the conversion of soluble protein (S) into an assembly competent state (A) that forms the nucleus (N) or seed. The formation of the nucleus from the soluble protein is the slow (rate-limiting) step in amyloid fibril formation. Fibril growth then proceeds via the addition of protein to the nucleus. Protein can bind to the nucleus in the soluble form (S) and then rearrange upon binding [173–175], or it can bind in the assembly competent (A) form [176] (Fig. 5.2).

Fig. 5.2 Nucleation and growth mechanisms. Key: S = soluble protein, S* = chemically modified protein, A = assembly competent state, N = nucleus, F = fibril.

a) Conformation hypothesis

1. $S \rightarrow A \rightarrow N + S \rightarrow F$
2. $S \rightarrow A \rightarrow N + A \rightarrow F$

Slow step is either $S \rightarrow A$ or $A \rightarrow N$

b) Chemical modification hypothesis

1. $S \rightarrow S^*\ (A) \rightarrow N + S \rightarrow F$
2. $S \rightarrow S^*\ (A) \rightarrow N + S^*\ (A) \rightarrow F$

Slow step is either $S \rightarrow S^*$ or $A \rightarrow N$

The assembly competent state (A) is generally believed to be a partially folded intermediate state [1–3, 177]. The destabilization of the native state, therefore, can increase the population of this intermediate and drive nucleus formation. Amyloidogenic mutants, for example, have been shown to be less stable than the wild-type proteins and it is proposed that the mutants populate the assembly competent intermediate state more readily [178]. Analogously, it is possible that post-translational chemical modifications can result in a protein (S*) that is less stable than the wild-type, readily populates the assembly competent intermediate state (i.e. $S^* \rightarrow A$ is fast) and facilitates nucleus formation (Fig. 5.2). The protein generated by post-translational chemical modification could even be the equivalent of the assembly competent protein (i.e. $S^* = A$).

Depending on the mechanism of protein addition to the nucleus, the growth of amyloid fibrils by the chemical modification mechanism could require either small or large amounts of modified protein. If the soluble protein adds directly to the nucleus, then only trace amounts of chemically modified proteins would be required to drive amyloid fibril formation. If, however, the assembly competent state is required for addition, then the majority of the protein in the fibrils would need to be chemically modified. Chemically modified proteins have been observed in *ex vivo* amyloid deposits in both major and minor quantities, suggesting that both mechanisms may occur *in vivo*.

Nucleus formation is inherently slow as a result of the entropically unfavorable association of the A-state protein. Increasing the concentration of the A-state protein, however, can help to overcome this barrier. The population of the A-state can be increased in two different ways; destabilization of the S-state could shift the equilibrium toward the A-state or the degradation of the A-state could be slowed. Chemical modifications of the primary structure likely affect both pathways.

The likelihood of amyloid fibril formation proceeding by the mechanism outlined in Fig. 5.2 b has been strongly supported by examples of proteins in which cross-seeding occurs. Cross-seeding experiments demonstrate that the nucleus does not need to be formed from the protein used in the growth process [179]. A nucleus formed from a fragment of a protein can, in many cases, nucleate fibril formation from the full-length protein (e.g. tau [180], lysozyme [181], β_2M

[182]). This suggests that cleavage of a protein could generate a highly amyloidogenic fragment which nucleates fibril formation of the full-length protein. Nuclei that contain sequences that are very similar (e.g. polyglutamine [183]), differ by only 1 amino acid (e.g. α-synuclein [184], lysozyme [185], protein G BI domain [186], Aβ [187]) or differ significantly (e.g. tau/α-synuclein [188], tubulin/α-synuclein [189], NAC/Aβ [190], Aβ/IAPP [187], Aβ/tau [191], Aβ, TTR fragments, IAPP, silk/serum amyloid A, reviewed in [192]) from the soluble protein have been shown to accelerate fibril formation. Cross-seeding, however, is not necessarily a reciprocal process. A nucleus composed of protein X may be able to seed protein Y, but a nucleus of protein Y may not be able to seed protein X [187, 193]. It is also important to note that there are cases in which a protein may bind to a nucleus but not seed additional fibril formation [194].

The discovery of heterologous seeding has implications for the purity of peptides and proteins used for *in vitro* experiments designed to understand fibril formation. The batch-to-batch variability of Aβ is well documented and may be a result of different impurities [123]. In some cases, only trace amounts of modified peptide are necessary to seed fibril formation from unmodified protein [195], indicating that experiments need to be performed with exceptionally pure materials to assess mechanistic implications of fibril formation. Furthermore, the protein deposited in the fibrils can also be characterized to determine the extent of modified protein in the amyloid fibrils [196].

The two pieces of evidence necessary to adequately assess the impact of a chemical modification on the kinetics of fibril formation *in vivo* are the demonstration that the chemical modification is a pre-fibrillogenic event and accelerates nucleus formation. One method to determine if the modification is pre-fibrillogenic *in vivo* is to determine if the modified protein is present in the circulation. Although this type of experiment is insightful, there are two potential obstacles. First, if modified protein is not observed in the circulation, it is possible that the rate of nucleus formation is so fast that the modified protein is not present in detectable levels. Second, if modified protein is observed in the circulation, it is possible that the protein originated from the dissolution of amyloid fibrils *in vivo* and the chemical modification was, in fact, a post-fibrillogenic event. Another method to assess if the modification is pre- or post-fibrillogenic is to examine the distribution of chemically modified protein in the fibrils. If a chemical modification is a post-fibrillogenic event, it is reasonable to assume that the distribution of chemically modified protein in the fibrils will be similar. However, it has been determined that many modifications of Aβ are not equally distributed and have temporal relationship to fibril deposition (see Section 5.2.2). Finally, the extent of chemically modified protein in the fibrils may provide an insight into whether or not the modification is pre- or post-fibrillogenic. If the chemical modification of a specific residue is a post-fibrillogenic event, over time all of the exposed residues will become modified. Thus, if 50% or more of these residues are modified in aged samples, this may suggest that this residue is surface exposed and may be modified after fibril formation. However, in cases in which only small amounts of modification are present, this chemical modification is either very slow or the residue is in the interior of the fibril. If the residue is in the core of the fibril, chemical modification is likely to be a pre-fibrillogenic event. New methods have recently been explored in which the extent of surface exposed residues of *in vitro* amyloid fibrils were elucidated by artificial aging of the fibrils [196]. It will be interesting to see if similar methods can be applied to *ex vivo* fibrils.

Another factor that is important to determine the effect of a chemical modification on the kinetics of fibril formation is the extent to which the modification accelerates fibril formation. *In vitro* studies have been used to determine that many chemical modifications of peptides lead to enhanced rates of fibril formation compared to wild-type as monitored by Thioflavin T fluorescence or Congo red binding. These rates are generally assessed by comparing equal concentrations of modified and wild-type peptides. These results, however, need to be examined in the context of the relative concentrations of modified and wild-type peptides *in vivo*. For example, Met oxidation of Aβ has been observed *in vivo* and *in vitro* and there is evidence that this peptide aggregates more slowly than Aβ *in vitro* (see Section 5.2.3). From this information, it is tempting to conclude that Met oxidation of Aβ has a protective effect by inhibiting Aβ fibril formation. However, if the Met-oxidized Aβ peptide is turned over at a significantly slower rate than Aβ, the Met-oxidized peptide could accumulate and nucleate fibril formation.

Pre- or post-fibrillogenic chemical modifications can play a role in amyloid fibril formation by altering the thermodynamic stability of the fibrils or the cytotoxicity. The stability can be enhanced by nucleating a different packing arrangement, the formation of covalent crosslinks (e.g. disulfides, AGEs, dityrosine), or decreasing the ability of the immune system to recognize or degrade the fibrils. One method to determine if the chemical modification can nucleate a different packing arrangement is to examine the fibril morphology. Phosphorylated tau, for example, results in fibrils that more closely resemble authentic PHFs than non-phosphorylated tau [170]. Fibrils can be examined for thermostability by comparing the measured solubility of modified and non-modified fibrils upon addition of denaturant.

Changes in thermodynamic stability and cytotoxicity could be closely related. Fibrils and small oligomers both appear cytotoxic, but oligomers may be able to diffuse through tissue and, therefore, be toxic to more cells. Therefore, increasing the thermodynamic stability of the fibrils may be both bad and good – bad because it will make the fibrils difficult to clear from the body, good because it may shift the equilibrium toward the fibrils which could decrease the production of cytotoxic oligomers.

5.4
Conclusions and Future Directions

The majority of amyloid deposits examined to date contain some protein that is post-translationally modified. The role of these chemical modifications in amyloid fibril formation, however, has remained underexplored. A major challenge for the future will be to establish a temporal relationship between chemical modifications and fibril formation, and to assess the impact of the modification on kinetics of fibril formation, thermodynamic stability and cytotoxicity. A clear understanding of the role of chemical modifications in amyloid fibril formation will not only expand our current knowledge of protein misfolding, but may also lead to new avenues for therapeutic intervention. One current strategy to inhibit amyloid fibril formation is to design molecules that can stabilize the native state of the protein. This strategy will also be applicable for cases in which chemical modification plays a role because modifications are less likely in regions of a protein that are not flexible. However, a new strategy to prevent amyloid fibril formation may be to inhibit aberrant protein chemical modifications.

Acknowledgments

I am grateful to Daniel Moriarty for a critical review of this manuscript, and to Carole Klapper and Randall Morrison for technical assistance in manuscript preparation.

References

1 Carrell, R.W. and D.A. Lomas. Conformational disease. *Lancet* **1997**, *350*, 134–138.

2 Kelly, J.W. The alternative conformations of amyloidogenic proteins and their multi-step assembly pathways. *Curr Opin Struct Biol* **1998**, *8*, 101–106.

3 Dobson, C.M. The structural basis of protein folding and its links with human disease. *Phil Trans R Soc Lond B Biol Sci* **2001**, *356*, 133–145.

4 Nalivaeva, N.N. and A.J. Turner. Post-translational modifications of proteins: acetylcholinesterase as a model system. *Proteomics* **2001**, *1*, 735–747.

5 Nagano, N., E.G. Hutchinson and J.M. Thornton. Barrel structures in proteins: automatic identification and classification including a sequence analysis of TIM barrels. *Protein Sci* **1999**, *8*, 2072–2084.

6 Damas, A.M. and M.J. Saraiva. Review: TTR amyloidosis – structural features leading to protein aggregation and their implications on therapeutic strategies. *J Struct Biol* **2000**, *130*, 290–299.

7 Clark, A., E.D. Koning, C. Baker, S. Charge and J. Morris. Localisation of N-terminal pro-islet amyloid polypeptide in β-cells of man and transgenic mice. *Diabetologia* **1993**, *36*, A136.

8 Westermark, P., U. Engstrom, G.T. Westermark, K.H. Johnson, J. Permerth and C. Betsholtz. Islet amyloid polypeptide (IAPP) and pro-IAPP immunoreactivity in human islets of Langerhans. *Diabetes Res Clin Pract* **1989**, *7*, 219–226.

9 Westermark, G.T., D.F. Steiner, S. Gebre-Medhin, U. Engstrom and P. Wes-

termark. Pro islet amyloid polypeptide (ProIAPP) immunoreactivity in the islets of Langerhans. *Ups J Med Sci* **2000**, *105*, 97–106.

10 Jaikaran, E.T., M.R. Nilsson and A. Clark. Pancreatic beta-cell granule peptides form heteromolecular complexes which inhibit islet amyloid polypeptide fibril formation. *Biochem J* **2004**, *377*, 709–716.

11 Park, K. and C.B. Verchere. Identification of a heparin binding domain in the N-terminal cleavage site of pro-islet amyloid polypeptide. Implications for islet amyloid formation. *J Biol Chem* **2001**, *276*, 16611–16616.

12 Sletten, K., P. Westermark and J.B. Natvig. Characterization of amyloid fibril proteins from medullary carcinoma of the thyroid. *J Exp Med* **1976**, *143*, 993–998.

13 Pucci, A., J. Wharton, E. Arbustini, M. Grasso, M. Diegoli, P. Needleman, M. Vigano and J. M. Polak. Atrial amyloid deposits in the failing human heart display both atrial and brain natriuretic peptide-like immunoreactivity. *J Pathol* **1991**, *165*, 235–241.

14 Goedert, M. Tau protein and the neurofibrillary pathology of Alzheimer's disease. *Trends Neurosci* **1993**, *16*, 460–465.

15 Robinson, N.E. and A.B. Robinson. Molecular clocks. *Proc Natl Acad Sci USA* **2001**, *98*, 944–949.

16 Baynes, J.W. and S.R. Thorpe. Glycoxidation and lipoxidation in atherogenesis. *Free Rad Biol Med* **2000**, *28*, 1708–1716.

17 Glenner, G.G., D. Ein, E.D. Eanes, H.A. Bladen, W. Terry and D.L. Page. Creation of "amyloid" fibrils from Bence Jones proteins *in vitro*. *Science* **1971**, *174*, 712–714.

18 Haggqvist, B., J. Naslund, K. Sletten, G.T. Westermark, G. Mucchiano, L.O. Tjernberg, C. Nordstedt, U. Engstrom, et al. Medin: an integral fragment of aortic smooth muscle cell-produced lactadherin forms the most common human amyloid. *Proc Natl Acad Sci USA* **1999**, *96*, 8669–8674.

19 Korvatska, E., H. Henry, Y. Mashima, M. Yamada, C. Bachmann, F.L. Munier and D.F. Schorderet. Amyloid and non-amyloid forms of 5q31-linked corneal dystrophy resulting from kerato-epithelin mutations at Arg-124 are associated with abnormal turnover of the protein. *J Biol Chem* **2000**, *275*, 11465–11469.

20 Berson, J.F., A.C. Theos, D.C. Harper, D. Tenza, G. Raposo and M.S. Marks. Proprotein convertase cleavage liberates a fibrillogenic fragment of a resident glycoprotein to initiate melanosome biogenesis. *J Cell Biol* **2003**, *161*, 521–533.

21 Chapman, M.R., L.S. Robinson, J.S. Pinkner, R. Roth, J. Heuser, M. Hammar, S. Normark and S.J. Hultgren. Role of *Escherichia coli* curli operons in directing amyloid fiber formation. *Science* **2002**, *295*, 851–855.

22 Cerini, C., V. Peyrot, C. Garnier, L. Duplan, S. Veesler, J.P. Le Caer, J.P. Bernard, H. Bouteille, et al. Biophysical characterization of lithostathine. Evidences for a polymeric structure at physiological pH and a proteolysis mechanism leading to the formation of fibrils. *J Biol Chem* **1999**, *274*, 22266–22267.

23 Duplan, L., B. Michel, J. Boucraut, S. Barthellemy, S. Desplat-Jego, V. Marin, D. Gambarelli, D. Bernard, et al. Lithostathine and pancreatitis-associated protein are involved in the very early stages of Alzheimer's disease. *Neurobiol Aging* **2001**, *22*, 79–88.

24 Jin, C.X., S. Naruse, M. Kitagawa, H. Ishiguro, T. Kondo, S. Hayakawa and T. Hayakawa. Pancreatic stone protein of pancreatic calculi in chronic calcified pancreatitis in man. *JOP: J Pancreas (Online)* **2002**, *3*, 54–61.

25 Sun, H.Q., M. Yamamoto, M. Mejillano and H.L. Yin. Gelsolin, a multifunctional actin regulatory protein. *J Biol Chem* **1999**, *274*, 33179–33182.

26 Maury, C.P. Gelsolin-related amyloidosis. Identification of the amyloid protein in Finnish hereditary amyloidosis as a fragment of variant gelsolin. *J Clin Invest* **1991**, *87*, 1195–1199.

27 de la Chapelle, A., R. Tolvanen, G. Boysen, J. Santavy, L. Bleeker-Wagemakers, C.P. Maury and J. Kere. Gelsolin-derived familial amyloidosis caused by asparagine or tyrosine substitution for aspartic acid at residue 187. *Nat Genet* **1992**, *2*, 157–160.

28 Ratnaswamy, G., M. E. Huff, A. I. Su, S. Rion and J. W. Kelly. Destabilization of Ca^{2+}-free gelsolin may not be responsible for proteolysis in Familial Amyloidosis of Finnish Type. *Proc Natl Acad Sci USA* **2001**, *98*, 2334–2339.

29 Chen, C. D., M. E. Huff, J. Matteson, L. Page, R. Phillips, J. W. Kelly and W. E. Balch. Furin initiates gelsolin familial amyloidosis in the Golgi through a defect in Ca^{2+} stabilization. *EMBO J* **2001**, *20*, 6277–6287.

30 Kangas, H., T. Paunio, N. Kalkkinen, A. Jalanko and L. Peltonen. *In vitro* expression analysis shows that the secretory form of gelsolin is the sole source of amyloid in gelsolin-related amyloidosis. *Hum Mol Genet* **1996**, *5*, 1237–1243.

31 Maury, C. P. and H. Rossi. Demonstration of a circulating 65K gelsolin variant specific for familial amyloidosis, Finnish type. *Biochem Biophys Res Commun* **1993**, *191*, 41–44.

32 Maury, C. P., K. Sletten, N. Totty, H. Kangas and M. Liljestrom. Identification of the circulating amyloid precursor and other gelsolin metabolites in patients with G654A mutation in the gelsolin gene (Finnish familial amyloidosis): pathogenetic and diagnostic implications. *Lab Invest* **1997**, *77*, 299–304.

33 Ghiso, J., O. Jensson and B. Frangione. Amyloid fibrils in hereditary cerebral hemorrhage with amyloidosis of Icelandic type is a variant of gamma-trace basic protein (cystatin C). *Proc Natl Acad Sci USA* **1986**, *83*, 2974–2978.

34 Olafsson, I., G. Gudmundsson, M. Abrahamson, O. Jensson and A. Grubb. The amino terminal portion of cerebrospinal fluid cystatin C in hereditary cystatin C amyloid angiopathy is not truncated: direct sequence analysis from agarose gel electropherograms. *Scand J Clin Lab Invest* **1990**, *50*, 85–93.

35 Staniforth, R. A., S. Giannini, L. D. Higgins, M. J. Conroy, A. M. Hounslow, R. Jerala, C. J. Craven and J. P. Waltho. Three-dimensional domain swapping in the folded and molten-globule states of cystatins, an amyloid-forming structural superfamily. *EMBO J* **2001**, *20*, 4774–4781.

36 Nagai, A., S. Kobayashi, K. Shimode, K. Imaoka, N. Umegae, S. Fujihara and M. Nakamura. No mutations in cystatin C gene in cerebral amyloid angiopathy with cystatin C deposition. *Mol Chem Neuropathol* **1998**, *33*, 63–78.

37 Bergstrom, J., C. Murphy, M. Eulitz, D. T. Weiss, G. T. Westermark, A. Solomon and P. Westermark. Codeposition of apolipoprotein A-IV and transthyretin in senile systemic (ATTR) amyloidosis. *Biochem Biophys Res Commun* **2001**, *285*, 903–908.

38 Westermark, P., G. Mucchiano, T. Marthin, K. H. Johnson and K. Sletten. Apolipoprotein A1-derived amyloid in human aortic atherosclerotic plaques. *Am J Pathol* **1995**, *147*, 1186–1192.

39 Mucchiano, G. I., B. Haggqvist, K. Sletten and P. Westermark. Apolipoprotein A-1-derived amyloid in atherosclerotic plaques of the human aorta. *J Pathol* **2001**, *193*, 27027–27035.

40 Nichols, W. C., F. E. Dwulet, J. Liepnieks and M. D. Benson. Variant apolipoprotein AI as a major constituent of a human hereditary amyloid. *Biochem Biophys Res Commun* **1988**, *156*, 762–768.

41 Husebekk, A., B. Skogen, G. Husby and G. Marhaug. Transformation of amyloid precursor SAA to protein AA and incorporation in amyloid fibrils *in vivo*. *Scand J Immunol* **1985**, *21*, 283–287.

42 Yamada, T., J. J. Liepnieks, B. Kluve-Beckerman and M. D. Benson. Cathepsin B generates the most common form of amyloid A (76 residues) as a degradation product from serum amyloid A. *Scand J Immunol* **1995**, *41*, 94–97.

43 Hatters, D. M. and G. J. Howlett. The structural basis for amyloid formation by plasma apolipoproteins: a review. *Eur Biophys J* **2002**, *31*, 2–8.

44 Baumann, M. H., T. Wisniewski, E. Levy, G. T. Plant and J. Ghiso. C-terminal fragments of alpha- and beta-tubulin form amyloid fibrils *in vitro* and associate with amyloid deposits of familial cerebral amyloid angiopathy, British type. *Biochem Biophys Res Commun* **1996**, *219*, 238–242.

45 Kim, S. H., R. Wang, D. J. Gordon, J. Bass, D. F. Steiner, G. Thinakaran, D. G. Lynn, S. C. Meredith, et al. Familial British dementia: expression and metabolism of BRI. *Ann NY Acad Sci* **2000**, *920*, 93–99.

46 Vidal, R., B. Frangione, A. Rostagno, S. Mead, T. Revesz, G. Plant and J. Ghiso. A stop-codon mutation in the BRI gene associated with familial British dementia. *Nature* **1999**, *399*, 776–781.

47 Schwab, C., M. Hosokawa, H. Akiyama and P. L. McGeer. Familial British dementia: colocalization of furin and ABri amyloid. *Acta Neuropathol (Berl)* **2003**, *106*, 278–284.

48 Kim, S. H., J. W. Creemers, S. Chu, G. Thinakaran and S. S. Sisodia. Proteolytic processing of familial British dementia-associated BRI variants: evidence for enhanced intracellular accumulation of amyloidogenic peptides. *J Biol Chem* **2002**, *277*, 1872–1877.

49 Bellotti, V., M. Stoppini, P. Mangione, M. Sunde, C. Robinson, L. Asti, D. Brancaccio and G. Ferri. Beta2-microglobulin can be refolded into a native state from ex vivo amyloid fibrils. *Eur J Biochem* **1998**, *258*, 61–67.

50 Linke, R. P., H. Hampl, H. Lobeck, E. Ritz, J. Bommer, R. Waldherr and M. Eulitz. Lysine-specific cleavage of beta 2-microglobulin in amyloid deposits associated with hemodialysis. *Kidney Int* **1989**, *36*, 675–681.

51 Esposito, G., R. Michelutti, G. Verdone, P. Viglino, H. Hernandez, C. V. Robinson, A. Amoresano, F. Dal Piaz, et al. Removal of the N-terminal hexapeptide from human beta2-microglobulin facilitates protein aggregation and fibril formation. *Protein Sci* **2000**, *9*, 831–845.

52 Saraiva, M. J. Transthyretin mutations in health and disease. *Hum Mutat* **1995**, *5*, 191–196.

53 Gustavsson, A., H. Jahr, R. Tobiassen, D. R. Jacobson, K. Sletten and P. Westermark. Amyloid fibril composition and transthyretin gene structure in senile systemic amyloidosis. *Lab Invest* **1995**, *73*, 703–708.

54 Gustavsson, A., U. Engstrom and P. Westermark. Normal transthyretin and synthetic transthyretin fragments form amyloid-like fibrils *in vitro*. *Biochem Biophys Res Commun* **1991**, *175*, 1159–1164.

55 MacPhee, C. E. and C. M. Dobson. Chemical dissection and reassembly of amyloid fibrils formed by a peptide fragment of transthyretin. *J Mol Biol* **2000**, *297*, 1203–1215.

56 Kelly, J. and P. L. Lansbury, Jr. A chemical approach to elucidate the mechanism of transthyretin and β-protein amyloid fibril formation. *Amyloid* **1994**, *1*, 186–205.

57 Lucking, C. B. and A. Brice. Alpha-synuclein and Parkinson's disease. *Cell Mol Life Sci* **2000**, *57*, 1894–1908.

58 Conway, K. A., J. D. Harper and P. T. Lansbury, Jr. Fibrils formed *in vitro* from alpha-synuclein and two mutant forms linked to Parkinson's disease are typical amyloid. *Biochemistry* **2000**, *39*, 2552–2563.

59 Culvenor, J. G., C. A. McLean, S. Cutt, B. C. Campbell, F. Maher, P. Jakala, T. Hartmann, K. Beyreuther, et al. Non-Abeta component of Alzheimer's disease amyloid (NAC) revisited. NAC and alpha-synuclein are not associated with Abeta amyloid. *Am J Pathol* **1999**, *155*, 1173–1181.

60 Baba, M., S. Nakajo, P. H. Tu, T. Tomita, K. Nakaya, V. M. Lee, J. Q. Trojanowski and T. Iwatsubo. Aggregation of alpha-synuclein in Lewy bodies of sporadic Parkinson's disease and dementia with Lewy bodies. *Am J Pathol* **1998**, *152*, 879–884.

61 Ueda, K., H. Fukushima, E. Masliah, Y. Xia, A. Iwai, M. Yoshimoto, D. A. Otero, J. Kondo, et al. Molecular cloning of cDNA encoding an unrecognized component of amyloid in Alzheimer disease. *Proc Natl Acad Sci USA* **1993**, *90*, 11282–11286.

62 Haass, C. and D. J. Selkoe. Cellular processing of beta-amyloid precursor protein and the genesis of amyloid beta-peptide. *Cell* **1993**, *75*, 1039–42.

63 Tjernberg, L. O., J. Naslund, J. Thyberg, S. E. Gandy, L. Terenius and C. Nordstedt. Generation of Alzheimer amyloid

beta peptide through nonspecific proteolysis. *J Biol Chem* **1997**, *272*, 1870–1875.

64 Lalowski, M., A. Golabek, C.A. Lemere, D.J. Selkoe, H.M. Wisniewski, R.C. Beavis, B. Frangione and T. Wisniewski. The "nonamyloidogenic" p3 fragment (amyloid beta17–42) is a major constituent of Down's syndrome cerebellar preamyloid. *J Biol Chem* **1996**, *271*, 33623–33631.

65 Saido, T.C., W. Yamao-Harigaya, T. Iwatsubo and S. Kawashima. Amino- and carboxyl-terminal heterogeneity of beta-amyloid peptides deposited in human brain. *Neurosci Lett* **1996**, *215*, 173–176.

66 Lemere, C.A., J.K. Blusztajn, H. Yamaguchi, T. Wisniewski, T.C. Saido and D.J. Selkoe. Sequence of deposition of heterogeneous amyloid beta-peptides and APO E in Down syndrome: implications for initial events in amyloid plaque formation. *Neurobiol Dis* **1996**, *3*, 16–32.

67 Jarrett, J.T., E.P. Berger and P.T. Lansbury, Jr. The carboxy terminus of the beta amyloid protein is critical for the seeding of amyloid formation: implications for the pathogenesis of Alzheimer's disease. *Biochemistry* **1993**, *32*, 4693–4697.

68 Chen, S.G., D.B. Teplow, P. Parchi, J.K. Teller, P. Gambetti and L. Autilio-Gambetti. Truncated forms of the human prion protein in normal brain and in prion diseases. *J Biol Chem* **1995**, *270*, 19173–19180.

69 Pan, T., R. Li, B.S. Wong, T. Liu, P. Gambetti and M. S. Sy. Heterogeneity of normal prion protein in two-dimensional immunoblot: presence of various glycosylated and truncated forms. *J Neurochem* **2002**, *81*, 1092–1101.

70 DiFiglia, M., E. Sapp, K.O. Chase, S.W. Davies, G.P. Bates, J.P. Vonsattel and N. Aronin. Aggregation of huntingtin in neuronal intranuclear inclusions and dystrophic neurites in brain. *Science* **1997**, *277*, 1990–1993.

71 McGowan, D.P., W. van Roon-Mom, H. Holloway, G.P. Bates, L. Mangiarini, G.J. Cooper, R.L. Faull and R.G. Snell. Amyloid-like inclusions in Huntington's disease. *Neuroscience* **2000**, *100*, 677–680.

72 Huang, C.C., P.W. Faber, F. Persichetti, V. Mittal, J.P. Vonsattel, M.E. MacDonald and J.F. Gusella. Amyloid formation by mutant huntingtin: threshold, progressivity and recruitment of normal polyglutamine proteins. *Somat Cell Mol Genet* **1998**, *24*, 217–233.

73 Scherzinger, E., A. Sittler, K. Schweiger, V. Heiser, R. Lurz, R. Hasenbank, G.P. Bates, H. Lehrach, et al. Self-assembly of polyglutamine-containing huntingtin fragments into amyloid-like fibrils: implications for Huntington's disease pathology. *Proc Natl Acad Sci USA* **1999**, *96*, 4604–4609.

74 Tarlac, V. and E. Storey. Role of proteolysis in polyglutamine disorders. *J Neurosci Res* **2003**, *74*, 406–416.

75 Cribbs, D.H., B.Y. Azizeh, C.W. Cotman and F.M. LaFerla. Fibril formation and neurotoxicity by a herpes simplex virus glycoprotein B fragment with homology to the Alzheimer's A beta peptide. *Biochemistry* **2000**, *39*, 5988–5994.

76 Lin, W.R., D. Shang, G.K. Wilcock and R.F. Itzhaki. Alzheimer's disease, herpes simplex virus type 1, cold sores and apolipoprotein E4. *Biochem Soc Trans* **1995**, *23*, 594S.

77 Jamieson, G.A., N.J. Maitland, G.K. Wilcock, C.M. Yates and R.F. Itzhaki. Herpes simplex virus type 1 DNA is present in specific regions of brain from aged people with and without senile dementia of the Alzheimer type. *J Pathol* **1992**, *167*, 365–368.

78 Robinson, A.B. and C.J. Rudd. Deamidation of glutaminyl and asparaginyl residues in peptides and proteins. *Curr Topics Cell Reg* **1974**, *8*, 247–295.

79 Bischoff, R. and H.V. Kolbe. Deamidation of asparagine and glutamine residues in proteins and peptides: structural determinants and analytical methodology. *J Chromatogr B Biomed Appl* **1994**, *662*, 261–278.

80 Wright, H.T. Sequence and structure determinants of the nonenzymatic deamidation of asparagine and glutamine residues in proteins. *Protein Eng* **1991**, *4*, 283–294.

81 Wright, H.T. Nonenzymatic deamidation of asparaginyl and glutaminyl resi-

dues in proteins. *Crit Rev Biochem Mol Biol* **1991**, *26*, 1–52.

82 Windisch, V., F. DeLuccia, L. Duhau, F. Herman, J.J. Mencel, S.Y. Tang and M. Vuilhorgne. Degradation pathways of salmon calcitonin in aqueous solution. *J Pharm Sci* **1997**, *86*, 359–364.

83 Hekman, C.M., W.S. DeMond, P.J. Kelley, S.F. Mauch and J.D. Williams. Isolation and identification of cyclic imide and deamidation products in heat stressed pramlintide injection drug product. *J Pharm Biomed Anal* **1999**, *20*, 763–772.

84 Brange, J., L. Langkjaer, S. Havelund and A. Volund. Chemical stability of insulin. 1. Hydrolytic degradation during storage of pharmaceutical preparations. *Pharm Res* **1992**, *9*, 715–726.

85 Michelitsch, M.D. and J.S. Weissman. A census of glutamine/asparagine-rich regions: implications for their conserved function and the prediction of novel prions. *Proc Natl Acad Sci USA* **2000**, *97*, 11910–11915.

86 Shimizu, T., A. Watanabe, M. Ogawara, H. Mori and T. Shirasawa. Isoaspartate formation and neurodegeneration in Alzheimer's disease. *Arch Biochem Biophys* **2000**, *381*, 225–234.

87 Lowenson, J.D. and S. Clarke. Recognition of D-aspartyl residues in polypeptides by the erythrocyte L-isoaspartyl/D-aspartyl protein methyltransferase. Implications for the repair hypothesis. *J Biol Chem* **1992**, *267*, 5985–5995.

88 Orru, S., L. Vitagliano, L. Esposito, L. Mazzarella, G. Marino and M. Ruoppolo. Effect of deamidation on folding of ribonuclease A. *Protein Sci* **2000**, *9*, 2577–2582.

89 Kato, A., S. Tanimoto, Y. Muraki, K. Kobayashi and I. Kumagai. Structural and functional properties of hen egg-white lysozyme deamidated by protein engineering. *Biosci Biotechnol Biochem* **1992**, *56*, 1424–1428.

90 Chiti, F., M. Calamai, N. Taddei, M. Stefani, G. Ramponi and C.M. Dobson. Studies of the aggregation of mutant proteins *in vitro* provide insights into the genetics of amyloid diseases. *Proc Natl Acad Sci USA* **2002**, *99 (Suppl 4)*, 16419–16426.

91 Hallak, M.E. and G. Bongiovanni. Posttranslational arginylation of brain proteins. *Neurochem Res* **1997**, *22*, 467–473.

92 La Spada, A.R., E.M. Wilson, D.B. Lubahn, A.E. Harding and K.H. Fischbeck. Androgen receptor gene mutations in X-linked spinal and bulbar muscular atrophy. *Nature* **1991**, *352*, 77–79.

93 Perutz, M.F. Glutamine repeats and neurodegenerative diseases: molecular aspects. *Trends Biochem Sci* **1999**, *24*, 58–63.

94 Chen, S., V. Berthelier, J.B. Hamilton, B. O'Nuallain and R. Wetzel. Amyloid-like features of polyglutamine aggregates and their assembly kinetics. *Biochemistry* **2002**, *41*, 7391–7399.

95 Violante, V., A. Luongo, I. Pepe, S. Annunziata and V. Gentile. Transglutaminase-dependent formation of protein aggregates as possible biochemical mechanism for polyglutamine diseases. *Brain Res Bull* **2001**, *56*, 169–172.

96 Sandmeier, E., P. Hunziker, B. Kunz, R. Sack and P. Christen. Spontaneous deamidation and isomerization of Asn108 in prion peptide 106–126 and in full-length prion protein. *Biochem Biophys Res Commun* **1999**, *261*, 578–583.

97 Qin, K., D.S. Yang, Y. Yang, M.A. Chishti, L.J. Meng, H.A. Kretzschmar, C.M. Yip, P.E. Fraser, et al. Copper(II)-induced conformational changes and protease resistance in recombinant and cellular PrP. Effect of protein age and deamidation. *J Biol Chem* **2000**, *275*, 19121–19131.

98 Miyata, T., Y. Iida, Y. Ueda, T. Shinzato, H. Seo, V.M. Monnier, K. Maeda and Y. Wada. Monocyte/macrophage response to beta 2-microglobulin modified with advanced glycation end products. *Kidney Int* **1996**, *49*, 538–550.

99 Odani, H., R. Oyama, K. Titani, H. Ogawa and A. Saito. Purification and complete amino acid sequence of novel beta 2-microglobulin. *Biochem Biophys Res Commun* **1990**, *168*, 1223–1229.

100 Kad, N.M., N.H. Thomson, D.P. Smith, D.A. Smith and S. E. Radford. $Beta_2$-microglobulin and its deamidated

variant, N17D form amyloid fibrils with a range of morphologies *in vitro*. *J Mol Biol* **2001**, *313*, 559–571.

101 Watanabe, A., K. Takio and Y. Ihara. Deamidation and isoaspartate formation in smeared tau in paired helical filaments. Unusual properties of the microtubule-binding domain of tau. *J Biol Chem* **1999**, *274*, 7368–7378.

102 Sandilands, A., A. M. Hutcheson, H. A. Long, A. R. Prescott, G. Vrensen, J. Loster, N. Klopp, R. B. Lutz, et al. Altered aggregation properties of mutant gamma-crystallins cause inherited cataract. *EMBO J* **2002**, *21*, 6005–6014.

103 Meehan, S., Y. Berry, B. Luisi, C. M. Dobson, J. A. Carver and C. E. MacPhee. Amyloid fibril formation by lens crystallin proteins and its implications for cataract formation. *J Biol Chem* **2004**, *279*, 3413–3419.

104 Takemoto, L., T. Emmons and D. Granstrom. The sequences of two peptides from cataract lenses suggest they arise by deamidation. *Curr Eye Res* **1990**, *9*, 793–797.

105 Takemoto, L. and D. Boyle. Increased deamidation of asparagine during human senile cataractogenesis. *Mol Vis* **2000**, *6*, 164–168.

106 Mori, H., K. Takio, M. Ogawara and D. J. Selkoe. Mass spectrometry of purified amyloid beta protein in Alzheimer's disease. *J Biol Chem* **1992**, *267*, 17082–17086.

107 Harigaya, Y., T. C. Saido, C. B. Eckman, C. M. Prada, M. Shoji and S. G. Younkin. Amyloid beta protein starting pyroglutamate at position 3 is a major component of the amyloid deposits in the Alzheimer's disease brain. *Biochem Biophys Res Commun* **2000**, *276*, 422–427.

108 Kuo, Y. M., M. R. Emmerling, A. S. Woods, R. J. Cotter and A. E. Roher. Isolation, chemical characterization, and quantitation of A beta 3-pyroglutamyl peptide from neuritic plaques and vascular amyloid deposits. *Biochem Biophys Res Commun* **1997**, *237*, 188–191.

109 Saido, T. C., T. Iwatsubo, D. M. Mann, H. Shimada, Y. Ihara and S. Kawashima. Dominant and differential deposition of distinct beta-amyloid peptide species, A beta N3(pE), in senile plaques. *Neuron* **1995**, *14*, 457–466.

110 He, W. and C. J. Barrow. The A beta 3-pyroglutamyl and 11-pyroglutamyl peptides found in senile plaque have greater beta-sheet forming and aggregation propensities *in vitro* than full-length A beta. *Biochemistry* **1999**, *38*, 10871–10877.

111 Szendrei, G. I., H. Fabian, H. H. Mantsch, S. Lovas, O. Nyeki, I. Schon and L. Otvos, Jr. Aspartate-bond isomerization affects the major conformations of synthetic peptides. *Eur J Biochem* **1994**, *226*, 917–924.

112 Roher, A. E., J. D. Lowenson, S. Clarke, C. Wolkow, R. Wang, R. J. Cotter, I. M. Reardon, H. A. Zurcher-Neely, et al. Structural alterations in the peptide backbone of beta-amyloid core protein may account for its deposition and stability in Alzheimer's disease. *J Biol Chem* **1993**, *268*, 3072–3083.

113 Shimizu, T., H. Fukuda, S. Murayama, N. Izumiyama and T. Shirasawa. Isoaspartate formation at position 23 of amyloid beta peptide enhanced fibril formation and deposited onto senile plaques and vascular amyloids in Alzheimer's disease. *J Neurosci Res* **2002**, *70*, 451–461.

114 Shin, Y., H. S. Cho, H. Fukumoto, T. Shimizu, T. Shirasawa, S. M. Greenberg and G. W. Rebeck. Abeta species, including IsoAsp23 Abeta, in Iowa-type familial cerebral amyloid angiopathy. *Acta Neuropathol (Berl)* **2003**, *105*, 252–258.

115 Fabian, H., G. I. Szendrei, H. H. Mantsch, B. D. Greenberg and L. Otvos, Jr. Synthetic post-translationally modified human A beta peptide exhibits a markedly increased tendency to form beta-pleated sheets *in vitro*. *Eur J Biochem* **1994**, *221*, 959–964.

116 Orpiszewski, J. and M. D. Benson. Induction of beta-sheet structure in amyloidogenic peptides by neutralization of aspartate: a model for amyloid nucleation. *J Mol Biol* **1999**, *289*, 413–428.

117 Orpiszewski, J., N. Schormann, B. Kluve-Beckerman, J. J. Liepnieks and

M. D. Benson. Protein aging hypothesis of Alzheimer disease. *FASEB J* **2000**, *14*, 1255–1263.

118 Bada, J. L. *In vivo* racemization in mammalian proteins. *Methods Enzymol* **1984**, *106*, 98–115.

119 Brown, R. Amino acid dating. *Origins* **1985**, *12*, 8–25.

120 De Miranda, J., A. Santoro, S. Engelender and H. Wolosker. Human serine racemase: molecular cloning, genomic organization and functional analysis. *Gene* **2000**, *256*, 183–188.

121 Janek, K., J. Behlke, J. Zipper, H. Fabian, Y. Georgalis, M. Beyermann, M. Bienert and E. Krause. Water-soluble beta-sheet models which self-assemble into fibrillar structures. *Biochemistry* **1999**, *38*, 8246–8252.

122 Shapira, R., G. E. Austin and S. S. Mirra. Neuritic plaque amyloid in Alzheimer's disease is highly racemized. *J Neurochem* **1988**, *50*, 69–74.

123 Zagorski, M.G., J. Yang, H. Shao, K. Ma, H. Zeng and A. Hong. Methodological and chemical factors affecting amyloid beta peptide amyloidogenicity. *Methods Enzymol* **1999**, *309*, 189–204.

124 Kaneko, I., K. Morimoto and T. Kubo. Drastic neuronal loss *in vivo* by beta-amyloid racemized at Ser(26) residue: conversion of non-toxic [D-Ser(26)]beta-amyloid 1–40 to toxic and proteinase-resistant fragments. *Neuroscience* **2001**, *104*, 1003–1011.

125 Kubo, T., Y. Kumagae, C. A. Miller and I. Kaneko. Beta-amyloid racemized at the Ser26 residue in the brains of patients with Alzheimer disease: implications in the pathogenesis of Alzheimer disease. *J Neuropathol Exp Neurol* **2003**, *62*, 248–259.

126 Lowenson, J. D. and S. Clarke. Structural elements affecting the recognition of L-isoaspartyl residues by the L-isoaspartyl/D-aspartyl protein methyltransferase. Implications for the repair hypothesis. *J Biol Chem* **1991**, *266*, 19396–19406.

127 Johnson, B. A., J. M. Shirokawa, J. W. Geddes, B. H. Choi, R. C. Kim and D. W. Aswad. Protein L-isoaspartyl methyltransferase in postmortem brains of aged humans. *Neurobiol Aging* **1991**, *12*, 19–24.

128 Yamamoto, A., H. Takagi, D. Kitamura, H. Tatsuoka, H. Nakano, H. Kawano, H. Kuroyanagi, Y. Yahagi, et al. Deficiency in protein L-isoaspartyl methyltransferase results in a fatal progressive epilepsy. *J Neurosci* **1998**, *18*, 2063–2074.

129 Lanthier, J., A. Bouthillier, M. Lapointe, M. Demeule, R. Beliveau and R. R. Desrosiers. Down-regulation of protein L-isoaspartyl methyltransferase in human epileptic hippocampus contributes to generation of damaged tubulin. *J Neurochem* **2002**, *83*, 581–591.

130 Smith, M. A., L. M. Sayre, V. M. Monnier and G. Perry. Radical AGEing in Alzheimer's disease. *Trends Neurosci* **1995**, *18*, 172–176.

131 Zhang, Q., E. T. Powers, J. Nieva, M. E. Huff, M. A. Dendle, J. Bieschke, C. G. Glabe, A. Eschenmoser, et al. Metabolite-initiated protein misfolding may trigger Alzheimer's disease. *Proc Natl Acad Sci USA* **2004**, *101*, 4752–4757.

132 Feughelman, M. and B. K. Willis. Thiol-disulfide interchange a potential key to conformational change associated with amyloid fibril formation. *J Theor Biol* **2000**, *206*, 313–315.

133 Takemoto, L. J. Oxidation of cysteine residues from alpha-A crystallin during cataractogenesis of the human lens. *Biochem Biophys Res Commun* **1996**, *223*, 216–220.

134 Uemichi, T., J. R. Murrell, S. Zeldenrust and M. D. Benson. A new mutant transthyretin (Arg 10) associated with familial amyloid polyneuropathy. *J Med Genet* **1992**, *29*, 888–891.

135 McCutchen, S. L. and J. W. Kelly. Intermolecular disulfide linkages are not required for transthyretin amyloid fibril formation *in vitro*. *Biochem Biophys Res Commun* **1993**, *197*, 415–421.

136 Heafield, M. T., S. Fearn, G. B. Steventon, R. H. Waring, A. C. Williams and S. G. Sturman. Plasma cysteine and sulphate levels in patients with motor neurone, Parkinson's and Alzheimer's disease. *Neurosci Lett* **1990**, *110*, 216–220.

137 Perry, T.L., M.G. Norman, V.W. Yong, S. Whiting, J.U. Crichton, S. Hansen and S.J. Kish. Hallervorden-Spatz disease: cysteine accumulation and cysteine dioxygenase deficiency in the globus pallidus. *Ann Neurol* **1985**, *18*, 482–489.

138 Lim, A., J. Wally, M.T. Walsh, M. Skinner and C.E. Costello. Identification and location of a cysteinyl posttranslational modification in an amyloidogenic kappa1 light chain protein by electrospray ionization and matrix-assisted laser desorption/ionization mass spectrometry. *Anal Biochem* **2001**, *295*, 45–56.

139 Lim, A., T. Prokaeva, M.E. McComb, L. H. Connors, M. Skinner and C.E. Costello. Identification of S-sulfonation and S-thiolation of a novel transthyretin Phe33Cys variant from a patient diagnosed with familial transthyretin amyloidosis. *Protein Sci* **2003**, *12*, 1775–1785.

140 Zhang, Q. and J.W. Kelly. Cys10 mixed disulfides make transthyretin more amyloidogenic under mildly acidic conditions. *Biochemistry* **2003**, *42*, 8756–8761.

141 Bellotti, V., P. Mangione and M. Stoppini. Biological activity and pathological implications of misfolded proteins. *Cell Mol Life Sci* **1999**, *55*, 977–991.

142 Stahl, N., M.A. Baldwin, D.B. Teplow, L. Hood, B.W. Gibson, A.L. Burlingame and S.B. Prusiner. Structural studies of the scrapie prion protein using mass spectrometry and amino acid sequencing. *Biochemistry* **1993**, *32*, 1991–2002.

143 Naslund, J., A. Schierhorn, U. Hellman, L. Lannfelt, A.D. Roses, L.O. Tjernberg, J. Silberring, S.E. Gandy, et al. Relative abundance of Alzheimer A beta amyloid peptide variants in Alzheimer disease and normal aging. *Proc Natl Acad Sci USA* **1994**, *91*, 8378–8382.

144 Kuo, Y.M., T.A. Kokjohn, T.G. Beach, L.I. Sue, D. Brune, J.C. Lopez, W.M. Kalback, D. Abramowski, et al. Comparative analysis of amyloid-beta chemical structure and amyloid plaque morphology of transgenic mouse and Alzheimer's disease brains. *J Biol Chem* **2001**, *276*, 12991–12998.

145 Snyder, S.W., U.S. Ladror, W.S. Wade, G.T. Wang, L.W. Barrett, E.D. Matayoshi, H.J. Huffaker, G.A. Krafft, et al. Amyloid-beta aggregation: selective inhibition of aggregation in mixtures of amyloid with different chain lengths. *Biophys J* **1994**, *67*, 1216–1228.

146 Varadarajan, S., J. Kanski, M. Aksenova, C. Lauderback and D.A. Butterfield. Different mechanisms of oxidative stress and neurotoxicity for Alzheimer's A beta(1–42) and A beta(25–35). *J Am Chem Soc* **2001**, *123*, 5625–5631.

147 Palmblad, M., A. Westlind-Danielsson and J. Bergquist. Oxidation of methionine 35 attenuates formation of amyloid beta-peptide 1–40 oligomers. *J Biol Chem* **2002**, *277*, 19506–19510.

148 Hou, L., I. Kang, R.E. Marchant and M.G. Zagorski. Methionine 35 oxidation reduces fibril assembly of the amyloid abeta-(1–42) peptide of Alzheimer's disease. *J Biol Chem* **2002**, *277*, 40173–40176.

149 Hokenson, M.J., V.N. Uversky, J. Goers, G. Yamin, L.A. Munishkina and A.L. Fink. Role of individual methionines in the fibrillation of methionine-oxidized alpha-synuclein. *Biochemistry* **2004**, *43*, 4621–4633.

150 Yamin, G., C.B. Glaser, V.N. Uversky and A.L. Fink. Certain metals trigger fibrillation of methionine-oxidized alpha-synuclein. *J Biol Chem* **2003**, *278*, 27630–27635.

151 Uversky, V.N., G. Yamin, P.O. Souillac, J. Goers, C.B. Glaser and A.L. Fink. Methionine oxidation inhibits fibrillation of human alpha-synuclein *in vitro*. *FEBS Lett* **2002**, *517*, 239–244.

152 Giasson, B.I., J.E. Duda, I.V. Murray, Q. Chen, J.M. Souza, H.I. Hurtig, H. Ischiropoulos, J.Q. Trojanowski, et al. Oxidative damage linked to neurodegeneration by selective alpha-synuclein nitration in synucleinopathy lesions. *Science* **2000**, *290*, 985–989.

153 Horiguchi, T., K. Uryu, B.I. Giasson, H. Ischiropoulos, R. Lightfoot, C. Bellmann, C. Richter-Landsberg, V.M. Lee,

et al. Nitration of tau protein is linked to neurodegeneration in tauopathies. *Am J Pathol* **2003**, *163*, 1021–1031.

154 Domigan, N.M., T.S. Charlton, M.W. Duncan, C.C. Winterbourn and A.J. Kettle. Chlorination of tyrosyl residues in peptides by myeloperoxidase and human neutrophils. *J Biol Chem* **1995**, *270*, 16542–16548.

155 Olszowski, S., E. Olszowska, T. Stelmaszynska, A. Krawczyk, J. Marcinkiewicz and N. Baczek. Oxidative modification of ovalbumin. *Acta Biochim Pol* **1996**, *43*, 661–672.

156 Hazen, S.L. and J.W. Heinecke. 3-Chlorotyrosine, a specific marker of myeloperoxidase-catalyzed oxidation, is markedly elevated in low density lipoprotein isolated from human atherosclerotic intima. *J Clin Invest* **1997**, *99*, 2075–2081.

157 Atwood, C.S., G. Perry, H. Zeng, Y. Kato, W.D. Jones, K.Q. Ling, X. Huang, R.D. Moir, et al. Copper mediates dityrosine cross-linking of Alzheimer's amyloid-beta. *Biochemistry* **2004**, *43*, 560–568.

158 Bucala, R. and A. Cerami. Advanced glycosylation: chemistry, biology and implications for diabetes and aging. *Adv Pharmacol* **1992**, *23*, 1–34.

159 Hunt, J.V. and S.P. Wolff. Oxidative glycation and free radical production: a causal mechanism of diabetic complications. *Free Rad Res Commun* **1991**, *12/13*, 115–123.

160 Nyhlin, N., Y. Ando, R. Nagai, O. Suhr, M. El Sahly, H. Terazaki, T. Yamashita, M. Ando, et al. Advanced glycation end product in familial amyloidotic polyneuropathy (FAP). *J Intern Med* **2000**, *247*, 485–492.

161 Kamalvand, G. and Z. Ali-Khan. Immunolocalization of lipid peroxidation/advanced glycation end products in amyloid A amyloidosis. *Free Rad Biol Med* **2004**, *36*, 657–664.

162 Ma, Z., P. Westermark and G.T. Westermark. Amyloid in human islets of Langerhans: immunologic evidence that islet amyloid polypeptide is modified in amyloidogenesis. *Pancreas* **2000**, *21*, 212–218.

163 Sasaki, N., R. Fukatsu, K. Tsuzuki, Y. Hayashi, T. Yoshida, N. Fujii, T. Koike, I. Wakayama, et al. Advanced glycation end products in Alzheimer's disease and other neurodegenerative diseases. *Am J Pathol* **1998**, *153*, 1149–1155.

164 Miyata, T., O. Oda, R. Inagi, Y. Iida, N. Araki, N. Yamada, S. Horiuchi, N. Taniguchi, et al. beta 2-Microglobulin modified with advanced glycation end products is a major component of hemodialysis-associated amyloidosis. *J Clin Invest* **1993**, *92*, 1243–1252.

165 Ledesma, M.D., P. Bonay and J. Avila. Tau protein from Alzheimer's disease patients is glycated at its tubulin-binding domain. *J Neurochem* **1995**, *65*, 1658–1664.

166 Munch, G., A.M. Cunningham, P. Riederer and E. Braak. Advanced glycation endproducts are associated with Hirano bodies in Alzheimer's disease. *Brain Res* **1998**, *796*, 307–310.

167 Vitek, M.P., K. Bhattacharya, J.M. Glendening, E. Stopa, H. Vlassara, R. Bucala, K. Manogue and A. Cerami. Advanced glycation end products contribute to amyloidosis in Alzheimer disease. *Proc Natl Acad Sci USA* **1994**, *91*, 4766–4770.

168 Kapurniotu, A., J. Bernhagen, N. Greenfield, Y. Al-Abed, S. Teichberg, R.W. Frank, W. Voelter and R. Bucala. Contribution of advanced glycosylation to the amyloidogenicity of islet amyloid polypeptide. *Eur J Biochem* **1998**, *251*, 208–216.

169 Askanas, V., W.K. Engel, M. Bilak, R.B. Alvarez and D.J. Selkoe. Twisted tubulofilaments of inclusion body myositis muscle resemble paired helical filaments of Alzheimer brain and contain hyperphosphorylated tau. *Am J Pathol* **1994**, *144*, 177–187.

170 Crowther, R.A., O.F. Olesen, M.J. Smith, R. Jakes and M. Goedert. Assembly of Alzheimer-like filaments from full-length tau protein. *FEBS Lett* **1994**, *337*, 135–138.

171 Smith, M.A., R.H. Nagaraj and G. Perry. Protocol for quantitative analysis of paired helical filament solubilization: a method applicable to insoluble amyloids and inclusion bodies. *Brain Res Brain Res Protoc* **1997**, *1*, 247–252.

172 Kelly, J. W. Mechanisms of amyloidogenesis. *Nat Struct Biol* **2000**, *7*, 824–826.

173 Griffith, J. S. Self-replication and scrapie. *Nature* **1967**, *215*, 1043–1044.

174 Serio, T. R., A. G. Cashikar, A. S. Kowal, G. J. Sawicki, J. J. Moslehi, L. Serpell, M. F. Arnsdorf and S. L. Lindquist. Nucleated conformational conversion and the replication of conformational information by a prion determinant. *Science* **2000**, *289*, 1317–1321.

175 Esler, W. P., E. R. Stimson, J. M. Jennings, H. V. Vinters, J. R. Ghilardi, J. P. Lee, P. W. Mantyh and J. E. Maggio. Alzheimer's disease amyloid propagation by a template-dependent dock-lock mechanism. *Biochemistry* **2000**, *39*, 6288–6295.

176 Harper, J. D. and P. T. Lansbury, Jr. Models of amyloid seeding in Alzheimer's disease and scrapie: mechanistic truths and physiological consequences of the time-dependent solubility of amyloid proteins. *Annu Rev Biochem* **1997**, *66*, 385–407.

177 Wetzel, R. For protein misassembly, it's the "I" decade. *Cell* **1996**, *86*, 699–702.

178 Takano, K., J. Funahashi and K. Yutani. The stability and folding process of amyloidogenic mutant human lysozymes. *Eur J Biochem* **2001**, *268*, 155–159.

179 Chien, P. and J. S. Weissman. Conformational diversity in a yeast prion dictates its seeding specificity. *Nature* **2001**, *410*, 223–227.

180 von Bergen, M., P. Friedhoff, J. Biernat, J. Heberle, E. M. Mandelkow and E. Mandelkow. Assembly of tau protein into Alzheimer paired helical filaments depends on a local sequence motif (^{306}VQIVYK^{311}) forming beta structure. *Proc Natl Acad Sci USA* **2000**, *97*, 5129–5134.

181 Krebs, M. R., D. K. Wilkins, E. W. Chung, M. C. Pitkeathly, A. K. Chamberlain, J. Zurdo, C. V. Robinson and C. M. Dobson. Formation and seeding of amyloid fibrils from wild-type hen lysozyme and a peptide fragment from the beta-domain. *J Mol Biol* **2000**, *300*, 541–549.

182 Kozhukh, G. V., Y. Hagihara, T. Kawakami, K. Hasegawa, H. Naiki and Y. Goto. Investigation of a peptide responsible for amyloid fibril formation of beta 2-microglobulin by achromobacter protease I. *J Biol Chem* **2002**, *277*, 1310–1315.

183 Chen, S., V. Berthelier, W. Yang and R. Wetzel. Polyglutamine aggregation behavior *in vitro* supports a recruitment mechanism of cytotoxicity. *J Mol Biol* **2001**, *311*, 173–182.

184 Wood, S. J., J. Wypych, S. Steavenson, J. C. Louis, M. Citron and A. L. Biere. Alpha-synuclein fibrillogenesis is nucleation-dependent. Implications for the pathogenesis of Parkinson's disease. *J Biol Chem* **1999**, *274*, 19509–19512.

185 Morozova-Roche, L. A., J. Zurdo, A. Spencer, W. Noppe, V. Receveur, D. B. Archer, M. Joniau and C. M. Dobson. Amyloid fibril formation and seeding by wild-type human lysozyme and its disease-related mutational variants. *J Struct Biol* **2000**, *130*, 339–351.

186 Ramirez-Alvarado, M. and L. Regan. Does the location of a mutation determine the ability to form amyloid fibrils? *J Mol Biol* **2002**, *323*, 17–22.

187 O'Nuallain, B., A. D. Williams, P. Westermark and R. Wetzel. Seeding specificity in amyloid growth induced by heterologous fibrils. *J Biol Chem* **2004**, *279*, 17490–17499.

188 Giasson, B. I., M. S. Forman, M. Higuchi, L. I. Golbe, C. L. Graves, P. T. Kotzbauer, J. Q. Trojanowski and V. M. Lee. Initiation and synergistic fibrillization of tau and alpha-synuclein. *Science* **2003**, *300*, 636–640.

189 Alim, M. A., M. S. Hossain, K. Arima, K. Takeda, Y. Izumiyama, M. Nakamura, H. Kaji, T. Shinoda, et al. Tubulin seeds alpha-synuclein fibril formation. *J Biol Chem* **2002**, *277*, 2112–2117.

190 Han, H., P. H. Weinreb and P. T. Lansbury, Jr. The core Alzheimer's peptide NAC forms amyloid fibrils which seed and are seeded by beta-amyloid: is

NAC a common trigger or target in neurodegenerative disease? *Chem Biol* **1995**, *2*, 163–169.

191 Giaccone, G., B. Pedrotti, A. Migheli, L. Verga, J. Perez, G. Racagni, M.A. Smith, G. Perry, et al. PP and Tau interaction. A possible link between amyloid and neurofibrillary tangles in Alzheimer's disease. *Am J Pathol* **1996**, *148*, 79–87.

192 Kisilevsky, R. Review: amyloidogenesis – unquestioned answers and unanswered questions. *J Struct Biol* **2000**, *130*, 99–108.

193 Balbirnie, M., R. Grothe and D.S. Eisenberg. An amyloid-forming peptide from the yeast prion Sup35 reveals a dehydrated beta-sheet structure for amyloid. *Proc Natl Acad Sci USA* **2001**, *98*, 2375–2380.

194 Nilsson, M.R. and C.M. Dobson. *In vitro* characterization of lactoferrin aggregation and amyloid formation. *Biochemistry* **2003**, *42*, 375–382.

195 Nilsson, M.R., M. Driscoll and D.P. Raleigh. Low levels of asparagine deamidation can have a dramatic effect on aggregation of amyloidogenic peptides: implications for the study of amyloid formation. *Protein Sci* **2002**, *11*, 342–349.

196 Nilsson, M.R. and C.M. Dobson. Chemical modification of insulin in amyloid fibrils. *Protein Sci* **2003**, *12*, 2637–2641.

6
Lipid Modulators of Protein Misfolding and Aggregation

Christopher A. MacRaild and Geoffrey J. Howlett

6.1
Introduction

Although at first glance the amyloid diseases are unified only by the characteristic presence of fibrous deposits of misfolded proteins, recent studies have revealed a number of other pathophysiologic features which appear to be common to many, if not all, amyloid-related conditions. One such common feature is the involvement of lipids at various stages of the disease etiology. For example, in Alzheimer's disease, a connection with lipid metabolism was first noted with the identification of the apolipoprotein E (ApoE) genotype as a major genetic risk factor for the disease [1, 2]. ApoE is a component of the plasma lipoproteins and plays a central role in plasma lipid metabolism. ApoE is also a component of the lipoproteins of the cerebrospinal fluid (CSF) and is presumed to have a similar role there. The association of ApoE with Alzheimer's disease remains poorly defined at the mechanistic level; however, investigations into this problem have revealed extensive links between Alzheimer's disease and lipid metabolism. Epidemiologically, increased levels of plasma cholesterol confer greater risk of disease [3]. Alzheimer's disease patients have elevated total and low-density lipoprotein (LDL) cholesterol levels [4], but lower high-density lipoprotein (HDL) levels than do healthy age-matched controls [5]. Similar alterations in plasma lipid levels are also associated with increased risk of heart disease and, indeed, patients with critical coronary artery disease also show cortical deposition of Aβ peptide similar to that seen in Alzheimer's brain [6]. Changes in brain phospholipid levels are also associated with Alzheimer's disease [7, 8]. These changes are likely the consequence of the up-regulation of a number of phospholipases, as well as increased oxidative modification of lipids in the Alzheimer's disease brain [9–11].

In other amyloid-related diseases, the evidence for lipid involvement is also building. ApoE is found co-localized to amyloid deposits associated with a number of disease types, including type II diabetes and several systemic amyloidoses [12–14]. In addition, the impact of amyloid fibrils on membrane integrity has been explored extensively recently: the amyloids or precursor proteins associated

Amyloid Proteins. The Beta Sheet Conformation and Disease. J. D. Sipe

ISBN: 3-527-31072-X

with Alzheimer's disease [15, 16], Type 2 diabetes [17], prion diseases [18], AA amyloidosis [19], senile cardiac amyloidosis [20] and dialysis-associated amyloidosis [21] have all been shown to form ion-conducting channels in lipid bilayers. Similar channels are observed in other proteins known to pathologically misfold such as the polyglutamine repeats associated with Huntington's disease [22] and α-synuclein, the protein which forms intracellular fibrillar aggregates in the Lewy bodies of Parkinson's disease [23].

Serum amyloid A (SAA) comprises a family of related acute-phase proteins, an N-terminal fragment of which forms amyloid deposits as a life-threatening complication to chronic inflammatory conditions [24]. During the acute-phase response to inflammation, circulating levels of SAA increase by up to 1000-fold and it becomes the predominant apolipoprotein of HDL particles, displacing the normal apolipoproteins of these particles [25]. Evidence from mutant forms of SAA indicates that the region in the N-terminal of the protein which is responsible for amyloid formation also contributes significantly to affinity for HDL, suggesting these two properties of SAA might be linked [26]. Interestingly, a number of other apolipoproteins also form amyloid, and it has been proposed that in these proteins, too, the properties of lipid affinity and amyloid formation share the same structural determinants [27].

The plasma thyroid hormone transporter transthyretin (TTR) forms amyloid deposits associated with senile cardiac amyloidosis and mutant forms of TTR are found in the deposits associated with some forms of familial amyloidotic polyneuropathy (FAP). Patients homozygous for several of these FAP-related mutants show decreased levels of ApoAII in plasma, apparently due to reduced affinity for HDL particles, and a subsequent increase in catabolism of this apolipoprotein [28]. *In vitro* studies of TTR aggregation suggest that lipoprotein particles enhance fibril formation [29].

We address one aspect of this relationship between amyloid formation and lipids, i.e. the effect of lipids and their catabolic and oxidative byproducts on the folding, unfolding and aggregation of amyloidogenic proteins. Other factors which are likely to play a role in this relationship, e.g. the effect of lipids on the production and processing of amyloidogenic precursors and the effect of amyloids on lipid peroxidation, have been reviewed elsewhere [22, 30–38] and are beyond the scope of the current work.

6.2
Protein Folding and Aggregation at Lipid Surfaces

Many proteins, and particularly many amyloidogenic proteins, are surface active and the interactions of these proteins with the surfaces of lipid membranes can have profound structural consequences. Lipid interactions can induce functionally important changes in protein structure. For example, the lipid bound forms of the apolipoproteins are conformationally and functionally distinct from the lipid-free form [39–42]. Similarly, the role of lipid in the folding and assembly of membrane proteins has been likened to that of a chaperone [43–45]. In contrast, surface-active globular proteins can be conformationally destabilized by interactions with lipid surfaces [46, 47], and these interactions can promote protein aggregation [48, 49] and amyloid formation [50].

Clearly then, the conformational consequences of protein–lipid interactions are complex, showing dependence not only on the nature of the protein, but also the physical and chemical properties of the lipid surface itself [47]. Not surprisingly, the effect of lipid surfaces on amyloid formation is also complex and multifactorial, reflecting the differing abilities of lipid surfaces to stabilize or destabilize a folded monomeric state and promote or inhibit protein–protein interactions. In what follows we will examine these effects in a number of different amyloidogenic systems.

6.2.1 The Aβ Peptide

The Aβ peptide is derived in part from the transmembrane domain of the amyloid precursor protein (AβPP) and thus its propensity for lipid interactions has long been a point of research interest. Furthermore, monomeric Aβ is largely unstructured in aqueous solution and aggregates into amyloid fibrils via an intermediate which is relatively rich in α-helical structure [51]. Thus, the effect of lipid surfaces on Aβ secondary structure is likely to manifest as an effect on the rate and extent of amyloid formation. The pathological relevance of such effects is suggested by a number of observations of Aβ aggregates associated with cell surface membranes in the brains of Alzheimer's disease patients, and in a number of animal and cell models [52–55].

Initial attempts to identify and quantify the effect of lipid interactions on the conformation and aggregation of Aβ were confounded by the range of apparently conflicting behaviors observed (Table 6.1). Aβ interacts with the surfaces of acidic lipid membranes via weak electrostatic interactions [56]. This interaction promotes a conformational transition in favor of the β-structure at low lipid:protein ratios, but at high lipid:protein ratios the α-helical structure is favored [57]. Solid-state nuclear magnetic resonance techniques reveal this interaction to be localized exclusively to the solvent–headgroup interface, with no significant insertion of the peptide into the lipid surface. Similarly, studies of lipid monolayers found insertion of Aβ only at lateral pressures well below those found in lipid bilayers [57]. In contrast, a number of studies have demonstrated a disruption of membrane integrity caused by Aβ which presumably implies the penetration of the peptide into the hydrophobic core of the bilayer [15, 16, 58]. Furthermore, Aβ has been observed to alter the fluidity of phospholipid membranes in a manner consistent with its insertion into the lipid bilayer, although the precise details of this effect have proved controversial [59–65].

More recent work to resolve the issues raised by these conflicting results has revealed that the interaction of Aβ with lipid surfaces can take several distinct modes dependent on a number of factors, including the aggregation state of

Table 6.1 Summary of selected interactions of lipids with amyloidogenic proteins.

Protein	Mode of lipid interaction	Effect on conformation and amyloid formation	References
Aβ peptide	electrostatic interactions with acidic lipid membranes; low protein : lipid ratio	favors β-sheet structure; promotes amyloid formation	56, 57, 74
	electrostatic interactions with acidic lipid membranes; high protein : lipid ratio	favors α-helical structure	57
	C-terminal insertion into membrane	favors α-helical structure; inhibits amyloid formation	74
	specific interaction with gangliosides in membrane	controversial: both promotion and inhibition of amyloid formation is reported	66, 67, 69, 70
	interaction with products of lipid oxidation	promotes amyloid formation	104, 105, 113
ApoAI	interaction with lipid surface via amphipathic helical regions	favors α-helical structure; inhibits amyloid formation	84, 86
α-Synuclein	native-like interaction with lipid surface via amphipathic helical regions	favors α-helical structure; inhibits amyloid formation	91, 92
	non-native interaction: low protein : lipid ratio or products of lipid damage	promotes aggregation and amyloid formation	93, 96–99
Apo CII	native-like interaction with lipid surface via amphipathic helical regions	favors α-helical structure; inhibits amyloid formation	27, 117, 140
	interaction with soluble lipid species	promotes amyloid formation	140

Aβ, the lipid composition and lateral pressure of the surface, the lipid : protein ratio, and the buffer conditions. For example, Kremer et al. have demonstrated an increase in exposed hydrophobic surface on aggregation of Aβ [59] and have correlated that with changes in membrane fluidity [59, 60], suggesting that Aβ insertion into lipid membranes might be dependent on aggregation. Furthermore, many recent studies of Aβ–lipid interactions have focused on the role of specific lipid species such as gangliosides and cholesterol or on the effect of the net charge of the lipid surfaces investigated [57, 58, 66–73].

These studies have not yet led to a comprehensive understanding of the structural details of the Aβ interaction with these various lipid species nor clearly identified the role of the interactions in terms of disease. However, they do reveal a diversity of lipid interactions which have implications for the conformation of Aβ, and the rate and mechanism of its aggregation. Exemplifying this fact is the study of Bokvist et al. which identifies two modes of interaction of

Aβ with acidic lipid membranes in a simple *in vitro* system [74]. The first of these modes involves a degree of peptide insertion into the membrane, and results in a peptide conformation with elements of both α-helical and β-sheet structure, while the second mode is based on superficial interactions between Aβ and the lipid headgroups, and shows little evidence of ordered peptide conformation. Over a period of several weeks, Aβ which interacts with the membrane in the first mode shows no conformation changes characteristic of aggregation, whereas Aβ superficially bound to the membrane surface undergoes a gradual increase in β-sheet content consistent with amyloid formation. In the Alzheimer's brain, a number of additional modes of Aβ-lipid interaction are likely to exist, mediated by specific interactions with gangliosides, cholesterol and other lipid species, and the simple two-state *in vitro* system described by Bokvist et al. [74] highlights the potential importance of this complex system of alternate interactions and conformational states in Aβ aggregation and disease progression.

6.2.2 ApoAI

ApoAI is the primary apolipoprotein of the HDLs and, like most apolipoproteins, ApoAI is helical when bound to the lipoprotein. Many mutant forms of ApoAI have been identified which are associated with amyloidogenic conditions, most resulting from the deposition of amyloid composed of 80- to 100-residue N-terminal fragments of the protein. This deposition is frequently systemic [75–78], although tissue-specific deposition has also been reported [79–81]. In addition, wild-type ApoAI and derivative peptides are a primary component within the amyloid associated with atherosclerotic plaques [82]. Many of the amyloid-promoting mutations result in a net gain of one charge leading to the hypothesis that an increase in charge is the cause of an increased propensity for amyloidosis either by a change in association of ApoAI with HDLs or a structural change that leads to altered catabolism and an increase in the concentration of the amyloidogenic N-terminal fragments [83]. In the lipid-free state, full-length ApoAI is partially α-helical, with extensive unstructured regions and a short stretch showing signs of β-structure [84, 85]. A conformational change is triggered by lipid interactions resulting in a large increase in helical structure in the disordered and β-sheet regions [84]. A similar conformational switching model is invoked to explain the aggregation of the amyloidogenic N-terminal fragment [86]. Here, the fragment is observed to be unstructured at neutral pH, but on acidification it forms a compact and transiently stable α-helical conformation, which subsequently self-associates and adopts the β-rich structure of the amyloid fibrils. The presence of lipid surfaces stabilizes the α-helical conformation and inhibits fibrillogenesis [86]. In this instance, the effect of lipid appears to be to favor the natively stabilizing interactions at the expense of the pathogenic interactions which lead to aggregation and fibril formation.

6.2.3
α-Synuclein

Amyloid fibrils composed of α-synuclein are the major components of the intraneuronal Lewy bodies found in Parkinson''s disease. The physiological role of α-synuclein is not yet established, although roles in synaptic plasticity and maintenance of synaptic vesicles have both been proposed [87]. α-Synuclein exists in the healthy brain in two states – a soluble cytosolic form, which has limited secondary structure, and an α-helical form bound to the lipid surface of synaptic vesicles [88, 89]. The N-terminal sequence of α-synuclein contains class A-like amphipathic α-helical domains [89] which are responsible for lipid binding and also corresponds to the amyloidogenic region [90].

The membrane interactions of α-synuclein are dependent on the lipid composition of the membrane and on membrane curvature, with a clear preference shown for negatively charged membranes and the highly curved surfaces of small vesicles. The implications of these interactions for aggregation and amyloid formation are again complex. Many reports have focused on an α-helical form of lipid-bound α-synuclein, which is presumed to correspond to the native, synaptic vesicle bound structure. In this context, lipid stabilizes helical structure and inhibits the transition to the β-sheet rich amyloid [91, 92]. However, there are several reports of accelerated aggregation of α-synuclein at lipid surfaces [93–95].

A number of lines of evidence point to the possibility that lipid-mediated aggregation occurs via a partially folded intermediate which is promoted by lipid interactions which are defective in some way, as compared to the native interaction. First, aggregation is accelerated in the presence of low lipid concentrations, but inhibited when the lipid concentration is increased [96]. Second, antioxidants reduce the aggregation of membrane-bound α-synuclein in brain extracts, suggesting that aggregation might be dependent on oxidative damage to the membrane [93]. Finally, aggregation is also accelerated by fatty acids and other products of lipid damage [97–99].

Two mutants of α-synuclein (A53T and A30P) are linked to early-onset Parkinson's disease and show enhanced oligomerization of amyloid intermediates [88]. In addition, these mutants have altered lipid-binding properties and it has been proposed that these differences may have implications for α-synuclein aggregation [90, 94, 100, 101].

6.2.4
Lipid Surfaces in Other Amyloidogenic Systems

There are reports of the involvement of lipid surfaces in a number of other amyloidogenic systems, including, for example, the islet amyloid polypeptide implicated in Type 2 diabetes [102]. One common factor to these reports and the studies outlined above is the involvement of acidic phospholipids. Indeed, one recent study of eight diverse proteins, including many without a link to disease, suggested that acidic lipid surfaces might act in a general manner to promote amyloid formation [103]. It is well established that otherwise stably folded and soluble proteins can often be induced to form amyloid by solutions conditions including low pH and reduced dielectricity. It is proposed, therefore, that the influence of negatively charged lipid surfaces might have a similar basis: proteins which associate peripherally with the membrane will be exposed to the high local proton concentration and reduced dielectric constant of the solution adjacent to the lipid surface.

6.3 Lipid Oxidation and Amyloid Formation

The association of amyloid and oxidative stress is well established and much research has focused on the role of amyloid in the formation of oxidative species. Conversely, a number of recent studies have examined the effect of oxidative products on protein misfolding and aggregation. Most of these effectors of amyloid formation are the byproducts of lipid oxidation. The products of copper and hydrogen peroxide oxidation of synthetic phospholipids accelerate β-sheet formation by Aβ42 to a similar extent to that observed for this peptide in the presence of ganglioside-enriched phosphatidylcholine bilayers [104]. Dye-binding assays and electron microscopy studies show that oxidized phospholipids increase the rate of amyloid fibril formation by Aβ42 [105], indicating a potential link between lipid oxidation and the pathogenesis of AD.

Amyloid has been associated with atherosclerotic plaques in 97% of patients over the age of 50 years [106]. Atherosclerotic plaques consist of foam cells, macrophages, and a necrotic core that accumulates oxidatively damaged lipids [107] and proteins [108] relative to normal plasma and arteries. We have proposed that oxidative processes may promote protein truncation or modification of apolipoproteins in a manner that perturbs their lipid binding properties and promotes amyloidogenic folding options [27]. The presence of oxidatively modified lipids [107] and the products of lipid hydrolysis such as fatty acids and lysolecithin [109] may also promote the formation of amyloid. For example, the presence of fatty acids promotes amyloid formation by tau protein *in vitro* [110], while oxidized fatty acids further promote amyloid formation by tau [111]. Similarly, α-synuclein aggregation is accelerated by the presence of long chain fatty acids [97, 98].

The discovery of ozone and the reaction product of ozone with cholesterol in excised atherosclerotic plaque tissue [112] has drawn attention to the potential role of ozonolysis oxidation products in amyloid fibril formation. The ozonolysis product of cholesterol (5,6-secosterol) and its structurally related aldolization product accelerate amyloid fibril formation by Aβ. The fatty acid oxidation metabolite, 4-hydroxynonenal, also accelerates Aβ fibril formation, whereas unmodified cholesterol and other derivatives lacking aldehyde groups have no effect [113].

6.4 Apolipoproteins and Amyloid

A significant number of apolipoproteins are implicated in amyloid formation or influence the amyloid formation of other proteins. These include the already introduced examples of ApoE and ApoAI, as well as others including ApoAII [114–116], ApoCII [117] and ApoB [118]. In addition, SAA and α-synuclein both share a number of structural and functional similarities with the apolipoproteins.

Insight into the common parameters that promote the formation of amyloid by apolipoproteins may be provided by considering the highly conserved nature of the apolipoproteins. This superfamily is thought to be derived from the duplication of an ancestral 22-amino-acid tandem repeat [119] and, consequently, these proteins share a number of significant structural features. They all contain a high proportion of class A amphipathic α-helical domains that mediate the binding to lipoprotein surfaces. Class A amphipathic α-helices are characterized by lysine residues on the hydrophilic face near the interface with the hydrophobic face and negatively charged residues at the outermost of the polar face [120]. This distribution of charged residues is believed to provide stabilizing interactions with the phosphate and choline moieties of the phospholipid head groups, thus anchoring the protein to the phospholipid surface [120]. These amphipathic helical domains bind parallel to the phospholipid surface in rapid equilibrium with unbound forms in contrast to the firm lipid-anchoring characteristic of transmembrane proteins. This mechanism of lipid binding permits switching between lipoprotein classes as part of the remodeling of lipoproteins that occurs during metabolism [120].

Another significant structural feature of apolipoproteins is their low conformational stability in the absence of lipid. While many apolipoproteins retain some α-helical structure in the absence of lipid, including ApoAI, ApoAII, ApoAIV and ApoE [121, 122], others have a diminished helical content under lipid-free conditions, including ApoCII [123] and ApoCIII [122]. With the possible exception of ApoE, all these apolipoproteins have limited conformational stability in the absence of lipid. Studies on ApoAI [124], ApoAII [125], ApoAIV [126] and the insect analog, apolipophorin III [127], indicate that even though these proteins retain α-helical structure, they have loosely packed tertiary structures. More specifically, thermal denaturation of ApoAI indicates a molten globular conformation in the absence of lipid [124], while ApoAIV retains a conformational stability energy near zero in the absence of lipid with a high content (54%) of α-helical structure [126]. ApoCI also has limited tertiary interactions in the absence of lipid, perhaps indicative of a conformationally flexible state, readily able to bind to lipid [128]. Even with ApoE, the C-terminal domain isolated by thrombin cleavage is α-helical in the absence of lipid and exhibits low conformational stability similar to that of the other apolipoproteins [129]. The C-terminal domain of ApoE is tetrameric and contains the majority of the predicted class A amphipathic α-helical domains in ApoE [130]. Evidence that loosely folded structures are important for lipid binding is provided by studies on apolipophorin III that show a correlation between lipid-binding kinetics and a molten globule conformation [127]. Such limited globular stability may reflect the requirement for conformational transformations of the apolipoproteins upon binding to lipid surfaces [84, 131].

The general lack of conformational stability of apolipoproteins in the absence of lipids suggests a mechanism for the propensity to form amyloid. Several globular proteins show a correlation between a partially folded state and amyloid formation. Mutants that promote molten globular conformations in lysozyme and TTR also promote amyloid formation [132, 133]. Similarly, conditions that partially denature proteins such as high concentrations of alcohols [134, 135] and extreme pH values [136] can promote amyloid formation leading to the conclusion that amyloid folding pathways are favored by destabilization of the native state [137]. In the context of apolipoproteins, we propose that events which destabilize the lipid-bound state (highly stable native complex), promote a partially folded conformation vulnerable to amyloid formation [27, 117].

6.5 The Effect of Lipids on the Stability of Apolipoproteins

We examined the effects of the short-chain phospholipid, dihexanoylphosphatidylcholine (DHPC), on the conformational stability of human ApoCII [27]. DHPC has the chemical structure of a phospholipid, but its short six-carbon acyl chains allow it to form submicellar and micellar species in aqueous solution, rather than forming the lipid bilayers characteristic of longer-chain phospholipids [138]. The motional and conformational characteristics of DHPC in micelles are similar to those in phospholipid bilayers [139]. ApoCII readily forms amyloid in the absence of lipid at physiological pH, salt concentrations and temperature, a process that is inhibited by the presence of micellar concentrations of DHPC. The conformational stability of ApoCII in the presence and absence of DHPC was determined by denaturation studies with guanidine hydrochloride (GuHCl) using circular dichroism to monitor the secondary structure. The results for ApoCII alone, over the concentration range of 0–5 M GuHCl, indicate little change in secondary structure. Parallel experiments were performed with ApoCII bound to DHPC micelles. The binding of this short-chain phospholipid to ApoCII in the absence of GuHCl induced α-helical structure. The unfolding profile for ApoCII in the presence of DHPC shows a cooperative unfolding event with a transition midpoint at approximately 1.5 M GuHCl, comparable to the results observed with the unfolding of globular proteins. Experiments were also performed with ApoCII bound to the longer-chain phospholipid, dimyristoylphosphatidylcholine (DMPC). In this case, GuHCl-induced unfolding also occurs, although there appears to be at least two transition phases indicating stable intermediate species in equilibrium. To further examine the unfolding, the samples were monitored for changes in the fluorescence properties of the single tryptophan residue in ApoCII. The results indicate that the changes in fluorescence correlate to the changes observed in the circular dichroism measurements. The shift of the average emission wavelength to a lower energy upon unfolding is consistent with the movement of the tryptophan from a non-polar environment to a more polar environment. The important conclusion is that there is little change in the fluorescence properties of ApoCII with increasing concentrations of GuHCl in the absence of lipid, suggesting that the protein in the lipid-free state is unordered. The results are consistent with a model where the binding of ApoCII to lipid confers conformational stability with native folding promoted by the hydrophobic core of the phospholipid.

Contrasting results on the affect of phospholipid on amyloid formation by human ApoCII are obtained using submicellar rather that micellar DHPC concentrations [140]. Turbidity measurements of the time course for amyloid formation by ApoCII show that micellar concentrations of DHPC (16 mM) reduce the rate of amyloid formation while submicellar concentrations of DHPC (below 10 mM) increase the rate by up to 6-fold. This suggests that the binding of single molecules of DHPC to ApoCII significantly reduces the activation energy for amyloid formation. Circular dichroism and fluorescence spectroscopy data show that submicellar concentrations of DHPC induce changes in secondary structure consistent with either the binding of individual DHPC molecules to DHPC to ApoCII or the induction of DHPC micelle formation by ApoCII. Similar observations have also been reported for α-synuclein where a series of simple alcohols induce changes in secondary structure that are characteristic of partially folded intermediates, coupled with markedly increased rates of amyloid fibril formation [141]. The ability of submicellar DHPC and short-chain alcohols to induce conformational changes and accelerate amyloid fibril formation identifies an additional mechanism for lipid modulation of amyloid fibril formation. A postulate arising from this mechanism is that lipid metabolites such as oxidized or truncated lipids, free fatty acids and lysolecithin exist as individual lipid effector molecules that promote the nucleation and elongation of amyloid fibrils.

6.6 Summary

The evidence for lipid involvement in amyloid-related diseases is building. Many amyloidogenic proteins are surface active and the interaction of these proteins with lipid surfaces can have profound structural consequences, as summarized in Fig. 6.1. Studies with several amyloid-forming systems indicate that proteins bind to negatively charged lipid surfaces in a reversible manner and that these interactions can promote amyloid fibril formation. However, proteins can also insert into membranes and form stable α-helical conformations that inhibit amyloid fibril formation. These competing mechanisms for the effects of lipids

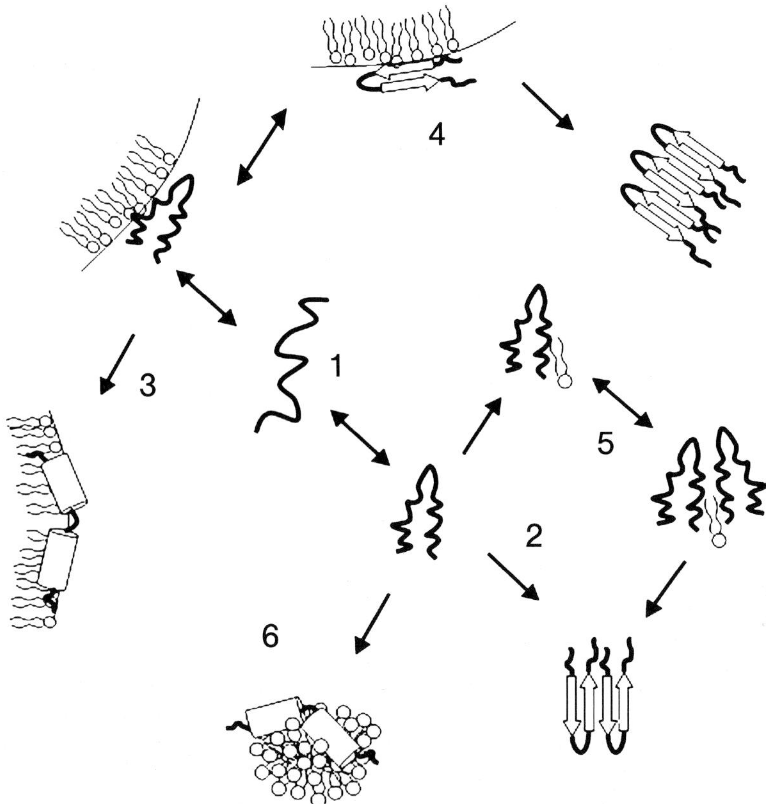

Fig. 6.1 Schematic of the proposed folding pathways for amyloidogenic proteins in the presence and absence of lipid. (1) Unfolded denatured state. (2) Pathway for amyloid formation by loosely folded, mostly unordered lipid-free conformation. (3) Lipid-bound α-helical conformation formed in the presence of negatively charged lipid bilayers under specific conditions. (4) Reversible binding to lipid bilayer surface favoring cross-β-sheet fibril formation. (5) Submicellar lipids promote formation of amyloid nucleus and promote fibril growth. (6) Micellar lipids interact with protein to stabilize α-helical conformation and inhibit fibril formation. Adapted from [140].

membranes can explain many of the apparently conflicting results pertaining to the modulating effects of lipids on amyloid fibril formation. In addition, individual lipid molecules (submicellar, oxidized and/or truncated lipids) can also influence amyloid fibril formation by exerting direct affects on protein folding and the assembly pathway.

Acknowledgments

We thank Danny Hatters for many valuable discussions of these topics.

References

1 Corder, E. H., A. M. Saunders, W. J. Strittmatter, D. E. Schmechel, P. C. Gaskell, G. W. Small, A. D. Roses, J. L. Haines and M. A. Pericak-Vance. Gene dose of apolipoprotein E type 4 allele and the risk of Alzheimer's disease in late onset families. *Science* **1993**, *261*, 921–923.

2 Strittmatter, W. J., A. M. Saunders, D. Schmechel, M. Pericak-Vance, J. Enghild, G. S. Salvesen and A. D. Roses. Apolipoprotein E: high-avidity binding to beta-amyloid and increased frequency of type 4 allele in late-onset familial Alzheimer disease. *Proc Natl Acad Sci USA* **1993**, *90*, 1977–1981.

3 Jarvik, G. P., E. M. Wijsman, W. A. Kukull, G. D. Schellenberg, C. Yu and E. B. Larson. Interactions of apolipoprotein E genotype, total cholesterol level, age, and sex in prediction of Alzheimer's disease: a case-control study. *Neurology* **1995**, *45*, 1092–1096.

4 Kuo, Y. M., M. R. Emmerling, C. L. Bisgaier, A. D. Essenburg, H. C. Lampert, D. Drumm and A. E. Roher. Elevated low-density lipoprotein in Alzheimer's disease correlates with brain abeta 1–42 levels. *Biochem Biophys Res Commun* **1998**, *252*, 711–715.

5 Fernandes, M. A., M. T. Proenca, A. J. Nogueira, L. M. Oliveira, B. Santiago, I. Santana and C. R. Oliveira. Effects of apolipoprotein E genotype on blood lipid composition and membrane platelet fluidity in Alzheimer's disease. *Biochim Biophys Acta* **1999**, *1454*, 89–96.

6 Sparks, D. L., J. C. Hunsaker, 3rd, S. W. Scheff, R. J. Kryscio, J. L. Henson and W. R. Markesbery. Cortical senile plaques in coronary artery disease, aging and Alzheimer's disease. *Neurobiol Aging* **1990**, *11*, 601–607.

7 Prasad, M. R., M. A. Lovell, M. Yatin, H. Dhillon and W. R. Markesbery. Regional membrane phospholipid alterations in Alzheimer's disease. *Neurochem Res* **1998**, *23*, 81–88.

8 Farooqui, A. A., S. I. Rapoport and L. A. Horrocks. Membrane phospholipid alterations in Alzheimer's disease: deficiency of ethanolamine plasmalogens. *Neurochem Res* **1997**, *22*, 523–527.

9 Pettegrew, J. W., J. Moossy, G. Withers, D. McKeag and K. Panchalingam. ^{31}P nuclear magnetic resonance study of the brain in Alzheimer's disease. *J Neuropathol Exp Neurol* **1988**, *47*, 235–248.

10 Pettegrew, J. W. Molecular insights into Alzheimer's disease. *Ann NY Acad Sci* **1989**, *568*, 5–28.

11 Farooqui, A. A., L. Liss and L. A. Horrocks. Increased activities of lipolytic enzymes in Alzheimer's disease. *Adv Neurol* **1990**, *51*, 127–129.

12 Gallo, G., T. Wisniewski, N. H. Choi-Miura, J. Ghiso and B. Frangione. Potential role of apolipoprotein-E in fibrillogenesis. *Am J Pathol* **1994**, *145*, 526–530.

13 Wisniewski, T. and B. Frangione. Apolipoprotein E: a pathological chaperone protein in patients with cerebral and systemic amyloid. *Neurosci Lett* **1992**, *135*, 235–238.

14 Charge, S. B., M. M. Esiri, C. A. Bethune, B. C. Hansen and A. Clark. Apolipoprotein E is associated with islet amyloid and other amyloidoses: implications for Alzheimer's disease. *J Pathol* **1996**, *179*, 443–447.

15 Arispe, N., H. B. Pollard and E. Rojas. Giant multilevel cation channels formed by Alzheimer disease amyloid beta-protein [A beta P-(1–40)] in bilayer membranes. *Proc Natl Acad Sci USA* **1993**, *90*, 10573–10577.

16 Hirakura, Y., M. C. Lin and B. L. Kagan. Alzheimer amyloid abeta1–42 channels: effects of solvent, pH, and Congo Red. *J Neurosci Res* **1999**, *57*, 458–466.

17 Mirzabekov, T. A., M. C. Lin and B. L. Kagan. Pore formation by the cytotoxic islet amyloid peptide amylin. *J Biol Chem* **1996**, *271*, 1988–1992.

18 Lin, M. C., T. Mirzabekov and B. L. Kagan. Channel formation by a neurotoxic prion protein fragment. *J Biol Chem* **1997**, *272*, 44–47.

19 Hirakura, Y., I. Carreras, J. D. Sipe and B. L. Kagan. Channel formation by serum amyloid A: a potential mechanism

for amyloid pathogenesis and host defense. *Amyloid* **2002**, *9*, 13–23.

20 Kourie, J. I., E. A. Hanna and C. L. Henry. Properties and modulation of alpha human atrial natriuretic peptide (alpha-hANP)-formed ion channels. *Can J Physiol Pharmacol* **2001**, *79*, 654–664.

21 Hirakura, Y. and B. L. Kagan. Pore formation by beta-2-microglobulin: a mechanism for the pathogenesis of dialysis associated amyloidosis. *Amyloid* **2001**, *8*, 94–100.

22 Kagan, B. L., Y. Hirakura, R. Azimov and R. Azimova. The channel hypothesis of Huntington's disease. *Brain Res Bull* **2001**, *56*, 281–284.

23 Lashuel, H. A., D. Hartley, B. M. Petre, T. Walz and P. T. Lansbury, Jr. Neurodegenerative disease: amyloid pores from pathogenic mutations. *Nature* **2002**, *418*, 291.

24 Sipe, J. D. Amyloidosis. *Annu Rev Biochem* **1992**, *61*, 947–975.

25 Coetzee, G. A., A. F. Strachan, D. R. van der Westhuyzen, H. C. Hoppe, M. S. Jeenah and F. C. de Beer. Serum amyloid A-containing human high density lipoprotein 3. Density, size, and apolipoprotein composition. *J Biol Chem* **1986**, *261*, 9644–9651.

26 Patel, H., J. Bramall, H. Waters, M. C. De Beer and P. Woo. Expression of recombinant human serum amyloid A in mammalian cells and demonstration of the region necessary for high-density lipoprotein binding and amyloid fibril formation by site-directed mutagenesis. *Biochem J* **1996**, *318*, 1041–1049.

27 Hatters, D. M. and G. J. Howlett. The structural basis for amyloid formation by plasma apolipoproteins: a review. *Eur Biophys J* **2002**, *31*, 2–8.

28 Nakamura, M., Y. Tanaka, Y. Ando, T. Yamashita, F. Salvi, A. Ferlini, P. Gobbi, C. Patrosso, M. Uchino and M. Ando. Decreased affinity of apolipoprotein AII to high-density lipoprotein in patients with transthyretin-related amyloidosis (Met30, Gln89, Pro36, and Thr34). *Biochem Biophys Res Commun* **1996**, *219*, 316–321.

29 Sun, X., Y. Ando, K. Haraoka, S. Katsuragi, T. Yamashita, S. Yamashita, M. Okajima, H. Terazaki and H. Okabe. Role of VLDL/chylomicron in amyloid formation in familial amyloidotic polyneuropathy. *Biochem Biophys Res Commun* **2003**, *311*, 344–350.

30 Kahn, S. E., S. Andrikopoulos and C. B. Verchere. Islet amyloid: a long-recognized but underappreciated pathological feature of type 2 diabetes. *Diabetes* **1999**, *48*, 241–253.

31 Eckert, G. P., C. Kirsch and W. E. Muller. Brain-membrane cholesterol in Alzheimer's disease. *J Nutr Health Aging* **2003**, *7*, 18–23.

32 Gibson Wood, W., G. P. Eckert, U. Igbavboa and W. E. Muller. Amyloid beta-protein interactions with membranes and cholesterol: causes or casualties of Alzheimer's disease. *Biochim Biophys Acta* **2003**, *1610*, 281–290.

33 Buchet, R. and S. Pikula. Alzheimer's disease: its origin at the membrane, evidence and questions. *Acta Biochim Pol* **2000**, *47*, 725–733.

34 Puglielli, L., R. E. Tanzi and D. M. Kovacs. Alzheimer's disease: the cholesterol connection. *Nat Neurosci* **2003**, *6*, 345–351.

35 Butterfield, D. A., A. Castegna, C. M. Lauderback and J. Drake. Evidence that amyloid beta-peptide-induced lipid peroxidation and its sequelae in Alzheimer's disease brain contribute to neuronal death. *Neurobiol Aging* **2002**, *23*, 655–664.

36 Kourie, J. I. Prion channel proteins and their role in vacuolation and neurodegenerative diseases. *Eur Biophys J* **2002**, *31*, 409–416.

37 Rocken, C. and A. Shakespeare. Pathology, diagnosis and pathogenesis of AA amyloidosis. *Virchows Arch* **2002**, *440*, 111–122.

38 Mahley, R. W., B. P. Nathan and R. E. Pitas. Apolipoprotein E. Structure, function, and possible roles in Alzheimer's disease. *Ann NY Acad Sci* **1996**, *777*, 139–145.

39 MacRaild, C. A., D. M. Hatters, G. J. Howlett and P. R. Gooley. NMR structure of human apolipoprotein C-II in the presence of sodium dodecyl sulfate. *Biochemistry* **2001**, *40*, 5414–5421.

40 Rozek, A., J.T. Sparrow, K.H. Weisgraber and R.J. Cushley. Conformation of human apolipoprotein C-I in a lipid-mimetic environment determined by CD and NMR spectroscopy. *Biochemistry* **1999**, *38*, 14475–14484.

41 Saito, H., P. Dhanasekaran, F. Baldwin, K.H. Weisgraber, S. Lund-Katz and M. C. Phillips. Lipid binding-induced conformational change in human apolipoprotein E. Evidence for two lipid-bound states on spherical particles. *J Biol Chem* **2001**, *276*, 40949–40954.

42 Borhani, D.W., D.P. Rogers, J.A. Engler and C.G. Brouillette. Crystal structure of truncated human apolipoprotein A-I suggests a lipid-bound conformation. *Proc Natl Acad Sci USA* **1997**, *94*, 12291–12296.

43 Zardeneta, G. and P.M. Horowitz. Micelle-assisted protein folding. Denatured rhodanese binding to cardiolipin-containing lauryl maltoside micelles results in slower refolding kinetics but greater enzyme reactivation. *J Biol Chem* **1992**, *267*, 5811–5816.

44 Bogdanov, M. and W. Dowhan. Lipid-assisted protein folding. *J Biol Chem* **1999**, *274*, 36827–36830.

45 Bogdanov, M., J. Sun, H.R. Kaback and W. Dowhan. A phospholipid acts as a chaperone in assembly of a membrane transport protein. *J Biol Chem* **1996**, *271*, 11615–11618.

46 Lo, Y.L. and Y.E. Rahman. Interaction between superoxide dismutase and dipalmitoylphosphatidylglycerol bilayers: a Fourier transform infrared (FT-IR) spectroscopic study. *Pharm Res* **1996**, *13*, 265–271.

47 Adams, S., A.M. Higgins and R.A.L. Jones. Surface-mediated folding and misfolding of proteins at lipid/water interfaces. *Langmuir* **2002**, *18*, 4854–4861.

48 Sluzky, V., J.A. Tamada, A.M. Klibanov and R. Langer. Kinetics of insulin aggregation in aqueous solutions upon agitation in the presence of hydrophobic surfaces. *Proc Natl Acad Sci USA* **1991**, *88*, 9377–9381.

49 Ball, A. and R.A.L. Jones. Conformational changes in adsorbed proteins. *Langmuir* **1995**, *11*, 3542–3548.

50 Sharp, J.S., J.A. Forrest and R.A. Jones. Surface denaturation and amyloid fibril formation of insulin at model lipid–water interfaces. *Biochemistry* **2002**, *41*, 15810–15819.

51 Kirkitadze, M.D., M.M. Condron and D.B. Teplow. Identification and characterization of key kinetic intermediates in amyloid beta-protein fibrillogenesis. *J Mol Biol* **2001**, *312*, 1103–1119.

52 Yamaguchi, H., M.L. Maat-Schieman, S.G. van Duinen, F.A. Prins, P. Neeskens, R. Natte and R.A. Roos. Amyloid beta protein (Abeta) starts to deposit as plasma membrane-bound form in diffuse plaques of brains from hereditary cerebral hemorrhage with amyloidosis-Dutch type, Alzheimer disease and nondemented aged subjects. *J Neuropathol Exp Neurol* **2000**, *59*, 723–732.

53 Torp, R., E. Head, N.W. Milgram, F. Hahn, O.P. Ottersen and C.W. Cotman. Ultrastructural evidence of fibrillar beta-amyloid associated with neuronal membranes in behaviorally characterized aged dog brains. *Neuroscience* **2000**, *96*, 495–506.

54 Oshima, N., M. Morishima-Kawashima, H. Yamaguchi, M. Yoshimura, S. Sugihara, K. Khan, D. Games, D. Schenk and Y. Ihara. Accumulation of amyloid β-protein in the low-density membrane domain accurately reflects the extent of β-amyloid deposition in the brain. *Am J Pathol* **2001**, *158*, 2209–2218.

55 Morishima-Kawashima, M. and Y. Ihara. The presence of amyloid beta-protein in the detergent-insoluble membrane compartment of human neuroblastoma cells. *Biochemistry* **1998**, *37*, 15247–15253.

56 Terzi, E., G. Holzemann and J. Seelig. Self-association of beta-amyloid peptide (1–40) in solution and binding to lipid membranes. *J Mol Biol* **1995**, *252*, 633–642.

57 Terzi, E., G. Holzemann and J. Seelig. Interaction of Alzheimer beta-amyloid peptide(1–40) with lipid membranes. *Biochemistry* **1997**, *36*, 14845–14852.

58 McLaurin, J. and A. Chakrabartty. Membrane disruption by Alzheimer beta-amyloid peptides mediated through spe-

cific binding to either phospholipids or gangliosides. Implications for neurotoxicity. *J Biol Chem* **1996**, *271*, 26482–26489.

59 Kremer, J. J., M. M. Pallitto, D. J. Sklansky and R. M. Murphy. Correlation of beta-amyloid aggregate size and hydrophobicity with decreased bilayer fluidity of model membranes. *Biochemistry* **2000**, *39*, 10309–10318.

60 Kremer, J. J., D. J. Sklansky and R. M. Murphy. Profile of changes in lipid bilayer structure caused by beta-amyloid peptide. *Biochemistry* **2001**, *40*, 8563–8571.

61 Muller, W. E., G. P. Eckert, K. Scheuer, N. J. Cairns, A. Maras and W. F. Gattaz. Effects of beta-amyloid peptides on the fluidity of membranes from frontal and parietal lobes of human brain. High potencies of A beta 1–42 and A beta 1–43. *Amyloid* **1998**, *5*, 10–15.

62 Muller, W. E., S. Koch, A. Eckert, H. Hartmann and K. Scheuer. *β*-Amyloid peptide decreases membrane fluidity. *Brain Res* **1995**, *674*, 133–136.

63 Avdulov, N. A., S. V. Chochina, U. Igbavboa, E. O. O'Hare, F. Schroeder, J. P. Cleary and W. G. Wood. Amyloid beta-peptides increase annular and bulk fluidity and induce lipid peroxidation in brain synaptic plasma membranes. *J Neurochem* **1997**, *68*, 2086–2091.

64 Chochina, S. V., N. A. Avdulov, U. Igbavboa, J. P. Cleary, E. O. O'Hare and W. G. Wood. Amyloid beta-peptide1-40 increases neuronal membrane fluidity: role of cholesterol and brain region. *J Lipid Res* **2001**, *42*, 1292–1297.

65 Mason, R. P., R. F. Jacob, M. F. Walter, P. E. Mason, N. A. Avdulov, S. V. Chochina, U. Igbavboa and W. G. Wood. Distribution and fluidizing action of soluble and aggregated amyloid beta-peptide in rat synaptic plasma membranes. *J Biol Chem* **1999**, *274*, 18801–18807.

66 Mandal, P. K. and J. W. Pettegrew. Alzheimer's disease: NMR studies of asialo (GM_1) and trisialo (GT_{1b}) ganglioside interactions with Abeta(1–40) peptide in a membrane mimic environment. *Neurochem Res* **2004**, *29*, 447–453.

67 Kakio, A., S. Nishimoto, K. Yanagisawa, Y. Kozutsumi and K. Matsuzaki. Interactions of amyloid beta-protein with various gangliosides in raft-like membranes: importance of GM_1 ganglioside-bound form as an endogenous seed for Alzheimer amyloid. *Biochemistry* **2002**, *41*, 7385–7390.

68 Kakio, A., S. Nishimoto, Y. Kozutsumi and K. Matsuzaki. Formation of a membrane-active form of amyloid beta-protein in raft-like model membranes. *Biochem Biophys Res Commun* **2003**, *303*, 514–518.

69 McLaurin, J., T. Franklin, P. E. Fraser and A. Chakrabartty. Structural transitions associated with the interaction of Alzheimer beta-amyloid peptides with gangliosides. *J Biol Chem* **1998**, *273*, 4506–4515.

70 ChooSmith, L. P. and W. K. Surewicz. The interaction between Alzheimer amyloid beta(1–40) peptide and ganglioside GM_1-containing membranes. *FEBS Lett* **1997**, *402*, 95–98.

71 Yip, C. M., E. A. Elton, A. A. Darabie, M. R. Morrison, J. McLaurin. Cholesterol, a modulator of membrane-associated Abeta-fibrillogenesis and neurotoxicity. *J Mol Biol* **2001**, *311*, 723–734.

72 Kirsch, C., G. P. Eckert and W. E. Muller. Cholesterol attenuates the membrane perturbing properties of beta-amyloid peptides. *Amyloid* **2002**, *9*, 149–159.

73 Gibson Wood, W., G. P. Eckert, U. Igbavboa and W. E. Muller. Amyloid beta-protein interactions with membranes and cholesterol: causes or casualties of Alzheimer's disease. *Biochim Biophys Acta* **2003**, *1610*, 281–290.

74 Bokvist, M., F. Lindstrom, A. Watts and G. Grobner. Two types of Alzheimer's beta-amyloid (1–40) peptide membrane interactions: aggregation preventing transmembrane anchoring versus accelerated surface fibril formation. *J Mol Biol* **2004**, *335*, 1039–1049.

75 Soutar, A. K., P. N. Hawkins, D. M. Vigushin, G. A. Tennent, S. E. Booth, T. Hutton, O. Nguyen, N. F. Totty, T. G. Feest, J. J. Hsuan and M. B. Pepys. Apolipoprotein AI mutation Arg-60 causes autosomal dominant amyloidosis. *Proc Natl Acad Sci USA* **1992**, *89*, 7389–7393.

76 Booth, D.R., S.Y. Tan, S.E. Booth, J.J. Hsuan, N.F. Totty, O. Nguyen, T. Hutton, D.M. Vigushin, G.A. Tennent, W.L. Hutchinson and M. Pepys. A new apolipoprotein Al variant, Trp50Arg, causes hereditary amyloidosis. *Q J Med* **1995**, *88*, 695–702.

77 Nichols, W.C., F.E. Dwulet, J. Liepnieks and M.D. Benson. Variant apolipoprotein AI as a major constituent of a human hereditary amyloid. *Biochem Biophys Res Commun* **1988**, *156*, 762–768.

78 Persey, M.R., D.R. Booth, S.E. Booth, R. van Zyl-Smit, B.K. Adams, A.B. Fattaar, G.A. Tennent, P.N. Hawkins and M.B. Pepys. Hereditary nephropathic systemic amyloidosis caused by a novel variant apolipoprotein A-I. *Kidney Int* **1998**, *53*, 276–281.

79 Hamidi Asl, K., J.J. Liepnieks, M. Nakamura, F. Parker and M.D. Benson. A novel apolipoprotein A-1 variant, Arg173Pro, associated with cardiac and cutaneous amyloidosis. *Biochem Biophys Res Commun* **1999**, *257*, 584–588.

80 Hamidi Asl, L., J.J. Liepnieks, K. Hamidi Asl, T. Uemichi, G. Moulin, E. Desjoyaux, R. Loire, M. Delpech, G. Grateau and M.D. Benson. Hereditary amyloid cardiomyopathy caused by a variant apolipoprotein A1. *Am J Pathol* **1999**, *154*, 221–227.

81 de Sousa, M.M., C. Vital, D. Ostler, R. Fernandes, J. Pouget-Abadie, D. Carles and M.J. Saraiva. Apolipoprotein AI and transthyretin as components of amyloid fibrils in a kindred with ApoAI Leu178His amyloidosis. *Am J Pathol* **2000**, *156*, 1911–1917.

82 Westermark, P., G. Mucchiano, T. Marthin, K.H. Johnson and K. Sletten. Apolipoprotein A1-derived amyloid in human aortic atherosclerotic plaques. *Am J Pathol* **1995**, *147*, 1186–1192.

83 Genschel, J., R. Haas, M.J. Propsting and H.H. Schmidt. Apolipoprotein A-I induced amyloidosis. *FEBS Lett* **1998**, *430*, 145–149.

84 Oda, M.N., T.M. Forte, R.O. Ryan and J.C. Voss. The C-terminal domain of apolipoprotein A-I contains a lipid-sensitive conformational trigger. *Nat Struct Biol* **2003**, *10*, 455–460.

85 Pownall, H.J., F.J. Hsu, M. Rosseneu, H. Peeters, A.M. Gotto and R.L. Jackson. Thermodynamics of lipid protein associations. Thermodynamics of helix formation in the association of high density apolipoprotein A-I (ApoA-I) to dimyristoyl phosphatidylcholine. *Biochim Biophys Acta* **1977**, *488*, 190–197.

86 Andreola, A., V. Bellotti, S. Giorgetti, P. Mangione, L. Obici, M. Stoppini, J. Torres, E. Monzani, G. Merlini and M. Sunde. Conformational switching and fibrillogenesis in the amyloidogenic fragment of apolipoprotein A-I. *J Biol Chem* **2003**, *278*, 2444–2451.

87 Clayton, D.F. and J.M. George. The synucleins: a family of proteins involved in synaptic function, plasticity, neurodegeneration and disease. *Trends Neurosci* **1998**, *21*, 249–254.

88 Conway, K.A., J.D. Harper and P.T. Lansbury. Accelerated *in vitro* fibril formation by a mutant alpha-synuclein linked to early-onset Parkinson disease. *Nat Med* **1998**, *4*, 1318–1320.

89 Davidson, W.S., A. Jonas, D.F. Clayton and J.M. George. Stabilization of alpha-synuclein secondary structure upon binding to synthetic membranes. *J Biol Chem* **1998**, *273*, 9443–9449.

90 Perrin, R.J., W.S. Woods, D.F. Clayton and J.M. George. Interaction of human alpha-synuclein and Parkinson's disease variants with phospholipids. Structural analysis using site-directed mutagenesis. *J Biol Chem* **2000**, *275*, 34393–34398.

91 Zhu, M. and A.L. Fink. Lipid binding inhibits alpha-synuclein fibril formation. *J Biol Chem* **2003**, *278*, 16873–16877.

92 Narayanan, V. and S. Scarlata. Membrane binding and self-association of alpha-synucleins. *Biochemistry* **2001**, *40*, 9927–9934.

93 Lee, H.J., C. Choi and S.J. Lee. Membrane-bound alpha-synuclein has a high aggregation propensity and the ability to seed the aggregation of the cytosolic form. *J Biol Chem* **2002**, *277*, 671–678.

94 Jo, E. A. A., Darabie, K. Han, A. Tandon, P. E. Fraser and J. McLaurin. α-Synuclein-synaptosomal membrane interactions: implications for fibrillogenesis. *Eur J Biochem* **2004**, *271*, 3180–3189.

95 Cole, N. B., D. D. Murphy, T. Grider, S. Rueter, D. Brasaemle and R. L. Nussbaum. Lipid droplet binding and oligomerization properties of the Parkinson's disease protein alpha-synuclein. *J Biol Chem* **2002**, *277*, 6344–6352.

96 Zhu, M., J. Li and A. L. Fink. The association of alpha-synuclein with membranes affects bilayer structure, stability, and fibril formation. *J Biol Chem* **2003**, *278*, 40186–40197.

97 Perrin, R. J., W. S. Woods, D. F. Clayton and J. M. George. Exposure to long chain polyunsaturated fatty acids triggers rapid multimerization of synucleins. *J Biol Chem* **2001**, *276*, 41958–41962.

98 Sharon, R., I. Bar-Joseph, M. P. Frosch, D. M. Walsh, J. A. Hamilton and D. J. Selkoe. The formation of highly soluble oligomers of alpha-synuclein is regulated by fatty acids and enhanced in Parkinson's disease. *Neuron* **2003**, *37*, 583–595.

99 Necula, M., C. N. Chirita and J. Kuret. Rapid anionic micelle-mediated alpha-synuclein fibrillization *in vitro*. *J Biol Chem* **2003**, *278*, 46674–46680.

100 Jo, E., J. McLaurin, C. M. Yip, P. St George-Hyslop and P. E. Fraser. α-Synuclein membrane interactions and lipid specificity. *J Biol Chem* **2000**, *275*, 34328–34334.

101 Jensen, P. H., M. S. Nielsen, R. Jakes, C. G. Dotti and M. Goedert. Binding of alpha-synuclein to brain vesicles is abolished by familial Parkinson's disease mutation. *J Biol Chem* **1998**, *273*, 26292–26294.

102 Knight, J. D. and A. D. Miranker. Phospholipid catalysis of diabetic amyloid assembly. *J Mol Biol* **2004**, *341*, 1175–1187.

103 Zhao, H., E. K. J. Tuominen and P. K. J. Kinnunen. Formation of amyloid fibres triggered by phosphatidylserine-containing membranes. *Biochemistry* **2004**, *42*, 10302–10307.

104 Koppaka, V. and P. H. Axelsen. Accelerated accumulation of amyloid beta proteins on oxidatively damaged lipid membranes. *Biochemistry* **2000**, *39*, 10011–10016.

105 Koppaka V., C. Paul, I. V. Murray and P. H. Axelsen. Early synergy between Abeta42 and oxidatively damaged membranes in promoting amyloid fibril formation by Abeta40. *J Biol Chem* **2003**, *278*, 36277–36284.

106 Mucchiano, G., G. G. Cornwell, 3rd and P. Westermark. Senile aortic amyloid. Evidence for two distinct forms of localized deposits. *Am J Pathol* **1992**, *140*, 871–877.

107 Suarna, C., R. T. Dean, J. May and R. Stocker. Human atherosclerotic plaque contains both oxidized lipids and relatively large amounts of alpha-tocopherol and ascorbate. *Arterioscler Thromb Vasc Biol* **1995**, *15*, 1616–1624.

108 Fu, S., M. J. Davies, R. Stocker and R. T. Dean. Evidence for roles of radicals in protein oxidation in advanced human atherosclerotic plaque. *Biochem J* **1998**, *333*, 519–525.

109 Williams, K. J. and I. Tabas. The response-to-retention hypothesis of early atherogenesis. *Arterioscler Thromb Vasc Biol* **1995**, *15*, 551–561.

110 Wilson, D. M. and L. I. Binder. Free fatty acids stimulate the polymerization of tau and amyloid beta peptides. *In vitro* evidence for a common effector of pathogenesis in Alzheimer's disease. *Am J Pathol* **1997**, *150*, 2181–2195.

111 Gamblin, T. C., M. E. King, J. Kuret, R. W. Berry and L. I. Binder. Oxidative regulation of fatty acid-induced tau polymerization. *Biochemistry* **2000**, *39*, 14203–14210.

112 Wentworth, P. Jr, J. Nieva, C. Takeuchi, R. Galve, A. D. Wentworth, R. B. Dilley, G. A. DeLaria, A. Saven, B. M. Babior, K. D. Janda, A. Eschenmoser and R. A. Lerner. Evidence for ozone formation in human atherosclerotic arteries. *Science* **2003**, *302*, 1053–1056.

113 Zhang, Q., E. T. Powers, J. Nieva, M. E. Huff, M. A. Dendle, J. Bieschke, C. G. Glabe, A. Eschenmoser, P. Wentworth, Jr, R. A. Lerner and J. W. Kelly. Metabo-

lite-initiated protein misfolding may trigger Alzheimer's disease. *Proc Natl Acad Sci USA* **2004**, *101*, 4752–4757.

114 Benson, M. D., J. J. Liepnieks, M. Yazaki, T. Yamashita, K. Hamidi Asl, B. Guenther and B. Kluve-Beckerman. A new human hereditary amyloidosis: the result of a stop-codon mutation in the apolipoprotein AII gene. *Genomics* **2001**, *72*, 272–277.

115 Higuchi, K., K. Kitagawa, H. Naiki, K. Hanada, M. Hosokawa and T. Takeda. Polymorphism of apolipoprotein A-II (ApoA-II) among inbred strains of mice. Relationship between the molecular type of ApoA-II and mouse senile amyloidosis. *Biochem J* **1991**, *279*, 427–433.

116 Higuchi, K., H. Naiki, K. Kitagawa, M. Hosokawa and T. Takeda. Mouse senile amyloidosis. ASSAM amyloidosis in mice presents universally as a systemic age-associated amyloidosis. *Virchows Arch B Cell Pathol* **1991**, *60*, 231–238.

117 Hatters, D. M., C. E. MacPhee, L. J. Lawrence, W. H. Sawyer and G. J. Howlett. Human apolipoprotein C-II forms twisted amyloid ribbons and closed loops. *Biochemistry* **2000**, *39*, 8276–8283.

118 O'Brien, K. D., K. L. Olin, C. E. Alpers, W. Chiu, M. Ferguson, K. Hudkins, T. N. Wight and A. Chait. Comparison of apolipoprotein and proteoglycan deposits in human coronary atherosclerotic plaques: colocalization of biglycan with apolipoproteins. *Circulation* **1998**, *98*, 519–527.

119 Li, W. H., M. Tanimura, C. C. Luo, S. Datta and L. Chan. The apolipoprotein multigene family: biosynthesis, structure, structure–function relationships, and evolution. *J Lipid Res* **1988**, *29*, 245–271.

120 Segrest, J. P., D. W. Garber, C. G. Brouillette, S. C. Harvey and G. M. Anantharamaiah. The amphipathic alpha helix: a multifunctional structural motif in plasma apolipoproteins. *Adv Protein Chem* **1994**, *45*, 303–369.

121 Perugini, M. A., P. Schuck and G. J. Howlett. Self-association of human apolipoprotein E3 and E4 in the presence and absence of phospholipid. *J Biol Chem* **2000**, *275*, 36758–36765.

122 Morrisett, J. D., R. L. Jackson and A. M. Gotto, Jr. Lipid–protein interactions in the plasma lipoproteins. *Biochim Biophys Acta* **1977**, *472*, 93–133.

123 Tajima, S., S. Yokoyama, Y. Kawai and A. Yamamoto. Behavior of apolipoprotein C-II in an aqueous solution. *J Biochem (Tokyo)* **1982**, *91*, 1273–1279.

124 Gursky, O. and D. Atkinson. Thermal unfolding of human high-density apolipoprotein A-1: implications for a lipid-free molten globular state. *Proc Natl Acad Sci USA* **1996**, *93*, 2991–2995.

125 Gursky, O. and D. Atkinson. High- and low-temperature unfolding of human high-density apolipoprotein A-2. *Protein Sci* **1996**, *5*, 1874–1882.

126 Weinberg, R. B. and M. S. Spector. Structural properties and lipid binding of human apolipoprotein A-IV. *J Biol Chem* **1985**, *260*, 4914–4921.

127 Soulages, J. L. and O. J. Bendavid. The lipid binding activity of the exchangeable apolipoprotein apolipophorin-III correlates with the formation of a partially folded conformation. *Biochemistry* **1998**, *37*, 10203–10210.

128 Gursky, O. and D. Atkinson. Thermodynamic analysis of human plasma apolipoprotein C-1: high-temperature unfolding and low-temperature oligomer dissociation. *Biochemistry* **1998**, *37*, 1283–1291.

129 Wetterau, J. R., L. P. Aggerbeck, S. C. Rall, Jr and K. H. Weisgraber. Human apolipoprotein E3 in aqueous solution. I. Evidence for two structural domains. *J Biol Chem* **1988**, *263*, 6240–6248.

130 Aggerbeck, L. P., J. R. Wetterau, K. H. Weisgraber, C. S. Wu and F. T. Lindgren. Human apolipoprotein E3 in aqueous solution. II. Properties of the amino- and carboxyl-terminal domains. *J Biol Chem* **1988**, *263*, 6249–6258.

131 Soulages, J. L., E. L. Arrese, P. S. Chetty and V. Rodriguez. Essential role of the conformational flexibility of helices 1 and 5 on the lipid binding activity of apolipophorin-III. *J Biol Chem* **2001**, *276*, 34162–34166.

132 Booth, D.R., M. Sunde, V. Bellotti, C.V. Robinson, W.L. Hutchinson, P.E. Fraser, P.N. Hawkins, C.M. Dobson, S.E. Radford, C.C. Blake and M.B. Pepys. Instability, unfolding and aggregation of human lysozyme variants underlying amyloid fibrillogenesis. *Nature* **1997**, *385*, 787–793.

133 Nettleton, E.J., M. Sunde, Z. Lai, J.W. Kelly, C.M. Dobson and C.V. Robinson. Protein subunit interactions and structural integrity of amyloidogenic transthyretins: evidence from electrospray mass spectrometry. *J Mol Biol* **1998**, *281*, 553–564.

134 Chiti, F., P. Webster, N. Taddei, A. Clark, M. Stefani, G. Ramponi and C. M. Dobson. Designing conditions for *in vitro* formation of amyloid protofilaments and fibrils. *Proc Natl Acad Sci USA* **1999**, *96*, 3590–3594.

135 Goda, S., K. Takano, Y. Yamagata, R. Nagata, H. Akutsu, S. Maki, K. Namba and K. Yutani. Amyloid protofilament formation of hen egg lysozyme in highly concentrated ethanol solution. *Protein Sci* **2000**, *9*, 369–375.

136 McParland, V.J., N.M. Kad, A.P. Kalverda, A. Brown, P. Kirwin-Jones, M.G. Hunter, M. Sunde and S.E. Radford. Partially unfolded states of beta(2)-microglobulin and amyloid formation *in vitro*. *Biochemistry* **2000**, *39*, 8735–8746.

137 Dobson, C.M. Protein misfolding, evolution and disease. *Trends Biochem Sci* **1999**, *24*, 329–332.

138 Tausk, R.J.M., J. Van Esch, J. Karmiggelt, G. Voordouw and J.Th.G. Overbeek. Physical chemical studies of short-chain lecithin homologues. II. Micellar weights of dihexanoyl- and diheptanoyllecithin. *Biophys Chem* **1974**, *1*, 184–203.

139 Burns, R.A. Jr and M.F. Roberts. Carbon-13 nuclear magnetic resonance studies of short-chain lecithins. Motional and conformational characteristics of micellar and monomeric phospholipid. *Biochemistry* **1980**, *19*, 3100–3106.

140 Hatters, D.M., L.J. Lawrence and G.J. Howlett. Sub-micellar phospholipid accelerates amyloid formation by apolipoprotein C-II. *FEBS Lett* **2001**, *494*, 220–224.

141 Munishkina, L.A., C. Phelan, V.N. Uversky and A.L. Fink. Conformational behavior and aggregation of alpha-synuclein in organic solvents: modeling the effects of membranes. *Biochemistry* **2003**, *42*, 2720–2730.

7
Extracellular Matrix Heparan Sulfate Proteoglycans

Peter J. Neame and John T. Gallagher

7.1
Introduction

Amyloid is a generic term for proteinaceous aggregates found in a variety of disease states, e.g. Alzheimer's disease, familial amyloid polyneuropathy, amyloid A (AA) amyloidosis and immunoglobulin-related light chain (AL) amyloidosis. The protein component varies considerably from disease to disease; there are presently at least 22 unrelated proteins that are considered to be amyloidogenic [1]. A common feature is the "cross-β" structure, in which β-strands align orthogonally to the long axis of the amyloid fiber [2]. *In vivo*, amyloid is frequently, or even always, associated with heparan sulfate glycosaminoglycans [3, 4] as well as other glycosaminoglycans such as chondroitin, dermatan and keratan sulfate [5]. Glycosaminoglycans are ubiquitous components of the cell surface and the extracellular matrix, and, with the exception of hyaluronic acid, are at least initially attached to a protein core.

Amyloid derives from innocuous proteins that are typically partially degraded or misfolded. The protein fragment or misfolded protein aggregates into fibrils which at least initially are often cytotoxic [6–9]. A nucleating center is probably critical for the deposition of a precipitated fibril and is the rate-limiting step in Alzheimer's amyloid fibril formation [10, 11]. Subsequent assembly of fibrils or amorphous deposits likely proceeds through a first-order addition of further monomers [12], although other mechanisms, such as progressive aggregation of protofibrils and protofilaments that each have different structures [13], have been proposed. There is an increasing body of evidence to suggest that proteins other than the "classic" amyloid-forming proteins can be induced to form amyloid-like structures *in vitro* (using circular dichroism, microscopy and Congo red binding as criteria). Amyloid formation may, therefore, be an inherent property of protein folding pathways, even though it is not usually a desirable outcome (reviewed by Ohnishi and Takano [14]).

The ubiquitous distribution, ordered structure and propensity for protein binding makes the glycosaminoglycan chains of extracellular proteoglycans potential candidate molecules for promotion of nucleation, although the mecha-

Amyloid Proteins. The Beta Sheet Conformation and Disease. J. D. Sipe

ISBN: 3-527-31072-X

nism of this is unclear. Once amyloid has formed, glycosaminoglycans are thought, in many cases, to provide a protective coat [15], increasing the stability of amyloid by reducing its opportunities for diffusion-mediated dispersal or protease-mediated destruction [16]. Prion proteins that are responsible for diseases such as Creutzfeldt-Jakob disease and scrapie also misfold into disease-causing, predominantly β-sheet, amyloid-like structures that may also be universally associated with glycosaminoglycans [17].

Glycosaminoglycans (or mucopolysaccharides as they used to be called) have been known to be associated with amyloid in the spleen [18] and other tissues [19] for over three decades. With the exception of hyaluronic acid, all glycosaminoglycans are, at least initially, attached to core proteins. Glycosaminoglycans comprise a family of polysaccharides that have as their distinguishing features a repeating disaccharide and extensive negative charge deriving from carboxyl and sulfate groups. Their roles are diverse; in cartilage, the predominant chondroitin/keratan sulfate proteoglycan largely functions to form a highly hydrated shock-absorbing extracellular matrix constrained by collagen fibrils. Other, smaller and more widespread chondroitin/dermatan/keratan sulfate proteoglycans with a leucine-rich repeat core protein are involved in control of collagen fibrillogenesis. Similar proteoglycans act to order the structure of the extracellular matrix in the cornea to ensure tissue transparency. Other heparan/chondroitin sulfate proteoglycans, either attached to the cell membrane or in the pericellular matrix, are critical for presentation of growth factors to their cell surface receptors and help to maintain blood fluidity on endothelial surfaces. Anticoagulant properties are vested in a specific pentasaccharide sequence in heparan sulfate and heparin that accelerates the rate of binding of antithrombin to thrombin and factor Xa [20]. Proteoglycans also assist in cell adhesion [21] and most proteoglycans have some role in ordering the extracellular matrix. For a more detailed overview, see Iozzo [22].

Our understanding of proteoglycans has increased enormously since the advent of modern molecular biology techniques, and dramatic advances have been made in understanding the nature of the core proteins (which fall into various unrelated protein families), and in defining the biosynthetic and degradative enzymes involved in the synthesis and processing of the glycosaminoglycan chains. Substantial improvements in chemical and physical analytical techniques have also enhanced our understanding of the structures of the very heterogeneous glycosaminoglycan chains. Even so, much remains to be learned about the detailed structure of proteoglycans, about the control of biosynthesis of glycosaminoglycan chains and about the role of the glycosaminoglycan chains in specific environments. It was once thought that the glycosaminoglycan chains of proteoglycans were capable of almost unlimited variety, but it is now clear that specific patterns of modification predominate, especially in heparan sulfate, in which the level of diversity is substantial but restricted to domains of variable sulfation [23]. A clear understanding of the role of proteoglycans in amyloid formation is likely to be dependent on careful analysis of the types of proteoglycan or glycosaminoglycan that are involved.

To visualize how glycosaminoglycans might be implicated in amyloid formation, it is instructive to consider the progression of pathological events within the context of the surroundings of the amyloid fibril. The initial step would be a globular precursor protein partially degraded by proteolysis, thus destabilizing its structure. Alternatively, the protein could be in a quasi-stable state due to a destabilizing mutation. Both of these situations would be likely to result in hydrophobic residues from the interior of the protein being exposed to solvent as a result of changes in conformation. If adequate solvent is present, then the protein fragment will likely refold to sequester the hydrophobic residues from solvent. However, if the protein concentration is high enough, such residues may be shielded from solvent by spontaneous and rapid aggregation with similar molecules – this process would bypass the refolding of the protein through intramolecular interactions. In this context, the concept of domain swapping has been proposed to be a mechanism for amyloid formation. Domain swapping involves the substitution of a domain in one protein for the equivalent domain in another [24]. While attractive as a process of fibril growth, it is unclear whether it is a viable mechanism in most cases of amyloid formation (reviewed by Liu and Eisenberg [25]).

A glycosaminoglycan-rich environment alters the surroundings of proteins or peptides substantially. Glycosaminoglycans have substantial negative charge at neutral pH as a result of large numbers of carboxyl and sulfate groups. Thus, water in the vicinity of the carbohydrate will tend to form a hydrogen-bonded ordered structure (this is the major function of the large chondroitin sulfate proteoglycan in cartilage). This has the net effect of decreasing free water and thus increasing the effective concentration of other molecules in the vicinity. As the glycosaminoglycans have a considerable overall negative charge, they are capable of binding to basic amino acids, thus temporarily "locking" a protein in a configuration that it would not ordinarily adopt.

Congo red dye, which is often used as one of the diagnostic tools for histochemical identification of amyloid (Fig. 7.1), has a superficial resemblance to glycosaminoglycans with two sulfate groups separated by unsaturated bonds that constrain the configuration [26]. While the sulfates in a glycosaminoglycan have considerably greater freedom of movement than those in Congo red, the regularly spaced charges are certainly analogous.

While heparan sulfate, by virtue of its ordered structure, appears to promote amyloid formation in some cases, low-molecular-weight mimics of heparan sulfate may act to inhibit fibrillogenesis, either through metabolic incorporation into the glycosaminoglycan or simply as competitors of glycosaminoglycan binding. Analogs of *N*-acetylglucosamine, incorporated into oligosaccharides, have proven to be inhibitory for amyloid formation [27, 28]. Other proteoglycan glycosaminoglycan chains (chondroitin, keratan and dermatan sulfates) have also been reported to be associated with amyloid by immunohistochemical studies [5]. The potential of glycosaminoglycan-like molecules to prevent amyloidogenesis has resulted in increased interest in the interaction between heparan sulfate and amyloid precursors. In the case of prion proteins, the transition from the

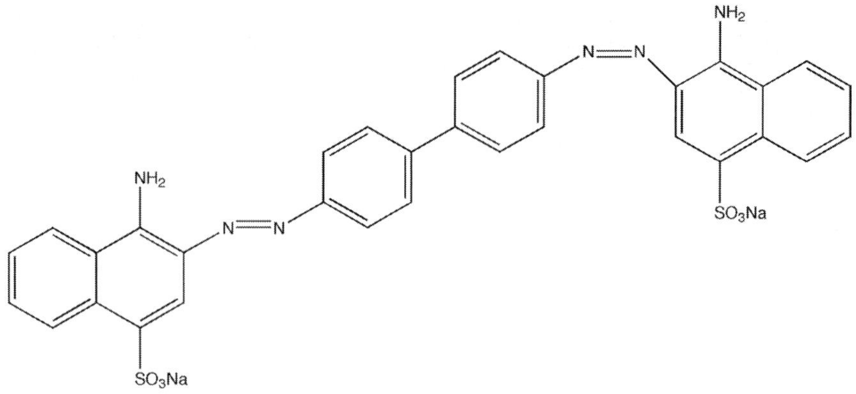

Fig. 7.1 Structure of Congo red. Congo red consists of a series of π-bonded residues with sulfate and amino groups conferring solubility. Superficially, the position of the sulfates could mimic the behavior of glycosaminoglycans [26], while the aromatic structure would intercolate itself between hydrophobic residues.

"good" prion protein (PrP^{C}), a primarily α-helical cell surface glycoprotein of unknown function, to the "bad", β-sheet-rich, prion (PrP^{Sc}) protein found in diseases such as Creutzfeldt-Jakob disease and scrapie is stimulated by cell surface heparan sulfate and this stimulation appears to depend on the level of sulfation of the glycosaminoglycan [29].

This review will briefly describe the heterogeneity of proteoglycans and discuss the nature of heparan sulfate–heparin interactions with proteins as far is it is understood. This will be further developed to discuss potential interactions with amyloid-causing proteins. Little detail is known about the ability of complex sulfated oligosaccharides to promote or inhibit amyloid formation; it is possible, however, to speculate about how sulfated carbohydrate and protein might interact to affect the process. The research in this area has largely been focused on heparin-derived oligosaccharides due to their commercial availability; it is to be hoped that future work will focus on other structures such as oversulfated chondroitin/dermatan sulfate.

7.2 Protein Folding and Glycosaminoglycans

Proteins fold into a defined structure as a result of interactions between the amino acids that comprise their primary structure or amino acid sequence. An early event in this process is the formation of secondary structure, either α-helix or β-sheet. The tertiary structure describes the final three-dimensional arrangement of the protein after the folding process has completed. The tertiary structure is not rigid; many proteins, notably enzymes, undergo structural changes as part of their normal function. Many forms of amyloid arise as a result of disruption of the tertiary structure of a protein. Amyloid thus develops from a variety of proteins, some of which [e.g. transthyretin (TTR)] are composed primarily of β-sheets and others [e.g. islet amyloid polypeptide (IAPP)] are composed primarily of α-helices.

Protein folding processes are largely driven by basic thermodynamics. Thus, entropy will tend to be maximized and free energy will be minimized. The reduction in free energy usually outweighs entropic changes, resulting in an ordered structure. Solvent-exposed hydrophobic residues and, to a lesser extent, solvent-exposed charges cause high levels of free energy and low entropy due to their influence on the surrounding water in biological systems. A simplistic view of protein folding is therefore the concept that it is driven by conformation changes that (1) neutralize exposed charge either by hydrogen bonding to water, to other structures or to other charged amino acids, and (2) bury hydrophobic side-chains, reducing their exposure to solvent and therefore the degree of ordering of the structure of the solvent. Protein folding is not a random thermodynamic process; the product would take much longer to form than is actually seen in practice. A folding "funnel" has been used as a tool to visualize the kinetic progression of a protein through its various folding intermediates, starting at a structure with high conformational entropy at the entrance of the funnel and proceeding to its most stable form with lowest free energy at the bottom of the funnel [30, 31].

A disruption to the protein structure can trigger refolding of the protein into a new structure that reduces solvent exposure by reducing the energy that needs to be added to the system to move to a state with overall lower free energy. Thus, proteolytic cleavage of a protein such as amyloid β precursor protein (AβPP) can cause a molecular rearrangement. Similarly, if a mutation changes a hydrophilic, solvent-exposed, residue to a hydrophobic residue, then this may cause a propensity to refold, as in TTR, for example (Section 7.7.5). Similar effects can be achieved through changes in pH, temperature or the presence of other molecules. If the individual protein concentrations are high enough, then enough monomers may be present to make the most rapid refolding the aggregation of hydrophobic sites into large multimolecular aggregates. A detailed discussion of amyloid formation from the point of view of protein folding has been developed by Ohnishi and Takano [14]. Proteoglycans might play a role in this process by binding to and stabilizing high-energy intermediates, effectively reducing their free energy.

The individual amino acid sequences that result in amyloid formation are almost impossible to determine. Conventional protein folding, as it is understood at present, is a unimolecular process that has a limited number of pathways and so is finite, whereas amyloid formation is multimolecular and can therefore proceed indefinitely. Intuitively, it would seem that amyloid would have little order. In fact, fibrillar amyloid seems to have a surprising degree of organized structure, exemplified by the cross-β structure consisting of β-strands arranged orthogonally to the long axis of the fiber [32]. In spite of the limited problem of folding of a single protein, it is still not possible to predict a protein structure based solely on the amino acid sequence. If the folding problem is expanded to multimolecular processes, then the difficulty of obtaining a solution expands to a level that is at present insoluble.

7.3 β-Sheets

The β-Sheets are the most common element of secondary structure for the proteins or peptides that comprise amyloid aggregates [16, 33–35]. The β-strand is the simplest and most flexible organized protein structure and is the conformation that a peptide will adopt if it is extended linearly. In a β-strand, amino acid side-chains extend in alternating directions at right angles to the peptide-bonded backbone and are approximately 3.5 Å apart (Fig. 7.2). The amide (N–H) and carbonyl groups (C=O) alternate, approximately at right angles to the side-chains, along the peptide backbone. The amide and carbonyl groups in separate β-strands hydrogen bond to each other in either a parallel or an antiparallel arrangement, resulting in β-sheets.

β-Sheets are not flat structures; adjacent strands typically have an approximately 25° twist with respect to each other (although parallel sheets have less twist and the degree of twist can vary significantly in the more flexible and more common antiparallel sheets) [36]. The β-sheet twist can be seen clearly in proteins such as TTR, lysozyme or any of the members of the immunoglobulin superfamily.

The structure of a β-strand is exemplified by the nuclear magnetic resonance (NMR)-derived structure of the TTR fragment found in TTR amyloid (Fig. 7.2) [37]. In this structure, one "side" of the strand consists of the amino acids Thr–Ala–Leu–Ser–Tyr (the side that is oriented towards solvent in the TTR [38]), while the other "side" is Ile–Ala–Leu–Pro and is oriented towards the interior of the protein. The amide and carbonyl groups hydrogen bond to adjacent β-strands to form, on one side, an antiparallel structure and, on the other, a parallel structure.

Ribosomal protein S6 from *Thermus thermophilus*, a 101-amino-acid protein, has been engineered by changing four hydrophilic amino acids to four hydrophobic amino acids (VEELGLRRLAYPIA → VAILGLMVLAYPIA) to contain a β-sheet structure that is very similar to that of Aβ. This structure has a pronounced propensity to aggregate [39] and may form a useful model for understanding the assembly of amyloid structures.

In principle, short β-strands would hydrogen bond to each other indefinitely to form enormous β-sheets; mechanical forces would be the major limitation on their assembly beyond a given size. It has been proposed that the β-strands on the outside edges of β-sheets in a protein would need to have structural features that would prevent this [40].

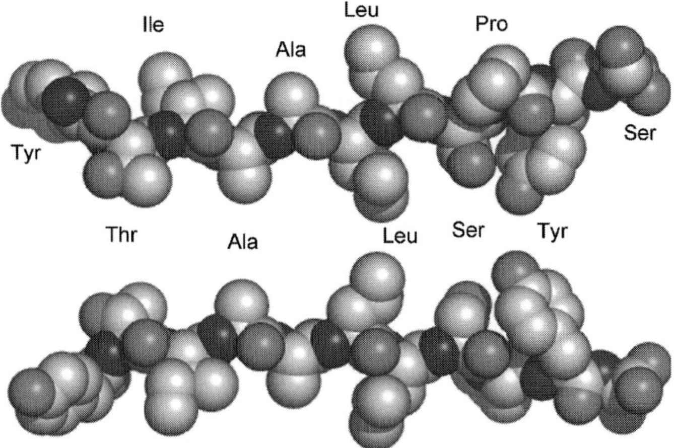

Fig. 7.2 The NMR-derived structure of a β-strand TTR fragment that forms amyloid *in vitro*. The peptide (Tyr–Thr–Ile–Ala–Ala–Leu–Leu–Ser–Pro–Tyr–Ser) derives from the penultimate β-sheet in intact TTR. Coordinates are from the PDB file 1RVS [37] and were rendered using MacPyMol. Individual amino acids are shown using their three-letter abbreviations. The viewing direction is in the plane of the β-sheet. The top and bottom views show opposite directions with the N-terminal on the left. Atoms are represented in standard CPK colors (red = oxygen, blue = nitrogen). Hydrogens are omitted for clarity. The locations of the carbonyl (CO) and amide (NH) groups that form the β-sheet-stabilizing hydrogen bonds are visible pointing towards the viewer on each side of the model. These residues hydrogen bond to adjacent β-strands to form the β-sheet. Note that the C-terminal amino acids (Ser–Pro–Tyr–Ser) begin to twist out of the plane of the sheet. In the native protein these residues form a turn and begin the next β-strand of the sheet.

De novo designed β-sheet structures often have severe solubility problems, as do β-strands synthesized or excised from otherwise soluble β-sheet proteins [41]. The β-strands at the edges of sheets typically have subtle differences from "ideal" β-sheets that preclude them from aggregating with other proteins [40]. Thus, when a β-sheet-containing protein is partially unfolded, it loses the "protection" of the edge strands and develops a propensity to aggregate.

Amyloid appears to result from a loss of the constraint imposed by the β-strands at the edge of the sheet and the net result is an aggregate of proteins or peptides that forms fibrils or other, more amorphous, aggregates. The structural chemistry and crystallographic data that leads to this model have been reviewed in detail by Serpell et al. [2]. At the level of amyloid fibril structure, there is no obvious reason why glycosaminoglycans might be associated with protein that has folded aberrantly. One explanation, discussed in Section 7.8, might be that the highly charged and hydrated glycosaminoglycans act as mischievous chaperones, stabilizing the protein or peptide precursor of amyloid so that intermolecular interactions rather than intramolecular interactions predominate.

7.4 Proteoglycans

Our understanding at present is that glycosaminoglycans, rather than the core protein, are capable of modulating amyloid formation or dissolution. Nevertheless, with the exception of hyaluronic acid which is synthesized at the cell membrane, glycosaminoglycans are initially assembled on a core protein and therefore their existence depends on the previous expression of the core protein. Proteoglycan biochemistry is therefore a complex field in which the core protein and the glycosaminoglycan chains can have different roles. Fig. 7.3 shows highly simplified diagrams of heparan sulfate proteoglycans. In agrin and perlecan, the core protein binds to other extracellular matrix components or to the cell surface. In the syndecans, the core proteins can aggregate, in part mediated by the intracellular domain, while the heparan sulfate chains can sequester cytokines or bind to extracellular matrix molecules.

The core protein structure is specified by the coding DNA sequence, but can vary due to alternative splicing. Post-translational modifications, such as proteolysis and *N*- and *O*-glycosylation, add another layer of structural variation. The true complexity of a proteoglycan lies in its glycosaminoglycan chains which are generally, but not always, attached to regions of the protein that adopt an extended configuration. Space precludes a general review of proteoglycan structure and this review will cover those that are most pertinent to amyloid formation – chondroitin/dermatan sulfate and heparin/heparan sulfate.

An example of proteoglycan diversity is AβPP, which can be alternatively spliced in astrocytes to yield a cryptic xylosyl transferase recognition site. A chondroitin sulfate chain is synthesized at this point to result in a proteoglycan, appican [42]. The chondroitin sulfate chains contain a heavily sulfated disaccharide, –GlcUA–GalNAc (4,6-*O*-disulfate)–, that has some of the properties of highly sulfated heparan sulfate in that it can bind cytokines and morphogens that are typically considered to be heparin binding [43].

In reviewing the literature for proteoglycan association with amyloid, it is important to remember that whenever the proteoglycan is identified, it is almost always the core protein that is being identified, either through immunoanalysis, through elevated mRNA levels or through direct analysis of the protein structure. The presence of the core protein does not, of itself, implicate glycosaminoglycan chains. The presence of a core protein and glycosaminoglycan in the same sample suggests, but does not confirm, the existence of intact proteoglycan. To infer the presence of the proteoglycan requires isolation of the intact molecule and identification of both the core protein and the glycosaminoglycan.

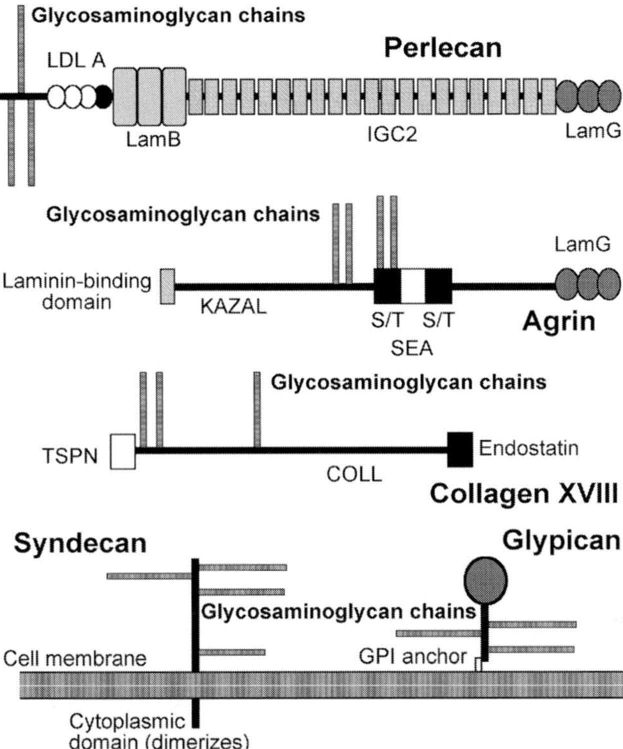

Fig. 7.3 The structures of heparan sulfate proteoglycans. Diagrams of the five major types of heparan sulfate proteoglycans showing the major features of their domain structure. Note that these diagrams are considerable simplifications and are not to scale. (Perlecan has approximately 4300 amino acids, agrin has 1900, collagen XVIII has 1500, but will have a more extended structure due to the collagen domains. The syndecans have between 200 and 350 amino acids, and the glypicans have in the region of 550 amino acids.) The major features and locations of the glycosaminoglycan chains in each type of proteoglycan are shown as bars at right angles to the core proteins. Perlecan has glycosaminoglycan chains near the N-terminal (left-hand side). The remainder of the proteoglycan has a variety of repeating motifs that fall into the low-density lipoprotein receptor, laminin and neural cell adhesion molecule (NCAM) families. Likewise, agrin consists primarily of repeating protein structural motifs. In agrin, the glycosaminoglycans are more centrally located. The locations of glycosaminoglycan attachment sites in collagen XVII have been determined in the chick protein and inferred in the human protein based on homology. The two membrane-associated proteoglycans, syndecan and glypican, are attached through a transmembrane domain (syndecan) and a phospholipid anchor (glypican) (reviewed in Bernfield et al. [67]). The domains of proteoglycans, while indicating their familial resemblance to other proteins, do not necessarily indicate a conserved function.

7.4.1 Basement Membrane-derived Heparan Sulfate Proteoglycans

Basement membranes are extracellular structures that delineate regions of different histological cell type, e.g. epithelial cells and connective tissue fibroblasts. The adhesion of epithelial and endothelial cells to basement membranes is essential for normal morphology and function. The basement membrane has a critical role in filtration in the kidney glomerulus and in nerve transmission at the neuromuscular synapse. Thus, basement membranes are near-ubiquitous structures, and are only really absent in cartilage and bone. Even within bone, they are present in the walls of blood vessels.

In molecular terms, basement membranes are not uniform entities and their composition varies from one region to another; however, in the vast majority of cases, collagen type IV, laminin and one or more extracellular heparan sulfate proteoglycan (perlecan, agrin and/or collagen type XVIII) are major constituents.

The heparan sulfate proteoglycans associated with basement membrane do not have any familial relationships, although they are all complex, elongated, multidomain proteins. Their multidomain nature enables them to interact with a broad range of extracellular and cell surface proteins that include other proteoglycans. Their glycosaminoglycan chains can be either of the heparan sulfate or the chondroitin/dermatan sulfate type (Section 7.5).

7.4.1.1 **Agrin**

Agrin is a multidomain proteoglycan with a 250-kDa core protein and is thought to have four glycolsaminoglycan chains. The intact molecule is approximately 400 kDa [44]. The glycosaminoglycan chains on the core protein are located in two clusters of serine/glycine-rich domains [45] (Fig. 7.3), but their precise position has not been determined. Agrin is important in the association of acetylcholine receptors in neuromuscular junction development; it is also found in the glomerular basement membrane where it is thought to be critical for glomerular filtration [46]. Agrin is found in the basal lamina of the microvasculature where its function is unclear, but may be associated with control of solute transport, contributing to maintenance of the blood–brain barrier [47].

Agrin has been detected in fibrillar and diffuse senile plaques [48, 49], and binds Aβ in a glycosaminoglycan dependent fashion. It accelerates fibril formation and protects Aβ from proteolysis *in vitro* [50], although it is unclear whether this is a property of the core protein or the intact proteoglycan as the proteoglycan has, to date, been primarily characterized by immunostaining.

7.4.1.2 **Perlecan (HSPG2)**

Perlecan, also known as HSPG2, is a large, multidomain heparan sulfate proteoglycan originally isolated from basement membrane [51]. Its name derives from the “beads-on-a-string” appearance of the molecule when visualized by rotary shadowing [52]. Perlecan is also found in other structures, most notably the growth plate of the developing skeleton [53]. Perlecan has a core protein of 400kDa and two or three heparan or chondroitin sulfate chains at the N-terminal (Fig. 7.3). There is some evidence that there may be additional glycosaminoglycans attached at the C-terminal [54]. Perlecan is present in all vascularized tissues [55] and has been identified associated with fibrillar amyloid plaques [56, 57]. While the proteoglycan was associated with diffuse plaques in the hippocampus, amygdala and neocortex, it was not found in diffuse plaques in the cerebellum [49, 57]. Perlecan has been shown to bind to Aβ via its heparan sulfate chains, whereupon it encourages and enhances Aβ fibril formation. The sulfate moiety on the glycosaminoglycan is required for perlecan-mediated Aβ fibril formation; removal of only *O*-sulfate or only *N*-sulfate reduced fibril formation, but not to the same extent as complete desulfation [58].

An HSPG2 single nucleotide polymorphism is correlated with a risk for Alzheimer's when present in carriers of an apolipoprotein E isoform (ApoE4) that has a cysteine-to-arginine change, but not when present by itself [59]. As this polymorphism is in an intron and therefore does not code for protein, its linkage is unlikely to be a direct indication of protein or glycosaminoglycan variability.

7.4.1.3 **Collagen XVIII**

Collagen XVIII [60] is found in basement membranes and associated with the basal surface of endothelial cells [61]. It is a heparan sulfate proteoglycan (and the only collagen that has HS chains) with approximately half its molecular mass being comprised of glycosaminoglycan [62]. The collagenous domain, which is broken into 10 discrete blocks, places it in the multiplexin family (*multiple* triple-helix domains and *in*terruptions) [63]. The glycosaminoglycan chains are located in the N-terminal half of the protein with the collagenous domains in the center of the protein [64] (Fig. 7.3). It has been shown to be associated with vascular Aβ and senile plaques [65]. The C-terminal of collagen XVIII can be cleaved off to yield endostatin, a protein that inhibits growth and promotes apoptosis of endothelial tissue. Endostatin has been shown to have a propensity to form cytotoxic amyloid structures on denaturation [33]. Endostatin is also found in amyloid plaques [66].

7.4.2 Cell Surface Heparan Sulfate Proteoglycans

Cell surface proteoglycans act to organize the extracellular environment around the cell, act as signal transducers, and mediate cell–cell and cell–matrix adhesion. Heparan sulfate proteoglycans are involved in all these areas. The external surface of the cell presents a very carbohydrate-rich environment and proteoglycans are a major component of this environment. The field of cell surface heparan sulfate proteoglycans has been comprehensively reviewed by Bernfield et al. [67].

Amyloid deposits in their early state are usually cytotoxic, although the mature fibrils are less so [68]; thus, the presence of a potentially amyloid-promoting macromolecule adjacent to the cell surface might potentiate this cytotoxicity. Importantly, heparan sulfate proteoglycans can also be cleaved from the cell surface through a variety of mechanisms (proteolysis, endoheparanase activity or, in the case of glypicans [69], phospholipase C activity) and so their rate of cleavage might be another factor in promoting or reducing amyloid formation.

7.4.2.1 **Glypicans**

There are six glypicans in the human genome. They all have common features (Fig. 7.3): (1) a C-terminal glycosyl-phosphatidyl-inositol link, (2) an approximately 50-kDa globular domain with 14 characteristic cysteine residues and (3) a stem region between these two domains that contains two to three glycosaminoglycan attachment motifs proximal to the cell surface [70]. They are heparan sulfate proteoglycans and the glycosaminoglycan chains are involved in sorting of the proteoglycan on the cell surface of epithelial cells [71] and in Wnt signaling [72].

Glypicans have been associated with fibrillar and diffuse senile plaques [49]. Glypican-1 binds to Aβ via glycosaminoglycan chains and is associated with the detergent-insoluble glycosphingolipid-enriched domains (lipid rafts) in the cell membrane [73] (Section |7.7.7|).

7.4.2.2 **Syndecans**

Like glypicans, syndecans are type I membrane proteins (i.e. the N-terminal is outside the cell), but are otherwise structurally the opposite of glypicans. Rather than consisting of extracellular molecules anchored to the membranes, syndecans have a transmembrane domain that is capable of interacting with intracellular structures such as the cytoskeleton (reviewed by Bernfield et al. [67] and Yoneda and Couchman [74]). In contrast to the glypicans, the heparan sulfate glycosaminoglycan chains are distal to the cell membrane and they can be shed, together with the extracellular domain, as a result of proteolytic activity [75] (Fig. 7.3). This process is regulated by a host of proteases and growth factors [76, 77].

There are five syndecans in the human genome with core proteins ranging in size from 22 to 43 kDa; two of these, syndecan-1 and syndecan-3, may have chondroitin sulfate chains adjacent to the cell surface. Syndecan has been associated with fibrillar senile plaques, but not with diffuse non-fibrillar senile plaques [49]. In common with glypicans, syndecans could conceivably immobilize amyloid or amyloid precursors near the cell membrane, enhancing their cytotoxic properties. Syndecans have a critical role in cell adhesion [21] and in the formation of dendritic spines [78]. The possibility exists, therefore, that cytotoxic effects mediated by amyloid might be a result of subversion of the signaling and adhesion roles of this family of cell surface proteoglycans through direct competition for binding sites on the glycosaminoglycan chains.

7.4.2.3 *β*-Glycan

β-Glycan is a "part-time" heparan sulfate proteoglycan, in that it does not always have attached glycosaminoglycans. It is a receptor for transforming growth factor (TGF)-*β* and is reminiscent of syndecans, in that it has an intracellular domain and is capable of signal transduction [79]. To date, it has not been implicated in amyloid biochemistry.

7.5 Heparin, Heparan Sulfate and Other Glycosaminoglycans

There are four subclasses of protein-attached glycosaminoglycans: chondroitin sulfate, dermatan sulfate (which can be considered to be an epimerized subset of chondroitin sulfate), keratan sulfate and heparan sulfate. While keratan sulfate has been reported to be associated with amyloid [5], the evidence for extensive involvement is sparse and we will not consider it here. Glycosaminoglycan chains have a strong negative charge as a result of carboxyl groups in each of the repeating disaccharides units and the variable but substantial sulfation of these units. Physically, proteoglycans all have the ability to bind substantial amounts of water by virtue of the extended structure of the glycosaminoglycan chain. The extended structure derives from the mutually repulsive effect of the negatively charged carboxylic acid and sulfate groups.

Given a glycosaminoglycan and a protein in solution in water, binding of the glycosaminoglycan to protein is energetically a more favorable outcome than binding to water, as the free energy is reduced. Thus, a positively charged area on a protein surface will be hydrogen bonded to water, as will a negatively charged group on the carbohydrate. The net effect of the two macromolecules binding to each other will be to eliminate two bound water molecules; glycosaminoglycans, therefore, "prefer" to bind to protein, if possible.

Glycosaminoglycans consist of repeating hexose disaccharides that contain a uronic acid (either D-glucuronic acid or its C5 isomer, L-iduronic acid) and an *N*-acetylated hexosamine (either *N*-acetyl glucosamine or its C4 isomer, *N*-acetyl galactosamine). With the exception of hyaluronic acid, which is synthesized on the plasma membrane, all glycosaminoglycans are synthesized in the endoplasmic reticulum and Golgi complex and are serine-*O*-linked to xylose through a linker tetrasaccharide, *N*-acetyl glucuronic acid–galactose–galactose–xylose. A further exception to the rule is keratan sulfate, which does not contain a uronic acid and may be *O*-linked via *N*-acetyl galactosamine or *N*-linked via *N*-acetyl glucosamine to the core protein. In the latter case, it derives from an *N*-linked oligosaccharide. The biosynthesis of glycosaminoglycans is reviewed in detail by Prydz and Dalen [80].

7.5.1
Chondroitin sulfate

Chondroitin sulfate is a widespread glycosaminoglycan that consists of repeating D-glucuronate β (1–3) linked to D-*N*-actyl galactosamine β (1–4) (Fig. 7.4). The galactosamine may be sulfated at the 4 or 6 positions, or sulfated at both

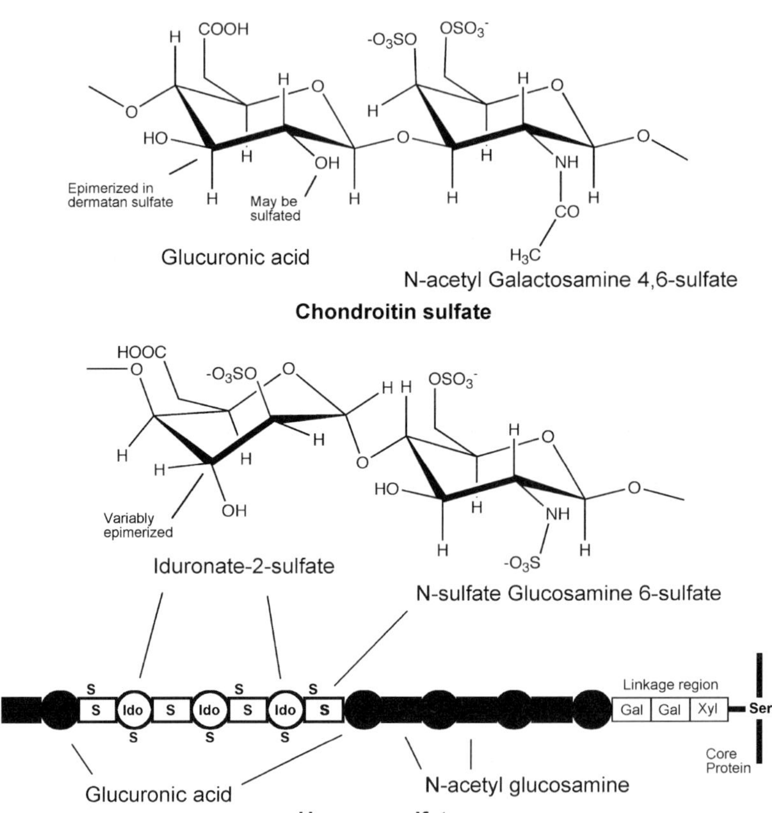

Fig. 7.4 Structures of heparan sulfate and chondroitin sulfate. The repeating disaccharides are shown in the upper structural representations, while the lower diagram illustrates how diversity develops in heparan sulfate. There are sulfate-poor regions (black objects), consequent to lack of modification of the *N*-acetyl glucosamine and there are sulfate-rich regions (white objects) where the glucosamine has been deacetylated and sulfated, resulting in epimerization and sulfation of the iduronic acid. Further subtle structural complexity (reviewed in [180, 181]) derives both from variation in the low- and high-sulfate regions, and from the ability of the iduronic acid in heparan sulfate to adopt a distorted skew-boat conformation rather than the chair conformation shown here.

positions. Additional sulfation can occur on the 2- or 3-positions of the glucuronic acid. Chondroitin sulfate is particularly prevalent in cartilage, where it forms the bulk of the abundant extracellular matrix. As many as 40–80 chondroitin sulfate chains can be attached to the core protein of the large proteoglycan aggrecan. Aggrecan also contains a number of *O*-linked keratan sulfate chains, depending on the species, and is non-covalently associated with hyaluronic acid to form extraordinarily large complexes with a net size in the range of 50–100 MDa. Related proteoglycans, versican, neurocan and brevican are prevalent in the central nervous system and critical for development and repair [81, 82].

Highly sulfated chondroitin sulfate has been found in the hybrid heparan/chondroitin sulfate proteoglycan, serglycin that is a prominent component of mast cell granules [83]. Recently, chondroitin/dermatan sulfate roles have been invoked in growth factor control [84, 85], mimicking the well-established behavior of heparan sulfate (Section 7.5.3). These topics are dealt with in a review by Sugahara et al. [86].

7.5.2 Dermatan Sulfate

Dermatan sulfate derives from chondroitin sulfate through C5 epimerization of the glucuronic acid to iduronic acid (Fig. 7.4) and is typically mixed with varying amounts of chondroitin sulfate on the same molecule, forming a continuum starting from sparsely epimerized chondroitin sulfate. Dermatan sulfate should, therefore, be considered together with chondroitin sulfate [86]. It is most frequently identified attached to the small, leucine-rich proteoglycans decorin and biglycan. The epimerized uronic acid derives from variable degrees of activity of chondroitin glucuronate 5-epimerase [87]. A small amount of the dermatan sulfate disaccharide unit is present in a novel proteoglycan, endocan, recently described in endothelial cells, that has a role in the function of hepatocyte growth factor/scatter factor [88]. Hepatocyte growth factor binds to an octasaccharide that has unsulfated iduronate residues in combination with 4-*O*-sulfated *N*-acetylgalactosamine [89].

The small, leucine-rich proteoglycans are strongly implicated in regulation of collagen fibril diameter regulation, but may have other functions. They include a number of keratan sulfate proteoglycans, derived through extension of *N*-linked oligosaccharides (reviewed by Funderburgh [90]). Biglycan and decorin have been shown to be present and associated with amyloid fibrils in glomeruli from diseased kidneys affected by renal AA amyloidosis [91]. Decorin and collagen were both found to be closely associated with amyloid in a second study [92], suggesting that this may be a result of increased extracellular matrix synthesis rather than specific association with amyloid.

7.5.3 Heparan Sulfate

Heparan sulfate proteoglycans are ubiquitous components of cell surface proteoglycans and of the extracellular matrix. Heparan sulfate proteoglycans consist of a protein core and one or more glycosaminoglycan chains, which may exclusively consist of heparan sulfate or may include other glycosaminoglycan chains, usually chondroitin/dermatan sulfate. Heparan sulfate proteoglycans are necessary for presentation of growth factors to their receptors. Arguably, heparan sulfate is the most complex glycosaminoglycan.

The heparan sulfate polymer is made up of a uronic acid–glucosamine repeat unit in which the uronate is β-D-glucuronate or α-L-iduronate and the glucosamine is *N*-acetylated or *N*-sulfated (Fig. 7.4). As in the case of dermatan sulfate, glucuronate and iduronate are present in the same polysaccharide chain. Epimerization of the glucuronic acid to iduronic acid occurs after the assembly of the repeating structure through the activity of the enzyme heparosan *N*-sulfate glucuronate 5-epimerase and is essentially irreversible [93]; the *N*-acetylation and *N*-sulfation of the adjacent hexose is a necessary step prior to epimerization. Thus the degree of epimerization of the glucuronic acid C5–C6 bond varies in proportion to the degree of *N*-sulfation of the glucosamine. Further complexity is added by optional sulfation of the glucosamine at C6 and the uronic acid (principally iduronic acid) C2. Sulfation also occurs, albeit rarely at the C3 position of *N*-sulfated glucosamine residues and is necessary for binding to antithrombin III (reviewed in [94]). In heparan sulfate the *N*- and *O*-sulfated sugars tend to occur on clusters (or domains) separated by *N*-acetyl rich regions of low or zero sulfation.

The length of the heparan sulfate chain is variable, resulting in glycosaminoglycan chain sizes from 10 to 100 kDa. The cellular control mechanism over the biosynthetic process is still unclear, but it has become increasingly apparent that multi-enzyme complexes are involved [95]. Different chains on the same core protein are unlikely to be identical [96].

Heparin is a subclass of heparan sulfate and is highly sulfated, the predominant disaccharide being L-iduronate-2-sulfate-α(1–4)-D-*N*-sulfoglucosamine-6-sulfate-α(1–4) [97]. Originally, heparin was considered to be exclusive to mast cells granules from which it was released as a response to antigen binding to IgE attached to IgE receptors on the cell surface. Heparin-like glycosaminoglycans have also been found in the liver [98] and brain [99], attached to extracellular matrix proteoglycans. Anticoagulant heparin typically contains glucosamine with a 3-*O*-sulfate, although not all the glucosamine is substituted in this way (Fig. 7.4). Nevertheless, heparin is more sulfated than heparan sulfate and the sulfation is more homogeneous along the glycosaminoglycan chain.

The diversity of heparan sulfate structures results in substantially different binding profiles. Even with the same heparan sulfate chain, there is room for structural variety, so that one region of a heparan sulfate chain may have different properties from another. There are distinct patterns of modification that occur along a heparan sulfate chain. The highly sulfated domains of heparan sulfate (S-domains) are heparin-like but only extend to six or seven disaccharide units and they are distributed along the chain at intervals of about 16 disaccharides. An unmodified domain [D-glucuronate-β(1–4)-D-*N*-acetyl glucosamine-α(1–4)] adjacent to the protein is typically found. In skin fibroblast heparan sulfate, this is 8–11 disaccharide units long [100], whereas in endothelial cells it is longer (10–17 disaccharides [101]). Complexity of sulfation in HS is further amplified by the presence of transition zones of alternating *N*-acetylated and *N*-sulfated disaccharides flanking the S-domains [23]. Mixed sequences such as this are not found in heparin and may define unique properties of heparan sulfate.

Heparan sulfate proteoglycan heterogeneity within the brain results in heparan sulfates that are unique for different regions; oligodendrocyte type 2 astrocyte progenitors, for example, synthesize heparin, rather than the more typical heparan sulfate [99] and so are considered to be highly unusual.

7.6 Heparin–Heparan Sulfate Interactions with Protein

Heparin and heparan sulfate have an established role in protein binding. It has become clear that heparan sulfate binds to a variety of cell effectors, such as members of the fibroblast growth factor (FGF) family, vascular endothelial growth factor (VEGF), members of the TGF family and bone morphogenetic proteins (BMPs), as well as other cytokines and chemokines [102]. Heparan sulfate can have two functions – it can be both necessary for presentation of a growth factor to its receptor (e.g. the FGF receptors) and it can also act as a reservoir for the growth factor in the extracellular matrix.

The majority of work performed to date on the three-dimensional structure of heparan sulfate interactions with proteins has been performed with heparin. This is primarily due to the ready availability of heparin in large amounts and the relative difficulty of obtaining adequate amounts of other homogeneous heparan sulfate oligosaccharides. At present, apart from conferring flexibility on the heparan sulfate polymer chain [103], the role of the poorly sulfated domains of heparan sulfate in biological systems is unclear.

Consensus sequences have been proposed for heparin-binding sites in proteins and peptides [104, 105]. The heparin-binding motif has been proposed to be XBBBXXBX or XBBXBX, where X is any non-basic amino acid and B is a basic amino acid, based on comparison of sequences that bind heparin *in vitro* nectin, ApoE and ApoB-100, and platelet factor 4 [105]. Overall, it was predicted that highly basic sequences would typically (and perhaps unsurprisingly) bind to heparin. In a different approach, proteins that were known to bind to heparin were analyzed for common structural motifs and 20-Å spacing for basic amino acids in α-helices was proposed [106]. However, as we discuss below, many heparin-binding proteins do not have defined secondary structure in their heparin-binding sites and so would not be constrained by a rigid protein backbone. The situation is, therefore, quite complex, reflecting the antiquity of heparan sulfate within the animal kingdom and there may not be a heparin-binding consensus sequence *per se* [107].

A number of structures of heparin associated with protein have now been determined (Table 7.1 shows the proteins complexed with heparin that are available in the PDB database). Common features of all the heparin binding sites in all these proteins are (1) the glycosaminoglycan binds to regions of protein that are so-called "random coil" (although there is nothing random about them; they have defined but irregular structures, although they may be somewhat flexible), in that they do not conform to β-strand or α-helix structures, and (2) they are dominated by the basic amino acids, lysine and arginine.

An alternative to the use of heparin in the structural analysis of heparin-binding domains has been the use of analogs of heparin such as sucrose octasulfate and D-myoinositol hexasulfate. These have been used in, for example, the analysis of heparin-binding domains in follistatin [108]. Perversely, the very high level of sulfation of these compounds results in less clear electron-density patterns, in all probability because there are several orientations of the ligand bound to the protein. The model further emphasizes the role of polar amino acids in the interaction of sulfated compounds with proteins [108]. These model compounds also tend to bind to basic amino acids in "random coil" regions of the protein.

It is tempting to assume that glycosaminoglycan chains form rigid, extended structures [103] although it is recognized that the 2-*O*-sulfated iduronic acid is capable of forming a "skew-boat" (also known as the $^{2}S_{O}$ form) [109, 110]. In Fig. 7.5, the IdoA 2-*O*-sulfate on the left can be seen to have this somewhat distorted structure. When bound to antithrombin III, the skew-boat conformation

Table 7.1 Known structures of heparin and heparan sulfate complexed with protein

PDB file	Proteins	Carbohydrate	Reference
2C4S		4-sulfated chondroitin sulfate tetramer	173
1HPN		heparin	174
1BFP	FGF2	heparin tetramer	
1BFC	FGF2	heparin hexamer	112
1AXM	FGF1 dimer	heparin decamer (only five hexoses are ordered)	175
2AXM	FGF1 dimer	heparin decamer (only five hexoses are ordered)	175
1AZX	antithrombin	heparin pentasaccharide	114
1NQ9	antithrombin	heparin pentasaccharide	176
1E0O	FGF1, FGFR2	heparin decasaccharide	177
1FQ9	FGF2, FGFR2	heparin decasaccharide	177
1G5N	Annexin	heparin tetrasaccharide	178
1GMN	NK1	heparin tetrasaccharide	179
1GMO	NK1	heparin tetrasaccharide	179

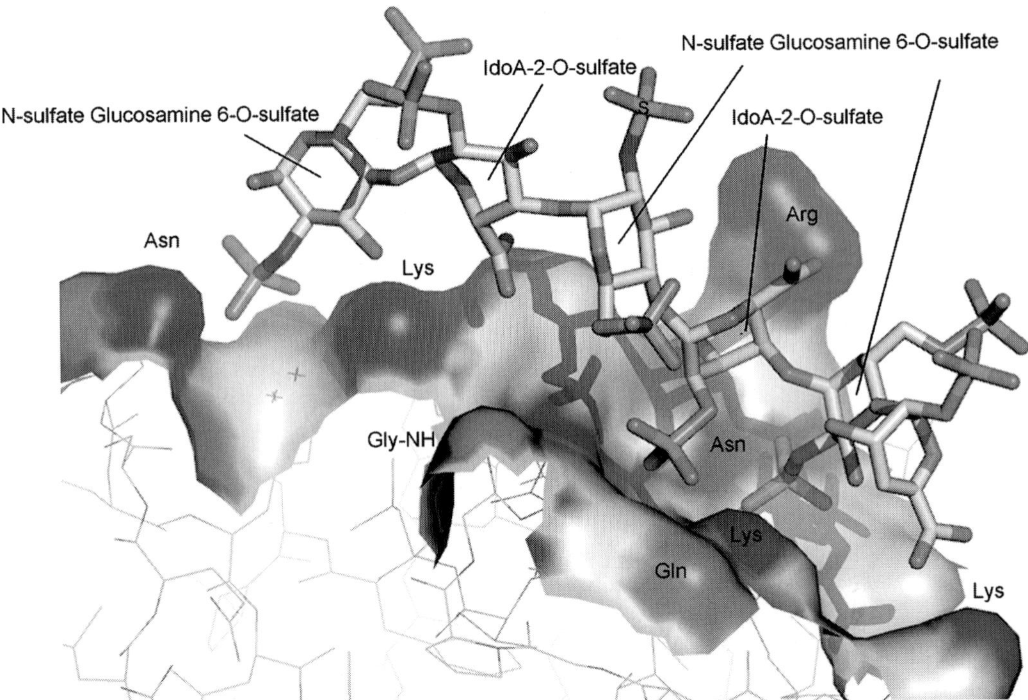

Fig. 7.5 The structure of a heparin hexamer (sticks) complexed with FGF2 (basic FGF). A surface representation of the region of FGF2 that is within 5 Å of heparin is shown. The coordinates are taken from PDB file 1BFC [112]. The individual hexoses (iduronic acid 2-sulfate and glucose-*N*-sulfate 6-*O*-sulfate) are labeled, as are the amino acid side-chains in the binding site. A backbone NH group (glycine) is also involved. Blue regions indicate amino groups, while red regions indicate carbonyl or hydroxyl groups. The sulfate residues in the oligosaccharide are shown in orange, oxygen atoms are red and nitrogen atoms are blue.

is the predominant form and transition to this structure may, in part, be driven by the interaction with the protein [111].

The structure of the heparin-binding site of basic FGF (FGF2) complexed with a heparin hexamer is shown in Fig. 7.5 and is based on the structural coordinates published by Faham et al. [112]. The binding site consists of five lysines, an arginine, a glutamine, two asparagines and a tyrosine, although not all these residues are intimately involved in the binding process. Two of the uronic acid residues in this case are in the skew-boat form while the third (the central one in Fig. 7.5) is in the chair configuration.

Heparin modifies the structure of antithrombin in its role as an inhibitor of coagulation [113]; the structure of the heparin-antithrombin complex has been solved by X-ray crystallography. The contact site for a heparan sulfate pentasaccharide in this case comprises basic amino acid side-chains from four separate regions of the protein (Lys11 and Arg 13, Arg46 and Arg47, Lys114, Arg125 and Arg 129) as well as two other charged amino acids (Asn45 and Glu113). These amino acids hydrogen bond to carboxyl and sulfate groups on the oligosaccharide [114].

Thus, while heparin/heparan-binding sites in peptides must, of necessity, consist of amino acids that are close to each other in the amino acid sequence, this is not necessarily the case in intact proteins. However, heparin/heparan-binding sites do primarily consist of the basic amino acids lysine and arginine. Oligosaccharide binding to protein can result in conformation changes in both components of the interaction.

7.7
Amyloid Proteins and Peptides

We will not attempt to survey all the amyloid-forming proteins and their interactions with heparan sulfate, but will focus on some illustrative examples. They divide conveniently into intact proteins [light chain amyloid (AL), β_2-microglobulin amyloid (Aβ_2M), TTR amyloid (ATTR)], small peptides/proteins [inflammation-associated amyloid (AA)] and peptides (IAPP and the Aβ peptide that assembles in amyloid fibrils in Alzheimer's disease].

7.7.1
Light Chain Amyloid (AL)

This amyloid fibril derives from fragments of an immunoglobulin light chain that is overexpressed, dimerizes and aggregates into a typical amyloid fibril whose structure has been solved in one case [115]. It is the causative factor in the lethal disease monoclonal, light chain (AL)-type systemic amyloidosis. Fibrils contain about 1–2% glycosaminoglycan, primarily in the form of heparan sulfate and dermatan sulfate chains [116]. It is unclear how or why glycosaminoglycans associate with this amyloid type; while it is predominately β-sheet, the individual units are fragments of typical immunoglobulin light chains with a size in the region of 20 kDa.

Fibril assembly proceeds by a first-order process, with consecutive addition of monomers to the extending fibril [12]. The same workers found that this process is accelerated by the presence of dermatan sulfate, although it was unclear whether this included the core protein or was entirely glycosaminoglycan dependant. It has been proposed that glycosaminoglycans (heparin or chondroitin sulfate) might bind to the opposite side from the antigen-binding site of light chain dimers, interacting with lysine residues in this region [117]. While the latter studies used quite high concentrations of glycosaminoglycans (above 1 mg/ml), causing some concern that the mechanism might be different *in vivo*, it is important to realize that the *in vivo* environment is considerably less hydrated than a typical *in vitro* experiment and that this is exacerbated by the presence of high concentrations of glycosaminoglycans.

7.7.2
Serum Amyloid P (SAP)

SAP is a pentameric protein that is a component of most amyloid deposits. Together with ApoE (also a common or even ubiquitous component of amyloid) it is thought that SAP mediates the tendency of amyloid fibrils to associate into tangles [118]. SAP acts to protect fibrillar, but not non-fibrillar, forms of Alzheimer's Aβ from proteolysis although it binds to both forms. This interaction can be inhibited by glycosaminoglycans and also by Congo red dye, providing some direct evidence that Congo red (Fig. 7.1) does mimic glycosaminoglycans to a certain extent [119].

SAP binds to heparan and dermatan sulfate in a calcium-dependant fashion, with a dissociation constant of 2×10^{-7} M [120] and can promote the aggregation of Alzheimer's Aβ peptide [121]. SAP also binds to other extracellular matrix components such as collagen, laminin and fibronectin.

7.7.3
Inflammation-associated AA

Inflammation-associated amyloidosis (AA amyloidosis) is a frequent complication in chronic inflammatory conditions. The causative agent is an apolipoprotein, ApoSAA (serum amyloid A) whose function is obscure but which is associated with high-density lipoprotein during the acute phase of an inflammatory response [122]. Heparin binds to the C-terminal [26]. Both the N-terminal 76 amino acids [123] and the full-length protein [124] have also been implicated in amyloid formation. The multimeric structure of SAA in solution has been inferred by sedimentation velocity analytical ultracentrifugation and found to be hexameric, primarily α-helical by circular dichroism, and possesses a central channel that can be visualized by electron microscopy [125].

The proteoglycans associated with SAA appear to be exclusively heparan and dermatan sulfates [116]. SAA is one of the few amyloid proteins to have a sequence of amino acids that does look as if it might bind to glycosaminoglycans. The C-terminal 21 amino acids of the three isoforms of human SAA have the sequence EWGRSG(KR)DP(DN)(HR)FRP(AD)GLPEKY and include five basic amino acids as well as five other amino acids capable of forming hydrogen bonds via their side-chains. There is no sign of the type of consensus sequence that has been proposed by some workers [105]. The PROF algorithm for structure prediction suggests that this region does not have a strongly β-sheet or α-helical component. Overall, the structure of SAA is predicted to have a helical N-terminal, a β-sheet domain and then another one or two 2 helical domains before the glycosaminoglycan-binding domain is reached [125].

The helical domain at the N-terminal of SAA, in common with α-helices in apolipoproteins in general, is likely to be amphipathic; thus, it will have one side consisting of hydrophobic residues and associating with the lipid in lipoproteins and the other consisting of hydrophilic residues that associate with solvent [126].

It is possible to suggest a role for extracellular matrix proteoglycans in the formation of amyloid A derived from the SAA precursor. The initial event would be binding of multiple SAA molecules to one or more glycosaminoglycans, restricting their lateral diffusion. This would promote aggregation of the remainder of the protein, allowing refolding. Finally, proteolytic degradation would release the aggregated N-terminals of SAA; it is noteworthy in this context that ApoSAA upregulates the expression of matrix metalloproteases in chondrocytes, extracellular proteases responsible in part for tissue remodeling [127], but which could mediate the cleavage of SAA to release the aggregated N-terminals.

7.7.4 β_2-Microglobulin (β_2M)

Patients on long-term hemodialysis often present with amyloid deposits associated with the skeletal system, as well as other organs. The major protein component of these deposits is β_2M, a member of the immunoglobulin superfamily whose protein folds are characterized by two β-sheets in which the β-strand arrays are antiparallel, stabilized by an inter-sheet disulfide bond. It is a component of the class I HLA complex. The protein-destabilizing event that promotes amyloid formation in this case may well be the disruption of the β-sheets at the edge of the protein, exposing hydrophobic residues and encouraging aggregation [128].

Depolymerization of β_2M amyloid is inhibited by the presence of glycosaminoglycans, perhaps by surface stabilization of the amyloid fibrils [129]. β_2M itself appears to have a general affinity for extracellular matrix, as it will bind to laminin and the major basement membrane collagen, collagen type IV [130]. However, heparan sulfate is markedly elevated at the sites of accumulation of amyloid, perhaps as a result of the local inflammatory response that amyloid deposits engender.

7.7.5 Transthyretin (TTR)

TTR (human) is a 127-amino-acid homo-tetrameric, predominantly β-sheet protein involved in thyroid hormone transport. It is the protein responsible for the most predominant form of familial amyloidotic polyneuropathy; over 40 disease-linked mutations are described in the National Institutes of Health OMIM (Online Mendelian Inheritance in Man) database and several of these have been found experimentally to promote amyloid formation [131, 132]. It is noteworthy that a cluster of residues on the outside β-strand of a β-sheet (amino acids 42–

47) when mutated, promote amyloid formation. This lends credibility to the idea that β-sheet outside strands have special properties (see Section 7.3) [40]. In contrast, the other mutation "hotspot", amino acid residues 68–71, is sandwiched between two β-sheets so this is clearly not a firm rule.

TTR has been shown to bind to perlecan, but the interaction is likely to be through the proteoglycan core protein, as inhibition of sulfation did not reduce the interaction [133]. Amyloid fibrils, while clearly in the β-sheet conformation as indicated by circular dichroism and X-ray diffraction analyses, are difficult to analyze rigorously as they possess considerable small-scale heterogeneity. A TTR-derived peptide that forms amyloid, the penultimate β-strand before the C-terminal of the intact protein is reached (Fig. 7.2), is primarily a β-sheet structure in the native protein; the only significant change in the structure of the peptide when formed into amyloid is that the C-terminal of the peptide straightens somewhat [37].

Ultrastructural analysis using electron microscopy and immunolabeling of the amyloid fibrils associated with familial amyloid neuropathy showed that a surface layer of heparan sulfate is associated with a relatively loose region that contains the β-sheet TTR filament [15]. However, the central region of the fibrils is proposed to consist of amyloid P surrounded by chondroitin sulfate. This model is not consistent with the cross-β structure derived from X-ray diffraction data proposed by Kirschner et al. [134] or Blake and Serpell [135] for the Aβ associated with Alzheimer's disease, suggesting that there may be a number of different processes responsible for amyloid formation. Blake and Serpell describe the structure of the protofilament as four β-sheets entwined around the fibril axis [135].

7.7.6
Islet Amyloid (AIAPP)

The amino acid sequence of the 37-amino-acid IAPP, also known as amylin, is KCNTATCATQ RLANFLVHSS NNFGAILSST NVGSNTY, and is derived from pro-IAPP by proteolytic removal of 11 amino acids from the N-terminal and 16 amino acids from the C-terminal, followed by removal of an additional 3 amino acids from the C-terminal. The two cysteines are disulfide bonded [136]. The N-terminal domain of pro-IAPP has been found in islet amyloid [137]. Perlecan is a component of AIAPP deposits in patients with islet amyloidosis, a complication in 90% of Type 2 diabetes cases [138]. The glycosaminoglycan chains and their sulfation state have been implicated in the process of amylin deposition [139].

The structure of the AIAPP nucleation site (SNNFGAILS) in detergent micelles, has been determined [140]; the first 4 amino acids form a turn while the remainder approximates an α-helix; it was suggested that the hydrophobic turn is the region that interacts with pancreatic cell β membranes.

It has been proposed that islet amyloid deposition is enhanced by defective processing of the precursor of IAPP, pro-IAPP, by β-cells and subsequent interaction with proteoglycans [141]. The defective products would bind to heparan sulfate in the basement membrane and act as a nucleation site. Both the N- and C-terminals of pro-IAPP have basic amino acids that could form glycosaminoglycan-binding sites. The N-terminal basic amino acids are required for heparin binding and may be masked from normal proteolytic processing by binding to glycosaminoglycan. Surprisingly, the C-terminal basic amino acids do not bind heparin [141].

In vitro, IAPP forms amyloid by a rapid transition to β-sheet and the ensuing structure has significant numbers of solvent-associated hydrophobic amino acid side-chains [142], encouraging further growth of the amyloid fibril.

7.7.7 Alzheimer's Aβ

The Aβ that is found in amyloid plaques and neurofibrillary tangles associated with Alzheimer's disease is a 39- to 43-amino-acid product of cleavage of AβPP also known as amyloid precursor protein (APP). As noted earlier, AβPP is also found as a proteoglycan, derived by alternative splicing that resolves into a glycosaminoglycan attachment site [143]. In non-polar solvents (20% water, 80% hexafluoroisopropanol), the Aβ monomer is largely helical [144], whereas in aqueous buffers, it is primarily β-sheet (reviewed in [145]) and forms aggregates. In the helical form, a potential glycosaminoglycan binding site [amino acid pattern Arg–His–(any 6 amino acids)–His] can be seen on one side of the helix. The other side of the helix has a predominantly negative charge. However, an alternative structure in sodium dodecylsulfate, which may better mimic the membrane-inserted state [146], does not have an N-terminal helical structure and seems unlikely to be able to bind glycosaminoglycans efficiently.

Aβ11–25 forms β-sheet structures with the peptide backbone orthogonal to the long axis of the amyloid fibril [147, 148]. The fibrils formed by Aβ1–40 are half the width (7 nm) of those that would be expected to form if the peptide was entirely an extended β-strand. Thus, it is likely that the structure of each individual peptide is a β-sheet, formed by two antiparallel β-strands connected through a hairpin turn [149]. The β-sheets between monomers are probably arranged in an antiparallel structure, although the evidence is equivocal (reviewed in [145]).

Proteoglycans, with or without glycosaminoglycans, are well documented to be preferentially associated with Alzheimer's amyloid plaques [56, 57, 150–155]. Whether this association results in promotion of fibril formation and enhancement of pathology or has the opposite effect is still equivocal. Glypican binds to fibrillar Aβ largely via its heparan sulfate chains [73]. Intriguingly, glypican is also associated with the same detergent-insoluble glycosphingolipid-enriched domains with which Aβ-42 is enriched, suggesting a possible direct role for glypican in Alzheimer's disease [73]. If glypican interacts with amyloid-forming proteins through its glycosaminoglycan chains, this would immobilize or promote the formation of the cytotoxic amyloid very close to the cell surface.

Evidence for direct involvement of glycosaminoglycans in Alzheimer's amyloid pathology has been obtained by investigating the interaction of heparan sulfate with the enzyme that is a crucial part of the processing of AβPP to yield Aβ peptide. It was previously shown that proteoglycans inhibit the degradation of fibrillar, but not non-fibrillar Aβ [156]. An earlier event in amyloid pathology is the action of β site AβPP cleaving enzyme (BACE)1. BACE1 (also known as Asp2 or memapsin 2; β-site APP-cleaving enzyme) is an aspartic protease that cleaves AβPP at the surface of the cell membrane. It binds to heparin oligosaccharides of at least 10 monosaccharides in length with the net result that cleavage of AβPP is inhibited at the β site (the N-terminal end of Aβ) [157]. There is significant structural complexity to this inhibition; *N*-acetyl heparin rather than *N*-sulfated heparin is a better inhibitor. 2- and 6-*O*-sulfation is a desirable trait for inhibition.

Heparan sulfate binds to the N-terminal of AβPP, likely to a basic region distal to the cell membrane [158]. As discussed by Scholefield et al. [157], this could sequester AβPP from BACE1, reducing the potential for forming Aβ. On the other hand, if AβPP is clustered prior to cleavage, the aggregation of Aβ could be enhanced. The situation becomes complex, however and would depend on the mix of syndecans (Section 7.4.2.2) and glypicans (Section 7.4.2.1) on the cell surface, as well as the nature of the glycosaminoglycan chains attached to them and the lipid rafts that these membrane associated proteins migrate within [159].

There is evidence that Alzheimer's disease Aβ can also form intracellularly; certainly all the ingredients, including glycosaminoglycans, will be present in the endoplasmic reticulum/Golgi complex [160, 161].

7.8 Heparan Sulfate and Amyloid

Heparan sulfate is frequently, perhaps always, associated with amyloid [3, 4]. However, heparan sulfate is ubiquitous and so it is important to compare the level of heparan sulfate in amyloid plaques with the levels in the surrounding tissues. When Alzheimer's amyloid plaque-containing material is compared with control tissue, it is sometimes unclear whether the presence of heparan sulfate glycosaminoglycan represents an accumulation in the plaque or not [162]. However, the heparan sulfate found in samples that included amyloid differed slightly after deaminative cleavage, suggesting that there may be selective association of particular heparan sulfate structures with amyloid.

AA amyloid (Section 7.7.3) has a distinct heparan sulfate profile. GlcA–GlcNSO_3 (6-*O*-sulfate) is increased approximately 2-fold in amyloid-containing samples of liver and spleen and IdoA (2-*O*-sulfate)-GlcNAc is reduced by 15% [163].

Castillo et al. have also shown that other glycosaminoglycans enhance amyloid deposition. In islet amyloid (Section 7.7.6) and Aβ amyloid (Section 7.7.7), heparin (the most sulfated form of heparan sulfate) is most effective, followed by heparan sulfate, and then chondroitin/dermatan sulfate [58, 139].

The amyloid β protein precursor binds to extracellular matrix and this binding is inhibited by 80% if glycosaminoglycan biosynthesis is competitively inhibited by 4-methylumbelliferyl-β-D-xyloside or if glycosaminoglycan sulfation is inhibited by chlorate; treatment of the extracellular matrix with heparatinase also caused appreciable inhibition of binding [164].

A variety of *N*-acetyl glucosamine analogs have been synthesized and evaluated for their potential as inhibitors of amyloid formation in a mouse transgenic model of Alzheimer's disease. The intent is to inhibit heparan sulfate synthesis in ways that reduce its potential for amyloid promotion without at the same time eliminating heparan sulfate itself (which would probably be developmentally lethal) [27]. For example, 2-acetamido-1,3,6-tri-*O*-acetyl-2,4-dideoxy-α and/or -β-D-xylo-hexopyranoses can be incorporated into heparan sulfate and reduce chain elongation; the same analogs inhibit amyloid deposition in a tissue culture system [165]. The same analog inhibits spleen and liver amyloid formation in a transgenic mouse model.

It is unclear to what extent glycosaminoglycans promote amyloid formation. When the role of glycosaminoglycans in the conversion of helical Aβ42 to β-sheet was analyzed using circular dichroism, chondroitin 6-sulfate, chondroitin 4-sulfate and dermatan sulfate promoted the conversion most efficiently [16]. Sulfate is critical for amyloid formation, as desulfated heparan sulfate is ineffective [58]. Sulfated monosaccharides compete with heparan sulfate for binding to Aβ, but for β-amyloid lateral association, disaccharides or larger carbohydrates are required [166].

Recent analysis of the interaction of 10 growth factors with an octasaccharide library shows that protein–glycosaminoglycan interactions can be quite specific [167] and are developmentally regulated [168]. It is still unclear whether the interaction of glycosaminoglycans with amyloid represents a specific interaction similar to that with growth factors, whether it is a consequence of the charge density and the effect of the charge density on the local ionic milieu (below, Section 7.8) or is a combination of both phenomena.

7.9 Conclusion

It is difficult, given the lack of potential glycosaminoglycan-binding basic amino acids in most amyloid β-sheets and the tendency for heparin, and presumably heparan sulfate, to bind to basic amino acids, to propose that heparan sulfate promotes amyloid formation simply by binding to and stabilizing the β-strands of the protein. Therefore, some other mechanism may exist. The glycosaminoglycan chains of proteoglycans are, in a gross sense, analogous to sponges. The most extreme form of these "sponges" is the large aggregating proteoglycan of cartilage, aggrecan. This proteoglycan has a core protein of approximately

300 kDa to which are attached approximately 100 20-kDa chondroitin sulfate chains. Several hundred of these proteoglycans are non-covalently bound to a hyaluronic acid molecule and the whole is trapped within a collagen matrix to give an underhydrated "biological shock absorber" [169]. While this is an extreme example of the mass properties of proteoglycans, it illustrates their essential property, that of providing a hydrated structure capable of hydrogen bonding to anything with a positive charge. Heparin/heparan sulfate is essentially a linear oligosaccharide that, by itself, will hydrogen bond to a considerable amount of water. The surrounding water will therefore be removed as a solvating agent in the region of the glycosaminoglycan, increasing the potential for proteins in the vicinity to adjust their folding in an environment that is more ordered and less hydrated than it would be without the glycosaminoglycan. If the "normal" protein structure is in a metastable state due to proteolysis, mutations or other factors such as an unusually high concentration, it might readily form alternative structures with a pronounced tendency to aggregate.

7.10 Future Directions

The majority of the work performed to date on the role of proteoglycans in amyloid formation and stabilization has suffered from a lack of consideration of the fine structure of the glycosaminoglycans. This is not to belittle the work that has been done, as it forms a valuable base and has developed the critical methodology from which it will be possible to advance the field. The analysis of the fine structure of glycosaminoglycans is difficult and time consuming. The difficulty of structurally characterizing glycosaminoglycans, as well as their heterogeneity, has limited the analysis of their role in most fields, with the notable exceptions of work performed to characterize the anticlotting structure in heparin and the growth factor-binding motifs in heparan sulfate.

An obvious question, based on the enormous diversity of glycosaminoglycan structures, is "are there glycosaminoglycan structures that promote or inhibit amyloid appearance?". At present, we are unable to answer this question, but the data suggest that a high level of sulfation is needed for protective effects. Therapeutically, small molecule glycosaminoglycan analogs have considerable promise, not least because they may, with modifications, be readily ingested and in some cases may be capable of transiting normal diffusional constraints. Chronic administration of low-molecular weight-heparan sulfate (e.g. certoparin, an approximately 12–20 hexose form of heparin, that is used for treatment of deep vein thrombosis and a 6-hexose fragment of certoparin) has been shown to inhibit cellular changes (tau-2 immunoreactivity) that normally arise in regions containing amyloid plaque, without affecting the extracellular accumulation of Aβ [170]. These glycosaminoglycan fragments were administered subcutaneously, suggesting that they, or their metabolites (disaccharides were ineffective and the 6-hexose fragment was less effective than certoparin), can cross the blood–brain barrier. Conceivably, this happens where the microvasculature has been damaged by amyloid deposition, although other workers have found that heparan sulfate disaccharides may be expected to cross the blood–brain barrier based on an *in vitro* model [171].

Alternative routes of intervention might be inhibition of specific enzymes in the glycosaminoglycan biosynthetic pathway that are responsible for amyloidogenic glycosaminoglycans. The large number of enzymes (sulfotransferases and glycosyltransferases) involved in glycosaminoglycan biosynthesis have mostly been cloned and now the difficult task of determining how they are controlled and how they control the structures presented by proteoglycans is being met [172]. The clinical difficulty with an enzyme inhibitory approach would be the rather unpredictable side effects. In the laboratory, however, it would be possible to use a variety of approaches to identify target enzymes that would be appropriate for use as pharmaceutical targets.

It is possible, or even likely, that just as some mutations of amyloid precursors result in a high risk of disease, subtle changes in the behavior of the biosynthetic machinery responsible for the glycosaminoglycans in the extracellular matrix might also promote amyloid-related diseases or conversely reduce the risk of accumulation of amyloid.

Acknowledgments

Molecular graphics were generated using PyMol (*http://www.pymol.org*).

References

1 Westermark, P., et al. Amyloid fibril protein nomenclature – 2002. *Amyloid* **2002**, *9*, 197–200.

2 Serpell, L. C., M. Sunde and C. C. Blake. The molecular basis of amyloidosis. *Cell Mol Life Sci* **1997**, *53*, 871–887.

3 Sipe, J. D. Amyloidosis. *Annu Rev Biochem* **1992**, *61*, 947–975.

4 Kisilevsky, R. and A. Snow. The potential significance of sulphated glycosaminoglycans as a common constituent of all amyloids: or, perhaps amyloid is not a misnomer. *Med Hypotheses* **1988**, *26*, 231–236.

5 Snow, A. D., D. Nochlin, R. Sekiguichi and S. S. Carlson. Identification in immunolocalization of a new class of proteoglycan (keratan sulfate) to the neuritic plaques of Alzheimer's disease. *Exp Neurol* **1996**, *138*, 305–317.

6 Urbanyi, Z., V. Lakics and S. L. Erdo. Serum amyloid P component-induced cell death in primary cultures of rat cerebral cortex. *Eur J Pharmacol* **1994**, *270*, 375–378.

7 Sopher, B. L., K. Fukuchi, A. C. Smith, K. A. Leppig, C. E. Furlong and G. M. Martin. Cytotoxicity mediated by conditional expression of a carboxyl-terminal derivative of the beta-amyloid precursor protein. *Brain Res Mol Brain Res* **1994**, *26*, 207–217.

8 Mirzabekov, T. A., M. C. Lin and B. L. Kagan. Pore formation by the cytotoxic islet amyloid peptide amylin. *J Biol Chem* **1996**, *271*, 1988–1992.

9 Yokota, M., T. C. Saido, E. Tani, I. Yamaura and N. Minami. Cytotoxic fragment of amyloid precursor protein accumulates in hippocampus after global

forebrain ischemia. *J Cereb Blood Flow Metab* **1996**, *16*, 1219–1223.

10 Jarrett, J.T. and P.T. Lansbury, Jr. Seeding "one-dimensional crystallization" of amyloid: a pathogenic mechanism in Alzheimer's disease and scrapie? *Cell* **1993**, *73*, 1055–1058.

11 Lomakin, A., D.B. Teplow, D.A. Kirschner and G.B. Benedek. Kinetic theory of fibrillogenesis of amyloid beta-protein. *Proc Natl Acad Sci USA* **1997**, *94*, 7942–7947.

12 Takahashi, N., K. Hasegawa, I. Yamaguchi, H. Okada, T. Ueda, F. Gejyo and H. Naiki. Establishment of a first-order kinetic model of light chain-associated amyloid fibril extension *in vitro*. *Biochim Biophys Acta* **2002**, **1601**, 110–120.

13 Ionescu-Zanetti, C., et al. Monitoring the assembly of Ig light-chain amyloid fibrils by atomic force microscopy. *Proc Natl Acad Sci USA* **1999**, *96*, 13175–13179.

14 Ohnishi, S. and K. Takano. Amyloid fibrils from the viewpoint of protein folding. *Cell Mol Life Sci* **2004**, *61*, 511–524.

15 Inoue, S., M. Kuroiwa, M.J. Saraiva, A. Guimaraes and R. Kisilevsky. Ultrastructure of familial amyloid polyneuropathy amyloid fibrils: examination with high-resolution electron microscopy. *J Struct Biol* **1998**, *124*, 1–12.

16 McLaurin, J., T. Franklin, X. Zhang, J. Deng and P.E. Fraser. Interactions of Alzheimer amyloid-beta peptides with glycosaminoglycans effects on fibril nucleation and growth. *Eur J Biochem* **1999**, *266*, 1101–1110.

17 Priola, S.A. and B. Caughey. Inhibition of scrapie-associated PrP accumulation. Probing the role of glycosaminoglycans in amyloidogenesis. *Mol Neurobiol* **1994**, *8*, 113–120.

18 Bitter, T. and H. Muir. Mucopolysaccharides of whole human spleens in generalized amyloidosis. *J Clin Invest* **1966**, *45*, 963–975.

19 Dalferes, E.R., Jr, B. Radhakrishnamurthy and G. S. Berenson. Acid mucopolysaccharides of amyloid tissue. *Arch Biochem Biophys* **1967**, *118*, 284–291.

20 Danielsson, A., E. Raub, U. Lindahl and I. Bjork. Role of ternary complexes, in which heparin binds both antithrombin and proteinase, in the acceleration of the reactions between antithrombin and thrombin or factor Xa. *J Biol Chem* **1986**, *261*, 15467–15473.

21 Woods, A. and J.R. Couchman. Syndecan-4 and focal adhesion function. *Curr Opin Cell Biol* **2001**, *13*, 578–583.

22 Iozzo, R.V. Matrix proteoglycans: from molecular design to cellular function. *Annu Rev Biochem* **1998**, *67*, 609–652.

23 Murphy, K.J., C.L. Merry, M. Lyon, J.E. Thompson, I.S. Roberts and J.T. Gallagher. A new model for the domain structure of heparan sulphate based on the novel specificity of K5 lyase. *J Biol Chem* **2004**, *279*, 27239–27245.

24 Taniuchi, H. and C.B. Anfinsen. Simultaneous formation of two alternative enzymology active structures by complementation of two overlapping fragments of staphylococcal nuclease. *J Biol Chem* **1971**, *246*, 2291–2301.

25 Liu, Y. and D. Eisenberg. 3D domain swapping: as domains continue to swap. *Protein Sci* **2002**, *11*, 1285–1299.

26 Ancsin, J.B. and R. Kisilevsky. The heparin/heparan sulfate-binding site on apo-serum amyloid A. Implications for the therapeutic intervention of amyloidosis. *J Biol Chem* **1999**, *274*, 7172–7181.

27 Kisilevsky, R. and W.A. Szarek. Novel glycosaminoglycan precursors as anti-amyloid agents, part II. *J Mol Neurosci* **2002**, *19*, 45–50.

28 Kisilevsky, R., W.A. Szarek, J. Ancsin, S. Bhat, Z. Li and S. Marone. Novel glycosaminoglycan precursors as anti-amyloid agents, part III. *J Mol Neurosci* **2003**, *20*, 291–297.

29 Ben-Zaken, O., S. Tzaban, Y. Tal, L. Horonchik, J.D. Esko, I. Vlodavsky and A. Taraboulos. Cellular heparan sulfate participates in the metabolism of prions. *J Biol Chem* **2003**, *278*, 40041–40049.

30 Leopold, P.E., M. Montal and J.N. Onuchic. Protein folding funnels: a kinetic approach to the sequence-structure relationship. *Proc Natl Acad Sci USA* **1992**, *89*, 8721–8725.

31 Onuchic, J.N. and P. G. Wolynes. Theory of protein folding. *Curr Opin Struct Biol* **2004**, *14*, 70–75.

32 Eanes, E. D. and G. G. Glenner. X-ray diffraction studies on amyloid filaments. *J Histochem Cytochem* **1968**, *16*, 673–677.

33 Kranenburg, O., L. M. Kroon-Batenburg, A. Reijerkerk, Y. P. Wu, E. E. Voest and M. F. Gebbink. Recombinant endostatin forms amyloid fibrils that bind and are cytotoxic to murine neuroblastoma cells *in vitro*. *FEBS Lett* **2003**, *539*, 149–155.

34 Cannon, M. J., A. D. Williams, R. Wetzel and D. G. Myszka. Kinetic analysis of beta-amyloid fibril elongation. *Anal Biochem* **2004**, *328*, 67–75.

35 Hamilton, J. A., L. K. Steinrauf, J. Liepnieks, M. D. Benson, G. Holmgren, O. Sandgren and L. Steen. Alteration in molecular structure which results in disease: the Met-30 variant of human plasma transthyretin. *Biochim Biophys Acta* **1992**, **1139**, 9–16.

36 Chothia, C. Coiling of beta-pleated sheets. *J Mol Biol* **1983**, *163*, 107–117.

37 Jaroniec, C. P., C. E. MacPhee, V. S. Bajaj, M. T. McMahon, C. M. Dobson and R. G. Griffin. High-resolution molecular structure of a peptide in an amyloid fibril determined by magic angle spinning NMR spectroscopy. *Proc Natl Acad Sci USA* **2004**, *101*, 711–716.

38 Hamilton, J. A. and M. D. Benson. Transthyretin: a review from a structural perspective. *Cell Mol Life Sci* **2001**, *58*, 1491–1521.

39 Otzen, D. E., O. Kristensen and M. Oliveberg. Designed protein tetramer zipped together with a hydrophobic Alzheimer homology: a structural clue to amyloid assembly. *Proc Natl Acad Sci USA* **2000**, *97*, 9907–9912.

40 Richardson, J. S. and D. C. Richardson. Natural beta-sheet proteins use negative design to avoid edge-to-edge aggregation. *Proc Natl Acad Sci USA* **2002**, *99*, 2754–2759.

41 Mattice, W. L. The beta-sheet to coil transition. *Annu Rev Biophys Biophys Chem* **1989**, *18*, 93–111.

42 Shioi, J., M. N. Pangalos, J. A. Ripellino, D. Vassilacopoulou, C. Mytilineou, R. U. Margolis and N. K. Robakis. The Alzheimer amyloid precursor proteoglycan (appican) is present in brain and is produced by astrocytes but not by neurons in primary neural cultures. *J Biol Chem* **1995**, *270*, 11839–11844.

43 Umehara, Y., S. Yamada, S. Nishimura, J. Shioi, N. K. Robakis and K. Sugahara. Chondroitin sulfate of appican, the proteoglycan form of amyloid precursor protein, produced by C6 glioma cells interacts with heparin-binding neuroregulatory factors. *FEBS Lett* **2004**, *557*, 233–238.

44 Tsen, G., W. Halfter, S. Kroger and G. J. Cole. Agrin is a heparan sulfate proteoglycan. *J Biol Chem* **1995**, *270*, 3392–3399.

45 Winzen, U., G. J. Cole and W. Halfter. Agrin is a chimeric proteoglycan with the attachment sites for heparan sulfate/chondroitin sulfate located in two multiple serine-glycine clusters. *J Biol Chem* **2003**, *278*, 30106–30114.

46 Groffen, A. J., et al. Agrin is a major heparan sulfate proteoglycan in the human glomerular basement membrane. *J Histochem Cytochem* **1998**, *46*, 19–27.

47 Barber, A. J. and E. Lieth. Agrin accumulates in the brain microvascular basal lamina during development of the blood–brain barrier. *Dev Dyn* **1997**, *208*, 62–74.

48 Berzin, T. M., et al. Agrin and microvascular damage in Alzheimer's disease. *Neurobiol Aging* **2000**, *21*, 349–355.

49 van Horssen, J., et al. Accumulation of heparan sulfate proteoglycans in cerebellar senile plaques. *Neurobiol Aging* **2002**, *23*, 537–545.

50 Cotman, S. L., W. Halfter and G. J. Cole. Agrin binds to beta-amyloid (Abeta), accelerates abeta fibril formation, and is localized to Abeta deposits in Alzheimer's disease brain. *Mol Cell Neurosci* **2000**, *15*, 183–198.

51 Hassell, J. R., P. G. Robey, H. J. Barrach, J. Wilczek, S. I. Rennard and G. R. Martin. Isolation of a heparan sulfate-containing proteoglycan from basement membrane. *Proc Natl Acad Sci USA* **1980**, *77*, 4494–4498.

52 Noonan, D. M., et al. The complete sequence of perlecan, a basement membrane heparan sulfate proteoglycan, reveals extensive similarity with laminin

A chain, low density lipoprotein-receptor, and the neural cell adhesion molecule. *J Biol Chem* **1991**, *266*, 22939–22947.

53 SundarRaj, N., D. Fite, S. Ledbetter, S. Chakravarti and J. R. Hassell. Perlecan is a component of cartilage matrix and promotes chondrocyte attachment. *J Cell Sci* **1995**, *108*, 2663–2672.

54 Tapanadechopone, P., J. R. Hassell, B. Rigatti and J. R. Couchman. Localization of glycosaminoglycan substitution sites on domain V of mouse perlecan. *Biochem Biophys Res Commun* **1999**, *265*, 680–690.

55 Murdoch, A. D., G. R. Dodge, I. Cohen, R. S. Tuan and R. V. Iozzo. Primary structure of the human heparan sulfate proteoglycan from basement membrane (HSPG2/perlecan). A chimeric molecule with multiple domains homologous to the low density lipoprotein receptor, laminin, neural cell adhesion molecules, and epidermal growth factor. *J Biol Chem* **1992**, *267*, 8544–8557.

56 Snow, A. D., et al. An important role of heparan sulfate proteoglycan (Perlecan) in a model system for the deposition and persistence of fibrillar A beta-amyloid in rat brain. *Neuron* **1994**, *12*, 219–234.

57 Snow, A. D., R. T. Sekiguchi, D. Nochlin, R. N. Kalaria and K. Kimata. Heparan sulfate proteoglycan in diffuse plaques of hippocampus but not of cerebellum in Alzheimer's disease brain. *Am J Pathol* **1994**, *144*, 337–347.

58 Castillo, G. M., W. Lukito, T. N. Wight and A. D. Snow. The sulfate moieties of glycosaminoglycans are critical for the enhancement of beta-amyloid protein fibril formation. *J Neurochem* **1999**, *72*, 1681–1687.

59 Iivonen, S., S. Helisalmi, A. Mannermaa, I. Alafuzoff, M. Lehtovirta, H. Soininen and M. Hiltunen. Heparan sulfate proteoglycan 2 polymorphism in Alzheimer's disease and correlation with neuropathology. *Neurosci Lett* **2003**, *352*, 146–150.

60 Rehn, M., E. Hintikka and T. Pihlajaniemi. Primary structure of the alpha 1 chain of mouse type XVIII collagen, partial structure of the corresponding gene, and comparison of the alpha 1 (XVIII) chain with its homologue, the alpha 1 (XV) collagen chain. *J Biol Chem* **1994**, *269*, 13929–13935.

61 Muragaki, Y., S. Timmons, C. M. Griffith, S. P. Oh, B. Fadel, T. Quertermous and B. R. Olsen. Mouse Col18a1 is expressed in a tissue-specific manner as three alternative variants and is localized in basement membrane zones. *Proc Natl Acad Sci USA* **1995**, *92*, 8763–8767.

62 Halfter, W., S. Dong, B. Schurer and G. J. Cole. Collagen XVIII is a basement membrane heparan sulfate proteoglycan. *J Biol Chem* **1998**, *273*, 25404–25412.

63 Oh, S. P., Y. Kamagata, Y. Muragaki, S. Timmons, A. Ooshima and B. R. Olsen. Isolation and sequencing of cDNAs for proteins with multiple domains of Gly-Xaa-Yaa repeats identify a distinct family of collagenous proteins. *Proc Natl Acad Sci USA* **1994**, *91*, 4229–4233.

64 Dong, S., G. J. Cole and W. Halfter. Expression of collagen XVIII and localization of its glycosaminoglycan attachment sites. *J Biol Chem* **2003**, *278*, 1700–1707.

65 van Horssen, J., M. M. Wilhelmus, R. Heljasvaara, T. Pihlajaniemi, P. Wesseling, R. M. de Waal and M. M. Verbeek. Collagen XVIII: a novel heparan sulfate proteoglycan associated with vascular amyloid depositions and senile plaques in Alzheimer's disease brains. *Brain Pathol* **2002**, *12*, 456–462.

66 Deininger, M. H., B. A. Fimmen, D. R. Thal, H. J. Schluesener and R. Meyermann. Aberrant neuronal and paracellular deposition of endostatin in brains of patients with Alzheimer's disease. *J Neurosci* **2002**, *22*, 10621–10626.

67 Bernfield, M., M. Gotte, P. W. Park, O. Reizes, M. L. Fitzgerald, J. Lincecum and M. Zako. Functions of cell surface heparan sulfate proteoglycans. *Annu Rev Biochem* **1999**, *68*, 729–777.

68 Bucciantini, M., et al. Inherent toxicity of aggregates implies a common mechanism for protein misfolding diseases. *Nature* **2002**, *416*, 507–511.

69 Campos, A., R. Nunez, C. S. Koenig, D. J. Carey and E. Brandan. A lipid-anchored heparan sulfate proteoglycan is present in the surface of differentiated skeletal muscle cells. Isolation and biochemical characterization. *Eur J Biochem* **1993**, *216*, 587–595.

70 Paine-Saunders, S., B. L. Viviano and S. Saunders. GPC6, a novel member of the glypican gene family, encodes a product structurally related to GPC4 and is colocalized with GPC5 on human chromosome 13. *Genomics* **1999**, *57*, 455–458.

71 Mertens, G., B. Van der Schueren, H. van den Berghe and G. David. Heparan sulfate expression in polarized epithelial cells: the apical sorting of glypican (GPI-anchored proteoglycan) is inversely related to its heparan sulfate content. *J Cell Biol* **1996**, *132*, 487–497.

72 De Cat, B., et al. Processing by proprotein convertases is required for glypican-3 modulation of cell survival, Wnt signaling, and gastrulation movements. *J Cell Biol* **2003**, *163*, 625–635.

73 Watanabe, N., W. Araki, D. H. Chui, T. Makifuchi, Y. Ihara and T. Tabira. Glypican-1 as an Abeta binding HSPG in the human brain: its localization in DIG domains and possible roles in the pathogenesis of Alzheimer's disease. *FASEB J* **2004**, *18*, 1013–1015.

74 Yoneda, A. and J. R. Couchman. Regulation of cytoskeletal organization by syndecan transmembrane proteoglycans. *Matrix Biol* **2003**, *22*, 25–33.

75 Kim, C. W., O. A. Goldberger, R. L. Gallo and M. Bernfield. Members of the syndecan family of heparan sulfate proteoglycans are expressed in distinct cell-, tissue-, and development-specific patterns. *Mol Biol Cell* **1994**, *5*, 797–805.

76 Subramanian, S. V., M. L. Fitzgerald and M. Bernfield. Regulated shedding of syndecan-1 and -4 ectodomains by thrombin and growth factor receptor activation. *J Biol Chem* **1997**, *272*, 14713–14720.

77 Fitzgerald, M. L., Z. Wang, P. W. Park, G. Murphy and M. Bernfield. Shedding of syndecan-1 and -4 ectodomains is regulated by multiple signaling pathways and mediated by a TIMP-3-sensitive metalloproteinase. *J Cell Biol* **2000**, *148*, 811–824.

78 Ethell, I. M., F. Irie, M. S. Kalo, J. R. Couchman, E. B. Pasquale and Y. Yamaguchi. EphB/syndecan-2 signaling in dendritic spine morphogenesis. *Neuron* **2001**, *31*, 1001–1013.

79 Lopez-Casillas, F., H. M. Payne, J. L. Andres and J. Massague. Betaglycan can act as a dual modulator of TGF-beta access to signaling receptors: mapping of ligand binding and GAG attachment sites. *J Cell Biol* **1994**, *124*, 557–568.

80 Prydz, K. and K. T. Dalen. Synthesis and sorting of proteoglycans. *J Cell Sci* **2000**, *113*, 193–205.

81 Morgenstern, D. A., R. A. Asher and J. W. Fawcett. Chondroitin sulphate proteoglycans in the CNS injury response. *Prog Brain Res* **2002**, *137*, 313–332.

82 Yamaguchi, Y. Lecticans: organizers of the brain extracellular matrix. *Cell Mol Life Sci* **2000**, *57*, 276–289.

83 Razin, E., R. L. Stevens, F. Akiyama, K. Schmid and K. F. Austen. Culture from mouse bone marrow of a subclass of mast cells possessing a distinct chondroitin sulfate proteoglycan with glycosaminoglycans rich in *N*-acetylgalactosamine-4,6-disulfate. *J Biol Chem* **1982**, *257*, 7229–7236.

84 Trowbridge, J. M., J. A. Rudisill, D. Ron and R. L. Gallo. Dermatan sulfate binds and potentiates activity of keratinocyte growth factor (FGF-7). *J Biol Chem* **2002**, *277*, 42815–42820.

85 Lyon, M., J. A. Deakin and J. T. Gallagher. The mode of action of heparan and dermatan sulfates in the regulation of hepatocyte growth factor/scatter factor. *J Biol Chem* **2002**, *277*, 1040–1046.

86 Sugahara, K., T. Mikami, T. Uyama, S. Mizuguchi, K. Nomura and H. Kitagawa. Recent advances in the structural biology of chondroitin sulfate and dermatan sulfate. *Curr Opin Struct Biol* **2003**, *13*, 612–620.

87 Tiedemann, K., T. Larsson, D. Heinegard and A. Malmstrom. The glucuronyl C5-epimerase activity is the limiting factor in the dermatan sulfate biosynthesis. *Arch Biochem Biophys* **2001**, *391*, 65–71.

88 Bechard, D., et al. Endocan is a novel chondroitin sulfate/dermatan sulfate proteoglycan that promotes hepatocyte growth factor/scatter factor mitogenic activity. *J Biol Chem* **2001**, *276*, 48341–48349.

89 Lyon, M., J.A. Deakin, H. Rahmoune, D.G. Fernig, T. Nakamura and J. T. Gallagher. Hepatocyte growth factor/scatter factor binds with high affinity to dermatan sulfate. *J Biol Chem* **1998**, *273*, 271–278.

90 Funderburgh, J.L. Keratan sulfate: structure, biosynthesis, and function. *Glycobiology* **2000**, *10*, 951–958.

91 Moss, J., I. Shore and D. Woodrow. An ultrastructural study of the colocalization of biglycan and decorin with AA amyloid fibrils in human renal glomeruli. *Amyloid* **1998**, *5*, 43–48.

92 Stokes, M.B., S. Holler, Y. Cui, K.L. Hudkins, F. Eitner, A. Fogo and C.E. Alpers. Expression of decorin, biglycan, and collagen type I in human renal fibrosing disease. *Kidney Int* **2000**, *57*, 487–498.

93 Hagner-McWhirter, A., J.P. Li, S. Oscarson and U. Lindahl. Irreversible glucuronyl C5-epimerization in the biosynthesis of heparan sulfate. *J Biol Chem* **2004**, *279*, 14631–14638.

94 Bourin, M.C. and U. Lindahl. Glycosaminoglycans and the regulation of blood coagulation. *Biochem J* **1993**, *289*, 313–330.

95 Sugahara, K. and H. Kitagawa. Heparin and heparan sulfate biosynthesis. *IUBMB Life* **2002**, *54*, 163–175.

96 Kato, M., H. Wang, M. Bernfield, J.T. Gallagher and J.E. Turnbull. Cell surface syndecan-1 on distinct cell types differs in fine structure and ligand binding of its heparan sulfate chains. *J Biol Chem* **1994**, *269*, 18881–18890.

97 Gallagher, J.T. and A. Walker. Molecular distinctions between heparan sulphate and heparin. Analysis of sulphation patterns indicates that heparan sulphate and heparin are separate families of *N*-sulphated polysaccharides. *Biochem J* **1985**, *230*, 665–674.

98 Lyon, M., J.A. Deakin and J.T. Gallagher. Liver heparan sulfate structure. A novel molecular design. *J Biol Chem* **1994**, *269*, 11208–11215.

99 Stringer, S.E., M. Mayer-Proschel, A. Kalyani, M. Rao and J.T. Gallagher. Heparin is a unique marker of progenitors in the glial cell lineage. *J Biol Chem* **1999**, *274*, 25455–25460.

100 Turnbull, J.E. and J.T. Gallagher. Sequence analysis of heparan sulphate indicates defined location of *N*-sulphated glucosamine and iduronate 2-sulphate residues proximal to the protein-linkage region. *Biochem J* **1991**, *277*, 297–303.

101 Lindblom, A., G. Bengtsson-Olivecrona and L.A. Fransson. Domain structure of endothelial heparan sulphate. *Biochem J* **1991**, *279*, 821–829.

102 Gallagher, J.T. Heparan sulfate: growth control with a restricted sequence menu. *J Clin Invest* **2001**, *108*, 357–361.

103 Mulloy, B. and M.J. Forster. Conformation and dynamics of heparin and heparan sulfate. *Glycobiology* **2000**, *10*, 1147–1156.

104 Sobel, M., D.F. Soler, J.C. Kermode and R.B. Harris. Localization and characterization of a heparin binding domain peptide of human von Willebrand factor. *J Biol Chem* **1992**, *267*, 8857–8862.

105 Cardin, A.D. and H.J. Weintraub. Molecular modeling of protein–glycosaminoglycan interactions. *Arteriosclerosis* **1989**, *9*, 21–32.

106 Margalit, H., N. Fischer and S.A. Ben-Sasson. Comparative analysis of structurally defined heparin binding sequences reveals a distinct spatial distribution of basic residues. *J Biol Chem* **1993**, *268*, 19228–19231.

107 Mulloy, B. and R.J. Linhardt. Order out of complexity – protein structures that interact with heparin. *Curr Opin Struct Biol* **2001**, *11*, 623–628.

108 Innis, C.A. and M. Hyvonen. Crystal structures of the heparan sulfate-binding domain of follistatin. Insights into ligand binding. *J Biol Chem* **2003**, *278*, 39969–39977.

109 Desai, U. R., H. M. Wang, T. R. Kelly and R. J. Linhardt. Structure elucidation of a novel acidic tetrasaccharide and hexasaccharide derived from a chemically modified heparin. *Carbohydr Res* **1993**, *241*, 249–259.

110 Yates, E. A., F. Santini, M. Guerrini, A. Naggi, G. Torri and B. Casu. ^{1}H and ^{13}C NMR spectral assignments of the major sequences of twelve systematically modified heparin derivatives. *Carbohydr Res* **1996**, *294*, 15–27.

111 Hricovini, M., M. Guerrini, A. Bisio, G. Torri, M. Petitou and B. Casu. Conformation of heparin pentasaccharide bound to antithrombin III. *Biochem J* **2001**, *359*, 265–272.

112 Faham, S., R. E. Hileman, J. R. Fromm, R. J. Linhardt and D. C. Rees. Heparin structure and interactions with basic fibroblast growth factor. *Science* **1996**, *271*, 1116–1120.

113 Olson, S. T., I. Bjork, R. Sheffer, P. A. Craig, J. D. Shore and J. Choay. Role of the antithrombin-binding pentasaccharide in heparin acceleration of antithrombin-proteinase reactions. Resolution of the antithrombin conformational change contribution to heparin rate enhancement. *J Biol Chem* **1992**, *267*, 12528–12538.

114 Jin, L., J. P. Abrahams, R. Skinner, M. Petitou, R. N. Pike and R. W. Carrell. The anticoagulant activation of antithrombin by heparin. *Proc Natl Acad Sci USA* **1997**, *94*, 14683–14688.

115 Schormann, N., J. R. Murrell, J. J. Liepnieks and M. D. Benson. Tertiary structure of an amyloid immunoglobulin light chain protein: a proposed model for amyloid fibril formation. *Proc Natl Acad Sci USA* **1995**, *92*, 9490–9494.

116 Nelson, S. R., M. Lyon, J. T. Gallagher, E. A. Johnson and M. B. Pepys. Isolation and characterization of the integral glycosaminoglycan constituents of human amyloid A and monoclonal light-chain amyloid fibrils. *Biochem J* **1991**, *275*, 67–73.

117 Jiang, X., E. Myatt, P. Lykos and F. J. Stevens. Interaction between glycosaminoglycans and immunoglobulin light chains. *Biochemistry* **1997**, *36*, 13187–13194.

118 MacRaild, C. A., et al. Non-fibrillar components of amyloid deposits mediate the self-association and tangling of amyloid fibrils. *J Biol Chem* **2004**, *279*, 21038–21045.

119 Gupta-Bansal, R. and K. R. Brunden. Congo red inhibits proteoglycan and serum amyloid P binding to amyloid beta fibrils. *J Neurochem* **1998**, *70*, 292–298.

120 Hamazaki, H. Ca^{2+}-mediated association of human serum amyloid P component with heparan sulfate and dermatan sulfate. *J Biol Chem* **1987**, *262*, 1456–1460.

121 Hamazaki, H. Amyloid P., component promotes aggregation of Alzheimer's beta-amyloid peptide. *Biochem Biophys Res Commun* **1995**, *211*, 349–353.

122 Benditt, E. P. and N. Eriksen. Amyloid protein SAA is associated with high density lipoprotein from human serum. *Proc Natl Acad Sci USA* **1977**, *74*, 4025–4028.

123 Husebekk, A., B. Skogen, G. Husby and G. Marhaug. Transformation of amyloid precursor SAA to protein AA and incorporation in amyloid fibrils *in vivo*. *Scand J Immunol* **1985**, *21*, 283–287.

124 Yamada, T., B. Kluve-Beckerman, J. J. Liepnieks and M. D. Benson. Fibril formation from recombinant human serum amyloid A. *Biochim Biophys Acta* **1994**, *1226*, 323–329.

125 Wang, L., H. A. Lashuel, T. Walz and W. Colon. Murine apolipoprotein serum amyloid A in solution forms a hexamer containing a central channel. *Proc Natl Acad Sci USA* **2002**, *99*, 15947–15952.

126 Segrest, J. P., H. J. Pownall, R. L. Jackson, G. G. Glenner and P. S. Pollock. Amyloid A: amphipathic helixes and lipid binding. *Biochemistry* **1976**, *15*, 3187–3191.

127 Vallon, R., et al. Serum amyloid A (apoSAA) expression is up-regulated in rheumatoid arthritis and induces transcription of matrix metalloproteinases. *J Immunol* **2001**, *166*, 2801–2807.

128 Trinh, C.H., D.P. Smith, A.P. Kalverda, S.E. Phillips and S.E. Radford. Crystal structure of monomeric human beta-2-microglobulin reveals clues to its amyloidogenic properties. *Proc Natl Acad Sci USA* **2002**, *99*, 9771–9776.

129 Yamaguchi, I., et al. Glycosaminoglycan and proteoglycan inhibit the depolymerization of beta2-microglobulin amyloid fibrils *in vitro*. *Kidney Int* **2003**, *64*, 1080–1088.

130 Ohashi, K. Pathogenesis of beta2-microglobulin amyloidosis. *Pathol Int* **2001**, *51*, 1–10.

131 McCutchen, S.L., W. Colon and J.W. Kelly. Transthyretin mutation Leu-55-Pro significantly alters tetramer stability and increases amyloidogenicity. *Biochemistry* **1993**, *32*, 12119–12127.

132 Kelly, J.W., et al. Transthyretin quaternary and tertiary structural changes facilitate misassembly into amyloid. *Adv Protein Chem* **1997**, *50*, 161–181.

133 Smeland, S., S.O. Kolset, M. Lyon, K.R. Norum and R. Blomhoff. Binding of perlecan to transthyretin *in vitro*. *Biochem J* **1997**, *326*, 829–836.

134 Kirschner, D.A., C. Abraham and D.J. Selkoe. X-ray diffraction from intraneuronal paired helical filaments and extraneuronal amyloid fibers in Alzheimer disease indicates cross-beta conformation. *Proc Natl Acad Sci USA* **1986**, *83*, 503–507.

135 Blake, C. and L. Serpell. Synchrotron X-ray studies suggest that the core of the transthyretin amyloid fibril is a continuous beta-sheet helix. *Structure* **1996**, *4*, 989–998.

136 Roberts, A.N., et al. Molecular and functional characterization of amylin, a peptide associated with type 2 diabetes mellitus. *Proc Natl Acad Sci USA* **1989**, *86*, 9662–9666.

137 Westermark, P., U. Engstrom, G.T. Westermark, K.H. Johnson, J. Permerth and C. Betsholtz. Islet amyloid polypeptide (IAPP) and pro-IAPP immunoreactivity in human islets of Langerhans. *Diabetes Res Clin Pract* **1989**, *7*, 219–226.

138 Young, I.D., L. Ailles, S. Narindrasorasak, R. Tan and R. Kisilevsky. Localization of the basement membrane heparan sulfate proteoglycan in islet amyloid deposits in type II diabetes mellitus. *Arch Pathol Lab Med* **1992**, *116*, 951–954.

139 Castillo, G.M., et al. Sulfate content and specific glycosaminoglycan backbone of perlecan are critical for perlecan's enhancement of islet amyloid polypeptide (amylin) fibril formation. *Diabetes* **1998**, *47*, 612–620.

140 Mascioni, A., F. Porcelli, U. Ilangovan, A. Ramamoorthy and G. Veglia. Conformational preferen of the amylin nucleation site in SDS micelles: an NMR study. *Biopolymers* **2003**, *69*, 29–41.

141 Park, K. and C.B. Verchere. Identification of a heparin binding domain in the N-terminal cleavage site of pro-islet amyloid polypeptide. Implications for islet amyloid formation. *J Biol Chem* **2001**, *276*, 16611–16616.

142 Kayed, R., J. Bernhagen, N. Greenfield, K. Sweimeh, H. Brunner, W. Voelter and A. Kapurniotu. Conformational transitions of islet amyloid polypeptide (IAPP) in amyloid formation *in vitro*. *J Mol Biol* **1999**, *287*, 781–796.

143 Pangalos, M.N., S. Efthimiopoulos, J. Shioi and N.K. Robakis. The chondroitin sulfate attachment site of appican is formed by splicing out exon 15 of the amyloid precursor gene. *J Biol Chem* **1995**, *270*, 10388–10391.

144 Crescenzi, O., S. Tomaselli, R. Guerrini, S. Salvadori, A.M. D'Ursi, P.A. Temussi and D. Picone. Solution structure of the Alzheimer amyloid beta-peptide (1–42) in an apolar microenvironment. Similarity with a virus fusion domain. *Eur J Biochem* **2002**, *269*, 5642–5648.

145 Serpell, L.C. Alzheimer's amyloid fibrils: structure and assembly. *Biochim Biophys Acta* **2000**, **1502**, 16–30.

146 Coles, M., W. Bicknell, A.A. Watson, D.P. Fairlie and D.J. Craik. Solution structure of amyloid beta-peptide (1–40) in a water-micelle environment. Is the membrane-spanning domain where we think it is? *Biochemistry* **1998**, *37*, 11064–11077.

147 Serpell, L. C., C. C. Blake and P. E. Fraser. Molecular structure of a fibrillar Alzheimer's A beta fragment. *Biochemistry* **2000**, *39*, 13269–13275.

148 Serpell, L. C. and J. M. Smith. Direct visualisation of the beta-sheet structure of synthetic Alzheimer's amyloid. *J Mol Biol* **2000**, *299*, 225–231.

149 Sikorski, P., E. D. Atkins and L. C. Serpell. Structure and texture of fibrous crystals formed by Alzheimer's beta (11–25) peptide fragment. *Structure (Camb)* **2003**, *11*, 915–926.

150 Perlmutter, L. S., H. C. Chui, D. Saperia and J. Athanikar. Microangiopathy and the colocalization of heparan sulfate proteoglycan with amyloid in senile plaques of Alzheimer's disease. *Brain Res* **1990**, *508*, 13–19.

151 Snow, A. D. and T. N. Wight. Proteoglycans in the pathogenesis of Alzheimer's disease and other amyloidoses. *Neurobiol Aging* **1989**, *10*, 481–497.

152 Snow, A. D., et al. The presence of heparan sulfate proteoglycans in the neuritic plaques and congophilic angiopathy in Alzheimer's disease. *Am J Pathol* **1988**, *133*, 456–463.

153 Snow, A. D., T. N. Wight, D. Nochlin, Y. Koike, K. Kimata, S. J. DeArmond and S. B. Prusiner. Immunolocalization of heparan sulfate proteoglycans to the prion protein amyloid plaques of Gerstmann–Straussler syndrome, Creutzfeldt-Jakob disease and scrapie. *Lab Invest* **1990**, *63*, 601–611.

154 Snow, A. D., H. Mar, D. Nochlin, R. T. Sekiguchi, K. Kimata, Y. Koike and T. N. Wight. Early accumulation of heparan sulfate in neurons and in the beta-amyloid protein-containing lesions of Alzheimer's disease and Down's syndrome. *Am J Pathol* **1990**, *137*, 1253–1270.

155 Su, J. H., B. J. Cummings and C. W. Cotman. Localization of heparan sulfate glycosaminoglycan and proteoglycan core protein in aged brain and Alzheimer's disease. *Neuroscience* **1992**, *51*, 801–813.

156 Gupta-Bansal, R., R. C. Frederickson and K. R. Brunden. Proteoglycan-mediated inhibition of A beta proteolysis. A potential cause of senile plaque accumulation. *J Biol Chem* **1995**, *270*, 18666–18671.

157 Scholefield, Z., E. A. Yates, G. Wayne, A. Amour, W. McDowell and J. E. Turnbull. Heparan sulfate regulates amyloid precursor protein processing by BACE1, the Alzheimer's beta-secretase. *J Cell Biol* **2003**, *163*, 97–107.

158 Rossjohn, J., et al. Crystal structure of the N-terminal, growth factor-like domain of Alzheimer amyloid precursor protein. *Nat Struct Biol* **1999**, *6*, 327–331.

159 Ehehalt, R., P. Keller, C. Haass, C. Thiele and K. Simons. Amyloidogenic processing of the Alzheimer beta-amyloid precursor protein depends on lipid rafts. *J Cell Biol* **2003**, *160*, 113–123.

160 Cook, D. G., et al. Alzheimer's A beta (1–42) is generated in the endoplasmic reticulum/intermediate compartment of NT2N cells. *Nat Med* **1997**, *3*, 1021–1023.

161 Hartmann, T., et al. Distinct sites of intracellular production for Alzheimer's disease A beta40/42 amyloid peptides. *Nat Med* **1997**, *3*, 1016–1020.

162 Lindahl, B., L. Eriksson and U. Lindahl. Structure of heparan sulphate from human brain, with special regard to Alzheimer's disease. *Biochem J* **1995**, *306*, 177–184.

163 Lindahl, B. and U. Lindahl. Amyloid-specific heparan sulfate from human liver and spleen. *J Biol Chem* **1997**, *272*, 26091–26094.

164 Small, D. H., V. Nurcombe, R. Moir, S. Michaelson, D. Monard, K. Beyreuther and C. L. Masters. Association and release of the amyloid protein precursor of Alzheimer's disease from chick brain extracellular matrix. *J Neurosci* **1992**, *12*, 4143–4150.

165 Kisilevsky, R., W. A. Szarek, J. B. Ancsin, E. Elimova, S. Marone, S. Bhat and A. Berkin. Inhibition of amyloid A amyloidogenesis *in vivo* and in tissue culture by 4-deoxy analogues of peracetylated 2-acetamido-2-deoxy-alpha- and beta-D-glucose: implications for the treatment of various amyloidoses. *Am J Pathol* **2004**, *164*, 2127–2137.

166 Fraser, P. E., A. A. Darabie and J. A. McLaurin. Amyloid-beta interactions with chondroitin sulfate-derived monosaccharides and disaccharides. implications for drug development. *J Biol Chem* **2001**, *276*, 6412–6419.

167 Ashikari-Hada, S., H. Habuchi, Y. Kariya, N. Itoh, A. H. Reddi and K. Kimata. Characterization of growth factor-binding structures in heparin/heparan sulfate using an octasaccharide library. *J Biol Chem* **2004**, *279*, 12346–12354.

168 Allen, B. L. and A. C. Rapraeger. Spatial and temporal expression of heparan sulfate in mouse development regulates FGF and FGF receptor assembly. *J Cell Biol* **2003**, *163*, 637–648.

169 Kiani, C., L. Chen, Y. J. Wu, A. J. Yee and B. B. Yang. Structure and function of aggrecan. *Cell Res* **2002**, *12*, 19–32.

170 Walzer, M., S. Lorens, M. Hejna, J. Fareed, I. Hanin, U. Cornelli and J. M. Lee. Low molecular weight glycosaminoglycan blockade of beta-amyloid induced neuropathology. *Eur J Pharmacol* **2002**, *445*, 211–220.

171 Leveugle, B., W. Ding, F. Laurence, M. P. Dehouck, A. Scanameo, R. Cecchelli and H. Fillit. Heparin oligosaccharides that pass the blood-brain barrier inhibit beta-amyloid precursor protein secretion and heparin binding to beta-amyloid peptide. *J Neurochem* **1998**, *70*, 736–744.

172 Nagai, N., H. Habuchi, J. D. Esko and K. Kimata. Stem domains of heparan sulfate 6-*O*-sulfotransferase are required for Golgi localization, oligomer formation and enzyme activity. *J Cell Sci* **2004**, *117*, 3331–3341.

173 Arnott, S., J. M. Guss, D. W. Hukins and M. B. Matthews. Dermatan sulfate and chondroitin 6-sulfate conformations. *Biochem Biophys Res Commun* **1973**, *54*, 1377–1383.

174 Mulloy, B., M. J. Forster, C. Jones and D. B. Davies. N.m.r. and molecular-modelling studies of the solution conformation of heparin. *Biochem J* **1993**, *293*, 849–858.

175 DiGabriele, A. D., I. Lax, D. I. Chen, C. M. Svahn, M. Jaye, J. Schlessinger and W. A. Hendrickson. Structure of a heparin-linked biologically active dimer of fibroblast growth factor. *Nature* **1998**, *393*, 812–817.

176 Johnson, D. J. and J. A. Huntington. Crystal structure of antithrombin in a heparin-bound intermediate state. *Biochemistry* **2003**, *42*, 8712–8719.

177 Pellegrini, L., D. F. Burke, F. von Delft, B. Mulloy and T. L. Blundell. Crystal structure of fibroblast growth factor receptor ectodomain bound to ligand and heparin. *Nature* **2000**, *407*, 1029–1034.

178 Capila, I., et al. Annexin V–heparin oligosaccharide complex suggests heparan sulfate-mediated assembly on cell surfaces. *Structure (Camb)* **2001**, *9*, 57–64.

179 Lietha, D., D. Y. Chirgadze, B. Mulloy, T. L. Blundell and E. Gherardi. Crystal structures of NK1-heparin complexes reveal the basis for NK1 activity and enable engineering of potent agonists of the MET receptor. *EMBO J* **2001**, *20*, 5543–5555.

180 Stringer, S. E., B. S. Kandola, D. A. Pye and J. T. Gallagher. Heparin sequencing. *Glycobiology* **2003**, *13*, 97–107.

181 Powell, A. K., E. A. Yates, D. G. Fernig and J. E. Turnbull. Interactions of heparin/heparan sulfate with proteins: Appraisal of structural factors and experimental approaches. *Glycobiology* **2004**, *14*, 17R–30R.

8
Serum Amyloid P Component

David C. Kilpatrick

8.1
Introduction to Pentraxins

The term pentraxin is given to a group of proteins possessing a distinctive quaternary structure that resembles a doughnut, with five identical polypeptide chains forming a ring with a hole approximately 20 Å (2 nm) in diameter [1]. The term was applied collectively, and virtually exclusively for many years, to two plasma proteins found in humans and other mammals, serum amyloid P component (SAP) and C-reactive protein (CRP). The former recognizes phosphoethanolamine, the latter recognizes phosphorylcholine and this has been the traditional difference between the two. Homologs of the human SAP and CRP pentraxins exist in other mammals, but murine CRP or lapine SAP are not so-called because of primary structural resemblance to the corresponding human pentraxins, but on account of their specificities towards phospholipids. It is not so surprising, therefore, that the sex-limited Syrian hamster pentraxin (female protein) cannot be readily classified: it binds phosphorylcholine, but its amino acid sequence more closely resembles that of human SAP and it is a constituent of hamster amyloid deposits [2].

Despite differing affinities for phospholipids, the two human pentraxins share about 50% amino acid sequence identity, as do the corresponding molecules in other species. One classical pentraxin CRP is typically an acute-phase reactant, but not necessarily the corresponding protein in different species. In humans, CRP can increase almost 1000-fold in response to a bacterial infection, while SAP is virtually unchanged. Under normal circumstances, SAP, at around 30 μg/ml in plasma, is typically 20 times more abundant in gravimetric terms than CRP. In mice, it is the phosphoethanolamine-binding pentraxin (SAP) which is the acute-phase reactant. In the guinea pig, however, neither pentraxin acts as a major acute-phase reactant [3].

CRP was discovered in 1930 as an antibody-like precipitin of pneumococcal C polysaccharide [4]. SAP, an apparently universal component of amyloid deposits, was probably first identified as an amyloid component in 1965 [5], although it was isolated independently several times before its relationship to CRP was es-

Amyloid Proteins. The Beta Sheet Conformation and Disease. J. D. Sipe

ISBN: 3-527-31072-X

tablished [6]. Many CRP/SAP homologs have been isolated from other vertebrate species and, additionally, related molecules have been described from a wide selection of invertebrates including horseshoe crabs [7] and tunicates [8]. Although both SAP and CRP have been found in all mammals investigated, only a single homologous pentraxin is found in teleost fish [9]. However, in the rather primitive horseshoe crab species, *Tachypleus tridentatus*, more than 20 different pentraxins have been found in the same hemolymph [10].

Even in humans, a third pentraxin-like molecule has been reported [11]. This molecule, expressed in cells of the monocyte/macrophage lineage and named PTX3, has a C-terminal domain homologous to CRP and SAP. The N-terminal domain, however, is unrelated to the other pentraxins. PTX3 is therefore sometimes referred to as a "long" pentraxin, in contrast to the classical or "short" pentraxins.

The horseshoe crab pentraxins have 12 subunits in a hexameric assembly. Nevertheless, all pentraxins have much in common and it is widely accepted that this group of doughnut-shaped molecules, the individual polypeptide subunits of which exhibit some primary structural homology (within a region of about 200 amino acids), a "jelly-roll" structural motif and a calcium-dependent carbohydrate binding site, constitute their own family of animal lectins [1]. The number of separate animal lectin families exceeds 12 and continues to grow year on year [12]. I personally consider a family requires at least two members, but if structurally unique lectins were considered to constitute one-member families, the number of so-defined animal lectin families would be far greater.

It may be helpful to digress briefly at this point to reflect on what is meant by an animal lectin. I have suggested elsewhere that "a carbohydrate binding protein not obviously involved in carbohydrate metabolism and which does not belong to any of the main classes of immunoglobulin" [1] is as good as definition as any and, certainly, any definition must be based on carbohydrate recognition. However, lectin classification based on sugar specificity makes no biological sense and, therefore, a structural classification seems much more logical. Classification based on primary structure was fine when most animal lectins could be considered C-type or S-type [13]; however, not only do we now have knowledge of so many more structural families, but it is clear many lectins belong to structural families most members of which are not lectins, and many lectins have specificities for various structures that are not carbohydrate in nature [12].

The pentraxins share this last characteristic with other lectin groups, as discussed in more detail below. In general, they seem to be protein-binding, lipid-binding and nucleic acid-binding proteins as well as carbohydrate-binding molecules. What, in equally general terms, are the functions of pentraxins?

Neither of the two classical human pentraxins is polymorphic and no deficiency state has been found for either. Any function therefore has to be inferred from circumstantial evidence. CRP is the quintessential acute-phase reactant, and has the ability to bind pathogens and activate complement; it is therefore reasonable to suppose CRP has a role in first-line defense against infection [1].

For the same reasons, SAP might participate in the initial stages of host defense, and since it has substantial DNA- and chromatin-binding ability, might also function to clear necrotic and apoptotic cell debris and thus have an anti-inflammatory role. CRP may also act at the resolution as well as at the initiation of inflammation, and therefore the two major human pentraxins may have related, complementary roles achieved by different molecular mechanisms or fine specificities.

Rodents naturally lacking pentraxins probably do not exist either, but, here, functional information can be acquired by genetic engineering. SAP and PTX3 knock-out mice, and CRP transgenic mice, have been described. Mice with targeted deletion of the SAP gene spontaneously develop antinuclear antibodies and glomerulonephritis [14] as will be discussed more fully later. Human CRP transgenic mice become more resistant to a challenge with pneumococci [15], while transgenic mice expressing high levels of lapine CRP were less susceptible to the harmful effects of bacterial lipopolysaccharide [16]. Mice genetically deficient in PTX3 were highly susceptible to the opportunistic fungal pathogen *Aspergillus fumigatus* [17]; in a disease model, all $Ptx3^{-/-}$ mice died within a few days, while all wild-type mice were still alive after 2 months. These findings were associated with defective recognition of fungal conidia by antigen-presenting cells; these defects could be corrected by addition of PTX3 *in vitro* and administration of purified PTX3 protected null mice from invasive pulmonary aspergillosis *in vivo*. PTX3 deficiency in these null mice also caused increase susceptibility to *Pseudomonas aeruginosa*, but not to *Listeria monocytogenes*. In both mice and humans, *A. fumigatus* conidia were able to induce production of PTX3 in bronchoalveolar lavage fluids and plasma. PTX3 has also been shown to bind to apoptotic cells (and necrotic cells to a lesser extent), shielding the dying cells from uptake by dendritic cells [18]. PTX3 might therefore be acting to prevent autoimmunity arising in inflamed tissues.

Less is known about the functions of horseshoe crab pentraxins, but limulin from *Limulus polyphemus* hemolymph possesses a cytotoxic activity compatible with an innate immune function [19]. The other invertebrate pentraxins are also presumed to function in host defense.

In so far as an overall function can be ascribed to pentraxins generally, it is that this family of lectins appears to regulate inflammation, in a pro-inflammatory manner by binding to pathogens and later in an anti-inflammatory manner by binding to damaged host cells and tissues to limit harm and speed the resolution of inflammation. It is noteworthy that the overall effect of CRP overexpressed in transgenic mice is anti-inflammatory [20].

It is against this family background that SAP will now be considered in more detail.

8.2 Structure of SAP

Synthesized and secreted by the liver, human SAP is a glycoprotein (12% carbohydrate; Fig. 8.1) component of basement membranes found in all types of amyloid deposits and is the major calcium-dependent DNA-binding protein of human blood plasma [21]. It is considered to have one of the most invariant carbohydrate moieties of any known glycoprotein [22]. Nevertheless, SAP in serum and after purification possesses at least five isoforms [23]. These isoforms do not reflect differences in sialic acid content, so some other post-translational modification (e.g. phosphorylation) may be responsible. Estimates of isoelectric point have varied from 4.1 to 5.7 [23–25]. Differences in the isoelectric focusing pattern and isoelectric points between purified SAP and SAP in serum have been attributed to ligand binding in the latter, but a recent investigation concluded that circulating SAP under normal circumstances does not bind any other protein ligand *in vivo* [26].

The gene encoding human SAP is located on chromosome 1 (band q2.1) and consists of two exons separated by an intron of 115 bp [27]. The first exon codes for a leader peptide and the first 2 amino acids of the mature polypeptide, while the second exon codes for a further 202 amino acids. The nucleotide sequence has 59% identity to that of the nearby gene for CRP and both human pentraxin genes map to a syntenic region in the mouse.

The full amino acid sequence of SAP isolated from the spleen of an amyloidosis patient was identical to that deduced from human SAP-specific cDNA clones [28, 29] (Fig. 8.2). The 204-amino-acid sequence contains only one methionine and has a single disulphide bridge linking Cys36 to Cys95. It is rich in glutamic and aspartic acids, leucine, glycine, serine and valine. The primary structure shows 52–54% identity to human CRP and 68–71% identity to the Syrian hamster SAP (female protein), and all have similarly positioned cysteinyl residues.

These M_r 24,000 subunits may arrange themselves as two disks interacting face-to-face, each disk comprised of a pentameric ring. The molecular weight of the native molecule under normal conditions was originally thought to be 240,000 [1]. However, the oligomeric state of human SAP in the presence or ab-

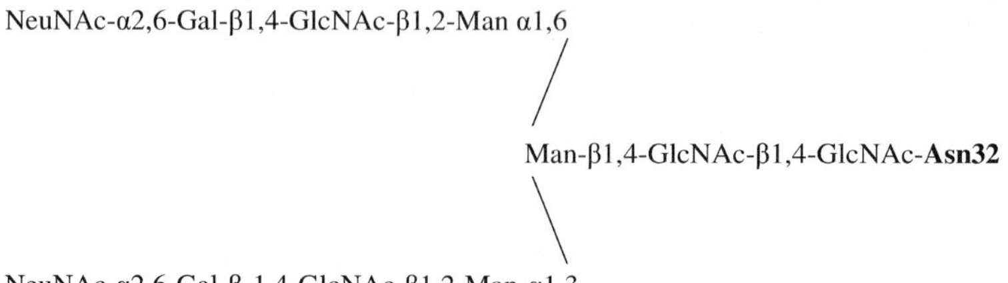

Fig. 8.1 Oligosaccharide moiety of human SAP. The carbohydrate moiety is attached to the polypeptide at Asn32.

His-Thr-Asp-Leu-Ser-Gly-Lys-Val-Phe-Val-Phe-Pro-Arg-Glu-Ser-Val-

20 ♣
Thr-Asp-His-Val-Asn-Leu-Ile-Thr-Pro-Leu-Glu-Lys-Pro-Leu-Gln-**Asn**-Phe-Thr-Leu-**Cys**-

40
Phe-Arg-Ala-Tyr-Ser-Asp-Leu-Ser-Arg-Ala-Tyr-Ser-Leu-Phe-Ser-Tyr-Asn-Thr-Gln-

60
Gly-Arg-Asp-Asn-Glu-Leu-Leu-Val-Tyr-Lys-Glu-Arg-Val-Gly-Glu-Tyr-Ser-Leu-Tyr-

80
Ile-Gly-Arg-His-Lys-Val-Thr-Pro-Lys-Val-Ile-Glu-Lys-Phe-Pro-Ala-Pro-Val-His-Ile-**Cys**-

100
Val-Ser-Trp-Glu-Ser-Ser-Ser-Gly-Ile-Ala-Glu-Phe-Trp-Ile-Asn-Gly-Thr-Pro-Leu-Val-Lys-

120
Lys-Gly-Leu-Arg-Gln-Gly-Tyr-Phe-Val-Glu-Ala-Gln-Pro-Lys-Ile-Val-Leu-Gly-Gln-Glu-

140
Gln-Asp-Ser-Tyr-Gly-Gly-Lys-Phe-Asp-Arg-Ser-Gln-Ser-Phe-Val-Gly-Glu-Ile-Gly-Asp-

160
Leu-Tyr-Met-Trp-Asp-Ser-Val-Leu-Pro-Pro-Glu-Asn-Ile-Leu-Ser-Ala-Tyr-Gln-Gly-Thr-

180
Pro-Leu-Pro-Ala-Asn-Ile-Leu-Asp-Trp-Gln-Ala-Leu-Asn-Tyr-Glu-Ile-Arg-Gly-Tyr-Val-

200
Ile-Ile-Lys-Pro-Leu-Val-Trp-Val

Fig. 8.2 Primary structure of human SAP. The single disulphide bridge and the position of attachment of the carbohydrate moiety (♣) are indicated.

sence of ligands has been investigated using mass spectrometry [30]. In the absence of calcium, SAP consists of pentameric and decameric forms. In the presence of physiological levels of calcium, however, SAP exists primarily as a pentamer. An interesting observation was that, while SAP and CRP form mixed decamers in Ca^{2+}-free buffers, this interaction was not observed in the presence of physiological concentrations of calcium nor was exchange of monomeric subunits observed between the SAP and CRP oligomers. This suggests that individual pentameric complexes are very stable in the presence of calcium.

The secondary structure of SAP contains about 50% β-pleated sheet and about 12% α-helix [31]. Emsley et al. [32] described the three-dimensional structure of pentameric SAP derived by X-ray diffraction at high (2 Å) resolution. A major finding was that each protomer has a tertiary fold constructed from antiparallel β-strands with a single long α-helix folded on top of the β-sheet. This "jelly-roll" structural motif is similar to that found in certain legume lectins and also in galectins-1 and -2. The β-sheets are separated by a hydrophobic core, one end of which is accessible to solvent. The five subunits are non-covalently assembled in a ring with a hole 20 Å in diameter and 35 Å deep. About 15% of the surfaces of the protomers are in contact with each other, maintained by three salt bridges, hydrogen bonds between the main-chain peptide groups and some hydrophobic contacts. The decamer, which is readily dissociated at pH 5.5, appears to be stabilized by ionic interactions involving carboxylate and/or imidazole groups, with the α-helical regions in contact with each other.

The ligand-binding site of each protomer contains two calcium ions located in the solvent-exposed region of the β-sheet. Calcium ion-1 is coordinated to the side-chains of Asp58, Asn59, Glu136, Asp138 and the main-chain carbonyl of Gln137. Calcium ion-2, which appears to be less firmly bound, is also coordinated to Glu136 and Asp138; an acetate ion (or perhaps Glu167) contributes to a bridge between the calcium ions, and the coordination of calcium ion-2 is completed by Gln148 and two water molecules.

Both calcium ions are implicated in ligand binding, although calcium does not seem to bind galactosides directly, but instead mediates binding indirectly by orientating side-chain amides. Binding to phosphoethanolamine involves a direct interaction between the phosphate group and both calcium ions. The sugar ligands of SAP interact with Asn59 and Gln148, while Glu66, Tyr74 and Lys79 interact with phosphoethanolamine. Binding to DNA and other polyanions also requires calcium and probably involves a closely related binding region.

The carbohydrate recognition site of SAP differs most notably from that of the true C-type lectins in being made up of amino acid residues from distinct loops within the sequence. There is, of course, no significant sequence homology between the primary structure of this calcium-dependent pentraxin lectin and the true C-type lectins [1, 12].

8.3 Lectin and Other Biological Activities of SAP

The ability of SAP to bind to agar/agarose has long been known, hence the ready acceptance of its status as a human lectin [33–35]. Indeed, agarose-based affinity chromatography is a ready means of purifying SAP from serum. SAP also binds other complex carbohydrates including zymosan [35–37], phosphomannans [37], galactans from snails [38], laminin [39], several glycosaminoglycans [40, 41] and sulfated glycolipids such as cerebroside-3-sulfate [42, 43]. All those interactions are calcium dependent, arguably specific and take place under physiological conditions (Tab. 8.1). However, a calcium-independent binding of SAP to glycosaminoglycans, which is optimal at pH 5.0, has also been reported [44].

The interaction of SAP and agarose has been studied in some detail [45]. This calcium-dependent binding of SAP varied with different sources of agarose and was found to correlate very strongly with the pyruvate content of agarose. Moreover, methylation of the carboxylic acid of the pyruvate residue abolished binding. A synthetic pyruvate acetyl at the 4,6 positions of galactose (4,6-*O*-[1-carboxyethylidene]-β-D-galactopyranose or M*O*βDG) inhibited SAP–agarose binding

Table 8.1 Some ligands of human serum amyloid P component.

Ligand	Reference
Agarose	33–35
C1q	62, 64
Collagen	72
DNA	69
Histones	61
Galactosides	38, 46
Glycosaminoglycans	40, 41, 44
Glycolipids	42
Laminin	39
Phosphoethanolamine	110
Zymosan	35–37

and could be used to elute SAP already bound to agarose gels. The precise specificity is therefore for the 4,6-pyruvate acetyl of β-D-galactopyranose present in trace amounts (less than 1%) in agarose (which is basically a linear galactan composed of repeating units of agarobiose). This finding is consistent with evidence from a different direction: Hamazaki [46] described an apparently new human lectin (later found to be SAP), the characteristics of which included specificity for β-galactosides involving the penultimate galactose residues in saccharide chains. This fine specificity explains why SAP is not inhibited by free galactose. Studies on the three-dimensional structure of SAP referred to earlier [32] confirm an interaction with a carboxylethylidene ring and explains the absence of binding to the non-cyclic acetyl of MOβDG.

SAP also binds to heparin [33], heparan sulfate [40, 41], dermatan sulfate [40, 41], and (with particularly high affinity) amyloid-associated galactosaminoglycans and large proteoglycans [41] in a calcium-dependent manner. Highest affinity binding to immobilized SAP was noted at 4 mM calcium. Cadmium (and, to a much lesser extent, zinc) ions partially substituted for calcium, but no binding to glycosaminoglycans was observed in the presence of a barium, copper, magnesium, manganese or strontium [40]. (Regrettably, nickel was not tested, as it was previously reported as having a high affinity for SAP [33].) The interactions of radiolabeled heparan and dermatan sulfate to SAP immobilized on Sepharose were strongly inhibited by heparan, heparin and dermatan sulfates, moderately so by chrondroitin-6-sulfate, and weakly so by chrondroitin-4-sulfate and hyaluronic acid, while keratin sulfate was without effect [40].

The calcium-dependent binding of SAP to zymosan has been localized to alkali-extractable ligands [36]. Since soluble zymosan extracts prepared by alkaline hydrolysis are rich in α1,3 and α1,6 mannosides, it was suggested that SAP has specificity for certain complex mannose-containing polysaccharides. This proposal is supported by a demonstration of SAP binding to mannose-terminated glycoproteins like ovalbumin and thyroglobulin but not to glycoproteins lacking terminal mannose like ovomucoid or glycophorin. Moreover, binding of SAP to zymosan extract absorbed with concanavalin A was significantly decreased. However, other researchers have reported that SAP has little affinity for yeast mannan [46, 47]. It is also curious that the binding of soluble SAP to immobilized zymosan extract ligands or to immobilized mannose-terminated glycoproteins is markedly suboptimal at calcium concentrations between 1 and 5 mM.

SAP can also bind to various bacteria including *Klebsiella rhinoscleromatis* [48], *Streptococcus pyogenes* [48], *Xanthomonas campestris* [48], *Salmonella enterica* [49], *Escherichia coli* [49, 50], *Haemophilus influenzae* [49] and *Neisseria meningitidis* [50]. The binding to *Klebsiella* and *S. pygenes* is clearly a lectin activity, as it was potently inhibited by MO*β*DG [48], but similar inhibition by saccharide does not appear to have been demonstrated for SAP–Gram-negative bacteria interactions and, indeed, the binding of SAP to lipopolysaccharide is via the lipid A part of the molecule [51]. Interestingly, SAP is the human plasma factor binding the dangerous toxin (Stx-2) from Shiga toxin-producing *E. coli* [52]; however, it appears to act *in vivo* as a carrier protein of Stx-2 rather than as a neutralizing entity [53].

SAP can bind to influenza A virus via the viral hemagglutinin [54, 55]. This seems to be a lectin activity since heparin, heparan sulfate and other saccharides reduced the binding, although a contribution from the viral lectin interacting with the carbohydrate moiety of SAP cannot be excluded.

A typical characteristic of lectins (plant or animal) is the ability to bind to and agglutinate cells and sometimes to stimulate or inhibit mitogenesis [1, 56]. SAP has been reported to bind autologous macrophages [57], while human SAP can bind human neutrophils [58], erythrocytes [38, 46] and mononuclear leukocytes [38].

The binding of [^{125}I]human SAP to human erythrocytes was detectable at less than 10 μg/ml and seemed to have reached a plateau at 50 μg/ml [38]. However, the ability of unlabeled SAP to agglutinate human red blood cells required a supraphysiological concentration of 60 μg/ml or above, as was also true for bovine, ovine and lapine erythrocytes [43]. Rat and equine erythrocytes could be agglutinated at 30 μg/ml. The best inhibitors were 2-*O*-*α*-D-glucopyranosyl-*O*-*β*-D-galactopyranosylhydroxylysine (Glc-Gal-Hyl), *N*-acetylated-Glc-Gal-Hyl and stachyose. Neither galactose nor any other monosaccharide tested had any effect. Human glycophorin was weakly inhibitory, but yeast mannan not at all [46].

SAP can also bind to human peripheral blood mononuclear cells, a calcium-dependent interaction that can be inhibited by galactans, but not by galactose [38]. Moreover, SAP abrogated the proliferative response of mononuclear cells to the mitogen, phytohemagglutinin, markedly inhibited their response to purified peptide derivative (PPD; a recall antigen) and significantly suppressed their pokeweed mitogen-induced polyclonal antibody production [38].

Another group has studied the binding of human SAP to human neutrophils [58]. The receptors involved were heterogeneous with respect to affinity, and could also interact with CRP. The binding was calcium-dependent and specific, but unfortunately inhibition by carbohydrate structures was not investigated, so this may not be a lectin activity. The previously reported binding of murine SAP to autologous peritoneal macrophages was dependent on the presence of Ca^{2+} or Mg^{2+} and could be inhibited by mannose or mannose phosphates [57]. This saccharide specificity, however, is not expected of SAP and, therefore, it seems more likely that in this instance SAP is a ligand for another lectin receptor. This possible explanation was strengthened by the observation that the partial removal of carbohydrate residues from SAP (with endoglycosidases) reduced the binding of SAP to macrophages.

Murine SAP can induce the production of serum colony-stimulating factors *in vivo* and lipopolysaccharide-independent production of colony-stimulating factors *in vitro* [59]. This property cannot be extrapolated to human SAP, however, especially as the corresponding human molecule could be considered to be CRP. Human SAP, however, does prevent the differentiation of human monocytes into fibrocytes *in vitro*, an activity that could be relevant to mechanisms promoting angiogenesis and wound healing [60].

It has been found that crosslinked SAP [61, 62], solid-phase histones complexed with SAP [61] and supraphysiological concentrations of SAP [63] can activate complement via the classical pathway. This may be related to its ability to bind the complement components C1q [62, 64], C3 [65] and C5b6 [66]. The SAP–C5b6 complex did not dissociate in the presence of EDTA. The other interactions with complement components may be calcium dependent, but there is no evidence of susceptibility to inhibition by saccharides. SAP can also bind to aggregated IgG [67] and, by implication, immune complexes, possibly via Fcγ receptors [68]. An early report of SAP interacting with Fabγ, but not Fcγ, seems not to have been confirmed [64].

Potentially the most important activity of SAP is its ability to bind chromatin [61, 63, 69–71]. While SAP shares this affinity for nuclear components with its fellow pentraxin CRP, the nature of the specificities involved is quite different.

Pepys and Butler [69] first reported that serum SAP, unlike CRP or any other serum protein, underwent specific calcium-dependent binding to single- or double-stranded DNA immobilized on Sepharose. Moreover, purified serum SAP in physiological buffers bound both native long chromatin and histone H1-stripped chromatin. Immobilized SAP specifically bound nucleosome core particles from solution. Breathnach et al. [70] also reported calcium-dependent binding of SAP to cell nuclei *in vitro*, with cryostat sections of normal human skin and cultured Hep-2 cells. Additionally their finding of SAP in globular dermal deposits of nuclear material from skin biopsies from two lupus patients was the first evidence of SAP binding to extracellular chromatin *in vivo* [70].

The first report of the binding of SAP to nucleoli [71] confirmed this calcium-dependent binding to chromatin and contrasted that with a comparative absence of chromatin–CRP interaction. In contrast, CRP was found to bind most obviously to small nuclear ribonucleoproteins.

Finally, it is notable that purified human SAP binding to native long chromatin selectively displaces histone H1 and thereby renders such chromatin soluble in physiological-buffered saline, in which the chromatin would otherwise precipitate completely [63]. In contrast, CRP had no effect on native chromatin or nucleosome core particles. Hicks et al. [61] had noted that SAP bound histones H1 and H2A, and had partly inhibited CRP binding to chromatin and to H1, but neither pentraxin inhibited the binding of the other to H2A. It seems that both SAP and CRP bind to nuclear components, including histones, but the two pentraxins largely recognize different nuclear materials, which could be an indication of functional complementarity rather than functional redundancy.

SAP binds to collagen, but this interaction is rather different. SAP interacts specifically with both human and murine type IV collagen [72], but not to collagen types I, II or III. Although this activity was calcium dependent (optimal at above 1 mM Ca^{2+}), it was not inhibited by phosphatidylethanolamine. It was, however, inhibited by C1q and CRP. It was suggested that SAP may bind to the triple helical region of collagen via a site distinct from its galactan-binding site, possibly in the region spanning residues 108–120 which is 60% identical to the CRP region involved in its binding to the collagen-like region of the C1q molecule [72]. It has also been suggested that SAP may bind C1q at its collagen-like region [64].

Other reported activities of human SAP include binding to immobilized calumenin [73], immobilized high-density lipoprotein and very-low-density lipoprotein (but not to low-density lipoprotein) [74], and to late apoptotic Jurkat cells [75]. Clearly SAP is capable of binding to a bewildering variety of molecules, and additional information is required to determine which interactions may be physiologically and/or pathologically relevant.

8.4 SAP: Its Physiological Role in Health

Based on the properties described above, two general functions have been proposed for SAP under normal physiological conditions. The first is that SAP participates in first-line host defense, in a collectin-like manner, by binding to pathogens (recognized as non-self), activating the complement system, and thereby promoting the elimination of microorganisms. The second proposed function, which is in no way mutually exclusive, is that SAP assists in the removal of apoptotic waste by binding to chromatin, etc., released by dying cells.

With regard to the first hypothesis, SAP can bind to certain bacteria as previously discussed, but we must be cautious about how we interpret this observation. Some plant lectins are mitogenic for human lymphocytes, after all, but this must be fortuitous, since plant lectins do not encounter human cells in their normal life circumstances. Similarly, aggregated or immobilized SAP can activate the complement system, but does it do so *in vivo*? The only evidence in support is the description of complexes between SAP and C3b/C3d or C4b/C4d in the circulation of healthy individuals [76]. However, far from being an inflammatory opsonin, SAP's binding to bacteria had a marked anti-opsonic effect *in vitro* and *in vivo*, inhibiting phagocytosis and killing [50]. Indeed the binding of pathogenic bacteria to SAP promoted virulence. Moreover, SAP knock-out mice survived potentially lethal infections with bacteria to which SAP binds. Also, SAP appeared to have no effect whatsoever in an experimental model of influenza virus infection concerning SAP knock-out mice transgenic for human SAP [77]. It seems unlikely therefore, that SAP recognizes and opsonizes microbial non-self, in the manner we attribute to collectins.

SAP binds chromatin and no other protein binds chromatin like SAP does [63, 69]. The binding of SAP to chromatin specifically does not activate complement [63]. Complexes of SAP and DNA exist in the serum of normal individuals and only a minor proportion of the SAP is bound to DNA. These complexes containing 5–50 ng/ml DNA required physiological calcium concentration and could be dissociated with heparin or heparan sulfate [78]. It seems plausible, therefore, that a major function of SAP is to remove chromatin from the bloodstream and other extracellular milieu when released as a result of apoptosis or tissue damage. Apoptosis is characterized by the translocation of potentially autoantigenic molecules from the nucleus to the cell surface where they are displayed to the extracellular environment as membranous blebs/apoptotic bodies [79].

Although caution must be exercised when extrapolating from murine SAP to human SAP, the observations on SAP knock-out mice [14] fit in perfectly with this explanation. These mice spontaneously developed high titer antinuclear antibodies (to DNA and histones) and had enhanced anti-double-stranded DNA responses to immunization with extrinsic chromatin. Moreover, degradation of long chromatin was retarded *in vitro* and *in vivo* in the presence of SAP [14]. It is therefore entirely plausible that SAP solubilizes and stabilizes chromatin *in vivo*, and may direct the catabolism of potential autoantigens to the liver, as hepatocytes are the only cells in which SAP is catabolized *in vivo* [80, 81]. Paul and Carroll [82], in discussing the SAP knock-out mice observations, suggest that in addition to sequestering nuclear antigens from the immune system, SAP sends a positively tolerogenic signal through the formation of SAP–C4 binding protein complexes [83] adhering to chromatin.

Many different molecules have been implicated in the non-inflammatory clearance of apoptotic cells [84, 85]. These include collectins, complement components and scavenger receptors. This implies a level of redundancy which could be taken as an indication of the physiological importance of clearing apoptotic cell debris. An alternative view might be that different molecules contribute to complementary but not identical processes. It may be significant, for example, that necrotic cell death is associated with early complement binding while apoptotic cell death is associated with later complement (and pentraxin) binding [86]. It may also be that cellular waste disposal is a much more complex process than is often assumed, with many optional mechanisms available. SAP can bind to complement components and to various cell types, after all, and multiple SAP interactions could be involved in apoptotic cell clearance not just the initial binding to chromatin. As others have pointed out, the relative importance of pentraxins, complement components and other molecules in the removal of apoptotic cells remains to be established, and the key question is the nature of the response triggered by clearance of opsonized apoptotic material [84]. Current thinking about the mechanisms involved in apoptotic cell clearance have been reviewed extensively elsewhere [85, 87].

The physiological role of SAP has not been convincingly determined. Presumably SAP and CRP arose from the duplication of a gene for a plasma proto-pentraxin and have since evolved separately. It is a reasonable educated guess that the proto-pentraxin had a role in inflammatory clearance of pathogens and in non- or anti-inflammatory clearance of self-waste in the form of apoptotic and perhaps necrotic cell debris. Perhaps both present day pentraxins have evolved more towards the second function, but have acquired separate specificities and separate receptors, although there may be still some degree of overlap with both substrates and receptors.

8.5
SAP: Its Role in Disease

Since SAP may play an important physiological role in the clearance of nuclear antigens, it has been suggested that SAP may be involved pathologically in diseases, such as systemic lupus erythematosus (SLE), characterized by antinuclear autoantibodies [81, 85]. This suggestion is superficially supported by the presence of glomerulonephritis in some SAP knock-out mice, principally affecting females [14]. A separate research group generated SAP deficient mice that produced high-titer antinuclear antibodies, but without severe glomerulonephritis [88]. The most recent findings in such animals are that transgenic expression of human SAP did not alter the autoimmune phenotype [89]. The reason for this is uncertain, but it is possible the autoimmunity is independent of SAP deficiency.

The striking fact remains that SAP deficiency has never been found in humans. Therefore SAP cannot be involved in the pathogenesis of SLE except by invoking fanciful conjectures like SAP gene polymorphism [82], glycosylation variants [82], less efficient binding to hypomethylated bacterial DNA [78] or SAP from lupus patients binding bacterial DNA differently from SAP in normal serum [78]. None of those speculations has any evidence to support it, and some have considerable evidence to the contrary. If pentraxin-mediated chromatin clearance is implicated in human SLE, the candidate pentraxin has to be CRP, not SAP, since the former is the acute phase reactant and, suggestively, the CRP response to lupus-induced inflammation is muted compared to other idiopathic rheumatic diseases such as rheumatoid arthritis [90]. Moreover, we should not forget that although human C1q deficiency appears to cause a form of SLE [91], the overwhelming majority of SLE patients do not have C1q deficiency. Much more evidence is clearly required before any credence can be given to a causal relationship between SAP function and SLE. Interestingly, elevated titers of both anti-SAP and anti-CRP autoantibodies are found in human lupus patients [92].

SAP is a universal non-fibrillar component of amyloid deposits. These are clearly pathological extracellular aggregates that disrupt tissue structure and function and are associated with diverse disorders including Alzheimer's disease, Type 2 diabetes mellitus and dialysis arthropathy as well as the systemic amyloidoses. Radiolabeled SAP can be used to locate these amyloid deposits and to monitor changes with time or in response to treatment [93, 94]. Since the original report of scintigraphic imaging with ^{123}I-labeled SAP [93], its usefulness in a variety of clinical circumstances has been demonstrated [95–99]. The applications of SAP scintigraphy have recently been reviewed [100].

The key question is whether SAP is involved in fibrillogenesis or fibril persistence. Statements like "...experiments, both *in vitro* and *in vivo*, suggest that SAP is important in the pathogenesis and persistence of amyloid..." [101] are common in reviews of amyloidosis, but the evidence to support them is thin. Certainly, the binding of SAP to amyloid fibrils *in vitro* protects the latter from degradation [102], but the *in vivo* evidence is confined to murine models with their inherent limitations. Amyloid deposition from experimental reactive AA amyloidosis was shown to be delayed and reduced in SAP knock-out mice [103]. These observations were confirmed independently by another group [104]. It seemed therefore that SAP accelerated the deposition of amyloid, although not essential to the process. However, a rather different inference was drawn from some later work [105]. Doubly transgenic mice, carrying both the human mutant transthyretin (responsible for familial amyloidotic polyneuropathy) and the human SAP gene, were generated. The age of onset, progression and distribution of amyloid deposits were similar in the doubly transgenic mice and in mice carrying only the human mutant transthyretin gene. Additionally, the regression of splenic AA amyloid fibrils in SAP knock-out and wild-type mice was studied. After induction of amyloid by daily injections of casein, the degree of amyloid deposition in the two groups was similar. These observations provide strong evidence that lack of SAP does not enhance regression of amyloid deposits *in vivo*. It must be stressed nevertheless that these interesting observations may not be relevant to natural human amyloid diseases.

Depletion of SAP is only one of many potential therapeutic strategies in amyloidosis [101, 106]. An interesting pharmacological approach has been the development of dimeric proline derivatives that can link SAP pentameters together [107, 108]. These compounds prevent SAP from binding to amyloid and, moreover, the drug–SAP complex is rapidly eliminated from the circulation. In animal models, the levels of SAP both in plasma and immobilized in amyloid deposits, were markedly decreased. Preliminary clinical trial results indicated that one of those drugs was well tolerated and effective at reducing circulating SAP. Further clinical trials are said to be in progress [109]. If efficacy can be demonstrated in treating established amyloidosis in well-controlled human trials, that will constitute more compelling evidence for the role of SAP in stabilizing and protecting amyloid deposits than could be obtained from the genetically modified murine models described above. Positive results would not necessarily benefit patients. It remains a theoretical possibility that such drugs used effectively enough to remove SAP from and disaggregate amyloid deposits might cause the induction of a lupus-like disease as well.

References

1 Kilpatrick, D.C. *Handbook of Animal Lectins: Properties and Biomedical Applications*. Wiley, Chichester, **2000**.

2 Coe, J.E. and Ross, M.J. Hamster female protein, a sex-limited pentraxin, is a constituent of Syrian hamster amyloid. *J Clin Invest* **1985**, *76*, 66–74.

3 Rubio, N., Sharpe, P.M., Rits, M., Zahedi, K. and Whitehead, A.S. Structure, expression and evolution of guinea pig serum amyloid P component and C-reactive protein. *J Biochem* **1993**, *113*, 277–284.

4 Tillett, W.S. and Francis, T. Serological reactions in pneumonia with a non-protein somatic fraction of pneumococcus. *J Exp Med* **1930**, *52*, 561–571.

5 Cathcart, E.S., Comerford, F.R. and Cohen, A.S. Immunological studies on a protein extracted from human secondary amyloid. *New Eng J Med* **1965**, *273*, 143–146.

6 Pepys, M.B. and Baltz, M.L. Acute phase proteins with special reference to C-reactive protein and related proteins (pentraxins) and serum amyloid A protein. *Adv Immunol* **1983**, *34*, 141–212.

7 Shrive, A.K., Metcalfe, A.M., Cartright, J.R. and Greenhough, T.J. C-reactive protein and SAP-like pentraxin are both present in *Limulus polyphemus* haemolymph: crystal structure of *Limulus* SAP. *J Mol Biol* **1999**, *290*, 997–1008.

8 Elola, M.T. and Vasta, G.R. Lectins from the colonial tunicate *Clavelina picta* are structurally related to acute phase reactants from vertebrates. *Ann NY Acad Sci* **1994**, *712*, 321–323.

9 Jensen, L.E., Hiney, M.P., Shields, D.C., Uhlar, C.M., Lindsay, A.J. and Whitehead, A.S. Acute phase proteins in salmonids. Evolutionary analyses and acute phase response. *J Immunol* **1997**, *158*, 384–392.

10 Iwaki, D., Osaki, T., Mizunoe, Y., Wai, S., Iwanaga, S. and Kawabata, S. Functional and structural diversities of C-reactive proteins present in horseshoe crab hemolymph plasma. *Eur J Biochem* **1999**, *264*, 314–326.

11 Alles, V.V., Bottazzi, B., Peri, G., Golay, J., Introna, M. and Mantovani, A. Inducible expression of PTX3, a new member of the pentraxin family, in human phagocytes. *Blood* **1994**, *84*, 3483–3493.

12 Kilpatrick, D.C. Animal lectins: a historical introduction and overview. *Biochim Biophys Acta* **2002**, *1572*, 187–197.

13 Drickamer, K. Two distinct classes of carbohydrate-recognition domains in animal lectins. *J Biol Chem* **1998**, *263*, 9557–9560.

14 Bickerstaff, M.C.M., Botto, M., Hutchinson, W.L., et al. Serum amyloid P component controls chromatin degradation and prevents antinuclear autoimmunity. *Nat Med* **1999**, *5*, 694–697.

15 Szalai, A.J., Briles, D.E. and Volanakis, J.E. Human C-reactive protein is protective against fatal *Streptococcus pneumoniae* infection in transgenic mice. *J Immunol* **1995**, *155*, 2557–2563.

16 Xia, D. and Samols, D. Transgenic mice expressing rabbit C-reactive protein are resistant to endotoxemia. *Proc Natl Acad Sci USA* **1997**, *94*, 2575–2580.

17 Garlanda, C., Hirsch, E., Botta, S., et al. Non-redundant role of the long pentraxin PTX3 in anti-fungal innate immune response. *Nature* **2002**, *420*, 182–186.

18 Rovere, P., Peri, G., Fazzini, F., et al. The long pentraxin PLX3 binds to apoptotic cells and regulates their clearance by antigen-presenting dendritic cells. *Blood* **2000**, *96*, 4300–4306.

19 Armstrong, P.B., Swarnaker, S., Srimal, S., et al. A cytolytic function for a sialic acid-binding lectin that is a member of the pentraxin family of proteins. *J Biol Chem* **1996**, *271*, 14717–14721.

20 Gabay, C. and Kushner, I. Acute-phase proteins and other systemic responses to inflammation. *New Eng J Med* **1999**, *340*, 448–454.

21 Tennant, G.A. and Pepys, M.B. Glycobiology of the pentraxins. *Biochem Soc Trans* **1994**, *22*, 74–79.

22 Pepys, M.B., Rademacher, T.W., Amatazakul-Chantler, S., et al. Human serum amyloid P component is an invariant constituent of amyloid deposits and has

a uniquely homogeneous glycostructure. *Proc Natl Acad Sci USA* **1994**, *91*, 5602–5606.

23 Nybo, M., Hackler, R., Kold, B., et al. Isoforms of murine and human serum amyloid P component. *Scand J Immunol* **1998**, *48*, 350–356.

24 Baltz, M. L., de Beer, F. C., Feinstein, A., et al. Phylogenetic aspects of C-reactive protein and related proteins. *Ann NY Acad Sci* **1982**, *389*, 49–75.

25 Ohkubo, I., Sahashi, W., Namikawa, C., et al. A procedure for large scale purification of human plasma amyloid P component. *Clin Chem Acta* **1986**, *157*, 95–102.

26 Sen, J. W. and Heegaard, N. H. Serum amyloid P component does not circulate in complex with C4-binding protein, fibronectin or any other major protein ligand. *Scand J Immunol* **2002**, *56*, 85–93.

27 Steel, D. M. and Whitehead, A. S. The major acute phase reactants: C-reactive protein and serum amyloid A protein. *Immunol Today* **1994**, *15*, 81–88.

28 Prelli, F., Pras, M. and Frangione, B. The primary structure of human tissue amyloid P component from a patient with primary idiopathic amyloidosis. *J Biol Chem* **1985**, *260*, 12895–12898.

29 Mantzouranis, E. C., Dowton, S. B., Whitehead, A. S., et al. Human serum amyloid P component. cDNA isolation, complete sequence of pre-serum amyloid P component, and localization of the gene to chromosome 1. *J Biol Chem* **1985**, *260*, 7752–7756.

30 Aquilina, J. A. and Robinson, C. V. Investigating interactions of the pentraxins serum amyloid P component and C-reactive protein by mass spectrometry. *Biochem J* **2003**, *375*, 323–328.

31 Dong, A., Caughey, W. S. and Du Clos, T. W. Effects of calcium, magnesium, and phosphorylcholine on secondary structures of human C-reactive protein and serum amyloid P component observed by infrared spectroscopy. *J Biol Chem* **1994**, *269*, 6424–6430.

32 Emsley, J., White, H. E., O'Hara, B. P., et al. Structure of pentameric human serum amyloid P component. *Nature* **1994**, *367*, 338–345.

33 Haupt, H., Heimburger, N., Kranz, T. and Baudner, S. Human serum proteins with a high affinity for carboxymethyl cellulose, III. *Hoppe-Seyler's Z Physiol Chem* **1972**, *353*, 1841–1849.

34 Pepys, M. B., Dash, A. C., Munn, E. A., et al. Isolation of amyloid P component (protein AP) from normal serum as a calcium-dependent binding protein. Lancet **1977**, *i*, 1029–1031.

35 Pepys, M. B., Dyck, R. F., de Beer, F. C., Skinner, M. and Cohen, A. S. Binding of serum amyloid P component (SAP) by amyloid fibrils. *Clin Exp Immunol* **1979**, *38*, 284–293.

36 Kubek, B. M., Potempa, L. A., Anderson, B., et al. Evidence that serum amyloid P component binds to mannose-terminated sequences of polysaccharides and glycoproteins. *Mol Immunol* **1988**, *25*, 851–858.

37 Potempa, L. A., Kubak, B. M. and Gewurz, H. Effect of divalent metal ions and pH upon the binding reactivity of human serum amyloid P component, a C-reactive protein homologue, for zymosan. *J Biol Chem* **1985**, *260*, 12142–12147.

38 Li, J. J., Pereira, M. E. A., DeLellis, R. A. and McAdam, P. W. J. Human amyloid P component: a circulating lectin that modulates immunological responses. *Scand J Immunol* **1984**, *19*, 227–236.

39 Zahedi, K. Characterization of the binding of serum amyloid P to laminin. *J Biol Chem* **1997**, *272*, 2143–2148.

40 Hamazaki, H. Ca^{2+}-mediated association of human serum amyloid P component with heparin sulfate and dermatin sulfate. *J Biol Chem* **1987**, *262*, 1456–1460.

41 Stenstad, T., Magnus, J. H., Syse, K. and Husby, G. On the association between amyloid fibrils and glycosaminoglycans; possible interactive role of Ca^{2+} and amyloid P-component. *Clin Exp Immunol* **1993**, *94*, 189–195.

42 Loveless, R. W., O'Sullivan, G. F., Raynes, J. G., Yuen, C.-T. and Feizi, T. Human serum amyloid P is a multispecific adhesive protein whose ligands include 6-phosphorylated mannose and the 3-sulfated saccharides galactose, *N*-

acetylgalactosamine and glucuronic acid. *EMBO J* **1992**, *11*, 813–819.

43 Wheeler, P. R., Raynes, J. G., O'Sullivan, G. M., Duggan, D. and McAdam, K. P. W. J. Sulfatide-binding properties are shared by serum amyloid P component and a polyreactive germ-line IgM autoantibody, the TH3 idiotype. *Clin Exp Immunol* **1988**, *112*, 262–269.

44 Danielsen, B., Sorenen, I. J., Nybo, M., et al. Calcium-dependent and -independent binding of the pentraxin serum amyloid P component to glycosaminoglycans and amyloid proteins: enhanced binding at slightly acid pH. *Biochim Biophys Acta* **1997**, *1339*, 73–78.

45 Hind, C. R. K., Collins, P. M., Renn, D., et al. Binding specificity of serum amyloid P component for the pyruvate acetal of galactose. *J Exp Med* **1984**, *159*, 1058–1069.

46 Hamazaki, H. Purification and characterization of a human lectin specific for penultimate galactose residues. *J Biol Chem* **1986**, *261*, 5455–5459.

47 Summerfield, J. A. and Taylor, M. E. Mannose-binding proteins in human serum: identification of mannose-specific immunoglobulins and a calcium-dependent lectin, of broader carbohydrate specificity, secreted by hepatocytes. *Biochim Biophys Acta* **1986**, *883*, 197–206.

48 Hind, C. R. K., Collins, P. M., Baltz, L. and Pepys, M. B. Human serum amyloid P component, a circulating lectin with specificity for the cyclic 4,6-pyruvate acetal of galactose. Interactions with various bacteria. *Biochem J* **1985**, *224*, 107–111.

49 De Haas, C. J. C., Van Leeuwen, E. M. M., Van Bommel, T., et al. Serum amyloid P component bound to gram-negative bacteria prevents lipopolysaccharide-mediated classical pathway complement activation. *Infect Immun* **2000**, *68*, 1753–1759.

50 Noursadeghi, M., Bickerstaff, M. C. M., Gallimore, J. R., Herbert, J., Cohen, J. and Pepys, M. B. Role of serum amyloid P component in bacterial infection: protection of the host or protection of the pathogen. *Proc Natl Acad Sci USA* **2000**, *97*, 14584–14589.

51 De Haas, C. J. C., Van der Tol, M. E., Van Kessel, K. P. M., Verhoef, J. and Van Strijp, J. A. G. A synthetic lipopolysaccharide-binding peptide based on amino acids 27–39 of serum amyloid P component inhibits lipopolysaccharide-induced responses in human blood. *J Immunol* **1998**, *161*, 3607–3615.

52 Kimura, T., Tani, S., Matsumoto, Y. and Takeda, T. Serum amyloid P component is the Shiga toxin-2 neutralising factor in human blood. *J Biol Chem* **2001**, *276*, 41576–41579.

53 Kimura, T., Tani, S., Motoki, M. and Matsumoto, Y. Role of sugar toxin 2 (Stx2) binding protein, human serum amyloid P component (HuSAP), in Shiga toxin-producing *Escherichia coli* infections: assumption from *in vitro* and *in vivo* study using HuSAP and anti-Stx2 humanised monoclonal antibody TMA-15. *Biochem Biophys Res Commun* **2003**, *305*, 1057–1060.

54 Andersen, O., Vilsgaard, R. K., Sorensen, J., Jonson, G. and Holm-Nielsen, E. Serum amyloid P component binds to influenza A virus haemagglutinin and inhibits the virus infection *in vitro*. *Scand J Immunol* **1997**, *46*, 331–337.

55 Horvath, A., Andesen, I., Junker, K., et al. Serum amyloid P component inhibits influenza A virus infections: *in vitro* and *in vivo* studies. *Antiviral Res* **2001**, *52*, 43–53.

56 Kilpatrick, D. C. Mechanisms and assessment of lectin-mediated mitogenesis. *Mol Biotechnol* **1999**, *11*, 55–65.

57 Siripont, J., Tebo, J. M. and Mortensen, R. F. Receptor-mediated binding of the acute-phase reactant mouse serum amyloid P component (SAP) to macrophages. *Cell Immunol* **1988**, *117*, 239–252.

58 Landsmann, P., Rosen, O., Pontet, M., et al. Binding of human serum amyloid P component (h. SAP) to human neutrophils. *Eur J Biochem* **1994**, *223*, 805–811.

59 Singh, S. and Singh, P. P. Serum amyloid P component-induced colony-stimulating factors. Production by macrophages. *Scand J Immunol* **2001**, *53*, 155–161.

60 Pilling, D., Buckley, C. D., Salmon, M. and Gomer, R. H. Inhibition of fibro-

cyte differentiation by serum amyloid P. *J Immunol* **2003**, *171*, 5537–5546.

61 Hicks, P., Saunero-Nava, L., Du Clos, T.W. and Mold, C. Serum amyloid P component binds to histones and activates the classical complement pathway. *J Immunol* **1992**, *149*, 3689–3694.

62 Ying, S.-C., Gewurz, A.T., Jiang, H. and Gewurz, H. Human serum amyloid P component oligomers bind and activate the classical complement pathway via residues 14–26 and 76–92 of the A chain collagen-like region of C1q. *J Immunol* **1993**, *150*, 169–176.

63 Butler, P.J.G., Tennant, G.A. and Pepys, M.B. Pentraxin–chromatin interactions: serum amyloid P component specifically displaces H1-type histones and solubilizes native long chromatin. *J Exp Med* **1990**, *172*, 13–18.

64 Bristow, C.L. and Boackle, R.J. Evidence for the binding of human serum amyloid P component to C1q and Faby. *Mol Immunol* **1986**, *23*, 1045–1052.

65 Hutchcraft, C.L., Gewurz, H., Hansen, B., Dyck, R.F. and Pepys, M.B. Agglutination of complement-coated erythrocytes by serum amyloid P component. *J Immunol* **1981**, *126*, 1217–1219.

66 Barbashov, S.F., Wang, C. and Nicholson-Weller, A. Serum amyloid P component forms a stable complex with human C5b6. *J Immunol* **1997**, *158*, 3830–3835.

67 Brown, M.R. and Anderson, B.E. Receptor–ligand interactions between serum amyloid P component and model soluble immune complexes. *J Immunol* **1993**, *151*, 2087–2095.

68 Mold, C., Baca, R. and Du Clos, T.W. Serum amyloid P component and C-reactive protein opsonise apoptotic cells for phagocytosis through Fc gamma receptors. *J Immunol* **2002**, *19*, 147–154.

69 Pepys, M.B. and Butler, P.J.G. Serum amyloid P component is the major calcium-dependent specific DNA-binding protein of the serum. *Biochem Biophys Res Commun* **1987**, *148*, 308–313.

70 Breathnach, S.M., Kofler, H., Sepp, N., et al. Serum amyloid P component binds to cell nuclei *in vitro* and to *in vivo* deposits of extracellular chromatin in systemic lupus erythematosus. *J Exp Med* **1989**, *170*, 1433–1438.

71 Pepys, M.B., Booth, S.E., Tennent, G.A., Butler, P.J.G. and Williams, D.G. Binding of pentraxins to different nuclear structures: C-reactive protein binds to small nuclear ribonucleoprotein particles, serum amyloid P component binds to chromatin and nucleoli. *Clin Exp Immunol* **1994**, *97*, 152–157.

72 Zahedi, K. Characterization of the binding of serum amyloid P to type IV collagen. *J Biol Chem* **1996**, *271*, 14897–14902.

73 Vorum, H., Jacobsen, C. and Honore, B. Calumenin interacts with serum amyloid P component. *FEBS Lett* **2000**, *465*, 129–134.

74 Li, X.A., Yutani, C. and Shimokado, K. Serum amyloid P component associates with high-density lipoprotein as well as very low density lipoprotein but not with low-density lipoprotein. *Biochem Biophys Res Commun* **1998**, *244*, 249–252.

75 Bijl, M., Horst, G., Bijzet, J., Bootsma, H., Limburg, P.C. and Kallenberg C.G. Serum amyloid P component binds to late apoptotic cells and mediates their uptake by monocyte derived macrophages. *Arthritis Rheum* **2003**, *48*, 248–254.

76 Familian, A., Van Mierlo, G.J. and Hack, C.E. Evidence for complement activation by serum amyloid P component (SAP) *in vivo*. Presented at the *Vth International Workshop on C1, the First Component of Complement and Collectins*, Seeheim, Germany, **2001**, abstr IV-2.

77 Herbert, J., Hutchinson, W.L., Carr, J., et al. Influenza virus infection is not affected by serum amyloid P component. *Mol Med* **2002**, *8*, 9–15.

78 Sorensen, I.J., Nielsen, E.H., Schroder, L., Voss, A., Horvath, L. and Svehag, S.-E. Complexes of serum amyloid P component and DNA in serum from healthy individuals and systemic lupus erythematosus patients. *J Clin Immunol* **2000**, *20*, 408–415.

79 Casciola-Rosen, L.A., Anhalt, G. and Rosen, A. Autoantigens targeted in systemic lupus erythematosus are clustered in two populations of surface structures

on apoptotic keratinocytes. *J Exp Med* **1994**, *179*, 1317–1330.

80 Hutchinson, W. L., Noble, G. E., Hawkins, P. N. and Pepys, M. B. The pentraxins, C-reactive protein and serum amyloid P component, are cleared and catabolized by hepatocytes *in vivo*. *J Clin Invest* **1994**, *94*, 1390–1396.

81 Gauthier, V. J., Tyler, L. N. and Mannik, M. Blood clearance kinetics and liver uptake of mononucleosomes in mice. *J Immunol* **1996**, *156*, 1151–1156.

82 Paul, E. and Carroll, M. C. SAP-less chromatin triggers systemic lupus erythematosus. *Nat Med* **1999**, *5*, 607–608.

83 Garcia de Frutos, P. and Dahlback, B. Interaction between serum amyloid P component and C4 b-binding protein associated with inhibition of factor I-mediated C4b degradation. *J Immunol* **1994**, *152*, 2430–2437.

84 Nauta, A. J., Daha, M. R., Van Kooten, C. and Roos, A. Recognition and clearance of apoptotic cells: a role for complement and pentraxins. *Trends Immunol* **2003**, *24*, 148–154.

85 Roos, A., Xu, W., Castellano, G., Nauta, A. J., Garred, P., Daha, M. R. and Van Kooten, C. A pivotal role for innate immunity in the clearance of apoptotic cells. *Eur J Immunol* **2004**, *34*, 921–929.

86 Gaipl, U. S., Kuenkele, S., Voll, R. E., et al. Complement binding is an early feature of necrotic and a rather late event during apoptotic cell death. *Cell Death Differ* **2001**, *8*, 327–334.

87 Du Clos, T. W. The interaction of C-reactive protein and serum amyloid P component with nuclear antigens. *Mol Biol Rep* **1996**, *23*, 253–260.

88 Soma, M., Tamaoki, T., Kawano, H., et al. Mice lacking serum amyloid P component do not necessarily develop severe auto-immune disease. *Biochem Biophys Res Commun* **2001**, *286*, 200–205.

89 Gillmore, J. D., Hutchinson, W. L., Herbert, J., et al. Autoimmunity and glomerulonephritis in mice with targeted deletion of the serum amyloid P component gene: SAP deficiency or strain combination? *Immunology* **2004**, *112*, 255–264.

90 Hassan, A. B., Ronnelid, J., Gunnarsson, I., Karlsson, G., Berg, L. and Lundberg, I. Increased serum levels of immunoglobulins, C-reactive protein, type 1 and type 2 cytokines in patients with mixed connective tissue disease. *J Autoimmunity* **1998**, *11*, 503–508.

91 Bowness, P., Davies, K. A., Norsworthy, P. J., et al. Hereditary C1q deficiency and systemic lupus erythematosus. *Q J Med* **1994**, *87*, 455–464.

92 Sjowall, C., Eriksson, P., Almer, S. and Skogh, T. Autoantibodies to C-reactive protein is a common finding in SLE, but not in primary Sjogren's syndrome, rheumatoid arthritis or inflammatory bowel disease. *J Autoimmunity* **2002**, *19*, 155–160.

93 Hawkins, P. N., Lavender, J. P. and Pepys, M. B. Evaluation of systemic amyloidosis by scintigraphy with ^{123}I-labeled serum amyloid P component. *New Eng J Med* **1990**, *323*, 508–513.

94 Hawkins, P. N., Aprile, C., Capri, G., et al. Scintigraphic imaging and turnover studies with iodine-131 labeled serum amyloid P component in systemic amyloidosis. *Eur J Nucl Med* **1998**, *25*, 701–708.

95 Nelson, S. R. and Hawkins, P. N. Imaging of haemodialysis-associated amyloidosis with ^{123}I-serum amyloid P component. *Lancet* **1991**, *338*, 335–339.

96 Rydh, A., Suhr, O., Hietala, S.-O., et al. Serum amyloid P component scintigraphy in familial amyloid polyneuropathy: regression of visceral amyloid following liver transplantation. *Eur J Nucl Med* **1998**, *25*, 709–713.

97 Jager, P. L., Hazenberg, B. P., Franssen, E. J., Limburg, P. C., Van Rijswijk, M. H. and Piers, D. A. Kinetic studies with iodine-123-labeled serum amyloid P component in patients with systemic AA and AL amyloidosis and assessment of clinical value. *J Nucl Med* **1998**, *39*, 699–706.

98 Tan, S. Y., Baillod, R., Brown, E., et al. Clinical, radiological and serum amyloid P component scintigraphic features of β_2-microglobulin amyloidosis associated with continuous ambulatory peritoneal dialysis. *Nephrol Dial Transplant* **1999**, *14*, 1467–1471.

99 Gillmore, J. D., Madhoo, S., Pepys, M. B. and Hawkins, P. N. Renal transplantation for amyloid end-stage renal failure – insights from serial serum amyloid P component scintigraphy. *Nucl Med Commun* **2000**, *21*, 735–740.

100 Hawkins, P. N. Serum amyloid P component scintigraphy for diagnosis and monitoring amyloidosis. *Curr Opin Nephrol Hyperten* **2002**, *11*, 649–655.

101 Hirschfield, G. M. Amyloidosis: a clinico-pathophysiological synopsis. *Semin Cell Dev Biol* **2004**, *15*, 39–44.

102 Tennant, G. A., Lovat, L. B. and Pepys, M. B. Serum amyloid P component prevents proteolysis of the amyloid fibrils of Alzheimer's disease and systemic amyloidosis. *Proc Natl Acad Sci USA* **1995**, *92*, 4299–4303.

103 Botto, M., Hawkins, P. N., Bickerstaff, M. C. M., et al. Amyloid deposition is delayed in mice with targeted deletion of the serum amyloid P component gene. *Nat Med* **1997**, *3*, 855–859.

104 Togashi, S., Lim, S.-K., Kawano, H., et al. Serum amyloid P component enhances induction of murine amyloidosis. *Lab Invest* **1997**, *77*, 525–531.

105 Maeda, S. Use of genetically altered mice to study the role of serum amyloid P component in amyloid deposition. *Amyloid: J Protein Folding Disord* **2003**, *10 (Suppl 1)*, 17–20.

106 Lachmann, H. J. and Hawkins, P. N. Novel pharmacological strategies in amyloidosis. *Nephron Clin Pract* **2003**, *94*, c85–c88.

107 Pepys, M. B., Herbert, J., Hutchinson, W. L., et al. Targeted pharmacological depletion of serum amyloid P component for treatment of human amyloidosis. *Nature* **2002**, *417*, 254–259.

108 Iversen, L. Small drugs lead the attack. *Nature* **2002**, *417*, 231–232.

109 Gertz, M. A., Lacy, M. Q. and Dispenzieri, A. Therapy for immunoglobulin light chain amyloidosis: the new and the old. *Blood Rev* **2004**, *18*, 17–37.

110 Hawkins, P. N., Tennant, G. A., Woo, P. and Pepys, M. B. Studies *in vivo* and *in vitro* of serum amyloid P component in normals and in a patient with AA amyloidosis. *Clin Exp Immunol* **1991**, *84*, 308–316.

9
Serum amyloid P Component – Structural Features and Amyloid Recognition

S. P. Wood and A. R. Coker

9.1
Introduction

The characteristic amyloid fibril-forming polypeptides and the amyloid P component (AP) are abundant constituents of amyloid deposits, and a growing body of evidence suggests that their interaction is an important feature of amyloid accumulation. When the mouse gene for serum amyloid P component (SAP), the circulating equivalent of AP, is inactivated or knocked-out, then the time course for deposition of amyloid after experimental induction is considerably retarded [1]. This strongly suggests that SAP is centrally involved in some way in fibrillogenesis and/or in stabilizing deposits against clearance. A recently discovered drug, CPHPC, induces rapid clearance of circulating SAP in transgenic mice expressing the human protein, and leads to a reduction in the amount of amyloid associated SAP and to a reduction in the amyloid load in animals where systemic amyloidosis has been induced [2]. These results suggest that SAP contributes to the stability and persistence of fibrils *in vivo.* Amyloid fibril formation and the calcium-dependent binding of SAP to the fibrils can be reproduced *in vitro,* and, in these *in vitro* conditions, it was found that bound SAP protects the fibrils against proteolysis and attack by phagocytic cells, reinforcing the stabilization hypothesis [3]. Physical/chemical stabilization of amyloid fibrils might also result from SAP binding. Amyloid formation *in vitro* is often driven by harsh protein destabilizing conditions and the process appears to be essentially irreversible. However, regression of deposits *in vivo* following depletion of circulating fibril precursor pools by immunization [4] or transplantation [5] suggests that they may be more dynamic structures. SAP binding might contribute to "locking in" materials to the deposit. However, the administration of radiolabeled SAP and whole-body scintigraphy, an established and effective procedure for locating amyloid laden organs, depends upon exchange of fibril-localized SAP with the circulating tracer pool and is further evidence of a dynamic aspect of amyloid deposits [6].

Although the interaction of SAP with amyloid fibrils appears to be a fundamental feature of amyloid disease pathology, the whole story of amyloid struc-

Amyloid Proteins. The Beta Sheet Conformation and Disease. J. D. Sipe

ISBN: 3-527-31072-X

ture is likely to be far more complex. A host of other components are found in association with amyloid deposits including the complement protein C1q, α_1-antichymotrypsin, apolipoprotein E, ubiquitin and glycosaminoglycans [7]. Their role is far from clear. There is evidence that SAP interacts with some of these constituents in other contexts, such as complement activation and basement membrane structure [8]. Some of the components may well be present in response to the local inflammatory cellular crisis at sites of amyloid deposition. However, the ability to reproduce amyloid recognition by SAP *in vitro* from pure components suggests that investigation of the interaction at a structural level is both a tractable and valid target. Here, we review what is known of the structure and properties of SAP, and how this is related to current views on the structure of amyloid and its formation, topics covered in greater depth elsewhere in this volume.

9.2
Amyloid Fibrils and their Formation

For many years, our understanding of fibril structure was limited to the interpretation of X-ray fibre diffraction patterns that reported the presence of cross-β structure. Technical advances in the use of synchrotron radiation for these investigations, the development of cryopreservation techniques and image analysis for electron microscopy and the use of multidimensional nuclear magnetic resonance (NMR) and more recently solid-state NMR are providing a clearer view of amyloid structure and potential motifs for SAP binding [9–13].

Amyloid fibrils are believed to be formed by a concentration dependent, nucleated growth pattern analogous to crystal growth [9]. Structurally perturbed precursors form stable intermolecular hydrogen bonding networks, perhaps involving domain swapping in some cases [14], that, together with any retained β-structure of the precursor, form the extended cross-β structure of the fibril. Strands are believed to run approximately normal to the fibril axis. This regular cross-β structure gives rise to characteristic reflections in the X-ray fibril diffraction pattern [15]. Fibril growth is probably far more complex than sequential monomer addition, as electron microscopy and atomic force microscopy have shown the presence of micelle-like components and beaded protofilaments during growth [16, 17]. Mature fibrils are often assembled as "rope-like" structures of entwined filaments of variable helical pitch and a hollow core [18] (Fig. 9.1a). Fibrils can be very long, but are usually in the range of 100–150 Å in diameter. The structural destabilization of precursors required to drive fibril growth may not always be extensive. It is clear that a number of mutated proteins associated with aggressive amyloid deposition exhibit reduced thermal stability [19, 20], and experimental conditions that promote fibril growth *in vitro* lead to partial denaturation and specific strand displacements from native folds. However, models for transthyretin amyloid (ATTR) that fit with much experimental evidence incorporate a good deal of the native β structure [21–24]. Crystal struc-

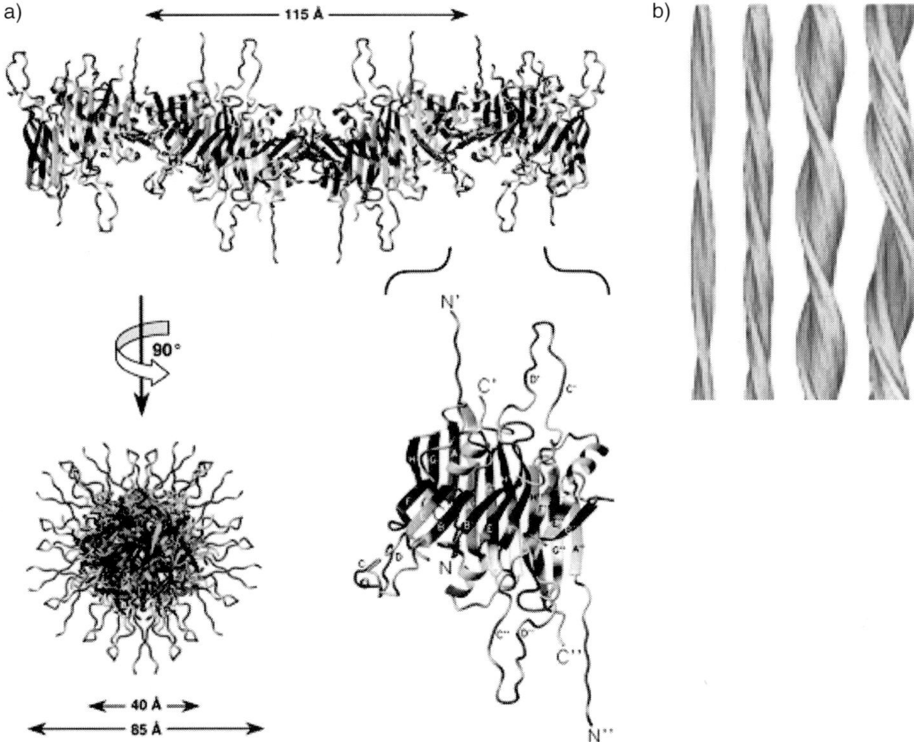

Fig. 9.1 (a) A model for ATTR derived from deuterium-exchange protection NMR and electron spin resonance data. Displacement of strands C and D from the native fold creates newly exposed A and B edge strands that together with a slightly modified form of the native intersubunit sheet provides a continuum of β-strands. Disordered loops and chain termini protrude from the fibril surface. (Reproduced from Oloffson et al. [27], with permission.) (b) Image processed electron microscopy images of insulin amyloid fibrils showing various degrees of intertwining of protofibrils. (Reproduced from Saibil et al. [10], with permission.)

tures of an amyloidogenic TTR mutant and of a κ-light chain variable region show extended β-helix arrangements, generated by the crystal symmetry, that emulate many of the features expected for amyloid fibrils [25, 26]. Recent solution NMR and mass spectrometry studies for TTR, Aβ and β_2-microglobulin (β_2M), using deuterium-exchange protection methods, suggest that considerable portions of the polypeptide chain, including chain termini and displaced loops, are probably disordered at the fibril surface providing an interaction surface distinct from that of the native proteins (Fig. 9.1 b) [27–29]. Monoclonal antibodies recognizing cryptic epitopes of TTR that are displayed by amyloid fibrils and by hyper-amyloidogenic mutants in solution, but not the native protein, reaffirm this view [30]. Solid-state NMR studies of Aβ1–40 and fibril proteolysis results suggest that the polar N-terminal region and turns are exposed at the surface of the fibril [31, 32] and that the N-terminal region is a primary epitope recognized by antibodies to Aβ fibrils [33]. The hydrogen bonding potential of the β-strands is likely to be fully satisfied by the cross-β structure, except at the fibril ends, and hydrophobic residues are probably buried, if indeed the fibril state represents a low-energy "primordial fold" [34]. Although many proteins, including those with no association with amyloid disease, can be driven to the fibril state by harsh destabilizing conditions, in general, protein structures have probably evolved with variable degrees of success to avoid exposure of sheet edge strands capable of repeating intermolecular interaction in physiological conditions [35].

On the order of 20–30 distinct proteins have been reported to produce amyloid deposits *in vivo*. Therefore, the universal calcium dependent binding of SAP to these fibrils is unlikely to be highly sequence specific, in view of the variety of the fibrillogenic proteins. Evidence for the existence of a generic amyloid motif, however, is provided by the identification of an antibody that recognizes amyloid derived from different source proteins [36]. In both cases, it is possible that there exists a conformation-dependent discontinuous epitope that obscures identification of important conserved patterns of residues. If the emerging structural models of amyloid fibrils formed by TTR, β_2M, Aβ1–40 are general in character, then we might expect the surface of the amyloid fibril to display turns between strands, displaced loops and flexible chain termini. These features are likely to be distributed in accord with the twist of the sheets and the inter-twining of the fibrils to produce a fuzz of polypeptide at the fibril surface within which motifs for SAP recognition must reside.

9.3
The Structure of SAP

The crystal structure of human SAP was first reported in 1994 and was determined by the classical method of isomorphous replacement [37]. Crystallization of the protein had been problematic due to its propensity to aggregate in the presence of calcium ions, a phenomenon that has continued to complicate many aspects of research into the behavior of SAP in solution [38].

SAP is a normal, relatively abundant (around 30 mg/l) plasma glycoprotein that is readily purified by virtue of its calcium-dependent affinity for immobilized phosphoethanolamine. Subsequent experimental manipulations are traditionally carried out in neutral buffers containing EDTA, where the protein exhibits the behavior of a highly soluble decamer of M_r 23,500 subunits. Each subunit comprises 204 amino acids and carries a single Asn32-linked biantennary oligosaccharide chain [39]. The isoelectric point is 5.5, giving the protein a net negative charge at physiological pH; however, this charge is not evenly distributed. Crystallization was carried out in the presence of high concentrations of calcium and acetate ions and at a pH close to the pI.

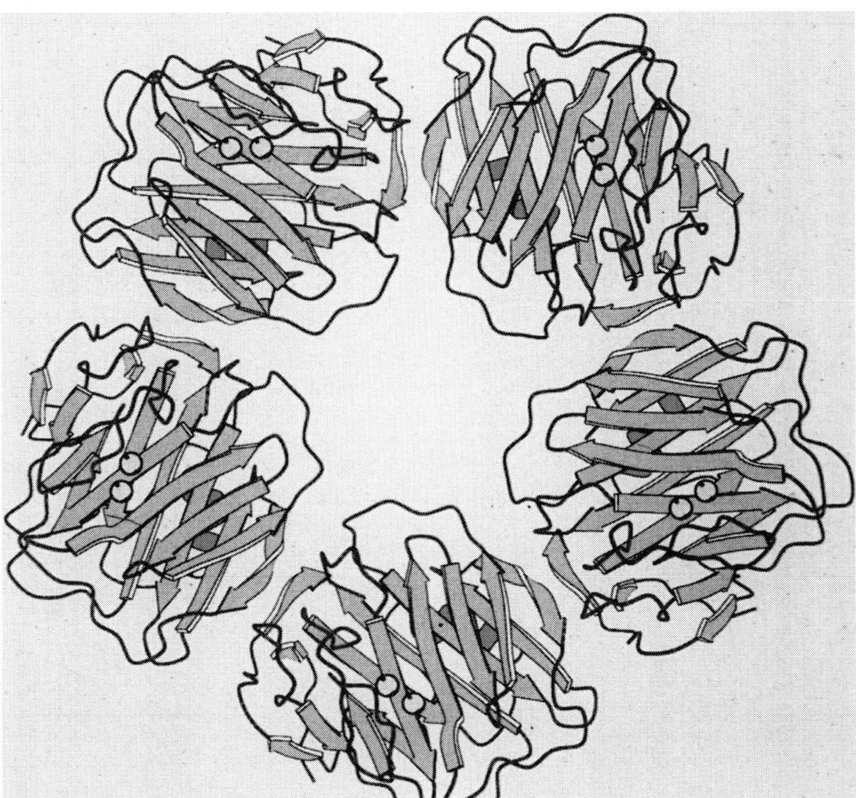

Fig. 9.2 The SAP pentamer viewed down the 5-fold symmetry axis showing two calcium ions (yellow) bound to each subunit.

The crystallographic structure revealed molecules with cyclic pentameric symmetry packed in a herring-bone fashion and containing only five subunits. These disc-like pentameric molecules are approximately 100 Å in diameter and 35 Å deep, and show a large 20 Å diameter hole through the center, in accord with previous negative stained electron microscopy (Fig. 9.2) [40]. The polypeptide chain in each subunit is organized as two layers of antiparallel β-strands, sandwiching hydrophobic residues of the core between them (Fig. 9.3). The first and last four strands sequentially visit the two layers in a "jelly-roll" topology, while the intervening six strands define two β-meanders of three strands at the same end of each layer. This architecture very closely resembles that of the plant lectins, concanavalin A and pea lectin [41, 42]. One of the sheets is flat and has a 10-residue α-helix positioned upon it. The other sheet is rather twisted and loops dipping into the concave surface contribute to the double calcium-binding site of each subunit. The loops connecting strands are generally short and this likely contributes to the remarkable stability of calcium bound SAP in the face of proteolytic digestion [43]. The amino acid sequence of SAP

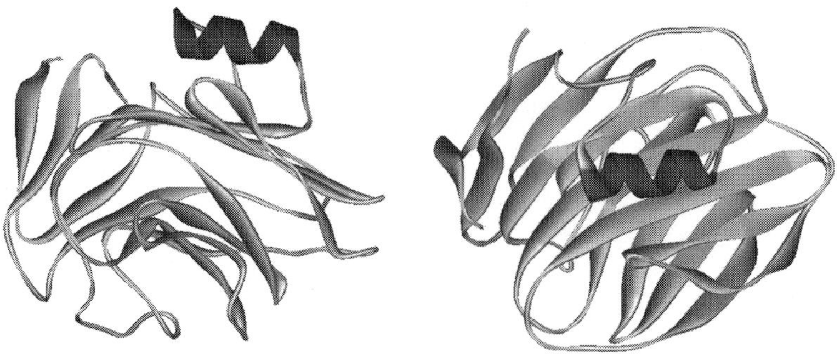

Fig. 9.3 Ribbon diagrams of orthogonal views of the SAP subunit showing the location of the A-face helix and the concavity of the B-face sheet.

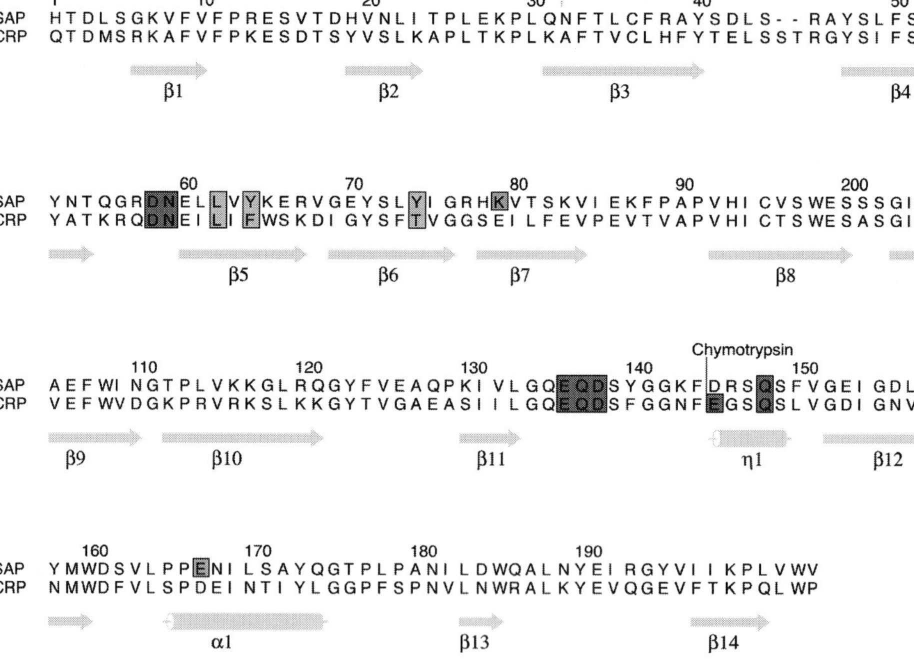

Fig. 9.4 Amino acid sequences for SAP and CRP with residues involved in calcium binding in red, residues of the hydrophobic pocket bound by MO*β*DG and CPHPC in blue. The oligosaccharide attachment site (CHO) and site of chymotrypsin cleavage are indicated, as is the distribution of secondary structure elements.

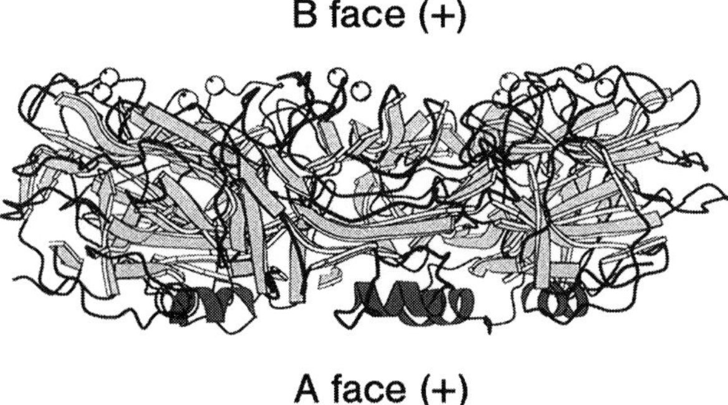

Fig. 9.5 Side view of the SAP pentamer showing the 10 bound calcium ions of the B face and five helices of the A face.

is provided in Fig. 9.4 in comparison with that of C-reactive protein (CRP), annotated with secondary structure features and with important residues in color.

All of the strands of each subunit approximately populate a plane normal to the 5-fold axis of the pentamer. The subunits interact via van der Waals contacts and hydrogen bonds through the open-core end of the β-sheet sandwich of one subunit and the N- and C-terminal strands of the adjacent subunit, to result in

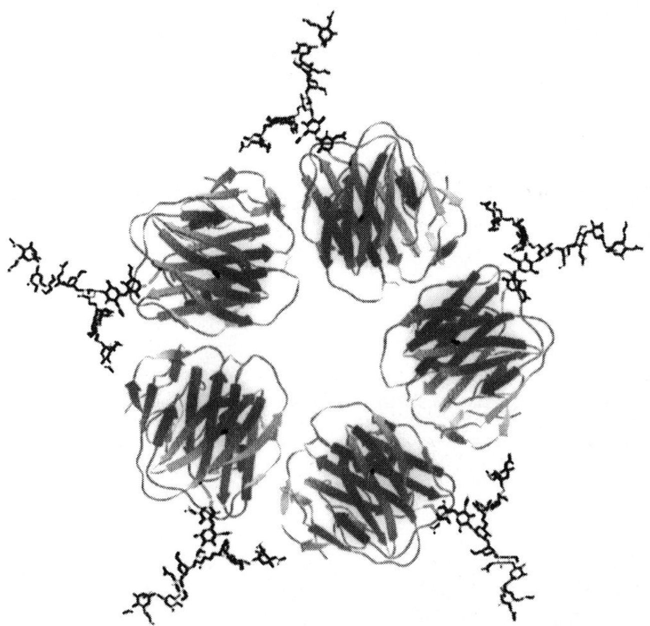

Fig. 9.6 SAP pentamer with fully extended oligosaccharide chains attached to each subunit, extending the effective radius of the molecule by around 40%.

the burying of 20% of the protomer surface on pentamer formation. The pentamer has a polar character, with five α-helices on one face (the A face) and five double calcium sites on the other (B face) (Fig. 9.5). Only the first ring of the oligosaccharide chain attached to Asn32 was visible in this and several subsequent crystal structure analysis. However, more recently, we have experimentally observed almost the entire length of a single sugar chain for one subunit (unpublished data). In this case the sugar chain was immobilized by crystal packing. If the oligosaccharide chains are fully extended, as suggested by solution X-ray and neutron scattering [44], then their area of influence around the pentamer is considerable, extending the effective molecular diameter from 100 to around 135 Å (Fig. 9.6).

9.4 The Calcium-binding Site

Each SAP subunit in the crystal structure contains two calcium ions bound very close together (4.2 Å) in the concavity formed by the buckled B-face sheet. This "tense" electrostatic arrangement is stabilized by a selection of carboxylate and amide protein side-chain ligands to the metals (Fig. 9.7). Unlike many other calcium-binding proteins, these side-chains come from different parts of the sequence and so it seems likely that the bound metals are important for stabilizing the structure in this region. For instance, Glu136 and Asp138 are positioned on an extended loop that dips into the site, with both side-chains forming bridging interactions with both calcium ions. The two metal ions have different total numbers of stabilizing ligands and are likely to be bound to the protein with

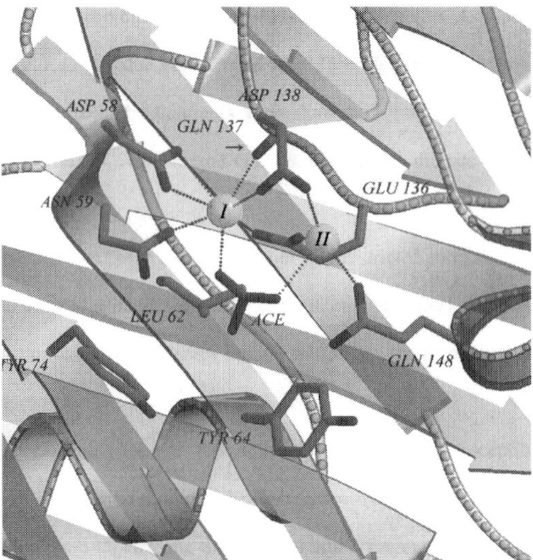

Fig. 9.7 The double calcium ion-binding site of SAP, showing the uneven distribution of amino acid side-chain ligands to the metals and a bound acetate ion.

quite different affinities. Furthermore, the ligand complement to the metals in this crystal form also included a bound acetate ion in four of the five subunits and an intermolecular interaction involving Glu167 from the prominent A-face helix of an adjacent pentamer. Acetate ions and other small molecules that bind this site inhibit the aggregation of SAP by blocking intermolecular interactions. There are many acidic groups exposed on the surface of the pentamer that could contribute to intermolecular contacts, but the Glu167 side-chain is of special interest as five copies project from the A face and could simultaneously interact with the five double calcium sites of another pentamer stacked on the same 5-fold symmetry axis. Such stacks would also be expected to be rather stable because the net charges on the contact surfaces are opposite. Limited stacks have been observed on the surface of *ex vivo* amyloid fibrils by electron microscopy [45], but the most impressive long stacks are reported in the presence of EDTA where, in the absence of calcium, the above stacking mechanism would not be expected to occur [46]. The presence of long strings of SAP pentamers in calcium-free solution seems unlikely as the protein is very soluble, monodisperse and generally well behaved in these conditions. It is possible that stack formation in the absence of calcium is the result of negative stain metal binding during sample preparation. Stack formation is a potentially important property of SAP as it might contribute to SAP accumulation at amyloid sites. Certainly the related troublesome solubility properties of SAP have provided major technical complications for studies of SAP binding to putative target macromolecules.

Calcium site I of each SAP subunit is coordinated to the side-chains of Asp58, Asn59, Glu136, Asp138 and the main chain carbonyl of Gln137. The acetate ion provides the seventh ligand to produce a pentagonal bipyramidal arrangement. Calcium site II is more open and has fewer protein ligands deriving from Glu136, Asp138 and Gln148 as well as the acetate ion and two water molecules. The site was found to release calcium easily, by washing crystals in calcium-free buffers, and indeed preferentially bound the much larger cerium ion. It is conceivable that metals larger than calcium, such as copper or zinc, might be preferred in site II in spite of their much lower availability in biological fluids. The integrity of the site probably also depends upon the presence of a bound ligand analogous to the acetate ion seen in the crystal, but no such ligand has been identified in SAP preparations. In the absence of bound metal ions SAP is readily cleaved by chymotrypsin at the 144–145 bond, suggesting that the loop carrying these residues, which is an integral part of the metal binding site, is much more accessible. The degree of protection conferred by calcium is concentration dependent [43]. Protection is complete at 5 mM and absent at 0.1 mM calcium. Protection is not entirely complete at 2 mM calcium and 35 mg/l SAP, conditions close to those expected *in vivo*.

9.5
Comparative studies of CRP

CRP is a close sequence and structural homologue of SAP and investigations of its structure and properties provide some intriguing comparisons with SAP, since CRP is not bound by amyloid fibrils. A number of crystal structures of CRP have been determined in the presence and absence of metal ions and ligands [47–50]. These show that the subunit fold of CRP is very similar to SAP. Two closely bound calcium ions reside on each subunit of a pentamer. The distribution of protein ligands to the two calcium ions is much more even in CRP and the metals are held with equal affinity [51]. In structures where no calcium ions are bound, the 140–150 loop region is no longer well ordered, and, in solution, the 145–146 and 146–147 peptide bonds are readily cleaved by proteolytic enzymes. Although no structure for metal-free SAP has been determined, it seems highly probable that an equivalent disorganization of the metal binding region occurs.

The organization of the pentamer of CRP is somewhat different from the SAP pentamer. Each subunit is tilted by 22° about an axis through its centre of mass towards the 5-fold axis of the pentamer such that the A-face helices are 5 Å closer to the axis while the five double calcium sites are displaced by an equivalent amount away from the axis. This reorganization may explain why CRP is not as susceptible to calcium-dependent aggregation and to stacking interactions as SAP, because the putative interacting groups are displaced to new pentamer radii differing by 10 Å. Perhaps it also contributes to the inability of CRP to bind motifs with a particular geometry at the surface of amyloid fibrils that are recognized by SAP. When an equivalent subunit rotation is modeled onto the SAP structure, surprisingly few steric clashes occur, suggesting that such rotations could be a dynamic feature of pentraxin structures.

The crystal structure of CRP in the complete absence of calcium ions shows two pentamers packed A face to A face on a common 5-fold axis, bringing the helices on each subunit close together. Although there is no evidence for the formation of CRP decamers in solution, this does illustrate a preferred packing arrangement for pentamers of CRP, albeit driven by the supersaturation of the crystallization cocktail.

9.6
SAP Structure in the Absence of Calcium

In all of the biological contexts in which SAP structure informs function, it is likely that metal ions are present and, thus, metal binding is only likely to be disrupted by proteolytic clipping of the 140–150 loop. In this respect, the decameric structure adopted by SAP in metal-free conditions appears to be of limited interest. However, in many investigations of the possible role of SAP in influencing fibril formation, sugar binding or protein folding (that we will visit later), metal-free conditions have often been employed. The value of observations made in these conditions is that they report on a functionality of the protein that persists when the metal site is severely disrupted. This might be attributed to a completely different part of the protein or to generation of a novel functionality for the disrupted region.

The decameric form of SAP that exists in solution in the absence of bound metals very likely involves a face-to-face interaction. The susceptibility of these decamers to proteolysis in the calcium site loop suggests that these loops are exposed and that the pentamers are interacting via their A faces, as they do in crystals of metal-free CRP. Binding of calcium however generates SAP pentamers. It is not clear how the ordering of the metal site would destabilize an A–A-face interaction some 35 Å away unless perhaps subunit rotation equivalent to that of CRP took place. It is conceivable that the SAP decamer involves interactions between pentamer B faces. The loop site of proteolysis is oriented to the outer edge of the pentamer and may not be completely obscured by a B–B-face interaction. Metal binding and ordering of the loop could disrupt a B–B-face decamer. Determination of the structure of metal-free SAP will clearly be informative.

9.7
Binding of Small Molecule Ligands to SAP

The calcium dependent binding of phosphoethanolamine (PE) and the 4,6-cyclic pyruvate acetal of β-D-galactose (MOβDG) to SAP are well-established observations [8]. Immobilized PE, linked to a matrix through the amino group, and MOβDG as a component of agarose are both effective affinity supports binding SAP from solutions containing calcium ions. The binding sites for the isolated ligands were identified by soaking SAP crystals in the compounds or by co-crystallization and the determination of crystal structures. Subsequently, both studies have been extended to higher resolution [52]. Both compounds were found to bind to the double calcium site of SAP via their acidic moieties.

The structure of bound MOβDG was of special interest, as this compound has been shown to inhibit the binding of SAP to amyloid fibrils *in vitro* [53], and, therefore, has potential both as a lead compound for drug design and as a guide to the nature of the amyloid interaction. The carboxylate of MOβDG interacts with the calcium ions of each SAP subunit and the bridge oxygens of the carboxyethylidine ring hydrogen bond with the side-chain amide nitrogens of Asn59 and Gln148 (Fig. 9.8). The side-chain carbonyl oxygens of the same residues also interact with the calcium ions. There are also hydrogen bonds between sugar oxygens of C1 and C3 to Gln148 and Lys79, respectively. The methyl group of the bound *R* isomer of MOβDG locates in a hydrophobic pocket formed by Tyr74, Phe64 and Leu62. In one subunit, crystal packing leaves no room for a molecule of MOβDG to bind and only one calcium ion is bound in site I. This suggests that a bound ligand of some sort is required to stabilize the filling of calcium site II.

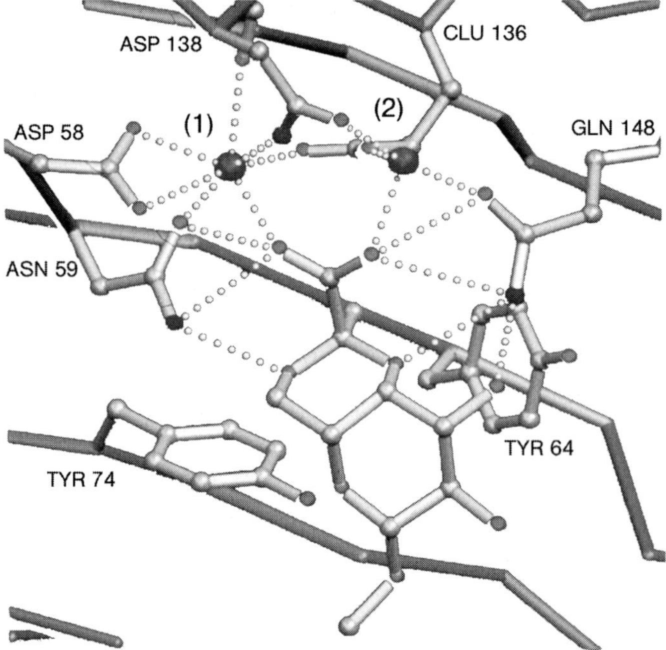

Fig. 9.8 The structure of the galactose analog (M$O\beta$DG) bound into the double calcium site of an SAP subunit, showing direct interaction with the calcium ions and hydrogen bonds to side-chains that also donate ligands to the metals.

Isothermal calorimetry measurements indicate a dissociation constant of 50 μM for the interaction of M$O\beta$DG with SAP [2], considerably (around 10^3-fold) weaker than that with amyloid fibrils. This may be due in large part to the likelihood that multiple attachment points and some associated cooperativity are involved in amyloid recognition. A tight β-turn between polypeptide strands and including aspartic acid at its apex would present a broadly similar array of binding groups, as M$O\beta$DG and such turns can be readily built into the binding site, an observation consistent with the supposition that exposed turns at the fibril surface are involved in SAP binding [8]. However, M$O\beta$DG would also be expected to interfere with the formation of SAP stacks. Inhibition of stack formation could account for much of the observed inhibition of SAP binding to fibrils in the presence of M$O\beta$DG and calcium ions.

The importance of cooperative binding effects was demonstrated particularly well for the first time in studies of the calcium-dependent binding of dAMP to SAP [54]. This ligand was identified as an effective inhibitor of SAP aggregation in the presence of calcium ions. The crystal structure of the SAP–dAMP complex revealed decamers of SAP. Each subunit bound a single dAMP molecule via its phosphate group to the double calcium site. The two pentamers were arranged B face to B face, but the dominant interaction stabilizing the complex was the non-covalent stacking of the adenine rings. This complex was stable during gel filtration even though the extent of the interactions at one double calcium site suggests a low-affinity binding, in the same range as MOβDG.

High-throughput screening approaches have identified a high-affinity ligand for SAP that clearly reinforces the importance of cooperative binding [2]. The screen selected compounds that inhibited binding of SAP to Aβ1–42 fibrils immobilized on a plastic surface. This approach identified CPHPC, a compound comprising two D-proline residues linked via a five-carbon spacer and that binds SAP with nanomolar affinity. The crystal structure of the SAP–CPHPC complex showed a decameric structure in which the two D-proline head groups of five molecules of the drug bound to the double calcium sites of subunits from two pentamers oriented B face to B face on a common 5-fold axis (Fig. 9.9). These decamers are stable in solution. The packing of the proline ring into the Leu62, Tyr74, Phe64 hydrophobic pocket and the interaction of the carboxylate with the calcium ions are the major stabilizing interactions. Binding of D-proline alone to SAP, where cooperative effects are not involved, is about 10^3-fold weaker, although the crystal structure of the *N*-acetyl D-Pro–SAP complex shows an identical set of interactions to the CPHPC head group complex. As noted earlier, the properties of this compound potentially provide an approach to treating amyloidoses. Although of considerable interest in a clinical context, it is not clear that the discovery of this compound directly aids our understanding of amyloid recognition. Multiple peptide chain C-termini at the fibril surface might bind SAP in a similar fashion to CPHPC and the side-chains of certain hydro-

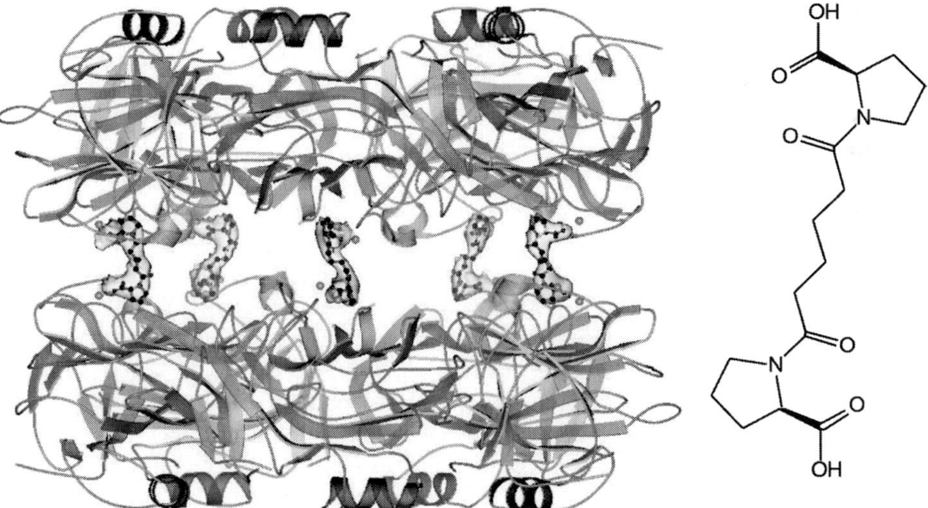

Fig. 9.9 Two pentamers of SAP (β-strands in green and helices in red) crosslinked by five molecules of the drug CPHPC (left).

phobic amino acids might fill the hydrophobic pocket, but they are unlikely to achieve the particularly snug fit observed with D-proline. L-proline binds considerably less tightly.

During studies of the chaperone activity of SAP, we discovered that high concentrations of chloride ions can effectively inhibit the aggregation of SAP in the presence of calcium ions. X-ray structure analysis of crystals grown in these conditions revealed that a single chloride ion can bind close to the calcium ions of each SAP subunit. Although the binding is weak, at sufficiently high concentrations it is capable of blocking the intermolecular interactions that lead to SAP precipitation with calcium ions. This provides a convenient solution condition for comparative investigations of the calcium dependent binding of SAP to other macromolecules (unpublished data).

In overview, structural investigations have defined the calcium-binding site of SAP in considerable detail. The most direct interpretation of the calcium dependence of fibril recognition by SAP suggests that this is the site involved although it is conceivable that the structural integrity of the site is required without direct metal bonding interactions as seen for sugar binding to structurally homologous lectins. The structural features of the binding of small molecule ligands provide clues to the nature of the binding mode of polypeptide ligands likely presented at the fibril surface. It is not clear how the architecture of the amyloid fibril might complement the 5-fold symmetric pattern of binding sites on SAP, although it seems that a multipoint attachment is required to explain the binding affinity observed for such a shallow recognition pocket. It is conceivable that the array of loops, turns and termini at the fibril surface satisfy the multipoint cooperative attachment by virtue of their density rather than their precise organization or sequence.

While we have detailed structural information for SAP and its likely amyloid recognition site, and a much clearer view of the structure of fibrils, difficult issues remain in bringing this information together into a coherent hypothesis for the form of decorated fibrils *in vivo*. The possible motifs for amyloid fibril recognition by SAP discussed above are likely to occur with high frequency due to the tight twisting together of protofilaments in mature fibrils and the relatively large number of repeats of the generally small protein constituents of most amyloid fibrils. A 150-Å diameter fibril could bind three SAP pentamers through a face interaction for each 140 Å of its length to fully coat the surface. However, the molar binding ratio of SAP pentamers to synthetic Aβ1–40 fibrils reaches a plateau at around 1:250, corresponding to around 12% of the mass of the fibril – roughly comparable to that found in *in vivo* amyloid fibrils [3]. Based on calculations of the average mass per unit length of mature synthetic Aβ1–40 fibrils [16], this would only give one SAP molecule every 370 Å along the fibril (Fig. 9.10). This low binding ratio of SAP to Aβ1–40 suggests that the amyloid fibrils in this model would be rather sparsely coated with SAP. For a fully coated surface providing maximum stabilization, the molar ratios would drop to around 1:20. An explanation for this might reside in the work of MacRaild [55]. Simultaneous attachment of SAP to more than one fibril, crosslinking and gen-

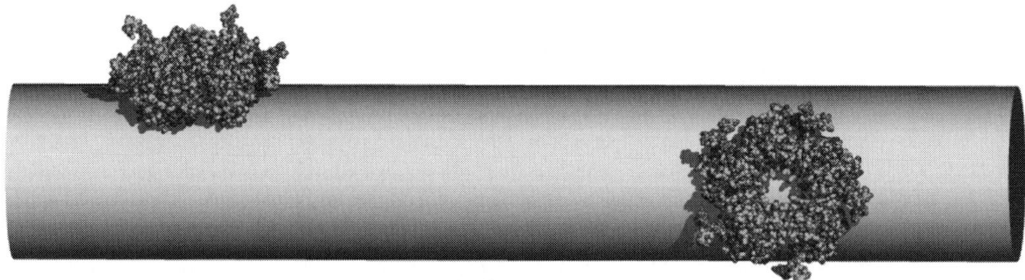

Fig. 9.10 Glycosylated SAP pentamers (around 140 Å diameter) attached through a B-face interaction with a 150-Å diameter fibril of Aβ would be around 370 Å apart if SAP made up around 12% of the mass of the complex.

erating fibril entanglement might limit the accessibility of potential fibril ligands to further molecules of SAP and provide stabilization and resistance towards proteolytic attack.

9.8 The Role of Glycosaminoglycans (GAGs)

Heparan sulfate and chondroitin sulfate proteoglycans (HSPG and CSPG) are well-documented constituents of amyloid deposits. *In vitro* experiments show that SAP interacts in a calcium-dependent fashion with dermatan and heparan sulfate polymers [56]. Kiselevsky et al. [57] have proposed on the basis of *in situ* electron microscopy that this interaction forms the foundation for the structure of amyloid fibrils *in vivo*, and that disruptive extraction procedures may explain discrepancies between this and other work. Their model suggests that stacked SAP molecules are coated by a coil of CSPG on top of which lie HSPG chains and surface-exposed 1-nm protofilaments. During water extraction these filaments are freed from the fibril and assemble into the familiar pattern of twisted filaments observed by others and reproduced in *in vitro* experiments. While all aspects of this model are not in accord with the views of many investigators, it is conceivable that the SAP stacks observed attached to *ex vivo* fibrils is a relic of a GAG bound structure that has survived fibril extraction from tissues. It is less easy to accept the proposed presence of single strand protofilaments enmeshed in the HSPG layer as a dominant form. TTR fibrils removed from aqueous humor with minimal disturbance clearly show the classical cross-β X-ray fibril diffraction pattern [10] and lend strong support to the concept that *in vivo* fibril assembly bears strong similarities to that *in vitro*.

9.9 SAP, Protein Folding and Amyloid Fibril Formation

SAP recognizes amyloid fibrils formed from a diverse array of source proteins and polypeptides, but there is no evidence to suggest that SAP has any preferential binding to these proteins in their native state. SAP has been reported to bind non-fibrillar β_2M, serum amyloid A protein (SAA) and Aβ, but the proteins were adsorbed to a plastic surface where some unfolding might be expected [58, 59]. Experiments designed to monitor the effects of SAP on fibril formation and protein folding are complicated by the calcium-induced aggregation of SAP and the technical difficulties arising from the need to employ harsh conditions to initiate fibril growth for some proteins. Synthetic Aβ has often been employed, as fibril growth proceeds spontaneously in dilute aqueous media. Amyloidogenic polypeptides like Aβ and SAA are likely to be metastable in aqueous solution, as their sequences are adapted to quite different environments. SAA preferentially associates with a high-density lipoprotein particle, while Aβ is a fragment of a larger protein and at least a portion of its sequence has evolved to reside within a lipid bilayer. Both materials have poor solubility and are rather difficult to work with. Precise solvent conditioning protocols are recommended by some investigators [29] to ensure reproducible behavior of synthetic Aβ, but these are not followed by all and this, together with the solubility difficulties of SAP, may go some way towards explaining diverging views on the effects of SAP on Aβ fibrillogenesis.

SAP has been reported to both enhance and to inhibit fibril formation by Aβ in different experimental conditions. Hamazaki [60] reported that SAP enhanced the production of sedimentable Aβ1–40 approximately 3-fold over 16 h in the presence of 1.5 mM calcium ions. Webster and Rodgers [61] reported a doubling of the Thioflavin T fluorescence for Aβ1–42 in the presence of SAP and 2 mM calcium ions over 24 h. In both cases, nanomolar concentrations of Aβ and SAP were employed, approximating to *in vivo* levels. Janciauskine et al. [62] found that in the absence of calcium ions, SAP completely inhibited the formation of fibrils by Aβ1–42. More recently, MacRaild et al. [55] have investigated the effects of SAP and apolipoprotein E on the state of amyloid fibrils using sedimentation velocity analysis, electron microscopy and rheology. They concluded that both proteins cause amyloid fibrils to associate more densely, becoming highly entangled. The additional stabilization of fibrils provided by calcium-dependent binding and/or crosslinking by SAP would be expected to favor fibril formation. Inhibition of fibril formation, however, suggests a different mode of interaction. There is considerable current interest in identifying small molecules that interfere with extension of the "crossed-β" structure [63–65] and peptide strand terminators offer a plausible guide to the way in which SAP might halt fibril growth. The disordered region made up of residues 140–150 is a unique feature of metal-free SAP and could contribute to this effect, donating a strand that cannot be propagated.

9.10 Perspective

The view that some disorganization of amyloidogenic proteins is required for them to restructure into a cross-β fold is well supported by experimental observations and the amyloidoses are often referred to as protein misfolding diseases [66]. The ability of SAP to recognize intermediate and end products of this process shows close parallels with the properties of molecular chaperones. These proteins interact with folding intermediates of other proteins in the cell to prevent aggregation and facilitate efficient production of correctly folded and biologically active materials [67, 68]. SAP has been shown to assist the refolding of chemically denatured lactate dehydrogenase (LDH) and to inhibit the turbulence induced inactivation of the same enzyme [69]. These effects were observed in the absence and presence of calcium ions, with the greatest effect being observed in the presence of the galactose analogue, MOβDG, which blocks amyloid binding. They are not inhibited by the presence of dAMP, which stabilizes a B–B-face interaction, suggesting that some other part of the SAP molecule is involved. The effects are saturable, possibly reflecting partitioning of refolding LDH intermediates between states that bind SAP and others that do not. Only a small proportion of molecules negotiate a direct route through the refolding energy landscape [70] in the absence of SAP, with the majority forming a misfolded and inactive product. SAP may bind misfolding intermediates that would otherwise inactivate correctly folded subunits of the dimeric enzyme and in the case of the turbulence inactivation assay, SAP may inhibit a nucleated aggregation process of inactivation analogous to amyloid formation. One might speculate that these effects of SAP reflect a surveillance role in detecting damaged proteins *in vivo*, which is overwhelmed in amyloid disease. A selection of other intracellular molecular chaperones have been identified by proteomic analysis of senile plaques, levels of heat-shock protein (Hsp) 90 being particularly elevated [7]. The extracellular chaperone clusterin is also found. This protein has been shown to bind tightly to Aβ and inhibit its aggregation and to inhibit the formation of amyloid fibrils by apolipoprotein CII *in vitro* [71, 72], effects reminiscent of those exhibited by calcium-free SAP on fibrillogenesis.

SAP binding to amyloid may be a pathological accident, perhaps a side-reaction to some other more general chaperone-like function, or SAP may have evolved as a safety mechanism to trap/localize misfolded proteins. A core role of SAP appears to be as an anti-opsonin in chromatin clearance and innate immunity [73] and masking the low level amyloid deposition that appears to be a common feature of ageing [74] may be a further element of this process.

References

1 Botto, M., et al. Amyloid deposition is delayed in mice with targeted deletion of the serum amyloid P component gene. *Nat Med* **1997**, *3*, 855–859.

2 Pepys, M. B., et al. Targeted pharmacological depletion of serum amyloid P component for treatment of human amyloidosis. *Nature* **2002**, *417*, 254–259.

3 Tennent, G. A., L. B. Lovat and M. B. Pepys. Serum amyloid P-component prevents proteolysis of the amyloid fibrils of Alzheimer-disease and systemic amyloidosis. *Proc Natl Acad Sci USA* **1995**, *92*, 4299–4303.

4 Schenk, D., et al. Immunization with amyloid-beta attenuates Alzheimer disease-like pathology in the PDAPP mouse. *Nature* **1999**, *400*, 173–177.

5 Holmgren, G., et al. Clinical improvement and amyloid regression after liver-transplantation in hereditary transthyretin amyloidosis. *Lancet* **1993**, *341*, 1113–1116.

6 Hawkins, P. N., et al. Scintigraphic imaging and turnover studies with iodine–131 labelled serum amyloid P component in systemic amyloidosis. *Eur J Nucl Med* **1998**, *25*, 701–708.

7 Liao, L., et al. Proteomic characterization of postmortem amyloid plaques isolated by laser capture microdissection. *J Biol Chem* **2004**, *279*, 37061–37068.

8 Pepys, M. B., D. R. Booth, W. L. Hutchinson, J. R. Gallimore, P. M. Collins and E. Hohenester. Amyloid P component. A critical review. *Amyloid: J Protein Fold Disorders* **1997**, *4*, 274–295.

9 Jarrett, J. T. and P. T. Lansbury. Seeding one-dimensional crystallization of amyloid – a pathogenic mechanism in Alzheimer's-disease and scrapie. *Cell* **1993**, *73*, 1055–1058.

10 Blake, C. and L. Serpell. Synchrotron X-ray studies suggest that the core of the transthyretin amyloid fibril is a continuous beta-sheet helix. *Structure* **1996**, *4*, 989–998.

11 Jimenez, J. L., J. L. Guijarro, E. Orlova, J. Zurdo, C. M. Dobson, M. Sunde and H. R. Saibil. Cryo-electron microscopy structure of an SH3 amyloid fibril and model of the molecular packing. *EMBO J* **1999**, *18*, 815–821.

12 Antzutkin, O. N., R. D. Leapman, J. J. Balbach and R. Tycko. Supramolecular structural constraints on Alzheimer's beta-amyloid fibrils from electron microscopy and solid-state nuclear magnetic resonance. *Biochemistry* **2002**, *41*, 15436–15450.

13 Antzutkin, O. N., J. J. Balbach and R. Tycko. Site-specific identification of non-beta-strand conformations in Alzheimer's beta-amyloid fibrils by solid-state NMR. *Biophys J* **2003**, *84*, 3326–3335.

14 Staniforth, R. A., et al. Three-dimensional domain swapping in the folded and molten-globule states of cystatins, an amyloid-forming structural superfamily. *EMBO J* **2001**, *20*, 4774–4781.

15 Serpell, L. C. Alzheimer's amyloid fibrils: structure and assembly. *Biochim Biophys Acta* **2000**, *1502*, 16–30.

16 Goldsbury, C. S., S. Wirtz, S. A. Muller, S. Sunderji, P. Wicki, U. Aebi and P. Frey. Studies on the *in vitro* assembly of A beta 1–40: implications for the search for A beta fibril formation inhibitors. *J Struct Biol* **2000**, *130*, 217–231.

17 Ionescu-Zanetti, C., et al. Monitoring the assembly of Ig light-chain amyloid fibrils by atomic force microscopy. *Proc Natl Acad Sci USA* **1999**, *96*, 13175–13179.

18 Jimenez, J. L., E. J. Nettleton, M. Bouchard, C. V. Robinson, C. M. Dobson and H. R. Saibil. The protofilament structure of insulin amyloid fibrils. *Proc Natl Acad Sci USA* **2002**, *99*, 9196–9201.

19 Booth, D. R., et al. Instability, unfolding and aggregation of human lysozyme variants underlying amyloid fibrillogenesis. *Nature* **1997**, *385*, 787–793.

20 Ferrao-Gonzales, A. D., L. Palmieri, M. Valory, J. L. Silva, H. Lashuel, J. W. Kelly and D. Foguel. Hydration and packing are crucial to amyloidogenesis as revealed by pressure studies on transthyretin variants that either protect or worsen amyloid disease. *J Mol Biol* **2003**, *328*, 963–974.

21 Inouye, H., F. S. Domingues, A. M. Damas, M. J. Saraiva, E. Lundgren, O.

Sandgren and D. A. Kirschner. Analysis of X-ray diffraction patterns from amyloid of biopsied vitreous humor and kidney of transthyretin (TTR) Met30 familial amyloidotic polyneuropathy (FAP) patients: axially arrayed TTR monomers constitute the protofilament. *Amyloid: Int J Exp Clin Invest* **1998**, *5*, 163–174.

22 Serag, A. A., C. Altenbach, M. Gingery, W. L. Hubbell and T. O. Yeates. Arrangement of subunits and ordering of beta-strands in an amyloid sheet. *Nat Struct Biol* **2002**, *9*, 734–739.

23 Serag, A. A., C. Altenbach, M. Gingery, W. L. Hubbell and T. O. Yeates. Arrangement of subunits and ordering of beta-strands in an amyloid sheet. *Nat Struct Biol* **2003**, *10*, 70.

24 Serag, A. A., C. Altenbach, M. Gingery, W. L. Hubbell and T. O. Yeates. Identification of a subunit interface in transthyretin amyloid fibrils: evidence for self-assembly from oligomeric building blocks. *Biochemistry* **2001**, *40*, 9089–9096.

25 Eneqvist, T., K. Andersson, A. Olofsson, E. Lundgren and A. E. Sauer-Eriksson. The beta-slip: a novel concept in transthyretin amyloidosis. *Mol Cell* **2000**, *6*, 1207–1218.

26 Steinrauf, L. K., M. Y. Chiang and D. Shiuan. Molecular structure of the amyloid-forming protein kappa I Bre. *J Biochem* **1999**, *125*, 422–429.

27 Olofsson, A., J. H. Ippel, S. S. Wijmenga, E. Lundgren and A. Ohman. Probing solvent accessibility of transthyretin amyloid by solution NMR spectroscopy. *J Biol Chem* **2004**, *279*, 5699–5707.

28 Hoshino, M., H. Katou, Y. Hagihara, K. Hasegawa, H. Naiki and Y. Goto. Mapping the core of the beta$_2$-microglobulin amyloid fibril by H/D exchange. *Nat Struct Biol* **2002**, *9*, 332–336.

29 Kheterpal, I., S. Zhou, K. D. Cook and R. Wetzel. A beta amyloid fibrils possess a core structure highly resistant to hydrogen exchange. *Proc Natl Acad Sci USA* **2000**, *97*, 13597–13601.

30 Goldsteins, G., et al. Exposure of cryptic epitopes on transthyretin only in amyloid and in amyloidogenic mutants. *Proc Natl Acad Sci USA* **1999**, *96*, 3108–3113.

31 Kheterpal, I., A. Williams, C. Murphy, B. Bledsoe and R. Wetzel. Structural features of the A beta amyloid fibril elucidated by limited proteolysis. *Biochemistry* **2001**, *40*, 11757–11767.

32 Tycko, R. Insights into the amyloid folding problem from solid-state NMR. *Biochemistry* **2003**, *42*, 3151–3159.

33 Town, T., J. Tan, N. Sansone, D. Obregon, T. Klein and M. Mullan. Characterization of murine immunoglobulin G antibodies against human amyloid-beta$_{1-42}$. *Neurosci Lett* **2001**, *307*, 101–104.

34 Dobson, C. M. Protein-misfolding diseases: getting out of shape. *Nature* **2002**, *418*, 729–730.

35 Richardson, J. S., D. C. Richardson. Natural beta-sheet proteins use negative design to avoid edge-to-edge aggregation. *Proc Natl Acad Sci USA* **2002**, *99*, 2754–2759.

36 O'Nuallain, B. and R. Wetzel. Conformational Abs recognizing a generic amyloid fibril epitope. *Proc Natl Acad Sci USA* **2002**, *99*, 1485–1490.

37 Emsley, J., et al. Structure of pentameric human serum amyloid-P component. *Nature* **1994**, *367*, 338–345.

38 Wood, S. P., et al. A pentameric form of human-serum amyloid-P component – crystallization, X-ray-diffraction and neutron-scattering studies. *J Mol Biol* **1988**, *202*, 169–173.

39 Pepys, M. B., et al. Human Serum amyloid-P component is an invariant constituent of amyloid deposits and has a uniquely homogeneous glycostructure. *Proc Natl Acad Sci USA* **1994**, *91*, 5602–5606.

40 Painter, R. H., I. Deescallon, A. Massey, L. Pinteric and S. B. Stern. The structure and binding characteristics of serum amyloid protein (9. 5S alpha-1-glycoprotein). *Ann NY Acad Sci* **1982**, *389*, 199–215.

41 Srinivasan, N., H. E. White, J. Emsley, S. P. Wood, M. B. Pepys and T L. Blundell. Comparative analyses of pentraxins – implications for protomer assembly and ligand-binding. *Structure* **1994**, *2*, 1017–1027.

42 Srinivasan, N., S. D. Rufino, M. B. Pepys, S. P. Wood and T. L. Blundell. A superfamily of proteins with the lectin fold.

Chemstracts Biochem Mol Biol **1996**, *6*, 149–164.

43 Kinoshita, C. M., et al. A protease-sensitive site in the proposed Ca^{2+}-binding region of human serum amyloid-P component and other pentraxins. *Protein Sci* **1992**, *1*, 700–709.

44 Ashton, A. W., M. K. Boehm, J. R. Gallimore, M. B. Pepys and S. J. Perkins. Pentameric and decameric structures in solution of serum amyloid P component by X-ray and neutron scattering and molecular modelling analyses. *J Mol Biol* **1997**, *272*, 408–422.

45 Shirahama, T. and A. S. Cohen. High-resolution electron microscopic analysis of the amyloid fibril. *J Cell Biol* **1967**, *33*, 679–708.

46 Pinteric, L. and R. H. Painter. Electron microscopy of serum amyloid protein in the presence of calcium; alternative forms of assembly of pentagonal molecules in two-dimensional lattices. *Can J Biochem* **1979**, *57*, 727–736.

47 Thompson, D., M. B. Pepys and S. P. Wood. The physiological structure of human C-reactive protein and its complex with phosphocholine. *Structure* **1999**, *7*, 169–177.

48 Shrive, A. K., A. W. Metcalfe, J. R. Cartwright and T. J. Greenhough. C-reactive protein and SAP-like pentraxin are both present in *Limulus polyphemus* haemolymph: crystal structure of *Limulus* SAP. *J Mol Biol* **1999**, *290*, 997–1008.

49 Ramadan, M. A. M., A. K. Shrive, D. Holden, D. A. A. Myles, J. E. Volanakis, L. J. DeLucas and T. J. Greenhough. The three-dimensional structure of calcium-depleted human C-reactive protein from perfectly twinned crystals. *Acta Crystallogr D Biol Crystallogr* **2002**, *58*, 992–1001.

50 Shrive, A. K., et al. Three dimensional structure of human C-reactive protein. *Nat Struct Biol* **1996**, *3*, 346–354.

51 Kinoshita, C. M., et al. Elucidation of a protease-sensitive site involved in the binding of calcium to C-reactive protein. *Biochemistry* **1989**, *28*, 9840–9848.

52 Thompson, D., M. B. Pepys, I. Tickle and S. Wood. The structures of crystalline complexes of human serum amyloid P component with its carbohydrate ligand, the cyclic pyruvate acetal of galactose. *J Mol Biol* **2002**, *320*, 1081–1086.

53 Hind, C. R. K., P. M. Collins, D. Renn, R. B. Cook, D. Caspi, M. L. Baltz and M. B. Pepys. Binding-specificity of serum amyloid-P component for the pyruvate acetal of galactose. *J Exp Med* **1984**, *159*, 1058–1069.

54 Hohenester, E., W. L. Hutchinson, M. B. Pepys and S. P. Wood. Crystal structure of a decameric complex of human serum amyloid P component with bound dAMP. *J Mol Biol* **1997**, *269*, 570–578.

55 MacRaild, C. A., et al. Non-fibrillar components of amyloid deposits mediate the self-association and tangling of amyloid fibrils. *J Biol Chem* **2004**, *279*, 21038–21045.

56 Hamazaki, H. Ca^{2+}-mediated association of human-serum amyloid-P component with heparan-sulfate and dermatan sulfate. *J Biol Chem* **1987**, *262*, 1456–1460.

57 Inoue, S., M. Kuroiwa and M. J. Saraiva, A. Guimaraes and R. Kisilevsky. Ultrastructure of familial amyloid polyneuropathy amyloid fibrils: examination with high-resolution electron microscopy. *J Struct Biol* **1998**, *124*, 1–12.

58 Danielsen, B., I. J. Sorensen, M. Nybo, E. H. Nielsen, B. Kaplan and S. E. Svehag. Calcium-dependent and -independent binding of the pentraxin serum amyloid P component to glycosaminoglycans and amyloid proteins: Enhanced binding at slightly acid pH. *Biochim Biophys Acta* **1997**, *1339*, 73–78.

59 Hamazaki, H. Ca^{2+}-dependent binding of human serum amyloid-P component to Alzheimer's beta-amyloid peptide. *J Biol Chem* **1995**, *270*, 10392–10394.

60 Hamazaki, H. Amyloid-P component promotes aggregation of Alzheimer's beta-amyloid peptide. *Biochem Biophys Res Commun* **1995**, *211*, 349–353.

61 Webster, S. and J. Rogers. Relative efficacies of amyloid beta peptide (A beta) binding proteins in A beta aggregation. *J Neurosci Res* **1996**, *46*, 58–66.

62 Janciauskiene, S., P. G. Defrutos, E. Carlemalm, B. Dahlback and S. Eriksson. Inhibition of Alzheimer beta-peptide fibril formation by serum amyloid-P com-

ponent. *J Biol Chem* **1995**, *270*, 26041–26044.

63 Lashuel, H.A., D.M. Hartley, D. Balakhaneh, A. Aggarwal, S. Teichberg and D.J.E. Callaway. New class of inhibitors of amyloid-beta fibril formation – implications for the mechanism of pathogenesis in Alzheimer's disease. *J Biol Chem* **2002**, *277*, 42881–42890.

64 Frenkel, D., B. Solomon and I. Benhar. Modulation of Alzheimer's beta-amyloid neurotoxicity by site-directed single-chain antibody. *J Neuroimmunol* **2000**, *106*, 23–31.

65 Soto, C., E.M. Sigurdsson, L. Morelli, R.A. Kumar, E.M. Castano and B. Frangione. Beta-sheet breaker peptides inhibit fibrillogenesis in a rat brain model of amyloidosis: implications for Alzheimer's therapy. *Nat Med* **1998**, *4*, 822–826.

66 Cohen, F.E. and J.W. Kelly. Therapeutic approaches to protein-misfolding diseases. *Nature* **2003**, *426*, 905–909.

67 Ellis, R.J. Roles of molecular chaperones in protein-folding. *Curr Opin Struct Biol* **1994**, *4*, 117–122.

68 Hartl, F.U. Molecular chaperones in cellular protein folding. *Nature* **1996**, *381*, 571–580.

69 Coker, A.R., A. Purvis, D. Baker, M.B. Pepys and S.P. Wood. Molecular chaperone properties of serum amyloid P component. *FEBS Lett* **2000**, *473*, 199–202.

70 Dill, K.A. and H.S. Chan. From Levinthal to pathways to funnels. *Nat Struct Biol* **1997**, *4*, 10–19.

71 Matsubara, E., C. Soto, S. Governale, B. Frangione and J. Ghiso. Apolipoprotein J and Alzheimer's amyloid beta solubility. *Biochem J* **1996**, *316*, 671–679.

72 Hatters, D.M., M.R. Wilson, S.B. Easterbrook-Smith and G. J. Howlett. Suppression of apolipoprotein C-II amyloid formation by the extracellular chaperone, clusterin. *Eur J Biochem* **2002**, *269*, 2789–2794.

73 Bickerstaff, M.C.M., et al. Serum amyloid P component controls chromatin degradation and prevents antinuclear autoimmunity. *Nat Med* **1999**, *5*, 694–697.

74 Haggqvist, B., et al. Medin: an integral fragment of aortic smooth muscle cell-produced lactadherin forms the most common human amyloid. *Proc Natl Acad Sci USA* **1999**, *96*, 8669–8674.

10
Apolipoprotein E: Structural and Functional Interactions with Amyloid β

W. Blaine Stine, Jr and Mary Jo LaDu

10.1
Introduction

Genetics, pathology and extensive experimental evidence suggest that apolipoprotein E (ApoE) and amyloid β (Aβ) interact structurally, and, thus, affect each other functionally. However, our understanding of the interactions between ApoE and Aβ is complicated by the experimental difficulty in determining and controlling the conformation of these two proteins. Thus, in evaluating experimental results, it is critical to consider whether the chemical form of ApoE and Aβ utilized is physiologically relevant. Is ApoE likely to exist *in vivo* in the same conformation as purified protein? Is the ApoE preparation to be used in a conformation that maintains ApoE receptor-binding affinity? Will large, triglyceride (TG)- and cholesteryl-ester (CE)-rich plasma ApoE-containing lipoproteins produce the same results as the small, lipid-poor, discoidal ApoE-containing lipoproteins secreted by glial cells? Are results using the Aβ fragments Aβ1–40 and Aβ1–42 comparable? Is the Aβ preparation structurally homogenous or is it likely to contain a variety of Aβ assemblies including small oligomers, fibrils and large globular aggregates? Will these different Aβ assemblies produce the same functional outcome? Thus, the thesis of this review is that *the structure of both ApoE and Aβ* profoundly affects their interactions and functions.

10.2
ApoE Background

10.2.1
Function in Plasma Lipid Metabolism

In the aqueous environment of the blood, neutral lipids circulate in complex with apolipoproteins and are thus packaged as lipoproteins. Lipoproteins are composed of a phospholipid and free cholesterol shell surrounding a TG and CE core, and are stabilized by surface apolipoproteins. Apolipoproteins also

Amyloid Proteins. The Beta Sheet Conformation and Disease. J. D. Sipe

ISBN: 3-527-31072-X

serve as cofactors for enzymatic reactions and ligands for lipoprotein receptors. Lipoproteins can be separated on the basis of size and density into four major classes – chylomicrons, very-low-density lipoproteins (VLDL), low-density lipoproteins (LDL) and high-density lipoproteins (HDL). The classes vary in their core TG/CE content and apolipoprotein composition. The apolipoprotein gene family encodes soluble proteins with amphipathic α-helical structures in the C-terminus that allow the proteins to exist at the water–lipid interface [1]. One such apolipoprotein is ApoE.

ApoE circulates in blood plasma associated with several classes of lipoproteins, including chylomicrons, VLDL and a subset of HDL. ApoE-containing lipoproteins are bound and internalized via receptor-mediated endocytosis involving a number of receptors in the LDL receptor (LDLR) family, including LDLR, LDLR-related protein (LRP), VLDL receptor (VLDLR), glycoprotein 330 (gp330) and ApoE receptor-2 (ER2) [2]. While the LDLR binds lipoproteins containing ApoB-100 and/or ApoE, the other receptors recognize only ApoE-containing particles or ApoJ for gp330. Thus, ApoE is a major contributor to the transport of lipid particles to and from the bloodstream, regulating plasma lipid and cholesterol metabolism by its involvement in the three primary pathways of lipid metabolism.

In the exogenous pathway, dietary lipids are packaged into chylomicrons and secreted by the intestine into the lymph, passing into the circulation via the thoracic duct. Lipoprotein lipase (LPL) anchored in the capillary endothelia hydrolyzes the TG from the chylomicron core, releasing free fatty acids that are then available for uptake by the surrounding tissue. The ApoE acquired by chylomicron remnants in the plasma allows these TG-poor particles to be cleared by ApoE receptors on the liver. In the similar endogenous pathway, ApoE-containing VLDL are produced by the liver and secreted into the circulation, where LPL again hydrolyzes their core TG. The resulting VLDL remnants are either taken up by hepatic receptors via recognition of ApoE or further modified to LDL. In reverse cholesterol transport, cholesterol from peripheral tissues is transported to the liver either directly by ApoE-containing HDL or via transfer of the cholesteryl esters to larger particles.

10.2.2
ApoE Structure

Human ApoE is a 299-amino-acid (34-kDa) glycosylated apolipoprotein that is encoded by a gene 3597 nucleotides in length with four exons and three introns [3]. The ApoE gene is located on the long arm of chromosome 19, and is part of a 45-kb multi-gene complex that includes ApoCI, ApoCII and ApoCIV, as well as a pseudo-ApoCI gene [4, 5]. This extensive DNA sequence is critical for the expression of ApoE, as tissue-specific expression sequences for the liver, macrophages and astrocytes, as well as several multi-enhancers, are located around 3–30 kb from the ApoE-coding region [6–9]. Three major ApoE isoforms exist in humans that differ at two residues; E2 (Cys112, Cys158), E3 (Cys112, Arg158) and E4 (Arg112, Arg158). The allelic variations that encode these isoforms have differing distributions in the general population. The $\varepsilon 3$ allele is the most common, with a gene frequency of around 77%. The frequencies of $\varepsilon 4$ (around 15%) and $\varepsilon 2$ (around8%) are lower, but they are still considered common variants [10]. Hepatic cells in the liver are the primary site of ApoE expression. However, the second largest source of ApoE is the brain, where it is primarily expressed by astrocytes and microglia (for review, see [11]). These two pools of ApoE are physically separated by the blood–brain barrier (BBB); evidence that the two pools remain segregated comes from studies of liver transplantation. The ApoE isoform present on plasma lipoproteins will change if the ApoE genotype of the donor liver is different from that of the recipient. However, this change does not affect the ApoE isoform present on the recipient's cerebrospinal fluid (CSF) lipoproteins, indicating that ApoE present in the brain is expressed locally and is not derived from, or exchangeable with, ApoE in the periphery [12].

The three-dimensional structure of full-length ApoE is comprised of independently folded N- and C-terminal domains joined by an unstructured flexible hinge region. The N-terminal domain contains several important structural features that functionally define ApoE. The LDLR binding region is located in the N-terminal domain on helix 4 (Fig. 10.1, yellow) encompassing residues 136–150 [13]. This region overlaps with a heparin-binding site between residues 142–147 [14]. X-ray crystallographic studies show that the N-terminal domain (residues 1–191) folds into a soluble 22-kDa amphipathic four-helix bundle (Fig. 10.1) [15]. The α-helices are oriented such that the hydrophobic faces are sequestered in the center of the bundle, creating a hydrophilic exterior. A conformational change in this region is thought to occur with lipid binding whereby the random coil region between helices 2 and 3 (Fig. 10.1, blue and green) serves as a hinge allowing the four-helix bundle to open in the presence of lipid, resulting in a receptor-competent conformation [13]. The N-terminal domain also contains the polymorphic sites that differentiate E2, E3 and E4 located on helices 3 and 4 (Fig. 10.1, green and yellow). Less is known about the structure of the hydrophobic C-terminal domain for which no high-resolution structural data exist. Protein modeling analysis suggests an extended α-helical region between residues 227 and 267 flanked by two shorter α helices. Isolated C-terminal ApoE also exhibits heparin-binding activity involving basic residues around Lys233, although this site appears to be masked in full-length ApoE [16]. Functionally, the C-terminal domain serves as the major lipid-binding region, anchoring ApoE to lipoprotein particles [17, 18].

Conformational changes that occur in the presence of lipid, within and between the N- and C-terminal domains, are influenced in part by the structure of the ApoE isoform [13]. ApoE is an arginine-rich protein containing several charged residues involved in inter- and intra-domain salt bridges. These interactions influence ApoE function and differ between isoforms [19, 20]. ApoE4 forms a unique inter-domain interaction between Arg61 on helix 2 (Fig. 10.1, blue) in the N-terminal domain and Glu255 in the C-terminal domain

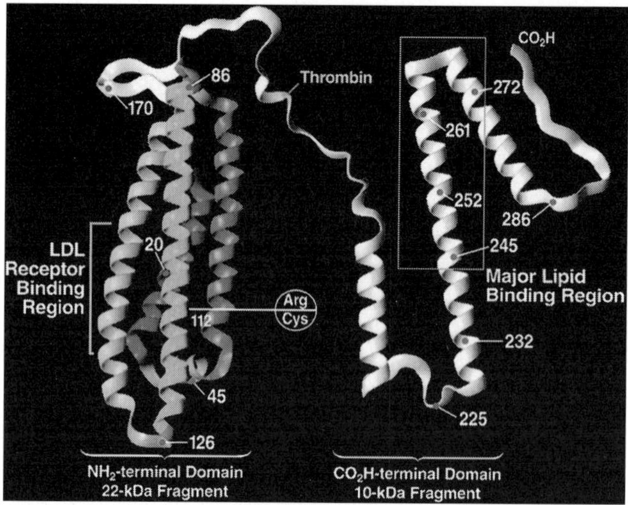

Fig. 10.1 ApoE structure. Model of N- and C-terminal domains connected by a protease-sensitive hinge region. The structure of the N-terminal domain is based on the X-ray crystal structure of non-lipidated N-terminal ApoE. It is predicted to fold as a four-helix bundle containing the LDLR-binding region, and the residues that determine ApoE isoform at positions 112 and 158. A flexible hinge region connects the two domains and is sensitive to thrombin proteolysis. The predicted structure of the C-terminal domain is based on computer modeling. It is predicted to be mostly α-helical and contains the major lipid binding region. (Reprinted from [244], with permission.)

(Fig. 10.1, white), resulting in a more compact structure [21], whereas ApoE3 Arg61 forms an intra-domain salt bridge with Asp65 on helix 2 [13].

As described above, a flexible unstructured "hinge" region (residues 165–215) that is sensitive to proteolysis [22, 23] divides ApoE into two structural domains. The 22-kDa N-terminal fragment contains the receptor-binding domain and the 10-kDa C-terminus contains several amphipathic α-helices that are important for the association of ApoE with lipoproteins. One functional consequence of these structural differences is that ApoE2 and ApoE3 preferentially bind to HDL, while ApoE4 binds to VLDL, although each is found in both classes of lipoproteins. While the C-terminus contains the lipid-binding domain, the N-terminus contains the residues that confer isoform specificity. Therefore, the preferential association of the ApoE isoforms with different classes of plasma lipoproteins is the result of an interaction between the two domains. Cleavage at the hinge region by either thrombin [24, 25] or a chymotrypsin-like serine protease [26] results in N- and C-terminal fragments that may function in a manner distinct from the intact holoprotein. Susceptibility to proteolysis has been reported to be isoform dependent with ApoE4 susceptibility greater than that of

ApoE3, suggesting differential stabilities between the ApoE isoforms. Thermal and chemical denaturation studies have provided additional evidence. The ApoE2 isoform is the most resistant to unfolding, with relative unfolding resistance of ApoE2 > ApoE3 > ApoE4 [27, 28]. These studies of resistance to unfolding also identified differences in the unfolding patterns between ApoE isoforms, where ApoE2 is thought to undergo a two-stage cooperative unfolding and ApoE4 to a significantly greater extent than ApoE3 undergoes non-cooperative unfolding. ApoE4 is associated with the population of a folding intermediate with the characteristics of a molten globule that may also influence lipid-binding properties in an isoform-specific manner [29]. In sum, these studies describe a structural basis for ApoE isoform-specific functional differences.

The structural differences that result from isoform differences in a single amino acid are also sufficient to influence plasma lipid metabolism, primarily by receptor-mediated clearance. ApoE2 is defective in binding to receptor, exhibiting only around 1% affinity for the LDLR [13] and around 40% for LRP [30]. Thus, individuals with the ApoE2/2 genotype are at risk for developing type III hyperlipoproteinemia, a disorder characterized by elevated plasma TG and cholesterol. ApoE2 differs from ApoE3 and ApoE in the salt bridge patterns within helices 3 and 4 that affect the electrostatic potential of the receptor binding region and result in defective LDLR binding [19]. While ApoE4 does not exhibit a decrease in binding to any known receptors in the LDLR family *in vitro*, it is associated with elevated plasma cholesterol and LDL, suggesting that its interaction with other aspects of the lipoprotein pathway differs from that of ApoE3 [31]. ApoE4/4 individuals have a small, but signifycant, increased risk for coronary heart disease [10] and a significantly greater risk for developing Alzheimer's disease [32]. The differences in receptor binding affinity may be due to conformational changes that result from the interaction with lipid. Purification of ApoE and the removal of lipid abolish its ability to bind LDLR [33], while reconstitution with lipid restores receptor-binding affinity [34], highlighting the functional dependency of ApoE on physiologically relevant conformations. In general, these observations illustrate that interactions between and within domains, and domain stability, are important features of ApoE. The subtleties of the effect of ApoE conformation, including extent and form of lipidation, on its function will be addressed further below.

10.3 ApoE and Aβ

10.3.1 Summary

Genetics, pathology and extensive experimental evidence suggest that ApoE and Aβ interact structurally and functionally. *In vivo*, ApoE, a component of senile plaques and peripheral Aβ deposits [35–37], influences amyloid deposition in humans [38, 39] and transgenic mice [40–42], and binds the soluble Aβ oligomers found in the brain, plasma and CSF [43, 44]. *In vitro*, ApoE forms stable complexes with Aβ [32, 45–53], alters the aggregation of various Aβ peptides [54–57], inhibits Aβ-induced neurotoxicity [58–62], modulates Aβ-induced inflammation, [63–70] and promotes clearance of Aβ, partially via ApoE receptors [38, 46, 71–78]. This section will focus on the effect of ApoE structure on its functional interactions with Aβ, specifically the effect of ApoE isoform and lipid-associated conformation. *Our hypothesis is that ApoE modulates the structure and function of two distinct assemblies of Aβ, soluble oligomers and amyloid.*

10.3.2
ApoE and Neurodegenerative Diseases

In 1993, Roses, Strittmatter and co-workers reported that $\varepsilon 4$ is a risk factor for Alzheimer's disease [32], including both sporadic [79] and familial Alzheimer's disease [80]. The risk for Alzheimer's disease is 3- and 8-fold greater in individuals with one or two copies of $\varepsilon 4$, compared to $\varepsilon 3$ homozygotes. Virtually all individuals who are homozygous for the $\varepsilon 4$ allele and live to be 80 years of age will develop Alzheimer's disease [80]. This genetic correlation has been confirmed in a variety of populations and by a number of statistical approaches [81–88], and expanded to include $\varepsilon 2$ as negatively correlated with Alzheimer's disease [89, 90]. Combining the increased risk of Alzheimer's disease due to the $\varepsilon 4$ allele and the reduced risk with $\varepsilon 2$, over 80% of all Alzheimer's disease cases can be attributed to genetic variance at the ApoE locus [80].

The increased risk of neurodegenerative disease with $\varepsilon 4$ appears to be confined to Alzheimer's disease and other conditions with amyloid deposition, including aging, Down's syndrome and cerebral amyloid angiopathy [38, 39, 91–97]. The $\varepsilon 4$ allele is also associated with poor cognitive recovery from cerebral insults (e.g. head injury, cerebral hemorrhage or stroke) compared to $\varepsilon 3$ [97–105]. $\varepsilon 4$ is not associated with other conditions of neuronal loss that lack cerebral amyloid, including amyotrophic lateral sclerosis, Parkinson's disease, Pick's disease, progressive supranuclear palsy and ischemic stroke [103, 106–110]. Although information is limited, ApoE2 offers cognitive protection from aging, as well as Alzheimer's disease [89, 111].

This genetic evidence has initiated extensive research efforts towards defining the functions of ApoE in the brain using various *in vivo* and *in vitro* models. Thus far, the potentially overlapping functions attributed to ApoE include lipid transport, maintenance of neuroplasticity, anti-inflammatory actions, modulation of the structure and function of tau, and modulation of the structure and function of Aβ, including amyloid deposition. In spite of this research effort, our understanding of the role of ApoE and ApoE receptors in the various processes that affect CNS function remains unresolved and often contradictory. In particular, an activity that accounts for the correlation between $\varepsilon 4$ and Alzheimer's disease has not been definitively identified. The remainder of this chapter will focus on the interactions between ApoE and Aβ.

10.3.3
Aβ: Oligomers and Amyloid

One mechanism by which ApoE may be involved in the pathology of Alzheimer's disease is modulation of the structure and function of Aβ. Aβ is derived by the proteolytic processing of Aβ precursor protein (AβPP) by β- and γ-secretase, resulting in a peptide predominantly 40 or 42 amino acids in length. Aβ is an amphipathic peptide with a hydrophilic N-terminus and a hydrophobic C-terminus with Aβ1–42 having two more hydrophobic residues than Aβ1–40 at the C-terminus. To date, AβPP and Aβ have no clearly defined physiological function (for review, see [112]).

Familial Alzheimer's disease is caused by mutations in AβPP that increase the amount of Aβ1–42. In addition, autosomal dominant mutations in the presenilins (PS1 and PS2) also appear to contribute to Alzheimer's disease by increasing the amount of Aβ1–42 [112]. Although this genetic evidence predicts that Aβ is a causative factor in Alzheimer's disease, its exact pathological function remains unclear. The "amyloid cascade hypothesis" maintains that the aggregation of Aβ into amyloid deposits, a pathologic hallmark of Alzheimer's disease, induces a toxic gain of function. The amyloid hypothesis gained considerable acceptance when initial reports indicated that fibrillar Aβ1–42 was cytotoxic *in vitro* [113–115]. However, amyloid plaques correlate poorly in number, tempo or distribution with the clinical progression of dementia [116–118], and several measures of neurodegenerative pathology are independent of amyloid deposition in APP transgenic mice [119–125]. Finally, *in vitro* inhibition of fibrillogenesis does not reduce toxicity [126–128]. Thus, recent research has focused on soluble oligomeric assemblies of Aβ as the proximate cause of neuronal injury, synaptic loss and the eventual dementia associated with Alzheimer's disease [129–133]. Small, stable oligomers of Aβ1–42, isolated from brain, plasma and CSF [43, 134, 135], correlate with the severity of neurodegeneration in Alzheimer's disease [136, 137].

In vitro data demonstrate that soluble Aβ oligomers and protofibrils are neurotoxic [138–141]. However, the relative contributions of fibrillar and oligomeric Aβ to the disease process remain unresolved. To directly assess the conformation-dependent differences among Aβ assemblies, we have developed protocols for the preparation of homogeneous unaggregated, oligomeric and fibrillar Aβ1–42 [142, 143]. These distinct assemblies are derived from chemically identical and structurally homogeneous starting material and are, thus, particularly suited for comparative structure–function studies. *In vitro*, oligomeric Aβ1–42 is around 10-fold more neurotoxic than the fibrillar (plaque-forming) assembly and around 40-fold more toxic than the unaggregated peptide [144]. *In vivo*, a diffusible form of Aβ directly injected into the brain of transgenic mice expressing human tau caused tangle formation [145]. In addition, while previous studies demonstrated that the chronic treatment of APP transgenic mice with an antibody to Aβ resulted in a reduction of amyloid burden [146], acute treatment with the same antibody resulted in a rapid reversal of memory impairment that preceded changes in amyloid deposition [147]. Based on this recent research, soluble Aβ was been identified as an early, causal factor in the pathogenesis Alzheimer's disease as the result of several revisions of the "amyloid hypothesis" [148–150].

Pathological mechanisms independent of frank protein aggregation are drawing attention in other neurodegenerative diseases, particularly those characterized by selective vulnerability. Acceleration of oligomerization, not fibril formation, is a key to α-synuclein mutations linked to early-onset Parkinson's disease [151–153]. In polyglutamine expansion diseases, including Huntington's disease, the number of protein aggregates can be reduced without the loss of cytotoxicity [154, 155]. In addition, there is a poor correlation between cell death and the number of Lewy bodies in the cortex of patients with diffuse Lewy body disease, again demonstrating a lack of correlation between neuronal damage and peptide aggregation [156]. Finally, expression of mutant tau in *Drosophila* induced neurodegeneration without aggregation of the protein into neurofibrillary tangles [157]. Thus, neurotoxicity in Alzheimer's disease and other neurodegenerative diseases may be the result of a change in protein conformation independent of the formation of large aggregates or amyloid.

In the following sections on ApoE and Aβ, the focus is on ApoE conformation. The oligomeric assembly state of Aβ undoubtedly effects its interactions with ApoE [158], as discussed briefly in this section. Because Aβ structure has not been fully addressed in previous functional studies, the resulting literature is complicated and controversial with respect to the function of Aβ *in vitro*. Indeed, specific commercial preparations of Aβ have been identified as "neurotoxic" and significant variability in biological activity between different lots of peptide has been described by several investigators [158–160]. These inconsistencies are the result of the difficulty in controlling and determining the conformational state of Aβ in a given preparation of fresh or aged Aβ1–40 or Aβ1–42. This comes about, in part, because conventional quality control methods focus primarily on chemical purity, not structural heterogeneity. Differences in secondary, tertiary or quaternary structure are frequently not taken into account. Thus, many Aβ functional studies do not fully address Aβ conformation. Although chapters 11–14 and 17 will address Aβ assembly and its effects on peptide function, it is another important aspect to consider when evaluating interactions with ApoE.

10.3.4
ApoE and Aβ Peptide

ApoE4 may contribute to Alzheimer's disease pathology by interaction with Aβ peptide, as well as promoting amyloid deposition, as will be discussed below. In separate investigations, ApoE was initially identified as an Aβ-binding protein in CSF [32, 44]. *In vitro*, ApoE interacts with Aβ to form a stable complex [32, 45–47, 49, 50]. From these published accounts, the effect of ApoE isoform on this interaction between ApoE and Aβ is unclear. At least a portion of this apparent discrepancy can be attributed to the source of ApoE. In biochemical assays, native ApoE2 and E3 (associated with lipid particles) form a sodium dodecylsulfate (SDS)-stable complex with Aβ1–40 that is more abundant than the ApoE4:Aβ complex [46–48, 50]. However, purified ApoE3 and E4 exhibit a comparable and lower affinity for the peptide than native ApoE3:Aβ [32, 47]. These results are consistent with the general observation that the functional activity of ApoE is affected by its conformation and the conformation of ApoE is largely determined by the presence or absence of lipid, as well as the nature of the lipoprotein particle.

In vivo, ApoE is secreted as a component of lipoprotein particles from a variety of cell types, including hepatocytes, endothelial cells and macrophages. In the brain, astrocytes secrete ApoE-containing lipoproteins that are small and discoidal, comparable to small HDL particles in plasma [11, 161, 162]. We have demonstrated ApoE isoform-specific SDS-stable ApoE:Aβ complex formation utilizing ApoE associated with VLDL, a large TG/CE-rich plasma lipoprotein [47], as well as the small, lipid-poor HDL-sized particles secreted by HEK cells stably transfected with cDNA encoding ApoE3 and E4 (HEK-ApoE) [46] (Fig. 10.2). While ApoE associated with two very different particles exhibited similar isoform-specific binding to Aβ, purified ApoE did not [46, 47, 163]. Whether *in vivo* as a component of plasma lipoproteins or *in vitro* secreted by primary glial cells or ApoE-transfected cell lines, ApoE is associated with lipid particles, indicating that lipid-associated ("native") preparations of ApoE may be a more physiologically relevant substrate for Aβ binding than purified ApoE. Like the receptor binding activity of ApoE, SDS-stable ApoE:Aβ complex formation is influenced by the conformation of ApoE.

Mouse [4] and rat [164] ApoE exist as single isoforms with an arginine at positions 112 and 158, comparable to ApoE4, and have around 70% sequence similarity with human ApoE. However, rat and mouse ApoE lack the arginine at position 61 necessary for the domain interactions present in human ApoE4 between Arg61 and Glu255. The conformational differences resulting from the loss of domain interaction with mouse and rat ApoE may be responsible for their preferential association with smaller HDL-sized lipoproteins, akin to human ApoE2 and E3 in plasma.

The lipoprotein context strongly influences the ability of rat ApoE to form SDS-stable complexes with Aβ. Rat ApoE on plasma VLDL, LDL or HDL does not form an SDS-stable complex with Aβ1–40 (Fig. 10.2). This behavior is comparable to human HEK-ApoE4 (Fig. 10.2). The only rat plasma apolipoprotein that binds to Aβ1–40 is ApoA-I on HDL, a complex that has also been detected in humans [165]. Purification of rat ApoE from plasma increases its affinity for Aβ1–40 as compared with plasma lipoprotein containing rat ApoE – a property previously described for human ApoE4 [47]. However, most surprising, is the observation that rat ApoE secreted by cultured glia on HDL-size lipoprotein particles [161] does form an abundant SDS-stable complex with Aβ1–40 (Fig. 10.2). Thus, it appears that astrocytic ApoE is an excellent substrate for Aβ in the formation of ApoE:Aβ complexes. These results are a compelling illustration of the

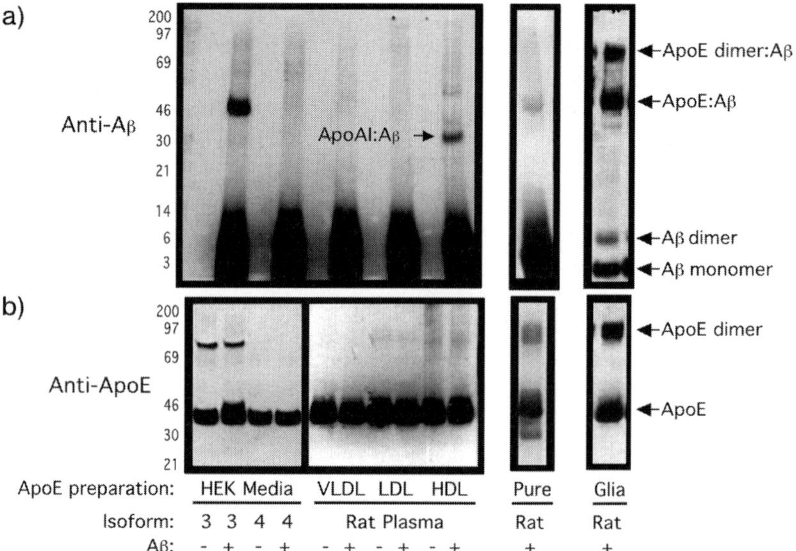

Fig. 10.2 Aβ binds rat ApoE associated with glial lipoproteins, but not rat ApoE associated with plasma lipoproteins. Western blots of 25 μg/ml (around 700 nM) ApoE incubated for 2 h at room temperature with either 5% DMSO control or 250 μM Aβ1–40, as indicated. Samples were run in non-reducing Laemmli buffer on 10–20% SDS–PAGE gels, transferred to Immobilon-P membranes and probed with 4G8 antibody (to Aβ residues 17–24) (A) or ApoE antisera (B). The source of ApoE was: HEK cells transfected with the cDNA encoding human ApoE3 or E4, fractions of VLDL, LDL and HDL from 1 ml of rat plasma fractionated by size exclusion chromatography using tandem Superose-6 columns, ApoE purified from rat plasma, and ApoE-containing lipoproteins isolated from cultured rat glial cells.

effect that the state of lipidation has on the function of ApoE, supporting the hypothesis that ApoE:Aβ interactions depend on specific ApoE conformations determined by isoform, domain structure and lipidation state. These results further support our hypothesis that, in the brain, astrocytic ApoE binds and clears oligomeric Aβ.

As an initial measure of the structural interactions between ApoE and Aβ1–42 oligomers and fibrils, we again used gel-shift assays to measure the formation of SDS-stable ApoE:Aβ complexes. HEK-ApoE3 formed a more abundant SDS-stable complex with both oligomeric and fibrillar Aβ1–42 than ApoE4. The order of degree of SDS-stable ApoE:Aβ1–42 complex formation is: ApoE3:oligomers > ApoE3:fibrils > ApoE4:oligomers > ApoE4:fibrils [143]. Again, these results are consistent with the hypothesis that ApoE3 preferentially binds oligomeric Aβ1–42.

The results measuring SDS-stable complex formation by gel-shift assay have been confirmed using a solid-phase binding assay [51] and upon using unpurified native recombinant ApoE [52]. The relevance of SDS-stable ApoE:Aβ complexes is further supported by *in vivo* data that demonstrated that nearly all of the Aβ in plaque-free, non-Alzheimer's disease brain samples is bound in SDS-stable complexes with ApoE. Alzheimer's disease brains contain significantly greater amounts of free Aβ and the ApoE:Aβ complexes present are more susceptible to proteolysis [166]. Thus, Alzheimer's disease brains may fail to degrade Aβ via an ApoE-mediated process, leading to neurotoxic levels of the peptide or a toxic aggregate. Our working hypothesis is that ApoE interacts in an isoform-specific manner with Aβ to form an SDS-stable complex that results in the differential clearance and/or deposition of the peptide (Fig. 10.3).

10.3.5
ApoE:Aβ Binding Domains

Attempts to identify the region(s) of Aβ involved in formation of SDS-stable ApoE:Aβ complexes have been limited; results using several different assay systems suggest that various regions of Aβ are involved [32, 167, 168]. Using HEK-ApoE, we reported that residues 13–28 of Aβ appear to be necessary, while complex formation is further enhanced by the presence of residues at the C-terminus of the peptide [169]. Using full-length Aβ1–40 and Aβ1–42, and a series of truncated peptides, ApoE2 and ApoE3, but not ApoE4, were observed to bind to the C-terminal region of Aβ and protect primary cortical neurons against neurotoxic effects of a non-fibrillar assembly of Aβ [62]. Both full-length and C-terminal fragments of Aβ form SDS-stable complexes with ApoE2 and ApoE3, but not with ApoE4 [62]. These findings are consistent with previous lipid-fusion experiments that suggest the participation of the C-terminal region of Aβ in the binding to ApoE [168]. Hydrophobic interactions are likely to dominate between ApoE and the C-terminus of Aβ which contains a cluster of hydrophobic residues. On the other hand, ApoE:Aβ complex stability is also likely to be enhanced by ionic interactions from the charged Aβ residues located closer to the N-terminus.

Previous observations suggest that three histidine residues located between residues 13–16 of Aβ are necessary for the binding of the peptide to microglia heparan sulfate proteoglycan (HSPG), resulting in activation of these cells [170]. Because the ApoE-binding domain of Aβ also appears to involve residues 13–16, ApoE may function as a competitive inhibitor of Aβ–HSPG binding, thereby attenuating the activity of the peptide. ApoE itself has two heparin-binding domains and HSPG binding facilitates uptake of ApoE by its receptors [171, 172]. Indeed, interactions with cell-surface proteoglycans may directly mediate the uptake of ApoE [173] and, in turn, could influence the activity of the Aβ, ApoE and/or an ApoE:Aβ complex.

Studies designed to identify the region(s) of ApoE involved in formation of SDS-stable ApoE:Aβ complexes have indicated that both ApoE domains can bind Aβ

Isolated N-terminal fragments of ApoE, produced by thrombin cleavage, formed SDS-stable complexes with Aβ, while isolated C-terminal ApoE fragments did not [174]. Additionally, if the C-terminus of ApoE alone mediates ApoE:Aβ complex formation, then VLDL particles known to bind the C-terminus of ApoE would be expected to compete for binding with Aβ. However, VLDL does not block ApoE:Aβ complex formation, providing further evidence that the interaction between ApoE and Aβ is primarily through the N-terminal domain [174]. The authors acknowledge that the interpretation of these experimental results must take into account the conformations of ApoE and Aβ – both of which influence complex formation. Isolated N-terminal ApoE3 (residues 1–201) retains a slightly greater affinity for Aβ1–40 than N-terminal fragments of ApoE4 [20]. However, native ApoE had significantly enhanced affinity for Aβ1–40, attributable to ionic interactions between the N- and C-terminal domains. Disruption of the Arg61–Glu255 salt bridge between these domains abolishes Aβ complex formation [20]. Therefore, both the structure of each domain and the interaction between domains influences SDS-stable complex formation with Aβ.

These interactions between the various functional domains of ApoE and Aβ are undoubtedly complex. Both ApoE and Aβ contain stretches of charged and hydrophobic residues that are implicated in ApoE:Aβ complex formation. It is unlikely that a simple explanation of complex formation focusing on a single region of either protein will suffice. Rather, it is likely that the formation of key contacts between charged and neutral residues that stabilize ApoE:Aβ complexes will largely be determined by the overall three dimensional structure of both proteins. This explanation will be a result of Aβ conformation, as determined by sequence and aggregation state, and ApoE conformation as determined by isoform and lipidation state.

10.3.6
ApoE and Amyloidosis

Amyloidosis is characterized by the accumulation of normally soluble proteins as insoluble deposits that are identified by staining with Congo red or Thioflavin T, indicating a β-pleated sheet secondary structure. ApoE co-localizes with both cerebral and systemic amyloid deposits, leading Wisniewski to call ApoE a "pathological chaperone" [36]. This label became specific to ApoE4, as ε4 correlates with amyloid burden in a dose-dependent manner in Alzheimer's disease patients [38, 39]. In the brain periphery, a variety of proteins can form amyloid, including amyloid A (AA) in patients with Mediterranean fever and amyloid L (AL) in patients with primary amyloidosis. AA and AL fibrils purified from these patients contained C-terminal fragments of ApoE [159, 175]. C-terminal fragments of ApoE also co-purified with Aβ extracted from the brain tissue of Alzheimer's disease patients [176, 177]. Aggregation of ApoE may actually contribute to amyloid plaque deposition. *In vitro*, C-terminal peptides of ApoE, ApoA-I, ApoA-II and serum AA protein form Congo red-positive fibrils [177]. These data indicate that ApoE aggregation may contribute to amyloid deposition [177].

The effect of ApoE on Aβ fibrillogenesis *in vitro* is more difficult to decipher from the published data, as the source of ApoE, the preparation and species of Aβ (1–40 versus 1–42), and the method of assessing fibril formation varies from report to report. For example, Aβ fibril formation is affected by the secondary structure of the peptide [178]. Early reports using purified ApoE suggested that ApoE enhances the formation of Aβ into fibrils [179], with ApoE4 a more potent catalyst than ApoE3 [54, 56, 180]. The type of Aβ fibril formed also appeared to depend on the presence of ApoE3 or E4 [55]. ApoE purified from plasma, primarily ApoE3, also increased the fibrillogenesis of synthetic peptides of gelsolin and AA [159]. However, later publications using lipidated forms of ApoE suggest that there is no isoform specificity or that both ApoE3 and E4 may actually delay the nucleation or seeding that precedes fibril formation [57, 181, 182]. Like ApoE receptor-binding and SDS-stable ApoE:Aβ complex formation described previously, purified and lipidated ApoE preparations appear to have different effects on fibril formation *in vitro*. The physiologic relevance of research results using purified ApoE remains questionable.

10.3.7
ApoE and Amyloid Deposition

ε4 correlates with amyloid burden in a dose-dependent manner in Alzheimer's disease patients [38, 39]. Further observations have been made in crosses between human ApoE and AβPP transgenic mice. ApoE-knock-out (KO) mice [183, 184] have been used to assess the role of ApoE in central nervous system function and are the background for several human ApoE transgenic mouse lines. Heterologous promoters have been used to drive the expression of human ApoE in neurons and/or glia, including glial fibrillary acidic protein (GFAP) [185–188], transferrin [189], platelet-derived growth factor (PDGF) [185] and neuron-specific enolase (NSE) [190]. The endogenous ApoE promoter has also been used, either with cosmid vectors containing the entire human ApoE multi-gene complex [191] or knock-in mice expressing human ApoE under the control of the endogenous mouse ApoE promoter [192, 193].

Transgenic mice expressing human AβPP with familial Alzheimer's disease mutations develop Aβ deposits in patterns similar to that seen in the Alzheimer's disease brain [194, 195]. Transgenic mice expressing human AβPP with the familial Alzheimer's disease mutation V717F (PDAPP mice) [194] crossed with ApoE-KO mice resulted in a dramatic overall reduction in Aβ deposition [40] as well as an altered deposition pattern [78]. Crosses of ApoE-KO/PDAPP mice with GFAP-human ApoE3 and E4 transgenic mice caused a further reduction of even diffuse Aβ deposits at young ages [42] although Aβ deposition increased as the mice aged, with ε4 correlating with greater levels of Aβ deposition than ε3 [41]. In addition, traumatic brain injury in these mice induced a significant increase in Aβ deposition specific to GFAP-ApoE4 mice [196]. Therefore, the net effect of expression of GFAP-human ApoE in PDAPP mice is an isoform-specific delay (E3 > E4) in Aβ deposition from 6–15 to 15–24 months.

Other reports demonstrate that in humans [116, 117] and PDAPP mice several measures of neurodegenerative pathology are independent of amyloid deposition [119–125]. Indeed, total levels of Aβ1–42 correlate with cognitive impairment even in the absence of detectable amyloid plaques [117, 122]. In PDAPP mice crossed with NSE-ApoE mice, ApoE3, but not E4, delays age- and Aβ-dependent synaptic and cholinergic deficits by a mechanism independent of plaque deposition [197]. Spatial memory is compromised in the presence of ApoE4 and not ApoE3 in these mice despite comparable concentrations of Aβ and the absence of detectable plaques [198]. These studies suggest that soluble Aβ rather than amyloid may produce the deficits reported and that ApoE may function both in the deposition and clearance of Aβ. However, as ApoE expression was driven by heterologous promotors, the impact of ApoE expression in its endogenous context on these results remains unknown. For example, glia, in particular astrocytes, are the primary cell type in the CNS that synthesize ApoE [199–208], making the relevance of overexpression of ApoE by neurons (NSE-ApoE mice) unclear.

10.3.8
ApoE Receptors in the CNS

ApoE receptors appear to mediate the effects of ApoE in neural cell processes, in general, and in the pathophysiology of Alzheimer's disease, in particular. First, neural cells express a variety of endocytic receptors in the LDLR family. LDLR is expressed by glia [38, 209], LRP is associated primarily with neurons, microglia and activated astrocytes [38, 210–212], ER2 is present in neurons [2, 213], and VLDLR is found primarily in neurons and activated microglia [214]. Second, ApoE enhances neurite outgrowth *in vitro* by a mechanism requiring ApoE receptors [186, 215–217]. Third, immunoreactivity for LRP and a number of its ligands including ApoE and α_2-macroglobulin (α_2M) is associated with senile plaques, and LRP expression is increased in Alzheimer's disease brains [212, 218]. Fourth, genetic evidence suggests that polymorphisms in either LRP or α_2M increase the risk of Alzheimer's disease [219, 220]. Finally, we have demonstrated that ApoE receptors are necessary for ApoE3-mediated protection against Aβ-induced neurotoxicity [60] and for Aβ-induced, glial-mediated inflammation [65]. Thus, the use of ApoE in a conformation that maintains ApoE receptor binding affinity may be critical to modulation of Aβ structure and function by ApoE, particularly if ApoE clears peptide intermediates via ApoE receptors.

10.3.9
ApoE and Aβ-induced Neurotoxicity

It has been difficult to isolate and determine the conformational species of Aβ responsible for its neural activity. Of the studies that have addressed the effect of ApoE on Aβ-induced activity in cell cultures, it is unclear what assembly of the peptide was responsible for the activity or what Aβ species interacted with

ApoE [59, 60]. For example, HEK-ApoE3, but not ApoE4, prevented Aβ-induced neurotoxicity in primary rat hippocampal cells [60]. SDS-stable complex formation may play a role in the ApoE3 protective effect as ApoE3 : Aβ complex in the media of these cultures increased over time, while ApoE4 : Aβ complex remained difficult to detect [60]. In addition, the ApoE3 inhibition of Aβ-induced neurotoxicity was dependent on ApoE receptor activity.

The effect of ApoE on oligomer- and fibril Aβ1–42-induced neurotoxicity was also examined using HEK-ApoE3 and E4. In the absence of ApoE, oligomeric Aβ1–42 is 10-fold more neurotoxic than fibril assemblies, as previously observed [144]. The addition of exogenous HEK-ApoE3, but not E4, attenuated the neurotoxicity of oligomeric Aβ1–42. Neither ApoE3 nor E4 affected the neurotoxicity induced by fibrillar Aβ [143]. Pillot et al. also demonstrated that ApoE2 and E3, but not E4, protects cortical neurons against apoptotic cell death induced by non-fibrillar Aβ, but has no effect on fibrillar-induced toxicity [62]. Further, the effect of ApoE2 and E3 was dependent on complex formation between ApoE and the C-terminus of Aβ. In addition, crossing ApoE2 transgenic mice with AβPP transgenic mice prevented soluble Aβ-induced dendritic spine loss [221]. Again, these *in vitro* results support the hypothesis that ApoE3 inhibits Aβ-induced neurotoxicity by binding oligomeric Aβ.

10.3.10 Conclusion

The exact mechanism by which ApoE and Aβ contribute to the pathogenesis of Alzheimer's disease is not yet entirely understood. However, a physical association between ApoE and Aβ may serve as either a means of clearing a soluble form of the peptide via the recognition of ApoE : Aβ complexes by ApoE receptors [38, 46] or ApoE may act as a pathological chaperone, facilitating the deposition of amyloid [36, 40, 56]. As illustrated in Fig. 10.3, our overall hypothesis is that ApoE may have two distinct interactions with Aβ, both reflecting isoform-specificity. First, ApoE interacts with oligomeric Aβ, ApoE3 > ApoE4, a process that may include binding and clearance of ApoE3:oligomer complexes by ApoE receptors. Second, ApoE facilitates the deposition of Aβ as amyloid, ApoE4 > ApoE3, as in humans [38, 39] and transgenic mice [41, 196, 222], ApoE4 correlates with amyloid burden. Together these interactions provide a more complete explanation of possible mechanisms responsible for the pathologies that characterize Alzheimer's disease.

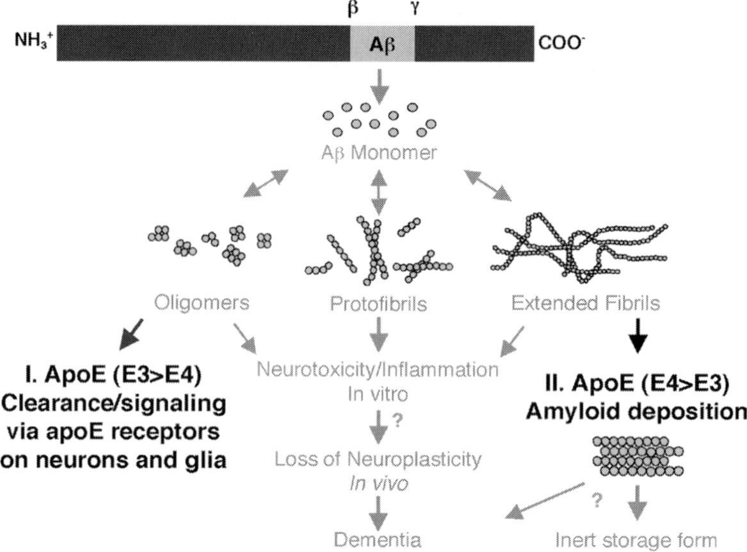

Fig. 10.3 Effect of ApoE on assembly of Aβ1–42. Our overall hypothesis is that ApoE has two general functions in relation to Aβ. First, ApoE interacts with oligomeric Aβ via an ApoE receptor-mediated processes to inhibit neurotoxicity and neuroinflammation, ApoE3 > ApoE4, a process possibly related to binding and clearance of ApoE3:oligomer complexes. Second, ApoE facilitates the deposition of Aβ as amyloid, ApoE4 > ApoE3. (Reprinted from [143], with permission.)

10.4 Other Aβ Binding Proteins

10.4.1 ApoJ

One of the more recently identified apolipoproteins is ApoJ, or clusterin. As an apolipoprotein, ApoJ was first identified by immunoaffinity chromatography localized to a HDL–3 [223]. ApoJ has several similarities to ApoE. The concentration of both proteins in human plasma is similar (around 50 μg/ml) and, although the peripheral circulation contains several classes of lipoproteins and numerous apolipoproteins, ApoE and ApoJ are the only apolipoproteins synthesized within the brain. ApoJ is expressed by glia, neurons and the ependymal cells lining the ventricles [224, 225]. We have previously reported that rat glial cells secrete HDL-like lipoprotein particles with ApoE and ApoJ as the primary protein component [161, 162]. *In vivo*, ApoJ immunoreactivity is localized to senile plaques [37, 226, 227], and soluble Aβ is found complexed to ApoJ-containing lipoproteins in plasma and CSF [45]. ApoJ also increases in response to different brain insults and is upregulated in Alzheimer's disease brain [224, 226].

The effect of ApoJ on amyloid deposition *in vivo* has been studied using PDAβPP transgenic mice [194]. Results from crossing ApoJ-KO with PDAβPP mice are similar to those observed with ApoE-KO/PDAβPP mice, a significant reduction in amyloid deposition [40, 228]. However, the simple interpretation that ApoE [229] and ApoJ [228] each promote amyloid deposition is called into question by the results obtained from ApoE-KO/ApoJ-KO/PDAβPP. If each protein were to independently function to promote the deposition of amyloid, then removal of *both* should result in even lower amyloid levels. In fact, the exact opposite result was reported. The loss of both ApoE and ApoJ expression in PDAβPP mice resulted in more amyloid deposition and at an earlier age [230]. Furthermore ApoE, in conjunction with ApoJ, were shown to regulate CNS Aβ levels. Therefore, while both ApoE and ApoJ are involved with amyloid deposition, they also appear to work in concert to promote the clearance of extracellular soluble forms of Aβ.

In vitro, ApoJ interacts with Aβ to form a stable complex [49] and alters the aggregation of Aβ to affect Aβ-induced neurotoxicity [49, 138, 139]. gp330, the only known receptor for mammalian ApoJ, is expressed by choroid plexus and ependyma, as well as brain capillary endothelial cells at the BBB [231–233]. Thus, while ApoJ is not likely to be important in receptor-mediated processes operating within the brain, ApoJ may play a role in transport of Aβ across the BBB as an ApoJ:Aβ complex ([234]; for review, see [235]).

10.4.2
α_2-macroglobulin (α_2M) and α_1-antichymotrypsin (ACT)

α_2M and ACT are two proteins that may interact with Aβ in a manner similar to ApoE. α_2M is a pan-protease inhibitor that functions in the clearance of several small molecules as a ligand for LRP [236, 237]. A polymorphism in an intronic region of the α_2M gene located on chromosome 12 has been shown to co-segregate with a late-onset sporadic Alzheimer's disease phenotype, suggesting that α_2M is a risk factor for Alzheimer's disease [220, 238]. *In vivo,* α_2M co-localizes with amyloid plaques [35, 212] and binds Aβ *in vitro* [239]. Functionally, α_2M may be involved in the clearance of soluble Aβ via LRP [237] whereby mutations in either α_2M or LRP affect Aβ levels contributing to Alzheimer's disease pathology.

Neuroinflammation is recognized as an important part of Alzheimer's disease pathology and release of inflammatory cytokines may contribute to neuronal dysfunction in Alzheimer's disease. ACT is an acute-phase inflammatory protein released from activated astrocytes and ACT also co-localizes with amyloid plaques (for review, see [240]). ACT binds Aβ *in vitro* [54] and has been shown to promote amyloid deposition *in vivo* [241], functioning as a "pathological chaperone" in a manner similar to ApoE [242]. However, like ApoE, other studies have demonstrated that ACT may inhibit Aβ fibril formation [243]. All these functions support the hypotheses of a physical interaction between these ApoJ/Aβ and a role in Alzheimer's disease pathogenesis.

Characterization of the components of amyloid plaques has fostered study of the interactions of several classes of proteins, although their mechanistic relationships are yet to be fully understood. Taking into account the importance of oligomeric forms of Aβ and the lack of evidence linking amyloid plaque formation and dementia, understanding the binding and clearance of soluble oligomeric Aβ involving ApoE, ApoJ, α_2M and ACT may provide a direct strategy for intervention in the pathogenesis of Alzheimer's disease.

Acknowledgments

The authors would like to thank Karl Weisgraber for defining the field of ApoE structure/function, Yadong Huang and Keith Crutcher for spirited discussion, Kate Reardon and Bill Rebeck for continuing input into our work, and Arlene Manelli for her collaborative efforts, including contributing Fig. 10.3. This research was supported by National Institute of Aging RO1 AG19121, RO1 AG20249 and PO1 AG021184, and the Charles Walgreen, Jr, Fund (M.J.L. D.).

References

1 Li, W.H., M. Tanimura, C.C. Luo, S. Datta and L. Chan. The apolipoprotein multigene family: biosynthesis, structure, structure–function relationships, and evolution. *J Lipid Res* **1988**, *29*, 245.

2 Kim, D.H., H. Iijima, K. Goto, J. Sakai, H. Ishii, H.J. Kim, H. Suzuki, H. Kondo, S. Saeki and T. Yamamoto. Human apolipoprotein E receptor 2. A novel lipoprotein receptor of the low density lipoprotein receptor family predominantly expressed in brain. *J Biol Chem* **1996**, *271*, 8373.

3 Das, H.K., J. McPherson, G.A. Bruns, S.K. Karathanasis and J.L. Breslow. Isolation, characterization, and mapping to chromosome 19 of the human apolipoprotein E gene. *J Biol Chem* **1985**, *260*, 6240.

4 Hoffer, M.J., M.H. Hofker, M.M. van Eck, L.M. Havekes and R.R. Frants. Evolutionary conservation of the mouse apolipoprotein e-c1-c2 gene cluster: structure and genetic variability in inbred mice. *Genomics* **1993**, *15*, 62.

5 Chang, D.J., Y.K. Paik, T.P. Leren, D.W. Walker, G.J. Howlett and J.M. Taylor. Characterization of a human apolipoprotein E gene enhancer element and its associated protein factors. *J Biol Chem* **1990**, *265*, 9496.

6 Grehan, S., E. Tse and J.M. Taylor. Two distal downstream enhancers direct expression of the human apolipoprotein E gene to astrocytes in the brain. *J Neurosci* **2001**, *21*, 812.

7 Mak, P.A., B.A. Laffitte, C. Desrumaux, S.B. Joseph, L.K. Curtiss, D.J. Mangelsdorf, P. Tontonoz and P.A. Edwards. Regulated expression of the ApoE/C-I/C-IV/C-II gene cluster in murine and human macrophages; a critical role for the nuclear receptors LXRalpha and LXRbeta. *J Biol Chem* **2002**, *24*, 24.

8 Simonet, W.S., N. Bucay, S.J. Lauer and J.M. Taylor. A far-downstream hepatocyte-specific control region directs expression of the linked human apolipoprotein E and C-I genes in transgenic mice. *J Biol Chem* **1993**, *268*, 8221.

9 Allan, C.M., D. Walker and J.M. Taylor. Evolutionary duplication of a hepatic control region in the human apolipoprotein E gene locus. Identification of a second region that confers high level and liver-specific expression of the hu-

man apolipoprotein E gene in transgenic mice. *J Biol Chem* **1995**, *270*, 26278.

10 Davignon, J., R.E. Gregg and C.F. Sing. Apolipoprotein E polymorphism and atherosclerosis. *Arteriosclerosis* **1988**, *8*, 1.

11 LaDu, M.J., C.A. Reardon, LJ. Van Eldik, A.M. Fagan, G. Bu, D. Holtzman and G S. Getz. Lipoproteins in the central nervous system. *Ann NY Acad Sci* **2000**, *903*, 167.

12 Linton, M.F., R. Gish, S.T. Hubl, E. Butler, C. Esquivel, W.I. Bry, J.K. Boyles, M.R. Wardell and S.G. Young. Phenotypes of apolipoprotein B and apolipoprotein E after liver transplantation. *J Clin Invest* **1991**, *88*, 270.

13 Weisgraber, K.H. Apolipoprotein E: structure–function relationships. *Adv Protein Chem* **1994**, *45*, 249.

14 Weisgraber, K.H., S.C. Rall, RW. Mahley, R.W. Milne, Y.L. Marcel and J.T. Sparrow. Human apolipoprotein E: determination of the heparin binding sites of apolipoprotein E3. *J Biol Chem* **1986**, *261*, 2068.

15 Wilson, C., M.R. Wardell, K.H. Weisgraber, R. Mahley, W. and D.A. Agard. Three-dimensional structure of the LDL-receptor-binding domain of human apolipoprotein E. *Science* **1991**, *252*, 1817.

16 Saito, H., P. Dhanasekaran, F. Baldwin, K.H. Weisgraber, M.C. Phillips and S. Lund-Katz. Effects of polymorphism on the lipid interaction of human apolipoprotein E. *J Biol Chem* **2003**, *278*, 40723.

17 Westerlund, J.A. and K.H. Weisgraber. Discrete carboxyl-terminal segments of apolipoprotein E mediate lipoprotein association and protein oligomerization. *J Biol Chem* **1993**, *268*, 15745.

18 Segrest, J.P., D.W. Garber, C.G. Brouillette, S.C. Harvey and G.M. Anantharamaiah. The amphipathic alpha helix: a multifunctional structural motif in plasma apolipoproteins. *Adv Protein Chem* **1994**, *45*, 303.

19 Dong, L.M., S. Parkin, S.D. Trakhanov, B. Rupp, T. Simmons, K.S. Arnold, Y. M. Newhouse, T.L. Innerarity and K.H. Weisgraber. Novel mechanism for defective receptor binding of apolipoprotein E2 in type III hyperlipoproteinemia. *Nat Struct Biol* **1996**, *3*, 718.

20 Bentley, N.M., M.J. Ladu, C. Rajan, G.S. Getz and C.A. Reardon. Apolipoprotein E structural requirements for the formation of SDS-stable complexes with beta-amyloid-(1–40): the role of salt bridges. *Biochem J* **2002**, *366*, 273.

21 Dong, L.M., C. Wilson, M.R. Wardell, T. Simmons, R.W. Mahley, K.H. Weisgraber and D.A. Agard. Human apolipoprotein E: role of arginine 61 in mediating the lipoprotein preferences of the E3 and E4 isoforms. *J Biol Chem* **1994**, *269*, 22358.

22 Wetterau, J.R., L.P. Aggerbeck, S.C. Rall, Jr and K. H. Weisgraber. Human apolipoprotein E3 in aqueous solution. I. Evidence for two structural domains. *J Biol Chem* **1988**, *263*, 6240.

23 Cho, H.S., B.T. Hyman, S.M. Greenberg and G.W. Rebeck. Quantitation of ApoE domains in Alzheimer disease brain suggests a role for ApoE in Abeta aggregation. *J Neuropathol Exp Neurol* **2001**, *60*, 342.

24 Tolar, M., M.A. Marques, J.A. Harmony and K. A. Crutcher. Neurotoxicity of the 22kDa thrombin-cleavage fragment of apolipoprotein E and related synthetic peptides is receptor-mediated. *J Neurosci* **1997**, *17*, 5678.

25 Tolar, M., J.N. Keller, S. Chan, M.P. Mattson, M.A. Marques and K.A. Crutcher. Truncated apolipoprotein E (ApoE) causes increased intracellular calcium and may mediate ApoE neurotoxicity. *J Neurosci* **1999**, *19*, 7100.

26 Harris, F.M., W. J. Brecht, Q. Xu, I. Tesseur, L. Kekonius, T. Wyss-Coray, J.D. Fish, E. Masliah, P.C. Hopkins, K. Scearce-Levie, K.H. Weisgraber, L. Mucke, R.W. Mahley and Y. Huang. Carboxyl-terminal-truncated apolipoprotein E4 causes Alzheimer's disease-like neurodegeneration and behavioral deficits in transgenic mice. *Proc Natl Acad Sci USA* **2003**, *100*, 10966.

27 Morrow, J.A., M.L. Segall, S. Lund-Katz, M.C. Phillips, M. Knapp, B. Rupp and K.H. Weisgraber. Differences in stability among the human apolipoprotein E iso-

forms determined by the amino-terminal domain. *Biochemistry* **2000**, *39*, 11657.

28 Acharya, P., M.L. Segall, M. Zaiou, J. Morrow, K.H. Weisgraber, M.C. Phillips, S. Lund-Katz and J. Snow. Comparison of the stabilities and unfolding pathways of human apolipoprotein E isoforms by differential scanning calorimetry and circular dichroism. *Biochim Biophys Acta* **2002**, *1584*, 9.

29 Morrow, J.A., D.M. Hatters, B. Lu, P. Hochtl, K.A. Oberg, B. Rupp and K.H. Weisgraber. Apolipoprotein E4 forms a molten globule. A potential basis for its association with disease. *J Biol Chem* **2002**, *277*, 50380.

30 Kowal, R.C., J. Herz, K.H. Weisgraber, R.W. Mahley, M.S. Brown and J.L. Goldstein. Opposing effects of apolipoproteins E and C on lipoprotein binding to low density lipoprotein receptor-related protein. *J Biol Chem* **1990**, *265*, 10771.

31 Mahley, R.W. Apolipoprotein E: cholesterol transport protein with expanding role in cell biology. *Science* **1988**, *240*, 622.

32 Strittmatter, W.J., A.M. Saunders, D. Schmechel, M. Pericak-Vance, J. Enghild, G.S. Salvesen and A. D. Roses. Apolipoprotein E: high avidity binding to β-amyloid and increased frequency of type 4 allele in late-onset familial Alzheimer disease. *Proc Natl Acad Sci USA* **1993**, *90*, 1977.

33 Innerarity, T.L. and R.W. Mahley. Enhanced binding by cultured human fibroblasts of apo-E-containing lipoproteins as compared with low density lipoproteins. *Biochemistry* **1978**, *17*, 1440.

34 Innerarity, T., D. Hui, T. Bersot and R.W. Mahley. Type III hyperlipoproteinemia: a focus on lipoprotein receptor–apolipoprotein E2 interactions. *Adv Exp Med Biol* **1986** *201*, 273.

35 Namba, Y., M. Tomonaga, H. Kawasaki, E. Otomo and K. Ikeda. Apolipoprotein E immunoreactivity in cerebral amyloid deposits and neurofibrillary tangles in Alzheimer's disease kuru plaque amyloid in Creutzfeldt-Jacob disease. *Brain Res* **1991**, *541*, 163.

36 Wisniewski, T. and B. Frangione. Apolipoprotein E: a pathological chaperone protein in patients with cerebral and systemic amyloid. *Neurosci Lett* **1992**, *135*, 235.

37 Choi-Miura, N.H., Y. Ihara, K. Fukuchi, M. Takeda, Y. Nakano, T. Tobe and M. Tomita. SP-40,40 is a constituent of Alzheimer's amyloid. *Acta Neuropathol* **1992**, *83*, 260.

38 Rebeck, G.W., J.S. Reiter, D.K. Strickland and B.T. Hyman. Apolipoprotein E in sporadic Alzheimer's disease: allelic variation and receptor interactions. *Neuron* **1993**, *11*, 575.

39 Schmechel, D.E., A.M. Saunders, W.J. Strittmatter, B.J. Crain, C.M. Hulette, S.H. Joo, M.A. Pericak-Vance, D. Goldgaber and A. D. Roses. Increased amyloid β-peptide deposition in cerebral cortex as a consequence of apolipoprotein genotype in late-onset Alzheimer disease. *Proc Natl Acad Sci USA* **1993**, *90*, 9649.

40 Bales, K.R., T. Verina, R.C. Dodel, Y. Du, L. Altstiel, M. Bender, P. Hyslop, E.M. Johnstone, S.P. Little, D.J. Cummins, P. Piccardo, B. Ghetti and S.M. Paul. Lack of apolipoprotein E dramatically reduces amyloid β-peptide deposition. *Nat Genet* **1997**, *17*, 263.

41 Holtzman, D.M., K.R. Bales, T. Tenkova, A.M. Fagan, M. Parsadanian, L.J. Sartorius, B. Macky, J. Olney, D. McKeel, D. Wozniak and S.M. Paul. Apolipoprotein E isoform-dependent amyloid deposition and neuritic degeneration in a mouse model of Alzheimer's disease. *Proc Natl Acad Sci USA* **2000**, *97*, 2892.

42 Holtzman, D.M., K.R. Bales, S. Wu, P. Bhat, M. Parsadanian, A.M. Fagan, L.K. Chang, Y. Sun and S.M. Paul. *In vivo* expression of apolipoprotein E reduces amyloid-β deposition in a mouse model of Alzheimer's disease. *J Clin Invest* **1999**, *103*, R15.

43 Kuo, Y.M., M.R. Emmerling, C. Vigo-Pelfrey, T.C. Kasunic, J.B. Kirkpatrick, G.H. Murdoch, M.J. Ball and A.E. Roher. Water-soluble Abeta (N-40, N-42) oligomers in normal and Alzheimer dis-

ease brains. *J Biol Chem* **1996**, *271*, 4077.

44 Wisniewski, T., A. Golabek, E. Matsubara, J. Ghiso and B. Frangione. Apolipoprotein E: binding to soluble Alzheimer's beta-amyloid. *Biochem Biophys Res Commun* **1993**, *192*, 359.

45 Ghiso, J., E. Matsubara, A. Koudinov, N.H. Choi-Miura, M. Tomita, T. Wisniewski and B. Frangione. The cerebrospinal-fluid form of Alzheimer's amyloid beta is complexed to SP-40,40 (apolipoprotein J), an inhibitor of the complement membrane-attack complex. *Biochem J* **1993**, *293*, 27.

46 LaDu, M.J., M.T. Falduto, A.M. Manelli, C.A. Reardon, G.S. Getz and D.E. Frail. Isoform-specific binding of apolipoprotein E to β-amyloid. *J Biol Chem* **1994**, *269*, 23404.

47 LaDu, M.J., T.M. Pederson, D.E. Frail, C.A. Reardon, G.S. Getz and M.T. Falduto. Purification of apolipoprotein E attenuates isoform-specific binding to β-amyloid. *J Biol Chem* **1995**, *270*, 9030.

48 LaDu, M.J., J.R. Lukens, C.A. Reardon and G.S. Getz. Association of human, rat, and rabbit apolipoprotein E with beta-amyloid. *J Neurosci Res* **1997**, *49*, 9.

49 Matsubara, E., B. Frangione and J. Ghiso. Characterization of apolipoprotein J-Alzheimer's Aβ interaction. *J Biol Chem* **1995**, *270*, 7563.

50 Aleshkov, S., C.R. Abraham and V.I. Zannis. Interaction of nascent ApoE2, ApoE3, and ApoE4 isoforms expressed in mammalian cells with amyloid peptide β(1–40): relevance to Alzheimer's disease. *Biochemistry* **1997**, *36*, 10571.

51 Tokuda, T., M. Calero, E. Matsubara, R. Vidal, A. Kumar, B. Permanne, B. Zlokovic, J. Smith, M.J. LaDu, A. Rostagno, B. Frangione and J. Ghiso. Lipidation of apolipoprotein E influences its isoform-specific interaction with Alzheimer's amyloid β-peptide. *Biochem J* **2000**, *348*, 359.

52 Yang, D.-S., J.D. Smith, Z. Zhou, S. Gandy and R.N. Martins. Characterization of the binding of amyloid-β peptide to cell culture-derived native apolipoprotein E2, E3, and E4 isoforms and to isoforms from human plasma. *J Neurochem* **1997**, *68*, 721.

53 Zhou, Z., J.D. Smith, P. Greengard and S. Gandy. Alzheimer amyloid-beta peptide forms denaturant-resistant complex with type epsilon 3 but not type epsilon 4 isoform of native apolipoprotein E. *Mol Med* **1996**, *2*, 175.

54 Ma, J., A. Yee, H.B. Brewer, S. Das and H. Potter. Amyloid-associated proteins alpha-1-antichymotrypsin and apolipoprotein E promote assembly of Alzheimer beta-protein into filaments. *Nature* **1994**, *372*, 92.

55 Sanan, D.A., K. H. Weisgraber, S.J. Russel, R.W. Mahley, D. Huang, A. Saunders, D. Schmechel, T. Wisniewksi, B. Frangione, B. Roses, A.D. Roses and W.J. Strittmatter. Apolipoprotein E associates with β amyloid peptide of Alzheimer's disease to form novel monofibrils. *J Clin Invest* **1994**, *94*, 860.

56 Wisniewski, T., E.M. Castano, A. Golabek, T. Vogel and B. Frangione. Acceleration of Alzheimer's fibril formation by apolipoprotein E *in vitro*. *Am J Path* **1994**, *145*, 1030.

57 Evans, K.C., E.P. Berger, C.G. Cho, K.H. Weisgraber and P. T. Lansbury. Apolipoprotein E is a kinetic but not a thermodynamic inhibitor of amyloid formation: implications for the pathogenesis and treatment of Alzheimer's disease. *Proc Natl Acad Sci USA* **1995**, *92*, 763.

58 Ma, J., H.B. Brewer, Jr and H. Potter. Alzheimer A beta neurotoxicity: promotion by antichymotrypsin, ApoE4; inhibition by A beta-related peptides. *Neurobiol Aging* **1996**, *17*, 773.

59 Miyata, M. and J.D. Smith. Apolipoprotein E allele-specific antioxidant activity and effects on cytotoxicity by oxidative insults and beta-amyloid peptides. *Nat Genet* **1996**, *14*, 55.

60 Jordán, J., M.F. Galindo, R.J. Miller, C. A. Reardon, G.S. Getz and M.J. LaDu. Isoform-specific effect of apolipoprotein E on cell survival and β-amyloid-induced toxicity in rat hippocampal pyramidal neuronal cultures. *J Neurosci* **1998**, *18*, 195.

61 Whitson, J.S., M.P. Mims, W.J. Strittmatter, T. Yamaki, J.D. Morrisett and S.H. Appel. Attenuation of the neurotoxic effect of A beta amyloid peptide by apolipoprotein E. *Biochem Biophys Res Commun* **1994**, *199*, 163.

62 Drouet, B., A. Fifre, M. Pincon-Raymond, J. Vandekerckhove, M. Rosseneu, J.L. Gueant, J. Chambaz and T. Pillot. ApoE protects cortical neurones against neurotoxicity induced by the non-fibrillar C-terminal domain of the amyloid-beta peptide. *J Neurochem* **2001**, *76*, 117.

63 Koistinaho, M., S. Lin, X. Wu, M. Esterman, D. Koger, J. Hanson, R. Higgs, F. Liu, S. Malkani, K.R. Bales and S.M. Paul. Apolipoprotein E promotes astrocyte colocalization and degradation of deposited amyloid-beta peptides. *Nat Med* **2004**, *10*, 719.

64 Guo, L., M.J. LaDu and L.J. Van Eldik. A dual role for apolipoprotein E in neuroinflammation: anti- and pro-inflammatory activity. *J Mol Neurosci* **2004**, *23*, 205.

65 LaDu, M.J., J.A. Shah, C.A. Reardon, G.S. Getz, G. Bu, J. Hu, L. Guo and L.J. Van Eldik. Apolipoprotein E receptors mediate the effects of beta-amyloid on astrocyte cultures. *J Biol Chem* **2000**, *275*, 33974.

66 Hu, J., M.J. LaDu and L.J. Van Eldik. Apolipoprotein E attenuates beta-amyloid-induced astrocyte activation. *J Neurochem* **1998**, *71*, 1626.

67 LaDu, M.J., J.A. Shah, C.A. Reardon, G. S. Getz, G. Bu, J. Hu, L. Guo and L.J. Van Eldik. Apolipoprotein E and apolipoprotein E receptors modulate A beta-induced glial neuroinflammatory responses. *Neurochem Int* **2001**, *39*, 427.

68 Laskowitz, D.T., D.M. Lee, D. Schmechel and H.F. Staats. Altered immune responses in apolipoprotein E-deficient mice. *J Lipid Res* **2000**, *41*, 613.

69 Lynch, J.R., W. Tang, H. Wang, M.P. Vitek, E.R. Bennett, P.M. Sullivan, D.S. Warner and D.T. Laskowitz. ApoE genotype and an ApoE-mimetic peptide modify the systemic and CNS inflammatory response. *J Biol Chem* **2003.**

70 Lynch, J.R., D. Morgan, J. Mance, W.D. Matthew and D.T. Laskowitz. Apolipoprotein E modulates glial activation and the endogenous central nervous system inflammatory response. *J Neuroimmunol* **2001**, *114*, 107.

71 Guillaume, D., P. Bertrand, D. Dea, J. Davignon and J. Poirier. Apolipoprotein E and low-density lipoprotein binding and internalization in primary cultures of rat astrocytes: isoform-specific alterations. *J Neurochem* **1996**, *66*, 2410.

72 Cole, G.M. and M.D. Ard. Influence of lipoproteins on microglial degradation of Alzheimer's amyloid beta-protein. *Microsc Res Tech* **2000**, *50*, 316.

73 Cole, G.M., W. Beech, S.A. Frautschy, J. Sigel, C. Glasgow and M.D. Ard. Lipoprotein effects on Abeta accumulation and degradation by microglia *in vitro*. *J Neurosci Res* **1999**, *57*, 504.

74 Urmoneit, B., I. Prikulis, G. Wihl, D. D'Urso, R. Frank, J. Heeren, U. Beisiegel and R. Prior. Cerebrovascular smooth muscle cells internalize Alzheimer amyloid beta protein via a lipoprotein pathway: implications for cerebral amyloid angiopathy. *Lab Invest* **1997**, *77*, 157.

75 Yang, D.S., D.H. Small, U. Seydel, J.D. Smith, J. Hallmayer, S. E. Gandy and R.N. Martins. Apolipoprotein E promotes the binding and uptake of beta-amyloid into Chinese hamster ovary cells in an isoform-specific manner. *Neuroscience* **1999**, *90*, 1217.

76 Winkler, K., H. Scharnagl, U. Tisljar, H. Hoschutzky, I. Friedrich, M.M. Hoffmann, M. Huttinger, H. Wieland and W. Marz. Competition of Abeta amyloid peptide and apolipoprotein E for receptor- mediated endocytosis. *J Lipid Res* **1999**, *40*, 447.

77 Beffert, U., N. Aumont, D. Dea, S. Lussier-Cacan, J. Davignon and P.J. Apolipoprotein E isoform-specific reduction of extracellular amyloid in neuronal cultures. *Mol Brain Res* **1999**, *68*, 181.

78 Irizarry, M.C., B.S. Cheung, G.W. Rebeck, S.M. Paul, K.R. Bales and B.T. Hyman. Apolipoprotein E affects the amount, form, and anatomical distribution of amyloid beta-peptide deposition in homozygous APP(V717F) transgenic mice. *Acta Neuropathol (Berl)* **2000**, *100*, 451.

79 Saunders, A.M., K. Schmader, J.C.S. Breitner, M.D. Benson, W.T. Brown, L. Goldfarb, D. Goldgaber, M.G. Manwaring, M.H. Szymanski, N. McCown, K.C. Dole, D.E. Schmechel, W.J. Strittmatter, M.A. Pericak-Vance and A.D. Roses. Apolipoprotein E ε4 allele distributions in late-onset Alzheimer's disease and in other amyloid-forming diseases. *Lancet* **1993**, *342*, 710.

80 Corder, E.H., S.M. Saunder, W.J. Strittmatter, D.E. Schmechel, P.C. Gaskell, G.W. Small, A.D. Roses, J.L. Haines and M.A. Pericak-Vance. Gene dose of apolipoprotein E type 4 allele and the risk of Alzheimer's disease in late onset families. *Science* **1993**, *261*, 921.

81 van Duijn, C.M., P. de Knijff, M. Cruts, A. Wehnert, L.M. Havekes, A. Hofman and C. Van Broeckhoven. Apolipoprotein E4 allele in a population-based study of early-onset Alzheimer's disease. *Nat Genet* **1994**, *7*, 74.

82 Houlden, H., R. Crook, K. Duff, C.J., R.P., M. Rossor and J. Hardy. Confirmation that the ApoE4 allele is associated with late onset, familial Alzheimer's disease. *Neurodegeneration* **1993**, *2*, 283.

83 Yu, C.E., H. Payami, J.M. Olson, M. Boehnke, E.M. Wijsman, H.T. Orr, W.A. Kukull, K.A. Goddard, E. Nemens, J.A. White, M.E. Alonso, T.D. Taylor, M.J. Ball, J. Kaye, J. Morris, H. Chui, A.D. Sadovnick, G.M. Martin, E.B. Larson, L.L. Heston, T.D. Bird and G.D. Schellenberg. The apolipoprotein E/CI/CII gene cluster and late-onset Alzheimer disease. *Am J Hum Genet* **1994**, *54*, 631.

84 Frisoni, G.B., S. Govoni, C. Geroldi, A. Bianchetti, L. Calabresi, G. Franceschini and M. Trabucchi. Gene dose of the epsilon 4 allele of apolipoprotein E and disease progression in sporadic late-onset Alzheimer's disease. *Ann Neurol* **1995**, *37*, 596.

85 Pericak-Vance, M.A., C.C. Johnson, J.B. Rimmler, A.M. Saunders, L.C. Robinson, E.G. D'Hondt, C.E. Jackson and J.L. Haines. Alzheimer's disease and apolipoprotein E-4 allele in an Amish population. *Ann Neurol* **1996**, *39*, 700.

86 Mak, Y.T., H. Chiu, J. Woo, R. Kay, Y.S. Chan, E. Hui, K.H. Sze, C. Lum, T. Kwok and C.P. Pang. Apolipoprotein E genotype and Alzheimer's disease in Hong Kong elderly Chinese. *Neurology* **1996**, *46*, 146.

87 Chartier-Harlin, M.C., M. Parfitt, S. Legrain, J. Perez-Tur, T. Brousseau, A. Evans, C. Berr, O. Vidal, P. Roques, V. Gourlet, J.C. Fruchart, A. Delacourte, M. Rossor and P. Amouyel. Apolipoprotein E, epsilon 4 allele as a major risk factor for sporadic early and late-onset forms of Alzheimer's disease: analysis of the 19q13.2 chromosomal region. *Hum Mol Genet* **1994**, *3*, 569.

88 Tsai, M.S., E.G. Tangalos, R.C. Petersen, G.E. Smith, D.J. Schaid, E. Kokmen, R.J. Ivnik and S.N. Thibodeau. Apolipoprotein E: risk factor for Alzheimer disease. *Am J Hum Genet* **1994**, *54*, 643.

89 Corder, E.H., S.M. Saunder, N.J. Risch, W.J. Strittmatter, D.E. Schmechel, P.C. Gaskell, Jr, J.B. Rimmler, P.A. Locke, P.M. Conneally, K.E. Schmader, G.W. Small, A.D. Roses, J.L. Haines and M.A. Pericak-Vance. Protective effect of apolipoprotein E type 2 allele for late onset Alzheimer disease. *Nature Genetics* **1994**, *7*, 180.

90 Royston, M.C., D. Mann, S. Pickering-Brown, F. Owen, R. Perry, R. Raghavan, C. Khin-Nu, S. Tyrer, K. Day, R. Crook, J. Hardy and G.W. Roberts. Apolipoprotein E epsilon 2 allele promotes longevity and protects patients with Down's syndrome from dementia. *Neuroreport* **1994**, *5*, 2583.

91 Greenberg, S.M., G.W. Rebeck, J.P. Vonsattel, T. Gomez-Isla and B.T. Hyman. Apolipoprotein E epsilon 4 and cerebral hemorrhage associated with amyloid angiopathy. *Ann Neurol* **1995**, *38*, 254.

92 Hyman, B.T., H.L. West, G.W. Rebeck, F. Lai and D.M. Mann. Neuropathological changes in Down's syndrome hippocampal formation. Effect of age and apolipoprotein E genotype. *Arch Neurol* **1995**, *52*, 373.

93 Nicoll, J. A. R., G. W. Roberts and D. I. Graham. ApoE E4 allele is associated with deposition of amyloid beta-protein following head injury. *Nat Med* **1995**, *1*, 135.

94 Polvikoski, T., R. Sulkava, M. Haltia, K. Kainulainen, A. Vuorio, A. Verkkoniemi, L. Niinisto, P. Halonen and K. Kontula. Apolipoprotein E, dementia and cortical deposition of beta-amyloid protein. *N Engl J Med* **1995**, *333*, 1242.

95 Lai, F., E. Kammann, G. W. Rebeck, A. Anderson, Y. Chen and R. A. Nixon. APOE genotype and gender effects on Alzheimer disease in 100 adults with Down syndrome. *Neurology* **1999**, *53*, 331.

96 Alonzo, N. C., B. T. Hyman, G. W. Rebeck and S. M. Greenberg. Progression of cerebral amyloid angiopathy: accumulation of amyloid-beta40 in affected vessels. *J Neuropathol Exp Neurol* **1998**, *57*, 353.

97 Bretsky, P., J. M. Guralnik, L. Launer, M. Albert and T. E. Seeman. The role of APOE-epsilon4 in longitudinal cognitive decline: MacArthur Studies of Successful Aging. *Neurology* **2003**, *60*, 1077.

98 Crawford, F. C., R. D. Vanderploeg, M. J. Freeman, S. Singh, M. Waisman, L. Michaels, L. Abdullah, D. Warden, R. Lipsky, A. Salazar and M. J. Mullan. APOE genotype influences acquisition and recall following traumatic brain injury. *Neurology* **2002**, *58*, 1115.

99 McCarron, M. O., C. J. Weir, K. W. Muir, K. L. Hoffmann, C. Graffagnino, J. A. Nicoll, K. R. Lees and M. J. Alberts. Effect of apolipoprotein E genotype on in-hospital mortality following intracerebral haemorrhage. *Acta Neurol Scand* **2003**, *107*, 106.

100 Liu, Y., M. P. Laakso, J. O. Karonen, R. L. Vanninen, J. Nuutinen, S. Soimakallio and H. J. Aronen. Apolipoprotein E polymorphism and acute ischemic stroke: a diffusion- and perfusion-weighted magnetic resonance imaging study. *J Cereb Blood Flow Metab* **2002**, *22*, 1336.

101 Tardiff, B. E., M. F. Newman, A. M. Saunders, W. J. Strittmatter, J. A. Blumenthal, M. D. White, N. D. Croughwell, R. D. J. Davis, A. D. Roses and J. G. Reves. Preliminary report of a genetic basis for cognitive decline after cardiac operations. The neurologic Outcome Research Group of the Duke Heart Center. *Ann Thorac Surg* **1997**, *64*, 715.

102 Teasdale, G. M., J. A. R. Nicoll, G. Murray and M. Fiddes. Association of apolipoprotein E polymorphism with outcome after head injury. *Lancet* **1997**, *350*, 1069.

103 Slooter, A. J. C., M. X. Tang and C. M. Vanduijn. Apolipoprotein E epsilon-4 and the risk of dementia with stroke-a population based investigation. *J Am Med Ass* **1997**, *277*, 818.

104 Friedman, G., P. Froom, L. Sazbon, I. Grinblatt, M. Shochina, J. Tsenter, S. Babaey, B. Yehuda and Z. Groswasser. Apolipoprotein E-epsilon4 genotype predicts a poor outcome in survivors of traumatic brain injury. *Neurology* **1999**, *52*, 244.

105 Sorbi, S., B. Nacmias, S. Piacentini, A. Repice, S. Latorraca, P. Forleo and L. Amaducci. ApoE as a prognostic factor for post-traumatic coma. *Nat Med* **1995**, *1*, 852.

106 Zarepesi, S., J. Kaye, R. Camicioli, H. Grimslid, B. Oken, M. Litt, J. Nutt, T. Bird, G. Schellenberg and H. Payami. Modulation of the age at onset of Parkinson's disease by apolipoprotein E genotype. *Ann Neurol* **1997**, *42*, 655.

107 Benjamin, R., A. Leake, J. A. Edwardson, I. G. McKeith, P. G. Ince, R. H. Perry and C. M. Morris. Apolipoprotein E genes in Lewy body and Parkinson's disease. *Lancet* **1994**, *343, 1565*

108 Mui, S., G. W. Rebeck, D. McKenna-Yasek, B. T. Hyman and R. H. Brown, Jr. Apolipoprotein E epsilon 4 allele is not associated with earlier age at onset in amyotrophic lateral sclerosis. *Ann Neurol* **1995**, *38*, 460.

109 Gomez-Isla, T., H. L. West, G. W. Rebeck, S. D. Harr, J. H. Growdon, J. J. Locascio, T. T. Perls, L. A. Lipsitz and B. T. Hyman. Clinical and pathological correlates of apolipoprotein E epsilon 4 in

Alzheimer's disease. *Ann Neurol* **1996**, *39*, 62.

110 Broderick, J., M. Lu, C. Jackson, A. Pancioli, B.C. Tilley, S.C. Fagan, R. Kothari, S.R. Levine, J.R. Marler, P.D. Lyden, E.C. Haley, Jr, T. Brott and J.C. Grotta. Apolipoprotein E phenotype and the efficacy of intravenous tissue plasminogen activator in acute ischemic stroke. *Ann Neurol* **2001**, *49*, 736.

111 Wilson, R.S., J.L. Bienias, E. Berry-Kravis, D.A. Evans and D.A. Bennett. The apolipoprotein E epsilon 2 allele and decline in episodic memory. *J Neurol Neurosurg Psychiatry* **2002**, *73*, 672.

112 Selkoe, D.J. Alzheimer's disease: genes, proteins, and therapy. *Physiol Rev* **2001**, *81*, 741.

113 Pike, C.J., D. Burdick, A.J. Walencewicz, C.G. Glabe and C.W. Cotman. Neurodegeneration induced by beta-amyloid peptides *in vitro*: the role of peptide assembly state. *J Neurosci* **1993**, *13*, 1676.

114 Pike, C.J., A.J. Walencewicz, C.G. Glabe and C. W. Cotman. *In vitro* aging of beta-amyloid protein causes peptide aggregation and neurotoxicity. *Brain Res* **1991**, *563*, 311.

115 Lorenzo, A. and B.A. Yankner. Beta-amyloid neurotoxicity requires fibril formation and is inhibited by Congo red. *Proc Natl Acad Sci USA* **1994**, *91*, 12243.

116 Cummings, B.J. and C.W. Cotman. Image analysis of beta-amyloid load in Alzheimer's disease and relation to dementia severity. *Lancet* **1995**, *346*, 1524.

117 Naslund, J., V. Haroutunian, R. Mohs, K.L. Davis, P. Davies, P. Greengard and J.D. Buxbaum. Correlation between elevated levels of amyloid beta-peptide in the brain and cognitive decline. *J Am Med Ass* **2000**, *283*, 1571.

118 Arriagada, P.V., J.H. Growdon, E.T. Hedley-Whyte and B.T. Hyman. Neurofibrillary tangles but not senile plaques parallel duration and severity of Alzheimer's disease. *Neurology* **1992**, *42*, 631.

119 Irizarry, M.C., M. McNamara, K. Fedorchak, K. Hsiao and B.T. Hyman. APPsw transgenic mice develop age-related Aβ deposits and neuropil abnormalities, but no neuronal loss in CA1. *J Neuropath Exp Neurol* **1997**, *56*, 965.

120 Irizarry, M.C., F. Soriano, M. McNamara, K.J. Page, D. Schenk, D. Games and B.T. Hyman. Aβ deposition is associated with neuropil changes, but not with overt neuronal loss in the human amyloid precursor protein V717F (PDAPP) transgenic mouse. *J Neurosci* **1997**, *17*, 7053.

121 Hsia, A.Y., E. Masliah, L. McConlogue, G.-Q. Yu, G. Tatsuno, K. Hu, D. Kholodenko, R.C. Malenka, R.A. Nicoll and L. Mucke. Plaque-independent disruption of neural circuits in Alzheimer's disease mouse models. *Proc Natl Acad Sci USA* **1999**, *96*, 3228.

122 Mucke, L., E. Masliah, G Q. Yu, M. Mallory, E.M. Rockenstein, G. Tatsuno, K. Hu, D. Kholodenko, K. Johnson-Wood and L. McConlogue. High-level neuronal expression of abeta 1–42 in wild-type human amyloid protein precursor transgenic mice: synaptotoxicity without plaque formation. *J Neurosci* **2000**, *20*, 4050.

123 Chui, D.H., H. Tanahashi, K. Ozawa, S. Ikeda, F. Checler, O. Ueda, H. Suzuki, W. Araki, H. Inoue, K. Shirotani, K. Takahashi, F. Gallyas and T. Tabira. Transgenic mice with Alzheimer presenilin 1 mutations show accelerated neurodegeneration without amyloid plaque formation. *Nat Med* **1999**, *5*, 560.

124 Holcomb, L., M.N. Gordon, E. McGowan, X. Yu, S. Benkovic, P. Jantzen, K. Wright, I. Saad, R. Mueller, D. Morgan, S. Sanders, C. Zehr, K. O'Campo, J. Hardy, C.-M. Prada, C. Eckman, S. Younkin, K. Hsaio and K. Duff. Accelerated Alzheimer-type phenotype in transgenic mice carrying both mutant *amyloid precursor protein* and *presenilin 1* transgenes. *Nat Med* **1998**, *4*, 97.

125 Moechars, D., I. Dewachter, K. Lorent, D. Reverse, V. Baekelandt, A. Naidu, I. Tesseur, K. Spittaels, C.V. Haute, F. Checler, E. Godaux, B. Cordell and F. Van Leuven. Early phenotypic changes in transgenic mice that overexpress different mutants of amyloid precursor protein in brain. *J Biol Chem* **1999**, *274*, 6483.

126 Aksenova, M.V., M.Y. Aksenov, D.A. Butterfield and J.M. Carney. alpha-1-antichymotrypsin interaction with A beta (1–40) inhibits fibril formation but does not affect the peptide toxicity. *Neurosci Lett* **1996**, *211*, 45.

127 Forloni, G., E. Lucca, N. Angeretti, P. Della Torre and M. Salmona. Amidation of beta-amyloid peptide strongly reduced the amyloidogenic activity without alteration of the neurotoxicity. *J Neurochem* **1997**, *69*, 2048.

128 Monji, A., K. Tashiro, I. Yoshida, H. Kaname, Y. Hayashi, K. Matsuda and N. Tashiro. Laminin inhibits both Abeta40 and Abeta42 fibril formation but does not affect Abeta40 or Abeta42-induced cytotoxicity in PC12 cells. *Neurosci Lett* **1999**, *266*, 85.

129 Klein, W.L., G.A. Krafft and C.E. Finch. Targeting small Abeta oligomers: the solution to an Alzheimer's disease conundrum? *Trends Neurosci* **2001**, *24*, 219.

130 Lansbury, P.T., Jr. Evolution of amyloid: what normal protein folding may tell us about fibrillogenesis and disease. *Proc Natl Acad Sci USA* **1999**, *96*, 3342.

131 Small, D.H. The Sixth International Conference on Alzheimer's disease, Amsterdam, The Netherlands, July 1998. The amyloid cascade hypothesis debate: emerging consensus on the role of A beta and amyloid in Alzheimer's disease. *Amyloid* **1998**, *5*, 301.

132 Terry, R.D. An honorable compromise regarding amyloid in Alzheimer disease. *Ann Neurol* **2001**, *49*, 684.

133 Haass, C. and H. Steiner. Protofibrils, the unifying toxic molecule of neurodegenerative disorders? *Nat Neurosci* **2001**, *4*, 859.

134 Roher, A.E., J. Baudry, M.O. Chaney, Y.M. Kuo, W. B. Stine and M. R. Emmerling. Oligomerization and fibril assembly of the amyloid-beta protein. *Biochim Biophys Acta* **2000**, *1502*, 31.

135 Roher, A.E., M.O. Chaney, Y.M. Kuo, S.D. Webster, W.B. Stine, L.J. Haverkamp, A.S. Woods, R.J. Cotter, J.M. Tuohy, G. A. Krafft, B. S. Bonnell and M.R. Emmerling. Morphology and toxicity of Abeta-(1–42) dimer derived from neuritic and vascular amyloid deposits of Alzheimer's disease. *J Biol Chem* **1996**, *271*, 20631.

136 Lue, L.F., Y.M. Kuo, A.E. Roher, L. Brachova, Y. Shen, L. Sue, T. Beach, J.H. Kurth, R.E. Rydel and J. Rogers. Soluble amyloid beta peptide concentration as a predictor of synaptic change in Alzheimer's disease. *Am J Pathol* **1999**, *155*, 853.

137 McLean, C.A., R.A. Cherny, F.W. Fraser, S.J. Fuller, M.J. Smith, K. Beyreuther, A.I. Bush and C.L. Masters. Soluble pool of Abeta amyloid as a determinant of severity of neurodegeneration in Alzheimer's disease. *Ann Neurol* **1999**, *46*, 860.

138 Oda, T., P. Wals, H.H. Osterburg, S.A. Johnson, G.M. Pasinetti, T.E. Morgan, I. Rozovsky, W.B. Stine, S.W. Snyder, T.F. Holzman, G.A. Krafft and C.E. Finch. Clusterin (apoJ) alters the aggregation of amyloid beta-peptide (A beta 1–42) and forms slowly sedimenting A beta complexes that cause oxidative stress. *Experimental Neurology* **1995**, *136*, 22.

139 Lambert, M.P., A.K. Barlow, B.A. Chromy, C. Edwards, R. Freed, M. Liosatos, T.E. Morgan, I. Rozovsky, B. Trommer, K.L. Viola, P. Wals, C. Zhang, C.E. Finch, G.A. Krafft and W.L. Klein. Diffusible, nonfibrillar ligands derived from Abeta1–42 are potent central nervous system neurotoxins. *Proc Natl Acad Sci USA* **1998**, *95*, 6448.

140 Walsh, D.M., D.M. Hartley, Y. Kusumoto, Y. Fezoui, M.M. Condron, A. Lomakin, G.B. Benedek, D.J. Selkoe and D.B. Teplow. Amyloid beta-protein fibrillogenesis. Structure and biological activity of protofibrillar intermediates. *J Biol Chem* **1999**, *274*, 25945.

141 Hartley, D.M., D.M. Walsh, C.P. Ye, T. Diehl, S. Vasquez, P.M. Vassilev, D.B. Teplow and D.J. Selkoe. Protofibrillar intermediates of amyloid beta-protein induce acute electrophysiological changes and progressive neurotoxicity

in cortical neurons. *J Neurosci* **1999**, *19*, 8876.

142 Stine, W. B., Jr, K. N. Dahlgren, G. K. Krafft and M. J. LaDu. *In vitro* characterization of conditions for amyloid-beta peptide oligomerization and fibrillogenesis. *J Biol Chem* **2003**, *278*, 11612.

143 Manelli, A. M., W. B. Stine, L. J. Van Eldik and M. J. LaDu. ApoE and Abeta1–42 interactions: effects of isoform and conformation on structure and function. *J Mol Neurosci* **2004**, *23*, 235.

144 Dahlgren, K. N., A. M. Manelli, W. B. Stine, Jr, L. K. Baker, G. A. Krafft and M. J. LaDu. Oligomeric and fibrillar species of amyloid-beta peptides differentially affect neuronal viability. *J Biol Chem* **2002**, *277*, 32046.

145 Gotz, J., F. Chen, J. van Dorpe and R. M. Nitsch. Formation of neurofibrillary tangles in P301l tau transgenic mice induced by Abeta 42 fibrils. *Science* **2001**, *293*, 1491.

146 DeMattos, R. B., K. R. Bales, D. J. Cummins, J. C. Dodart, S. M. Paul and D. M. Holtzman. Peripheral anti-A beta antibody alters CNS and plasma A beta clearance and decreases brain A beta burden in a mouse model of Alzheimer's disease. *Proc Natl Acad Sci USA* **2001**, *98*, 8850.

147 Dodart, J. C., K. R. Bales, K. S. Gannon, S. J. Greene, R. B. DeMattos, C. Mathis, C. A. DeLong, S. Wu, X. Wu, D. M. Holtzman and S. M. Paul. Immunization reverses memory deficits without reducing brain Abeta burden in Alzheimer's disease model. *Nat Neurosci* **2002**, *5*, 452.

148 Hardy, J. and D. J. Selkoe. The amyloid hypothesis of Alzheimer's disease: progress and problems on the road to therapeutics. *Science* **2002**, *297*, 353.

149 Selkoe, D. J. Alzheimer's disease is a synaptic failure. *Science* **2002**, *298*, 789.

150 Golde, T. E. Alzheimer disease therapy: can the amyloid cascade be halted? *J Clin Invest* **2003**, *111*, 11.

151 Conway, K. A., S. J. Lee, J. C. Rochet, T. T. Ding, R. E. Williamson and P. T. Lansbury, Jr. Acceleration of oligomerization, not fibrillization, is a shared property of both alpha-synuclein mutations linked to early-onset Parkinson's disease: implications for pathogenesis and therapy. *Proc Natl Acad Sci USA* **2000**, *97*, 571.

152 Volles, M. J. and P. T. Lansbury, Jr. Zeroing in on the pathogenic form of alpha-synuclein and its mechanism of neurotoxicity in Parkinson's disease. *Biochemistry* **2003**, *42*, 7871.

153 Rochet, J. C., T. F. Outeiro, K. A. Conway, T. T. Ding, M. J. Volles, H. A. Lashuel, R. M. Bieganski, S. L. Lindquist and P. T. Lansbury. Interactions among alpha-synuclein, dopamine, and biomembranes: some clues for understanding neurodegeneration in Parkinson's disease. *J Mol Neurosci* **2004**, *23*, 23.

154 Saudou, F., S. Finkbeiner, D. Devys and M. E. Greenberg. Huntingtin acts in the nucleus to induce apoptosis but death does not correlate with the formation of intranuclear inclusions. *Cell* **1998**, *95*, 55.

155 Klement, I. A., P. J. Skinner, M. D. Kaytor, H. Yi, S. M. Hersch, H. B. Clark, H. Y. Zoghbi and H. T. Orr. Ataxin-1 nuclear localization and aggregation: role in polyglutamine- induced disease in SCA1 transgenic mice [see Comments]. *Cell* **1998**, *95*, 41.

156 Tompkins, M. M. and W. D. Hill. Contribution of somal Lewy bodies to neuronal death. *Brain Res* **1997**, *775*, 24.

157 Wittmann, C. W., M. F. Wszolek, J. M. Shulman, P. M. Salvaterra, J. Lewis, M. Hutton and M. B. Feany. Tauopathy in *Drosophila*: neurodegeneration without neurofibrillary tangles. *Science* **2001**, *293*, 711.

158 Golabek, A. A., C. Soto, T. Vogel and T. Wisniewski. The interaction between apolipoprotein E and Alzheimer's amyloid β-peptide is dependent on β-peptide conformation. *J Biol Chem* **1996**, *271*, 10602.

159 Soto, C., E. M. Castano, F. Prelli, R. A. Kumar and M. Baumann. Apolipoprotein E increases the fibrillogenic potential of synthetic peptides derived from Alzheimer's, gelsolin and AA amyloids. *FEBS Lett* **1995**, *371*, 110.

160 Howlett, D.R., K.H. Jennings, D.C. Lee, M.S. Clark, F. Brown, R. Wetzel, S.J. Wood, P. Camilleri and G.W. Roberts. Aggregation state and neurotoxic properties of Alzheimer beta-amyloid peptide. *Neurodegeneration* **1995**, *4*, 23.

161 LaDu, M.J., S.M. Gilligan, S.R. Lukens, V.G. Cabana, C.A. Reardon, L.J. Van Eldik and D.M. Holtzman. Nascent astrocyte particles differ from lipoproteins in CSF. *J Neurochem* **1998**, *70*, 2070.

162 Fagan, A.M., D.M. Holtzman, G. Munson, T. Mathur, D. Schneider, L.K. Chang, G.S. Getz, C.A. Reardon, J. Lukens, J.A. Shah and M.J. LaDu. Unique lipoproteins secreted by primary astrocytes from wild type, ApoE$^{-/-}$, and human ApoE transgenic mice. *J Biol Chem* **1999**, *274*, 30001.

163 Strittmatter, W.J., K.H. Weisgraber, D.Y. Huang, L.-Y. Dong, G.S. Salvesen, M. Pericak-Vance, D. Schmechel, A.M. Saunders, D. Goldgaber and A.D. Roses. Binding of human apolipoprotein E to synthetic amyloid β peptide: isoform specific-effects and implications for late-onset Alzheimer disease. *Proc Natl Acad Sci USA* **1993**, *90*, 8098.

164 Fung, W.P., G.J. Howlett and G. Schreiber. Structure and expression of the rat apolipoprotein E gene. *J Biol Chem* **1986**, *261*, 13777.

165 Koudinov, A., E. Matsubara, B. Frangione and J. Ghiso. The soluble form of Alzheimer's amyloid beta protein is complexed to high density lipoprotein 3 and very high density lipoprotein in normal human plasma. *Biochem Biophys Res Commun* **1994**, *205*, 1164.

166 Russo, C., G. Angelini, D. Dapino, A. Piccini, G. Piombo, G. Schettini, S. Chen, J.K. Teller, D. Zaccheo, P. Gambetti and M. Tabaton. Opposite roles of apolipoprotein E in normal brains and in Alzheimer's disease. *Proc Natl Acad Sci USA* **1998**, *95*, 15598.

167 Haas, C., P. Cazorla, C.D. Miguel, F. Valdivieso and J. Vazquez. Apolipoprotein E forms stable complexes with recombinant Alzheimer's disease beta-amyloid precursor protein. *Biochem J* **1997**, *325*, 169.

168 Pillot, T., M. Goethals, B. Vanloo, L. Lins, R. Brasseur, J. Vandekerckhove and M. Rosseneu. Specific modulation of the fusogenic properties of the Alzheimer beta-amyloid peptide by apolipoprotein E isoforms. *Euro J Biochem* **1997**, *243*, 650.

169 Munson, G.W., A.E. Roher, Y.M. Kuo, S.M. Gilligan, C.A. Reardon, G.S. Getz and M.J. LaDu. SDS-stable complex formation between native apolipoprotein E3 and beta-amyloid peptides. *Biochemistry* **2000**, *39*, 16119.

170 Giulian, D., L.J. Haverkamp, J. Yu, W. Karshin, D. Tom, J. Li, A. Kazanskaia, J. Kirkpatrick and A.E. Roher. The HHQK domain of beta-amyloid provides a structural basis for the immunopathology of Alzheimer's disease. *J Biol Chem* **1998**, *273*, 29719.

171 Ji, A.-S., S. Fazio, Y.L. Lee and R.W. Mahley. Secretion capture role for apolipoprotein E in remnant lipoprotein metabolism involving cell surface heparan sulfate proteoglycans. *J Biol Chem* **1994**, *269*, 2764.

172 Mahley, R.W., Z.S. Ji, W.J. Brecht, R.D. Miranda and D. He. Role of heparan sulfate proteoglycans and the LDL receptor-related protein in remnant lipoprotein metabolism. *Ann NY Acad Sci* **1994**, *737*, 39.

173 Ji, Z.S., R.E. Pitas and R.W. Mahley. Differential cellular accumulation/retention of apolipoprotein E mediated by cell surface heparan sulfate proteoglycans. Apolipoproteins E3 and E2 greater than E4. *J Biol Chem* **1998**, *273*, 13452.

174 Golabek, A.A., E. Kida, M. Walus, C. Perez, T. Wisniewski and C. Soto. Sodium dodecyl sulfate-resistant complexes of Alzheimer's amyloid beta-peptide with the N-terminal, receptor binding domain of apolipoprotein E. *Biophys J* **2000**, *79*, 1008.

175 Castano, E.M., F. Prelli, M. Pras and B. Frangione. Apolipoprotein E carboxy-terminal fragments are complexed to amyloids A and L. *J Biol Chem* **1995**, *270*, 17610.

176 Naslund, J., J. Thyberg, L.O. Tjernberg, C. Wernstedt, A. R. Karlstrom, N. Bog-

danovic, S.E. Gandy, L. Lannfelt, L. Terenius and C. Nordstedt. Characterization of stable complexes involving apolipoprotein E and the amyloid β peptide in Alzheimer's disease brain. *Neuron* **1995**, *15*, 219.

177 Wisniewski, T., M. Lalowski, A.A. Golabek, T. Vogel and B. Frangione. Is Alzheimer's disease an apolipoprotein E amyloidosis? *Lancet* **1995**, *345*, 956.

178 Soto, C., E.M. Castano, R.A. Kumar, R.C. Beavis and B. Frangione. Fibrillogenesis of synthetic amyloid-beta peptides is dependent on their initial secondary structure. *Neurosci Lett* **1995**, *200*, 105.

179 Berg, M.J., R. Durrie, V.S. Sapirstein and N. Marks. Composition of white matter bovine brain coated vesicles: evidence that several components influence beta-amyloid peptide to form oligomers and aggregates *in vitro*. *Brain Res* **1997**, *752*, 72.

180 Castano, E.M., F. Prelli, T. Wisniewski, A. Golabek, R.A. Kumar, C. Soto and B. Frangione. Fibrillogenesis in Alzheimer's disease of amyloid beta peptides and apolipoprotein E. *Biochem J* **1995**, *306*, 599.

181 Chauhan, A., T. Pirttilä, V.P. Chauhan, P. Mehta and H.M. Wisniewski. Aggregation of amyloid beta-protein as function of age and apolipoprotein E in normal and Alzheimer's serum. *J Neurol Sci* **1998**, *154*, 159.

182 Olesen, O.F. and L. Dago. High density lipoprotein inhibits assembly of amyloid beta-peptides into fibrils. *Biochem Biophys Res Commun* **2000**, *270*, 62.

183 Piedrahita, J.A., S.H. Zhang, J.R. Hagaman, P.M. Oliver and N. Maeda. Generation of mice carrying a mutant apolipoprotein E gene inactivated by gene targeting in embryonic stem cells. *Proc Natl Acad Sci USA* **1992**, *89*, 4471.

184 Plump, A.S., J.D. Smith, T. Hayek, K. Aalto-Setala, A. Walsh, J.G. Verstuyft, E.M. Rubin and J.L. Breslow. Severe hypercholesterolemia and atherosclerosis in apolipoprotein E-deficient mice created by homologous recombination in ES cells. *Cell* **1992**, *71*, 343.

185 Tesseur, I., J. Van Dorpe, K. Bruynseels, F. Bronfman, R. Sciot, A. Van Lommel and F. Van Leuven. Prominent axonopathy and disruption of axonal transport in transgenic mice expressing human apolipoprotein E4 in neurons of brain and spinal cord. *Am J Pathol* **2000**, *157*, 1495.

186 Sun, Y., S. Wu, G. Bu, M.K. Onifade, S.N. Patel, M.J. LaDu, A.M. Fagan and D.M. Holtzman. GFAP-ApoE transgenic mice: astrocyte specific expression and differing biological effects of astrocyte-secreted ApoE3 and ApoE4 lipoproteins. *J Neurosci* **1998**, *18*, 3261.

187 Smith, J.D., J. Sikes and J.A. Levin. Human apolipoprotein E allele-specific brain expressing transgenic mice. *Neurobiol Aging* **1998**, *19*, 407.

188 Huber, G., W. Marz, J.R. Martin, P. Malherbe, J.G. Richards, N. Sueoka, T. Ohm and M.M. Hoffmann. Characterization of transgenic mice expressing apolipoprotein E4(C112R) and apolipoprotein E4(L28P; C112R). *Neuroscience* **2000**, *101*, 211.

189 Bowman, B.H., L. Jansen, F. Yang, G.S. Adrian, M. Zhao, S.S. Atherton, J.M. Buchanan, R. Greene, C. Walter, D.C. Herbert, F.J. Weaker, L.K. Chiodo, K. Kagan-Hallet and J.E. Hixson. Discovery of a brain promoter from the human transferrin gene and its utilization for development of transgenic mice that express human apolipoprotein E alleles. *Proc Natl Acad Sci USA* **1995**, *92*, 12115.

190 Raber, J., D. Wong, M. Buttini, M. Orth, S. Bellosta, R.E. Pitas, R.W. Mahley and L. Mucke. Isoform-specific effects of human apolipoprotein E on brain function revealed in ApoE knockout mice: increased susceptibility of females. *Proc Natl Acad Sci USA* **1998**, *95*, 10914.

191 Xu, P.-T., D. Schmechel, T. Rothrock-Christian, D.S. Burkhart, H.-L. Qiu, B. Popko, P. Sullivan, N. Maeda, A.M. Saunders, A.D. Roses and J.R. Gilbert. Human apolipoprotein E2, E3, and E4 isoform-specific transgenic mice: human-like pattern of glial and neuronal immunoreactivity in central nervous

system not observed in wild-type mice. *Neurobiol Dis* **1996**, *3*, 229.

192 Hamanaka, H., Y. Katoh-Fukui, K. Suzuki, M. Kobayashi, R. Suzuki, Y. Motegi, Y. Nakahara, A. Takeshita, M. Kawai, K. Ishiguro, M. Yokoyama and S.C. Fujita. Altered cholesterol metabolism in human apolipoprotein E4 knock-in mice. *Hum Mol Genet* **2000**, *9*, 353.

193 Sullivan, P.M., H. Mezdour, Y. Aratani, C. Knouff, J. Najib, R.L. Reddick, S.H. Quarfordt and M.N. Targeted replacement of the mouse apolipoprotein E gene with the common human APOE3 allele enhances diet-induced hypercholesterolemia and atherosclerosis. *J Biol Chem* **1997**, *272*, 17972.

194 Games, D., D. Adams, R. Alessandrini, R. Barbour, P. Berthelette, C. Blackwell, T. Carr, J. Clemens, T. Donaldson, F. Gillespie, T. Guido, S. Hagopian, K. Johnson-Wood, K. Khan, M. Lee, P. Leibowitz, I. Lieberburg, S. Little, E. Masliah, L. McConlogue, Montoya-Zavala, L. Mucke, L. Paganini, E. Penniman, M. Power, D. Schenk, P. Seubert, B. Snyder, F. Soriano, H. Tan, J. Vitale, S. Wadsworth, B. Wolozin and J. Zhao. Alzheimer-type neuropathology in transgenic mice overexpressing V717F β-amyloid precursor protein. *Nature* **1995**, *373*, 523.

195 Hsiao, K., P. Chapman, S. Nilsen, C. Eckman, Y. Harigaya, S. Youkin, F. Yang and G. Cole. Correlative memory deficits, Aβ elevation, and amyloid plaques in transgenic mice. *Science* **1996**, *274*, 99.

196 Hartman, R.E., H. Laurer, L. Longhi, K.R. Bales, S.M. Paul, T.K. McIntosh and D.M. Holtzman. Apolipoprotein E4 influences amyloid deposition but not cell loss after traumatic brain injury in a mouse model of Alzheimer's disease. *J Neurosci* **2002**, *22*, 10083.

197 Buttini, M., G.Q. Yu, K. Shockley, Y. Huang, B. Jones, E. Masliah, M. Mallory, T. Yeo, F.M. Longo and L. Mucke. Modulation of Alzheimer-like synaptic and cholinergic deficits in transgenic mice by human apolipoprotein E depends on isoform, aging and overexpression of amyloid beta peptides but not on plaque formation. *J Neurosci* **2002**, *22*, 10539.

198 Raber, J., D. Wong, G.Q. Yu, M. Buttini, R.W. Mahley, R.E. Pitas and L. Mucke. Apolipoprotein E and cognitive performance. *Nature* **2000**, *404*, 352.

199 Boyles, J.K., R.E. Pitas, E. Wilson, R.W. Mahley and J.M. Taylor. Apolipoprotein E associated with astrocytic glia of the central nervous system and with nonmyelinating glia of the peripheral nervous system. *J Clin Invest* **1985**, *76*, 1501.

200 Poirier, J., M. Hess, P.C. May and C.E. Finch. Astrocytic apolipoprotein E mRNA and GFAP mRNA in hippocampus after entorhinal cortex lesioning. *Mol Brain Res* **1991**, *11*, 97.

201 Diedrich, J.F., H. Minnigan, R.I. Carp, J.N. Whitaker, R. Race, W. Frey and A.T. Haase. Neuropathological changes in scrapie and Alzheimer's disease are associated with increased expression of apolipoprotein E and cathepsin D in astrocytes. *J Virol* **1991**, *65*, 4759.

202 Mufson, E.J., W.C. Benzing, G.M. Cole, H. Wang, D.F. Emerich, J.R. Sladek, Jr, J.H. Morrison and J.H. Kordower. Apolipoprotein E-immunoreactivity in aged rhesus monkey cortex: colocalization with amyloid plaques. *Neurobiol Aging* **1994**, *15*, 621.

203 Gearing, M., G.W. Rebeck, B.T. Hyman, J. Tigges and S.S. Mirra. Neuropathology and apolipoprotein E profile of aged chimpanzees: implications for Alzheimer disease. *Proc Natl Acad Sci USA* **1994**, *91*, 9382.

204 Pitas, R.E., J.K. Boyles, S.H. Lee, D. Foss and R.W. Mahley. Astrocytes synthesize apolipoprotein E and metabolize apolipoprotein E-containing lipoproteins. *Biochim Biophys Acta* **1987**, *917*, 148.

205 Nakai, M., T. Kawamata, K. Maeda and C. Tanaka. Expression of ApoE mRNA in rat microglia. *Neurosci Lett* **1996**, *211*, 41.

206 Stone, D.J., I. Rozovsky, T.E. Morgan, C.P. Anderson, H. Hajian and C.E. Finch. Astrocytes and microglia respond to estrogen with increased ApoE

mRNA *in vivo* and *in vitro*. *Exp Neurol* **1997**, *143*, 313.

207 Stoll, G. and H.W. Muller. Macrophages in the peripheral nervous system and astroglia in the central nervous system of rat commonly express apolipoprotein E during development but differ in their response to injury. *Neurosci Lett* **1986**, *72*, 233.

208 Fujita, S.C., K. Sakuta, R. Tsuchiya and H. Hamanaka. Apolipoprotein E is found in astrocytes but not in microglia in the normal mouse brain. *Neurosci Res* **1999**, *35*, 123.

209 Pitas, R. E., J.K. Boyles, S.H. Lee, D.Hui and K.H. Weisgraber. Lipoproteins and their receptors in the central nervous system. *J Biol Chem* **1987**, *262*, 14352.

210 Bu, G., E.A. Maksymovitch, J.M. Nerbonne and A.L. Schwartz. Expression and function of the low density lipoprotein receptor-related protein (LRP) in mammalian central neurons. *J Biol Chem* **1994**, *269*, 18521.

211 Marzolo, M.P., R. von Bernhardi, G. Bu and N.C. Inestrosa. Expression of alpha$_2$-macroglobulin receptor/low density lipoprotein receptor-related protein (LRP) in rat microglial cells. *J Neurosci Res* **2000**, *60*, 401.

212 Rebeck, G.W., S.D. Harr, D.K. Strickland and B.T. Hyman. Multiple, diverse senile plaque-associated proteins are ligands of an apolipoprotein receptor, the alpha 2-macroglobulin receptor/low-density-lipoprotein receptor-related protein. *Ann Neurol* **1995**, *37*, 211.

213 Clatworthy, A.E., W. Stockinger, R.H. Christie, W.J. Schneider, J. Nimpf, B.T. Hyman and G.W. Rebeck. Expression and alternative splicing of apolipoprotein E receptor 2 in brain. *Neuroscience* **1999**, *90*, 903.

214 Christie, R.H., H. Chung, G.W. Rebeck, D. Strickland and B.T. Hyman. Expression of the very low-density lipoprotein receptor (VLDL-r), an apolipoprotein-E receptor, in the central nervous system and in Alzheimer's disease. *J Neuropathol Exp Neurol* **1996**, *55*, 491.

215 Holtzman, D.M., R.E. Pitas, J. Kilbridge, B. Nathan, R.W. Mahley, G. Bu and A.L. Schwartz. LRP mediates apolipoprotein E-dependent neurite outgrowth in a CNS-derived neuronal cell line. *Proc Natl Acad Sci USA* **1995**, *92*, 9480.

216 Bellosta, S., B.P. Nathan, M. Orth, L. M. Dong, R.W. Mahley and R.E. Pitas. Stable expression and secretion of apolipoproteins E3 and E4 in mouse neuroblastoma cells produces differential effects on neurite outgrowth. *J Biol Chem* **1995**, *270*, 27063.

217 Veinbergs, I., E. Van Uden, M. Mallory, M. Alford, C. McGiffert, R. DeTeresa, R. Orlando and E. Masliah. Role of apolipoprotein E receptors in regulating the differential *in vivo* neurotrophic effects of apolipoprotein E. *Exp Neurol* **2001**, *170*, 15.

218 Qiu, Z., D.K. Strickland, B.T. Hyman and G.W. Rebeck. Elevation of LDL receptor-related protein levels via ligand interactions in Alzheimer disease and *in vitro*. *J Neuropathol Exp Neurol* **2001**, *60*, 430.

219 Kang, D.E., T. Saitoh, X. Chen, Y. Xia, E. Masliah, L.A. Hansen, R.G. Thomas, L.J. Thal and R. Katzman. Genetic association of the low-density lipoprotein receptor-related protein gene (LRP), an apolipoprotein E receptor, with late-onset Alzheimer's disease. *Neurology* **1997**, *49*, 56.

220 Blacker, D., M.A. Wilcox, N.M. Laird, L. Rodes, S.M. Horvath, R.C.P. Go, R. Perry, B. Watson, S.S. Bassett, M.G. Mcinnis, M.S. Albert, B.T. Hyman and R.E. Tanzi. Alpha-2 macroglobulin is genetically associated with Alzheimer's disease. *Nat Genet* **1998**, *19*, 357.

221 Lanz, T. A., D.B. Carter and K.M. Merchant. Dendritic spine loss in the hippocampus of young PDAPP and Tg2576 mice and its prevention by the ApoE2 genotype. *Neurobiol Dis* **2003**, *13*, 246.

222 Carter, D.B., E. Dunn, D.D. McKinley, N.C. Stratman, T.P. Boyle, S L. Kuiper, J.A. Oostveen, R.J. Weaver, J.A. Boller and M. E. Gurney. Human apolipoprotein E4 accelerates beta-amyloid deposi-

tion in APPsw transgenic mouse brain. *Ann Neurol* **2001**, *50*, 468.

223 de Silva, H.V., W.D. Stuart, Y.B. Park, S.J.T. Mao, C.M. Gil, J.R. Wetterau, S.J. Busch and J.A.K. Harmony. Purification and characterization of apolipoprotein J. *J Biol Chem* **1990**, *265*, 14292.

224 May, P.C. and C.E. Finch. Sulfated glycoprotein 2: new relationships of this multifunctional protein to neurodegeneration. *Trends Neurol Sci* **1992**, *15*, 391.

225 Aronow, B.J., S.D. Lund, T.L. Brown, J.A.K. Harmony and D.P. Witte. Apolipoprotein J expression at fluid-tissue interfaces: Potential role in barrier cytoprotection. *Proc Natl Acad Sci USA* **1993**, *90*, 725.

226 May, P.C., M. Lampert-Etchells, S.A. Johnson, J. Poirier, J.N. Masters and C.E. Finch. Dynamics of gene expression for a hippocampal glycoprotein elevated in Alzheimer's disease and in response to experimental lesions in rat. *Neuron* **1990**, *5*, 831.

227 Calero, M., A. Rostagno, E. Matsubara, B. Zlokovic, B. Frangione and J. Ghiso. Apolipoprotein J (clusterin) and Alzheimer's disease. *Microsc Res Tech* **2000**, *50*, 305.

228 DeMattos, R.B., A. O'Dell M, M. Parsadanian, J.W. Taylor, J.A. Harmony, K.R. Bales, S.M. Paul, B.J. Aronow and D.M. Holtzman. Clusterin promotes amyloid plaque formation and is critical for neuritic toxicity in a mouse model of Alzheimer's disease. *Proc Natl Acad Sci USA* **2002**, *99*, 10843.

229 Bales, K.R., T. Verina, D.J. Cummins, Y. Du, R.C. Dodel, J. Saura, C.E. Fishman, C.A. DeLong, P. Piccardo, V. Petegnief, B. Ghetti and S.M. Paul. Apolipoprotein E is essential for amyloid deposition in the APP(V717F) transgenic mouse model of Alzheimer's disease. *Proc Natl Acad Sci USA* **1999**, *96*, 15233.

230 DeMattos, R.B., J.R. Cirrito, M. Parsadanian, P.C. May, M.A. O'Dell, J.W. Taylor, J.A. Harmony, B.J. Aronow, K.R. Bales, S.M. Paul and D.M. Holtzman. ApoE and clusterin cooperatively suppress Abeta levels and deposition. Evidence that ApoE regulates extracellular Abeta metabolism *in vivo*. *Neuron* **2004**, *41*, 193.

231 Kounnas, M.Z., C.C. Haudenschild, D.K. Strickland and W.S. Argraves. Immunological localization of glycoprotein 330, low density lipoprotein receptor related protein and 39 kDa receptor associated protein in embryonic mouse tissues. *In Vivo* **1994**, *8*, 343.

232 Kounnas, M.Z., E.B. Loukinova, S. Stefansson, J.A. Harmony, B.H. Brewer, D.K. Strickland and W.S. Argraves. Identification of glycoprotein 330 as an endocytic receptor for apolipoprotein J/clusterin. *J Biol Chem* **1995**, *270*, 13070.

233 Zlokovic, B.V., C.L. Martel, E. Matsubara, J.G. McComb, G. Zheng, R.T. McCluskey, B. Frangione and J. Ghiso. Glycoprotein 330/megalin: probable role in receptor-mediated transport of apolipoprotein J alone and in a complex with Alzheimer disease amyloid beta at the blood–brain and blood–cerebrospinal fluid barriers. *Proc Natl Acad Sci USA* **1996**, *93*, 4229.

234 Zlokovic, B.V. Cerebrovascular transport of Alzheimer's amyloid beta and apolipoproteins J and E: possible anti-amyloidogenic role of the blood–brain barrier. *Life Sci* **1996**, *59*, 1483.

235 Zlokovic, B.V. Clearing amyloid through the blood–brain barrier. *J Neurochem* **2004**, *89*, 807.

236 Narita, M., G. Bu, D.M. Holtzman and A.L. Schwartz. The low density lipoprotein receptor-related protein (LRP), a multifunctional ApoE receptor, modulates hippocampal neurite development. *J Neurochem* **1997**, *68*, 587.

237 Qiu, Z., D.K. Strickland, B.T. Hyman and G. W. Rebeck. Alpha2-macroglobulin enhances the clearance of endogenous soluble beta-amyloid peptide via low-density lipoprotein receptor-related protein in cortical neurons. *J Neurochem* **1999**, *73*, 1393.

238 Liao, A., R.M. Nitsch, S.M. Greenberg, U. Finckh, D. Blacker, M. Albert, G.W. Rebeck, T. Gomez-Isla, A. Clatworthy, G. Binetti, C. Hock, T. Mueller-Thom-

sen, U. Mann, K. Zuchowski, U. Beisiegel, H. Staehelin, J. H. Growdon, R. E. Tanzi and B. T. Hyman. Genetic association of an alpha2-macroglobulin (Val1000lle) polymorphism and Alzheimer's disease. *Hum Mol Genet* **1998**, *7*, 1953.

239 Du, Y., B. Ni, M. Glinn, R. C. Dodel, K. R. Bales, Z. Zhang, P. Hyslop and S. M. Paul. $\alpha 2$ macroglolbulin as a β-amyloid peptide-binding plasma protein. *J Neurochem* **1997**, *69*, 299.

240 Abraham, C. R. Reactive astrocytes and alpha1-antichymotrypsin in Alzheimer's disease. *Neurobiol Aging* **2001**, *22*, 931.

241 Nilsson, L. N., K. R. Bales, G. DiCarlo, M. N. Gordon, D. Morgan, S. M. Paul and H. Potter. Alpha-1-antichymotrypsin promotes beta-sheet amyloid plaque deposition in a transgenic mouse model of Alzheimer's disease. *J Neurosci* **2001**, *21*, 1444.

242 Potter, H., I. M. Wefes and L. N. Nilsson. The inflammation-induced pathological chaperones ACT and apo-E are necessary catalysts of Alzheimer amyloid formation. *Neurobiol Aging* **2001**, *22*, 923.

243 Inouye, H., P. Fraser and D. Kirschner. Structure of beta-crystallite assemblies formed by Alzheimer beta-amyloid protein analogues: analysis by X-ray diffraction. *Biophys J* **1993**, *64*, 502.

244 Huang, Y., X. Q. Liu, T. Wyss-Coray, W. J. Brecht, D. A. Sanan and R. W. Mahley. Apolipoprotein E fragments present in Alzheimer's disease brains induce neurofibrillary tangle-like intracellular inclusions in neurons. *Proc Natl Acad Sci USA* **2001**, *98*, 8838.

Part IV
Pathway to Amyloid Fibril Formation

Amyloid Proteins. The Beta Sheet Conformation and Disease. J. D. Sipe

ISBN: 3-527-31072-X

11
Pathways to Amyloid Fibril Formation: Partially Folded Intermediates in the Fibrillation of Natively Unfolded Proteins

Vladimir N. Uversky and Anthony L. Fink

11.1
Introduction

The realization that many proteins are intrinsically disordered, or natively unfolded, under physiological conditions (for the purified protein) is a relatively recent one. In fact, current estimates suggest that between 25 and 33% of the human genome encoding proteins may code for intrinsically disordered regions [1–4]! Thus, rather than being rarities, such proteins may be rather common. Several hypotheses have been proposed for the existence of natively unfolded proteins; probably the most widely accepted is that the lack of ordered structure permits greater flexibility in binding partners, a valuable characteristic in a protein that must react with multiple binding partners in its normal function, as is true for many signaling molecules and gene regulators. It is likely that in their normal physiological milieu intrinsically unstructured proteins are not unfolded, due to interactions with other proteins, nucleic acids, membranes or small ligands. This would, at least in part, explain their stability to intracellular proteases. Natively unfolded, or intrinsically disordered, proteins are specifically localized within a unique region of charge–hydrophobicity phase space, and are characterized by a combination of low overall hydrophobicity and large net charge [5].

As will be discussed, a variety of conditions can bring about folding of natively unfolded proteins, resulting in either relatively non-compact, partially folded conformations or relatively compact, tightly folded conformations. An informative example is the effect of increasing trimethylamine *N*-oxide (TMAO) on α-synuclein. TMAO is an osmolyte, and is known to stabilize native states and induce structure in disordered states. When the structure of α-synuclein was examined as a function of increasing TMAO, the initially natively unfolded conformation adopted a partially folded intermediate conformation in the vicinity of 1 M TMAO, and became compact and helical in 3 M TMAO.

The term conformational disease or protein deposition disease has been given to a number of pathological states in which a specific protein or protein fragment changes from its natural soluble form into stable, insoluble, ordered fila-

Amyloid Proteins. The Beta Sheet Conformation and Disease. J. D. Sipe

ISBN: 3-527-31072-X

mentous protein aggregates, commonly referred to as amyloid fibrils [6–11]. Given the prevalence of natively unfolded proteins it should not be surprising that of the more than two dozen proteins known so far to be involved in protein deposition diseases, more than 50% are either completely unfolded or contain significant unfolded regions under physiological conditions (Tables 11.1 and 11.2). For many years it has been generally assumed that the ability to form amyloid fibrils is limited to a relatively small number of proteins, essentially those found in the diseases, and that these proteins possess specific sequence motifs encoding the unique structure of amyloid core. However, recent studies have suggested that the ability of proteins to form amyloid fibrils is probably very common and may be a general feature of the polypeptide chain [7, 12, 13]. In fact, many proteins that are not associated with diseases have been shown to form fibrils *in vitro* (reviewed in [14]). Overall, there is an increasing belief that the ability to form fibrils is an inherent property of the polypeptide chain, i.e. many proteins, perhaps all, are potentially able to form amyloid fibrils under appropriate conditions [7, 15–17]. Amyloid fibrils are highly organized structures (similar to one-dimensional crystals) in which the essential features of the structure are determined by the physicochemical properties of the polypeptide chain [18]. Furthermore, it has been pointed out that, as with other highly organized compounds (including crystals) whose structures are based on repetitive long-range interactions, the most stable structures are those consisting of a single type of peptide or protein where self-interactions can be optimized [18, 19]. Obviously, this can explain the remarkable specificity of protein aggregation in protein deposition diseases.

Although amyloidogenic polypeptides may be rich in β-sheet, α-helix, β-helix or natively unfolded [5, 20–27], the resulting fibrils display many common properties including a core cross-β-sheet structure in which continuous β-sheets are formed with β-strands running perpendicular to the long axis of the fibril [28]. All fibrils have similar morphologies, with mature fibrils usually having a twisted, rope-like structure, which usually involves two to six unbranched protofilaments around 2–3 nm in diameter associated laterally or twisted together to form fibrils of around 10 nm diameter (see, e.g. [29–31]).

Since all fibrils, regardless of the original structure of the given amyloidogenic protein, have a common cross-β structure, considerable conformational rearrangements have to occur in order for fibril formation to happen. The molecular basis of amyloid fibril formation from globular proteins is described in Chapter 12; briefly, fibril formation can only occur when the rigid native structure is *destabilized,* favoring *partial unfolding* and the formation of an amyloidogenic *partially unfolded conformation* [6–11, 32–35]. This is because the structural rearrangements required for the formation of fibrils cannot take place within the tightly packed native globular protein, due to the constraints of its tertiary structure. This situation, obviously, is not applicable for natively unfolded proteins, as they are devoid of rigid structure in their starting state. Thus, in contrast to globular proteins, the primary step of fibrillogenesis of natively unfolded proteins is *partial folding* rather than partial unfolding. The details of the pro-

Table 11.1 Some natively unfolded or significantly disordered amyloidogenic proteins and the corresponding amyloid-based clinical disorders.

Amyloidogenic protein	Type of structure	Disease
Prion protein and its fragments	N-terminal fragment (23–121) is natively unfolded; C-terminal domain (121–230) is α-helical (predominantly)	Creutzfeldt-Jacob disease (CJD) Gerstmann-Straussler-Schneiker syndrome (GSS) fatal familial insomnia (FFI) kuru bovine spongiform encephalopathy (BSE) and scrapie
Aβ and its fragments	natively unfolded	Alzheimer's disease (AD) Dutch hereditary cerebral hemorrhage with amyloidosis (HCHWA, also known as cerebrovascular amyloidosis) Congophilic angiopathy
ABri	natively unfolded	familial British dementia
Huntingtin	exon 1 is unfolded and forms fibrils	Huntington disease
Androgen receptor protein	ligand-binding and DNA-binding domains are α-helical; N-terminal domain is natively unfolded	spinal and bulbar muscular atrophy (SBMA)
Ataxin-1	unknown (natively unfolded)	spinocerebellar ataxia (SCA) neuronal intranuclear inclusion disease (NIID)
DRPLA protein (atrophin-1)	unknown (probably natively unfolded)	hereditary dentatorubral-pallidoluysian atrophy (DRPLA)
IAPP (amylin)	natively unfolded	pancreatic islet amyloidosis in late-onset diabetes (type II diabetes mellitus)
Calcitonin	natively unfolded	medullary carcinoma of the thyroid (MCT)
α-Synuclein	natively unfolded	Parkinson's disease (PD) diffuse Lewy bodies disease (DLBD) Lewy bodies variant of Alzheimer's disease (LBVAD) dementia with Lewy bodies (DLB) multiple system atrophy (MSA) Hallervorden-Spatz disease
Tau protein	natively unfolded	Alzheimer disease (AD) Pick's disease progressive supranuclear palsy (PSP)

Table 11.2 Non-disease-related amyloidogenic proteins and peptides.

Protein (peptide)	Type of structure
Prothymosin *α* [164]	natively unfolded
Human complement receptor 1, 18–34 fragment [165]	unfolded
GAGA factor [166]	natively unfolded
Yeast prion Ure2p [167]	*α*-helical/unfolded
ApoCII [156, 157, 160]	Natively unfolded
Cold shock protein B, 1–22 fragment [168]	Unfolded
Core histones [163]	Natively unfolded
Soluble homopolypeptides [169]:	
poly(L-lysine)	
poly(L-glutamic acid)	
poly(L-threonine)	

cess will be illustrated using results from a detailed analysis of the aggregation of human *α*-synuclein, a protein whose fibrillogenesis *in vitro* has been studied extensively. The generality of the results will be emphasized with a variety of other natively unfolded proteins that assemble into amyloid fibrils.

11.2 Molecular Mechanisms of Amyloid Fibril Formation by a Natively Unfolded Protein: *α*-Synuclein

11.2.1 *α*-Synuclein in Parkinson's Disease and other Neurodegenerative Disorders

Parkinson's disease is a progressive disorder resulting from loss of neurons of the substantia nigra, a small area of cells in the mid-brain. Gradual degeneration of these dopaminergic neurons causes a reduction in the release of the neurotransmitter dopamine in the striatum. This, in turn, can produce one or more of the classic signs of Parkinson's disease: resting tremor on one (or both) side(s) of the body, generalized slowness of movements (bradykinesia), stiffness of limbs (rigidity) and gait or balance problems (postural dysfunction). The precise mechanisms of neuronal death are unknown as yet. Some surviving nigral dopaminergic neurons contain cytosolic filamentous inclusions known as Lewy bodies (LBs) and Lewy neurites (LNs) [36, 37]. LBs are also neurophathological hallmarks of several other diseases collectively termed synucleinopathies.

There is a strong link between *α*-synuclein aggregation and the pathogenesis of Parkinson's disease, including three different missense mutations in the *α*-synuclein gene, corresponding to A30P, E46K and A53T substitutions in *α*-synuclein protein, in a small number of kindreds with autosomal-dominantly inherited, early-onset Parkinson's disease [38–40]; triplication of the α-synuclein gene locus causes familial autosomal dominant Parkinson's disease with early onset [41]; overexpression of wild-type (WT) α-synuclein in transgenic mice [42] or of WT, A30P and A53T in transgenic flies [43] leads to motor deficits and neuronal inclusions reminiscent of Parkinson's disease.

11.2.2
Key Structural Properties of α-Synuclein: A Natively Unfolded Protein

α-Synucleins from different organisms exhibit a high degree of sequence conservation: human α-synuclein consists of 140 amino acid residues and can be divided into the N-terminal region, residues 1–95, containing six 11-amino-acid imperfect repeats with a highly conserved hexamer motif (KTKEGV), and the C-terminal region, residues 96–140, which is enriched in acidic residues and prolines, suggesting that it adopts a disordered conformation.

α-Synuclein is a typical intrinsically unstructured or natively unfolded protein (for recent reviews, see [13, 21, 23, 26, 44, 45]). The sequence of intrinsically disordered proteins is characterized by amino acid compositional bias and the existence of highly predictable flexibility [46]. Further, the majority of the intrinsically disordered proteins, are substantially depleted in I, L, V, W, F, Y, C and N, and enriched in E, K, R, G, Q, S, P and A [21]. These features account for the low hydrophobicity and high net charge of the intrinsically unstructured proteins, a property that allows discrimination of globular (folded) and intrinsically unstructured proteins based solely on their amino acid composition [5].

Several detailed studies show that purified α-synuclein possesses little ordered structure under physiological conditions [45, 47–49] and is characterized by far-UV circular dichroism (CD) and Fourier transform IR (FTIR) spectra typical of a substantially unfolded polypeptide pH [49]. The hydrodynamic properties of α-synuclein show that the protein is quite expanded, but is slightly more compact than expected for a random coil and lacks a tightly packed globular structure [45, 49, 50]. The results of pulsed-field gradient nuclear magnetic resonance (NMR; which allows an estimation of the hydrodynamic radius) confirm that α-synuclein is unfolded, but slightly collapsed [51], and a high-resolution NMR analysis revealed that α-synuclein is largely unfolded in solution, but exhibits a region between residues 6 and 37 with a preference for helical conformation [48]. Finally, Raman optical activity spectra suggested that α-synuclein may contain some poly(L-proline) II helical conformation [52].

One of the characteristic features of natively unstructured proteins is a reverse response to changes in their environment compared with that of globular proteins. For example, intrinsically unstructured proteins gain rather than lose ordered structure at extremes of pH or high temperatures. The structure-forming effects of low pH on α-synuclein [45, 49, 50] were attributed to the minimization of the large net negative charge present at neutral pH, thereby decreasing intramolecular charge–charge repulsion and permitting hydrophobic-driven collapse to the partially folded conformation. The effect of elevated temperatures was attributed to increased strength of the hydrophobic interaction at higher temperatures, leading to a stronger hydrophobic driving force for folding [49]. In fact, any changes in the environment of a natively unfolded protein leading to an increase in protein hydrophobicity and/or decrease in its net charge are expected to lead to partial folding.

11.2.3
Major Structural Characteristics of Partially Folded α-Synuclein

As the pH is decreased (or temperature increased) changes were observed in the shape of the CD spectrum for α-synuclein. Fig. 11.1 (a) shows that the minimum at 196 nm becomes less intense, whereas the negative intensity of the spectrum around 222 nm increases, reflecting pH-induced formation of secondary structure. Similarly, the FTIR spectrum of α-synuclein at pH 7.5 was typical of an unfolded polypeptide (Fig. 11.1 b), whereas a decrease in pH leads to significant spectral changes, indicating an increase in ordered secondary structure. The most evident change is the appearance of a new β-sheet band in the vicinity of 1626 cm^{-1}. Thus, at acidic pH natively unfolded α-synuclein is transformed into a partially folded conformation with a significant amount of β-structure. Furthermore, Fig. 11.1 (c) shows that a decrease in pH leads to a considerable increase in 1-anilino-8-naphthalene sulfonate (ANS) fluorescence intensity and a large blue shift of the ANS fluorescence maximum (from around 515 to around 475 nm), reflecting the pH-induced transformation of the natively unfolded α-synuclein to a partially folded conformation with solvent-exposed hydrophobic regions. Hydrodynamic methods revealed that this partially folded conformation is accompanied by a substantial decrease in hydrodynamic dimensions ($R_S = 27.9 \pm 0.4$ and $R_g = 30 \pm 1$ Å). Moreover, changes in the profile of the small-angle X-ray scattering Kratky plot at pH 3 were consistent with the development of a tightly packed core (Fig. 11.1 D). This means that protonation of α-synuclein results in the transformation of this natively unfolded protein into a more compact conformation with a significant amount of ordered secondary structure, affinity for ANS and the beginnings of a tightly packed core, all hallmarks of a partially folded intermediate [49]. Comparable structural changes are induced in α-synuclein by high temperatures [49].

The analysis of these data suggests that the partially folded conformation induced in α-synuclein by low pH or high temperature resembles the pre-molten globule state, an intermediate, preceding the molten globule in the refolding of globular proteins. In fact, it has been shown that the pre-molten globule is a denatured state, i.e. a conformation where the protein lacks a rigid three-dimensional structure. It is characterized by some secondary structure, although much less than that of the molten globule or native globular protein. A protein in the pre-molten globule state is considerably less compact than in the molten globule state and does not have a tightly packed globular structure, but it is more compact than the corresponding random coil. Furthermore, the pre-molten globule can interact with the hydrophobic fluorescent probe ANS, although more weakly than a molten globular protein. This means that at least some hy-

a)
pH 7.5
95°C
pH 3.0
[θ] (deg cm2 dmol-1)
0
-5000
-10000
-15000
190
200
210
220
230
240
250
Wavelength (nm)
b)
Fibril
pH 7.5
pH 3.0
Normalized absorbance
1700
1650
1600
Wavenumbers (cm-1)
c)
pH 3.0
pH 7.5
ANS fluorescence
500x10³
250x10³
0
400
450
500
550
600
Wavelength (nm)

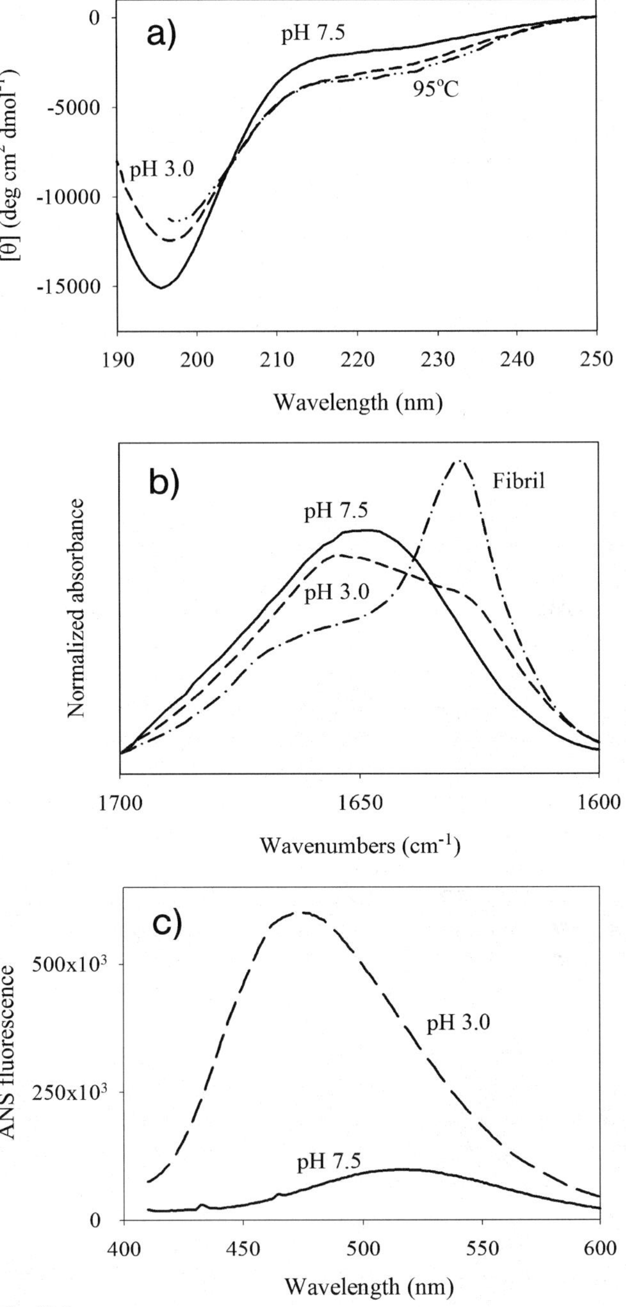

Fig. 11.1 a–c

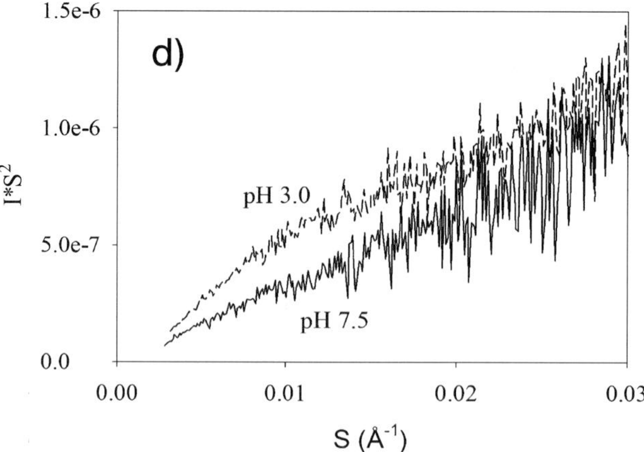

Fig. 11.1 Major structural characteristics of the natively unfolded and the partially folded pre-molten globule-like conformation induced in human α-synuclein by low pH. (a) Far-UV CD spectra measured under different conditions. (b) FTIR spectra in the amide I region measured for natively unfolded, partially folded and fibrillar forms of human α-synuclein. (c) ANS spectra measured under different experimental conditions. (d) Kratky plot representation of the results of small angle X-ray scattering analysis of α-synuclein at different experimental conditions.

drophobic clusters are already formed in the pre-molten globule state, although there is no globular structure [53–55].

11.2.4 Fibril Formation by α-Synuclein and the Partially Folded Amyloidogenic Conformation

The fibrillogenesis of α-synuclein *in vitro* has been studied extensively (for recent reviews, see [45, 56]). Although α-synuclein is an intrinsically unstructured protein, it forms fibrils of highly organized structure. Fig. 11.2 shows that both decreasing the pH and increasing the temperature result in accelerated fibrillation of α-synuclein [49]; thus, there is an excellent correlation between the intramolecular conformational change described above and the increased kinetics of fibril formation, suggesting that the partially structured conformation is a key intermediate on the fibril-forming pathway [49]. In contrast to an unfolded polypeptide chain, a partially folded conformation is anticipated to have contiguous hydrophobic patches on its surface, which might foster self-association and fibrillation. Thus, factors that shift the equilibrium in favor of this monomeric partially folded conformation will favor fibril formation.

An increase in protein concentration obviously increases the absolute concentration of the intermediate and, thus, is predicted to accelerate fibrillation, as is ob-

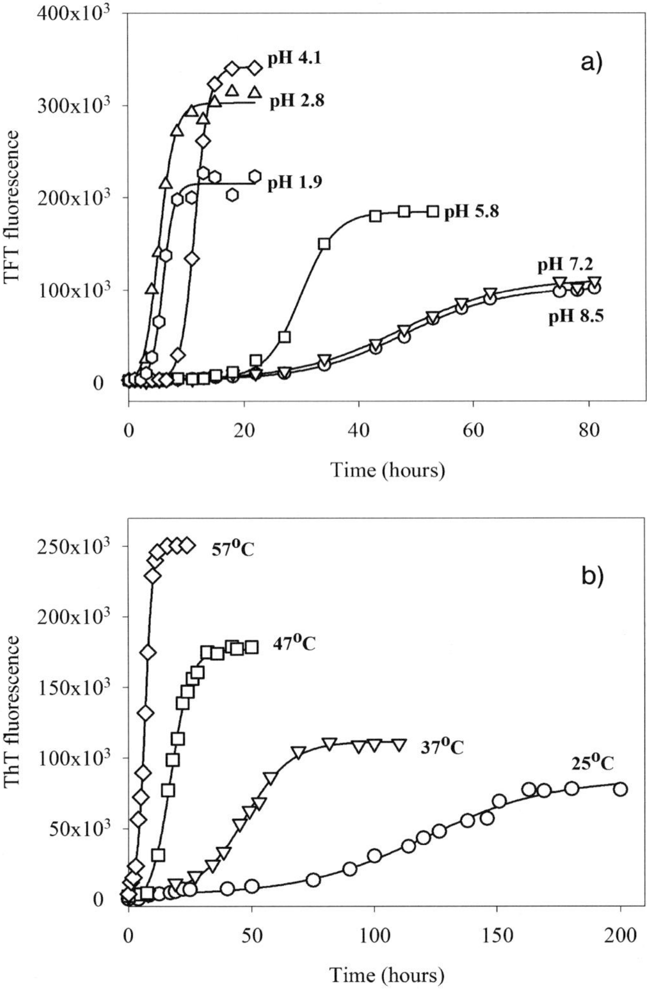

Fig. 11.2 Effect of pH (A) and temperature (B) on fibril formation of human α-synuclein, monitored by Thioflavin T fluorescence. Decreasing pH and increasing temperature accelerate fibrillation, due to decreased net charge and increased hydrophobicity, respectively.

served (Fig. 11.3). There is an inverse linear correlation between the logarithm of α-synuclein concentration and the duration of the lag time (Fig. 11.3B). This reflects an important aspect of the kinetics of nucleus formation, i.e. the formation of the aggregation-prone partially folded intermediate represents the rate-limiting step in α-synuclein nucleation and this process is probably responsible for the observed first-order kinetics. An alternative way to view this is that association to form fibrils involves the association of a rare conformer, the amyloidogenic par-

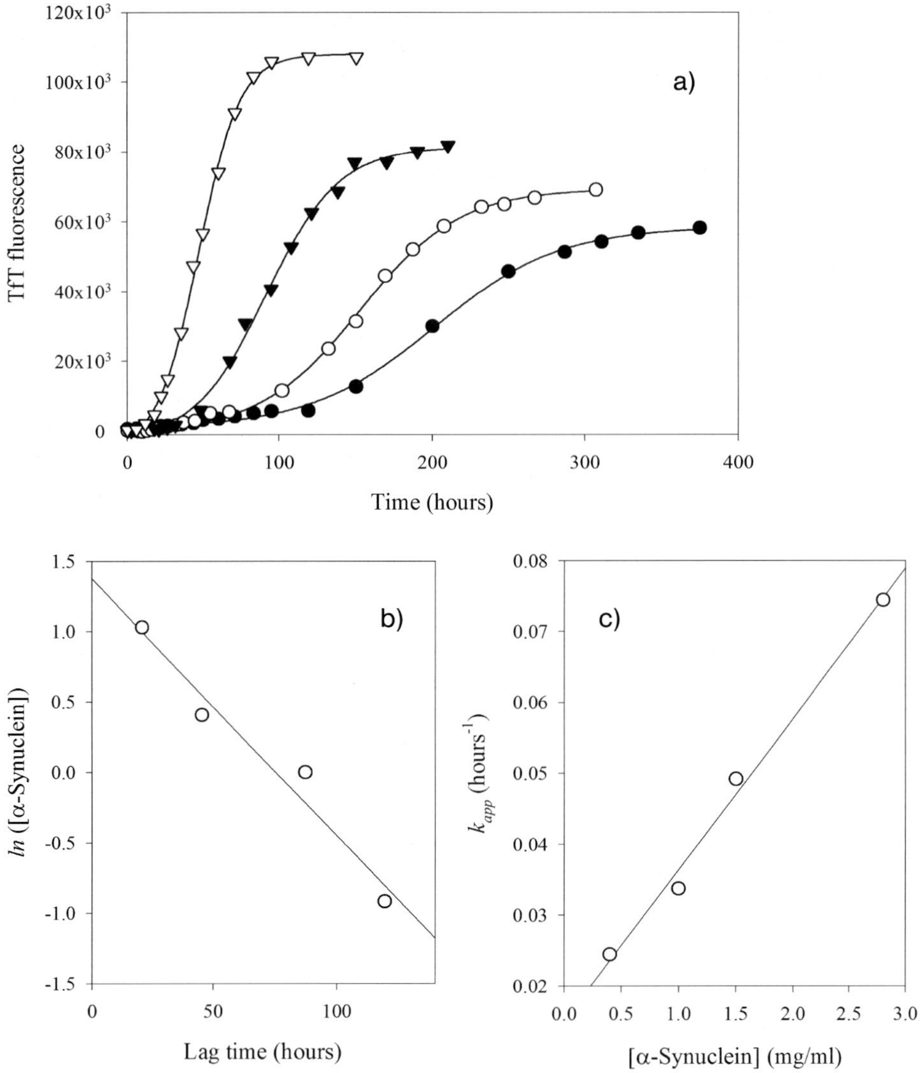

Fig. 11.3 The effect of protein concentration on the kinetics of fibril formation of human recombinant α-synuclein. (a) Kinetics of fibrillation at different α-synuclein concentrations. Protein concentrations were 21 (solid circles), 70 (open circles), 105 (solid triangles) and 190 μM (open triangles). (b) Inverse linear dependence of the logarithm of α-synuclein concentration as a function of lag time. (c) Linear dependence of the rate constant k_{app} for fibril growth on the α-synuclein concentration.

tially folded intermediate. The data suggest that the key partially folded intermediate, once formed, oligomerizes rapidly to form fibrils. Fig. 11.3(C) shows that there is also a linear dependence of the first-order rate constant for fibril growth (elongation) on protein concentration. The kinetics of elongation have been shown to follow first-order kinetics for some other proteins, including Aβ and insulin [57–60]. The simplest explanation for this fact is that increasing protein concentration leads to increasing numbers of fibrils, due to increasing concentration of nuclei and to increasing concentration of monomeric protein (either in its native-like form or as a partially folded intermediate, see below).

It has been assumed that the aggregation of α-synuclein, leading to the development of Parkinson's disease and other synucleinopathies, arises from various factors that would significantly increase the concentration of the critical partially folded intermediate [49]. For example, a number of factors have been shown to induce partial folding of α-synuclein and accelerated fibrillation *in vitro*: these include some common pesticides and herbicides [61–63], metal ions [63, 64], moderate concentrations of osmolytes such as TMAO [65], and some alcohols [66]. Under all these conditions α-synuclein was shown to form fibrils rapidly, thus confirming the correlation between the partially folded conformation and fibril formation (see Figs 11.2 B and 11.4 A). In contrast, Fig. 11.4 (B) shows that the process of fibril formation can be slowed or completely inhibited under conditions favoring formation of more folded conformations [65, 66] or by stabilization of off-pathway oligomers, e.g. via nitration of tyrosines [67] or methionine oxidation [68], or interaction with β- or γ-synucleins [50]. All conditions that populated the aggregation-prone partially folded conformation accelerated both the nucleation and elongation stages of fibril assembly. This means that the partially folded intermediate is likely involved in both the formation of the nucleus and in the subsequent propagation of fibrils.

It is worth noting that the formation of amyloid-like fibrils is not the only pathological outcome of protein deposition diseases and in several disorders (as well as in numerous *in vitro* experiments) protein deposits are composed of amorphous aggregates, without local order, e.g. light chain deposition disease, which involves deposition of immunoglobulin light chains. Similarly, soluble oligomers represent another alternative final product of the aggregation process. The physiological significance of oligomers is that they may be the toxic species. Fig. 11.5 represents a simplified model of α-synuclein aggregation, demonstrating the fact that aggregation is a very complex process, which can be divided into three major steps. The model shows that the first stage of the aggregation process is the structural transformation of a soluble natively unfolded protein, U_N, into the "sticky" aggregation-prone precursor or intermediate (I), which plays a central role in the entire aggregation process. This species can partition kinetically between a minimum of three distinct pathways, leading to fibrils, amorphous aggregates and soluble oligomers. The formation of a nucleus leading to both fibrils and amorphous aggregates is a kinetically disfavored event and is responsible for the lag period preceding significant formation of aggregates. Once a critical nucleus has been generated, the conditions change in favor of a rapid increase in aggregate size.

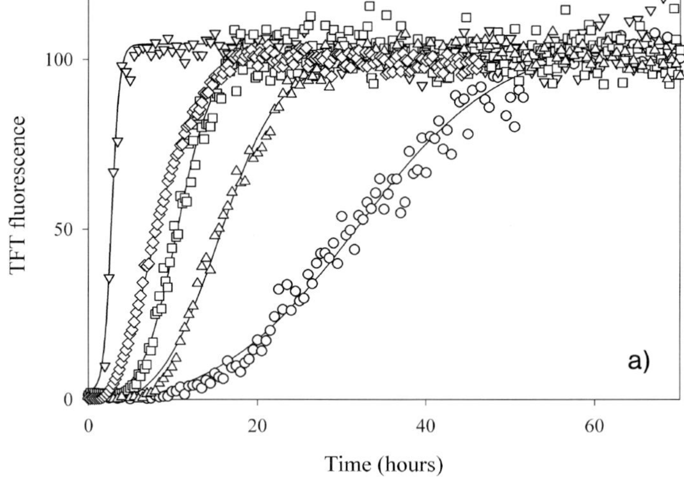

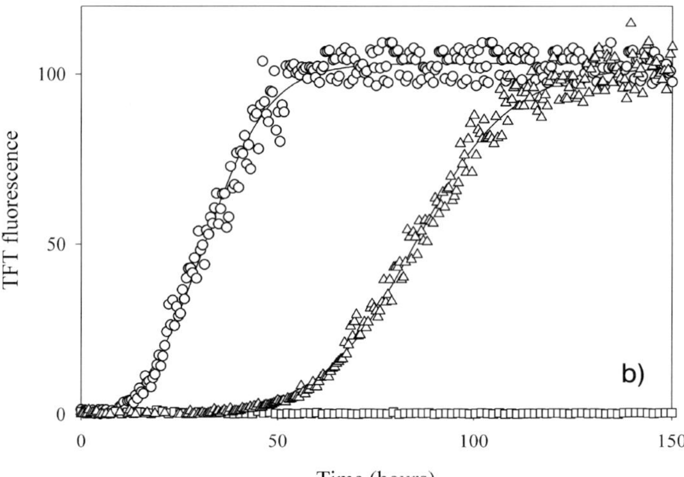

Fig. 11.4 Effect of different environmental factors on fibrillation of human α-synuclein. (a) Factors accelerating protein fibrillation: control (circles), 1 μM TMAO (inverse triangles), 10% methanol (diamonds), 500 M paraquat (squares) and 500 μM $HgCl_2$ (triangles). (b) Factors inhibiting fibrillation of α-synuclein: control (circles), nitrated α-synuclein (squares) and 1:1 mixture of α-/β-synucleins (triangles).

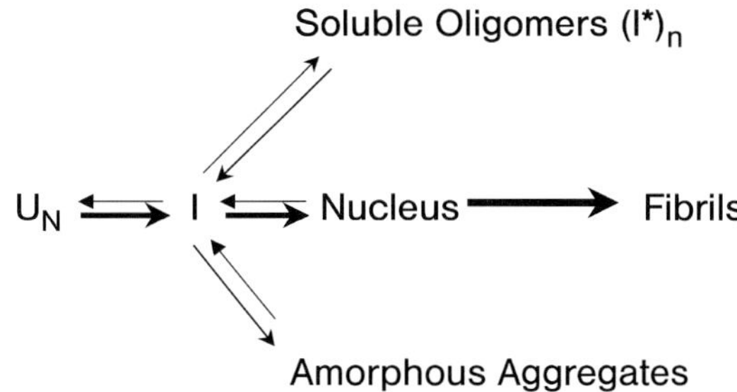

Fig. 11.5 Schematic overview of the aggregation of α-synuclein, showing the central position of the partially folded intermediate and the different final aggregated products.

11.3 Fibrillogenesis of Natively Unfolded Proteins Requires Partial Folding

11.3.1 Fibril Formation by Proteins Involved in Conformational Disorders

Data presented in Tables 11.1 and 11.2 show that many of the known amyloidogenic proteins are natively unfolded. It is reasonable to assume that such proteins are well suited for amyloidogenesis, as they lack significant secondary and tertiary structure. In the absence of such conformational constraints they would be expected to be substantially more conformationally mobile, and thus more able to polymerize in the form of β-sheets than tightly packed globular proteins. However, this is not always the case and many natively unfolded proteins form fibrils *in vitro* with similar rates to those of native globular proteins (i.e. within a time frame of a few hours to several days). The delay in the fibril formation of these proteins has been attributed to the requirement for considerable structural rearrangement within the unfolded polypeptide chain, giving rise to a partially folded conformation. The following section considers illustrative examples of fibril formation of natively unfolded proteins.

11.3.2 Amyloid β protein (Aβ)

Alzheimer's disease is the most prevalent age-dependent dementia, and is characterized by the accumulation of extracellular amyloid deposits, senile plaques, in the cerebral cortex and of intracellular neurofibrillary tangles. Senile plaques contain Aβ, which is a 40- to 42-residue peptide produced by endoproteolytic cleavage of the Aβ protein precursor (AβPP), whereas neurofibrillary tangles are assembled from the protein tau (see below). Many lines of evidence support the crucial role of Aβ in Alzheimer's disease, in which the aggregation of Aβ is associated with the development of the cascade of neuropathogenetic events, ending with the appearance of cognitive and behavioral features typical of Alzheimer's disease. NMR studies have shown that monomers of Aβ1–40 or Aβ1–42 are unfolded under physiological conditions [69]. However, partial folding has been detected at the earliest stages of Aβ fibril formation [69] and Aβ aggregation fits the "standard" pattern of fibril formation, i.e. native(ly unfolded) to partially folded to oligomers to fibrillar species. Many recent studies have shown that Aβ forms small soluble oligomers, in which the polypeptide is partially folded and that appear to be the toxic species [70–78].

11.3.3 Tau protein

Tau, a microtubule assembly protein [79–85], represents a family of isoforms with one, two, three or four repeats in the C-terminal region [86]. Post-translational phosphorylation of tau is an additional source of microheterogeneity [87]. Tau is a significantly disordered protein. Interest in tau dramatically increased with the discovery of its aggregation in neuronal cells in the progress of Alzheimer's disease and various other neurodegenerative disorders, especially frontotemporal dementia [88, 89]. In these cases specific tau-containing neurofibrillary tangles [paired helical filaments (PHFs)] are formed [90]. Hyperphosphorylation was shown to be a common characteristic of pathological tau [91], and hyperphosphorylated tau isolated from patients with Alzheimer's disease is unable to bind to microtubules and promote microtubule assembly. However, both of these activities are restored after enzymatic dephosphorylation of tau [92–96].

During brain development, tau is phosphorylated at many residues and by several kinases [97, 98]. Most of the *in vitro* phosphorylation sites of tau are located within the microtubule interacting region (repeat domain) and sequences flanking the repeat domain. Many of these sites are also phosphorylated in PHF-tau [99, 100]. In fact, 10 major phosphorylation sites have been identified in tau isolated from PHFs from patients with Alzheimer's disease [99]. All of these sites are located in regions flanking tau's repeat domain and constitute recognition sites for several Alzheimer's disease diagnostic antibodies, which may point to an important role for these phosphorylation sites for Alzheimer's disease pathogenesis.

Hyperphosphorylation is accompanied by transformation from the unfolded state of tau into a partially folded conformation [101, 102], dramatically accelerating the self-assembly into paired helical filaments *in vitro* [93]. To analyze the potential role of tau hyperphosphorylation in tauopathies, mutated tau proteins have been produced, in which all 10 serine/threonine residues known to be highly phosphorylated in PHF-tau were substituted for negatively charged residues, thus producing a model for a defined and permanent hyperphosphorylation-like state [103]. Analogous to hyperphosphorylation, glutamate substitutions induce compact structure elements and sodium dodecylsulfate (SDS)-resistant conformational domains in tau, as well as dramatic acceleration of its fibrillation [103].

11.3.4 Islet Amyloid Polypeptide (IAPP) or Amylin

In addition to insulin, pancreatic islet cells β produce a peptide called amylin or IAPP [104]. Dysfunction of amylin due to mutation and/or amyloid fibril formation is associated with the development of non-insulin-dependent diabetes mellitus (Type 2 diabetes) [105–107]. Amylin is an unstructured peptide hormone of 37 amino acid residues, which forms fibrils under physiological conditions *in vitro*. The process of polymerization is relatively fast (lag times of 100 and 50 min for full-length amylin and its 8–37 fragment, respectively) and results in the appearance of typical amyloid fibrils [108]. Interestingly, both peptides showed formation of a partially folded intermediate early in the fibrillation process. It takes around 90 min for full-length amylin to form such an intermediate, whereas this period was almost half as long for the truncated peptide, showing excellent correlation with the fibrillation lag-times [108].

11.3.5
Prion Protein

Prion diseases, collectively referred to as the transmissible spongiform encephalopathies (TSEs), are characterized by unique infectious prion particles. TSEs include Creutzfeldt-Jakob disease, scrapie, bovine spongiform encephalopathy, and chronic wasting disease of mule deer and elk [109]. The characteristic pathological features of TSEs are spongiform degeneration of the brain and accumulation of the abnormal, protease-resistant prion protein isoform in the central nervous system, which sometimes forms amyloid-like plaques.

The N-terminal region of about 100 amino acids in PrP^{C} is unstructured in the isolated molecule in solution in the absence of copper. The C-terminal domain is folded into a largely α-helical conformation (three α-helices and a short antiparallel β-sheet) and stabilized by a single disulphide bond linking helices 2 and 3 [110]. The central event in the pathogenesis of prion diseases is a major conformational change of the C-terminal region of the prion protein (PrP) from an α-helical (PrP^{C}) to a β-sheet-rich isoform (PrP^{Sc}) and PrP^{Sc} propagates itself by causing the conversion of PrP^{C} to PrP^{Sc}. Although unstructured in the isolated molecule, the N-terminal region contains tight binding sites for Cu^{2+} ions and acquires structure following copper binding [111–115]. Investigations of the steps required for prion propagation and neurodegeneration in transgenic mice expressing chimeric mouse–hamster–mouse or mouse–human–mouse PrP transgenics indicated that the last 50 residues in the disordered N-terminal region play a particularly important role in the interaction of PrP^{C} with PrP^{Sc} leading to the conversion of the former to the latter [116, 117]. Those residues are largely unordered or weakly helical in the full-length PrP^{C} [118, 119], but are predicted to be β-structure in PrP^{Sc} [120]. These observations emphasize a crucial role of the disordered N-terminal region in the modulation of prion protein aggregation. Several kinetics studies have revealed the existence of partially folded intermediates for the prion protein [121–123] and is reasonable to assume that fibrillation requires partial unfolding of the C-terminal domain prior to self-association.

11.3.6
Polyglutamine Repeat Diseases

Currently there are eight known hereditary diseases, including Huntington's disease, in which the expansion of a CAG repeat in the gene leads to neurodegeneration [124, 125]. The neurotoxicity in these diseases is due to the expansion of the CAG_n-encoded polyglutamine [poly(Gln)] repeat. The mechanistic hypothesis linking CAG repeat expansion to toxicity involves the tendency of longer poly(Gln) sequences, regardless of protein context, to form insoluble aggregates [126–134]. To help evaluate various possible mechanisms, the biophysical properties of a series of simple poly(Gln) peptides have been analyzed. The CD spectra of poly(Gln) peptides with repeat lengths of 5, 15, 28 and 44 residues were shown to be nearly identical, and were consistent with a high degree of random coil structure, suggesting that the length dependence of disease is not related to a conformational change in the monomeric states of expanded poly(Gln) sequences [132]. In contrast, there was a dramatic acceleration in the spontaneous formation of ordered, amyloid-like aggregates for poly(Gln) peptides with repeat lengths of greater than 37 residues. Several studies have revealed the existence of partially folded intermediates of poly(Gln)-repeat proteins as key species in fibrillation [133, 135, 136] and these are assumed to be precursors of oligomeric aggregates [132].

11.4
Fibrillation of Proteins Unrelated to Conformational Disease

11.4.1
Yeast Prions

There are at least two natural genetic elements, the [PSI^+] and [*URE3*] factors, in the budding yeast *Saccharomyces cerevisiae* that exhibit non-Mendelian inheritance, but can be "cured" by treatment of the cells with low concentrations of the protein denaturant guanidine hydrochloride [137]. The [PSI^+] factor is a self-perpetuating, conformationally altered form of a cellular protein [138–140]. The inheritance of [PSI^+] from mother to daughter cells is based on the transmission of conformational information from ordered non-functional Sup35p aggregates (amyloid-like fibrils) to the soluble, functional form of Sup35p, a subunit of the polypeptide chain release complex that is essential for translation termination [141, 142]. *In vitro*, the prion-determining region of Sup35p, NM, is initially disordered, but slowly converts to a partially folded structure and then fibrils [143–145]. In particular, factors such as elevated temperatures, chemical chaperones and certain mutations that increased the amount of structure increased the amount of aggregation [145]. The acceleration of Sup35p fibril formation is determined by the acceleration of slow conformational changes rather than by providing stable nuclei. Strikingly, inhibitory mutations map exclusively within a short glutamine/asparagine-rich region of Sup35p and all but one occur at polar residues. Even after replacement of this region with poly(Gln), Sup35p retains its ability to form fibrils [140]. This suggests similarities between the prion-like propagation of [*PSI*$^+$] and poly(Gln)-mediated pathogenesis of several neurodegenerative diseases (see above).

The [*URE3*] element is another factor of *S. cerevisiae* that propagates by a prion-like mechanism and corresponds to the loss of function of the cellular protein Ure2 [146, 147]. The N-terminal region of the protein is flexible and unstructured, while its C-terminal region is compact and folded [148–150]. The overexpression of full-length Ure2p in wild-type *S. cerevisiae* strains induced a 20- to 200-fold increase in the frequency with which [*URE3*] arose. On the other hand, expression of just the N-terminal 65 residues of Ure2p increased the frequency of [*URE3*] induction 6000-fold [148]. These observations emphasize a key role of the disordered N-terminal domain in the fibrillation of Ure2 protein. A dimeric intermediate is populated transiently during refolding and populated under conditions correlating with fibril formation. Interestingly, the native dimer and dimeric intermediate are significantly more stable than either of their monomeric counterparts. Stabilization of the native state disfavors amyloid nucleation [151].

11.4.2 Prothymosin α

This is a very acidic protein, containing around 50% aspartic and glutamic acid, no aromatic or cysteine residues, and very few large hydrophobic aliphatic amino acids [152]. Because of these features, prothymosin α adopts a random coil-like conformation with no regular secondary structure at neutral pH ([152, 153], and Tables 11.1 and 11.2). However, at acidic pH prothymosin α folds into a partially folded conformation [153]. Interestingly, it has been recently shown that at low pH (below pH 3, i.e. under conditions favoring the formation of the partially folded conformation [153]), prothymosin α is capable of relatively fast formation (lag time around 100 min) of regular elongated fibrils with a flat ribbon-like structure 4–5 nm in height and 12–13 nm in width [154].

11.4.3 Apolipoprotein CII (ApoCII)

Human ApoCII is a plasma protein consisting of 79 amino acid residues, whose function is to activate lipoprotein lipase. The structure of ApoCII in the presence of the lipid mimetic, SDS, reveals three regions of well-defined amphipathic α-helix with loosely defined intervening regions that may reflect flexible hinge regions [155]. In the absence of lipid or detergent, human ApoCII lacks ordered structure and forms amyloid ribbons after incubation for several days [156–162]. Fibril formation was dramatically accelerated by the addition of phospholipids in submicellar concentrations, which were shown to induce partial folding of the protein into a pre-molten globule-like intermediate [156]. In contrast, the fibrillation of ApoCII was completely inhibited in the presence of micellar phospholipids, i.e. under conditions favoring α-helical conformation [156].

11.4.4 Histones

The bovine core histones are natively unfolded proteins in solutions of low ionic strength, due to their high net positive charge at pH 7.5 [163]. Analysis of the structural properties of core histones as a function of pH, ionic strength and organic solvent concentration demonstrated a correlation between conformational change and propensity to aggregate. Overall, the histones are able to adopt at least five different relatively stable conformations: the intrinsically unstructured form (pH 2.0, low protein concentration, low ionic strength), a fully-unfolded conformation (6 M urea), partially folded α-helical dimers (low ionic strength, pH 2.0, high protein concentrations), helical monomers (low ionic strength, pH 2.0, 10% trifluoroethanol) and molten globule-like α-helical oligomers (pH 2.0, high salt) [163]. Under most of the conditions studied the histones formed amyloid fibrils with typical morphology as seen by electron microscopy. In contrast to most aggregation/amyloidogenic systems, the kinetics of fibrillation showed an inverse dependence on histone concentration, which has been attributed to partitioning to a faster pathway leading to non-fibrillar self-associated aggregates at higher protein concentrations. In keeping with the hypothesis that partial folding is required of intrinsically disordered proteins for aggregation, the rate of fibril formation correlated with conditions favoring the stabilization of the partially folded conformation [163].

11.5 Conclusions

Intrinsically disordered or natively unfolded proteins represent a high fraction of naturally occurring amyloidogenic proteins. This is most likely due to the ease with which they can form the β-sheet topology required in amyloid fibrils, in contrast to folded globular proteins, for which the native conformation places major constraints for the required topological rearrangement. In contrast to tightly folded globular proteins, the first, and critical, step in the aggregation of intrinsically disordered proteins or polypeptides is partial folding. This results in a partially folded conformation/intermediate with hydrophobic surface patches that favors self-association. As with such intermediates formed from the partial folding of globular proteins, such aggregate-prone intermediates can polymerize to form fibrillar or amorphous aggregates, or soluble oligomers.

Acknowledgments

This work was supported by grants INTAS 2001-2347 (V. N. U.), and NIH NS39985, DK55675 and NS43778 (A. L. F.).

References

1 Pandey, N., M. Ganapathi, K. Kumar, D. Dasgupta, S. D. Sutar and D. Dash. Comparative analysis of protein unfoldedness in human housekeeping and non-housekeeping proteins. *Bioinformatics* **2004**, 20, 2904–2910.

2 Dunker, A. K., Z. Obradovic, P. Romero, E. C. Garner and C. J. Brown. Intrinsic protein disorder in complete genomes. *Genome Inform Ser Workshop Genome Inform* **2000**, *11*, 161–171.

3 Wright, P. E. and H. J. Dyson. Intrinsically unstructured proteins: re-assessing the protein structure–function paradigm. *J Mol Biol* **1999**, *293*, 321–331.

4 Radivojac, P., Z. Obradovic, C. J. Brown and A. K. Dunker. Improving sequence alignments for intrinsically disordered proteins. *Pacific Symp Biocomput* **2002**, 589–600.

5 Uversky, V. N., J. R. Gillespie and A. L. Fink. Why are "natively unfolded" proteins unstructured under physiologic conditions? *Proteins* **2000**, *41*, 415–427.

6 Kelly, J. W. The alternative conformations of amyloidogenic proteins and their multi-step assembly pathways. *Curr Opin Struct Biol* **1998**, *8*, 101–106.

7 Dobson, C. M. Protein misfolding, evolution and disease. *Trends Biochem Sci* **1999**, *24*, 329–332.

8 Bellotti, V., P. Mangione and M. Stoppini. Biological activity and pathological implications of misfolded proteins. *Cell Mol Life Sci* **1999**, *55*, 977–991.

9 Uversky, V. N., A. Talapatra, J. R. Gillespie and A. L. Fink. Protein deposits as the molecular basis of amyloidosis. I. Systemic amyloidoses. *Med Sci Monitor* **1999**, *5*, 1001–1012.

10 Uversky, V. N., A. Talapatra, J. R. Gillespie and A. L. Fink. Protein deposits as the molecular basis of amyloidosis. II. Localized amyloidosis and neurodegenerative disorders. *Med Sci Monitor* **1999**, *5*, 1238–1254.

11 Rochet, J. C. and P. T. Lansbury, Jr. Amyloid fibrillogenesis: themes and variations. *Curr Opin Struct Biol* **2000**, *10*, 60–68.

12 Chiti, F., P. Webster, N. Taddei, A. Clark, M. Stefani, G. Ramponi and C. M. Dobson. Designing conditions for *in vitro* formation of amyloid protofilaments and fibrils. *Proc Natl Acad Sci USA* **1999**, *96*, 3590–3594.

13 Uversky, V. N. Protein folding revisited. A polypeptide chain at the folding–misfolding–nonfolding cross-roads: which way to go? *Cell Mol Life Sci* **2003**, *60*, 1852–1871.

14 Uversky, V. N. and A. L. Fink. Conformational constraints for amyloid fibrillation: the importance of being unfolded. *Biochim Biophys Acta* **2004**, *1698*, 131–153.

15 Dobson, C. M. Protein folding and its links with human disease. *Biochem Soc Symp* **2001**, 1–26.

16 Fandrich, M., M. A. Fletcher and C. M. Dobson. Amyloid fibrils from muscle myoglobin – even an ordinary globular protein can assume a rogue guise if conditions are right. *Nature* **2001**, *410*, 165–166.

17 Pertinhez, T. A., M. Bouchard, E. J. Tomlinson, R. Wain, S. J. Ferguson, C. M. Dobson and L. J. Smith. Amyloid fibril formation by a helical cytochrome. *FEBS Lett* **2001**, *495*, 184–186.

18 Dobson, C. M. Principles of protein folding, misfolding and aggregation. *Semin Cell Dev Biol* **2004**, *15*, 3–16.

19 Stefani, M. and C. M. Dobson. Protein aggregation and aggregate toxicity: new insights into protein folding, misfolding diseases and biological evolution. *J Mol Med* **2003**, *81*, 678–699.

20 Dyson, H. J. and P. E. Wright. Elucidation of the protein folding landscape by NMR. *Methods Enzymol* **2005**, *394*, 299–321.

21 Dunker, A. K., J. D. Lawson, C. J. Brown, R. M. Williams, P. Romero, J. S. Oh, C. J. Oldfield, A. M. Campen, C. M. Ratliff, K. W. Hipps, J. Ausio, M. S. Nissen, R. Reeves, C. Kang, C. R. Kissinger, R. W. Bailey, M. D. Griswold, W. Chiu, E. C. Garner and Z. Obradovic. Intrinsically disordered protein. *J Mol Graph Model* **2001**, *19*, 26–59.

22 Uversky, V. N. What does it mean to be natively unfolded? *Eur J Biochem* **2002**, *269*, 2–12.
23 Uversky, V. N. Natively unfolded proteins: a point where biology waits for physics. *Protein Sci* **2002**, *11*, 739–756.
24 Dunker, A. K., C. J. Brown, J. D. Lawson, L. M. Iakoucheva and Z. Obradovic. Intrinsic disorder and protein function. *Biochemistry* **2002**, *41*, 6573–6582.
25 Dunker, A. K., C. J. Brown and Z. Obradovic. Identification and functions of usefully disordered proteins. *Adv Protein Chem* **2002**, *62*, 25–49.
26 Tompa, P. Intrinsically unstructured proteins. *Trends Biochem Sci* **2002**, *27*, 527–533.
27 Dyson, H. J. and P. E. Wright. Coupling of folding and binding for unstructured proteins. *Curr Opin Struct Biol* **2002**, *12*, 54–60.
28 Sunde, M., L. C. Serpell, M. Bartlam, P. E. Fraser, M. B. Pepys and C. C. Blake. Common core structure of amyloid fibrils by synchrotron X-ray diffraction. *J Mol Biol* **1997**, *273*, 729–739.
29 Shirahama, T., M. D. Benson, A. S. Cohen and A. Tanaka. Fibrillar assemblage of variable segments of immunoglobulin light chains: an electron microscopic study. *J Immunol* **1973**, *110*, 21–30.
30 Shirahama, T. and A. S. Cohen. High-resolution electron microscopic analysis of the amyloid fibril. *J Cell Biol* **1967**, *33*, 679–708.
31 Jimenez, J. L., J. I. Guijarro, E. Orlova, J. Zurdo, C. M. Dobson, M. Sunde and H. R. Saibil. Cryo-electron microscopy structure of an SH3 amyloid fibril and model of the molecular packing. *EMBO J* **1999**, *18*, 815–821.
32 Lansbury, P. T., Jr. Evolution of amyloid: what normal protein folding may tell us about fibrillogenesis and disease. *Proc Natl Acad Sci USA* **1999**, *96*, 3342–3344.
33 Fink, A. L. Protein aggregation: folding aggregates, inclusion bodies and amyloid. *Fold Des* **1998**, *3*, 9–15.
34 Dobson, C. M. The structural basis of protein folding and its links with human disease. *Philos Trans R Soc Lond B Biol Sci* **2001**, *356*, 133–145.
35 Zerovnik, E. Amyloid-fibril formation. Proposed mechanisms and relevance to conformational disease. *Eur J Biochem* **2002**, *269*, 3362–3371.
36 Lewy, F. H. Paralysis Agitans. Pathologische Anatomie. In *Handbuch der Neurologie*, Lewandowski, M. (ed.). Springer, Berlin, **1912**.
37 Forno, L. S. Neuropathology of Parkinson's disease. *J Neuropathol Exp Neurol* **1996**, *55*, 259–272.
38 Polymeropoulos, M. H., C. Lavedan, E. Leroy, S. E. Ide, A. Dehejia, A. Dutra, B. Pike, H. Root, J. Rubenstein, R. Boyer, E. S. Stenroos, S. Chandrasekharappa, A. Athanassiadou, T. Papapetropoulos, W. G. Johnson, A. M. Lazzarini, R. C. Duvoisin, G. Di Iorio, L. I. Golbe and R. L. Nussbaum. Mutation in the alpha-synuclein gene identified in families with Parkinson's disease. *Science* **1997**, *276*, 2045–2047.
39 Kruger, R., W. Kuhn, T. Muller, D. Woitalla, M. Graeber, S. Kosel, H. Przuntek, J. T. Epplen, L. Schols and O. Riess. Ala30 Pro mutation in the gene encoding alpha-synuclein in Parkinson's disease. *Nat Genet* **1998**, *18*, 106–108.
40 Zarranz, J. J., J. Alegre, J. C. Gomez-Esteban, E. Lezcano, R. Ros, I. Ampuero, L. Vidal, J. Hoenicka, O. Rodriguez, B. Atares, V. Llorens, T. E. Gomez, T. del Ser, D. G. Munoz and J. G. de Yebenes. The new mutation, E46K, of alpha-synuclein causes Parkinson and Lewy body dementia. *Ann Neurol* **2004**, *55*, 164–173.
41 Singleton, A. B., M. Farrer, J. Johnson, A. Singleton, S. Hague, J. Kachergus, M. Hulihan, T. Peuralinna, A. Dutra, R. Nussbaum, S. Lincoln, A. Crawley, M. Hanson, D. Maraganore, C. Adler, M. R. Cookson, M. Muenter, M. Baptista, D. Miller, J. Blancato, J. Hardy and K. Gwinn-Hardy. α-Synuclein locus triplication causes Parkinson's disease. *Science* **2003**, *302*, 841.
42 Masliah, E., E. Rockenstein, I. Veinbergs, M. Mallory, M. Hashimoto, A. Takeda, Y. Sagara, A. Sisk and L. Mucke. Dopaminergic loss and inclusion body formation in alpha-synuclein mice: im-

plications for neurodegenerative disorders. *Science* **2000**, *287*, 1265–1269.

43 Feany, M. B. and W. W. Bender. A *Drosophila* model of Parkinson's disease [see Comments]. *Nature* **2000**, *404*, 394–398.

44 Uversky, V. N. What does it mean to be natively unfolded? *Eur J Biochem* **2002**, *269*, 2–12.

45 Uversky, V. N. A protein-chameleon: conformational plasticity of α-synuclein, a disordered protein involved in neurodegenerative disorders. *J Biomol Struct Dyn* **2003**, *21*, 211–234.

46 Dunker, A. K., E. Garner, S. Guilliot, P. Romero, K. Albercht, J. Hart, Z. Obradovic, C. Kissinger and J. E. Villafranca. Protein disorder and the evolution of molecular recognition: theory, predictions and observations. *Pacific Symp Biocomput* **1998**, *3*, 473–484.

47 Weinreb, P. H., W. Zhen, A. W. Poon, K. A. Conway and P. T. Lansbury, Jr. NACP, a protein implicated in Alzheimer's disease and learning, is natively unfolded. *Biochemistry* **1996**, *35*, 13709–13715.

48 Eliezer, D., E. Kutluay, R. Bussell, Jr and G. Browne. Conformational properties of alpha-synuclein in its free and lipid-associated states. *J Mol Biol* **2001**, *307*, 1061–1073.

49 Uversky, V. N., J. Li and A. L. Fink. Evidence for a partially folded intermediate in alpha-synuclein fibril formation. *J Biol Chem* **2001**, *276*, 10737–10744.

50 Uversky, V. N., J. Li, P. Souillac, I. S. Millett, S. Doniach, R. Jakes, M. Goedert and A. L. Fink. Biophysical properties of the synucleins and their propensities to fibrillate – inhibition of alpha-synuclein assembly by beta- and gamma-synucleins. *J Biol Chem* **2002**, *277*, 11970–11978.

51 Morar, A. S., A. Olteanu, G. B. Young and G. J. Pielak. Solvent-induced collapse of alpha-synuclein and acid-denatured cytochrome *c*. *Protein Sci* **2001**, *10*, 2195–2199.

52 Syme, C. D., E. W. Blanch, C. Holt, R. Jakes, M. Goedert, L. Hecht and L. D. Barron. A Raman optical activity study of rheomorphism in caseins, synucleins and tau. New insight into the structure and behaviour of natively unfolded proteins. *Eur J Biochem* **2002**, *269*, 148–156.

53 Uversky, V. N. and O. B. Ptitsyn. Further evidence on the equilibrium "pre-molten globule state": Four-state guanidinium chloride-induced unfolding of carbonic anhydrase B at low temperature. *J Mol Biol* **1996**, *255*, 215–228.

54 Uversky, V. N. Diversity of equilibrium compact forms of denatured globular proteins. *Protein Peptide Lett* **1997**, *4*, 355–367.

55 Karnoup, A. S. and V. N. Uversky. Sequential compaction of a random copolymer of hydrophilic and hydrophobic amino acid residues. *Macromolecules* **1997**, *30*, 7427–7434.

56 Uversky, V. N. and A. L. Fink. Biophysical properties of human α-synuclein and its role in Parkinson's disease. In *Recent Research Developments in Proteins*, Pandalai, S. G. (ed.). Transworld Research Network, Kerala, **2002**.

57 Lomakin, A., D. S. Chung, G. B. Benedek, D. A. Kirschner and D. B. Teplow. On the nucleation and growth of amyloid beta-protein fibrils: detection of nuclei and quantitation of rate constants. *Proc Natl Acad Sci USA* **1996**, *93*, 1125–1129.

58 Naiki, H. and K. Nakakuki. First-order kinetic model of Alzheimer's beta-amyloid fibril extension *in vitro*. *Lab Invest* **1996**, *74*, 374–383.

59 Esler, W. P., E. R. Stimson, J. R. Ghilardi, Y. A. Lu, A. M. Felix, H. V. Vinters, P. W. Mantyh, J. P. Lee and J. E. Maggio. Point substitution in the central hydrophobic cluster of a human beta-amyloid congener disrupts peptide folding and abolishes plaque competence. *Biochemistry* **1996**, *35*, 13914–13921.

60 Nielsen, L., R. Khurana, A. Coats, S. Frokjaer, J. Brange, S. Vyas, V. N. Uversky and A. L. Fink. Effect of environmental factors on the kinetics of insulin fibril formation: elucidation of the molecular mechanism. *Biochemistry* **2001**, *40*, 6036–6046.

61 Uversky, V. N., J. Li and A. L. Fink. Pesticides directly accelerate the rate of alpha-synuclein fibril formation: a possi-

ble factor in Parkinson's disease. *FEBS Lett* **2001**, *500*, 105–108.

62 Manning-Bog, A. B., A. L. McCormack, J. Li, V. N. Uversky, A. L. Fink and D. A. Di Monte. The herbicide paraquat causes up-regulation and aggregation of alpha-synuclein in mice. *J Biol Chem* **2002**, *277*, 1641–1644.

63 Uversky, V. N., J. Li, K. Bower and A. L. Fink. Synergistic effects of pesticides and metals on the fibrillation of α-synuclein: implications for Parkinson's disease. *Neurotoxicology* **2002**, *23*, 527–536.

64 Uversky, V. N., J. Li and A. L. Fink. Metal-triggered structural transformations, aggregation and fibrillation of human alpha-synuclein. A possible molecular link between Parkinson's disease and heavy metal exposure. *J Biol Chem* **2001**, *276*, 44284–44296.

65 Uversky, V. N., J. Li and A. L. Fink. Trimethylamine-*N*-oxide-induced folding of alpha-synuclein. *FEBS Lett* **2001**, *509*, 31–35.

66 Munishkina, L. A., C. Phelan, V. N. Uversky and A. L. Fink. Conformational behavior and aggregation of alpha-synuclein in organic solvents: modeling the effects of membranes. *Biochemistry* **2003**, *42*, 2720–2730.

67 Yamin, G., V. N. Uversky and A. L. Fink. Nitration inhibits fibrillation of human alpha-synuclein *in vitro* by formation of soluble oligomers. *FEBS Lett* **2003**, *542*, 147–152.

68 Uversky, V. N., G. Yamin, P. O. Souillac, J. Goers, C. B. Glaser and A. L. Fink. Methionine oxidation inhibits fibrillation of human alpha-synuclein *in vitro*. *FEBS Lett* **2002**, *517*, 239–244.

69 Kirkitadze, M. D., M. M. Condron and D. B. Teplow. Identification and characterization of key kinetic intermediates in amyloid beta-protein fibrillogenesis. *J Mol Biol* **2001**, *312*, 1103–1119.

70 Kheterpal, I., H. A. Lashuel, D. M. Hartley, T. Wlaz, P. T. Lansbury and R. Wetzel. A beta protofibrils possess a stable core structure resistant to hydrogen exchange. *Biochemistry* **2003**, *42*, 14092–14098.

71 Kayed, R., Y. Sokolov, B. Edmonds, T. M. McIntire, S. C. Milton, J. E. Hall, and C. G. Glabe. Permeabilization of lipid bilayers is a common conformation-dependent activity of soluble amyloid oligomers in protein misfolding diseases. *J Biol Chem* **2004**, *279*, 46363–46366.

72 Yong, W., A. Lomakin, M. D. Kirkitadze, D. B. Teplow, S. H. Chen and G. B. Benedek. Structure determination of micelle-like intermediates in amyloid beta-protein fibril assembly by using small angle neutron scattering. *Proc Natl Acad Sci USA* **2002**, *99*, 150–154.

73 Zerovnik, E. Amyloid-fibril formation. Proposed mechanisms and relevance to conformational disease. *Eur J Biochem* **2002**, *269*, 3362–3371.

74 Bitan, G., A. Lomakin and D. B. Teplow. Amyloid β-protein oligomerization. I. Prenucleation interactions revealed by photo-induced cross-linking of unmodified proteins. *J Biol Chem* **2001**, *276*, 35176–35184.

75 Chromy, B. A., R. J. Nowak, M. P. Lambert, K. L. Viola, L. Chang, P. T. Velasco, B. W. Jones, S. J. Fernandez, P. N. Lacor, P. Horowitz, C. E. Finch, G. A. Krafft and W. L. Klein. Self-assembly of Abeta(1–42) into globular neurotoxins. *Biochemistry* **2003**, *42*, 12749–12760.

76 El Agnaf, O. M. and G. B. Irvine. Aggregation and neurotoxicity of alpha-synuclein and related peptides. *Biochem Soc Trans* **2002**, *30*, 559–565.

77 Hoshi, M., M. Sato, S. Matsumoto, A. Noguchi, K. Yasutake, N. Yoshida and K. Sato. Spherical aggregates of beta-amyloid (amylospheroid) show high neurotoxicity and activate tau protein kinase I/glycogen synthase kinase-3beta. *Proc Natl Acad Sci USA* **2003**, *100*, 6370–6375.

78 Stine, W. B., Jr, K. N. Dahlgren, G. A. Krafft and M. J. LaDu. *In vitro* characterization of conditions for amyloid-beta peptide oligomerization and fibrillogenesis. *J Biol Chem* **2003**, *278*, 11612–11622.

79 Cleveland, D. W., S. Y. Hwo and M. W. Kirschner. Purification of tau, a microtubule-associated protein that induces assembly of microtubules from purified tubulin. *J Mol Biol* **1977**, *116*, 207–225.

80 Cleveland, D.W., S.Y. Hwo and M.W. Kirschner. Physical and chemical properties of purified tau factor and role of tau in microtubule assembly. *J Mol Biol* **1977**, *116*, 227–247.

81 Cleveland, D.W., J.A. Connolly, V.I. Kalnins, B.M. Spiegelman and M.W. Kirschner. Physical-properties and cellular localization of tau, a microtubule associated protein which induces assembly of purified tubulin. *J Cell Biol* **1977**, *75*, A283–A283.

82 Drechsel, D.N., A.A. Hyman, M.H. Cobb and M.W. Kirschner. Modulation of the dynamic instability of tubulin assembly by the microtubule-associated protein tau. *Mol Biol Cell* **1992**, *3*, 1141–1154.

83 Brandt, R. and G. Lee. The balance between tau proteins microtubule growth and nucleation activities – implications for the formation of axonal microtubules. *J Neurochem* **1993**, *61*, 997–1005.

84 Brandt, R. and G. Lee. Functional-organization of microtubule-associated protein-tau – identification of regions which affect microtubule growth, nucleation and bundle formation *in vitro*. *J Biol Chem* **1993**, *268*, 3414–3419.

85 Binder, L.I., A. Frankfurter and L.I. Rebhun. The distribution of tau in the mammalian central nervous system. *J Cell Biology* **1985**, *101*, 1371–1378.

86 Himmler, A. Structure of the bovine tau-gene – alternatively spliced transcripts generate a protein family. *Mol Cell Biol* **1989**, *9*, 1389–1396.

87 Kenessey, A. and S.H.C. Yen. The extent of phosphorylation of fetal-tau is comparable to that of PHF-tau from Alzheimer paired helical filaments. *Brain Res* **1993**, *629*, 40–46.

88 Crowther, R.A. and M. Goedert. Abnormal tau-containing filaments in neurodegenerative diseases. *J Struct Biol* **2000**, *130*, 271–279.

89 Goedert, M. The significance of tau and alpha-synuclein inclusions in neurodegenerative diseases. *Curr Opin Genet Dev* **2001**, *11*, 343–351.

90 Delacourte, A. and L. Buee. Normal and pathological tau proteins as factors for microtubule assembly. *Int Rev Cytol* **1997**, *171*, 167–224.

91 Vulliet, R., S.M. Halloran, R.K. Braun, A.J. Smith and G. Lee. Proline-directed phosphorylation of human tau protein. *J Biol Chem* **1992**, *267*, 22570–22574.

92 Lu, Q. and J.G. Wood. Functional studies of Alzheimers disease tau protein. *J Neurosci* **1993**, *13*, 508–515.

93 Alonso, A.C., I. Grundkeiqbal and K. Iqbal. Alzheimer's disease hyperphosphorylated tau sequesters normal tau into tangles of filaments and disassembles microtubules. *Nat Med* **1996**, *2*, 783–787.

94 Alonso, A.D., I. Grundkeiqbal and K. Iqbal. Abnormally phosphorylated tau from Alzheimer disease brain depolymerizes microtubules. *Neurobiol Aging* **1994**, *15*, S37–S37.

95 Alonso, A.D., T. Zaidi, I. Grundkeiqbal and K. Iqbal. Role of abnormally phosphorylated tan in the breakdown of microtubules in Alzheimer disease. *Proc Natl Acad Sci USA* **1994**, *91*, 5562–5566.

96 Iqbal, K., T. Zaidi, C. Bancher and I. Grundkeiqbal. Alzheimer paired helical filaments – restoration of the biological-activity by dephosphorylation. *FEBS Lett* **1994**, *349*, 104–108.

97 Watanabe, A., M. Hasegawa, M. Suzuki, K. Takio, M. Morishimakawashima, K. Titani, T. Arai, K.S. Kosik and Y. Ihara. *In vivo* phosphorylation sites in fetal and adult rat tau. *J Biol Chem* **1993**, *268*, 25712–25717.

98 Billingsley, M.L. and R.L. Kincaid. Regulated phosphorylation and dephosphorylation of tau protein: effects on microtubule interaction, intracellular trafficking and neurodegeneration. *Biochem J* **1997**, *323*, 577–591.

99 Morishimakawashima, M., M. Hasegawa, K. Takio, M. Suzuki, H. Yoshida, A. Watanabe, K. Titani and Y. Ihara. Hyperphosphorylation of Tau in PHF. *Neurobiol Aging* **1995**, *16*, 365–371.

100 Morishimakawashima, M., M. Hasegawa, K. Takio, M. Suzuki, H. Yoshida, K. Titani and Y. Ihara. Proline-directed and non-proline-directed phosphorylation of PHF-tau. *J Biol Chem* **1995**, *270*, 823–829.

101 Uversky, V. N., S. Winter, O. V. Galzitskaya, L. Kittler and G. Lober. Hyperphosphorylation induces structural modification of tau protein. *FEBS Lett* **1998**, *439*, 21–25.

102 Hagestedt, T., B. Lichtenberg, H. Wille, E. M. Mandelkow and E. Mandelkow. Tau-protein becomes long and stiff upon phosphorylation – correlation between paracrystalline structure and degree of phosphorylation. *J Cell Biol* **1989**, *109*, 1643–1651.

103 Eidenmuller, J., T. Fath, A. Hellwig, J. Reed, E. Sontag and R. Brandt. Structural and functional implications of tau hyperphosphorylation: information from phosphorylation-mimicking mutated tau proteins. *Biochemistry* **2000**, *39*, 13166–13175.

104 Cooper, G. J., A. C. Willis, A. Clark, R. C. Turner, R. B. Sim and K. B. Reid. Purification and characterization of a peptide from amyloid-rich pancreases of type 2 diabetic patients. *Proc Natl Acad Sci USA* **1987**, *84*, 8628–8632.

105 Higham, C. E., E. T. Jaikaran, P. E. Fraser, M. Gross and A. Clark. Preparation of synthetic human islet amyloid polypeptide (IAPP) in a stable conformation to enable study of conversion to amyloid-like fibrils. *FEBS Lett* **2000**, *470*, 55–60.

106 Jaikaran, E. T., C. E. Higham, L. C. Serpell, J. Zurdo, M. Gross, A. Clark and P. E. Fraser. Identification of a novel human islet amyloid polypeptide beta-sheet domain and factors influencing fibrillogenesis. *J Mol Biol* **2001**, *308*, 515–525.

107 Jaikaran, E. T. and A. Clark. Islet amyloid and type 2 diabetes: from molecular misfolding to islet pathophysiology. *Biochim Biophys Acta* **2001**, *1537*, 179–203.

108 Goldsbury, C., K. Goldie, J. Pellaud, J. Seelig, P. Frey, S. A. Muller, J. Kistler, G. J. S. Cooper and U. Aebi. Amyloid fibril formation from full-length and fragments of amylin. *J Struct Biol* **2000**, *130*, 352–362.

109 Prusiner, S. B. Shattuck lecture – neurodegenerative diseases and prions. *N Engl J Med* **2001**, *344*, 1516–1526.

110 Riek, R., S. Hornemann, G. Wider, M. Billeter, R. Glockshuber and K. Wuthrich. NMR structure of the mouse prion protein domain PrP(121–231). *Nature* **1996**, *382*, 180–182.

111 Collinge, J. Prion diseases of humans and animals: their causes and molecular basis. *Annu Rev Neurosci* **2001**, *24*, 519–550.

112 Aronoff-Spencer, E., C. S. Burns, N. I. Avdievich, G. J. Gerfen, J. Peisach, W. E. Antholine, H. L. Ball, F. E. Cohen, S. B. Prusiner and G. L. Millhauser. Identification of the Cu^{2+} binding sites in the N-terminal domain of the prion protein by EPR and CD spectroscopy. *Biochemistry* **2000**, *39*, 13760–13771.

113 Burns, C. S., E. Aronoff-Spencer, C. M. Dunham, P. Lario, N. I. Avdievich, W. E. Antholine, M. M. Olmstead, A. Vrielink, G. J. Gerfen, J. Peisach, W. G. Scott and G. L. Millhauser. Molecular features of the copper binding sites in the octarepeat domain of the prion protein. *Biochemistry* **2002**, *41*, 3991–4001.

114 Burns, C. S., E. Aronoff-Spencer, G. Legname, S. B. Prusiner, W. E. Antholine, G. J. Gerfen, J. Peisach and G. L. Millhauser. Copper coordination in the full-length, recombinant prion protein. *Biochemistry* **2003**, *42*, 6794–6803.

115 Millhauser, G. L. Copper binding in the prion protein. *Acc Chem Res* **2004**, *37*, 79–85.

116 Scott, M., D. Groth, D. Foster, M. Torchia, S. L. Yang, S. J. DeArmond and S. B. Prusiner. Propagation of prions with artificial properties in transgenic mice expressing chimeric PrP genes. *Cell* **1993**, *73*, 979–988.

117 Telling, G. C., M. Scott, J. Mastrianni, R. Gabizon, M. Torchia, F. E. Cohen, S. J. DeArmond and S. B. Prusiner. Prion propagation in mice expressing human and chimeric PrP transgenes implicates the interaction of cellular PrP with another protein. *Cell* **1995**, *83*, 79–90.

118 Donne, D. G., J. H. Viles, D. Groth, I. Mehlhorn, T. L. James, F. E. Cohen, S. B. Prusiner, P. E. Wright and H. J. Dyson. Structure of the recombinant full-length hamster prion protein

PrP(29–231): the N terminus is highly flexible. *Proc Natl Acad Sci USA* **1997**, *94*, 13452–13457.

119 James, T.L., H. Liu, N.B. Ulyanov, S. Farr-Jones, H. Zhang, D.G. Donne, K. Kaneko, D. Groth, I. Mehlhorn, S.B. Prusiner and F.E. Cohen. Solution structure of a 142-residue recombinant prion protein corresponding to the infectious fragment of the scrapie isoform. *Proc Natl Acad Sci USA* **1997**, *94*, 10086–100891.

120 Prusiner, S.B., M.R. Scott, S.J. DeArmond and F.E. Cohen. Prion protein biology. *Cell* **1998**, *93*, 337–348.

121 Apetri, A.C., K. Surewicz and W.K. Surewicz. The effect of disease-associated mutations on the folding pathway of human prion protein. *J Biol Chem* **2004**, *279*, 18008–18014.

122 Apetri, A.C. and W.K. Surewicz. Kinetic intermediate in the folding of human prion protein. *J Biol Chem* **2002**, *277*, 44589–44592.

123 Martins, S.M., A. Chapeaurouge and S.T. Ferreira. Folding intermediates of the prion protein stabilized by hydrostatic pressure and low temperature. *J Biol Chem* **2003**, *278*, 50449–50455.

124 Cummings, C.J. and H.Y. Zoghbi. Trinucleotide repeats: mechanisms and pathophysiology. *Annu Rev Genomics Hum Genet* **2000**, *1*, 281–328.

125 Cummings, C.J. and H.Y. Zoghbi. Fourteen and counting: unraveling trinucleotide repeat diseases. *Hum Mol Genet* **2000**, *9*, 909–916.

126 Zoghbi, H.Y. and H.T. Orr. Polyglutamine diseases: protein cleavage and aggregation. *Curr Opin Neurobiol* **1999**, *9*, 566–570.

127 Ross, C.A., J.D. Wood, G. Schilling, M.F. Peters, F.C. Nucifora, Jr, J.K. Cooper, A.H. Sharp, R.L. Margolis and D.R. Borchelt. Polyglutamine pathogenesis. *Phil Trans R Soc Lond B Biol Sci* **1999**, *354*, 1005–1011.

128 Preisinger, E., B.M. Jordan, A. Kazantsev and D. Housman. Evidence for a recruitment and sequestration mechanism in Huntington's disease. *Phil Trans R Soc Lond B Biol Sci* **1999**, *354*, 1029–1034.

129 Wanker, E.E. Protein aggregation and pathogenesis of Huntington's disease: mechanisms and correlations. *Biol Chem* **2000**, *381*, 937–942.

130 McCampbell, A., J.P. Taylor, A.A. Taye, J. Robitschek, M. Li, J. Walcott, D. Merry, Y. Chai, H. Paulson, G. Sobue and K.H. Fischbeck. CREB-binding protein sequestration by expanded polyglutamine. *Hum Mol Genet* **2000**, *9*, 2197–2202.

131 McCampbell, A. and K.H. Fischbeck. Polyglutamine and CBP: fatal attraction? *Nat Med* **2001**, *7*, 528–530.

132 Chen, S., V. Berthelier, W. Yang and R. Wetzel. Polyglutamine aggregation behavior *in vitro* supports a recruitment mechanism of cytotoxicity. *J Mol Biol* **2001**, *311*, 173–182.

133 Chen, S., V. Berthelier, J.B. Hamilton, B. O'Nuallain and R. Wetzel. Amyloid-like features of polyglutamine aggregates and their assembly kinetics. *Biochemistry* **2002**, *41*, 7391–7399.

134 Perutz, M.F., B.J. Pope, D. Owen, E.E. Wanker and E. Scherzinger. Aggregation of proteins with expanded glutamine and alanine repeats of the glutamine-rich and asparagine-rich domains of Sup35 and of the amyloid beta-peptide of amyloid plaques. *Proc Natl Acad Sci USA* **2002**, *99*, 5596–5600.

135 Chow, M.K., H.L. Paulson and S.P. Bottomley. Destabilization of a non-pathological variant of ataxin-3 results in fibrillogenesis via a partially folded intermediate: a model for misfolding in polyglutamine disease. *J Mol Biol* **2004**, *335*, 333–341.

136 Poirier, M.A., H. Li, J. Macosko, S. Cai, M. Amzel and C.A. Ross. Huntingtin spheroids and protofibrils as precursors in polyglutamine fibrilization. *J Biol Chem* **2002**, *277*, 41032–41037.

137 Lindquist, S. Mad cows meet psi-chotic yeast: the expansion of the prion hypothesis. *Cell* **1997**, *89*, 495–498.

138 Patino, M.M., J.J. Liu, J.R. Glover and S. Lindquist. Support for the prion hypothesis for inheritance of a phenotypic trait in yeast. *Science* **1996**, *273*, 622–626.

139 Paushkin, S.V., V.V. Kushnirov, V.N. Smirnov and M.D. Ter Avanesyan. Propagation of the yeast prion-like [psi^+] determinant is mediated by oligomerization of the SUP35-encoded polypeptide chain release factor. *EMBO J* **1996**, *15*, 3127–3134.

140 DePace, A.H., A. Santoso, P. Hillner and J.S. Weissman. A critical role for amino-terminal glutamine/asparagine repeats in the formation and propagation of a yeast prion. *Cell* **1998**, *93*, 1241–1252.

141 Serio, T.R. and S.L. Lindquist. [PSI^+]: an epigenetic modulator of translation termination efficiency. *Annu Rev Cell Dev Biol* **1999**, *15*, 661–703.

142 Serio, T. R., A. G. Cashikar, J. J. Moslehi, A. S. Kowal and S. L. Lindquist. Yeast prion [psi^+] and its determinant, Sup35p. *Methods Enzymol* **1999**, *309*, 649–673.

143 Glover, J.R., A.S. Kowal, E.C. Schirmer, M.M. Patino, J.J. Liu and S. Lindquist. Self-seeded fibers formed by Sup35, the protein determinant of [PSI^+], a heritable prion-like factor of *S. cerevisiae. Cell* **1997**, *89*, 811–819.

144 Serio, T.R., A.G. Cashikar, A.S. Kowal, G.J. Sawicki, J.J. Moslehi, L. Serpell, M.F. Arnsdorf and S.L. Lindquist. Nucleated conformational conversion and the replication of conformational information by a prion determinant. *Science* **2000**, *289*, 1317–1321.

145 Scheibel, T. and S.L. Lindquist. The role of conformational flexibility in prion propagation and maintenance for Sup35p. *Nat Struct Biol* **2001**, *8*, 958–962.

146 Wickner, R.B. [*URE3*] as an altered URE2 protein: evidence for a prion analog in *Saccharomyces cerevisiae. Science* **1994**, *264*, 566–569.

147 Thual, C., L. Bousset, A.A. Komar, S. Walter, J. Buchner, C. Cullin and R. Melki. Stability, folding, dimerization and assembly properties of the yeast prion Ure2p. *Biochemistry* **2001**, *40*, 1764–1773.

148 Masison, D.C. and R.B. Wickner. Prion-inducing domain of yeast Ure2p and protease resistance of Ure2p in prion-containing cells. *Science* **1995**, *270*, 93–95.

149 Masison, D.C., M.L. Maddelein and R.B. Wickner. The prion model for [*URE3*] of yeast: spontaneous generation and requirements for propagation. *Proc Natl Acad Sci USA* **1997**, *94*, 12503–12508.

150 Thual, C., A.A. Komar, L. Bousset, E. Fernandez-Bellot, C. Cullin and R. Melki. Structural characterization of *Saccharomyces cerevisiae* prion-like protein Ure2. *J Biol Chem* **1999**, *274*, 13666–13674.

151 Zhu, L., X.J. Zhang, L.Y. Wang, J.M. Zhou and S. Perrett. Relationship between stability of folding intermediates and amyloid formation for the yeast prion Ure2p: a quantitative analysis of the effects of pH and buffer system. *J Mol Biol* **2003**, *328*, 235–254.

152 Gast, K., H. Damaschun, K. Eckert, K. Schulze-Forster, H.R. Maurer, M. Muller-Frohne, D. Zirwer, J. Czarnecki and G. Damaschun. Prothymosin alpha: a biologically active protein with random coil conformation. *Biochemistry* **1995**, *34*, 13211–13218.

153 Uversky, V.N., J.R. Gillespie, I.S. Millett, A.V. Khodyakova, A.M. Vasiliev, T.V. Chernovskaya, R.N. Vasilenko, G.D. Kozovskaya, D.A. Dolgikh, A.L. Fink, S. Doniach and V.M. Abramov. Natively unfolded human prothymosin alpha adopts partially folded collapsed conformation at acidic pH. *Biochemistry* **1999**, *38*, 15009–15016.

154 Pavlov, N.A., D.I. Cherny, G. Heim, T.M. Jovin and V. Subramaniam. Amyloid fibrils from the mammalian protein prothymosin alpha. *FEBS Lett* **2002**, *517*, 37–40.

155 MacRaild, C.A., D.M. Hatters, G.J. Howlett and P.R. Gooley. NMR structure of human apolipoprotein C-II in the presence of sodium dodecyl sulfate. *Biochemistry* **2001**, *40*, 5414–5421.

156 Hatters, D.M., L.J. Lawrence and G.J. Howlett. Sub-micellar phospholipid accelerates amyloid formation by apolipoprotein C-II. *FEBS Lett* **2001**, *494*, 220–224.

157 Hatters, D.M., C.E. MacPhee, L.J. Lawrence, W.H. Sawyer and G.J. Howlett. Human apolipoprotein C-II forms

twisted amyloid ribbons and closed loops. *Biochemistry* **2000**, *39*, 8276–8283.

158 Hatters, D. M., R. A. Lindner, J. A. Carver and G. J. Howlett. The molecular chaperone, alpha-crystallin, inhibits amyloid formation by apolipoprotein C-II. *J Biol Chem* **2001**, *276*, 33755–33761.

159 Hatters, D. M., A. P. Minton and G. J. Howlett. Macromolecular crowding accelerates amyloid formation by human apolipoprotein C-II. *J Biol Chem* **2002**, *277*, 7824–7830.

160 Hatters, D. M. and G. J. Howlett. The structural basis for amyloid formation by plasma apolipoproteins: a review. *Eur Biophys J* **2002**, *31*, 2–8.

161 Hatters, D. M., M. R. Wilson, S. B. Easterbrook-Smith and G. J. Howlett. Suppression of apolipoprotein C-II amyloid formation by the extracellular chaperone, clusterin. *Eur J Biochem* **2002**, *269*, 2789–2794.

162 Pham, C. L., D. M. Hatters, L. J. Lawrence and G. J. Howlett. Cross-linking and amyloid formation by N- and C-terminal cysteine derivatives of human apolipoprotein C-II. *Biochemistry* **2002**, *41*, 14313–14322.

163 Munishkina, L. A., A. L. Fink and V. N. Uversky. Conformational prerequisites for formation of amyloid fibrils from histones. *J Mol Biol* **2004**, *342*, 1305–1324.

164 Pavlov, N. A., D. I. Cherny, G. Heim, T. M. Jovin and V. Subramaniam. Amyloid fibrils from the mammalian protein prothymosin alpha. *FEBS Lett* **2002**, *517*, 37–40.

165 Pertinhez, T. A., M. Bouchard, R. A. G. Smith, C. M. Dobson and L J. Smith. Stimulation and inhibition of fibril formation by a peptide in the presence of different concentrations of SDS. *FEBS Lett* **2002**, *529*, 193–197.

166 Agianian, B., K. Leonard, E. Bonte, H. Van der Zandt, P. B. Becker and P. A. Tucker. The glutamine-rich domain of the *Drosophila* GAGA factor is necessary for amyloid fibre formation *in vitro*, but not for chromatin remodelling. *J Mol Biol* **1999**, *285*, 527–544.

167 Bousset, L., N. H. Thomson, S. E. Radford and R. Melki. The yeast prion Ure2p retains its native alpha-helical conformation upon assembly into protein fibrils *in vitro*. *EMBO J* **2002**, *21*, 2903–2911.

168 Wilkins, D. K., C. M. Dobson and M. Gross. Biophysical studies of the development of amyloid fibrils from a peptide fragment of cold shock protein B. *Eur J Biochem* **2000**, *267*, 2609–2616.

169 Fandrich, M. and C. M. Dobson. The behaviour of polyamino acids reveals an inverse side chain effect in amyloid structure formation. *EMBO J* **2002**, *21*, 5682–5690.

12
Structural Intermediates of Globular Proteins as Precursors to Amyloid Formation

Daniel F. Moriarty and Wilfredo Colón

12.1
Introduction

Many proteins and peptides have been found capable of amyloid formation *in vivo* and/or *in vitro*. Because most peptides lack a defined and stable structure, it has been easier to accept the idea that peptides are inherently susceptible to amyloid formation. In the case of proteins it was at first difficult to understand how a stably folded protein could form amyloid fibrils *in vivo* and, therefore, proteolysis was viewed as a prerequisite. Over the past 15 years, the number of proteins linked to amyloid diseases has increased significantly (Table 12.1) and this number is likely to grow. Remarkably, there is no unique structural feature among proteins that form amyloid. This suggests that the ability to form amyloid fibrils is not confined to any given structural property and this concept has been confirmed by the observation that many proteins can form amyloid under destabilizing conditions [1]. Thus, amyloid fibril formation is a property of many globular proteins that surfaces as a result of a misfolding event. Most of our understanding about the mechanism by which proteins form amyloid comes from *in vitro* studies, where the intrinsic stability and folding kinetics, as well as the effect of mutation on these properties, can be easily investigated. A wealth of data suggests that partially folded protein intermediates are the precursor species that upon misfolding, initiate the aberrant self-association process that results in amyloid formation and disease. In this chapter, we will review the evidence that partially folded intermediates are usually the precursors of protein aggregation and will describe their role in the mechanism of amyloid formation. We will conclude by discussing examples from the literature that suggest that blocking the formation of amyloidogenic intermediates offers one of the most promising approaches for therapeutic intervention.

Amyloid Proteins. The Beta Sheet Conformation and Disease. J. D. Sipe

ISBN: 3-527-31072-X

12.2 Protein Folding

The native state for most proteins is usually the most thermodynamically stable structure that is kinetically accessible under physiological conditions. Despite over five decades of research, the mechanism by which the amino acid sequence of a protein results in a unique three-dimensional structure remains poorly understood and is generally referred to as the "protein folding problem" [2, 3]. Understanding how a protein is able to fold into its functional structure is critical because the function of a given protein is determined by its three-dimensional structure. In the late 1950s, it was shown that all the information required for a protein to fold into its native state is encoded in the primary sequence [4]. The process of protein folding is driven by the thermodynamic "need" of the system to sequester non-polar amino acid side-chains from the aqueous environment, a process known as the hydrophobic effect. The folding pathway of a protein depends on the folding energy landscape, which is determined by the amino acid sequence of the protein. It has been very useful to describe this folding landscape in terms of a funnel shape model, to explain how a protein can fold rapidly into its thermodynamically stable native state [5]. In this analogy, like the trajectory of a drop of water placed anywhere at the top of a cone, an unfolded protein anywhere at the top (high free energy) of the energy landscape can quickly reach a unique three-dimensional structure (low free energy). Interestingly, the funnel-like energy landscape of protein folding is not cone-like smooth, but rather is rough like the slope of a mountain. It is this "roughness" of the energy landscape that determines not only the rate of folding, but also whether the protein transiently populates states of intermediate structure, commonly known as "folding intermediates".

Although many proteins can fold *in vitro* without any help, protein folding in the extremely crowded environment of cells is a complex and perilous situation. The macromolecular concentration inside the cell exceed 300 mg/ml [6, 7] and therefore the probability for unwanted interactions before folding is complete is very high [8–10]. As a protein is released from the ribosome, the nascent polypeptide chain begins to fold into its native structure or perhaps may encounter folding problems in which the hydrophobic surface may become exposed and act as a template for undesired interactions, including self-association (aggregation). To combat this problem, there are several chaperone systems in place to help guide proteins along the process of folding [11]. These chaperones generally fall into two categories. Smaller chaperones such as hsp70 and hsp40 bind to the polypeptide chain as it comes off the ribosome by interacting with exposed hydrophobic patches, thus blocking potential aggregation, as well as delaying the folding process until the protein is ready to achieve its structure. Larger chaperones, such as the GroEL/GroES complex, function in a very different way by interacting with proteins that are fully synthesized, but are not in their proper conformation [7]. The profound role of the protein quality control machinery in aiding protein folding in the cell [12] makes it clear that correct protein folding in any living organism cannot be taken for granted and that the potential for protein misfolding is a problem that has been addressed by nature. An improperly folded protein that cannot be rescued from a misfolded conformation represents not only a loss of cellular resources and energy, but also a "waste management" problem. Fortunately, the proteosome and lysosomes can degrade defective proteins, and in cases of excess insoluble aggregate formation cells are able to sequester them into cellular inclusion bodies called aggresomes [13]. However, if the quality control system is overwhelmed, the accumulation of misfolded protein may cause their aggregation into toxic species before they can be degraded, leading to cellular dysfunction and disease.

12.3 Folding Intermediates as Precursors to Protein Aggregation

During the process of folding, most small proteins (roughly 100 residues or less) tend to fold by an apparent two-state process where only the native and denatured states are populated [14]. In these cases, intermediates cannot be directly observed, although mutations may alter the folding energetics such that they may become populated during the folding process [15]. Larger proteins tend to have folding intermediates that are populated, but are short lived because they are destabilized relative to the native state. These intermediates may be necessary for the protein to achieve its native structure or they may be off-pathway traps from which the protein must escape by partially unfolding before it can reach the proper conformation. In either case, if an intermediate is present at high enough concentrations for a long enough period of time, it may be possible for aberrant self-association to take place.

Our understanding of the relationship between partially folded intermediates and aggregation began in the 1970s, primarily with work on lactic dehydrogenase [16, 17], tryptophanase [18] and chymotrypsinogen A [19, 20]. During refolding of lactic dehydrogenase, competition was observed between the formation of the native tetrameric structure and aggregation of a stable folding intermediate. Aggregates were formed from subunits that contained some of the native secondary structure, but none of the quaternary contacts [17]. A similar competition was observed with tryptophanase, where both the native tetramer and aggregates were formed during renaturation, and the amount of aggregation varied with the concentration of urea. Remarkably, the association was specific to the tryptophanase monomer, as other denatured proteins introduced into the solution were not incorporated into the aggregate [18]. This study demonstrated that protein aggregation involves specific interactions of partially folded intermediates. In the case of the monomeric chymotrypsinogen A, a number of intermediates were present during refolding, some of which contained non-native disulfide bonds and were prone to irreversible aggregation [19, 20]. At very low concentrations of guanidine–HCl, upon triggering refolding by dilution, aggregation was favored due to the population of an aggregation-prone intermediate. However, if the concentration of guanidine–HCl was slowly increased, the intermediate was destabilized and most of the protein was able to refold correctly [19]. Coincidently, during this period when the connection between partially folded intermediates in the process of aggregation was being made, a protein component of an amyloid fibril was positively identified for the first time, i.e. the N-terminal variable fragment of an immunoglobulin light chain [21].

The role of partially structured intermediates in protein aggregation was firmly established by the 1980s. The thermostable P22 tailspike protein provided one of the best examples of the interplay between folding and aggregation. During folding, the P22 tailspike protein rapidly assumes a β-helical conformation that self-assembles into a native-like trimeric intermediate, which further rearranges to form the native state [22–28]. Several mutations were identified which had no effect on the native state of the protein, but drastically affected the folding efficiency both *in vivo* and *in vitro* [29–31]. The mutant proteins could not assemble into the correct trimeric structure, but instead would form inclusion bodies. It was later shown that these mutations favored the formation of a partially folded and aggregation-prone monomeric intermediate [22–28]. Another protein, rhodanese, was found to aggregate during denaturation studies at low concentrations of the guanidine–HCl (below 1.3 M), where the protein maintained almost all of its native secondary structure but lacked a well-defined and compact tertiary structure [32]. In another striking example, denaturation of bovine growth hormone with guanidine–HCl resulted in the formation of a well-populated intermediate that was prone to aggregation due to an exposed amphipathic helix [33, 34].

In the early part of the 1990s a direct connection between partial denaturation and amyloid formation was first reported based on studies involving the protein transthyretin (TTR) [35]. As the decade progressed, other examples emerged, including immunoglobulin light chain V_L domain [36, 37], lysozyme [38] and cystatin C [39]. Towards the mid to late 1990s, it became widely accepted that amyloid formation in globular proteins is a conformational misfolding event that is preceded by partially folded intermediates [40–42]. Later, it was discovered that the SH3 domain of the p85 α subunit of bovine phosphatidylinositol-3′-kinase (PI3-SH3), which is not associated with any known amyloid disease, is able to form amyloid fibrils upon acid-induced partial denaturation [43]. Other examples of amyloid formation (Table 12.1) by proteins not linked to disease have been reported, strongly supporting the idea that amyloid formation may be a generic property of proteins [44].

12.4
Structural Intermediates in Amyloid Formation

In this section we will discuss various examples of proteins involved in amyloidosis where intermediates have been shown to be precursors of amyloid formation. The focus will be on the evidence for intermediates, since a more in-depth

Table 12.1 Amyloid-forming globular proteins [a)]

Precursor protein	Clinical syndrome	Native structure	References
SAA	reactive amyloidosis	mainly helix	120, 121
TTR	senile systemic amyloidosis; familial amyloid polyneuropathy	mainly β	122, 123
Immunoglobulin light chain	antibody light chain-related amyloidosis	mainly β	124–126
α_1-Antitrypsin	progressive juvenile cirrhosis	mainly β	127–129
Lactadherin	aortic medial amyloid	unknown	130
Lysozyme	non-neuropathic systemic amyloidosis	mixed α/β	38, 63
SP-C	pulmonary alveolar proteinosis	mainly helical	85, 131
Superoxide dismutase	familial amyotrophic lateral sclerosis	mainly β	132, 133
Gelsolin	Finnish hereditary systemic amyloidosis	mixed α/β	134–136
Cystatin C	hereditary cerebral amyloid angiopathy	mainly β	137, 138
β_2M	hemodialysis-related amyloidosis	mainly β	55
Prion	spongiform encephallpathies	mainly helix	139–142
Fibrinogen	hereditary renal amyloidosis	mixed α/β	143, 144
Insulin	injection-localized amyloidosis	mainly helix	145–147
ApoAI	familial amyloid polyneuropathy (III)	mainly helix	148, 149
β-Lactoglobulin	none known	mainly β	150, 151
α-Lactalbumin	none known	mainly helix	152
β-Helix domain of P22 tailspike	none known	β-helix	153
Yeast phosphoglycerate kinase	none known	mixed α/β	154, 155
Apomyoglobin	none known	mainly helix	106, 156
Cytochrome *c*(552)	none known	mainly helix	157
SH3 domain [b)]	none known	mainly β	43
B1 domain of protein G	none known	mainly β	158
Acylphosphatase	none known	mixed α/β	159
Core domain of tumor suppressor P53	none known	mainly β	160
Monellin	none known	mainly β	161, 162
Acidic fibroblast growth factor	none known	mainly β	163–165

a) Small peptides and natively unfolded proteins involved in amyloid diseases were omitted from this list.

b) Of the p85α subunit of bovine PI3-SH3.

discussion of most of these proteins and their related disease is found in later chapters of this book. The proteins chosen illustrate the general role of intermediates in amyloid formation despite the diversity in protein structure.

12.4.1
TTR

TTR is a homotetrameric β-sheet protein containing subunits of 127 residues [45]. TTR is the main protein component of amyloid in individuals with familial amyloid polyneuropathy (FAP) and senile systemic amyloidosis (SSA) [46]. In its native form, TTR does not form amyloid fibrils. However, at low pH, the tetrameric wild-type and FAP-related TTR mutants dissociate into a monomeric intermediate that misfolds and self-assembles into amyloid fibrils [35, 47]. In the case of FAP-associated mutations, many of them lead to a destabilized form of the tetramer or a reduced kinetic barrier towards unfolding [48], thereby facilitating the population of the amyloidogenic monomeric intermediate. Two mutants, L55P and V30M, are less resistant to acid denaturation than the wild-type protein [49–51]. In fact, even under physiological conditions, the L55P mutant can dissociate into a monomeric species which is capable of forming protofilaments [49–51]. Remarkably, there is a natural mutation, T119M, which makes TTR more resistant to amyloid formation and suppresses FAP in individuals carrying the pathogenic V30M TTR mutation on the second TTR allele [52]. *In vitro* studies have shown that the T119M mutation increases the energetic barrier towards TTR dissociation and therefore essentially inhibits the formation of the amyloidogenic intermediate [53].

12.4.2
β_2-Microglobulin (β_2M)

β_2M is a seven-stranded β-sandwich protein consisting of 99 residues and containing a single disulfide bond linking two sheets [54]. In individuals undergoing hemodialysis treatment, the concentration of β_2M increases, sometimes leading to amyloid formation and hemodialysis-related amyloidosis [55]. *In vitro,* β_2M forms amyloid under a variety of conditions that destabilize the native state [56–62], including low pH, the removal of the N-terminal six residues and the binding of copper at neutral pH. At pH 3.6 the A and G outer strands of β_2M become destabilized while the remaining strands retain a stable structure and therefore the protein aggregates, although mostly protofibrils (non-fibrillar oligomers that are intermediates in amyloid fibril formation) are formed under these conditions. Lowering the pH to below 3 leads to a more destabilized structure that is able to form mature fibrils. The native state can also be destabilized through the addition of Cu^{2+} [58, 60], where the interaction of Cu^{2+} with histidine residues in a partially unfolded β_2M state is mainly responsible for the increased flexibility and decreased stability that leads to fibril formation.

12.4.3
Lysozyme

Lysozyme is a monomeric α/β protein of 129 residues containing four disulfide bonds and two domains (the α and β domains). Lysozyme missense mutations (D67H, I56T, F57I and W64R) are associated with non-neuropathic systemic amyloidosis [63–65]. Studies have shown that the folding kinetics and thermodynamic stabilities of I56T and D67H are altered compared to the wild-type protein due to disruption of interactions in the β-domain. While the α-domain remains largely unchanged, the β-domain (along with the C-helix from the α-domain) of the D67H mutant undergoes local cooperative unfolding, exposing portions of the protein under amyloid-forming conditions (pH 5) [38, 66, 67]. As expected, in this region there is also an accelerated rate of hydrogen exchange compared to the wild-type protein [68]. The amyloidogenic mutants of lysozyme were shown to be less thermostable than the wild-type protein under amyloid-forming conditions and to have faster unfolding rates [38, 66, 67]. These data point to the existence of a partially unfolded intermediate for the destabilized amyloidogenic lysozyme mutants.

12.4.4
Cystatin C

Human cystatin C is a monomeric protein of 120 residues comprised of two disulfide bonds, a five-stranded anti-parallel β-sheet that curves around a central helix, and three loops containing many hydrophobic residues [69]. A mutation in cystatin C, L68Q, leads to amyloid formation in the brain of young adults, leading to hereditary cerebral amyloid angiopathy. The L68Q mutant is active *in vitro* initially, but begins to dimerize over time at neutral pH and 37 °C, a process that is extremely sensitive to temperature. The wild-type form is more stable, but can form inactive dimers when exposed to elevated temperature, high concentration of guanidine–HCl or low pH. These conditions help generate a destabilized monomeric intermediate which still contains most of the native secondary and tertiary contacts [70–72].

The crystal structure of human cystatin C shows that it can irreversibly form a three-dimensional domain-swapped dimer [73]. The secondary structure and general fold of the dimer is largely intact when compared with the monomer, and nuclear magnetic resonance experiments show a symmetrical dimer [74, 75]. The formation of the dimer is essentially irreversible and more thermodynamically stable than the monomer [76, 77]. The loss of stability of the monomeric state of L68Q leads to a partially folded intermediate more susceptible to forming three-dimensional domain-swapped dimers and amyloid formation, the latter presumably through an open-ended swapping mechanism [73]. It has been shown that prevention of domain swapping inhibits amyloid fibril formation of cystatin C [78]. This is consistent with the proposal that domain swapping may be a common mechanism by which many partially folded proteins may form amyloid [79, 80]

12.4.5
Serum Amyloid A (SAA)

SAA is a major acute-phase reactant protein of 104 residues that is found predominantly in the serum bound to high-density lipoproteins (HDL). In cases of chronic inflammation, SAA may deposit into amyloid fibrils, leading to the disease of reactive amyloidosis. Recently, it was shown that murine SAA2.2, a non-amyloidogenic isoform *in vivo*, forms a hexamer *in vitro* containing a putative central channel [81]. Further studies revealed that upon incubation at physiological temperature and pH, SAA2.2 irreversibly dissociates to a monomeric species that undergoes a significant α to β conformational change. Within a few days of incubation under these conditions the misfolded SAA2.2 species forms amyloid fibrils [82]. Considering the marginal stability and high amyloidogenicity of SAA, it is intriguing to understand the *in vivo* factors that usually protect this protein from forming amyloid formation. We speculate that avoiding the build up of monomeric SAA via HDL binding to the monomer or ligand-mediated hexamer stabilization may play an important role in protecting against SAA amyloid formation.

12.5
Factors that Favor the Formation of Amyloidogenic Intermediates

Proteins are marginally stable and changes in the environment, such as a decrease in pH, increase in temperature, changes in solvent composition or changes to the protein itself, including mutations or post-translational modifications, can result in a protein with increased propensity for populating amyloidogenic intermediate states. *In vivo*, missense mutations have been found to be a common factor responsible for amyloid diseases, although other risk factors clearly play a role, including molecular and cellular changes resulting from the aging process. Thus, a fundamental issue is to understand how mutations and other factors may affect the population of amyloidogenic intermediates.

12.5.1
Thermodynamic versus Kinetic Stability Effects

The native state of a protein is separated from partially and globally unfolded states by a transition state (TS) energy barrier. The height of this barrier is determined by the free energy of the native state and the free energy of the TS (Fig. 12.1). A decrease in the TS energy leads to a reduction in kinetic stability, a term associated with proteins that are resistant to unfolding by being "trapped" in their native conformations by an energy barrier. The unfolding process can be illustrated as a simple reaction where the native (N) and denatured (D) states are separated by a higher energy TS (Fig. 12.1) that determines the rate of protein unfolding. Changes in protein structure as a result of mutation

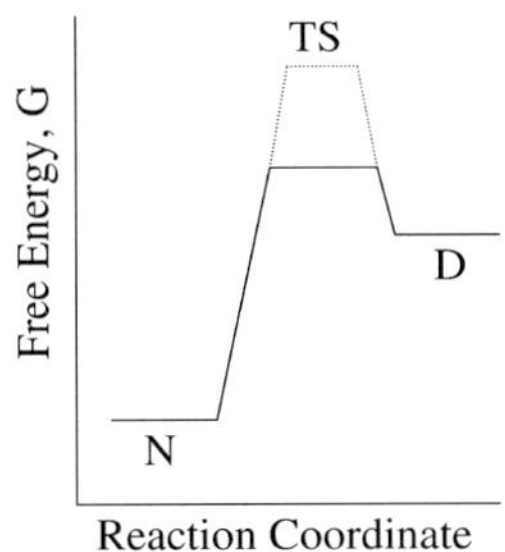

Fig. 12.1 Free energy diagram of protein folding illustrating the concept of "kinetic stability". Free energy, also known as Gibbs energy, is a thermodynamic property of a system that is defined as *H–TS*, where *H* is the enthalpy, *S* the entropy and *T* the temperature. The diagram shows that two proteins that have the same thermodynamic stability (G_D–G_N=ΔG) may have very different kinetic stability. The dashed line represents the higher transition state free energy and therefore slower unfolding, for a kinetically stable protein.

or other factors (e.g. oxidative damage) could increase the population of a denatured state (e.g. amyloidogenic intermediate) after the protein has folded, either by decreasing the stability of the native state (thermodynamic effect), by decreasing the unfolding TS energy (kinetic effect) or both (Fig. 12.2).

Although a decrease in native state stability can lead to an increased population of aggregation-prone intermediate structures, a mutation could alter the kinetic stability of a protein and make it more susceptible to aggregation even in the absence of a thermodynamic effect. Thus, thermodynamic stability alone may not be able to protect proteins that are susceptible to aggregation due to partially unfolded states that become transiently populated under physiological conditions [83]. Therefore, the presence of a high unfolding energy barrier for some proteins may serve to protect them against aberrant self-association.

12.5.2 The Effect of Aging on Amyloid Formation

Amyloidogenic intermediates could potentially populate during protein folding after their synthesis. However, under these conditions it appears that it is not likely that aggregation-prone species will build to any significant extent if the quality control machinery of the cell (e.g. chaperones, proteosomes, unfolded protein response in the endoplasmic reticulum) is working efficiently. However, it is likely that a cell with a compromised quality control system will be inefficient in its ability to avoid the build up and aggregation of amyloidogenic intermediates that may be formed during and after protein synthesis. Thus, a decreased efficiency of protein folding quality control as a result of the aging process may be the main determinant for the late onset of many protein misfolding diseases, even those that are familial. This has therapeutic implications, as it suggests that if the quality control machinery could be modestly protected from the effects of aging, it may be possible to significantly delay the onset of protein misfolding diseases.

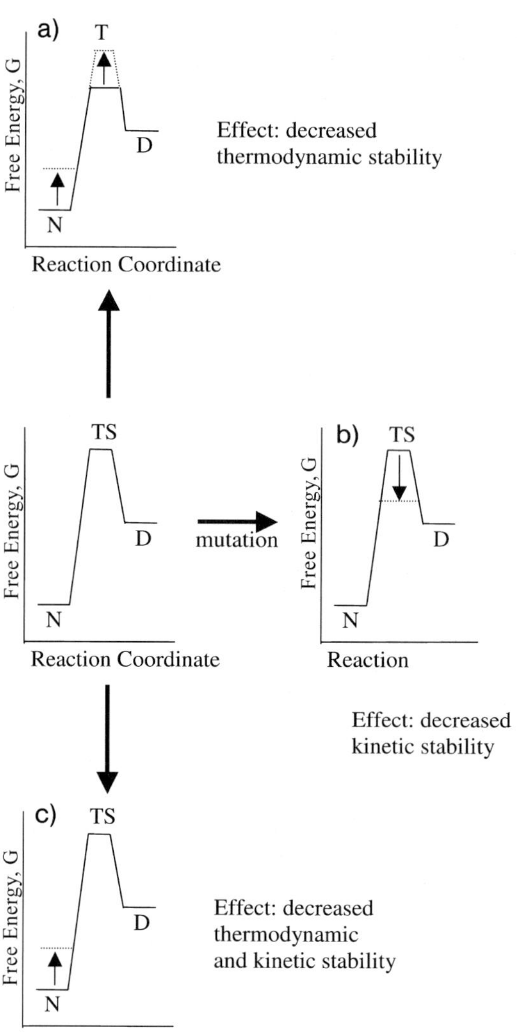

Fig. 12.2 Potential adverse effect of mutation on thermodynamic and kinetic stability. Mutations may increase a protein's access to partially or globally unfolded denatured (D) states by decreasing its thermodynamic stability (A), decreasing its kinetic stability (B) or by decreasing both (C). The arrows in the diagram illustrate the relative change in native (N) or transition state (TS) free energy as a result of mutation.

12.5.3
From α-Helix to β-Sheet: "Jekyll and Hyde Sequences"

The process of amyloid formation often involves a significant α-helix to β-sheet conformational change. This is consistent with studies on peptides demonstrating that secondary structure formation is highly dependent on its environment, including pH, temperature, ionic strength, ligands, etc. [84]. A protein's ability to undergo an α to β conformational change may be facilitated by containing amino acid regions that adopt an α-helical conformation within the native structure, while having a higher statistical propensity for the β structure. An example that illustrates the role of environment and secondary structure propensity in amyloid formation is the lung surfactant protein C (SP-C).

Pulmonary alveolar proteinosis (PAP) is characterized by amyloid deposits in the cell-free fraction of bronchoalveolar lavage fluid [85]. The principal component of the fibrils is the 35-residue SP-C, a lipopeptide palmitoylated on consecutive cysteine residues near the N-terminus that forms a transmembrane helix (residues 9–34). The primary sequence of SP-C is extremely hydrophobic [86, 87] and, remarkably, there is a high percentage of valine residues (12 out of 35 residues), which are more likely to be found in a β-sheet than in an α-helix. The presence of helical regions with sequences predicted to favor the β structure has been observed in other amyloid-forming proteins [88]. However, in the case of SP-C, this propensity for forming β-sheets is suppressed by the lipid environment where it normally resides in the α-helical conformation [89].

The SP-C helix is stable *in vitro* in the presence of micelles, but less structured in mixed organic solvents, where it misfolds into the preferred β structure and forms amyloid fibrils [90]. Aggregation of SP-C is inhibited at neutral pH when containing a 12-residue N-terminal propeptide flanker sequence that interacts and stabilizes the helix. By decreasing the pH, the propeptide–helix interaction is weakened and the helix begins to unwind, misfolding to the preferred β structure and eventually forming amyloid fibrils [91]. Removal of one or both of the palmitoyl groups has a similar effect on the stability and amyloidogenicity of the SP-C helix [92]. Interestingly, the helix can be stabilized due to improper cleavage of the propeptide sequence, but at the cost of loss of functionality, which leads to a fatal (though not amyloid-related) disease [93]. Thus, SP-C illustrates the existence *in vivo* of sequences in peptides or within a protein that may possess dual "conformational personality" and may represent an inherent risk for amyloidosis.

12.6 Mechanism of Amyloid Formation

12.6.1 Nucleation-dependent Amyloid Fibril Formation

Most of the mechanistic understanding concerning the process of amyloid formation comes from *in vitro* studies using purified proteins. Excellent reviews have been written on the mechanism of amyloid formation [94, 95] and, therefore, we will focus here on the formation of the intermediate species that precedes amyloid formation. Just as experimental evidence suggests that amyloid formation is a common property of many proteins, the mechanism by which a protein forms amyloid is also thought to be somewhat generic. In the case of globular proteins it is clear that partially folded intermediates populated from the native state or during folding is the first critical step of the pathway [40–42]. Upon misfolding, if present at a critical concentration, these partially folded in-

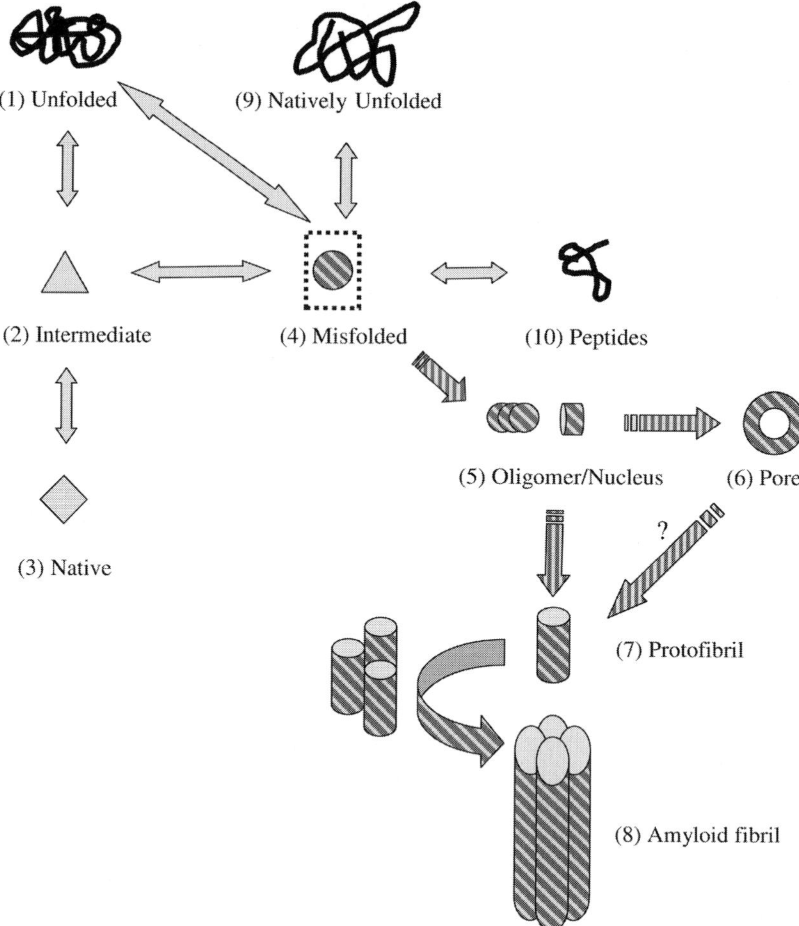

Fig. 12.3 Model illustrating the role of intermediates in amyloid formation. During the normal process of protein folding, a protein will go from a predominately unstructured state (1) to its unique native conformation (3). This may or may not occur through a partially folded intermediate (2). Various factors, including mutation, post-translational modification, etc., may induce the formation of an amyloidogenic misfolded intermediate (4) usually via a native-like partially folded intermediate. If the misfolded state is stable enough and/or the concentration is high enough, oligomers (5) may form and eventually a nucleus will be produced. This leads to the formation of pores (6) or protofibrils (7), which then may aggregate further to form mature amyloid fibrils (8). The formation of an amyloidogenic intermediate may also arise from natively unfolded proteins (9) and largely unstructured peptides (10).

termediates slowly assemble into a nucleus, which then rapidly grows into soluble aggregates, such as protofibrils, and then into fibrils (Fig, 12.3) [96, 97]. Below the critical concentration, the monomers remain the predominant form because the enthalpic gain from molecular interactions is outweighed by the entropic cost of forming the nucleus. The amount of time that is necessary to form the nucleus is known as the lag time. This delay between the achievement of the critical concentration of the amyloidogenic intermediate and formation of fibrils is usually very sensitive to concentration, with small changes in protein concentration leading to large changes in the rate of amyloid formation. The lag time of the system can be abolished by addition of preformed seeds, which present an ordered template for the fast addition of monomers.

If all amyloid fibrils and amyloidogenic intermediates had a common structural motif, then it could be expected that amyloid fibrils from one protein would be able to seed another protein and increase the rate of amyloid formation. In a few instances this has been shown to be the case, but usually heterologous seeding does not occur [96, 98, 99]. The fact that the seeding effect is usually highly protein specific indicates that structural uniqueness exist not only in the structure of the intermediate, but is also preserved in the seed and fibril. It is interesting that the side-chains, which perhaps are not the determining factor for a protein's ability to form amyloid, are most likely the main determinant of the seeding specificity. This seems to be another mechanism by which cells are protected by the devastating effects that would surely occur from non-specific heterologous seeding.

12.6.2
Partial versus Global Unfolding

Since the formation of amyloid fibrils involves the self-association of β-sheet structural elements, the conformational changes required vary depending on the structure of the protein. For an intermediate to successfully form amyloid *in vivo*, it must be different enough from the native conformation so that naturally occurring negative design elements for blocking interstrand propagation [54, 100–105] will not be effective, yet perhaps not so different that it will be recognized and eliminated by the cell's quality control mechanisms. Thus, it is not surprising that relatively few proteins have been associated with amyloid formation *in vivo*. The fact that amyloid fibrils contain a high degree of β-sheet structure regardless of the native secondary structure content points to a large structural rearrangement in some proteins. For proteins rich in β-sheet structure like TTR, β_2M and immunoglobulin light chain, the extent of conformational change needed to form an amyloidogenic intermediate need not be extensive; however, for other proteins that contain significant amounts of helical structure (e.g. lysozyme, prion, acylphosphatase and apomyoglobin), a major conformational change accompanied with significant misfolding is required.

Unlike proper protein folding, which is dictated largely by side-chain interactions, amyloid fibril formation relies heavily on backbone chain interactions.

This helps explain why many different proteins and peptides can form amyloid. The α-helical proteins in their native state are less likely to undergo the large structural changes needed for amyloid formation, whereas for predominately β-sheet proteins little structural change would be required. Thus, one could say that the β-sheet structure is an intrinsic risk factor for amyloid formation. It seems that nature has taken this into account. For example, formation of β-barrels avoids the exposure of edge strands and, therefore, chain propagation cannot take place without first breaking the barrel. For those β-sheets proteins that do expose edge strands, these usually contain negative designed features to prevent them from aggregation; these include the presence of proline or charged residues in the center of the strands, shortening the edge strand to minimize favorable contacts between proteins, surrounding the edge strand with loops that can act as a barrier for other proteins or formation of a strained structure to distort the interface [54, 100–105].

For proteins with a large content of α-helical structure, a significant degree of unfolding and misfolding is likely to be required for amyloid formation. Lysozyme has been shown to contain a fair amount of residual structure in its amyloidogenic intermediate [68]. In contrast, α-lactalbumin, which is a member of the lysozyme family, has not been linked to amyloid formation *in vivo* and can only form amyloid fibrils *in vitro* when it is essentially unfolded. A similar finding was shown for the α-helical protein apomyoglobin. When the tertiary structure was disrupted, but the native secondary structure remained intact, no fibril formation occurred. At a lower pH, apomyoglobin was shown to partially unfold and aggregate, even while retaining some helical structure. However, these aggregates contained no amyloid-like structure. In conditions where fibril growth was favorable, there was no evidence of β-sheet structure before fibril formation and all of the intermediates that were observed contained various amounts of α-helical structure. Binding of the heme group prevented fibril formation under conditions that otherwise promoted it for apomyoglobin [106]. It is important to note that for apomyoglobin and α-lactalbumin, the predominantly unfolded state still contained some residual, but not necessarily native structure. There is a growing body of evidence that proteins in environments which favor the unfolded state may slightly favor extended conformations that could contain some non-random structure [107–109]. This has implications for proteins such as α-synuclein and tau, which possess little stable secondary structure, but can form amyloid fibrils, or for peptides like Aβ and islet amyloid polypeptide (IAPP; amylin), which are highly amyloidogenic. Therefore, for those proteins and peptides that lack a well-ordered structure, amyloid formation seem to involve an amyloidogenic species formed through partial folding rather than partial denaturation.

12.7 An "Eye" for an "I": Inhibiting the Formation of Intermediates

Various therapeutic strategies for amyloid diseases have been discussed over the years. The widely held belief that protein aggregation plays a causative role in amyloid diseases leads to the obvious conclusion that an effective therapeutic drug would either inhibit aggregation or dissolve the toxic aggregates. Advances made in understanding the general mechanism of amyloid formation have provided various targets for drug design, including the fibrils, protofibrils, soluble aggregates and monomeric intermediate species. The strategy aimed at dissolving protein aggregates is complicated by the realization in recent years that the most toxic species may not be the amyloid fibrils themselves, but rather smaller misfolded monomeric or oligomeric forms of the protein [110, 111]. Rational drug design is further hampered by our limited understanding of the detailed molecular structure of the relevant aggregated species. Therefore, the use of "nature-designed" monoclonal antibodies seems the most promising approach for targeting protein aggregates [112]. In contrast, the use of rational drug design to inhibit protein misfolding and aggregation seems like a very promising strategy, especially for misfolding diseases involving globular proteins of known three-dimensional structure. The basic rationale is that stabilizing the native state of an amyloidogenic protein should inhibit the population of the intermediate prone to misfolding, thereby blocking misfolding, oligomerization, and aggregation. Recent reviews have discussed different therapeutic approaches for protein misfolding diseases [42, 95, 113]. Here, we will focus on several examples where proteins and small molecules have been used to stabilize the native state, leading to inhibition of amyloid fibril formation *in vitro*.

12.7.1 Native State Stabilization via Binding of Small Molecules

Most TTR mutations leading to FAP destabilize the native tetrameric structure of TTR in favor of an amyloidogenic monomeric intermediate [48, 114]. Small molecules have been developed that bind to native TTR with nanomolar dissociation constants and stabilize its native structure, thereby inhibiting the tetramer to monomer transition that is a prerequisite for amyloid formation [53, 115, 116]. These inhibitors function by binding to the tetramer at the thyroxine-binding sites located in the central channel of the protein. Upon ligand binding, the tetramer dissociation rates were significantly decreased, thus explaining its amyloid inhibition effect. It is important to point out that small molecule inhibitors need not block TTR dissociation to have an impact on amyloid formation, as even a small decrease in the concentration of the amyloidogenic monomeric intermediate has a significant effect on the *in vitro* rate of amyloid formation.

12.7.2
Native State Stabilization via Binding of Protein Molecules

Because of their potentially exquisite binding properties, proteins could also serve as therapeutic agents to inhibit the formation of amyloidogenic intermediates. However, unlike as is the case with small molecules, it is extremely difficult to rationally design *de novo* a protein that will bind selectively and tightly to another protein. This problem can be dealt with by using specific antibodies or by using a mutated version of the protein in question to bind and suppress the formation of the amyloidogenic intermediate. The former strategy can be applied to any protein, whereas the latter one is most applicable to oligomeric proteins. Here, we will discuss recent examples where these approaches have been used to stabilize the native states of the amyloidogenic proteins lysozyme and TTR.

The formation of amyloid fibrils by the pathogenic lysozyme mutant, D67H, was inhibited by a camelid antibody raised against the wild type protein [117]. Interestingly, the antibody did not bind to the mutated region or to the region destabilized by the mutation. Biophysical studies demonstrated that the antibody achieved its inhibitory effect by stabilizing the native state and by restoring the folding cooperativity of lysozyme. Thermal denaturation studies revealed that the temperature mid-point (T_m) of unfolding of the variant protein was raised by 15 °C, which overcompensates for the 10 °C decrease in T_m as a result of the mutation. Even more important was the observation that the increase in stability was accompanied by an increase in unfolding cooperativity. Unlike the D67H mutant, which partially unfolds to populate the amyloidogenic intermediate, the antibody-bound protein exhibited a single cooperative two-state transition. Thus, it is clear that the camelid antibody inhibits lysozyme amyloid formation by inhibiting the partial unfolding event that leads to the formation of the amyloidogenic intermediate.

Another protein-based strategy for inhibiting the formation of amyloidogenic intermediates involves *trans*-suppression [52]. Unlike *cis*-suppression, where a second mutation within the same protein chain can suppress the effect of a primary mutation [118], *trans*-suppression involves the intermolecular interaction of a mutated protein with another molecule of the same protein containing the suppressor mutation. *Trans*-suppression has been observed naturally with heterozygote individuals who carry the pathogenic V30M mutation and the FAP suppressor T119M mutation on different alleles [119]. The suppression effect appears to be caused by an increased stability of T119M TTR, thereby protecting V30M TTR subunits present in mixed T119M/V30M tetramers from forming the monomeric amyloidogenic intermediates [52]. Interestingly, the high free energy of the TS in heterotetramers resulting in very slow dissociation rates is not caused by a stabilization of the native tetramer, but rather by a destabilization of the dissociation TS [53]. Thus, unlike the effect of small molecules and camelids, *trans*-suppression of TTR amyloid formation is caused by kinetic stabilization.

12.7.3
Therapeutic Potential

The studies described above show that inhibiting the formation of amyloidogenic intermediates is a very effective means for blocking amyloid formation *in vitro*. The advantage of this strategy is that it is not necessary to know the identity of the pathogenic species because the aggregation process is blocked upstream of misfolding and oligomerization. Another advantage is that even a modest decrease in the formation of the aggregation-prone intermediate is likely to have a dramatic effect in nucleation and aggregation rate. The availability and advantages of the various (small molecules, *trans*-suppression and antibody) approaches for inhibiting amyloidogenic intermediates makes this a promising approach for treating misfolding diseases.

12.8
Conclusion

A large number of misfolding diseases have been discovered to date and we predict that many more will be discovered in the future. In misfolding diseases where globular proteins aggregate into toxic species, there is clear evidence that partially folded intermediate are the precursors that initiate the aberrant self-assembly. In most cases, the disease appears to be associated with a mutation that through a kinetic or thermodynamic effect facilitates the population of a partially folded intermediate prone to misfolding. The assembly of this intermediate into a nucleus is often the rate determining step in amyloid formation and the relatively slow rate of this process seems to be an advantage for the cell, as it allows the quality control machinery to do its work. However, if the concentration of the amyloidogenic species increases (e.g. by mutation, oxidative damage, etc.) or the efficiency of the quality control machinery is compromised (e.g. by aging), the onset of disease may occur. Despite the uncertainty of the nature of the toxic species in amyloid diseases and the lack of structural information along the amyloid formation pathway, the stabilization of the native state of amyloidogenic proteins seems a very promising approach for developing therapeutic treatment for these usually fatal diseases.

References

1 Dobson, C.M. Principles of protein folding, misfolding and aggregation. *Semin Cell Dev Biol* **2004**, *15*, 3–16.

2 Grantcharova, V., E.J. Alm, D. Baker and A.L. Horwich. Mechanisms of protein folding. *Curr Opin Struct Biol* **2001**, *11*, 70–82.

3 Levinthal, C. Are there pathways for protein folding? *J Chem Phys* **1968**, *34*, 1963–1974.

4 Anfinsen, C.B. Principles that govern the folding of protein chains. *Science* **1973**, *181*, 223–230.

5 Bryngelson, J.D., J.N. Onuchic, N.D. Socci and P.G. Wolynes. Funnels, pathways and the energy landscape of protein folding: a synthesis. *Proteins Struct Funct Genet* **1995**, *21*, 167–195.

6 Zimmerman, S.B. and S.O. Trach. Estimation of macromolecule concentrations and excluded volume effects for the cytoplasm of *Escherichia coli*. *J Mol Biol* **1991**, *222*, 599–620.

7 Ellis, R.J. Molecular chaperones: avoiding the crowd. *Curr Biol* **1997**, *7*, R531–533.

8 van den Berg, B., R.J. Ellis and C.M. Dobson. Effects of macromolecular crowding on protein folding and aggregation. *EMBO J* **1999**, *18*, 6927–6933.

9 Minton, A.P. Protein folding: thickening the broth. *Curr Biol* **2000**, *10*, R97–99.

10 Kinjo, A.R. and S. Takada. Competition between protein folding and aggregation with molecular chaperones in crowded solutions: insight from mesoscopic simulations. *Biophys J* **2003**, *85*, 3521–3531.

11 Ranson, N.A., N.J. Dunster, S.G. Burston and A. R. Clarke. Chaperonins can catalyse the reversal of early aggregation steps when a protein misfolds. *J Mol Biol* **1995**, *250*, 581–586.

12 Ellis, R.J. and F.U. Hartl. Principles of protein folding in the cellular environment. *Curr Opin Struct Biol* **1999**, *9*, 102–110.

13 Kopito, R.R. Aggresomes, inclusion bodies and protein aggregation. *Trends Cell Biol* **2000**, *10*, 524–530.

14 Jackson, S.E. How do small single-domain proteins fold? *Fold Des* **1998**, *3*, 81–91.

15 Hobart, S.A., D.W. Meinhold, R. Osuna and W. Colón. From two-state to three-state: effect of P61A mutation on the dynamics and stability of the factor for inversion stimulation results in an altered equilibrium denaturation mechanism. *Biochemistry* **2002**, *41*, 13744–13754.

16 Rudolph, R., G. Zettlmeissl and R. Jaenicke. Reconstitution of lactic dehydrogenase. Noncovalent aggregation vs. reactivation. 2. Reactivation of irreversibly denatured aggregates. *Biochemistry* **1979**, *18*, 5572–5575.

17 Zettlmeissl, G., R. Rudolph and R. Jaenicke. Reconstitution of lactic dehydrogenase. Noncovalent aggregation vs. reactivation. 1. Physical properties and kinetics of aggregation. *Biochemistry* **1979**, *18*, 5567–5571.

18 London, J., C. Skrzynia and M.E. Goldberg. Renaturation of *Escherichia coli* tryptophanase after exposure to 8 M urea. Evidence for the existence of nucleation centers. *Eur J Biochem* **1974**, *47*, 409–415.

19 Orsini, G., C. Skrzynia and M.E. Goldberg. The renaturation of reduced polyalanyl-chymotrypsinogen and chymotrypsinogen. *Eur J Biochem* **1975**, *59*, 433–440.

20 Orsini, G. and M.E. Goldberg. The renaturation of reduced chymotrypsinogen A in guanidine HCl. Refolding versus aggregation. *J Biol Chem* **1978**, *253*, 3453–3458.

21 Glenner, G.G., J. Harbaugh, J.I. Ohma, M. Harada and P. Cuatrecasas. An amyloid protein: the amino-terminal variable fragment of an immunoglobulin light chain. *Biochem Biophys Res Commun* **1970**, *41*, 1287–1289.

22 King, J. and C. Haase-Pettingell. Aggregate formation from thermolabile intermediates in the maturation of the thermostable P22 tailspike. *Biochem Soc Trans* **1988**, *16*, 105–108.

23 Mitraki, A. and J. King. Protein folding intermediates and inclusion body formation. *Bio/Technology* **1989**, *7*, 690–697.

24 Speed, M. A., D. I. Wang and J. King. Multimeric intermediates in the pathway to the aggregated inclusion body state for P22 tailspike polypeptide chains. *Protein Sci* **1995**, *4*, 900–908.

25 King, J. Genetic analysis of protein folding pathways. *Bio/Technology* **1986**, *4*, 297–303.

26 Seckler, R., A. Fuchs, J. King and R. Jaenicke. Reconstitution of the thermostable trimeric phage P22 tailspike protein from denatured chains *in vitro*. *J Biol Chem* **1989**, *264*, 11750–11753.

27 King, J. and M.H. Yu. Mutational analysis of protein folding pathways: the P22 tailspike endorhamnosidase. *Methods Enzymol* **1986**, *131*, 250–266.

28 Brunschier, R., M. Danner and R. Seckler. Interactions of phage P22 tailspike protein with GroE molecular chaperones during refolding *in vitro*. *J Biol Chem* **1993**, *268*, 2767–2772.

29 Sturtevant, J.M., M.H. Yu, C. Haase-Pettingell and J. King. Thermostability of temperature-sensitive folding mutants of the P22 tailspike protein. *J Biol Chem* **1989**, *264*, 10693–10698.

30 Yu, M.H. and J. King. Single amino acid substitutions influencing the folding pathway of the phage P22 tail spike endorhamnosidase. *Proc Natl Acad Sci USA* **1984**, *81*, 6584–6588.

31 Goldenberg, D.P. and J. King. Temperature-sensitive mutants blocked in the folding or subunit of the bacteriophage P22 tail spike protein. II. Active mutant proteins matured at 30 degrees C. *J Mol Biol* **1981**, *145*, 633–651.

32 Horowitz, P. and N.L. Criscimagna. Low concentrations of guanidinium chloride expose apolar surfaces and cause differential perturbation in catalytic intermediates of rhodanese. *J Biol Chem* **1986**, *261*, 15652–15658.

33 Havel, H.A., E.W. Kauffman, S.M. Plaisted and D.N. Brems. Reversible self-association of bovine growth hormone during equilibrium unfolding. *Biochemistry* **1986**, *25*, 6533–6538.

34 Brems, D.N., S.M. Plaisted, E.W. Kauffman and H.A. Havel. Characterization of an associated equilibrium folding intermediate of bovine growth hormone. *Biochemistry* **1986**, *25*, 6539–6543.

35 Colón, W. and J. W. Kelly. Partial denaturation of transthyretin is sufficient for amyloid fibril formation *in vitro*. *Biochemistry* **1992**, *31*, 8654–8660.

36 Hurle, M.R., L.R. Helms, L. Li, W. Chan and R. Wetzel. A role for destabilizing amino acid replacements in light-chain amyloidosis. *Proc Natl Acad Sci USA* **1994**, *91*, 5446–5450.

37 Khurana, R., J.R. Gillespie, A. Talapatra, L.J. Minert, C. Ionescu-Zanetti, I. Millet, and A.L. Fink. Partially folded intermediates as critical precursors of light chain amyloid fibrils and amorphous aggregates. *Biochemistry* **2001**, *40*, 3525–3535.

38 Booth, D.R., M. Sunde, V. Bellotti, C.V. Robinson, W.L. Hutchinson, P.E. Fraser, P.N. Hawkins, C.M. Dobson, S.E. Radford, C.C. Blake and M.B. Pepys. Instability, unfolding and aggregation of human lysozyme variants underlying amyloid fibrillogenesis. *Nature* **1997**, *385*, 787–793.

39 Gerhartz, B., I. Ekiel and M. Abrahamson. Two stable unfolding intermediates of the disease-causing L68Q variant of human cystatin C. *Biochemistry* **1998**, *37*, 17309–17317.

40 Wetzel, R. For protein misassembly, it's the "I" decade. *Cell* **1996**, *86*, 699–702.

41 Kelly, J.W. The alternative conformations of amyloidogenic proteins and their multi-step assembly pathways. *Curr Opin Struct Biol* **1998**, *8*, 101–106.

42 Horwich, A. Protein aggregation in disease: a role for folding intermediates forming specific multimeric interactions. *J Clin Invest* **2002**, *110*, 1221–1232.

43 Guijarro, J.I., M. Sunde, J.A. Jones, I.D. Campbell and C. M. Dobson. Amyloid fibril formation by an SH3 domain. *Proc Natl Acad Sci USA* **1998**, *95*, 4224–4228.

44 Dobson, C.M. Protein misfolding, evolution and disease. *Trends Biochem Sci* **1999**, *24*, 329–332.

45 Blake, C.C., M.J. Geisow, S.J. Oatley, B. Rerat and C. Rerat. Structure of prealbumin: secondary, tertiary and quaternary interactions determined by Fourier refinement at 1.8 Å. *J Mol Biol* **1978**, *121*, 339–356.

46 Benson, M.D. Familial amyloidotic polyneuropathy. *Trends Neurosci* **1989**, *12*, 88–92.

47 Lai, Z., W. Colón and J.W. Kelly. The acid-mediated denaturation pathway of transthyretin yields a conformational intermediate that can self-assemble into amyloid. *Biochemistry* **1996**, *35*, 6470–6482.

48 Kelly, J.W., W. Colón, Z. Lai, H.A. Lashuel, J. McCulloch, S.L. McCutchen, G.J. Miroy and S.A. Peterson. Transthyretin quaternary and tertiary structural changes facilitate misassembly into amyloid. *Adv Protein Chem* **1997**, *50*, 161–181.

49 McCutchen, S.L., W. Colón and J.W. Kelly. Transthyretin mutation Leu-55-Pro significantly alters tetramer stability and increases amyloidogenicity. *Biochemistry* **1993**, *32*, 12119–12127.

50 Lashuel, H.A., Z. Lai and J. W. Kelly. Characterization of the transthyretin acid denaturation pathways by analytical ultracentrifugation: implications for wild-type, V30M and L55P amyloid fibril formation. *Biochemistry* **1998**, *37*, 17851–17864.

51 Lashuel, H.A., C. Wurth, L. Woo and J.W. Kelly. The most pathogenic transthyretin variant, L55P, forms amyloid fibrils under acidic conditions and protofilaments under physiological conditions. *Biochemistry* **1999**, *38*, 13560–13573.

52 Hammarstrom, P., F. Schneider and J.W. Kelly. *Trans*-suppression of misfolding in an amyloid disease. *Science* **2001**, *293*, 2459–2462.

53 Hammarstrom, P., R.L. Wiseman, E.T. Powers and J.W. Kelly. Prevention of transthyretin amyloid disease by changing protein misfolding energetics. *Science* **2003**, *299*, 713–716.

54 Trinh, C.H., D.P. Smith, A.P. Kalverda, S.E. Phillips and S.E. Radford. Crystal structure of monomeric human beta-2-microglobulin reveals clues to its amyloidogenic properties. *Proc Natl Acad Sci USA* **2002**, *99*, 9771–9776.

55 Floege, J. and G. Ehlerding. Beta-2-microglobulin-associated amyloidosis. *Nephron* **1996**, *72*, 9–26.

56 Esposito, G., R. Michelutti, G. Verdone, P. Viglino, H. Hernandez, C.V. Robinson, A. Amoresano, F. Dal Piaz, M. Monti, P. Pucci, P. Mangione, M. Stoppini, G. Merlini, G. Ferri and V. Bellotti. Removal of the N-terminal hexapeptide from human beta2-microglobulin facilitates protein aggregation and fibril formation. *Protein Sci* **2000**, *9*, 831–845.

57 Chiti, F., E. De Lorenzi, S. Grossi, P. Mangione, S. Giorgetti, G. Caccialanza, C.M. Dobson, G. Merlini, G. Ramponi and V. Bellotti. A partially structured species of beta 2-microglobulin is significantly populated under physiological conditions and involved in fibrillogenesis. *J Biol Chem* **2001**, *276*, 46714–46721.

58 Morgan, C.J., M. Gelfand, C. Atreya and A.D. Miranker. Kidney dialysis-associated amyloidosis: a molecular role for copper in fiber formation. *J Mol Biol* **2001**, *309*, 339–345.

59 Kad, N.M., N.H. Thomson, D.P. Smith, D.A. Smith and S. E. Radford. Beta$_2$-microglobulin and its deamidated variant, N17D form amyloid fibrils with a range of morphologies *in vitro*. *J Mol Biol* **2001**, *313*, 559–571.

60 Villanueva, J., M. Hoshino, H. Katou, J. Kardos, K. Hasegawa, H. Naiki and Y. Goto. Increase in the conformational flexibility of β2-microglobulin upon copper binding: a possible role for copper in dialysis-related amyloidosis. *Protein Sci* **2004**, *13*, 797–809.

61 Jones, S., D.P. Smith and S.E. Radford. Role of the N and C-terminal strands of beta 2-microglobulin in amyloid formation at neutral pH. *J Mol Biol* **2003**, *330*, 935–941.

62 Kozhukh, G.V., Y. Hagihara, T. Kawakami, K. Hasegawa, H. Naiki and Y. Goto. Investigation of a peptide responsible for amyloid fibril formation of beta 2-microglobulin by achromobacter pro-

tease I. *J Biol Chem* **2002**, *277*, 1310–1315.

63 Pepys, M.B., P.N. Hawkins, D.R. Booth, D.M. Vigushin, G.A. Tennent, A.K. Soutar, N. Totty, O. Nguyen, C.C. Blake, C.J. Terry, T.G. Feest, A.M Zalin and J.J. Hsuan. Human lysozyme gene mutations cause hereditary systemic amyloidosis. *Nature* **1993**, *362*, 553–557.

64 Valleix, S., S. Drunat, J.B. Philit, D. Adoue, J.C. Piette, D. Droz, B. MacGregor, D. Canet, M. Delpech and G. Grateau. Hereditary renal amyloidosis caused by a new variant lysozyme W64R in a French family. *Kidney Int* **2002**, *61*, 907–912.

65 Yazaki, M., S.A. Farrell and M.D. Benson. A novel lysozyme mutation Phe57Ile associated with hereditary renal amyloidosis. *Kidney Int* **2003**, *63*, 1652–1657.

66 Canet, D., M. Sunde, A.M. Last, A. Miranker, A. Spencer, C.V. Robinson and C.M. Dobson. Mechanistic studies of the folding of human lysozyme and the origin of amyloidogenic behavior in its disease-related variants. *Biochemistry* **1999**, *38*, 6419–6427.

67 Takano, K., J. Funahashi and K. Yutani. The stability and folding process of amyloidogenic mutant human lysozymes. *Eur J Biochem* **2001**, *268*, 155–159.

68 Canet, D., A.M. Last, P. Tito, M. Sunde, A. Spencer, D.B. Archer, C. Redfield, C.V. Robinson and C.M. Dobson. Local cooperativity in the unfolding of an amyloidogenic variant of human lysozyme. *Nat Struct Biol* **2002**, *9*, 308–315.

69 Bode, W., R. Engh, D. Musil, U. Thiele, R. Huber, A. Karshikov, J. Brzin, J. Kos and V. Turk. The 2.0 Å X-ray crystal structure of chicken egg white cystatin and its possible mode of interaction with cysteine proteinases. *EMBO J* **1988**, *7*, 2593–2599.

70 Palsdottir, A., M. Abrahamson, L. Thorsteinsson, A. Arnason, I. Olafsson, A. Grubb and O. Jensson. Mutation in cystatin C gene causes hereditary brain haemorrhage. *Lancet* **1988**, *2*, 603–604.

71 Abrahamson, M. and A. Grubb. Increased body temperature accelerates aggregation of the Leu-68 → Gln mutant cystatin C, the amyloid-forming protein in hereditary cystatin C amyloid angiopathy. *Proc Natl Acad Sci USA* **1994**, *91*, 1416–1420.

72 Ekiel, I. and M. Abrahamson. Folding-related dimerization of human cystatin C. *J Biol Chem* **1996**, *271*, 1314–1321.

73 Janowski, R., M. Kozak, E. Jankowska, Z. Grzonka, A. Grubb, M. Abrahamson and M. Jaskolski. Human cystatin C, an amyloidogenic protein, dimerizes through three-dimensional domain swapping. *Nat Struct Biol* **2001**, *8*, 316–320.

74 Ekiel, I., M. Abrahamson, D.B. Fulton, P. Lindahl, A.C. Storer, W. Levadoux, M. Lafrance, S. Labelle, Y. Pomerleau, D. Groleau, L. LeSauteur and K. Gehring. NMR structural studies of human cystatin C dimers and monomers. *J Mol Biol* **1997**, *271*, 266–277.

75 Staniforth, R.A., S. Giannini, L.D. Higgins, M.J. Conroy, A.M. Hounslow, R. Jerala, C.J. Craven and J.P. Waltho. Three-dimensional domain swapping in the folded and molten-globule states of cystatins, an amyloid-forming structural superfamily. *Embo J.* **2001**, *20*, 4774–4781.

76 Engh, R.A., T. Dieckmann, W. Bode, E.A. Auerswald, V. Turk, R. Huber and H. Oschkinat. Conformational variability of chicken cystatin. Comparison of structures determined by X-ray diffraction and NMR spectroscopy. *J Mol Biol* **1993**, *234*, 1060–1069.

77 Staniforth, R.A., J.L. Dean, Q. Zhong, E. Zerovnik, A.R. Clarke and J.P. Waltho. The major transition state in folding need not involve the immobilization of side chains. *Proc Natl Acad Sci USA* **2000**, *97*, 5790–5795.

78 Nilsson, M., X. Wang, S. Rodziewicz-Motowidlo, R. Janowski, V. Lindstrom, P. Onnerfjord, G. Westermark, Z. Grzonka, M. Jaskolski and A. Grubb. Prevention of domain swapping inhibits dimerization and amyloid fibril formation of cystatin C: use of engineered disulfide bridges, antibodies and carboxymethylpapain to stabilize the mono-

meric form of cystatin C. *J Biol Chem* **2004**, *279*, 24236–24245.

79 Bennett, M.J., M.P. Schlunegger and D. Eisenberg. 3D domain swapping: a mechanism for oligomer assembly. *Protein Sci* **1995**, *4*, 2455–2468.

80 Jaskolski, M. 3D domain swapping, protein oligomerization and amyloid formation. *Acta Biochim Pol* **2001**, *48*, 807–827.

81 Wang, L., H.A. Lashuel, T. Walz and W. Colón. Murine apolipoprotein serum amyloid A in solution forms a hexamer containing a central channel. *Proc Natl Acad Sci USA* **2002**, *99*, 15947–15952.

82 Wang, L., H.A. Lashuel and W. Colón. From hexamer to amyloid: marginal stability of apolipoprotein SAA2.2 leads to *in vitro* fibril formation at physiological temperature. *Amyloid* **2005**, in press.

83 Plaza del Pino, I.M., B. Ibarra-Molero and J.M. Sanchez-Ruiz. Lower kinetic limit to protein thermal stability: a proposal regarding protein stability *in vivo* and its relation with misfolding diseases. *Proteins* **2000**, *40*, 58–70.

84 Waterhous, D.V. and W.C. Johnson, Jr. Importance of environment in determining secondary structure in proteins. *Biochemistry* **1994**, *33*, 2121–2128.

85 Gustafsson, M., J. Thyberg, J. Naslund, E. Eliasson and J. Johansson. Amyloid fibril formation by pulmonary surfactant protein C. *FEBS Lett* **1999**, *464*, 138–142.

86 Johansson, J. Structure and properties of surfactant protein C. *Biochim Biophys Acta* **1998**, *1408*, 161–172.

87 Wang, Z., O. Gurel, J.E. Baatz and R.H. Notter. Acylation of pulmonary surfactant protein-C is required for its optimal surface active interactions with phospholipids. *J Biol Chem* **1996**, *271*, 19104–19109.

88 Kallberg, Y., M. Gustafsson, B. Persson, J. Thyberg and J. Johansson. Prediction of amyloid fibril-forming proteins. *J Biol Chem* **2001**, *276*, 12945–12950.

89 Li, S.C. and C.M. Deber. A measure of helical propensity for amino acids in membrane environments. *Nat Struct Biol* **1994**, *1*, 558.

90 Szyperski, T., G. Vandenbussche, T. Curstedt, J.M. Ruysschaert, K. Wuthrich and J. Johansson. Pulmonary surfactant-associated polypeptide C in a mixed organic solvent transforms from a monomeric alpha-helical state into insoluble beta-sheet aggregates. *Protein Sci* **1998**, *7*, 2533–2540.

91 Li, J., W. Hosia, A. Hamvas, J. Thyberg, H. Jornvall, T.E. Weaver and J. Johansson. The N-terminal propeptide of lung surfactant protein C is necessary for biosynthesis and prevents unfolding of a metastable alpha-helix. *J Mol Biol* **2004**, *338*, 857–862.

92 Gustafsson, M., W.J. Griffiths, E. Furusjo and J. Johansson. The palmitoyl groups of lung surfactant protein C reduce unfolding into a fibrillogenic intermediate. *J Mol Biol* **2001**, *310*, 937–950.

93 Whitsett, J.A. and T.E. Weaver. Hydrophobic surfactant proteins in lung function and disease. *N Engl J Med* **2002**, *347*, 2141–2148.

94 Zerovnik, E. Amyloid-fibril formation. Proposed mechanisms and relevance to conformational disease. *Eur J Biochem* **2002**, *269*, 3362–3371.

95 Dobson, C.M. The structural basis of protein folding and its links with human disease. *Phil Trans R Soc Lond B Biol Sci* **2001**, *356*, 133–145.

96 Jarrett, J.T. and P.T. Lansbury, Jr. Seeding "one-dimensional crystallization" of amyloid: a pathogenic mechanism in Alzheimer's disease and scrapie? *Cell* **1993**, *73*, 1055–1058.

97 Harper, J. D. and P.T. Lansbury, Jr. Models of amyloid seeding in Alzheimer's disease and scrapie: mechanistic truths and physiological consequences of the time-dependent solubility of amyloid proteins. *Annu Rev Biochem* **1997**, *66*, 385–407.

98 Morozova-Roche, L. A., J. Zurdo, A. Spencer, W. Noppe, V. Receveur, D.B. Archer, M. Joniau and C. M. Dobson. Amyloid fibril formation and seeding by wild-type human lysozyme and its disease-related mutational variants. *J Struct Biol* **2000**, *130*, 339–351.

99 O'Nuallain, B., A.D. Williams, P. Westermark and R. Wetzel. Seeding specificity in amyloid growth induced by heterologous fibrils. *J Biol Chem* **2004**, *279*, 17490–17499.

100 Richardson, J. S. and D. C. Richardson. Natural beta-sheet proteins use negative design to avoid edge-to-edge aggregation. *Proc Natl Acad Sci USA* **2002**, *99*, 2754–2759.

101 Wang, W. and M. H. Hecht. Rationally designed mutations convert de novo amyloid-like fibrils into monomeric beta-sheet proteins. *Proc Natl Acad Sci USA* **2002**, *99*, 2760–2765.

102 Wood, S. J., R. Wetzel, J. D. Martin and M. R. Hurle. Prolines and amyloidogenicity in fragments of the Alzheimer's peptide beta/A4. *Biochemistry* **1995**, *34*, 724–730.

103 Moriarty, D. F. and D. P. Raleigh. Effects of sequential proline substitutions on amyloid formation by human amylin 20–29. *Biochemistry* **1999**, *38*, 1811–1818.

104 Williams, A. D., E. Portelius, I. Kheterpal, J. T. Guo, K. D. Cook, Y. Xu and R. Wetzel. Mapping abeta amyloid fibril secondary structure using scanning proline mutagenesis. *J Mol Biol* **2004**, *335*, 833–842.

105 Otzen, D. E., O. Kristensen and M. Oliveberg. Designed protein tetramer zipped together with a hydrophobic Alzheimer homology: a structural clue to amyloid assembly. *Proc Natl Acad Sci USA* **2000**, *97*, 9907–9912.

106 Fandrich, M., V. Forge, K. Buder, M. Kittler, C. M. Dobson and S. Diekmann. Myoglobin forms amyloid fibrils by association of unfolded polypeptide segments. *Proc Natl Acad Sci USA* **2003**, *100*, 15463–15468.

107 Smith, L. J., K. A. Bolin, H. Schwalbe, M. W. MacArthur, J. M. Thornton and C. M. Dobson. Analysis of main chain torsion angles in proteins: prediction of NMR coupling constants for native and random coil conformations. *J Mol Biol* **1996**, *255*, 494–506.

108 Shortle, D. and M. S. Ackerman. Persistence of native-like topology in a denatured protein in 8 M urea. *Science* **2001**, *293*, 487–489.

109 Narayan, M., E. Welker and H. A. Scheraga. Characterizing the unstructured intermediates in oxidative folding. *Biochemistry* **2003**, *42*, 6947–6955.

110 Kirkitadze, M. D., G. Bitan and D. B. Teplow. Paradigm shifts in Alzheimer's disease and other neurodegenerative disorders: the emerging role of oligomeric assemblies. *J Neurosci Res* **2002**, *69*, 567–577.

111 Caughey, B. and P. T. Lansbury. Protofibrils, pores, fibrils and neurodegeneration: separating the responsible protein aggregates from the innocent bystanders. *Annu Rev Neurosci* **2003**, *26*, 267–298.

112 Solomon, B. Anti-aggregation antibodies, a new approach towards treatment of conformational diseases. *Curr Med Chem* **2002**, *9*, 1737–1749.

113 Cohen, F. E. and J. W. Kelly. Therapeutic approaches to protein-misfolding diseases. *Nature* **2003**, *426*, 905–909.

114 Colón, W., Z. Lai, S. L. McCutchen, G. J. Miroy, C. Strang and J. W. Kelly. FAP mutations destabilize transthyretin facilitating conformational changes required for amyloid formation. *Ciba Found Symp* **1996**, *199*, 228–238; discussion 239–242.

115 Miroy, G. J., Z. Lai, H. A. Lashuel, S. A. Peterson, C. Strang and J. W. Kelly. Inhibiting transthyretin amyloid fibril formation via protein stabilization. *Proc Natl Acad Sci USA* **1996**, *93*, 15051–15016.

116 Miller, S. R., Y. Sekijima and J. W. Kelly. Native state stabilization by NSAIDs inhibits transthyretin amyloidogenesis from the most common familial disease variants. *Lab Invest* **2004**, *84*, 545–552.

117 Dumoulin, M., A. M. Last, A. Desmyter, K. Decanniere, D. Canet, G. Larsson, A. Spencer, D. B. Archer, J. Sasse, S. Muyldermans, L. Wyns, C. Redfield, A. Matagne, C. V. Robinson and C. M. Dobson. A camelid antibody fragment inhibits the formation of amyloid fibrils by human lysozyme. *Nature* **2003**, *424*, 783–788.

118 Mitraki, A., B. Fane, C. Haase-Pettingell, J. Sturtevant and J. King. Global suppression of protein folding defects and inclusion body formation. *Science* **1991**, *253*, 54–58.

119 Coelho, T. Compound heterozygotes of transthyretin Met30 and transthyretin Met119 are protected from the devastating effects of familial amyloid polyneuropathy. *Neuromuscular Disorders* **1996**, *6*, S20.

120 Westermark, P. and K. Sletten. A serum AA-like protein as a common constituent of secondary amyloid fibrils. *Clin Exp Immunol* **1982**, *49*, 725–731.

121 Gillmore, J. D., L. B. Lovat, M. R. Persey, M. B. Pepys and P. N. Hawkins. Amyloid load and clinical outcome in AA amyloidosis in relation to circulating concentration of serum amyloid A protein. *Lancet* **2001**, *358*, 24–29.

122 Saraiva, M. J. Transthyretin mutations in health and disease. *Hum Mutat* **1995**, *5*, 191–196.

123 Westermark, P., K. Sletten, B. Johansson and G. G. Cornwell. Fibril in senile systemic amyloidosis is derived from normal transthyretin. *Proc Natl Acad Sci USA* **1990**, *87*, 2843–2845.

124 Glenner, G. G., J. Harbaugh, J. I. Ohma, M. Harada and P. Cuatrecasas. An amyloid protein: the amino-terminal variable fragment of an immunoglobulin light chain. *Biochem Biophys Res Commun* **1970**, *41*, 1287–1289.

125 Ferri, G., M. Stoppini, P. Iadarola, V. Bellotti and G. Merlini. Structural characterization of kappa II Inc, a new amyloid immunoglobulin. *Biochim Biophys Acta* **1989**, *995*, 103–108.

126 Wetzel, R. Domain stability in immunoglobulin light chain deposition disorders. *Adv Protein Chem* **1997**, *50*, 183–242.

127 Janciauskiene, S., E. Carlemalm and S. Eriksson. *In vitro* amyloid fibril formation from alpha 1-antitrypsin. *Biol Chem Hoppe Seyler* **1995**, *376*, 103–109.

128 Lomas, D. A., D. L. Evans, J. T. Finch and R. W. Carrell. The mechanism of Z alpha 1-antitrypsin accumulation in the liver. *Nature* **1992**, *357*, 605–607.

129 Carrell, R. W. and D. A. Lomas. Conformational disease. *Lancet* **1997**, *350*, 134–138.

130 Haggqvist, B., J. Naslund, K. Sletten, G. T. Westermark, G. Mucchiano, L. O. Tjernberg, C. Nordstedt, U. Engstrom and P. Westermark. Medin: an integral fragment of aortic smooth muscle cell-produced lactadherin forms the most common human amyloid. *Proc Natl Acad Sci USA* **1999**, *96*, 8669–8674.

131 Seymour, J. F. and J. J. Presneill. Pulmonary alveolar proteinosis: progress in the first 44 years. *Am J Respir Crit Care Med* **2002**, *166*, 215–235.

132 DiDonato, M., L. Craig, M. E. Huff, M. M. Thayer, R. M. Cardoso, C. J. Kassmann, T. P. Lo, C. K. Bruns, E. T. Powers, J. W. Kelly, E. D. Getzoff and J. A. Tainer. ALS mutants of human superoxide dismutase form fibrous aggregates via framework destabilization. *J Mol Biol* **2003**, *332*, 601–615.

133 Elam, J. S., A. B. Taylor, R. Strange, S. Antonyuk, P. A. Doucette, J. A. Rodriguez, S. S. Hashein, L. J. Hayward, J. S. Valentine, T. O. Yeates, and P. J. Hart. Amyloid-like filaments and water-filled nanotubes formed by SODI mutant proteins linked to familial ALS. *Nat Struct Biol* **2003**, *10*, 461–467.

134 Maury, C. P. Gelsolin-related amyloidosis. Identification of the amyloid protein in Finnish hereditary amyloidosis as a fragment of variant gelsolin. *J Clin Invest* **1991**, *87*, 1195–1199.

135 Benyamini, H., K. Gunasekaran, H. Wolfson and R. Nussinov. Conservation and amyloid formation: a study of the gelsolin-like family. *Proteins* **2003**, *51*, 266–282.

136 Huff, M. E., L. J. Page, W. E. Balch and J. W. Kelly. Gelsolin domain 2 Ca^{2+} affinity determines susceptibility to furin proteolysis and familial amyloidosis of Finnish type. *J Mol Biol* **2003**, *334*, 119–127.

137 Ghiso, J., B. Pons-Estel and B. Frangione. Hereditary cerebral amyloid angiopathy: the amyloid fibrils contain a protein which is a variant of cystatin C, an inhibitor of lysosomal cysteine proteases. *Biochem Biophys Res Commun* **1986**, *136*, 548–554.

138 Greenberg, S. M. Cerebral amyloid angiopathy: prospects for clinical diagnosis and treatment. *Neurology* **1998**, *51*, 690–694.

139 Kocisko, D.A., J.H. Come, S.A. Priola, B. Chesebro, G.J. Raymond, P.T. Lansbury and B. Caughey. Cell-free formation of protease-resistant prion protein. *Nature* **1994**, *370*, 471–474.

140 Prusiner, S.B. The prion diseases. *Brain Pathol* **1998**, *8*, 499–513.

141 Cohen, F.E. Protein misfolding and prion diseases. *J Mol Biol* **1999**, *293*, 313–320.

142 Morillas, M., D.L. Vanik and W.K. Surewicz. On the mechanism of alpha-helix to beta-sheet transition in the recombinant prion protein. *Biochemistry* **2001**, *40*, 6982–6987.

143 Hamidi Asl, L., V. Fournier, C. Billerey, E. Justrabo, D. Chevet, D. Droz, C. Pecheux, M. Delpech and G. Grateau. Fibrinogen A alpha chain mutation (Arg554 Leu) associated with hereditary renal amyloidosis in a French family. *Amyloid* **1998**, *5*, 279–284.

144 Benson, M.D., J. Liepnieks, T. Uemichi, G. Wheeler and R. Correa. Hereditary renal amyloidosis associated with a mutant fibrinogen alpha-chain. *Nat Genet* **1993**, *3*, 252–255.

145 Burke, M.J. and M.A. Rougvie. Cross-beta protein structures. I. Insulin fibrils. *Biochemistry* **1972**, *11*, 2435–2439.

146 Nilsson, M.R. and C.M. Dobson. Chemical modification of insulin in amyloid fibrils. *Protein Sci* **2003**, *12*, 2637–2641.

147 Dische, F.E., C. Wernstedt, G.T. Westermark, P. Westermark, M.B. Pepys, J.A. Rennie, S.G. Gilbey and P.J. Watkins. Insulin as an amyloid-fibril protein at sites of repeated insulin injections in a diabetic patient. *Diabetologia* **1988**, *31*, 158–161.

148 Wisniewski, T., A.A. Golabek, E. Kida, K.E. Wisniewski and B. Frangione. Conformational mimicry in Alzheimer's disease. Role of apolipoproteins in amyloidogenesis. *Am J Pathol* **1995**, *147*, 238–244.

149 Westermark, P., G. Mucchiano, T. Marthin, K.H. Johnson and K. Sletten. Apolipoprotein A1-derived amyloid in human aortic atherosclerotic plaques. *Am J Pathol* **1995**, *147*, 1186–1192.

150 Hamada, D. and C.M. Dobson. A kinetic study of beta-lactoglobulin amyloid fibril formation promoted by urea. *Protein Sci* **2002**, *11*, 2417–2426.

151 Carrotta, R., R. Bauer, R. Waninge and C. Rischel. Conformational characterization of oligomeric intermediates and aggregates in beta-lactoglobulin heat aggregation. *Protein Sci* **2001**, *10*, 1312–1318.

152 Goers, J., S.E. Permyakov, E.A. Permyakov, V.N. Uversky and A.L. Fink. Conformational prerequisites for alpha-lactalbumin fibrillation. *Biochemistry* **2002**, *41*, 12546–12551.

153 Schuler, B., R. Rachel and R. Seckler. Formation of fibrous aggregates from a non-native intermediate: the isolated P22 tailspike beta-helix domain. *J Biol Chem* **1999**, *274*, 18589–18596.

154 Damaschun, G., H. Damaschun, K. Gast and D. Zirwer. Proteins can adopt totally different folded conformations. *J Mol Biol* **1999**, *291*, 715–725.

155 Damaschun, G., H. Damaschun, H. Fabian, K. Gast, R. Krober, M. Wieske and D. Zirwer. Conversion of yeast phosphoglycerate kinase into amyloid-like structure. *Proteins* **2000**, *39*, 204–211.

156 Fandrich, M., M.A. Fletcher and C.M. Dobson. Amyloid fibrils from muscle myoglobin. *Nature* **2001**, *410*, 165–166.

157 Pertinhez, T.A., M. Bouchard, E.J. Tomlinson, R. Wain, S.J. Ferguson, C.M. Dobson and L.J. Smith. Amyloid fibril formation by a helical cytochrome. *FEBS Lett* **2001**, *495*, 184–186.

158 Ramirez-Alvarado, M., J.S. Merkel and L. Regan. A systematic exploration of the influence of the protein stability on amyloid fibril formation *in vitro*. *Proc Natl Acad Sci USA* **2000**, *97*, 8979–8984.

159 Chiti, F., N. Taddei, P. Webster, D. Hamada, T. Fiaschi, G. Ramponi and C. M. Dobson. Acceleration of the folding of acylphosphatase by stabilization of local secondary structure. *Nat Struct Biol* **1999**, *6*, 380–387.

160 Ishimaru, D., L.R. Andrade, L.S. Teixeira, P.A. Quesado, L.M. Maiolino, P.M. Lopez, Y. Cordeiro, L.T. Costa, W.

M. Heckl, G. Weissmuller, D. Foguel and J. L. Silva. Fibrillar aggregates of the tumor suppressor p53 core domain. *Biochemistry* **2003**, *42*, 9022–9027.

161 Konno, T. Multistep nucleus formation and a separate subunit contribution of the amyloidgenesis of heat-denatured monellin. *Protein Sci* **2001**, *10*, 2093–2101.

162 Konno, T., K. Murata and K. Nagayama. Amyloid-like aggregates of a plant protein: a case of a sweet-tasting protein, monellin. *FEBS Lett* **1999**, *454*, 122–126.

163 Srisailam, S., T. K. Kumar, D. Rajalingam, K. M. Kathir, H. S. Sheu, F. J. Jan, P. C. Chao and C. Yu. Amyloid-like fibril formation in an all beta-barrel protein. Partially structured intermediate state(s) is a precursor for fibril formation. *J Biol Chem* **2003**, *278*, 17701–17709.

164 Srisailam, S., H. M. Wang, T. K. Kumar, D. Rajalingam, V. Sivaraja, H. S. Sheu, Y. C. Chang and C. Yu. Amyloid-like fibril formation in an all beta-barrel protein involves the formation of partially structured intermediate(s). *J Biol Chem* **2002**, *277*, 19027–19036.

165 Samuel, D., T. K. Kumar, T. Srimathi, H. Hsieh and C. Yu. Identification and characterization of an equilibrium intermediate in the unfolding pathway of an all beta-barrel protein. *J Biol Chem* **2000**, *275*, 34968–34975.

13
Computational Approaches and Tools for Establishing Structural Models for Short Amyloid-forming Peptides

Nurit Haspel, David Zanuy, Hui-Hsu (Gavin) Tsai, Buyong Ma, Haim Wolfson and Ruth Nussinov

13.1
Introduction

The process of protein misfolding and fibrillar aggregation is associated with a large number of pathologically unrelated human diseases such as Alzheimer's, Parkinson's and Huntington's diseases [1–4], all sharing the presence of insoluble fibrillar deposits of proteins or deposits of proteins that can readily be converted to amyloid fibrils [5]. The proteins involved in these processes are unrelated in sequence or in function. Despite this unrelatedness, the general similarity in morphology has led to many attempts to obtain generalized models that would explain and characterize amyloid fiber formation [6, 7]. It is generally assumed that the amyloid fibril organization is compatible with a cross-β structure [8]. However, the details of the organization and the interactions are not always clear, since usually no direct three-dimensional information is available. Understanding amyloid ultrastructure organization should yield clues into the chemical mechanisms that lead to the formation of these invasive fibers [9]. It is the insolubility of the fibrils that prevents the use of standard structure determination techniques such as X-ray diffraction or nuclear magnetic resonance (NMR). Currently available experimental techniques can provide only either short-range distances or mesoscopic levels of organization, but are usually unable to provide high-resolution data. Recently, solid-state NMR studies were able, in some cases, to provide information regarding the backbone and side-chain arrangement of a single amyloid strand or sheet [10, 11]. However, this type of information is usually not available for amyloid systems. From the computational standpoint, achieving the goal of finding a structural model for a molecular system is a challenging task. The increase in computational capabilities now allows enlarging the scope of systems that can be studied by atomistic molecular simulations. Within this context, molecular dynamics (MD) simulations have become an important tool for understanding conformational and aggregation phenomena at the molecular level [12, 13]. MD simulation is the major computational tool used to help define the structure of many molecular systems, amyloid fibrils included. Of particular note, using MD simulations, the first kinetic stages of β-sheet formation are currently being intensively explored [12, 14].

Amyloid Proteins. The Beta Sheet Conformation and Disease. J. D. Sipe

ISBN: 3-527-31072-X

The major problem, however, with MD is the computational time. For example, a MD simulation of 4 ns for a fully atomistic model of several thousands of atoms may take a number of days on a powerful cluster of 16 central processing units. Since this method is extremely time consuming and the structural starting point is unknown, many trials may need to be done until a plausible structural model is obtained. A method that performs initial structural prediction of amyloid fibrils may be very useful and save much time, since the number of false trials can be substantially reduced.

The general "protein folding problem", which is the problem of predicting protein structure given its amino acid sequence, is extremely difficult computationally and the chances of a successful prediction are small. The specific case of resolving the amyloid fibril structure, while also very difficult, could become more tractable, since there exists some *a priori* information regarding the structure, i.e. the secondary structure components and the fact that it is usually a repetitive system, so only the basic repeating unit needs to be probed. Therefore, it could be possible to develop an algorithm that uses heuristics, based on that information, and it may be very efficient.

In this chapter we describe the efforts to build structural models for specific short amyloidogenic peptides and a study of the interactions underlying the amyloid structure, using MD. In parallel, we develop computationally efficient algorithms for the initial prediction of the internal arrangement of amyloid fibrils. The prediction includes side-chain conformations, and arrangement of strands between and within the sheets. An efficient prediction algorithm for biologically stable conformations is not aimed at substituting the MD methods, but it may save considerable simulation time of testing *a priori* unstable conformations.

13.2
Computational Tools in the Service of Amyloid Structure Prediction

The structure of amyloid fibrils usually cannot be determined experimentally, due to their insolubility and lack of crystal formation that prevent the use of classical experimental structure determination methods. In addition, even if some structural features can be determined (the fibrillar structure on a higher level or, in some cases, the inter-sheet arrangement), the experimental processes take up substantial time and computational methods can help by eliminating improbable conformations in advance. Computational methods are now an inseparable part of amyloid research, both combined with experimental work or standalone [15–18].

MD is a well-established computational method that is used to simulate the conformational behavior of molecular systems over a time interval, based on the computation of the internal energy of the system. As mentioned earlier, the computational time is a major problem in MD, due to the huge amount of calculations that are required. Furthermore, in the case of constructing a model for an amyloid system, not all the physical parameters computed by the simulation (e.g. velocities) are needed and therefore considerable unnecessary calculations are performed. For this reason, there is a need for developing computational algorithms that will focus on the specific characteristics necessary for determining the structural stability of amyloid fiber models. The goal of an efficient algorithm for predicting biologically stable conformations is not to substitute for the MD methods or experimental techniques; instead, it may save simulation time by *a priori* filtering out unstable conformations. These algorithms can provide important clues about the amyloid-forming processes and the interactions necessary to create and stabilize amyloid fibers.

To date, algorithms aimed at solving the problem of amyloid structure prediction and the lateral association of amyloid sheets have not been developed. The application of existing general structure prediction and docking algorithms is not efficient because the algorithms are either highly inefficient in terms of computational time and may consume days or weeks of runtime, or may not consider sufficient biological and chemical features. In the case of amyloid fibrils, this is critical, since β-sheets have flat surfaces and, therefore, the main driving force in their association is side-chain interactions. In order to predict the structure of amyloid fibrils, we are now in the process of developing efficient computational algorithms, based on geometrical and biological considerations. Preliminary results with these algorithms yield promising results in predicting the inter-sheet arrangement and side-chain conformations of amyloids.

13.3 Constructing Amyloid Models

Considerable efforts have been devoted to studies of the kinetic stages of amyloid formation. Our group has been studying the amyloid problem from a different starting point. We assume that the final fibril micro-organization holds clues to the amyloid aggregation. Experimentally, it is well established that fibril growth is initiated by a minimal pre-formed, soluble, self-assembled seed complex that must retain its organization in water in order for the seed to assume the role of a structural template for the final growth process. Thus, rather than sampling many micro-states of aggregation of a very small number of molecular assemblies, we explore possible final arrangements of molecular complexes and evaluate their relative stabilities [19–24]. Due to computational limitations, we do not study the final shape and size, but, instead, a number of molecular species that could mimic the minimal organization requirements. Once we have built different combinations, through high-temperature MD simulations, we establish the relative stabilities of the models. Those models that are able to retain some degree of self-organization for the time simulated would be good candidates to study the structural properties of the amyloid fibers formed by a given sequence. This strategy is based on systematic exploration of the molecular self-assembly of a system consisting of a small number of copies of the same peptide.

Based on the algorithm-suggested models, we have been able to interpret some experimental results and to suggest an elongation mechanism for the protofilament of a prion peptide segment (residues 113–120) [20]. Then, we rationalized the observed structural preferences of Aβ peptide segments [21, 25] and predicted the micro-structural differences of the islet amyloid polypeptide segments 22–27 and 22–29 [19, 22], and the role of some critical amino acids in the peptide segment 22–27 [19]. In all cases, the organization of the smallest ordered structure always implied lateral association, either of sheets in the shorter peptides, or of peptide segments of the longer sequences [19].

Here, we describe our research of the structural and chemical properties of the amyloid forming peptide fragment DFNKF [24] [residues 15–19 of human calcitonin (hCT)] and some of its tetra- and pentapeptide single point mutations. hCT [26] is a 32-amino-acid polypeptide hormone, produced by the thyroid C-cells. It is related to calcium homeostasis. Amyloid fibril formation of hCT is associated with medullary carcinoma of the thyroid [27, 28]. It was shown that, while hCT has little secondary structure at room temperature, hCT amyloid fibrils were found to be highly ordered, consisting of both α-helix and β-sheet elements. Experimentally, the pentapeptide fragment DFNKF is known to form fibrils similar to those formed by the entire 32-residue hormone [29, 30]. Studies performed on truncated segments of DFNKF showed that the DFNK tetrapeptide was also capable of forming fibrils, although bifurcated and much shorter than the ones obtained for the pentapeptide. On the other hand, the FNKF segment did not form detectable amyloid fibrils.

Our previous research on amyloid fibril-forming systems revealed many possible arrangements of β-sheet fibril complexes. The β-strands that associate to form the β-sheets can be parallel or antiparallel and the sheets, in turn, usually associate with each other laterally in a parallel or antiparallel fashion to create the basic protofibril units, which are the subject of our studies. Based on our studies of the other amyloid systems, we assumed that, due to its short length, the DFNKF protofilament also had to consist of at least two laterally associated sheets. We tested DFNKF models with practically all the potential arrangements of parallel/antiparallel strands within sheets and of sheets with respect to each other. Our extensive efforts spanned 18 MD explicit water simulations with over 72 ns. Nevertheless, neither models with two sheets of four strands in each nor those with three sheets of three strands per sheet had sufficient stability to explain the experimental results. Detailed examination of the more stable multisheet systems within these, illustrated that for those models that contained parallel-stranded sheets, while the sheets separated from each other, the sheets themselves remained largely intact. Hence, we concluded that the most probable organization for the protofilament structure of the DFNKF peptide is a single-sheet layer. This has led us to test a long, parallel, nine-stranded single-layer β-sheet. The model consisting of a single β-sheet with only a small hydrophobic core maintained the highest degree of bulk β-sheet organization and preserved the largest amount of backbone and side-chain interactions during the simulation. Moreover, our preliminary results (H.-H. Tsai et al., preliminary results) of parallel tempering simulations over a temperature range of 300–600 K have further confirmed that a single sheet with parallel strands is the stable organization for this short peptide. This raised the interesting question of the origin of the chemical interactions stabilizing this single layer β-sheet in water.

In the next sections we present a summary of the conformational and structural properties of the model we recently constructed for residues 15–19 of the hCT hormone. We also characterize its validation using the available experimental data. We further focus on the utilization and the relevance of computational tools in establishing structural models for fibril systems.

13.4 The Calcitonin Pentapeptide System: Bulk Organization and Interactions

Our initially simulated DFNKF models were multi-sheet systems. They consisted of two four-stranded sheet octamers and three three-stranded sheet nonamers. Some of the initial two and three sheet models are shown in Fig. 13.1. Octapeptides and nonapeptides are good candidates for simulations, since their small size makes them suitable for atomic studies. The two- and three-sheet structure enables some strands to be hidden from water and some to be exposed, thus allowing us to study the effect of different surroundings of both the peptide and the solvent on the association process. The basic set of models differed in the organization of the strands within the sheets (parallel or antiparallel) and of the sheets with respect to one another (again, parallel or antiparallel). The main difference between a computational simulation and an experimental protocol is that here one can design a specific organization of β-sheet arrangement and simulate its behavior in solution, knowing that the results obtained will pertain to this organization only. In addition, one can run many simulations in parallel, each testing a different organization of β-sheets. In our case, each model was simulated for 4 ns at 350 K. To quantitatively assess the structural bulk stability of the models and to give a general view of the stability, we used geometric criteria, the retention of the inter-strand and inter-sheet distances during the simulation. We previously developed these criteria [19, 22, 23] since using existing criteria such as RMSD cannot give us an absolute assessment of the structural stability, only a measurement relative to a known structure, which is unavailable in our case. We also measured the percentage of main-chain hydrogen bonds at any stage of the simulation to estimate the organization of the β-sheet bulk structure.

At an early stage of the study we realized that the strands should probably be parallel within the sheets. Several indications pointed to this conclusion: The interstrand distance parameters remained much more constant for the parallel models than for the antiparallel models. In addition, in the antiparallel models, the ability to retain the main-chain hydrogen bonds appeared to be seriously compromised compared to the parallel models. This issue raises a problem of MD. Despite the fact that MD enables us to design models of specific bulk orga-

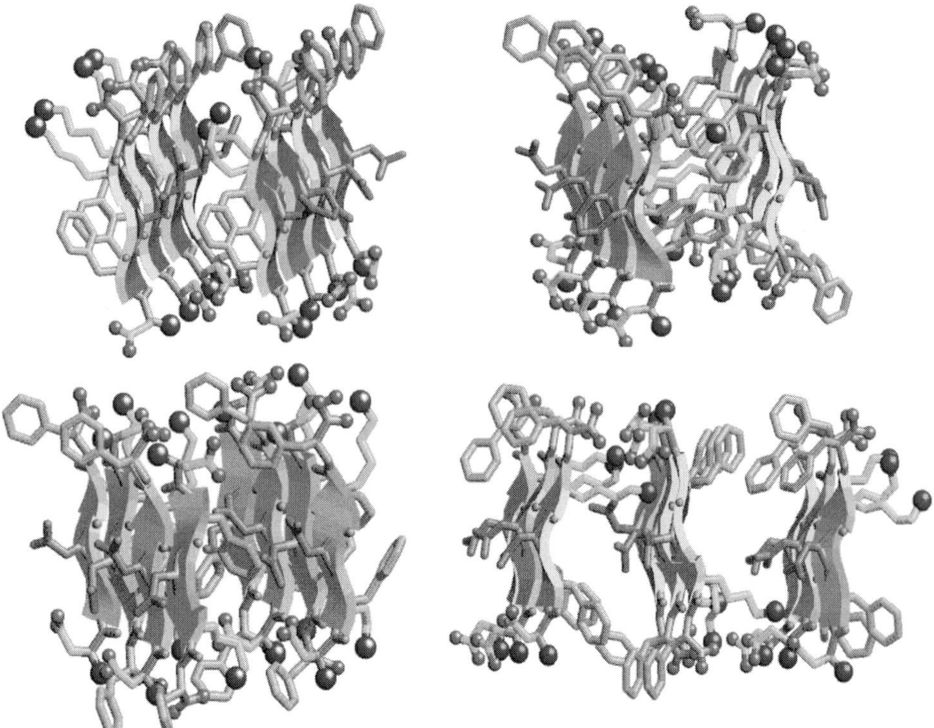

Fig. 13.1 Atomistic representation of some of the two- and three-sheet models. In all representations hydrogen atoms have been deleted for clarity. Main-chain atoms are depicted in gray while side-chains are colored depending on their chemical nature. Asp is represented in green, Phe in orange, Asn in magenta, Lys in cyan, N-termini and positive charge centers in blue spacefill, and C-termini and negative charge centers in red spacefill.

nization and test them, the computation time poses a severe limitation, since in order to reach the conclusion that antiparallel β-sheets are structurally unstable, many days of simulations need to be performed. Having an efficient computational algorithm that is based on simpler heuristics and a smaller number of calculations could enable us to detect the large energetic differences between the parallel and antiparallel strand models within a matter of minutes or less. Using this algorithm, we could simulate only the most plausible structural models and save considerable time.

The parallel stranded models showed an interesting phenomenon – at a very early stage of the simulation the two sheets were separated from one another due to steric hindrance and each sheet continued to be stable on its own, like a self-containing system. The geometry of each sheet became self-optimized at a very early stage of the simulation, with the strands of each sheet twisting with respect to one another (a classical characteristic of a cross-β structure), remaining twisted throughout the entire simulation. As a consequence, each sheet separately exhibits the characteristic interactions of the single-sheet model.

Due to the fact that none of the two-sheet models remained sufficiently stable throughout the simulation, we decided to test models composed of three sheets with three strands in each. One advantage of this arrangement is that there is one sheet (the middle one) which is completely buried between the other two, shielded from the solvent. Since most residues have hydrophilic side-chains, the feature of being buried is mainly relevant to the aromatic rings of phenylalanine. Having at least one buried sheet allows us to study the possible hydrophobic interactions taking place in the system. Further, in three-sheet models, the complex nature of amyloid formation may illustrate some differences in the relative stabilities when the number of associated sheets in the MD models is changed. Our three-sheet models had parallel strands within the sheets and most had antiparallel orientation between the sheets. Here, as in the two-sheet case, the inter-sheet geometrical and chemical magnitudes were less constant than the intra-sheet ones, showing again a clear tendency to a looser organization between sheets.

Based on the results of the two- and three-sheet systems, we realized that each sheet formed its own system of interactions, which led us to conclude that the system may be a one-sheet system. We decided, based on this observation, to simulate a long one-sheet system of parallel strands. The one-sheet model we simulated contained nine parallel strands stacked upon each other. Fig. 13.2 (a) presents an atomistic representation of the minimized model. Initially, the strands have no twist between them. A nine-stranded system was selected based on the assumption that a sheet of this size provides sufficient information about the side-chain interactions underlying the system, while saving computational time that may be very long for a larger system with more strands.

The one-sheet parallel-stranded model was very stable throughout the entire simulation time, compared to the other models. The only significant change in the organization of this single-sheet model was the spontaneous twist of the strands with respect to one another after less than 100 ps of simulation. The strands remained twisted thereafter, with the β-sheet organization being very stable and maintaining a high degree of structural order. The average twist angle was approximately 18° and changed only a little during the simulation. The conformation of the DFNKF system after 1 and 2 ns of simulation is shown in Fig. 13.3. Looking at the snapshots, we see that the twist between the strands enables the formation of hydrophobic interactions between the side-chains. These hydrophobic interactions enable the Phe rings to associate and stack.

Thus, the computational tools allow us to evaluate the conformational stability of a molecular system by obtaining snapshots of its configurations at very short time intervals and, based on the information we get, conclude whether the system is structurally stable or not. In terms of bulk structure, the structural analysis provides clear indications to the high organization of the protofilament. The main structural parameter used to initially evaluate the organization of the one-

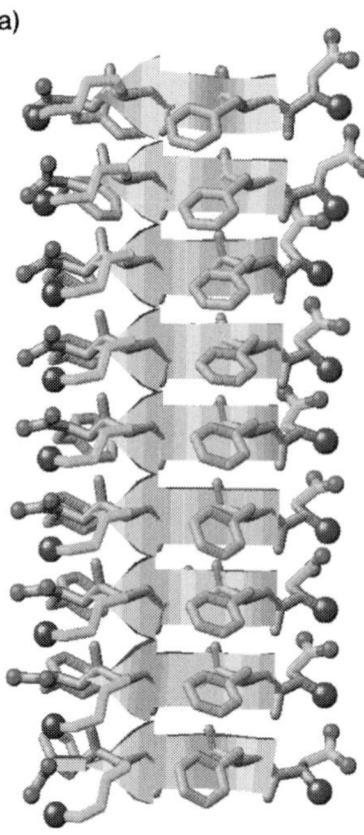

Fig. 13.2 Representation of the one-sheet mutant models. The color coding is similar to the one used for Fig. 13.1, plus the following: Ala in yellow and Glu in dark green. From the top to the bottom: DFNKF, Mono1 (a), DFNK (b), DANKF (c), DFAKF (d) and EFNKF (e).

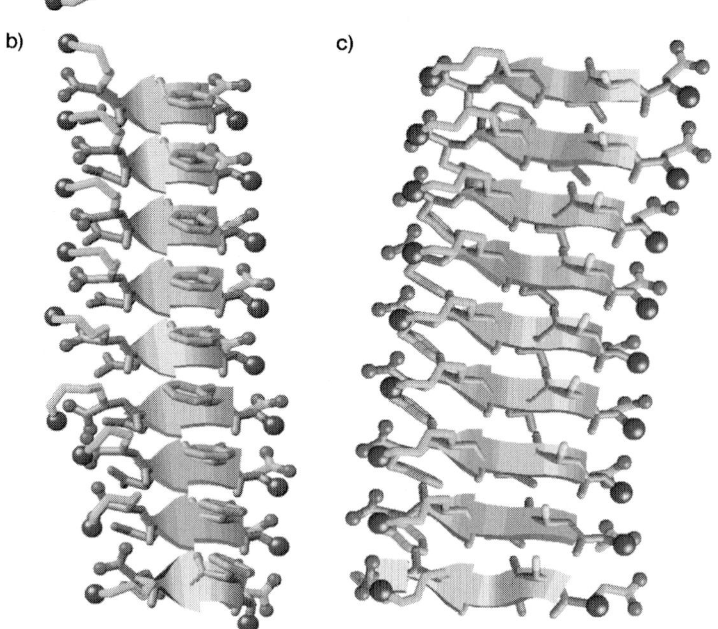

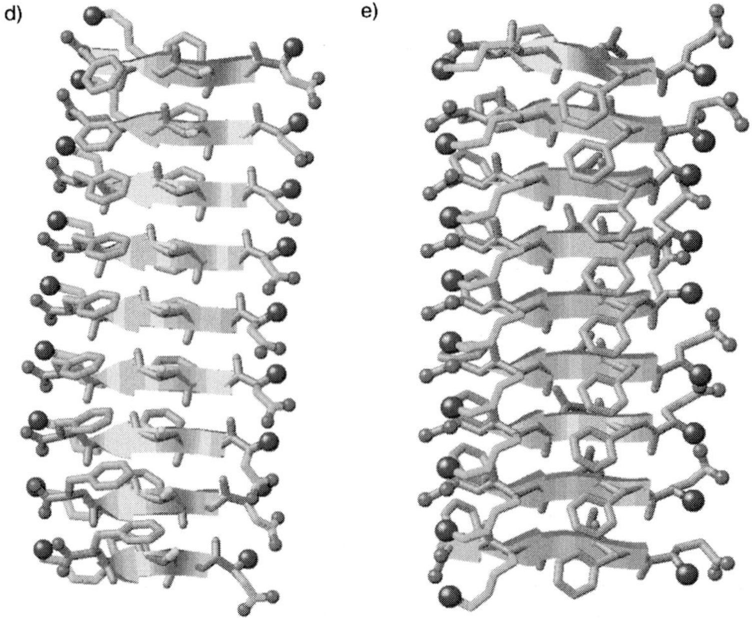

Fig. 13.2 d–e

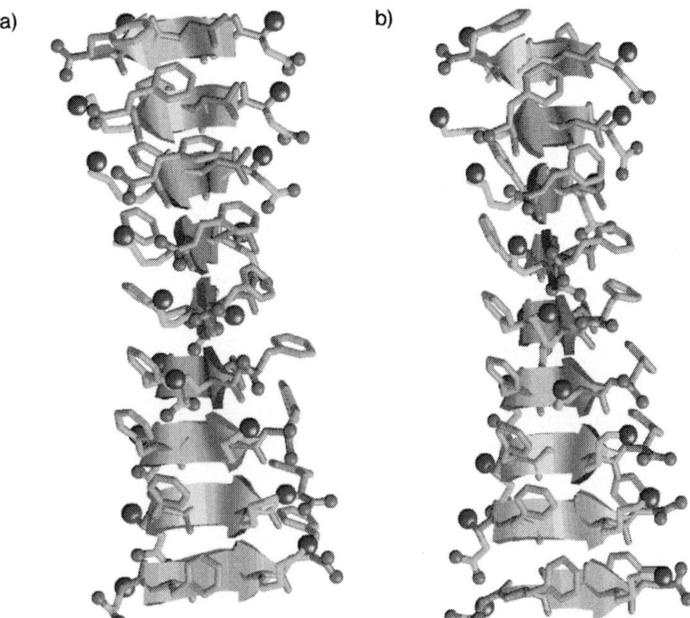

Fig. 13.3 The conformational evolution of model Mono1 after (a) 1 and (b) 2 ns. For the color coding, see Fig. 13.1.

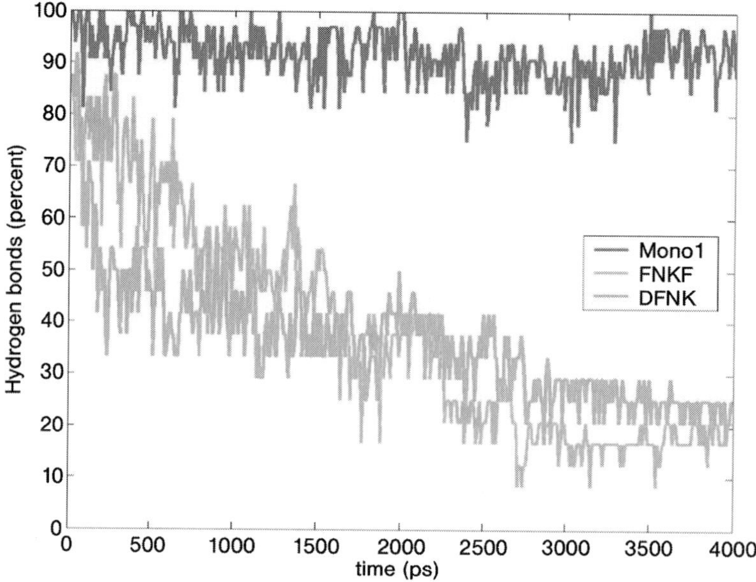

Fig. 13.4 Analysis of the main-chain interactions during the simulation time of some of the one-sheet models: The fraction of main-chain hydrogen bond. The percentage is with respect to the minimized stage. Each model is represented by a different color (see inset legend).

sheet model was the percentage of main-chain hydrogen bonds (with respect to the minimized structure) retained by the system. The more hydrogen bonds retained during the simulation, the more stable the bulk organization is. As seen in Fig. 13.4, model Mono1 retained at least 90% of its hydrogen bonds throughout the entire simulation.

Analyzing the conformations obtained during the simulation also enables us to gain insight into the side-chain interactions that take place in the system and stabilize it. This too cannot be done experimentally in this case, since NMR or crystal data not are available. In the DFNKF case we observe three main types of side-chain interactions. The structure is stabilized mainly by a network of side-chain–main-chain electrostatic interactions. Also important is a network of hydrogen bonds between adjacent Asn side-chains. This interaction functions as a pivot that holds the system together. Phe-mediated interactions may also provide another key to the high structural organization of the DFNKF single-sheet layer model.

13.5 Calcitonin Mutation Study: Simulation and Prediction of Specific Changes in Amino Acids

In an attempt to validate our results, we explore single point mutations of DFNKF for which experimental data are available. The mutant study further highlights the role played by each amino acid in the sequence and assesses the interconnection of experimental and computational procedures. We first simulated the two truncated tetrapeptide filaments, FNKF and DFNK. Experimentally, the FNKF does not create fibrils, while DFNK does, albeit short and less ordered than DFNKF [30]. Both our models consisted of a single β-sheet with nine parallel strands, similar in the macroscopic level to the original DFNKF system. Our assumption was that the fibrils created by DFNK are similar to those of DFNKF, i.e. a single-sheet layer with parallel strands. On the other hand, we note that this assumption for the DFNK model may possibly bias our results for this truncated mutant.

The minimized DFNK system can be seen in Fig. 13.2 (b). Consistent with experiment, the FNKF system was less stable than the DFNK and the DFNKF pentapeptide. The FNKF lost the β-sheet organization at a very early stage of the simulation with the strands becoming an amorphous aggregate. However, DFNK also lost many of its original interactions and became less organized. As demonstrated, the removal of each of the terminal amino acids disrupts the ability of the sequence to form ordered fibrils and to maintain the bulk organization. However, the removal of Asp from the sequence has a larger destabilizing effect on the organization of the β-sheet structure than the removal of the Phe at position 5 from the sequence. The DFNK system did not result in highly ordered fibrils neither in experiment nor in the simulations. However, it is also conceivable that its topology may be different from that of the DFNKF used in the simulations.

The next stage in the mutation study was replacing amino acids in the sequence to study the specific role of each amino acid. Using computational algorithms we can perform a fast exploration of initial conformational states of the mutants. We can explore not only different possible arrangements, but also many possible mutations, efficiently obtaining insight into the role of each specific amino acid we "mutated". The replacement may affect the bulk organization of the structure by changing the chemical interactions taking place in the system. Therefore, we have further simulated pentapeptidic mutated filaments, DANKF, DFNKA, EFNKF and DFAKF. Each mutant is simulated with a nine-stranded, single-layer β-sheet. Fig. 13.2 provides the atomistic representation of some of the minimized mutants: (b) shows the DANKF system, (c) shows the DFAKF system and (d) shows the EFNKF system.

As expected, both DANKF and DFNKA were structurally more stable than the DFNK and FNKF tetrapeptides simulated above. Both were able to maintain a larger β-sheet core and a more organized structure. In addition, they preserved a larger percentage of their original side-chain and main-chain interactions.

However, both were considerably less stable than the original wild-type DFNKF. The EFNKF mutant was more stable than the Phe–Ala mutants. In the EFNKF system, the β-sheet structure maintained a considerable amount of structural organization. However, it lost more of their bulk organization and initial interactions than DFNKF. In agreement with experimental data, the DFAKF mutants failed to create ordered fibrils and the β-structure fell apart at a very early stage of the simulation. The reason for this is that the spine of side-chain hydrogen bonds, which helped maintain the β-sheet structure in the original peptide, is missing here. This simulation highlights the important role played by Asn in DFNKF fibril formation, and further shows how it is possible to use computational methods to study the chemical and biological behavior of a system based on structural predictions.

An additional study in our group [24] has recently observed that DFNKA is more likely to create ordered fibrils than DANKF.

13.6
DFNKF Amyloid Seed and its Stability and Dynamics [21]

It is generally believed that the formation of an amyloid fibril follows a nucleation mechanism and proceeds via a conformational change. The amyloid fibril grows very fast once the critical nucleus seed is formed. In order to understand the behavior of amyloid nucleus seed, we studied the stabilities and dynamics of the DFNKF dimer, trimer and tetramer in detail, again using MD simulations. The results show that the crucial nucleus seed size cannot be very large, since we observed that the DFNKF trimer and tetramer are relatively stable with their parallel in-register packing for a sufficient time during the 350 K MD simulation. The relatively small size of the nucleus seed of DFNKF can explain the experimental observation of its rapid formation of amyloid fibrils. The simulation results also show that the ordered β-strand acts as a β-sheet "template" in prompting other strands to bind non-cooperatively in a parallel arrangement. In general, the residues near the N/C-termini are found to be more flexible. In contrast, the interior residues are relatively more stable. The sequence variant studies indicate that the side-chain–side-chain interactions in the DFNKF oligomers are particularly important. In particular, the Asn side-chain hydrogen bond was found to be crucial in stabilizing the parallel DFNKF oligomers.

13.7
Conclusions

The problem of defining the exact three-dimensional arrangement of amyloid fibrils and the interactions underlying amyloid systems is one of great importance, since amyloid fibrils are formed in many known and common diseases. Understanding the structural features and the kinetic process of their formation may help in blocking fibril formation and, thus, preventing or curing disease. In this work, we show the challenges inherent to the building of models of amyloid fibers, using our latest work with residues 15–19 of the hCT hormone, DFNKF, for which only limited experimental data are available. In the DFNKF case, our MD simulations revealed an interesting three-dimensional organization. We presented a highly stable single-layer, parallel-stranded β-sheet model with a very small hydrophobic core. This structure is consistent with the available experimental data.

MD is an important, well-established and powerful computational tool used to simulate the behavior of molecular systems over time. Using MD, it is possible to construct structural models for molecular systems whose lack of solubility or crystallinity does not enable experimental determination of their structures. Thus, it is possible to alter the initial or final conditions of the system at will, and study it in different temperatures, pressures and concentrations.

However, since MD simulations require a huge amount of calculations even to obtain a simulation of a few nanoseconds, our group is currently developing computational methods that may help in the initial prediction of amyloid structure in shorter times by saving many of the computations performed by MD. These algorithms use simpler heuristics than those used by MD. This way less information is obtained; however, at the next stage, potential promising candidate organizations can be tested rigorously and, thus, overall the necessary information can still be gained in much shorter computational time.

Acknowledgments

The computation times are provided by the National Cancer Institute's Frederick Advanced Biomedical Supercomputing Center and by the NIH Biowulf. This project has been funded in whole or in part with Federal funds from the National Cancer Institute, National Institutes of Health, under contract NO1-CO-12400. The content of this publication does not necessarily reflect the views or policies of the Department of Health and Human Services nor does mention of trade names, commercial products or organizations imply endorsement by the US Government.

References

1 Wanker, E. E. Protein aggregation in Huntington's and Parkinson's disease: implications for therapy. *Mol Med Today* **2000**, *6*, 387–391.

2 Dobson, C. M. Protein misfolding, evolution and disease. *Trends Biochem Sci* **1999**, *24*, 329–332.

3 Dobson, C. M. Protein folding and disease: a view from the first Horizon symposium. *Nat Rev Drug Disc* **2003**, *2*, 154–160.

4 Harper, J. D. and Lansbury, P. T., Jr. Models of amyloid seeding in Alzheimer's disease and Scrapie: mechanistic truths and physiological consequences of the

time-dependent solubility of amyloid proteins. *Annu Rev Biochem* **1997**, *66*, 385–407.

5 Kisilevsky, R. Amyloidogenesis: unquestioned answers and unanswered questions. *J Struct Biol* **2000**, *130*, 99–108.

6 Gazit, E. Possible role of p-stacking in self-assembly of amyloid fibrils. *FASEB J* **2002**, *16*, 77–83.

7 Fandrich, M. and Dobson, C. M. The behavior of polyamino acids reveals an inverse side chain effect in amyloid structure formation. *EMBO J* **2002**, *21*, 5682–5690.

8 Sunde, M. and Blake, C. F. F. From the globular to the fibrous state: protein structure and structural conversion in amyloid formation. *Q Rev Biophys* **1998**, *31*, 1–39.

9 Serrano, L. The relationship between sequence and structure in elementary folding units. *Adv Protein Chem* **2000**, *53*, 49–85

10 Tycko, R. Insights into amyloid folding problem from solid-state NMR. *Biochemistry* **2003**, *42*, 3151–3159.

11 Jaroniec, C. P., MacPhee, C. E., Bajaj, V. S., McMahon, M. T., Dobson, C. M. and Griffin, R. G. High-resolution molecular structure of a peptide in an amyloid fibril determined by magic angle spinning NMR spectroscopy. *Proc Natl Acad Sci USA* **2004**, *101*, 711–716.

12 Klimov, D. and Thirumalai, D. Dissecting the assembly of $A\beta_{16-22}$ amyloid peptides into antiparallel β sheets. *Structure* **2003**, *11*, 295–307.

13 Gnanakaran, S., Nymeyer, H., Portman, J., Sanbonmatsu, K. Y. and Garcia, A. E. Peptide folding simulations. *Curr Opin Struct Biol* **2003**, *13*, 168–174.

14 Gsopner, J., Haberthür, U. and Caflisch, A. The role of side-chain interactions in the early steps of aggregation: molecular dynamics simulations of an amyloid-forming peptide from the yeast prion Sup 35. *Proc Natl Acad Sci USA* **2003**, *100*, 5154–5159.

15 Koide, S., Huang, X., Link, K., Koide, A., Bu, Z. and Engelman, D. M. Design of single-layer β-sheets without a hydrophobic core. *Nature* **2000**, *403*, 456–460.

16 Hosia, W., Bark, N., Liepinsh, E., Tjernberg, A., Persson, B., Hallén, D., Thyberg, J., Johansson, J. and Tjernberg, L. Folding into a β-hairpin can prevent amyloid fibril formation. *Biochemistry* **2004**, *43*, 4655–4661.

17 O'Nuallain, B., Williams, A. D., Westermark, P. and Welzel, R. Seeding specificity in amyloid growth induced by heterologous fibrils. *J Biol Chem* **2004**, *279*, 17490–17499.

18 Kajava, A. V., Baxa, U., Wickner, R. B. and Steven, A. C. A model for Ure2p prion filaments and other amyloids: the parallel superpleated β-structure. *Proc Natl Acad Sci* **2004**, *101*, 7885–7890.

19 Zanuy, D., Ma B. and Nussinov, R. Short peptide amyloid organization: stabilities and conformations of the Islet amyloid peptide NFGAIL. *Biophys J* **2003**, *84*, 1–11.

20 Ma, B. and Nussinov, R. Molecular dynamics simulations of alanine rich β-sheet oligomers: insight into amyloid formation. *Protein Sci* **2002**, *11*, 2335–2350.

21 Ma, B. and Nussinov, R. Stabilities and conformations of Alzheimer's β-amyloid peptide oligomers ($A\beta_{16-22}$, $A\beta_{16-35}$, $A\beta_{10-35}$): sequence effects. *Proc Natl Acad Sci USA* **2002**, *99*, 14126–14131.

22 Zanuy, D. and Nussinov, R. The sequence dependence of fiber organization: a comparative molecular dynamics study of the Islet amyloid polypeptide segments 22–27 and 22–29. *J Mol Biol* **2003**, *329*, 565–584.

23 Zanuy, D., Porat, Y., Gazit, E. and Nussinov, R. Peptide sequence and amyloid formation: Molecular simulations and experimental study of a human Islet amyloid polypeptide fragment and its analogs. *Structure* **2004**, *12*, 439–455.

24 Tsai, H. H., Zanuy, D., Haspel, N., Gunasekaran, K., Ma, B., Tsai, C. J. and Nussinov, R. The stability and dynamic of the human calcitonin amyloid peptide DFNKF. *Biophys J* **2004**, *87*, 146–158.

25 Sikorski, P., Atkins, E. D. T. and Serpell, L. C. Structure and texture of fibrous crystals formed by Alzheimer's $A\beta$ (11–25) peptide fragment. *Structure* **2003**, *11*, 915–926.

26 Zaidi, M., Inzerillo, A.M., Moonga, B.S., Bevis, P.J. and Huang, C.L. Forty years of calcitonin – where are we now? A tribute to the work of Iain Macintyre, FRS. *Bone* **2002**, *30*, 655–663.

27 Butler, M. and Khan, S. Immunoreactive calcitonin in amyloid fibrils of medullary carcinoma of the thyroid gland. An immunogold staining technique. *Arch Pathol Lab Med* **1986**, *110*, 647–6499.

28 Berger, G., Berger, N., Guillaud, M.H., Trouillas, J. and Vauzelle, J.L. Calcitonin-like immunoreactivity of amyloid fibrils in medullary thyroid carcinomas. An immunoelectron microscope study. *Virchows Arch A Pathol Anat Histopathol* **1988**, *412*, 543–551.

29 Arvinte, T., Cudd, A. and Drake, A.F. The Structure and mechanism of formation of human calcitonin fibrils. *J Biol Chem* **1993**, *268*, 6415–6422.

30 Reches, M., Porat, Y. and Gazit, E. Amyloid fibril formation by pentapeptide and tetrapeptide segment of human calcitonin. *J Biol Chem* **2002**, *277*, 35475–35480.

Part V
Pathophysiology of Amyloid Fibril Formation

Amyloid Proteins. The Beta Sheet Conformation and Disease. J. D. Sipe

ISBN: 3-527-31072-X

14
Oligomers and Cellular Toxicity

Bruce Kagan

14.1
Introduction

"Amyloid" was first described by Virchow in the 19th century as an amorphous starch-like deposit in tissues that stained with iodine, indicating the presence of carbohydrate. Investigators in the 20th century concluded that amyloid deposits were largely proteinaceous and exhibited a characteristic staining with the dye Congo red. It also became clear that many different proteins could form amyloid deposits and that these deposits were composed of fibrils with a characteristic structure (for review, see Sipe and Cohen, 2000). Fibrils are 80–100 Å in width and often of very extended length. At least 21 known proteins and peptides have been reported to deposit as amyloid (Table 14.1) and are associated with a wide variety of human illnesses (Merlini and Westermark, 2004). Although amyloid-forming proteins exhibit no amino acid sequence homology, they all adopt a characteristic β-sheet structure within amyloid fibrils.

Amyloid deposits also contain a variety of common components, such as glycosaminoglycans, proteoglycans and the pentraxin serum amyloid P. The structural role of these components in amyloid fibrils and deposits is unclear. Although β-sheet conformation is at the essence of amyloid fibril formation, its role in determining pathology has been less clear. Amyloid deposits were often found to be associated with disease, but were also found in otherwise apparently healthy tissue. Subsequently, studies began to show that amyloid deposits are associated with disease and, perhaps, play an etiologic role (e.g. Hardy and Alsop, 1991). However, the experimental evidence for a correlation of clinical illness with amyloid burden has been irregular and unpersuasive. Recent studies have suggested that smaller, soluble aggregates of proteins, many of which go on to form amyloid fibrils, may be responsible for cellular dysfunction and death. These early aggregates will be referred to as oligomers in this chapter, with the understanding that the term oligomer might refer to species containing as few as two or as many as thousands of molecules. More precise numbers may come from future studies, but the important distinction is that these oligomers exist in solution (at least for a time), as opposed to fibrils which precipitate out of solution.

Amyloid Proteins. The Beta Sheet Conformation and Disease. J. D. Sipe

ISBN: 3-527-31072-X

Table 14.1 Diseases of protein misfolding: amyloidoses

Disease	Protein	Abbreviation
Alzheimer's disease Down's Syndrome (trisomy 21) Heredity cerebral angiopathy (Dutch)	amyloid β precursor protein (Aβ1–42)	AβPP (Aβ1–42)
Kuru Gerstmann-Sträussler-Scheinker syndrome (GSS) Creutzfeldt-Jacob disease Scrapie (sheep) Bovine spongiform encephalopathy ("mad cow")	prion protein	PrP^{C}/PrP^{Sc}
Type II diabetes mellitus (adult onset)	islet amyloid polypeptide (amylin)	IAPP
Dialysis-associated amyloidosis	β_2-microglobulin	β_2M
Senile cardiac amyloidosis	atrial natriuretic protein	ANP
Familial amyloid polyneuropathy	transthyretin	TTR
Reactive amyloidosis (familial Mediterranean fever)	serum amyloid A	SAA
Familial amyloid polyneuropathy (Finnish)	gelsolin	AGel
Macroglobulinemia	λ1 heavy chain	AH
Primary systemic amyloidoses	Igλ, Igκ	AL
Familial Polyneuropathy – Iowa (Irish)	apolipoprotein AI	ApoAI
Hereditary cerebral myopathy – Iceland	cystatin C	ACys
Non-neuropathic hereditary amyloid with renal disease	fibrinogen α	AFibA
Non-neuropathic hereditary amyloid with renal disease	lysozyme	ALys
Familial British dementia	FBDP	ABri

A number of common, significant, costly, and devastating diseases are amyloidoses. These include Alzheimer's disease, Type 2 diabetes mellitus and the spongiform encephalopathies [prion diseases, e.g. scrapie, "mad cow" disease, new variant Creutzfeldt-Jakob disease (Table 14.1)]. The aging of the population in the developed world has caused an epidemic of these age-related diseases to begin, making the search for an understanding of disease pathogenesis more critical than ever. Although a great deal of research has strongly implicated amyloid proteins in the pathogenesis of these illnesses, the molecular mechanisms by which these misfolded proteins cause dysfunction and/or toxicity remains elusive (for review, see Merlini and Bellotti, 2003). In this chapter, I will first review the evidence that oligomers (as opposed to fibrils) of amyloid peptides are the pathogenic species. I will go on to review the current evidence as to how

Table 14.2 Diseases of protein misfolding: non-amyloidoses

Disease	Protein	Abbreviation
Diffuse Lewy body disease	α-synuclein	AS
Parkinson's disease		
Fronto-temporal dementia	tau	tau
Amyotrophic lateral sclerosis	superoxide dismutase-1	SoD-1
Triplet-repeat diseases:	polyglutamine tracts in the following proteins:	PG
Huntington's	huntingtin	
spinal and bulbar muscular atrophy	androgen receptor	
spinocerebellar ataxias	ataxins	
spinocerebellar ataxia 17	TATA box-binding protein	

these oligomers cause cellular dysfunction and/or toxicity. It has been also reported that protein misfolding and aggregation is linked to the pathophysiology of a number of other, non-amyloid diseases (Table 14.2). Furthermore, it has been reported that non-disease-related proteins can form amyloid oligomers with toxicity. Here, the links amongst these various disease-associated phenomena will be considered.

14.2 Aggregation

The early findings that amyloid burden in patients did not correlate well with clinical severity of illness was troubling to investigators who were thinking along the lines that amyloid fibrils and deposits play an etiologic role in amyloidoses. Subsequently, the discovery of familial forms of amyloidoses linked to mutations in amyloid proteins gave new impetus to the notion that these proteins "caused" the illness (e.g. Pepys et al., 1993; Samaia et al., 1997). Further research, initially with the Aβ1–42 peptide of Alzheimer's disease, showed that amyloid peptides could demonstrate cytotoxicity in isolated cell systems and that these peptides were toxic to the cell types damaged in diseases (Yankner et al., 1990). Thus, Aβ and PrP106–126 (Forloni et al., 1994) were shown to be toxic to neurons, and IAPP was shown to be toxic to β cells from the islets of Langerhans in the pancreas (Lorenzo et al., 1994). Further experiments showed that amyloid fibrils themselves usually lacked toxicity. Monomers, also, seemed to lack cytotoxicity. In early experiments, however, cytotoxicty was poorly reproducible, especially for the Aβ peptides. This was eventually shown to be due to differences in the aggregation status of Aβ under varying experimental conditions (Pike et al., 1993). The aggregation of amyloid fibril-forming peptides was found to be a highly complex and variable process (Harper and Lansbury, 1997). The initial phases of protein aggregation from monomers into small oligomers progressed quite slowly. However, once a critical mass was reached, aggregation proceeded quite rapidly. Thus, growth of fibrils could be "nucleated" by the addition of pre-formed "seeds" to a solution of peptide. The seeds greatly accelerated the aggregation of the amyloid protein. However, it was soon discovered that, while promotion of amyloid peptide aggregation (by acidic pH, high concentration, "aging" or seeding) was *necessary* to achieve toxicity, extensive aggregation could actually lead to a decline in cytotoxicty (Hirakura et al., 1998). It appeared that an intermediate state of aggregation was associated with the cytotoxicity of amyloid fibril-forming peptides.

The notion of amyloid peptide-induced change also broadened. While cytotoxicity was clearly relevant to diseases in which cells died, animal models of Alzheimer's disease created with mice transgenic for the Aβ precursor protein (AβPP) or Aβ developed amyloid deposits, learning and memory deficits, and synaptic loss, but no frank neuronal loss. Although some criticized the models as inadequate, or attributed the lack of nerve loss to differences between mice and men, others began to think that, perhaps in the early stages of Alzheimer's disease, there were characteristic dysfunctions that precede neuronal loss. Developmental investigations of transgenic mouse models showed that memory deficits actually *preceded* amyloid deposits and synaptic loss. Further studies showed that Aβ peptides could inhibit long-term potentiation, a model for memory (Chen et al., 2000). Additionally, only oligomeric forms of Aβ could inhibit long-term potentiation, not monomeric or fibrillar Aβ (Walsh et al., 2002). These observations led to the implication of oligomers in cellular dysfunction relevant to clinical symptoms and to cytotoxicity.

In vitro, it was shown that fibroblasts could be killed by Aβ that was "freshly prepared" (and thus less highly aggregated) and that appeared to be "globular" by atomic force microscopy. Toxicity was blocked by antibodies or the channel-blocking zinc ion, but not by antioxidants (Zhu et al., 2002).

A similar story emerged in Huntington's disease, where early deficits in learning and memory could be linked to the presence of aggregates of the disease related protein huntingtin with a long PG tract. These deficits appear well before any frank neuronal loss in the brains of affected transgenic mice. Electrophysiologic abnormalities accompanied these deficits and could be detected even prior to the appearance of huntingtin aggregates, suggesting that the large, visible aggregates may not be responsible for the early cellular effects of huntingtin with extended PG tracts (Cepeda et al., 2003). It has also been shown that PG aggregates targeted to the nucleus can be directly cytotoxic (Yang et al., 2002).

Oligomers have also been implicated in the cytotoxicity of AS. Mutations in AS which promote "protofibril" (oligomers) formation also enhance cytotoxicity (Lashuel et al., 2002). Similarly, the "arctic" mutation of Aβ, which enhances oligomer formation, shows increased cytotoxicity in comparison to native Aβ peptide (Lashuel et al., 2003). Toxic oligomers of IAPP were shown by light scattering to contain between 25 and 6000 IAPP molecules (Janson et al., 1999). Once again, fibrils and monomers were shown to lack toxicity.

An antibody preparation to the "soluble oligomer" conformation was generated, which recognized several different amyloid peptides, but only in this precise conformation (Kayed et al., 2003). This antibody could block cytotoxicity of a range of amyloid peptides, strongly suggesting that oligomers were the toxic species and that the toxic mechanism (or at least a part of it) was common to all the amyloids.

Biological dogma has maintained that the three-dimensional structure and function of proteins is dictated by their primary amino acid sequence. Our knowledge of amyloid diseases has led to the recognition that natively folded proteins may unfold at times and adopt non-native conformations. Proteins may refold to their native conformation (with or without the assistance of chaperones) or they may be targeted for degradation via the ubiquitin–proteasome pathway. Alternately, some partially unfolded or misfolded proteins may aggregate. This aggregation may be driven thermodynamically by new hydrogen bonding possibilities (Fernandez and Berg, 2003) or by the hydrophobic effect, the need to shield hydrophobic parts of proteins from the aqueous environment. Proteins that form amyloid appear to have a propensity to misfold into β-sheet structures. These β-sheets tend to self-aggregate, forming *inter*molecular bonds that aggregate the proteins or peptides. Proteins that possess hydrogen-bonding defects are more likely to interact with lipid bilayers (Fernandez and Berg, 2003).

It is still not well understood why these complexes are not handled by the cell's usual defense mechanisms, i.e. refolding to native structures through chaperone assistance or targeting for degradation through ubiquitin–proteasomes. It may be that these processes are at work, but are merely overwhelmed by the amount of misfolded protein they must handle. This might be the place where disease begins. Alternatively, it is possible that the nature of these β-sheet aggregates leads them to attack or insert into cellular membranes. This would render them inaccessible to chaperones and to the ubiquitin–proteasome pathway. Another possibility is that, once seeding or nucleation has occurred, the fibrillation process proceeds so rapidly that the aggregates become too large to be handled by the cell's defenses. It has been suggested that amyloid fibril formation, with or without vacuolization, is actually a cellular defense mechanism to isolate the misfolded protein and keep it separate from the cell's vital machinery. Support for this idea comes from experiments showing that amyloid fibrils have little, if any, toxicity to cells, while smaller aggregates appear to possess considerable cytotoxicity.

The initial misfolding of a protein may be driven by mutations, metal interactions, environmental conditions such as temperature, acidic pH, oxidation and proteolysis or a simple increase in concentration. All these phenomena can potentially increase the presence of misfolded conformations and increase the probability of aggregation. The presence of membranes may also play a key role in this process, as lipid bilayers have been shown to influence both the conformation of amyloid peptides and the propensity of amyloid peptides to aggregate (McLaurin et al., 1998). Thus, the intracellular component in which an amyloid protein finds itself may significantly affect its ultimate conformational fate.

The many amyloid peptides exhibit no primary sequence homology, but they all possess the possibility of converting to relatively high β-sheet content. Sometimes this β-sheet content is unpredictable, as in the prion protein where sequences which would usually be predicted to be α-helical adopt a β-sheet conformation when synthesized (De Gioia et al., 1994).

While the earliest stages of amyloid peptide aggregation cannot be easily observed with current imaging techniques, the results of recent electron and atomic force microscopic studies have shown the presence of spherical or globular aggregates early in the process of amyloid fibril formation. The smallest particle appears to aggregate into chains or annular rings referred to as "protofibrils". These structures appear to be the precursors of fibrils that exist in mature amyloid deposits. The protofibrillar structures possess significant toxicity in comparison to fibrils themselves. It is unclear whether the annular structures or rings are more toxic than their linear brethren, but it is of interest that pathogenic mutations of Aβ amyloid and AS lead to increased formation of these annular structures (Lashuel et al., 2002).

14.3
Cellular Mechanisms of Oligomeric Toxicity

Recently, it has been reported that proteins and peptides associated with amyloid disease are not unique in their ability to form amyloid. Aggregation and amyloid fibril formation have been achieved for a number of non-disease associated proteins (Table 14.3) (Fandrich et al., 2003). It has been proposed that under appropriate conditions all proteins might be capable of forming amyloid fibrils. Beyond the structural similarity of these non-disease-associated amyloid proteins to actual amyloid proteins, there are functional similarities as well. For example, HypF is able to aggregate, and the aggregates can kill cells and permeabilize liposomes (Relini et al., 2001). Fibrils lack these properties, but aggregation of HypF is required for toxicity and permeabilization, thus implicating smaller oligomers as the cytotoxic, membrane-permeabilizing species. Preceding

Table 14.3 Non-disease related amyloid-forming proteins/peptides

SH3 domain p 85	fibronectin type III
Phosphatidylinositol-3-kinase	phosphoglycerate kinase acylphosphatase
HypF N-terminal domain (*E. coli*)	amphoterin (human)
Apomyoglobin (equine)	apocytochrome *c*
Endostatin (human)	Met aminopeptidase
Stefin B (human)	ADA2H
Fibroblast growth factor (*Notophthalmus viridescens*)	apolipoprotein CII
VI domain (murine)	B1 domain of IgG-binding protein
Curlin CgsA subunit	monellin

cytotoxicity, HypF oligomers, but not fibrils, raised intercellular Ca^{2+} levels and levels of reactive oxygen species (Bucciantini et al., 2004). These intracellular effects were reversible, and they were prevented with the omission of Ca^{2+} from the media or the addition of reducing agents. These physiologic effects are strikingly similar to those observed with disease related amyloid peptides (Schubert et al., 1995; Kawahara et al., 2000). These results are also consistent with the findings that antibodies raised against soluble pre-fibrillar oligomers of various amyloid peptides recognize that state in other amyloid peptides and can prevent cytotoxicity (Kayed et al., 2003). This suggests that amyloid peptides share an oligomeric conformation critical to cytotoxicity and independent of amino acid sequence.

Thus, a wide variety of structural, biochemical, and physiologic studies suggest that pre-fibrillar amyloid peptide oligomers act to damage and/or kill cells, especially neurons, and that they appear to share a common mechanism of action. As we discuss below, a growing body of biophysical evidence implicates channel formation by amyloid peptides in cellular membranes as the molecular mechanism of this damage.

14.4 Loss of Function Hypothesis

This view postulates that cytotoxicity results from a loss of native protein function due to misfolding of native protein and aggregation. This idea seems untenable in light of the fact that so many different, structurally and functionally unrelated proteins are involved in amyloid diseases. Transgenic mouse models in which an amyloid protein-encoding gene is inserted and disease results also argue against this idea. An alternative form of this concept is that aggregated amyloid proteins might bind and inactivate essential cellular proteins, such as transcription factors. However, there is little evidence to suggest that physiologically significant amounts of critical cellular proteins can be found in amyloid deposits or inclusion bodies. Substantial evidence has now accumulated to show that amyloid proteins possess toxic activity toward cells. The molecular mechanism of this toxicity remains unknown.

14.5 Receptors for Advanced End-products of Glycation (RAGE) Receptors

It has been proposed that RAGE bind various amyloid oligomers, inducing cellular stress and activation of NF-κB (Bucciarelli et al., 2002). While some experimental evidence supports this idea, it has not been clearly documented that this mechanism accounts for tissue damage in amyloid diseases. Further investigation is warranted.

14.6 Oxidative Stress

Another hypothesis for the pathogenesis of amyloidosis and amyloid-like diseases is that oxidative stress is induced directly by amyloid peptides, leading to free radical production, mitochondrial dysfunction and death (Butterfield and Bush, 2004). Although many elements of oxidative stress pathology occur in amyloid induced cytotoxicity, it does not explain *how* oxidative stress and/or free radicals are generated by amyloid peptides. Initial reports that the peptide itself could generate free radicals have not been confirmed. Oxidative stress clearly occurs in cells affected with amyloid toxicity. It remains to be established where in the chain of toxicity oxidation plays its role.

14.7 The Channel Hypothesis

A large body of research has focused on the ability of amyloid peptides to interact with membranes. A series of provocative studies has shown that many (if not all) amyloid peptides can aggregate into β-sheet oligomers capable of spontaneously inserting into lipid membranes. The peptides form relatively permanent ion-permeable channel structures across the cell membrane. These channels are reported to have properties that would damage or kill most cells (Table 14.4). The channels are reported to be (a) large, (b) non-selective, (c) heterogeneous, (d) voltage independent, (e) irreversible, (f) inhibited by agents that prevented aggregation such as Congo red and (g) blocked by zinc ions. These channels would likely cause a leakage pathway in plasma, mitochondrial, endoplasmic reticulum, lysosomal or other membranes. These leaks could damage cells by (1) disrupting membrane potentials and ion gradients, (2) causing loss of vital intracellular ions such as K^+ and Mg^{2+}, (3) allowing influx of toxic ions such as Ca^{2+}, (4) running down energy stores by forcing ion pumps to work harder, (5) disrupting mitochondrial membrane potential and initiating apoptosis by allowing cytochrome *c* to leak out of mitochondria, and (6) allowing toxic enzymes and other factors to leak out of lysosomes and peroxisomes. In the sections that follow, the evidence for channel formation by the amyloid peptides which have been most studied by these techniques is reviewed.

14.8 Aβ

The "channel hypothesis" was first proposed by Arispe et al. (1993 a) after they reported that Aβ1–40 could form cation-selective, calcium permeable channels of various conductances in planar lipid bilayer membranes (BLMs). The channels were formed by Aβ1–42 as well and were large, voltage independent and

Table 14.4 Channel properties of amyloid peptides

Peptide	Ring diameter (inner/outer, in Å, by microscopy)	Single-channel conductance (pS)	Ion selectivity (permeability ratio)	Blockade by zinc	Inhibition by Congo red	Reference
Aβ25–35		10–400	cation ($P_K/P_{Cl}=1.6$)	+	+	Mirzabekov et al. (1994); Lin and Kagan (2002)
Aβ1–40		10–2000	cation ($P_K/P_{Cl}=1.8$)	+		Hirakura et al. (1999)
Aβ1–40		50–4000	cation ($P_K/P_{Cl}=11.1$)	+		Arispe et al. (1993a,b, 1996)
Aβ1–40 ARC (E22G)	15–20/70–100					Lashuel et al. (2003)
Aβ1–42		10–2000	cation ($P_K/P_{Cl}=1.8$)	+	+	Hirakura et al. (1999)
CT105 (C-terminal fragment of amyloid precursor protein (APP)		120	cation	+	+	Kim et al. (1999)
Islet amyloid polypeptide (Amylin)		7.5	cation ($P_K/P_{Cl}=1.9$)	+	+	Mirzabekov et al. (1996)
PrP106–126		10–400	cation ($P_K/P_{Cl}=2.5$)	+	+	Lin et al. (1997)
PrP106–126		140, 900, 1444	cation ($P_K/P_{Cl} > 10$)			Kourie and Culverson (2000)
PrP82–146			cation (variable)			Bahadi et al. (2003)
SAA		10–1000	cation ($P_K/P_{Cl}=2.9$)	+	+	Hirakura et al. (2002)
SAA 2.2 (murine hexamer)	25/80					Wang et al. (2002)
C-type natriuretic peptide		21, 63	cation ($P_K/P_{Cl} > 10$)	+	+	Kourie (1989)
Atrial natriuretic factor		68, 160, 273	variable			Kourie et al. (2001)
$Beta_2$-microglobulin		0.5–120	non-selective	+	+	Hirakura and Kagan (2001)
Transthyretin		variable	cation (variable)	+	+	Hirakura et al. (2001)
Polyglutamine (average molecular weight = 6000)		19–220	non-selective	–	–	Hirakura et al. (2000)

Table 14.4 (continued)

Peptide	Ring diameter (inner/outer, in Å, by microscopy)	Single-channel conductance (pS)	Ion selectivity (permeability ratio)	Blockade by zinc	Inhibition by Congo red	Reference
Polyglutamine 40		17	cation			Monoi et al. (2000)
NAC (AS65–95)		10–300	variable	+	+	Azimova et al. (2003)
AS A30P/A53T	20–25/80–120					Lashuel et al. (2002)
CT (human/salmon)		12/580	non-selective			Stipani et al. (2001)
Lysozyme 87–114	permeable to *β*-galactosidase (molecular weight 116 kD)		non-selective			Ibrahim et al. (2001)
Cu/Zn SoD	50/190					Chung et al. (2003)

blocked by tromethamine ($Tris^+$) and aluminum (Arispe, 1993b). The largest channels observed (4 nS) could potentially change the interior [Na] of a cell by as much as 10 μM/s. They proposed that ionic leaking of Na^+, K^+ and Ca^{2+} could disrupt membrane potential and ionic regulation within a few seconds.

While these findings were not immediately confirmed by other laboratories, due to problems with irregular aggregation of A*β* (e.g. Mirzabekov et al., 1994), eventually a series of studies found A*β* peptides to be capable of forming channels in BLMs (Hirakura et al., 1999; Kourie et al., 2001), liposomes (Lin et al., 1999), neurons and oocytes (Fraser et al., 1997), and fibroblasts (Zhu et al., 2000). The state of A*β* aggregation is critical not only to its cytotoxicity (Pike et al., 1993; Hirakura et al., 1998), but also to its channel-forming abilities. Indeed, monomers and fibrils of A*β* that are non-toxic fail to form ion channels (Hirakura et al., 1999). Oligomeric species of A*β* cause a variety of channel entities, which can be distinguished by their single-channel conductance, ionic selectivity, kinetics and other channel properties (Kourie et al., 2002). It was also found that conditions (aging, acidic pH, etc.) that favor the aggregation of monomers into oligomers led to an increase in channel activity (Hirakura et al., 1999). Exposure to organic solvents, which promote monomeric A*β*, led to loss of channel activity. However, channel activity could be recovered by allowing the peptide to "age" in aqueous solution.

Although A*β*1–40 and 1–42 are the primary forms of A*β* peptides found *in vivo,* other A*β* fragments have been of experimental interest. A*β*25–35, a cytotoxic peptide not found *in vivo,* is a voltage-dependent, non-selective channel former (Mirzabekov et al., 1994). Studies using variants of A*β*25–35 showed that channel formation was *necessary* for cytotoxicity, but not sufficient, i.e. all cytotoxic species formed channels, but there were two channel forming variants of Aβ25–35 that did not kill cells (Lin, 1996). Channel activity could be enhanced by lipids carrying a net negative surface charge and this effect could be countered by high salt concentrations. Addition of cholesterol, which stiffens membranes, decreased Aβ25–35 channel activity (Lin and Kagan, 2002). Aβ25–35 variants could not form channels if they were not at least 10 residues long, indicating a minimum bilayer spanning length of about 30 Å, a result consistent with the β-sheet span lengths of the known channels generated by Staphylococcal α toxin and anthrax toxin (Song et al., 1996; Petosa et al., 1997). More recently, an extremely short channel-forming Aβ variant (31–35) has been reported (Qi and Qiao, 2001). Whether this peptide might form hemi-channels to span the bilayer similar to the peptide Gramicidin is unknown. *In vivo*, Aβ1–40 or 1–42 can induce currents in rat cortical neurons (Weiss et al., 1994; Furukawa et al., 1994), HNT cells (Sanderson et al., 1997) and gonadotrophin-releasing hormone secreting neurons (Kawahara et al., 1997). The channels observed *in vivo* seem indistinguishable in their properties from those observed *in vitro*.

Aβ1–40 or 1–42 can also kill fibroblasts in a manner inhibited by antibodies, tromethamine or zinc, but not by antioxidants, suggesting that channel formation is the mechanism of cytotoxicity. The freshly prepared Aβ used in these studies appeared "globular:" consistent with an early stage of aggregation (Bhatia et al., 2000; Zhu et al., 2000). It has also been shown that the cholesterol content of plasma membranes affects a cell's vulnerability to Aβ1–40 and 1–42 (Arispe and Doh, 2002), suggesting that the membrane plays a critical role in Aβ cytotoxicity. Also, it has been reported that Aβ can directly induce cytochrome *c* with release from mitochondria. This could occur through the action of Aβ to decrease mitochondrial membrane potential or even through Aβ channel mediated release of cytochrome *c* (Kim et al., 2002).

14.9 PrP106–126

The prion protein (PrP) has at least two distinct tertiary conformations, PrP^C and PrP^{Sc}, the latter of which results in a transmissible neurodegenerative disease known as a spongiform encephalopathy. Prion diseases include scrapie in sheep and "mad cow" disease as well as Creutzfeldt-Jakob disease, Gerstmann-Sträussler-Scheinker syndrome and fatal familial insomnia in humans. These illnesses may be sporadic, infectious or hereditary. The familial versions are associated with mutations in the prion protein (for review, see DeArmond and Prusiner, 2003). PrP^{Sc} deposits in the brains of afflicted organisms in a form that is readily converted in amyloid fibrils *in vitro*. A critical step in the conformational transition from PrP^C to PrP^{Sc} is the conversion of α-helical and random coil regions of PrP to β-sheets (Pan et al., 1993). One region predicted to be α-helical, PrP106–126, actually forms β-sheets when chemically synthesized and self-aggregates into amyloid fibrils (Gasset et al., 1992). The β-sheet-rich form of PrP106–126 binds to membranes, unfolds and ultimately disrupts the bilayer (Kazlauskaite et al., 2003). Forloni et al. (1993) demonstrated that PrP106–126 was toxic to neurons in culture. Lin et al. (1997) reported that PrP106–126 could form ion-permeable channels in planar lipid bilayer membranes at neurotoxic concentrations. PrP106–126 channels were irreversibly associated with the membranes, demonstrated a multiplicity of single channel conductances (10–400 pS in 0.1 M NaCl) and had relatively long lifetimes (seconds to minutes). Ionic selectivity of the channels was meager, with significant permeability being shown to Na^{+}, K^{+}, Cl^{-} and Ca^{+} ($P_{Na}/P_{Cl}=2.5$). Channel activity could be enhanced dramatically by "aging" of the peptide in aqueous solution, a procedure which promotes aggregation and increases neurotoxicity. Incubation of PrP106–126 at acidic pH also enhanced channel activity by nearly 100 times and shifted the distribution of observed single-channel conductances to higher conductance levels. It has also been reported that acidic pH converts α-helical PrP106–126 to the β-sheet conformation (De Gioia et al., 1994). Kourie and Culverson (2000) characterized three distinct channel types formed by PrP106–126. These included: (1) a dithiodipyin sensitive channel of 40 pS with slow kinetic behavior, (2) a giant channel, 900–1500 pS, exhibiting five separate subconductance states and (3) a tetraethyl-ammonium chloride (TEA)-sensitive channel of 140 pS with rapid kinetics. Manunta et al. (2000) were unable to observe PrP106–126 neurotoxicity or channel formation, but this may have been a result of the highly variable aggregation state of the PrP106–126 peptide, reminiscent of the variability seen with aggregation of Aβ peptides.

Bahadi et al. (2003) have reported that PrP82–146, a peptide found in the PrP^{Sc} brains of patients with Gerstmann-Sträussler-Scheinker syndrome, can also form ion channels. Scrambling the amino acid sequence of the 106–126 region of this longer peptide abolishes the ability to form ion channels, whereas scrambling the 127–146 region has no effect, thus implicating the 106–126 region as key in channel-forming ability. The electrophysiologic properties of PrP82–145 are very similar to those of PrP106–126. Channel activity could be decreased by the antibiotic rifampicin which had previously been shown to decrease aggregation and toxicity of Aβ peptides. The association of amyloid deposits with prion diseases is variable. Intriguingly, in one prion disease where amyloid fibrils are not found, the mutant prion protein adapts a transmembrane conformation (Hegde et al., 1998). It is tempting to speculate that this transmembrane protein may be causing ionic leakage across the cell membrane.

Lysosomotropic agents have been reported to inhibit PrP^{Sc} accumulation in neuroblastoma cells (Doh-Ura et al., 2000). One of these agents, quinacrine, has been reported to block PrP106–126 channels (Farrelly et al., 2003). Quinacrine is also able to repair the impaired functioning of N-type calcium channels in prion-infected neurons (Sandberg et al., 2004). Thus, it seems likely that channel blockers, such as quinacrine, may be useful as potential therapeutic agents in prion related diseases. Indeed, there is at least one report of quinacrine improving the clinical status of four patients with Creutzfeldt-Jakob disease (Nakajima et al., 2004). Other acridine derivatives and tricyclic compounds may have even better anti-prion efficiency (Korth et al., 2001). Congo red can inhibit channel formation, block PrP106–126 cytotoxicity and inhibit the development of scrapie (Ingrosso et al., 1995; Hirakura et al., 2000a). It remains to be seen whether the anti-channel blocking or anti-aggregation activities of quinacrine are key to its anti-prion effects.

It has also been reported that a peptide PrP170–175 bearing a prion protein mutation related to schizoaffective disorder (N171S; Samaia et al., 1997) increases the permeability of planar lipid bilayers and forms channels with conductance of 8–26 pS in 0.5 M KCl. The native PrP170–175 does not form channels in membranes. This result suggests that yet another segment of PrP may be capable of influencing toxicity via channel formation and that this may be directly relevant to human disease.

14.10 IAPP

IAPP (amylin) is a 37-residue amyloidogenic hormone which is co-secreted with insulin from β cells in the islets of Langerhans in the pancreas. Amyloid deposits comprising IAPP are found in the islets of patients with Type 2 diabetes, and are positively correlated with β cell loss and clinical insulin requirements (Westermark and Wilander, 1978; Butler et al., 2003). IAPP is cytotoxic to β cells in culture (Lorenzo et al., 1994).

Although IAPP is α-helical in aqueous solution, exposure to lipid membranes induced a transition to the β-sheet structure (McLean and Balasubramaniam, 1992). Human IAPP formed ion-permeable channels in planar lipid membranes at cytotoxic concentrations (Mirzabekov et al., 1996). Rat IAPP, which differs from human IAPP at five amino acid positions, and is non-amyloidogenic and non-toxic, did not form channels. Human IAPP channels could be inserted into membranes irrespective of voltage; however, once inserted, channels rapidly opened at negative voltages and rapidly inactivated at positive voltages (voltages being relative to the IAPP-containing side). Inactivation faded gradually over a time course of several minutes. Open IAPP channels were ohmic and exhibited a single-channel conductance of 7.5 pS in 0.1 M KCl. Channels were permanently associated with the membrane and showed lifetimes of seconds to minutes depending on voltages. Increasing concentrations of net negatively charged lipids in the membrane led to an increase in IAPP channel activity. Increasing salt concentrations in the aqueous solution decreased channel activity. Anguiano et al. (2002) showed that liposomes could be permeabilized by IAPP in a graded fashion, allowing Ca^{2+} to cross the membrane, while not allowing fura-2 (molecular weight=832) or FITC–dextran (molecular weight=4400) to escape. IAPP has also been reported to disrupt Ca^{2+} homeostasis in cells in a manner similar to Aβ and prion-related peptides (Kawahara et al., 2000).

Large fibrils of IAPP have been reported to be non-toxic, whereas smaller aggregates are associated with cytotoxicity (Janson et al., 1999). The aggregates, but not fibrils, could disrupt planar lipid bilayers. Light scattering showed these oligomers to range in size from 25 to 6000 IAPP molecules. Hirakura et al. (2000a) showed that Congo red incubation with IAPP, Aβ or PrP106–126, prior to membrane exposure, could inhibit channel formation. They also reported that Zn^{2+} could reversibly block these channels. The concurrence of channel-forming properties, physiologic effects and cytotoxicity strongly suggests a common mechanism for the channel-forming action for these three amyloid peptides.

14.11 ANP

A family of hormones, C-type natriuretic peptide (CNP), ANP and brain-derived natriuretic peptide (BNP), helps to regulate fluid and ionic balance. As people age, increasingly large amyloid deposits of ANP are found in their hearts. These fibril-containing deposits are thought to play a deleterious role, perhaps leading to atrial fibrillation and other cardiac pathology (McCarthy and Kasper, 1998). Channels in lipid bilayers have been reported for ANP1–28 (Kourie et al., 2001), CNP-22 and OaC-type natriuretic peptide(18–39) from platypus (Kourie, 1999). ANP channels exhibited a multiplicity of single channel conductances, but all were cation selective. The channels could be divided into three types: (1) Ba^{2+}-sensitive, fast kinetics with three modes (spike, burst and open), 68 pS, (2) a large conductance channel possessing subconductance states, time dependent inactivation, 273 pS, and (3) transient activation, 160 pS. CNP channels were weakly cation selective, with a high open probability and large single-channel conductance (546 pS). The physiologic properties of these peptides were believed to be concordant with known pathological effects described in animal models and human disease. Although the peptides act through a well-known receptor and second messenger signaling system, it has been proposed that the channel-forming activity of the peptides could play a role in physiologic events. ANP channels would likely hyperpolarize muscle cell membrane and inhibit depolarization driven contractions. Large conductance channels were postulated to degrade membrane potential and ionic balance.

14.12 SAA

SAA refers to a family of related apolipoproteins. During states of infection or inflammation the acute-phase isoforms of SAA can increase their levels in serum by as much as 3 orders of magnitude. AA fibrils most commonly comprising the N-terminal 76 residues of SAA are found as amyloid deposits in various organs such as spleen, kidney and liver. Patients with chronic infections, such as tuberculosis, or inflammatory diseases, such as rheumatoid arthritis, are particularly at risk. Patients with cancer, arteriosclerosis and Alzheimer's disease have also been reported to have elevated SAA concentrations (Sipe, 2000).

One commercially available, recombinant-generated acute-phase isoform, SAAp, has been reported to form ion-permeable channels in planar lipid bilayer membranes at physiologically relevant concentrations (Hirakura et al., 2002). A wide variety of single-channel conductances (10–1000 pS) were observed, consistent with a peptide aggregated into multiple oligomeric states. SAA channels were permeable to most physiologic ions including Na^+, K^+, Ca^{2+} and Cl^-, exhibiting only a weak preference for cations over anions. Channel formation could be inhibited by pre-incubation of SAA with Congo red, but addition of Congo red after channel formation had no effect. Channels could be reversibly blocked by 100 μM Zn^{2+}.

The naturally occurring acute phase isoform SAA1 was reported to lyse bacterial cells when expressed in *Escherichia coli*, whereas expression of the constitutive isoform SAA4 did not. SAA1 and SAA4 differ in their sequences at approximately 50% of residues. SAA1 has a greater concentration of hydrophobic residues in the N-terminal region. The resemblance of these results to the properties of channel forming toxins such as colicins (Schein et al., 1978), yeast killer toxins (Kagan, 1983), defensins (Kagan et al., 1990) and protegrins (Sokolov et al., 1999) led to the proposal that SAA might play a role in host defense against microbes. Electron microscopic analysis revealed that murine SAA 2.2 can exist as an annular hexamer with a central "pore-like" region (Wang et al., 2002). Although membranes were not present, the observed pore diameter of 25 Å was consistent with the physiologic findings of Hirakura et al. (2002).

14.13 AS

Parkinson's disease is a progressive neurodegenerative disorder characterized by tremor, rigidity and brachykinesia. The hallmark lesion of Parkinson's disease is the Lewy body, an inclusion body in dopaminergic neurons consisting largely of non-amyloid component (NAC; residues 66–95 of AS). Incorrectly named, NAC is actually a fibril-forming amyloid peptide. Mutations in AS are associated with familial Parkinson's disease and implicate AS in the pathophysiology of the illnessβ (Baptista et al., 2004). NAC is also found in Alzheimer's disease amyloid deposits, which suggests a link between the two amyloid diseases. This notion is supported by the clinical overlap in these illnesses. Alzheimer's disease patients are commonly found to have motor abnormalities. Parkinson's disease patients frequently have cognitive problems, such as dementia and depression. Intermediate syndromes such as dementia with Lewy bodies also implicate AS in damage to neurons outside the dopaminergic system (McKeith et al., 2004).

It has been reported that AS can increase the permeability of liposomes in a graded manner to substances of increasing size. This "sieving" action is characteristic of "pore-like" transport systems (Volles and Lansbury, 2002). Pathogenic mutations in AS such as A30P and A53T have been reported to accelerate the formation of oligomers (protofibrils) capable of permeabilizing activity (Volles and Lansbury, 2002). Electron microscopy revealed that AS formed annular, pore-like oligomers and that the Parkinson's disease-related mutations enhanced the formation of these structures (Lashuel et al., 2002 a). The pathogenic "arctic" mutation of Aβ also showed a similar enhancing effect on these annular structures (Lashuel et al., 2002 b). Electrophysiologic studies of NAC have confirmed the formation of ion-permeable channels in lipid bilayers (Azimova et al., 2003). These channels have properties strikingly similar to those of other amyloidogenic peptides. Single-channel conductances are heterogeneous. Ionic selectivity is weak. Channels are irreversible and have extended lifetimes. Channel formation is inhibited by Congo red and channels are blocked by Zn^{2+}.

14.14 β_2M

β_2M forms amyloid deposits in bones and joints of patients on hemodialysis or peritoneal dialysis, a syndrome referred to as "dialysis-associated amyloidosis". This 99-residue peptide belongs to the MHC class I complex which is involved in the presentation of foreign antigens to lymphocytes. β_2M levels can rise 100-fold during states of renal failure (Drüeke, 1998). Renal transplantation can result in lower β_2M levels and clinical symptoms. β_2M's physiologic effects include the induction of Ca^{2+}, efflux from calvariae, bone resorption and increasing collagenase production (Brinckerhoff et al., 1989; Moe and Sprague, 1992; Peterson and Kang, 1994). β_2M is mainly found in amyloid deposits as full-length native protein. In contrast to other amyloid protein "misfolding" diseases, the misfolding here appears to be solely a function of increased protein concentration in the serum, rather than mutation or proteolysis.

Channel formation by β_2M was reported by Hirakura and Kagan (2001). A multiplicity of single channel conductances were observed, ranging from 0.5 to 120 pS, with 90 pS being the most commonly observed size in 0.1 M KCl. Channel lifetimes were typically extended and ionic selectivity was poor. β_2M associated irreversibly with the membrane. Incubation of β_2M with Congo red inhibited formation of channels. Zn^{2+} could reversibly block inserted channels. Open β_2M channels exhibited a slight degree of rectification with *trans* positive currents being somewhat larger than *trans* negative currents. Channel formation could be accelerated by acidic pH, compatible with the idea that the acidotic/uremic state of renal failure could enhance the generation of pathologic oligomers of β_2M.

14.15
AL Amyloidosis

AL (light chain) amyloidosis is characterized by fibrillar deposits of the variable domain of immunoglobulin light chains. Fibril assembly is dependent on environmental conditions, e.g. the process may be different on surfaces versus in solution (Zhu et al., 2000). AL deposits are frequently found on surfaces such as arterial walls and basement membranes. A recent study of AL aggregation using atomic force microscopy found that AL protein of the variable domain SMA could form annular aggregates similar to those seen with AS (see Section 14.13). The SMA annular aggregates were significantly larger, however. Acidic pH was critical to the formation of these aggregates. It was suggested that these annular species might form pores in membranes (Zhu et al., 2004).

14.16
PG

Although triplet-repeat diseases are not classic "amyloidoses", their pathology seems to involve protein misfolding and the accumulation of toxic protein aggregates. Huntington's disease is the most common and best known of these hereditary illnesses, which are caused by an expansion of the codon CAG which codes for glutamine. In Huntington's disease, PG tracts larger than 37 residues cause disease, although this number varies amongst the different triplet-repeat illnesses. Several different proteins are associated with the various triplet-repeat diseases, but all are affected by an extended PG tract with a minimum threshold length for causing clinical symptoms. In Huntington's disease, the best-studied triplet-repeat disease, the presence of amyloid-like neuronal aggregates of huntingtin, the PG expanded protein, correlates with disease progression in transgenic mice (for review, see Li and Li, 2004). The toxicity of huntingtin is proportional to the PG tract repeat length. Age of onset of illness is inversely proportional to PG repeat length. However, a PG repeat length = 12 cell line shows vulnerability to apoptosis without *visible* aggregates of huntingtin (PG repeat length = 150). Thus, visible deposit or aggregates may not be necessary to cause dysfunction or cell death. Indeed, some transgenic mice show electrophysiologic abnormalities in striatum before aggregates are visible (Cepeda et al., 2003.)

Channel formation by PG was reported almost simultaneously by Hirakura et al. (2000 b) and Monoi et al. (2000). The former group reported channels that were long lived, non-selective and heterogeneous in single channel conductance size, ranging from 19 to 220 pS in 0.1 M KCl. Channel formation was increased by acidic pH. Unlike classic amyloid peptides, Congo red pre-incubation did not inhibit channel activity nor was Zn^{2+} able to block PG channels. These distinct properties indicate that PG aggregates are clearly different structurally from classic amyloids.

Monoi et al. (2000) reported that PG40 could form cation-selective, long-lived channels with a single-channel conductance of 17 pS. PG29 could not form channels, consistent with the 37-residue cut-off for clinical illness. The investigators also proposed a structural model, the μ-helix, for PG channels, a model which just spans the bilayer hydrophobic core at a length of 37 residues, again in agreement with the clinical data.

Further evidence of possible channel formation by PG in mitochondria was reported by Panov et al. (2002), who showed that, in Huntington's disease, mitochondria exhibited decreased membrane potential and were depolarized at lower Ca^{2+} levels than control mitochondria. Mitochondria from brain of transgenic mice expressing huntingtin with a pathogenic PG tract exhibited a similar dysfunction. Electron microscopy revealed mutant huntingtin localized to mitochondrial membranes. Most strikingly, the mitochondrial defects could be reproduced by a fusion protein with a long PG repeat. These results suggest that PG tracts are toxic to mitochondria and likely act via a channel-forming mechanism, depolarizing mitochondria and leaving them more vulnerable to apoptosis or metabolic insults. These data are consistent with a report that Aβ peptide can directly induce cytochrome *c* release in isolated mouse brain mitochondria, by directly inducing a permeability increase in the mitochondrial membrane (Rodrigues et al., 2000; Kim et al., 2002).

14.17 HypF

The recent studies of non-disease related amyloid proteins has led to a reexamination of the nature of amyloid structure with respect to disease. It has been reported that a wide variety of non-disease proteins can form amyloid fibrils under the "appropriate" conditions (Stefani and Dobson, 2003). It is a curiosity that the aggregation process of at least one such non-disease-related amyloid protein, HypF, can lead to formation of oligomeric structures and permeabilization of lipid membranes, without forming amyloid fibrils (Relini et al., 2004). These oligomers are cytotoxic as well. These data strongly suggest that it is the oligomer β-sheet conformation itself that leads to membrane insertion, channel formation, cellular dysfunction and, eventually, cytotoxicity. It has been proposed that these are latent properties of all polypeptide chains (see Chapter 3).

14.18 Calcitonin (CT)

Human calcitonin is (hCT) is a 32-residue peptide hormone involved in the regulation of calcium and phosphorous metabolism. It is produced by the C cells of the thyroid gland and is found in fibrillar amyloid deposits in patients with medullary carcinoma of the thyroid. The calcitonin peptide and fragments as small as four residues can form amyloid fibrils when incubated in aqueous solution. Calcitonin has pharmacological use in humans as a treatment for Paget's disease and osteoporosis. The tendency of hCT to form fibrils limits its utility and has favored the use of salmon calcitonin (sCT), which is less fibrillogenic. A peptide corresponding to residues 16–21 of hCT form antiparallel β-sheets in methanol water and peptides consisting of residues 15–19 are highly amyloidogenic (Reches et al., 2002). Fibril formation is accompanied by a conformational charge from α-helix to β-sheet in this central region or a change from random coil to β-sheet in the C-terminal region (Kawahara et al., 2000). CT from several species, including human, salmon, eel and porcine, can form ion-permeable channels in lipid bilayer membranes (Stipani et al., 2001). These channels were long lived, voltage dependent, non-selective between cations and anions, permeable to Ca^{2+}, and enhanced by the presence of negatively charged phospholipids. It is possible that the channel-forming activity of hCT could have other physiologic effects on signal transduction, particularly in light of its ability to allow Ca^{2+} across membranes. Further investigation is needed to investigate the potential physiological role of these channels.

14.19 Lysozyme

Lysozyme is an enzyme that exhibits antimicrobial activity by cleaving bonds in the outer wall of Gram-positive bacteria. Lysozyme also forms toxic amyloid deposits in human. Mutations or partial protein unfolding leads to aggregation, oligomer formation and fibrilization of lysozyme. Partial unfolding of lysozyme also leads to the development of a non-enzymatic, broad-spectrum antimicrobial activity and a membrane-permeabilizing activity (Ibrahim et al., 2001). These activities were localized to a helix–loop–helix peptide at the upper lip of the active site cleft (87–114 of hen lysozyme). Similar peptides from human and chicken lysozyme possessed these activities as well. These results suggest that the unfolding of lysozyme leads to fundamental shifts in protein function and activity. The aggregation of lysozyme monomers into oligomers appears to create a membrane-penetrating, antimicrobial complex which can aggregate into amyloid rings as seen by imaging (Malisauskas et al., 2003).

References

Anguiano M, Nowak RJ, Lansbury PT, Jr. Protofibrillar islet amyloid polypeptide permeabilizes synthetic vesicles by a pore-like mechanism that may be relevant to type II diabetes. *Biochemistry* **2002**, *41*, 11338–11343.

Arispe N, Doh M. Plasma membrane cholesterol controls the cytotoxicity of Alzheimer's disease AβP (1–40) and (1–42) peptides. *FASEB J* **2002**, *16*, 1526–1536.

Arispe N, Pollard HB, Rojas E. Giant multilevel cation channels formed by Alzheimer disease amyloid β-protein [AβP-(1–40)] in bilayer membranes. *Proc Natl Acad Sci USA* **1993a**, *90*, 10573–10577.

Arispe N, Rojas E, Pollard HB. Alzheimer disease amyloid *β* protein forms calcium channels in bilayer membranes: blockade by tromethamine and aluminum. *Proc Natl Acad Sci USA* **1993b**, *90*, 567–571.

Azimova, RK, Kagan BL. Ion channels formed by a fragment of *α*-synuclein (NAC) in lipid membranes. *Biophys J* **2003**, *84*, 53a.

Bahadi R, Farrelly PV, Kenna BL, Kourie JI, Tagliavini F, Forloni G, Salmona M. Channels formed with a mutant prion protein PrP(82–146) homologous to a 7-kDa fragment in diseased brain of GSS patients. *Am J Physiol Cell Physiol* **2003**, *285*, C862–C872.

Baptista MJ, Cookson MR, Miller DW. Parkin and *α*-synuclein: opponent actions in the pathogenesis of Parkinson's disease. *Neuroscientist* **2004**, *10*, 63–72.

Bhatia R, Lin H, Lal R. Fresh and globular amyloid *β* protein (1–42) induces rapid cellular degeneration: evidence for A*β*P channel-mediated cellular toxicity. *FASEB J* **2000**, *14*, 1233–1243.

Brinckerhoff CE, Mitchell TI, Karmilowicz MJ, Kluve-Beckerman B, Benson MD. Autocrine induction of collagenase by serum amyloid A-like and β_2-microglobulin-like proteins [see Comments]. *Science* **1989**, *243*, 655–657.

Bucciarelli LG, Wendt T, Rong L, Lalla E, Hofmann MA, Goova MT, Taguchi A, Yan SF, Yan SD, Stern DM, Schmidt AM. RAGE is a multiligand receptor of the immunoglobulin superfamily: implications for homeostasis and chronic disease. *Cell Mol Life Sci* **2002**, *59*, 1117–1128.

Bucciantini M, Calloni G, Chiti F, Formigli L, Nosi D, Dobson CM, Stefani M. Prefibrillar amyloid protein aggregates share common features of cytotoxicity. *J Biol Chem* **2004**, *279*, 31374–31382.

Butler AE, Janson J, Soeller WC, Butler PC. Increased *β*-cell apoptosis prevents adaptive increase in *β*-cell mass in mouse model of type 2 diabetes: evidence for role of islet amyloid formation rather than direct action of amyloid. *Diabetes* **2003**, *52*, 2304–2314.

Butterfield DA, Bush AI. Alzheimer's amyloid *β*-peptide (1–42): involvement of methionine residue 35 in the oxidative stress and neurotoxicity properties of this peptide. *Neurobiol Aging* **2004**, *25*, 563–568.

Cepeda C, Hurst RS, Calvert CR, Hernandez-Echeagaray E, Nguyen OK, Jocoy E, Christian LJ, Ariano MA, Levine MS. Transient and progressive electrophysiological alterations in the corticostriatal pathway in a mouse model of Huntington's disease. *J Neurosci* **2003**, *23*, 961–969.

Chen QS, Kagan BL, Hirakura Y, Xie CW. Impairment of hippocampal long-term potentiation by Alzheimer amyloid *β*-peptides. *J Neurosci Res* **2000**, *60*, 65–72.

DeArmond SJ, Prusiner SB. Perspectives on prion biology, prion disease pathogenesis, and pharmacologic approaches to treatment. *Clin Lab Med* **2003**, *23*, 1–41.

De Gioia L, Selvaggini C, Ghibaudi E, Diomede L, Bugiani O, Forloni G, Tagliavini F, Salmona M. Conformational polymorphism of the amyloidogenic and neurotoxic peptide homologous to residues 106–126 of the prion protein. *J Biol Chem* **1994**, *269*, 7859–7862.

Doh-Ura K, Iwaki T, Caughey B. Lysosomotropic agents and cysteine protease inhibitors inhibit scrapie-associated prion protein accumulation. *J Virol* **2000**, *74*, 4894–4897.

Drüeke TB. Dialysis-related amyloidosis. *Nephrol Dial Transplant* **1998**, *13 (Suppl 1)*, 58–64.

Fandrich M, Forge V, Buder K, Kittler M, Dobson CM, Diekmann S. Myoglobin forms amyloid fibrils by association of unfolded polypeptide segments. *Proc Natl Acad Sci USA* **2003**, *100*, 15463–15468.

Farrelly PV, Kenna BL, Laohachai KL, Bahadi R, Salmona M, Forloni G, Kourie JI. Quinacrine blocks PrP (106–126)-formed channels. *J Neurosci* Res **2003**, *74*, 934–941.

Fernandez A, Berry RS. Proteins with H-bond packing defects are highly interactive with lipid bilayers: implications for amyloidogenesis. *Proc Natl Acad Sci USA* **2003**, *100*, 2391–2396.

Forloni G, Chiesa R, Smiroldo S, Verga L, Salmona M, Tagliavini F, Angeretti N.

Apoptosis mediated neurotoxicity induced by chronic application of *β* amyloid fragment 25–35. *Neuroreport* **1993**, *4*, 523–526.

Fraser SP, Suh YH, Djamgoz MB. Ionic effects of the Alzheimer's disease *β*-amyloid precursor protein and its metabolic fragments. *Trends Neurosci* **1997**, *20*, 67–72.

Furukawa K, Abe Y, Akaike N. Amyloid *β* protein-induced irreversible current in rat cortical neurones. *Neuroreport* **1994**, *5*, 2016–2018.

Gasset M, Baldwin MA, Lloyd DH, Gabriel JM, Holtzman DM, Cohen F, Fletterick R, Prusiner SB. Predicted *α*-helical regions of the prion protein when synthesized as peptides form amyloid. *Proc Natl Acad Sci USA* **1992**, *89*, 10940–10944.

Hardy J, Allsop D. Amyloid deposition as the central event in the aetiology of Alzheimer's disease. *Trends Pharmacol Sci* **1991**, *12*, 383–388.

Harper JD, Lansbury PT, Jr. Models of amyloid seeding in Alzheimer's disease and scrapie: mechanistic truths and physiological consequences of the time-dependent solubility of amyloid proteins. *Annu Rev Biochem* **1997**, *66*, 385–407.

Hegde RS, Mastrianni JA, Scott MR, DeFea KA, Tremblay P, Torchia M, DeArmond SJ, Prusiner SB, Lingappa VR. A transmembrane form of the prion protein in neurodegenerative disease. *Science* **1998**, *279*, 827–834.

Hirakura Y, Kagan BL. Pore formation by *β*-2-microglobulin: a mechanism for the pathogenesis of dialysis associated amyloidosis. *Amyloid* **2001**, *8*, 94–100.

Hirakura Y, Satoh Y, Hirashima N, Suzuki T, Kagan BL, Kirino Y. Membrane perturbation by the neurotoxic Alzheimer amyloid fragment *β*25–35 requires aggregation and *β*-sheet formation. *Biochem Mol Biol Int* **1998**, *46*, 787–794.

Hirakura Y, Lin MC, Kagan BL. Alzheimer amyloid a*β*1–42 channels: effects of solvent, pH, and Congo Red. *J Neurosci Res* **1999**, *57*, 458–466.

Hirakura Y, Yiu WW, Yamamoto A, Kagan BL. Amyloid peptide channels: blockade by zinc and inhibition by Congo red (amyloid channel block). *Amyloid* **2000a**, *7*, 194–199.

Hirakura Y, Azimov R, Azimova R, Kagan BL. Polyglutamine-induced ion channels: a possible mechanism for the neurotoxicity of huntington and other CAG repeat diseases. *J Neurosci Res* **2000b**, *60*, 490–494.

Hirakura Y, Carreras I, Sipe JD, Kagan BL. Channel formation by serum amyloid A: a potential mechanism for amyloid pathogenesis and host defense. *Amyloid* **2002**, *9*, 13–23.

Ibrahim HR, Thomas U, Pellegrini A. A helix–loop–helix peptide at the upper lip of the active site cleft of lysozyme confers potent antimicrobial activity with membrane permeabilization action. *J Biol Chem* **2001**, *276*, 43767–43774.

Ingrosso L, Ladogana A, Pocchiari M. Congo red prolongs the incubation period in scrapie-infected hamsters. *J Virol* **1995**, *69*, 506–508.

Janson J, Ashley RH, Harrison D, McIntyre S, Butler PC. The mechanism of islet amyloid polypeptide toxicity is membrane disruption by intermediate-sized toxic amyloid particles. *Diabetes* **1999**, *48*, 491–498.

Kagan BL. Mode of action of yeast killer toxins: channel formation in lipid bilayer membranes. *Nature* **1983**, *302*, 709–711.

Kagan BL, Selsted ME, Ganz T, Lehrer RI. Antimicrobial defensin peptides form voltage-dependent ion-permeable channels in planar lipid bilayer membranes. *Proc Natl Acad Sci USA* **1990**, *87*, 210–214.

Kagan BL, Hirakura Y, Azimov R, Azimova R. The channel hypothesis of Huntington's disease. *Brain Res Bull* **2001**, *56*, 281–284.

Kawahara M, Arispe N, Kuroda Y, Rojas E. Alzheimer's disease amyloid *β*-protein forms Zn^{2+}-sensitive, cation-selective channels across excised membrane patches from hypothalamic neurons. *Biophys J* **1997**, *73*, 67–75.

Kawahara M, Kuroda Y, Arispe N, Rojas E. Alzheimer's *β*-amyloid, human islet amylin, and prion protein fragment evoke intracellular free calcium elevations by a common mechanism in a hypothalamic GnRH neuronal cell line. *J Biol Chem* **2000**, *275*, 14077–14083.

Kayed R, Head E, Thompson JL, McIntire TM, Milton SC, Cotman CW, Glabe CG.

Common structure of soluble amyloid oligomers implies common mechanism of pathogenesis. *Science* **2003**, *300*, 486–489.

Kazlauskaite J, Sanghera N, Sylvester I, Venien-Bryan C, Pinheiro TJ. Structural changes of the prion protein in lipid membranes leading to aggregation and fibrillization. *Biochemistry* **2003**, *42*, 3295–3304.

Kim HS, Lee JH, Lee JP, Kim EM, Chang KA, Park CH, Jeong SJ, Wittendorp MC, Seo JH, Choi SH, Suh YH. Amyloid β peptide induces cytochrome C release from isolated mitochondria. *Neuroreport* **2002**, *13*, 1989–1993.

Korth C, May BC, Cohen FE, Prusiner SB. Acridine and phenothiazine derivatives as pharmacotherapeutics for prion disease. *Proc Natl Acad Sci USA* **2001**, *98*, 9836–9841.

Kourie JI. Characterization of a C-type natriuretic peptide (CNP-39)-formed cation-selective channel from platypus (*Ornithorhynchus anatinus*) venom. *J Physiol* **1999**, *518*, 59–69.

Kourie JI, Culverson A. Prion peptide fragment PrP[106–126] forms distinct cation channel types. *J Neurosci Res* **2000**, *62*, 120–133.

Kourie JI, Hanna EA, Henry CL. Properties and modulation of α human atrial natriuretic peptide (α-hANP)-formed ion channels. *Can J Physiol Pharmacol* **2001**, *79*, 654–664.

Kourie JI, Henry CL, Farrelly P. Diversity of amyloid β protein fragment [1–40]-formed channels. *Cell Mol Neurobiol* **2001 b**, *21*, 255–284.

Kourie JI, Culverson AL, Farrelly PV, Henry CL, Laohachai KN. Heterogeneous amyloid-formed ion channels as a common cytotoxic mechanism: implications for therapeutic strategies against amyloidosis. *Cell Biochem Biophys* **2002**, *36*, 191–207.

Lashuel HA, Petre BM, Wall J, Simon M, Nowak RJ, Walz T, Lansbury PT, Jr. Alpha-synuclein, especially the Parkinson's disease-associated mutants, forms pore-like annular and tubular protofibrils. *J Mol Biol* **2002 a**, *322*, 1089–1102.

Lashuel HA, Hartley D, Petre BM, Walz T, Lansbury PT, Jr. Neurodegenerative disease: amyloid pores from pathogenic mutations. *Nature* **2002 b**, *418*, 291.

Lashuel HA, Hartley DM, Petre BM, Wall JS, Simon MN, Walz T, Lansbury PT, Jr. Mixtures of wild-type and a pathogenic (E22G) form of Aβ40 *in vitro* accumulate protofibrils, including amyloid pores. *J Mol Biol* **2003**, *332*, 795–808.

Li SH, Li XJ. Huntingtin–protein interactions and the pathogenesis of Huntington's disease. *Trends Genet* **2004**, *20*, 146–154.

Lin H, Zhu YJ, Lal R. Amyloid β protein (1–40) forms calcium-permeable, Zn^{2+}-sensitive channel in reconstituted lipid vesicles. *Biochemistry* **1999**, *38*, 11189–11196.

Lin MC. Channel formation by amyloidogenic neurotoxic and neurodegenerative disease related peptides. *PhD Dissertation*, Division of Neuroscience, UCLA, **1996**.

Lin MC, Kagan BL. Electrophysiologic properties of channels induced by Aβ25–35 in planar lipid bilayers. *Peptides* **2002**, *23*, 1215–1228.

Lin MC, Mirzabekov T, Kagan BL. Channel formation by a neurotoxic prion protein fragment. *J Biol Chem* **1997**, *272*, 44–47.

Lorenzo A, Razzaboni B, Weir GC, Yankner BA. Pancreatic islet cell toxicity of amylin associated with type-2 diabetes mellitus. *Nature* **1994**, *368*, 756–760.

Malisauskas M, Zamotin V, Jass J, Noppe W, Dobson CM, Morozova-Roche LA. Amyloid protofilaments from the calcium-binding protein equine lysozyme: formation of ring and linear structures depends on pH and metal ion concentration. *J Mol Biol* **2003**, *330*, 879–890.

Manunta M, Kunz B, Sandmeier E, Christen P, Schindler H. Reported channel formation by prion protein fragment 106–126 in planar lipid bilayers cannot be reproduced [Letter]. *FEBS Lett* **2000**, *474*, 255–256.

McCarthy RE, 3rd, Kasper EK. A review of the amyloidoses that infiltrate the heart. *Clin Cardiol* **1998**, *21*, 547–552.

McKeith IG, Mosimann UP. Dementia with Lewy bodies and Parkinson's disease. *Parkinsonism Relat Disord* **2004**, *10 (Suppl 1)*, S15–S18.

McLaurin J, Franklin T, Fraser PE, Chakrabartty A. Structural transitions associated with the interaction of Alzheimer β-amyloid peptides with gangliosides. *J Biol Chem* **1998**, *273*, 4506–4515.

McLean LR, Balasubramaniam A. Promotion of β-structure by interaction of diabetes associated polypeptide (amylin) with phosphatidylcholine. *Biochim Biophys Acta* **1992**, *1122*, 317–320.

Merlini G, Bellotti V. Molecular mechanisms of amyloidosis. *N Engl J Med* **2003**, *349*, 583–596.

Merlini G, Westermark P. The systemic amyloidoses: clearer understanding of the molecular mechanisms offers hope for more effective therapies. *J Intern Med* **2004**, *255*, 159–178.

Mirzabekov T, Lin MC, Yuan WL, Marshall PJ, Carman M, Tomaselli K, Lieberburg I, Kagan BL. Channel formation in planar lipid bilayers by a neurotoxic fragment of the β-amyloid peptide. *Biochem Biophys Res Commun* **1994**, *202*, 1142–1148.

Mirzabekov TA, Lin MC, Kagan BL. Pore formation by the cytotoxic islet amyloid peptide amylin. *J Biol Chem* **1996**, *271*, 1988–1992.

Moe SM, Sprague SM. Beta 2-microglobulin induces calcium efflux from cultured neonatal mouse calvariae. *Am J Physiol* **1992**, *263*, F540–F545.

Monoi H, Futaki S, Kugimiya S, Minakata H, Yoshihara K. Poly-L-glutamine forms cation channels: relevance to the pathogenesis of the polyglutamine diseases [see Comments]. *Biophysical J* **2000**, *78*, 2892–2899.

Nakajima M, Yamada T, Kusuhara T, Furukawa H, Takahashi M, Yamauchi A, Kataoka Y. Results of quinacrine administration to patients with Creutzfeldt–Jakob disease. *Dement Geriatr Cogn Disorders* **2004**, *17*, 158–163.

Pan KM, Baldwin M, Nguyen J, Gasset M, Serban A, Groth D, Mehlhorn I, Huang Z, Fletterick RJ, Cohen FE, Prusiner SB. Conversion of α-helices into β-sheets features in the formation of the scrapie prion proteins. *Proc Natl Acad Sci USA* **1993**, *90*, 10962–10966.

Panov AV, Gutekunst CA, Leavitt BR, Hayden MR, Burke JR, Strittmatter WJ, Greenamyre JT. Early mitochondrial calcium defects in Huntington's disease are a direct effect of polyglutamines. *Nat Neurosci* **2002**, *5*, 731–736.

Pepys MB, Hawkins PN, Booth DR, Vigushin DM, Tennent GA, Soutar AK, Totty N, Nguyen O, Blake CC, Terry CJ, Feest TG, Zalin AM, Hsuan JJ. Human lysozyme gene mutations cause hereditary systemic amyloidosis. *Nature* **1993**, *362*, 553–557.

Petersen J, Kang MS. *In vivo* effect of β_2-microglobulin on bone resorption. *Am J Kidney Dis* **1994**, *23*, 726–730.

Petosa C, Collier RJ, Klimpel KR, Leppla SH, Liddington RC. Crystal structure of the anthrax toxin protective antigen. *Nature* **1997**, *385*, 833–838.

Pike CJ, Burdick D, Walencewicz AJ, Glabe CG, Cotman CW. Neurodegeneration induced by β-amyloid peptides *in vitro*: the role of peptide assembly state. *J Neurosci* **1993**, *13*, 1676–1687.

Qi JS, Qiao JT. Amyloid β-protein fragment 31–35 forms ion channels in membrane patches excised from rat hippocampal neurons. *Neuroscience* **2001**, *105*, 845–852.

Reches M, Porat Y, Gazit E. Amyloid fibril formation by pentapeptide and tetrapeptide fragments of human calcitonin. *J Biol Chem* **2002**, *277*, 35475–35480.

Relini A, Torrassa S, Rolandi R, Gliozzi A, Rosano C, Canale C, Bolognesi M, Plakoutsi G, Bucciantini M, Chiti F, Stefani M. Monitoring the process of HypF fibrillization and liposome permeabilization by protofibrils. *J Mol Biol* **2004**, *338*, 943–957.

Rodrigues CM, Sola S, Brites D. Bilirubin induces apoptosis via the mitochondrial pathway in developing rat brain neurons. *Hepatology* **2002**, *35*, 1186–1195.

Samaia HB, Mari JJ, Vallada HP, Moura RP, Simpson AJ, Brentani RR. A prion-linked psychiatric disorder. *Nature* **1997**, *390*, 241.

Sandberg MK, Wallen P, Wikstrom MA, Kristensson K. Scrapie-infected GT1-1 cells show impaired function of voltage-gated N-type calcium channels (Ca(v) 2.2) which is ameliorated by quinacrine treatment. *Neurobiol Dis* **2004**, *15*, 143–151.

Sanderson KL, Butler L, Ingram VM. Aggregates of a β-amyloid peptide are required to induce calcium currents in neuron-like human teratocarcinoma cells: relation to Alzheimer's disease. *Brain Res* **1997**, *744*, 7–14.

Schein SJ, Kagan BL, Finkelstein A. Colicin K acts by forming voltage-dependent channels in phospholipid bilayer membranes. *Nature* **1978**, *276*, 159–163.

Schubert D, Behl C, Lesley R, Brack A, Dargusch R, Sagara Y, Kimura H. **1995**, Amyloid peptides are toxic via a common oxidative mechanism. *Proc Natl Acad Sci USA 92*, 1989–1993.

Simmons MA, Schneider CR. Amyloid β peptides act directly on single neurons. *Neurosci Lett* **1993**, *150*, 133–136.

Sipe JD. Serum amyloid A: from fibril to function. Current status. *Amyloid* **2000**, *7*, 10–12.

Sipe JD, Cohen AS. Review: history of the amyloid fibril. *J Struct Biol* **2000**, *130*, 88–98.

Sokolov Y, Mirzabekov T, Martin DW, Lehrer RI, Kagan BL. Membrane channel formation by antimicrobial protegrins. *Biochim Biophys Acta* **1999**, *1420*, 23–29.

Song L, Hobaugh MR, Shustak C, Cheley S, Bayley H, Gouaux JE. Structure of staphylococcal *α*-hemolysin, a heptameric transmembrane pore. *Science* **1996**, *274*, 1859–1866.

Stefani M, Dobson CM. Protein aggregation and aggregate toxicity: new insights into protein folding, misfolding diseases and biological evolution. *J Mol Med* **2003**, *81*, 678–699.

Stipani V, Gallucci E, Micelli S, Picciarelli V, Benz R. Channel formation by salmon and human calcitonin in black lipid membranes. *Biophys J* **2001**, *81*, 3332–3338.

Volles MJ, Lansbury PT, Jr. Vesicle permeabilization by protofibrillar *α*-synuclein is sensitive to Parkinson's disease-linked mutations and occurs by a pore-like mechanism. *Biochemistry* **2002**, *41*, 4595–4602.

Walsh DM, Klyubin I, Fadeeva JV, Cullen WK, Anwyl R, Wolfe MS, Rowan MJ, Selkoe DJ. Naturally secreted oligomers of amyloid *β* protein potently inhibit hippocampal long-term potentiation *in vivo*. *Nature* **2002**, *416*, 483–484.

Wang L, Lashuel HA, Walz T, Colon W. Murine apolipoprotein serum amyloid A in solution forms a hexamer containing a central channel. *Proc Natl Acad Sci USA* **2002**, *99*, 15947–15952.

Weiss JH, Pike CJ, Cotman CW. Ca^{2+} channel blockers attenuate *β*-amyloid peptide toxicity to cortical neurons in culture. *J Neurochem* **1994**, *62*, 372–375.

Westermark P, Wilander E. The influence of amyloid deposits on the islet volume in maturity onset diabetes mellitus. *Diabetologia* **1978**, *15*, 417–421.

Yang AJ, Knauer M, Burdick DA, Glabe C. Intracellular A *β*1–42 aggregates stimulate the accumulation of stable, insoluble amyloidogenic fragments of the amyloid precursor protein in transfected cells. *J Biol Chem* **1995**, *270*, 14786–14792.

Yang W, Dunlap JR, Andrews RB, Wetzel R. Aggregated polyglutamine peptides delivered to nuclei are toxic to mammalian cells. *Hum Mol Genet* **2002**, *11*, 2905–2917.

Yankner BA, Duffy LK, Kirschner DA. Neurotrophic and neurotoxic effects of amyloid *β* protein: reversal by tachykinin neuropeptides. *Science* **1990**, *250*, 279–282.

Zhu M, Souillac PO, Ionescu-Zanetti C, Carter SA, Fink AL. Surface-catalyzed amyloid fibril formation. *J Biol Chem* **2002**, *277*, 50914–50922.

Zhu M, Han S, Zhou F, Carter SA, Fink AL. Annular oligomeric amyloid intermediates observed by *in situ* atomic force microscopy. *J Biol Chem* **2004**, *279*, 24452–24459.

Zhu YJ, Lin H, Lal R. Fresh and nonfibrillar amyloid *β* protein(1–40) induces rapid cellular degeneration in aged human fibroblasts: evidence for A*β*P-channel-mediated cellular toxicity. *FASEB J* **2000**, *14*, 1244–1254.

15
The Future of Molecular Diagnostics and Targeted Therapeutics in the Amyloidoses

David C. Seldin

15.1
Introduction

The chapters in this book document the explosion of knowledge, over the last few years, in the understanding of the chemistry of the amyloid proteins and their diverse roles in human diseases. The challenge ahead is to convert this understanding into an improvement in the diagnostics and therapy for the amyloidoses. Today, the diagnostic and therapeutic tools for the amyloid diseases are of insufficient specificity and limited efficacy. Amyloid diseases are rarely diagnosed until organ damage is well advanced; it is difficult to distinguish the type of amyloid (testing by a specialized amyloidosis center is often required); staging is difficult and relies on indirect assessment of disease burden; and, finally, treatment options are few for the majority of amyloidoses, and may require heroic and dangerous interventions for those cases that are treatable. In this chapter, we will discuss some of the new findings and concepts that relate to the amyloid diseases which promise to rectify this situation in the future. Let us begin by considering what would be the ideal strategy and examine approaches that lead to the implementation of such as strategy.

Ideally, amyloid diseases would be detected at a preclinical stage. The extent of disease would be determined in a non-invasive manner. The type of amyloid would be determined precisely. Treatment would be undertaken that would be specific for the amyloid disease protein and would not damage normal cells or tissues, or interfere with normal protein function. Eventually, one would be able to predict who is prone to the development of amyloid disease and make use of preventative treatment. In the field of amyloidosis treatment and research, these ideas sound like science fiction, yet one can take encouragement from paradigms in other fields. Consider, for example, the current state of diagnostics and therapeutics in breast cancer. Breast cancer can be detected at a preclinical stage, by mammography, before there is any gross palpable tumor and such microscopic carcinomas *in situ* can often be easily cured with surgery. Patients with breast cancer are routinely staged by computerized tomography (CT) scanning, magnetic resonance imaging (MRI), bone scanning and now by positron

Amyloid Proteins. The Beta Sheet Conformation and Disease. J. D. Sipe

ISBN: 3-527-31072-X

emission tomography (PET) scanning, so the extent of disease can be accurately assessed. The specific type of breast cancer can be determined by immunohistochemical and molecular analyses – techniques which even provide information about the molecular pathogenesis of the disease. Tumor-specific immunotherapy is now widely used, with anti-HER2/*neu* receptor antibodies [Herceptin (trastuzumab)] supplementing cytotoxic chemotherapy and radiation. Lastly, genetic testing can identify patients carrying mutant breast cancer susceptibility (*BRCA*) genes who are at very high risk of developing cancer, which can be largely prevented by prophylactic mastectomy.

Obviously, these advances in breast cancer diagnostics and treatment have not yet eliminated breast cancer morbidity and mortality, but they have clearly saved lives, identifying patients at earlier curable stages of disease and prolonging life even for patients with metastatic disease. In the amyloid field, we should target research efforts towards similar goals.

15.2 Early Diagnosis of Amyloid Diseases

For breast cancer, the development of imaging techniques, such as mammography, and breast examination by a physician and self-examination by women have provided non-invasive methods that are reasonably sensitive and economical for early diagnosis of breast cancer. Prospective studies have demonstrated the usefulness of these techniques and their ability to save lives. The cost-effectiveness of this approach is, in large part, due to the commonness of the disease: if one of eight women will develop breast cancer in their lifetime, unnecessary annual testing of seven others is an acceptable cost to society. It does not appear that any amyloid disease in the general population is likely to achieve an incidence that would justify widespread testing of normal individuals. Thus, one would anticipate that such testing will only be carried out in families known to have a genetic predisposition to a systemic or neurodegenerative amyloidosis (see below).

Early diagnosis of sporadic disease in people with no family history of amyloid disease is likely to be based upon timely clinical ascertainment. Historically, this has depended upon the individual physician's acumen and has been suboptimal in the diagnosis of rare diseases, as physicians are taught that "when you hear hoof beats, think of horses, not zebras". In the amyloidoses, it is the stampeding zebras or other rare hoofed beasts that cause disease. To diagnose amyloidosis, physicians must become more attuned to unusual causes of common syndromes. Differential diagnosis can be aided by computer-assisted algorithms, but in the current environment of cost-containment in medicine, unfortunately, physicians often lack the time and resources for reflection and in-depth research. Thus, it is worthwhile to review some of the syndromes that should initiate a diagnostic workup for amyloidosis.

The Alzheimer's, neuroserpinopathies and other forms of dementia can be diagnosed when subtle indications of decline in cognition and memory are investigated with neuropsychological testing. Parkinson's disease, Huntington's disease and other movement disorders can be diagnosed upon careful evaluation of early signs of lack of coordination, falls and stiffness. The systemic amyloid diseases can be detected through investigation of symptoms and signs related to the target organs, including cardiomyopathy with diastolic dysfunction, nephrotic renal insufficiency, and autonomic and peripheral neuropathy in the absence of other identifiable causes. These early and subtle clinical indications of disease should be aggressively pursued, with appropriate tissue biopsy for the systemic processes and serial neurological studies for the neurodegenerative processes. Table 15.1 identifies some of the common clinical syndromes and diagnostic testing for the systemic amyloidoses. Fig. 15.1 illustrates the current diagnostic process.

Examples in schematic form for the kinetics of onset of amyloid diseases are presented in Fig. 15.2. In the hereditary amyloidoses such as amyloidogenic transthyretin (ATTR), the mutant amyloidogenic gene defect is present at birth. The patient enjoys a long preclinical period (light gray), when he or she carries

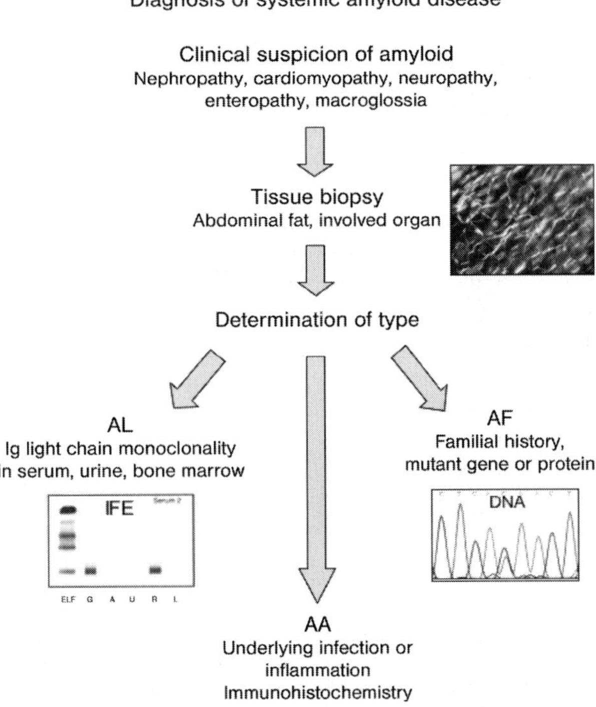

Fig. 15.1 Flow chart for the current diagnosis of the systemic amyloidoses. See text for details and new methodology under development.

Table 15.1 The systemic amyloidoses

Type	Fibril composition	Clinical features	Laboratory diagnostic studies	Body fluid markers
AL, primary	immunoglobulin light chain (κ or λ)	multisystem disease can include proteinuria, cardiomyopathy, hepatomegaly, macroglossia, orthostatic hypotension, peripheral neuropathy, cutaneous ecchymoses	monoclonal protein in serum and/or urine; monoclonal bone marrow plasmacytosis; elevated free light chain in serum	proteinuria, low serum albumin; elevated alkaline phosphatase; elevated BNP; depressed clotting factor X
ATTR, familial	TTR	mid-life onset of peripheral and autonomic neuropathy, cardiomyopathy, vitreous opacities; family history	mutant TTR protein in serum or mutant TTR gene sequence	elevated BNP
Other familial	ApoAI	nephropathy, cardiomyopathy	disease-associated DNA mutation	proteinuria, low serum albumin
	ApoAII	nephropathy		
	fibrinogen	nephropathy		
	lysozyme	nephropathy		
	gelsolin	cranial neuropathy, nephropathy	disease-associated DNA mutation	proteinuria, low serum albumin
AA, secondary	amyloid A protein	underlying inflammatory disease; multisystem disease including proteinuria, renal insufficiency, hepatomegaly, splenomegally, orthostatic hypotension	Permanganate-sensitive Congo red staining of tissue biopsy; immunohistochemical staining of tissue biopsy with anti-AA antiserum	proteinuria, low serum albumin; elevated alkaline phosphatase
Aβ_2M, dialysis associated	β_2M	long-term hemodialysis or chronic ambulatory peritoneal dialysis; carpal tunnel syndrome or musculoskeletal disorders of shoulder or hip	elevated serum β_2M; immunohistochemical staining of tissue biopsy with anti β_2M antiserum	none

Table 15.1 (continued)

Type	Fibril composition	Clinical features	Laboratory diagnostic studies	Body fluid markers
Senile cardiac amyloidosis	TTR (wild-type)	cardiomyopathy	exclusion of other systemic amyloidoses	elevated brain natiuretic peptide

β_2M = β_2-microglobulin; Apo = apolipoprotein;
BNP = brain natiuretic peptide.

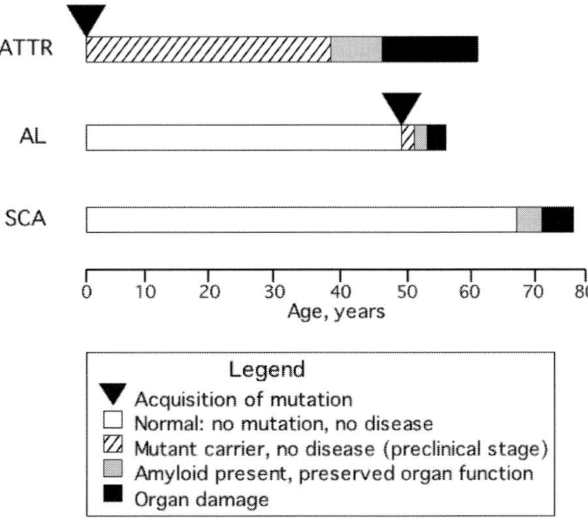

Fig. 15.2 Schematic of the kinetics of disease onset in the systemic amyloidoses. In an hereditary amyloidosis such as ATTR, the mutation is present at birth, but the disease develops much later in life, after an asymptomatic carrier state. In AL amyloidosis, the mutation is acquired during somatic immunoglobulin gene rearrangement in a plasma cell and leads to disease more rapidly. In senile cardiac amyloidosis, the wild-type protein causes disease, without any identifiable mutation.

the TTR mutation, but does not have any amyloid deposition or organ disease. In adult life, perhaps due to aging, oxidative stress or environmenttal influence, the amyloid precursor protein, TTR, undergoes some alteration and the process of amyloid deposition begins. Over a period of years, the amyloid deposits are laid down, but organ function remains normal (medium gray). Finally, neuropathy and/or cardiomyopathy develop(s), leading to an untimely death in the absence of treatment (see below).

In AL amyloidosis, the amyloid precursor protein is an immunoglobulin light chain produced by a B cell whose light chain genes undergo somatic rearrangement during maturation of the immune system and clonal expansion. Prior to such an event, the patient is biochemically, genetically and clinically normal. In the case of amyloidosis resulting from amyloid fibril formation by light chains, there is more likely a much shorter preclinical phase than for ATTR, as the process of amyloid deposition and organ failure appears to progress very quickly in the AL disease and the median survival of untreated patients is only about 1 year.

In senile cardiac amyloidosis, there is no hereditary or acquired gene mutation. Rather, the aging process and (hypothetically) post-translational modification of the wild-type TTR amyloid precursor protein or alteration in the host milieu leads to misfolding, amyloid deposition and cardiac failure. Progression seems to be somewhat slower than in AL cardiomyopathy, but, currently, there is no treatment at all.

15.3
Accurate Classification of Amyloid Diseases

In the example of breast cancer, the classification of the type of tumor rests upon the histological appearance and the immunohistochemical features with respect to estrogen, progesterone and Her2/*neu* receptor expression. Similarly, in the amyloidoses, we hope and expect that sophisticated histochemical techniques will aid in the classification of disease. However, such diagnosis will rest upon the availability of biopsy material. This will not likely ever be routinely available in pre-mortem specimens for patients with neurodegenerative diseases. For the systemic amyloidoses, biopsies are performed of abdominal fat or involved organs and stained with Congo red to identify the presence of amyloid fibrils. Immunohistochemistry can be done on these specimens, but can be unreliable – because light chains comprise the AL amyloid deposits, the secondary reagents may give non-specific reactivity leading to false positive results. The specificity is better with immunoelectron microscopy, in which the coincidence of gold-labeled primary antibodies and the fibrils themselves can be observed in electron micrographs. This technique is available in specialized centers only. In addition, sensitive biochemical techniques can be applied to identify the fibril components, e.g. protein extraction and mass spectrometry. Here, also, appropriate controls and cautious interpretation are essential: all human biopsy materials are obtained with some degree of blood contamination, and therefore contamination with circulating immunoglobulins and other serum proteins. Presumably, over time, mass spectrometry will become more sensitive, accurate and automated, and a combination of immunochemical and biochemical techniques will lead to precise identification of the fibril components.

In cancer diagnostics, molecular techniques are being used to analyze chromosomal abnormalities in tumors and gene expression patterns. Chromosomal rearrangements are useful in identifying those leukemias, lymphomas and other cancers in which characteristic non-random chromosomal translocations are associated with the disease. These, however, are uncommon. In contrast, all cancers have dysregulated gene expression manifest as alterations in mRNA expression compared with normal cells. These can be detected using so-called “microarrays”, in which mRNAs from the tumor tissue are isolated and hybridized to oligonucleotides spotted on a glass slide. The “oligos” represent known and unknown genes; within a few years, oligonucleotides for all expressed genes in the human genome will be available on a single inexpensive such chip. The pattern of mRNA expression reflects the network of transcription factors and other regulatory elements present and provides a “fingerprint” that can identify the cell of origin and or the underlying molecular defect, and for some cancers can provide prognostic information. It is quite possible that such fingerprints will eventually be established for the amyloidoses.

15.4 Non-invasive Staging of Amyloid Diseases

Currently, staging of the amyloidoses depends upon indirect clinical assessment of organ involvement with disease. In the neurodegenerative diseases, as noted above, we depend upon neurological evaluation of brain and nervous system function. In the systemic amyloidoses, we depend upon serum evaluation of kidney and liver function, and non-invasive testing of cardiac function. There is an urgent need for imaging techniques that provide semiquantitative assessment of organ amyloid burden. A number of such techniques are in development. One that has had some clinical success is the serum amyloid P component (SAP) scan developed in the UK. In this technique, SAP is purified, labeled with ^{131}I and injected into patients. SAP binds to most fibrillar amyloid deposits and whole-body scintigraphy gives an assessment of disease burden [1]. Amyloid scintigraphy is not very quantitative and gives false positives, due to blood pooling in the heart. Particularly, given the clinical importance of cardiac disease, this approach is not ideal. Additionally, SAP is a purified serum component that carries some theoretical risk of viral contamination.

There are other approaches to assessment of amyloid disease burden; one, for which there is some recent limited clinical experience, is the technetium-labeled aprotinin scan [2]. Aprotinin, or Trasylol, a serine protease inhibitor, has been purified from bovine lung and binds to proteases associated with amyloid fibrils. It is excreted in the urine, and therefore gives false positive signals in kidney and bladder; also, anaphylactic reactions have been reported. Thioflavin T and Congo red derivatives have been labeled and used for fibril imaging; the most promising to date is a benzothiazole derivative Pittsburgh compound B (PIB), that can be detected by PET scanning and does appear to identify amyloid deposits in Alzheimer's patients [3], see Fig. 15.3. These reagents should be investigated as soon as possible in a broad range of systemic and neurological amyloid diseases. PET scanning based on uptake of [^{18}F]fluorodeoxyglucose in hypermetabolic tumor cells is proving to be extremely useful in identifying

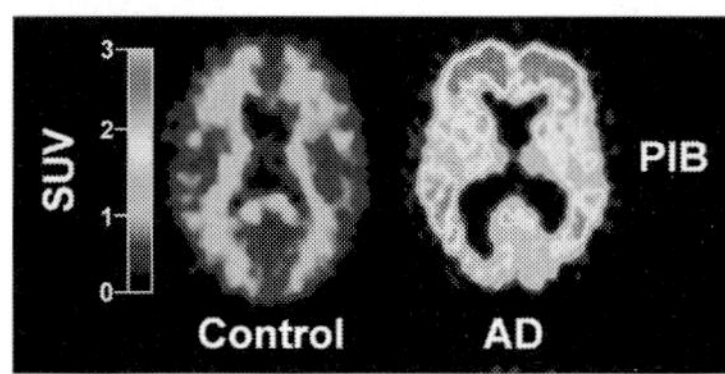

Fig. 15.3 PIB is retained in the brain of an Alzheimer's patient (right) compared with a control patient (left). (Reprinted from [3], with permission.)

metastases in breast cancer and other solid tumors. Refinement of other techniques such as magnetic resonance imaging may also eventually become useful in non-invasive detection of amyloid burden.

15.5
Targeted Therapeutics of Amyloid Diseases

Modern cancer therapeutics can be summarized as two types of approaches: non-specific cytotoxic treatment and targeted therapeutics. Chemotherapy and radiotherapy fall into the former category, killing malignant cells by DNA-damage-induced apoptosis and necrosis by processes that also affect normal cells. This treatment has a small "therapeutic index", i.e. the differential killing of cancer cells versus normal cells is not great.

In contrast, modern cancer therapeutics involves targeted therapeutics that usually have a much higher specificity for the malignant cells. The number of examples of such treatments currently available is small, but there are promising new agents in preclinical and clinical development. Examples include imatinib (Gleevec), a relatively specific inhibitor of the *bcr–abl* chimeric fusion kinase present in chronic myelogenous leukemia, and gefitinib, an inhibitor of the epidermal growth factor receptor kinase that is effective in a subset of patients with lung cancer. In addition to these kinase inhibitors, monoclonal antibody therapy is targeted and efficacious, and these include Rituxan (rituximab; anti-CD20) in lymphomas and, returning to our breast cancer example, Herceptin, an anti-HER2/*neu* antibody.

Similarly, targeted therapeutics are being developed for the amyloidoses. The agents that have entered clinical trials are targeted at accessory molecules in fibrillogenesis, as discussed in other chapters in this volume (see Chapters 8 and 9 about SAP). The compound CPHPC promotes SAP dimer formation and accelerates SAP degradation [4]. The rationale was the hope to "denude" amyloid deposits of their SAP, a component of most amyloid deposits, and accelerate fibril degradation in extracellular tissues. Phase I clinical trials appear to demonstrate safety, but, so far there, are no reports of the efficacy of this strategy.

Another common component of amyloid deposits is proteoglycan, macromolecules made up of a protein core and glycosaminoglycan side-chains, which are found associated with amyloid fibrils. Fibrillex, a synthetic glycosaminoglycan mimetic, blocks binding of proteoglycans to amyloid fibrils. Fibrillex has been through phase III multicenter trials for AA amyloidosis, with the results scheduled to be announced in early 2005.

Chemicals which bind to amyloid fibrils have the potential to interfere with aggregation and assembly into fibrils. This has been tested in experiments demonstrating that Congo red can protect cells and tissues from the toxicity of aggregation of Huntington's disease polyglutamine repeat peptides [5]. While Congo red is unlikely to be useful clinically, this is proof of principle for the approach and will, undoubtedly, serve as the foundation of future studies.

Biophysical studies are beginning to identify stable non-amyloidogenic protein conformations and unstable amyloid-prone conformations. For TTR, the ligand-bound tetramer is thermodynamically more stable than the free tetramer. Compounds have been identified that bind to the ligand-binding site, but lack the biological effects of excess thyroid hormone or retinoids. A relatively safe non-steroidal anti-inflammatory agent, diflunisal, exhibits such activity [6] and will soon be entering clinical trials. Diflunisal will be the prototype of conformational modifying or stabilizing drugs for the amyloidoses.

Immunotherapy is another approach to targeted treatment of cancer. In addition to passive immunization with the Herceptin and Rituxan types of antibody reagents, there is active research on dendritic cell vaccines in cancer, aimed to present tumor antigens to cytotoxic cells of the immune system. Amyloid fibrils are made up of proteins that are identical or nearly identical to normal proteins and are likely to be poorly immunogenic. However, there are techniques being developed to "break tolerance" to promote responses against endogenous proteins. The price of this response would be the potential for autoimmune disease, which could be as deleterious as the amyloid disease, if the antigen is an essential cellular protein. Alternatively, it is conceivable that conformation-specific antigens corresponding to the β-sheet structures will stimulate fibril-specific immune responses without autoimmunity.

15.6 Amyloid Disease Prevention

In cancer, prevention strategies rely upon three approaches: (1) avoidance of co-carcinogens (asbestos, tobacco, chemicals, radiation, etc.), (2) early detection of pre-malignant or early malignant lesions (colonoscopy, mammography, etc.) and (3) increased screening or organ ablation in highly susceptible populations.

Are there environmental influences on amyloid disease? The amyloidoses are diseases of aging. The neurodegenerative amyloidoses, Alzheimer's disease and the pathogenetically related Parkinson's and Huntington's diseases, appear in older adults. When a clear genetic predisposition is inherited, the clock is moved up considerably, but the disease still takes years to appear. Likewise, the systemic amyloidoses manifest in adulthood, even the hereditary diseases; AL amyloidosis, the plasma cell disease, is also uncommon before the sixth decade of life. These epidemiologic data strongly speak to a contribution of the environment on disease. It is possible that the environment that is important is the endogenous tissue environment, in which oxidant stress, cellular senescence and other manifestations of aging promote protein misfolding and amyloidosis. It is also possible that external environmental influences contribute to this process.

The second strategy for disease prevention, early diagnosis, was discussed above. The third point addresses the management of disease-susceptible subpopulations. In the last 35 years, it has been recognized that a genetic predisposition to cancer can be inherited. Usually, the predisposition occurs through inheritance of a defective tumor suppressor gene, providing the first hit in a two-hit tumor suppressor gene inactivation process. This has been characterized as "Knudson's hypothesis", developed in response to observations about tumor onset sporadic and familial retinoblastoma patients [7]. Following this prediction of the *Rb* tumor suppressor gene, Li and Fraumeni identified cancer-susceptible families, leading to the identification of the *p53* tumor suppressor gene [8], and many others have followed. This is of particular importance in breast cancer, where two major tumor suppressor genes, *BRCA1* and *BRCA2* [9, 10], were identified through heroic mapping and linkage analysis efforts in the pre-genome era. Patients bearing heterozygous mutations in these genes have a vastly increased lifetime risk of breast cancer, approaching 80–90%. They are strongly urged to have intensive breast cancer surveillance with examinations and mammography, and to consider prophylactic mastectomy, which markedly reduces, but does not totally eliminate, the risk of breast cancer.

Patients bearing the mutations in amyloid proteins, amyloid precursor proteins and amyloid-related proteins discussed in other chapters, presenilins, α-synuclein, huntintin, transthyretin, etc., are at markedly elevated risk of developing the related disease, probably approaching 100%. The first challenge is to identify patients carrying such mutations. For such genetic diseases, perhaps half occur in an obvious hereditary context (i.e. multiple family members have the disease and Mendelian inheritance of susceptibility can be identified), but roughly half occur with new mutations, where a patient is the first to develop the mutation, which can then be passed on to offspring. Thus, there should ideally be screening of both obvious familial cases and apparently non-hereditary sporadic cases of disease for gene mutations.

Once such mutations are identified, they have two consequences. When disease is preventable, early diagnosis and treatment strategies can be implemented. Alternatively, when disease is highly morbid and not preventable, consideration can be given to prenatal diagnosis and pregnancy termination.

How can amyloidosis be prevented? The best example is that of the hereditary systemic amyloidoses, in which organ-based "gene therapy" can be used: orthotopic liver transplantation removes the major factory of the abundant serum proteins that cause these diseases. In AL amyloidosis, the removal is accomplished with high-dose anti-plasma cell chemotherapy and stem cell transplantation. These are expensive and morbid treatments, like breast removal for women with *BRCA* mutations or total colectomy in patients with hereditary polyposis of the colon at risk for colon cancer. It would be preferable to have small molecules that interfere with amyloid fibril formation from which non-toxic drugs could be developed that could dramatically delay the onset or prevent disease. Such drugs could be useful for the neurodegenerative diseases, which are not amenable to organ ablation approaches. The conformation stabilization drug strategy described above would be promising for this approach.

15.7 Conclusions

The diagnosis and treatment of the amyloidoses are entering a promising and exciting new phase, based upon the insights into the molecular basis of these diseases discussed in other chapters of this book. Building upon the basic research, a revolution is beginning that will lead to early diagnosis, accurate classification, non-invasive disease staging, targeted therapeutics and disease prevention for the amyloidoses. To accomplish this, basic scientists, clinical investigators, clinicians and patients need to work hand in hand. Their interactions and communications are facilitated by the rate at which information can be disseminated through the Internet. Patients and doctors can use the Internet to seek assistance for early diagnosis of disease through expert help at specialized centers of research. This is a common practice for patients with rare diseases or unusual symptoms and the Internet has provided a unique forum for patients to communicate with each other around the globe, and to communicate with physicians and researchers. This facilitates interactions, promotes knowledge and state-of-the-art patient care and research. Interactions of patients and researchers make patient material more readily available for study and facilitate patient participation in clinical trials. Moving this process ahead quickly and efficiently will be the most rapid way of bringing new treatments to clinical use, and the support of the public and public funding of research is essential.

References

1 Hawkins, P. N., Lavender, J. P. and Pepys, M. B. Evaluation of systemic amyloidosis by scintigraphy with ^{123}I-labeled serum amyloid P component. *N Engl J Med* **1990**, *323*, 508–513.

2 Aprile, C., Marinone, G., Saponaro, R., Bonino, C. and Merlini, G. Cardiac and pleuropulmonary AL amyloid imaging with technetium-99m labelled aprotinin. *Eur J Nucl Med* **1995**, *22*, 1393–1401.

3 Klunk, W. E., Engler, H., Nordberg, A., Wang, Y., Blomqvist, G., Holt, D. P., Bergstrom, M., Savitcheva, I., Huang, G. F., Estrada, S., Ausen, B., Debnath, M. L., Barletta, J., Price, J. C., Sandell, J., Lopresti, B. J., Wall, A., Koivisto, P., Antoni, G., Mathis, C. A. and Langstrom, B. Imaging brain amyloid in Alzheimer's disease with Pittsburgh Compound-B. *Ann Neurol* **2004**, *55*, 306–319.

4 Pepys, M. B., Herbert, J., Hutchinson, W. L., Tennent, G. A., Lachmann, H. J., Gallimore, J. R., Lovat, L. B., Bartfai, T., Alanine, A., Hertel, C., Hoffmann, T., Jakob-Roetne, R., Norcross, R. D., Kemp, J. A., Yamamura, K., Suzuki, M., Taylor, G. W., Murray, S., Thompson, D., Purvis, A., Kolstoe, S., Wood, S. P. and Hawkins,

P. N. Targeted pharmacological depletion of serum amyloid P component for treatment of human amyloidosis. *Nature* **2002**, *417*, 254–259.

5 Sanchez, I., Mahlke, C. and Yuan, J. Pivotal role of oligomerization in expanded polyglutamine neurodegenerative disorders. *Nature* **2003**, *421*, 373–379.

6 Adamski-Werner, S. L., Palaninathan, S. K., Sacchettini, J. C. and Kelly, J. W. Diflunisal analogues stabilize the native state of transthyretin. Potent inhibition of amyloidogenesis. *J Med Chem* **2004**, *47*, 355–374.

7 Knudson, A. G., Jr. Mutation and cancer: statistical study of retinoblastoma. *Proc Natl Acad Sci USA* **1971**, *68*, 820–823.

8 Malkin, D., Li, F. P., Strong, L. C., Fraumeni, J. F., Jr, Nelson, C. E., Kim, D. H., Kassel, J., Gryka, M. A., Bischoff, F. Z. and Tainsky, M. A. Germ line *p53* mutations in a familial syndrome of breast cancer, sarcomas, and other neoplasms. *Science* **1990**, *250*, 1233–1238.

9 Friedman, L. S., Ostermeyer, E. A., Szabo, C. I., Dowd, P., Lynch, E. D., Rowell, S. E. and King, M. C. Confirmation of *BRCA1* by analysis of germline mutations linked to breast and ovarian cancer in ten families. *Nat Genet* **1994**, *8*, 399–404.

10 Schubert, E. L., Mefford, H. C., Dann, J. L., Argonza, R. H., Hull, J. and King, M. C. *BRCA1* and *BRCA2* mutations in Ashkenazi Jewish families with breast and ovarian cancer. *Genet Test* **1997**, *1*, 41–46.

16
Brain Dysfunction Associated with Amyloid Fibrils and Other Aggregated Proteins

Giorgio Giaccone, Mario Salmona, Fabrizio Tagliavini and Gianluigi Forloni

16.1
Introduction

The presence of protein aggregates has been proposed as a unifying feature of neurodegenerative disorders. The aggregates may consist of fibrils with a β-pleated-sheet conformation, termed amyloid, or be made up of misfolded proteins without the staining and ultrastructural properties of amyloid. This abnormal material can accumulate at the intra- or extracellular levels, but is invariably associated with neuronal degeneration. In some instances, new categories of disorders could be identified on the basis of the misfolded protein. The formation of inclusions may represent an early event or the end stage of a molecular cascade of several steps and the pathogenic role of the aggregates might be variable in different diseases. For several neurodegenerative disorders, genetic variants assist in explaining the pathogenesis of the more common sporadic forms, and in developing mouse and other models. Our understanding of the pathways involved in protein aggregation and in the molecular mechanisms of cellular toxicity is growing rapidly [1, 2].

The presence of extracellular amyloid deposits and of intracellular neurofibrillary tangles is the main neuropathological feature of Alzheimer's disease. The β amyloid (Aβ) protein, a peptide of 40–42 amino acids, is the major component of amyloid, while neurofibrillary tangles are made up of hyperphosphorylated tau protein. Based on the presence of abnormal hyperphosphorylation and aggregation of tau without other specific neuropathological abnormalities, a heterogeneous group of neurodegenerative disorders clinically characterized by dementia and/or motor syndromes is now unified under the term of "tauopathies". The group includes Pick's disease, progressive supranuclear palsy (PSP) and corticobasal degeneration (CBD). The discovery that mutations of protein tau are associated with familial forms of frontotemporal dementia and parkinsonism (FTDP-17) provides strong evidence that the derangement of tau metabolism contributes to neurodegeneration.

The accumulation of a misfolded form of the prion protein characterizes prion diseases and plays an essential role also in their transmissibility. Molecu-

Amyloid Proteins. The Beta Sheet Conformation and Disease. J. D. Sipe

ISBN: 3-527-31072-X

lar genetic investigations allowed the identification of α-synuclein aggregates as the main component of Lewy bodies, the intracellular inclusions that characterize the degenerating cells in Parkinson's disease. α-Synuclein is also found in the intraneural inclusions of other neurodegenerative disorders, such as dementia with Lewy bodies (DLB) and multiple system atrophy, which are all now commonly described as "α-synucleinopathies".

The pathogenetic role of protein aggregates found in other groups of neurodegenerative diseases is less defined. Ubiquitinated cytoplasmatic inclusions have been detected in amyotrophic lateral sclerosis, and distinguished morphologically in skein inclusions, hyaline Lewy body-like inclusions and Bunina bodies. The composition of these aggregates is varied: in the familial forms associated with copper–zinc superoxide dismutase (SOD1) mutations, the mutated form of the enzyme was consistently found in the aggregates. Recent studies have proposed a direct involvement of SOD1 misfolding and aggregation in the pathogenesis of this disease. According to this hypothesis, Stathopulos et al. [3] have shown that SOD1 mutants undergo conformational changes that facilitate their polymerization. However, in terms of temporal development of the neurodegenerative process investigated in animal models, the appearance of intracellular inclusions is not an early event. Furthermore, the heterogeneous composition of intracellular inclusions in sporadic cases did not support a seminal role of protein aggregation in the pathogenesis of amyotrophic lateral sclerosis [4–6].

The presence of a trinucleotide (CAG) repeat expansion is characteristic of many neurological disorders including Huntington's disease, spinobulbar muscular atrophy, dentatorubro-pallidoluysian atrophy and different forms of spinocerebellar ataxia. The expanded polyglutamine [poly(Q)] tract lies in functionally unrelated genes and the accumulation of poly(Q) sequences is a common pathological hallmark of these disorders. In most cases, the accumulation occurs at the level of neuronal nuclei (intranuclear inclusions). The molecular basis of the toxicity of poly(Q) has been explained by two alternative concepts. One implies that the expansion leads to conformational changes creating a misfolded structure with cytotoxic properties [7]; the other that the long poly(Q) sequence increases the capacity of the protein to form amyloid aggregates, according to the hypothesis of nucleation [8]. However, the central role of protein aggregates in poly(Q) disorders has been questioned by researchers who consider them to be a neuronal byproduct that, in some cases, may have protective functions [9, 10].

16.2 Neuropathology

16.2.1 Alzheimer's Disease

The neuropathologic hallmark of Alzheimer's disease is the co-occurrence of extracellular deposits of amyloid made up of straight fibrils of around 10 nm whose main component is the Aβ polypeptide and the intracellular build-up of abnormal twisted filaments [paired helical filaments (PHFs)], mainly composed of hyperphosphorylated tau protein. Aβ deposition takes place in the neuropil as well as in leptomeningeal and parenchymal vessel walls, whereas PHF formation occurs within perykaria and neurites of selected neuronal populations. Therefore, Alzheimer's disease is a unique protein misfolding disease, since it is characterized by misfolding of two unrelated proteins, Aβ and tau, that cause two distinct histopathologic changes. This peculiarity, together with the fact that Aβ deposition and neurofibrillary changes can occur independently of each other, is the basis of several questions regarding the pathogenesis, and, in particular, which one of the two lesions is more strictly associated with the development of diffuse neuronal loss and dysfunction, synaptic rarefaction, and the appearance of the clinical signs of dementia. The issue is further complicated by the fact that Aβ deposition and tau pathology, although with differences in severity and topographic distribution, may also be present in non-demented elderly subjects [11]. A main distinction is that, in non-demented subjects, neurofibrillary changes, when present, are less severe and confined to restricted brain areas (the mesial temporal structures), while Aβ deposits may be widespread in the neocortex, in some instances in an amount similar to Alzheimer's disease. This is in keeping with the concept that neurofibrillary changes evolve in an anatomical stereotypical hierarchical fashion, with the entorhinal cortex being the earliest area affected, followed by the hippocampal formation. With increasing disease progression, neurofibrillary changes occur in the association cortex, while the primary sensory and motor cortices are spared until the very advanced stages of the disease [12, 13] (Fig. 16.1 A and B).

Aβ is a 40- to 42-residue peptide derived by proteolytic cleavage of a 695- to 770-amino-acid Aβ protein precursor (AβPP) [14–16], a transmembrane glycoprotein encoded on chromosome 21, in which Aβ is positioned partly in the extracellular domain and partly in the transmembrane domain (see also Chapter 17). The Aβ deposits in the vessel wall take the form of congophilic (amyloid) angiopathy, while in the neuropil, the deposits determine the formation of the senile plaques (Fig. 16.2). These are complex lesions whose morphogenesis is influenced by the participation of cellular elements, either neuronal (neurites and synaptic terminals) or glial (reactive astrocytes and activated microglia), as well as by the association of Aβ amyloid with other "chaperone" molecules. These may be associated to all forms of amyloid, such as P component, complement factors, apolipoprotein (Apo) E and J, or more specific for Aβ deposits as

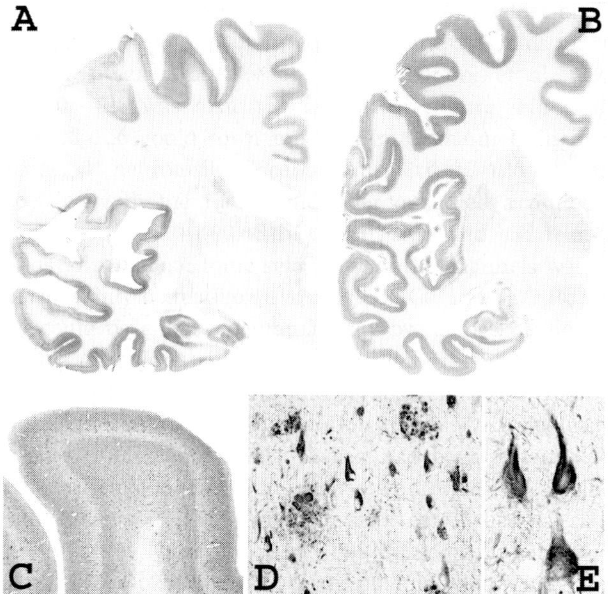

Fig. 16.1 Alzheimer's disease. Immunohistochemistry with monoclonal antibody AT8 to phosphorylated tau (immunoreactivity corresponds to the brown reaction product) revealed severe involvement of the cerebral cortex by neurofibrillary changes that can spare the primary motor and sensory areas (A) or involve all cortical fields (B). AT8 labeled cell bodies and a network of neuronal processes dispersed in the neuropil of the cerebral cortex (neuropil threads) (C). Clustering of AT8-immunoreactive profiles occurred in direct correspondence to senile plaques (D). Labeled neurons often showed pyramidal morphology, with AT8 immunostaining extending to the apical dendrite, reproducing the picture of flame-shaped NFTs (E).

α_1-antichymotrypsin [17]. The introduction of immunohistochemical techniques using antibodies against Aβ revealed a far higher density and wider distribution of Aβ deposits in Alzheimer's disease than had been appreciated by the use of classic silver impregnation methods or by applying specific staining for amyloid, such as Congo red or Thioflavin S. This was because immunohistochemistry could also recognize Aβ deposits that lacked the staining and ultrastructural properties of amyloid. These amorphous appearing, non-fibrillar plaques were referred to as "diffuse plaques" or "pre-amyloid deposits", since they might represent an early stage in the morphogenesis of senile plaques, also considering that they are not associated with neuritic or glial changes [18–21] (Fig. 16.3). This hypothesis is supported by studies on patients with Down's syndrome (trisomy 21) who invariably develop Alzheimer's neuropathological changes, starting with pre-amyloid deposits in their teenage years and displaying fully-developed

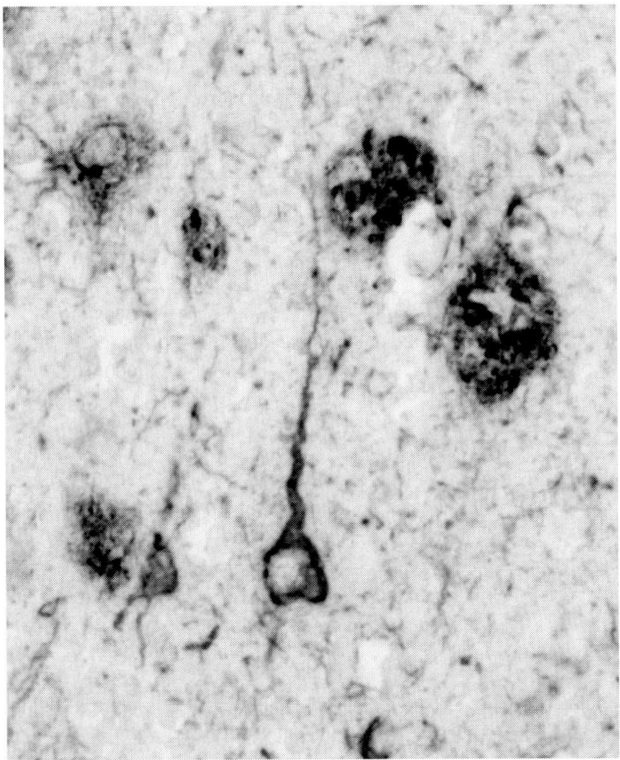

Fig. 16.2 Alzheimer's disease. Double immunohistochemical detection of phosphorylated tau (blue) and Aβ protein (brown). Note the intermingling of Aβ deposition with tau-immunoreactive neuronal profiles, configuring the senile plaque.

senile plaques with neuritic/glial changes and neurofibrillary tangles (NFTs) about two decades later [22, 23]. Moreover, by immunohistochemistry, it has been demonstrated that, in Alzheimer's disease, Aβ deposition also affects brain regions previously considered to be spared by the pathologic process, such as the cerebellum, striatum and brainstem [24, 25]. Interestingly, non-fibrillar pre-amyloid deposits are the predominant form of Aβ deposition in these regions. The concept of the C-terminal heterogeneity of the Aβ peptides deposited in Alzheimer's disease brains provided a further distinguishing element between pre-amyloid deposits and senile plaques, since the former are made up almost exclusively by Aβ1–42, while in the latter contain a mixture of Aβ1–40 and Aβ1–42. Aβ1–40 is the main constituent of amyloid in the vessel walls [26, 27].

NFTs are caused by the intracellular accumulation of PHFs, abnormal twisted filaments 80–200 nm thick, whose main component is an abnormally phosphorylated form of tau protein [28–30]. Tau is a microtubule-associated protein, encoded by a gene on chromosome 17 [31, 32], highly expressed in the central

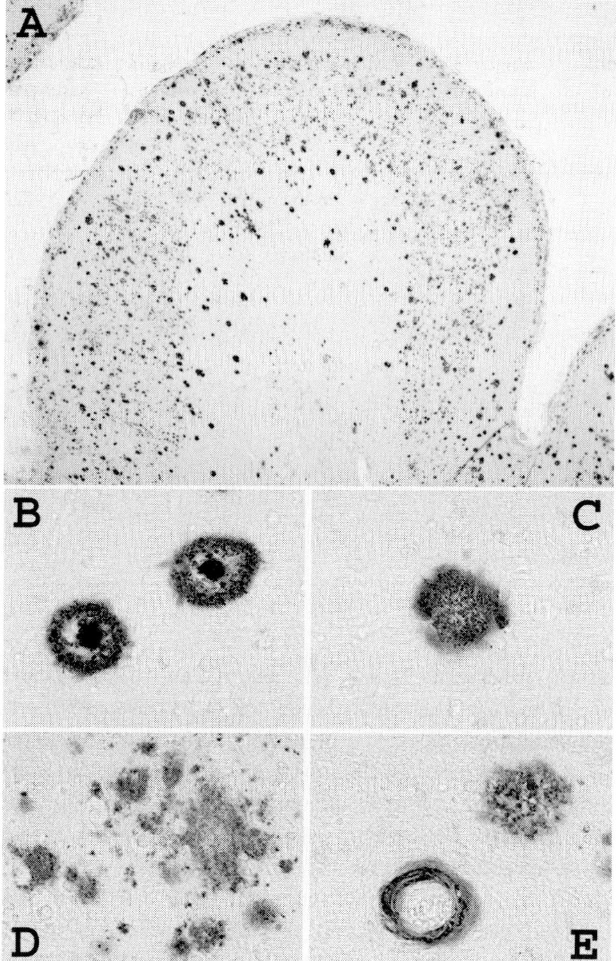

Fig. 16.3 Alzheimer's disease. Immunohistochemistry with antibodies to Aβ (immunoreactivity corresponds to the brown reaction product) revealed the high density of Aβ deposits in the cerebral cortex (low magnification, A) and the different morphologies of Aβ deposition: classic senile plaques with a dense core and a more dispersed peripheral halo (B), amorphous amyloid deposit (C), pre-amyloid deposits (D) and amyloid angiopathy close to a senile plaque (E).

nervous system (CNS). In normal conditions, tau is present exclusively in axons, while in Alzheimer's disease, it accumulates in an insoluble, hyperphosphorylated form ectopically in the perikaria and dendrites. The main biologic function of tau is to bind microtubules via the C-terminal binding domains, promoting their polymerization and stabilization. Tau therefore plays an important role in maintaining neuronal integrity, axonal transport and polarity [33, 34]. Highly phosphorylated tau has fewer tendencies to bind microtubules [35–38]. NFTs and degenerating neurites of plaques are consistently immunoreactive with anti-phosphorylated tau antibodies (Fig. 16.1 D). Moreover, immunohistochemistry with these antibodies is much more sensitive than classic silver impregnation methods in revealing neurofibrillary changes and enabled recognition of an additional form of hyperphosphorylated tau (Fig. 16.1 C–E). These are the "neuropil threads", a meshwork of randomly oriented, tau-immunoreactive neurites, dispersed in the neuropil, that represent a consistent part of the total burden of neurofibrillary pathology in Alzheimer's disease [39]. NFT and degenerating neurites of plaques are also consistently immunoreactive for ubiquitin. The molecular mechanism of PHF formation has not been clarified: several kinases and phosphatases have been implicated [40, 41], but it remains to be determined whether hyperphosphorylation is the primary event that makes tau insoluble and resistant to degradation or is a consequence of the aggregation process.

In summary, several lines of evidence indicate the cerebral deposition of Aβ amyloid is the seminal event that initiates a cascade of biochemical and cellular changes leading to tau-related alteration of the neuronal cytoskeleton and to neurodegeneration. This scenario – known as the "amyloid cascade hypothesis" [42, 43] – is supported by the observation that quantitative modification of the total levels of Aβ or of the ratio between long forms and short forms of this peptide (i.e. Aβ1–42 versus Aβ1–40) are sufficient to induce the whole neuropathologic phenotype of the disease. The first situation takes place in Down's syndrome, in which one extra copy of the gene coding for the AβPP is present as a consequence of the chromosome 21 trisomy, while the second corresponds to the familiar cases of Alzheimer's disease associated with mutations of the genes for AβPP, presenilin1 or presenilin2 (see below) [43].

NFT composed of tau aggregates that are biochemically similar to those of Alzheimer's disease have been described in many neurodegenerative diseases in which Aβ is absent. These are referred as tauopathies, and include Pick's disease, CBD and PSP. A subset of tauopathies is familial and mutations of the tau gene have been demonstrated in a percentage of them [44–47].

16.2.2 Tauopathies

Starting from 1994, an autosomal-dominantly inherited form of frontotemporal dementia with parkinsonism was described in several families and linked to the region of chromosome 17 that contains the tau gene, resulting in the denomination "frontotemporal dementia and parkinsonism linked to chromosome 17" (FTDP-17) [44, 48]. The neuropathologic picture of FTDP-17 is characterized by tau-positive inclusions that occur in neurons, but may also be numerous in glial cells. The now widely used term "tauopathy" was introduced to highlight the abundance of tau accumulation, in the absence of Aβ aggregates [45]. In 1998, the first mutations in the tau gene (MAPT) were reported in FTDP-17 families and now the total of the known mutations is more than 30 [46, 47, 49].

Tau protein is encoded by a single gene of 16 exons from which different forms are produced via alternative mRNA splicing. In adult human brain, six tau forms are generated ranging from 352 to 441 amino acids, differing with respect to the presence of three (3R-tau) or four (4R-tau) C-terminal repeat sequences of 31–32 residues. Three different forms of each (3R-tau and 4R-tau) exist, as a result of the presence of none, one or two N-terminal inserts of unknown function [41]. The complex structure of the tau protein is the basis of the variability in the chemicophysical characteristics of the insoluble tau filaments in different tauopathies. The partial elucidation of the relationship between the site of the mutation on the MAPT gene, the tau isoforms engaged in the misfolding process, and the biochemical and structural characteristics of the tau aggregates provided very important insights into the role of tau abnormalities in neurodegenerative diseases, and more generally into the pathogenesis of protein misfolding diseases.

The most important implication derived from studies of patients with FDTP-17 is that the mutation of the tau gene is sufficient to cause intracellular tau deposition and to induce neurodegeneration. Moreover, the studies of these rare inherited forms of tauopathies have shed light on the pathogenesis of other more common, sporadic neurodegenerative diseases characterized by tau inclusions (Pick's disease, CBD and PSP), that can be considered as primary tauopathies, due to the absence of other disease-specific neuropathological abnormalities. In contrast, other brain diseases are known in which tau inclusions are likely secondary to metabolic (Niemann–Pick type C), infective (subacute sclerosing panencephalitis) or traumatic (dementia pugilistica) brain lesions.

PSP is characterized clinically by supranuclear gaze palsy and by postural instability [50]. Neuropathologically, the main features are atrophy of the basal ganglia and brainstem, with neuronal loss and gliosis. In these brain areas, there is a high density of tau pathology, including NFTs in neurons and consistent tau aggregates in glial cells, both astrocytes ("tufted astrocytes") and oligodendrocytes ("coiled bodies") [51] (Fig. 16.4 C and D). In contrast to Alzheimer's disease, ultrastructurally these aggregates are made up straight fibrils of 15–18 nm [52] and biochemically are mainly composed of 4R-tau isoforms [41].

CBD is a progressive neurodegenerative disease involving the cerebral cortex and deep grey and white matter structures such as the striatum, thalamus, substantia nigra, capsula interna, subcortical frontal white matter and brainstem [53]. The neuropathological features are depigmentation of substantia nigra and asymmetrical frontoparietal atrophy. In affected regions, neuronal loss, gliosis and tau-immunoreactive glial and neuronal inclusions are prominent, and surviving nerve cells are often swollen ("achromatic" or "ballooned" neurons). The neuronal tau inclusions are morphologically pleomorphic. In some neurons they appear as dispersed globose NFTs; in others as small compact inclusion bodies and occasionally tau-immunoreactivity diffusely fills the cytoplasm. A typical finding of CBD is the prominent deposition of abnormal tau in neuronal processes

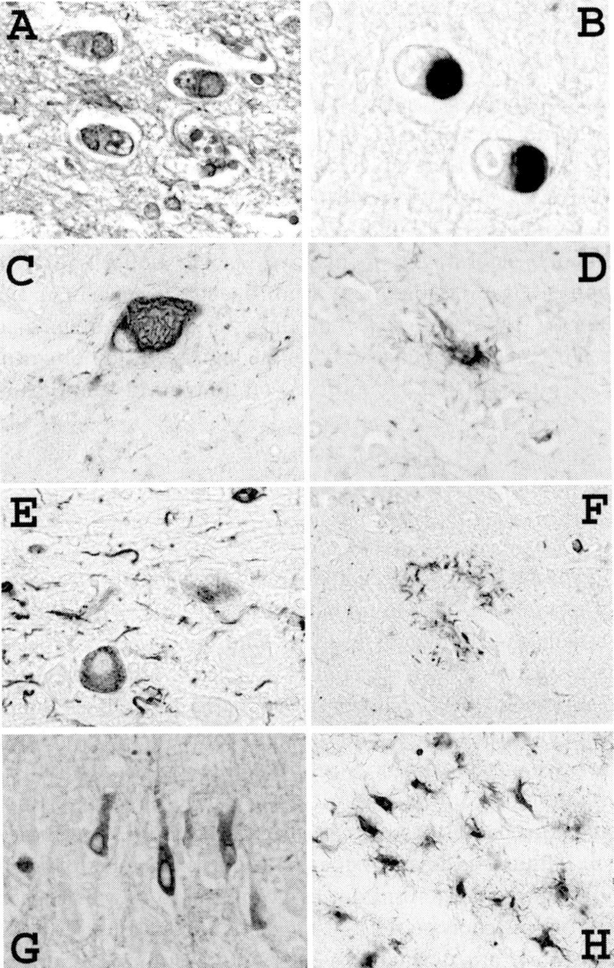

Fig. 16.4 Tauopathies. In Pick's disease the characteristic neuronal, round, well-demarcated inclusions (Pick bodies) are revealed by Bodian silver impregnation (A) and by immunohistochemistry with monoclonal antibody to phosphorylated tau (AT8, immunoreactivity corresponds to the brown reaction product) (B). In PSP the tau pathology is stained by monoclonal antibody AT8 to phosphorylated tau, and includes NFTs in neurons (C) and tau deposition in glial cells, as the "tufted astrocytes" (D). In CBD the deposition of phosphorylated tau disclosed by AT8 immunohistochemistry is prominent in neuronal perikaria (E), but also in neuritic processes (E). The "astrocytic plaque" (F) is considered typical of this disease. FTDP-17 is characterized by the presence of phosphorylated tau deposition revealed by AT8 immunohistochemistry in neurons (often diffuse in the cytoplasm, G) and in glial cells (H).

("thread-like neuritic profiles") and in glial cells, with consistent involvement of the white matter. In the cerebral cortex, peculiar lesions consisting of annular clusters of tau-immunoreactive processes ("astrocytic plaques") are considered typical of CBD [54] (Fig. 16.4 E and F). The abnormal filaments include both PHF-like and straight filaments [55]. The biochemical profile of insoluble tau is similar to that of PSP, being mainly composed of 4R-tau isoforms [41].

Pick's disease is now a term that should be restricted to forms of frontotemporal degeneration with tau-immunoreactive lesions and, in particular, with the characteristic neuronal, round, well-demarcated argyrophilic inclusions (Pick bodies) [56] (Fig. 16.4 A and B). The neuropathologic picture is that of severe frontotemporal atrophy with marked neuronal loss, gliosis and neuropil vacuolization. Pick bodies are abundant in the dentate gyrus of the hippocampus and focally in layers II, III and VI of the cerebral cortex. They are detected by silver impregnation and by immunohistochemistry with antibodies to hyperphosphorylated tau that reveal additional cytoskeletal changes, including neuritic profiles and astrocytic and oligodendroglial inclusions [57]. Biochemically, the tau aggregates are predominantly made up by the 3R-tau isoforms [41]. It should be recalled that the majority of patients with the clinical features of FTD do not show the neuropathology of Pick's disease, but a more unspecific picture of atrophy of the frontal and temporal lobes with neuronal loss, gliosis and microvacuolar neuropil degeneration, in the absence of tau-immunoreactive changes.

The neuropathology of FTDP-17 is heterogeneous with respect to the morphologic characteristic of the lesions as well as their severity and topographic distribution. It can overlap with the scenario of Pick's disease, PSP and CBD – the invariable change being the presence of tau deposits in neurons and glial cells in several grey and white matter structures of the CNS (Fig. 16.4 G and H). The heterogeneity of the clinical/pathologic phenotypes partially reflects the position of the mutations in the tau gene. They can be separated in two broad categories: (1) coding region missense mutations, and (2) intronic mutations and silent mutations of the coding region.

The first group of tau mutations is the ordinary one, in which a nucleotide substitution in the coding region of the MAPT gene changes the corresponding amino acidic residue. Most of the reported missense mutations occur in the highly conserved region within or near the microtubule-binding domains. The effects of these mutations seem to be 2-fold: (1) they reduce the ability of tau protein to bind to microtubules and to promote microtubule assembly *in vitro* [58, 59], and (2) they increase the tendency of tau to aggregate in insoluble filaments [60, 61]. The structure and biochemical characteristics of the misfolded tau aggregates are dependent on the exon of the mutation. While mutations located outside exon 10 modify all six tau isoforms, those within exon 10 affect only the three tau isoforms bearing four microtubule-binding domains (4R-tau), since the supplementary repeat of this portion of the molecule is coded by exon 10.

The second class of mutations is more puzzling. These genetic defects are single base pair substitutions occurring within intron 10 or silent nucleotide changes in the adjacent exon 10. At first it was difficult to explain how mutations that do not affect the sequence of the tau protein could provoke its misfolding. The explanation came from biochemical studies demonstrating that these mutations increase the levels of 4R forms of tau, likely due to a higher proportion of mRNAs in which exon 10 is retained [62, 63], as a consequence of the disruption of a regulatory stem–loop, "splicing enhancer" structure [64]. The mechanisms by which changes in the ratio of 3R-tau to 4R-tau lead to neuronal dysfunction is unclear, but it seems established that a ratio of 3R-tau forms to 4R-tau forms of about unity is necessary for the normal structure and function of microtubules. It has been suggested that some missense mutations in exon 10 close to the downstream intron may also act with this mechanism.

16.2.3 Prion Diseases

Prion diseases are fatal, rapidly progressive, neurodegenerative disorders of humans and animals. They are a unique category of diseases, since may be sporadic or inherited in origin and can be transmitted. They are therefore also called "transmissible spongiform encephalopathies", a term that underlines their infectious character and the classic main neuropathological hallmark of neuropil vacuolation. The transmissible agent, the prion, is devoid of informational nucleic acid and consists only of protein [65] (see below). Several lines of evidence indicate that prions are composed of an abnormal, pathogenic isoform of the prion protein (PrP). The normal form of PrP (PrP^C) is widely expressed in neurons and glia in the CNS, and little is known about its function(s). The human cellular gene which encodes PrP^C has been called PRNP, whose open reading frame is in a single exon. The pathogenic isoform (PrP^{Sc}) results when the normal form undergoes a conformational change, converting α-helical regions to β-sheet motifs, and possesses abnormal physicochemical properties such as detergent insolubility and protease resistance [66]. Therefore, prion diseases are "transmissible protein misfolding diseases", in which the abnormally folded protein may induce the conversion of normal molecules into the misfolded status.

Animal prion diseases include scrapie in sheep and goats and bovine spongiform encephalopathy (BSE) in cattle. In humans, the most common prion disease is Creutzfeldt–Jakob disease (CJD), which can be inherited as autosomal dominant disease associated with mutations of the PRNP gene, acquired iatrogenically through exposure to material contaminated with prions or arise sporadically for no obvious reason. Possible causes of sporadic CJD include spontaneous formation of PrP^{Sc} as a rare stochastic event, somatic mutations of PRNP or unrecognized prion exposure. A variant CJD (vCJD) form was reported in 1996 [67], characterized by clustering in the UK and peculiar neuropathological and biochemical features. Geographic and temporal association, as well as transmission studies revealing strong analogies between the vCJD and BSE agents, lead to the conclusion that vCJD is acquired by humans by exposure to BSE contaminated material.

The detection of the deposition of PrP^{Sc} is now possible not only biochemically by immunoblot analysis, but also in histological specimens following the introduction of effective pre-treatments that make immunohistochemistry with anti-PrP antibodies sensitive and specific [68–70]. These techniques revealed how PrP^{Sc} accumulates in the CNS of CJD patients with variable topographic distribution and different patterns of immunostaining – synaptic, perivacuolar, perineuronal and plaque-like (Fig. 16.5 A–D). The accumulation of PrP^{Sc} in the CNS is accompanied by neuronal and glial changes. The triad of spongiform changes, neuronal loss and gliosis, in the absence of an inflammatory response, is the classic neuropathological hallmark of CJD. Spongiform change is relatively specific to CJD, but may differ in severity in different patients and from regions to regions of the CNS. It is characterized by small, round or oval vacuoles in the neuropil of the cerebral cortex and other grey structures (Fig. 16.5 E). Glial changes consist of reactive astrogliosis and microglial activation (Fig. 16.5 F). Neuronal loss in the affected cortical and subcortical regions is often severe, sometimes with complete depletion of some neuronal populations [71]. While in classic CJD cases the formation of true amyloid deposits is a rare event and is restricted to specific brain regions, in Gerstmann–Sträussler–Scheinker (GSS) disease this phenomenon is relevant, and PrP amyloid deposits are abundant and widespread in the CNS [72] (Fig. 16.5 G). GSS is determined by specific mutations in the PRNP gene and has a longer disease duration than CJD. Amyloid PrP deposition is also abundant in vCJD, in which the typical lesion is the florid plaques – an amyloid core surrounded by vacuoles of spongiosis (Fig. 16.5 H).

The neuropathology of CJD is heterogeneous and it has been shown that this may be related to variations in the tertiary structure of PrP^{Sc}, resulting in different conformers of the abnormal protein having distinct physicochemical and pathogenic properties [73]. This is confirmed by the observation of molecular size differences in the protease-resistant core of PrP^{Sc} that are due to different sites of proteolytic cleavage reflecting different tertiary structures. In particular, the clinico-pathological heterogeneity of sporadic CJD has been linked to two types of PrP^{Sc}, termed type 1 and type 2, having a molecular weight of 21 and 19 kDa. Evidence suggests that the PrP^{Sc} type in combination with the genotype at codon 129 of the PRNP gene – a common polymorphic site encoding methionine or valine – is a major determinant of deposition pattern (i.e. diffuse or focal) and brain regional distribution of PrP^{Sc}. The view that different conformers of PrP^{Sc} may possess distinct pathogenic properties is supported by a study on CJD cases with both type 1 and type 2 PrP^{Sc} in the same brain. This event is relatively common, involving about 25% of the sporadic patients with a close relationship between PrP^{Sc} type, pattern of PrP immunoreactivity and severity of spongiform changes [74].

Further support for the prion hypothesis has been provided by transgenic animal studies; in particular, the finding that mice in which the PrP gene has been ablated (PrP knock-out mice) neither develop neuropathological features of prion diseases nor accumulate PrP^{Sc} or propagate the disease after inoculation

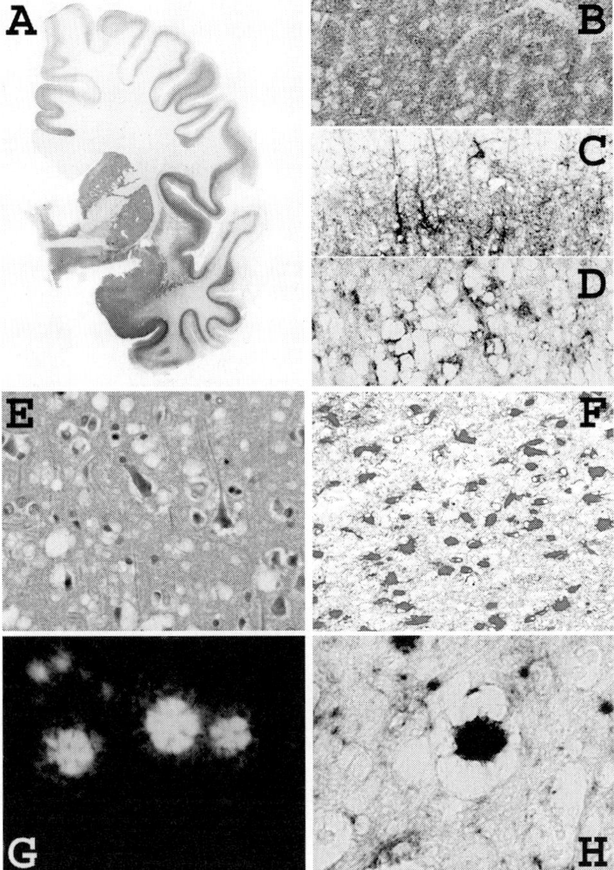

Fig. 16.5 Prion diseases. In CJD, most grey structures exhibit PrP^{Sc} immunoreactivity (corresponding to the brown reaction product) (A) that may assume different patterns: diffuse synaptic type (B), perineuronal (C) and perivacuolar (D), and is associated with spongiform changes (hematoxylin & eosin, E) and astrogliosis (anti-glial fibrillary acid immunohistochemistry, F). PrP amyloid deposits are consistent in GSS (Thioflavin S, G) and vCJD (PrP immunohistochemistry, H).

of prions, demonstrating that the absence of PrP^C expression prevents prion spread and neuronal dysfunction [75].

16.2.4 Synucleinopathies

Synucleinopathies represent a heterogeneous group of neurodegenerative disorders characterrized by the presence of cytoplasmic neuronal and glial inclusions. The synucleinopathies can be divided into Lewy body disorders (Parkinson's disease and DLB) and disorders with intracytoplasmic inclusions in glial cells (multiple system atrophy).

Parkinson's disease is the second most common neurodegenerative disease after Alzheimer's disease. Parkinson's disease is characterized clinically by resting tremor, rigidity and bradykinesia, resulting from the progressive and selective loss of dopaminergic neurons in the pars compacta of the substantia nigra. Histopathologically, it is characterized by the degeneration of specific nerve cell populations that develop filamentous inclusions in the form of Lewy bodies and dilated neurites. The onset of clinical symptoms occurs when the loss of dopaminergic neurons is over 50%. Lewy bodies are found not only in the substantia nigra, but also in other brain structures [76]. The classical Lewy body is a spherical, cytoplasmatic inclusion with a diameter of about 10–30 μm. After hematoxylin & eosin staining, the Lewy body assumes the classical "target shape" with a central core more intensely stained and a more faded external portion (Fig. 16.6 A). At the ultrastructural level, the central portion is formed by tightly packed filamentous and granular material, while in the external part filaments of 7–20 nm are associated with electron-dense material. Other round-shaped inclusions ("pale" bodies) are also identified in Parkinson's disease, and appear as granular and eosinophilic material displacing neuromelanin of pigmented neurons, without the classical "target shape" of Lewy bodies. Pale bodies are considered precursors of Lewy bodies. The identification of familial forms of Parkin-

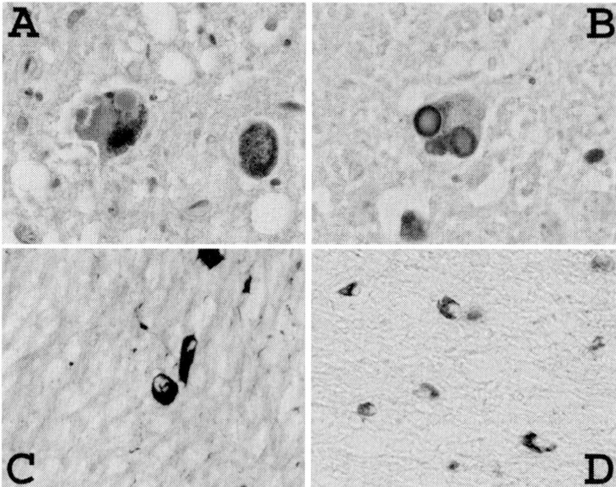

Fig. 16.6 Synucleinopathies. In Parkinson's disease, Lewy bodies appear as target-like, spherical, cytoplasmic inclusions (hematoxylin & eosin, A) intensely immunoreactive with anti-SYN antibodies (immunoreactivity corresponds to the brown reaction product, B). In MSA, GCIs appear black in sections stained with the Gallyas silver impregnation (C) and are immunolabeled by anti-SYN antibodies (D).

son's disease caused by missense mutations in the α-synuclein gene [77] prompted immunohistochemical studies showing that Lewy bodies are consistently and intensely immunoreactive with anti-SYN antibodies [78] (Fig. 16.6 B). Previously, Lewy bodies were recognized by means of anti-ubiquitin antibodies [79]. Furthermore, parkin, another protein genetically associated to familial Parkinson's disease, was invariably found in Lewy bodies [80]. SYN immunoreactivity was found not only in perykarial inclusions but also in neurites (Lewy neurites).

DLB defines a specific clinical syndrome characterized by progressive dementia with fluctuation of cognitive deficits and hallucinations, mainly visual, associated with the neuropathological finding of Lewy bodies. However, the threshold of Lewy bodies for a DLB diagnosis has not been defined yet. This could lead to the undetermined condition of subjects with a clinical diagnosis of dementia where the typical Alzheimer's disease pathology (Aβ deposits and NFTs) co-exists with the presence of Lewy bodies. The problem is that the relative contribution to the clinical symptoms of the individual pathological components is difficult to define. Other DLB cases are "pure", in which Alzheimer-type pathology is absent or minimal. Based on the localization and density of Lewy bodies, three categories have been identified: Lewy bodies at predominance in the brainstem, in the limbic cortex and in the neocortex. Unexpectedly, DLB is not necessary associated with the third category (Lewy bodies at predominance in neocortex). Cortical Lewy bodies after hematoxylin & eosin show different shapes: round, oval and bean-like, lacking the target shape typical of nigral Lewy bodies. On the other hand, the antigenic characteristics of cortical Lewy bodies are essentially the same of the classical Lewy bodies, as they are immunopositive for SYN, ubiquitin and parkin.

Multiple system atrophy (MSA) is clinically associated to disautonomic disturbances, parkinsonism and cerebellar signs. Neuropathologically, it is characterized by intracellular glial inclusions originally recognized by the silver impregnation method of Gallyas [81]. They appear as oval, or triangle-, flame- or sickle-shaped glial cytoplasmic inclusions (GCIs or Papp–Lantos inclusions) localized in oligodendrocytes (Fig. 16.6 C). The size is variable, but the inclusions often completely fill the cytoplasm, confining the nucleus at the periphery. GCIs are immunoreactive with antibodies against SYN [82] (Fig. 16.6 D). At the ultrastructural level, GCIs are composed of tubules or filaments having a diameter of 20–40 nm in association with granular material. In MSA subjects, SYN-immunoreactive inclusions can be found in glial and neuronal cells nuclei and in neurites. GCIs are prominent in the motor and sensory primary cortices, in the putamen, globus pallidus, internal and external capsula, corpus callosum, and several other structures of the brainstem and spinal cord.

16.3
The Neurotoxic Proteins

16.3.1
Alzheimer's Disease

The pathogenic role of Aβ in Alzheimer's disease has been put forward on the basis of genetic, experimental and biochemical evidence, and an essential contribution to this hypothesis derived from the demonstration that Aβ peptides can exert neurotoxic effects *in vitro* [42]. The toxicity of Aβ peptides is mediated by apoptosis [83, 84] and was initially associated with the polymerization into fibrils of the peptide [85]. The biological active fragment was found to be the Aβ 25–35 sequence. The fibrillogenic capacity, tested *in vitro*, increases progressively from peptide Aβ1–40 to Aβ1–42 and Aβ25–35. Aβ peptides containing 40 or 42 residues are those found in the brains of Alzheimer's disease patients in senile plaques and originate from the precursor of Aβ, AβPP. In the amyloidogenic pathway of catabolism, AβPP is cleaved sequentially by β- and γ-secretases at the N- and C-terminus of the Aβ region (see also Chapter 17). β-Secretase cleaves between residues 671 (Met) and 672 (Asp), while the cleavage at the C-terminal by γ-secretases can occur either at residue 711 (Val) or 713 (Iso) to generate Aβ1–40 or Aβ1–42 [86–88]. The amyloid-like fibrils spontaneously formed *in vitro* by these peptides have morphological, staining and ultrastructural features similar to the amyloid deposits isolated from Alzheimer's disease brains. The amyloid fibrils are composed of protofilaments that are hydrogen-binding β-sheet structures with the β-strands running perpendicular to the long fibril axis. It is not clear whether the conformational alterations in the oligomers are a consequence of the oligomer assembling or whether the misfolded monomers induce the formation of the aggregates. An intermediate possibility is that an initial, unstable, misfolded conformation of peptide monomer can be stabilized by interaction with other proteins to form oligomers that lead the formation of fibrils and aggregates [89]. In this case, the conformational changes are not triggered by oligomerization, but their stable conformation is indissolubly associated with the oligomers. The formation of aggregates in Alzheimer's disease brains is thought to be seeded by Aβ1–42, which is more prone to form aggregates, whereas Aβ1–40 accumulates subsequently. The essential role of Aβ1–42 is supported by evidence that the common molecular change associated with the mutations of three different genes linked to familial Alzheimer's disease (AβPP, presenilin1 and presenilin2) is the increase of Aβ1–42 production [90].

The close association between the neurotoxicity and the aggregation of the Aβ peptide was subsequently reconsidered, and the hypothesis that Aβ deposition is a primary factor in the pathogenesis of Alzheimer's disease has been challenged on the basis of studies showing that the degree of cortical synapse loss and the quantity of neurofibrillary changes predict the severity of dementia more accurately than the density of Aβ deposits [91, 92]. Recently, new data have emerged that lead to reconsideration of the amyloid cascade hypothesis;

they support as the central pathogenic element the soluble oligomers of Aβ rather than the fibrillar aggregates of the peptide [93, 94]. Using amidation at the C-terminal of β25–35 to reduce the fibrillogenic capacity of the peptide, we demonstrated that the neurotoxic activity was independent of the aggregation state of the peptide [95]. Accordingly, several other studies suggested that neuronal death better correlates with the presence of oligomer species rather than with aggregates of Aβ peptides [96, 97]. It is possible that senile plaques represent the reservoir of aggregated Aβ that can continuously release diffusible oligomers and protofibrils to induce injury not only in the surrounding cells, but also at a certain distance of the senile plaques. Other theories advance the notion that neuronal death is triggered by intracellular events that occur during AβPP processing. In this regard, a role of intraneuronal Aβ in Alzheimer's disease pathogenesis is suggested by the study of triple transgenic mice (AβPP/PS1/Tau). In the hippocampus of these animals, the intraneuronal accumulation of Aβ is an early event that precedes plaque formation and correlates with synaptic dysfunction [98]. A good correlation between total content of Aβ1–42 (biochemically determined in specific cerebral regions) and cognitive decline was recently found in Alzheimer's disease, in keeping with the closer association of intellectual deterioration with Aβ production rather than with the density of plaques [99].

16.3.2
Prion Diseases

As mentioned earlier, the pathogenic mechanism of prion diseases is based on the conformational conversion of the cellular prion protein (PrP^{C}) into the disease-specific species (PrP^{Sc}). The "protein only" hypothesis on the nature of the infectious agent (prion) indicates that it is essentially constituted of PrP^{Sc} [65] that is able to induce a "conformational transmission" via the host PrP^{C} [100]. It has been postulated that other element(s) (called protein X) facilitate the conversion of PrP^{C} to PrP^{Sc} [101].

Although the pathogenetic mechanisms are not fully clarified, the direct role of PrP^{Sc} in neuronal degeneration and glial activation is largely accepted. However, based on the evidence obtained in transgenic mice, Chiesa et al. [102] have demonstrated that the disease-associated, infectious form of the prion protein differs from the neurotoxic species. To evaluate the toxicity of PrP^{Sc}, in principle, one could directly apply the purified protein to neurons in culture. Although there have been several reports of these experiments [103], they are difficult to interpret because of uncertainties about the physical state of the PrP^{Sc}, since detergents that are required to maintain the protein in solution have to be removed prior to application to cell cultures. An alternative strategy has been to analyze the effect on cultured neurons of synthetic peptides derived from the PrP sequence [104]. The concept that misfolding of PrP causes a transmissible neurodegenerative disorder has prompted studies aimed at identifying polypeptide segments that are central to the conversion process. In particular, a synthetic peptide corresponding to the human PrP region 106–126 (PrP106–126) recapitulates several chemicophysical characteristics of PrP^{Sc}, including the propensity to form β-sheet-rich, insoluble and protease-resistant fibrils similar to those found in prion diseases [104, 105]. Similar to the studies with synthetic Aβ peptides, 10 years ago PrP106–126 was proposed as a model with which to investigate the biological effects of PrP^{Sc} [106]. This approach was successful with respect to the capacity of the synthetic peptide to mimic *in vitro* several aspects of the disease associated to the presence of PrP^{Sc}, including neurodegeneration, glial activation and alteration of membrane fluidity. Although PrP106–126 is not normally found in the brain of individuals with prion diseases, the various N- and C-terminal truncated fragments of PrP produced in the CNS in the course of sporadic and inherited prion diseases invariably contain the 106–126 sequence, suggesting that this region of the protein may possess the ability to trigger a fundamental pathogenic mechanism. The neurotoxic effect of PrP106–126 was abolished or reduced in neurons derived from PrP knock-out mice that are resistant to prion infectivity, arguing for a role of the cellular prion protein in the neurotoxic cascade activated by PrP106–126 and underscoring the relevance of this model [107, 108].

Numerous other peptides homologous to PrP fragments were synthesized in wild-type and mutated forms, and their chemicophysical characteristics and biological effects investigated [109]. In some cases, the introduction of missense mutations associated to familial prion diseases (i.e. D178N) increased the fibrillogenic capacity of the PrP peptides [110].

The PrP106–126 peptide is highly fibrillogenic, but its *in vitro* toxicity was independent of this feature, while the ability to induce astrogliosis seems to be associated with the aggregation state of the peptide [110]. The amidation at the C-terminal combined with the acetylation at the N-terminal strongly reduced the fibrillogenic capacity of PrP106–126, but did not affect the toxicity of the peptide [111]. These results are in agreement with the neuropathological evidence that PrP^{Sc} is deposited as amyloid fibrils only in particular conditions, as in GSS and vCJD, while in sporadic CJD PrP^{Sc} is assembled in non-fibrillar form, most likely as oligomers, strongly supporting the concept that soluble oligomers rather than mature amyloid fibrils are actually the pathogenetic species that causes neurodegeneration. It is noteworthy, in this regard, that an antibody recently developed and recognizing soluble oligomers of Aβ [112] prevented the toxicity induced by Aβ oligomers. Surprisingly, the antibody also recognized oligomer aggregates of SYN, islet amyloid polypeptide, polyglutamine, insulin and PrP peptides. As for Aβ, the antibody did not react with the monomeric form or the fibrillar version of these proteins, but only with the oligomers, indicating that the antibody recognizes a common structural epitope independent of the amino acid sequence.

16.3.3 Synucleinopathies

Epidemiological studies suggest that, in the etiology of synucleinopathies, environmental factors are more relevant than genetic ones. Nevertheless, recent data about the existence of rare familiar forms of Parkinson's disease have been collected, and a pathogenetic role has been demonstrated for several genes, as SYN, parkin and DJ-1, whose mutations have been linked to familiar forms of Parkinson's disease [113].

Two different missense mutations in the SYN gene (A30P and A53T) cause rare familial forms of Parkinson's disease [77, 114] and recently a family with a new mutation (E46K) has been described [115]. Based on this evidence, SYN was identified as a major component of Lewy bodies [78]. On the other hand, a recessive form of juvenile Parkinson's disease was associated with mutations in the gene encoding parkin [116].

Several pieces of evidence, most of which are based on the analysis of transfected cells, support the hypothesis that accumulation of SYN in Parkinson's disease is a consequence of an alteration of the ubiquitin/proteasome system (UPS) [117, 118]. The self-aggregation capacity of SYN [119, 120] and the ubiquitin ligase activity attributed to parkin [121, 122] suggest that protein aggregation and dysfunction of the UPS might play a causal role in the development of sporadic and familial Parkinson's disease. SYN is a highly conserved 140-amino-acid protein widespread in the CNS. It interacts with other cerebral proteins (14-3-3, synphilin-1, parkin, tyrosine hydroxylase, dopamine transporter) and it is involved in dopamine vesicle trafficking [123]. In aqueous solution, free SYN shows a natively unfolded conformation that favors protein self-aggregation and toxicity [124]. Experimental injuries reproducing possible Parkinson's disease triggers (toxins, oxidative stress, proteasome impairment) increase SYN aggregation [125–127]. Like the other amyloidogenic proteins, SYN aggregation follows a multistep process, starting from SYN monomers that form oligomers (protofibrils) whose coalescence generates fibrils that, in turn, can aggregate in inclusion bodies. Recent data suggest that the protofibrils are the toxic intermediate, while the final inclusions may have a protective value [128]. Several *in vitro* studies have confirmed that SYN mutations predispose to protein aggregation and the overexpression of the mutated forms of the protein are toxic for the cell [129, 130]. According to this hypothesis, the accumulation of SYN in the Lewy bodies is a toxic event that eventually triggers neuronal death [131]. We found that the fragment corresponding to the residues 61–95 [also known as non-amyloid component of plaques (NAC)] was specifically neurotoxic for the dopaminergic cells and the pre-aggregation of the peptide amplified this effect [132]. Conversely, other studies demonstrated that SYN can exert neuroprotective activity against various cellular stresses like oxidative stress, serum deprivation and apoptosis [133, 134]. In our investigation, we have also noted that PC12 transfected with SYN were more resistant to oxidative stress. This evidence was derived mainly by *in vitro* overexpression of the wild-type protein whose protective role was reduced by the mutations A30P and A53T. Accordingly, the loss of SYN protective function, rather than a gain of toxic function, may determine the cell death in Parkinson's disease. Nevertheless, the situation is not completely clear, as other authors have reported a protective role also for the mutated forms of the protein [134]. It is possible that the two apparent opposing actions of SYN are not mutually exclusive. SYN may be at first neuroprotective, but, if the protein self-aggregation becomes relevant, the disease occurs. In this regard, it was recently reported, using an *in vitro* model that the extracellular administration of wild-type SYN on the nanomolar scale is protective against oxidative stress, while the overexpression of the protein or its extracellular administration on the micromolar scale is toxic [135]. We have confirmed these data using a chimeric SYN associated with the TAT [136] sequence (TAT–SYN) to translocate the protein inside the cells. Mutated (A30P, A53T) and wild-type TAT–SYN sequences equally protected against oxidative stress induced by hydrogen peroxide in PC12 and the effect was mediated by the induction of HSP70 protein [137]. The protective effect by TAT–SYN in SK-NBE neuroblastoma cells was efficacious also against the dopamine-specific neurotoxin 6-hydroxydopamine (6OHDA).

A third important point that has been intensively investigated for Parkinson's disease is the so called "selective vulnerability", i.e. why only a particular region of the brain (substantia nigra pars compacta) is affected even though SYN is widely expressed in brain and the factors that probably trigger the disease are not region specific. Recent studies have pointed out that a first important factor is dopamine synthesis and secretion that are peculiar features of the neurons of the affected areas [138]. In fact, dopamine is a molecule that is prone to oxidation, which may contribute to generate reactive molecules potentially able to damage cellular components like proteins or lipids. From this point of view several experimental data indicate that SYN is implicated in dopamine metabolism and trafficking [139, 140] suggesting a relationship between protein function, dopamine homeostasis, oxidative stress and neuronal damage.

The development of animal models has provided further insight about SYN pathogenetic mechanisms. Transgenic mice overexpressing human SYN (wild-type or mutated) showed intraneuronal inclusions, but the associated phenotype was quite heterogeneous as for the neuronal populations involved and no model recapitulated completely the human pathology. For instance, transgenic mice expressing human SYN gene carrying the A53T mutation under Thy-1 promoter control [141] developed a progressive motor impairment starting from 40 days post-birth. Many telencephalic and brainstem neurons and the neurons of the spinal cord show a strong immunoreactivity to SYN antibodies in cellular bodies and dendrites; this picture differs from the axonal and presynaptic distribution of endogenous SYN in control mice. Brainstem neurons and motor neurons seem particularly vulnerable: motor neurons alterations comprise axonal damage and neuromuscular junction denervation, suggesting that an increased SYN expression may interfere with synaptic trophic mechanisms. Otherwise, transgenic mice with an elevated human SYN expression under platelet-derived growth factor promoter control showed a dopaminergic neuronal loss, and SYN- and ubiquitin-positive inclusions formation in different cerebral areas [142].

16.4 Conclusions

Although the presence of aggregates made up of misfolded proteins is a common feature of many neurodegenerative disorders, the biological processes responsible for this phenomenon are variable and its significance in the pathogenesis of the diseases is different. As illustrated in this chapter, the neuropathological manifestations may be extremely heterogeneous even within disorders characterized by aggregates made up by the same protein. In some conditions, like Alzheimer's disease, prion diseases and tauopathies, the evidence supporting a pathogenic role of the aggregates is quite compelling, although the molecular mechanisms triggering the protein accumulation remain to be established. In the synucleinopathies, the responsibility of the pathologic inclusions in determining the degenerative process is more uncertain. In ALS and diseases with CAG repeat expansions the role played by the aggregates is still elusive, although the number of studies focused on their potential pathogenic role is growing. Finally, the consensus around the neurotoxic role of oligomer species rather than fibrils themselves is vast and propose new therapeutic approaches at these diseases.

Acknowledgments

Supported by the Italian Ministry of Health, Department of Social Services (PS-03-4 and PS-03-10) and by the European Community (LSHM-CT-2004-503039 and Neuroprion FOOD-CT-2004-506579).

References

1 Taylor, J. P., Hardy, J. and Fischbeck, J. H. Toxic proteins in neurodegenerative disease. *Science* **2002**, *296*, 1991–1995.

2 Soto, C. Unfolding the role of protein misfolding in neurodegenerative diseases. *Nat Rev Neurosci* **2003**, *4*, 49–60.

3 Stathopulos, P. B., Rumfeldt, J. A., Scholz, G. A., Irani, R. A., Frey, H. E., Hallewell, R. A., Lepock, J. R. and Meiering, E. M. Cu/Zn superoxide dismutase mutants associated with amyotrophic lateral sclerosis show enhanced formation of aggregates *in vitro*. *Proc Natl Acad Sci USA* **2003**, *100*, 7021–7026.

4 Bendotti, C. and Carri, M. T. Lessons from models of SOD1-linked familial ALS. *Trends Mol Med* **2004**, *10*, 393–400.

5 van Welsem, M. E., Hogenhuis, J. A., Meininger, V., Metsaars, W. P., Hauw, J. J. and Seilhean, D. The relationship between Bunina bodies, skein-like inclusions and neuronal loss in amyotrophic lateral sclerosis. *Acta Neuropathol* **2002**, *10*, 583–589

6 Piao, Y. S., et al. Neuropathology with clinical correlations of sporadic amyotrophic lateral sclerosis: 102 autopsy cases examined between 1962 and 2000. *Brain Pathol* **2003**, *13*, 10–22.

7 Busch, A., Engemann, S., Lurz, R., Okazawa, H., Lehrach, H. and Wanker, E. E. Mutant huntingtin promotes the fibrillogenesis of wild-type huntingtin: a potential mechanism for loss of huntingtin function in Huntington's disease. *J Biol Chem* **2003**, *278*, 41452–41461.

8 Scherzinger, E., et al. Self-assembly of polyglutamine-containing huntingtin fragments into amyloid-like fibrils: implications for Huntington's disease pathology. *Proc Natl Acad Sci USA* **1999**, *96*, 4604–4609.

9 Lee, H. G., Petersen, R. B., Zhu, X., Honda, K., Aliev, G., Smith, M. A. and Perry, G. Will preventing protein aggregates live up to its promise as prophylaxis against neurodegenerative diseases? *Brain Pathol* **2003**, *13*, 630–638.

10 Michalik, A. and Van Broeckhoven, C. Pathogenesis of polyglutamine disorders: aggregation revisited. *Hum Mol Genet* **2003**, *12*, R173–R186.

11 Price, J. L., Davis, P. B., Morris, J. C. and White, D. L. The distribution of tangles, plaques, and related immunohistochemical markers in healthy aging and Alzheimer's disease. *Neurobiol Aging* **1991**, *12*, 295–312.

12 Braak, H. and Braak, E. Neuropathological stageing of Alzheimer-related changes. *Acta Neuropathol* **1991**, *82*, 239–259.

13 Delacourte, A., et al. The biochemical pathway of neurofibrillary degeneration in aging and Alzheimer's disease. *Neurology* **1999**, *52*, 1158–1165.

14 Glenner, G. G. and Wong C. W. Alzheimer's disease: initial report of the purification and characterization of a novel cerebrovascular amyloid protein. *Biochem Biophys Res Commun* **1984**, *120*, 885–890.

15 Masters, C. L., Simms, G., Weinman, N. A., Multhaup, G., McDonald, B. L. and Beyreuther, K. Amyloid plaque core protein in Alzheimer disease and Down syndrome. *Proc Natl Acad Sci USA* **1985**, *82*, 4245–4249.

16 Kang, J., et al. The precursor of Alzheimer's disease amyloid A4 protein resembles a cell-surface receptor. *Nature* **1987**, *325*, 733–736.

17 Muchowski, P. J. and Wacker, J. L. Modulation of neurodegeneration by molecular chaperones. *Nat Rev Neurosci* **2005**, *6*, 11–22.

18 Tagliavini, F., Giaccone, G., Frangione, B. and Bugiani, O. Preamyloid deposits in the cerebral cortex of patients with Alzheimer's disease and nondemented individuals. *Neurosci Lett* **1988**, *93*, 191–196.

19 Yamaguchi, H., Hirai, S., Morimatsu, M., Shoji, M, and Ihara, Y. Diffuse type of senile plaques in the brains of Alzheimer-type dementia. *Acta Neuropathol* **1988**, *77*, 113–119.

20 Verga, L., Frangione, B., Tagliavini, F., Giaccone, G., Migheli, A. and Bugiani, O. Alzheimer patients and Down patients: cerebral preamyloid deposits differ ultrastructurally and histochemically from the amyloid of senile plaques. *Neurosci Lett* **1989**, *105*, 294–299.

21 Bugiani, O., Tagliavini, F. and Giaccone, G. Preamyloid deposits, amyloid deposits, and senile plaques in Alzheimer's disease, Down syndrome, and aging. *Ann NY Acad Sci* **1991**, *640*, 122–128.

22 Giaccone, G., Tagliavini, F., Linoli, G., Bouras, C., Frigerio, L., Frangione, B. and Bugiani, O. Down patients: extracellular preamyloid deposits precede neuritic degeneration and senile plaques. *Neurosci Lett* **1989**, *97*, 232–238.

23 Mann, D. and Esiri, M. The pattern of acquisition of plaques and tangles in the brains of patients under 50 years of age with Down's syndrome. *J Neurol Sci* **1989**, *89*, 169–179.

24 Bugiani, O., Giaccone, G., Frangione, B., Ghetti, B. and Tagliavini, F. Alzheimer patients: preamyloid deposits are more widely distributed than senile plaques throughout the central nervous system. *Neurosci Lett* **1989**, *103*, 263–268.

25 Joachim, C. L., Morris, J. H. and Selkoe, D. J. Diffuse senile plaques occur commonly in the cerebellum in Alzheimer's disease. *Am J Pathol* **1989**, *135*, 309–319.

26 Iwatsubo, T., Odaka, A., Suzuki, N., Mizusawa, H., Nukina, N. and Ihara, Y. Visualization of Aβ42(43) and Aβ40 in senile plaques with end-specific A monoclonals: evidence that an initially deposited species is Aβ42(43). *Neuron* **1994**, *13*, 45–53.

27 Fukumoto, H., Asami-Odaka, A., Suzuki, N., Shimada, H., Ihara, Y. and Iwatsubo, T. Amyloid beta-protein deposition in normal aging has the same characteristics as that in Alzheimer's: predominance of Aβ42(43) and association of Aβ40 with cored plaques. *Am J Pathol* **1996**, *148*, 259–265.

28 Kidd, M. Paired helical filaments in electron microscopy of Alzheimer's disease. *Nature* **1963**, *197*, 192–194.

29 Crowther, R.A. and Wischik, C.M. Image reconstruction of the Alzheimer paired helical filament. *EMBO J* **1985**, *4*, 3661–3665.

30 Goedert, M., Wischik, C.M. Crowther, R.A., Walker, J. E. and Klug, A. Cloning and sequencing of the cDNA encoding a core protein of the paired helical filament of Alzheimer disease: identification as the microtubule associated protein tau. *Proc Natl Acad Sci USA* **1988**, *85*, 4051–4055.

31 Neve, R.L., Harris, P., Kosik, K.S., Kurnit, D.M. and Donlon, T.A. Identification of cDNA clones for the human microtubule-associated protein tau and chromosomal localization of the genes for tau and microtubule-associated protein 2. *Brain Res* **1986**, *387*, 271–280.

32 Andreadis, A., Brown, W.M. and Kosik, K.S. Structure and novel exons of the human tau gene. *Biochemistry* **1992**, *31*, 10626–10633.

33 Weingarten, M.D., Lockwood, A.H., Hwo, S.Y. and Kirschner, M.W. A protein factor essential for microtubule assembly. *Proc Natl Acad Sci USA* **1975**, *72*, 1858–1862.

34 Cleveland, D.W., Hwo, S.Y. and Kirschner, M.W. Purification of tau, a microtubule-associated protein that induces assembly of microtubules from purified tubulin. *J Mol Biol* **1977**, *116*, 207–225.

35 Drechsel, D.N., Hyman, A.A., Cobb, M.H. and Kirschner, M.W. Modulation of the dynamic instability of tubulin assembly by the microtubule-associated protein tau. *Mol Biol Cell* **1992**, *3*, 1141–1154.

36 Biernat, J., Gustke, N., Drewes, G., Mandelkow, E.M. and Mandelkow, E. Phosphorylation of Ser262 strongly reduces binding of tau to microtubules: distinction between PHF-like immunoreactivity and microtubule binding. *Neuron* **1993**, *11*, 153–163.

37 Bramblett, G.T., Goedert, M., Jakes, R., Merrick, S.E., Trojanowski, J.Q. and Lee, V.M. Abnormal tau phosphorylation at Ser396 in Alzheimer's disease recapitulates development and contributes to reduced microtubule binding. *Neuron* **1993**, *10*, 1089–1099.

38 Yoshida, H. and Ihara, Y. Tau in paired helical filaments is functionally distinct from fetal tau: assembly incompetence of paired helical filament-tau. *J Neurochem* **1993**, *61*, 1183–1186.

39 Braak, H., Braak, E., Grundke-Iqbal, I. and Iqbal, K. Occurrence of neuropil threads in the senile human brain and in Alzheimer's disease: a third location of paired helical filaments outside of neurofibrillary tangles and neuritic plaques. *Neurosci Lett* **1986**, *65*, 351–355.

40 Lee, V.M.Y., Goedert, M. and Trojanowski, J.Q. Neurodegenerative Tauopathies. *Annu Rev Neurosci* **2001**, *24*, 1121–1159.

41 Buée, L., Bussiére, T., Buée-Scherrer, V., Delacourte, A. and Hof, P.R. Tau protein isoforms, phosphorylation and role in neurodegenerative disorders. *Brain Res Rev* **2000**, *33*, 95–130.

42 Hardy, J.A. and Higgins, G.A. Alzheimer's disease: the amyloid cascade hypothesis. *Science* **1992**, *256*, 184–185.

43 Selkoe, D. Alzheimer's disease: genes, proteins, and therapy. *Physiol Rev* **2001**, *81*, 741–766.

44 Foster, N.L., Wilhelmsen, K., Sima, A.A., Jones, M.Z., D'Amato, C.J. and Gilman, S. Frontotemporal dementia and parkinsonism linked to chromosome 17: a consensus conference. *Ann Neurol* **1997**, *41*, 706–715.

45 Spillantini, M.G., Goedert, M., Crowther, R.A. Murrell, J.R., Farlow, M.R. and Ghetti, B. Familial multiple system tauo-

pathy with presenile dementia: a disease with abundant neuronal and glial tau filaments. *Proc Natl Acad Sci USA* **1997**, *94*, 4113–4118.

46 Hutton, M., et al. Association of missense and 5′-splice-site mutations in tau with the inherited dementia FTDP-17. *Nature* **1998**, *393*, 702–705.

47 Poorkaj, P., et al. Tau is a candidate gene for chromosome 17 frontotemporal dementia. *Ann Neurol* **1998**, *43*, 815–825.

48 Wilhelmsen, K.C., Lynch, T., Pavlou, E., Higgins, M. and Nygaard, T.G. Localization of disinhibition–dementia–parkinsonism–amyotrophic complex to 17q21–22. *Am J Hum Genet* **1994**, *55*, 1159–1165.

49 Ghetti, B., Hutton, M.L. and Wszolek, Z. Frontotemporal dementia and parkinsonism linked to chromosome 17 associated with tau gene mutations (FTDP-17T). In *Neurodegeneration: The Molecular Pathology of Dementia and Movement Disorders*, D. Dickson (ed.). ISN Neuropath Press, Basel, **2003**.

50 Steele, J.C., Richardson, J.C. and Olszewski, J. Progressive supranuclear palsy. *Arch Neurol* **1964**, *10*, 333–359.

51 Hauw, J.J., Verny, M., Delaere, P., Cervera, P., He, Y. and Duyckaerts, C. Constant neurofibrillary changes in the neucortex in progressive supranuclear palsy. Basic differences with Alzheimer's disease and aging. *Neurosci Lett* **1990**, *119*, 182–186.

52 Tellez-Nagel, I. and Wisniewski, H.M. Ultrastructure of neurofibrillary tangles in Steele–Richardson–Olszewski syndrome. *Arch Neurol* **1973**, *29*, 324–327.

53 Rebeiz, J.J., Kolodny, E.H. and Richardson, E.P., Jr. Corticodentatonigral degeneration with neuronal achromasia. *Arch Neurol* **1968**, *18*, 20–33.

54 Feany, M.B. and Dickson, D.W. Widespread cytoskeletal pathology characterizes corticobasal degeneration. *Am J Pathol* **1995**, *146*, 1388–1396.

55 Ksiezak-Reding, H., Morgan, K., Mattiace, L.A., Davies, P., Liu, W.K., Yen, S.H., Weidenheim, K. and Dickson, D.W. Ultrastructure and biochemical composition of paired helical filaments in corticobasal degeneration. *Am J Pathol* **1994**, *145*, 1496–1508.

56 Constantinidis, J., Richard, J. and Tissot, R. Pick's disease. Histological and clinical correlations. *Eur Neurol* **1974**, *11*, 208–217.

57 Feany, M.B., Mattiace, L.A. and Dickson, D.W. Neuropathologic overlap of progressive supranuclear palsy, Pick's disease and corticobasal degeneration. *J Neuropathol Exp Neurol* **1996**, *55*, 53–67.

58 Hong, M., et al. Mutation-specific functional impairments in distinct tau isoforms of hereditary FTDP-17. *Science* **1998**, *282*, 1914–1917.

59 Hasegawa, M., Smith, M.J. and Goedert, M. Tau proteins with FTDP-17 mutations have a reduced ability to promote microtubule assembly. *FEBS Lett* **1998**, *437*, 207–210.

60 Arrasate, M., Perez, M., Armas-Portela, R. and Avila, J. Polymerization of tau peptides into fibrillary structures. The effect of FTDP-17 mutations. *FEBS Lett* **1999**, *446*, 199–202.

61 Nacharaju, P., Lewis, J., Easson, C., Yen, S., Hackett, J., Hutton, M. and Yen, S.H. Accelerated filaments formation from tau protein with specific FTDP-17 missense mutations. *FEBS Lett* **1999**, *447*, 195–199.

62 D'Souza, I., Poorkaj, P., Hong, M., Nochlin, D., Lee, V.M., Bird, T.D. and Schellenberg, G.D. Missense and silent tau gene mutations cause frontotemporal dementia with parkinsonism-chromosome 17 type, by affecting multiple alternative RNA splicing regulatory elements. *Proc Natl Acad Sci USA*, **1999**, *96*, 5598–5603.

63 Grover, J., et al. 5′ splice site mutations in tau associated with the inherited dementia FTDP-17 affect a stem–loop structure that regulates alternative splicing of exon 10. *J Biol Chem* **1999**, *274*, 15134–15143.

64 Jiang, Z., Cote, J., Kwon, J.M., Goate, A.M. and Wu, J.Y. Aberrant splicing of tau pre-RNA caused by intronic mutations associated with the inherited frontotemporal dementia with parkinsonism linked to chromosome 17. *Mol Cell Biol* **2000**, *20*, 4036–4048.

65 Prusiner, S. B. Novel proteinaceous infectious particles cause scrapie. *Science*, **1982**, *216*, 136–144.

66 Prusiner, S. B. Prions. *Proc Natl Acad Sci USA* **1998**, *95*, 13363–13383.

67 Will, R. G., et al. A new variant of Creutzfeldt–Jakob disease in the UK. *Lancet* **1996**, *347*, 921–925.

68 Budka, H., et al. Neuropathological diagnostic criteria for Creutzfeldt–Jakob disease (CJD) and other human spongiform encephalopathies (prion diseases). *Brain Pathol* **1995**, *5*, 459–466.

69 Kretzschmar, H. A., Ironside, J. W., DeArmond, S. J. and Tateishi, J. Diagnostic criteria for sporadic Creutzfeldt–Jakob disease. *Arch Neurol* **1996**, *53*, 913–920.

70 Giaccone, G., et al. Creutzfeldt–Jakob disease: Carnoy's fixative improves the immunohistochemistry of the proteinase-K resistant prion protein. *Brain Pathol* **2000**, *10*, 31–37.

71 Budka, H., Head, M. W., Ironside, J. W., Gambetti, P., Parchi, P., Zeidler, M. and Tagliavini, F. Sporadic Creutzfeldt–Jakob disease. In *Neurodegeneration: The Molecular Pathology of Dementia and Movement Disorders*, D. Dickson (ed.). ISN Neuropath Press, Basel, **2003**.

72 Ghetti, B., Dlouhy, S. R., Giaccone, G., Bugiani, O., Frangione, B., Farlow, M. and Tagliavini, F. Gerstmann–Sträussler–Scheinker disease and the Indiana kindred. *Brain Pathol* **1995**, *5*, 61–75.

73 Parchi, P., et al. Classification of sporadic Creutzfeldt–Jakob disease based on molecular and phenotypic analysis of 300 subjects. *Ann Neurol* **1999**, *46*, 224–233.

74 Puoti, G., Giaccone, G., Rossi, G., Canciani, B., Bugiani, O. and Tagliavini, F. Sporadic Creutzfeldt–Jakob disease: co-occurrence of different types of PrP^{Sc} in the same brain. *Neurology* **1999**, *53*, 2173–2176.

75 Bueler, H., Aguzzi, A., Sailer, A., Greiner, R. A., Autenried, P., Aguet, M. and Weissmann C. Mice devoid of PrP are resistant to scrapie. *Cell* **1993**, *73*, 1339–1347.

76 Forno, L. S. Neuropathology of Parkinson's disease. *J Neuropathol Exp Neurol* **1996**, *55*, 259–272.

77 Polymeropoulos, M. H., et al. Mutation in the α-synuclein gene identified in families with Parkinson's disease. *Science* **1997**, *276*, 2045–2077.

78 Spillantini, M. G., Schmidt, M. L., Lee, V. M., Trojanowski, J. Q., Jakes, R. and Goedert, M. α-synuclein in Lewy bodies. *Nature* **1997**, *388*, 839–840.

79 Kazuhara, S., Mori, H., Izumiyama, N., Yoshimura, M. and Ihara, Y. Lewy bodies are ubiquitinated. A light and electron microscopic immunohistochemical study. *Acta Neuropathol* **1988**, *75*, 345–353.

80 Schlossmacher, M. G., et al. Parkin localizes to the Lewy bodies of Parkinson disease and dementia with Lewy bodies. *Am J Pathol* **2002**, *160*, 1655–1667.

81 Papp, M. I., Kahn, J. E. and Lantos, P. L. Glial cytoplasmic inclusions in the CNS of patients with multiple system atrophy (striatonigral degeneration, olivopontocerebellar atrophy and Shy–Drager syndrome). *J Neurol Sci* **1989**, *94*, 79–100.

82 Arima, K., et al. NACP/-synuclein immunoreactivity in fibrillary components of neuronal and oligodendroglial cytoplasmic inclusion in the pontine nuclei in multiple system atrophy. *Acta Neuropathol* **1998**, *96*, 439–444.

83 Forloni, G., Chiesa, R., Smiroldo, S., Verga, L., Salmona, M., Tagliavini, F. and Angeretti, N. Apoptosis mediated neurotoxicity induced by chronic application of beta amyloid fragment 25–35. *Neuroreport* **1993**, *4*, 523–526.

84 Loo, D. T., Copani, A., Pike, C. J., Whittemore, E. R., Walencewicz, A. J. and Cotman, C. W. Apoptosis is induced by beta-amyloid in cultured central nervous system neurons. *Proc Natl Acad Sci USA* **1993**, *90*, 7951–7955.

85 Pike, C. J., Walencewicz, A. J., Glabe, C. G. and Cotman, C. W. *In vitro* aging of beta-amyloid protein causes peptide aggregation and neurotoxicity. *Brain Res* **1991**, *563*, 311–314.

86 Luo, Y., et al. Mice deficient in BACE1, the Alzheimer's beta-secretase, have normal phenotype and abolished beta-amyloid generation. *Nat Neurosci* **2001**, *4*, 231–232.

87 Dewachter, I. and Van Leuven, F. Secretases as targets for the treatment of Alz-

heimer's disease: the prospects. *Lancet Neurol* **2002**, *1*, 409–416.

88 Selkoe, D. J. The cell biology of beta-amyloid precursor protein and presenilin in Alzheimer's disease. *Trends Cell Biol* **1998**, *8*, 447–453.

89 Caughey, B. and Lansbury, P. T. Protofibrils, pores, fibrils and neurodegeneration: separating the responsible protein aggregates from the innocent bystanders. *Annu Rev Neurosci* **2003**, *26*, 267–298.

90 Forloni, G., et al. Protein misfolding in Alzheimer's and Parkinson's disease: genetics and molecular mechanisms. *Neurobiol Aging* **2002**, *23*, 957–976.

91 Arriagada, P. V., Growdon, J. H., Hedley-Whyte, E. T. and Hyman, B. T. Neurofibrillary tangles but not senile plaques parallel duration and severity of Alzheimer's disease. *Neurology* **1992**, *42*, 631–639.

92 Bennett, D. A., Cochran, E. J., Saper, C. B., Leverenz, J. B., Gilley, D. W. and Wilson, R. S. Pathological changes in frontal cortex from biopsy to autopsy in Alzheimer's disease. *Neurobiol Aging* **1993**, *14*, 589–596.

93 Klein, W. L., Krafft, G. A. and Finch, C. E. Targeting small Abeta oligomers: the solution to an Alzheimer's disease conundrum? *Trends Neurosci* **2001**, *24*, 219–224.

94 Hardy, J. and Selkoe, D. J. The amyloid hypothesis of Alzheimer's disease: progress and problems on the road to therapeutics. *Science* **2002**, *297*, 353–356.

95 Forloni, G., Lucca, E., Angeretti, N., Della Torre, P. and Salmona, M. Amidation of beta-amyloid peptide strongly reduced the amyloidogenic activity without alteration of the neurotoxicity. *J Neurochem* **1997**, *69*, 2048–2054.

96 Hartley, D. M., Walsh, D. M., Ye, C. P., Diehl, T., Vasquez, S., Vassilev, P. M., Teplow, D. B. and Selkoe, D. J. Protofibrillar intermediates of amyloid beta-protein induce acute electrophysiological changes and progressive neurotoxicity in cortical neurons. *J Neurosci* **1999**, *19*, 8876–8884.

97 Cleary, J. P., Walsh, D. M., Hofmeister, J. J., Shankar, G. M., Kuskowski, M. A., Selkoe, D. J. and Ashe, K. H. Natural oligomers of the amyloid-beta protein specifically disrupt cognitive function. *Nat Neurosci* **2005**, *8*, 79–84.

98 Oddo, S., et al. Triple-transgenic model of Alzheimer's disease with plaques and tangles: intracellular Abeta and synaptic dysfunction. *Neuron* **2003**, *39*, 409–421.

99 Naslund, J., Haroutunian, V., Mohs, R., Davis, K. L., Davies, P., Greengard, P. and Buxbaum, J. D. Correlation between elevated levels of amyloid beta-peptide in the brain and cognitive decline. *J Am Med Ass* **2000**, *283*, 1571–1577.

100 Cohen, F. E. and Prusiner, S. B. Pathologic conformations of prion proteins. *Annu Rev Biochem* **1998**, *67*, 793–819.

100 Soto, C. and Saborio, G. P. Prions: disease propagation and disease therapy by conformational transmission. *Trends Mol Med* **2001** *7*, 109–114.

102 Chiesa, R., Piccardo, P., Quaglio, E., Drisaldi, B., Si-Hoe, S. L., Takao, M., Ghetti, B. and Harris, D. A. Molecular distinction between pathogenic and infectious properties of the prion protein. *J Virol* **2003**, *77*, 7611–7622.

103 Hetz, C., Russelakis-Carneiro, M., Maundrell, K., Castilla, J. and Soto, C. Caspase-12 and endoplasmic reticulum stress mediate neurotoxicity of pathological prion protein. *EMBO J* **2003**, *22*, 5435–5445.

104 Forloni, G., Angeretti, N., Chiesa, R., Monzani, E., Salmona, M., Bugiani, O. and Tagliavini, F. Neurotoxicity of a prion protein fragment. *Nature* **1993**, *362*, 543–546.

105 Tagliavini, F., et al. Synthetic peptides homologous to prion protein residues 106–147 form amyloid-like fibrils *in vitro*. *Proc Natl Acad Sci USA* **1993**, *90*, 9678–9682.

106 Forloni, G., Tagliavini, F., Bugiani, O. and Salmona, M. Amyloid in Alzheimer's disease and prion-related encephalopathies: studies with synthetic peptides. *Prog Neurobiol* **1996**, *49*, 287–315.

107 Brown, D. R., Herms, J. and Kretzschmar, H. A. Mouse cortical cells lacking cellular PrP survive in culture with a neurotoxic PrP fragment *Neuroreport* **1994**, *27*, 2057–2060.

108 Fioriti, L., Quaglio, E., Harris, D., Forloni, G. and Chiesa, R. The neurotoxicity of prion protein (PrP) peptide 106–126 is independent of the expression level of PrP and is not mediated by abnormal PrP species. *Cell Mol Neurosci* **2005**, *28*, 165–176.

109 Tagliavini, F., Forloni, G., D'Ursi, P., Bugiani, O. and Salmona, M. Studies on peptide fragments of prion. *Adv Protein Chem* **2001**, *57*, 171–201.

110 Forloni, G., Angeretti, N., Malesani, P., Peressini, E., Rodriguez Martin, T., Della Torre, P. and Salmona, M. Influence of mutations associated with familial prion-related encephalopathies on biological effects of PrP peptides. *Ann Neurol* **1999**, *45*, 489–494.

111 Forloni, F., Tagliavini, F., Bugiani, O. and Salmona, M. Apoptosis-mediated neurotoxicity induced by beta-amyloid and PrP fragments. *Mol Chem Neuropathol* **1996**, *28*, 163–171.

112 Kayed, R., Head, E., Thompson, J. L., McIntire, T. M., Milton, S. C., Cotman, C. W. and Glabe, C. G. Common structure of soluble amyloid oligomers implies common mechanism of pathogenesis. *Science* **2003**, *300*, 486–489.

113 Bonifati, V., Oostra, B. A. and Heutink, P. Unraveling the pathogenesis of Parkinson's disease – the contribution of monogenic forms. *Cell Mol Life Sci* **2004**, *61*, 1729–1750.

114 Kruger, R., et al. Ala30Pro mutation in the gene encoding alpha-synuclein in Parkinson's disease. *Nat Genet* **1998**, *18*, 106–108.

115 Zarranz, J. J., et al. The new mutation, E46K, of alpha-synuclein causes Parkinson and Lewy body dementia. *Ann Neurol* **2004**, *55*, 164–173.

116 Kitada, T., et al. Mutations in the parkin gene cause autosomal recessive juvenile parkinsonism. *Nature* **1998**, *392*, 605–608.

117 Biasini, E., Fioriti, L., Ceglia, I., Invernizzi, R., Bertoli, A., Chiesa, R. and Forloni, G. Proteasome inhibition and aggregation in Parkinson's disease: a comparative study in untransfected and transfected cells *J Neurochem* **2004**, *88*, 545–553.

118 Petrucelli, L. and Dawson, T. M. Mechanism of neurodegenerative disease: role of the ubiquitin proteasome system. *Ann Med* **2004**, *36*, 315–320.

119 el-Agnaf, O. M. and Irvine, G. B. Aggregation and neurotoxicity of alpha-synuclein and related peptides. *Biochem Soc Trans* **2000**, *30*, 559–565.

120 Wood, S. J., Wypych, J., Steavenson, S., Louis, J. C., Citron, M. and Biere, A. L. Synuclein fibrillogenesis is nucleation-dependent. Implications for the pathogenesis of Parkinson's disease. *J Biol Chem* **1999**, *274*, 19509–19512.

121 Shimura, H., et al. Familial Parkinson disease gene product, parkin, is a ubiquitin-protein ligase. *Nat Genet* **2000**, *25*, 302–305.

122 Yamamoto, A., Friedlein, A., Imai, Y., Takahashi, R., Kahle, P. J. and Haass, C. Parkin phosphorylation and modulation of its E3 ubiquitin ligase activity. *J Biol Chem* **2005**, *280*, 3390–3399.

123 Wersinger, C., Prou, D., Vernier, P., Niznik, H. B. and Sidhu, A. Mutations in the lipid-binding domain of alpha-synuclein confer overlapping, yet distinct, functional properties in the regulation of dopamine transporter activity. *Mol Cell Neurosci* **2003**, *24*, 91–103.

124 Munishkina, L. A., Henriques, J., Uversky, V. N. and Fink, A. L. Role of protein–water interactions and electrostatics in alpha-synuclein fibril formation. *Biochemistry* **2004**, *43*, 3289–3300.

125 Hashimoto, M., Hsu, L. J., Xia, Y., Takeda, A., Sisk, A., Sundsmo, M. and Masliah, E. Oxidative stress induces amyloid-like aggregate formation of NACP/alpha-synuclein *in vitro*. *Neuroreport* **1999**, *10*, 717–721.

126 Norris, E. H., Giasson, B. I., Ischiropoulos, H. and Lee, V. M. Effects of oxidative and nitrative challenges on alpha-synuclein fibrillogenesis involve distinct mechanisms of protein modifications. *J Biol Chem* **2003**, *278*, 27230–27240.

127 Sherer, T. B., Kim, J. H., Betarbet, R. and Greenamyre, J. T. Subcutaneous rotenone exposure causes highly selective dopaminergic degeneration and alpha-synuclein aggregation. *Exp Neurol* **2003**, *179*, 9–16.

128 Conway, K.A., Lee, S.J., Rochet, J.C., Ding, T.T., Williamson, R.E. and Lansbury, P.T., Jr. Acceleration of oligomerization, not fibrillization, is a shared property of both alpha-synuclein mutations linked to early-onset Parkinson's disease: implications for pathogenesis and therapy. *Proc Natl Acad Sci USA* **2000**, *97*, 571–576.

129 Conway, K.A., Harper, J.D. and Lansbury, P.T. Accelerated *in vitro* fibril formation by a mutant alpha-synuclein linked to early-onset Parkinson disease. *Nat Med* **1998**, 4, 1318–1320.

130 Greenbaum, E.A., Graves, C.L., Mishizen-Eberz, A.J., Lupoli, M.A., Lynch, D.R., Englander, S.W., Axelsen, P.H. and Giasson, B.I. The E46K mutation in alpha-synuclein increases amyloid fibril formation. *J Biol Chem* **2005**, Jan 4 [Epub ahead of print]

131 Volles, M.J. and Lansbury, P.T., Jr. Zeroing in on the pathogenic form of alpha-synuclein and its mechanism of neurotoxicity in Parkinson's disease. *Biochemistry* **2003**, *42*, 7871–7878.

132 Forloni, G., Bertani, I., Calella, A.M., Thaler, F. and Invernizzi, R. Alpha-synuclein and Parkinson's disease: selective neurodegenerative effect of alpha-synuclein fragment on dopaminergic neurons *in vitro* and *in vivo*. *Ann Neurol* **2000**, *47*, 632–640.

133 Manning-Bog, A.B., McCormack, A.L., Purisai, M.G., Bolin, L.M. and Di Monte, D.A. Alpha-synuclein overexpression protects against paraquat-induced neurodegeneration. *J Neurosci* **2002**, *23*, 3095–3099.

134 Ancolio, K., Alves da Costa, C., Ueda, K. and Checler, F. Alpha-synuclein and the Parkinson's disease-related mutant Ala53Thr-alpha-synuclein do not undergo proteasomal degradation in HEK293 and neuronal cells. *Neurosci Lett* **2000**, *285*, 79–82.

135 Seo, J.H., et al. Alpha-synuclein regulates neuronal survival via Bcl-2 family expression and PI3/Akt kinase pathway. *FASEB J* **2002**, *16*, 1826–1828

136 Becker-Hapak, M., McAllister, S.S. and Dowdy, S. F. TAT-mediated protein transduction into mammalian cells. *Methods* **2001**, *24*, 247–256.

137 Albani, D., Peverelli, E., Rametta, R., Batelli, S., Veschini, L., Negro, A. and Forloni, G. Protective effect of TAT-delivered alpha-synuclein: relevance of the C-terminal domain and involvement of HSP70. *FASEB J* **2004**, *18*, 1713–1715.

138 Xu, J., Kao, S.Y., Lee, F.J., Song, W., Jin, L.W. and Yankner, B.A. Dopamine-dependent neurotoxicity of alpha-synuclein: a mechanism for selective neurodegeneration in Parkinson disease. *Nat Med* **2002**, *8*, 600–606.

139 Lee, F.J., Liu, F., Pristupa, Z. B. and Niznik, H.B. Direct binding and functional coupling of alpha-synuclein to the dopamine transporters accelerate dopamine-induced apoptosis. *FASEB J* **2001**, *15*, 916–922.

140 Perez, R.G., Waymire, J.C., Lin, E., Liu, J.J., Guo, F. and Zigmond, M.J. A role for alpha-synuclein in the regulation of dopamine biosynthesis. *J Neurosci* **2002**, *22*, 3090–3099.

141 van der Putten, H., et al. Neuropathology in mice expressing human alpha-synuclein. *J Neurosci* **2000**, *20*, 6021–6029.

142 Masliah, E, et al. Dopaminergic loss and inclusion body formation in alpha-synuclein mice: implications for neurodegenerative disorders. *Science* **2000**, *287*, 1265–1269.

Amyloid Proteins

Volume 2

edited by Jean D. Sipe

Further Titles of Interest

Johannes Buchner, Thomas Kiefhaber (eds.)

Protein Folding Handbook

(5 Volumes)
2005
ISBN: 3-527-30784-2

Knut H. Nierhaus, Daniel N. Wilson (eds.)

Protein Synthesis and Ribosome Structure
Translating the Genome

2004
ISBN: 3-527-30638-2

Giovanni Cesareni, Mario Gimona, Marius Sudol, Michael Yaffe (eds.)

Modular Protein Domains

2004
ISBN: 3-527-30813-X

Jean-Charles Sanchez, Garry L. Corthals, Denis F. Hochstrasser (eds.)

Biomedical Applications of Proteomics

2004
ISBN : 3-527-30807-5

R. John Mayer, Aaron J. Ciechanover, Martin Rechsteiner (eds.)

Protein Degradation
Vol. 1: Ubiquitin and the Chemistry of Life

2005
ISBN: 3-527-30837-7

Amyloid Proteins

The Beta Sheet Conformation and Disease

edited by Jean D. Sipe

Volume 2

WILEY-VCH Verlag GmbH & Co. KGaA

Editor

Jean D. Sipe
8406 North Brook Lane
Bethesda, MD 20814-2615
USA

This book was carefully produced. Nevertheless, authors, editors, and publisher do not warrant the information contained therein to be free of errors. Readers are advised to keep in mind that statements, data, illustrations, procedural details or other items may inadvertently be inaccurate.

Library of Congress Card No.: applied for

British Library Cataloguing-in-Publication Data: A catalogue record for this book is available from the British Library.

Die Deutsche Bibliothek – CIP Cataloguing-in-Publication Data: A catalogue record for this publication is available from Die Deutsche Bibliothek

Printed on acid-free paper

Composition K+V Fotosatz GmbH, Beerfelden
Printing Strauss GmbH, Mörlenbach
Bookbinding Litges & Dopf Buchbinderei GmbH, Heppenheim
Cover Design Gunther Schulz, Fußgönheim

Cover illustration includes images kindly provided by David Teplow, Department of Neurology, University of California Los Angeles, Los Angeles, CA, USA

Printed in the Federal Republic of Germany

ISBN-13: 978-3-527-31072-2
ISBN-10: 3-527-31072-X

Contents

Volume 1

Amyloid Proteins. The Beta Sheet Conformation and Disease. J. D. Sipe

ISBN: 3-527-31072-X

Preface

The folding of proteins into unique three dimensional structures is integral to their specific biological functions within the body. Of the tens of thousands of proteins encoded within the human genome, fewer than thirty are known to share the feature of being susceptible to increased folding of the polypeptide backbone into the beta sheet conformation and assembly into amyloid fibrils. In each case, the increased beta sheet folding is associated with a clinically distinct disease or disorder, one of the amyloidoses. The pathological consequences of amyloid fibril formation are implicated in a wide range of both common and rare diseases, including Alzheimer's disease and other brain disorders, adult onset (type II) diabetes mellitus, plasma B-cell dsycrasias, long term hemodialysis, hereditary polyneuropathies and hereditary periodic fever syndromes.

During the latter half of the twentieth century, chemical and physical studies of amyloid fibrils that had been isolated and purified from amyloid laden tissues and organs led to the recognition that there is a unique association between the chemical identity of the fibril forming protein and the pattern of localized or systemic clinical symptoms. However, despite their biochemical and clinical differences, each of the amyloidoses shares the common pathophysiologic features of an amyloidogenic protein precursor, permissive host genetic background, abnormalities in proteolysis that permit accumulation of precursor protein and fibril intermediates, and alteration in the chemistry of the extracellular matrix. Each of the more than 20 chemically distinct types of amyloid deposits contains a common set of extracellular matrix constituents, glycosaminoglycans and non-fibril forming proteins, serum amyloid P component and apolipoprotein E.

Until now, to some extent, clinical studies and basic studies have proceeded in parallel. Now, we are poised to integrate and utilize our knowledge of protein structure, physiology and pathology to forestall or ameliorate the clinical consequences of the tendency of amyloid fibril forming proteins to undergo increased folding into the beta sheet conformation. Outside the body, using conditions that alter protein folding, amyloid fibrils have been created from many more than 30 proteins; this is an indication of the key role of the local tissue environment in triggering amyloid fibril formation within the body.

Amyloid Proteins. The Beta Sheet Conformation and Disease. J. D. Sipe

ISBN: 3-527-31072-X

These volumes bring together preeminent amyloid clinicians and basic scientists to consider our present knowledge in terms of those structural and thermodynamic features which, over time, lead to amyloid fibril formation, deposition and disease. The authors present an overview of amyloidosis and amyloid proteins today, including the history of amyloid investigation, the internationally accepted nomenclature and the anatomic and clinical clues as to why amyloid fibrils form within the body. Protein folding, unfolding and refolding are considered in terms of thermodynamics, posttranslational modification and lipid association, and the influence of the extracellular matrix, serum amyloid P component and apolipoprotein E. Pathways to amyloid fibril formation are considered in terms of folding of natively unfolded proteins and unfolding of natively folded, globular proteins. The use of computational approaches to derive potential structures of amyloid fibril intermediates is presented. How the process of amyloid fibril formation causes damage to organs and tissues of the body is considered in terms of oligomeric fibril forming intermediates and cellular toxicity and brain dysfunction. Most of the amyloid proteins are considered individually in terms of current knowledge of structure, function and metabolism. Some of the more recently identified forms of amyloid, including medin, lactoferrin, apoA-IV or keratoepithelin (Table 1.1) have not been considered in detail here. It is anticipated they will be the subject of greater study in the future and that additional chemical forms of amyloid, particularly localized, will be identified in future studies. There is also within this volume the call for development of molecular diagnostics and targeted therapeutics in the amyloidoses. It is to that end that this volume is dedicated, to the acceleration of progress in understanding the contribution of the beta sheet conformation to the etiology and pathophysiology of disease, in order to enable prevention and better informed treatment of the amyloidoses.

I would like to express my gratitude to the contributing authors, to Frank Weinreich, Wiley-VCH, for his invitation to edit this volume and for his continuous support and encouragement, to Waltraud Wüst, Wiley-VCH for her able and amiable management of the publication process and to Byron Caughey and Margaret Johns for helpful discussions about the content of this volume.

Bethesda, Maryland
April, 2005

Jean D. Sipe

List of Contributors

Prof. Dr. Ilia V. Baskakov
Chapter 4
University of Maryland
Biotechnology Institute
Medical Biotechnology Center
725 W. Lombard St.
Baltimore, MD 21201
USA

Prof. Dr. Vittorio Bellotti
Chapter 24
Dipartimento di Biochimica
Universita degli Studi di Pavia
Via Taramelli 3b
I-27100 Pavia
Italy

Prof. Dr. Wilfredo Colon
Chapter 12
Department of Chemistry
& Chemical Biology
Rensselaer Polytechnic Institute
Cogswell Building, Room 126
110 Eighth Street
Troy, NY 12180
USA

Prof. Dr. Ana Margarida Damas
Chapter 21
IBMC-Molecular Structure Unit
Universidade do Porto
Rua do Campo Alegre, 823
4150-180 Porto
Portugal

Dr. Philippe Derreumaux
Chapter 18
Laboratoire de Biochimie Theorique –
UPR 9080 CNRS
Institut de Biologie Physico-Chimique
11 rue Pierre et Marie Curie
75005 Paris
France

Prof. Dr. Anthony L. Fink
Chapter 11
Department of Chemistry &
Biochemistry
University of California, Santa Cruz
1156 High Street
Santa Cruz, CA 95064-1099
USA

Prof. Dr. Jorge A. Ghiso
Chapter 19
Departments Pathology & Psychiatry
New York University School
of Medicine
550 First Avenue, TH-432
New York, NY 10016-6402
USA

Prof. Dr. Gilles Grateau
Chapter 25
Pierre et Marie Curie University
Service de medicine interne
Hôpital Hotel-Dieu
1, place du parvis Notre Dame
75181 Paris cedex 04
France

Amyloid Proteins. The Beta Sheet Conformation and Disease. J. D. Sipe

ISBN: 3-527-31072-X

Prof. Dr. Geoffrey Howlett
Chapter 6
Department of Biochemistry &
Molecular Biology
University of Melbourne, Parkville
Melbourne, Victoria 3010
Australia

Prof. Dr. Mariusz Jaskolski
Chapter 27
Department of Crystallography
Faculty of Chemistry
Adam Mickiewicz University
Grunaldzka 6
60-780 Poznan
Poland

Prof. Dr. Bruce Kagan
Chapter 14
Department of Psychiatry
Neuropsychiatric Institute
University of California, Los Angeles
760 Westwood Plaza
Los Angeles, CA 90024
USA

Prof. Dr. David C. Kilpatrick
Chapter 8
Scottish National Blood Transfusion
Service
National Science Laboratory
Ellen's Glen Road
Edinburgh, EH17 7QT
Scotland, UK

Prof. Dr. Robert Kisilevsky
Chapter 2
Department of Pathology
& Molecular Medicine
Queens University
Kingston, ON K7L 3N6
Canada

Prof. Dr. Barbara Kluve-Beckerman
Chapter 22
Department of Pathology &
Laboratory of Medicine
Indiana University
School of Medicine
635 Barnhill Drive, MS-128
Indianapolis, IN 46202
USA

Prof. Dr. Mary Jo LaDu
Chapter 10
ENH Research Institute
Northwestern University
1801 Maple Avenue, Suite 6240
Evanston, IL 60201
USA

Prof. Dr. Peter J. Neame
Chapter 10
Department of Biochemistry
& Molecular Biology
University of South Florida
3500 East Fletcher Avenue, Suite 302
Tampa, FL 33613
USA

Prof. Dr. Melanie R. Nilsson
Chapter 5
Department of Chemistry
McDaniel College
Eaton Hall, 2 College Hill
Westminster, MD 21157
USA

Dr. Ruth Nussinov
Chapters 13, 23
National Cancer Institute
Laboratory of Experimental &
Computational Biology
SAIC Frederick Inc.
P.O. Box B, Bldg. 469, Rm 149
Frederick, MD 21702-1201
USA

Prof. Dr. Sheena E. Radford
Chapter 26
School of Biochemistry &
Molecular Biology
University of Leeds
Mt. Preston St.
Leeds, LS2 9JT
UK

Prof. Dr. David C. Seldin
Chapter 15
Department of Medicine
Boston University Medical Center
EBRC420
650 Albany Street
Boston, MA 02118
USA

Prof. Dr. Jean D. Sipe
Chapter 2
Department of Biochemistry
Boston University
School of Medicine
8406 North Brook Lane
Bethesda, MD 20814-2615
USA

Dr. Fred J. Stevens
Chapter 30
Biosciences Division
Argonne National Laboratory
9700 Cass Avenue
Argonne, IL 60439
USA

Prof. Dr. Fabrizio Tagliavini
Chapter 16
Istituto Nazionale Neurologico
Carlo Besta
Via Celoria 11
20133 Milano
Italy

Prof. Dr. David Teplow
Chapter 17
Department of Neurology
University of California, Los Angeles
710 Westwood Plaza
(Reed CII 1A)
Los Angeles, CA 90095
USA

Prof. Dr. Gunilla T. Westermark
Chapter 28
Department of Cell Biology
University Hospital
Linköping University
58185 Linköping
Sweden

Prof. Dr. Per Westermark
Chapter 1
Departments of Genetics & Pathology
University Hospital
Uppsala University
75185 Uppsala
Sweden

Prof. Dr. Steve P. Wood
Chapter 9
Division of Biochemistry
and Molecular Biology
School of Biological Sciences
University of Southampton
Bassett Crescent East
Southampton SO16 7PX
UK

Part VI
Amyloid Proteins

Amyloid Proteins. The Beta Sheet Conformation and Disease. J. D. Sipe

ISBN: 3-527-31072-X

Brain

17
The Amyloid β-Protein

Noel D. Lazo, Samir K. Maji, Erica A. Fradinger, Gal Bitan and David B. Teplow

17.1
Introduction

The amyloid β-protein (Aβ) may be the most clinically important and the least well understood of all the amyloid proteins. Aβ actually is a peptide that is produced normally and ubiquitously in the human body throughout life [1, 2]. The primary structure of Aβ is presented in Fig. 17.1. There are no cysteines and thus no intra- or intermolecular disulfide pairing is possible. The peptide is not constitutively modified post-translationally. The sequence is amphipathic. Polar amino acids occur only within the first 28 positions. The last 14 amino acids are apolar. Considering the rather unremarkable structure of Aβ, one may ask "why is this peptide so important?" In this chapter we seek to answer this question. In doing so, the reader will be provided a broad perspective on Aβ, including its history and clinical relevance, physiology, genetics, cell and molecular biology, and structural biology. Particular emphasis is placed on Aβ as an etiologic agent in Alzheimer's disease (AD). However, at a more basic level, Aβ also may be viewed as an archetype of natively unfolded proteins with amyloidogenic potential. Studies of Aβ folding and assembly thus hold the promise of providing insight into fundamental aspects of protein science as well as advancing efforts towards curing amyloidoses.

DAEFRHDSGYEVHHQKLVFFAEDVGSNKGAIIGLMVGGVVIA
1 5 10 15 20 25 30 35 40

Fig. 17.1 Amino acid sequence of the Aβ protein.

Amyloid Proteins. The Beta Sheet Conformation and Disease. J. D. Sipe

ISBN: 3-527-31072-X

17.2
Aβ, AD and Amyloid

Aβ is an important focus of study in the biomedical community because it is linked causally to AD [3–5]. This linkage was established a century ago when the Bavarian neurologist Alois Alzheimer reported histologic findings of an "unusual" case of dementia [6]. The two prominent findings were agyrophilic extracellular plaques and intraneuronal tangles in the brain [6]. Plaques and tangles remain pathognomonic for AD [7, 8]. At the time of Alzheimer's report, the proteinaceous nature of amyloid plaques was not known. Deposits of this type were thought to be starch-like, hence the name *amyloid* [from the words *amylum* (Latin) and *amylom* (Greek), both of which mean *starch*]. Protein chemical studies performed almost 80 years later by Glenner and Wong [9, 10] demonstrated the proteinaceous nature of amyloid plaques and provided the first primary structure information on Aβ. The modern era of AD research had begun.

In 1987, four groups reported the cloning of a gene encoding a large precursor protein, the amyloid β-protein precursor (AβPP), within which the sequence for Aβ was found [11–14]. AβPP is a type I integral membrane glycoprotein [11]. AβPP is produced in three predominant forms, 695, 751 or 770 amino acids long (Fig. 17.2). Aβ is produced through post-translational endoproteolysis of AβPP (Section 17.3.1). Two predominant forms of Aβ are produced *in vivo*, Aβ40 (residues 1–40; Fig. 17.1) and Aβ42 (residues 1–42; Fig. 17.1) [The pre-

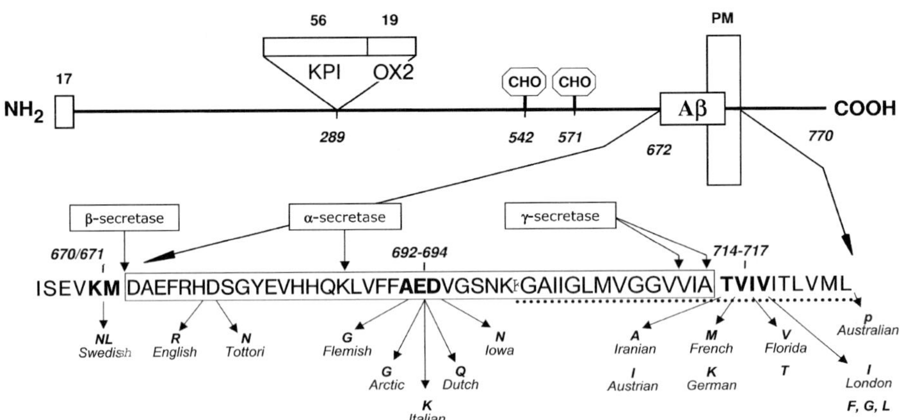

Fig. 17.2 Schematic representation of AβPP. The top diagram shows the longest variant of AβPP (770 residues), containing a 17-residue signal peptide, two alternatively spliced exons at residue 289 (KPI and OX-2), sites of glycosylation (CHO) and a transmembrane domain (PM, residues 700–723). The bottom diagram shows the locations of the cleavage sites of the β, α and γ-secretases. Pathogenic mutations and their names are indicated by arrows below the amino acid sequence (Section 17.5.1). KPI = Kunitz-type serine protease inhibitor motif.

dominant forms of full-length Aβ are denoted Aβ40 and Aβ42. Fragments are denoted by Aβ(x–y), where x is the position of the N-terminal amino acid and y is the position of the C-terminal amino acid, relative to the full-length peptide.] The cloning of the structural gene for AβPP enabled genetic, cell, molecular biological and immunological studies. The AβPP gene is located on chromosome 21, the same chromosome that is triplicated in Down's syndrome [13]. Down's syndrome patients invariably develop Alzheimer-type pathology should they live into their fifth decade [15] and amyloid deposits have been detected in some patients as young as 12 [16]. This observation suggested that this pathology was an effect of gene dosage, and therefore that the AβPP gene and AD may be linked [14]. Subsequent studies have provided strong evidence supporting this hypothesis (Section 17.5.1). Nineteen kindreds now have been identified in which mutations occur within or immediately adjacent to the Aβ region of the AβPP gene (Fig. 17.2). Invariably, the phenotypic effect of these mutations is to: (1) increase Aβ concentrations, (2) increase the Aβ42/Aβ40 concentration ratio or (3) produce Aβ alloforms with increased propensities to aggregate into amyloid fibrils (Section 17.5.1). Histopathological, molecular biological and genetic studies thus all support a central role for Aβ in AD.

What is the mechanistic basis for the linkage of qualitative and quantitative changes in Aβ metabolism to AD? Clues were available in Alzheimer's time, i.e., the presence of intra- and extracellular protein deposits and dead and dying neurons [6]. Continued investigation has revealed that the amyloid plaque is a dynamic structure [17] comprising: (1) an accretion of amyloid fibrils, within and around which are found dystrophic (swollen) neurites [18, 19], (2) a variety of macromolecules, including complement components, serum amyloid P (SAP), apolipoprotein E (ApoE) and proteoglycans (PGs) [20], (3) an infiltrate of phagocytic cells that includes astrocytes and microglia [21], and (4) dead and dysmorphic neurons [19]. These observations suggested that plaques were neurotoxic. The next step was to determine if Aβ fibrils were the neurotoxic component of the plaques. They were [22, 23]. These data, and work by others, led Hardy and Higgins in 1992 to propose the "amyloid cascade hypothesis" [3]. The central tenet of this hypothesis is that AD is a "direct result of Aβ deposition".

In the last decade, tests of the amyloid cascade hypothesis have weakened its foundation. The temporal and anatomical development of plaques is inconsistent with the course of disease in humans [24], and with the progression of neuronal dysfunction and neuron loss in transgenic animals [25]. Neurofibrillary tangle formation (intracellular fibrillar deposits of tau protein) is a better indicator of disease status [26]. However, correlations have been made between the concentration of soluble, non-fibrillar Aβ42 assemblies in brain, plasma and cerebrospinal fluid (CSF) [27, 28] and the severity of AD [29, 30]. In addition, studies in transgenic mice expressing human AβPP and producing Aβ, and in whom extensive amyloid deposition occurs in adulthood and late life, reveal neurological deficits prior to the observation of deposits [31]. A paradigm shift thus has occurred away from the primacy of fibrils in the causation of AD towards the primacy of fibril precursors and oligomeric assemblies [32–38]. This shift now is reflected in a revised amyloid cascade hypothesis [5] that emphasizes the role of oligomeric forms of Aβ in AD.

In vitro studies of the assembly and biological activity of Aβ have provided compelling evidence supporting the aforementioned paradigm shift. Basic studies of Aβ self-association have revealed an increasing number of pre-fibrillar assemblies, including multiple monomer conformers [39], different types of oligomers [40–42], Aβ-derived diffusible ligands (ADDLs) [43, 44], protofibrils [45, 46], fibrils [47] and spheroids [48, 49]. These assemblies are potent neurotoxins [40, 43, 44, 50, 51] and the toxicity of some can be greater than that of the corresponding fibrils [52].

An impediment to curing AD and other amyloidoses is the identification of the proximate toxic agent(s). In the century since Alzheimer's case presentation [6], a broad (and increasing) foundation of information has been produced, supporting work towards overcoming this barrier. The essence of this work, as expressed in the title of this book, is linking the biophysical behavior of the Aβ peptide with its clinical effects. Doing so provides the targets necessary for the development of therapeutic agents. In the sections that follow, we discuss in greater detail the current status and future directions of efforts to understand the role of Aβ in AD and to develop efficacious therapies.

17.3
Pathogenetic Process – Biology

17.3.1
Aβ Metabolism and AD

The production and accumulation of Aβ in the central nervous system (CNS) are seminal events in the development of AD [53]. Levels of Aβ in the brain are regulated by the activity of enzymes involved in its production, degradation and clearance. Although AD is not linked to a single gene, genetic studies have revealed that mutations in the AβPP gene in or near the Aβ region, and in the genes encoding presenilin (PS)-1 and -2, proteases involved in the production of Aβ, are linked to early-onset familial AD. It should be noted, however, that the majority of AD cases are "sporadic" – they have no known genetic cause, although it is possible that they involve genes that regulate the metabolism of Aβ. Environmental factors, including diet, could also contribute to sporadic cases, but reports in the literature have been inconsistent [54].

Aβ is derived through endoproteolytic processing of AβPP [55]. The cleavage sites are termed α, β, γ, δ, ε and ζ (Fig. 17.3 and Table 17.1). In this discussion, for simplicity, the locations of these sites are defined relative to the Aβ sequence itself, beginning with Asp1. In the non-amyloidogenic pathway, AβPP first is cleaved after Lys16 by α-secretase, resulting in the release of a soluble N-terminal AβPP fragment, $A\beta PP_{s\alpha}$, and a membrane-associated C-terminal fragment,

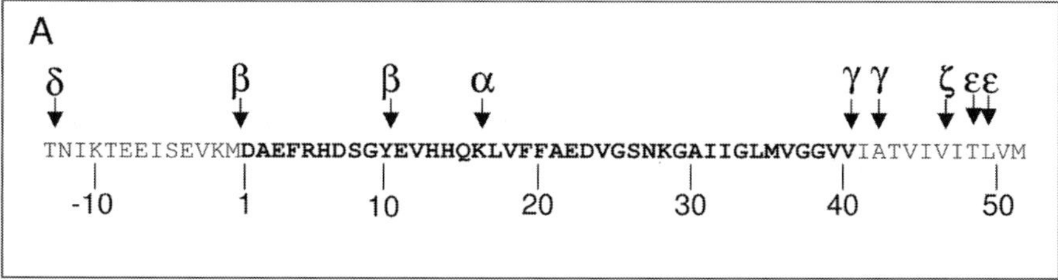

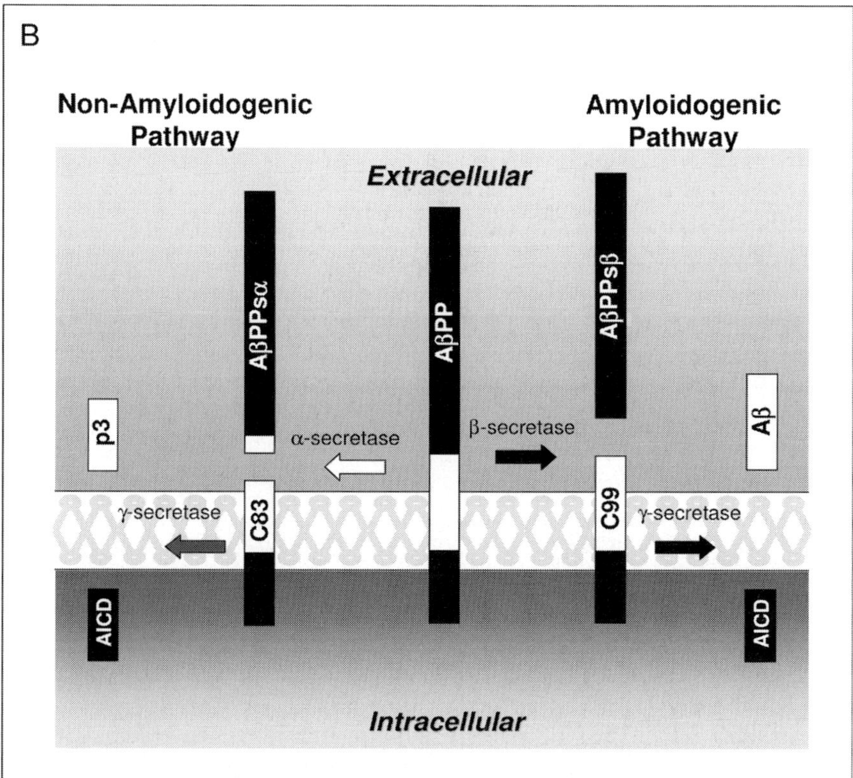

Fig. 17.3 Proteolytic processing of AβPP. (A) The Aβ region (boxed) and flanking sequences in AβPP. Sites of α, β, γ, ε, δ and ζ cleavage are indicated. (B) Predominant processing pathways and their products (C = C-terminal fragment). Aβ is produced through the amyloidogenic pathway.

C83. The C83 fragment is cleaved at the ε-cleavage site (either after Thr48 or Leu49), releasing the AβPP intracellular domain (AICD), and subsequently at the γ-cleavage site (after either Val40 or Ala42) to produce p3 [56–59]. Cleavage at the α-site is mediated by α-secretase, a protease thought to be a disintegrin and metalloprotease (ADAM) family member [60, 61]. α-Secretase cleavage pre-

Table 17.1 Localization of AβPP processing

Organelle	Secretases	AβPP products
ER/IC [78–80]	β, γ	$A\beta PP_{s\beta}$, C99, Aβ42, AICD
Trans-Golgi [785]	α, β, γ	$A\beta PP_{s\alpha/\beta}$, C83, C99, p3, Aβ, AICD
Secretory vesicles [786–788]	α, β	$A\beta PP_{s\alpha/\beta}$, C83, C99
Plasma membrane [75]	α, γ	$A\beta PP_{s\alpha}$, C83, C99, p3, Aβ, AICD
Endosomes [74, 212, 789]	β, γ	$A\beta PP_{s\beta}$, C99, Aβ, AICD

cludes formation of full-length Aβ. γ-Cleavage is mediated by the γ-secretase complex, which is made up of Aph-1, Pen-2, nicastrin and PS-1 [62]. Studies in AβPP/PS-1 overexpressing cell lines have revealed that γ-secretase inhibitors and PS-1 mutants also inhibit ε- and ζ-cleavage, indicating that these cleavages involve PS-1 and γ-secretase-like activity [57, 63]. In the amyloidogenic pathway, AβPP first is cleaved by β-secretase to generate soluble $A\beta PP_{s\beta}$ and membrane-bound C-terminal fragment C99. Subsequently, C99 is cleaved by γ-secretase to produce Aβ and the AICD. γ-Secretase actually is active at multiple cleavage sites in the C99 fragment, thus Aβ alloforms ranging from 39 to 43 amino acids in length may be generated. However, the 40 and 42 amino acid forms of Aβ are the predominant alloforms [47]. β-Site cleavage of AβPP is mediated by a transmembrane aspartyl protease, β-site AβPP-cleaving enzyme (BACE1), also referred to as Asp2 or memapsin2 [64–67]. AβPP also may be cleaved at the δ-cleavage site (after Thr584 of AβPP695) resulting in an N-terminally extended peptide [68], the ζ-cleavage site (after Val46) or the ε-cleavage sites (after Thr48 and Leu49) [57, 63]). The ζ-cleavage site corresponds to the site of the London mutation (Fig. 17.3) and involves γ-secretase-like activity [63]. In most cells, the non-amyloidogenic pathway is the predominant AβPP processing pathway and stimulation of α-secretase results in a significant decrease in Aβ production [55]. However, in neural cells, both non-amyloidogenic and amyloidogenic pathways normally function [1, 2, 69, 70]. In AD patients, a 6-fold increase in soluble Aβ concentration and a 12-fold increase in Aβ42 concentration have been observed [27]. Shifts in processing pathway choice, resulting in increased levels of Aβ, thus have been postulated to cause AD [55].

One mechanism to regulate α- or β-cleavage of AβPP is membrane trafficking. Lipid rafts, composed of cholesterol and sphingolipids, are membrane microdomains that can selectively incorporate or exclude proteins, thereby influencing protein–protein interactions in the membrane [71]. Studies have shown that both the β-secretase and the γ-secretase complex are localized to lipid rafts [72–74]. BACE1 was found to be associated with detergent-insoluble membranes, and cholesterol depletion inhibited β-cleavage of AβPP in N2a cells [73]. In contrast, α-secretase activity was not associated with lipid rafts, and cholesterol depletion resulted in an increase in α-secretase cleavage [73]. Trafficking of AβPP to specific membrane microdomains thus may be one mechanism regulating the production of Aβ.

Post-translational processing of AβPP also occurs within the secretory pathway, at the cell surface and in the endosomal pathway, resulting in the release of AβPP derivatives into vesicles or the extracellular space (Fig. 17.4). Pulse–chase studies have revealed that α- and ε-cleavage occur late in the secretory pathway in the *trans*-Golgi network (TGN) and secretory vesicles, at the plasma membrane, and in the endocytic pathway [57, 75, 76]. Both β- and γ-secretase cleavage can occur early in the secretory pathway, as well as in the late secretory pathway, at the plasma membrane and in the endosomal pathway. Cellular studies using brefeldin A, a compound that sequesters protein in the endoplasmic reticulum (ER) by blocking protein transport to the Golgi complex and redistribution of the *cis*-Golgi into the ER [77], have demonstrated that AβPP$_{s\beta}$ and Aβ42 are produced in the ER/intermediate compartment (ER/IC) [78–80]. Therefore, Aβ40 and Aβ42 produced later in the secretory pathway is released extracellularly, whereas Aβ42 produced in the ER/IC accumulates intracellularly [81]. Aβ42 is the major alloform found in the ER/IC and endosomes. Aβ40 is the major alloform produced in the late secretory pathway and plasma membrane

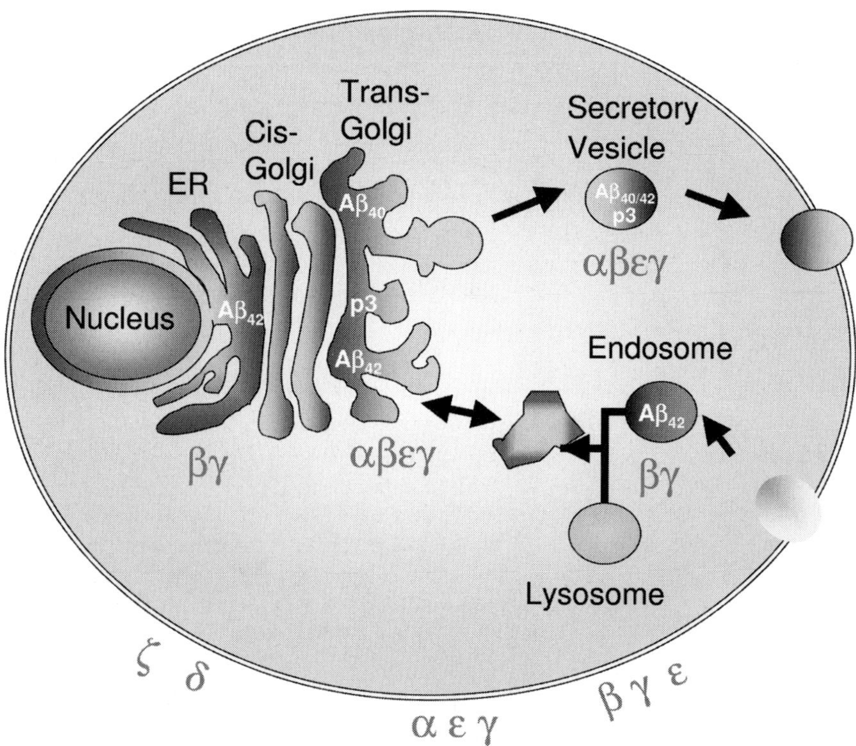

Fig. 17.4 Cellular localization of AβPP processing. Aβ42 is the predominant product in the ER and endosome. Aβ40 is predominantly produced in the late secretory pathway and at the plasma membrane.

and Aβ40 production may be linked to ε- and γ-secretase activity. Understanding the mechanisms underlying the generation of Aβ42 is critical to our understanding of AD because Aβ42 has a higher propensity to self-associate than does Aβ40 and Aβ42 appears to be more neurotoxic [47].

The level of Aβ in the brain is maintained through a metabolic balance between anabolic and catabolic processes. Therefore, although early-onset AD is linked to increased production of Aβ, the development of late-onset AD, the most common form of the disease, may be due to impaired Aβ clearance. Cerebral Aβ may be cleared through receptor-mediated transport across the blood–brain barrier (BBB), enzymatic digestion, or microglial phagocytosis (Fig. 17.5). The first evidence for the involvement of clearance pathways in the development of AD came from the discovery that ApoE, the major cholesterol transporter in the brain [82], was linked to the development of AD [83]. ApoE is encoded by the *APOE* gene, which has three alleles, $\varepsilon 2$, $\varepsilon 3$ and $\varepsilon 4$ [82]. The $\varepsilon 4$ allele is a strong risk factor for AD [83, 84] (Section 17.5.3). ApoE binds to Aβ and can facilitate aggregation [85] or promote receptor-mediated clearance [86, 87]. Alternatively, the $\varepsilon 4$ allele may increase the risk of AD indirectly through alteration of cholesterol levels, promoting the generation of Aβ from the AβPP precursor [88, 89]. ApoE significantly increased levels of Aβ and cholesterol in the brain of *APOE*-knock-in/AβPP transgenic mice [90]. Scavenger receptors, such as the

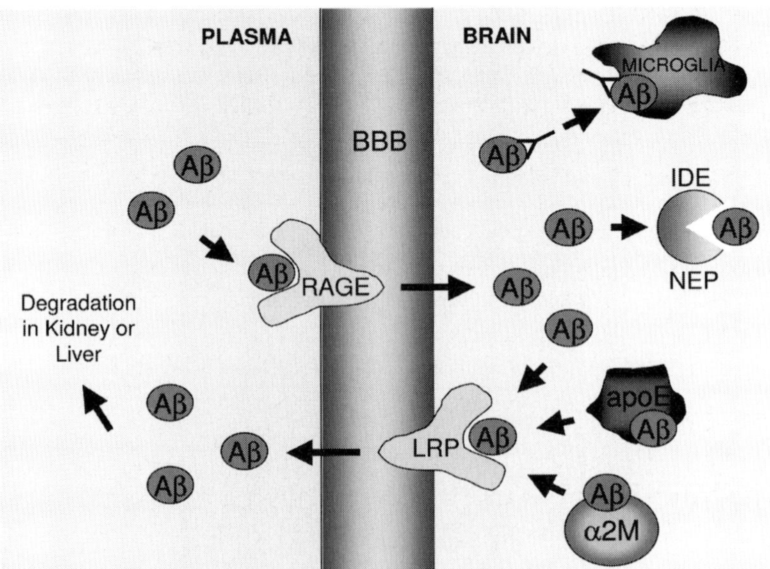

Fig. 17.5 Clearance pathways for the removal of Aβ from the brain include: (1) microglial phagocytosis, (2) enzymatic digestion by IDE or NEP, (3) receptor-mediated clearance across the BBB via LRP or RAGE and (4) chaperone-assisted (ApoE or α_2-macroglobulin) clearance across the BBB.

low-density lipoprotein (LDL) receptor-related protein (LRP) and the receptor for advanced glycation end-products (RAGE), are involved in the clearance of Aβ across the BBB [91]. LRP appears to be involved primarily in the movement of Aβ40 from the CNS to the periphery for degradation in the liver and kidney [92]. In mice injected with radiolabeled Aβ40, LRP antagonists reduced Aβ clearance from the brain by 90% [87]. Additionally, knock-out of the LRP chaperone receptor-associated protein (RAP) resulted in increased cerebral Aβ levels [93]. LRP-mediated clearance may occur through direct binding of Aβ to the receptor or through binding of chaperones such as ApoE and α_2-macroglobulin [92]. In contrast, RAGE appears to be involved in the transport of peripheral Aβ back into the CNS [92].

In addition to clearance of holoAβ, direct proteolytic degradation of Aβ appears to be an important catabolic pathway (Fig. 17.6). A variety of enzymes have been shown to degrade Aβ *in vitro*, including insulin-degrading enzyme (IDE), neprilysin (NEP), endothelin-converting enzyme (ECE), angiotensin-converting enzyme (ACE) and plasmin [94–96]. *In vivo* evidence indicates that IDE and NEP are the primary enzymes involved in the degradation of Aβ. IDE is a cytosolic zinc metalloprotease involved in the hydrolysis of regulatory peptides including insulin and glucagon [96]. Evidence for the involvement of IDE in the degradation of Aβ includes genetic linkage to AD [97], IDE-dependent Aβ degradation in cultured cells [98] and IDE mouse models [99, 100]. IDE knock-out mice have increased endogenous brain Aβ levels and overexpression in IDE/AβPP double-transgenics resulted in a 50% decrease in Aβ [99, 100]. The zinc metalloprotease NEP is a type II integral membrane protein localized to the extracellular faces of pre-synaptic and axonal neural membranes that functions to turn off neuropeptide signals at the synapse [94, 101, 102]. NEP is a member of the M13 family of metalloproteases, which also includes ECE. Although both NEP and ECE are able to degrade Aβ, NEP appears to be the major Aβ-degrading enzyme in the brain [103]. Endogenous NEP was able to degrade exogenously administered Aβ in the rat brain and thiorphan, a specific NEP inhibitor, abolished this activity, leading to Aβ deposition [104]. NEP knock-out mice have

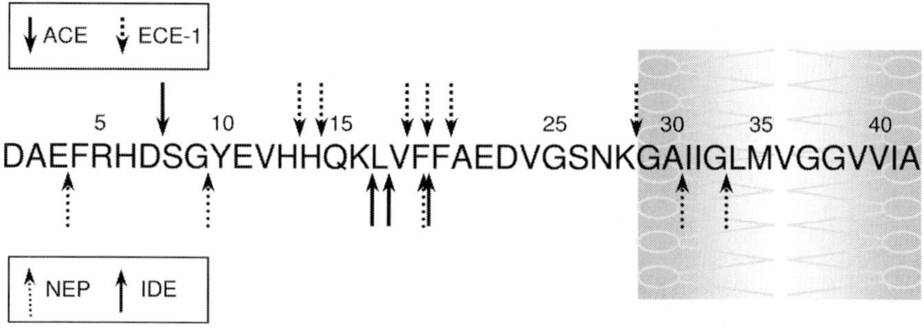

Fig. 17.6 Aβ sequence showing the cleavage sites for enzymes involved in the catabolism of Aβ (see text).

impaired ability to degrade exogenously administered Aβ, and have elevated levels of endogenous Aβ40 and Aβ42 [104]. Additionally, NEP/AβPP transgenic mice have decreased levels of Aβ [105]. The differential localization of IDE in the cytosol and NEP on the extracellular face of the plasma membrane may indicate that these enzymes are involved in the catabolism of different pools of Aβ. Both IDE and NEP can degrade monomeric Aβ, and NEP can also degrade oligomeric Aβ [106]. However, elimination of larger assemblies of Aβ may require the activity of phagocytic pathways.

Microglia and astrocytes are cells of the CNS that maintain neural cell health and are involved in the immune response in reaction to CNS injuries. In microglial cultures from AD and non-demented elderly human brains, Aβ42 exposure results in microglial activation causing expression of major histocompatibility complex type II antigens, pro-inflammatory cytokines, chemokines, reactive oxygen species (ROS) and complement proteins [107, 108]. Microglia express scavenger receptors, including formyl peptide receptor, macrophage scavenger receptor and RAGE, that can bind Aβ and induce a phagocytic response [108, 109]. Both soluble and insoluble assemblies of Aβ may be cleared from the brain by microglial phagocytosis [110, 111], and cultured rat microglia can clear Aβ from serum-free medium through phagocytosis [112]. Microglial phagocytosis is, in fact, the proposed Aβ clearance mechanism in AβPP-overexpressing transgenic mice immunized with Aβ [113]. Astrocytes have been found to internalize and degrade non-fibrillar Aβ in AD brains and in Aβ-containing brain sections from AβPP transgenic mice [114–116]. NEP is localized in astrocytes and may be the primary enzyme involved in the degradation of Aβ in these cells. A connection between astrocytes and Aβ degradation is also suggested by immunohistochemical studies demonstrating that the astrocyte-specific cell surface protein glial fibrillary acidic protein (GFAP) co-localizes with some Aβ deposits [117]. These deposits, termed diffuse plaques, are nebulous deposits of Aβ [116, 118] and are distinct from senile plaques, which are smaller and more densely packed with Aβ [116, 118]. It has been suggested that GFAP-positive diffuse plaques may be derived from the lysis of astrocytes that contain large amounts of Aβ, but were unable to metabolize it [116, 118]. In summary, the levels of Aβ in the brain are regulated through the interplay of anabolic and catabolic processes, the complexity of which suggests multiple points at which dysfunction may cause AD as well as where therapeutic intervention may be targeted.

17.3.2
Mechanisms of Aβ-induced Neuronal Injury

In the AD brain, and in mouse models of AD, evidence exists both for necrotic [119–121] and apoptotic [122–125] neuronal cell death. The damaging effects of Aβ are associated with synaptic dysfunction, mitochondrial dysfunction, oxidative stress, NO radical generation, disruption of Ca^{2+} homeostasis and inflammation (Fig. 17.7). Although neuronal cell loss is a pathological hallmark of AD, it is preceded by cognitive impairment due to neuronal dysfunction and synapse

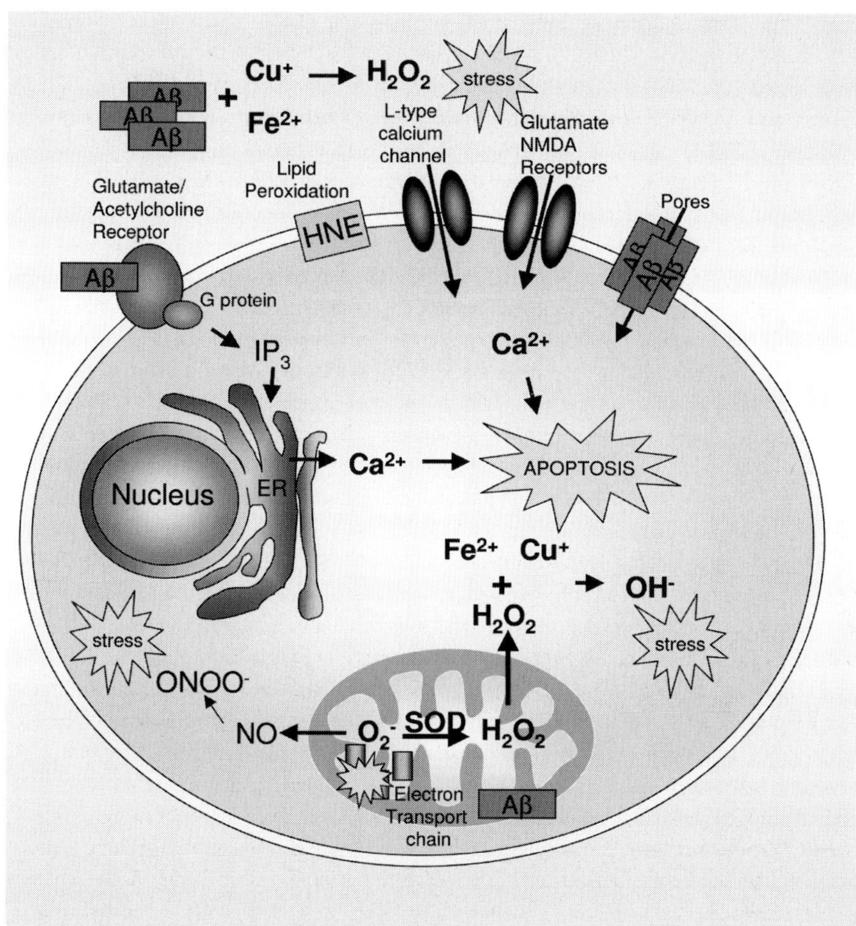

Fig. 17.7 Aβ-mediated neurotoxic mechanisms involving oxidative stress and calcium (Ca^{2+}) disruption. Aβ can form pores in the plasma membrane leading to Ca^{2+} disruption. Aβ aggregation in the presence of Cu^{+} and Fe^{2+} produces hydrogen peroxide (H_2O_2), resulting in membrane-associated oxidative stress that causes: (1) lipid peroxidation and the generation of HNE, and (2) oxidative modification of G-protein-coupled receptors (glutamate or acetycholine receptors), ion channels (Ca^{2+}) and voltage-dependent channels (NMDA) that leads to Ca^{2+} disruption. Aβ also can cause mitochondrial dysfunction leading to the production of superoxide ($O_2^{-}\cdot$) that can: (1) interact with nitric oxide (NO) to produce peroxynitrite ($ONOO^{-}$) and (2) convert to H_2O_2 through the action of SOD. H_2O_2 interacts with Cu^{+} and Fe^{2+} producing hydroxyl radicals ($OH\cdot$) that can cause membrane-associated oxidative stress or dysfunction of the ER.

loss. The phenomenon of long-term potentiation (LTP) is thought to mediate learning and memory and is widely used as a model of activity-dependent synaptic plasticity. LTP is a long-lasting enhancement of synaptic efficacy in response to high-frequency afferent stimulation. Soluble Aβ oligomers can inhibit hippocampal LTP *in vivo* [40, 126–128], and AβPP transgenic mice display reduced basal synaptic transmission and inhibition of LTP before plaque deposition [129]. The cellular mechanism through which Aβ inhibits LTP is poorly understood. Studies have investigated the role of Ca^{2+}, cholinergic [130, 131] and glutamatergic [132, 133] pathways [134]. Glutamate mediates fast excitatory synaptic responses in the CNS via both ionotropic and metabotropic receptors. The *N*-methyl-D-aspartate (NMDA) receptors are ionotropic glutamate receptors localized to excitatory post-synaptic membranes that regulate synaptic plasticity and LTP through the mediation of Ca^{2+} influx into the cell [135]. The NMDA receptor is sensitive to oxidizing or reducing agents in the extracellular milieu that can act on the sulfhydryl redox subsite to either decrease or increase its function, respectively. NMDA receptors are heteromultimeric protein complexes and the differential assembly of NR1 or NR2 subunit proteins modulates the pharmacological properties of the receptor [135, 136]. Although there is no clear link between the NMDA receptor and AD, expression of the $NR1_{OXX}$ subunit appears to be positively correlated with age of disease onset [135, 137]. Aβ can induce L-glutamate toxicity via excessive activation of the NMDA receptor [138, 139]. Additionally, Aβ42 was found to bind to the α_7-acetylcholine receptor, blocking Ca^{2+} influx and resulting in cellular vulnerability to glutamate toxicity [131, 140, 141]. Although the causative relationship between Aβ and disruption of these pathways remains unclear, it may involve Aβ-induced oxidative stress and disruption in calcium homeostasis.

In AD, neurons adjacent to plaques and neurons bearing neurofibrillary tangles exhibit elevated levels of free-radical-mediated damage, including oxidatively modified proteins [142], lipids [143–145] and nucleic acids [146–149]. Reduced energy metabolism, superoxide and redox-active metals have been implicated in the oxidative stress found in AD. Neurons are particularly vulnerable to radicals due to their high rates of oxygen metabolism. Mitochondria produce free radicals, predominantly superoxide ($O_2\cdot^-$), as byproducts of the electron transport chain. Superoxide dismutase (SOD) converts the $O_2\cdot^-$ radical to hydrogen peroxide (H_2O_2), which diffuses freely through the mitochondrial membrane into the cytoplasm where catalase (CAT) converts H_2O_2 to H_2O and O_2. AD brains show decreased energy metabolism [150], decreased activity of mitochondrial enzymes, including cytochrome *c* oxidase, the pyruvate dehydrogenase complex and α-ketoglutarate dehydrogenase [151, 152], and altered SOD/CAT ratios [153]. These deficiencies are thought to increase $O_2\cdot^-$ levels in the mitochondria leading to increased cytoplasmic levels of H_2O_2 [151, 154]. In support of this hypothesis, elevated levels of cytoplasmic H_2O_2 are found in AD patients and AβPP transgenic mice [155]. Transition metals such as Fe^{2+} and Cu^+ can react with H_2O_2 (Fenton chemistry) to produce hydroxyl radicals ($OH\cdot$) that cause extensive cellular damage. Additionally, $O_2\cdot^-$ may react with nitric oxide ($NO\cdot$) to generate peroxinitrite, a highly reactive nitrating agent. Nitration of tyrosine has been associated with AD [156] and AβPP-expressing PC12 cells have increased NO· levels [157]. Extracellularly, the aggregation of Aβ in the presence of Fe^{2+} and Cu^{+} produces ROS that impair membrane proteins involved in Ca^{2+} regulation [151].

Ca^{2+} plays a fundamental role in learning, memory and neuronal survival. In AD, perturbed Ca^{2+} regulation appears to be involved in neuronal dysfunction and death [158]. Aβ may disrupt Ca^{2+} regulation through the generation of ROS, causing lipid peroxidation and the impairment of receptors coupled to Ca^{2+}, including voltage-dependent Ca^{2+} channels and ionotropic glutamate receptors. Aβ-induced lipid peroxidation produces 4-hydroxy-2,3-nonenal (HNE), an aldehyde that covalently modifies proteins. The end-products of HNE modification are increased in AD patients [143]. HNE has been shown to inhibit the dephosphorylation of tau [159] and to kill cholinergic neurons when administered intraparenchymally in rats [160]. In addition to oxidative stress-mediated mechanisms, Aβ may have a direct effect on Ca^{2+} homeostasis. In cultured neurons, Aβ can potentiate the activation of voltage-dependent Ca^{2+} channels [161]. Alternatively, *in vitro* studies with Aβ(25–35) have indicated that soluble Aβ oligomers can form pores in the membrane [162, 163] leading to Ca^{2+} influx into the cell [164, 165]. Aβ also has been shown to disrupt Ca^{2+} regulation in the ER and mitochondria. Familial AD (FAD)-linked PS-1 mutations can alter Ca^{2+} release from the ER in response to membrane depolarization, inositol triphosphate activation and glutamate signaling [166].

Recent studies provide evidence that neuronal cell death in AD may be due to ER stress arising from perturbation of Ca^{2+} homeostasis, inhibition of protein glycosylation or reduction of disulfide bonds resulting in the accumulation of unfolded proteins. A specific signaling pathway, the unfolded protein response (UPR), is present in cells to ensure that the protein-folding capacity of the ER is not overwhelmed. This pathway is activated by the accumulation of misfolded proteins in the ER. Activation of the UPR results in increased transcription of genes involved in reducing protein translation and encoding chaperones that aid in the refolding of misfolded proteins or target them for degradation. However, prolonged activation of the UPR leads to activation of the apoptotic cell death pathway [167]. FAD-linked PS mutations increase neuronal susceptibility to ER stress through down-regulation of the UPR [168]. PS mutations decrease expression of ER molecular chaperones GRP78/Bip [169], disrupt activation of ER stress transducers IRE1, PERK and ATF6 [170, 171], and impede translational attenuation [172], leading to apoptotic cell death.

Neurons are not the only cells affected by Aβ. Astrocytes, oligodendrocytes and microglia undergo changes after exposure to Aβ. Activated microglia and reactive astrocytes are associated with amyloid deposits in the AD brain [173]. In astrocytes, Aβ induces cytokine release, perturbs Ca^{2+} regulation and impairs glutamate transport via lipid peroxidation. The impairment of glutamate metabolism causes synaptic dysfunction [174]. Aβ can induce changes in Ca^{2+} homeostasis in oligodendrocytes making them vulnerable to glutamate toxicity [175]. In microglia, Aβ induces the production of inflammatory cytokines, tumor necrosis factor-α and interleukin-1β, and NO· [176] through a Ca^{2+}-dependent pathway [177]. The activation of toxic cytokines contributes to neuronal degeneration.

17.4
Normal Physiologic Function of AβPP and Aβ

17.4.1
AβPP Structure

AβPP comprises a heterogeneous group of ubiquitously expressed proteins of mass 110–135 kDa [178] (Fig. 17.2). The function of AβPP is not well understood. The longer variants of AβPP, comprising 751 or 770 amino acid residues, are widely expressed in non-neuronal cells and also occur in neurons. The shorter isoform, containing 695 residues, is expressed at high levels in neurons and at very low levels in other cell types [179]. The longer forms contain a 56-residue, Kunitz-type serine protease inhibitor (KPI) motif, suggesting that one function of AβPP is protease inhibition [180]. The longest form of AβPP, comprising 770 amino acids, also contains a 19-residue OX-2 domain. In addition to KPI and OX-2 motifs, the large, N-terminal, extracellular domain of AβPP contains heparin- and collagen-binding domains, copper- and zinc-binding sites, acidic (Glu/Asp-rich) and Thr-rich domains, multiple disulfide bonds, and sites of glycosylation, sulfonation and phosphorylation. ("Extracellular" and "intracellular" refer to AβPP attached to the plasma membrane. These designations are used for convenience and do not reflect the actual cellular distribution of AβPP, which to a large extent reveals AβPP in association with intracellular membranes such as the ER [181, 182].) The C-terminal, intracellular tail contains a $G_{o\alpha}$-binding site, cleavage sites for caspase-3, -6, -8 and -9, and phosphorylation sites. AβPP is a member of a family of type I integral membrane proteins that contains at least two other homologs, known as the amyloid precursor-like proteins 1 and 2 (APLP1 and APLP2) [183–185]. APLP1 and APLP2 share most of their structural domains with AβPP, including the transmembrane domain, *N*-glycosylation sites, and copper- and zinc-binding domains. Like AβPP, APLP2, but not APLP1, also contains a KPI motif. These similarities suggest that the APLPs have functions overlapping those of AβPP. Interestingly, APLP1 and APLP2 do not contain an Aβ domain.

17.4.2
AβPP and Aβ Function

In the CNS, an increase in AβPP expression during development coincides with a peak of neuronal differentiation and neurite outgrowth [186, 187], suggesting a role for AβPP in these functions. However, AβPP is not an essential protein. AβPP-deficient mice are viable and fertile. These mice show decreased locomotor activity and reduced strength [188], their brains display reduced brain weight and size of forebrain commissures [189], and they develop reactive gliosis by 14 weeks of age [188]. In comparison, transgenic mice expressing human AβPP have been reported to show improved resistance to oxidative stress and excitotoxicity relative to control mice [190]. However, a different study reported that transgenic mice overexpressing AβPP had increased ischemic brain damage relative to control mice [191]. Cultured neurons from AβPP-deficient mice are similar to neurons from wild-type mice in their sensitivity to oxidative stress and Aβ neurotoxicity, suggesting that the neuroprotective functions of AβPP may be compensated for by APLPs [192]. In particular, APLP2 appears to share function with AβPP. APLP2-defficient mice are healthy and fertile, and do not display any gross functional problems. However, 80% of mice in which both AβPP and APLP2 are knocked-out die in the first week of their life, and the survivors display substantial growth and development deficiencies [193, 194]. A neuroprotective role for AβPP is supported by the observations that AβPP expression is increased by neurotrophic factors [195] and in response to neuronal injury [196].

Functions of AβPP have been attributed to the membrane-bound holoprotein, its secreted fragments, including $A\beta PP_{s\alpha}$, $A\beta PP_{s\beta}$ and Aβ, and its intracellular C-terminal fragments. Cell surface AβPP stimulates neurite growth in cultured neurons [197, 198], promotes cell adhesion [198, 199] and protects neurons against various insults, including Aβ toxicity [200]. N-terminal $A\beta PP_s$ fragments are released from presynaptic terminals in response to electrical activity, and regulate neuronal excitability, synaptic plasticity, and learning and memory [201]. In particular $A\beta PP_{s\alpha}$ has been shown to be involved in neuritogenesis and neuroprotection in primary neuronal cultures [202–206] and neuronal stem cells [207, 208].

As discussed in Section 17.3.1, $A\beta PP_{s\alpha}$ and $A\beta PP_{s\beta}$ are released by α- and β-secretases, leaving behind the membrane-anchored, C-terminal, AβPP fragments C83 and C99, respectively. These fragments are further cleaved by γ-secretase, releasing the AICD. AICD contains a 7-amino-acid recognition motif, ^{756}GYENPTY, which is conserved in all species and mediates clathrin-coated pit internalization [209–212]. Another recognition motif in AICD is ^{728}YTSI, which has been shown to be important for basolateral sorting of AβPP in polarized epithelial cells and for degradation of AβPP in lysosomes [213]. The GYENPTY motif also mediates binding of AβPP to adaptor proteins that contain a phosphotyrosine-binding (PTB) domain, such as those in the X11, Fe65 and JIP families [214]. X11α interacts with AβPP and with the APLPs [215–217] in a manner that requires the presence of the YENP tetrapeptide (the central part of the GYENPTY motif) [215, 218]. The interaction of X11α with AβPP has been implicated in inhibition of γ-secretase cleavage. The basis for this inhibition is thought to be diminution in AβPP trafficking to cellular compartments containing active γ-secretase [219]. In contrast, interaction of AβPP with Fe65 family members facilitates γ-secretase cleavage of C99 and release of AICD and Aβ [220, 221]. Based on work in transfected COS cells, it has been suggested that binding to AβPP may anchor Fe65 in the cytoplasm, impairing nuclear translocation [222]. However, further work in neuronal and non-neuronal cells showed that the AICD–Fe65 complex interacts with the histone acetyltransferase Tip60, translocates into the nucleus [223, 224], and functions in the transcriptional regulation of genes including glycogen synthase kinase–3 [225] and AβPP itself [226]. Fe65 forms similar transcription complexes with the APLPs [227]. c-Jun N-terminal kinase (JNK)-interacting proteins (JIPs), particularly JIP-1b, also interact with AβPP through the YENPTY motif [228–230]. The resulting complex translocates into the nucleus and activates gene expression. However, the mechanism of action of the AICD–JIP-1b complex differs from that of AICD–Fe65. Fe65 enters the nucleus in the absence of full-length AβPP, but JIP-1b does not. In addition, the translocation is Tip60 independent and occurs with AβPP, but not with APLPs [231]. In addition, JIP-1b serves as a link between AβPP (but not APLP2) and kinesin light chain-1 [230, 232], mediating fast axonal transport of vesicles containing AβPP, PS-1 and BACE1 [233]. Other adaptor proteins may interact with the C-terminal part of AβPP (for a recent review, see [214]), but their biological function remains to be determined. Because APLP cannot substitute for AβPP in all cases, an interesting question is how AβPP-null mice compensate for the absence of AICD. Taken together, studies of AβPP processing and physiological activity clearly establish a role for the AICD in gene regulation.

What about the Aβ fragment of AβPP? What is its normal physiologic function? This is one of the most interesting, important and puzzling questions in the field. Aβ is produced as part of normal metabolism [69, 70], but despite massive efforts to characterize its biological function, a clear answer has not been obtained. In view of the important roles of the N-terminal and C-terminal fragments of AβPP, one hypothesis is that the Aβ sequence serves merely as a linker between these two parts and is destined for endosomal/lysosomal degradation. However, a myriad of Aβ activities have been reported. Initial indication of possible neurogenic/neurotrophic effects of Aβ came from observations in AD brains and experiments in rats treated intracerebrally with Aβ that showed neurite sprouting near plaque-derived lesions [234, 235]. When added to primary rat hippocampal neurons at nanomolar concentrations, Aβ(1–28) was found to enhance cell survival [236]. A similar effect on neuron survival was found for submicromolar concentrations of Aβ40 and Aβ42, which also promoted neurite growth [22, 237, 238]. Sequence homology between the region Aβ(31–35) (Ile-Ile-Gly-Leu-Met) and the C-terminus of tachykinins (Phe-X-Gly-Leu-Met-NH_2, where X is a hydrophobic amino acid), a family of neuropeptides with known cytotrophic characteristics [239], led to the hypothesis that the trophic effects displayed by Aβ may be mediated by similar mechanisms to those used by the tachykinins. The fragment Aβ(25–35) displayed neurotrophic effects at submicromolar concentrations but was neurotoxic at higher concentrations [238]. Recently, a sequence-specific neurogenic effect on the development of neuronal stem cells was shown for Aβ42, but not Aβ40 or Aβ(25–35) [793]. In a study using rat and human neurons in culture, treatment with secretase inhibitors, preventing Aβ production, or capturing Aβ using an antibody, led to a substantial decrease in cell viability. This effect was not observed with non-neuronal cells. Neuronal viability could be rescued by picomolar concentrations of Aβ40, but Aβ(25–35) or Aβ42 had minimal protective effects [240]. The tachykinin hypothesis thus has not been supported experimentally.

Aβ is a high-affinity chelator of transition metal ions such as Cu^{2+}, Fe^{3+} and Zn^{2+} [241]. Conceivably, Aβ can bind these ions *in vivo*, thereby preventing them from reacting with metalloproteins that mediate redox reactions and may be involved in oxidative stress responses. Aβ thus may function as an antioxidant [242, 243]. Oxidative stress promotes Aβ production (see references 31–38 in [243]), supporting this hypothesis. Production and aggregation of Aβ thus have been suggested to occur secondary to neuronal stress and to function as a protective adaptation to the disease rather than causing it [244]. However, others have shown that Aβ assembly causes oxidative stress, presumably through the action of activated microglia and astrocytes, leading to a cycle in which more Aβ is produced, increasing the concentration of ROS and leading to neuronal damage [245].

A role for Aβ as an apolipoprotein mediating cholesterol homeostasis and synaptic plasticity has been suggested [246]. In support of this view, Aβ40 increases LTP and NMDA receptor-mediated synaptic transmission in rat hippocampus [247, 248], and inhibition of cholesterol synthesis abolishes this effect [249].

In summary, a normal physiologic role for Aβ is possible, but remains to be established clearly. Although Aβ may be critical for survival of neurons in culture [240], Aβ is not an essential protein *in vivo* [188], suggesting that specific inhibition of its production, enhancement of its clearance or prevention of its assembly are viable routes for treatment of AD.

17.5 Genetic Evidence for a Role of Aβ in AD

Over 100 candidate genes for AD have been reported, but only four have been linked convincingly to AD – *Aβ*PP, *PSEN1* (encoding PS-1), *PSEN2* (encoding PS-2) and *APOE* (Table 17.2). Mutations in these genes result in increased pro-

Table 17.2 Causative genes for AD

Gene	Protein name	Protein sequence [a]	Chromosomal location	Mode of inheritance	Pathogenic Mutations
*Aβ*PP	AβPP	P05067	21q21.2	autosomal dominant	19 [b]
PSEN1	PS-1	P49768	14q24.3	autosomal dominant	140 [c]
PSEN2	PS-2	P49810	1q31–42	autosomal dominant	10 [c]
APOE	ApoE	P02649	19q13.32	complex	NA [d]

a) SwissProt accession number.
b) As of November 2004, see http://www.alzforum.org/res/com/mut/app/.
c) As of November 2004, see htpp://www.molgen.ua.ac.be/ADMutations/.
d) NA, not applicable.

duction of Aβ, an increased Aβ42/Aβ40 concentration ratio, alterations in the biophysical or biological properties of Aβ, or altered metabolism of the peptide.

17.5.1 Mutations in AβPP

The first pathogenic missense mutation in AβPP was reported in 1990. Since then, 18 other pathogenic mutations have been discovered. These mutations comprise two general classes – those located inside or outside of the region encoding Aβ.

17.5.1.1 Mutations Inside the Aβ Region of AβPP

There are seven pathogenic missense mutations within Aβ (Table 17.3). The effects of these mutations on the biochemistry of AβPP and on the metabolism and biophysical properties of Aβ vary. For some mutations, production of Aβ was not affected. However, differences in the biophysical properties of mutant and wild-type peptides were observed (Table 17.3). This emphasizes the importance of biophysical investigations for elucidating potential disease mechanisms.

H6R (English Mutation)

This mutation was discovered in one of two 55-year old English siblings [250]. Post-mortem neuropathologic analyses of both siblings were consistent with a diagnosis of AD. The patient with the H6R mutation did not have the ε4 allele of ApoE (see below) nor a PS mutation. However, because nothing is known about the previous generation of the proband's family, it is possible that the patient

Table 17.3 Pathogenic mutations within the Aβ-coding region of AβPP

Mutation a)	Name b)	Phenotype	Aβ production c)	Aβ aggregation d)
H677R/H6R	English	AD	Aβ40 =; Aβ42 =	fibrils: ↑; oligomers: NA
D678N/D7N	Tottori	AD	Aβ40 =; Aβ42 =	fibrils: ↑; oligomers: NA
A692G/A21G	Flemish	CAA	Aβ40 ↑; Aβ42 ↑	fibrils: ↓; oligomers: ↓
E693G/E22G	Arctic	EOAD	Aβ40 ↓; Aβ42 ↓	fibrils: ↑; oligomers: ↑
E693G/E22Q	Dutch	HCHWA	Aβ40 =; Aβ42 ↓	fibrils: ↑; oligomers: ↑
E693K/E22K	Italian	HCHWA	unknown	fibrils: ↑; oligomers: NA
D694N/D23N	Iowa	CAA	Aβ40 =; Aβ42 =	fibrils: ↑; oligomers: NA

a) Sites of mutations are given for AβPP (relative to the 770-amino-acid form)/Aβ.
b) Mutations are named according to the ethnicities of the proband's kindred.
c) Relative to wild-type AβPP: ↑, increased; ↓ decreased; =, no effect.
d) Kinetics of assembly relative to wild-type Aβ: ↑, increased; ↓, decreased; =, no effect; NA, not available.

had sporadic AD, with the H6R mutation being a rare non-pathogenic polymorphism [250]. Recent *in vitro* studies of Aβ fibril assembly, however, suggest that the English form of Aβ elongates more rapidly than does wild-type Aβ [251].

D7N (Tottori Mutation)

The proband is a 72-year-old woman living in the Japanese city of Tottori who showed signs of AD at about 60 years old [252]. Diffuse and severe cortical atrophy in the internal portion of the temporal lobes was observed in magnetic resonance imaging of the proband's brain at age 71. Expression of Tottori AβPP in COS-1 or N2a cells resulted in production of Aβ40 and Aβ42 in amounts equivalent to that observed in wild-type AβPP-expressing cells [251]. Whether a similar lack of effect on AβPP processing occurs *in vivo* is not known, but could be addressed by measuring Aβ levels in plasma of patients harboring the D7N mutation. Biophysical studies of Tottori Aβ reveal increased rates of fibril elongation relative to wild-type Aβ, as observed for the adjacent English mutation [251].

E22Q (Dutch Mutation)

The Dutch mutation was the first AβPP mutation reported. It was discovered in Dutch patients manifesting a severe form of hereditary cerebral hemorrhage with amyloidosis (HCHWA-Dutch type; HCHWA-D), a form of cerebral amyloid angiopathy (CAA) [253, 254]. This disease involves extensive deposition of Aβ40 in arterioles and small cerebral vessels, but only limited senile plaque formation. Plaques that do form are composed largely of Aβ42 or Aβ(1–43) [255]. HCHWA-D patients suffer recurrent strokes and dementia leading to premature (around 50 years old) death. The amount of CAA, as determined by computerized morphometry, is highly correlated with dementia, but is independent of plaque density, age or neurofibrillary pathology [256].

Three mechanisms may be involved in the pathogenesis of HCHWA-Dutch: (1) altered processing of AβPP, (2) increased propensity of Dutch Aβ to form protofibrils and fibrils, and (3) reduced clearance of Dutch Aβ from the brain. Bornebroek et al. measured the concentration of Aβ40 and Aβ42 in the plasma of Dutch mutation carriers and non-carriers [257]. The ratio of Aβ40 to Aβ42 was significantly higher in carriers than in controls. In CHO-K1 [258] and HEK293 [259] cells transfected with the Dutch AβPP gene, the Aβ40/Aβ42 concentration ratios also were higher relative to wild-type controls. In transgenic mice expressing human Dutch AβPP, the Aβ40/Aβ42 ratio before amyloid deposition was detected was approximately twice that found in mice expressing wild-type AβPP [260]. Taken together, these results suggest that the Dutch mutation favors the production of Aβ40, which then leads to increased vascular amyloid deposition.

In vitro studies of the aggregation of Dutch Aβ indicate that it has a greater tendency to form β-sheet containing assemblies, including protofibrils and fibrils, than does wild-type Aβ [45, 261–263]. This has been attributed to a decreased propensity of Dutch Aβ to adopt an α-helical structure, resulting in an increase in β-sheet formation [264]. Fibril formation by Dutch Aβ *in vivo* may be facilitated by transglutaminase, which covalently crosslinks proteins through glutamine residues [265], or by increased binding of the peptide to vascular wall components, such as perlecan, accelerating vascular amyloid formation [266]. Dutch Aβ40, but not wild-type Aβ40 or Aβ40 with the charge-preserving E22D substitution, assembled into fibrils on human cerebrovascular smooth muscle cells, leading to cell degeneration [267, 268]. Dutch Aβ40 fibrils are also toxic to cultured human brain pericytes [269] and human BBB endothelial cells [270].

Using a ventriculo-cisternal perfusion technique in guinea pigs, Monro et al. demonstrated that the clearance of Dutch Aβ40 from the CNS to blood was reduced significantly, relative to wild-type Aβ40. Dutch Aβ40 is also more resistant to proteolysis by neprilysin [271] and insulin-degrading enzyme [272].

A21G (Flemish Mutation)

In 1992 Hendriks et al. identified a family whose members displayed cerebral hemorrhage and presenile dementia due to CAA [273]. Post-mortem examinations of the brains of carriers of this mutation show amyloid angiopathy and unusually large senile plaques [274, 275]. Intracellular neurofibrillary tangles also were observed [258, 274]. The senile plaques comprised predominately Aβ40 [275].

Expression of Flemish AβPP in HEK-293 [276] or CHO-K1 [258] cells revealed increased production of Aβ40 and Aβ42, and an increase in the Aβ42/Aβ40 concentration ratio. The A692G mutation is four residues from the α-secretase cleavage site at Lys16, suggesting that the Flemish mutation alters processing of AβPP by α-secretase [258, 276]. The processing of AβPP thus is shifted toward β-secretase processing, resulting in elevated Aβ levels. The precise mechanism through which this effect occurs is unknown.

Transgenic mice overexpressing AβPP with the Flemish mutation showed behavioral disturbances and died prematurely [277]. Examination of their brain tissues revealed apoptotic neurons and microspongiosis in white matter. No amyloid deposits were observed, suggesting that amyloid formation is not required for the observed phenotype.

The effects of the Flemish mutation on the biophysical and biological properties of Aβ have been studied. Flemish Aβ40 showed increased solubility and decreased rates of fibril formation relative to wild-type Aβ40 [278]. The mutant peptide also formed oligomers that were more stable to sodium dodecylsulfate (SDS) [278]. Fibrils from Flemish Aβ40 are as toxic to cultured primary rat cortical neurons as fibrils from wild-type Aβ [278].

E22G (Arctic Mutation)

This mutation was identified in a Swedish family living in northern Sweden, above the Arctic Circle [259]. Carriers of this mutation show clinical features of early-onset AD (EOAD), but without the cerebral hemorrhaging observed in patients carrying other mutations at position 693 of AβPP. Plasma levels of Aβ in carriers of the Arctic mutation were lower than those in non-carriers, suggesting decreased production of Aβ [259]. This also was observed in *in vitro* studies in which cells overexpressing Arctic AβPP produced lower Aβ levels relative to those expressing wild-type AβPP. These observations are the opposite of those observed in studies of other AD mutations in which increased production of Aβ, particularly Aβ42, is observed. It has been reported that the Arctic mutation interferes with the processing of AβPP [279]. However, how this leads to decreased Aβ production is not known.

In vitro biophysical studies of Arctic Aβ provided insights into how carriers of the mutation develop EOAD [259]. While studies of fibril formation indicate no difference in overall kinetics, Arctic Aβ formed protofibrils at a much higher rate and in greater amounts than wild-type Aβ. The enhanced production of protofibrils, which have been shown to be neurotoxic [50, 51], suggests that the Arctic form of AD is caused by protofibrils.

E22K (Italian Mutation)

An E693K mutation in AβPP, causing an E22K change in Aβ, has been found in several members of three Italian kindreds [280, 281]. In contrast to carriers of the Dutch mutation, the Italian patients had recurrent strokes much later in life (around 60–70 years of age). HCHWA-Italian is characterized by extensive Aβ deposits in leptomeningeal and cortical vessels, some amyloid plaques in cerebral cortex, and the absence of neurofibrillary tangles [280, 281]. The vascular deposits comprise primarily Aβ40, whereas the parenchymal deposits contain predominately Aβ42.

The effects of the Italian mutation on the processing of AβPP are not known. However, the biological and biophysical properties of Italian Aβ have been reported. Italian Aβ40 and Aβ42 aggregate faster and are more toxic to PC12 cells than is wild-type Aβ [263]. Italian Aβ40 binds to the surface of cultured human cerebrovascular smooth muscle cells, where it aggregates [268]. Italian Aβ40, at a concentration of 25 μM, did not cause apoptosis in cerebral endothelial cells, whereas an equivalent concentration of Dutch Aβ40 did [262].

D23N (Iowa Mutation)

This mutation was discovered in an Iowa family that presented with dementia in their seventh and eighth decades of life [282]. Severe CAA was observed, with plaques consisting mostly of Aβ40. In contrast to CAA caused by other Aβ mutations, widespread neurofibrillary tangles were observed.

Van Nostrand et al. showed that the Iowa mutation does not affect the processing of AβPP expressed in human H4 neuroglioma cells as indicated by comparative quantitation of levels of $A\beta PP_{sa}$, C83 or Aβ40 produced by cells expressing Iowa or wild-type AβPP [283]. However, Iowa Aβ did form fibrils faster than did wild-type Aβ and also displayed increased pathogenicity to cultured human cerebrovascular smooth muscle cells.

17.5.1.2 Mutations Outside the Aβ Region of AβPP

Table 17.4 shows pathogenic mutations in AβPP found outside the Aβ coding region. These mutations occur at or near the β- and γ-secretase cleavage sites, suggesting that the mutations affect AβPP processing. In fact, the KM670/671NL Swedish double mutation found at the P2/P1 site of β-secretase leads to the production of 6- to 8-fold higher levels of Aβ than produced by wild-type AβPP [284]. The mechanism for this increase may involve more efficient processing of AβPP by β-secretase. How this occurs is not entirely understood. The other mutations in Table 17.4 are found near the γ-secretase cleavage sites. These mutations increase the Aβ42/Aβ40 concentration ratio. Increased Aβ42 concentrations are linked strongly to AD [285]. This may be related to the relatively high propensity of Aβ42 to self-associate, resulting in increased formation of fibrils and of oligomeric assemblies with neurotoxic properties [32, 33]. Lichtenthaler et al. have suggested that the γ-secretase site exists within an AβPP transmembrane helix [286]. In this model, the cleavage sites for Aβ40 and Aβ42 are on opposite sides of a helical turn. Mutations may disrupt the structure of this turn facilitating cleavage after Ala42 in the Aβ region.

Table 17.4 Pathogenic mutations outside the Aβ-coding region of AβPP

Mutation [a)]	Name [b)]	Neuropathology	Aβ production [c)]
K670N/M671L	Swedish	typical AD	Aβ ↑
T714A	Iranian	typical AD	NA
T714I	Austrian	diffuse plaques from N-truncated Aβ42 with no Aβ40	Aβ42/Aβ40 ratio ↑
V715M	French	typical AD	Aβ42/Aβ40 ratio ↑
V715A	German	typical AD	Aβ42/Aβ40 ratio ↑
I716V	Florida	typical AD	Aβ42/Aβ40 ratio ↑
I716T	None	typical AD	NA
V717I	London	typical AD, Lewy bodies	Aβ42/Aβ40 ratio ↑
V717L	Indiana	typical AD	Aβ42/Aβ40 ratio ↑
V717F	none	typical AD	Aβ42/Aβ40 ratio ↑
V717G	none	typical AD	Aβ42/Aβ40 ratio ↑
L723P	Australian	typical AD	Aβ42/Aβ40 ratio ↑

a) Sites of mutations are given for AβPP (relative to the 770 amino acid form).
b) Mutations are named according to the ethnicities of the proband's kindred.
c) Relative to wild-type AβPP: ↑, increased; ↓, decreased; =, no effect; NA, not available.

17.5.2 Mutations in *PSEN1* and *PSEN2*

The majority of EOAD appears to be caused by mutations in *PSEN1*. To date, 140 pathogenic mutations in *PSEN1* have been reported (Table 17.2). Carriers of these mutations present with AD symptoms in their sixth decade and die in their seventh decade. Fewer mutations have been identified in *PSEN2* (Table 17.2) and the clinical effects of these mutations generally are less severe than are those of the *PSEN1* mutations.

Mutations in *PSEN1* and *PSEN2* result in an increased Aβ42/Aβ40 concentration ratio. Scheuner et al. showed that media from fibroblasts or from plasma of carriers of *PSEN1*, *PSEN2* or AβPP (Swedish and London) mutations had elevated levels of Aβ42 and Aβ(1–43) relative to controls [201]. They noted that the data from plasma were particularly meaningful because they showed that the mutations increased Aβ42 and Aβ(1–43) *in vivo*. The increased production of Aβ42 has been confirmed in transfected cells, transgenic mice, and in human brains. Citron et al. studied HEK cells and transgenic mice co-expressing human AβPP and PS-1 [287]. In both models, the expression of mutant PS resulted in increased production of Aβ42. Xia et al. observed similar results using doubly transfected CHO cells [288]. In addition, Aβ42 oligomers were observed in the cell culture supernates, presumably due to accelerated Aβ oligomerization resulting from the increased steady-state concentrations of Aβ42. In transgenic mice co-expressing mutant *PSEN1* and AβPP, Borchelt et al. showed that amyloid deposits developed much earlier than in age-matched control mice [289]. Lemere et al. examined the neuropathological phenotype of four AD patients from a Colombian kindred carrying a *PSEN1* mutation (E280A) [290]. Massive brain deposition of Aβ42 was noted. Quantification of Aβ burden in frontal, temporal and occipital cortices, and in cerebellum, showed significantly greater deposition of Aβ42 than Aβ40 relative to normal controls.

The PS protein likely is the catalytic component of γ-secretase [291], which comprises at least three other proteins, Aph-1, Pen-2 and nicastrin [62]. PS undergoes endoproteolysis in cells to yield N- and C-terminal fragments that associate as heterodimers. Heterodimeric PS appears to be the functional form of the protein [292]. Compounds designed to inhibit γ-secretase bind specifically to PS heterodimers [293]. De Strooper et al. showed that gene deletion of PS-1 dramatically reduced γ-secretase activity [294]. Wolfe et al. showed that mutation of two conserved intramembrane aspartates in PS-1 interfered with γ-secretase activity [295]. Furthermore, under appropriate conditions, PS is co-immunoprecipitated with full-length AβPP in cell preparations [296, 297], as might be expected for an enzyme and its substrate. Mutations in *PSEN1* and *PSEN2* that result in increased production of Aβ42 thus may act by altering the native fold of the PS proteins, which affects substrate specificity.

17.5.3
APOE is an AD risk factor

ApoE is the major brain lipoprotein. Three human isoforms exist, ApoE2, ApoE3 and ApoE4. These proteins are encoded by the *ε2*, *ε3* and *ε4* alleles, respectively, of the *APOE* gene. The three proteins each contain 299 amino acids and have identical primary structure, except at positions 112 and 158 [298, 299], at which the amino acids Cys and Cys, respectively, occur in ApoE2, Cys and Arg occur in ApoE3, and Arg and Arg occur in ApoE4 (Table 17.5). The presence of Cys in ApoE2 and ApoE3 allows formation of disulfide-linked homo- and hetero-oligomers. Disulfide-mediated multimerization of ApoE4 is not possible. The most common ApoE genotype, *ε3/ε3*, occurs in 50–70% of the population. In 1993, Corder et al. provided evidence that carriers of either one or both *ε4* alleles are at significantly increased risk for AD [83]. Saunders et al. presented clinical data indicating that 40% of autopsy-documented AD patients have at least one *ε4* allele [300]. Subsequent studies have confirmed that *APOEε4* is a risk factor for AD (for a recent review, see [301]). It should be emphasized that *APOEε4*, although a risk factor, is not a predeterminant of AD. Possession of one or both *APOEε4* alleles does not predict that an individual will develop AD, in contrast to the other mutations discussed earlier that have absolute or high penetrance.

Expression of ApoE4 is associated with enhanced deposition of Aβ [302–309] (see also Section 17.6.4.1), tau hyperphosphorylation and formation of neurofibrillary tangles [310–313], inhibition of the outgrowth of neurites [314], potentiation of Aβ-induced lysosomal leakage and cell death in neuronal cells [315], disruption of cytoskeletal structure and function [316], and induction of cognitive decline [317–319]. Expression of ApoE3, in contrast, appears to be protective. ApoE3 effects include potentiation of Aβ clearance [320, 321], protection of tau against phosphorylation [310], stimulation of the outgrowth of neurites [314, 322, 323], protection against oxidative insults [324, 325] and protection from cognitive decline [317].

Structural differences between ApoE3 and ApoE4 underlie their distinct effects on Aβ. ApoE has an N-terminal domain, residues 1–191, that contains the LDL receptor-binding region. In ApoE4, the N- and C-terminal domains interact through a salt bridge between Arg61 and Glu255, leading to a more compact structure and preferential binding to triglyceride-rich LDLs [326–328]. Domain

Table 17.5 ApoE isoforms

Allele	Residue 112	Residue 158
ε2	Cys	Cys
ε3	Cys	Arg
ε4	Arg	Arg

interaction has been implicated in increased susceptibility of ApoE4 to proteolysis leading to production of neurotoxic fragments (Y. Huang, personal communication). Domain interaction is lacking in ApoE3, resulting in a more open structure and preferential binding to phospholipid-rich high-density lipoproteins (HDLs) [326–328]. Guanidine and temperature denaturation studies have shown that ApoE4 is less stable than ApoE3 [329, 330]. Furthermore, in contrast to ApoE3, ApoE4 does not unfold cooperatively, but forms a partially folded, molten globule-like intermediate [330]. This intermediate contains β structure, which may account for the ability of ApoE4 to enhance the deposition of Aβ. It also has been suggested that the molten globule form of ApoE4 may interact with membranes, potentiating Aβ-induced lysosomal leakage [330]. Potential mechanisms by which ApoE4 enhances Aβ deposition are discussed in Section 17.6.4.1.

17.5.4
Other Genetic Factors

One of the strongest genetic risk factors for AD is the possession of a first-degree relative with the disease [331–333]. This suggests that there exist other genes, as yet unidentified, that are linked to AD [333]. Bertram and Tanzi have noted that studies of the molecular genetics of AD published in 2003 have revealed, on average, 10 genes per month with either positive or negative linkage to AD [334]. Unfortunately, the statistical significance of the positive association of the majority of these genes could not be proven reliably, thus additional studies are required if associations akin to that of ApoE are to be established.

17.6
Pathogenetic Process – Biophysics

The folding and assembly of nascent Aβ monomers into neurotoxic conformers, oligomers and higher-order assemblies is thought to underlie the pathogenesis of AD. Structure–activity studies have shown that unstructured Aβ monomers are non-toxic in *in vitro* neurotoxicity assays. However, monomer folding results in the production of toxic structures. This has made understanding Aβ folding and assembly important.

The folding landscape [335] of Aβ is complex, both with respect to the number of different stable conformational and assembly states that the peptide may populate, and with respect to the pathways through which these states may be reached. This observation is important, because it suggests that efforts to define single pathways of Aβ assembly and to classify particular conformers or assemblies as "on" or "off" pathway will not be possible or meaningful. If one seeks to understand the structural biology of Aβ and its various assemblies for the purpose of developing therapeutic approaches for AD, the primary goal(s) must be the correlation of structure and pathogenicity. As will be discussed below, a myriad of Aβ conformers and assemblies have been described, most of which are neurotoxic. This may necessitate the development and clinical testing of agents targeting many different toxic species to determine whether single or combination therapies are efficacious. Whichever approach is pursued, the starting point will be the identification of an appropriate target. In this section, we discuss Aβ assembly dynamics, and the identification of the stable and metastable Aβ structures comprising the assembly landscape. We begin with fibrils, proceed to oligomers and "end with the beginning," i.e., Aβ monomer folding. Each section provides a brief historical perspective, followed by discussions of the biophysics and biology associated with each structural entity.

17.6.1 Aβ Folding and Assembly – From Fibrils Back to Monomers

17.6.1.1 Fibrils

Discovery

The fibrillar nature of amyloid deposits was first established in 1959 by Cohen and Calkins in electron microscopy (EM) studies of renal and subcutaneous amyloids [336]. The existence of fibrils in amyloid plaques in AD brain was first demonstrated 5 years later in EM studies of Terry et al. [337] and Kidd [338]. However, it was not until two decades later that the Aβ protein that forms the fibrils in AD brain was isolated, purified and partially sequenced by Glenner and Wong [9].

Structure

EM studies of amyloid fibrils at high magnification indicates a hierarchical subfibrillar organization. End-on views of amyloid fibrils from brains of AD and Down's syndrome patients revealed filaments, each of which comprised five subunits [339]. Similar cross-sections were observed in EM studies of fibrils formed from synthetic fragments of Aβ, including Aβ(11–28), Aβ(1–28) [340, 341] and Aβ(6–25) [340]. The subunits, each having a diameter of 2.5–3 nm, were named protofilaments or subfibrils. Analysis of X-ray diffraction patterns of aligned fibrils of Aβ also led to the conclusion that the fibrils are composed of five or six protofilaments [342].

Initial insights into the types of secondary structure present in senile plaque amyloid fibrils came from investigations of the tinctorial properties of amyloid deposits in brain parenchyma and vasculature. When stained with Congo red and examined using polarized light, senile plaques exhibit apple green birefringence [10]. The binding of Congo red, as well as the birefringence, is due to the presence of ordered protein assemblies exhibiting β-sheet structure [343]. Fiber X-ray diffraction studies of cores of senile plaques by Kirschner et al. provided direct experimental evidence for the presence of cross-β structure [344]. The cross-β motif produces a characteristic diffraction pattern with a meridional reflection at ~4.76 Å and an equatorial reflection at ~10.6 Å. This pattern was observed in earlier studies by Eanes and Glenner [345] and Bonar et al. [346] of human hepatic and splenic amyloids. The diffraction pattern is consistent with the Pauling and Corey cross-β-sheet model first proposed to explain the structure of silk [347]. In this model, extended β-sheets are aligned parallel to the fibril axis. The β-strands comprising the β-sheet lie perpendicular to the fibril axis. The reflection at 4.76 Å corresponds to interstrand hydrogen bonding and the reflection at 10.6 Å corresponds to the distance between β-sheets.

Insights into Aβ fibril structure, including identification of residues involved in β-strands, arrangement of the strands in the β-sheets and other secondary structure elements, have been provided by solid-state nuclear magnetic resonance (NMR) (Table 17.6). The orientation of strands appears to depend on the length of the Aβ monomer. In fibrils formed by short fragments of Aβ, including Aβ(34–42) [348], Aβ(16–22) [349] and Aβ(11–25), [350], the β-strands are arranged in an antiparallel fashion. Full-length Aβ40 [351] and Aβ42 [352], and the long fragment Aβ(10–35) [353], form parallel β-strands. Based on ^{13}C chemical shifts and resonance linewidths, the N-terminus of Aβ40 was concluded to be disordered [351]. A Gly37–Gly38 turn, stabilized by intermolecular hydrogen bonding and hydrophobic interactions between the side-chains of Leu34, Val36, Val38 and Ile41, was postulated to exist in fibrils formed by Aβ(34–42) [354]. However, no turns were found in Aβ(10–35) fibrils [353]. A bend in the region Gly25–Gly29 in Aβ40 fibrils has been suggested [351]. This structure would bring the central hydrophobic cluster (CHC; Leu17–Ala21) and C-terminus into contact, allowing hydrophobic side-chain interactions, a necessary step in fibril formation. Fibrils from Aβ42 have not been investigated to the same extent as Aβ40 (Table 17.6). It would be informative to determine if the Gly37–Gly38 turn and the Gly25–Gly29 bend are present in Aβ42.

Structural constraints from solid-state NMR experiments, and fibril dimensions and mass-per-length data from EM studies, have been used to develop a model for an Aβ40 protofilament [351] (Fig. 17.8). This model is consistent with

Table 17.6 β-Sheet organization in Aβ fibrils determined by solid-state NMR

Fibril precursor	Location of β-strands	β-Sheet organization	Other secondary structure present
Aβ(34–42) [348]	Leu34 → Val36; Val39 → Val42	antiparallel	Gly37–Gly38 turn [354]
Aβ(16–22) [349]	Leu15 → Ala21	antiparallel	none
Aβ(10–35) [353]	Tyr10 → µMet35	in-register, parallel	none
Aβ(11–25) [350]	Leu17 → Ala21	antiparallel	none
Aβ40 [351]	Gly9 → Val24; Ala30 → Val40	parallel	Asp1–Ser8, disordered; Gly25–Gly29, bend
Aβ42 [352]	not determined	in-register, parallel	not determined

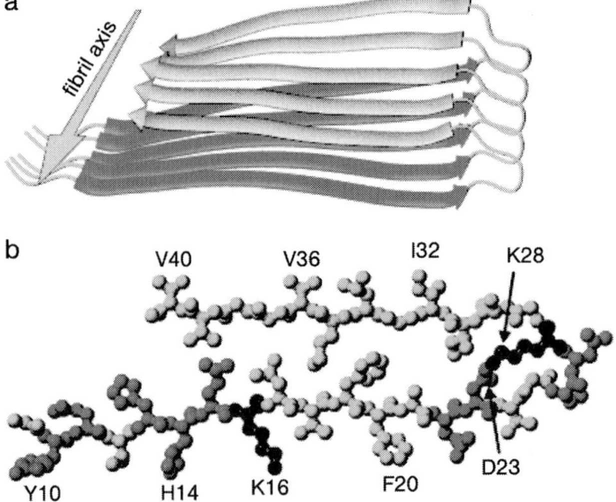

Fig. 17.8 Model of a protofilament from Aβ40 based on constraints obtained by solid-state NMR [351]. (a) Residues 9–40 form two β-strands, separated by a bend. The β-strands form parallel β-sheets in a cross-β arrangement. (b) Residues 9–40, colored according to type of side-chain (green, hydrophobic; magenta, polar; red, negatively charged; and blue, positively charged). A salt bridge between Asp23 and Lys28 helps stabilize the bend region. (Reprinted with permission from National Academy of Sciences, USA)

the supposition that hydrophobic collapse of Aβ monomers initiates fibril formation. The model also predicts that electrostatic interactions are involved in fibril assembly through formation of a salt bridge between Asp23 and Lys28 which stabilizes the Gly25–Gly29 bend.

Other biophysical methods have elucidated structural features of the Aβ fibril not amenable to NMR studies. Limited proteolysis of Aβ40 fibrils using trypsin and chymotrypsin showed that the peptide bonds Phe4–Arg5, Arg5–His6 and Tyr10–Glu11 are cleaved from fibrils as rapidly as from the monomers [355]. Other scissile peptide bonds are cleaved much slower. Hydrogen–deuterium exchange monitored by mass spectrometry [356] showed that at least three classes of backbone amide protons exist in fibrils: (1) those that are solvent exposed and not protected by hydrogen bonds and thus exchange rapidly, (2) those that are partially exposed or hydrogen bonded and exchange at intermediate rates, and (3) those that are buried and fully protected, and thus do not exchange even after prolonged incubation at neutral pH. Approximately 50% of the amide protons of Aβ40 do not exchange. Taken together, the limited proteolysis and hydrogen exchange data suggest that the N-terminus of Aβ40 is disordered, whereas the C-terminus is buried and presumably involved in β-sheet formation. Electron paramagnetic resonance (EPR) measurements on spin-labeled fully-hydrated Aβ40 and Aβ42 fibrils support an in-register, parallel alignment of β-strands and a disordered N-terminus [357].

Other models of amyloid fibril structure include polar zippers and β-helices. Perutz et al. showed that synthetic poly-L-glutamine forms β-sheets stabilized by an array of hydrogen bonds between the side-chain amide and carbonyl groups [358]. This array was termed a "polar zipper" and has been incorporated in models of amyloid fibrils from glutamine- or asparagine-rich proteins [359]. Experimental evidence for this model is lacking, but could be provided by solid-state NMR studies of fibrils. Modeling based on established examples of β-helical conformation, and on hydrophobic and hydrogen bonding effects on protein folding, led Lazo and Downing to propose that amyloid fibrils contain β-helices [360]. Experimental evidence for this model was provided by studies of model peptides [361] and re-examination of the solid-state NMR data on Aβ(34–42) fibrils [354]. The β-helix model has been incorporated in models of fibrils from scrapie prion protein [362] and Aβ(15–36) [363]. The model shown in Fig. 17.8 might be considered a β-helix if the N-terminus of one Aβ40 molecule is closer to the C-terminus of the adjacent molecule in the protofilament [364].

Kinetics

Jarrett et al. probed the kinetics of polymerization of Aβ(1–39), Aβ40, Aβ42, Aβ(26–39), Aβ(26–40), Aβ(26–42) and Aβ(26–43) using turbidity measurements [365]. They observed that the development of turbidity by peptides ending at Val39 or Val40 commenced after a lag time, whereas no lag time was observed for the longer peptides. The lag time was attributed to fibril nucleation and could be eliminated by the addition of preformed fibrils to the Aβ solution. These observations are consistent with a nucleation-dependent polymerization mechanism, characterized by three features [366]: (1) a critical protein concentration below which no aggregation occurs, (2) a lag time, at protein concentrations at or above the critical concentration, prior to polymerization, and (3) elimination or diminution of the lag time by the addition of preformed nuclei. It should be noted that turbidity measurements are not capable of monitoring nucleation directly, as the nuclei formed are too small to scatter measurable amounts of light [47]. The lag times observed by Jarrett et al. actually represent the time required for the production of Aβ aggregates large enough to produce measurable scattering. Nucleation characteristics can only be studied by establishing appropriate constants of proportionality relating nucleation time to aggregate size.

Naiki and Nakakuki used Thioflavin T, a dye that fluoresces after binding to β-sheets, to monitor the growth of fibrils from Aβ40 [367]. They showed that the rate of fibril extension is affected by seed concentration, Aβ concentration, pH and temperature. They proposed a first-order kinetic model that postulates that fibril elongation occurs by reversible addition of monomers to fibril ends. This model is appealing because of its simplicity. It does not, however, consider nucleation.

Tomski and Murphy used quasielastic light scattering spectroscopy (QLS) and static light scattering (SLS) to investigate the kinetics of assembly of Aβ40 in phosphate-buffered saline [368]. They hypothesized that monomers of Aβ self-associate into octamers that then stack to form short fibrils. Long, amyloid-type fibrils formed by irreversible end-to-end association of the short fibrils. A mathematical model, based on diffusion-limited aggregation, was derived that accounted for time-dependent changes in fibril length. The model assumed complete conversion of monomers to oligomers and thus could not provide a mechanism whereby monomers added to fibril ends. Lomakin et al. [369] also used QLS to investigate assembly of Aβ(1–40). Their experiments were done in 0.1 N HCl, a milieu in which reproducible fibril growth was achieved that was sufficiently slow to allow monitoring of the entire assembly process, including nucleation. Lomakin et al. showed that above a certain critical concentration, Aβ formed micelles (see below) from which fibril nuclei emanated. Fibril elongation proceeded by monomer addition to fibril ends. The experimental data provided the basis for development of a mathematical formulation for the temporal evolution of the Aβ fibril length distribution that included terms for nucleation and elongation rates, micelle size, and critical concentration [370].

Studies of Aβ fibril assembly kinetics also have been done using *ex situ* amyloid plaques in tissue sections obtained from AD brain [371]. "Plaque growth" assays were performed by addition of radioiodinated Aβ to the sections followed by autoradiography and scintillation counting. Fibril growth rates in this system, as in strict solution systems, also displayed first-order dependence on Aβ concentration [371].

The development of systems in which controlled Aβ folding and fibril assembly occurred allowed studies of the thermodynamics of the assembly process. Kusumoto et al. showed that Aβ monomer addition to fibril ends occurred coincident with substantial conformational rearrangement [372]. Esler et al. probed the deposition activity of Aβ congeners designed to possess similar structures but have different energies of activation for plaque growth. They showed that [cyclo-H14K–E22]Aβ(10–35)-NH_2, designed with a covalent bridge preventing conformational rearrangement, was inactive whereas [Glu22Gln]Aβ(10–35) (the Dutch peptide) had an enhanced deposition rate relative to wild-type peptide. Studies of the Teplow and Maggio groups formed the basis for development of an "energy landscape model" of Aβ folding and fibril assembly [373].

Biological Activity

Work done over a decade ago showed that Aβ fibrils were neurotoxic [22, 23], an observation that has been validated many times [111, 374, 375]. To correlate Aβ assembly state and neurotoxicity, Lorenzo and Yankner showed that amorphous aggregates of Aβ42, produced by dissolving the peptide in phosphate-buffered saline (pH 7.3), were non-toxic, whereas Aβ42 fibrils were neurotoxic [376]. Similar results were obtained by Howlett et al. for Aβ40 [374]. These and other data led to the conclusion that Aβ neurotoxicity required fibril formation. However, recent clinical, biological and biophysical data have challenged this assumption [32, 33, 36]. Lue et al. showed that levels of Aβ deposition, quantified by immunohistochemistry, Thioflavin T binding and immunochemistry (ELISA), do not distinguish AD patients from controls [29]. Levels of soluble Aβ, in contrast, did distinguish AD patients and correlated strongly with synapse loss. The authors suggested that the results were reasonable "when one considers that soluble Aβ has the potential to affect neurons and neurites over a much wider area than insoluble Aβ, which is essentially pinned to fixed points in the neuropil in the form of plaques" [29]. Transgenic mice expressing wild-type and FAD-associated AβPP exhibit neurological deficits prior to the appearance of fibrils [31, 377]. In addition, levels of soluble Aβ correlated strongly with synapse loss, akin to results from AD patients [31]. As will be discussed in greater detail below, non-fibrillar Aβ oligomers have been found to be neurotoxic both *in vitro* and *in vivo* [32–34].

17.6.1.2 **Protofibrils**

Discovery

In 1997, biophysical studies done by two research groups working independently revealed an amyloid fibril intermediate that was termed a "protofibril" [45, 46]. Walsh et al. used size-exclusion chromatography (SEC), QLS and EM to characterize intermediates in the formation of fibrils by Aβ40 and Aβ42 [45]. Incubation of Aβ for periods of hours to days resulted in a chromatographic profile characterized by the appearance of a void volume peak that increased in size over time. EM showed that the material in this peak comprised short, curly structures 6–8 nm in diameter and generally less than 150 nm in length. These structures were called protofibrils to distinguish them from the longer, straighter fibrils normally found in amyloid plaques or produced after prolonged (days to weeks) incubation of Aβ. The kinetics of protofibril appearance/disappearance and fibril formation was consistent with protofibrils being intermediates in Aβ fibrillization. Protofibril formation also was observed in experiments using Dutch Aβ40. Harper et al. used atomic force microscopy (AFM) to monitor fibril formation by Aβ40 and Aβ42 [46]. Early in the fibrillization process, small elongated oligomers were observed with diameters of around 3–4 nm and lengths ranging from 20 to over 100 nm. These assemblies grew slowly and then disappeared concurrent with the appearance of fibrils. The approaches taken by the two groups produced data that were entirely consistent. Subsequent studies have revealed that protofibril formation is a common property of amyloid proteins. At least eight other proteins associated with disease form protofibrils, including α-synuclein [378, 379], prions [380], ABri peptide [381], β_2-microglobulin [382], huntingtin [383], islet amyloid polypeptide [384], immunoglobulin light chain [385] and transthyretin [386].

Biophysics

Structure studies have shown that protofibrils bind significant amounts of Congo red and Thioflavin T and that this binding is Aβ concentration dependent, consistent with the presence of significant amounts of β-sheet [45]. Circular dichroism (CD) spectroscopy showed that protofibrils contain significant (45–50%) β structure (β-sheet and β-turn) [45]. Päiviö et al. also have reported significant amounts of β structure [387].

Molecular mass estimates of protofibrils using dextran standards indicate a size distribution ranges of 100–3000 kDa or 24–700 molecules, respectively [387]. Hydrogen-exchange studies have shown that protofibrils possess an ordered core structure [388]. Approximately 40% of the backbone amide protons of wild-type or Arctic Aβ40 protofibrils are protected from exchange even after 2 days of incubation. By comparison, around 60% of the backbone amides of mature fibrils are protected [356]. This suggests that protofibrils are a relatively mature assembly with certain secondary and tertiary structure characteristics in common with fibrils.

Studies of protofibril formation have shown that amino acid residues at the N-terminus, Met35 and Glu22 all affect assembly. For example, Aβ40 containing the dipeptide substitution Asp1Glu2 → Glu1Val2 (designated EV40) formed abundant protofibrils, whereas wild-type Aβ40 formed long fibrillar aggregates [389]. Aβ peptides lacking the N-terminal 11 residues, including Aβ(12–40) and [E22G]Aβ(12–40), do not form protofibrils, but display increased rates of fibril formation [387]. Oxidation of Met35 in Aβ42 prevents the formation of protofibrils [390]. The rank order of protofibril formation from variants of Aβ40 containing substitutions at Glu22 correlates with the hydrophobicity of the substituent, i.e., Glu22Val > Glu22Ala >> Glu22Gly > Glu22Gln >> Glu22 [387]. Italian Aβ40, containing the Glu22Lys substitution, does not produce protofibrils within the same time frame examined for the other variants.

The conversion of protofibrils to fibrils may involve both inter- and intrafibrillary changes [45, 46]. End-to-end annealing, lateral association and end-to-end annealing following lateral association are simple mechanisms in which protofibrils might grow to form fibrils. An *in situ* AFM study of Aβ fibrillization showed that the growth of protofibrils is bidirectional [391]. Protofibrils from Aβ40 were directed along the elongation pathway by addition of Aβ40 monomers, aggregates or protofibrils [391], or along the lateral association pathway by the addition of NaCl [392]. The conversion of protofibrils into fibrils is seeded by preformed fibrils [393]. In contrast, co-incubation of Arctic Aβ40 with wild-type Aβ40 leads to kinetic stabilization of protofibrils [394].

In principle, the solid-state NMR methods and labeling schemes that have been used to obtain insights into the structure of Aβ fibrils [364, 395] could be used to probe the structure of protofibrils. However, to do so, protofibrils must be stabilized to prevent their dissociation into monomers and smaller oligomers, and their maturation into fibrils.

Biological Activity

Protofibrils, like fibrils, are potent neurotoxins. Incubation of rat cortical neurons with protofibrils has been shown to cause decreased metabolism of redox-active dyes such as MTT, indicating detrimental effects on neuronal metabolism [50]. Protofibrils also altered the electrical activity of cultured rat cortical neurons [51]. In neocortical neurons, protofibrils from Aβ42 inhibited K^+ currents and induced reversible, Ca^{2+}-dependent increases in spontaneous action potentials and membrane depolarizations [396]. This result suggests that protofibrils bind to specific channels or membrane proteins. Taken together, the rapid alterations in neuronal metabolism caused by protofibrils is consistent with *in vivo* alteration in normal neuronal function and synaptic activity linked to AD. Strong support for this hypothesis has come from study of the Arctic (Glu693Gly) AβPP mutation (see Section 17.5.1.1 and [259, 397]), which results in the production of a form of Aβ with high propensity to form protofibrils. The Arctic form of FAD may be the first "protofibril disease".

17.6.1.3 **α-Helical Intermediate**

Discovery

CD studies of protofibril formation revealed what appeared to be the transitory formation of significant levels of α-helix [50]. Extending this work, Kirkitadze et al. sought to determine if helix formation was a general feature of Aβ fibrillogenesis and, if so, what factors affected its formation [42]. To do so, CD was used to monitor the fibrillogenesis of 18 different Aβ peptides comprising all biologically relevant Aβ alloforms known at that time. Wild-type Aβ40 and Aβ42 were studied first. Immediately after dissolution, Aβ40 was largely unstructured. However, over a period of 21 days, spectra were observed consistent with conformational transitions from random coil $\rightarrow$ α-helix $\rightarrow$ β-sheet. The α-helix content was maximal at 11–12 days, a point at which β-sheet content increased exponentially and then decreased to essentially undetectable levels when fibril formation was completed. Aβ42 displayed an identical random coil $\rightarrow$ α-helix $\rightarrow$ β-sheet transition, but with accelerated kinetics. All the 16 other peptides also exhibited time-dependent conformational changes similar to those observed for wild-type Aβ40 and Aβ42. These data supported the postulation that transitory formation of an α-helix-containing intermediate occurs during Aβ fibrillogenesis.

Biophysics

Filtration studies by Kirkitadze et al. revealed that the intermediate had a molecular mass above 100,000 Da, indicating helix formation is associated with oligomerization [42]. Asp23 and His13 exerted significant control over the kinetics of α-helix formation. The kinetics was most rapid in the pH regime in which the β-carboxyl group of Asp23 was ionized and the imidazole ring of His13 was protonated. Additional experimental evidence for a helix-containing intermediate has come from solid-state NMR studies of mixtures of spherical aggregates and fibrils from Arctic Aβ40 [398]. ^{1}H and ^{13}C magic-angle spinning spectra showed peaks with chemical shifts consistent with the presence of α-helices. These peaks were not observed in populations of pure fibrils, demonstrating that the observed α-helix resonances were produced by the spherical aggregates. Molecular dynamics simulations of Aβ(16–22) oligomerization also support the existence of an on-pathway, α-helix-containing intermediate [399].

Biological Activity

The biological activity of the α-helical intermediate has not been studied.

17.6.1.4 Micelles

Discovery

Aβ is an amphipathic peptide. The first 28 residues comprise a mixture of both polar and apolar residues, but the C-terminal 12–14 residues are apolar (Fig. 17.1). This structure suggested that the peptide might exhibit properties of surfactants. Soreghan et al. investigated the effect of Aβ on the surface tension of water [400]. For peptides of 36 or more residues, a progressive decrease in surface tension was observed up to a concentration of 25 μM, above which little additional decrease occurred. In addition, at concentrations above 25 μM, Soreghan et al. noted fluorescence in mixtures of Aβ and 1,6-diphenyl-1,3,5-hexatriene, a dye that fluoresces strongly in hydrophobic milieus. The existence of a critical concentration and a hydrophobic environment is characteristic of micellar phases. This led Soreghan et al. to suggest an axial, amphipathic organization of Aβ within "micellar" intermediates in Aβ fibrillization [400]. Lomakin et al. used QLS to monitor the concentration dependence of the Aβ fibril elongation rate [369]. Below an Aβ concentration of 100 μM, the fibril elongation rate was proportional to peptide concentration. However, above an Aβ concentration of 100 μM, the elongation rate was constant. These observations also suggest a micellization phenomenon, in this case occurring at a critical concentration of 100 μM. Consistent with this interpretation, addition of the surfactant *n*-dodecylhexaoxyethylene glycol monoether ($C_{12}E_6$) to the fibril formation reaction altered both nucleation and fibril elongation phases in a concentration-dependent manner. These data formed the basis for a comprehensive kinetic theory of Aβ fibrillogenesis [370].

Biophysics

The micelles observed by Lomakin et al. had diameters of around 14 nm, and existed in dynamic equilibrium with monomers and small oligomers [369]. Kinetic data suggested they were sites of Aβ fibril nucleation. To further examine the structure of these micellar intermediates, small angle neutron scattering (SANS) was performed [401]. Deconvolution of the SANS data produced a spherocylindrical model of the Aβ micelle with a diameter of ~4.8 nm and a length of ~11 nm. The micelles were composed of 30–50 monomers. Changes in the Aβ concentration at which the micelles formed did not alter their geometry, suggesting that a non-repetitive internal organization of Aβ monomers spanning the entire length of the structures existed.

Biological Activity

The biological activity of Aβ micelles has not been determined.

17.6.1.5 ADDLs

Discovery

In 1995, Oda et al. showed that ApoJ, an apolipoprotein found in the brain and also known as clusterin, blocked fibril formation by Aβ42 [43]. However, in the presence of ApoJ, Aβ42 formed slowly sedimenting, multimeric complexes that were found to be more neurotoxic to PC12 cells than were Aβ42 fibrils. Oda et al. inferred that ApoJ was part of the Aβ42 complexes based on earlier studies that showed 1:1 complexes of Aβ and ApoJ in human serum and CSF. Three years later, Lambert et al. demonstrated that the Aβ42 oligomers were non-fibrillar, diffused readily and could be made in the absence of ApoJ [44]. These oligomers were termed "Aβ-derived diffusible ligands" (ADDLs). Importantly, Gong et al. later showed that ADDLs could be detected in the brains of AD patients and that their concentrations were, on average, greater than an order of magnitude higher than in normal, age-matched controls [402].

Biophysics

ADDLs form in ApoJ-free solutions upon incubation at low temperatures (4–8°C) or in F12 tissue culture medium at low (50 nM) concentrations [44]. AFM studies showed that ADDLs are small globular structures ~5–6 nm in height. ADDLs from synthetic Aβ can occur in several sizes, but the dominant brain-derived ADDL is a dodecamer. The secondary structure composition of ADDLs is not known. However, it has been shown that synthetic and brain-derived ADDLs possess similar conformations because they both are recognized by ADDL-specific antibodies [402]. It is noteworthy that ADDLs have heights identical to the diameters of protofibrils. However, the structural relationships between ADDLs, protofibrils and other assembly intermediates remain to be determined. In principle, if the structure of ADDLs could be stabilized through freezing, chemically crosslinking or other means, NMR methods could be used to determine ADDL structure. ADDLs are formed by Aβ42, but not by Aβ40, indicating that Ile41 and Ala42 are necessary for their formation [44]. This unique ability of Aβ42 to form ADDLs provides one explanation for the strong clinical linkage of Aβ42 with AD.

Despite a lack of structural data, inhibitors of ADDL formation have been reported. Yao et al. showed that Ginkgo biloba extracts rescued PC12 cells from Aβ-induced death by stopping the formation of ADDLs [403]. A β-cyclodextrin (per-6-substituted-per-6-deoxy-β-cyclodextrin) has been found to inhibit the formation of synthetic ADDLs [404].

Biological Activity

The biological activity of ADDLs has been reviewed recently [33] and is summarized here. ADDLs have been found in AD brain [402] and are neurologically active [33]. ADDLs inhibit LTP [33, 44, 405], providing a basis for the memory-specific nature of AD. The molecular basis of LTP inhibition may involve ADDL-dependent damage of memory-relevant synapses [33, 406], disrupting the signal transduction necessary for the formation of memory [33]. Toxicity studies using neuroblastoma N2A cells indicated that ADDLs are more toxic than Aβ42 fibrils [407]. In mice, water maze experiments suggest that when ADDL levels increase, memory functions decrease [408]. It has been suggested that ADDLs bind to specific proteins enriched in synaptosomes [33]. Disrupting this interaction could be a potential AD therapy. Recently, a localized surface plasmon resonance nanosensor based on the optical properties of silver nanotriangles has been used to demonstrate the sensitivity of ADDL-functionalized nanoparticles to anti-ADDL antibodies [409]. This could lead to a sensitive method for the detection of ADDLs in body fluids and a method for the diagnosis of AD.

17.6.1.6 Paranuclei

Discovery

To study the earliest phases of Aβ oligomerization, one must be able to observe assembly events that produce small numbers of metastable structures with short lifetimes. Bitan et al. applied the method of photo-induced crosslinking of unmodified proteins (PICUP) [410], which covalently stabilizes oligomers, to determine Aβ oligomer size distributions [41, 411–413]. Aβ40 formed a mixture of monomer, dimer, trimer and tetramer, in rapid equilibrium. In contrast, Aβ42 preferentially formed pentamer/hexamer units as well as dodecamers, octadecamers, and larger assemblies. The hexamer-like multiplicity of the larger oligomers suggested that a hexameric building block self-associated to form the large-order structures. For this reason, the pentamer/hexamer units were termed "paranuclei" [41]. As with ADDLs, Aβ42 formed paranuclei, but Aβ40 did not – an observation relevant to efforts to understand why the longer form of Aβ is linked particularly strongly to AD [41, 413].

Biophysics

To determine the secondary structure composition of paranuclei, Aβ was studied using CD before and after crosslinking [41]. No substantial conformational differences were observed either with Aβ40 or Aβ42. However, Aβ42 had slightly higher ($\sim$15 versus $\sim$10%) β-turn/sheet content than did Aβ40. Whether this difference or the additional two amino acids in Aβ42 contribute to paranucleus formation is unclear [41].

EM was used to determine the morphology of paranuclei [41]. Low-molecular-weight (LMW) Aβ40 formed nebulous string-like structures that increased in size after PICUP crosslinking. LMW Aβ42, in contrast, formed spheroidal particles with an average diameter of 5 nm that were indistinguishable from ADDLs.

Studies of the primary structure dependence of paranucleus formation showed that Ile41 was critical [412]. However, in the absence of Ala42, paranucleus self-association did not occur. Systematic alteration of side-chains at position 41 showed that paranucleus formation also required a side-chain of sufficient size and hydrophobicity. For example, the Ile41 $\rightarrow$ Gly substitution blocked paranucleus formation whereas Ile41 $\rightarrow$ Val and Ile41 $\rightarrow$ Leu substitutions resulted in abundant paranuclei [414]. The size of the side-chain at position 42 was not critical, indicating that it was the length of the peptide rather than the nature of residue 42 that was important for paranucleus self-association [414]. An important additional structural question was how oxidation of Met35 affected oligomerization. This question evolves from *in vivo* and *in vitro* studies suggesting that redox chemistry is involved in AD [415, 416]. Oxidation of Met35 to the corresponding sulfoxide or sulfone had no effect on the oligomerization of Aβ40 but completely abolished paranucleus formation by Aβ42 [412].

Biological Activity

The biological activity of paranuclei has not been reported.

17.6.2
Aβ Monomer Folding

At the most basic level, formation of all of the structures discussed above requires Aβ monomer folding. Ideally, one would like to determine a high-resolution tertiary structure of the Aβ monomer and how this structure changes during self-assembly. However, achieving this goal has been difficult. Aβ has not been crystallized, and thus X-ray crystallography, which accounts for around 85% of all protein structures in the Protein Data Bank, cannot be used. Use of the other major structure determination method, solution-state NMR, also has been problematic because of the natural propensity of Aβ to aggregate. Nevertheless, structural information about Aβ monomer has been obtained using NMR as well as a number of lower resolution techniques.

Table 17.7 summarizes NMR studies conducted on full-length, unmodified Aβ. To prevent aggregation, organic co-solvents such as trifluoroethanol (TFE)

Table 17.7 Summary of solution-state NMR studies of full-length, unmodified Aβ

Peptide	Conditions	Major structural features [a)]
Aβ40 [418]	2.5 mM Aβ, 40% TFE/H_2O, pH 2.8, 25 °C	α-helix: Gln15–Asp23; Ile31–Met35
Aβ40 [421]	1 mM Aβ, micellar SDS/H_2O, pH 5.1, 25 °C	α-helix: Gln15–Val24; Lys28–Val36; kink: Gly25–Asn27
Aβ40/Aβ42 [420]	1 mM Aβ, micellar SDS/H_2O/HFIP, pH 7.2, 20 °C	α-helix: Tyr10–C terminus; loop: Gly25–Asn27
Aβ42 [419]	2 mM Aβ, 80% HFIP/H_2O, 27 °C	α-helix: Ser8–Gly25; Lys28–Gly38; type I β-turn between the helices
Aβ40/Aβ42 [423]	0.2–0.8 mM Aβ, H_2O, pH 7.2, 5 °C	turn- or bend-like structure: Asp7–Glu11 and Phe20–Ser26

a) Unspecified areas exist as extended chain or disordered structures.

[417, 418] and hexafluoroisopropanol (HFIP) [419] were used in some studies. SDS, at concentrations above its critical micellar concentration, has also been used to mimic the membrane environment and to block aggregation [420, 421]. In these milieus, Aβ monomer exists in a predominantly α-helical form. The location of this structure in Aβ has been used to obtain insights into the membrane-spanning domain of AβPP [421]. Another approach to obtain NMR data for Aβ monomer in an aqueous solvent was to acquire data at near freezing temperature, a condition under which Aβ aggregation is inhibited [422]. Hou et al. studied Aβ40 and Aβ42 at 5 °C and found that both peptides possessed predominantly random, extended-chain conformations with turn- or bend-like local structures at Asp7–Glu11 and Phe20–Ser26 [423]. An insufficient number of constraints were observed to allow complete structure determination, suggesting that Aβ, like α-synuclein [424], is largely disordered in the "native" state.

Reductionist approaches to structure determination also have been implemented in studies of Aβ. To do so, peptide fragments have been synthesized chemically and then NMR studies have been performed. A number of local structural features have been determined in this way (Table 17.8). In the presence of SDS or lithium dodecylsulfate, Aβ(1–28) and Aβ(25–35) were largely helical. Aβ(10–35) in water possessed an unusual "collapsed coil" structure with two turn-like regions. Aβ(25–35) in water displayed an N-terminal turn-like structure with the C-terminal region disordered. Aβ(21–30) in water forms a turn-like structure centered at Val24–Lys28. Structures obtained from NMR studies of Aβ fragments have been used as starting structures in molecular dynamic simulations of conformational transitions involved in Aβ assembly [425]. However, three caveats must be considered in evaluating data produced in studies of Aβ fragments: (1) how the presence of co-solvents or solvent additives affects the structure of the peptide, (2) how the truncation of full-length Aβ, and the consequent alteration in cooperative folding interactions, affects the folding of the remain-

Table 17.8 Summary of NMR studies of fragments Aβ which include structure calculations

Peptide	Conditions	Major structural features [a)]
Aβ(1–28) [790]	2 mM Aβ, micellar SDS/H_2O, pH 3, 25 °C	α-helix, with bend centered at Val12
Aβ(25–35) [791]	2 mM Aβ, micellar LiDS/H_2O, pH 4, 25 °C	α-helix: Lys28–Met35
Aβ(10–35) [649]	0.25 mM Aβ, H_2O, pH 5.7, 25 °C	collapsed coil centered at Leu17–Ala21; turn-like regions: Asp7–Glu11 and Phe20–Ser26
Aβ(25–35) [792]	1 mM Aβ, micellar SDS/H_2O, pH 4.2, 25 °C	α-helix: Lys28–Leu34
Aβ(25–35) [792]	1 mM Aβ, 20% HFIP/H_2O, pH 4.2, 25 °C	type I β-turn: Ser26–Asn27
Aβ(21–30) [426]	1 mM Aβ, H_2O, pH 6, 10 °C	turn- or loop-like structure: Val24–Lys28

a) Unspecified areas exist as extended chain or disordered structures.

ing peptide region, and (3) how consideration of the first two caveats bears on the biological relevance of the results.

Recently, limited proteolysis, mass spectrometry and NMR were combined to monitor Aβ monomer folding [426]. Limited proteolysis is a powerful method to probe conformational dynamics in proteins that are partially unfolded [427] and have a propensity to aggregate [355, 428, 429]. The technique is performed in the absence of denaturants, which provides the means to identify folded (protease resistant) versus unfolded (protease sensitive) regions of a protein. Seven endoproteases were used to digest Aβ40 and Aβ42. Liquid chromatography/ mass spectrometry of the resulting digests showed, for both proteins, that the segment Ala21–Ala30 was protease resistant. The decapeptide Aβ(21–30) showed similar protease resistance and thus was studied by solution-state NMR. Structure determination using NMR-derived constraints revealed that Aβ(21–30) populated two conformational families, both of which contained a turn-like structure at Val24–Lys28 and were stabilized in part by hydrophobic interactions between the propyl side-chain of Val24 and the butyl side-chain of Lys28. The difference between the families was a Coulombic interaction between Lys28 and either Glu22 (Family I) or Asp23 (Family II). It was postulated that turn formation at Val24–Lys28 nucleated the intramolecular folding of Aβ monomer – the initial event in Aβ assembly (Fig. 17.9).

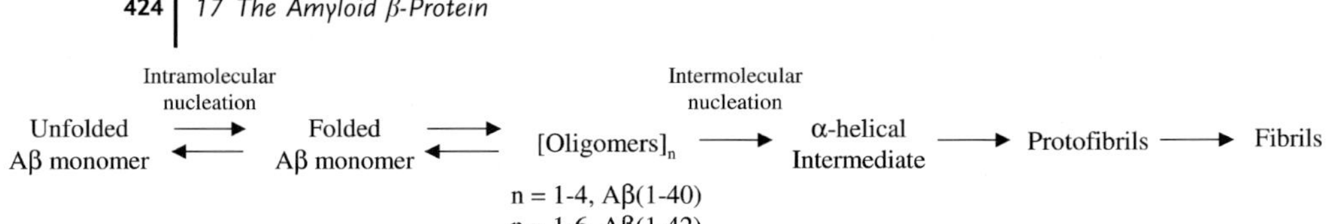

Fig. 17.9 Aβ folding and assembly. Several nucleation processes occur during Aβ fibrillogenesis. "Intramolecular nucleation" refers to structure formation in an initially unfolded Aβ monomer. The partially folded conformers then oligomerize in an alloform-specific manner, producing distinct oligomer distributions displayed by Aβ40 and Aβ42 [41]. "Intermolecular nucleation" refers to conformational transformations associated with the formation of higher-order oligomers.

17.6.3 Other Aβ Assemblies

17.6.3.1 Channels

Discovery

Arispe et al. showed that when Aβ40 was dissolved in water and incorporated into lipid bilayers, a Ca^{2+}-selective ion channel could be formed [430]. The channel could be blocked with tromethamine and Al^{3+}. The lifetime of the channel extended from minutes to hours. These data were consistent with the hypothesis that Aβ channels could kill neurons by allowing external Ca^{2+} entry into the cell [430]. This hypothesis was met initially with skepticism, because the results of Arispe et al. could not be reproduced [165]. Subsequent work has shown that the lack of reproducibility could stem from the unpredictable aggregation of Aβ [165]. Results from Aβ40 have been reproduced [431, 432] and others have shown that channels can be formed by Aβ42 [433], Aβ(25–35) [434] and other amyloid proteins, including islet amyloid polypeptide [435], transthyretin [165] and β_2-microglobulin [436].

Biophysics

Hirakura et al. showed that the formation of channels is increased by acidic pH and inhibited by Congo red, indicating that the channels are formed by aggregates of Aβ [437]. This is consistent with results obtained by Lashuel et al. who showed that a subpopulation of protofibrils from Arctic Aβ40 forms annular pores displaying outer diameters of 7–10 nm, inner diameters of 1.5–2 nm and masses of 150–250 kDa, corresponding to 40–60 Aβ monomers [162]. The pore-like morphology of these protofibrils could explain the channel-like properties of Aβ aggregates [162].

The structure of the Aβ channel in biological membranes is not known. High-resolution AFM of Aβ42 channels in planar lipid bilayers showed doughnut-shaped structures that protruded ~1 nm above the surface of the bilayer [438]. The structures possessed outer diameters of 8–12 nm and contained a centralized pore-like depression. The AFM images revealed both tetrameric and hexameric structures. Only theoretical structures have been proposed for the Aβ40 channel [439]. These models show oligomeric structures in which the monomer is organized into two α-helices (Lys16–Val24 and Lys28–Val40) and a β-hairpin (Glu3–His13). Hydrophilic amino acid residues found in the N-terminus are oriented in the lumen and a ring of His residues are found around the entrance of the channel. Experimental support for this model has been indirect and has come from studies of the interaction of small molecules with Aβ channels. Zn^{2+} ions, which bind to His, have been shown to block the channel [440]. Peptides that match the sequence of residues in the lumen block the channel, whereas other peptides have no effect [441].

Biological Activity

Aβ has been shown to form channels in rat cortical neurons [442], membrane patches from hypothalamic neurons [443] and neuron-like human teratocarcinoma cells [444]. These channels transport Ca^{2+} ions into cells, disrupting Ca^{2+} homeostasis [165], a mechanism that has been proposed to account for neuronal death in AD (see above).

17.6.3.2 βamy Balls

At high Aβ concentrations (300–600 μM), Westlind-Danielsson and Arnerup showed that Aβ40 forms fibrils that spontaneously assemble into spheres with diameters of around 20–200 μm. These spherical structures were given the name βamy balls [49]. βamy balls were highly stable at 30 °C and not affected by dilution (1:2) in cell culture media. They withstood freezing and thawing. Interestingly, Aβ42 did not form βamy balls but accelerated Aβ40 βamy ball formation at low Aβ42/Aβ40 ratios (below 10% w/w). The secondary structure composition present in and the toxicity of βamy balls have not been determined.

17.6.3.3 Amylospheroids (ASPDs)

ASPDs are spherical assemblies of Aβ40 and Aβ42 that form during slow rotation of Aβ solutions [48]. Examination of ASPD by AFM and EM revealed perfect spheres. ASPDs with diameters above 10 nm were toxic to primary neuronal cultures from rat brain, whereas those with diameters below 10 nm were non-toxic. ASPDs from Aβ42 formed more rapidly and at lower peptide concentration than did Aβ40 ASPDs. In addition, the Aβ42 ASPDs were approximately 100-fold more toxic than were the Aβ40 ASPDs. The formation of ASPD was not inhibited by "β-sheet breaker" peptides, including KLVFF and LPFFD, indicating that the mechanism of formation of ASPDs differs from that of amyloid assembly intermediates and fibrils.

17.6.4 Modulators of Aβ Folding and Assembly

Aβ folding and assembly are affected by many factors, including peptide concentration and primary structure, pH, ionic strength, and cellular and extracellular milieu [20, 42, 47, 366]. Nascent Aβ *in vivo* exists in an extremely complex environment in which interactions with many different chemical moieties and macromolecules may occur. Proteins, lipids and metal ions, for example, all may affect Aβ aggregation or Aβ-mediated neurotoxicity [20]. Interactions with these factors may explain why Aβ, which is present in the body at average concentrations in the picomolar and nanomolar range (far below critical concentrations for assembly observed *in vitro*), can self-assemble and deposit as amyloid plaques [445]. Factors that may interact in a clinically important manner with Aβ are discussed below.

17.6.4.1 Proteins

Many plaque-associated proteins have been reported [20]. Detailed *in vitro* biophysical studies have been performed to demonstrate and characterize their interactions with Aβ. These studies have demonstrated that different proteins may inhibit or enhance Aβ fibrillogenesis. In fact, depending on the experimental system employed, the same protein has been observed to exhibit either characteristic. We discuss below a number of proteins found to interact with Aβ. The reader is cautioned that: (1) co-localization of proteins, or other molecules, with amyloid plaques is a correlative observation that does not prove that Aβ assembly and deposition result from interaction with the co-localized molecule/protein, and (2) *in vitro* studies of binary interactions (e.g., Aβ with another protein) do not account for ternary and higher-order interactions nor for the effects of the complex cellular and extracellular milieus in which the interactions occur *in vivo*, and therefore data thus derived must be considered suggestive, but not formally demonstrative, of a biologically relevant interaction.

α_1-Antichymotrypsin (ACT)

A link between ACT, a serine protease inhibitor, and AD was established by the observation that ACT is found in senile plaques [446]. *In vitro* studies have shown that ACT can accelerate or inhibit Aβ aggregation in a concentration-dependent manner [305, 447–450]. ACT was thought to interact with the tripeptide sequence Asp-Ser-Gly (residues 7–9) in Aβ, based upon the similarity of this sequence to the catalytic domain of serine proteases [446]. However, subsequent studies showed that both the amino and carboxyl termini of Aβ undergo strand insertion within two different β-sheets of ACT [449, 451]. This bimodal insertion of Aβ into ACT β-sheets provides a mechanistic explanation for the concentration-dependence of the effect of ACT on fibrillogenesis. The mechanism also predicts that the concentration ratio of ACT to Aβ determines whether and how this interaction contributes to Aβ aggregation and to AD.

Laminin

Laminin is a major basement membrane glycoprotein ($M_r \sim 850{,}000$) widely distributed in the extracellular matrix (see also Chapter 7). The three individual chains (α, β and γ) of this protein self-assemble into disulfide-bonded heterotrimers arranged in a cross-shaped structure [452–454]. The C-terminal globular domains of the α-chain are critical for correct assembly of the glycoprotein [455]. This region also is important for driving laminin polymerization and heparin binding [456, 457]. Laminin accumulates in senile plaques, and also has been found to regulate AβPP processing in fibroblasts and neuronal cells [458–461]. Studies of the interaction of laminin with Aβ40, Aβ42, and the Dutch variant of Aβ40, have revealed that laminin inhibits fibril formation [462–465]. Laminin also has been shown to disaggregate pre-formed Aβ fibrils [463, 466, 467]. Inhibition of fibril formation by laminin-1 or laminin-2, which differ in their α-chain, but share the same β and γ-chains, protects cultured rat cortical neurons [468], but not PC12 cells [469], from Aβ-mediated toxicity. Efforts to define the structural determinants responsible for the inhibitory effects of laminin have shown that the α_1-chain-derived peptide Tyr-Phe-Gln-Arg-Tyr-Leu-Ile inhibits fibril formation as well as the holoprotein [467]. Laminin was not effective as an inhibitor of islet amyloid polypeptide fibril formation, suggesting that laminin's amyloid inhibitory effects were peptide specific [470]. The C-terminal globular domain of the laminin α-chain apparently contains an active motif mediating interaction with Aβ [470]. The inhibitory effects of laminin and laminin-derived peptides have led to the suggestion that these may be effective therapeutic agents for AD [471].

Acetylcholinesterase (AChE)

It has been suggested that AChE, an enzyme involved in the hydrolysis of the neurotransmitter acetylcholine [472], may be involved in the development of AD [473, 474]. AChE co-localizes with amyloid deposits and has been suggested to enhance AβPP processing and to promote Aβ aggregation into fibrils [475–479].

CD, turbidity, EM and Thioflavin T-binding studies all show, regardless of the source of AChE (i.e., human, bovine or mouse) or its quaternary structure (monomer, dimer or tetramer), that it accelerates amyloidogenic conformational transitions in Aβ and fibril formation [473, 480–484]. AChE also promotes the aggregation of full-length Aβ with a Val18Ala substitution and of several amyloidogenic fragments of Aβ [473, 480]. Amyloid fibril–AChE complexes may form at the growing tips of amyloid fibrils, facilitating addition of Aβ monomers [480]. AChE has been shown to form stable complexes with senile plaque components through its peripheral anionic site [473]. This site is located close to the rim of the active site of the enzyme and might be involved in the acceleration of fibril formation [484]. A monoclonal antibody that binds to the peripheral anionic site blocks AChE enhancement of fibril formation, supporting the involvement of the site in Aβ binding [485]. Studies of interactions between different fragments of Aβ and AChE showed that residues 12–28 of Aβ were necessary.

This region includes the cationic amino acid segment His13-His14-Gln15-Lys16 and the CHC (Leu17-Val18-Phe19-Phe20-Val21). Both segments of Aβ apparently were necessary for the interaction of Aβ with the anionic site of AChE [486]. Finally, toxicity studies of Aβ–AChE complexes formed with Aβ40 or Aβ42 have shown that the complexes are more toxic than the peptide itself [487].

Serum Amyloid P (SAP)

The plasma glycoprotein SAP, a member of the pentraxin family of proteins, is an invariant component of all amyloid deposits *in vivo* (see also Chapters 8 and 9). Pentraxins are composed of five identical non-covalently bound subunits and possess Ca^{2+}-binding properties [488, 489]. SAP is found in senile plaques and amorphous Aβ deposits as well as in neurofibrillary tangles [490–497]. Botto et al. have shown that induction of reactive amyloidogenesis is retarded in mice with targeted deletion of the SAP gene, supporting a role for SAP in amyloid formation [498]. *In vitro,* SAP undergoes reversible calcium-dependent binding to all types of amyloid fibrils [499]. Although the normal physiological role of SAP is unknown, it has been suggested that SAP might protect amyloid from degradation *in vivo* by masking the abnormal fibrillar conformation that would otherwise be expected to trigger phagocytic clearance mechanisms [500, 501]. SAP isolated from amyloid deposits is identical in structure to circulating SAP [502] and therefore is not an enzyme inhibitor itself. The protein inhibits proteolysis of fibrils from Aβ40 by pronase [503]. However, when SAP was prevented from binding to the fibrils by the addition of methyl-4,6-*O*-[(*R*)-1-carboxyethylidene]-β-D-galactopyranoside, the inhibitory effect of SAP on the proteolysis of the fibrils was vitiated [503]. Using analytical ultracentrifugation, EM and rheological measurements, MacRaild et al. demonstrated that SAP caused soluble fibrils to condense into localized fibrillar aggregates with a greatly enhanced local density of fibril entanglements [504]. Taken together, the work on SAP suggests that compounds blocking SAP–Aβ interactions could have therapeutic value [489].

ApoE

ApoE is a 299-residue (34-kDa) plasma protein that transports lipids and cholesterol [505]. Numerous studies support the hypothesis that the ApoE4 form of the protein is a risk factor for AD (see Section 17.5.4, Chapter 10 and [83, 300, 301, 303, 304, 506–511]). Brain tissue from AD patients expressing ApoE4 contains more amyloid than brain tissue from AD patients with other ApoE genotypes, implying a direct role for ApoE in cerebral amyloidogenesis [302, 303, 512]. *In vitro* experiments have shown that both ApoE3 and ApoE4 bind Aβ and form SDS-stable complexes, implying avid intermolecular interactions [320, 513]. Importantly, Aβ–ApoE complexes have been isolated from AD brain extracts and shown to be stable after treatment with denaturants that include 10% SDS, 10 M urea, 6 M guanidine–HCl, 90% phenol and 70–100% formic acid

[514]. Brain-derived Aβ–ApoE complexes examined by EM existed as tightly packed amyloid fibrils morphologically similar to non-complexed Aβ fibrils [514].

To understand how ApoE affects assembly of Aβ, interactions between ApoE and Aβ have been monitored *in vitro* using turbidity and dye-binding (Thioflavin T and Congo red) assays [305, 306, 515–522]. Both stimulation and inhibition of fibril formation have been reported. It has been suggested that ApoE acts as a pathological chaperone, promoting conformational transitions in Aβ from a soluble form into β-sheet-rich, toxic aggregates [305, 520, 523–526]. ApoE also facilitated fibrillogenesis of [Ala18]Aβ40, an Aβ alloform that is not normally fibrillogenic [527]. Golabek et al. suggested that the fibril-promoting effect of ApoE depended upon the initial secondary structure of Aβ – the ApoE-mediated promotion of fibrillogenesis was slight with Aβ already containing β-sheets and large if the peptide was substantially disordered [526]. The importance of the ApoE–Aβ interaction has been supported by studies demonstrating that a synthetic peptide that can block ApoE–Aβ interaction reduced amyloid-induced toxicity and fibril formation *in vitro*, and β-amyloid plaque load in transgenic mice [528].

Evans et al. showed that ApoE3 and ApoE4 both inhibited fibril nucleation at physiological concentrations, but that ApoE4 was less effective than was ApoE3 [306]. They also showed that the inhibitory activity of a proteolytic fragment of ApoE3 containing the N-terminal 191 amino acids was comparable to that of the native protein, whereas the C-terminal fragment had no activity. Neither ApoE3 nor ApoE4 inhibited seeded growth of fibrils or affected fibril solubility or structure. Studies by Wood et al. showed that all three ApoE isoforms are equally potent at inhibiting both nucleation and seeded growth of Aβ40 fibrils [517, 518]. However, Esler et al. suggested that Aβ amyloid formation was faster in the presence of ApoE4 than ApoE3, while growth of existing plaques was unaffected by either isoform [529]. They also showed that the amyloid promoting effect of ApoE3 was mediated through interaction with oligomeric intermediates rather than through specific contacts with monomeric Aβ [529].

Data obtained from *in vivo* studies of the effects of ApoE in transgenic mice are complex. Studies indicate that murine ApoE facilitates the development of Thioflavin S-positive Aβ deposits [511, 530–534]. In contrast, human ApoE suppresses Aβ deposition and may be involved in Aβ clearance [531, 535].

Binding studies of human ApoE and various truncated Aβ peptides suggest that the region Val12–Lys28 in Aβ is important for ApoE binding [536, 537]. The domain of ApoE responsible for Aβ binding is unclear. Evans et al. reported that the N-terminus of ApoE binds Aβ [306], whereas others have reported the involvement the ApoE C-terminus [514, 520, 536, 538, 539]. The contradictory conclusions about the effects of ApoE on Aβ assembly likely have their bases in four areas: (1) the mode of preparation and initial conformational and assembly state of Aβ, (2) the mode of preparation of ApoE, especially with respect to whether lipidated or delipidated protein was used, (3) conditions (solvent, pH, temperature, buffer composition, protein concentrations) under which the assembly process was monitored, and (4) the phenomenon measured (nucleation, elongation, secondary structure changes, etc.). Genetic, histopathologic and clinical studies clearly link the *APOEε4* genotype to AD. Additional studies will be required before a well-supported mechanistic explanation for this linkage emerges.

Proteoglycans (PGs)

PGs are glycoproteins composed of a core protein to which at least one, but frequently many more (tens to hundreds), glycosaminoglycan (GAG) chains are attached (see also Chapter 7). Four different PGs, containing the GAG subunits heparan sulfate, keratan sulfate, dermatan sulfate or chondroitin sulfate, have been found associated with plaques in AD [540–544]. Growing evidence implicates PGs in the pathophysiology of amyloid [545–548]. GAGs are responsible for PG binding to Aβ, accelerate fibril formation and stabilize fibrils once formed [547–549].

Aβ assembly reactions monitored by CD show that amyloidogenic (random coil → β-sheet) folding of Aβ occurs more rapidly in the presence of GAGs [550]. Secondary structure changes in Aβ42 occurred immediately upon addition of GAGs, whereas changes were not immediately apparent with Aβ40. In the presence of chondroitin-4/6-sulfates, Aβ40 converted from random coil to β-sheet within 4 h. This change was not observed in the absence of these GAGs. The chemical nature of the GAG affects its ability to alter Aβ assembly. Chondroitin-6-sulfate more effectively accelerates conformational changes in Aβ42 than does chondroitin-4-sulfate [550]. Ionization state is another variable. Aβ nucleation occurs more rapidly at pH 7 in the presence of heparan sulfate than it does in the presence of the more highly charged heparin [550]. The importance of sulfate groups in facilitating Aβ aggregation was confirmed by demonstrating that desulfation of heparan sulfate decreased the rate of GAG-induced Aβ fibril formation [551]. These results, as well as those obtained with dermatan and chondroitin sulfates, suggest that the position and distribution of the active sulfate groups on the GAG backbone are defining elements in Aβ–GAG interactions.

EM studies have shown that GAGs also affect fibril morphology. In the presence of chondroitin-4/6-sulfates and dermatan sulfate, Aβ formed more densely packed fibrils due to enhanced lateral aggregation [549]. It has been suggested that GAG binding to Aβ is mediated by the basic peptide segment His-His-Gln-Lys (residues 13–16) [552–554]. However, once fibrils which contain a predominantly β-sheet structure are formed, the binding does not depend on the His-His-Gln-Lys sequence alone, but may involve other charged residues and nonionic interactions, including hydrophobic interactions [554]. The requirement of a β-sheet motif in Aβ–GAG interactions is supported further by affinity co-electrophoresis studies of Aβ and heparin [266]. Results showed that: (1) fibrillar Aβ binds heparin, whereas non-fibrillar Aβ does not, (2) Congo red, which binds to β-sheets, blocks the Aβ–heparin interaction, and (3) incubation of Dutch Aβ in a solvent that promotes α-helix formation reversed the ability of the peptide to bind heparin [266].

The cumulative data on GAG–Aβ interactions indicate that GAGs accelerate amyloidogenesis by facilitating fibril nucleation and elongation. In addition, EM data support the suggestion that GAGs promote fibril–fibril interactions and accretion of suspended fibrils onto existing plaques, thereby accelerating plaque growth [549, 554]. Compounds that block Aβ–GAG interactions thus could have therapeutic value [555].

17.6.4.2 **Lipids**

Amyloid fibrils are intimately associated with neuronal, microglial and endothelial membranes. Lipid rafts, and the component lipid cholesterol, have been found to modulate AβPP processing and therefore affect AD risk [556, 557]. The interactions between Aβ and membrane lipids have received special attention due to speculation that these interactions may be a primary mechanism of amyloid-mediated toxicity [558–561]. For example, as discussed previously (Section 17.6.3.1), one pathogenetic mechanism for Aβ-mediated cytotoxicity is the formation of cation-selective membrane channels that disrupt ion homeostasis and lead to cellular injury and death [430, 431, 438, 441, 443, 562]. Alterations in membrane lipid composition that have been observed in AD patients may promote these and other types of pathogenic Aβ–membrane interactions [563–565]. In this section we discuss the linkage of sphingolipids (gangliosides), phospholipids and cholesterol to AD.

Gangliosides

Gangliosides are complex neutral oligoglycosphingolipids. The involvement of gangliosides in AD was suggested by the identification, in human brains exhibiting early pathological changes consistent with AD, of a novel Aβ species characterized by tight binding to GM_1 ganglioside (GM_1) [566, 567]. The GM_1-bound Aβ exhibited altered immunoreactivity and a strong tendency to aggregate, suggesting that GM_1 binding caused conformational changes in Aβ and contributed to peptide aggregation by producing Aβ–GM_1 complexes that could seed fibril formation [568]. EM studies have shown that both Aβ40 and Aβ42 form amyloid fibrils in the presence of gangliosides [568, 569]. Support for the involvement of GM_1 in altering Aβ folding and assembly has come from CD studies that showed that binding of monomeric, random coil Aβ40 to mixed gangliosides or purified GM_1-containing membranes induced α-helix or β-sheet formation, depending upon pH and ganglioside concentration [569–573]. Questions remain about the relative roles of GM_1 and other gangliosides (GM_2, GM_3, GD_{1a} and GD_{1b}) in binding Aβ and influencing its conformational transitions [569, 570, 574, 575]. What is clear is that differences in the number and positions of sialic acid residues in the carbohydrate backbone significantly affect Aβ conformation [570, 571, 575].

Phospholipids

A link between defective phospholipid synthesis and AD has been suggested [563], thus experiments have been done to determine how phospholipids affect Aβ assembly *in vitro*. Acidic phospholipids were found to induce random coil → β-sheet transition in Aβ40, Aβ42 and Aβ(25–35) [560, 576, 577]. Extrapolating these data to the *in vivo* milieu, it was suggested that vesicular Aβ–phospholipid interactions may increase local peptide concentration or accelerate Aβ folding, leading to enhanced fibril formation. Aβ is thought to interact electrostatically with phospholipid head groups [576–578]. Different phospholipids, including phosphatidylinositol, phosphatidylinositol-4-phosphate and phosphatidylinositol-4,5-diphosphate, affect the structural transition of Aβ to β-sheets differently, and both the charged groups and the sugar moiety itself appear to be involved [576]. Structure–activity studies using inositol and Aβ42 have shown that proper orientation of hydroxyl groups in inositol are necessary for Aβ–inositol complexation [579]. Phospholipid–Aβ interactions also are mediated by non-electrostatic interactions, as NaCl-mediated Debye shielding of charged groups does not block phospholipid-induced β-sheet formation by Aβ42 [576].

Cholesterol

Cholesterol is a major structural lipid component of eukaryotic plasma membranes, and forms bile acids and hormones. Elevated plasma cholesterol concentration is a risk factor for cardiovascular disease and emerging evidence suggests that cholesterol also plays a role in the pathogenesis of AD [556, 557, 580, 581]. Epidemiological studies of patients using statins, cholesterol-lowering agents, reveal a highly significant decrease in AD risk [556]. The mechanism of this effect may be the ability of statins to cause reduced production of Aβ42 and Aβ40 [582, 583]. Plasma membrane cholesterol has been found to modulate AβPP processing *in vivo* and *in vitro* [88, 556, 584–589].

Cholesterol may affect not only Aβ generation, but also Aβ aggregation and neurotoxicity as well [556, 590–594]. For example, fibrillogenesis of Aβ42 is accelerated in the presence of cholesterol and cholesterol derivatives, including cholesteryl acetate, micellar cholesterol polyoxyethyl cholesteryl sebacate/cholesteryl PEG 600 sebacate and cholesterol–sphingomyelin liposomes [590]. This potentiation of Aβ42 polymerization may involve hydrophobic interactions among Aβ amino acid side-chains and the tetracyclic sterol nucleus. Zhang et al. recently showed that cholesterol ozonolysis accelerated Aβ aggregation *in vitro* [595]. The ozonolysis products contain aldehyde groups attached to a large hydrocarbon. Acylation of Aβ side-chain amines (Lys16 and Lys18), and the N-terminus thus could produce modified peptides of increased hydrophobicity and altered folding properties, causing increased aggregation. Importantly, the cholesterol ozonolysis products identified *in vitro* also are found in human brain.

17.6.4.3 Metal Ions

Disregulation of metal ion homeostasis has been proposed to cause AD [596–604]. Cationic transition metals such as copper, iron and zinc exist at high concentration within the healthy brain neocortex, but increased concentrations of these metal ions are detected in AD, where they are concentrated in amyloid plaques [600, 605–607]. In addition to its importance as a component of metalloproteins, zinc is abundant in the pre-synaptic vesicles of specialized neurons termed "zinc-containing" neurons and is released into the extracellular environment during normal neuronal activity [608]. Extracellular zinc concentrations are normally low (below 1 μM), but during intense neuronal activity zinc concentrations of around 100 μM may be observed [608]. This concentration is sufficient to promote Aβ aggregation [609]. Metal ions may be released from metalloproteins under slightly acidic conditions and during inflammatory responses [610–614]. The correlation of increased metal ion levels with AD and studies of the role of metals in fibril stability and Aβ-mediated toxicity suggest that metal ions modulate Aβ aggregation [615–618]. In fact, Aβ40 and Aβ42 have been found to bind metal ions avidly, which causes rapid peptide aggregation in a pH-dependent manner [609, 617, 619, 620]. The metal ion binding sites appear to be identical in both peptides and involve His13 and His14 [621–623]. Studies of Zn^{2+} binding study using Aβ(1–28) fragments suggest that His13 is the most important metal binding site [624]. Substitution of His13 with uncharged residues or Arg (found at position 13 in rodent Aβ) eliminates the zinc-induced random coil $\rightarrow$ β-sheet transition as well as fibril aggregation [624]. Rodent and human Aβ sequences differ at three sites, i.e., 5, 10 and 13 [625]. Rodent Aβ40 (Arg5 $\rightarrow$ Gly, Tyr10 $\rightarrow$ Phe and His13 $\rightarrow$ Arg) and histidine-modified (*N*-carbethoxyhistidine) human Aβ40 did not aggregate in the presence of metal ions at physiological concentration, further supporting the involvement of His residues in metal binding and aggregation [617].

Based on *in vitro* biophysical studies and correlative *in vivo* results, metals have been targeted for AD therapy (see Section 17.8.2.3). Metal chelators have been shown to solubilize aggregated Aβ from AD brain [626, 627]. Chelators also attenuate cerebral amyloid burden in transgenic mice expressing human AβPP and Aβ [628]. Lee et al. reported marked decreases in Aβ deposition in the brains of Tg2576 mice lacking the synaptic ZnT3 zinc transporter [629], a result supporting the hypothesis that metals released during synaptic activity may have an etiologic role in AD.

17.7 Identification of Therapeutic Targets

Amyloid fibril formation is pathognomonic for AD. However, continuing study of Aβ folding and assembly, and resulting structure–activity correlations, have revealed a myriad of neurotoxic assemblies, including activated monomers, oligomers, protofibrils and fibrils [32–34]. These findings suggest that inhibition

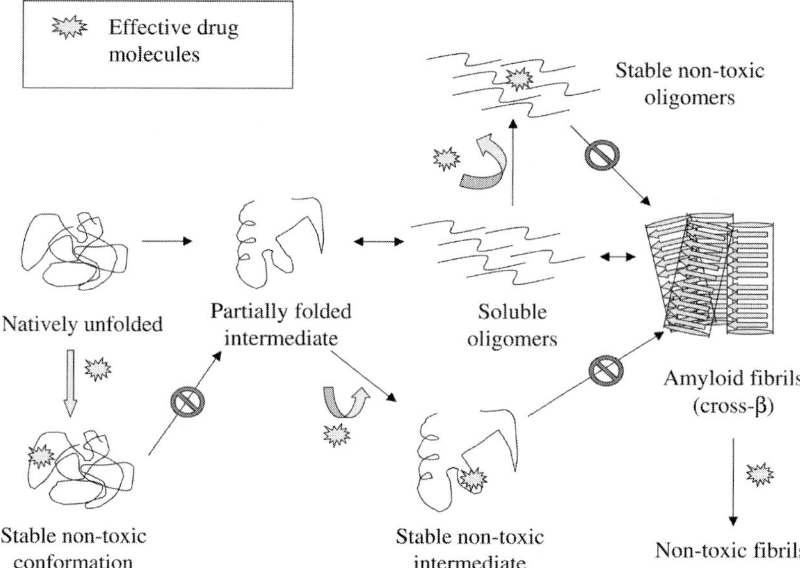

Fig. 17.10 Schematic representation of Aβ folding, aggregation and possible therapeutic strategies to block amyloidogenesis.

of amyloidogenic conformational transitions in the Aβ monomer, blocking oligomerization and higher-order assembly, and dissociation of preformed fibrils have potential therapeutic value. Which conformers or assemblies are most relevant remains unclear. However, two general strategic approaches may be envisioned: (1) blocking production of toxic conformers or assemblies and (2) stabilizing non-toxic structures. We discuss here the current state of knowledge about potential targets for AD therapy (Fig. 17.10).

17.7.1 Fibrils

For the last century, the involvement of fibrils in AD has been axiomatic [4]. Although many new hypotheses about the etiology of AD have been promulgated, therapies targeting fibrils remain important. Aβ fibrils are potent neurotoxins, thus much prior effort has been focused on understanding fibril assembly and identifying small molecules able to inhibit fibrillogenesis or disassemble pre-existing fibrils [630–633]. A number of small molecule inhibitors, including Congo red, peptides, benzofurans, 4,5-dianilinophthalimide, apomorphine and nitrophenols, have been shown to block either Aβ aggregation or the toxicity of Aβ fibrils. Leu-Pro-Phe-Phe-Asp, which includes the β-sheet-disrupting amino acid Pro, has been shown to inhibit Aβ fibrillogenesis, disassemble preformed fibrils and prevent neuronal cell death induced by amyloid fibrils *in vitro*. This peptide also blocks amyloid fibril formation *in vivo* [634]. Two caveats exist with strategies targeting fibrils: (1) if more than one neurotoxic assembly is involved in AD, targeting a single assembly may not produce significant therapeutic effects, and (2) dissociation of pre-existing fibrils may produce protofibrils or other fibril intermediates that are equally or more toxic than fibrils. The former caveat is true of all strategies targeting Aβ and its assemblies. Evaluation of the latter caveat will require clinical trials.

17.7.2 Protofibrils

Protofibrils, relatively short (<150 nm), flexible, narrow (~5 nm) fibrillar precursors of amyloid-type fibrils, are neurotoxic and are linked to a familial form of AD (Section 17.6.1.2). Inhibiting protofibril formation, facilitating protofibril dissociation or converting protofibrils to non-toxic forms thus could have clinical value. One feature of protofibrils that complicates development of therapeutic agents is protofibril metastability. However, recent work by Qahwash et al. showed that substitution of the first two amino acids (Asp–Ala) of Aβ by Glu–Val significantly increases Aβ40 protofibril stability and produces protofibrils that are more cytotoxic than those produced by wild-type Aβ40 [389]. This modified peptide could prove useful both for basic structural and functional studies of protofibrils and as a target for the design and testing of lead compounds.

17.7.3 α-Helix-rich Intermediate

For most natively folded proteins, formation of amyloid fibrils requires a partially unfolded intermediate. Aβ fibril assembly is a mirror image of this process in that the natively disordered Aβ conformer must partially fold to assemble [42, 635]. Experimental studies of the conformational transitions of Aβ40, Aβ42 and their disease-related congeners during fibril formation showed that a transient, α-helix-containing oligomer was an obligate intermediate in the process [42]. Studies of the thermodynamics of formation of this intermediate showed that helix stability was a key factor controlling assembly [635]. Addition of low concentrations (20% or less, v/v) of the helix-permissive co-solvent TFE accelerated formation of the intermediate and subsequent fibril formation. However, higher TFE concentrations blocked fibril formation, presumably by increasing the activation energy necessary for the subsequent required α-helix$\rightarrow\beta$-sheet transition. TFE cannot be used *in vivo* as it was *in vitro*. However, development of TFE substitutes that could alter the thermodynamics of formation of the α-helix-containing intermediate is worthy of exploration. In fact, the thermodynamic principles of stabilizing non-toxic structures and destabilizing toxic structures have been applied elegantly and successfully to the problem of transthyretin amyloidosis by Hammerström et al. [636].

17.7.4 Oligomers

Biophysical studies of Aβ oligomerization have revealed an increasing number of oligomeric assemblies, including ADDLs (Section 17.6.1.5), paranuclei (Section 17.6.1.6), βamy balls (Section 17.6.3.2) and ASPDs (Section 17.6.3.3). ADDLs have been studied in most detail. They are powerful neurotoxins that are found in significantly higher amounts in AD patients versus normal controls. These facts support the relevance of efforts to develop ADDL-specific therapeutic agents, and, in fact, small organic compounds have recently been synthesized that have the ability to block ADDL formation and subsequent toxicity [404, 637]. βamy balls and ASPDs form at supraphysiologic concentrations (above 300 μM), thus their relevance to AD remains unproven.

17.7.5 Targeting Aβ Conformation

17.7.5.1 Stabilization of Native Conformation

Aβ monomers prepared under carefully controlled conditions that preclude self-association are non-neurotoxic. However, monomer folding and self-assembly produce neurotoxins. Stabilization of the native (irregular) structure of the Aβ monomer thus is one strategy to inhibit amyloidogenic (toxic) conformational transitions. Theoretically, this strategy could be implemented through the design of drug-like molecules blocking Aβ monomer folding. This strategy has been used in efforts to block amyloid fibril formation of natively folded amyloid proteins that must undergo partial unfolding to self-assemble [638, 639]. Recent work by Lazo et al. has suggested that Aβ monomer folding is nucleated by the formation of a turn-like structure in the C-terminal half of the peptide (Section 17.6.2). NMR studies of this region have produced models of moderate resolution, sufficient to allow rational design of binding compounds. Active compounds would either block formation of the turn-like element or prevent the conformational rearrangement (maturation) of the region necessary for β-sheet formation.

17.7.5.2 (De)stabilization of Specific Conformers

During Aβ assembly, the monomer undergoes a series of conformational transitions, including random coil $\rightarrow$ α-helix $\rightarrow$ β-sheet. These secondary structure changes occur concurrently with quaternary structure changes, including formation of paranuclei, α-helix-rich oligomers and protofibrils. If toxic conformers or assemblies were destabilized, or non-toxic structures stabilized, therapeutic benefit could result. Soto et al. introduced a single amino acid substitution in Aβ (Val18Ala) and showed that this change caused a significant increase in α-helix content [264]. The result was a significant diminution in fibril assembly, assessed using turbidity, Thioflavin T binding, Congo red staining and EM. Soto et al. proposed that the α-helical conformer was unable to form amyloid [264], a suggestion supported by structure–activity studies of Dutch and wild-type Aβ [264]. In these studies, helix content and fibrillogenesis rates were correlated. In the presence of 20% (v/v) TFE co-solvent, Dutch Aβ(1–40) possessed around 6% helix, whereas wild-type Aβ possessed around 17% helix. Dutch Aβ assembled significantly faster than did wild-type Aβ, a negative correlation consistent with the hypothesis that stabilization of helix content can block Aβ assembly [264, 635].

McLaurin et al. studied the effects of gangliosides on Aβ assembly [569]. They observed that monomeric Aβ40 and Aβ42, in the presence of mixed gangliosides or GM_1-containing vesicles, formed stable α-helical conformers that did not transform into fibrils. Gangliosides could not induce formation of the α-helical conformers if the Aβ already existed as a β-sheet-rich aggregate. Similar results were observed in studies of Aβ42. Aβ40, Aβ42, and disease-related congeners all have been found to form an α-helix-rich intermediate during fibril assembly (Section 17.6.1.3). The data of Soto, McLaurin and others [264, 569, 635], support the idea that blocking Aβ assembly by stabilization of helical conformers is a strategy worthy of further exploration.

17.7.5.3 **Aβ Monomer Subregions and Residues**

Fine-structure analysis of Aβ folding and assembly have revealed much about the involvement of specific regions of the holopeptide, and thus provided potential therapeutic targets. We discuss here these peptide segments, as well as the involvement of specific amino acids in Aβ assembly. We emphasize that conclusions derived from the experimental systems employed to study these segments/residues are scientifically valid, but cannot be extrapolated directly to the organismal level without additional investigation. Regions and residues found to be important for Aβ folding and aggregation *in vitro* are obvious drug targets, but target validation will require *in vivo* studies of relevance and efficacy.

N-terminus

The strong linkage of Aβ42 to AD, relative to Aβ40, has focused research efforts on the C-terminus of Aβ; however, evidence is accumulating that the N-terminus has important effects on Aβ assembly and thus must be considered a potential target for therapy. Qahwash et al. have reported that [Glu1–Val2]Aβ40 preferentially forms protofibrils and that these protofibrils are more cytotoxic than fibrils formed by wild-type peptide [389]. Bitan et al. showed that N-terminal truncations alter significantly the oligomerization of Aβ40 [414]. New studies of the oligomerization of Tyr-substituted Aβ alloforms ([Tyr1]Aβ40 and [Tyr1]-Aβ42) show that the oligomer distributions of the modified peptides differ from those of the respective wild-type peptides. This observation demonstrates that the N-terminus does not exist solely as a solvent-exposed coil, but must at least partially populate an organized conformational state. This latter conclusion is consistent with the increasing recognition that the N-terminus mediates monomer packing within quaternary Aβ assemblies, especially fibrils [351].

Cationic Tetrad (His-His-Gln-Lys)

A cluster of basic amino acids is found in the Aβ N-terminus, His13-His14-Gln15-Lys16. The imidazole ring of His endows it with a number of activities, including metal binding, hydrogen bonding and salt bridge formation. These activities are important for Aβ folding and assembly as well as Aβ-mediated toxicity [341, 423, 554, 603, 640, 641]. Early structural studies on Aβ(1–28) produced a tetramer model of the β-sheet region comprising the fibrils and showed that this structure was stabilized in part by a His13–Asp23 salt bridge [341]. Hilbich et al. studied Aβ(10–42) and reported that substitution of His13–His14 by Ala13–Ala14 increased fibrillogenesis and decreased solubility [641]. However, McLaurin et al. showed that a His13Ala substitution in Aβ(1–28) produced a disordered Aβ conformer that did not undergo subsequent conformational changes or fibril assembly, even after lengthy incubation [554]. Wild-type Aβ(1–28), in contrast, readily transformed into fibrils. Interestingly, [Ala14]Aβ(1–28) did form fibrils, suggesting that His14 was not necessary for fibril formation. However, the length and surface density of fibrils formed by [Ala14]Aβ(1–28) were lower those found in wild-type Aβ, indicating that the side-chain of residue 14 is involved in controlling assembly kinetics and possibly fibril packing. The disparate results of Hilbich et al. and McLaurin et al. emphasize the complexity of the Aβ folding and assembly process. Identical amino acid substitutions in different peptides, e.g., Aβ40 or Aβ42, can produce significantly different biophysical and biological effects. Kirschner et al. have speculated that the His-His-Gln-Lys tetrad is involved in amyloid fibril packing [341]. Consistent with this suggestion are data from studies of Aβ in which the tetrad was replaced by the uncharged Gly-Gly-Gln-Gly tetrad [640]. The substituted peptide displayed reduced β-sheet stability and formed only short fibrils with diameters of 35–40 Å that displayed little or no lateral aggregation [640]. McLaurin et al. studied the effects of single Ala substitutions of the three basic amino acids as well that of triple substitution [554]. [Ala16]Aβ(1–28) behaved similarly to [Ala14]Aβ(1–28) in that it underwent a random coil → β-sheet transition and formed short, dispersed fibrils and globular aggregates. Unlike single Ala substitution, the triply-substituted peptides [Ala13, Ala14, Ala16]Aβ(1–28) showed β-sheet structure without incubation. EM studies of samples incubated for 3 days showed numerous fibrils with decreased lengths compared to wild-type. Hou et al. studied side-chain NMR chemical shifts of Aβ40 and Aβ42 during aggregation [423]. Significant up-field chemical shifts were observed for the ^{2}H aromatic chemical shifts of His13 and His14 during aggregation. This suggests that the His13–His14 region is important for β-sheet formation and might be involved in a salt bridge interaction with Asp residues [423]. Recently, Stephenson et al. reported successful development of dianionic compounds that bind in this region and inhibit fibril formation [642].

Histidines also are important for metal binding. Disregulation of metal ion homeostasis correlates with AD [596–604], and *in vitro* studies have shown that Aβ40 and Aβ42 avidly bind metal ions, a process resulting in rapid aggregation [609, 617, 619, 620]. His13 is particularly important in metal binding [624].

In addition to the effect of the basic tetrad on peptide folding, Giulian et al. have proposed that His-His-Gln-Lys in Aβ42 is responsible for binding to microglial cells [643]. Microglial activation, and subsequent inflammatory effects, has been linked to the pathogenesis of AD [644], thus agents that interfered with the His-His-Gln-Lys "activation domain" would be useful. Small peptides containing His-His-Gln-Lys inhibited binding of Aβ42 to microglial cells as well as the induction of neurotoxicity by microglial cells *in vivo*. The His-His-Gln-Lys tetrapeptide also reduced rat brain inflammation elicited after infusion of Aβ peptides or implantation of native plaque fragments *in vivo* [643].

17.7.5.4 **Central Hydrophobic Cluster (CHC)**

The Aβ CHC (Leu17-Val18-Phe19-Phe20-Ala21) is an important mediator of Aβ aggregation. Wood et al. have shown that Pro substitution anywhere within the region Leu17–Asp23 blocks fibril formation [645]. Intrinsic fluorescence studies of Aβ40 have shown that Aβ assembly alters the local environment of Phe20 [646]. The necessity for structural rearrangement within the CHC of Aβ40 is one explanation for the slower kinetics of Aβ40 assembly relative to Aβ42 assembly. One aspect of structural change may be the conversion of helical segments within the oligomeric fibril intermediate to β-sheet (see Section 17.6.1.3). Such conversions have been observed both experimentally and observed in simulations *in silico* [399, 647]. A random coil → β-sheet CHC conversion has also been described, but in an Aβ fragment, Aβ(12–28) [648]. Aqueous solution NMR studies of Aβ(10–35) have revealed an unusual "collapsed coil" structure centered at the CHC [649]. Stabilization of this structure within the monomer prevented the conformational transformations necessary for fibril formation [650], presumably by creating a kinetic barrier to amyloidogenic folding. This result adds additional support to the "thermodynamic approach" to inhibitor targeting and design exemplified by studies of transthyretin amyloid formation [636]. An intriguing approach to CHC inhibitor design has been the synthesis of peptides containing portions of the domain itself [634, 651–658]. In addition to strict sequence mimetics, these inhibitor peptides have included D-amino acids, Pro and oligolysine as β-sheet breakers [634, 651–658]. Peptidomimetics with *N*-methylated amino acids and other modifications have been reported to inhibit Aβ fibril formation [659–663]. A CHC-based mimetic also has been implemented *in vivo* against AD [634]. Using Leu-Val-Phe-Phe as a template, Soto et al. designed a β-sheet breaker peptide, Leu-Pro-Phe-Phe-Asp, incorporating the β-sheet-disrupting amino acid Pro. The activity of this peptide was demonstrated using three systems. *In vitro*, the peptide inhibited Aβ fibrillogenesis and disassembled preformed fibrils from Aβ40 and Aβ42. In cell culture, it prevented neuronal cell death induced by Aβ42 fibrils. *In vivo*, it blocked amyloid fibril formation in rat brain [634].

C-terminus

Despite the small primary structural difference between Aβ40 and Aβ42, i.e., the C-terminal dipeptide Ile41–Ala42, the clinical impact and biophysical behavior of the two Aβ alloforms are distinct. This fact, and early work on Aβ fibril nucleation [365, 664], has focused attention on the Aβ C-terminus. Solid-state NMR studies of Aβ40 suggest that the segment Ala30–Val40 adopts a β-strand conformation and participates in forming the core of the amyloid fibril [351]. However, other contributions of the C-terminus to Aβ folding and the assembly of oligomeric and protofibrillar intermediates remain a subject of active study. For example, it is unclear how the Ile41–Ala42 dipeptide alters local structure at the Aβ C-terminus, whether structural alteration have allosteric effects, or whether both occur. Solution NMR study of Aβ variants [Met(O)35]Aβ40 and [Met(O)35]Aβ42 revealed that neither peptide folded into a globular structure and that both deviated from random coil behavior by local conformational preferences of a number of short segments [665]. The comparison of ^{15}N–^{1}H nuclear Overhauser enhancements (NOEs), which provides insight into dynamics, showed that residues 30–40 of [Met(O)35]Aβ42 had higher values of ^{15}N–^{1}H NOEs compared to [Met(O)35]Aβ40, indicating that the C-terminus of Aβ42 is less flexible than that of Aβ40 [665]. Hou et al. also monitored [Met(O)35]Aβ40 and [Met(O)35]Aβ42 structure, and those of wild-type Aβ40 and Aβ42, by NMR [423]. They reported that the C-terminal segment Val39-Val40-Ile41 in both [Met(O)35]Aβ42 and wild-type Aβ42 had a tendency to form β-strand structure. In contrast, the C-terminal segment Gly29–Val40 in both [Met(O)35]Aβ40 and wild-type Aβ40 did not have a strong tendency for β-strand structure [423]. Intrinsic fluorescence studies of Aβ40 and Aβ42 in a mixed solvent (dimethylsulfoxide–water) system revealed that residues 30 and 42 are solvent-shielded in Aβ42, whereas only the C-terminus of Aβ40 is shielded [646].

The most detailed insights into Aβ folding and oligomerization have been recently through molecular dynamics simulations [666]. Contact maps (two-dimensional representations of binary interactions between amino acids) from these studies have shown a folded Aβ42 monomer contains a turn centered at Gly37–Gly38 that is not found in Aβ40. This result is consistent with results obtained using limited proteolysis [426]. The contact maps also showed a turn-like region from Val24–Lys28, but this forms later than the Gly37–Gly38 turn [666]. The pentamer from Aβ42 was stabilized primarily by hydrophobic interactions involving Leu17–Ala21, Ala30–Met35 and Val40–Ala42. These results were consistent with *in vitro* experimental studies of Aβ oligomerization (Section 17.6.1.6). A powerful feature of *in silico* approaches is the ability to visualize structures that exist in infinitesimally small numbers and for infinitesimally small time periods. Nuclei and transition-state structures are but two examples of the types of structures that may be determined *in silico*, and are of utmost importance in amyloid formation and chemical biology. The ability to determine these structures facilitates drug design and the dynamic possibilities offered through simulation provide the means to study inhibitor–target interactions.

Met35

The single methionine residue of Aβ, located at position 35 (Met35), contributes to Aβ folding and has been suggested to be involved in AD-associated redox chemistry. For example, endogenous anti-oxidant activity could be provided through oxidation of Met35 to its corresponding sulfoxide. In post-mortem AD brain, [Met(O)35]Aβ has been reported to comprise 10–50% of total brain Aβ [667]. Raman spectroscopy also has revealed oxidized Met35 in AD brain [607]. Whether these data represent agonal effects or physiologically relevant information is unclear. Biophysically, oxidation of Met35 has a profound effect on Aβ aggregation – it is inhibited substantially or blocked [390, 412, 668, 669]. Solution NMR studies of wild-type Aβ and [Met(O)35]Aβ reveal little regular structure in either [423]. Hou et al. have suggested that Met35 oxidation inhibits amyloid assembly by preventing early, site-specific hydrophobic and electrostatic associations [423]. Bitan et al. showed that the distinct oligomerization behaviors of Aβ40 and Aβ42 are eliminated by Met35 oxidation [412]. [Met(O)35]Aβ42 and [Met(O_2)35]Aβ42 both oligomerized indistinguishably from Aβ40 [412]. Theoretical analysis of the impact of side-chain van der Waals radii and polarity suggested that the primary effect of Met oxidation was to increase the Met35 solvation free energy, thereby disfavoring the burial of the Met side-chain in the apolar core of an Aβ aggregate [412]. Controlling the redox state and activity of Met35 may be strategies to control both Aβ assembly and physiologic activity.

Ile41

In addition to its effects in the context of the C-terminal dipeptide Ile41–Ala42, specific effects of Ile41 can be recognized. Ile41 is the more hydrophobic of the two amino acids and the most hydrophobic within Aβ. It possesses a higher propensity for β-strand formation than does Ala42. In simulations of Aβ folding, Ile41 is organized within the core of Aβ aggregates [666]. Studies of Aβ oligomerization have revealed that Ile41 is necessary and sufficient for paranucleus formation [41]. The pentapeptide amide H-Gly-Val-Val-Ile-Ala-NH_2, a proxy for the Aβ segment Gly38–Ala42, binds to Aβ42 and can prevent its aggregation and toxicity [670]. Docking experiments show that Ile41 in the Aβ holopeptide is one of the important sites for binding the pentapeptide ligand [670]. Experimental and theoretical bases for targeting compounds to Ile41 thus exist.

Charged Residues (His6, Asp7, Lys16, Glu22, Asp23, Lys28)

The hydrophobic effect [671] is a significant driving force in Aβ assembly and is especially relevant with respect to the involvement of the CHC and C-terminus [641, 664, 672–674]. However, electrostatics also contribute to the overall Aβ folding process, as demonstrated by the profound effect of pH [417, 675, 676]. Structural and model building studies have suggested the importance of charged residues in controlling Aβ folding and assembly [341, 677]. One of the most important observations supporting the importance of electrostatics in the structural biology and function of Aβ is the identification of six kindreds in which mutations resulting in amino acid substitutions at His6, Asp7, Glu22 or Asp23 cause FAD or CAA (Section 17.5.1.1). These mutations affect both AβPP processing and the biophysical and biological activities of the resulting peptides (Section 17.5.1.1). The latter effects, in particular, support an interventional strategy targeting the altered amino acid or toxic conformers formed as a result of the amino acid substitution.

Detailed study of early phases of Aβ assembly and of fibril formation suggest that substitutions at Glu22 and Asp23 produce different effects depending on whether they occur in Aβ40 or Aβ42. Quantitative determinations of oligomer size distributions of the mutant Aβ40 peptides showed that all four substitutions (Glu22Gln, Dutch; Glu22Gly, Arctic; Glu22Lys, Italian; Asp23Asn, Iowa) produced oligomer distributions extending to relatively high-order oligomers, relative to wild-type Aβ40. In addition, the gel mobilities of tetramers and higher-order oligomers of Aβ40 analogues containing substitutions at Glu22 were larger than that of the respective wild-type Aβ40 oligomers. This suggests that oligomers of the mutant peptides were relatively compact. In contrast, Aβ42 peptides containing the disease-associated substitutions produced oligomer distributions virtually indistinguishable from wild-type peptides [414]. Studies of the kinetics of formation of the α-helix-rich Aβ assembly intermediate showed a similar effect [42]. The Arctic and Dutch substitutions in Aβ40 significantly accelerated formation of the intermediate formation whereas no change in kinetics was observed with Aβ42 peptides [42]. Salt bridge formation/disruption may be a mechanistic explanation for these effects [351]. Recent studies of the nucleation of Aβ monomer folding also postulate important roles for Glu22, Asp23 and Lys28 in salt bridge formation and stabilization of a folding nucleus [426]. Drugs that could disrupt salt bridge formation would be expected to inhibit Aβ folding and aggregation. The differences in behavior of Aβ40 and Aβ42 peptides bearing the FAD- or CAA-lined mutations do not exclude a salt bridge mechanism. Instead, the differences likely reflect the relative contribution of salt bridge formation/disruption to the overall thermodynamics of the system. As discussed earlier, the addition of Ile41–Ala42 to Aβ40 appears to alter the relative contributions of non-C-terminal peptide segments, relative to the C-terminus, to the control of Aβ folding and assembly. These considerations emphasize the difficulty of selecting therapeutic targets when local changes in structure produce different effects in different assemblies.

Lys16 is one of the basic amino acids within the basic His13–Lys16 tetrapeptide segment of Aβ. As discussed above, these amino acids have been shown to be important in the folding and aggregation of Aβ. Lys16 itself exerts significant control over β-sheet formation and Aβ assembly, presumably by ion pairing with negatively charged residues (Asp, Glu) [341, 640, 677]. Substitution of Lys16 by Ala in the peptide fragment [Ala16]Aβ(1–28) did not block peptide self-assembly, but resulted in the formation of small numbers of thin protofilaments without significant lateral aggregation and globular aggregates ~7–10 nm diameter size. This observation suggested that Lys16 was involved in formation of mature, amyloid-type fibrils. The effect of Lys16 on fibril nucleation appeared minimal and its effects in full-length Aβ were not studied.

17.8 Current Therapies for AD

AD is a progressive, fatal disorder without cure. The mechanism(s) of the disease is not entirely understood, impeding development of effective therapy. Current treatments provide short-term alleviation of some symptoms, but generally have poor efficacy and can be accompanied by moderate to severe side-effects [678–680].

The "amyloid cascade hypothesis" [3, 53] posits that self-assembly of Aβ initiates a cascade of events leading to changes in cell membrane conductivity, production of ROS [245], tau hyperphosphorylation [681], glutamatergic excitotoxicity [682], deficits in neurotransmitters, including acetylcholine [683], norepinephrine [684] and serotonin [685], inflammation [686], and apoptosis [687]. These events cause impairment of neuronal function and neuron death, producing the symptoms of AD. Other hypotheses ascribe etiologic primacy to oxidative stress [244] or related metabolic changes [243, 249] resulting from aging and argue that Aβ-mediated effects occur secondarily to these initial insults.

Wide acceptance of the amyloid cascade hypothesis has focused therapeutic efforts on inhibiting formation, or facilitating elimination, of neurotoxic Aβ assemblies. Leading approaches are Aβ immunization [688], inhibition of Aβ production [689, 690], enhancement of Aβ clearance by specific proteases or general proteolytic mechanisms [691] and prevention of Aβ assembly using a wide variety of chemicals (Section 17.7). Other approaches include the use of neuroprotective agents [692], antioxidants [693], anti-inflammatory agents [694], hormone replacement therapy [695], cholinesterase inhibitors [696] and NMDA receptor antagonists [794].

17.8.1 Approved Drugs

17.8.1.1 AChE Inhibitors (AChEIs)

The most commonly used drugs for treatment of AD are cholinesterase inhibitors. During the disease, degeneration of cholinergic nuclei localized in the basal forebrain occurs. Impairment of the cholinergic system, which projects into large areas of the limbic system and the neocortex, is followed by disturbance of attentional processes and cognitive decline. In the absence of better therapeutic alternatives, treatment of AD-induced cholinergic deficiency was considered an attractive and reasonable strategy in the 1980s [697]. By 1993, execution of the strategy resulted in FDA approval of the AChEI Tacrine, the first drug specifically approved for the treatment of AD [698]. Tacrine provides modest symptomatic relief in mild to moderate cases of AD, but like other AChEIs, it can cause severe side-effects, including nausea, vomiting, diarrhea, constipation, headache, dizziness and sleep disturbance [699, 700]. New AChEIs that cause less severe side-effects have been developed in the last decade [679], including Donepezil [701], Galantamine [702], Rivastigmine [703] and Metrifonate [704].

Mechanistically, AChEI increase non-amyloidogenic AβPP processing and production of AβPP$_{s\alpha}$ [203, 705–707], an activity that likely contributes to their beneficial influence on patients with AD and which is mediated through modulation of several downstream signaling pathways [708–711]. Nevertheless, the effects of AChEIs on memory and cognitive abilities are modest and limited in time to 12–36 months, after which efficacy is lost [696, 712].

17.8.1.2 Memantine

Memantine is a low-to-moderate affinity, uncompetitive NMDA receptor antagonist that appears to block pathological, but not physiological, activation of NMDA receptors [713, 714]. The drug may interfere with glutamatergic excitotoxicity and provide temporary relief to AD patients with moderate-to-severe disease. Memantine has been shown to cause modest improvement in clinical symptoms in severe stages of AD and may retard disease progression. Side-effects of memantine are minimal. Strong voltage dependency and rapid blocking/unblocking kinetics are thought to be the basis for memantine's clinical tolerability. Clinical studies demonstrate positive effects of memantine in AD both as a monotherapy and in combination with AChEI treatment [715]. Memantine offers a complementary therapeutic approach to AChEI, which may slow down the progress of AD, but like AChEI it is not curative.

17.8.1.3 Antioxidants

Excessive accumulation of ROS contributes to neuronal loss and dysfunction, and has been implicated in many studies as one pathological mechanism of AD [716, 717]. Based on these observations, vitamin E, a potent antioxidant and neuroprotective agent, is often prescribed as a treatment for AD. A clinical trial of vitamin E in patients with moderately advanced AD was conducted by the Alzheimer's Disease Cooperative Study [718]. The results indicated that vitamin E may slow functional brain deterioration, confirming results of previous trials [719]. The selective monoamine oxidase inhibitor selegiline also slowed down AD progression, but its combination with vitamin E did not improve the outcome relative to each drug alone [719].

17.8.2 Clinical Studies

17.8.2.1 Immunotherapy

Based on initial data from *in vitro* experiments [720, 721] followed by studies in rodents [722], immunization with Aβ has been a very active area of research. Active and passive immunization strategies have been shown to reduce AD-like pathology and restore cognitive deficits in transgenic mice, rabbits, guinea pigs and monkeys [688, 723, 724]. These results evoked substantial optimism. Unfortunately, phase IIa clinical trials, in which patients with AD were immunized with Aβ(1–42) in adjuvant QS-21 (preparation AN-1792) were halted because a small but significant number of patients developed meningoencephalitis [725, 726]. Post-mortem examination of two treated patients revealed few plaques in the neocortex and no dystrophic neurites or reactive astrocytes (as compared with unimmunized controls). Reactive microglia associated with areas devoid of plaques were not observed [727]. These results suggest that an effective immune response was generated in these patients, resulting in clearance of Aβ plaques. In addition, patients who produced antibodies in response to immunization with AN-1792 exhibited slower rates of cognitive decline [728]. Additional studies now must be performed to understand the mechanistic basis of the iatrogenic problems and to develop safer vaccines for future use [729].

17.8.2.2 **Statins**

Statins are molecules that lower plasma levels of LDL and cholesterol, and increase HDL levels. Statins have been in use for many years in the treatment of cardiovascular and other diseases, and are considered safe medications [730]. Recent studies now suggest that statins may be of use for the prevention of AD. It should be noted that these studies involved relatively small numbers of subjects and therefore that the data extant are insufficient to justify the use of statins in the general, non-demented population without hyperlipidemia [731]. Large-scale, placebo-controlled clinical trials will be required to establish the efficacy and safety of statins in prevention and treatment of AD.

17.8.2.3 **Chelation Therapy**

In vitro and *in vivo* data have suggested that formation of neurotoxic assemblies of Aβ is mediated by interaction of Aβ with ions of transition metals such as copper and zinc [603, 732]. This hypothesis led to *in vitro* studies of the effects of chelators on Aβ fibril formation and dissociation [733, 734]. These studies showed that chelators could block fibril formation *in vitro* [615], dissociate *ex vivo* amyloid [627] and inhibit Aβ accumulation *in vivo* [628, 735]. Based on these results, clinical trials were initiated to determine if chelators could be used to treat AD [736]. The chelator chosen for study was clioquinol, a copper and zinc-chelating agent that had been in use for more than 70 years in the treatment of amoebic dysentery. The drug had been in disfavor since the 1970s because many patients developed blindness and paralysis following treatment [737, 738]. Further investigation revealed that these severe adverse effects resulted from deficiency in vitamin B_{12} [739] and could be prevented by vitamin B_{12} supplementation. Recently, a phase II clinical trial of clioquinol with vitamin B_{12} supplementation showed that the cognitive abilities of treated patients stabilized and their plasma Aβ(1–42) levels were reduced relative to a placebo group [736]. These results suggest that clioquinol, or other chelators, may hold promise as therapeutic agents for AD.

17.8.2.4 Hormone Replacement Therapy

Epidemiologic studies have shown a higher prevalence of AD in women than in men, suggesting a link between gonadal hormone levels and AD [740]. In view of these studies, and of evidence supporting a role for estrogen in brain regions involved in learning and memory and in the protection and regulation of cholinergic neurons, hormone replacement therapy has been studied with respect to its ability to decrease the risk for, or delay the onset of, AD in post-menopausal women. Unfortunately, recent trials have suggested that estrogen treatment has no significant effect on the clinical course of AD in elderly women with the disease [741]. Moreover, the Women's Health Initiative study of estrogen plus medroxyprogesterone acetate showed an increased risk of dementia among post-menopausal women who showed no cognitive deficits before entering the trial and were in the active-treatment group [742, 743]. Hormone replacement therapy thus does not appear to be a viable approach for the treatment or prevention of AD.

17.8.2.5 Anti-inflammatory Drugs

Interest in anti-inflammatory drugs has been driven by observations of inflammation in brain regions affected by AD [744, 745] and by studies showing that certain non-steroidal anti-inflammatory drugs (NSAIDs) reduce the likelihood of developing AD [746, 747]. One mechanism by which the NSAID effect is mediated has been suggested by the finding that a subset of NSAIDs specifically reduces Aβ42 levels [748, 749], possibly through interaction with the γ-secretase complex [750] and/or the small G-protein Rho [751]. Long-term use of NSAIDs has been found to reduce AD risk and greater risk reduction was observed among patients with longer NSAID treatment histories [752]. However, other studies have failed to show clear beneficial effects of NSAID treatment [753–755]. These conflicting data suggest that simple anti-inflammatory treatment of AD patients may be ineffective and that better understanding of the mechanisms by which NSAIDs affect AβPP processing will be required if the approach is to be successful.

17.8.2.6 Natural Products

Several herbal remedies and dietary supplements have been promoted for the treatment of AD and related disorders. Claims about the safety and effectiveness of these products are based largely on testimonials, tradition and a small body of scientific research [756]. These products include coenzyme Q_{10} (ubiquinone), Ginko biloba, the moss extract huperzine A, phosphatidylserine and "coral" calcium. The latter product differs from pure calcium salts only in the composition (unknown) of trace contaminants. Clinical studies are being conducted with the former four compounds [757–760]. Thus far, data exist supporting a small beneficial effect of Ginko biloba [761, 762].

17.8.3 Pre-clinical Studies

Strategies to prevent Aβ production through inhibition of β- or γ-secretases are being pursued actively [689, 690, 763, 764]. Challenges with respect to bioavailability, BBB penetration and selectivity of the inhibitors currently prevent the use of such compounds in clinical trials. Other strategies attempt to enhance α-secretase activity [60, 765] or Aβ clearance [92]. Because Aβ may have a normal physiologic role, attempts to lower its levels may produce undesirable side-effects. Targeting assembly of neurotoxic Aβ assemblies, a wholly pathologic process, does not raise this problem.

17.8.4 Accelerating Progress Toward a Cure

Although AD has no cure, current clinical and pre-clinical research efforts hold promise for the eventual development of efficacious therapies for this devastating disease. This process would be accelerated if effective diagnostic and prognostic tests were available. AD diagnosis has been an ongoing challenge. To date, definitive diagnosis of AD has required post-mortem neuropathologic observation of extracellular amyloid plaques and intracellular neurofibrillary tangles. To enable the prevention and treatment of AD, the disease must be diagnosed at the earliest stage possible, preferably no later than the onset of mild cognitive impairment (MCI) [766]. The diagnosis method must have high specificity and sensitivity, and preferably be non-invasive. Studies show that MCI typically leads to development of AD; however, the symptoms of MCI are difficult to distinguish from normal aging and from other dementing diseases [767].

Imaging techniques may become particularly valuable in AD diagnosis. Currently, magnetic resonance imaging (MRI) and computed tomography are used primarily to rule out dementia due to causes other than AD, such as brain tumors, infarcts, damage from head trauma or accumulation of CSF [768–771]. Detection of brain atrophy using these techniques suggests a neurodegenerative condition and atrophy in particular brain regions may help determine the cause of dementia. However, some atrophy may exist in older age even in the absence of disease and an individual may have AD or a related disorder with no noticeable sign of atrophy. Functional imaging, using functional MRI (fMRI), single-photon emission computed tomography (SPECT) or positron emission tomography (PET), provides images of brain function and assists in the differential diagnosis of dementia [772–776]. Several studies have demonstrated differences between structural images of brains in people with and without AD [777, 778]. These images also may help confirm a diagnosis when symptoms are subtle.

Despite advances in imaging techniques, reliable, early-stage diagnosis of AD remains difficult. Amyloid imaging, in combination with behavioral and cognitive testing, may provide an improved, non-invasive diagnostic method for AD [779, 780]. A relatively unexplored diagnostic and prognostic area is the quantitation of neurotoxic Aβ oligomers. Detection of elevated levels of these oligomers in plasma or CSF may provide highly valuable information about disease status. This approach is the object of active study [781–784].

17.9 Concluding Remarks

A remarkable amount of knowledge has been produced in studies of the relationship of Aβ to AD. For the first time since Alois Alzheimer's case presentation in 1906 [6], therapeutic strategies are being developed targeting what may be the proximate neuropathogenetic processes. Clinical trials will establish whether these approaches are efficacious. In addition, and independent of the results of these trials, biochemical and biophysical studies of the folding and assembly of Aβ have contributed significantly to the elucidation of fundamental principles of protein self-assembly. These studies also support the hypothesis that formation of β-sheet structures is an intrinsic property of proteins that is evolutionarily conserved. In fact, recent work in bacterial, yeast and mammalian systems shows that amyloid fibril formation can be a normal, physiologically beneficial process (for a brief review, see [795]). We expect that continued, rigorous, unbiased study of the structural and functional dynamics of the β-sheet conformation will produce additional important and unexpected outcomes.

References

1 Seubert, P., Vigo-Pelfrey, C., Esch, F., Lee, M., Dovey, H., Davis, D., Sinha, S., Schlossmacher, M. G., et al. Isolation and quantitation of soluble Alzheimer's β-peptide from biological fluids. *Nature* **1992**, *359*, 325–327.

2 Shoji, M., Golde, T. E., Ghiso, J., Cheung, T. T., Estus, S., Shaffer, L. M., Cai, X., McKay, D. M., et al. Production of the Alzheimer amyloid protein by normal proteolytic processing. *Science* **1992**, *258*, 126–129.

3 Hardy, J. A. and Higgins, G. A. Alzheimer's disease: the amyloid cascade hypothesis. *Science* **1992**, *256*, 184–185.

4 Selkoe, D. J. Alzheimer's disease: genes, proteins, and therapy. *Physiol Rev* **2001**, *81*, 741–766.

5 Hardy, J. and Selkoe, D. J. The amyloid hypothesis of Alzheimer's disease: progress and problems on the road to therapeutics. *Science* **2002**, *297*, 353–356.

6 Alzheimer, A. Über einen eigenartigen schweren Erkrankungsprozeß der Hirnrinde. *Neurologisches Centralblatt* **1906**, *23*, 1129–1136.

7 Selkoe, D. J. Alzheimer disease: mechanistic understanding predicts novel therapies. *Ann Intern Med* **2004**, *140*, 627–638.

8 Hardy, J. The relationship between amyloid and tau. *J Mol Neurosci* **2003**, *20*, 203–206.

9 Glenner, G. G. and Wong, C. W. Alzheimer's disease: initial report of the purification and characterization of a novel cerebrovascular amyloid protein. *Biochem Biophys Res Commun* **1984**, *120*, 885–890.

10 Glenner, G. G. and Wong, C. W. Alzheimer's disease and Down's syndrome: sharing of a unique cerebrovascular amyloid fibril protein. *Biochem Biophys Res Commun* **1984**, *122*, 1131–1135.

11 Kang, J., Lemaire, H.-G., Unterbeck, A., Salbaum, J. M., Masters, C. L., Grzeschik, K.-H., Multhaup, G., Beyreuther, K., et al. The precursor of Alzheimer's disease amyloid A4 protein resembles a cell-surface receptor. *Nature* **1987**, *325*, 733–736.

12 Goldgaber, D., Lerman, M. I., McBridge, O. W. V. S. and Gajdusek, D. C. Characterization and chromosomal localization of a cDNA encoding brain amyloid of Alzheimer's disease. *Science* **1987**, *235*, 877–880.

13 Robakis, N. K., Ramakrishna, N., Wolfe, G. and Wisniewski, H. M. Molecular cloning and characterization of a cDNA encoding the cerebrovascular and the neuritic plaque amyloid peptides. *Proc Natl Acad Sci USA* **1987**, *84*, 4190–4194.

14 Tanzi, R. E., Gusella, J. F., Watkins, P. C., Bruns, G. A. B., St George-Hyslop, P. H., Van Keuren, M. L., Patterson, D., Pagan, S., et al. Amyloid β-protein gene: cDNA, mRNA distribution, and genetic linkage near the Alzheimer locus. *Science* **1987**, *235*, 880–884.

15 Oliver, C. and Holland, A. J. Down's syndrome and Alzheimer's disease: a review. *Psychol Med* **1986**, *16*, 307–322.

16 Lemere, C. A., Blusztajn, J. K., Yamaguchi, H., Wisniewski, T., Saido, T. C. and Selkoe, D. J. Sequence of deposition of heterogeneous amyloid β-peptides and Apo E in Down syndrome – implications for initial events in amyloid plaque formation. *Neurobiol Dis* **1996**, *3*, 16–32.

17 Hyman, B. T. Molecular and anatomical studies in Alzheimer's disease. *Neurologia* **2001**, *16*, 100–104.

18 Benzing, W. C., Mufson, E. J. and Armstrong, D. M. Alzheimer's disease-like dystrophic neurites characteristically associated with senile plaques are not found within other neurodegenerative diseases unless amyloid β-protein deposition is present. *Brain Res* **1993**, *606*, 10–18.

19 Spires, T. L. and Hyman, B. T. Neuronal structure is altered by amyloid plaques. *Rev Neurosci* **2004**, *15*, 267–278.

20 McLaurin, J., Yang, D., Yip, C. M. and Fraser, P. E. Review: modulating factors in amyloid-β fibril formation. *J Struct Biol* **2000**, *130*, 259–270.

21 D'Andrea, M. R., Cole, G. M. and Ard, M. D. The microglial phagocytic role with specific plaque types in the Alzheimer disease brain. *Neurobiol Aging* **2004**, *25*, 675–683.

22 Pike, C. J., Walencewicz, A. J., Glabe, C. G. and Cotman, C. W. *In vitro* aging of β-amyloid protein causes peptide aggregation and neurotoxicity. *Brain Res* **1991**, *563*, 311–314.

23 Roher, A. E. β-amyloid from Alzheimer disease brains inhibits sprouting and survival of sympathetic neurons. *Biochem Biophys Res Commun* **1991**, *174*, 572–579.

24 Näslund, J., Haroutunian, V., Mohs, R., Davis, K. L., Davies, P., Greengard, P. and Buxbaum, J. D. Correlation between elevated levels of amyloid β-peptide in the brain and cognitive decline. *J Am Med Ass* **2000**, *283*, 1571–1577.

25 Chui, D. H., Tanahashi, H., Ozawa, K., Ikeda, S., Checler, F., Ueda, O., Suzuki, H., Araki, W., et al. Transgenic mice with Alzheimer presenilin 1 mutations show accelerated neurodegeneration without amyloid plaque formation. *Nat Med* **1999**, *5*, 560–564.

26 Giannakopoulos, P., Herrmann, F. R., Bussiere, T., Bouras, C., Kovari, E., Perl, D. P., Morrison, J. H., Gold, G., et al. Tangle and neuron numbers, but not amyloid load, predict cognitive status in Alzheimer's disease. *Neurology* **2003**, *60*, 1495–1500.

27 Kuo, Y. M., Emmerling, M. R., Vigo-Pelfrey, C., Kasunic, T. C., Kirkpatrick, J. B., Murdoch, G. H., Ball, M. J. and Roher, A. E. Water-soluble Aβ (N-40, N-42) oligomers in normal and Alzheimer disease brains. *J Biol Chem* **1996**, *271*, 4077–4081.

28 Roher, A. E., Chaney, M. O., Kuo, Y. M., Webster, S. D., Stine, W. B., Haverkamp, L. J., Woods, A. S., Cotter, R. J., et al. Mor-

phology and toxicity of Aβ(1–42) dimer derived from neuritic and vascular amyloid deposits of Alzheimer's disease. *J Biol Chem* **1996**, *271*, 20631–20635.

29 Lue, L.F., Kuo, Y.M., Roher, A.E., Brachova, L., Shen, Y., Sue, L., Beach, T., Kurth, J.H., et al. Soluble amyloid β peptide concentration as a predictor of synaptic change in Alzheimer's disease. *Am J Pathol* **1999**, *155*, 853–862.

30 McLean, C.A., Cherny, R.A., Fraser, F.W., Fuller, S.J., Smith, M.J., Beyreuther, K., Bush, A.I. and Masters, C.L. Soluble pool of Aβ amyloid as a determinant of severity of neurodegeneration in Alzheimer's disease. *Ann Neurol* **1999**, *46*, 860–866.

31 Mucke, L., Masliah, E., Yu, G. Q., Mallory, M., Rockenstein, E.M., Tatsuno, G., Hu, K., Kholodenko, D., et al. High-level neuronal expression of Aβ(1–42) in wild-type human amyloid protein precursor transgenic mice: synaptotoxicity without plaque formation. *J Neurosci* **2000**, *20*, 4050–4058.

32 Kirkitadze, M.D., Bitan, G. and Teplow, D.B. Paradigm shifts in Alzheimer's disease and other neurodegenerative disorders: the emerging role of oligomeric assemblies. *J Neurosci Res* **2002**, *69*, 567–577.

33 Klein, W.L., Stine, W.B., Jr and Teplow, D.B. Small assemblies of unmodified amyloid β-protein are the proximate neurotoxin in Alzheimer's disease. *Neurobiol Aging* **2004**, *25*, 569–580.

34 Walsh, D.M. and Selkoe, D.J. Deciphering the molecular basis of memory failure in Alzheimer's disease. *Neuron* **2004**, *44*, 181–193.

35 Hardy, J. and Selkoe, D.J. Medicine – the amyloid hypothesis of Alzheimer's disease: progress and problems on the road to therapeutics. *Science* **2002**, *297*, 353–356.

36 Klein, W.L., Krafft, G.A. and Finch, C.E. Targeting small Aβ oligomers: the solution to an Alzheimer's disease conundrum? *Trends Neurosci* **2001**, *24*, 219–224.

37 Small, D.H. The Sixth International Conference on Alzheimer's disease, Amsterdam, The Netherlands, July 1998. The amyloid cascade hypothesis debate: emerging consensus on the role of Aβ and amyloid in Alzheimer's disease. *Amyloid* **1998**, *5*, 301–304.

38 Haass, C. and Steiner, H. Protofibrils, the unifying toxic molecule of neurodegenerative disorders? *Nat Neurosci* **2001**, *4*, 859–860.

39 Taylor, B.M., Sarver, R.W., Fici, G., Poorman, R.A., Lutzke, B.S., Molinari, A., Kawabe, T., Kappenman, K., et al. Spontaneous aggregation and cytotoxicity of the β-amyloid $A\beta^{1-40}$: a kinetic model. *J Protein Chem* **2003**, *22*, 31–40.

40 Walsh, D.M., Klyubin, I., Fadeeva, J.V., Cullen, W.K., Anwyl, R., Wolfe, M.S., Rowan, M.J. and Selkoe, D.J. Naturally secreted oligomers of amyloid β protein potently inhibit hippocampal long-term potentiation *in vivo*. *Nature* **2002**, *416*, 535–539.

41 Bitan, G., Kirkitadze, M.D., Lomakin, A., Vollers, S.S., Benedek, G.B. and Teplow, D.B. Amyloid β-protein (Aβ) assembly: Aβ40 and Aβ42 oligomerize through distinct pathways. *Proc Natl Acad Sci USA* **2003**, *100*, 330–335.

42 Kirkitadze, M.D., Condron, M.M. and Teplow, D.B. Identification and characterization of key kinetic intermediates in amyloid β-protein fibrillogenesis. *J Mol Biol* **2001**, *312*, 1103–1119.

43 Oda, T., Wals, P., Osterburg, H.H., Johnson, S.A., Pasinetti, G.M., Morgan, T.E., Rozovsky, I., Stine, W.B., et al. Clusterin (ApoJ) alters the aggregation of amyloid β-peptide ($A\beta_{1-42}$) and forms slowly sedimenting Aβ complexes that cause oxidative stress. *Exp Neurol* **1995**, *136*, 22–31.

44 Lambert, M.P., Barlow, A.K., Chromy, B.A., Edwards, C., Freed, R., Liosatos, M., Morgan, T.E., Rozovsky, I., et al. Diffusible, nonfibrillar ligands derived from $A\beta_{1-42}$ are potent central nervous system neurotoxins. *Proc Natl Acad Sci USA* **1998**, *95*, 6448–6453.

45 Walsh, D.M., Lomakin, A., Benedek, G.B., Condron, M.M. and Teplow, D.B. Amyloid β-protein fibrillogenesis – detection of a protofibrillar intermediate. *J Biol Chem* **1997**, *272*, 22364–22372.

46 Harper, J.D., Wong, S.S., Lieber, C.M. and Lansbury, P.T. Observation of metastable Aβ amyloid protofibrils by atomic force microscopy. *Chem Biol* **1997**, *4*, 119–125.

47 Teplow, D. B. Structural and kinetic features of amyloid β-protein fibrillogenesis. *Amyloid: Int J Exp Clin Invest* **1998**, *5*, 121–142.

48 Hoshi, M., Sato, M., Matsumoto, S., Noguchi, A., Yasutake, K., Yoshida, N. and Sato, K. Spherical aggregates of β-amyloid (amylospheroid) show high neurotoxicity and activate tau protein kinase I/glycogen synthase kinase-3β. *Proc Natl Acad Sci USA* **2003**, *100*, 6370–6375.

49 Westlind-Danielsson, A. and Arnerup, G. Spontaneous *in vitro* formation of supramolecular β-amyloid structures, "βamy balls", by β-amyloid 1–40 peptide. *Biochemistry* **2001**, *40*, 14736–14743.

50 Walsh, D. M., Hartley, D. M., Kusumoto, Y., Fezoui, Y., Condron, M. M., Lomakin, A., Benedek, G. B., Selkoe, D. J., et al. Amyloid β-protein fibrillogenesis – structure and biological activity of protofibrillar intermediates. *J Biol Chem* **1999**, *274*, 25945–25952.

51 Hartley, D. M., Walsh, D. M., Ye, C. P. P., Diehl, T., Vasquez, S., Vassilev, P. M., Teplow, D. B. and Selkoe, D. J. Protofibrillar intermediates of amyloid β-protein induce acute electrophysiological changes and progressive neurotoxicity in cortical neurons. *J Neurosci* **1999**, *19*, 8876–8884.

52 Dahlgren, K. N., Manelli, A. M., Stine, W. B., Jr, Baker, L. K., Krafft, G. A. and LaDu, M. J. Oligomeric and fibrillar species of amyloid-β peptides differentially affect neuronal viability. *J Biol Chem* **2002**, *277*, 32046–32053.

53 Selkoe, D. J. Toward a comprehensive theory for Alzheimer's disease. Hypothesis: Alzheimer's disease is caused by the cerebral accumulation and cytotoxicity of amyloid β-protein. *Ann NY Acad Sci* **2000**, *924*, 17–25.

54 Luchsinger, J. A. and Mayeux, R. Dietary factors and Alzheimer's disease. *Lancet Neurol* **2004**, *3*, 579–587.

55 Ling, Y., Morgan, K. and Kalsheker, N. Amyloid precursor protein (APP) and the biology of proteolytic processing: relevance to Alzheimer's disease. *Int J Biochem Cell Biol* **2003**, *35*, 1505–1535.

56 Haass, C., Hung, A. Y., Schlossmacher, M. G., Teplow, D. B. and Selkoe, D. J. β-amyloid peptide and a 3-kDa fragment are derived by distinct cellular mechanisms. *J Biol Chem* **1993**, *268*, 3021–3024.

57 Weidemann, A., Eggert, S., Reinhard, F. B. M., Vogel, M., Paliga, K., Baier, G., Masters, C. L., Beyreuther, K., et al. A novel ε-cleavage within the transmembrane domain of the Alzheimer amyloid precursor protein demonstrates homology with notch processing. *Biochemistry* **2002**, *41*, 2825–2835.

58 Kametani, F. Secretion of long Aβ-related peptides processed at ε-cleavage site is dependent on the α-secretase pre-cutting. *FEBS Lett* **2004**, *570*, 73–76.

59 Funamoto, S., Morishima-Kawashima, M., Tanimura, Y., Hirotani, N., Saido, T. C. and Ihara, Y. Truncated carboxyl-terminal fragments of β-amyloid precursor protein are processed to amyloid β-proteins 40 and 42. *Biochemistry* **2004**, *43*, 13532–13540.

60 Allinson, T. M., Parkin, E. T., Turner, A. J. and Hooper, N. M. ADAMs family members as amyloid precursor protein α-secretases. *J Neurosci Res* **2003**, *74*, 342–352.

61 Turner, A. J. and Hooper, N. M. Role for ADAM-family proteinases as membrane protein secretases. *Biochem Soc Trans* **1999**, *27*, 255–259.

62 Kimberly, W. T., LaVoie, M. J., Ostaszewski, B. L., Ye, W., Wolfe, M. S. and Selkoe, D. J. γ-secretase is a membrane protein complex comprised of presenilin, nicastrin, Aph-1, and Pen-2. *Proc Natl Acad Sci USA* **2003**, *100*, 6382–6387.

63 Zhao, G., Mao, G., Tan, J., Dong, Y., Cui, M. Z., Kim, S. H. and Xu, X. Identification of a new presenilin-dependent ζ-cleavage site within the transmembrane domain of amyloid precursor protein. *J Biol Chem* **2004**, *279*, 50647–50650.

64 Hussain, I., Powell, D., Howlett, D. R., Tew, D. G., Week, T. D., Chapman, C., Gloger, I. S., Murphy, K. E., et al. Identification of a novel aspartic protease (Asp 2) as β-secretase. *Mol Cell Neurosci* **1999**, *14*, 419–427.

65 Sinha, S., Anderson, J. P., Barbour, R., Basi, G. S., Caccavello, R., Davis, D.,

Doan, M., Dovey, H. F., et al. Purification and cloning of amyloid precursor protein β-secretase from human brain. *Nature* **1999**, *402*, 537–540.

66 Vassar, R., Bennett, B. D., Babu-Khan, S., Kahn, S., Mendiaz, E. A., Denis, P., Teplow, D. B., Ross, S., et al. β-secretase cleavage of Alzheimer's amyloid precursor protein by the transmembrane aspartic protease BACE. *Science* **1999**, *286*, 735–741.

67 Yan, R. Q., Bienkowski, M. J., Shuck, M. E., Miao, H. Y., Tory, M. C., Pauley, A. M., Brashler, J. R., Stratman, N. C., et al. Membrane-anchored aspartyl protease with Alzheimer's disease β-secretase activity. *Nature* **1999**, *402*, 533–537.

68 Simons, M., de Strooper, B., Multhaup, G., Tienari, P. J., Dotti, C. G. and Beyreuther, K. Amyloidogenic processing of the human amyloid precursor protein in primary cultures of rat hippocampal neurons. *J Neurosci* **1996**, *16*, 899–908.

69 Haass, C., Schlossmacher, M. G., Hung, A. Y., Vigo-Pelfrey, C., Mellon, A., Ostaszewski, B. L., Lieberburg, I., Koo, E. H., et al. Amyloid β-peptide is produced by cultured cells during normal metabolism. *Nature* **1992**, *359*, 322–325.

70 Busciglio, J., Gabuzda, D. H., Matsudaira, P. and Yankner, B. A. Generation of β-amyloid in the secretory pathway in neuronal and nonneuronal cells. *Proc Natl Acad Sci USA* **1993**, *90*, 2092–2096.

71 Ikonen, E. Roles of lipid rafts in membrane transport. *Curr Opin Cell Biol* **2001**, *13*, 470–477.

72 Wahrle, S., Das, P., Nyborg, A. C., McLendon, C., Shoji, M., Kawarabayashi, T., Younkin, L. H., Younkin, S. G., et al. Cholesterol-dependent γ-secretase activity in buoyant cholesterol-rich membrane microdomains. *Neurobiol Dis* **2002**, *9*, 11–23.

73 Ehehalt, R., Keller, P., Haass, C., Thiele, C. and Simons, K. Amyloidogenic processing of the Alzheimer β-amyloid precursor protein depends on lipid rafts. *J Cell Biol* **2003**, *160*, 113–123.

74 Vetrivel, K. S., Cheng, H., Lin, W., Sakurai, T., Li, T., Nukina, N., Wong, P. C., Xu, H., et al. Association of γ-secretase with lipid rafts in post-Golgi and endosome membranes. *J Biol Chem* **2004**, *279*, 44945–44954.

75 Sisodia, S. S. β-amyloid precursor protein cleavage by a membrane-bound protease. *Proc Natl Acad Sci USA* **1992**, *89*, 6075–6079.

76 Parvathy, S., Hussain, I., Karran, E. H., Turner, A. J. and Hooper, N. M. Cleavage of Alzheimer's amyloid precursor protein by α-secretase occurs at the surface of neuronal cells. *Biochemistry* **1999**, *38*, 9728–9734.

77 Klausner, R. D., Donaldson, J. G. and Lippincott-Schwartz, J. Brefeldin A: insights into the control of membrane traffic and organelle structure. *J Cell Biol* **1992**, *116*, 1071–1080.

78 Chyung, A. S. C., Greenberg, B. D., Cook, D. G., Doms, R. W. and Lee, V. M. Y. Novel β-secretase cleavage of β-amyloid precursor protein in the endoplasmic reticulum intermediate compartment of NT2N cells. *J Cell Biol* **1997**, *138*, 671–680.

79 Skovronsky, D. M., Pijak, D. S., Doms, R. W. and Lee, V. M. Y. A distinct ER/IC γ-secretase competes with the proteasome for cleavage of AβPP. *Biochemistry* **2000**, *39*, 810–817.

80 Greenfield, J. P., Tsai, J., Gouras, G. K., Hai, B., Thinakaran, G., Checler, F., Sisodia, S. S., Greengard, P., et al. Endoplasmic reticulum and *trans*-Golgi network generate distinct populations of Alzheimer β-amyloid peptides. *Proc Natl Acad Sci USA* **1999**, *96*, 742–747.

81 Wilson, C. A., Doms, R. W., Zheng, H. and Lee, V. M. Presenilins are not required for Aβ42 production in the early secretory pathway. *Nat Neurosci* **2002**, *29*, 29.

82 Puglielli, L., Tanzi, R. E. and Kovacs, D. M. Alzheimer's disease: the cholesterol connection. *Nat Neurosci* **2003**, *6*, 345–351.

83 Corder, E. H., Saunders, A. M., Strittmatter, W. J., Schmechel, D. E., Gaskell, P. C., Small, G. W., Roses, A. D., Haines, J. L., et al. Gene dose of apolipoprotein E type 4 allele and the risk of Alzheimer's disease in late onset families. *Science* **1993**, *261*, 921–923.

84 Tanzi, R. E. and Bertram, L. New frontiers in Alzheimer's disease genetics. *Neuron* **2001**, *32*, 181–184.

85 Ladu, M.J., Pederson, T.M., Frail, D.E., Reardon, C.A., Getz, G.S. and Falduto, M.T. Purification of Apolipoprotein E attenuates isoform-specific binding to β-amyloid. *J Biol Chem* **1995**, *270*, 9039–9042.

86 Kang, D.E., Pietrzik, C.U., Baum, L., Chevallier, N., Merriam, D.E., Kounnas, M.Z., Wagner, S.L., Troncoso, J.C., et al. Modulation of amyloid β-protein clearance and Alzheimer's disease susceptibility by the LDL receptor-related protein pathway. *J Clin Invest* **2000**, *106*, 1159–1166.

87 Shibata, M., Yamada, S., Kumar, S.R., Calero, M., Bading, J., Frangione, B., Holtzman, D.M., Miller, C.A., et al. Clearance of Alzheimer's amyloid-β(1–40) peptide from brain by LDL receptor-related protein-1 at the blood–brain barrier. *J Clin Invest* **2000**, *106*, 1489–1499.

88 Refolo, L.M., Pappolla, M.A., Malester, B., LaFrancois, J., Bryant-Thomas, T., Wang, R., Tint, G.S., Sambamurti, K., et al. Hypercholesterolemia accelerates the Alzheimer's amyloid pathology in a transgenic mouse model. *Neurobiol Dis* **2000**, *7*, 321–331.

89 Kuo, Y.M., Emmerling, M.R., Bisgaier, C.L., Essenburg, A.D., Lampert, H.C., Drumm, D. and Roher, A.E. Elevated low-density lipoprotein in Alzheimer's disease correlates with brain Aβ1–42 levels. *Biochem Biophys Res Commun* **1998**, *252*, 711–715.

90 Mann, K.M., Thorngate, F.E., Katoh-Fukui, Y., Hamanaka, H., Williams, D.L., Fujita, S. and Lamb, B.T. Independent effects of APOE on cholesterol metabolism and brain Aβ levels in an Alzheimer disease mouse model. *Hum Mol Genet* **2004**, *13*, 1959–1968.

91 Zlokovic, B.V. Clearing amyloid through the blood–brain barrier. *J Neurochem* **2004**, *89*, 807–811.

92 Tanzi, R.E., Moir, R.D. and Wagner, S.L. Clearance of Alzheimer's Aβ peptide: the many roads to perdition. *Neuron* **2004**, *43*, 605–608.

93 Van Uden, E., Mallory, M., Veinbergs, I., Alford, M., Rockenstein, E. and Masliah, E. Increased extracellular amyloid deposition and neurodegeneration in human amyloid precursor protein transgenic mice deficient in receptor-associated protein. *J Neurosci* **2002**, *22*, 9298–9304.

94 Turner, A.J. Exploring the structure and function of zinc metallopeptidases: old enzymes and new discoveries. *Biochem Soc Trans* **2003**, *31*, 723–727.

95 Carson, J.A. and Turner, A.J. β-Amyloid catabolism: roles for neprilysin (NEP) and other metallopeptidases? *J Neurochem* **2002**, *81*, 1–8.

96 Selkoe, D.J. Clearing the brain's amyloid cobwebs. *Neuron* **2001**, *32*, 177–180.

97 Prince, J.A., Feuk, L., Gu, H.F., Johansson, B., Gatz, M., Blennow, K. and Brookes, A.J. Genetic variation in a haplotype block spanning IDE influences Alzheimer disease. *Hum Mutat* **2003**, *22*, 363–371.

98 Qiu, W.Q., Walsh, D.M., Ye, Z., Vekrellis, K., Zhang, J.M., Podlisny, M.B., Rosner, M.R., Safavi, A., et al. Insulin-degrading enzyme regulates extracellular levels of amyloid β-protein by degradation. *J Biol Chem* **1998**, *273*, 32730–32738.

99 Farris, W., Mansourian, S., Chang, Y., Lindsley, L., Eckman, E.A., Frosch, M.P., Eckman, C.B., Tanzi, R.E., et al. Insulin-degrading enzyme regulates the levels of insulin, amyloid β-protein, and the β-amyloid precursor protein intracellular domain *in vivo*. *Proc Natl Acad Sci USA* **2003**, *100*, 4162–4167.

100 Leissring, M.A., Farris, W., Chang, A.Y., Walsh, D.M., Wu, X., Sun, X., Frosch, M.P. and Selkoe, D.J. Enhanced proteolysis of β-amyloid in AβPP transgenic mice prevents plaque formation, secondary pathology, and premature death. *Neuron* **2003**, *40*, 1087–1093.

101 Iwata, N., Mizukami, H., Shirotani, K., Takaki, Y., Muramatsu, S., Lu, B., Gerard, N.P., Gerard, C., et al. Presynaptic localization of neprilysin contributes to efficient clearance of amyloid-β peptide in mouse brain. *J Neurosci* **2004**, *24*, 991–998.

102 Fukami, S., Watanabe, K., Iwata, N., Haraoka, J., Lu, B., Gerard, N.P., Gerard, C., Fraser, P., et al. Aβ-degrading endopeptidase, neprilysin, in mouse brain: synaptic and axonal localization inversely correlating with Aβ pathology. *Neurosci Res Suppl* **2002**, *43*, 39–56.

103 Shirotani, K., Tsubuki, S., Iwata, N., Takaki, Y., Harigaya, W., Maruyama, K., Kiryu-Seo, S., Kiyama, H., et al. Neprilysin degrades both amyloid β peptides 1–40 and 1–42 most rapidly and efficiently among thiorphan- and phosphoramidon-sensitive endopeptidases. *J Biol Chem* **2001**, *276*, 21895–21901.

104 Iwata, N., Tsubuki, S., Takaki, Y., Shirotani, K., Lu, B., Gerard, N. P., Gerard, C., Hama, E., et al. Metabolic regulation of brain Aβ by neprilysin. *Science* **2001**, *292*, 1550–1552.

105 Marr, R. A., Rockenstein, E., Mukherjee, A., Kindy, M. S., Hersh, L. B., Gage, F. H., Verma, I. M. and Masliah, E. Neprilysin gene transfer reduces human amyloid pathology in transgenic mice. *J Neurosci* **2003**, *23*, 1992–1996.

106 Kanemitsu, H., Tomiyama, T. and Mori, H. Human neprilysin is capable of degrading amyloid β peptide not only in the monomeric form but also the pathological oligomeric form. *Neurosci Lett* **2003**, *350*, 113–116.

107 Tuppo, E. E. and Arias, H. R. The role of inflammation in Alzheimer's disease. *Int J Biochem Cell Biol* **2005**, *37*, 289–305.

108 Rogers, J., Strohmeyer, R., Kovelowski, C. J. and Li, R. Microglia and inflammatory mechanisms in the clearance of amyloid β peptide. *Glia* **2002**, *40*, 260–269.

109 Bard, F., Barbour, R., Cannon, C., Carretto, R., Fox, M., Games, D., Guido, T., Hoenow, K., et al. Epitope and isotype specificities of antibodies to β-amyloid peptide for protection against Alzheimer's disease-like neuropathology. *Proc Natl Acad Sci USA* **2003**, *100*, 2023–2028.

110 Frautschy, S. A., Cole, G. M. and Baird, A. Phagocytosis and deposition of vascular β-amyloid in rat brains injected with Alzheimer β-amyloid. *Am J Pathol* **1992**, *140*, 1389–1399.

111 Weldon, D. T., Rogers, S. D., Ghilardi, J. R., Finke, M. P., Cleary, J. P., Ohare, E., Esler, W. P., Maggio, J. E., et al. Fibrillar β-amyloid induces microglial phagocytosis, expression of inducible nitric oxide synthase, and loss of a select population of neurons in the rat CNS *in vivo*. *J Neurosci* **1998**, *18*, 2161–2173.

112 Ard, M. D., Cole, G. M., Wei, J., Mehrle, A. P. and Fratkin, J. D. Scavenging of Alzheimer's amyloid β-protein by microglia in culture. *J Neurosci Res* **1996**, *43*, 190–202.

113 Bard, F., Cannon, C., Barbour, R., Burke, R. L., Games, D., Grajeda, H., Guido, T., Hu, K., et al. Peripherally administered antibodies against amyloid β-peptide enter the central nervous system and reduce pathology in a mouse model of Alzheimer disease. *Nat Med* **2000**, *6*, 916–919.

114 Wyss-Coray, T., Loike, J. D., Brionne, T. C., Lu, E., Anankov, R., Yan, F. R., Silverstein, S. C. and Husemann, J. Adult mouse astrocytes degrade amyloid-β *in vitro* and *in situ*. *Nat Med* **2003**, *9*, 453–457.

115 Guenette, S. Y. Astrocytes: a cellular player in Aβ clearance and degradation. *Trends Mol Med* **2003**, *9*, 279–280.

116 Nagele, R. G., Wegiel, J., Venkataraman, V., Imaki, H. and Wang, K. C. Contribution of glial cells to the development of amyloid plaques in Alzheimer's disease. *Neurobiol Aging* **2004**, *25*, 663–674.

117 Vehmas, A. K., Kawas, C. H., Stewart, W. F. and Troncoso, J. C. Immune reactive cells in senile plaques and cognitive decline in Alzheimer's disease. *Neurobiol Aging* **2003**, *24*, 321–331.

118 Nagele, R. G., D'Andrea, M. R., Lee, H., Venkataraman, V. and Wang, H. Y. Astrocytes accumulate Aβ42 and give rise to astrocytic amyloid plaques in Alzheimer disease brains. *Brain Res* **2003**, *971*, 197–209.

119 Lucassen, P. J., Chung, W. C., Kamphorst, W. and Swaab, D. F. DNA damage distribution in the human brain as shown by *in situ* end labeling; area-specific differences in aging and Alzheimer disease in the absence of apoptotic morphology. *J Neuropathol Exp Neurol* **1997**, *56*, 887–900.

120 Guo, Q., Sebastian, L., Sopher, B. L., Miller, M. W., Ware, C. B., Martin, G. M. and Mattson, M. P. Increased vulnerability of hippocampal neurons from presenilin-1 mutant knock-in mice to amyloid β-peptide toxicity: central roles of superoxide production and caspase activation. *J Neurochem* **1999**, *72*, 1019–1029.

121 Behl, C., Davis, J. B., Klier, F. G. and Schubert, D. Amyloid β peptide induces necrosis rather than apoptosis. *Brain Res* **1994**, *645*, 253–264.

122 Su, J. H., Anderson, A. J., Cummings, B. J. and Cotman, C. W. Immunohistochemical evidence for apoptosis in Alzheimer's disease. *Neuroreport* **1994**, *5*, 2529–2533.

123 Kitamura, Y., Shimohama, S., Kamoshima, W., Ota, T., Matsuoka, Y., Nomura, Y., Smith, M. A., Perry, G., et al. Alteration of proteins regulating apoptosis, bcl-2, bcl-x, bax, bak, bad, ich-1 and cpp32, in Alzheimer's disease. *Brain Res* **1998**, *780*, 260–269.

124 Troy, C. M., Rabacchi, S. A., Friedman, W. J., Frappier, T. F., Brown, K. and Shelanski, M. L. Caspase-2 mediates neuronal cell death induced by β-amyloid. *J Neurosci* **2000**, *20*, 1386–1392.

125 Troy, C. M., Rabacchi, S. A., Xu, Z. H., Maroney, A. C., Connors, T. J., Shelanski, M. L. and Greene, L. A. β-Amyloid-induced neuronal apoptosis requires c-Jun N-terminal kinase activation. *J Neurochem* **2001**, *77*, 157–164.

126 Klyubin, I., Walsh, D. M., Cullen, W. K., Fadeeva, J. V., Anwyl, R., Selkoe, D. J. and Rowan, M. J. Soluble Arctic amyloid β protein inhibits hippocampal long-term potentiation *in vivo*. *Eur J Neurosci* **2004**, *19*, 2839–2846.

127 Chen, Q. S., Kagan, B. L., Hirakura, Y. and Xie, C. W. Impairment of hippocampal long-term potentiation by Alzheimer amyloid β-peptides. *J Neurosci Res* **2000**, *60*, 65–72.

128 Nalbantoglu, J., Tiradosantiago, G., Lahsaini, A., Poirier, J., Goncalves, O., Verge, A., Momoli, F., Welner, S. A., et al. Impaired learning and LTP in mice expressing the carboxy terminus of the Alzheimer amyloid precursor protein. *Nature* **1997**, *387*, 500–505.

129 Rowan, M. J., Klyubin, I., Cullen, W. K. and Anwyl, R. Synaptic plasticity in animal models of early Alzheimer's disease. *Phil Trans R Soc Lond B* **2003**, *358*, 821–828.

130 Auld, D. S., Kornecook, T. J., Bastianetto, S. and Quirion, R. Alzheimer's disease and the basal forebrain cholinergic system: relations to β-amyloid peptides, cognition, and treatment strategies. *Prog Neurobiol* **2002**, *68*, 209–245.

131 Kihara, T. and Shimohama, S. Alzheimer's disease and acetylcholine receptors. *Acta Neurobiol Exp (Wars)* **2004**, *64*, 99–105.

132 Kim, J. H., Anwyl, R., Suh, Y. H., Djamgoz, M. B. A. and Rowan, M. J. Use-dependent effects of amyloidogenic fragments of β-amyloid precursor protein on synaptic plasticity in rat hippocampus *in vivo*. *J Neurosci* **2001**, *21*, 1327–1333.

133 Chen, Q. S., Wei, W. Z., Shimahara, T. and Xie, C. W. Alzheimer amyloid β-peptide inhibits the late phase of long-term potentiation through calcineurin-dependent mechanisms in the hippocampal dentate gyrus. *Neurobiol Learn Mem* **2002**, *77*, 354–371.

134 Rowan, M. J., Klyubin, I., Wang, Q. and Anwyl, R. Mechanisms of the inhibitory effects of amyloid β-protein on synaptic plasticity. *Exp Gerontol* **2004**, *39*, 1661–1667.

135 Hynd, M. R., Scott, H. L. and Dodd, P. R. Selective loss of NMDA receptor NR1 subunit isoforms in Alzheimer's disease. *J Neurochem* **2004**, *89*, 240–247.

136 Hynd, M. R., Scott, H. L. and Dodd, P. R. Differential expression of *N*-methyl-D-aspartate receptor NR2 isoforms in Alzheimer's disease. *J Neurochem* **2004**, *90*, 913–919.

137 Hynd, M. R., Scott, H. L. and Dodd, P. R. Glutamate (NMDA) receptor NR1 subunit mRNA expression in Alzheimer's disease. *J Neurochem* **2001**, *78*, 175–182.

138 Wu, J., Anwyl, R. and Rowan, M. J. β-Amyloid selectively augments NMDA receptor-mediated synaptic transmission in rat hippocampus. *Neuroreport* **1995**, *6*, 2409–2413.

139 Miguel-Hidalgo, J. J., Alvarez, X. A., Cacabelos, R. and Quack, G. Neuroprotec-

tion by memantine against neurodegeneration induced by β-amyloid(1–40). *Brain Res* **2002**, *958*, 210–221.

140 Wang, H.Y., Lee, D.H.S., D'Andrea, M.R., Peterson, P.A., Shank, R.P. and Reitz, A.B. β-amyloid (1–42) binds to α7 nicotinic acetylcholine receptor with high affinity – implications for Alzheimer's disease pathology. *J Biol Chem* **2000**, *275*, 5626–5632.

141 Wang, H.Y., Lee, D.H., Davis, C.B. and Shank, R.P. Amyloid peptide Aβ(1–42) binds selectively and with picomolar affinity to α7 nicotinic acetylcholine receptors. *J Neurochem* **2000**, *75*, 1155–1161.

142 Castegna, A., Aksenov, M., Aksenova, M., Thongboonkerd, V., Klein, J.B., Pierce, W.M., Booze, R., Markesbery, W.R., et al. Proteomic identification of oxidatively modified proteins in Alzheimer's disease brain. Part I: creatine kinase BB, glutamine synthase, and ubiquitin carboxy-terminal hydrolase L-1. *Free Rad Biol Med* **2002**, *33*, 562–571.

143 Sayre, L.M., Zelasko, D.A., Harris, P.L., Perry, G., Salomon, R.G. and Smith, M.A. 4-Hydroxynonenal-derived advanced lipid peroxidation end products are increased in Alzheimer's disease. *J Neurochem* **1997**, *68*, 2092–2097.

144 Sun, A.Y., Draczynska-Lusiak, B. and Sun, G.Y. Oxidized lipoproteins, β amyloid peptides and Alzheimer's disease. *Neurotox Res* **2001**, *3*, 167–178.

145 Butterfield, D.A. and Lauderback, C.M. Lipid peroxidation and protein oxidation in Alzheimer's disease brain: potential causes and consequences involving amyloid β-peptide-associated free radical oxidative stress. *Free Rad Biol Med* **2002**, *32*, 1050–1060.

146 Mecocci, P., MacGarvey, U. and Beal, M.F. Oxidative damage to mitochondrial DNA is increased in Alzheimer's disease. *Ann Neurol* **1994**, *36*, 747–751.

147 Shan, X., Tashiro, H. and Lin, C.L.G. The identification and characterization of oxidized RNAs in Alzheimer's disease. *J Neurosci* **2003**, *23*, 4913–4921.

148 Lyras, L., Perry, R.H., Perry, E.K., Ince, P.G., Jenner, A., Jenner, P. and Halliwell, B. Oxidative damage to proteins, lipids, and DNA in cortical brain regions from patients with dementia with Lewy bodies. *J Neurochem* **1998**, *71*, 302–312.

149 Butterfield, D.A., Drake, J., Pocernich, C. and Castegna, A. Evidence of oxidative damage in Alzheimer's disease brain: central role for amyloid β-peptide. *Trends Mol Med* **2001**, *7*, 548–554.

150 Blass, J.P. Brain metabolism and brain disease: is metabolic deficiency the proximate cause of Alzheimer dementia? *J Neurosci Res* **2001**, *66*, 851–856.

151 Marlatt, M., Lee, H.G., Perry, G., Smith, M.A. and Zhu, X. Sources and mechanisms of cytoplasmic oxidative damage in Alzheimer's disease. *Acta Neurobiol Exp (Wars)* **2004**, *64*, 81–87.

152 Casley, C.S., Canevari, L., Land, J.M., Clark, J.B. and Sharpe, M.A. β-Amyloid inhibits integrated mitochondrial respiration and key enzyme activities. *J Neurochem* **2002**, *80*, 91–100.

153 Gsell, W., Conrad, R., Hickethier, M., Sofic, E., Frolich, L., Wichart, I., Jellinger, K., Moll, G., et al. Decreased catalase activity but unchanged superoxide dismutase activity in brains of patients with dementia of Alzheimer type. *J Neurochem* **1995**, *64*, 1216–1223.

154 Behl, C., Davis, J.B., Lesley, R. and Schubert, D. Hydrogen peroxide mediates amyloid β protein toxicity. *Cell* **1994**, *77*, 817–827.

155 Takahashi, M., Dore, S., Ferris, C.D., Tomita, T., Sawa, A., Wolosker, H., Borchelt, D.R., Iwatsubo, T., et al. Amyloid precursor proteins inhibit heme oxygenase activity and augment neurotoxicity in Alzheimer's disease. *Neuron* **2000**, *28*, 461–473.

156 Castegna, A., Thongboonkerd, V., Klein, J. B., Lynn, B., Markesbery, W.R. and Butterfield, D.A. Proteomic identification of nitrated proteins in Alzheimer's disease brain. *J Neurochem* **2003**, *85*, 1394–1401.

157 Keil, U., Bonert, A., Marques, C.A., Scherping, I., Weyermann, J., Strosznajder, J.B., Muller-Spahn, F., Haass, C., et al. Amyloid β-induced changes in nitric oxide production and mitochondrial activity lead to apoptosis. *J Biol Chem* **2004**, *279*, 50310–50320.

158 Mattson, M. P. and Chan, S. L. Calcium orchestrates apoptosis. *Nat Cell Biol* **2003**, *5*, 1041–1043.

159 Mattson, M. P., Fu, W., Waeg, G. and Uchida, K. 4-Hydroxynonenal, a product of lipid peroxidation, inhibits dephosphorylation of the microtubule-associated protein tau. *Neuroreport* **1997**, *8*, 2275–2281.

160 Bruce-Keller, A. J., Li, Y. J., Lovell, M. A., Kraemer, P. J., Gary, D. S., Brown, R. R., Markesbery, W. R. and Mattson, M. P. 4-Hydroxynonenal, a product of lipid peroxidation, damages cholinergic neurons and impairs visuospatial memory in rats. *J Neuropathol Exp Neurol* **1998**, *57*, 257–267.

161 Price, S. A., Held, B. and Pearson, H. A. Amyloid β protein increases Ca^{2+} currents in rat cerebellar granule neurones. *Neuroreport* **1998**, *9*, 539–545.

162 Lashuel, H. A., Hartley, D., Petre, B. M., Walz, T. and Lansbury, P. T. Neurodegenerative disease – amyloid pores from pathogenic mutations. *Nature* **2002**, *418*, 291.

163 Kayed, R., Sokolov, Y., Edmonds, B., MacIntire, T. M., Milton, S. C., Hall, J. E. and Glabe, C. G. Permeabilization of lipid bilayers is a common conformation-dependent activity of soluble amyloid oligomers in protein misfolding diseases. *J Biol Chem* **2004**, *279*, 46363–46366.

164 Kawahara, M., Kuroda, Y., Arispe, N. and Rojas, E. Alzheimer's β-amyloid, human islet amylin, and prion protein fragment evoke intracellular free calcium elevations by a common mechanism in a hypothalamic GnRH neuronal cell line. *J Biol Chem* **2000**, *275*, 14077–14083.

165 Kagan, B. L., Hirakura, Y., Azimov, R., Azimova, R. and Lin, M. C. The channel hypothesis of Alzheimer's disease: current status. *Peptides* **2002**, *23*, 1311–1315.

166 LaFerla, F. M. Calcium dyshomeostasis and intracellular signalling in Alzheimer's disease. *Nat Rev Neurosci* **2002**, *3*, 862–872.

167 Forman, M. S., Lee, V. M. and Trojanowski, J. Q. "Unfolding" pathways in neurodegenerative disease. *Trends Neurosci* **2003**, *26*, 407–410.

168 Imaizumi, K., Miyoshi, K., Katayama, T., Yoneda, T., Taniguchi, M., Kudo, T. and Tohyama, M. The unfolded protein response and Alzheimer's disease. *Biochim Biophys Acta Mol Basis Dis* **2001**, *31*, 2–3.

169 Katayama, T., Imaizumi, K., Sato, N., Miyoshi, K., Kudo, T., Hitomi, J., Morihara, T., Yoneda, T., et al. Presenilin-1 mutations downregulate the signalling pathway of the unfolded-protein response. *Nat Cell Biol* **1999**, *1*, 479–485.

170 Katayama, T., Imaizumi, K., Honda, A., Yoneda, T., Kudo, T., Takeda, M., Mori, K., Rozmahel, R., et al. Disturbed activation of endoplasmic reticulum stress transducers by familial Alzheimer's disease-linked presenilin-1 mutations. *J Biol Chem* **2001**, *276*, 43446–43454.

171 Steiner, H., Winkler, E., Shearman, M. S., Prywes, R. and Haass, C. Endoproteolysis of the ER stress transducer ATF6 in the presence of functionally inactive presenilins. *Neurobiol Dis* **2001**, *8*, 717–722.

172 Yasuda, Y., Kudo, T., Katayama, T., Imaizumi, K., Yatera, M., Okochi, M., Yamamori, H., Matsumoto, N., et al. FAD-linked presenilin-1 mutants impede translation regulation under ER stress. *Biochem Biophys Res Commun* **2002**, *296*, 313–318.

173 Meda, L., Baron, P. and Scarlato, G. Glial activation in Alzheimer's disease: the role of Aβ and its associated proteins. *Neurobiol Aging* **2001**, *22*, 885–893.

174 McGeer, P. L. and McGeer, E. G. Inflammation, autotoxicity and Alzheimer disease. *Neurobiol Aging* **2001**, *22*, 799–809.

175 Pak, K., Chan, S. L. and Mattson, M. P. Presenilin-1 mutation sensitizes oligodendrocytes to glutamate and amyloid toxicities, and exacerbates white matter damage and memory impairment in mice. *Neuromol Med* **2003**, *3*, 53–64.

176 Benveniste, E. N., Nguyen, V. T. and O'Keefe, G. M. Immunological aspects of microglia: relevance to Alzheimer's disease. *Neurochem Int* **2001**, *39*, 381–391.

177 Combs, C. K., Johnson, D. E., Cannady, S. B., Lehman, T. M. and Landreth, G. E. Identification of microglial signal transduction pathways mediating a neurotoxic response to amyloidogenic fragments of -amyloid and prion proteins. *J Neurosci* **1999**, *19*, 928–939.

178 Selkoe, D. J., Podlisny, M. B., L, J. C., Vickers, E. A., Lee, G., Fritz, L. C. and Oltersdorf, T. β-amyloid precursor protein of Alzheimer disease occurs as 110–135 kilodalton membrane-associated proteins in neural and nonneural tissues. *Proc Natl Acad Sci USA* **1988**, *85*, 7341–7345.

179 Haass, C., Hung, A. Y. and Selkoe, D. J. Processing of β-amyloid precursor protein in microglia and astrocytes favors a localization in internal vesicles over constitutive secretion. *J Neurosci* **1991**, *11*, 3783–3793.

180 Smith, R. P., Higuchi, D. A. and Broze, G. J., Jr. Platelet coagulation factor XIa-inhibitor, a form of Alzheimer amyloid precursor protein. *Science* **1990**, *248*, 1126–1128.

181 Tarassishin, L., Yin, Y. I., Bassit, B. and Li, Y. M. Processing of Notch and amyloid precursor protein by γ-secretase is spatially distinct. *Proc Natl Acad Sci USA* **2004**, *101*, 17050–17055.

182 Chyung, J. H., Raper, D. M. and Selkoe, D. J. γ-Secretase exists on the plasma membrane as an intact complex that accepts substrates and effects intramembrane cleavage. *J Biol Chem* **2005**, *280*, 4383–4392.

183 Wasco, W., Bupp, K., Magendantz, M., Gusella, J., Tanzi, R. E. and Solomon, F. Identification of a mouse brain cDNA that encodes a protein related to the Alzheimer disease-associated amyloid β-protein precursor. *Proc Natl Acad Sci USA* **1992**, *89*, 10758–10762.

184 Sprecher, C. A., Grant, F. J., Grimm, G., O'Hara, P. J., Norris, F., Norris, K. and Foster, D. C. Molecular cloning of the cDNA for a human amyloid precursor protein homolog: evidence for a multigene family. *Biochemistry* **1993**, *32*, 4481–4486.

185 Slunt, H. H., Thinakaran, G., Von Koch, C., Lo, A. C., Tanzi, R. E. and Sisodia, S. S. Expression of a ubiquitous, cross-reactive homologue of the mouse β-amyloid precursor protein (AβPP). *J Biol Chem* **1994**, *269*, 2637–2644.

186 Hung, A. Y., Koo, E. H., Haass, C. and Selkoe, D. J. Increased expression of β-amyloid precursor protein during neuronal differentiation is not accompanied by secretory cleavage. *Proc Natl Acad Sci USA* **1992**, *89*, 9439–9443.

187 Salbaum, J. M. and Ruddle, F. H. Embryonic expression pattern of amyloid protein precursor suggests a role in differentiation of specific subsets of neurons. *J Exp Zool* **1994**, *269*, 116–127.

188 Zheng, H., Jiang, M. H., Trumbauer, M. E., Sirinathsinghji, D. J. S., Hopkins, R., Smith, D. W., Heavens, R. P., Dawson, G. R., et al. β-Amyloid precursor protein-deficient mice show reactive gliosis and decreased locomotor activity. *Cell* **1995**, *81*, 525–531.

189 Magara, F., Muller, U., Li, Z. W., Lipp, H. P., Weissmann, C., Stagljar, M. and Wolfer, D. P. Genetic background changes the pattern of forebrain commissure defects in transgenic mice underexpressing the β-amyloid-precursor protein. *Proc Natl Acad Sci USA* **1999**, *96*, 4656–4661.

190 Mucke, L., Abraham, C. R. and Masliah, E. Neurotrophic and neuroprotective effects of hAPP in transgenic mice. *Ann NY Acad Sci* **1996**, *777*, 82–88.

191 Zhang, F., Eckman, C., Younkin, S., Hsiao, K. K. and Iadecola, C. Increased susceptibility to ischemic brain damage in transgenic mice overexpressing the amyloid precursor protein. *J Neurosci* **1997**, *17*, 7655–7661.

192 White, A. R., Zheng, H., Galatis, D., Maher, F., Hesse, L., Multhaup, G., Beyreuther, K., Masters, C. L., et al. Survival of cultured neurons from amyloid precursor protein knock-out mice against Alzheimer's amyloid-β toxicity and oxidative stress. *J Neurosci* **1998**, *18*, 6207–6217.

193 von Koch, C. S., Zheng, H., Chen, H., Trumbauer, M., Thinakaran, G., van der Ploeg, L. H., Price, D. L. and Sisodia, S. S. Generation of APLP2 KO mice and early postnatal lethality in APLP2/AβPP double KO mice. *Neurobiol Aging* **1997**, *18*, 661–669.

194 Heber, S., Herms, J., Gajic, V., Hainfellner, J., Aguzzi, A., Rulicke, T., von Kretzschmar, H., von Koch, C., et al. Mice with combined gene knock-outs reveal essential and partially redundant functions of amyloid precursor protein family members. *J Neurosci* **2000**, *20*, 7951–7963.

195 Ohyagi, Y. and Tabira, T. Effect of growth factors and cytokines on expression of amyloid β protein precursor mRNAs in cultured neural cells. *Brain Res Mol Brain Res* **1993**, *18*, 127–132.

196 Nakamura, Y., Takeda, M., Niigawa, H., Hariguchi, S. and Nishimura, T. Amyloid β-protein precursor deposition in rat hippocampus lesioned by ibotenic acid injection. *Neurosci Lett* **1992**, *136*, 95–98.

197 Milward, E. A., Papadopoulos, R., Fuller, S. J., Moir, R. D., Small, D., Beyreuther, K. and Masters, C. L. The amyloid protein precursor of Alzheimer's disease is a mediator of the effects of nerve growth factor on neurite outgrowth. *Neuron* **1992**, *9*, 129–137.

198 Qiu, W. Q., Ferreira, A., Miller, C., Koo, E. H. and Selkoe, D. J. Cell-surface β-amyloid precursor protein stimulates neurite outgrowth of hippocampal neurons in an isoform-dependent manner. *J Neurosci* **1995**, *15*, 2157–2167.

199 Schubert, D., Jin, L.-W., Saitoh, T. and Cole, G. The regulation of amyloid β protein precursor secretion and its modulatory role in cell adhesion. *Neuron* **1989**, *3*, 689–694.

200 Schubert, D. and Behl, C. The expression of amyloid β protein precursor protects nerve cells from β-amyloid and glutamate toxicity and alters their interaction with the extracellular matrix. *Brain Res* **1993**, *629*, 275–282.

201 Scheuner, D., Eckman, C., Jensen, M., Song, X., Citron, M., Suzuki, N., Bird, T. D., Hardy, J., et al. Secreted amyloid β-protein similar to that in the senile plaques of Alzheimer's disease is increased *in vivo* by the Presenilin 1 and 2 and APP mutations linked to familial Alzheimer's disease. *Nat Med* **1996**, *2*, 864–870.

202 Araki, W., Kitaguchi, N., Tokushima, Y., Ishii, K., Aratake, H., Shimohama, S., Nakamura, S. and Kimura, J. Trophic effect of β-amyloid precursor protein on cerebral cortical neurons in culture. *Biochem Biophys Res Commun* **1991**, *181*, 265–271.

203 Salvietti, N., Cattaneo, E., Govoni, S. and Racchi, M. Changes in β amyloid precursor protein secretion associated with the proliferative status of CNS derived progenitor cells. *Neurosci Lett* **1996**, *212*, 199–203.

204 Mattson, M., Culwell, A. R., Esch, F. S., Lieberburg, I. and Rydel, R. Evidence for neuroprotective and intraneuronal calcium-regulating roles for secreted forms of the β-amyloid precursor protein. *Neuron* **1993**, *10*, 243–254.

205 Ohsawa, I., Hirose, Y., Ishiguro, M., Imai, Y., Ishiura, S. and Kohsaka, S. Expression, purification, and neurotrophic activity of amyloid precursor protein-secreted forms produced by yeast. *Biochem Biophys Res Commun* **1995**, *213*, 52–58.

206 Ohsawa, I., Takamura, C. and Kohsaka, S. The amino-terminal region of amyloid precursor protein is responsible for neurite outgrowth in rat neocortical explant culture. *Biochem Biophys Res Commun* **1997**, *236*, 59–65.

207 Ohsawa, I., Takamura, C., Morimoto, T., Ishiguro, M. and Kohsaka, S. Amino-terminal region of secreted form of amyloid precursor protein stimulates proliferation of neural stem cells. *Eur J Neurosci* **1999**, *11*, 1907–1913.

208 Caille, I., Allinquant, B., Dupont, E., Bouillot, C., Langer, A., Muller, U. and Prochiantz, A. Soluble form of amyloid precursor protein regulates proliferation of progenitors in the adult subventricular zone. *Development* **2004**, *131*, 2173–2181.

209 Essalmani, R., Macq, A. F., Mercken, L. and Octave, J. N. Missense mutations associated with familial Alzheimer's disease in Sweden lead to the production of the amyloid peptide without internalization of its precursor. *Biochem Biophys Res Commun* **1996**, *218*, 89–96.

210 Koo, E. H. and Squazzo, S. L. Evidence that production and release of amyloid β-protein involves the endocytic pathway. *J Biol Chem* **1994**, *269*, 17386–17389.

211 Leblanc, A. C. and Gambetti, P. Production of Alzheimer 4kda β-amyloid peptide requires the C-terminal cytosolic domain of the amyloid precursor protein. *Biochem Biophys Res Commun* **1994**, *204*, 1371–1380.

212 Perez, R. G., Soriano, S., Hayes, J. D., Ostaszewski, B., Xia, W. M., Selkoe, D. J., Chen, X. H., Stokin, G. B., et al. Mutagenesis identifies new signals for β-amyloid precursor protein endocytosis, turnover, and the generation of secreted fragments, including Aβ42. *J Biol Chem* **1999**, *274*, 18851–18856.

213 Lai, A., Gibson, A., Hopkins, C. R. and Trowbridge, I. S. Signal-dependent trafficking of β-amyloid precursor protein-transferrin receptor chimeras in Madin-Darby canine kidney cells. *J Biol Chem* **1998**, *273*, 3732–3739.

214 King, G. D. and Scott Turner, R. Adaptor protein interactions: modulators of amyloid precursor protein metabolism and Alzheimer's disease risk? *Exp Neurol* **2004**, *185*, 208–219.

215 Borg, J. P., Ooi, J., Levy, E. and Margolis, B. The phosphotyrosine interaction domains of X11 and Fe65 bind to distinct sites on the YENPTY motif of amyloid precursor protein. *Mol Cell Biol* **1996**, *16*, 6229–6241.

216 McLoughlin, D. M. and Miller, C. C. J. The intracellular cytoplasmic domain of the Alzheimer's disease amyloid precursor protein interacts with phosphotyrosine-binding domain proteins in the yeast two-hybrid system. *FEBS Lett* **1996**, *397*, 197–200.

217 Tomita, S., Ozaki, T., Taru, H., Oguchi, S., Takeda, S., Yagi, Y., Sakiyama, S., Kirino, Y., et al. Interaction of a neuron-specific protein containing PDZ domains with Alzheimer's amyloid precursor protein. *J Biol Chem* **1999**, *274*, 2243–2254.

218 King, G. D., Perez, R. G., Steinhilb, M. L., Gaut, J. R. and Turner, R. S. X11α modulates secretory and endocytic trafficking and metabolism of amyloid precursor protein: mutational analysis of the YENPTY sequence. *Neuroscience* **2003**, *120*, 143–154.

219 King, G. D., Cherian, K. and Turner, R. S. X11α impairs γ- but not β-cleavage of amyloid precursor protein. *J Neurochem* **2004**, *88*, 971–982.

220 Chang, Y., Tesco, G., Jeong, W. J., Lindsley, L., Eckman, E. A., Eckman, C. B., Tanzi, R. E. and Guenette, S. Y. Generation of the β-amyloid peptide and the amyloid precursor protein C-terminal fragment γ are potentiated by FE65L1. *J Biol Chem* **2003**, *278*, 51100–51107.

221 Tanahashi, H. and Tabira, T. Characterization of an amyloid precursor protein-binding protein Fe65L2 and its novel isoforms lacking phosphotyrosine-interaction domains. *Biochem J* **2002**, *367*, 687–695.

222 Minopoli, G., de Candia, P., Bonetti, A., Faraonio, R., Zambrano, N. and Russo, T. The β-amyloid precursor protein functions as a cytosolic anchoring site that prevents Fe65 nuclear translocation. *J Biol Chem* **2001**, *276*, 6545–6550.

223 Cao, X. and Sudhof, T. C. A transcriptionally active complex of AβPP with Fe65 and histone acetyltransferase Tip60. *Science* **2001**, *293*, 115–120.

224 Kimberly, W. T., Zheng, J. B., Guenette, S. Y. and Selkoe, D. J. The intracellular domain of the β-amyloid precursor protein is stabilized by Fe65 and translocates to the nucleus in a Notch-like manner. *J Biol Chem* **2001**, *276*, 40288–40292.

225 Kim, H. S., Kim, E. M., Lee, J. P., Park, C. H., Kim, S., Seo, J. H., Chang, K. A., Yu, E., et al. C-terminal fragments of amyloid precursor protein exert neurotoxicity by inducing glycogen synthase kinase-3β expression. *FASEB J* **2003**, *17*, 1951–1953.

226 von Rotz, R. C., Kohli, B. M., Bosset, J., Meier, M., Suzuki, T., Nitsch, R. M. and

Konietzko, U. The APP intracellular domain forms nuclear multiprotein complexes and regulates the transcription of its own precursor. *J Cell Sci* **2004**, *117*, 4435–4448.

227 Scheinfeld, M.H., Ghersi, E., Laky, K., Fowlkes, B.J. and D'Adamio, L. Processing of β-amyloid precursor-like protein-1 and -2 by γ-secretase regulates transcription. *J Biol Chem* **2002**, *277*, 44195–44201.

228 Matsuda, S., Yasukawa, T., Homma, Y., Ito, Y., Niikura, T., Hiraki, T., Hirai, S., Ohno, S., et al. c-Jun N-terminal kinase (JNK)-interacting protein-1b/islet-brain-1 scaffolds Alzheimer's amyloid precursor protein with JNK. *J Neurosci* **2001**, *21*, 6597–6607.

229 Scheinfeld, M.H., Roncarati, R., Vito, P., Lopez, P.A., Abdallah, M. and D'Adamio, L. Jun NH_2-terminal kinase (JNK) interacting protein 1 (JIP1) binds the cytoplasmic domain of the Alzheimer's β-amyloid precursor protein (AβPP). *J Biol Chem* **2002**, *277*, 3767–3775.

230 Inomata, H., Nakamura, Y., Hayakawa, A., Takata, H., Suzuki, T., Miyazawa, K. and Kitamura, N. A scaffold protein JIP-1b enhances amyloid precursor protein phosphorylation by JNK and its association with kinesin light chain 1. *J Biol Chem* **2003**, *278*, 22946–22955.

231 Scheinfeld, M. H., Matsuda, S. and D'Adamio, L. JNK-interacting protein-1 promotes transcription of Aβ protein precursor but not Aβ precursor-like proteins, mechanistically different than Fe65. *Proc Natl Acad Sci USA* **2003**, *100*, 1729–1734.

232 Matsuda, S., Matsuda, Y. and D'Adamio, L. Amyloid beta protein precursor (AβPP), but not AβPP-like protein 2, is bridged to the kinesin light chain by the scaffold protein JNK-interacting protein 1. *J Biol Chem* **2003**, *278*, 38601–38606.

233 Kamal, A., Almenar-Queralt, A., LeBlanc, J.F., Roberts, E.A. and Goldstein, L.S.B. Kinesin-mediated axonal transport of a membrane compartment containing β-secretase and presenilin-1 requires APP. *Nature* **2001**, *414*, 643–648.

234 Geddes, J.W., Anderson, K.J. and Cotman, C.W. Senile plaques as aberrant sprout-stimulating structures. *Exp Neurol* **1986**, *94*, 767–776.

235 Hyman, B.T., Kromer, L.J. and Van Hoesen, G.W. Reinnervation of the hippocampal perforant pathway zone in Alzheimer's disease. *Ann Neurol* **1987**, *21*, 259–267.

236 Whitson, J.S., Selkoe, D.J. and Cotman, C.W. Amyloid β protein enhances the survival of hippocampal neurons *in vitro*. *Science* **1989**, *243*, 1488–1490.

237 Whitson, J.S., Glabe, C.G., Shintani, E., Abcar, A. and Cotman, C.W. β-Amyloid protein promotes neuritic branching in hippocampal cultures. *Neurosci Lett* **1990**, *110*, 319–324.

238 Yankner, B.A., Duffy, L.K. and Kirschner, D.A. Neurotrophic and neurotoxic effects of amyloid β protein: reversal by tachykinin neuropeptides. *Science* **1990**, *250*, 279–282.

239 Reynolds, P.N., Holmes, M.D. and Scicchitano, R. Role of tachykinins in bronchial hyper-responsiveness. *Clin Exp Pharmacol Physiol* **1997**, *24*, 273–280.

240 Plant, L.D., Boyle, J.P., Smith, I.F., Peers, C. and Pearson, H.A. The production of amyloid β peptide is a critical requirement for the viability of central neurons. *J Neurosci* **2003**, *23*, 5531–5535.

241 Atwood, C.S., Robinson, S.R. and Smith, M.A. Amyloid-β: redox-metal chelator and antioxidant. *J Alzheimer's Dis* **2002**, *4*, 203–214.

242 Zou, K., Gong, J.S., Yanagisawa, K. and Michikawa, M. A novel function of monomeric amyloid β-protein serving as an antioxidant molecule against metal-induced oxidative damage. *J Neurosci* **2002**, *22*, 4833–4841.

243 Atwood, C.S., Obrenovich, M.E., Liu, T., Chan, H., Perry, G., Smith, M.A. and Martins, R.N. Amyloid-β: a chameleon walking in two worlds: a review of the trophic and toxic properties of amyloid-β. *Brain Res Brain Res Rev* **2003**, *43*, 1–16.

244 Lee, H.G., Casadesus, G., Zhu, X., Takeda, A., Perry, G. and Smith, M.A. Challenging the amyloid cascade hypothesis: senile plaques and amyloid-β as protective adaptations to Alzheimer disease. *Ann NY Acad Sci* **2004**, *1019*, 1–4.

245 Squier, T.C. Oxidative stress and protein aggregation during biological aging. *Exp Gerontol* **2001**, *36*, 1539–1550.

246 Koudinov, A.R., Berezov, T.T. and Koudinova, N.V. The levels of soluble amyloid-β in different high density lipoprotein subfractions distinguish Alzheimer's and normal aging cerebrospinal fluid: implication for brain cholesterol pathology? *Neurosci Lett* **2001**, *314*, 115–118.

247 Wu, J., Anwyl, R. and Rowan, M.J. β-Amyloid selectively augments NMDA receptor-mediated synaptic transmission in rat hippocampus. *Neuroreport* **1995**, *6*, 2409–2413.

248 Wu, J., Anwyl, R. and Rowan, M.J. β-Amyloid-(1–40) increases long-term potentiation in rat hippocampus *in vitro*. *Eur J Pharmacol* **1995**, *284*, R1-R3.

249 Koudinov, A.R. and Berezov, T.T. Alzheimer's amyloid-β (Aβ) is an essential synaptic protein, not neurotoxic junk. *Acta Neurobiol Exp (Wars)* **2004**, *64*, 71–79.

250 Janssen, J.C., Beck, J.A., Campbell, T.A., Dickinson, A., Fox, N.C., Harvey, R.J., Houlden, H., Rossor, M.N., et al. Early onset familial Alzheimer's disease – Mutation frequency in 31 families. *Neurology* **2003**, *60*, 235–239.

251 Hashimoto, M., Hori, Y., Yamada, K., Wakutani, Y., Condron, M.C., Tsubuki, S., Saido, T.C., Teplow, D.B., et al. APP (H6R and D7N) mutations linked to familial Alzheimer's disease alter Aβ amyloid assembly but not APP processing. *Soc Neurosci Abstr* **2004**, 218.-214.

252 Wakutani, Y., Watanabe, K., Adachi, Y., Wada-Isoe, K., Urakami, K., Ninomiya, H., Saido, T.C., Hashimoto, T., et al. Novel amyloid precursor protein gene missense mutation (D678N) in probable familial Alzheimer's disease. *J Neurol Neurosurg Psychiatry* **2004**, *75*, 1039–1042.

253 Levy, E., Carman, M.D., Fernandez-Madrid, I.J., Power, M.D., Lieberburg, I., van Duinen, S.G., Bots, G.T.A.M., Luyendijk, W., et al. Mutation of the Alzheimer's disease amyloid gene in hereditary cerebral hemorrhage, Dutch-type. *Science* **1990**, *248*, 1124–1126.

254 van Broeckhoven, C., Haan, J., Bakker, E., Hardy, J.A., Hul, W.V., Vegter-Van Der Vlis, M. and Roos, R.A.C. Amyloid β-protein precursor gene and hereditary cerebral hemorrhage with amyloidosis (Dutch). *Science* **1990**, *248*, 1120–1122.

255 Mann, D.M.A., Iwatsubo, T., Ihara, Y., Cairns, N.J., Lantos, P.L., Bogdanovic, N., Lannfelt, L., Winblad, B., et al. Predominant deposition of amyloid-β(42(43)) in plaques in cases of Alzheimer's disease and hereditary cerebral hemorrhage associated with mutations in the amyloid precursor protein gene. *Am J Pathol* **1996**, *148*, 1257–1266.

256 Natte, R., Maat-Schieman, M.L.C., Haan, J., Bornebroek, M., Roos, R.A.C. and van Duinen, S.G. Dementia in hereditary cerebral hemorrhage with amyloidosis-Dutch type is associated with cerebral amyloid angiopathy but is independent of plaques and neurofibrillary tangles. *Ann Neurol* **2001**, *50*, 765–772.

257 Bornebroek, M., De Jonghe, C., Haan, J., Kumar-Singh, S., Younkin, S., Roos, R. and Van Broeckhoven, C. Hereditary cerebral hemorrhage with amyloidosis Dutch type (AβPP 693): decreased plasma amyloid-β 42 concentration. *Neurobiol Dis* **2003**, *14*, 619–623.

258 De Jonghe, C., Zehr, C., Yager, D., Prada, C.M., Younkin, S., Hendriks, L., Van Broeckhoven, C. and Eckman, C.B. Flemish and Dutch mutations in amyloid β precursor protein have different effects on amyloid secretion. *Neurobiol Dis* **1998**, *5*, 281–286.

259 Nilsberth, C., Westlind-Danielsson, A., Eckman, C.B., Condron, M.M., Axelman, K., Forsell, C., Stenh, C., Luthman, J., et al. The 'Arctic' APP mutation (E693G) causes Alzheimer's disease by enhanced Aβ protofibril formation. *Nat Neurosci* **2001**, *4*, 887–893.

260 Herzig, M.C., Winkler, D.T., Burgermeister, P., Pfeifer, M., Kohler, E.,

Schmidt, S. D., Danner, S., Abramowski, D., et al. Aβ is targeted to the vasculature in a mouse model of hereditary cerebral hemorrhage with amyloidosis. *Nat Neurosci* **2004**, *7*, 954–960.

261 Clements, A., Allsop, D., Walsh, D. M. and Williams, C. H. Aggregation and metal-binding properties of mutant forms of the amyloid Aβ peptide of Alzheimer's disease. *J Neurochem* **1996**, *66*, 740–747.

262 Miravalle, L., Tokuda, T., Chiarle, R., Giaccone, G., Bugiani, O., Tagliavini, F., Frangione, B. and Ghiso, J. Substitutions at codon 22 of Alzheimer's Aβ peptide induce diverse conformational changes and apoptotic effects in human cerebral endothelial cells. *J Biol Chem* **2000**, *275*, 27110–27116.

263 Murakami, K., Irie, K., Morimoto, A., Ohigashi, H., Shindo, M., Nagao, M., Shimizu, T. and Shirasawa, T. Neurotoxicity and physicochemical properties of Aβ mutant peptides from cerebral amyloid angiopathy: implication for the pathogenesis of cerebral amyloid angiopathy and Alzheimer's disease. *J Biol Chem* **2003**, *278*, 46179–46187.

264 Soto, C., Castaño, E. M., Frangione, B. and Inestrosa, N. C. The α-helical to β-strand transition in the amino-terminal fragment of the amyloid β-peptide modulates amyloid formation. *J Biol Chem* **1995**, *270*, 3063–3067.

265 Dudek, S. M. and Johnson, G. Transglutaminase facilitates the formation of polymers of the β-amyloid peptide. *Brain Res* **1994**, *651*, 129–133.

266 Watson, D. J., Lander, A. D. and Selkoe, D. J. Heparin-binding properties of the amyloidogenic peptides Aβ and amylin. Dependence on aggregation state and inhibition by Congo red. *J Biol Chem* **1997**, *272*, 31617–31624.

267 Van Nostrand, W. E., Melchor, J. P. and Ruffini, L. Pathologic amyloid β-protein cell surface fibril assembly on cultured human cerebrovascular smooth muscle cells. *J Neurochem* **1998**, *70*, 216–223.

268 Melchor, J. P., McVoy, L. and Van Nostrand, W. E. Charge alterations of E22 enhance the pathogenic properties of the amyloid β-protein. *J Neurochem* **2000**, *74*, 2209–2212.

269 Verbeek, M. M., Dewaal, R. M. W., Schipper, J. J. and Vannostrand, W. E. Degeneration of cultured human brain pericytes by amyloid β protein. *J Neurochem* **1997**, *68*, 1135–1141.

270 Eisenhauer, P. B., Johnson, R. J., Wells, J. M., Davies, T. A. and Fine, R. E. Toxicity of various amyloid β peptide species in cultured human blood–brain barrier endothelial cells: increased toxicity of Dutch-type mutant. *J Neurosci Res* **2000**, *60*, 804–810.

271 Tsubuki, S., Takaki, Y. and Saido, T. C. Dutch, Flemish, Italian, and Arctic mutations of APP and resistance of Aβ to physiologically relevant proteolytic degradation. *Lancet* **2003**, *361*, 1957–1958.

272 Morelli, L., Llovera, R., Gonzalez, S. A., Affranchino, J. L., Prelli, F., Frangione, B., Ghiso, J. and Castano, E. M. Differential degradation of amyloid β genetic variants associated with hereditary dementia or stroke by insulin-degrading enzyme. *J Biol Chem* **2003**, *278*, 23221–23226.

273 Hendriks, L., van Duijn, C. M., Cras, P., Cruts, M., Van Hul, W., van Harskamp, F., Warren, A., McInnis, M. G., et al. Presenile dementia and cerebral haemorrhage linked to a mutation at codon 692 of the β-amyloid precursor protein gene. *Nat Genet* **1992**, *1*, 218–221.

274 Cras, P., van Harskamp, F., Hendriks, L., Ceuterick, C., van Duijn, C. M., Stefanko, S. Z., Hofman, A., Kros, J. M., et al. Presenile Alzheimer dementia characterized by amyloid angiopathy and large amyloid core type senile plaques in the APP 692Ala → Gly mutation. *Acta Neuropathol (Berl)* **1998**, *96*, 253–260.

275 Kumar-Singh, S., Cras, P., Wang, R., Kros, J. M., van Swieten, J., Lubke, U., Ceuterick, C., Serneels, S., et al. Dense-core senile plaques in the Flemish variant of Alzheimer's disease are vasocentric. *Am J Pathol* **2002**, *161*, 507–520.

276 Haass, C., Hung, A. Y., Selkoe, D. J. and Teplow, D. B. Mutations associated with a locus for familial Alzheimer's disease result in alternative processing of amyloid β-protein precursor. *J Biol Chem* **1994**, *269*, 17741–17748.

277 Kumar-Singh, S., Dewachter, I., Moechars, D., Lubke, U., De Jonghe, C., Ceuterick, C., Checler, F., Naidu, A., et al. Behavioral disturbances without amyloid deposits in mice overexpressing human amyloid precursor protein with Flemish (A692G) or Dutch (E693Q) mutation. *Neurobiol Dis* **2000**, *7*, 9–22.

278 Walsh, D.M., Hartley, D.M., Condron, M.M., Selkoe, D.J. and Teplow, D.B. *In vitro* studies of amyloid β-protein fibril assembly and toxicity provide clues to the aetiology of Flemish variant ($Ala^{692} \rightarrow Gly$) Alzheimer's disease. *Biochem J* **2001**, *355*, 869–877.

279 Stenh, C., Nilsberth, C., Hammarback, J., Engvall, B., Naslund, J. and Lannfelt, L. The Arctic mutation interferes with processing of the amyloid precursor protein. *Neuroreport* **2002**, *13*, 1857–1860.

280 Bugiani, O., Padovani, A., Magoni, M., Andora, G., Sgarzi, M., Savoiardo, M., Bizzi, A., Giaccone, G., et al. An Italian type of HCHWA. *Neurobiol Aging* **1998**, *19*, S238.

281 Tagliavini, F., Rossi, G., Padovani, A., Magoni, M., Andora, G., Sgarzi, M., Bizzi, A., Savioardo, M., et al. A new βpp mutation related to hereditary cerebral hemorrhage. *Alzheimer's Rep* **1999**, *2 (Suppl)*, S28.

282 Grabowski, T.J., Cho, H.S., Vonsattel, J.P.G., Rebeck, G.W. and Greenberg, S.M. Novel amyloid precursor protein mutation in an Iowa family with dementia and severe cerebral amyloid angiopathy. *Ann Neurol* **2001**, *49*, 697–705.

283 Van Nostrand, W.E., Melchor, J.P., Cho, H.S., Greenberg, S.M. and Rebeck, G.W. Pathogenic effects of D23N Iowa mutant amyloid β-protein. *J Biol Chem* **2001**, *276*, 32860–32866.

284 Citron, M., Oltersdorf, T., Haass, C., McConlogue, L., Hung, A.Y., Seubert, P., Vigo-Pelfrey, C., Lieberburg, I., et al. Mutation of the β-amyloid precursor protein in familial Alzheimer's disease increases β-protein production. *Nature* **1992**, *360*, 672–674.

285 Younkin, S.G. Evidence that Aβ42 is the real culprit in Alzheimer's disease. *Ann Neurol* **1995**, *37*, 287–288.

286 Lichtenthaler, S.F., Wang, R., Grimm, H., Uljon, S.N., Masters, C.L. and Beyreuther, K. Mechanism of the cleavage specificity of Alzheimer's disease γ-secretase identified by phenylalanine-scanning mutagenesis of the transmembrane domain of the amyloid precursor protein. *Proc Natl Acad Sci USA* **1999**, *96*, 3053–3058.

287 Citron, M., Westaway, D., Xia, W.M., Carlson, G., Diehl, T., Levesque, G., Johnsonwood, K., Lee, M., et al. Mutant presenilins of Alzheimer's disease increase production of 42-residue amyloid β-protein in both transfected cells and transgenic mice. *Nat Med* **1997**, *3*, 67–72.

288 Xia, W.M., Zhang, J.M., Kholodenko, D., Citron, M., Podlisny, M.B., Teplow, D.B., Haass, C., Seubert, P., et al. Enhanced production and oligomerization of the 42-residue amyloid β-protein by Chinese hamster ovary cells stably expressing mutant presenilins. *J Biol Chem* **1997**, *272*, 7977–7982.

289 Borchelt, D.R., Ratovitski, T., Vanlare, J., Lee, M.K., Gonzales, V., Jenkins, N. A., Copeland, N.G., Price, D.L., et al. Accelerated amyloid deposition in the brains of transgenic mice coexpressing mutant presenilin 1 and amyloid precursor proteins. *Neuron* **1997**, *19*, 939–945.

290 Lemere, C.A., Lopera, F., Kosik, K.S., Lendon, C.L., Ossa, J., Saido, T.C., Yamaguchi, H., Ruiz, A., et al. The E280A presenilin 1 Alzheimer mutation produces increased Aβ42 deposition and severe cerebellar pathology. *Nat Med* **1996**, *2*, 1146–1150.

291 Wolfe, M.S. and Selkoe, D.J. Perspectives: biochemistry – intramembrane proteases – mixing oil and water. *Science* **2002**, *296*, 2156–2157.

292 Thinakaran, G., Borchelt, D.R., Lee, M.K., Slunt, H.H., Spitzer, L., Kim, G., Ratovitsky, T., Davenport, F., et al. Endoproteolysis of presenilin 1 and accumulation of processed derivatives *in vivo. Neuron* **1996**, *17*, 181–190.

293 Esler, W. P., Kimberly, W. T., Ostaszewski, B. L., Diehl, T. S., Moore, C. L., Tsai, J. Y., Rahmati, T., Xia, W. M., et al. Transition-state analogue inhibitors of γ-secretase bind directly to presenilin-1. *Nat Cell Biol* **2000**, *2*, 428–434.

294 De Strooper, B., Saftig, P., Craessaerts, K., Vanderstichele, H., Guhde, G., Annaert, W., Von Figura, K. and Van Leuven, F. Deficiency of presenilin-1 inhibits the normal cleavage of amyloid precursor protein. *Nature* **1998**, *391*, 387–390.

295 Wolfe, M. S., Xia, W. M., Ostaszewski, B. L., Diehl, T. S., Kimberly, W. T. and Selkoe, D. J. Two transmembrane aspartates in presenilin-1 required for presenilin endoproteolysis and γ-secretase activity. *Nature* **1999**, *398*, 513–517.

296 Weidemann, A., Paliga, K., Durrwang, U., Czech, C., Evin, G., Masters, C. L. and Beyreuther, K. Formation of stable complexes between two Alzheimer's disease gene products – presenilin-2 and β-amyloid precursor protein. *Nat Med* **1997**, *3*, 328–332.

297 Xia, W. M., Zhang, J. M., Perez, R., Koo, E. H. and Selkoe, D. J. Interaction between amyloid precursor protein and presenilins in mammalian cells – implications for the pathogenesis of Alzheimer's disease. *Proc Natl Acad Sci USA* **1997**, *94*, 8208–8213.

298 Weisgraber, K. H., Rall, S. C., Jr and Mahley, R. W. Human E apoprotein heterogeneity. Cysteine-arginine interchanges in the amino acid sequence of the apo-E isoforms. *J Biol Chem* **1981**, *256*, 9077–9083.

299 Rall, S. C., Jr, Weisgraber, K. H. and Mahley, R. W. Human apolipoprotein E. The complete amino acid sequence. *J Biol Chem* **1982**, *257*, 4171–4178.

300 Saunders, A. M., Strittmatter, W. J., Schmechel, D., George-Hyslop, P. H., Pericak-Vance, M. A., Joo, S. H., Rosi, B. L., Gusella, J. F., et al. Association of apolipoprotein E allele $\varepsilon 4$ with late-onset familial and sporadic Alzheimer's disease. *Neurology* **1993**, *43*, 1467–1472.

301 Raber, J., Huang, Y. and Ashford, J. W. ApoE genotype accounts for the vast majority of AD risk and AD pathology. *Neurobiol Aging* **2004**, *25*, 641–650.

302 Schmechel, D. E., Saunders, A. M., Strittmatter, W. J., Crain, B. J., Hulette, C. M., Joo, S. H., Pericak-Vance, M. A., Goldgaber, D., et al. Increased amyloid β-peptide deposition in cerebral cortex as a consequence of apolipoprotein E genotype in late-onset Alzheimer disease. *Proc Natl Acad Sci USA* **1993**, *90*, 9649–9653.

303 Rebeck, G. W., Reiter, J. S., Strickland, D. K. and Hyman, B. T. Apolipoprotein E in sporadic Alzheimer's disease: allelic variation and receptor interactions. *Neuron* **1993**, *11*, 575–580.

304 Strittmatter, W. J., Saunders, A. M., Schmechel, D., Perciak-Vance, M., Enghild, J., Salvesen, G. S. and Roses, A. D. Apolipoprotein E: high-avidity binding to β-amyloid and increased frequency of type 4 allele in late-onset familial Alzheimer disease. *Proc Natl Acad Sci USA* **1993**, *90*, 1977–1981.

305 Ma, J. Y., Yee, A., Brewer, H. B., Das, S. and Potter, H. Amyloid-associated proteins α_1-antichymotrypsin and apolipoprotein E promote assembly of Alzheimer β-protein into filaments. *Nature* **1994**, *372*, 92–94.

306 Evans, K. C., Berger, E. P., Cho, C. G., Weisgraber, K. H. and Lansbury, P. T. Apolipoprotein E is a kinetic but not a thermodynamic inhibitor of amyloid formation – implications for the pathogenesis and treatment of Alzheimer disease. *Proc Natl Acad Sci USA* **1995**, *92*, 763–767.

307 Bales, K. R., Verina, T., Cummins, D. J., Du, Y. S., Dodel, T. C., Saura, J., Fishman, C. E., DeLong, C. A., et al. Apolipoprotein E is essential for amyloid deposition in the AβPP(V717F) transgenic mouse model of Alzheimer's disease. *Proc Natl Acad Sci USA* **1999**, *96*, 15233–15238.

308 Holtzman, D. M., Bales, K. R., Tenkova, T., Fagan, A. M., Parsadanian, M., Sartorius, L. J., Mackey, B., Olney, J., et al. Apolipoprotein E isoform-dependent amyloid deposition and neuritic degeneration in a mouse model of Alzheimer's disease. *Proc Natl Acad Sci USA* **2000**, *97*, 2892–2897.

309 Irizarry, M.C., Cheung, B.S., Rebeck, G.W., Paul, S.M., Bales, K.R. and Hyman, B.T. Apolipoprotein E affects the amount, form, and anatomical distribution of amyloid β-peptide deposition in homozygous AβPP(V717F) transgenic mice. *Acta Neuropathol (Berl)* **2000**, *100*, 451–458.

310 Strittmatter, W.J., Saunders, A.M., Goedert, M., Weisgraber, K.H., Dong, L.M., Jakes, R., Huang, D.Y., Pericak-vance, M., et al. Isoform-specific interactions of apolipoprotein E with microtubule-associated protein Tau: implications for Alzheimer disease. *Proc Natl Acad Sci USA* **1994**, *91*, 11183–11186.

311 Tesseur, I., Van Dorpe, J., Spittaels, K., Van den Haute, C., Moechars, D. and Van Leuven, F. Expression of human apolipoprotein E4 in neurons causes hyperphosphorylation of protein tau in the brains of transgenic mice. *Am J Pathol* **2000**, *156*, 951–964.

312 Huang, Y.D., Liu, X.Q., Wyss-Coray, T., Brecht, W.J., Sanan, D.A. and Mahley, R.W. Apolipoprotein E fragments present in Alzheimer's disease brains induce neurofibrillary tangle-like intracellular inclusions in neurons. *Proc Natl Acad Sci USA* **2001**, *98*, 8838–8843.

313 Harris, F.M., Brecht, W.J., Xu, Q., Tesseur, I., Kekonius, L., Wyss-Coray, T., Fish, J.D., Masliah, E., et al. Carboxyl-terminal-truncated apolipoprotein E4 causes Alzheimer's disease-like neurodegeneration and behavioral deficits in transgenic mice. *Proc Natl Acad Sci USA* **2003**, *100*, 10966–10971.

314 Nathan, B.P., Bellosta, S., Sanan, D.A., Weisgraber, K.H., Mahley, R.W. and Pitas, R.E. Differential effects of apolipoproteins E3 and E4 on neuronal growth *in vitro*. *Science* **1994**, *264*, 850–852.

315 Ji, Z.S., Miranda, R.D., Newhouse, Y.M., Weisgraber, K.H., Huang, Y.D. and Mahley, R.W. Apolipoprotein E4 potentiates amyloid β peptide-induced lysosomal leakage and apoptosis in neuronal cells. *J Biol Chem* **2002**, *277*, 21821–21828.

316 Nathan, B.P., Chang, K.C., Bellosta, S., Brisch, E., Ge, N.F., Mahley, R.W. and Pitas, R.E. The inhibitory effect of apolipoprotein E4 on neurite outgrowth is associated with microtubule depolymerization. *J Biol Chem* **1995**, *270*, 19791–19799.

317 Raber, J., Wong, D., Yu, G.Q., Buttini, M., Mahley, R.W., Pitas, R.E. and Mucke, L. Apolipoprotein E and cognitive performance. *Nature* **2000**, *404*, 352–354.

318 Raber, J., Wong, D., Buttini, M., Orth, M., Bellosta, S., Pitas, R.E., Mahley, R.W. and Mucke, L. Isoform-specific effects of human apolipoprotein E on brain function revealed in ApoE knockout mice: increased susceptibility of females. *Proc Natl Acad Sci USA* **1998**, *95*, 10914–10919.

319 Raber, J., Bongers, G., LeFevour, A., Buttini, M. and Mucke, L. Androgens protect against apolipoprotein E4-induced cognitive deficits. *J Neurosci* **2002**, *22*, 5204–5209.

320 Ladu, M.J., Falduto, M.T., Manelli, A.M., Reardon, C.A., Getz, G.S. and Frail, D.E. Isoform-specific binding of apolipoprotein E to β-amyloid. *J Biol Chem* **1994**, *269*, 23403–23406.

321 DeMattos, R.B., Cirrito, J.R., Parsadanian, M., May, P.C., O'Dell, M.A., Taylor, J.W., Harmony, J.A., Aronow, B.J., et al. ApoE and clusterin cooperatively suppress Aβ levels and deposition. Evidence that ApoE regulates extracellular Aβ metabolism *in vivo*. *Neuron* **2004**, *41*, 193–202.

322 Holtzman, D.M., Pitas, R.E., Kilbridge, J., Nathan, B., Mahley, R.W., Bu, G.J. and Schwartz, A.L. Low density lipoprotein receptor-related protein mediates Apolipoprotein E-dependent neurite outgrowth in a central nervous system-derived neuronal sell line. *Proc Natl Acad Sci USA* **1995**, *92*, 9480–9484.

323 DeMattos, R.B., Curtiss, L.K. and Williams, D.L. A minimally lipidated form of cell-derived apolipoprotein E exhibits isoform-specific stimulation of neurite outgrowth in the absence of exogenous lipids or lipoproteins. *J Biol Chem* **1998**, *273*, 4206–4212.

324 Miyata, M. and Smith, J. D. Apolipoprotein E allele-specific antioxidant activity and effects on cytotoxicity by oxidative insults and β-amyloid peptides. *Nat Genet* **1996**, *14*, 55–61.

325 Lauderback, C. M., Kanski, J., Hackett, J. M., Maeda, N., Kindy, M. S. and Butterfield, D. A. Apolipoprotein E modulates Alzheimer's Aβ(1–42)-induced oxidative damage to synaptosomes in an allele-specific manner. *Brain Res* **2002**, *924*, 90–97.

326 Dong, L. M., Wilson, C., Wardell, M. R., Simmons, T., Mahley, R. W., Weisgraber, K. H. and Agard, D. A. Human apolipoprotein E. Role of arginine 61 in mediating the lipoprotein preferences of the E3 and E4 isoforms. *J Biol Chem* **1994**, *269*, 22358–22365.

327 Dong, L. M. and Weisgraber, K. H. Human Apolipoprotein E4 domain interaction – arginine 61 and glutamic acid 255 interact to direct the preference for very low density lipoproteins. *J Biol Chem* **1996**, *271*, 19053–19057.

328 Xu, Q., Brecht, W. J., Weisgraber, K. H., Mahley, R. W. and Huang, Y. Apolipoprotein E4 domain interaction occurs in living neuronal cells as determined by fluorescence resonance energy transfer. *J Biol Chem* **2004**, *279*, 25511–25516.

329 Morrow, J. A., Segall, M. L., Lund-Katz, S., Phillips, M. C., Knapp, M., Rupp, B. and Weisgraber, K. H. Differences in stability among the human apolipoprotein E isoforms determined by the amino-terminal domain. *Biochemistry* **2000**, *39*, 11657–11666.

330 Morrow, J. A., Hatters, D. M., Lu, B., Hochtl, P., Oberg, K. A., Rupp, B. and Weisgraber, K. H. Apolipoprotein E4 forms a molten globule. A potential basis for its association with disease. *J Biol Chem* **2002**, *277*, 50380–50385.

331 Bergem, A. L., Engedal, K. and Kringlen, E. The role of heredity in late-onset Alzheimer disease and vascular dementia. A twin study. *Arch Gen Psychiatry* **1997**, *54*, 264–270.

332 Rao, V. S., Cupples, A., van Duijn, C. M., Kurz, A., Green, R. C., Chui, H., Duara, R., Auerbach, S. A., et al. Evidence for major gene inheritance of Alzheimer disease in families of patients with and without apolipoprotein E ε4. *Am J Hum Genet* **1996**, *59*, 664–675.

333 Warwick Daw, E., Payami, H., Nemens, E. J., Nochlin, D., Bird, T. D., Schellenberg, G. D. and Wijsman, E. M. The number of trait loci in late-onset Alzheimer disease. *Am J Hum Genet* **2000**, *66*, 196–204.

334 Bertram, L. and Tanzi, R. E. Alzheimer's disease: one disorder, too many genes? *Hum Mol Genet* **2004**, *13 (Spec 1)*, R135–R141.

335 Dobson, C. M. Protein folding and misfolding. *Nature* **2003**, *426*, 884–890.

336 Cohen, A. S. and Calkins, E. Electron microscopic observation on a fibrous component in amyloid of diverse origins. *Nature* **1959**, *183*, 1202–1203.

337 Terry, R. D., Gonatas, N. K. and Weiss, M. Ultrastructural studies in Alzheimer's presenile dementia. *Am J Pathol* **1964**, *44*, 269–297.

338 Kidd, M. Alzheimer's disease – an electron microscopical study. *Brain* **1964**, *87*, 307–320.

339 Miyakawa, T., Watanabe, K. and Katsuragi, S. Ultrastructure of amyloid fibrils in Alzheimer's disease and Down's syndrome. *Virchows Arch B Cell Pathol Incl Mol Pathol* **1986**, *52*, 99–106.

340 Fraser, P. E., Duffy, L. K., O'Malley, M. B., Nguyen, J., Inouye, H. and Kirschner, D. A. Morphology and antibody recognition of synthetic β-amyloid peptides. *J Neurosci Res* **1991**, *28*, 474–485.

341 Kirschner, D. A., Inouye, Y., Duffy, L. K., Sinclair, A., Lind, M. and Selkoe, D. J. Synthetic peptide homologous to β protein from Alzheimer disease forms amyloid-like fibrils *in vitro*. *Proc Natl Acad Sci USA* **1987**, *84*, 6953–6957.

342 Inouye, H., Fraser, P. E. and Kirschner, D. A. Structure of β-crystallite assemblies formed by Alzheimer β-amyloid protein analogues: analysis by X-ray diffraction. *Biophys J* **1993**, *64*, 502–519.

343 Glenner, G. G., Eanes, E. D. and Page, D. L. The relation of the properties of

Congo red-stained amyloid fibrils to the β-conformation. *J Histochem Cytochem* **1972**, *20*, 821–826.

344 Kirschner, D.A., Abraham, C. and Selkoe, D.J. X-ray diffraction from intraneuronal paired helical filaments and extraneuronal amyloid fibers in Alzheimer's disease indicates cross-β conformation. *Proc Natl Acad Sci USA* **1986**, *83*, 503–507.

345 Eanes, E.D. and Glenner, G.G. X-ray diffraction studies on amyloid filaments. *J Histochem Cytochem* **1968**, *16*, 673–677.

346 Bonar, L., Cohen, A.S. and Skinner, M.M. Characterization of the amyloid fibril as a cross-β protein. *Proc Soc Exp Biol Med* **1969**, *131*, 1373–1375.

347 Pauling, L. and Corey, R. Configuration of polypeptide chains with favoured orientation around single bonds: two new pleated sheets. *Proc Natl Acad Sci USA* **1951**, *37*, 729–739.

348 Lansbury, P.T., Jr, Costa, P.R., Griffiths, J.M., Simon, E.J., Auger, M., Halverson, K.J., Kocisko, D.A., Hendsch, Z.S., et al. Structural model for the β-amyloid fibril based on interstrand alignment of an antiparallel-sheet comprising a C-terminal peptide. *Nature Struct Biol* **1995**, *2*, 990–998.

349 Balbach, J.J., Ishii, Y., Antzutkin, O.N., Leapman, R.D., Rizzo, N.W., Dyda, F., Reed, J. and Tycko, R. Amyloid fibril formation by Aβ(16–22), a seven-residue fragment of the Alzheimer's β-amyloid peptide, and structural characterization by solid state NMR. *Biochemistry* **2000**, *39*, 13748–13759.

350 Petkova, A.T., Buntkowsky, G., Dyda, F., Leapman, R.D., Yau, W.M. and Tycko, R. Solid state NMR reveals a pH-dependent antiparallel β-sheet registry in fibrils formed by a β-amyloid peptide. *J Mol Biol* **2004**, *335*, 247–260.

351 Petkova, A.T., Ishii, Y., Balbach, J.J., Antzutkin, O.N., Leapman, R.D., Delaglio, F. and Tycko, R.A structural model for Alzheimer's β-amyloid fibrils based on experimental constraints from solid state NMR. *Proc Natl Acad Sci USA* **2002**, *99*, 16742–16747.

352 Antzutkin, O.N., Leapman, R.D., Balbach, J.J. and Tycko, R. Supramolecular structural constraints on Alzheimer's β-amyloid fibrils from electron microscopy and solid-state nuclear magnetic resonance. *Biochemistry* **2002**, *41*, 15436–15450.

353 Benzinger, T.L.S., Gregory, D.M., Burkoth, T.S., Miller-Auer, H., Lynn, D.G., Botto, R.E. and Meredith, S.C. Two-dimensional structure of β-amyloid(10–35) fibrils. *Biochemistry* **2000**, *39*, 3491–3499.

354 Lazo, N.D. and Downing, D.T. Fibril formation by amyloid-β proteins may involve β-helical protofibrils. *J Peptide Res* **1999**, *53*, 633–640.

355 Kheterpal, I., Williams, A., Murphy, C., Bledsoe, B. and Wetzel, R. Structural features of the Aβ amyloid fibril elucidated by limited proteolysis. *Biochemistry* **2001**, *40*, 11757–11767.

356 Kheterpal, I., Zhou, S., Cook, K.D. and Wetzel, R. Aβ amyloid fibrils possess a core structure highly resistant to hydrogen exchange. *Proc Natl Acad Sci USA* **2000**, *97*, 13597–13601.

357 Torok, M., Milton, S., Kayed, R., Wu, P., McIntire, T., Glabe, C. and Langen, R. Structural and dynamic features of Alzheimer's Aβ peptide in amyloid fibrils studied by site-directed spin labeling. *J Biol Chem* **2002**, *277*, 40810–40815.

358 Perutz, M.F., Johnson, T., Suzuki, M. and Finch, J.T. Glutamine repeats as polar zippers: their possible role in inherited neurodegenerative diseases. *Proc Natl Acad Sci USA* **1994**, *91*, 5355–5358.

359 Perutz, M.F., Pope, B.J., Owen, D., Wanker, E.E. and Scherzinger, E. Aggregation of proteins with expanded glutamine and alanine repeats of the glutamine-rich and asparagine-rich domains of Sup35 and of the amyloid β-peptide of amyloid plaques. *Proc Natl Acad Sci USA* **2002**, *99*, 5596–5600.

360 Lazo, N.D. and Downing, D.T. Amyloid fibrils may be assembled from β-helical protofibrils. *Biochemistry* **1998**, *37*, 1731–1735.

361 Lazo, N.D. and Downing, D.T. β-Helical fibrils from a model peptide. *Biochem Biophys Res Commun* **1997**, *235*, 675–679.

362 Wille, H., Michelitsch, M. D., Guenebaut, V., Supattapone, S., Serban, A., Cohen, F. E., Agard, D. A. and Prusiner, S. B. Structural studies of the scrapie prion protein by electron crystallography. *Proc Natl Acad Sci USA* **2002**, *99*, 3563–3568.

363 Guo, J. T., Wetzel, R. and Xu, Y. Molecular modeling of the core of Aβ amyloid fibrils. *Proteins* **2004**, *57*, 357–364.

364 Tycko, R. Progress towards a molecular-level structural understanding of amyloid fibrils. *Curr Opin Struct Biol* **2004**, *14*, 1–8.

365 Jarrett, J. T., Berger, E. P. and Lansbury, P. T., Jr. The carboxy terminus of the β amyloid protein is critical for the seeding of amyloid formation: implications for the pathogenesis of Alzheimer's disease. *Biochemistry* **1993**, *32*, 4693–4697.

366 Harper, J. D. and Lansbury, P. T., Jr. Models of amyloid seeding in Alzheimer's disease and scrapie: mechanistic truths and physiological consequences of the time-dependent solubility of amyloid proteins. *Annu Rev Biochem* **1997**, *66*, 385–407.

367 Naiki, H. and Nakakuki, K. First-order kinetic model of Alzheimer's β-amyloid fibril extension *in vitro*. *Lab Invest* **1996**, *74*, 374–383.

368 Tomski, S. J. and Murphy, R. M. Kinetics of aggregation of synthetic β-amyloid peptide. *Arch Biochem Biophys* **1992**, *294*, 630–638.

369 Lomakin, A., Chung, D. S., Benedek, G. B., Kirschner, D. A. and Teplow, D. B. On the nucleation and growth of amyloid β-protein fibrils: detection of nuclei and quantitation of rate constants. *Proc Natl Acad Sci USA* **1996**, *93*, 1125–1129.

370 Lomakin, A., Teplow, D. B., Kirschner, D. A. and Benedek, G. B. Kinetic theory of fibrillogenesis of amyloid β-protein. *Proc Natl Acad Sci USA* **1997**, *94*, 7942–7947.

371 Esler, W. P., Stimson, E. R., Ghilardi, J. R., Vinters, H. V., Lee, J. P., Mantyh, P. W. and Maggio, J. E. *In vitro* growth of Alzheimer's disease β-amyloid plaques displays first-order kinetics. *Biochemistry* **1996**, *35*, 749–757.

372 Kusumoto, Y., Lomakin, A., Teplow, D. B. and Benedek, G. B. Temperature dependence of amyloid β-protein fibrillization. *Proc Natl Acad Sci USA* **1998**, *95*, 12277–12282.

373 Massi, F. and Straub, J. E. Energy landscape theory for Alzheimer's amyloid β-peptide fibril elongation. *Proteins* **2001**, *42*, 217–229.

374 Howlett, D. R., Jennings, K. H., Lee, D. C., Clark, M. S., Brown, F., Wetzel, R., Wood, S. J., Camilleri, P., et al. Aggregation state and neurotoxic properties of Alzheimer β-amyloid peptide. *Neurodegeneration* **1995**, *4*, 23–32.

375 Forloni, G., Bugiani, O., Tagliavini, F. and Salmona, M. Apoptosis-mediated neurotoxicity induced by β-amyloid and PrP fragments. *Mol Chem Neuropathol* **1996**, *28*, 163–171.

376 Lorenzo, A. and Yankner, B. A. β-Amyloid neurotoxicity requires fibril formation and is inhibited by Congo red. *Proc Natl Acad Sci USA* **1994**, *91*, 12243–12247.

377 Hsia, A. Y., Masliah, E., McConlogue, L., Yu, G. Q., Tatsuno, G., Hu, K., Kholodenko, D., Malenka, R. C., et al. Plaque-independent disruption of neural circuits in Alzheimer's disease mouse models. *Proc Natl Acad Sci USA* **1999**, *96*, 3228–3233.

378 Conway, K. A., Harper, J. D. and Lansbury, P. T. Fibrils formed *in vitro* from α-synuclein and two mutant forms linked to Parkinson's disease are typical amyloid. *Biochemistry* **2000**, *39*, 2552–2563.

379 Ding, T. T., Lee, S. J., Rochet, J. C. and Lansbury, P. T. Annular α-synuclein protofibrils are produced when spherical protofibrils are incubated in solution or bound to brain-derived membranes. *Biochemistry* **2002**, *41*, 10209–10217.

380 Perutz, M. F., Finch, J. T., Berriman, J. and Lesk, A. Amyloid fibers are water-filled nanotubes. *Proc Natl Acad Sci USA* **2002**, *99*, 5591–5595.

381 Srinivasan, R., Jones, E. M., Liu, K., Ghiso, J., Marchant, R. E. and Zagorski, M. G. pH-dependent amyloid and protofibril formation by the ABri peptide

of familial British dementia. *J Mol Biol* **2003**, *333*, 1003–1023.

382 Kad, N. M., Myers, S. L., Smith, D. P., Smith, D. A., Radford, S. E. and Thomson, N. H. Hierarchical assembly of β_2-microglobulin amyloid *in vitro* revealed by atomic force microscopy. *J Mol Biol* **2003**, *330*, 785–797.

383 Poirier, M. A., Li, H. L., Macosko, J., Cai, S. W., Amzel, M. and Ross, C. A. Huntingtin spheroids and protofibrils as precursors in polyglutamine fibrilization. *J Biol Chem* **2002**, *277*, 41032–41037.

384 Goldsbury, C., Goldie, K., Pellaud, J., Seelig, J., Frey, P., Muller, S. A., Kistler, J., Cooper, G. J. S., et al. Amyloid fibril formation from full-length and fragments of amylin. *J Struct Biol* **2000**, *130*, 352–362.

385 Ionescu-Zanetti, C., Khurana, R., Gillespie, J. R., Petrick, J. S., Trabachino, L. C., Minert, L. J., Carter, S. A. and Fink, A. L. Monitoring the assembly of Ig light-chain amyloid fibrils by atomic force microscopy. *Proc Natl Acad Sci USA* **1999**, *96*, 13175–13179.

386 Quintas, A., Vaz, D. C., Cardoso, I., Saraiva, M. J. M. and Brito, R. M. M. Tetramer dissociation and monomer partial unfolding precedes protofibril formation in amyloidogenic transthyretin variants. *J Biol Chem* **2001**, *276*, 27207–27213.

387 Päiviö, A., Jarvet, J., Graslund, A., Lannfelt, L. and Westlind-Danielsson, A. Unique physicochemical profile of β-amyloid peptide variant Aβ1–40E22G protofibrils: conceivable neuropathogen in Arctic mutant carriers. *J Mol Biol* **2004**, *339*, 145–159.

388 Kheterpal, I., Lashuel, H. A., Hartley, D. M., Walz, T., Lansbury, P. T., Jr and Wetzel, R. Aβ protofibrils possess a stable core structure resistant to hydrogen exchange. *Biochemistry* **2003**, *42*, 14092–14098.

389 Qahwash, I., Weiland, K. L., Lu, Y. F., Sarver, R. W., Kletzien, R. F. and Yan, R. Q. Identification of a mutant amyloid peptide that predominantly forms neurotoxic protofibrillar aggregates. *J Biol Chem* **2003**, *278*, 23187–23195.

390 Hou, L., Kang, I., Marchant, R. E. and Zagorski, M. G. Methionine 35 oxidation reduces fibril assembly of the amyloid β-(1–42) peptide of Alzheimer's disease. *J Biol Chem* **2002**, *277*, 40173–40176.

391 Blackley, H. K. L., Sanders, G. H. W., Davies, M. C., Roberts, C. J., Tendler, S. J. B. and Wilkinson, M. J. *In situ* atomic force microscopy study of β-amyloid fibrillization. *J Mol Biol* **2000**, *298*, 833–840.

392 Nichols, M. R., Moss, M. A., Reed, D. K., Lin, W. L., Mukhopadhyay, R., Hoh, J. H. and Rosenberry, T. L. Growth of β-amyloid(1–40) protofibrils by monomer elongation and lateral association. Characterization of distinct products by light scattering and atomic force microscopy. *Biochemistry* **2002**, *41*, 6115–6127.

393 Harper, J. D., Lieber, C. M. and Lansbury, P. T. Atomic force microscopic imaging of seeded fibril formation and fibril branching by the Alzheimer's disease amyloid-β protein. *Chem Biol* **1997**, *4*, 951–959.

394 Lashuel, H. A., Hartley, D. M., Petre, B. M., Wall, J. S., Simon, M. N., Walz, T. and Lansbury, P. T. Mixtures of wild-type and a pathogenic (E22G) form of Aβ40 *in vitro* accumulate protofibrils, including amyloid pores. *J Mol Biol* **2003**, *332*, 795–808.

395 Tycko, R. Insights into the amyloid folding problem from solid-state NMR. *Biochemistry* **2003**, *42*, 3151–3159.

396 Ye, C. P., Selkoe, D. J. and Hartley, D. M. Protofibrils of amyloid β-protein inhibit specific K+ currents in neocortical cultures. *Neurobiol Dis* **2003**, *13*, 177–190.

397 Nilsberth, C., Westlind-Danielsson, A., Eckman, C. B., Forsell, C., Axelman, K., Luthman, J., Younkin, S. G. and Näslund, J. The Arctic APP mutation (E693G) causes Alzheimer's disease through a novel mechanism: Increased amyloid β protofibril formation and decreased amyloid β levels in plasma and conditioned media. *Neurobiol Aging* **2000**, *21*, S58.

398 Antzutkin, O. N. Amyloidosis of Alzheimer's Aβ peptides: solid-state neu-

clear magnetic resonance, electron paramagnetic resonance, transmission electron microscopy, scanning transmission electron microscopy and atomic force microscopy studies. *Magn Reson Chem* **2004**, *42*, 231–246.

399 Klimov, D. K. and Thirumalai, D. Dissecting the assembly of Aβ(16–22) amyloid peptides into antiparallel β sheets. *Structure* **2003**, *11*, 295–307.

400 Soreghan, B., Kosmoski, J. and Glabe, C. Surfactant properties of Alzheimer's Aβ peptides and the mechanism of amyloid aggregation. *J Biol Chem* **1994**, *269*, 28551–28554.

401 Yong, W., Lomakin, A., Kirkitadze, M. D., Teplow, D. B., Chen, S. H. and Benedek, G. B. Structure determination of micelle-like intermediates in amyloid β-protein fibril assembly by using small angle neutron scattering. *Proc Natl Acad Sci USA* **2002**, *99*, 150–154.

402 Gong, Y., Chang, L., Viola, K. L., Lacor, P. N., Lambert, M. P., Finch, C. E., Krafft, G. A. and Klein, W. L. Alzheimer's disease-affected brain: presence of oligomeric Aβ ligands (ADDLs) suggests a molecular basis for reversible memory loss. *Proc Natl Acad Sci USA* **2003**, *100*, 10417–10422.

403 Yao, Z. X., Drieu, K. and Papadopoulos, V. The Ginkgo biloba extract EGb 761 rescues the PC12 neuronal cells from β-amyloid-induced cell death by inhibiting the formation of β-amyloid-derived diffusible neurotoxic ligands. *Brain Res* **2001**, *889*, 181–190.

404 Wang, Z., Chang, L., Klein, W. L., Thatcher, G. R. and Venton, D. L. Per–6-substituted-per–6-deoxy β-cyclodextrins inhibit the formation of β-amyloid peptide derived soluble oligomers. *J Med Chem* **2004**, *47*, 3329–3333.

405 Wang, H. W., Pasternak, J. F., Kuo, H., Ristic, H., Lambert, M. P., Chromy, B., Viola, K. L., Klein, W. L., et al. Soluble oligomers of β amyloid (1–42) inhibit long-term potentiation but not long-term depression in rat dentate gyrus. *Brain Res* **2002**, *924*, 133–140.

406 Lacor, P. N., Viola, K. L., Fernandez, S., Velasco, P. T., Bigio, E. H. and Klein, W. L. Targeting of synapses and memory-linked IEG protein Arc by ADDLs. *Soc Neurosci Abstr* **2003**, *29*.

407 Manelli, A., Dahlgren, K., Krafft, G. A. and LaDu, M. J. Neurotoxicity induced by oligomeric versus fibrillar Aβ1–42. *Soc Neurosci Abstr* **2001**, *27*, 322–312.

408 Westerman, M. A., Chang, L., Frautschy, S., Kotilinek, L., Cole, G. and Klein, W. L. Ibuprofen reverses memory loss in transgenic mice modeling Alzheimer's disease. *Soc Neurosci Abstr* **2002**, *28*, 690–694.

409 Haes, A. J., Hall, W. P., Chang, L., Klein, W. L. and Van Duyne, R. P. A localized surface plasmon resonance biosensor: First steps toward an assay for Alzheimer's disease. *Nano Lett* **2004**, *4*, 1029–1034.

410 Fancy, D. A. and Kodadek, T. Chemistry for the analysis of protein–protein interactions: rapid and efficient cross-linking triggered by long wavelength light. *Proc Natl Acad Sci USA* **1999**, *96*, 6020–6024.

411 Bitan, G., Lomakin, A. and Teplow, D. B. Amyloid β-protein oligomerization: prenucleation interactions revealed by photo-induced cross-linking of unmodified proteins. *J Biol Chem* **2001**, *276*, 35176–35184.

412 Bitan, G., Tarus, B., Vollers, S. S., Lashuel, H. A., Condron, M. M., Straub, J. E. and Teplow, D. B. A molecular switch in amyloid assembly: Met^{35} and amyloid β-protein oligomerization. *J Am Chem Soc* **2003**, *125*, 15359–15365.

413 Bitan, G. and Teplow, D. B. Rapid photochemical cross-linking – a new tool for studies of metastable, amyloidogenic protein assemblies. *Acc Chem Res* **2004**, *37*, 357–364.

414 Bitan, G., Vollers, S. S. and Teplow, D. B. Elucidation of primary structure elements controlling early amyloid β-protein oligomerization. *J Biol Chem* **2003**, *278*, 34882–34889.

415 Butterfield, D. A. and Kanski, J. Methionine residue 35 is critical for the oxidative stress and neurotoxic properties of Alzheimer's amyloid β-peptide 1–42. *Peptides* **2002**, *23*, 1299–1309.

416 Schöneich, C. Redox processes of methionine relevant to β-amyloid oxidation and Alzheimer's disease. *Arch Biochem Biophys* **2002**, *397*, 370–376.

417 Barrow, C. J. and Zagorski, M. G. Solution structures of β peptide and its constituent fragments: relation to amyloid deposition. *Science* **1991**, *253*, 179–182.

418 Sticht, H., Bayer, P., Willbold, D., Dames, S., Hilbich, C., Beyreuther, K., Frank, R. W. and Rosch, P. Structure of amyloid A4-(1–40)-peptide of Alzheimer's disease. *Eur J Biochem* **1995**, *233*, 293–298.

419 Crescenzi, O., Tomaselli, S., Guerrini, R., Salvadori, S., D'Ursi, A. M., Temussi, P. A. and Picone, D. Solution structure of the Alzheimer amyloid β-peptide (1–42) in an apolar microenvironment – similarity with a virus fusion domain. *Eur J Biochem* **2002**, *269*, 5642–5648.

420 Shao, H. Y., Jao, S. C., Ma, K. and Zagorski, M. G. Solution structures of micelle-bound amyloid β-(1–40) and β-(1–42) peptides of Alzheimer's disease. *J Mol Biol* **1999**, *285*, 755–773.

421 Coles, M., Bicknell, W., Watson, A. A., Fairlie, D. P. and Craik, D. J. Solution structure of amyloid β-peptide(1–40) in a water-micelle environment – is the membrane-spanning domain where we think it is? *Biochemistry* **1998**, *37*, 11064–11077.

422 Gursky, O. and Aleshkov, S. Temperature-dependent β-sheet formation in β-amyloid Aβ(1–40) peptide in water: uncoupling β-structure folding from aggregation. *Biochim Biophys Acta Prot Struct Mol Enzymol* **2000**, *1476*, 93–102.

423 Hou, L., Shao, H., Zhang, Y., Li, H., Menon, N. K., Neuhaus, E. B., Brewer, J. M., Byeon, I.-J. L., et al. Solution NMR studies of the Aβ(1–40) and Aβ(1–42) peptides establish that the Met35 oxidation state affects the mechanism of amyloid formation. *J Am Chem Soc* **2004**, *126*, 1992–2005.

424 Conway, K. A., Harper, J. D. and Lansbury, P. T. Accelerated *in vitro* fibril formation by a mutant α-synuclein linked to early-onset Parkinson disease. *Nat Med* **1998**, *4*, 1318–1320.

425 Straub, J. E., Guevara, J., Huo, S. H. and Lee, J. P. Long time dynamic simulations: Exploring the folding pathways of an Alzheimer's amyloid Aβ-peptide. *Acc Chem Res* **2002**, *35*, 473–481.

426 Lazo, N. D., Grant, M. A., Condron, M. C., Rigby, A. C. and Teplow, D. B. On the nucleation of amyloid β-protein monomer folding. *Protein Sci* **2005**, in press.

427 Fontana, A., Polverino de Laureto, P., De Filippis, V., Scaramella, E. and Zambonin, M. Probing the partly folded states of proteins by limited proteolysis. *Fold Des* **1997**, *2*, R17–R26.

428 Polverino de Laureto, P., Taddei, N., Frare, E., Capanni, C., Costantini, S., Zurdo, J., Chiti, F., Dobson, C. M., et al. Protein aggregation and amyloid fibril formation by an SH3 domain probed by limited proteolysis. *J Mol Biol* **2003**, *334*, 129–141.

429 Monti, M., Garolla di Bard, B. L., Calloni, G., Chiti, F., Amoresano, A., Ramponi, G. and Pucci, P. The regions of the sequence most exposed to the solvent within the amyloidogenic state of a protein initiate the aggregation process. *J Mol Biol* **2004**, *336*, 253–262.

430 Arispe, N., Pollard, H. B. and Rojas, E. Giant multilevel cation channels formed by Alzheimer disease amyloid β-protein [AβP-(1–40)] in bilayer membranes. *Proc Natl Acad Sci USA* **1993**, *90*, 10573–10577.

431 Lin, H., Zhu, Y. W. J. and Lal, R. Amyloid β protein (1–40) forms calcium-permeable, Zn^{2+}-sensitive channel in reconstituted lipid vesicles. *Biochemistry* **1999**, *38*, 11189–11196.

432 Hirakura, Y., Yiu, W. W., Yamamoto, A. and Kagan, B. L. Amyloid peptide channels: blockade by zinc and inhibition by Congo red (amyloid channel block). *Amyloid* **2000**, *7*, 194–199.

433 Bhatia, R., Lin, H. and Lal, R. Fresh and globular amyloid β-protein (1–42) induces rapid cellular degeneration: evidence for Aβ channel-mediated cellular toxicity. *FASEB J* **2000**, *14*, 1233–1243.

434 Lin, M. C. A. and Kagan, B. L. Electrophysiologic properties of channels induced by Aβ 25–35 in planar lipid bilayers. *Peptides* **2002**, *23*, 1215–1228.

435 Mirzabekov, T.A., Lin, M.C. and Kagan, B.L. Pore formation by the cytotoxic islet amyloid peptide amylin. *J Biol Chem* **1996**, *271*, 1988–1992.

436 Hirakura, Y. and Kagan, B. L. Pore formation by β-2-microglobulin: a mechanism for the pathogenesis of dialysis associated amyloidosis. *Amyloid: J Protein Fold Disord* **2001**, *8*, 94–100.

437 Hirakura, Y., Lin, M.C. and Kagan, B.L. Alzheimer amyloid Aβ 1–42 channels: Effects of solvent, pH, and congo red. *J Neurosci Res* **1999**, *57*, 458–466.

438 Lin, H., Bhatia, R. and Lal, R. Amyloid β protein forms ion channels: implications for Alzheimer's disease pathophysiology. *FASEB J* **2001**, *15*, 2433–2444.

439 Durell, S.R., Guy, H.R., Arispe, N., Rojas, E. and Pollard, H.B. Theoretical models of the ion channel structure of amyloid β-protein. *Biophys J* **1994**, *67*, 2137–2145.

440 Arispe, N., Pollard, H.B. and Rojas, E. Zn^{2+} interaction with Alzheimer amyloid β protein calcium channels. *Proc Natl Acad Sci USA* **1996**, *93*, 1710–1715.

441 Arispe, N. Architecture of the Alzheimer's AβP ion channel pore. *J Membr Biol* **2004**, *197*, 33–48.

442 Furukawa, K., Abe, Y. and Akaike, N. Amyloid β protein-induced irreversible current in rat cortical neurones. *Neuroreport* **1994**, *5*, 2016–2018.

443 Kawahara, M., Arispe, N., Kuroda, Y. and Rojas, E. Alzheimer's-disease amyloid β-protein forms Zn^{2+}-sensitive, cation-selective channels across excised membrane patches from hypothalamic neurons. *Biophys J* **1997**, *73*, 67–75.

444 Sanderson, K.L., Butler, L. and Ingram, V.M. Aggregates of a β-amyloid peptide are required to induce calcium currents in neuron-like human teratocarcinoma cells – relation to Alzheimer's disease. *Brain Res* **1997**, *744*, 7–14.

445 Selkoe, D.J. Normal and abnormal biology of the β-amyloid precursor protein. *Annu Rev Neurosci* **1994**, *17*, 489–517.

446 Abraham, C.R., Selkoe, D.J. and Potter, H. Immunochemical identification of the serine protease inhibitor, α_1-antichymotrypsin in the brain amyloid deposits of Alzheimer's disease. *Cell* **1988**, *52*, 487–501.

447 Fraser, P.E., Nguyen, J.T., McLachlan, D.R., Abraham, C.R. and Kirschner, D.A. α_1-Antichymotrypsin binding to Alzheimer Aβ peptides is sequence specific and induces fibril disaggregation *in vitro*. *J Neurochem* **1993**, *61*, 298–305.

448 Eriksson, S., Janciauskiene, S. and Lannfelt, L. α_1-Antichymotrypsin regulates Alzheimer β-amyloid peptide fibril formation. *Proc Natl Acad Sci USA* **1995**, *92*, 2313–2317.

449 Lukacs, C.M. and Christianson, D.W. Is the binding of β-amyloid protein to antichymotrypsin in Alzheimer plaques mediated by a β-strand insertion? *Proteins* **1996**, *25*, 420–424.

450 Janciauskiene, S., Eriksson, S. and Wright, H.T. A specific structural interaction of Alzheimer's peptide Aβ 1–42 with α_1-antichymotrypsin. *Nat Struct Biol* **1996**, *3*, 668–671.

451 Janciauskiene, S., Rubin, H., Lukacs, C.M. and Wright, H.T. Alzheimer's peptide Aβ(1–42) binds to two β-sheets of α_1-antichymotrypsin and transforms it from inhibitor to substrate. *J Biol Chem* **1998**, *273*, 28360–28364.

452 Timpl, R. Macromolecular organization of basement membranes. *Curr Opin Cell Biol* **1996**, *8*, 618–624.

453 Beck, K., Hunter, I. and Engel, J. Structure and function of laminin: anatomy of a multidomain glycoprotein. *FASEB J* **1990**, *4*, 148–160.

454 Nomizu, M., Otaka, A., Utani, A., Roller, P.P. and Yamada, Y. Assembly of synthetic laminin peptides into a triple-stranded coiled-coil structure. *J Biol Chem* **1994**, *269*, 30386–30392.

455 Utani, A., Nomizu, M., Timpl, R., Roller, P.P. and Yamada, Y. Laminin chain assembly. Specific sequences at the C terminus of the long arm are required for the formation of specific double- and triple-stranded coiled-coil structures. *J Biol Chem* **1994**, *269*, 19167–19175.

456 Yurchenco, P.D., Tsilibary, E.C., Charonis, A.S. and Furthmayr, H. Laminin polymerization *in vitro*. Evidence for a

two-step assembly with domain specificity. *J Biol Chem* **1985**, *260*, 7636–7644.

457 Yurchenco, P.D., Cheng, Y.S. and Schittny, J.C. Heparin modulation of laminin polymerization. *J Biol Chem* **1990**, *265*, 3981–3991.

458 Bronfman, F.C., Soto, C., Tapia, L., Tapia, V. and Inestrosa, N.C. Extracellular matrix regulates the amount of the β-amyloid precursor protein and its amyloidogenic fragments. *J Cell Physiol* **1996**, *166*, 360–369.

459 Murtomaki, S., Risteli, J., Risteli, L., Koivisto, U.M., Johansson, S. and Liesi, P. Laminin and its neurite outgrowth-promoting domain in the brain in Alzheimer's disease and Down's syndrome patients. *J Neurosci Res* **1992**, *32*, 261–273.

460 Perlmutter, L.S. and Chui, H.C. Microangiopathy, the vascular basement membrane and Alzheimer's disease: a review. *Brain Res Bull* **1990**, *24*, 677–686.

461 Perlmutter, L.S. Microvascular pathology and vascular basement membrane components in Alzheimer's disease. *Mol Neurobiol* **1994**, *9*, 33–40.

462 Bronfman, F.C., Garrido, J., Alvarez, A., Morgan, C. and Inestrosa, N.C. Laminin inhibits amyloid-β-peptide fibrillation. *Neurosci Lett* **1996**, *218*, 201–203.

463 Bronfman, F.C., Alvarez, A., Morgan, C. and Inestrosa, N.C. Laminin blocks the assembly of wild-type Aβ and the Dutch variant peptide into Alzheimer's fibrils. *Amyloid: Int J Exp Clin Invest* **1998**, *5*, 16–23.

464 Monji, A., Tashiro, K., Yoshida, I., Hayashi, Y. and Tashiro, N. Laminin inhibits Aβ42 fibril formation *in vitro*. *Brain Res* **1998**, *788*, 187–190.

465 Monji, A., Tashiro, K., Hayashi, Y., Yoshida, I. and Tashiro, N. The inhibitory effect of laminin 1 and synthetic peptides deduced from the sequence in the laminin α–1 chain on Aβ–40 fibril formation *in vitro*. *Neurosci Lett* **1998**, *251*, 65–68.

466 Morgan, C. and Garrido, J. Laminin disaggregates amyloid β-fibrils. *Neurobiol Aging* **1998**, *19*, abstr 185.

467 Morgan, C., Bugueno, M.P., Garrido, J. and Inestrosa, N.C. Laminin affects polymerization, depolymerization and neurotoxicity of Aβ peptide. *Peptides* **2002**, *23*, 1229–1240.

468 Drouet, B., Pincon-Raymond, M., Chambaz, J. and Pillot, T. Laminin 1 attenuates β-amyloid peptide Aβ(1–40) neurotoxicity of cultured fetal rat cortical neurons. *J Neurochem* **1999**, *73*, 742–749.

469 Monji, A., Tashiro, K., Yoshida, I., Kaname, H., Hayashi, Y., Matsuda, K. and Tashiro, N. Laminin inhibits both Aβ40 and Aβ42 fibril formation but does not affect Aβ40 or Aβ42-induced cytotoxicity in PC12 cells. *Neurosci Lett* **1999**, *266*, 85–88.

470 Castillo, G.M., Lukito, W., Peskind, E., Raskind, M., Kirschner, D.A., Yee, A.G. and Snow, A.D. Laminin inhibition of β-amyloid protein (Aβ) fibrillogenesis and identification of an Aβ binding site localized to the globular domain repeats on the laminin A chain. *J Neurosci Res* **2000**, *62*, 451–462.

471 Morgan, C. and Inestrosa, N.C. Interactions of laminin with the amyloid β peptide. Implications for Alzheimer's disease. *Braz J Med Biol Res* **2001**, *34*, 597–601.

472 Fernandez, H.L., Moreno, R.D. and Inestrosa, N.C. Tetrameric (G4) acetylcholinesterase: structure, localization, and physiological regulation. *J Neurochem* **1996**, *66*, 1335–1346.

473 Inestrosa, N.C., Alvarez, A., Perez, C.A., Moreno, R.D., Vicente, M., Linker, C., Casanueva, O.I., Soto, C., et al. Acetylcholinesterase accelerates assembly of amyloid-β peptides into Alzheimer's fibrils: possible role of the peripheral site of the enzyme. *Neuron* **1996**, *16*, 881–891.

474 Geula, C. and Mesulam, M.M. Cortical cholinergic fibers in aging and Alzheimer's disease: a morphometric study. *Neuroscience* **1989**, *33*, 469–481.

475 Geula, C. and Mesulam, M. Special properties of cholinesterases in the cerebral cortex of Alzheimer's disease. *Brain Res* **1989**, *498*, 185–189.

476 Gomez-Ramos, P., Mufson, E.J. and Moran, M.A. Ultrastructural localization of acetylcholinesterase in neurofibrillary tangles, neuropil threads and

senile plaques in aged and Alzheimer's brain. *Brain Res* **1992**, *569*, 229–237.

477 Wright, C.I., Guela, C. and Mesulam, M.M. Protease inhibitors and indoleamines selectively inhibit cholinesterases in the histopathologic structures of Alzheimer disease. *Proc Natl Acad Sci USA* **1993**, *90*, 683–686.

478 Struble, R.G., Cork, L.C., Whitehouse, P.J. and Price, D.L. Cholinergic innervation in neuritic plaques. *Science* **1982**, *216*, 413–415.

479 Carson, K.A., Geula, C. and Mesulam, M.M. Electron microscopic localization of cholinesterase activity in Alzheimer brain tissue. *Brain Res* **1991**, *540*, 204–208.

480 Alvarez, A., Opazo, C., Alarcon, R., Garrido, J. and Inestrosa, N.C. Acetylcholinesterase promotes the aggregation of amyloid-β peptide fragments by forming a complex with the growing fibrils. *J Mol Biol* **1997**, *272*, 348–361.

481 Alvarez, A., Bronfman, F., Perez, C.A., Vicente, M., Garrido, J. and Inestrosa, N.C. Acetylcholinesterase, a senile plaque component, affects the fibrillogenesis of Amyloid-β peptides. *Neurosci Lett* **1995**, *201*, 49–52.

482 Bartolini, M., Bertucci, C., Cavrini, V. and Andrisano, V. β-Amyloid aggregation induced by human acetylcholinesterase: inhibition studies. *Biochem Pharmacol* **2003**, *65*, 407–416.

483 Inestrosa, N.C., Alvarez, A. and Calderon, F. Acetylcholinesterase is a senile plaque component that promotes assembly of amyloid β-peptide into Alzheimer's filaments. *Mol Psychiatry* **1996**, *1*, 359–361.

484 De Ferrari, G.V., Canales, M.A., Shin, I., Weiner, L.M., Silman, I. and Inestrosa, N.C. A structural motif of acetylcholinesterase that promotes amyloid β-peptide fibril formation. *Biochemistry* **2001**, *40*, 10447–10457.

485 Reyes, A.E., Perez, D.R., Alvarez, A., Garrido, J., Gentry, M.K., Doctor, B.P. and Inestrosa, N.C. A monoclonal antibody against acetylcholinesterase inhibits the formation of amyloid fibrils induced by the enzyme. *Biochem Biophys Res Commun* **1997**, *232*, 652–655.

486 Alvarez, A., Opazo, C., Alarcon, R., Garrido, J. and Inestrosa, N.C. Acetylcholinesterase promotes the aggregation of amyloid-β-peptide fragments by forming a complex with the growing fibrils. *J Mol Biol* **1997**, *272*, 348–361.

487 Munoz, F.J. and Inestrosa, N.C. Neurotoxicity of acetylcholinesterase amyloid β-peptide aggregates is dependent on the type of Aβ peptide and the AChE concentration present in the complexes. *FEBS Lett* **1999**, *450*, 205–209.

488 Emsley, J., White, H.E., O'Hara, B.P., Oliva, G., Srinivasan, N., Tickle, I.J., Blundell, T.L., Pepys, M.B., et al. Structure of pentameric human serum amyloid P component. *Nature* **1994**, *367*, 338–345.

489 Pepys, M.B., Herbert, J., Hutchinson, W.L., Tennent, G.A., Lachmann, H.J., Gallimore, J.R., Lovat, L.B., Bartfai, T., et al. Targeted pharmacological depletion of serum amyloid P component for treatment of human amyloidosis. *Nature* **2002**, *417*, 254–259.

490 Coria, F., Castano, E., Prelli, F., Larrondo-Lillo, M., van Duinen, S., Shelanski, M.L. and Frangione, B. Isolation and characterization of amyloid P component from Alzheimer's disease and other types of cerebral amyloidosis. *Lab Invest* **1988**, *58*, 454–458.

491 Duong, T., Pommier, E.C. and Scheibel, A.B. Immunodetection of the amyloid P component in Alzheimer's disease. *Acta Neuropathol (Berl)* **1989**, *78*, 429–437.

492 Kalaria, R.N. and Grahovac, I. Serum amyloid P immunoreactivity in hippocampal tangles, plaques and vessels: implications for leakage across the blood–brain barrier in Alzheimer's disease. *Brain Res* **1990**, *516*, 349–353.

493 Kalaria, R.N. Serum amyloid P and related molecules associated with the acute-phase response in Alzheimer's disease. *Res Immunol* **1992**, *143*, 637–641.

494 Kalaria, R.N., Galloway, P.G. and Perry, G. Widespread amyloid P component immunoreactivity in cortical amyloid deposits of Alzheimer's disease and other degenerative disorders. *Neuropath Appl Neurobiol* **1991**, *17*, 189–201.

495 Kalaria, R. N., Golde, T. E., Cohen, M. L. and Younkin, S. G. Serum amyloid P in Alzheimer's disease. *Ann NY Acad Sci* **1991**, *640*, 145–148.

496 Kalaria, R. N. and Kroon, S. N. Serum proteins in neurofibrillary pathology of Alzheimer's disease: antithrombin-III-like immunoreactivity. *Soc Neurosci Abstr* **1991**, *17*, 725.

497 Akiyama, H., Yamada, T., Kawamata, T. and McGeer, P. L. Association of amyloid P component with complement proteins in neurologically diseased brain tissue. *Brain Res* **1991**, *548*, 349–352.

498 Botto, M., Hawkins, P. N., Bickerstaff, M. C. M., Herbert, J., Bygrave, A. E., McBride, A., Hutchinson, W. L., Tennent, G. A., et al. Amyloid deposition is delayed in mice with targeted deletion of the serum amyloid P component gene. *Nat Med* **1997**, *3*, 855–859.

499 Hamazaki, H. Ca^{2+}-dependent binding of human serum amyloid P component to Alzheimer's β-amyloid peptide. *J Biol Chem* **1995**, *270*, 10392–10394.

500 Hind, C. R., Collins, P. M., Caspi, D., Baltz, M. L. and Pepys, M. B. Specific chemical dissociation of fibrillar and non-fibrillar components of amyloid deposits. *Lancet* **1984**, *2*, 376–378.

501 Pepys, M. B. Amyloidosis: some recent developments. *Q J Med* **1988**, *67*, 283–298.

502 Pepys, M. B., Rademacher, T. W., Amatayakul-Chantler, S., Williams, P., Noble, G. E., Hutchinson, W. L., Hawkins, P. N., Nelson, S. R., et al. Human serum amyloid P component is an invariant constituent of amyloid deposits and has a uniquely homogeneous glycostructure. *Proc Natl Acad Sci USA* **1994**, *91*, 5602–5606.

503 Tennent, G. A., Lovat, L. B. and Pepys, M. B. Serum amyloid P component prevents proteolysis of the amyloid fibrils of Alzheimer disease and systemic amyloidosis. *Proc Natl Acad Sci USA* **1995**, *92*, 4299–4303.

504 MacRaild, C. A., Stewart, C. R., Mok, Y.-F., Gunzburg, M. J., Perugini, M. A., Lawrence, L. J., Tirtaatmadja, V., Cooper-White, et al. Non-fibrillar components of amyloid deposits mediate the self-association and tangling of amyloid fibrils. *J Biol Chem* **2004**, *279*, 21038–21045.

505 Weisgraber, K. H. Apolipoprotein E: structure–function relationships. *Adv Protein Chem* **1994**, *45*, 249–302.

506 Strittmatter, W. J., Weisgraber, K. H., Huang, D. Y., Dong, L. M., Salvesen, G. S., Pericak-Vance, M., Schmechel, D., Saunders, A. M., et al. Binding of human apolipoprotein E to synthetic amyloid β-peptide: isoform-specific effects and implications for late-onset Alzheimer disease. *Proc Natl Acad Sci USA* **1993**, *90*, 8098–8102.

507 Ueki, A., Kawano, M., Namba, Y., Kawakami, M. and Ikeda, K. A high frequency of apolipoprotein E4 isoprotein in Japanese patients with late-onset nonfamilial Alzheimer's disease. *Neurosci Lett* **1993**, *163*, 166–168.

508 Namba, Y., Tomonaga, M., Kawasaki, H., Otomo, E. and Ikeda, K. Apolipoprotein E immunoreactivity in cerebral amyloid deposits and neurofibrillary tangles in Alzheimer's disease and kuru plaque amyloid in Creutzfeldt–Jakob disease. *Brain Res* **1991**, *541*, 163–166.

509 Wisniewski, T. and Frangione, B. Apolipoprotein E: a pathological chaperone protein in patients with cerebral and systemic amyloid. *Neurosci Lett* **1992**, *135*, 235–238.

510 Mayeux, R., Stern, Y., Ottman, R., Tatemichi, T. K., Tang, M. X., Maestre, G., Ngai, C., Tycko, B., et al. The apolipoprotein ε4 allele in patients with Alzheimer's disease. *Ann Neurol* **1993**, *34*, 752–754.

511 Holtzman, D. M. *In vivo* effects of ApoE and clusterin on amyloid-β metabolism and neuropathology. *J Mol Neurosci* **2004**, *23*, 247–254.

512 Berr, C., Hauw, J. J., Delaere, P., Duyckaerts, C. and Amouyel, P. Apolipoprotein E allele ε4 is linked to increased deposition of the amyloid β-peptide (Aβ) in cases with or without Alzheimer's disease. *Neurosci Lett* **1994**, *178*, 221–224.

513 Strittmatter, W. J., Weisgraber, K. H., Huang, D. Y., Dong, L. M., Salvesen, G. S., Pericakvance, M., Schmechel, D., Saunders, A. M., et al. Binding of human apolipoprotein-E to synthetic amyloid β peptide – isoform-specific effects and implications for late-onset Alzheimer disease. *Proc Natl Acad Sci USA* **1993**, *90*, 8098–8102.

514 Naslund, J., Thyberg, J., Tjernberg, L. O., Wernstedt, C., Karlstrom, A. R., Bogdanovic, N., Gandy, S. E., Lannfelt, L., et al. Characterization of stable complexes involving apolipoprotein E and the amyloid β-peptide in Alzheimer's disease brain. *Neuron* **1995**, *15*, 219–228.

515 Wisniewski, T., Castano, E. M., Golabek, A., Vogel, T. and Frangione, B. Acceleration of Alzheimer's fibril formation by apolipoprotein E *in vitro*. *Am J Pathol* **1994**, *145*, 1030–1035.

516 Wisniewski, T., Golabek, A. A., Kida, E., Wisniewski, K. E. and Frangione, B. Conformational mimicry in Alzheimer's disease – role of apolipoproteins in amyloidogenesis. *Am J Pathol* **1995**, *147*, 238–244.

517 Wood, S. J., Chan, W. and Wetzel, R. Seeding of Aβ fibril formation is inhibited by all three isotypes of apolipoprotein E. *Biochemistry* **1996**, *35*, 12623–12628.

518 Wood, S. J., Chan, W. and Wetzel, R. An ApoE–Aβ inhibition complex in Aβ fibril extension. *Chem Biol* **1996**, *3*, 949–956.

519 Castano, E. M., Prelli, F., Wisniewski, T., Golabek, A., Kumar, R. A., Soto, C. and Frangione, B. Fibrillogenesis in Alzheimer's disease of amyloid β peptides and apolipoprotein E. *Biochem J* **1995**, *306*, 599–604.

520 Castano, E. M., Prelli, F., Pras, M. and Frangione, B. Apolipoprotein E carboxyl-terminal fragments are complexed to amyloids A and L. Implications for amyloidogenesis and Alzheimer's disease. *J Biol Chem* **1995**, *270*, 17610–17615.

521 Soto, C., Castano, E. M., Prelli, F., Kumar, R. A. and Baumann, M. Apolipoprotein E increases the fibrillogenic potential of synthetic peptides derived from Alzheimer's, gelsolin and AA amyloids. *FEBS Lett* **1995**, *371*, 110–114.

522 Castaño, E. M., Prelli, F. C. and Frangione, B. Apolipoprotein E and amyloidogenesis. *Lab Invest* **1995**, *73*, 457–460.

523 Wisniewski, T. and Frangione, B. Apolipoprotein E: a pathological chaperone protein in patients with cerebral and systemic amyloid. *Neurosci Lett* **1992**, *135*, 235–238.

524 Wisniewski, T., Castaño, E. M., Golabek, A., Vogel, T. and Frangione, B. Acceleration of Alzheimer's fibril formation by apolipoprotein E *in vitro*. *Am J Pathol* **1994**, *145*, 1030–1035.

525 Sanan, D. A., Weisgraber, K. H., Russell, S. J., Mahley, R. W., Huang, D., Saunders, A., Schmechel, D., Wisniewski, T., et al. Apolipoprotein E associates with β-amyloid peptide of Alzheimer's disease to form novel monofibrils. Isoform apoE4 associates more efficiently than apoE3. *J Clin Invest* **1994**, *94*, 860–869.

526 Golabek, A. A., Soto, C., Vogel, T. and Wisniewski, T. The interaction between apolipoprotein E and Alzheimer's amyloid β-peptide is dependent on β-peptide conformation. *J Biol Chem* **1996**, *271*, 10602–10606.

527 Castaño, E. M., Prelli, F., Wisniewski, T., Golabek, A., Kumar, R. A., Soto, C. and Frangione, B. Fibrillogenesis in Alzheimer's disease of amyloid β-peptides and apolipoprotein E. *Biochem J* **1995**, *306*, 599–604.

528 Sadowski, M., Pankiewicz, J., Scholtzova, H., Ripellino, J. A., Li, Y., Schmidt, S. D., Mathews, P. M., Fryer, J. D., et al. A synthetic peptide blocking the apolipoprotein E/β-amyloid binding mitigates β-amyloid toxicity and fibril formation *in vitro* and reduces β-amyloid plaques in transgenic mice. *Am J Pathol* **2004**, *165*, 937–948.

529 Esler, W. P., Marshall, J. R., Stimson, E. R., Ghilardi, J. R., Vinters, H. V., Mantyh, P. W. and Maggio, J. E. Apolipoprotein E affects amyloid formation but not amyloid growth *in vitro*: mechanistic implications for ApoE4 enhanced amyloid burden and risk for

Alzheimer's disease. *Amyloid* **2002**, *9*, 1–12.

530 Holtzman, D.M., Fagan, A.M., Mackey, B., Tenkova, T., Sartorius, L., Paul, S.M., Bales, K., Ashe, K.H., et al. Apolipoprotein E facilitates neuritic and cerebrovascular plaque formation in an Alzheimer's disease model. *Ann Neurol* **2000**, *47*, 739–747.

531 Holtzman, D.M. Role of apoE/Aβ interactions in the pathogenesis of Alzheimer's disease and cerebral amyloid angiopathy. *J Mol Neurosci* **2001**, *17*, 147–155.

532 Costa, D.A., Nilsson, L.N., Bales, K.R., Paul, S.M. and Potter, H. Apolipoprotein is required for the formation of filamentous amyloid, but not for amorphous Aβ deposition, in an AβPP/PS double transgenic mouse model of Alzheimer's disease. *J Alzheimer's Dis* **2004**, *6*, 509–514.

533 Potter, H., Wefes, I.M. and Nilsson, L.N.G. The inflammation-induced pathological chaperones ACT and apo-E are necessary catalysts of Alzheimer amyloid formation. *Neurobiol Aging* **2001**, *22*, 923–930.

534 Nilsson, L.N., Arendash, G.W., Leighty, R.E., Costa, D.A., Low, M.A., Garcia, M.F., Cracciolo, J.R., Rojiani, A., et al. Cognitive impairment in PDAβPP mice depends on ApoE and ACT-catalyzed amyloid formation. *Neurobiol Aging* **2004**, *25*, 1153–1167.

535 Koistinaho, M., Lin, S., Wu, X., Esterman, M., Koger, D., Hanson, J., Higgs, R., Liu, F., et al. Apolipoprotein E promotes astrocyte colocalization and degradation of deposited amyloid-β peptides. *Nat Med* **2004**, *10*, 719–726.

536 Strittmatter, W.J., Saunders, A.M., Schmechel, D., Pericak-Vance, M., Enghild, J., Salvesen, G.S. and Roses, A.D. Apolipoprotein E: high-avidity binding to β-amyloid and increased frequency of type 4 allele in late-onset familial Alzheimer disease. *Proc Natl Acad Sci USA* **1993**, *90*, 1977–1981.

537 Munson, G.W., Roher, A.E., Kuo, Y.M., Gilligan, S.M., Reardon, C.A., Getz, G.S. and LaDu, M.J. SDS-stable complex formation between native apolipoprotein E3 and β-amyloid peptides. *Biochemistry* **2000**, *39*, 16119–16124.

538 Cho, H.S., Hyman, B.T., Greenberg, S.M. and Rebeck, G.W. Quantitation of apoE domains in Alzheimer disease brain suggests a role for apoE in Aβ aggregation. *J Neuropathol Exp Neurol* **2001**, *60*, 342–349.

539 Pillot, T., Goethals, M., Najib, J., Labeur, C., Lins, L., Chambaz, J., Brasseur, R., Vandekerckhove, J., et al. β-amyloid peptide interacts specifically with the carboxy-terminal domain of human apolipoprotein E: relevance to Alzheimer's disease. *J Neurochem* **1999**, *72*, 230–237.

540 Snow, A.D. and Castillo, G.M. Specific proteoglycans as potential causative agents and relevant targets for therapeutic intervention in Alzheimer's disease and other amyloidoses. *Amyloid: Int J Exp Clin Invest* **1997**, *4*, 135–141.

541 Snow, A.D., Mar, H., Nochlin, D., Kimata, K., Kato, M., Suzuki, S., Hassell, J. and Wight, T.N. The presence of heparan sulfate proteoglycans in the neuritic plaques and congophilic angiopathy in Alzheimer's disease. *Am J Pathol* **1988**, *133*, 456–463.

542 Snow, A.D., Nochlin, D., Sekiguchi, R. and Carlson, S.S. Identification and immunolocalization of a new class of proteoglycan (keratan sulfate) to the neuritic plaques of Alzheimer's disease. *Exp Neurol* **1996**, *138*, 305–317.

543 Snow, A. D., Mar, H., Nochlin, D., Kresse, H. and Wight, T. N. Peripheral distribution of dermatan sulfate proteoglycans (decorin) in amyloid-containing plaques and their presence in neurofibrillary tangles of Alzheimer's disease. *J Histochem Cytochem* **1992**, *40*, 105–113.

544 DeWitt, D.A., Silver, J., Canning, D.R. and Perry, G. Chondroitin sulfate proteoglycans are associated with the lesions of Alzheimer's disease. *Exp Neurol* **1993**, *121*, 149–152.

545 Sipe, J.D. Amyloidosis. *Annu Rev Biochem* **1992**, *61*, 947–975.

546 Fraser, P.E., Nguyen, J.T., Chin, D.T. and Kirschner, D.A. Effects of sulfate ions on Alzheimer β/A4 peptide assemblies: implications for amyloid fi-

bril-proteoglycan interactions. *J Neurochem* **1992**, *59*, 1531–1540.
547 Guptabansal, R., Frederickson, R. C. A. and Brunden, K. R. Proteoglycan-mediated inhibition of Aβ proteolysis – a potential cause of senile plaque accumulation. *J Biol Chem* **1995**, *270*, 18666–18671.
548 Castillo, G. M., Ngo, C., Cummings, J., Wight, T. N. and Snow, A. D. Perlecan binds to the β-amyloid proteins (Aβ) of Alzheimer's disease, accelerates Aβ fibril formation, and maintains Aβ fibril stability. *J Neurochem* **1997**, *69*, 2452–2465.
549 McLaurin, J., Franklin, T., Zhang, X., Deng, J. and Fraser, P. E. Interactions of Alzheimer amyloid-β peptides with glycosaminoglycans effects on fibril nucleation and growth. *Eur J Biochem* **1999**, *266*, 1101–1110.
550 McLaurin, J., Franklin, T., Zhang, X. Q., Deng, J. P. and Fraser, P. E. Interactions of Alzheimer amyloid-β peptides with glycosaminoglycans – effects on fibril nucleation and growth. *Eur J Biochem* **1999**, *266*, 1101–1110.
551 Castillo, G. M., Lukito, W., Wight, T. N. and Snow, A. D. The sulfate moieties of glycosaminoglycans are critical for the enhancement of β-amyloid protein fibril formation. *J Neurochem* **1999**, *72*, 1681–1687.
552 Kisilevsky, R. Theme and variations on a string of amyloid. *Neurobiol Aging* **1989**, *10*, 499–500; discussion 510–492.
553 Brunden, K. R., Richter-Cook, N. J., Chaturvedi, N. and Frederickson, R. C. pH-dependent binding of synthetic β-amyloid peptides to glycosaminoglycans. *J Neurochem* **1993**, *61*, 2147–2154.
554 McLaurin, J. and Fraser, P. E. Effect of amino-acid substitutions on Alzheimer's amyloid-β peptide–glycosaminoglycan interactions. *Eur J Biochem* **2000**, *267*, 6353–6361.
555 Kisilevsky, R., Lemieux, L. J., Fraser, P. E., Kong, X. Q., Hultin, P. G. and Szarek, W. A. Arresting amyloidosis *in vivo* using small-molecule anionic sulphonates or sulphates – implications for Alzheimer's disease. *Nat Med* **1995**, *1*, 143–148.
556 Wolozin, B. Cholesterol and the biology of Alzheimer's disease. *Neuron* **2004**, *41*, 7–10.
557 Raffai, R. L. and Weisgraber, K. H. Cholesterol: from heart attacks to Alzheimer's disease. *J Lipid Res* **2003**, *44*, 1423–1430.
558 Hertel, C., Terzi, E., Hauser, N., Jakob-Rotne, R., Seelig, J. and Kemp, J. A. Inhibition of the electrostatic interaction between β-amyloid peptide and membranes prevents β-amyloid-induced toxicity. *Proc Natl Acad Sci USA* **1997**, *94*, 9412–9416.
559 Kremer, J. J. and Murphy, R. M. Kinetics of adsorption of β-amyloid peptide Aβ(1–40) to lipid bilayers. *J Biochem Biophys Methods* **2003**, *57*, 159–169.
560 Terzi, E., Holzemann, G. and Seelig, J. Interaction of Alzheimer β-amyloid peptide(1–40) with lipid membranes. *Biochemistry* **1997**, *36*, 14845–14852.
561 Talaga, P. and Quere, L. The plasma membrane: a target and hurdle for the development of anti-Aβ drugs? *Curr Drug Targets CNS Neurol Disord* **2002**, *1*, 567–574.
562 Mirzabekov, T., Lin, M. C., Yuan, W. L., Marshall, P. J., Carman, M., Tomaselli, K., Lieberburg, I. and Kagan, B. L. Channel formation in planar lipid bilayers by a neurotoxic fragment of the β-amyloid peptide. *Biochem Biophys Res Commun* **1994**, *202*, 1142–1148.
563 Nitsch, R. M., Blusztajn, J. K., Pittas, A. G., Slack, B. E., Growdon, J. H. and Wurtman, R. J. Evidence for a membrane defect in Alzheimer disease brain. *Proc Natl Acad Sci USA* **1992**, *89*, 1671–1675.
564 Soderberg, M., Edlund, C., Alafuzoff, I., Kristensson, K. and Dallner, G. Lipid composition in different regions of the brain in Alzheimer's disease/senile dementia of Alzheimer's type. *J Neurochem* **1992**, *59*, 1646–1653.
565 Svennerholm, L. and Gottfries, C. G. Membrane lipids, selectively diminished in Alzheimer brains, suggest synapse loss as a primary event in early-onset form (type I) and demyelination in late-onset form (type II). *J Neurochem* **1994**, *62*, 1039–1047.

566 Yanagisawa, K., Odaka, A., Suzuki, N. and Ihara, Y. Gm1 Gangloside-bound amyloid-β-protein (Aβ) – a possible form of preamyloidin Alzheimer's disease. *Nat Med* **1995**, *1*, 1062–1066.

567 Yanagisawa, K. and Ihara, Y. GM_1 ganglioside-bound amyloid β-protein in Alzheimer's-disease brain. *Neurobiol Aging* **1998**, *19*, S 65–S 67.

568 Hayashi, H., Kimura, N., Yamaguchi, H., Hasegawa, K., Yokoseki, T., Shibata, M., Yamamoto, N., Michikawa, M., et al. A seed for Alzheimer amyloid in the brain. *J Neurosci* **2004**, *24*, 4894–4902.

569 McLaurin, J., Franklin, T., Fraser, P. E. and Chakrabartty, A. Structural transitions associated with the interaction of Alzheimer β-amyloid peptides with gangliosides. *J Biol Chem* **1998**, *273*, 4506–4515.

570 Matsuzaki, K. and Horikiri, C. Interactions of amyloid β-peptide (1–40) with ganglioside-containing membranes. *Biochemistry* **1999**, *38*, 4137–4142.

571 Kakio, A., Nishimoto, S., Yanagisawa, K., Kozutsumi, Y. and Matsuzaki, K. Interactions of amyloid β-protein with various gangliosides in raft-like membranes: Importance of GM_1 ganglioside-bound form as an endogenous seed for Alzheimer amyloid. *Biochemistry* **2002**, *41*, 7385–7390.

572 Kakio, A., Nishimoto, S., Yanagisawa, K., Kozutsumi, Y. and Matsuzaki, K. Cholesterol-dependent formation of GM_1 ganglioside-bound amyloid β-protein, an endogenous seed for Alzheimer amyloid. *J Biol Chem* **2001**, *276*, 24985–24990.

573 Tashima, Y., Oe, R., Lee, S., Sugihara, G., Chambers, E. J., Takahashi, M. and Yamada, T. The effect of cholesterol and monosialoganglioside (GM_1) on the release and aggregation of amyloid β-peptide from liposomes prepared from brain membrane-like lipids. *J Biol Chem* **2004**, *279*, 17587–17595.

574 Choo-Smith, L. P., Garzon-Rodriguez, W., Glabe, C. G. and Surewicz, W. K. Acceleration of amyloid fibril formation by specific binding of Aβ-(1–40) peptide to ganglioside-containing membrane vesicles. *J Biol Chem* **1997**, *272*, 22987–22990.

575 Ariga, T., Kobayashi, K., Hasegawa, A., Kiso, M., Ishida, H. and Miyatake, T. Characterization of high-affinity binding between gangliosides and amyloid β-protein. *Arch Biochem Biophys* **2001**, *388*, 225–230.

576 McLaurin, J. and Chakrabartty, A. Characterization of the interactions of Alzheimer β-amyloid peptides with phospholipid membranes. *Eur J Biochem* **1997**, *245*, 355–363.

577 Terzi, E., Holzemann, G. and Seelig, J. Alzheimer β-amyloid peptide 25–35: electrostatic interactions with phospholipid membranes. *Biochemistry* **1994**, *33*, 7434–7441.

578 McLaurin, J., Franklin, T., Chakrabartty, A. and Fraser, P. E. Phosphatidylinositol and inositol involvement in Alzheimer amyloid-β fibril growth and arrest. *J Mol Biol* **1998**, *278*, 183–194.

579 McLaurin, J., Golomb, R., Jurewicz, A., Antel, J. P. and Fraser, P. E. Inositol stereoisomers stabilize an oligomeric aggregate of Alzheimer amyloid β peptide and inhibit Aβ-induced toxicity. *J Biol Chem* **2000**, *275*, 18495–18502.

580 Wellington, C. L. Cholesterol at the crossroads: Alzheimer's disease and lipid metabolism. *Clin Genet* **2004**, *66*, 1–16.

581 Reiss, A. B., Siller, K. A., Rahman, M. M., Chan, E. S., Ghiso, J. and de Leon, M. J. Cholesterol in neurologic disorders of the elderly: stroke and Alzheimer's disease. *Neurobiol Aging* **2004**, *25*, 977–989.

582 Fassbender, K., Simons, M., Bergmann, C., Stroick, M., Lutjohann, D., Keller, P., Runz, H., Kuhl, S., et al. Simvastatin strongly reduces levels of Alzheimer's disease β-amyloid peptides Aβ42 and Aβ40 *in vitro* and *in vivo*. *Proc Natl Acad Sci USA* **2001**, *98*, 5856–5861.

583 Wolozin, B. Cholesterol and Alzheimer's disease. *Biochem Soc Trans* **2002**, *30*, 525–529.

584 Jarvik, G. P., Wijsman, E. M., Kukull, W. A., Schellenberg, G. D., Yu, C. and Larson, E. B. Interactions of apolipoprotein E genotype, total cholesterol level, age, and sex in prediction of Alzheimer's disease: a case-control study. *Neurology* **1995**, *45*, 1092–1096.

585 Sparks, D. L., Scheff, S. W., Hunsaker, J. C., 3rd, Liu, H., Landers, T. and Gross, D. R. Induction of Alzheimer-like β-amyloid immunoreactivity in the brains of rabbits with dietary cholesterol. *Exp Neurol* **1994**, *126*, 88–94.

586 Shie, F. S., Jin, L. W., David, G. C., Leverenz, J. B. and LeBoeuf, R. C. Diet-induced hypercholesterolemia enhances brain Aβ accumulation in transgenic mice. *Neuroreport* **2002**, *13*, 455–459.

587 Simons, M., Keller, P., De Strooper, B., Beyreuther, K., Dotti, C. G. and Simons, K. Cholesterol depletion inhibits the generation of β-amyloid in hippocampal neurons. *Proc Natl Acad Sci USA* **1998**, *95*, 6460–6464.

588 Frears, E. R., Stephens, D. J., Walters, C. E., Davies, H. and Austen, B. M. The role of cholesterol in the biosynthesis of β-amyloid. *Neuroreport* **1999**, *10*, 1699–1705.

589 Kirsch, C., Eckert, G. P., Koudinov, A. R. and Muller, W. E. Brain cholesterol, statins and Alzheimer's Disease. *Pharmacopsychiatry* **2003**, *36 (Suppl 2)*, S113–119.

590 Harris, J. R. *In vitro* fibrillogenesis of the amyloid β(1–42) peptide: cholesterol potentiation and aspirin inhibition. *Micron* **2002**, *33*, 609–626.

591 Sponne, I., Fifre, A., Koziel, V., Oster, T., Olivier, J. L. and Pillot, T. Membrane cholesterol interferes with neuronal apoptosis induced by soluble oligomers but not fibrils of amyloid-β peptide. *FASEB J* **2004**, *18*, 836–838.

592 Sambamurti, K., Granholm, A. C., Kindy, M. S., Bhat, N. R., Greig, N. H., Lahiri, D. K. and Mintzer, J. E. Cholesterol and Alzheimer's disease: clinical and experimental models suggest interactions of different genetic, dietary and environmental risk factors. *Curr Drug Targets* **2004**, *5*, 517–528.

593 Yanagisawa, K. and Matsuzaki, K. Cholesterol-dependent aggregation of amyloid β-protein. *Ann NY Acad Sci* **2002**, *977*, 384–386.

594 McLaurin, J., Darabie, A. A. and Morrison, M. R. Cholesterol, a modulator of membrane-associated Aβ fibrillogenesis. *Ann NY Acad Sci* **2002**, *977*, 376–383.

595 Zhang, Q., Powers, E. T., Nieva, J., Huff, M. E., Dendle, M. A., Bieschke, J., Glabe, C. G., Eschenmoser, A., et al. Metabolite-initiated protein misfolding may trigger Alzheimer's disease. *Proc Natl Acad Sci USA* **2004**, *101*, 4752–4757.

596 Hershey, C. O., Hershey, L. A., Varnes, A., Vibhakar, S. D., Lavin, P. and Strain, W. H. Cerebrospinal fluid trace element content in dementia: clinical, radiologic, and pathologic correlations. *Neurology* **1983**, *33*, 1350–1353.

597 Ehmann, W. D., Markesbery, W. R., Alauddin, M., Hossain, T. I. and Brubaker, E. H. Brain trace elements in Alzheimer's disease. *Neurotoxicology* **1986**, *7*, 195–206.

598 Thompson, C. M., Markesbery, W. R., Ehmann, W. D., Mao, Y. X. and Vance, D. E. Regional brain trace-element studies in Alzheimer's disease. *Neurotoxicology* **1988**, *9*, 1–7.

599 Basun, H., Forssell, L. G., Wetterberg, L. and Winblad, B. Metals and trace elements in plasma and cerebrospinal fluid in normal aging and Alzheimer's disease. *J Neural Transm Park Dis Dement Sect* **1991**, *3*, 231–258.

600 Deibel, M. A., Ehmann, W. D. and Markesbery, W. R. Copper, iron, and zinc imbalances in severely degenerated brain regions in Alzheimer's disease: possible relation to oxidative stress. *J Neurol Sci* **1996**, *143*, 137–142.

601 Cornett, C. R., Markesbery, W. R. and Ehmann, W. D. Imbalances of trace elements related to oxidative damage in Alzheimer's disease brain. *Neurotoxicology* **1998**, *19*, 339–345.

602 Gonzalez, C., Martin, T., Cacho, J., Brenas, M. T., Arroyo, T., Garcia-Berrocal, B., Navajo, J. A. and Gonzalez-Buitrago, J. M. Serum zinc, copper, insulin and lipids in Alzheimer's disease ε4 apolipoprotein E allele carriers. *Eur J Clin Invest* **1999**, *29*, 637–642.

603 Bush, A. I. The metallobiology of Alzheimer's disease. *Trends Neurosci* **2003**, *26*, 207–214.

604 Lynch, T., Cherny, R.A. and Bush, A.I. Oxidative processes in Alzheimer's disease: the role of Aβ metal interactions. *Exp Gerontol* **2000**, *35*, 445–451.

605 Lovell, M.A., Robertson, J.D., Teesdale, W.J., Campbell, J.L. and Markesbery, W.R. Copper, iron and zinc in Alzheimer's disease senile plaques. *J Neurol Sci* **1998**, *158*, 47–52.

606 Danscher, G., Jensen, K.B., Frederickson, C.J., Kemp, K., Andreasen, A., Juhl, S., Stoltenberg, M. and Ravid, R. Increased amount of zinc in the hippocampus and amygdala of Alzheimer's-diseased brains – a proton-induced X-ray emission spectroscopic analysis of cryostat sections from autopsy material. *J Neurosci Methods* **1997**, *76*, 53–59.

607 Dong, J., Atwood, C.S., Anderson, V.E., Siedlak, S.L., Smith, M.A., Perry, G. and Carey, P.R. Metal binding and oxidation of amyloid-β within isolated senile plaque cores: Raman microscopic evidence. *Biochemistry* **2003**, *42*, 2768–2773.

608 Assaf, S.Y. and Chung, S.H. Release of endogenous Zn^{2+} from brain tissue during activity. *Nature* **1984**, *308*, 734–736.

609 Huang, X.D., Atwood, C.S., Moir, R.D., Hartshorn, M.A., Vonsattel, J.P., Tanzi, R.E. and Bush, A.I. Zinc-induced Alzheimer's Aβ1–40 aggregation is mediated by conformational factors. *J Biol Chem* **1997**, *272*, 26464–26470.

610 Owen, C.A., Jr. Uptake of ^{67}Cu by ceruloplasmin *in vitro*. *Proc Soc Exp Biol Med* **1975**, *149*, 681–682.

611 Lestas, A.N. The effect of pH upon human transferrin: selective labelling of the two iron-binding sites. *Br J Haematol* **1976**, *32*, 341–350.

612 Brieland, J.K. and Fantone, J.C. Ferrous iron release from transferrin by human neutrophil-derived superoxide anion: effect of pH and iron saturation. *Arch Biochem Biophys* **1991**, *284*, 78–83.

613 Lamb, D.J. and Leake, D.S. Acidic pH enables caeruloplasmin to catalyse the modification of low-density lipoprotein. *FEBS Lett* **1994**, *338*, 122–126.

614 Lamb, D.J. and Leake, D.S. Iron released from transferrin at acidic pH can catalyse the oxidation of low density lipoprotein. *FEBS Lett* **1994**, *352*, 15–18.

615 Atwood, C.S., Scarpa, R.C., Huang, X.D., Moir, R.D., Jones, W.D., Fairlie, D.P., Tanzi, R.E. and Bush, A.I. Characterization of copper interactions with Alzheimer amyloid β peptides: identification of an attomolar-affinity copper binding site on amyloid β 1–42. *J Neurochem* **2000**, *75*, 1219–1233.

616 Bush, A.I., Pettingell, W.H., Multhaup, G., Paradis, M.D., Vonsattel, J.P., Gusella, J.F., Beyreuther, K., Masters, C.L., et al. Rapid induction of Alzheimer Aβ amyloid formation by zinc. *Science* **1994**, *265*, 1464–1467.

617 Atwood, C.S., Moir, R.D., Huang, X.D., Scarpa, R.C., Bacarra, N.M.E., Romano, D.M., Hartshorn, M.K., Tanzi, R.E., et al. Dramatic aggregation of Alzeimer Aβ by Cu(II) is induced by conditions representing physiological acidosis. *J Biol Chem* **1998**, *273*, 12817–12826.

618 Morgan, D.M., Dong, J.J., Jacob, J., Lu, K., Apkarian, R.P., Thiyagarajan, P. and Lynn, D.G. Metal switch for amyloid formation: Insight into the structure of the nucleus. *J Am Chem Soc* **2002**, *124*, 12644–12645.

619 Bush, A.I., Pettingell, W.H., Paradis, M.D. and Tanzi, R.E. Modulation of A-β adhesiveness and secretase site cleavage by zinc. *J Biol Chem* **1994**, *269*, 12152–12158.

620 Bush, A.I., Pettingell, W.H., Multhaup, G., Paradis, M.D., Vonsattel, J.P., Gusella, J.F., Beyreuther, K., Masters, C.L., et al. Rapid Induction of Alzheimer A-β amyloid formation by zinc. *Science* **1994**, *265*, 1464–1467.

621 Yang, D. S., McLaurin, J., Qin, K. F., Westaway, D. and Fraser, P. E. Examining the zinc binding site of the amyloid-β peptide. *Eur J Biochem* **2000**, *267*, 6692–6698.

622 Miura, T., Suzuki, K., Kohata, N. and Takeuchi, H. Metal binding modes of Alzheimer's amyloid β-peptide in insoluble aggregates and soluble complexes. *Biochemistry* **2000**, *39*, 7024–7031.

623 Curtain, C.C., Ali, F., Volitakis, I., Cherny, R.A., Norton, R.S., Beyreuther,

K., Barrow, C. J., Masters, C. L., et al. Alzheimer's disease amyloid-β binds copper and zinc to generate an allosterically ordered membrane-penetrating structure containing superoxide dismutase-like subunits. *J Biol Chem* **2001**, *276*, 20466–20473.

624 Liu, S. T., Howlett, G. and Barrow, C. J. Histidine-13 is a crucial residue in the zinc ion-induced aggregation of the Aβ peptide of Alzheimer's disease. *Biochemistry* **1999**, *38*, 9373–9378.

625 Yamada, T., Sasaki, H., Furuya, H., Miyata, T., Goto, I. and Sakaki, Y. Complementary DNA for the mouse homolog of the human amyloid beta protein precursor. *Biochem Biophys Res Commun* **1987**, *149*, 665–671.

626 Cherny, R. A., Legg, J. T., McLean, C. A., Fairlie, D. P., Huang, X. D., Atwood, C. S., Beyreuther, K., Tanzi, R. E., et al. Aqueous dissolution of Alzheimer's disease Aβ amyloid deposits by biometal depletion. *J Biol Chem* **1999**, *274*, 23223–23228.

627 Cherny, R. A., Barnham, K. J., Lynch, T., Volitakis, I., Li, Q. X., McLean, C. A., Multhaup, G., Beyreuher, K., et al. Chelation and intercalation: complementary properties in a compound for the treatment of Alzheimer's disease. *J Struct Biol* **2000**, *130*, 209–216.

628 Cherny, R. A., Atwood, C. S., Xilinas, M. E., Gray, D. N., Jones, W. D., McLean, C. A., Barnham, K. J., Volitakis, I., et al. Treatment with a copper–zinc chelator markedly and rapidly inhibits β-amyloid accumulation in Alzheimer's disease transgenic mice. *Neuron* **2001**, *30*, 665–676.

629 Lee, J. Y., Cole, T. B., Palmiter, R. D., Suh, S. W. and Koh, J. Y. Contribution by synaptic zinc to the gender-disparate plaque formation in human Swedish mutant AβPP transgenic mice. *Proc Natl Acad Sci USA* **2002**, *99*, 7705–7710.

630 Cohen, F. E. and Kelly, J. W. Therapeutic approaches to protein-misfolding diseases. *Nature* **2003**, *426*, 905–909.

631 Mason, J. M., Kokkoni, N., Stott, K. and Doig, A. J. Design strategies for anti-amyloid agents. *Curr Opin Struct Biol* **2003**, *13*, 526–532.

632 Findeis, M. A. Peptide inhibitors of β-amyloid aggregation. *Curr Top Med Chem* **2002**, *2*, 417–423.

633 Lansbury, P. T. Inhibition of amyloid formation – a strategy to delay the onset of Alzheimer's disease. *Curr Opin Chem Biol* **1997**, *1*, 260–267.

634 Soto, C., Sigurdsson, E. M., Morelli, L., Kumar, R. A., Castano, E. M. and Frangione, B. β-sheet breaker peptides inhibit fibrillogenesis in a rat brain model of amyloidosis – implications for Alzheimer's therapy. *Nat Med* **1998**, *4*, 822–826.

635 Fezoui, Y. and Teplow, D. B. Kinetic studies of amyloid β-protein fibril assembly – differential effects of α-helix stabilization. *J Biol Chem* **2002**, *277*, 36948–36954.

636 Hammarström, P., Wiseman, R. L., Powers, E. T. and Kelly, J. W. Prevention of transthyretin amyloid disease by changing protein misfolding energetics. *Science* **2003**, *299*, 713–716.

637 De Felice, F. G., Vieira, M. N., Saraiva, L. M., Figueroa-Villar, J. D., Garcia-Abreu, J., Liu, R., Chang, L., Klein, W. L., et al. Targeting the neurotoxic species in Alzheimer's disease: inhibitors of Aβ oligomerization. *FASEB J* **2004**, *18*, 1366–1372.

638 Chiti, F., Taddei, N., Stefani, M., Dobson, C. M. and Ramponi, G. Reduction of the amyloidogenicity of a protein by specific binding of ligands to the native conformation. *Protein Sci* **2001**, *10*, 879–886.

639 Peterson, S. A., Klabunde, T., Lashuel, H. A., Purkey, H., Sacchettini, J. C. and Kelly, J. W. Inhibiting transthyretin conformational changes that lead to amyloid fibril formation. *Proc Natl Acad Sci USA* **1998**, *95*, 12956–12960.

640 Fraser, P. E., McLachlan, D. R., Surewicz, W. K., Mizzen, C. A., Snow, A. D., Nguyen, J. T. and Kirschner, D. A. Conformation and fibrillogenesis of Alzheimer Aβ peptides with selected substitution of charged residues. *J Mol Biol* **1994**, *244*, 64–73.

641 Hilbich, C., Kisters-Woike, B., Reed, J., Masters, C. L. and Beyreuther, K. Aggregation and secondary structure of synthetic amyloid βA4 peptides of Alzheimer's disease. *J Mol Biol* **1991**, *218*, 149–163.

642 Stephenson, V.C. and Weaver, D.F. Halting amyloidosis with dianionic compounds: A comprehensive molecular modelling and *in vitro* investigation. *Soc Neurosci Abstr* **2004**, 673–623.

643 Giulian, D., Haverkamp, L.J., Yu, J.H., Karshin, M., Tom, D., Li, J., Kazanskaia, A., Kirkpatrick, J., et al. The HHQK domain of β-amyloid provides a structural basis for the immunopathology of Alzheimer's disease. *J Biol Chem* **1998**, *273*, 29719–29726.

644 Streit, W.J. Microglia and Alzheimer's disease pathogenesis. *J Neurosci Res* **2004**, *77*, 1–8.

645 Wood, S.J., Wetzel, R., Martin, J.D. and Hurle, M.R. Prolines and amyloidogenicity in fragments of the Alzheimer's peptide β/A4. *Biochemistry* **1995**, *34*, 724–730.

646 Maji, S.K., Amsden, J.J., Rothschild, K.J., Condron, M.M. and Teplow, D.B. Assembly-associated conformational transitions of Aβ40 and Aβ42 probed using intrinsic fluorescence. *Submitted for publication* 2005.

647 Tarus, B., Straub, J.E. and Thirumalai, D. Probing the initial stage of aggregation of the $A\beta_{10-35}$-protein: assessing the propensity for peptide dimerization. *J Mol Biol* **2005**, *345*, 1141–1156.

648 Jarvet, J., Damberg, P., Bodell, K., Eriksson, L. E. and Graslund, A. Reversible random coil to β-sheet transition and the early stage of aggregation of the Aβ(12–28) fragment from the Alzheimer peptide. *J Am Chem Soc* **2000**, *122*, 4261–4268.

649 Zhang, S., Iwata, K., Lachenmann, M.J., Peng, J.W., Li, S., Stimson, E.R., Lu, Y., Felix, A.M., et al. The Alzheimer's peptide Aβ adopts a collapsed coil structure in water. *J Struct Biol* **2000**, *130*, 130–141.

650 Esler, W.P., Felix, A.M., Stimson, E.R., Lachenmann, M.J., Ghilardi, J.R., Lu, Y.A., Vinters, H.V., Mantyh, P.W., et al. Activation barriers to structural transition determine deposition rates of Alzheimer's disease Aβ amyloid. *J Struct Biol* **2000**, *130*, 174–183.

651 Tjernberg, L.O., Naslund, J., Lindqvist, F., Johansson, J., Karlstrom, A.R., Thyberg, J., Terenius, L. and Nordstedt, C. Arrest of β-amyloid fibril formation by a pentapeptide ligand. *J Biol Chem* **1996**, *271*, 8545–8548.

652 Tjernberg, L.O., Lilliehook, C., Callaway, D.J.E., Naslund, J., Hahne, S., Thyberg, J., Terenius, L. and Nordstedt, C. Controlling amyloid β-peptide fibril formation with protease-stable ligands. *J Biol Chem* **1997**, *272*, 12601–12605.

653 Soto, C., Kindy, M.S., Baumann, M. and Frangione, B. Inhibition of Alzheimer's amyloidosis by peptides that prevent β-sheet conformation. *Biochem Biophys Res Commun* **1996**, *226*, 672–680.

654 Adessi, C. and Soto, C. β-sheet breaker strategy for the treatment of Alzheimer's disease. *Drug Dev Res* **2002**, *56*, 184–193.

655 Adessi, C., Frossard, M.J., Boissard, C., Fraga, S., Bieler, S., Ruckle, T., Vilbois, F., Robinson, S.M., et al. Pharmacological profiles of peptide drug candidates for the treatment of Alzheimer's disease. *J Biol Chem* **2003**, *278*, 13905–13911.

656 Ghanta, J., Shen, C.L., Kiessling, L.L. and Murphy, R.M. A strategy for designing inhibitors of β-amyloid toxicity. *J Biol Chem* **1996**, *271*, 29525–29528.

657 Lowe, T.L., Strzelec, A., Kiessling, L.L. and Murphy, R.M. Structure–function relationships for inhibitors of β-amyloid toxicity containing the recognition sequence KLVFF. *Biochemistry* **2001**, *40*, 7882–7889.

658 Chalifour, R.J., McLaughlin, R.W., Lavoie, L., Morissette, C., Tremblay, N., Boule, M., Sarazin, P., Stea, D., et al. Stereoselective interactions of peptide inhibitors with the β-amyloid peptide. *J Biol Chem* **2003**, *278*, 34874–34881.

659 Gordon, D.J., Sciarretta, K.L. and Meredith, S.C. Inhibition of β-amyloid(40) fibrillogenesis and disassembly of β-amyloid(40) fibrils by short β-amyloid congeners containning *N*-methyl amino acids at alternate residues. *Biochemistry* **2001**, *40*, 8237–8245.

660 Gordon, D.J., Tappe, R. and Meredith, S.C. Design and characterization of a membrane permeable *N*-methyl amino acid-containing peptide that inhibits Aβ(1–40) fibrillogenesis. *J Pept Res* **2002**, *60*, 37–55.

661 Gordon, D.J. and Meredith, S.C. Probing the role of backbone hydrogen bonding in β-amyloid fibrils with inhibitor peptides containing ester bonds at alternate positions. *Biochemistry* **2003**, *42*, 475–485.

662 Findeis, M.A., Musso, G.M., Arico-Muendel, C.C., Benjamin, H.W., Hundal, A.M., Lee, J.J., Chin, J., Kelley, M., et al. Modified-peptide inhibitors of amyloid β-peptide polymerization. *Biochemistry* **1999**, *38*, 6791–6800.

663 Findeis, M.A., Lee, J.J., Kelley, M., Wakefield, J.D., Zhang, M.H., Chin, J., Kubasek, W. and Molineaux, S.M. Characterization of Cholyl-Leu-Val-Phe-Phe-Ala-OH as an inhibitor of amyloid β-peptide polymerization. *Amyloid* **2001**, *8*, 231–241.

664 Jarrett, J.T., Berger, E.P. and Lansbury, P.T., Jr. The C-terminus of the β protein is critical in amyloidogenesis. *Ann NY Acad Sci* **1993**, 144–148.

665 Riek, R., Guntert, P., Döbeli, H., Wipf, B. and Wüthrich, K. NMR studies in aqueous solution fail to identify significant conformational differences between the monomeric forms of two Alzheimer peptides with widely different plaque-competence, Aβ(1–40)^{ox} and Aβ(1–42)^{ox}. *Eur J Biochem* **2001**, *268*, 5930–5936.

666 Urbanc, B., Cruz, L., Yun, S., Buldyrev, S.V., Bitan, G., Teplow, D.B. and Stanley, H. E. *In silico* study of amyloid β-protein folding and oligomerization. *Proc Natl Acad Sci USA* **2004**, *101*, 17345–17350.

667 Kuo, Y.M., Kokjohn, T.A., Beach, T.G., Sue, L.I., Brune, D., Lopez, J.C., Kalback, W.M., Abramowski, D., et al. Comparative analysis of amyloid-β chemical structure and amyloid plaque morphology of transgenic mouse and Alzheimer's disease brains. *J Biol Chem* **2001**, *276*, 12991–12998.

668 Watson, A.A., Fairlie, D.P. and Craik, D.J. Solution structure of methionine-oxidized amyloid β-peptide (1–40) – does oxidation affect conformational switching? *Biochemistry* **1998**, *37*, 12700–12706.

669 Palmblad, M., Westlind-Danielsson, A. and Bergquist, J. Oxidation of methionine 35 attenuates formation of amyloid β-peptide 1–40 oligomers. *J Biol Chem* **2002**, *277*, 19506–19510.

670 Hetenyi, C., Szabo, Z., Klement, T., Datki, Z., Kortvelyesi, T., Zarandi, M. and Penke, B. Pentapeptide amides interfere with the aggregation of β-amyloid peptide of Alzheimer's disease. *Biochem Biophys Res Commun* **2002**, *292*, 931–936.

671 Tanford, C. *The Hydrophobic Effect: Formation of Micelles and Biological Membranes.* Wiley, New York, 1973.

672 Hilbich, C., Kisters-Woike, B., Reed, J., Masters, C.L. and Beyreuther, K. Substitutions of hydrophobic amino acids reduce the amyloidogenicity of Alzheimer's disease βA4 peptides. *J Mol Biol* **1992**, *228*, 460–473.

673 Halverson, K., Fraser, P.E., Kirschner, D.A. and Lansbury, P.T., Jr. Molecular determinants of amyloid deposition in Alzheimer's disease: conformational studies of synthetic β-protein fragments. *Biochemistry* **1990**, *29*, 2639–2644.

674 Jarrett, J.T. and Lansbury, P.T., Jr. Seeding "one-dimensional crystallization" of amyloid: A pathogenic mechanism in Alzheimer's disease and scrapie? *Cell* **1993**, *73*, 1055–1058.

675 Fraser, P.E., Nguyen, J.T., Surewicz, W.K. and Kirschner, D.A. pH-dependent structural transitions of Alzheimer amyloid peptides. *Biophys J* **1991**, *60*, 1190–1201.

676 Fraser, P.E., Nguyen, J.T., Inouye, H., Surewicz, W. K., Selkoe, D.J., Podlisny, M.B. and Kirschner, D.A. Fibril formation by primate, rodent, and Dutch-hemorrhagic analogues of Alzheimer amyloid β-protein. *Biochemistry* **1992**, *31*, 10716–10723.

677 Tjernberg, L.O., Callaway, D.J.E., Tjernberg, A., Hahne, S., Lilliehook, C., Terenius, L., Thyberg, J. and Nordstedt, C. A molecular model of Alzheimer amyloid β-peptide fibril formation. *J Biol Chem* **1999**, *274*, 12619–12625.

678 Cummings, J. L. Alzheimer's disease. *N Engl J Med* **2004**, *351*, 56–67.

679 Racchi, M., Mazzucchelli, M., Porrello, E., Lanni, C. and Govoni, S. Acetylcholinesterase inhibitors: novel activities of old molecules. *Pharmacol Res* **2004**, *50*, 441–451.

680 Bullock, R. Future directions in the treatment of Alzheimer's disease. *Expert Opin Invest Drugs* **2004**, *13*, 303–314.

681 Morishima-Kawashima, M. and Hara, Y. Alzheimer's disease: β-amyloid protein and tau. *J Neurosci Res* **2002**, *70*, 392–401.

682 Smith-Swintosky, V. L. and Mattson, M. P. Glutamate, β-amyloid precursor proteins, and calcium mediated neurofibrillary degeneration. *J Neural Transm Suppl* **1994**, *44*, 29–45.

683 Giacobini, E. Cholinesterase inhibitors stabilize Alzheimer's disease. *Ann NY Acad Sci* **2000**, *920*, 321–327.

684 Fu, W. M., Luo, H., Parthasarathy, S. and Mattson, M. P. Catecholamines potentiate amyloid β-peptide neurotoxicity – involvement of oxidative stress, mitochondrial dysfunction, and perturbed calcium homeostasis. *Neurobiol Dis* **1998**, *5*, 229–243.

685 Gonzalo-Ruiz, A., Gonzalez, I. and Sanz-Anquela, J. M. Effects of β-amyloid protein on serotoninergic, noradrenergic, and cholinergic markers in neurons of the pontomesencephalic tegmentum in the rat. *J Chem Neuroanat* **2003**, *26*, 153–169.

686 Weiner, H. L. and Selkoe, D. J. Inflammation and therapeutic vaccination in CNS diseases. *Nature* **2002**, *420*, 879–884.

687 Cotman, C. W. and Anderson, A. J. A potential role for apoptosis in neurodegeneration and Alzheimer's disease. *Mol Neurobiol* **1995**, *10*, 19–45.

688 Monsonego, A. and Weiner, H. L. Immunotherapeutic approaches to Alzheimer's disease. *Science* **2003**, *302*, 834–838.

689 Tomita, T. and Iwatsubo, T. The inhibition of γ-secretase as a therapeutic approach to Alzheimer's disease. *Drug News Perspect* **2004**, *17*, 321–325.

690 Citron, M. Strategies for disease modification in Alzheimer's disease. *Nat Rev Neurosci* **2004**, *5*, 677–685.

691 LeVine, H., 3rd. The amyloid hypothesis and the clearance and degradation of Alzheimer's β-peptide. *J Alzheimer's Dis* **2004**, *6*, 303–314.

692 Mattson, M. P. Emerging neuroprotective strategies for Alzheimer's disease: dietary restriction, telomerase activation, and stem cell therapy. *Exp Gerontol* **2000**, *35*, 489–502.

693 Behl, C. and Moosmann, B. Antioxidant neuroprotection in Alzheimer's disease as preventive and therapeutic approach. *Free Rad Biol Med* **2002**, *33*, 182–191.

694 Halliday, G., Robinson, S. R., Shepherd, C. and Kril, J. Alzheimer's disease and inflammation: a review of cellular and therapeutic mechanisms. *Clin Exp Pharmacol Physiol* **2000**, *27*, 1–8.

695 Bieber, E. J. and Cohen, D. P. Estrogens and hormone replacement therapy: is there a role in the preservation of cognitive function? *Int J Fertil Womens Med* **2001**, *46*, 206–209.

696 Giacobini, E. Long-term stabilizing effect of cholinesterase inhibitors in the therapy of Alzheimer's disease. *J Neural Transm Suppl* **2002**, 181–187.

697 Price, D. L. New perspectives on Alzheimer's disease. *Annu Rev Neurosci* **1986**, *9*, 489–512.

698 Whitehouse, P. J. Cholinergic therapy in dementia. *Acta Neurol Scand Suppl* **1993**, *149*, 42–45.

699 Knapp, M. J., Knopman, D. S., Solomon, P. R., Pendlebury, W. W., Davis, C. S. and Gracon, S. I. A 30-week randomized controlled trial of high-dose tacrine in patients with Alzheimer's disease. The Tacrine Study Group. *J Am Med Ass* **1994**, *271*, 985–991.

700 Wagstaff, A. J. and McTavish, D. Tacrine. A review of its pharmacodynamic and pharmacokinetic properties, and therapeutic efficacy in Alzheimer's disease. *Drugs Aging* **1994**, *4*, 510–540.

701 Kelly, C. A., Harvey, R. J. and Cayton, H. Drug treatments for Alzheimer's disease raise clinical and ethical problems. *Br Med J* **1997**, *314*, 693–694.

702 Scott, L. J. and Goa, K. L. Galantamine: a review of its use in Alzheimer's disease. *Drugs* **2000**, *60*, 1095–1122.

703 Gottwald, M. D. and Rozanski, R. I. Rivastigmine, a brain-region selective acetylcholinesterase inhibitor for treating Alzheimer's disease: review and current status. *Expert Opin Investig Drugs* **1999**, *8*, 1673–1682.

704 Ringman, J. M. and Cummings, J. L. Metrifonate: update on a new antidementia agent. *J Clin Psychiatry* **1999**, *60*, 776–782.

705 Nitsch, R. M., Slack, B. E., Wurtman, R. J. and Growdon, J. H. Release of Alzheimer amyloid precursor derivatives stimulated by activation of muscarinic acetylcholine receptors. *Science* **1992**, *258*, 304–307.

706 Wolf, B. A., Wertkin, A. M., Jolly, Y. C., Yasuda, R. P., Wolfe, B. B., Konrad, R. J., Manning, D., Ravi, S., et al. Muscarinic regulation of Alzheimer's disease amyloid precursor protein secretion and amyloid β-protein production in human neuronal nt2n cells. *J Biol Chem* **1995**, *270*, 4916–4922.

707 Nitsch, R. M., Farber, S. A., Growdon, J. H. and Wurtman, R. J. Release of amyloid β-protein precursor derivatives by electrical depolarization of rat hippocampal slices. *Proc Natl Acad Sci USA* **1993**, *90*, 5191–5193.

708 Racchi, M., Sironi, M., Caprera, A., Konig, G. and Govoni, S. Short- and long-term effect of acetylcholinesterase inhibition on the expression and metabolism of the amyloid precursor protein. *Mol Psychiatry* **2001**, *6*, 520–528.

709 Slack, B. E., Breu, J., Petryniak, M. A., Srivastava, K. and Wurtman, R. J. Tyrosine phosphorylation-dependent stimulation of amyloid precursor protein secretion by the M3 muscarinic acetylcholine receptor. *J Biol Chem* **1995**, *270*, 8337–8344.

710 Slack, B. E. The m3 muscarinic acetylcholine receptor is coupled to mitogen-activated protein kinase via protein kinase C and epidermal growth factor receptor kinase. *Biochem J* **2000**, *348*, 381–387.

711 Haring, R., Fisher, A., Marciano, D., Pittel, Z., Kloog, Y., Zuckerman, A., Eshhar, N. and Heldman, E. Mitogen-activated protein kinase-dependent and protein kinase C-dependent pathways link the m1 muscarinic receptor to β-amyloid precursor protein secretion. *J Neurochem* **1998**, *71*, 2094–2103.

712 Giacobini, E. Cholinesterase inhibitors stabilize Alzheimer's disease. *Neurochem Res* **2000**, *25*, 1185–1190.

713 Parsons, C. G., Danysz, W. and Quack, G. Memantine is a clinically well tolerated *N*-methyl-D-aspartate (NMDA) receptor antagonist – a review of preclinical data. *Neuropharmacology* **1999**, *38*, 735–767.

714 Ferris, S. H. Evaluation of memantine for the treatment of Alzheimer's disease. *Expert Opin Pharmacother* **2003**, *4*, 2305–2313.

715 Mobius, H. J., Stoffler, A. and Graham, S. M. Memantine hydrochloride: pharmacological and clinical profile. *Drugs Today* **2004**, *40*, 685–695.

716 Rao, A. V. and Balachandran, B. Role of oxidative stress and antioxidants in neurodegenerative diseases. *Nutr Neurosci* **2002**, *5*, 291–309.

717 Multhaup, G., Ruppert, T., Schlicksupp, A., Hesse, L., Beher, D., Masters, C. L. and Beyreuther, K. Reactive oxygen species and Alzheimer's disease. *Biochem Pharmacol* **1997**, *54*, 533–539.

718 Grundman, M. Vitamin E and Alzheimer disease: the basis for additional clinical trials. *Am J Clin Nutr* **2000**, *71*, 630S–636S.

719 Sano, M., Ernesto, C., Thomas, R. G., Klauber, M. R., Schafer, K., Grundman, M., Woodbury, M., Growdon, P., et al. A controlled trial of selegiline, α-tocopherol, or both as treatments for Alzheimer's disease. The Alzheimer's Disease Cooperative Study. *New Eng J Med* **1997**, *36*, 1216–1222.

720 Solomon, B., Koppel, R., Hanan, E. and Katzav, T. Monoclonal antibodies inhibit *in vitro* fibrillar aggregation of the Alzheimer β-amyloid peptide. *Proc Natl Acad Sci USA* **1996**, *93*, 452–455.

721 Solomon, B., Koppel, R., Frankel, D. and Hanan-Aharon, E. Disaggregation

of Alzheimer β-amyloid by site-directed mAb. *Proc Natl Acad Sci USA* **1997**, *94*, 4109–4112.

722 Schenk, D., Barbour, R., Dunn, W., Gordon, G., Grajeda, H., Guido, T., Hu, K., Huang, J. P., et al. Immunization with amyloid-β attenuates Alzheimer disease-like pathology in the PDAβPP mouse. *Nature* **1999**, *400*, 173–177.

723 Holtzman, D. M., Bales, K. R., Paul, S. M. and DeMattos, R. B. Aβ immunization and anti-Aβ antibodies: potential therapies for the prevention and treatment of Alzheimer's disease. *Adv Drug Deliv Rev* **2002**, *54*, 1603–1613.

724 Gelinas, D. S., DaSilva, K., Fenili, D., St George-Hyslop, P. and McLaurin, J. Immunotherapy for Alzheimer's disease. *Proc Natl Acad Sci USA* **2004**, *101 (Suppl 2)*, 14657–14662.

725 Senior, K. Dosing in phase II trial of Alzheimer's vaccine suspended. *Lancet Neurol* **2002**, *1*, 3.

726 Orgogozo, J. M., Gilman, S., Dartigues, J. F., Laurent, B., Puel, M., Kirby, L. C., Jouanny, P., Dubois, B., et al. Subacute meningoencephalitis in a subset of patients with AD after Aβ42 immunization. *Neurology* **2003**, *61*, 46–54.

727 Nicoll, J. A. R., Wilkinson, D., Holmes, C., Steart, P., Markham, H. and Weller, R. O. Neuropathology of human Alzheimer disease after immunization with amyloid-β peptide: a case report. *Nat Med* **2003**, *9*, 448–452.

728 Hock, C., Konietzko, U., Streffer, J. R., Tracy, J., Signorell, A., Muller-Tillmanns, B., Lemke, U., Henke, K., et al. Antibodies against β-amyloid slow cognitive decline in Alzheimer's disease. *Neuron* **2003**, *38*, 547–554.

729 Lemere, C. A., Spooner, E. T., Leverone, J. F., Mori, C., Iglesias, M., Bloom, J. K. and Seabrook, T. J. Amyloid-β immunization in Alzheimer's disease transgenic mouse models and wildtype mice. *Neurochem Res* **2003**, *28*, 1017–1027.

730 Hartmann, T. Cholesterol, Aβ and Alzheimer's disease. *Trends Neurosci* **2001**, *24*, S45–S48.

731 Miller, L. J. and Chacko, R. The role of cholesterol and statins in Alzheimer's disease. *Ann Pharmacother* **2004**, *38*, 91–98.

732 Barnham, K. J., Masters, C. L. and Bush, A. I. Neurodegenerative diseases and oxidative stress. *Nat Rev Drug Discov* **2004**, *3*, 205–214.

733 Cuajungco, M. P., Faget, K. Y., Huang, X., Tanzi, R. E. and Bush, A. I. Metal chelation as a potential therapy for Alzheimer's disease. *Ann NY Acad Sci* **2000**, *920*, 292–304.

734 Gnjec, A., Fonte, J. A., Atwood, C. and Martins, R. N. Transition metal chelator therapy – a potential treatment for Alzheimer's disease? *Front Biosci* 1016, *7*, D1016–D1023.

735 Dedeoglu, A., Cormier, K., Payton, S., Tseitlin, K. A., Kremsky, J. N., Lai, L., Li, X., Moir, R. D., et al. Preliminary studies of a novel bifunctional metal chelator targeting Alzheimer's amyloidogenesis. *Exp Gerontol* **2004**, *39*, 1641–1649.

736 Ritchie, C. W., Bush, A. I., Mackinnon, A., Macfarlane, S., Mastwyk, M., MacGregor, L., Kiers, L., Cherny, R., et al. Metal–protein attenuation with iodochlorhydroxyquin (clioquinol) targeting Aβ amyloid deposition and toxicity in Alzheimer disease – a pilot phase 2 clinical trial. *Arch Neurol* **2003**, *60*, 1685–1691.

737 Aoyama, H., Oira, M., Ota, T., Yoshioka, S. and Yoshida, T. [Epidemiological studies on subacute myelo-optico neuropathy (SMON) in the town of Yubara, Japan. 3]. *Nippon Eiseigaku Zasshi* **1970**, *25*, 468–471.

738 Oakley, G. P., Jr. The neurotoxicity of the halogenated hydroxyquinolines. A commentary. *J Am Med Ass* **1973**, *225*, 395–397.

739 Ricoy, J. R., Ortega, A. and Cabello, A. Subacute myelo-optic neuropathy (SMON). First neuro-pathological report outside Japan. *J Neurol Sci* **1982**, *53*, 241–251.

740 Fillit, H. M. The role of hormone replacement therapy in the prevention of Alzheimer disease. *Arch Intern Med* **2002**, *162*, 1934–1942.

741 Henderson, V. W., Paganini-Hill, A., Miller, B. L., Elble, R. J., Reyes, P. F., Shoupe, D., McCleary, C. A., Klein,

R. A., et al. Estrogen for Alzheimer's disease in women – randomized, double-blind, placebo-controlled trial. *Neurology* **2000**, *54*, 295–301.

742 Shumaker, S. A., Legault, C., Rapp, S. R., Thal, L., Wallace, R. B., Ockene, J. K., Hendrix, S. L., Jones, B. N., 3rd, et al. Estrogen plus progestin and the incidence of dementia and mild cognitive impairment in postmenopausal women: the Women's Health Initiative Memory Study: a randomized controlled trial. *J Am Med Ass* **2003**, *289*, 2651–2662.

743 Henderson, V. W. Hormone therapy and Alzheimer's disease: benefit or harm? *Expert Opin Pharmacother* **2004**, *5*, 389–406.

744 Aisen, P. S. Inflammation and Alzheimer disease. *Mol Chem Neuropathol* **1996**, *28*, 83–88.

745 Cooper, N. R., Bradt, B. M., O'Barr, S. and Yu, J. X. Focal inflammation in the brain: role in Alzheimer's disease. *Immunol Res* **2000**, *21*, 159–165.

746 Breitner, J. C. S. The role of anti-inflammatory drugs in the prevention and treatment of Alzheimer's disease. *Annu Rev Med* **1996**, *47*, 410–411.

747 Aisen, P. S. Anti-inflammatory agents in Alzheimer's disease. *Curr Neurol Neurosci Rep* **2002**, *2*, 405–409.

748 Weggen, S., Eriksen, J. L., Das, P., Sagi, S. A., Wang, R., Pietrzik, C. U., Findlay, K. A., Smith, T. E., et al. A subset of NSAIDs lower amyloidogenic Aβ42 independently of cyclooxygenase activity. *Nature* **2001**, *414*, 212–216.

749 Yan, Q., Zhang, J. H., Liu, H. T., Babu-Khan, S., Vassar, R., Biere, A. L., Citron, M. and Landreth, G. Anti-inflammatory drug therapy alters β-amyloid processing and deposition in an animal model of Alzheimer's disease. *J Neurosci* **2003**, *23*, 7504–7509.

750 Eriksen, J. L., Sagi, S. A., Smith, T. E., Weggen, S., Das, P., McLendon, D. C., Ozols, V. V., Jessing, K. W., et al. NSAIDs and enantiomers of flurbiprofen target γ-secretase and lower Aβ42 *in vivo*. *J Clin Invest* **2003**, *112*, 440–449.

751 Zhou, Y., Su, Y., Li, B., Liu, F., Ryder, J. W., Wu, X., Gonzalez-DeWhitt, P. A., Gelfanova, V., et al. Nonsteroidal anti-inflammatory drugs can lower amyloidogenic Aβ42 by inhibiting Rho. *Science* **2003**, *302*, 1215–1217.

752 Zandi, P. P., Anthony, J. C., Hayden, K. M., Mehta, K., Mayer, L. and Breitner, J. C. S. Reduced incidence of AD with NSAID but not H-2 receptor antagonists – the Cache County Study. *Neurology* **2002**, *59*, 880–886.

753 Tabet, N. and Feldmand, H. Ibuprofen for Alzheimer's disease. *Cochrane Database Syst Rev* **2003**, CD004031.

754 Aisen, P. S., Schafer, K. A., Grundman, M., Pfeiffer, E., Sano, M., Davis, K. L., Farlow, M. R., Jin, S., et al. Effects of rofecoxib or naproxen vs placebo on Alzheimer disease progression: a randomized controlled trial. *J Am Med Ass* **2003**, *289*, 2819–2826.

755 Reines, S. A., Block, G. A., Morris, J. C., Liu, G., Nessly, M. L., Lines, C. R., Norman, B. A. and Baranak, C. C. Rofecoxib: no effect on Alzheimer's disease in a 1-year, randomized, blinded, controlled study. *Neurology* **2004**, *62*, 66–71.

756 Alzheimer's Association. *Alternative Treatments for Alzheimer's*. Alzheimer's Association, Chicago, IL. Available: http://www.alz.org/AboutAD/Treatment/Alternative.asp.

757 Shults, C. W. Coenzyme Q10 in neurodegenerative diseases. *Curr Med Chem* **2003**, *10*, 1917–1921.

758 van Dongen, M., van Rossum, E., Kessels, A., Sielhorst, H. and Knipschild, P. Ginkgo for elderly people with dementia and age-associated memory impairment: a randomized clinical trial. *J Clin Epidemiol* **2003**, *56*, 367–376.

759 [Anonymous] Huperzine A. *Drugs R D* **2004**, *5*, 44–45.

760 McDaniel, M. A., Maier, S. F. and Einstein, G. O. "Brain-specific" nutrients: a memory cure? *Nutrition* **2003**, *19*, 957–975.

761 Le Bars, P. L., Katz, M. M., Berman, N., Itil, T. M., Freedman, A. M. and Schatzberg, A. F. A placebo-controlled, double-blind, randomized trial of an extract of Ginkgo biloba for dementia. North American EGb Study Group. *J Am Med Ass* **1997**, *278*, 1327–1332.

762 Oken, B.S., Storzbach, D.M. and Kaye, J.A. The efficacy of Ginkgo Biloba on cognitive function in Alzheimer disease. *Arch Neurol* **1998**, *55*, 1409–1415.

763 Harrison, T., Churcher, I. and Beher, D. γ-Secretase as a target for drug intervention in Alzheimer's disease. *Curr Opin Drug Discov Dev* **2004**, *7*, 709–719.

764 Vassar, R. BACE1: the β-secretase enzyme in Alzheimer's disease. *J Mol Neurosci* **2004**, *23*, 105–114.

765 Hooper, N.M. and Turner, A.J. The search for α-secretase and its potential as a therapeutic approach to Alzheimer's disease. *Curr Med Chem* **2002**, *9*, 1107–1119.

766 Morris, J.C., Storandt, M., Miller, J.P., McKeel, D.W., Price, J.L., Rubin, E.H. and Berg, L. Mild cognitive impairment represents early-stage Alzheimer disease. *Arch Neurol* **2001**, *58*, 397–405.

767 Collie, A., Maruff, P., Shafiq-Antonacci, R., Smith, M., Hallup, M., Schofield, P.R., Masters, C.L. and Currie, J. Memory decline in healthy older people: implications for identifying mild cognitive impairment. *Neurology* **2001**, *56*, 1533–1538.

768 Scheltens, P., Weinstein, H.C. and Leys, D. Neuro-imaging in the diagnosis of Alzheimer's disease. I. Computer tomography and magnetic resonance imaging. *Clin Neurol Neurosurg* **1992**, *94*, 277–289.

769 Laakso, M.P., Soininen, H., Partanen, K., Lehtovirta, M., Hallikainen, M., Hanninen, T., Helkala, E.L., Vainio, P., et al. MRI of the hippocampus in Alzheimer's disease: sensitivity, specificity, and analysis of the incorrectly classified subjects. *Neurobiol Aging* **1998**, *19*, 23–31.

770 Chui, H. and Zhang, Q. Evaluation of dementia: a systematic study of the usefulness of the American Academy of Neurology's practice parameters. *Neurology* **1997**, *49*, 925–935.

771 Knopman, D.S., DeKosky, S.T., Cummings, J.L., Chui, H., Corey-Bloom, J., Relkin, N., Small, G.W., Miller, B., et al. Practice parameter: diagnosis of dementia (an evidence-based review). Report of the Quality Standards Subcommittee of the American Academy of Neurology. *Neurology* **2001**, *56*, 1143–1153.

772 Weinstein, H.C., Scheltens, P., Hijdra, A. and van Royen, E.A. Neuro-imaging in the diagnosis of Alzheimer's disease. II. Positron and single photon emission tomography. *Clin Neurol Neurosurg* **1993**, *95*, 81–91.

773 Zamrini, E., De Santi, S. and Tolar, M. Imaging is superior to cognitive testing for early diagnosis of Alzheimer's disease. *Neurobiol Aging* **2004**, *25*, 685–691.

774 Almkvist, O. Functional brain imaging as a looking-glass into the degraded brain: reviewing evidence from Alzheimer disease in relation to normal aging. *Acta Psychol (Amst)* **2000**, *105*, 255–277.

775 Rosen, A.C., Bokde, A.L., Pearl, A. and Yesavage, J.A. Ethical, and practical issues in applying functional imaging to the clinical management of Alzheimer's disease. *Brain Cogn* **2002**, *50*, 498–519.

776 Small, G.W. What does imaging add to the management of Alzheimer's disease? *CNS Spectr* **2004**, *9*, 20–23.

777 De Toledo-Morrell, L., Goncharova, I., Dickerson, B., Wilson, R.S. and Bennett, D.A. From healthy aging to early Alzheimer's disease: *in vivo* detection of entorhinal cortex atrophy. *Ann NY Acad Sci* **2000**, *911*, 240–253.

778 Pantel, J., Schroder, J., Essig, M., Jauss, M., Schneider, G., Eysenbach, K., von Kummer, R., Baudendistel, K., et al. *In vivo* quantification of brain volumes in subcortical vascular dementia and Alzheimer's disease. An MRI-based study. *Dement Geriatr Cogn Disord* **1998**, *9*, 309–316.

779 Clark, C.M. and Karlawish, J.H. Alzheimer disease: current concepts and emerging diagnostic and therapeutic strategies. *Ann Intern Med* **2003**, *138*, 400–410.

780 Soininen, H.S. and Scheltens, P. Early diagnostic indices for the prevention of Alzheimer's disease. *Ann Med* **1998**, *30*, 553–559.

781 Pitschke, M., Prior, R., Haupt, M. and Riesner, D. Detection of single amyloid β-protein aggregates in the cerebrospinal fluid of Alzheimer's patients by fluorescence correlation spectroscopy. *Nat Med* **1998**, *4*, 832–834.

782 Kayed, R., Head, E., Thompson, J. L., McIntire, T. M., Milton, S. C., Cotman, C. W. and Glabe, C. G. Common structure of soluble amyloid oligomers implies common mechanism of pathogenesis. *Science* **2003**, *300*, 486–489.

783 Chang, L., Bakhos, L., Wang, Z., Venton, D. L. and Klein, W. L. Femtomole immunodetection of synthetic and endogenous amyloid-β oligomers and its application to Alzheimer's disease drug candidate screening. *J Mol Neurosci* **2003**, *20*, 305–313.

784 El-Agnaf, O. M., Walsh, D. M. and Allsop, D. Soluble oligomers for the diagnosis of neurodegenerative diseases. *Lancet Neurol* **2003**, *2*, 461–462.

785 Xu, H. X., Sweeney, D., Wang, R., Thinakaran, G., Lo, A. C. Y., Sisodia, S. S., Greengard, P., and Gandy, S. Generation of Alzheimer β-amyloid protein in the trans-Golgi network in the apparent absence of vesicle formation. *Proc Natl Acad Sci USA* **1997**, *94*, 3748–3752.

786 Sambamurti, K., Shioi, J., Anderson, J. P., Pappolla, M. A., and Robakis, N. K. Evidence for intracellular cleavage of the Alzheimer's amyloid precusor in PC12 cells. *J Neurosci Res* **1992**, *33*, 319–329.

787 De Strooper, B., Umans, L., Van Leuven, F., and Van Den Berghe, H. Study of the synthesis and secretion of normal and artificial mutants of murine amyloid precursor protein (APP): cleavage of APP occurs in a late compartment of the default secretion pathway. *J Cell Biol* **1993**, *121*, 295–304.

788 Haass, C., Lemere, C. A., Capell, A., Citron, M., Seubert, P., Schenk, D., Lannfelt, L., and Selkoe, D. J. The Swedish mutation causes early-onset Alzheimer's disease by β-secretase cleavage within the secretory pathway. *Natl Med* **1995**, *1*, 1291–1296.

789 Huse, J. T., Pijak, D. S., Leslie, G. J., Lee, V. M., and Doms, R. W. Maturation and endosomal targeting of β-site amyloid precursor protein-cleaving enzyme. The Alzheimer's disease β-secretase. *J Biol Chem* **2000**, *275*, 33729–33737.

790 Talafous, J., Marcinowski, K. J., Klopman, G., and Zagorski, M. G. Solution structure of residues 1–28 of the amyloid β-peptide. *Biochemistry* **1994**, *33*, 7788–7796.

791 Kohno, T., Kobayashi, K., Maeda, T., Sato, K., and Takashima, A. Three-dimensional structure of amlyoid β-peptide (25–35) in membrane-mimicking environment. *Biochemistry* **1996**, *35*, 16094–16104.

792 D'Ursi, A. M., Armenante, M. R., Guerrine, R., Salvadori, S., Sorrentino, G., and Picone, D. Solution structure of amyloid β-peptide (25–35) in different media. *J Med Chem* **2004**, *17*, 4231–4238.

793 López-Toledano, M. A. and Shelanski, M. L. Neurogenic effect of β-amyloid peptide in the development of neural stem cells. *J Neurosci* **2004**, *26*, 5439–5444.

794 Farlow, M. R. NMDA receptor anatagonists. A new therapeutic approach for Alzheimer's disease. *Geriatrics* **2004**, *59*, 22–27.

795 Kelly, J. W. and Balch, W. E. Amyloid as natural product. *J Cell Biol* **2003**, *161*, 461–462.

18
Prion Protein

Philippe Derreumaux

18.1
Introduction

The deposition of amyloid fibrils sharing a common cross-β-sheet structure with the β-strands perpendicular to the fiber axis is a hallmark of several fatal diseases [1–3]. These disorders associated with the failure of proteins to fold correctly can affect the brain, the central nervous system, and various organs such as the liver and heart in humans and animals. The list of fatal diseases includes, among others, primary and secondary systemic amyloidosis, Alzheimer's, Parkinson's and Huntington's diseases, diabetes Type 2, and transmissible spongiform encephalopathies (TSE). The first TSE symptom observed is pronounced astrogliosis [4], followed by spongiform degeneration caused by the formation of vacuoles in neuronal processes and astrocytes [5]. The late step in the TSE disease often involves the formation of amyloid plaques containing PrP^{Sc} [6], the abnormal (or scrapie) isoform of the cellular prion protein (PrP^{C}) encoded by the PrP gene and found predominantly on the outer surface of neurons.

The "protein-only" hypothesis states that a single prion (proteinaceous infectious particle) protein, lacking a small nucleic acid, can give rise to multiple isolates or strains with varying infectivity and incubation time, and is able to convert PrP^{C} to PrP^{Sc} in an autocatalytic manner [7–9]. This theory is at variance with the virus [10] and virino [11] hypotheses which postulate that the infectious particle consists of a small nucleic acid coated by PrP^{Sc}. The original translation product consists of 253 amino acids, but region 1–22 is cleaved as signal peptide during trafficking and region 232–253 is replaced by a glycosylphosphatidylinositol (GPI) anchor at position 231. As seen in Fig. 18.1, mammalian PrP^{C} is a highly conserved secretory cell surface glycoprotein of approximately 210 amino acids (residues 23–231). PrP^{C} is also characterized by two glycosylation sites at N181 and N197, and a single disulfide bond between the cysteines C179 and C214. PrP^{C} and PrP^{Sc} share the same primary structure – no sequence or post-translational differences have been detected [12] – but have distinct biophysical and biochemical properties. PrP^{C} is monomeric, soluble in detergents and sen-

Amyloid Proteins. The Beta Sheet Conformation and Disease. J. D. Sipe

ISBN: 3-527-31072-X

sitive to proteinase K, while PrP^{Sc} is oligomeric, insoluble and partially resistant to proteinase K [13, 14]. In addition, the conversion of PrP^{C} to PrP^{Sc} involves a large variation in secondary structure as determined by IR spectroscopy: PrP^{C} has 42% α-helix and 3–5% β-sheet, whereas PrP^{Sc} has 30% α-helix and 43% β-sheet, and PrP^{Sc} 27–30 (i.e. PrP^{Sc} with removal of residues 23–90) has 25% α-helix and 48–54% β-sheet [15–18].

Despite decades of study and recent promising discoveries, scientists do not know how to prevent prion diseases or what leads prion proteins to become misfolded and toxic. No purified recombinant PrP has been successfully con-

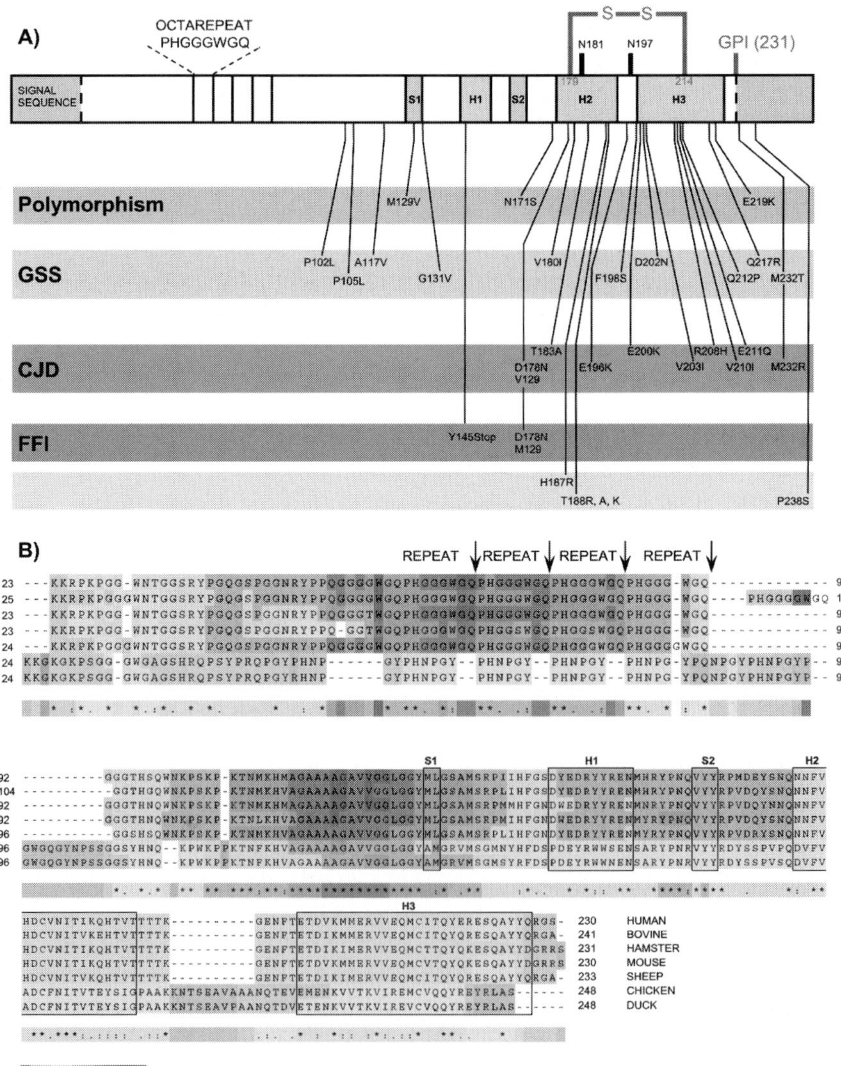

verted *in vitro* to infectious PrP^{Sc}. Yet, scrapie in sheep and goats was first recognized in the 18th century, and mad cow disease or bovine spongiform encephalopathy (BSE) was discovered in 1986. In humans, different types of TSE were identified during the 20th century. Human TSEs are essentially sporadic (80%) with the modes of natural transmission remaining undetermined; around 15% are inherited [familial forms of Creutzfeldt–Jakob disease (CJD), fatal familial insomnia (FFI) and Gerstmann–Sträussler syndrome (GSS)] by mutations in the human PrP gene within chromosome 20. Sporadic CJD occurs at a rate of 0.5–1 per million population per year and GSS, which is the most common familial TSE, generally occurs in the third or fourth decade of life. New pathogenic mutations in the human prion protein continue to be discovered and the current identified list is given in Fig. 18.1. Finally, around 5% of human TSEs are infectious through CJD-infected surgical equipment or tissue transplants; kuru, recognized in 1957, results from exposure to contaminated human tissues during endocannibalistic rituals, whereas new variant CJD (vCJD), identified in 1996, is believed to be spread by dietary exposure to the BSE agent [19] and might also be transmitted by blood transmission [20].

In this chapter, we review our current understanding of the underlying factors that may promote the topological change of mammalian prion proteins. We have selected to focus here on the structural, thermodynamic and dynamic aspects of the *in vitro* and *in vivo* conversion. Reviews on the link between misfolding, pathogenesis and neurotoxicity on the pathological implications of prion glycosylation, on the differences between mammalian and yeast prions, and, finally, on the possible destinations taken by PrP^{C} can be found elsewhere [21–24].

Fig. 18.1 Species variations and mutations of the prion protein gene. (A) The octa-repeats within the tail are represented by white boxes, the regular secondary structures (helices: H1, H2 and H3; β-strands: S1 and S2) within the NMR PrP^{C} structure [28] are also shown. The disulfide bond (S–S) between Cys179 and Cys214 is drawn as a red line; the GPI anchor at position 231 and the two glycosylation sites N181 and N197 are also shown. Polymorphisms and point mutations associated with known human GSS, CJD or FFI are given. The mutations H187R, T188R, T188A, T188K and P238S are still unclassified [137]. (B) Aligned amino acid sequences of human, bovine, hamster, mouse, sheep, chicken and duck PrP. The multiple alignment was carried using T-Coffee [136]. The reliability of the alignment is given in a color code (e.g. red means very reliable). In the last line below each aligned sequences, the asterisk indicates the strictly conserved residues. Note that the fragment spanning residues 106–126 is fully identical between the seven species.

18.2
Conformations of PrP^C and PrP^{Sc}

Knowledge of the three-dimensional structure of PrP in its protease-sensitive and protease-resistant forms is important for our understanding of prion replication. However, both the PrP^C and PrP^{Sc} structures are still unknown at atomic resolution. Amyloid fibrils are non-crystalline and insoluble, and therefore not amenable to X-ray crystallography and nuclear magnetic resonance (NMR) spectroscopy. Furthermore, there is accumulating evidence that the PrP^{Sc} structure is not unique and differences in the conformation of the infectious protein determine prion strain variation [25, 26]. Based on far-UV circular dichroism (CD) and limited proteolysis experiments, PrP^C and recombinant PrPs in membranes (lacking glycosylation sites) or in solution (lacking both GPI anchor and glycosylation sites) seem to adopt similar conformations [27], but no detailed structure is available for exact comparison.

Thus far, only the solution structure of recombinant PrP from various mammalian species was characterized by NMR spectroscopy and X-ray diffraction. All NMR solution structures consist of a disordered tail (residues 23–123), and a globular domain with three α-helices and a short antiparallel β-sheet (Fig. 18.2). The β-strands S1 and S2 comprise residues 128–131 and 161–164, respectively; helix H1 spans residues 144–153, helix H2 residues 172–194 and

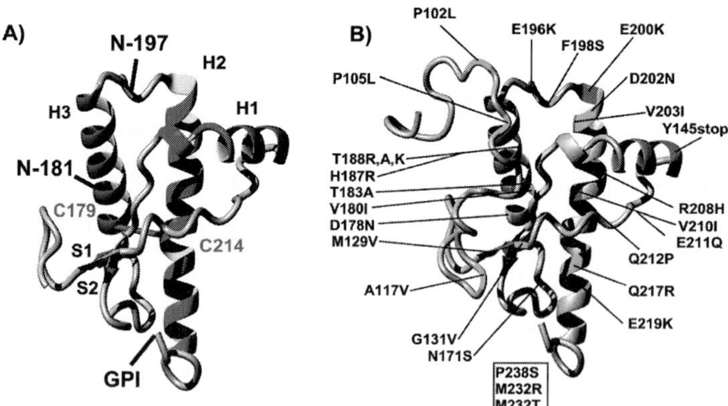

Fig. 18.2 Mutations of the PrP gene are associated with prion diseases. (A) The NMR recombinant Syrian hamster PrP^C NMR structure [28] with the three α-helices H1, H2 and H3, the two β-strands S1 and S2, the disulfide bond between the cysteines C179 and C214, the GPI anchor, and the two glycosylation sites N181 and N197. For clarity, only residues 120–231 are shown. (B) The positions of all polymorphisms and disease-causing mutations as discussed in Fig. 18.1. The mutations leading to GSS are given in green; those leading to CJD are in red. Here, all residues spanning PrP90–231 are shown.

helix H3 residues 200–225 (numbering in Syrian hamster [28]). Although the NMR structures from mouse PrP121–231 [29], bovine PrP23–230 [30], Syrian hamster PrP23–231 [28] and human PrP121–230 at pH 7 and 4.5 [31, 32] superpose well, conformational changes are observed within the loops connecting the helices and even in the length of the helices. The disorder in the loops is an intrinsic property of PrP as deduced by hydrogen–deuterium exchange measurements [33] and is independent of the oligomeric (monomeric or dimeric) state of PrP, as deduced by molecular dynamics (MD) simulations [34]. Helix H3 spans residues 200–222 in mouse PrP121–231 versus 200–228 in human PrP90–231 and helix H2 is 12 residues shorter (173–182) in the human PrP125–228(S170N) variant [35].

Strikingly, the crystal PrP structures do not always superpose on the NMR conformations. The crystal structure of recombinant human PrP119–226 shows a covalently dimer with an intermolecular disulfide bridge, swapping of helix H3 and formation of an interchain antiparallel β-sheet through residues 190–194, i.e. in helix H2 within the monomeric NMR structure [36]. In contrast to human PrP, the crystal structure of recombinant sheep PrP123–230 points to a monomeric state with an intramolecular disulfide bridge, superposing within 1.7 Å r.m.s. deviation from the NMR human PrP structure [37]. This X-ray structure also shows intermolecular or lattice contacts between the strands S1, allowing for the formation of a four-stranded intermolecular β-sheet and the possibility that the residues 190–194 could easily shift from an α-helix to a β-strand [37]. Whether the dimeric form with swapping of helix H3 and intermolecular disulfide bridge is an artifact of the crystallization protocol or indicates a physiological state remains to be determined.

An essential aspect of the recombinant PrP structure is that the entire tail encompassing residues 23 to around 120 is highly flexible and largely disordered for a wide range of pH values. This is not totally surprising since the tail spanning residues 23–90 contains around 40% of glycine residues. Region 90–120 is, however, critical in prion propagation. Residues 108–111 influence prion replication efficiency [38]. PrP with the region 113–120 deleted is not converted to a protease-resistant form when expressed in scrapie-infected neuroblastoma cells [39]. Mice expressing PrP with a deletion of residues 32–106 are not susceptible to infection *in vivo* [40]. In addition, the mutations P102L and P105L lead to GSS syndrome and the mutation P101L in murine PrP (P102L in human numbering) alters TSE incubation time across three species barriers [41].

The flexible tail contains the octapeptide PHGGGWGQ repeated 4 times from residues 60 to 90 which binds copper within the physiological range [42]. Using PrP-derived peptides, full-length Syrian hamster PrP and electron paramagnetic resonance measurements, copper was found to interact not only with each HGGGW segment of the tail, but also with the PrP92–96 segment GGGTH through the His96 imidazole group and the carboxyl oxygen atom of Gly94 [43]. The 106–126 peptide, cytotoxic *in vivo* [44] and containing the palindromic sequence VAGAAAAGAV, was shown to form fibrils with parallel β-sheets [45], but the extrapolation of this result to PrP90–231 remains to be determined. Recent NMR studies of recombinant human PrP23–230 at pH 6.2 showed that the octapeptide repeats are structured, with the segments HGGGW and GWGQ adopting a loop conformation and a β-turn, respectively [46]. This result, along with the crystal structure of the copper-binding pentapeptide HGGGW-Cu^{2+} [47], suggests that the conformations of the HGGGW loop depend on both pH and copper binding.

Three PrP^{Sc} models were first constructed on the basis of immunological studies and CD spectra. In the first model, the region between residues 90 and 145 was modeled by two consecutive β-hairpins (four β-strands) with β-strand 1 parallel to β-strand 3 [48]. In the second model, the secondary H1, S1 and S2 structural elements were replaced by a Greek key consisting of four adjacent antiparallel β-strands [49], whereas in the last model, PrP^{Sc} adopted β-helical conformations [50]. More recently, three other PrP^{Sc} models were proposed on the basis of specific criteria and electron microscopy data at 7 Å resolution from *in vitro* infectious Syrian and mouse PrP two-dimensional protofibril crystals. Fig. 18.3 shows the fluctuations of the secondary structures within these predicted PrP^{Sc} models. Wille et al. proposed a hexamer with each monomer featuring β-helices in the region 90–170 and α-helices H2 and H3 formed using threading techniques onto known β-helical structures. Nevertheless, they did not exclude the possibility that PrP^{Sc} could adopt a novel protein fold [51]. Mornon et al., using sequence analysis, identified the TATA box-binding protein fold as a template for PrP^{Sc} which allowed for the construction of another hexameric PrP^{Sc} model [52, 53]. In their model, the S1H1S2 motif and the N-terminal end of helix H3 are preserved, whereas the helix H2 and its neighboring residues are transformed into three new β-strands. Finally, De Marco and Daggett used MD simulations of the Syrian hamster PrP109–219 (D147N) at low pH to generate β-rich conformations and then docking between the units to build an alter-

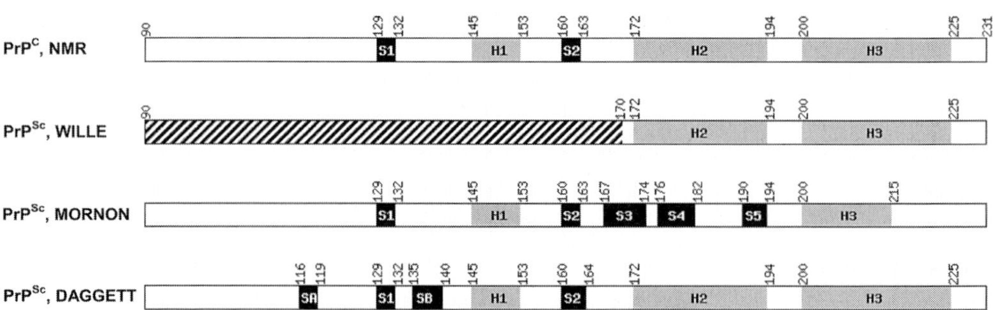

Fig. 18.3 Schematic representation of the secondary structures of wild-type recombinant Syrian hamster PrP^{C} by NMR [28], and the three PrP^{Sc} models proposed by Wille et al. [51], Mornon et al. [53] and De Marco and Daggett [54]. Helices H1, H2 and H3 are presented by grey boxes; S1, S2 and the new β-strands by black boxes; β-helical conformations by cross-hatched boxes.

native hexameric PrP^{Sc} model [54]. In this model, the three helices are preserved, and the core of the β structure consists of an isolated and new β-strand preceding helix H1, and a new β-strand spanning residues 116–119 packed against the S1 and S2 strands.

Since it is possible to generate several PrP^{Sc} models within the electron microscopy envelope, other experimental data are needed to discriminate the solution from false positives. These can include differential proteinase K, vibrational spectroscopies and antibodies which recognize selectively PrP^C or PrP^{Sc}. The model of Wille et al. [51] with β-helices was recently questioned using vibrational Raman optical activity measurements, which pointed to strong similarities between the spectra of PrP^{Sc} and flat β-sheet proteins, but not β-helical proteins [55]. The model of Mornon et al. is consistent with a recently discovered antibody recognizing the YYR motif within the strand S2 of PrP^{Sc} [56] and with the antibody V5B2, raised against residues 214–226, which recognizes PrP^{Sc}, but not PrP^C, indicating a structural rearrangement of the C-terminus [57]. The model of De Marco and Daggett [54] is consistent with peptide-binding studies of PrP^{Sc} [58, 59] potently inhibiting the PrP-res induced cell-free conversion of PrP-sen to the protease-resistant state. This model is also consistent with the monoclonal antibody 15B3, which recognizes PrP^{Sc} at positions 142–148 (N-terminal half of helix H1 in the normal protein), 162–170 and 214–226, but not PrP^C [49]. However, neither Mornon's or De Marco's model discuss the conformations of the fragment 90–109, and fail to reproduce, in their present forms, the experimental percentage of β structure. Fourier transform IR studies suggest that 48% of residues (or about 68 residues) [18] are in β-strands within PrP^{Sc}90–231, whereas for instance the De Marco's model has 37 residues in β-strands within the region 109–219. These differences require study in more detail.

18.3
Stability and Unfolding/Folding of PrP^C *In Vitro*

An important step towards understanding the conformational conversion of PrP^C to PrP^{Sc} is to characterize the energy landscape and unfolding/folding pathways of the wild-type and mutant prion proteins *in vitro*. It is well established that the recombinant prion protein can be folded either to a monomeric α-helical topology or β-rich oligomers and that folding to the PrP^C isoform is under kinetic control [60]. Starting from disordered conformations, PrP folds to its PrP^C isoform at pH 7, whereas under acidic pH conditions, PrP avoids the kinetic trap and folds to the β-rich isoform. The energy barrier separating the PrP^C state and the PrP β-rich oligomers was estimated to be 35–45 kcal/mol for wild-type mouse PrP [60]. There is also strong evidence that the full tertiary context of the protein is necessary to stabilize the terminal α-helices and the β-strands within PrP^C. Biophysical studies showed that the helices H2 and H3 are largely disordered in the PrP fragment spanning helices H2 and H3 [61].

Monte Carlo simulations suggested that the sequence PrP127–164 can adopt two distinct tertiary folds with different secondary structures [62].

Initial experiments on recombinant mouse PrP121–231 at 298 K, based on stopped-flow fluorescence using the variant F175W which has the same overall structure and stability as wild-type mouse PrP121–231, suggested a two-state folding model at neutral pH, i.e. PrP either unfolded or native. The mouse PrP121–231 fragment was also found to fold very rapidly to PrP^C with a half-life of 170 μs [63]. Then, other kinetic studies on human PrP90–231 and mouse PrP121–231 suggested that the change from neutral to acid pH conditions shifted the two-state kinetic to a three-state kinetic by stabilizing a monomeric intermediate with β-sheet structure in equilibrium with the native helical state. An intermediate state was detected at pH 4.0 and 1–1.5 M Gdn–HCl [64], pH 3.6 and 0.75–1.75 M Gdn–HCl [65], pH 5.0 and 2–2.5 M Gdn–HCl [66], and 3.5–4.5 M urea [67]. These β-sheet-rich intermediates, generated under acidic pH conditions (pH below 5), are, however, oligomeric rather than monomeric in character [68–71]. More recently, by following the kinetics of folding and unfolding reactions of human PrP90–231 at 278 K and pH 7, it was again proposed that the prion protein folds by a three-state mechanism involving a monomeric intermediate [72]. Since then, the existence of a folding intermediate, PrP*, has been advocated by a number of studies. These include hydrogen exchanges monitored by NMR [73] and high-pressure NMR experiments on Syrian hamster PrP90–231 [74] – the population of this intermediate was found to be 1% at pH 5.2, 303 K and 1 bar – or mouse PrP121–231 [75].

The point mutations associated with inherited human TSEs are also of interest for characterizing the energy landscape. It was initially believed that mutations in PrP gene could promote the conformational conversion by destabilizing the native structure of PrP^C [76]. Fig. 18.2 shows the positions of polymorphisms and disease-causing mutations within the NMR human PrP structure. The most frequent polymorphism, modulating disease susceptibility and onset, is located at codon 129. For instance, the M129/N178 allele segregates with FFI, whereas the V129/N178 allele leads to CJD [77]. The codon stop mutation at position 145 which generates an atypical GSS variant is also given [78]. As seen in Fig. 18.2, most familial mutations are located within the globular domain 124–226: strand S1 (G131V), helix H2 (D178N, V180I, T183A, H187R, T188A or T188K or T188R), loop between H2 and H3 (E196K and F198S), and helix H3 (E200K, D202N, V203I, R208H, V210I, E211Q, Q212P and Q217R). However, three mutations also occur within the disordered N-terminal tail: P102L, P105L and A117V. Although destabilization of the PrP^C structure upon mutations has been confirmed for D178N, T183A, F198S, R208H and Q217R mutants by equilibrium unfolding studies in urea [79–81], it is not a general mechanism underlying the formation of PrP^{Sc}. The structure of the E200K variant of human PrP superposes well on the wild-type sequence structure [82]. The M129V and P102L variants of human PrP show PrP^C-like structural properties from CD analysis [79, 83]. The recently discovered G131V mutation does not lead to any identifiable effects on PrP^C secondary structure, as deduced by MD simulations

[84]. Clearly, other mechanisms rather than changes in the thermodynamic stability of PrP^{C} are thus important for conversion.

One solution is that single-point mutations reduce the energy barrier by stabilizing the transition state. This effect has not been fully explored yet. Another possibility is that the point mutations change the electrostatic energy surface of PrP^{C} and thus alter the interaction (binding or conversion step) with PrP^{Sc}. This effect has been proposed for the E200K mutant [82]. Finally, it is possible that the point mutations stabilize a partially folded intermediate. Recent kinetic experiments on a number of human PrP90–231 variants carrying mutations associated with familial forms of prion disease suggest the existence of partially structured intermediates on the refolding pathway of the following PrP variants: F198S, Q217R, V180I, V210I, R208H, D178N/M129 and D178N/V129. In each case, the partially folded state was found to be at least an order of magnitude more populated than the fully unfolded state. The strongest effect was seen for the variant F198S where the ratio of intermediate states relative to the native state is 1/350 versus 1/42,000 for wild-type PrP [81]. In contrast, the P102L mutation did not lead to any increase in the population of the intermediate, leaving unanswered the question of its impact on the energy surface. While it is not a definitive proof for the existence of a monomeric PrP* intermediate, it is interesting that such an intermediate has not been detected for Doppel, a homolog of PrP that does not form infectious prions [73].

Little is still known on the structure of the monomeric PrP* intermediate. High-pressure NMR experiments point to an intermediate with the helices H2 and H3 disordered and helix H1 formed at pH 5.2, 303 K and 1 bar [74]. MD simulations of PrP90–231 with either the strand S1 or the strand S2 deleted – these variants do not inhibit prion propagation in a cell-free assay system [85] – suggest partially unfolded intermediates of molten globule type with the helices H2 and H3 disordered, and the helix H1 either fully formed or partially disordered at its C-terminal end [86]. Clearly, the structural features of PrP*, if this intermediate exists in monomeric form, need to be investigated in more detail. Nonetheless, based on MD simulations coupled with Φ analysis on globular proteins, it would not be too surprising that PrP* consists of an ensemble of distinct conformations having different secondary structure compositions [87].

18.4 Mechanisms of Prion Replication *In Vivo*

Within the protein-only hypothesis, a detailed mechanism for the conformational transition is still unclear. A number of facts are, however, well accepted. Interactions between PrP^{C} and PrP^{Sc} provide the main forces to propagate the topological change [60]. The conversion is promoted by partially denaturing conditions and in the presence of low concentrations of urea, i.e. under conditions expected to increase the population of an intermediate. Irrespective of the kinetic model (see below), transition from PrP^{C} to PrP^{Sc} is a two-step process, which begins with binding between the two PrP isoforms, followed by conversion of the cellular form to the pathogenic form [88]. By designing specific synthetic peptides which inhibit binding and conversion reaction, it was found that the binding surfaces include residues 119–138 (region preceding helix H1), 165–174 (loop between S2 and H2) and 206–223 (helix H3), while conversion to PrP^{Sc} is strongly influenced by residues 139, 155 and 170 [58, 59, 88]. The species barrier, which is the most intriguing feature of prion propagation, limits transmission of prions across species. Mouse scrapie prion can be transmitted to Syrian hamster with an incubation time of 380 days, but transmission in mice is 130 days [89]. At present, two main factors have been identified to contribute to the species barrier from studies on transgenic animals: the conformations of individual strains of PrP^{Sc}, and the difference in PrP sequences between the donor and acceptor species [90, 91]. For instance, six positions have been identified in affecting prion transmission between mice and humans, i.e. residues 138, 143, 145, 155 and 166 [29].

Several kinetic models have been proposed to explain why spontaneous formation is an extremely rare event and how infection with PrP^{Sc} promotes the conversion of the cellular prion protein (Fig. 18.4). The template-assisted or heterodimer model [92] postulates that PrP^{Sc} is thermodynamically more stable than PrP^{C}, but kinetically inaccessible. PrP^{C} exists in equilibrium with a conformational intermediate PrP* which is able to interact with PrP^{Sc}. Binding of PrP* to PrP^{Sc} lowers the high activation energy barrier and allows the formation of a PrP^{Sc} heterodimer, but other cellular factors such as molecular chaperones or prion-disease causing mutations might also contribute by reducing the energy barriers or increasing the stability of the partially folded intermediate species. In this model, the rate-limiting step is the conformational conversion between PrP* and PrP^{Sc} (Fig. 18.4 A). The second kinetic model is nucleated polymerization [71, 93]. In this model, PrP^{C} and PrP^{Sc} are in equilibrium in solution, but PrP^{Sc} is rare and unstable. The rate-limiting step is the formation of a stable PrP^{Sc} nucleus or seed. This is discerned by the observation of a lag phase in polymer growth. Once a seed is present, molecular association facilitates the conformational change of PrP^{C} at a rapid rate (Fig. 18.4 B).

Within these models, two variants have also emerged recently. The β-nucleation model proposes that unfolding of helix H1 in PrP^{C}, catalyzed by the PrP^{Sc} aggregate, is the key event in PrP^{Sc} propagation, and that intermolecular salt bridges between the helix H1 D and R residues of adjacent molecules are critical for the conversion [94]. This kinetic model, partially supported by MD simulations of wild-type and its D178N and E200K variants [95], was, however, not validated by cell-free conversion data [96]. The nucleated conformational conversion model, which is a dynamic version of nucleated polymerization, assumes that structurally fluid oligomeric complexes are crucial intermediates and rapid assembly occurs when these complexes conformationally convert upon association with nuclei [97]. At present, computer simulations have not been able to distinguish these kinetic models, independently of the protein models used, because the size of the oligomer is too small [98–102], the energy surface is biased towards the native state [103, 104] or the simulation conditions are not appropri-

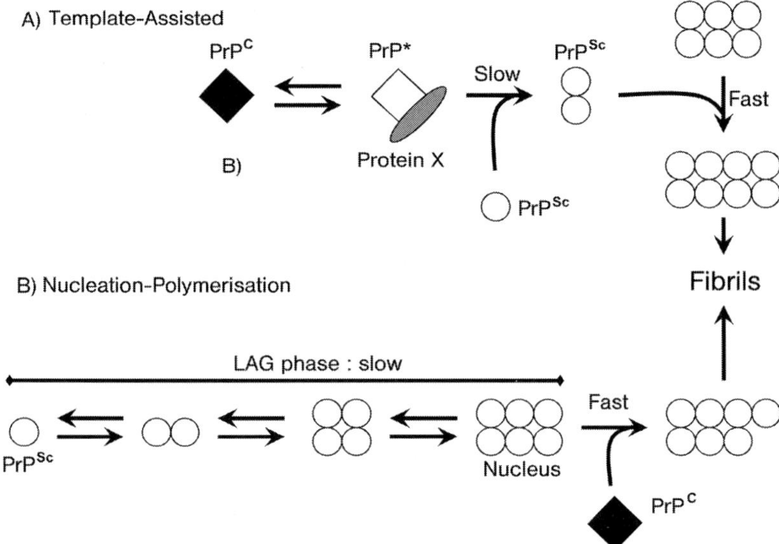

Fig. 18.4 Kinetic models for the conformational conversion of PrP^C to PrP^{Sc}. (A) The template-assisted or heterodimer model [92]. Here, the rate-limiting step during assembly is the conformational conversion between the partially folded PrP intermediate (PrP*) and a newly formed PrP^{Sc}. Binding of PrP* to PrP^{Sc} under the control of the putative protein X allows the formation of a PrP^{Sc} heterodimer. (B) The polymerization nucleation model [71, 93]. Here the rate-limiting step is the formation of a stable PrP^{Sc} nucleus or seed. This is discerned by the observation of a lag phase. The exact size of the nucleus is unknown. PrP^{Sc} monomers are stabilized by joining the nucleus, as in seeded crystallization.

ate [105]. Regardless of the kinetic model, template-assisted or nucleation models, several important questions remain to be answered.

(1) What is the role played by PrP^C plasticity in the conformational change? Based on the experimental percentage of β-sheet within PrP^{Sc} [15], both the tail and the protein core undergo a structural change, but does one region initiate the conformational change? To address this issue, several MD simulations focused on the tail [84, 106–109]. In particular, Alonso et al. performed 10-ns MD simulations of Syrian hamster PrP109–219 at 300 K under neutral and low pH conditions [106]. Their results showed that acidic conditions favor the formation of a three-stranded antiparallel β-sheet spanning residues 109–131, whereas neutral conditions maintain the N-terminal region in a disordered state. Santini et al., by using MD simulations on various Syrian hamster PrP90–231 sequences, provided evidence that the rate-limiting factor for the formation of a three-stranded antiparallel β-sheet within the tail is thermodynamic rather than kinetic in character at neutral pH [84]. This β-sheet within the tail, which is marginally populated in both PrP90–231 M129V and G131V mutants (30% of β-sheet versus 70% of random conformations) and not populated in wild-type PrP90–231 or its P102L variant, might be stabilized upon binding to the PrP^{Sc} template.

In parallel, other studies focused on the possible role of the structured core. By using MD studies of PrP in solution [106, 110], molecular mechanics calculations of PrP fragments in monomeric, dimeric and tetrameric forms [94], and NMR structures of sheep PrP142–166 in solution [111], unfolding of helix H1 was suggested to provide the first internal driving force. However, other studies pointed to the high stability of helix H1. These include CD and NMR studies of human and murine prion peptides encompassing helix 1 and flanking sequences under various pH conditions [112–114], high-pressure NMR experiments on Syrian hamster PrP90–231 [74], MD simulations of PrP at low pH and high temperature conditions [115], and MD simulations of two PrP deletion variants at pH 7 and 300 K [86]. Alternatively, other theoretical investigations led to the proposal that helix H2 was the key region in the conformational change [52, 53, 116], because helix H2 is frustrated in the monomeric PrP^C structure [117, 118]. Frustration in secondary structure elements is defined as the incompatibility between the predicted (e.g. by using neural networks) secondary structure and the experimentally determined structure. Or they concluded using biased MD simulations that there are two populated unfolding routes, the helices H2 and H3 being disordered first with the S1H1S2 motif formed and vice versa [119]. Clearly, this diversity of unfolding and folding routes for PrP is not surprising, but it remains to simulate the dynamics of PrP^C in interaction with a PrP^{Sc} model to reflect experimental reality.

(2) What is the exact role of the β-rich oligomers formed by non reduced recombinant PrP proteins in the conversion? Recent studies have shown that these oligomers, sensitive to digestion by proteinase K, are apparently not on the pathway to amyloid fibril formation [69, 70]. However, it remains possible that these β-rich oligomers may offer a pool of pre-aggregated material for further structural reorganization through alternative pathways. Furthermore, these oligomers have been observed in other neurodegenerative diseases and there is accumulating evidence that both soluble Aβ oligomers (early aggregates) and fibrils are toxic in cell cultures [120].

(3) What is the minimal size of the infectious PrP^{Sc} particle? Ionizing irradiation experiments have suggested that the minimally infectious PrP^{Sc} is a dimer [121], and protease-resistant PrP dimers were observed in hamster brain [122]. On the other hand, recent studies based on size-exclusion chromatography, electrospray mass spectrometry and dynamic light scattering showed that the β-rich oligomers consist of octamers [60, 69, 123], whereas electron crystallography experiments have deduced an hexamer from *in vitro* infectious Syrian and mouse PrP two-dimensional protofibril crystals [51]. Taken together, these data suggest that infectivity may be associated with several sizes of aggregates under specific external conditions such as pH, temperature, concentration.

(4) Finally, what is the role played by copper in the interconversion process and is our inability to generate infectious PrP^{Sc} in test tubes due to the absence of cellular factors? There is strong genetic evidence supporting for the importance of the copper-binding domain, since modifications in the number of octa-repeats cause familial prion diseases [124]. The exact role of copper in the interconversion

process is far from being elucidated. It is currently believed that the binding of copper ions induces a conformational transition that presumably modulates PrP aggregation [46], thereby enhancing its infectivity [125].

Prior studies on the transmission of human prions to transgenic mice have suggested the role of a cofactor, protein X, in the formation of PrP^{Sc}. Protein X appears to bind to the side-chains of residues that form a discontinuous epitope: some amino acids are in the loop composed of residues 165–171 and at the end of helix H2 (Q168 and Q172), whereas others are on the surface of helix H3 (T215 and Q219) [126]. It is intriguing that the protein X epitope coincide with the binding surfaces between PrP^{Sc} and PrP^{C} and the disordered regions within the NMR PrP^{C} structures of human, mouse and Syrian hamster. However, all physical attempts to identify the protein X have failed thus far.

Recent studies have also indicated that PrP^{C} can associate with various molecular complexes [127]. For instance, Edenhofer et al. found interactions between the molecular chaperone Hsp60 and recombinant glutathione S-transferase-fusion PrP^{C} peptides [128]. In addition, *in vitro* conversion experiments using a variant of the protein-misfolding cyclic amplification method [129] have shown that specific RNA molecules stimulate PrP^{Sc} formation [130]. This raises the possibility that RNA molecules are cellular cofactors and are involved in generating strain diversity. Clearly, advances in mass spectrometry should help clarify the nature of the cellular factors.

18.5 Perspectives

Since 1997 when Stanley Prusiner was awarded the Nobel Prize for Medicine, important steps have been made in understanding prion replication at the structural level, but many questions remain to be answered. Thus far, many molecules have been identified to be effective for inhibiting prion replication in cell culture essays. For instance, Soto et al. partly reversed *in vitro* PrP^{Sc} to PrP^{C} using β-sheet breaker peptide spanning PrP115–122 [131]. Caughey et al. inhibited the conversion reaction *in vitro* using diferoylmethane [132]. However, all these inhibitors failed on sick animals. The design of efficient inhibitors blocking prion propagation in mammals requires, at least, a combined effort in three directions:

1. *PrP functional role.* Several studies have pointed to the possible role of PrP in copper transport or metabolism [125], signal transduction [133] and apoptosis [134], to name a few, but the physiological function of PrP remains unknown. Yet, PrP is highly conserved across species, and is expressed in most adult tissues and at high levels in the central nervous system. Thus, PrP is likely to be very useful.
2. *Cellular factors.* Advances in experimental techniques are needed to identify the cellular factors because no purified recombinant PrP has yet been successfully converted *in vitro* to infectious PrP^{Sc}.

3. *Structural characterization of the soluble PrP oligomers and insoluble PrP^{Sc} fibrils.* A detailed characterization of the soluble oligomeric intermediates is very difficult because the intermediates are typically short lived and are present in a wide range of conformations and degrees of aggregation; however, these atomic models would greatly facilitate the design of new drugs by computers.

Such a combined effort should not only be useful in prion science but also in Alzheimer's or Huntington's diseases because soluble oligomers apparently share a unique common structural feature [135], and, eventually, in understanding the generic rules between amino acid sequences, folded and misfolded structures for all protein coding genes.

References

1 Sipe, J. D. Amyloidosis. *Annu Rev Biochem* **1992**, *61*, 947–975.
2 Kelly, J. F. The alternative conformations of amyloidogenic proteins and their multistep assembly pathways. *Curr Opin Struct Biol* **1998**, *8*, 101–106.
3 Dobson, C. M. Protein-misfolding diseases: getting out of shape. *Nature* **2002**, *418*, 729–730.
4 Diedrich, J. F., P. E. Bendheim, Y. S. Kim, R. I. Carp and A. T. Haase. Scrapie-associated prion protein accumulates in astrocytes during scrapie infection. *Proc Natl Acad Sci USA* **1991**, *88*, 375–379.
5 Jeffrey, M., J. R. Scott, A. Williams and H. Fraser. Ultrastructural features of spongiform encephalopathy transmitted to mice from three species of bovidae. *Acta Neuropathol* **1992**, *84*, 559–569.
6 Bendheim, P. E., R. A. Barry, S. J. DeArmond, D. P. Stites and S. B. Prusiner. Antibodies to a scrapie prion protein. *Nature* **1984**, *310*, 418–421.
7 Alper, T., W. A. Cramp, D. A. Haig and M. C. Clarke. Does the agent of scrapie replicate without nucleic acid? *Nature* **1967**, *214*, 764–766.
8 Griffith, J. S. Self-replication and scrapie. *Nature* **1967**, *215*, 1043–1044.
9 Prusiner, S. B. Novel proteinaceous infectious particles cause scrapie. *Science* **1982**, *216*, 136–144.
10 Rohwer, R. G. Scrapie infectious agent is virus-like in size and susceptibility to inactivation. *Nature* **1984**, *308*, 658–662.
11 Dickinson, A. G. and G. W. Outram. Genetic aspects of unconventional virus infections: the basis of the virino hypothesis. *Ciba Found Symp* **1988**, *135*, 63–83.
12 Stahl, N., M. A. Baldwin, D. B. Teplow, L. Hood, B. W. Gibson, A. L. Burlingame and S. B. Prusiner. Structural studies of the scrapie prion protein using mass spectrometry and amino acid sequencing. *Biochemistry* **1993**, *32*, 1991–2002.
13 Prusiner, S. B. Prions. *Proc Natl Acad Sci USA* **1998**, *95*, 13363–13383.
14 Kocisko, D. A., J. H. Come, S. A. Priola, B. Chesbro, G. J. Raymond, P. T. Lansbury and B. Caughey. Cell-free formation of protease-resistant prion protein. *Nature* **1994**, *370*, 471–474.
15 Caughey, B. W., A. Dong, K. Bhat, D. Ernst, S. Hayes and W. Caughey. Secondary structure analysis of the scrapie-associated protein PrP 27–30 in water by infrared spectroscopy. *Biochemistry* **1991**, *30*, 7672–7680.
16 Gasset, M., M. A. Baldwin, R. J. Fletterick and S. B. Prusiner. Perturbation of the secondary structure of the scrapie prion protein under conditions that alter infectivity. *Proc Natl Acad Sci USA* **1993**, *90*, 1–5.
17 Pan, K. M., M. Baldwin, J. Nguyen, M. Gasset, A. Serban, D. Groth, I. Mehlhorn, Z. Huang, et al. Conversion of alpha-helices into beta-sheets features in the formation of the scrapie prion pro-

teins. *Proc Natl Acad Sci USA* **1993**, *90*, 10962–10966.

18 Wille, H., G.F. Zhang, M.A. Baldwin, F.E. Cohen and S.B. Prusiner. Separation of scrapie prion infectivity from PrP amyloid polymers. *J Mol Biol* **1996**, *259*, 608–621.

19 Bruce, M.E., R.G. Will, J.W. Ironside, I. McConnell, D. Drummond, A. Suttie, L. McCardle, A. Chree, et al. Transmissions to mice indicate that "new variant" CJD is caused by the BSE agent. *Nature* **1997**, *389*, 498–501.

20 Llewelyn, C.A., P.E. Hewitt, R.S. Knight, K. Amar, S. Cousens, J. Mackenzie and R.G. Will. Possible transmission of variant Creutzfeldt-Jakob disease by blood transfusion. *Lancet* **2004**, *363*, 417–421.

21 Bousset, L. and R. Melki. Similar and divergent features in mammalian and yeast prions. *Microbes Infect* **2002**, *4*, 461–469.

22 Ermonval, M., S. Mouillet-Richard, P. Codogno, O. Kellermann and J. Botti. Evolving views in prion glycosylation: functional and pathological implications. *Biochimie* **2003**, *85*, 33–45.

23 Hetz, C. and C. Soto. Protein misfolding and disease: the case of prion disorders. *Cell Mol Life Sci* **2003**, *60*, 133–143.

24 Prado, M.A., J. Alves-Silva, A.C. Magalhaes, V.F. Prado, R. Linden, V.R. Martins and B.R. Brentani. PrP^C on the road: trafficking of the cellular prion protein. *J Neurochem* **2004**, *88*, 769–781.

25 Tanaka, M., P. Chien, N. Naber, R. Cooke and J.S. Weissman. Conformational variations in an infectious protein determine prion strain differences. *Nature* **2004**, *428*, 323–328.

26 King, C.Y. and R. Diaz-Avalos. Protein-only transmission of three yeast prion strains. *Nature* **2004**, *428*, 319–323.

27 Eberl, H., P. Tittmann and R. Glockshuber. Characterization of recombinant, membrane-attached full-length prion protein. *J Biol Chem* **2004**, *279*, 25058–25065.

28 Donne, D.G., J.H. Viles, D. Groth, I. Mehlhorn, T.L. James, F.E. Cohen, S.B. Prusiner, P.E. Wright, et al. Structure of the recombinant full-length hamster prion protein PrP(29–231): the N terminus is highly flexible. *Proc Natl Acad Sci USA* **1997**, *94*, 13452–13457.

29 Riek, R., S. Hornemann, G. Wider, M. Billeter, R. Glockshuber and K. Wüthrich. NMR structure of the mouse prion protein domain PrP(121–321). *Nature* **1996**, *382*, 180–182.

30 Lopez Garcia, F., R. Zahn, R. Riek and K. Wüthrich. NMR structure of the bovine prion protein. *Proc Natl Acad Sci USA* **2000**, *97*, 8334–8339.

31 Zahn, R., A. Liu, T. Luhrs, R. Riek, C. von Schroetter, F. Lopez Garcia, M. Billeter, L. Calzolai, et al. NMR solution structure of the human prion protein. *Proc Natl Acad Sci USA* **2000**, *97*, 145–150.

32 Calzolai, L. and R. Zahn. Influence of pH on NMR structure and stability of the human prion protein globular domain. *J Biol Chem* **2003**, *278*, 35592–35596.

33 Hosszu, L.L., N.J. Baxter, G.S. Jackson, A. Power, A.R. Clarke, J.P. Waltho, C.J. Craven and J. Collinge. Structural mobility of the human prion protein probed by backbone hydrogen exchange. *Nat Struct Biol* **1999**, *6*, 740–743.

34 Sekijima, M., C. Motono, S. Yamasaki, K. Kaneko and Y. Akiyama. Molecular dynamics simulation of dimeric and monomeric forms of human prion protein: insight into dynamics and properties. *Biophys J* **2003**, *85*, 1176–1185.

35 Calzolai, L., D.A. Lysek, P. Guntert, C. von Schroetter, R. Riek, R. Zahn and K. Wüthrich. NMR structures of three single-residue variants of the human prion protein. *Proc Natl Acad Sci USA* **2000**, *97*, 8340–8345.

36 Knaus, K.J., M. Morillas, W. Swietnicki, M. Malone, W.K. Surewicz and V.C. Yee. Crystal structure of the human prion protein reveals a mechanism for oligomerization. *Nat Struct Biol* **2001**, *8*, 770–774.

37 Haire, L.F., S.M. Whyte, N. Vasisht, A.C. Gill, C. Verma, E.J. Dodson, G.G. Dodson and P.M. Bayley. The crystal structure of the globular domain of sheep prion protein. *J Mol Biol* **2004**, *336*, 1175–1183.

38 Supattapone S., T. Muramoto, G. Legname, I. Mehlhorn, F.E. Cohen, S.J. DeArmond, S.B. Prusiner and M.R. Scott. Identification of two prion protein regions that modify scrapie incubation time. *J Virol* **2001**, *75*, 1408–1413.

39 Holscher, C., H. Delius and A. Burkle. Overexpression of nonconvertible PrP^{C}delta 114–121 in scrapie-infected mouse neuroblastoma cells leads to trans-dominant inhibition of wild-type PrP^{Sc} accumulation. *J Virol* **1998**, *72*, 1153–1159.

40 Weissmann, C. Molecular genetics of transmissible spongiform encephalopathies. *J Biol Chem* **1999**, *274*, 3–6.

41 Barron, R.M., V. Thomson, E. Jamieson, D.W. Melton, J. Ironside, R. Will and J.C. Manson. Changing a single amino acid in the N-terminus of murine PrP alters TSE incubation time across three species barriers. *EMBO J* **2001**, *20*, 5070–5078.

42 Kramer, M.L., H.D. Kratzin, B. Schmidt, A. Romer, O. Windl, S. Liemann, S. Hornemann and H. Kretzschmar. Prion protein binds copper within the physiological concentration range. *J Biol Chem* **2001**, *276*, 16711–16719.

43 Burns, C.S., E. Aronoff-Spencer, G. Legname, S.B. Prusiner, W.E. Antholine, G.J. Gerfen, J. Peisach and G.L. Millhauser. Copper coordination in the full-length, recombinant prion protein *Biochemistry* **2003**, *42*, 6794–6803.

44 Ettaiche, M., R. Pichot, J.P. Vincent and J. Chabry. *In vivo* cytotoxicity of the prion protein fragment 106–126. *J Biol Chem* **2000**, *275*, 36487–36490.

45 Kuwata, K., T. Matumoto, H. Cheng, K. Nagayama, T.L. James and H. Roder. NMR-detected hydrogen exchange and molecular dynamics simulations provide structural insight into fibril formation of prion protein fragment 106–126. *Proc Natl Acad Sci USA* **2003**, *100*, 14790–14795.

46 Zahn, R. The octapeptide repeats in mammalian prion protein constitute a pH-dependent folding and aggregation site. *J Mol Biol* **2003**, *334*, 477–488.

47 Burns, C.S., E. Aronoff-Spencer, C.M. Dunham, P. Lario, N.I. Avdievich, W.E. Antholine, M.M. Olmstead, A. Vrielink, et al. Molecular features of the copper binding sites in the octarepeat domain of the prion protein. *Biochemistry* **2002**, *41*, 3991–4001.

48 Huang, Z., S.B. Prusiner and F.E. Cohen. Scrapie prions: a three-dimensional model of an infectious fragment. *Fold Design* **1996**, *1*, 13–19.

49 Korth, C., B. Stierli, P. Streit, M. Moser, O. Schaller, R. Fischer, W. Schulz-Schaeffer, H. Kretzschmar, et al. Prion (PrP^{Sc})-specific epitope defined by a monoclonal antibody. *Nature* **1997**, *390*, 74–77.

50 Downing, D.T. and N.D. Lazo. Molecular modelling indicates that the pathological conformations of prions might be β-helical. *Biochem J* **1999**, *343*, 453–460.

51 Wille, H., M.D. Michelitsch, V. Guenebaut, S. Supattapone, A. Serban, F.E. Cohen, D.A. Agard and S.B. Prusiner. Structural studies of the scrapie prion protein by electron crystallography. *Proc Natl Acad Sci USA* **2002**, *99*, 3563–3568.

52 Mornon, J.P., K. Prat, F. Dupuis and I. Callebaut. Structural features of prions explored by sequence analysis. I. Sequence data. *Cell Mol Life Sci* **2002**, *59*, 1366–1376.

53 Mornon, J.P., K. Prat, F. Dupuis, N. Boisset and I. Callebaut. Structural features of prions explored by sequence analysis. II. A PrP^{Sc} model. *Cell Mol Life Sci* **2002**, *59*, 2144–2154.

54 DeMarco, M.L. and V. Daggett. From conversion to aggregation: protofibril formation of the prion protein. *Proc Natl Acad Sci USA* **2004**, *101*, 2293–2298.

55 McColl, I.H., E.W. Blanch, A.C. Gill, A.G. Rhie, M.A. Ritchie, L. Hecht, K. Nielsen and L.D. Barron. A new perspective on beta-sheet structures using vibrational Raman optical activity: from poly(L-lysine) to the prion protein. *J Am Chem Soc* **2003**, *125*, 10019–10026.

56 Paramithiotis, E., M. Pinard, T. Lawton, S. LaBoissiere, V.L. Leathers, W.Q. Zou, L.A. Estey, J. Lamontagne, et al. A prion protein epitope selective for the pathologically misfolded conformation. *Nat Med* **2003**, *9*, 893–899.

57 Serbec, C.V., M. Bresjanac, M. Popovic, P.K. Hartman, V. Galvani, R. Rupreht,

M. Cernilec, T. Vranac, et al. Monoclonal antibody against a peptide of human prion protein discriminates between Creutzfeldt–Jacob's disease-affected and normal brain tissue. *J Biol Chem* **2003**, *279*, 3694–3698.

58 Chabry, J., B. Caughey and B. Chesebro. Specific inhibition of *in vitro* formation of protease-resistant prion protein by synthetic peptides. *J Biol Chem* **1998**, *273*, 13203–13207.

59 Horiuchi, M., G. S. Baron, L. W. Xiong and B. Caughey. Inhibition of interactions and interconversions of prion protein isoforms by peptide fragments from the C-terminal folded domain. *J Biol Chem* **2001**, *276*, 15489–15497.

60 Baskakov, I. V., G. Legname, S. B. Prusiner and F. E. Cohen. Folding of prion protein to its native alpha-helical conformation is under kinetic control. *J Biol Chem* **2001**, *276*, 19687–19690.

61 Eberl, H. and R. Glockshuber. Folding and intrinsic stability of deletion variants of PrP(121–231), the folded C-terminal domain of the prion protein. *Biophys Chem* **2002**, *96*, 293–303.

62 Derreumaux, P. Evidence that the 127–164 region of prion proteins has two equienergetic conformations with beta or alpha features. *Biophys J* **2001**, *81*, 1657–1665.

63 Wildegger, G., S. Liemann and R. Glockshuber. Extremely rapid folding of the C-terminal domain of the prion protein without kinetic intermediates. *Nat Struct Biol* **1999**, *6*, 550–553.

64 Jackson, G. S., L. L. Hosszu, A. Power, A. F. Hill, J. Kenney, H. Saibil, C. J. Craven, J. P. Waltho, et al. Reversible conversion of monomeric human prion protein between native and fibrilogenic conformations. *Science* **1999**, *283*, 1935–1937.

65 Swietnicki, W., R. Petersen, P. Gambetti and W. K. Surewicz. pH-dependent stability and conformation of the recombinant human prion protein PrP(90–231). *J Biol Chem* **1997**, *272*, 27517–27520.

66 Zhang, H., J. Stockel, I. Mehlhorn, D. Groth, M. A. Baldwin, S. B. Prusiner, T. L. James and F. E. Cohen. Physical studies of conformational plasticity in a recombinant prion protein. *Biochemistry* **1997**, *36*, 3543–3553.

67 Hornemann, S. and R. Glockshuber. A scrapie-like unfolding intermediate of the prion protein domain PrP(121–231) induced by acidic pH. *Proc Natl Acad Sci USA* **1998**, *95*, 6010–6014.

68 Morillas, M., D. L. Vanik and W. K. Surewicz. On the mechanism of alpha-helix to beta-sheet transition in the recombinant prion protein. *Biochemistry* **2001**, *40*, 6982–6987.

69 Baskakov, I. V., G. Legname, M. A. Baldwin, S. B. Prusiner and F. E. Cohen. Pathway complexity of prion protein assembly into amyloid. *J Biol Chem* **2002**, *277*, 21140–21148.

70 Lu, B. Y. and J. Y. Chang. Isolation and characterization of a polymerized prion protein. *Biochem J* **2002**, *364*, 81–87.

71 Baskakov, I. V., G. Legname, Z. Gryczynski and S. B. Prusiner. The peculiar nature of unfolding of the human prion protein. *Protein Sci* **2004**, *13*, 586–595.

72 Apetri, A. C. and W. K. Surewicz. Kinetic intermediate in the folding of human prion protein. *J Biol Chem* **2002**, *277*, 44589–44592.

73 Nicholson, E. M., H. Mo, S. B. Prusiner, F. E. Cohen and S. Marqusee. Differences between the prion protein and its homolog Doppel: a partially structured state with implications for scrapie formation. *J Mol Biol* **2002**, *316*, 807–815.

74 Kuwata, K., H. Li, Y. Yamada, G. Legname, S. B. Prusiner, K. Akasaka and T. L. James. Locally disordered conformer of the hamster prion protein: a crucial intermediate to PrP^{Sc}? *Biochemistry* **2002**, *41*, 12277–12283.

75 Martins, S. M., A. Chapeaurouge and S. T. Ferreira. Folding intermediates of the prion protein stabilized by hydrostatic pressure and low temperature. *J Biol Chem* **2003**, *278*, 50449–50455.

76 Cohen, F. E., K. Pan, Z. Huang, M. Baldwin, R. Fletterick and S. B. Prusiner. Structural clues to prion replication. *Science* **1994**, *264*, 530–531.

77 Goldfarb, L. G., R. B. Petersen, M. Tabaton, P. Brown, A. C. LeBlanc, P. Montagna, P. Cortelli, J. Julien, et al. Fatal familial insomnia and familial Creutzfeldt-Jakob disease: disease phenotype determined by a DNA polymorphism. *Science* **1992**, *258*, 806–808.

78 Kitamoto, T., R. Iizuka and J. Tateishi. An amber mutation of prion protein in Gerstmann–Sträussler syndrome with mutant PrP plaques. *Biochem Biophys Res Commun* **1993**, *192*, 525–531.

79 Liemann, S. and R. Glockshuber. Influence of amino acid substitutions related to inherited human prion diseases on the thermodynamic stability of the cellular prion protein. *Biochemistry* **1999**, *38*, 3258–3267.

80 Vanik, D. L. and W. K. Surewicz. Disease-associated F198S mutation increases the propensity of the recombinant prion protein for conformational conversion to scrapie-like form. *J Biol Chem* **2002**, *277*, 49065–49070.

81 Apetri, A. C., K. A. Surewicz and W. K. Surewicz. The effect of disease-associated mutations on the folding pathway of human prion protein. *J Biol Chem* **2004**, *279*, 18008–18014.

82 Zhang, Y., W. Swietnicki, M. G. Zagorski, W. K. Surewicz and F. D. Sonnichsen. Solution structure of the E200K variant of human prion protein. *J Biol Chem* **2000**, *275*, 33650–33654.

83 Swietnicki, W., R. Petersen, P. Gambetti and W. Surewicz. Familial mutations and the thermodynamic stability of the recombinant human prion protein. *J Biol Chem* **1998**, *273*, 31048–31052.

84 Santini, S., J.-B. Claude, S. Audic and P. Derreumaux. Impact of the tail and mutations G131V and M129V on prion protein flexibility. *Proteins* **2003**, *51*, 258–265.

85 Vorberg, I., K. Chan and A. Priola. Deletion of β-strand and α-helix secondary structure in normal prion protein inhibits its formation of its protease-resistant isoform. *J Virol* **2001**, *75*, 10024–10032.

86 Santini, S. and P. Derreumaux. The helix H1 of the prion protein is rather stable against environmental perturbations: molecular dynamics of mutation and deletion variants of PrP(90–231). *Cell Mol Life Sci* **2004**, *61*, 951–960.

87 Paci, E., M. Vendruscolo, C. M. Dobson and M. Karplus. Determination of a transition state at atomic resolution from protein engineering data. *J Mol Biol* **2002**, *324*, 151–163.

88 Horiuchi, M., S. A. Priola, J. Cabry and B. Caughey. Interactions between heterologous forms of prion protein: binding, inhibition of conversion and species barriers. *Proc Natl Acad Sci USA* **2000**, *97*, 5836–5841.

89 Kimberlin, R. H., S. Cole and C. A. Walker. Temporary and permanent modifications to a single strain of mouse scrapie on transmission to rats and hamsters. *J Gen Virol* **1987**, *68*, 1875–1881.

90 Prusiner, S. B., M. Scott, D. Foster, K. Pan, D. Groth, C. Mirenda, M. Torchia, S. L. Yang, et al. Transgenetic studies implicate interactions between homologous PrP isoforms in scrapie prion replication. *Cell* **1990**, *63*, 673–686.

91 Asante, E. A., J. M. Linehan, M. Desbruslais, S. Joiner, I. Gowland, A. L. Wood, J. Welch, A. F. Hill, et al. BSE prions propagate as either variant CJD-like or sporadic CJD-like prion strains in transgenic mice expressing human prion protein. *EMBO J* **2002**, *21*, 6358–6366.

92 Cohen, F. E. and S. B. Prusiner. Pathologic conformations of prion proteins. *Annu Rev Biochem* **1998**, *67*, 793–819.

93 Jarrett, J. T. and P. T. Lansbury Jr. Seeding one dimensional crystallization of amyloid: a pathogenic mechanism in Alzheimer's disease and scrapie? *Cell* **1993**, *73*, 1055–1058.

94 Morrissey, M. P. and E. I. Shakhnovich. Evidence for the role of PrP^C helix 1 in the hydrophilic seeding of prion aggregates. *Proc Natl Acad Sci USA* **1999**, *96*, 11293–11298.

95 Levy, Y. and O. M. Becker. Conformational polymorphism of wild-type and mutant prion proteins: energy landscape analysis. *Proteins* **2002**, *47*, 458–468.

96 Speare, J. O., T. S. Rush 3rd, M. E. Bloom and B. Caughey. The role of helix 1 aspartates and salt bridges in the stability and conversion of prion protein. *J Biol Chem* **2003**, *278*, 12522–12529.

97 Serio, T. R., A. G. Cashikar, A. S. Kowal, G. J. Sawicki, J. J. Moslehi, L. Serpell, M. F. Arnsdorf and S. L. Lindquist. Nucleated conformational conversion and the replication of conformational information by a prion determinant. *Science* **2000**, *289*, 1317–1321.

98 Harrison, P. M., H. S. Chan, S. B. Prusiner and F. E. Cohen. Conformational propagation with prion-like characteristics in a simple model of protein folding. *Protein Sci* **2001**, *10*, 819–835.

99 Gsponer, J., U. Haberthur and A. Caflisch. The role of side-chain interactions in the early steps of aggregation: molecular dynamics simulations of an amyloid-forming peptide from the yeast prion Sup35. *Proc Natl Acad Sci USA* **2003**, *100*, 5154–5159.

100 Klimov, D. K. and D. Thirumalai. Dissecting the assembly of $A\beta^{16-22}$ amyloid peptides into antiparallel beta sheets. *Structure* **2003**, *11*, 295–307.

101 Santini, S., G. Wei, N. Mousseau and P. Derreumaux. Exploring the folding pathways of proteins through energy landscape sampling: application to Alzheimer's beta-amyloid peptide. *Internet Electron J Mol Des* **2003**, *2*, 564–577.

102 Santini, S., G. Wei, N. Mousseau and P. Derreumaux. Pathway complexity of Alzheimer's -amyloid A^{1622} peptide assembly. *Structure* **2004**, *12*, 1245–1255.

103 Ding, F., N. V. Dokholyan, S. V. Buldyrev, H. E. Stanley and E. I. Shakhnovich. Molecular dynamics simulation of the SH3 domain aggregation suggests a generic amyloidogenesis mechanism. *J Mol Biol* **2002**, *324*, 851–857.

104 Jang, H., C. K. Hall and Y. Zhou. Assembly and kinetic folding pathways of a tetrameric beta-sheet complex: molecular dynamics simulations on simplified off-lattice protein models. *Biophys J* **2004**, *86*, 31–49.

105 Dima, R. I. and D. Thirumalai. Exploring protein aggregation and self-propagation using lattice models: phase diagram and kinetics. *Protein Sci* **2002**, *11*, 1036–1049.

106 Alonso, D. O., S. J. DeArmond, F. E. Cohen and V. Daggett. Mapping the early steps in the pH-induced conformational conversion of the prion protein. *Proc Natl Acad Sci USA* **2001**, *98*, 2985–2989.

107 Zuegg, J. and J. E. Gready. Molecular dynamics simulations of human prion protein: importance of correct treatment of electrostatic interactions. *Biochemistry* **1999**, *38*, 13862–13876.

108 Okimoto, N., K. Yamanaka, A. Suenaga, M. Hata and T. Hoshino. Computational studies on prion proteins: effect of Ala^{117}–Val mutation. *Biophys J* **2002**, *82*, 2746–2757.

109 Parchment, O. G. and J. W. Essex. Molecular dynamics of mouse and Syrian hamster PrP: implications for activity. *Proteins* **2000**, *38*, 327–340.

110 Gsponer J., P. Ferrara and A. Caflisch. Flexibility of the murine prion protein and its Asp178Asn mutant investigated by molecular dynamics simulations. *J Mol Graph Model* **2001**, *20*, 169–182.

111 Kozin S. A., G. Bertho, A. K. Mazur, H. Rabesona, J. P. Girault, T. Haertle, M. Takahashi, P. Debey, et al. Sheep prion protein synthetic peptide spanning helix 1 and beta-strand 2 (residues 142–166) shows beta-hairpin structure in solution. *J Biol Chem* **2001**, *276*, 46364–46370.

112 Liu, A., R. Riek, R. Zahn, S. Hornemann, R. Glockshuber and K. Wüthrich. Peptides and proteins in neurodegenerative disease: helix propensity of a polypeptide containing helix 1 of the mouse prion protein studied by NMR and CD spectroscopy. *Biopolymers* **1999**, *51*, 145–152.

113 Ziegler, J., H. Sticht, U. C. Marx, W. Muller, P. Rosch and S. Schwarzinger. CD and NMR-studies of prion protein helix 1: novel implications for its role in the PrP^{C} to PrP^{Sc} conversion process. *J Biol Chem* **2003**, *278*, 50175–50181.

114 Jamin, N., Y. M. Coic, C. Landon, L. Ovtracht, F. Baleux, J. M. Neumann and A. Sanson. Most of the structural elements of the globular domain of murine prion protein form fibrils with predominant beta-sheet structure. *FEBS Lett* **2002**, *529*, 256–260.

115 Gu, W., T. Wang, J. Zhu, Y. Shi, H. Liu. Molecular dynamics simulation of the unfolding of the human prion protein domain under low pH and high temperature conditions. *Biophys Chem* **2003**, *104*, 79–94.

116 Gilis, D. and M. Rooman. PoPMuSiC, an algorithm for predicting protein mutant stability changes: application to prion proteins. *Protein Eng* **2000**, *13*, 849–856.

117 Kallberg, Y., M. Gustafsson, B. Persson, J. Thyberg and J. Johansson. Prediction of amyloid fibril-forming proteins. *J Biol Chem* **2001**, *276*, 12945–12950.

118 Dima, R. I. and D. Thirumalai. Exploring the propensities of helices in PrP^C to form beta sheet using NMR structures and sequence alignments. *Biophys J* **2002**, *83*, 1268–1280.

119 Settanni, G., T.X. Hoang, C. Micheletti and A. Maritan. Folding pathways of prion and doppel. *Biophys J* **2002**, *83*, 3533–3541.

120 Walsh, D.M., I. Klyubin, J.V. Fadeeva, W.K. Cullen, R. Anwyl, M.S. Wolfe, M.J. Rowan and D.J. Selkoe. Naturally secreted oligomers of amyloid beta protein potently inhibit hippocampal long-term potentiation *in vivo*. *Nature* **2002**, *416*, 483–484.

121 Bellinger-Kawahara, C.G., E. Kempner, D. Groth, R. Gabizon and S.B. Prusiner. Scrapie prion liposomes and rods exhibit target sizes of 55,000 Da. *Virology* **1988**, *164*, 537–541.

122 Turk, E., D.B. Teplow, L.E. Hood, S.B. Prusiner. Purification and properties of the cellular and scrapie hamster prion proteins. *Eur J Biochem* **1988**, *176*, 21–30.

123 Sokolowski, F., A.J. Modler, R. Masuch, D. Zirwer, M. Baier, G. Lutsch, D.A. Moss, K. Gast, et al. Formation of critical oligomers is a key event during conformational transition of recombinant Syrian hamster prion protein. *J Biol Chem* **2003**, *278*, 40481–40492.

124 Capellari, S., P. Parchi, J. Campbell, D.M. Posey, R.B. Petersen and P. Gambetti. Creutzfeldt–Jakob disease associated with a deletion of two repeats in the prion protein gene. *Neurology* **2002**, *59*, 1628–1630.

125 Sigurdsson, E.M., D.R. Brown, M.A. Alim, H. Scholtzova, R. Carp, H.C. Meeker, F. Prelli, B. Frangione, et al. Copper chelation delays the onset of prion disease. *J Biol Chem* **2003**, *278*, 46199–46202.

126 Kaneko, K., L. Zulianello, M. Scott, C.M. Cooper, A.C. Wallace, T.L. James, F.E. Cohen and S.B. Prusiner. Evidence for protein X binding to a discontinuous epitope on the cellular prion protein during scrapie prion propagation. *Proc Natl Acad Sci USA* **1997**, *94*, 10069–10074.

127 Lee, K.S., R. Linden, M.A. Prado, R.R. Brentani and V.R. Martins. Towards cellular receptors for prions. *Rev Med Virol* **2003**, *13*, 399–408.

128 Edenhofer, F., R. Rieger, M. Famulok, W. Wendler, S. Weiss and E.L. Winnacker. Prion protein PrP^C interacts with molecular chaperones of the Hsp60 family. *J Virol* **1996**, *70*, 4724–4728.

129 Saborio, G.P., B. Permanne and C. Soto. Sensitive detection of pathological prion protein by cyclic amplification of protein misfolding. *Nature* **2001**, *411*, 810–813.

130 Deleault, N.R., R.W. Lucassen, S. Supattapone. RNA molecules stimulate prion protein conversion. *Nature* **2003**, *425*, 717–720.

131 Soto, C., R.J. Kascsak, G.P. Saborio, P. Aucouturier, T. Wisniewski, F. Prelli, R. Kascsak, E. Mendez, et al. Reversion of prion protein conformational changes by synthetic -sheet breaker peptides. *Lancet* **2000**, *355*, 192–197.

132 Caughey, B., L.D. Raymond, G.J. Raymond, L. Maxson, J. Silveira and G.S. Baron. Inhibition of protease-resistant prion protein accumulation *in vitro* by curcumin. *J Virol* **2003**, *77*, 5499–5502.

133 Mouillet-Richard, S., M. Ermonval, C. Chebassier, J.L. Laplanche, S. Lehmann, J.M. Launay and O. Kellermann. Signal transduction through prion protein. *Science* **2000**, *289*, 1925–1928.

134 Schatzl, H.M., L. Laszlo, D.M. Holtzman, J. Tatzelt, S.J. DeArmond, R.I. Weiner, W.C. Mobley and S.B. Prusiner. A hypothalamic neuronal cell line persistently infected with scrapie prions exhibits apoptosis. *J Virol* **1997**, *71*, 8821–8831.

135 Kayed, R., E. Head, J.L. Thompson, T.M. McIntire, S.C. Milton, C.W. Cotman and C.G. Glabe. Common structure of soluble amyloid oligomers implies common mechanism of pathogenesis. *Science* **2003**, *300*, 486–489.

136 Notredame, C., D.G. Higgins and J. Heringa. T-Coffee: a novel method for fast and accurate multiple sequence alignment. *J Mol Biol* **2000**, *302*, 205–217.

137 Kovacs, G.G., G. Trabattoni, J.A. Hainfellner, J.W. Ironside, R.S. Knight and H. Budka. Mutations of the prion protein gene phenotypic spectrum. *J Neurol* **2002**, *249*, 1567–1582.

19
Familial British and Danish Dementias

Jorge Ghiso, Agueda Rostagno, Yasushi Tomidokoro, Tammaryn Lashley, Janice L. Holton, Gordon Plant, Tamas Revesz and Blas Frangione

19.1
Introduction

Systemic and cerebral amyloid diseases are considered to be part of an emerging complex group of chronic and progressive entities collectively known as "protein folding disorders", among them Alzheimer's disease, polyglutamine repeat disorders, cataracts, amyotropic lateral sclerosis, Parkinson's disease and other synucleinopathies, tauopathies, prion diseases, and Type 2 diabetes. For reasons that are poorly understood, normally soluble proteins that are found in biological fluids undergo a change in their conformation and form insoluble structures that accumulate in the form of intra- and extracellular aggregates. In the case of amyloid, the aggregates consist of fibrillar structures. So far, eight out of the 25 amyloid subunits known to produce amyloid diseases in humans have been associated with lesions in the central nervous system [1], with those related to Aβ being the most common. The deposition of amyloid fibrils in the cerebrovasculature, generically referred to as cerebral amyloid angiopathy (CAA), is one of the major neuropathological hallmarks of the aging brain and many neurodegenerative disorders, including sporadic and familial Alzheimer's disease, hereditary cerebral hemorrhage with amyloidosis – Icelandic type (HCHWA-I), and the recently described chromosome 13 dementias, familial British and Danish dementias (FBD and FDD, respectively).

19.2
FBD and FDD

19.2.1
Clinical Presentation

FBD, a neurodegenerative disorder described in members of three British pedigrees [2, 3], is clinically characterized by progressive dementia, cerebellar ataxia and spastic paraparesis. Disease onset occurs typically in the fourth to fifth de-

Amyloid Proteins. The Beta Sheet Conformation and Disease. J. D. Sipe

ISBN: 3-527-31072-X

cade of life and runs a course of some 10 years until death occurs. FBD was first described in 1933 by Worster-Drought et al. as a familial pre-senile dementia with spastic paralysis [4, 5]. Patients with the disease have marked memory impairment progressing to global dementia, with personality changes being the earliest clinical manifestation. They develop spastic paralysis that is far more severe than that seen in atypical Alzheimer's disease or in Gerstmann-Sträussler-Sheinker disease. Patients progress to a chronic vegetative state becoming mute, unresponsive, incontinent and quadriplegic.

FDD is also a neurodegenerative disease observed, at the moment, only in a single small family in Denmark. Although many clinical features are similar to those in FBD, patients with FDD also have unique characteristics. The earliest manifestation in the disease is the presence of cataracts before the age of 30 followed by sensory hearing loss about 10 years later. By the mid-40s, patients develop cerebellar ataxia, psychosis and dementia, leading to death between the ages of 50 and 60 years [6–8].

19.2.2
Neuropathology

The neuropathology in both diseases is remarkably similar to Alzheimer's disease, including cerebral amyloid angiopathy, pre-amyloid lesions, amyloid plaques of various types and neurofibrillary tangles (NFTs), ultrastructurally composed of classical paired-helical filaments (Fig. 19.1). NFTs in FBD and FDD brains are indistinguishable from those found in Alzheimer's disease. They are recognized by a panel of anti-hyperphosphorylated tau antibodies in immunohistochemical analysis and, when extracted from the lesions and evaluated by electrophoresis and Western blot analysis, they show identical patterns of abnormally hyperphosphorylated tau immunoreactivity (Fig. 19.2) [2, 9–12]. Activated microglia expressing the major histocompatibility class II antigens, characteristic of inflammatory processes, abound around amyloid plaques and dystrophic neurites in both diseases [9, 12]. Complement activation products from both the classical and the alternative pathways co-localize also with FBD and FDD parenchymal plaques and cerebrovascular amyloid deposits, mainly associated with Congo red/Thioflavin-positive amyloid deposits rather than Congo red/Thioflavin-negative parenchymal pre-amyloid lesions [13]. Also closely resembling the findings in Alzheimer's disease, multiple amyloid-associated proteins, among them apolipoproteins J and E, as well as serum amyloid P component are associated with the lesions [14–16].

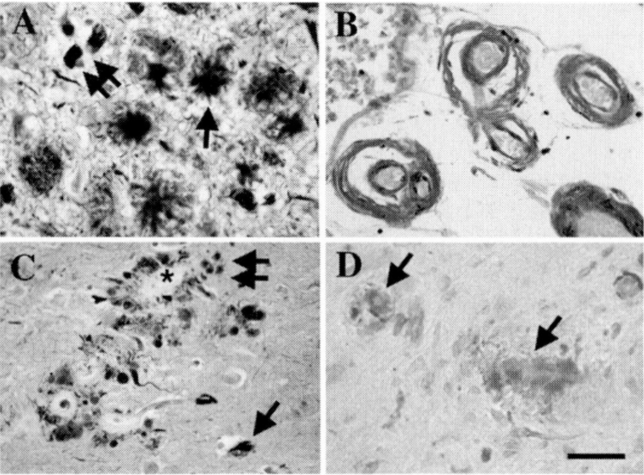

Fig. 19.1 Parenchymal and vascular amyloid lesions in FBD and FDD. (A) Silver stain demonstrates numerous argyrophilic plaques (arrow) co-localizing with NFTs (double arrow) in a case of FBD. (B) Widespread severe cerebral amyloid angiopathy is also characteristic in FBD. (C) Silver stain reveals NFTs (arrow) and abnormal neurites (double arrow) around amyloid-laden blood vessels (asterisk), but no amyloid plaques in an FDD case. (D) Severe amyloid deposition, similar to FBD, affects both leptomeningeal and parenchymal blood vessels throughout the central nervous system in FDD. (A, C) Bielschowsky's silver impregnation; (B, D) periodic acid–Schiff. Bar on (D) represents 50 μm on (A)–(D).

19.3 A Novel Gene *BRI2*

The *BRI* gene, organized into six exons of varied length and extending more than 20 kb, belongs to a multigene family comprising at least three homologs in both mice and humans, *BRI1*, *BRI2* and *BRI3* (also commonly referred to as *ITM2A*, *ITM2B* and *ITM2C* or *E25A*, *E25B* and *E25C*, respectively) [17–20]. Gene expression of the human isoforms is strikingly dissimilar. Whereas *BRI2* is highly ubiquitous, *BRI1* shows a limited expression profile mostly restricted to osteo- and chondrogenic tissues [18], and *BRI3* exclusively localizes to the brain [20, 21].

The *BRI2* gene (Fig. 19.3 A), located on the long arm of chromosome 13 [18, 19], is of particular interest, since it is associated with the chromosome 13 dementias FBD and FDD (see below). The *BRI2* gene is broadly expressed in peripheral organs including the kidney, pancreas and placenta, as well as in the brain, in which it is more abundant in the hippocampus and cerebellum compared with the cerebral cortex [19]. *In situ* hybridization studies of human cell cultures showed that, within the different cerebral cell populations, BRI2 mRNA is ubiquitously present in neurons, astrocytes, smooth muscle and cerebral endothelial cells (Fig. 19.3 B).

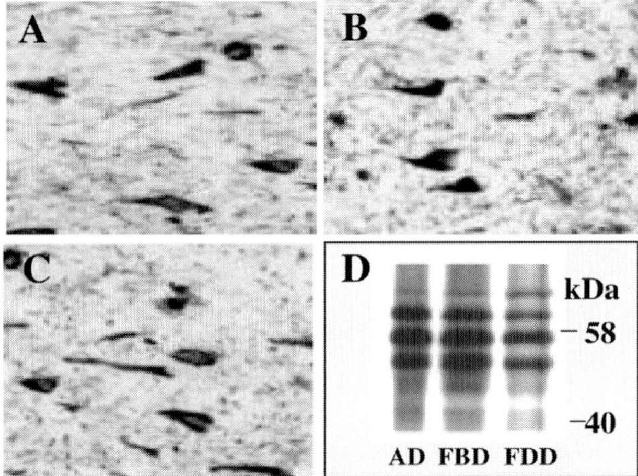

Fig. 19.2 Neurofibrillary tangles in FBD and FDD in comparison with Alzheimer's disease. Immunohistochemical and biochemical identification of hyperphosphorylated tau (AT8 antibody) in Alzheimer's disease (A and D, lane 1), FBD (B and D, lane 2) and FDD (C and D, lane 3). Note the similarities in immunostaining as well as in the electrophoretic pattern among the three diseases.

The protein product of *BRI2* is a 266-amino-acid protein containing a single glycosylation site, and a predicted structure similar to the type II transmembrane glycoproteins in which the extracellular portion is the C-terminal region and the N-terminus is intracytoplasmic [19]. A furin-like processing at peptide bond 243–244 releases a 23-residues peptide with still unknown biological func-

Fig. 19.3 BRI2 gene, protein expression and genetic defects in FBD and FDD. (A) Schematic representation of the multiexonic *BRI2* gene located on chromosome 13. (B) *In situ* hybridization for BRI2 mRNA in brain cells in culture. (C) Schematic representation of the protein product of the *BRI2* gene. Protein BRI contains a single transmembrane domain spanning from residues 52 to 74 and a single *N*-glycosylation site at position 170 (shaded diamond). (D) Immunocytochemical detection of protein BRI in brain cultured cells via fluorescence microscopy. (E) A stop-to-arg mutation in the *BRI2* gene translates in a longer BRI protein; the 34 C-terminal amino acids ABri, released by a furin-like proteolytic processing, is the main component of amyloid deposits in FBD. (F) A 10-nucleotide duplication-insertion in the *BRI2* gene translates in a longer BRI protein; the 34 C-terminal amino acids ADan, released by a furin-like proteolytic processing, is the main component of amyloid deposits in FDD.

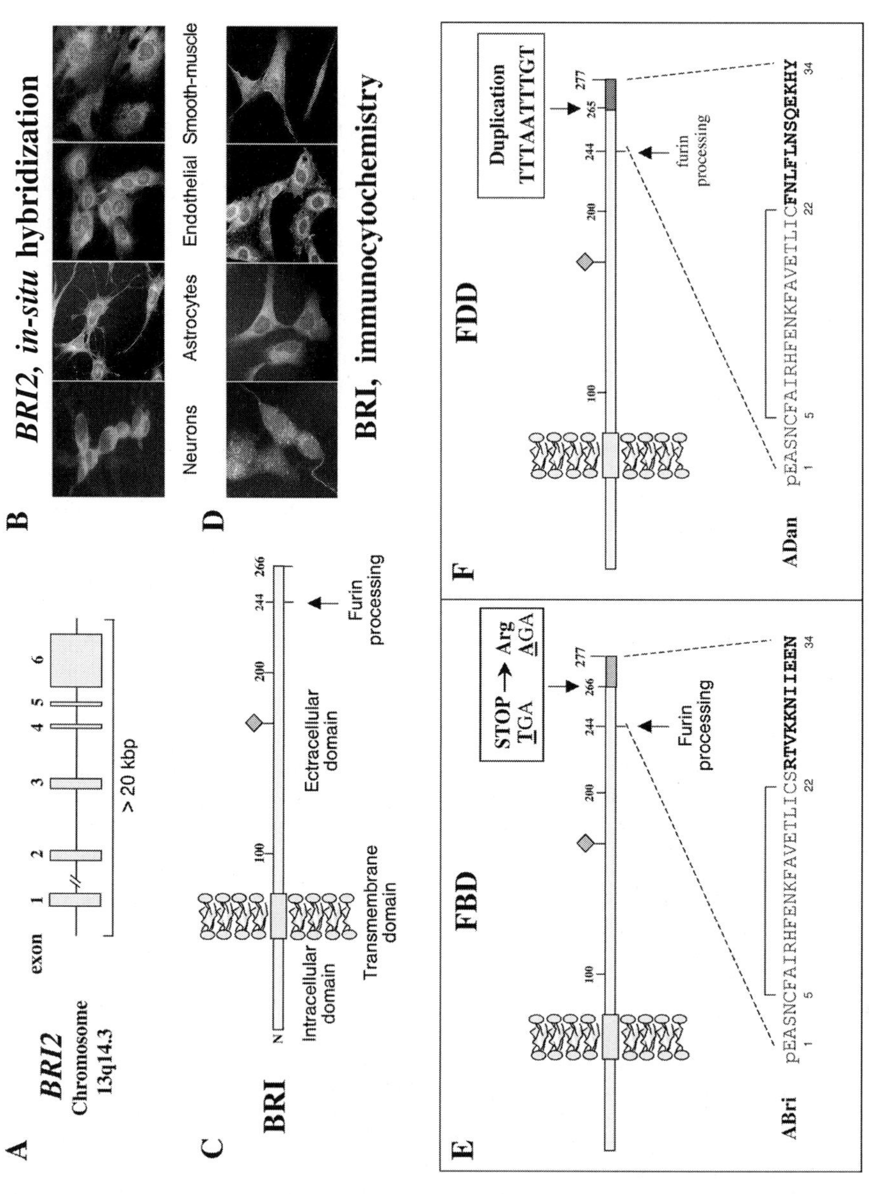
A
BRI2
Chromosome
13q14.3
exon
1
2
3
4
5
6
> 20 kbp
B
BRI2, in-situ hybridization
Neurons
Astrocytes
Endothelial
Smooth-muscle
C
BRI
N
100
200
244
266
Intracellular domain
Transmembrane domain
Ectracellular domain
Furin processing
D
BRI, immunocytochemistry
E
FBD
STOP → Arg
TGA AGA
100
200
244
266
277
Furin processing
ABri
pEASNCFAIRHFENKFAVETLICSRTVKKNIIEEN
1 5 22 34
F
FDD
Duplication
TTTAATTTGT
100
200
244
265
277
furin processing
ADan
pEASNCFAIRHFENKFAVETLICFNLFLNSQEKHY
1 5 22 34

tion [22] (Fig. 19.3 C). Within the brain, protein BRI was detected as fine granules in the neuronal cytoplasm, particularly in pyramidal neurons of the CA3 and CA4 layers in the hippocampus as well as in Purkinje cells in the cerebellum [23]. Immunocytochemical analysis of cerebral cells in culture employing polyclonal antibodies raised versus the C-terminal region of BRI (residues 222–233) demonstrated that, consistent with the mRNA data, BRI protein is present in neurons, both in the cellular body and axons, as well as in astrocytes, cerebral smooth muscle and endothelial cells (Fig. 19.3 D).

19.4
BRI2 Mutations Generate Two New Amyloid Subunits, ABri and ADan

In the autosomal dominant diseases FBD and FDD, two distinct genetic defects occur in the *BRI2* gene, i.e. a stop-to-arg mutation in FBD [19] (Fig. 19.3 E) and a 10-nucleotide duplication-insertion immediately before the stop codon in FDD [24] (Fig. 19.3 F). As a result, and regardless of the nucleotide changes, the final outcome is common to both diseases: the ordinarily occurring stop codon is not in frame causing the genesis of an extended precursor featuring a C-terminal piece that does not exist in normal conditions. The 34-amino-acid peptides that deposit in both disorders (ABri in FBD and ADan in FDD) originate from the above-mentioned furin-like proteolytic cleavage at peptide bond 243–244 of the respective mutated longer precursor protein [22, 25]. As a result, ABri and ADan share 100% homology only on their first 22 residues (Fig. 19.3 E and F; non-homologous sequences highlighted in bold). These particular differences at the C-terminus, and the fact that both segments are generated *de novo* and are not present in the normal population, prompted the generation of specific antibodies that allowed the topographical characterization of the deposited proteins [19, 24]. As illustrated in Fig. 19.4 (A), antibodies specific for ABri (Ab338) [19]

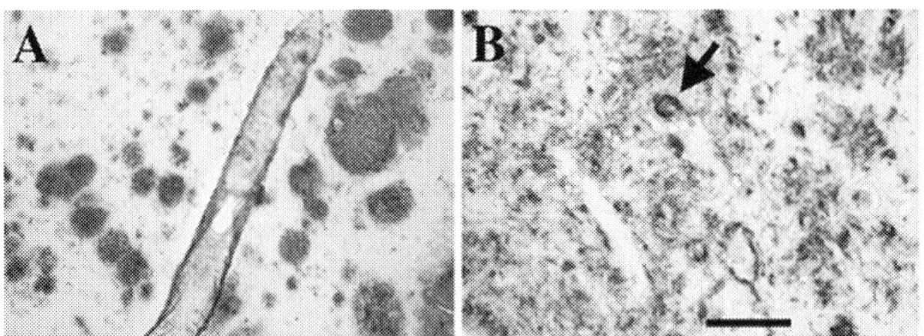

Fig. 19.4 Immunohistochemical analysis of FBD and FDD amyloid lesions. (A) Amyloid plaques and a blood vessel with cerebral amyloid angiopathy (center) are strongly immunoreactive for ABri (Ab338). (B) Deposition of ADan can be demonstrated in blood vessels affected by amyloid deposition (arrow) and also in diffuse, pre-amyloid parenchymal plaques (Ab5282). Bar represents 50 μm on (A) and (B).

recognize parenchymal plaques and cerebrovascular deposits in FBD, whereas antibodies specific for ADan (Ab5282) [24] highlight vascular amyloid deposits and pre-amyloid lesions in FDD (Fig. 19.4 B).

19.5 Biochemical Properties of Amyloid Subunits ABri and ADan

Deposited ABri and ADan, as stated above, are both 34 amino acids long, identical in their first 22 residues, but with different C-termini. Both peptides contain two cysteine residues that, in patients with FBD and FDD, are oxidized, forming a single intrachain disulfide bond between residues 5 and 22. Although some degree of N- and C-terminal heterogeneity was observed in the deposited materials, both amyloid subunits were primarily post-translationally modified at their N-terminus, featuring pyroglutamate instead of glutamate [19, 24]. Both peptides showed a high tendency to form oligomers and aggregate, *in vivo* and *in vitro*. As observed in every known amyloid, subunits isolated from tissue deposits showed a variable degree of oligomerization. For ABri and ADan, the process of oligomerization seems to be directly related to the degree of peptide insolubility. From FBD and FDD cerebral tissue homogenates, only monomeric ABri and ADan species together with a small proportion (less than 10%) of dimeric and trimeric forms were extracted with water-based buffers. In contrast, detergent (sodium dodecylsulfate)-solubilized fractions, thought to mainly represent pre-amyloid deposits, exhibited a similar composition, although there was some enrichment (less than 40%) in dimeric and trimeric forms that, in some cases, appeared in an equivalent proportion to the monomeric species. Lastly, the most insoluble fractions, representing fibrillar deposits, could be solubilized in formic acid and exhibited a pattern of heavy oligomerization with the simultaneous appearance of a broad spectrum of species ranging from monomeric forms to high-molecular-mass aggregates (above 70 kDa) [26].

In vitro studies using synthetic homologues of full-length ABri and ADan are able to reproduce the oligomerization propensity seen with the naturally derived peptides, forming spontaneous β-sheet-rich structures that exhibit fast aggregation kinetics, with an even higher tendency than the Alzheimer's associated Aβ42 peptide to form high-order oligomers. In an extremely short time (less than 1 h), under physiologic conditions of pH, ionic strength and temperature, both peptides form oligomers to a high degree, with ABri exhibiting faster kinetics. It is interesting that, under these conditions, fibril formation seems to proceed very slowly (weeks). In fact, enhanced aggregation kinetics and protofibril formation by ABri is favored by slightly acidic pH [27], consistent with the behavior of other amyloid-forming proteins [28] and supporting the premise that common mechanism(s) may be involved in protein misfolding and amyloidosis. ABri formed protofibrils as intermediate structures during maturation into fibrils at pH 4.9 and, similar to previous findings with Aβ, fibril seeds are required to initiate amyloid formation at neutral pH [27]. As previously observed with the Alzheimer's disease-associated Aβ, synthetic peptides homologous to ABri and ADan [29, 30] exhibited cytotoxic properties to neuronal cell lines.

19.6 Soluble Forms of ABri and ADan in Biological Fluids

In Alzheimer's disease, soluble forms of the Aβ peptide circulate in blood and are present in other biological fluids, such as cerebrospinal fluid (CSF) and urine (reviewed in [1]). Similarly, soluble forms of ABri occur in serum, CSF and urine of both FBD patients and pre-symptomatic carriers of the stop-to-arg mutation [31], whereas the peptide is consistently absent in non-carrier family members. In the case of FDD, soluble ADan species are also found in the circulation [26]. Using a combination of immunoprecipitation and mass spectrometry analysis, we demonstrated that, in contrast to the deposited species that are post-translationally modified at the N-terminus (pyroglutamate), both sABri and sADan feature an N-terminal glutamic acid, strongly suggesting that the conversion of glutamate to pyroglutamate occurs at the site of deposition [26, 31]. Since the reaction to form pyroglutamate is practically irreversible, sABri and sADan in circulation are not likely to represent ABri or ADan species being cleared from the deposits.

19.7 Unique Features of FBD or FDD

Although FBD and FDD share striking similarities, both disorders also show significant differences.

19.7.1 FBD is a Systemic Disorder

Immunohistochemical studies carried out with organs obtained at autopsy identified ABri amyloid and pre-amyloid deposits in tissues other than the brain (e.g. cardiac muscle and pancreas), with the vast majority of Congo red-positive amyloid lesions found in the form of vascular deposits [11, 31]. Biochemical analysis of the extracted systemic amyloid indicates the presence of the same ABri species found in brain deposits, including the presence of the internal disulphide bond and the N-terminally modified pyroglutamate [31], clearly indicating that the enzyme(s) involved in these transitions/modifications are not exclusive of cerebral origin, but also exist in organs throughout the body. It is not yet known up to date whether similar systemic deposits outside the brain also occur in FDD.

19.7.2
FDD is not a Single Amyloid Disease

Biochemical and immunohistochemical analysis of FDD amyloid laden tissues revealed an unexpected feature of this disease: the presence of Aβ co-localizing with ADan deposits [12, 26]. Some of the cerebral capillaries, arterioles and veins also contain Aβ peptide, although the number of vessels affected by ADan deposition exceeds those with Aβ deposits. Whereas some of the blood vessels are affected with solely Aβ or ADan deposition, in a considerable number, co-localization of both peptides can be established by confocal microscopy [12]. Aβ amyloid angiopathy is most severe in the hippocampus, limbic structures, neocortex, thalamus and cerebellum, while there is little involvement of the brain stem and spinal cord. In the CNS parenchyma subpial peptide deposits, perivascular plaques and diffuse parenchymal deposits also contain Aβ in addition to ADan. The hippocampus, limbic structures and neocortex are most severely and consistently affected by parenchymal Aβ deposits. Electron microscopy after colloidal gold labeling of ADan and Aβ revealed the presence of both components within the same fibrillar deposits. Immunoprecipitation, in conjunction with Western blot and mass spectrometry analysis, revealed Aβ peptides ending at position 42 as the major components whereas Aβ ending at residue 40 was almost non-existent. Of the different Aβ42 species present, Aβ4–42 was the most relevant [26]. The presence of Aβ42 in vascular lesions was also an unexpected finding that contrasts with previous findings in Aβ-related disorders (familial and sporadic Alzheimer's disease, Down's syndrome, sporadic CAA, and normal aging) in which the major Aβ species in vascular deposits is consistently 40 residues long; peptides 42 residues in length are mainly associated with parenchymal deposits. Which mechanisms are responsible for the formation of the double amyloid deposits and why Aβ42 is the predominant species are still not known. Since several mutations in the AβPP gene are known to produce vasculotropic variants and mutations in presenilin 1 (PS1) increase the ratio of Aβ42/Aβ40, we investigated the possibility of nucleotide changes in these molecules as one of the potential causes of Aβ42 deposition in FDD cerebral vessels. It is worth noting that analysis of the nucleotide sequences of exons 16 and 17 of AβPP as well as all exons in PS1 did not reveal any mutations in FDD patients.

19.8
Potential Implications of FBD and FDD for Alzheimer's Disease

Different genetic defects in the *BRI2* gene generate ABri and ADan peptides structurally unrelated to any known amyloid proteins, including those deposited in the brain. Both ABri and ADan peptides share an identical N-terminus and feature different C-termini. In spite of the structural differences in the amyloid subunits, amyloid lesions in both diseases co-localize with NFTs, exhibiting an electrophoretic pattern of abnormally hyperphosphorylated tau indistinguishable from that observed in Alzheimer's disease. The data indicate that amyloid peptides distinct from Aβ can trigger similar neuropathological changes, leading to the same scenario, neuronal loss and dementia. Therefore, we propose that FBD and FDD constitute alternative models with which to study the mechanisms of neurofibrillar degeneration, neuronal cell death and amyloid formation in the brain.

Acknowledgments

Supported by the Alzheimer's Association, the American Health Assistant Foundation and the National Institutes of Health (grants AG05891, AG08721 and NS38777).

References

1 Ghiso, J. and B. Frangione. Amyloidosis and Alzheimer's disease. *Adv Drug Del Rev* **2002**, *54*, 1539–1551.
2 Plant, G., T. Revesz, R. Barnard, A. Harding and P. Gautier-Smith. Familial cerebral amyloid angiopathy with non-neuritic plaque formation. *Brain* **1990**, *113*, 721–747.
3 Mead, S., M. J. Galton, T. Revesz, R. B. Doshi, G. Harwood, E. L. Pan, J. Ghiso, B. Frangione, et al. Familial British dementia with amyloid angiopathy: Early clinical, neuropsychological and imaging findings. *Brain* **2000**, *123*, 975–986.
4 Worster-Drought, C., T. Hill and W. McMenemey. Familial presenile dementia with spastic paralysis. *J Neurol Psychopathol* **1933**, *14*, 27–34.
5 Worster-Drought, C., J. G. Greenfield and W. McMenemey. A form of familial presenile dementia with spastic paralysis (including the pathological examination of a case). *Brain* **1940**, *63*, 237–254.
6 Strömgren, E., A. Dalby, M. A. Dalby and B. Ranheim. Cataract, deafness, cerebellar ataxia, psychosis and dementia: a new syndrome. *Acta Neurol Scand* **1970**, *46*, 261–262.
7 Strömgren, E. Heredopathia ophthalmo-oto-encephalica. In *Handbook of Clinical Neurology*, Vinken, P. J. and Bruyn, G. W. (eds). North Holland, Amsterdam, **1981**.
8 Bek, T. Ocular changes in heredo-oto-ophthalmo-encephalopathy. *Br J Ophtalmol* **2000**, *84*, 1298h–1302.
9 Holton, J., J. Ghiso, T. Lashley, A. Rostagno, C. Guerin, G. Gibb, H. Houlden, H. Ayling, et al. Regional distribution of fibrillar and non-fibrillar ABri deposition and its association with neurofibrillary degeneration in Familial British Dementia. *Am J Pathol* **2001**, *158*, 515–526.
10 Revesz, T., J. L. Holton, B. Doshi, B. H. Anderton, F. Scaravilli and G. Plant. Cytoskeletal pathology in familial cerebral amyloid angiopathy (British type) with non-neuritic amyloid plaque formation. *Acta Neuropathol* **1999**, *97*, 170–176.
11 Holton, J., J. Ghiso, T. Lashley, M. Ganguly, K. Strand, A. Rostagno, G. Plant, B. Frangione, et al. Familial British Dementia (FBD): a cerebral amyloidosis with systemic amyloid deposition. *Neuropathol Appl Neurobiol* **2002**, *28*, 148.
12 Holton, J., T. Lashley, J. Ghiso, H. Braendgaard, R. Vidal, C. Guerin, G. Gibb, D. P. Hanger, et al. Familial Danish Dementia: a novel form of cerebral amyloidosis associated with deposition of both amyloid-Dan and amyloid-beta. *J*

Neuropathol Exp Neurol **2002**, *61*, 254–267.

13 Rostagno, A., T. Revesz, T. Lashley, Y. Tomidokoro, L. Magnotti, H. Braendgaard, M. Bojsen-Moller, J. Holton, et al. Complement activation in Chromosome 13 dementias: similarities with Alzheimer's Disease. *J Biol Chem* **2003**, *277*, 49782–49790.

14 Revesz, T., J. Ghiso, T. Lashley, G. Plant, A. Rostagno, B. Frangione and J. Holton. Cerebral amyloid angiopathies: a pathologic, biochemical and genetic view. *J Neuropathol Exp Neurol* **2003**, *62*, 885–898.

15 Rostagno, A., R. McGinty, D. Ng, T. Lashley, J. Holton, B. Frangione, T. Revesz and J. Ghiso. P-component in familial British and Danish dementias. *Soc Neurosci* **2003**, 203.3 A.

16 Calero, M., A. Rostagno, B. Frangione and J. Ghiso. Clusterin and Alzheimer's disease. In *Subcellular Biochemistry: Alzheimer's Disease*, Harris, R. (ed.). Plenum Press/Kluwer, Berlin, **2005**, *38*, 273–298.

17 Deleersnijder, W., G. Hong, R. Cortvrindt, C. Poirier, P. Tyzanowski, K. Pittois, E. Van Marck and J. Merregaert. Isolation of markers for chondro-osteogenic differentiation using cDNA library subtraction. *J Biol Chem* **1996**, *271*, 19475–19482.

18 Pittois, K., W. Deleersnijder and J. Merregaert. cDNA sequence analysis, chromosomal assignment and expression pattern of the gene coding for integral membrane protein 2B. *Gene* **1998**, *217*, 141–149.

19 Vidal, R., B. Frangione, A. Rostagno, S. Mead, T. Revesz, G. Plant and J. Ghiso. A stop-codon mutation in the *BRI* gene associated with familial British dementia. *Nature* **1999**, *399*, 776–781.

20 Vidal, R., M. Calero, T. Revesz, G. Plant, J. Ghiso and B. Frangione. Sequence, genomic structure and tissue expression of human BRI3, a member of the BRI gene family. *Gene* **2001**, *266*, 95–102.

21 Choi, S.C., J. Kim, T.H. Kim, S.Y. Cho, S.S. Park, K.D. Kim and S.H. Lee. Cloning and characterization of a type II integral transmembrane protein gene, Itm2c, that is highly expressed in the mouse brain. *Mol Cells* **2001**, *12*, 391–397.

22 Kim, S.-H., R. Wang, D.J. Gordon, J. Bass, D. Steiner, D.G. Lynn, G. Thinakaran, S. Meredith, et al. Furin mediates enhanced production of fibrillogenic ABri peptides in familial British dementia. *Nat Neurosci* **1999**, *2*, 984–988.

23 Akiyama, H., H. Kondo, T. Arai, K. Ikeda, M. Kato, E. Iseki, C. Schwab and P. L. McGeer. Expression of BRI, the normal precursor of the amyloid protein of familial British dementia, in human brain. *Acta Neuropathol* **2004**, *107*, 53–58.

24 Vidal, R., J. Ghiso, T. Revesz, A. Rostagno, E. Kim, J. Holton, T. Bek, M. Bojsen-Moller, et al. A decamer duplication in the 3′ region of the *BRI* gene originates a new amyloid peptide that is associated with dementia in a Danish kindred. *Proc Natl Acad Sci USA* **2000**, *97*, 4920–4925.

25 Kim, S.-H., J.W. Creemers, S. Chu, G. Thinakaran and S.S. Sisodia. Proteolytic processing of familial British dementia-associated BRI variants: evidence for enhanced intracellular accumulation of amyloidogenic peptides. *J Biol Chem* **2002**, *277*, 1872–1877.

26 Tomidokoro, Y., S. Fleire, A. Rostagno, B. Frangione, J. Ghiso, T. Lashley, J. Holton, H. Houlden, et al. Co-existence of amyloid ADan and amyloid β in familial Danish dementia. *Neurobiol Aging* **2002**, *23*, S455.

27 Srinivasan, R., E.M. Jones, K. Liu, J. Ghiso, R.E. Marchant and M.G. Zagorsky. pH-dependent amyloid and protofibrils formation by the ABri peptide of familial British dementia. *J Mol Biol* **2003**, *333*, 1003–1023.

28 Rostagno, A., R. Vidal, B. Kaplan, J. Chuba, A. Kumar, J.I. Elliott, B. Frangione, G. Gallo, et al. pH-dependent fibrillogenesis of a VκIII Bence Jones protein. *Br J Haematol* **1999**, *107*, 835–843.

29 El-Agnaf, O.M.A., S. Nagala, B.P. Patel and B.M. Austen. Non-fibrillar oligomeric species of the amyloid ABri peptide, implicated in Familial British Dementia, are more potent at inducing apoptotic cell death than protofibrils or mature fibrils. *J Mol Biol* **2001**, *310*, 157–168.

30 Austen, B.M., O.M.A. El-Agnaf, S. Nagala, B.P. Patel, N. Gunasekera, M. Lee and V. Lelyveld. Properties of neurotoxic peptides related to the *BRI* gene. *Biochem Soc Trans* **2002**, *30*, 557–559.

31 Ghiso, J., J. Holton, L. Miravalle, M. Calero, T. Lashley, R. Vidal, H. Houlden, N. Wood, et al. Systemic amyloid deposits in Familial British Dementia. *J Biol Chem* **2001**, *276*, 43909–43914.

Systemic

20
Immunoglobulin

Fred J. Stevens

20.1
Introduction

Antibody light chains were first documented by Bence Jones [1, 2], a pioneering clinical chemist [3–5], more than 150 years ago as a urinary substance produced by a patient with a disease now known to be the cancer multiple myeloma. The material exhibited the peculiar behavior of precipitating as the urine was heated, yet going back into solution as the temperature was further increased. This precipitable material, which became known as Bence Jones protein, was much later identified as the antibody light chain [6] product of a monoclonal proliferation of antibody-producing cells during multiple myeloma and other plasma cell dyscrasias, including monoclonal gammopathy of undetermined significance (MGUS) [7], Waldenstrom's macroglobulinemia [8], lymphoma [9] and others [10–16].

Bence Jones proteins, which are often produced in large quantities by patients with multiple myeloma, proved to be an important resource from which was garnered the earliest insights into the structural basis of antibody diversity and specificity. The protein was relatively easily purified from the urine of patients who had healthy kidneys, thus limiting the passage of other, larger serum proteins into the urine. This availability enabled successful application of the laborious amino acid sequencing techniques of the day, which consumed significant quantities of protein. The work of many groups [17–28] quickly established two classes of light chain, which came to be known as κ and λ. Each class was further divided by similarity into several subgroups, implying the existence of multiple variable domain genes. Wu and Kabat [29, 30] demonstrated that the

Amyloid Proteins. The Beta Sheet Conformation and Disease. J. D. Sipe

ISBN: 3-527-31072-X

amino acid variations were not uniformly distributed within the variable domain, but were concentrated in three segments, which they termed the hypervariable segments or complementarity-determining regions (CDRs). The remainder of the domain was designated as framework (FR). It was predicted that CDRs clustered together to form the antigen-combining site, a prediction that was born out by structural characterization of the λ class Bence Jones protein, Mcg [31–34] and of an Fab [35].

Although the immunochemistry of Bence Jones proteins was the driving force behind studies of primary and tertiary structures, the potential clinical relevance of structural variation was also recognized by Solomon [36–41] and other clinical researchers [42–46] in extensive biochemical and immunochemical studies. Over time, the immunochemical aspects of light chains would fade as a motivator of primary and tertiary structural studies. The motivating role would be taken over by the fatal complications related to the Bence Jones proteins produced by some patients.

Antibody light chains consist of two β-type domains termed the variable domain and constant domain [17–21] on the basis of extensive and limited amino acid variation, respectively. It was early realized that the light chain was not the product of a single gene [47]. With the currently known exceptions of the antibodies produced by camelids and certain sharks [48, 49], light chains are paired with heavy chains consisting of an N-terminal variable domain and three to four constant domains. Light chains contribute in at least three ways to the formation of a functional antibody. First, the combinatorial association of variable domains encoded by approximately 40 germline genes (15 κ and 25 λ) is a major contributor to diversity of specificity of the antibody repertoire. Second, light chains provide a mechanism for receptor editing [50–55] by which many potentially autoimmune antibodies are deleted from the immune response. Third, light chain variable domains effectively serve as chaperones for often unstable heavy chain variable domains [56–63] and *vice versa*; the light chain/heavy chain complex is more stable than its components in isolation. The earliest available Fab structure, New [35], exhibited a deletion of residues 55–61 in the light chain [64]; this light chain would be expected to be very unstable. Likewise, since light chain variable domains self-associate to form transient dimers, it is probable that variable domains can serve as autochaperones in which the domain in the complex is effectively more stable [65] than in a free form.

Two sporadic features of light chains may have implications in both their immunological roles and be relevant to their properties during plasma cell disease, but detailed discussions are beyond the scope of this chapter. Many light chains are glycosylated [66–81] via the presence of an *N*-linked glycosylation site generated by somatic mutation. The contribution of *N*-glycosylation to the potential disease state of the protein is unclear. Although the presence of an *N*-glycosylation site on κ chains was found to be a "risk factor" for amyloidosis, no such correlation has emerged for glycosylation of λ light chains. No light chain germline gene encodes an *N*-linked glycosylation site. All light chain variable and constant domains are characterized by numerous serine and threonine residues on their non-interface surfaces. Walker et al. [82] reported that all Bence Jones proteins and myeloma light chains were glycosylated, on the basis that all tested samples reacted with the lectin, concanavalin A. They inferred exploitation of the hydroxyl group of serine and threonine as providing the basis for *O*-glycosylation. However, it is certain that not all Bence Jones proteins and myeloma-produced light chains are *O*-glycosylated. Nevertheless, it is probable that a sufficient fraction of light chains are *O*-glycosylated to render all samples reactive with concanavalin A.

Another interesting property of some light chains is low levels of catalytic activity (see Chapter 21). Paul and others [83–92] have demonstrated proteolytic activity in several Bence Jones proteins. It is not known if this property has physiological significance, either in the context of normal functional antibodies, of overexpressed myeloma-related antibodies or as Bence Jones proteins. Catalytic efficiency is low; however, when the light chain is produced in gram quantities daily during myeloma, it is possible to envision that some patients might experience a large "proteolytic burden".

A general review of the detailed structure of the light chain variable domain was authored by Schiffer [93] and need not be reprised here. Our major emphasis in this chapter is on the relationship between normal primary structure variation and its impact on the integrity of structure, leading to various pathological assembly and aggregation processes. A second objective is to demonstrate that an understanding of the linkages between light chain primary structure and pathogenesis has significance beyond proteins produced during multiple myeloma or primary amyloidosis. The antibody light chain variable domain has numerous paralogs, i.e. homologs encoded by the human genome. Using the concept of "immunoproteomics" we suggest that insight gained by study of the structural basis of light chain pathology will contribute in varying degrees to understanding abnormal and normal function in many of these paralogs.

20.2 Amyloidosis (AL)

Among the myriad array of proteins that are responsible for clinical development of amyloidoses [94], understanding the underlying cause of primary amyloidosis (AL) or amyloidosis associated with multiple myeloma arguably has presented the largest challenge [95]. Most examples of physiological amyloidoses can be attributed to the overproduction of a normal protein (islet amyloid polypeptide, gelsolin, cystatin) or generation of a truncated version of a normal protein (amyloid β precursor protein, serum amyloid A protein). Mutated forms or inherited polymorphic variations of human lysozyme and transthyretin (TTR) show enhanced fibrillogenic capability, due to destabilizing effects of single amino acid variations to which pathogenesis can be directly attributed. Unmutated TTR is also prone to fibrillogenesis, usually in elderly individuals, resulting in senile cardiac amyloidosis [96, 97], thus posing the question of why the (presumably) same protein does not generate amyloid at an earlier age.

The complications posed by amyloidogenic antibody light chains arise from at least three sources. First, antibody light chains are pathogenically pluripotent; in addition to fibril formation, light chains are capable of forming tubular casts [98–100] in the kidney, depositing amorphously in tissues [light chain deposition disease (LCDD)] [101–104] or can be produced in large quantities without clinical complication. In some patients, the protein crystallizes as is frequently observed in acquired Fanconi's syndrome [16, 105–108] or in crystal-storing histiocytosis [16, 109–111]. Second, no "wild-type" light chain has been observed clinically – every sequenced protein produced by a patient with primary amyloidosis, multiple myeloma or Waldenstrom's macroglobulinemia has undergone somatic mutation during the normal process of immune system diversification. However, the assumption that fibrillogenic capability is a gain of function that originates in somatic mutation may be in error in some cases, such as is likely with proteins of the $\lambda 6$ subgroup [112–114]. Finally, it is not possible to directly compare any two light chain variable domain sequences and infer the particular variations that are responsible for amyloid formation. Typically, light chains derived from the same germline gene exhibit 5–30 amino acid substitutions, whereas those that originate from different germline genes may vary at 20–60 positions.

Development of methods to relate the primary structure of the light chain variable domain to propensity for fibril formation is potentially significant in several ways. The most obvious is that one would be able to diagnose probable amyloidosis from the primary structure of the protein produced by a patient. In current practice, the development of amyloidosis is a presenting factor for the patient. However, it is conceivable that if future improvement in methods of detection of plasma cell diseases lead to significantly earlier diagnosis, prior to the development of extensive amyloid burden, then analysis of sequence, perhaps from plasma cell DNA, would allow anticipation of possible amyloid development and implementation of prophylactic measures.

However, the more immediate significance, and a possible precursor to the establishment of potential preventative or therapeutic methods alluded to above, is that the ability to recognize amyloidogenic potential from sequence would reflect or enable a comprehensive understanding of the light chain fibril assembly process. This understanding involves three phases of the disease process. The first involves the identification of amino acid substitutions that *enable* the protein to escape its native conformation and achieve a partially folded or alternative conformation that is amenable to self-assembly in the form of fibrils. Second, what is the process of fibril *assembly*, a mechanism that is highly degenerate based on the multiplicity of proteins of unrelated sequence that are able to form fibrils? Third, how do fibrils, again degenerately, physiologically interact with serum amyloid P component (SAP) [115, 116] and glycosaminoglycans [117–121] resulting in the heteromolecular complex that provides the clinical challenge?

The first issue, *enablement*, has been largely resolved, for the physiological amyloidoses, according to three major mechanisms: (1) In some cases, amyloidosis results from the overaccumulation of normal proteins or peptides such as found with β_2-microglobulin [122–125] during long-term dialysis treatment of kidney failure, the islet amyloid polypeptide [126–128] during adult onset diabetes or the prion proteins involved in spongiform encephalopathies [129–135]. (2) In the cases of secondary amyloidosis (SAA) and Alzheimer's disease, cleavage of a precursor protein is generally involved. Amyloid formation by SAA is also dependent upon overproduction of the protein in response to inflammation and it is not clear that cleavage of the protein is absolutely essential [136]. In Alzheimer's disease, separation of the Aβ peptide from the cell bound amyloid precursor protein would appear to be essential. (3) The third mechanism involves genetic variation, either inherited or sporadic. Huntington's disease is linked to the presence of an expanded string of CAG codons encoding a polyglutamine-repeat region of the protein, huntingtin [137–140], that may mediate oligomerization [141, 142]. Amino acid substitutions determine the probability of amyloid formation by other proteins, including apolipoprotein AI (ApoAI) [143], lysozyme [144–146], TTR [147–149] and antibody light chains [10, 95, 112, 150–155]. Amyloidogenic variants of ApoAI, lysozyme and TTR almost invariably differ from the wild-type proteins at a single amino acid position; amyloidosis occurs at normal physiological expression levels. In contrast, all amyloid-related light chain variable domains have accrued multiple amino acid variations relative to the germline sequences of variable domain exon, joining segment exon, the exon junction site and addition of non-germline-encoded nucleotides by deoxynucleotidyl transferase (TdT) [156]. Whereas most individuals do not produce highly amyloidogenic forms of ApoAI, TTR, and lysozyme, it can be assumed that every human continually produces numerous potentially amyloidogenic light chains. It is only when a free light chain of suitable sequence is significantly overproduced by detectable plasma cell tumors during plasma cell dyscrasias or by cryptic tumors in primary amyloidosis that the pathogenic potential is realized.

The second issue, *fibril architecture and assembly*, is the subject of intense activity by several groups using various model systems [157, 158]. Fink et al. [159–167] have been the center of activity on studies of the structural architecture of fibrils formed by recombinant light chain variable domains [168]. A β-helical model for prion fibril formation has recently been proposed by Govaerts et al. [169].

The third issue, the mechanisms by which apparently all fibrils are able to associate with SAP, glycosaminoglycans and the receptor for advanced glycation end-products (RAGE) [170–175], largely remains to be addressed. A key question is posed by the word *all*. The ability of all amyloid fibrils to bind Congo red resulting in birefringent staining is essentially a part of the definition of amyloid [176]. Typically, non-congophilic fibrillar protein deposits are considered "amyloid-like" [177]. Likewise, the apparent existence of a universal binding site for another small molecule, Thioflavin T, has evoked little curiosity; indeed, Thioflavin T has been considered to be specific for fibrils [178, 179]. The binding of several glycosaminoglycans [117, 121, 176, 180–185] might also be considered rather non-specific at the level the identity of the protein. All proteins have basic amino acids; it is highly probable, therefore, that any fibril-like array of protein would serve as a nanoscale anion-exchange column suited for capture of polysulfonates like glycosaminoglycans. In this manner, interaction between fibrils and proteoglycans would be a consequence of the metaspecific interaction with the glycosaminoglycan moieties. On the other hand, proteoglycans contain domains that are homologs of the variable domain itself; it is conceivable that interaction of proteoglycans with fibrils is mediated all, or in part, through the protein core.

The binding of both SAP and RAGE to fibrils formed by unrelated proteins is less easy to dismiss, although SAP, a pentraxin, could fortuitously bind through an avidity enhancement provided by its five subunits and a highly repeated low-affinity ligand on the fibril. Presumably this putative low-affinity ligand must be present on all fibrils, despite primary structure disparity. RAGE is not a pentraxin and, therefore, does not benefit from multiple subunits. As a result, the affinity of the RAGE domain for a particular fibril must be significant. It is therefore difficult to rationalize RAGE pan-specificity based on a conventional interaction between receptor and ligand. Moreover, antibodies raised against fibrils [186] or fibril subunits [187, 188] have demonstrated interactions with xenogenic fibrils. Such antibodies may interact with fibrils in a manner similar to that of SAP and RAGE.

Understanding the interactions that fibrils undertake with multiple proteins should provide insight into the assembly of fibrils themselves. The primary structure diversity of antibody light chains, which long obscured the structural correlates of AL amyloidogenesis, may prove a useful resource in the context of developing general concepts of fibril formation.

20.3 Physicochemistry of Antibody Light Chains

20.3.1 Self-association of Variable Domains

All light chain variable domains, κ or λ, self-associate to varying degrees. Tetrameric light chains were early reported in serum [189–195] and were demonstrated by *in vitro* experiments as well [196]. Tetramers are likely to be formed as non-covalent dimers of covalent dimers, i.e. two light chains that are linked via the terminal (κ) or penultimate (λ) cysteine of the constant domain. A hexameric complex of λ chain dimers in serum was recently reported by Abraham et al. [197] in a patient with a high serum level of light chain, but relatively low production of Bence Jones protein. Tetrameric and hexameric complexes are both too large to be filtered through the glomerulus, resulting in serum retention. This feature may have contributed to the lack of correlation between concentrations of free light chain in the serum and urine in comparisons between patients as seen by Abraham et al. [198], although a good correlation was observed for multiple samples from individual patients.

Oligomers are more likely to be formed by variable domains that have low affinities of self-association. A covalently linked pair of light chains can be viewed as occupying a sphere of radius of approximately 7 nm, or a volume of roughly 3.5×10^{-27} m^3. Since two molecules occupy this sphere, the local concentration can be estimated as approximately 100 mM. The relationship between monomer concentration (C_m), total protein concentration (C) and dimerization constant (K_D) is given by the relationship $C_m=[-1+(8CK_D+1)^{1/2}]/4K_D=[-1+(0.8K_D+1)^{1/2}]/4K$, assuming a protein concentration of 100 mM. This suggests minimal free variable domain available for tetramer formation for $K_D>10^3\ M^{-1}$. However, the term K_D should probably be replaced with K_D^*, in which the asterisk acknowledges that the interactions between two variable domains that are physically constrained do not conform to the free diffusion assumptions of the law of mass action. Therefore, the apparent K_D as measured experimentally is unlikely to equal K_D^*. However, the consequences of the constraining disulfide link should be the same in all cases. As a result, one would expect a correlation between measured affinity of dimerization and the preferential formation of tetramers. Unstable dimer formation by linked light chains could lead to stable tetramers by avidity; tetramer dissociation requires concurrent dissociation of two independent interacting molecules. The effective "avidity constant" is the square of the intrinsic affinity constant. Although the avidity enhancement also applies to covalent dimers governed by higher affinity constants, the very low concentration of "open" dimers is likely to be insufficient to support tetramer formation. However, the relatively small number of reported tetramers may not accurately reflect degree of rareness.

The self-association kinetics of the variable domain are characterized by fast rates of association and dissociation. Bence Jones protein Au (κ) was characterized by Maeda et al. [199] to have an association rate of $9\times10^6\ M^{-1}\ s^{-1}$ and a dissociation rate of $1.5\times10^2\ s^{-1}$. No evidence of dimerization of the constant domains was found [200]. The ratio of the two variable domain forward and reverse kinetic rates provides an equilibrium constant of $6\times10^4\ M^{-1}$. These numbers can be crudely interpreted as indicating that the average lifetime of a variable domain dimer is less than 1 s. However, at a free monomer concentration of 0.1 mM, approximately half of all monomers would participate in at least one dimerization event. Comparable rate constants were obtained by Azuma et al. [201] with additional Bence Jones proteins of the λ class.

Bence Jones proteins of the λ class are typically found as mostly covalent dimers, whereas the κ class is usually a mixture of monomer and covalent dimer [40]. The reason for this has never been determined. Dimerization constants for κ light chains from 17 patients were roughly estimated from size-exclusion gel chromatography profiles [202] and exhibited affinities ranging from 10^3 to $10^6\ M^{-1}$. Tetramer formation was not rare. Although there was an apparent correlation between K_D and the relative abundance of covalent dimer, it did not appear to exhibit a three orders of magnitude span of difference. It can be hypothesized that, if λ light chains have higher self-association constants than κ light chains, increased self-association of λ variable domains would bring the constant domains into proximity and enhance the probability of disulfide bond formation. However, it is also possible that dimerization of variable domains impairs disulfide bond formation. Whereas crystallographic structures of several λ Bence Jones proteins have been obtained for covalent dimers: Mcg (λ2, PDB [203] identifier 1DCL), Loc (λ1, 4BJL [204]), Cle (λ3, 1LIL [205]), Mcg/Weir (λ2/λ2, 1MCW [206]), no crystallographic data characterizing covalently linked κ light chains have been published to date. The crystal structure for one (virtually) intact κ chain (Del; 1B6D) was determined for a protein associated with acquired Fanconi syndrome [207]. The structure was obtained from a non-covalent dimer in which the C-terminus was apparently truncated, removing the terminal cysteine; covalent dimers from the same patient did not crystallize. Perhaps a disulfide bond between κ chain constant domains and a well-defined variable domain interaction are not concurrently possible. The κ chain J segments are 1 amino acid shorter than those of λ chains; the shorter length may place constraints on the relative positions of κ variable and constant domains not present for λ chains. On the other hand, it may be simply possible that insufficient efforts have been made, to date, to crystallize disulfide-bonded κ chains.

The conventional variable domain dimer usually emulates the interactions between and light chain and heavy chain in an Fab. These interactions have been observed in intact light chains (above) and variable domain dimers: Rei (κ1, 1REI [208]), Bre (κ1, 1QP1 [209]), Wat (κ1, 1WTL [210]), Len (κ4, 1LVE [211]), Rec (κ4, 1EK3), Rhe (λ1, 2RHE [212]), Jto (λ6, 1CD0 [113]) and Wil (λ6, 2CD0 [210]). Of particular note is position 38, in which glutamine is invariant in all light chain germline genes. With the exception of the dimer structure of protein Rhe, a major component of the interaction between variable domains is contributed by a hydrogen bond between the side chains of Gln38. This corresponds to the interaction between the Gln38 side-chain and Gln39 in heavy chains.

The self-interaction of Gln38 suggested that changing this neutral residue to any ionic residue would generate variable domains that could not self-associate [213]. All four ionic amino acid variants were constructed; three of the four mutants exhibited little or no self-association capability; a control variant, Q38A showed the expected decrease in affinity. However, one mutant, Q38E, dimerized with affinity at least twice that of the wild-type domain. To determine the basis for this anomaly, the Q38E construct was crystallized and its structure determined [214].

Fig. 20.1 compares the quaternary structures of the dimer formed by the Len variable domain with that obtained by the Q38E variant. The α carbons of the domains to the left in each dimer were superimposed (r.m.s. deviation, 0.72 Å). The interactions between variable domains in the Q38E variant involve the same interface that is active in the typical variable domain dimer, but the mechanism of interaction has been completely reinvented. Whereas the previously observed dimers could be described as having a "parallel" arrangement of domains, the self-association of Q38E is "anti-parallel". One domain is flipped relative to its orig-

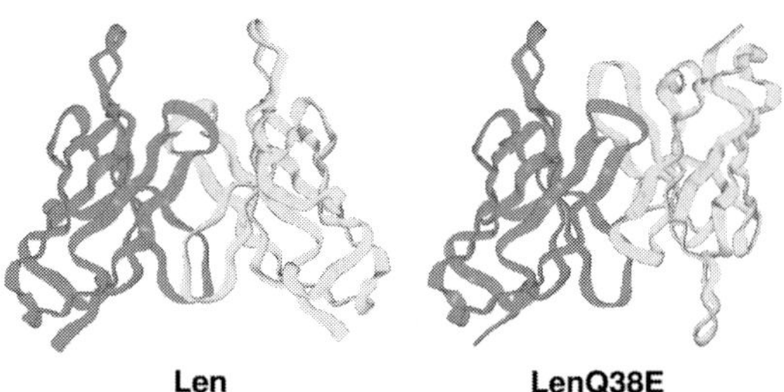

Fig. 20.1 Two forms of light chain dimer. (Left) Typical variable domain dimer, which emulates the interaction between the light and heavy chain in an antibody Fab; the CDRs in both monomers are arranged at the top of the image (1LVE). (Right) Structure obtained following the mutation of Gln38 to glutamic acid. The domains shown in dark have been superimposed; the domain in light has rotated approximately 180° from its original orientation and has assumed . The interaction seen on the right is the mode of interaction found in related cell adhesion molecules such as MPZ.

inal orientation; this quaternary arrangement is reminiscent of that seen in the structure [215] of the cell adhesion protein, myelin protein zero (MPZ).

Similar domain arrangements have been observed for other single-amino-acid variants of Len, including K30T (4LVE [214]) and Q89L (1QAC [216]). It is thought that the domain rearrangement in the Q38E and K30T mutations is the result of increasing the ionic disparity between the two "poles" of the domain. The mechanism of the Q89L reorientation is somewhat more subtle. Structural analysis of the Q38E and K30T mutants revealed the Q89 had unfulfilled hydrogen-bonding capacity, a condition that is destabilizing [217]. Thus, the domain shift was energetically favorable despite generating a unfavorable location for Q38. The replacement of Q38 with either alanine or leucine removed the hydrogen bond penalty imposed by the inverted orientation of the wild-type and were sufficient to allow a domain flip. Because three additional methylene groups were buried in the dimerization of Q89L, its dimerization affinity (above 10^8 M^{-1}) is much higher than that obtained with the alanine replacement (4×10^6 M^{-1}) and is the highest dimerization affinity observed for a variable domain, to our knowledge.

20.3.2 Variable Domain Stability

The linkage between light chain amyloid formation and the intrinsic stability of the variable domain was first suggested by Wetzel [218, 219], as had been previously demonstrated in the familial amyloidoses of TTR [220, 221]. Fig. 20.2 illustrates the correlation between variable domain stability and ability to form fibrils *in vitro*. Results obtained for approximately 75 variants of the κ4 protein Len are sorted on their ability (F+) or inability (F–) under a standard fibril for-

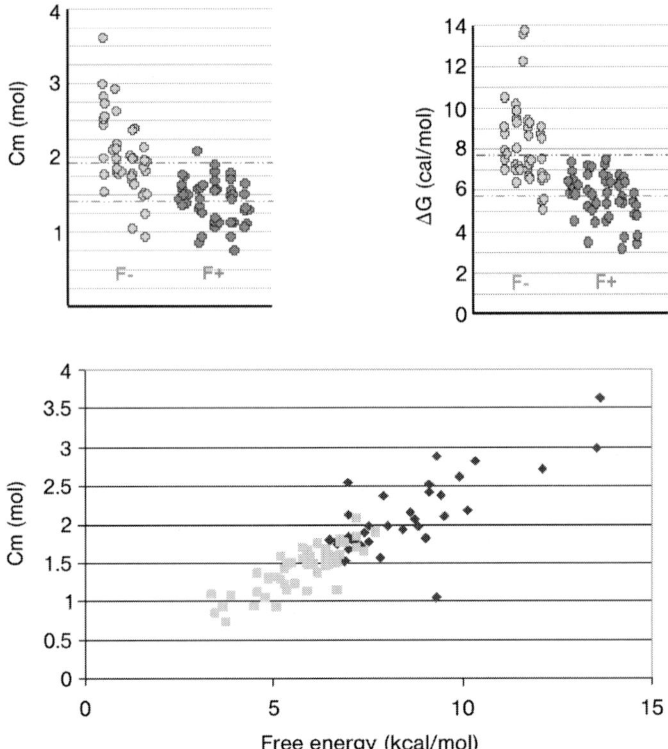

Fig. 20.2 Correlation of fibril formation and domain stability. (Upper left) The plot sorts the results of *in vitro* fibril-forming assays involving a number of site-specific mutants of the recombinant Len variable domain as a function of apparent stability as assessed by C_m, the concentration of denaturant required for 50% unfolding. Light grey symbols represent non-fibril formers; dark grey represents fibril forming constructs. Horizontal light grey and dark grey dashed lines indicate the mean C_m values for two populations. (Upper right) Results are sorted by the calculated stability, ΔG. (Lower) Correlation plot of C_m versus ΔG. Imperfect segregation is the result of different cooperativity of folding/unfolding exhibited by various constructs.

mation assay, involving protein at 1 mg/ml in standard phosphate buffer, pH 7, and incubated with vigorous shaking at 37 °C. Samples that did not form fibrils within 1 week were scored as F–.

In earlier work [222], a sharp demarcation was found between the thermodynamic stabilities of F+ and F– single amino acid variants of recombinant Len variable domain. This observation hinted at the feasibility of establishing a reasonably definitive amyloidosis threshold that might have diagnostic relevance. Fig. 20.2 (upper left) distinguishes fibril-forming capability on the basis of the concentration (C_m) of denaturant (guanidine hydrochloride) required to achieve equal representation of native and unfolded form as evidenced by reversal of quenching of fluorescence of Trp35 when exposed to solvent or displaced from the proximity of the disulfide bond linking Cys23 and Cys88 [223, 224]. Fig. 20.2 (upper right) compares F+ and F− on the basis of ΔG, the apparent change in free energy of unfolding. Both measures indicate a correlation between fibril formation and low stability; however, by each measure there are several instances of constructs with stabilities higher than the non-fibrillogenic Len ($C_m = 1.78$ M Gdn; $\Delta G = 7.5$ kcal/mol).

As seen in Fig. 20.2 (lower), the two measures of stability are discordant in a number of cases. ΔG is derived from fitting the fractional fluorescence observed in the presence of a series of denaturation concentrations. The method of Santoro and Bolen is typically used [225, 226] fitting six parameters to a data set that includes baseline values obtained for denaturant concentrations below and above the unfolding transition range (Fig. 20.3). One of the fitted parameters is m, a measure of the cooperativity of folding given in units of kcal/mol-M that is the slope of the unfolding curve at the midpoint concentration. ΔG is calculated as the product of the denaturant concentration that yields 50% unfolding and m, the cooperativity factor.

It is possible that trivial explanations exist to account for at least some of the apparently ambiguous stability/fibril relationships. A few unstable mutant constructs failed to form fibrils because they first precipitated as an amorphous aggregate. In other cases, the apparently contradictory results could be a quirk of a noisy data point skewing the fitting process. However, it is likely that many of the observations are real reflections of the stability/fibril relationship and may point to consequences of the mutation to the folding pathway(s) as well as to the total free energy of folding. This may provide an unanticipated linkage between propensity to form fibrils under physiologically relevant conditions to folding kinetics as well as the equilibrium between native and partially unfolded forms. If so, it may be difficult to estimate risk of fibril formation from measurement of thermodynamic stability alone, although it is probable that identification of high- and low-risk domains is possible.

The vast majority of stability/fibrillogenesis data is from work in this laboratory with $\kappa 4$ variable domains (currently approaching 200 constructs). Limited work has been done on identifying determinants of stability in $\kappa 1$ [60, 61, 151, 218, 227–232]. No mutational work has been reported for $\kappa 2$ and $\kappa 3$ proteins; Ewert et al. [60] have recently indicated that $\kappa 3$ variable domains may be the most stable, based on comparative properties exhibited by domains engineered on the basis of consensus germline sequence analysis. Ewert et al. [60] found their $\kappa 1$ construct to be the least stable of the κ domain constructs and to exhibit the poorest yield in expression of recombinant protein; this may correlate with the $\kappa 1$ subgroup being the most prevalent amyloid-forming κ protein. Little work to date has been done with λ proteins [114, 233]; a rare non-amyloidogenic $\lambda 6$ protein was found to have improved thermodynamic stability introduced by two fortuitous somatic mutations that introduced a salt bridge be-

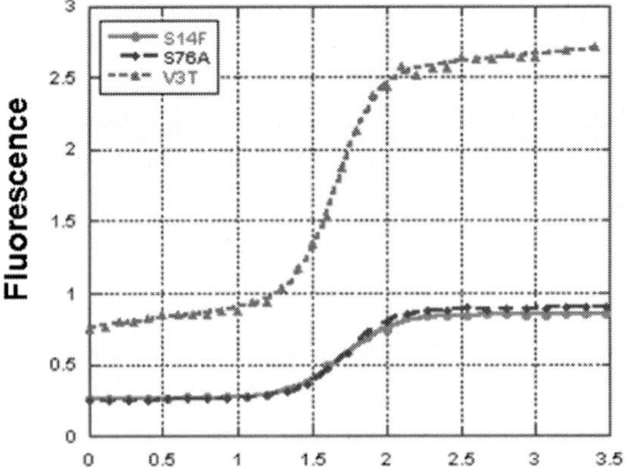

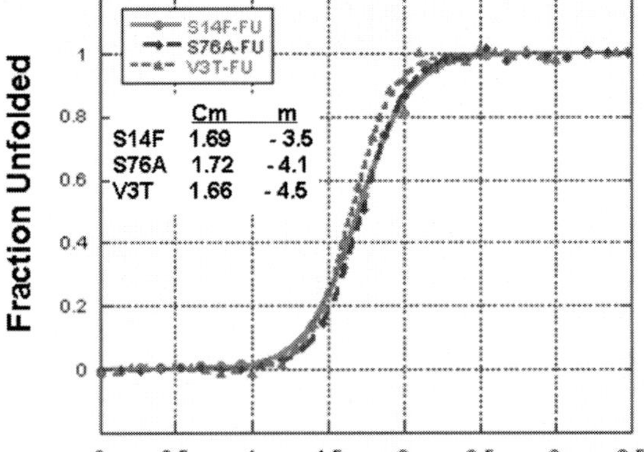

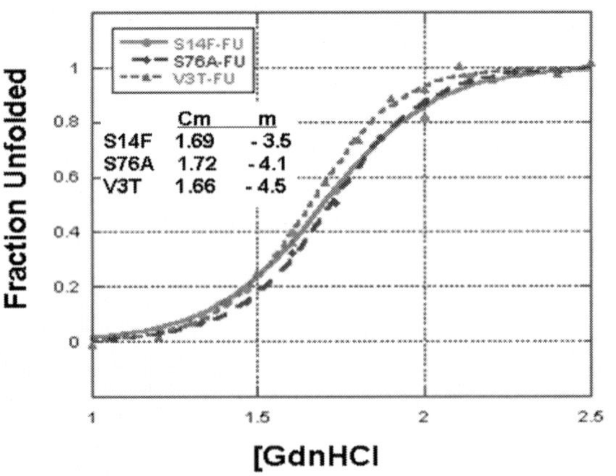

Fig. 20.3 Illustration of the method for determining variable domain stability. (Upper) Panel depicts raw fluorescence data obtained for three variants of Len. Protein concentrations were 1.5 μM in phosphate buffer, pH 7.5 to which was added guanidine hydrochloride to provide a denaturation profile from 0 to 3.5 M Gdn in steps of 0.1 M. Data were collected at 350 nm with excitation at 295 nm after overnight incubation of the protein in denaturant to assure equilibrium. Displacement of V3T data is due to use of different instrument settings. (Middle) Data have been normalized and fitted to the Santoro and Bolen model [225]. The inset summarizes the results for the three constructs; the C_m value are quite close: 1.69 ± 0.03 M. However, the m (cooperativity) value is widely spread, ranging from –3.5 to –4.5 kcal/(mol·mol). As a consequence, the calculated ΔG for folding, which is given by $m \cdot C_m$ varies: S14F, –5.9 kcal/mol; S76A, –7.0 kcal/mol; V3T, –7.5 kcal/mol. V3T has the same calculated ΔG as Len, but forms fibrils; the other constructs do not form fibrils during the typical scoring period (1 week). (Lower) Expanded plot of fitted data to illustrate the different slopes of the unfolding curves, corresponding to the different cooperativity value (slope).

tween as aspartic acid at CDR1 position 29 and an arginine at framework position 68 [113].

Humans are thermally mesophilic organisms; as a result most of our proteins exhibit intermediate characteristics of stability. Proteins that have evolved to be very short lived (e.g. transcription factors) or long lived (e.g. crystallines in the eye) may be effectively psychrophilic or thermotolerant, but most human proteins, such as the light chain variable domain, have substantial leeway for "improvement". The consensus amino acid at most positions is not necessarily the optimal from the perspective of thermodynamic stability. For instance, at position 4 in the κ sequence, all germline genes, with three exceptions, encode methionine. The exceptions include the rarely observed κ1f subgroup (germline gene L18 [234]), which is represented only four times in our database of primary structures (see below), and the κ3b and κ3c subgroups (germline genes A27 and L6, respectively). Of 19 observations of κ3b and κ3c subgroup representatives, only two are associated with amyloidosis. In contrast, eight of 17 samples derived from the κ3a subgroup, which encodes the consensus methionine at position 4 (germline gene L12–L16), were from amyloidogenic proteins. Replacement of Met4 with Leu has been shown to improve stability by 1.1–1.3 kcal/mol [168, 235], approaching a 10-fold improvement in stability in terms of the equilibrium ratio of native fold to unfolded form, and perhaps contributing to the finding of Ewert et al. [60] of the superior stability of the κ3 consensus structure.

Although net destabilization of the variable domain enhances the probability of fibril formation, the somatic mutations exhibited by each patient's protein is likely to include amino acid replacements that enhance stability as well as diminish it. Pokkuluri et al. [235] combined three amino acid changes Met4Leu, Tyr27dAsp and Thr94His to increase the stability of a κ4 variable domain by a factor of 10,000, in terms of the ratio of native and unfolded forms. All of these mutations have been observed in amyloid forming light chains.

Tables 20.1 and 20.2 summarize mutations observed at positions that are invariant in κ and λ germline genes. Positions at which three or fewer mutations are not included; such positions include hyperconserved sites Gln6, Cys23, Trp35, Cys88, Arg61 and Asp82 in λ. These positions, with the exception of Arg61 in κ chains, are seldom seen mutated in soluble proteins but have been observed mutated in functional antibodies. The limited numbers of observations of mutations that have been made at these sites reflect improvements in technologies to sequence minute amounts of deposited protein extracted from tissue [236] or fixed tissue [237, 238] and the broad use of DNA sequencing approaches [152, 195, 239–246].

Because the positions listed in Tables 20.1 and 20.2 are conserved within the κ and λ classes, respectively, and several of the positions are conserved in all light chains, it is reasonable to expect that they are structurally or functionally important. If so, then these may be expected to be common sites for amyloid related mutations. Interestingly, only one κ position shows a strong mutational correlation with amyloidogenesis. Loss of arginine at position 61 is found in 20 amyloidogenic proteins (upper number) and in four proteins not known to be

Table 20.1 Amyloid association with invariant κ germline positions

			A	C	D	E	F	G	H	I	K	L	M	N	P	Q	R	S	T	V	W	Y	
2	I	218					0					1								7			8
							1					2								2			5
16	G	250															1				1		2
																	2				0		2
26	S	266	0		1			0						2			1		0				4
			1		0			1						5			2		2				10
27	Q	265				4	0		3			0					1						8
						2	1		4			3					1						10
33	M	264	0				0													5			5
			1				3													4			8
38	Q	263					0		1		2	2					1			0			6
							1		5		0	1					1			1			8
39	K	265				1			1	0		0		0		1	5	1	0				9
						0			0	1		2		1		1	18	0	3				26
40	P	265	2									3						2		1			8
			3									0						2		0			5
41	G	266				1					0						0						1
						2					1						2						5
47	L	261								3							0			3			6
										1							2			1			4
48	I	261					1			1		2	2							2			8
							2			1		2	6							2			13
49	Y	260			1		6					0						2					9
					0		7					1						8					16
52	S	262	2				2		2								0		3		0	1	10
			2				4		0								1		2		1	1	11
59	P	264	2													1		4	1				8
			0													0		3	2				5
61	R	264						1			7		0	9					3				20
								0			1		1	0					2				4
63	S	259								1							5		2	0			8
										2							3		5	1			11
64	G	263	1		0										0								1
			2		1										1								4
65	S	261						8															8
								5															5
66	G	263	2			1				0	0		0			1	2						5
			2			1				1	1		1			0	0						7
67	S	263	0				1					1								1		3	6
			1				4					0								0		2	6
68	G	263	1		0	0											0	0					1
			1		1	4											1	1					8
69	T	263	0								1						0	2					3
			3								0						1	1					5
71	F	263																		0		3	3
																				1		2	3

Table 20.1 (cont.)

			A	C	D	E	F	G	H	I	K	L	M	N	P	Q	R	S	T	V	W	Y	
72	T	263								6				2	0			5				1	14
										3				2	1			3				0	9
75	I	260					1						0					0		0			1
							0						1					1		3			5
76	S	262	0		2		0			1				7	1				7			0	17
			4		3		1			1				9	0				13			1	32
82	D	261				0				0	0	1		0									1
						1				1	1	0		1									4
86	Y	265					2				0												2
							1				1												2
87	Y	263					7		3									2					12
							14		1									2					17
90	Q								4		0	0					0						4
									9		1	3					3						16

amyloidogenic (lower number). The category of "not known to be amyloidogenic" includes non-pathological proteins, proteins that exhibit non-amyloid pathologies and proteins for which no clinical information is available. No clinical information was provided for the four position 61 mutants that are not known to be amyloidogenic. Mutation of Arg61 is one of four structural "risk factors" [247] that were previously described and are associated with for a majority of the amyloidogenic proteins of the κ class. Arg61 forms an energetically important salt bridge with Asp82; both amino acids are absolutely conserved in both κ and λ germline repertoires.

Since approximately one-third of the known κ sequences are derived from amyloid proteins (directly or indirectly from cDNA), it is probable that mutations that are associated with amyloidosis in 60% of its observations represent reasonable hypotheses of being linked. These sites include Ile2, Pro40, Leu47, Pro59, Ser65 and Thr72. Replacement of isoleucine with valine creates a destabilizing void in the core of the domain. Replacement of Pro40 with hydrophobic residues was previously recognized [168, 248] as potentially linked to fibril formation. Substitution of isoleucine for Leu47 without a compensating mutation elsewhere is not a conservative replacement; likewise, replacement by valine introduces a large destabilizing void. Pro59 helps stabilize the turn that positions Arg61 for interaction with Asp82; thus, a mutation of this unique amino acid is an indirect mutation of Arg61. Ser65 does not have an obvious critical structural role; all mutations are to glycine, which usually has a destabilizing effect on stability due to increasing the entropy penalty for folding by increasing the number of conformations available to the unfolded polypeptide. Mutation of Thr65 to isoleucine introduces a large hydrophobic residue on the solvent-exposed surface of the domain. The potential significance of replacement by serine is less

Table 20.2 Amyloid association with invariant κ germline positions

			A	C	D	E	F	G	H	I	K	L	M	N	P	Q	R	S	T	V	W	Y	
5	T	277	2							2	0			0				1		2			7
			2							0	1			1				1		1			7
16	G	295			1	1								0			4						5
					0	1								1			0						3
37	Q	297				0			0		1	2					2	1	1				7
						1			1		0	1					5	0	0				8
38	Q	297				1			27		1	5					1			0			35
						1			7		0	3					0			1			11
40	P	297	8									2				0	4		2				16
			2									0				1	5		0				8
49	Y	287			1		21		3		1	1		2				4					33
					1		6		3		0	0		1				2					13
55	P	296	7															3					10
			3															2					5
56	S	288	2						1			1			0				0			1	5
			1						0			0			1				1			0	3
57	G	293	1		2	2				1					1		3			2	0		11
			0		0	2				0					0		0			0	1		4
62	F	293								2		0	1					1		2			6
										5		1	0					1		0			7
63	S	291	5				1												0				6
			2				0												1				3
64	G	291	4												0			0	1	1			6
			8												1			1	0	1			11
65	S	290	2				1				1				0							0	4
			0				3				0				1							1	5
67	S	289	2				1								1				0			1	3
			0				1								1				2			0	4
75	I	293																		8			8
																				8			8
83	E	296	1		2			1			1					0			1				6
			0		0			0			0					1			0				1
84	A	295						6										1	1				8
								1										2	2				5
85	D	296	0			8		0	4	1			1	1				1	0	3		4	22
			1			3		1	3	0			0	0				0	1	2		1	12
86	Y	296			1		4											1					6
					0		1											0					1
87	Y	296		2			15		5									2	0				24
				0			12		1									1	1				15

clear; however, it is noted that threonines are more common than serines on inner strands of β-sheets.

Two-thirds of the data characterizing sites of germline invariance in λ light chains are associated with amyloidogenic proteins. Therefore, formal statistically significant correlations with amyloidosis are rare. Six out of seven observations of Glu83 mutations and six out of seven mutations of Tyr86 are associated with amyloidosis. As seen in several λ Bence Jones protein structures, the side-chain of Glu83 interacts with the backbone amide of position 107 and forms an ionic interaction with a lysine in the constant domain. In the Bence Jones dimer structure, only one of the light chains forms this arrangement. It is also observed in λ chain Fabs, such as the human tetanus toxoid binding Fab recently described by Faber et al. [249]. Perhaps this interaction between variable and constant domains contributes to λ chain stability and its loss facilitates fibril formation (Fig. 20.4).

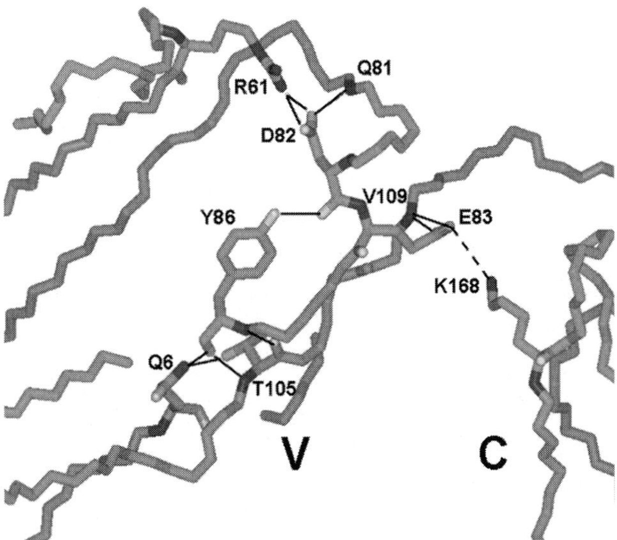

Fig. 20.4 A map of a portion of the hydrogen-bonding network that is a major contributor to cooperativity of folding. Arg61 forms a partially buried salt bridge with Asp82. The side-chain of Asp82 forms a hydrogen bond to the backbone nitrogen of the residue at position 79. The carbonyl of Asp82 is involved in a buried hydrogen bond to the hydroxyl of Tyr86, forming the tyrosine corner. The backbone carbonyl of Tyr86 forms buried hydrogen bonds to the side-chain of Gln6 and to the backbone nitrogen of Thr105; the backbone carbonyl of Thr105 forms a buried hydrogen bond to the backbone nitrogen of Tyr86. Adjacent to Asp82, the side-chain of Glu83 forms hydrogen bonds to the backbone nitrogen of position 109; it also exhibits a 3.7-Å ionic interaction with Lys168 in the constant domain. Hydrogen bonds are shown as solid lines; the ionic interaction as a dashed line. The structure depicted is that of a human λ light chain in a tetanus toxoid-binding Fab (1AQK [249]).

The reason that mutation of Tyr86 is significant is less ambiguous. Tyr86 generates a tyrosine corner [250, 251], forming a buried hydrogen bond to the carboxyl oxygen of Asp82. This hydrogen bond, by being protected from the solvent, is itself a major contributor to domain stability, as is the buried hydrogen bond between Gln6 and Thr105. In addition, by anchoring the backbone of Asp82, it augments the Arg61–Asp82 salt bridge.

Also found in Table 20.2 are a number of positions that appear to be highly mutated in λ proteins. Although the frequency of mutations is not significantly different in the amyloid and non-amyloid sets, the ratio of amyloid to non-amyloid sequences is about 2:1 in the database. However, the recent emphasis on sequencing amyloidogenic λ light chains should result in overrepresentation of mutations associated with amyloidosis, even in the absence of statistical significance. Therefore, although conclusions cannot be drawn directly from most of the observations, it is reasonable to use the appearance of frequently mutated sites as a guide for site-directed mutagenesis. One particularly intriguing mutation is that of Gln38 to histidine, which has been observed 34 times with at least 27 occurrences in amyloidogenic proteins. Since Gln38 contributes little to domain stability, if this apparent relationship to amyloidogenesis is meaningful, it might be related to interference with the association of the light chain to the heavy chain.

20.4 Database of Dyscrasia-related Variable Domain Sequences

The observations reviewed above were made possible by the compilation of a database of the primary structures of variable domains of light chains produced by plasma cell disease. The vast majority of the sequences are from multiple myeloma and primary amyloidosis patients. The pathological distribution of 267 κ and 299 λ sequences is provided in Table 20.3. The apparent distribution of several of the pathological end-points is effectively inverted in the database relative to the distribution in clinical cases. For instance, most light chains are nonpathogenic or precipitate as nephritic casts if produced at high enough levels [252–258]. This process may simply reflect supersaturation of the urine during

Table 20.3 Light chain distribution

Pathology	κ	λ
Non-pathogenic	21	8
Amyloid	98	196
LCDD	20	7
Cast	8	7
Fanconi	10	1
Unknown	110	80
Total	267	299

water recovery by the kidney or may be augmented by interactions of the protein with Tamm–Horsfall protein [259–263]. However, only 8% of the sequence data are from this source. Slightly more than half of the sequence data are from amyloidogenic proteins, while amyloidosis affects approximately 10% of multiple myeloma patients and an approximately equal number of primary amyloidosis patients. The distribution of data is an understandable consequence of the perceived lack of novelty of "normal" light chains and the interest in identifying the structural basis for light chain amyloidosis. The consequence, however, is that as more than half the data are from amyloidogenic light chains, and two-thirds of the remainder are of unknown pathology, statistical analyses lose robustness. The dataset that is labeled unknown should probably be divided into two cohorts, unknown-A and unknown-B with the latter set representing sequences determined from cDNA. Unknown-B has a significantly higher probability of including cryptic amyloidogenic sequences than unknown-A, for which the general assumption of lack of pathogenesis is more reasonable based on their derivation from urinary protein.

Table 20.4 shows the distribution of pathogenic light chains broken down by germline gene of origin. In general terms, λ chains have a higher risk of fibril formation than κ, whereas the opposite relationship holds true for LCDD. It is clear that a few germline genes exhibit elevated or lowered risk of pathogenesis, while most germline gene products exhibit pathogenicity frequencies similar to the overall average. It appears likely the germline λ6 gene probably encodes a light chain that is inherently amyloidogenic. Of the 53 entries, 51 (96%) characterize amyloidogenic proteins. The single example of a documented non-amyloidogenic representative involved, as noted earlier, fortuitous improvement in stability resulting from a salt bridge introduced by two somatic mutations [113, 233, 264]. The rare λ3b subgroup has the lowest rate of amyloidosis (38%). Among κ light chains, the κ1e sub-subgroup has an 81% rate of amyloidogenesis; this group constitutes 6% of the κ data. According to Girschick and Lipsky [265], about 1% of peripheral adult IgM^+ B cells exhibit κ1e usage. In contrast, only 13% of the κ3b entries represent amyloidogenic proteins. Products of the A27 germline gene constitute 6% of the κ dataset, but appear to be the most frequently used (14%) κ gene [265].

The germline of origin may also influence the site of fibril deposition. Evidence of germline-directed tropism is described by Comenzo et al. [152, 239] in findings of the preferential involvement of λ6 proteins with the kidney, while other λ proteins were preferentially associated with the heart and other organs. An independent study by Abraham et al. [246] supported this finding and extended it by noting that renal involvement correlated with extended patient survival. Thus, extraction of the amino acid sequence of each patient's light chain could eventually contribute to determination of prognosis. Indeed, it is perhaps appropriate to consider determining the primary structure of the light chain produced by all multiple myeloma and primary AL patients. To do so would generate a sequence database unbiased by the novelty criterion of publication and linked to detailed clinical data. Coupling this database with an in-depth un-

Table 20.4 Germline distribution of pathological light chains

SSG [a)]	Name [b)]	Amyloid [c)]	Cast	LCDD	Fanconi [d)]	None [e)]	Unknown [f)]	Total
1 a	L12a	8	1	4	0	2	16	31
1 b	O8-O18	30	2	3	3	8	25	71
1 c	O12-O2	14	2	0	3	0	16	35
1 d	A30	1	0	0	0	0	5	6
1 e	L1	13	0	0	0	2	1	16
1 f	L18	1	0	0	0	0	1	2
1 g	L5	0	0	2	0	0	5	7
2 a	O11-O1	0	0	0	0	0	2	2
2 b	A19-A3	4	1	0	0	3	6	14
2 c	A2	0	1	0	0	0	2	3
3 a	L2-L16	8	1	3	1	1	3	17
3 b	A27	2	0	0	2	1	10	15
3 c	L6	0	0	0	1	0	3	4
4	B3	17	0	8	0	3	14	42
5	B2	0	0	0	0	1	1	2
Σ (κ)		98	8	20	10	21	110	267
1 a	humlv114	13	0	0	0	0	6	19
1 b	humlv112	26	1	2	0	2	13	44
1 c	IGLV1S2	9	0	0	0	0	2	11
2 a	VL2.1	25	0	1	0	1	11	38
2 b	DPL12	13	0	2	0	1	11	27
3 a	humlv318	10	2	1	0	1	10	24
3 b	hsigg1150	5	0	1	0	1	6	13
3 c	VIII.1	32	1	0	1	2	10	46
4	humlv418	8	0	0	0	0	7	15
6	IGLV6S1	51	1	0	0	0	1	53
7	DPL18	0	0	0	0	0	1	1
8	humlv801	4	2	0	0	0	2	8
Σ (λ)		196	7	7	1	8	81	299
$\Sigma\,\Sigma$ [g)]		294	15	27	11	29	191	566

a) Subgroup.
b) Representative germline gene name (not necessarily unique).
c) Number of sequences of amyloidogenic proteins.
d) Number of sequences related to acquired Fanconi's syndrome.
e) Number of sequences from proteins documented to be non-pathogenic.
f) Unknown pathology, no clinical information published.
g) Grand totals.

derstanding of the physicochemical implications of each amino acid variation should ultimately yield substantial new insight into diagnosis of the implications encoded in each patient's unique protein.

The existing database made it possible to identify overrepresented amino acids in κ amyloid proteins, and enabled the identification of 80% of amyloidogenic κ proteins on the basis of four elements, mutation of Arg61, introduction of aspartic acid at position 31, the presence of valine at position 27b and the presence of an *N*-linked glycosylation site anywhere in the molecule. Addition of information from other positions in the sequence improves the identification rate. Similar results have been obtained from an artificial intelligence approach [266]. However, little progress has been made to date in identification of amyloidogenic λ light chains on the basis of sequence. The reasons for this are unclear, but may suggest an increased propensity for fibril formation. Analysis may also be complicated by the overrepresentation of amyloid data in the λ database.

20.5 Amyloidosis (AH)

The possibility that heavy chains might be capable of fibril formation was suggested by the early work of Pruzanski [267] who demonstrated *in vitro* fibril formation by a $\gamma 3$ product of heavy chain disease. In contrast to the hundreds of primary structures that have been determined for amyloidogenic light chains and the extensive clinical characterization of light chain-related amyloidosis, the literature describing immunoglobulin heavy chain-related amyloidosis appears to currently total seven publications accrued since 1990. The six AH studies involved are found in Table 20.5. AH appears to be promiscuous in its protein constituency; samples have been derived from both IgM and IgG involving variable domain components and, as has been found in AL, constant domains [268, 269]. The production of heavy chain fragments is also a characteristic of heavy chain deposition disease and may reflect synthesis of the incomplete chains or/and proteolytic degradation [270]. There are few biophysical studies on constant domains of any type. In a series of studies, Hamaguchi et al. [224, 271–275]

Table 20.5 Cases of heavy chain amyloidosis (AH)

Case	H chain	L chain	Organ	Reference
1	IgG1 (VDJ–C_H3)	κ	systemic	309
2	IgG (VD)	κ	kidney; spleen	310
3	IgG1/4 (C_H3)	–	eye	311, 312
4	IgM (not reported)	λ	kidney	313
5	IgG (not reported)	–	kidney	314
6	IgG (VDJ?)	λ	kidney	315

a) Sequence data consistent with either IgG1 or IgG4.

found κ and λ constant domains to have guanidine hydrochloride midpoints of unfolding at 0.9 and 1.2 M, respectively, corresponding to ΔG values of 5.2 and 5.7 kcal/mol. These values, if applied to light chains, would be consistent with a high probability of fibril formation.

Thies et al. [276–278] have demonstrated that the C_H3 domain (C-terminal domain of IgG) is able to fold and dimerize even when the intradomain disulfide bond is reduced; in contrast, typical light chain variable domains precipitate when the disulfide link is removed. They reported a free energy of folding of –66.5 kJ/mol, which corresponds to –15.8 kcal/mol. If a direct comparison is possible, this free energy value suggests that C_H3 is approximately 10^6 times more stable than the non-amyloidogenic light chain, Len. However, mid-point transitions were strongly concentration dependent, suggesting improved stability by dimer formation and were 1.0 M guanidine or less. It appears that stability was calculated for the dimer, rather than the monomer. This suggests that the free energy of folding for the monomeric constant domain was no more than –33.2 kJ/mol (–7.9 kcal/mol). As the observed free energy change was highly concentration dependent, the intrinsic stability of C_H3 may be comparable to light chain constant domains.

Interestingly, Thies et al. [277] reported that under acidic conditions (pH 2) and in the presence of anions such as $KClO_4$, preferentially, but also NaCl and KCl, C_H3 self-assembles into large oligomers. The self-assembly followed an undefined conformational change. Like fibrils, the assemblies were very stable. Unlike fibrils, the oligomer was in equilibrium with free monomers and dimers, and the process was reversible, but not instantaneous, when the pH was raised. No image of the oligomeric assembly appears to have been published. Similar observations were made for a reduced and oxidized Fab under acidic conditions [279]. Thus, it is clear that, although extensive characterization of AH has not yet been undertaken, the stability properties of the heavy chain variable domain and at least one of the heavy chain constant domains are compatible with fibril formation by a mechanism comparable to that of the light chain variable domain.

20.6 Immunoproteomics

Williams [280, 281] introduced the concept of the immunoglobulin superfamily, which at that time consisted of antibodies, class I and class II MHC antigens, the poly Ig receptor, and the Thy-1 antigen. He wondered "whether other immunoglobulin-like molecules that function in various cell interactions remain to be discovered". Although numerous variable domain homologs were identified by the end of the previous century [282, 283], several hundred examples have emerged by genomic study. According to one analysis of the human genome, the immunoglobulin domain is the most prolific, accounting to about 2% of the currently estimated number of human genes [284].

The notion of immunoproteomics is based on the idea that progress in understanding the biophysical basis for amyloid formation might be found by defocusing, to some degree, and studying basic properties of appropriately selected homologs. Likewise, study of light chain amyloidosis might have implications for understanding properties of the many variable domain homologs that embody other functions as receptors or cell adhesion molecules. An example of this was provided by the myelin sheath coat protein, MPZ. The structure of MPZ (1NEU [215]) exhibited an inverted dimer interaction, reminiscent of the interaction seen by a mutant of Len, seen in Fig. 20.1. More than 60 inherited polymorphic variations associated with familial neuropathies, Charcot-Marcot-Tooth disease and the more severe Dejerine–Sottas syndrome (DSS) have been documented.

Several missense mutations of Arg69 in MPZ have been causally linked to DSS. Arg69 in MPZ corresponds to Arg61 in the antibody light chain variable domain. In both proteins, the arginine makes a salt bridge to a complementary aspartic acid located at comparable sites [285]. Hurle et al. [218] demonstrated that replacement of Arg61 in a $\kappa 1$ variable domain resulted in a loss of stability of 2.9 kcal/mol, i.e. a 100-fold increase in the relative concentration of unfolded domain. On the basis of database analysis, we had pointed out the apparent connection between Arg61 mutations and amyloidogenesis [286]. However, as noted earlier, all instances of Arg61 mutations in light chains were accompanied by other mutations, so no firm conclusions as to its significance could be drawn. The description of DSS mutations of Arg69 in MPZ was published in 1996 [287]. This information would have substantially strengthened support for the hypothesis that mutation of Arg61 was a high risk factor [288] for amyloidosis by demonstrating physiological significance of breaking the salt bridge in the absence of multiple other amino acid variations. Unfortunately, this link was not made until 2000, within 1 h on the Internet after the similarity of quaternary structures of MPZ and LenQ38E was noted by my crystallographic colleagues, M. Schiffer and P. Pokkuluri (personal communication). There are many other variable domain paralogs for which disease-related mutations have been described.

Study of distant homologs may also provide insight into the identification of the most critical structural features that contribute to the maintenance of the variable domain's thermodynamic stability or folding pathways. Table 20.6 lists 40 homologs of MPZ found in the third round of Psi-BLAST (National Center for Biotechnology Information, National Library of Medicine, National Institutes of Health). The homologs include functionally diverse representatives from humans, other mammals, chicken, fish and an amphibian. Most of the homologs have no statistically significant similarity to MPZ when compared directly by BLAST. It is expected that searches for homologs will be improved by the implementation of Psi-BLAST substitution matrices that have been optimized for the immunoglobulin fold [289]. The sequence alignments are shown in Fig. 20.5. As expected with the majority of these sequences having no significant similarity to MPZ, extensive variability of sequence is observed. However, a few positions exhibit near-perfect conservation. Shown in red, these positions correspond to light chain positions Gly16, Cys23, Trp35, Arg61, Asp82, Tyr86 and Cys88. In addition, position 86 is highly conserved as a glycine, but is usually

Table 20.6 Distant homologs of MPZ

gi	Protein	Organism	E3 a)	E1 b)
4505243	MPZ	*Homo sapiens*	0.0	0.0
10048456	cortical thymocyte receptor	*Xenopus laevis*	1×10^{-14}	0.22
628017	carcinoma-associated antigen pE4	*Rattus norvegicus*	1×10^{-14}	NS
29477102	similar to adhesion molecule AMICA	*Mus musculus*	3×10^{-14}	0.29
			3×10^{-12}	0.49
47213396	unnamed protein	*Tetradon nigroviridis*	3×10^{-13}	0.057
			3.6	NS
5031561	transmembrane glycoprotein A33	*Homo sapiens*	1×10^{-12}	0.22
46015553	Car D1 Domain	*Homo sapiens*	1×10^{-12}	0.044
21703782	TAP-binding protein-like	*Mus musculus*	2×10^{-11}	1.4
35215304	adipose tissue transmembrane protein	*Rattus norvegicus*	3×10^{-10}	2.4
2118920	PRR2α	*Homo sapiens*	5×10^{-10}	NS
12656130	sialoadhesin	*Homo sapiens*	3×10^{-9}	0.37
6855343	butyrophilin	*Homo sapiens*	7×10^{-9}	1.8
11990415	signal regulatory protein (SIRP-β1)	*Homo sapiens*	4×10^{-8}	NS
4071286	Ig λ light chain	*Monodelphis domestica*	2×10^{-7}	NS
30519900	immune co-stimulatory protein B7-H4	*Mus musculus*	3×10^{-7}	NS
543085	proteoglycan link protein	*Equus caballus*	5×10^{-7}	NS
18426911	protein tyrosine phosphatase	*Homo sapiens*	8×10^{-7}	NS
9800454	T cell receptor V_α	*Takifugu rubripes*	3×10^{-6}	NS
243055	Ig κ light chain	*Mus musculus*	3×10^{-6}	NS
565161	T cell receptor β	*Homo sapiens*	3×10^{-6}	NS
30584087	CD7 antigen	*Homo sapiens*	4×10^{-6}	NS
1620937	Ig heavy chain V region	*Oncorhynchus mykiss*	5×10^{-6}	NS
26006825	OX-2 membrane glycoprotein	*Mus musculus*	6×10^{-6}	NS
14714574	PVRL4 protein	*Homo sapiens*	7×10^{-6}	NS
2136629	T cell receptor δ	*Sus scrofa*	9×10^{-6}	NS
5381424	CD86 antigen	*Felis catus*	1×10^{-5}	NS
3273918	inhibitory receptor SHPS-1	*Mus musculus*	2×10^{-5}	NS
21307651	NKp30S	*Macaca mulatta*	2×10^{-5}	NS
17148679	TIM1	*Mus musculus*	3×10^{-5}	NS
104730	*B-G antigen 14/8*	*Gallus gallus*	3×10^{-5}	*NS*
2827462	hepatitis A virus cellular receptor	*Cercopi aethiops*	5×10^{-5}	NS
28277261	Necl1-pending-pro protein	*Xenopus laevis*	6×10^{-5}	NS
47550825	dermacan	*Danio rerio*	7×10^{-5}	NS
2677624	CD80 antigen	*Bos taurus*	8×10^{-5}	NS
18996289	mSIRP-1α	*Mus musculus*	9×10^{-5}	NS
1218052	zipper protein	*Gallus gallus*	7×10^{-4}	NS
3776056	tapasin	*Gallus gallus*	0.002	NS
1143285	*brevican core protein*	*Rattus norvegicus*	*0.004*	*NS*
9506521	chondroitin sulfate proteoglycans	*Mus musculus*	0.004	NS
2623877	1C7	*Homo sapiens*	0.005	NS
6754720	myelin oligodendrocyte glycoprotein	*Mus musculus*	0.008	NS

Italics indicate that the sequence is not included in Fig. 20.5.
a) E3, expect value in the third round of Psi-BLAST.
b) E1, expect value by direct BLAST comparison; NS, no significant similarity.

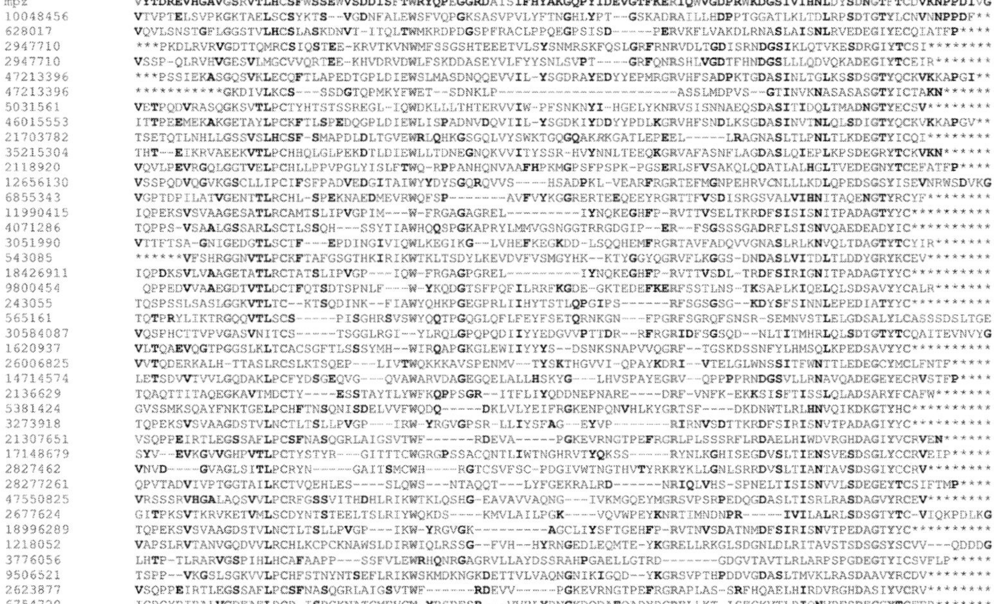

Fig. 20.5 Aligned sequences of human MPZ and 40 homologs. Alignments are as produced in the third round of Psi-BLAST; gi numbers are provided in the left column. Descriptions (protein and species) for each entry are provided in Table 20.6. Highly conserved residues are shown in bold, red letters. Matches to MPZ are shown in bold, black letters. Dashes indicate gaps; asterisks indicate that Psi-BLAST provided no alignment. Unique inserts were ignored since emphasis was identification of elements of conservation.

an alanine in the antibody and T cell proteins. It is Ser86 in the human protein 1C7, a receptor on natural killer T cells, perhaps demonstrating a link between the innate and acquired immune systems.

Although the positions highlighted in red are extremely well conserved, only Cys23 and Cys88 are perfectly conserved among this sample of 41 homologs. Interestingly, MPZ is the only example presented here in which the tyrosine corner is lost. The Arg61–Asp82 salt bridge is not present in the second V domain in gi|47213396, an unnamed protein from the puffer fish. The absence of both the arginine and the aspartic acid assure that the lack of an apparent salt bridge is not an artifact of misalignment. This unnamed protein has homology to a human protein of unknown function that serves as a receptor for coxsackie virus and adenovirus. Its structure (from mouse sialoadhesin) has been determined (1KAC) and is shown in the upper right corner of Fig. 20.6. Cys88 is missing in gi|12656130, the N-terminal domain of human sialoadhesin [290]. However, a cysteine residue appears at light chain variable domain position 69. Indeed, the three dimensional structure of human sialoadhesin (1QFO [291]) re-

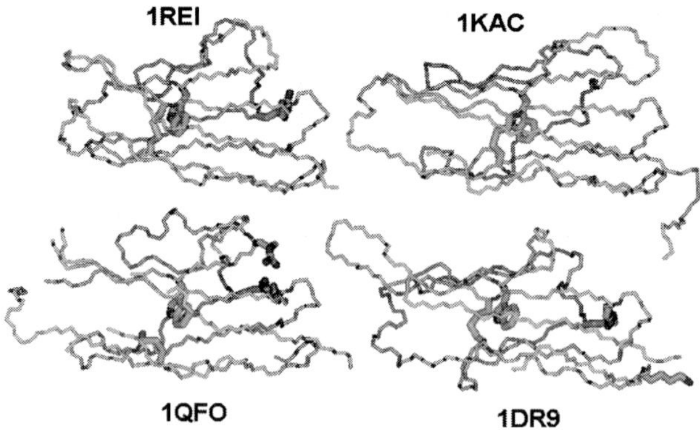

Fig. 20.6 Exceptions to conformational conservation. 1REI illustrates the conventional variable domain backbone; highlighted are the juxtaposition of Trp35 to the disulfide bond of Cys23 and Cys88, as well as the salt bridge between Arg61 and Asp82. This salt bridge is lost in 1KAC (virus receptor). The salt bridge is present in 1QFO (N-terminal domain of sialoadhesin) and is augmented by a second arginine interaction. In addition, the disulfide bridge from Cys23 now involves a cysteine at position 69 rather than 88, yielding an intra-sheet link rather than an inter-sheet linkage. A glycine at position 16 has been replaced by lysine in CD80 (1DR9).

veals a disulfide bond between these positions. This effectively exchanges an inter-sheet disulfide for an intra-sheet disulfide. The consequences of this trade are unknown. In addition, sialoadhesin exhibits two arginine residues forming salt bridges to Asp82. The structure of sialoadhesin indicates that the Psi-BLAST alignment depicted in Fig 20.5 is incorrect for gi|12656130 and that the arginine shown in the aligned position corresponds to position 65; the conserved arginine is at the initial site in the string: $_{61}$RFRGR$_{65}$. The conservation of glycine at position 16 is generally rationalized that the absence of a side-chain contributes to formation of a tight β turn. However, the CD80 (B7-1) antigen (gi|2677624) replaces glycine with the bulky lysine without apparent structural consequences as shown in 1DR9 [292]. No close structural representative of gi|30584087, the CD7 antigen lacking Trp35, was found. Similarly, no structural representative of gi|21703782, annotated as TAP-binding protein like, was found. The Psi-BLAST alignment indicates that Arg61 is not present in this structure. However, Asp82 is found, which suggests that an Arg61 equivalent is present and that there is an insertion of four residues between the two contributors of the salt bridge, or that Asp82 has found an alternative partner.

20.7 Concluding Remarks

Understanding amyloid formation by immunoglobulin molecules has proven to be a major challenge due to the extreme sequence diversity of these molecules. However, as we understand better the structural determinants of immunoglobulin stability, patterns are beginning to emerge and more specific hypotheses of fibril development have become possible. Demonstration of *in vitro* antagonism of fibril formation presents at least the feasibility of eventual pharmaceutical control of fibril formation for antibody light chain amyloidosis. The fact that a peptide has been observed to block fibril formation by both κ and λ light chains, as well as by superoxide dismutase (SOD) and reduced ribonuclease A (Gu and Stevens, unpublished results) suggests the possibility that the amyloidoses may no longer be orphan diseases in that it may be conceivable that a relatively small number of drugs that cross disease classifications may treat large numbers of patients.

Finally, research on the bases of the amyloidoses of immunoglobulins has evolved from the earliest days of protein sequence and structure determination to today's era of potential comprehensive knowledge of human's genetic and molecular structure resources. Human heavy and light chain variable domains come in thousands of sequence variations introduced by somatic mutation and other mechanisms; all of these proteins are potentially flawed, having bypassed scrutiny by the trials of evolution. The most badly flawed are subject to censorship by the quality control systems in the endoplasmic reticulum [293–308]. The most highly conserved sites in human light chain variable domains and their 200–300 homologs are very seldom found mutated, although those mutations must occur. In other protein families, such as SOD, sites of high conservation can be associated with familial Lou Gehrigs's disease in which virtually every one of the known disease-related mutations occur at such positions (Stevens, unpublished results). The non-antibody and non-T cell homologs of light and heavy chain variable domains all represent different solutions to the problem of constructing a variable domain. In principle, all of these structures have been tested and optimized by evolution. Understanding how all of the functionally diverse sequences represented in Table 20.6 and Fig. 20.5, the remaining functional prototypes that were not entered and their innumerable homologs that have also been optimized by evolution would contribute extensively to a final understanding of immunoglobulin amyloidosis and, perhaps, to understanding all forms of amyloidosis.

Acknowledgments

This work was supported by the US Department of Energy, Office of Biological and Environmental Research under contract W-31-100-Eng-38 and NIH grants DK43957 and AG18001.

References

1 Bence Jones, H. Papers on chemical pathology: prefaced by the Gulstonian lectures, read at the Royal College of Physicians, 1846. Lecture III. *Lancet* **1847**, *2*, 88–92.

2 Bence Jones, H. On a new substance occurring in the urine of a patient with mollities ossium. *Phil Trans R Soc Lond* **1847**, 55–62.

3 Coley, N.G. Early blood chemistry in Britain and France. *Clin Chem* **2001**, *47*, 2166–2178.

4 Coley, N.G. Henry Bence Jones, M.D., F.R.S. (1813–1873). *Notes Rec R Soc Lond* **1973**, *28*, 31–56.

5 Kyle, R.A. Henry Bence Jones – physician, chemist, scientist and biographer: a man for all seasons. *Br J Haematol* **2001**, *115*, 13–18.

6 Edelman, G.M. and J.A. Gally. The nature of Bence Jones proteins. Chemical similarities to polypeptide chains of myeloma globulins and normal gamma-globulins. *J Exp Med* **1962**, *116*, 207–227.

7 Kyle, R.A. and S.V. Rajkumar. Monoclonal gammopathies of undetermined significance: a review. *Immunol Rev* **2003**, *194*, 112–139.

8 Stone, M.J. Myeloma and macroglobulinemia: what are the criteria for diagnosis? *Clin Lymphoma* **2002**, *3*, 23–25.

9 Hummel, M. and H. Stein. Clinical relevance of immunoglobulin mutation analysis. *Curr Opin Oncol* **2000**, *12*, 395–402.

10 Stevens, F.J., A. Solomon and M. Schiffer. Bence Jones proteins: a powerful tool for fundamental study of protein chemistry and pathophysiology. *Biochemistry* **1991**, *30*, 6803–6805.

11 Ma, C.X., M.Q. Lacy, J.F. Rompala, A. Dispenzieri, S.V. Rajkumar, P.R. Greipp, R. Fonseca, R.A. Kyle, et al. Acquired Fanconi syndrome is an indolent disorder in the absence of overt multiple myeloma. *Blood* **2004**, *104*, 40–42.

12 Leung, N., D.J. Lager, M.A. Gertz, K. Wilson, S. Kanakiriya and F.C. Fervenza. Long-term outcome of renal transplantation in light-chain deposition disease. *Am J Kidney Dis* **2004**, *43*, 147–153.

13 Buxbaum, J. and G. Gallo. Nonamyloidotic monoclonal immunoglobulin deposition disease. Light-chain, heavy-chain, and light- and heavy-chain deposition diseases. *Hematol Oncol Clin North Am* **1999**, *13*, 1235–1248.

14 Aucouturier, P., M. Bauwens, A.A. Khamlichi, L. Denoroy, S. Spinelli, G. Touchard, J.-L. Preud'homme and M. Cogne. Monoclonal Ig L chain and L chain V domain fragment crystallization in myeloma-associated Fanconi's syndrome. *J Immunol* **1993**, *150*, 3561–3568.

15 Messiaen, T., S. Deret, B. Mougenot, F. Bridoux, P. Dequiedt, J.J. Dion, R. Makdassi, F. Meeus, et al. Adult Fanconi syndrome secondary to light chain gammopathy. Clinicopathologic heterogeneity and unusual features in 11 patients. *Medicine (Baltimore)* **2000**, *79*, 135–154.

16 Jones, D., V.K. Bhatia, T. Krausz and G.S. Pinkus. Crystal-storing histiocytosis: a disorder occurring in plasmacytic tumors expressing immunoglobulin kappa light chain. *Hum Pathol* **1999**, *30*, 1441–1448.

17 Titani, K. and F.W. Putnam. Immunoglobulin structure: amino- and carboxyl-terminal peptides of Type I Bence Jones proteins. *Science* **1965**, *147*, 1304–1305.

18 Titani, K., E. Whitley, Jr, L. Avogardo and F.W. Putnam. Immunoglobulin structure: partial amino acid sequence of a Bence Jones protein. *Science* **1965**, *149*, 1090–1092.

19 Putnam, F.W., K. Titani and E. Whitley, Jr. Chemical structure of light chains: amino acid sequence of type K Bence Jones proteins. *Proc R Soc Lond B Biol Sci* **1966**, *166*, 124–137.

20 Hilschmann, N. and L.C. Craig. Countercurrent-distribution studies with Bence Jones protein in a dissociating system. *Biochemistry* **1965**, *10*, 5–11.

21 Hilschmann, N. and L.C. Craig. Amino acid sequence studies with Bence Jones proteins. *Proc Natl Acad Sci USA* **1965**, *53*, 1403–1409.

22 Milstein, C. Variations in amino-acid sequence near the disulphide bridges of

Bence Jones proteins. *Nature* **1966**, *209*, 370–373.

23 Hood, L. E., W. R. Gray and W. J. Dreyer. On the mechanism of antibody synthesis: a species comparison of L-chains. *Proc Natl Acad Sci USA* **1966**, *55*, 826–832.

24 Baglioni, C. and D. Cioli. A study of immunoglobulin structure. II. The comparison of Bence Jones proteins by peptide mapping. *J Exp Med* **1966**, *124*, 307–330.

25 Cioli, D. and C. Baglioni. Origin of structural variation in Bence Jones proteins. *J Mol Biol* **1966**, *15*, 385–388.

26 Tischendorf, F. W. and E. F. Osserman. [Differences in the invariant (C-terminal) peptide composition of human type L-Bence Jones protein]. *Z Naturforsch B* **1967**, *22*, 1337–1339.

27 Tischendorf, F. W. and E. F. Osserman. Two antigenic subtypes of human lambda immunoglobulin chains. *J Immunol* **1969**, *102*, 172–178.

28 Tischendorf, F. W., M. M. Tischendorf and E. F. Osserman. Subgroup-specific antigenic marker on immunoglobulin lambda chains: identification of three subtypes of the variable region. *J Immunol* **1970**, *105*, 1033–1035.

29 Wu, T. T. and E. A. Kabat. An analysis of the sequences of the variable regions of Bence Jones proteins and myeloma light chains and their implications for antibody complementarity. *J Exp Med* **1970**, *132*, 211–250.

30 Kabat, E. A. and T. T. Wu. Attempts to locate complementarity-determining residues in the variable positions of light and heavy chains. *Ann NY Acad Sci* **1971**, *190*, 382–393.

31 Edmundson, A. B., M. K. Wood, M. Schiffer, K. D. Hardman, C. F. Ainsworth and K. R. Ely. A crystallographic investigation of a human IgG immunoglobulin. *J Biol Chem* **1970**, *245*, 2763–2764.

32 Schiffer, M., K. D. Hardman, M. K. Wood, A. B. Edmundson, M. E. Hood, K. R. Ely and H. F. Deutsch. A preliminary crystallographic investigation of human L-type Bence Jones protein. *J Biol Chem* **1970**, *245*, 728–730.

33 Edmundson, A. B., M. Schiffer, K. R. Ely and M. K. Wood. Structure of a lambda-type Bence Jones protein at 6-Å resolution. *Biochemistry* **1972**, *11*, 1822–1827.

34 Schiffer, M., R. L. Girling, K. R. Ely and A. B. Edmundson. Structure of a lambda-type Bence Jones protein at 3.5-Å resolution. *Biochemistry* **1973**, *12*, 4620–4631.

35 Poljak, R. J., L. M. Amzel, H. P. Avey and L. N. Becka. Structure of Fab′ New at 6 Å resolution. *Nat New Biol* **1972**, *235*, 137–140.

36 Solomon, A. and C. L. McLaughlin. Bence Jones proteins and light chains of immunoglobulins. I. Formation and characterization of amino-terminal (variant) and carboxyl-terminal (constant) halves. *J Biol Chem* **1969**, *244*, 3393–3404.

37 Caggiano, V., J. Cuttner and A. Solomon. Myeloma proteins, Bence Jones proteins and normal immunoglobulins in multiple myeloma. *Blood* **1967**, *30*, 265–287.

38 Solomon, A., T. A. Waldmann, J. L. Fahey and A. S. McFarlane. Metabolism of Bence Jones proteins. *J Clin Invest* **1964**, *43*, 103–117.

39 McLaughlin, C. L. and A. Solomon. A hidden antigenic site localized to the constant region of light chains of immunoglobulins. *Science* **1973**, *179*, 580–582.

40 Solomon, A. Bence Jones proteins and light chains of immunoglobulins [First of two parts]. *N Engl J Med* **1976**, *294*, 17–23.

41 Solomon, A. Bence Jones proteins and light chains of immunoglobulins [Second of two parts]. *N Engl J Med* **1976**, *294*, 91–98.

42 Wolfenstein-Todel, C., E. C. Franklin and R. A. Rudders. Similarities of the light chains and the variable regions of the heavy chains of the IgG2 lambda and IgA1 lambda myeloma proteins from a single patient. *J Immunol* **1974**, *112*, 871–876.

43 Franklin, E. C. Structure and function of immunoglobulins. *Acta Endocrinol Suppl* **1975**, *194*, 77–95.

44 Zolla, S., J. Buxbaum, E. C. Franklin and M. D. Scharff. Synthesis and assembly of immunoglobulins by malignant human plasmacytes. I. Myelomas producing gamma-chains and light chains. *J Exp Med* **1970**, *132*, 148–162.

45 Buxbaum, J., S. Zolla, M. D. Scharff and E. C. Franklin. Synthesis and assembly of immunoglobulins by malignant human plasmacytes and lymphocytes. II. Heterogeneity of assembly in cells producing IgM proteins. *J Exp Med* **1971**, *133*, 1118–1130.

46 Weitzman, S., D. H. Margulies and M. D. Scharff. Mutations in mouse myeloma cells: implications for human multiple myeloma and the production of immunoglobulins. *Ann Intern Med* **1976**, *85*, 110–116.

47 Hood, L. and D. Ein. Immunoglobulin lambda chain structure: two genes, one polypeptide chain. *Nature* **1968**, *220*, 764–767.

48 Conrath, K. E., U. Wernery, S. Muyldermans and V. K. Nguyen. Emergence and evolution of functional heavy-chain antibodies in *Camelidae*. *Dev Comp Immunol* **2003**, *27*, 87–103.

49 Dottorini, T., C. K. Vaughan, M. A. Walsh, P. LoSurdo and M. Sollazzo. Crystal structure of a human V_H: requirements for maintaining a monomeric fragment. *Biochemistry* **2004**, *43*, 622–628.

50 Radic, M. Z., J. Erikson, S. Litwin and M. Weigert. B lymphocytes may escape tolerance by revising their antigen receptors. *J Exp Med* **1993**, *177*, 1165–1173.

51 Chen, C., E. L. Prak and M. Weigert. Editing disease-associated autoantibodies. *Immunity* **1997**, *6*, 97–105.

52 Li, H., Y. Jiang, E. L. Prak, M. Radic and M. Weigert. Editors and editing of anti-DNA receptors. *Immunity* **2001**, *15*, 947–957.

53 Li, Y., Y. Louzoun and M. Weigert. Editing anti-DNA B cells by V_2x. *J Exp Med* **2004**, *199*, 337–346.

54 Goodnow, C. C., J. Crosbie, H. Jorgensen, R. A. Brink and A. Basten. Induction of self-tolerance in mature peripheral B lymphocytes. *Nature* **1989**, *342*, 385–391.

55 Nemazee, D. A. and K. Burki. Clonal deletion of B lymphocytes in a transgenic mouse bearing anti-MHC class I antibody genes. *Nature* **1989**, *337*, 562–566.

56 Benhar, I. and I. Pastan. Identification of residues that stabilize the single-chain Fv of monoclonal antibody B3. *J Biol Chem* **1995**, *270*, 23373–23380.

57 Davies, J. and L. Riechmann. Antibody V_H domains as small recognition units. *Biotechnology* **1995**, *13*, 475–479.

58 Tan, P. H., B. M. Sandmaier and P. S. Stayton. Contributions of a highly conserved V_H/V_L hydrogen bonding interaction to scFv folding stability and refolding efficiency. *Biophys J* **1998**, *75*, 1473–1482.

59 Worn, A. and A. Pluckthun. Stability engineering of antibody single-chain Fv fragments. *J Mol Biol* **2000**, *305*, 989–1010.

60 Ewert, S., T. Huber, A. Honegger and A. Pluckthun. Biophysical properties of human antibody variable domains. *J Mol Biol* **2003**, *325*, 531–553.

61 Ewert, S., A. Honegger and A. Pluckthun. Structure-based improvement of the biophysical properties of immunoglobulin V_H domains with a generalizable approach. *Biochemistry* **2003**, *42*, 1517–1528.

62 Worn, A. and A. Pluckthun. Different equilibrium stability behavior of ScFv fragments: identification, classification, and improvement by protein engineering. *Biochemistry* **1999**, *38*, 8739–8750.

63 Worn, A. and A. Pluckthun. Mutual stabilization of V_L and V_H in single-chain antibody fragments, investigated with mutants engineered for stability. *Biochemistry* **1998**, *37*, 13120–13127.

64 Chen, B. L. and R. J. Poljak. Amino acid sequence of the (lambda) light chain of a human myeloma immunoglobulin (IgG New). *Biochemistry* **1974**, *13*, 1295–1302.

65 Dubnovitsky, A. P., E. G. Kapetaniou and A. C. Papageorgiou. Expression, purification, crystallization and preliminary crystallographic analysis of phosphoserine aminotransferase from *Bacillus alcalophilus*. *Acta Crystallogr D* **2003**, *59*, 2319–2321.

66 Garver, F. A. and N. Hilschmann. The primary structure of a monoclonal human lambda-type immunoglobulin L-chain of subgroup II (Bence Jones protein NEI). *Eur J Biochem* **1972**, *26*, 10–32.

67 Bergman, L.W. and W.M. Kuehl. Temporal relationship of translation and glycosylation of immunoglobulin heavy and light chains. *Biochemistry* **1978**, *17*, 5174–5180.

68 Tveteraas, T., K. Sletten and P. Westermark. The amino acid sequence of a carbohydrate-containing immunoglobulin-light-chain-type amyloid-fibril protein. *Biochem J* **1985**, *232*, 183–190.

69 Fykse, E.-M., K. Sletten, G. Husby and G.G. Cornwell. The primary structure of the variable region of an immunoglobulin IV light-chain amyloid-fibril protein (AL GIL). *Biochem J* **1988**, *256*, 973–980.

70 Cogne, M., J.-L. Preud'homme, M. Bauwens, G. Touchard and P. Aucouturier. Structure of a monoclonal κ chain of the V_{κ}IV subgroup in the kidney and plasma cells in light chain deposition disease. *J Clin Invest* **1991**, *87*, 2186–2190.

71 Aucouturier, P., A.A. Khamlichi, J.L. Preud'homme, M. Bauwens, G. Touchard and M. Cogne. Complementary DNA sequence of human amyloidogenic immunoglobulin light-chain precursors. *Biochem J* **1992**, *285*, 149–152.

72 Renouf, D.V. and E.F. Hounsell. Molecular modelling of glycoproteins by homology with non-glycosylated protein domains, computer simulated glycosylation and molecular dynamics. *Adv Exp Med Biol* **1995**, *376*, 37–45.

73 Leung, S.O., A.S. Dion, M.C. Pellegrini, M.J. Losman, R.C. Grebenau, D.M. Goldenberg and H.J. Hansen. Effect of V_{κ} framework-1 glycosylation on the binding affinity of lymphoma-specific murine and chimeric LL2 antibodies and its potential use as a novel conjugation site. *Int J Cancer* **1995**, *60*, 534–538.

74 Decourt, C., M. Cogne and A. Rocca. Structural peculiarities of a truncated V_{κ}III immunoglobulin light chain in myeloma with light chain deposition disease. *Clin Exp Immunol* **1996**, *106*, 357–361.

75 Kagimoto, T., H. Nakakuma, H. Hata, M. Hidaka, K. Horikawa, T. Kawaguit, S. Nagakura, N. Iwamoto, et al. Differential glycosylation of Bence Jones protein and kidney impairment in patients with plasma cell dyscrasia. *J Lab Clin Med* **1997**, *129*, 217–223.

76 Foss, G.S., R. Nilsen, G.C. Cornwell, G. Husby and K. Sletten. A glycosylated Bence Jones protein and its autologous amyloid light chain containing potentially amyloidogenic residues. *Scand J Immunol* **1998**, *47*, 348–354.

77 Nilsen, T., K. Sletten and P. Westermark. AL612, a glycosylated lambda light chain of subgroup I. *Amyloid* **2000**, *7*, 51–53.

78 Omtvedt, L.A., G. Husby, G.G. Cornwell, R.A. Kyle and K. Sletten. The amino acid sequence of the glycosylated amyloid immunoglobulin light chain protein AL MS. *Scand J Immunol* **1997**, *45*, 551–556.

79 Omtvedt, L.A., S. Haavik, E.F. Hounsell, H. Barsett and K. Sletten. The carbohydrate structure of the amyloid immunoglobulin light chain EPS. *Amyloid* **1995**, *2*, 150–158.

80 Omtvedt, L.A., D. Bailey, D.V. Renouf, M.J. Davies, N.A. Paramonov, S. Haavik, G. Husby, K. Sletten, et al. Glycosylation of immunoglobulin light chains associated with amyloidosis. *Amyloid* **2000**, *7*, 227–244.

81 Karimi, M., K. Sletten and P. Westermark. Biclonal systemic AL-amyloidosis with one glycosylated and one nonglycosylated AL-protein. *Scand J Immunol* **2003**, *57*, 319–323.

82 Walker, M.R., J. Lee and R. Jefferis. Immunogenicity and antigenicity of immunoglobulins: detection of human immunoglobulin light-chain carbohydrate, using concanavalin A. *Biochim Biophys Acta* **1987**, *915*, 314–320.

83 Sun, M., Q.S. Gao, L. Li and S. Paul. Proteolytic activity of an antibody light chain. *J Immunol* **1994**, *153*, 5121–5126.

84 Paul, S., L. Li, R. Kalaga, P. Wilkins-Stevens, F.J. Stevens and A. Solomon. Natural catalytic antibodies: peptide-hydrolyzing activities of Bence Jones proteins and V_L fragment. *J Biol Chem* **1995**, *270*, 15257–15261.

85 Paul, S. Natural catalytic antibodies. *Mol Biotechnol* **1996**, *5*, 197–207.

86 Tyutyulkova, S., Q.S. Gao, A. Thompson, S. Rennard and S. Paul. Efficient vasoactive intestinal polypeptide hydrolyzing

autoantibody light chains selected by phage display. *Biochim Biophys Acta* **1996**, *1316*, 217–223.

87 Sun, M. and S. Paul. Altered cleavage site preference of a proteolytic antibody light chain induced by denaturation. *FEBS Lett* **1997**, *407*, 289–290.

88 Sun, M., Q. S. Gao, L. Kirnarskiy, A. Rees and S. Paul. Cleavage specificity of a proteolytic antibody light chain and effects of the heavy chain variable domain. *J Mol Biol* **1997**, *271*, 374–385.

89 Thiagarajan, P., R. Dannenbring, K. Matsuura, A. Tramontano, G. Gololobov and S. Paul. Monoclonal antibody light chain with prothrombinase activity. *Biochemistry* **2000**, *39*, 6459–6465.

90 Rangan, S. K., R. Liu, D. Brune, S. Planque, S. Paul and M. R. Sierks. Degradation of beta-amyloid by proteolytic antibody light chains. *Biochemistry* **2003**, *42*, 14328–14334.

91 Mitsuda, Y., E. Hifumi, K. Tsuruhata, H. Fujinami, N. Yamamoto and T. Uda. Catalytic antibody light chain capable of cleaving a chemokine receptor CCR-5 peptide with a high reaction rate constant. *Biotechnol Bioeng* **2004**, *86*, 217–225.

92 Hifumi, E., Y. Mitsuda, K. Ohara and T. Uda. Targeted destruction of the HIV-1 coat protein gp41 by a catalytic antibody light chain. *J Immunol Methods* **2002**, *269*, 283–298.

93 Schiffer, M. Molecular anatomy and the pathological expression of antibody light chains. *Am J Pathol* **1996**, *148*, 1339–1344.

94 Westermark, P., M. D. Benson, J. N. Buxbaum, A. S. Cohen, B. Frangione, S. Ikeda, C. L. Masters, G. Merlini, et al. Amyloid fibril protein nomenclature – 2002. *Amyloid* **2002**, *9*, 197–200.

95 Bellotti, V., P. Mangione and G. Merlini. Review: immunoglobulin light chain amyloidosis – the archetype of structural and pathogenic variability. *J Struct Biol* **2000**, *130*, 280–289.

96 Kyle, R. A., P. C. Spittell, M. A. Gertz, C. Y. Li, W. D. Edwards, L. J. Olson and S. N. Thibodeau. The premortem recognition of systemic senile amyloidosis with cardiac involvement. *Am J Med* **1996**, *101*, 395–400.

97 Westermark, P., J. Bergstrom, A. Solomon, C. Murphy and K. Sletten. Transthyretin-derived senile systemic amyloidosis: clinicopathologic and structural considerations. *Amyloid* **2003**, *10 (Suppl 1)*, 48–54.

98 Markowitz, G. S. Dysproteinemia and the kidney. *Adv Anat Pathol* **2004**, *11*, 49–63.

99 Paueksakon, P., M. P. Revelo, R. G. Horn, S. Shappell and A. B. Fogo. Monoclonal gammopathy: significance and possible causality in renal disease. *Am J Kidney Dis* **2003**, *42*, 87–95.

100 Basic-Jukic, N., P. Kes and B. Labar. Myeloma kidney: pathogenesis and treatment. *Acta Med Croatica* **2001**, *55*, 169–175.

101 Buxbaum, J. N., J. V. Chuba, G. C. Hellman, A. Solomon and G. R. Gallo. Monoclonal immunoglobulin deposition disease: light chain and light and heavy chain deposition diseases and their relation to light chain amyloidosis. Clinical features, immunopathology, and molecular analysis. *Ann Intern Med* **1990**, *112*, 455–464.

102 Randall, R. E., W. C. Williamson, F. Mullinax, M. Y. Tung and W. J. S. Still. Manifestation of systemic light chain deposition. *Am J Med* **1976**, *60*, 293–299.

103 Vidal, R., F. Goni, F. Stevens, P. Aucouturier, A. Kumar, B. Frangione, J. Ghiso and G. Gallo. Somatic mutations of the L12a gene in V-kappa(1) light chain deposition disease: potential effects on aberrant protein conformation and deposition. *Am J Pathol* **1999**, *155*, 2009–2017.

104 Teppo, A. M. and C. P. Maury. Urinary protein excretion patterns in reactive (secondary) systemic amyloidosis. *Rheumatol Int* **1988**, *8*, 213–217.

105 Decourt, C., A. Rocca, F. Bridoux, F. Vrtovsnik, J. L. Preud'homme, M. Cogne and G. Touchard. Mutational analysis in murine models for myeloma-associated Fanconi's syndrome or cast myeloma nephropathy. *Blood* **1999**, *94*, 3559–3566.

106 Deret, S., L. Denoroy, M. Lamarine, R. Vidal, B. Mougenot, B. Frangione, F. J.

Stevens, P. M. Ronco, et al. Kappa light chain-associated Fanconi's syndrome: molecular analysis of monoclonal immunoglobulin light chains from patients with and without intracellular crystals. *Protein Eng* **1999**, *12*, 363–369.

107 Orfila, C., J. C. Lepert, A. Modesto, P. Bernadet and J. M. Suc. Fanconi's syndrome, kappa light-chain myeloma, non-amyloid fibrils and cytoplasmic crystals in renal tubular epithelium. *Am J Nephrol* **1991**, *11*, 345–349.

108 Lacy, M. Q. and M. A. Gertz. Acquired Fanconi's syndrome associated with monoclonal gammopathies. *Hematol Oncol Clin North Am* **1999**, *13*, 1273–1280.

109 Terashima, K., K. Takahashi, M. Kojima, Y. Imai, S. Tsuchida, S. Migita, S. Ebina and C. Itoh. Kappa-type light chain crystal storage histiocytosis. *Acta Pathol Jpn* **1978**, *28*, 111–138.

110 Sethi, S., B. P. Cuiffo, G. S. Pinkus and H. G. Rennke. Crystal-storing histiocytosis involving the kidney in a low-grade B-cell lymphoproliferative disorder. *Am J Kidney Dis* **2002**, *39*, 183–188.

111 Coupland, S. E., H. D. Foss, M. Hummel and H. Stein. Extranodal marginal zone B-cell lymphoma of the lacrimal gland associated with crystal-storing histiocytosis. *Ophthalmology* **2002**, *109*, 105–110.

112 Solomon, A., B. Frangione and E. C. Franklin. Bence Jones proteins and light chains of immunoglobulins. Preferential association of the V-lambda-VI subgroup of human light chains with amyloidosis AL (lambda). *J Clin Invest* **1982**, *70*, 453–460.

113 Pokkuluri, P. R., A. Solomon, D. T. Weiss, F. J. Stevens and M. Schiffer. Tertiary structure of human lambda 6 light chains. *Amyloid* **1999**, *6*, 165–171.

114 Wall, J., M. Schell, C. Murphy, R. Hrncic, F. J. Stevens and A. Solomon. Thermodynamic instability of human lambda 6 light chains: correlation with fibrillogenicity. *Biochemistry* **1999**, *38*, 14101–14108.

115 Pepys, M. B., R. F. Dyck, F. C. de Beer, M. Skinner and A. S. Cohen. Binding of serum amyloid P-component (SAP) by amyloid fibrils. *Clin Exp Immunol* **1979**, *38*, 284–293.

116 Pepys, M. B., J. Herbert, W. L. Hutchinson, G. A. Tennent, H. J. Lachmann, J. R. Gallimore, L. B. Lovat, T. Bartfal, et al. Targeted pharmacological depletion of serum amyloid P component for treatment of human amyloidosis. *Nature* **2002**, *417*, 254–259.

117 Stevens, F. J. and R. Kisilevsky. Immunoglobulin light chains, glycosaminoglycans, and amyloid. *Cell Mol Life Sci* **2000**, *57*, 441–449.

118 Kisilevsky, R. Amyloids: tombstones or triggers? *Nat Med* **2000**, *6*, 633–634.

119 Kisilevsky, R., L. J. Lemieux, P. E. Fraser, X. Kong, P. G. Hultin and W. A. Szarek. Arresting amyloidosis *in vivo* using small-molecule anionic sulphonates or sulphates: implications for Alzheimer's disease. *Nat Med* **1995**, *1*, 143–148.

120 Kisilevsky, R., W. A. Szarek, J. Ancsin, S. Bhat, Z. Li and S. Marone. Novel glycosaminoglycan precursors as anti-amyloid agents, part III. *J Mol Neurosci* **2003**, *20*, 291–297.

121 Snow, A. D. and R. Kisilevsky. Temporal relationship between glycosaminoglycan accumulation and amyloid deposition during experimental amyloidosis. A histochemical study. *Lab Invest* **1985**, *53*, 37–44.

122 Rosano, C., S. Zuccotti, P. Mangione, S. Giorgetti, V. Bellotti, F. Pettirossi, A. Corazza, P. Viglino, et al. Beta2-microglobulin H31Y variant 3D structure highlights the protein natural propensity towards intermolecular aggregation. *J Mol Biol* **2004**, *335*, 1051–1064.

123 Borysik, A. J., S. E. Radford and A. E. Ashcroft. Co-populated conformational ensembles of beta 2-microglobulin uncovered quantitatively by electrospray ionisation mass spectrometry. *J Biol Chem* **2004**, 27069–27077.

124 Villanueva, J., M. Hoshino, H. Katou, J. Kardos, K. Hasegawa, H. Naiki and Y. Goto. Increase in the conformational flexibility of beta 2-microglobulin upon copper binding: a possible role for cop-

per in dialysis-related amyloidosis. *Protein Sci* **2004**, *13*, 797–809.
125 Dzido, G. and S. M. Sprague. Dialysis-related amyloidosis. *Minerva Urol Nefrol* **2003**, *55*, 121–129.
126 Sumner Makin, O. and L. C. Serpell. Structural characterisation of islet amyloid polypeptide fibrils. *J Mol Biol* **2004**, *335*, 1279–1288.
127 Green, J. D., C. Goldsbury, J. Kistler, G. J. Cooper and U. Aebi. Human amylin oligomer growth and fibril elongation define two distinct phases in amyloid formation. *J Biol Chem* **2004**, *279*, 12206–12212.
128 Westermark, P. Quantitative studies on amyloid in the islets of Langerhans. *Uppsala J Med Sci* **1972**, *77*, 91–94.
129 Carrell, R. W. and B. Gooptu. Conformational changes and disease – serpins, prions, and Alzheimer's. *Curr Opin Struct Biol* **1998**, *8*, 799–809.
130 Liao, Y. C., R. V. Lebo, G. A. Clawson and E. A. Smuckler. Human prion protein cDNA: molecular cloning, chromosomal mapping, and biological implications. *Science* **1986**, *233*, 364–367.
131 Gajdusek, D. C. and C. J. Gibbs, Jr. Slow, latent and temperate virus infections of the central nervous system. *Res Publ Ass Res Nerv Ment Dis* **1968**, *44*, 254–280.
132 Diener, T. O. Viroids: the smallest known agents of infectious disease. *Annu Rev Microbiol* **1974**, *28*, 23–39.
133 Prusiner, S. B. Novel proteinaceous infectious particles cause scrapie. *Science* **1982**, *216*, 136–144.
134 Heikenwalder, M., M. Prinz, F. L. Heppner and A. Aguzzi. Current concepts and controversies in prion immunopathology. *J Mol Neurosci* **2004**, *23*, 3–12.
135 Scheibel, T. Amyloid formation of a yeast prion determinant. *J Mol Neurosci* **2004**, *23*, 13–22.
136 Ericsson, L. H., N. Eriksen, K. A. Walsh and E. P. Benditt. Primary structure of duck amyloid protein A. The form deposited in tissues may be identical to its serum precursor. *FEBS Lett* **1987**, *218*, 11–16.
137 Kimura, Y. and A. Kakizuka. Polyglutamine diseases and molecular chaperones. *IUBMB Life* **2003**, *55*, 337–345.
138 Singhrao, S. K., P. Thomas, J. D. Wood, J. C. MacMillan, J. W. Neal, P. S. Harper and A. L. Jones. Huntingtin protein colocalizes with lesions of neurodegenerative diseases: an investigation in Huntington's, Alzheimer's, and Pick's diseases. *Exp Neurol* **1998**, *150*, 213–222.
139 Hirakura, Y., R. Azimov, R. Azimova and B. L. Kagan. Polyglutamine-induced ion channels: a possible mechanism for the neurotoxicity of Huntington and other CAG repeat diseases. *J Neurosci Res* **2000**, *60*, 490–494.
140 Perutz, M. F. Glutamine repeats and inherited neurodegenerative diseases: molecular aspects. *Curr Opin Struct Biol* **1996**, *6*, 848–858.
141 Sanchez, I. and J. Yuan. A convoluted way to die. *Neuron* **2001**, *29*, 563–566.
142 Sanchez, I., C. Mahlke and J. Yuan. Pivotal role of oligomerization in expanded polyglutamine neurodegenerative disorders. *Nature* **2003**, *421*, 373–379.
143 Mucchiano, G. I., B. Haggqvist, K. Sletten and P. Westermark. Apolipoprotein A-1-derived amyloid in atherosclerotic plaques of the human aorta. *J Pathol* **2001**, *193*, 270–275.
144 Pepys, M. B., P. N. Hawkins, D. R. Booth, D. M. Vigushin, G. A. Tennent, A. K. Soutar, N. Totty, O. Nguyen, et al. Human lysozyme gene mutations cause hereditary systemic amyloidosis. *Nature* **1993**, *362*, 553–557.
145 Canet, D., M. Sunde, A. M. Last, A. Miranker, A. Spencer, C. V. Robinson and C. M. Dobson. Mechanistic studies of the folding of human lysozyme and the origin of amyloidogenic behavior in its disease-related variants. *Biochemistry* **1999**, *38*, 6419–6427.
146 Booth, D. R., M. Sunde, V. Bellotti, C. V. Robinson, W. L. Hutchinson, P. E. Fraser, P. N. Hawkins, C. M. Dobson, et al. Instability, unfolding and aggregation of human lysozyme variants underlying amyloid fibrillogenesis. *Nature* **1997**, *385*, 787–793.

147 Connors, L.H., A.M. Richardson, R. Theberge and C.E. Costello. Tabulation of transthyretin (TTR) variants as of 1/1/2000. *Amyloid* **2000**, *7*, 54–69.

148 Fitch, N.J.S., M.T. Akbari and D.B. Ramsden. An inherited non-amyloidogenic transthyretin variant, [Ser6]-TTR, with increased thyroxine-binding affinity, characterized by DNA sequencing. *J Endocrinol* **1991**, *129*, 309–313.

149 Haltia, M., J. Ghiso, F. Prelli, G. Gallo, S. Kiuru, H. Somer, J. Palo and B. Frangione. Amyloid in familial amyloidosis, Finnish type, is antigenically and structurally related to gelsolin. *Am J Pathol* **1990**, *136*, 1223–1228.

150 Perfetti, V., P. Ubbiali, M.C. Vignarelli, R.F.M. Diegoli, M. Stoppini, A. Lisa, P. Mangione, L. Obici, et al. Evidence that amyloidogenic light chains undergo antigen-driven selection. *Blood* **1998**, *91*, 2948–2954.

151 Helms, L.R. and R. Wetzel. Specificity of abnormal assembly in immunoglobulin light chain deposition disease and amyloidosis. *J Mol Biol* **1996**, *257*, 77–86.

152 Comenzo, R.L., J. Wally, G. Kica, J. Murray, T. Ericsson, M. Skinner and Y. Zhang. Clonal immunoglobulin light chain variable region germline gene use in AL amyloidosis: association with dominant amyloid-related organ involvement and survival after stem cell transplantation. *Br J Haematol* **1999**, *106*, 744–751.

153 Glenner, G.G., W. Terry, M. Harada, C. Isersky and D. Page. Amyloid fibril proteins: proof of homology with immunoglobulin light chains by sequence analyses. *Science* **1971**, *172*, 1150–1151.

154 Stevens, P.W., R. Raffen, D.K. Hanson, Y.L. Deng, M. Berrios-Hammond, F.A. Westholm, C. Murphy, M. Eulitz, et al. Recombinant immunoglobulin variable domains generated from synthetic genes provide a system for *in vitro* characterization of light-chain amyloid proteins. *Protein Sci* **1995**, *4*, 421–432.

155 Husby, G. Immunoglobulin-related (AL) amyloidosis. *Clin Exp Rheumatol* **1983**, *1*, 353–358.

156 Burrows, P.D., J.F. Kearney, H.W. Schroeder, Jr and M.D. Cooper. Normal B lymphocyte differentiation. *Baillieres Clin Haematol* **1993**, *6*, 785–806.

157 Tycko, R. Progress towards a molecular-level structural understanding of amyloid fibrils. *Curr Opin Struct Biol* **2004**, *14*, 96–103.

158 Ohnishi, S. and K. Takano. Amyloid fibrils from the viewpoint of protein folding. *Cell Mol Life Sci* **2004**, *61*, 511–524.

159 Souillac, P.O., V.N. Uversky, I.S. Millett, R. Khurana, S. Doniach and A.L. Fink. Effect of association state and conformational stability on the kinetics of immunoglobulin light chain amyloid fibril formation at physiological pH. *J Biol Chem* **2002**, *277*, 12657–12665.

160 Souillac, P.O., V.N. Uversky, I.S. Millett, R. Khurana, S. Doniach and A.L. Fink. Elucidation of the molecular mechanism during the early events in immunoglobulin light chain amyloid fibrillation. Evidence for an off-pathway oligomer at acidic pH. *J Biol Chem* **2002**, *277*, 12666–12679.

161 Ionescu-Zanetti, C., R. Khurana, J.R. Gillespie, J.S. Petrick, L.C. Trabachino, L.J. Minert, S.A. Carter and A.L. Fink. Monitoring the assembly of Ig light-chain amyloid fibrils by atomic force microscopy. *Proc Natl Acad Sci USA* **1999**, *96*, 13175–13179.

162 Khurana, R., J.R. Gillespie, A. Talapatra, L.J. Minert, C. Ionescu-Zanetti, I. Millett and A.L. Fink. Partially folded intermediates as critical precursors of light chain amyloid fibrils and amorphous aggregates. *Biochemistry* **2001**, *40*, 3525–3535.

163 Zhu, M., P.O. Souillac, C. Ionescu-Zanetti, S.A. Carter and A.L. Fink. Surface-catalyzed amyloid fibril formation. *J Biol Chem* **2002**, *277*, 50914–50922.

164 Souillac, P.O., V.N. Uversky and A.L. Fink. Structural transformations of oligomeric intermediates in the fibrillation of the immunoglobulin light chain LEN. *Biochemistry* **2003**, *42*, 8094–8104.

165 Khurana, R., C. Ionescu-Zanetti, M. Pope, J. Li, L. Nielson, M. Ramirez-Al-

varado, L. Regan, A.L. Fink, et al. A general model for amyloid fibril assembly based on morphological studies using atomic force microscopy. *Biophys J* **2003**, *85*, 1135–1144.

166 Khurana, R., P.O. Souillac, A.C. Coats, L. Minert, C. Ionescu-Zanetti, S.A. Carter, A. Solomon and A.L. Fink. A model for amyloid fibril formation in immunoglobulin light chains based on comparison of amyloidogenic and benign proteins and specific antibody binding. *Amyloid* **2003**, *10*, 97–109.

167 Zhu, M., S. Han, F. Zhou, S.A. Carter and A.L. Fink. Annular oligomeric amyloid intermediates observed by *in-situ* AFM. *J Biol Chem* **2004**, in press.

168 Wilkins Stevens, P., R. Raffen, D.K. Hanson, Y.-L. Deng, M. Berrios-Hammond, F.A. Westholm, C. Murphy, M. Eulitz, et al. Recombinant immunoglobulin variable domains generated from synthetic genes provide a system for *in vitro* characterization of light-chain amyloid proteins. *Protein Sci* **1995**, *4*, 421–432.

169 Govaerts, C., H. Wille, S.B. Prusiner and F.E. Cohen. Evidence for assembly of prions with left-handed beta-helices into trimers. *Proc Natl Acad Sci USA* **2004**, *101*, 8342–8347.

170 Rocken, C., R. Kientsch-Engel, S. Mansfeld, B. Stix, K. Stubenrauch, B. Weigle, F. Buhling, M. Schwan, et al. Advanced glycation end products and receptor for advanced glycation end products in AA amyloidosis. *Am J Pathol* **2003**, *162*, 1213–1220.

171 Bucciarelli, L.G., T. Wendt, L. Rong, E. Lalla, M.A. Hofmann, M.T. Goova, A. Taguchi, S.F. Yan, et al. RAGE is a multiligand receptor of the immunoglobulin superfamily: implications for homeostasis and chronic disease. *Cell Mol Life Sci* **2002**, *59*, 1117–1128.

172 Sousa, M.M., S. Du Yan, R. Fernandes, A. Guimaraes, D. Stern and M.J. Saraiva. Familial amyloid polyneuropathy: receptor for advanced glycation end products-dependent triggering of neuronal inflammatory and apoptotic pathways. *J Neurosci* **2001**, *21*, 7576–7586.

173 Hou, F.F., D.N. Reddan, W.K. Seng and W.F. Owen, Jr. Pathogenesis of beta$_2$-microglobulin amyloidosis: role of monocytes/macrophages. *Semin Dial* **2001**, *14*, 135–139.

174 Sousa, M.M., S.D. Yan, D. Stern and M.J. Saraiva. Interaction of the receptor for advanced glycation end products (RAGE) with transthyretin triggers nuclear transcription factor κB (NF-κB) activation. *Lab Invest* **2000**, *80*, 1101–1110.

175 Miyata, T., O. Hori, J. Zhang, S.D. Yan, L. Ferran, Y. Iida and A.M. Schmidt. The receptor for advanced glycation end products (RAGE) is a central mediator of the interaction of AGE-beta2 microglobulin with human mononuclear phagocytes via an oxidant-sensitive pathway. Implications for the pathogenesis of dialysis-related amyloidosis. *J Clin Invest* **1996**, *98*, 1088–1094.

176 Sipe, J.D. and A.S. Cohen. Review: history of the amyloid fibril. *J Struct Biol* **2000**, *130*, 88–98.

177 Hvala, A., D. Ferluga, A. Vizjak and M. Koselj-Kajtna. Fibrillary noncongophilic renal and extrarenal deposits: a report on 10 cases. *Ultrastruct Pathol* **2003**, *27*, 341–347.

178 LeVine, H., 3rd. Thioflavine T interaction with synthetic Alzheimer's disease beta-amyloid peptides: detection of amyloid aggregation in solution. *Protein Sci* **1993**, *2*, 404–410.

179 Ban, T., D. Hamada, K. Hasegawa, H. Naiki and Y. Goto. Direct observation of amyloid fibril growth monitored by thioflavin T fluorescence. *J Biol Chem* **2003**, *278*, 16462–16465.

180 Kisilevsky, R. Anti-amyloid drugs: potential in the treatment of diseases associated with aging. *Drugs Aging* **1996**, *8*, 75–83.

181 Jiang, X., E. Myatt, P. Lykos and F.J. Stevens. Interaction between glycosaminoglycans and immunoglobulin light chains. *Biochemistry* **1997**, *36*, 13187–13194.

182 Snow, A.D., J. Willmer and R. Kisilevsky. Sulfated glycosaminoglycans: a common constituent of all amyloids? *Lab Invest* **1987**, *56*, 120–123.

183 Kisilevsky, R. Heparan sulfate proteoglycans in amyloidogenesis: an epiphenomenon, a unique factor, or the tip of a more fundamental process? *Lab Invest* **1990**, *63*, 589–591.

184 Kisilevsky, R. Proteoglycans, glycosaminoglycans, amyloid-enhancing factor, and amyloid deposition. *J Intern Med* **1992**, *232*, 515–516.

185 Kisilevsky, R. and P. Fraser. Proteoglycans and amyloid fibrillogenesis. *Ciba Found Symp* **1996**, *199*, 58–67; discussion 68–72, 90–103.

186 O'Nuallain, B. and R. Wetzel. Conformational Abs recognizing a generic amyloid fibril epitope. *Proc Natl Acad Sci USA* **2002**, *99*, 1485–1490.

187 Solomon, A., D.T. Weiss and J.S. Wall. Immunotherapy in systemic primary (AL) amyloidosis using amyloid-reactive monoclonal antibodies. *Cancer Biother Radiopharm* **2003**, *18*, 853–860.

188 Solomon, A., D.T. Weiss and J.S. Wall. Therapeutic potential of chimeric amyloid-reactive monoclonal antibody 11-1F4. *Clin Cancer Res* **2003**, *9*, 3831S–3838S.

189 Grey, H.M. and P.F. Kohler. A case of tetramer Bence Jones proteinaemia. *Clin Exp Immunol* **1968**, *3*, 277–285.

190 Parr, D.M., W. Pruzanski, J.G. Scott and D.M. Mills. Primary amyloidosis with plasmacytic dyscrasia and a tetramer of Bence Jones type lambda globulin in the serum and urine. *Blood* **1971**, *37*, 473–484.

191 Sato, K., Y. Imai, M. Takayasu, M. Tamaki and T. Tange. [An autopsied case of multiple myeloma with dimer and tetramer of Bence Jones protein in the serum and urine]. *Rinsho Ketsueki* **1975**, *16*, 620–627.

192 Kozuru, M., H. Benoki, H. Sugimoto, K. Sakai and H. Ibayashi. A case of lambda type tetramer Bence Jones proteinemia. *Acta Haematol* **1977**, *57*, 359–365.

193 Ibayashi, H., M. Kozuru, H. Sugimoto, M. Enjoji and H. Taira. [Multiple mass formation throughout the body (immunoelectrophoresis) – partial excision of epidural tumors: myeloma–lambda chain tetramer type Bence Jones proteinemia]. *Nippon Rinsho* **1977**, *35 (Suppl 2)*, 3152–3153, 3158–3159.

194 Gallango, M.L., R. Suinaga and M. Ramirez. Bence Jones myeloma with a tetramer of kappa-type globulin in serum. *Clin Chem* **1980**, *26*, 1741–1744.

195 Kosaka, M., Y. Iishi, K. Okagawa, S. Saito, J. Sugihara and Y. Muto. Tetramer Bence Jones protein in the immunoproliferative diseases. Angioimmunoblastic lymphadenopathy, primary amyloidosis, and multiple myeloma. *Am J Clin Pathol* **1989**, *91*, 639–646.

196 Berggard, I. and P.A. Peterson. Polymeric forms of free normal kappa and lambda chains of human immunoglobulin. *J Biol Chem* **1969**, *244*, 4299–4307.

197 Abraham, R.S., M.C. Charlesworth, B.A. Owen, L.M. Benson, J.A. Katzmann, C.B. Reeder and R.A. Kyle. Trimolecular complexes of lambda light chain dimers in serum of a patient with multiple myeloma. *Clin Chem* **2002**, *48*, 1805–1811.

198 Abraham, R.S., R.J. Clark, S.C. Bryant, J.F. Lymp, T. Larson, R.A. Kyle and J.A. Katzmann. Correlation of serum immunoglobulin free light chain quantification with urinary Bence Jones protein in light chain myeloma. *Clin Chem* **2002**, *48*, 655–657.

199 Maeda, H., J. Engel and H.J. Schramm. Kinetics of dimerization of the variable fragment of the Bence Jones protein Au. *Eur J Biochem* **1976**, *69*, 133–139.

200 Maeda, H., E. Steffen and J. Engel. Kinetics of dimerization of the Bence Jones protein AU. *Biophys Chem* **1978**, *9*, 57–64.

201 Azuma, T., K. Hamaguchi and S. Migita. Interactions between immunoglobulin polypeptide chains. *J Biochem (Tokyo)* **1974**, *76*, 685–693.

202 Stevens, F.J., F.A. Westholm, A. Solomon and M. Schiffer. Self-association of human immunoglobulin kappa I light chains: role of the third hypervariable region. *Proc Natl Acad Sci USA* **1980**, *77*, 1144–1148.

203 Berman, H. M., J. Westbrook, Z. Feng, G. Gilliland, T. N. Bhat, H. Weissig, I. N. Shindyalov and P. E. Bourne. The Protein Data Bank. *Nucleic Acids Res* **2000**, *28*, 235–242.

204 Huang, D. B., C. F. Ainsworth, F. J. Stevens and M. Schiffer. Three quaternary structures for a single protein. *Proc Natl Acad Sci USA* **1996**, *93*, 7017–7021.

205 Huang, D.-B., C. Ainsworth, A. Solomon and M. Schiffer. Pitfalls of molecular replacement: the structure determination of an immunoglobulin light chain dimer. *Acta Crystallogr D* **1996**, *52*, 1058–1066.

206 Ely, K. R., J. N. Herron and A. B. Edmundson. Three-dimensional structure of a hybrid light chain dimer: protein engineering of a binding cavity. *Mol Immunol* **1990**, *27*, 101–114.

207 Roussel, A., S. Spinelli, S. Deret, J. Navaza, P. Aucouturier and C. Cambillau. The structure of an entire noncovalent immunoglobulin kappa light-chain dimer reveals a weak and unusual constant domains association. *Eur J Biochem* **1999**, *260*, 192–199.

208 Epp, O., E. E. Lattman, M. Schiffer, R. Huber and W. Palm. The molecular structure of a dimer composed of the variable portions of the Bence Jones protein REI refined at 2.0 Å resolution. *Biochemistry* **1975**, *14*, 4943–4952.

209 Steinrauf, L. K., M. Y. Chiang and D. Shiuan. Molecular structure of the amyloid-forming protein kappa I Bre. *J Biochem (Tokyo)* **1999**, *125*, 422–429.

210 Huang, D. B., C. H. Chang, C. Ainsworth, A. T. Brunger, M. Eulitz, A. Solomon, F. J. Stevens and M. Schiffer. Comparison of crystal structures of two homologous proteins: structural origin of altered domain interactions in immunoglobulin light-chain dimers. *Biochemistry* **1994**, *33*, 14848–14857.

211 Huang, D. B., C. H. Chang, C. Ainsworth, G. Johnson, A. Solomon, F. J. Stevens and M. Schiffer. Variable domain structure of kappaIV human light chain Len: high homology to the murine light chain McPC603. *Mol Immunol* **1997**, *34*, 1291–1301.

212 Furey, W., B. C. Wang, C. S. Yoo and M. Sax. Structure of a novel Bence Jones protein (Rhe) fragment at 1.6 Å resolution. *J Mol Biol* **1983**, *167*, 661–692.

213 Raffen, R., P. W. Stevens, C. Boogaard, M. Schiffer and F. J. Stevens. Reengineering immunoglobulin domain interactions by introduction of charged residues. *Protein Eng* **1998**, *11*, 303–309.

214 Pokkuluri, P. R., D. B. Huang, R. Raffen, X. Cai, G. Johnson, P. W. Stevens, F. J. Stevens and M. Schiffer. A domain flip as a result of a single amino-acid substitution. *Structure* **1998**, *6*, 1067–1073.

215 Shapiro, L., J. P. Doyle, P. Hensley, D. R. Colman and W. A. Hendrickson. Crystal structure of the extracellular domain from P0, the major structural protein of peripheral nerve myelin. *Neuron* **1996**, *17*, 435–449.

216 Pokkuluri, P. R., X. Cai, G. Johnson, F. J. Stevens and M. Schiffer. Change in dimerization mode by removal of a single unsatisfied polar residue located at the interface. *Protein Sci* **2000**, *9*, 1852–1855.

217 McDonald, I. K. and J. M. Thornton. Satisfying hydrogen bonding potential in proteins. *J Mol Biol* **1994**, *238*, 777–793.

218 Hurle, M. R., L. R. Helms, L. Li, W. Chan and R. Wetzel. A role for destabilizing amino acid replacements in light-chain amyloidosis. *Proc Natl Acad Sci USA* **1994**, *91*, 5446–5450.

219 Wetzel, R. Domain stability in immunoglobulin light chain deposition disorders. *Adv Protein Chem* **1997**, *50*, 183–242.

220 McCutchen, S. L., W. Colon and J. W. Kelly. Transthyretin mutation Leu-55-Pro significantly alters tetramer stability and increases amyloidogenicity. *Biochemistry* **1993**, *32*, 12119–12127.

221 McCutchen, S. L., Z. Lai, G. J. Miroy, J. W. Kelly and W. Colon. Comparison of lethal and nonlethal transthyretin variants and their relationship to amyloid disease. *Biochemistry* **1995**, *34*, 13527–13536.

222 Raffen, R., L. J. Dieckman, M. Szpunar, C. Wunschl, P. R. Pokkuluri, P. Dave,

P. Wilkins Stevens, X. Cai, et al. Physicochemical consequences of amino acid variations that contribute to fibril formation by immunoglobulin light chains. *Protein Sci* **1999**, *8*, 509–517.

223 Cowgil, R. Fluorescence and the structure of proteins. XII. Pancreatic trypsin inhibitor. *Biochim Biophys Acta* **1967**, *140*, 37–44.

224 sunenaga, M., Y. Goto, Y. Kawata and K. Hamaguchi. Unfolding and refolding of a type kappa immunoglobulin light chain and its variable and constant fragments. *Biochemistry* **1987**, *26*, 6044–6051.

225 Santoro, M.M. and D.W. Bolen. Unfolding free energy changes determined by the linear extrapolation method. 1. Unfolding of phenylmethanesulfonyl alpha-chymotrypsin using different denaturants. *Biochemistry* **1988**, *27*, 8063–8068.

226 Bolen, D.W. and M.M. Santoro. Unfolding free energy changes determined by the linear extrapolation method. 2. Incorporation of delta G degrees N–U values in a thermodynamic cycle. *Biochemistry* **1988**, *27*, 8069–8074.

227 Frisch, C., H. Kolmar and H.J. Fritz. A soluble immunoglobulin variable domain without a disulfide bridge: construction, accumulation in the cytoplasm of *E. coli*, purification and physicochemical characterization. *Biol Chem Hoppe-Seyler* **1994**, *375*, 353–356.

228 Frisch, C., H. Kolmar, A. Schmidt, G. Kleeman, A. Reinhard, E. Pohl, I. Uson, T.R. Schneider, et al. Contribution of the intramolecular disulfide bridge to the folding stability of REIv, the variable domain of a human immunoglobulin kappa light chain. *Fold Design* **1996**, *1*, 431–440.

229 Kolmar, H., C. Frisch, K. Gotze and H.-J. Fritz. Immunoglobulin mutant library genetically screened for folding stability exploiting bacterial signal transduction. *J Mol Biol* **1995**, *251*, 471–476.

230 Uson, I., E. Pohl, T.R. Schneider, Z. Dauter, A. Schmidt, H.-J. Fritz and G.M. Sheldrick. 1.7 Å structure of the stabilized REIv mutant T39K. Application of local NCS restraints. *Acta Crystallogr D* **1999**, *55*, 1158–1167.

231 Helms, L.R. and R. Wetzel. Destabilizing loop swaps in the CDRs of an immunoglobulin V_L domain. *Protein Sci* **1995**, *4*, 2073–2081.

232 Chan, W., L.R. Helms, I. Brooks, G. Lee, S. Ngola, D. McNulty, B. Maleef, P. Hensley, et al. Mutational effects on inclusion body formation in the periplasmic expression of the immunoglobulin V_L domain REI. *Fold Design* **1996**, *1*, 77–89.

233 Wall, J.S., V. Gupta, M. Wilkerson, M. Schell, R. Loris, P. Adams, A. Solomon, F. Stevens, et al. Structural basis of light chain amyloidogenicity: comparison of the thermodynamic properties, fibrillogenic potential and tertiary structural features of four $V_{\lambda}6$ proteins. *J Mol Recognit* **2004**, *17*, 323–231.

234 Klein, R., R. Jaenichen and H.G. Zachau. Expressed human immunoglobulin genes and their hypermutation. *Eur J Immunol* **1993**, *23*, 3248–3271.

235 Pokkuluri, P.R., R. Raffen, L. Dieckman, C. Boogaard, F.J. Stevens and M. Schiffer. Increasing protein stability by polar surface residues: domain-wide consequences of interactions within a loop. *Biophys J* **2002**, *82*, 391–398.

236 Kaplan, B., C.L. Murphy, V. Ratner, M. Pras, D.T. Weiss and A. Solomon. Micro-method to isolate and purify amyloid proteins for chemical characterization. *Amyloid* **2001**, *8*, 22–29.

237 Murphy, C.L., M. Eulitz, R. Hrncic, K. Sletten, P. Westermark, T. Williams, S.D. Macy, C. Wooliver, et al. Chemical typing of amyloid protein contained in formalin-fixed paraffin-embedded biopsy specimens. *Am J Clin Pathol* **2001**, *116*, 135–142.

238 Kaplan, B., R. Hrncic, C.L. Murphy, G. Gallo, D.T. Weiss and A. Solomon. Microextraction and purification techniques applicable to chemical characterization of amyloid proteins in minute amounts of tissue. *Methods Enzymol* **1999**, *309*, 67–81.

239 Comenzo, R.L., Y. Zhang, C. Martinez, K. Osman and G.A. Herrera. The tropism of organ involvement in primary systemic amyloidosis: contributions of Ig V_L germ line gene use and clonal plasma cell burden. *Blood* **2001**, *98*, 714–720.

240 Perfetti, V., S. Casarini, G. Palladini, M.C. Vignarelli, C. Kiersy, M. Diegoli, E. Ascari and G. Merlini. Analysis of V_λ–J_λ expression in plasma cells from primary (AL) amyloidosis and normal bone marrow identifies *3r* (λIII) as a new amyloid-associated germline gene segment. *Blood* **2002**, *100*, 948–953.

241 Perfetti, V., P. Ubbiali, M.C. Vignarelli, M. Diegoli, R. Fasani, M. Stoppini, A. Lisa, P. Mangione, et al. Evidence that amyloidogenic light chains undergo antigen-driven selection. *Blood* **1998**, *91*, 2948–2954.

242 Stevens, F.J. Analysis of protein–protein interaction by simulation of small-zone size-exclusion chromatography: application to an antibody–antigen association. *Biochemistry* **1986**, *25*, 981–993.

243 Stevenson, F., S. Sahota, D. Zhu, C. Ottensmeier, C. Chapman, D. Oscier and T. Hamblin. Insight into the origin and clonal history of B-cell tumors as revealed by analysis of immunoglobulin variable region genes. *Immunol Rev* **1998**, *162*, 247–259.

244 Kosmas, C., K. Stamatopoulos, T. Papadaki, C. Belessi, X. Yataganas, D. Anagnostou and D. Loukopoulos. Somatic hypermutation of immunoglobulin variable region genes: focus on follicular lymphoma and multiple myeloma. *Immunol Rev* **1998**, *162*, 281–292.

245 Kosmas, C., K. Stamatopoulos, N. Stavroyianni, K. Zoi, C. Belessi, N. Viniou, P. Kollia and X. Yataganas. Origin and diversification of the clonogenic cell in multiple myeloma: lessons from the immunoglobulin repertoire. *Leukemia* **2000**, *14*, 1718–1726.

246 Abraham, R.S., S.M. Geyer, T.L. Price-Troska, C. Allmer, R.A. Kyle, M.A. Gertz and R. Fonseca. Immunoglobulin light chain variable (V) region genes influence clinical presentation and outcome in light chain-associated amyloidosis (AL). *Blood* **2003**, *101*, 3801–3808.

247 Stevens, F.J. Four structural risk factors identify most fibril-forming kappa light chains. *Amyloid* **2000**, *7*, 200–211.

248 Stevens, F.J., E.A. Myatt, C.H. Chang, F.A. Westholm, M. Eulitz, D.T. Weiss, C. Murphy, A. Solomon, et al. A molecular model for self-assembly of amyloid fibrils: immunoglobulin light chains. *Biochemistry* **1995**, *34*, 10697–10702.

249 Faber, C., L. Shan, Z. Fan, L.W. Guddat, C. Furebring, M. Ohlin, C.A. Borrebaeck and A.B. Edmundson. Three-dimensional structure of a human Fab with high affinity for tetanus toxoid. *Immunotechnology* **1998**, *3*, 253–270.

250 Hemmingsen, J.M., K.M. Gernert, J.S. Richardson and D.S. Richardson. The tyrosine corner: a feature of most Greek key beta-barrel proteins. *Protein Sci* **1994**, *3*, 1927–1933.

251 Hamill, S.J., E. Cota, C. Chothia and J. Clarke. Conservation of folding and stability within a protein family: the tyrosine corner as an evolutionary cul-de-sac. *J Mol Biol* **2000**, *295*, 641–649.

252 Coleman, R.F., C.H. Lupton, Jr and J.F. McManus. Renal lesions in inbred mice with plasma-cell tumors. A morphological characterization and histochemical study of the "myeloma kidney" type of lesions which have recently been found to occur in C3H and BALB/C mice carrying transmissible plasma-cell tumors. *Arch Pathol* **1962**, *74*, 6–15.

253 Levi, D.F., R.C. Williams, Jr and F.D. Lindstrom. Immunofluorescent studies of the myeloma kidney with special reference to light chain disease. *Am J Med* **1968**, *44*, 922–933.

254 Zlotnick, A. and E. Rosenmann. Renal pathologic findings associated with monoclonal gammopathies. *Arch Intern Med* **1975**, *135*, 40–45.

255 Melato, M., G. Falconieri, E. Pascali and A. Pezzoli. Amyloid casts within renal tubules: a singular finding in myelomatosis. *Virchows Arch A Pathol Anat Histol* **1980**, *387*, 133–145.

256 Kyle, R.A. Monoclonal gammopathies and the kidney. *Annu Rev Med* **1989**, *40*, 53–60.

257 Sanders, P.W. Pathogenesis and treatment of myeloma kidney. *J Lab Clin Med* **1994**, *124*, 484–488.

258 Teng, J., W.J. Russell, X. Gu, J. Cardelli, M.L. Jones and G.A. Herrera. Different types of glomerulopathic light chains interact with mesangial cells using a common receptor but exhibit different intracellular trafficking patterns. *Lab Invest* **2004**, *84*, 440–451.

259 Huang, Z.Q. and P.W. Sanders. Biochemical interactions between Tamm-Horsfall glycoprotein and Ig light chains in the pathogenesis of cast nephropathy. *Lab Invest* **1995**, *73*, 810–817.

260 Huang, Z.Q. and P.W. Sanders. Localization of a single binding site for immunoglobulin light chains on human Tamm-Horsfall glycoprotein. *J Clin Invest* **1997**, *99*, 732–736.

261 Huang, Z.Q., K.A. Kirk, K.G. Connelly and P.W. Sanders. Bence Jones proteins bind to a common peptide segment of Tamm-Horsfall glycoprotein to promote heterotypic aggregation. *J Clin Invest* **1993**, *92*, 2975–2983.

262 Sanders, P.W. and B.B. Booker. Pathobiology of cast nephropathy from human Bence Jones proteins. *J Clin Invest* **1992**, *89*, 630–639.

263 Ying, W.Z. and P.W. Sanders. Mapping the binding domain of immunoglobulin light chains for Tamm-Horsfall protein. *Am J Pathol* **2001**, *158*, 1859–1866.

264 Wall, J., M. Schell, C. Murphy, R. Hrncic, F.J. Stevens and A. Solomon. Thermodynamic instability of human lambda 6 light chains: correlation with fibrillogenicity. *Biochemistry* **1999**, *38*, 14101–14108.

265 Girschick, H.J. and P.E. Lipsky. The kappa gene repertoire of human neonatal B cells. *Mol Immunol* **2002**, *38*, 1113–1127.

266 Zavaljevski, N., F.J. Stevens and J. Reifman. Support vector machines with selective kernel scaling for protein classification and identification of key amino acid positions. *Bioinformatics* **2002**, *18*, 689–696.

267 Pruzanski, W., A. Katz, S.C. Nyburg and M.H. Freedman. *In vitro* production of an amyloid-like substance from gamma 3 heavy chain disease protein. *Immunol Commun* **1974**, *3*, 469–476.

268 Solomon, A., D.T. Weiss, C.L. Murphy, R. Hrncic, J.S. Wall and M. Schell. Light-chain associated amyloid deposits comprised of a novel kappa constant domain. *Proc Natl Acad Sci USA* **1998**, *95*, 9547–9551.

269 Olsen, K.E., K. Sletten and P. Westermark. Fragments of the constant region of immunoglobulin light chains are constituents of AL-amyloid proteins. *Biochem Biophys Res Commun* **1998**, *251*, 642–647.

270 Buxbaum, J.N. Abnormal immunoglobulin synthesis in monoclonal immunoglobulin light chain and light and heavy chain deposition disease. *Amyloid* **2001**, *8*, 84–93.

271 Goto, Y. and K. Hamaguchi. Role of amino-terminal residues in the folding of the constant fragment of the immunoglobulin light chain. *Biochemistry* **1987**, *26*, 1879–1884.

272 Goto, Y. and K. Hamaguchi. Conformation and stability of the constant fragment of the immunoglobulin light chain containing an intramolecular mercury bridge. *Biochemistry* **1986**, *25*, 2821–2828.

273 Goto, Y. and K. Hamaguchi. Unfolding and refolding of the constant fragment of the immunoglobulin light chain containing an intramolecular mercury bridge. *J Biochem (Tokyo)* **1986**, *99*, 1501–1511.

274 Goto, Y. and K. Hamaguchi. Unfolding and refolding of the reduced constant fragment of the immunoglobulin light chain. Kinetic role of the intrachain disulfide bond. *J Mol Biol* **1982**, *156*, 911–926.

275 Goto, Y. and K. Hamaguchi. Unfolding and refolding of the constant fragment of the immunoglobulin light chain. *J Mol Biol* **1982**, *156*, 891–910.

276 Thies, M.J., F. Talamo, M. Mayer, S. Bell, M. Ruoppolo, G. Marino and J. Buchner. Folding and oxidation of the antibody domain C_H3. *J Mol Biol* **2002**, *319*, 1267–1277.

277 Thies, M.J., R. Kammermeier, K. Richter and J. Buchner. The alternatively folded state of the antibody C_H3 domain. *J Mol Biol* **2001**, *309*, 1077–1085.

278 Thies, M.J., J. Mayer, J.G. Augustine, C.A. Frederick, H. Lilie and J. Buchner. Folding and association of the antibody domain C_H3: prolyl isomerization precedes dimerization. *J Mol Biol* **1999**, *293*, 67–79.

279 Buchner, J., R. Rudolph and H. Lilie. Intradomain disulfide bonds impede formation of the alternatively folded state of antibody chains. *J Mol Biol* **2002**, *318*, 829–836.

280 Williams, A.F. The immunoglobulin superfamily takes shape. *Nature* **1984**, *308*, 12–13.

281 Williams, A.F. and A.N. Barclay. The immunoglobulin superfamily – domains for cell surface recognition. *Annu Rev Immunol* **1988**, *6*, 381–405.

282 Halaby, D.M. and J.P.E. Mornon. The immunoglobulin superfamily: an insight on its tissular, species, and functional diversity. *J Mol Evol* **1998**, *46*, 389–400.

283 Halaby, D.M., A. Poupon and J.-P. Mornon. The immunoglobulin fold family: sequence analysis and 3D structure comparisons. *Protein Eng.* **1999**, *12*, 563–571.

284 International Human Genome Sequencing Consortium. Initial sequencing and analysis of the human genome. *Nature* **2001**, *409*, 860–921.

285 Stevens, F.J., P.R. Pokkuluri and M. Schiffer. Protein conformation and disease: pathological consequences of analogous mutations in homologous proteins. *Biochemistry* **2000**, *39*, 15291–15296.

286 Stevens, F.J., E.A. Myatt, C.-H. Chang, F.A. Westholm, E. Eutlitz, D.T. Weiss, C. Murphy, A. Solomon, et al. A molecular model for the formation of amyloid: immunoglobulin light chains. *Biochemistry* **1995**, *34*, 10697–10702.

287 Rouger, H., F. LeGuern, R. Gouider, S. Tardieu, N. Birouk, M. Gugenheim, P. Bouch, Y. Agid, et al. High frequency of mutations in codon 98 of the peripheral myelin protein P0 protein in 20 French CMT1 patients. *Am J Hum Genet* **1996**, *58*, 638–641.

288 Stevens, F.J. Four structural risk factors identify most fibril-forming kappa light chains. *Amyloid* **2000**, *7*, 200–211.

289 Vilim, R.B., R.M. Cunningham, B. Lu, P. Kheradpour and F.J. Stevens. Fold-specific substitution matrices for protein classification. *Bioinformatics* **2004**, *20*, 847–853.

290 Kelm, S., R. Schauer and P.R. Crocker. The sialoadhesins – a family of sialic acid-dependent cellular recognition molecules within the immunoglobulin superfamily. *Glycoconj J* **1996**, *13*, 913–926.

291 May, A.P., R.C. Robinson, M. Vinson, P.R. Crocker and E.Y. Jones. Crystal structure of the N-terminal domain of sialoadhesin in complex with 3′ sialyllactose at 1.85 Å resolution. *Mol Cell* **1998**, *1*, 719–728.

292 Ikemizu, S., R.J. Gilbert, J.A. Fennelly, A.V. Collins, K. Harlos, E.Y. Jones, D.I. Stuart and S.J. Davis. Structure and dimerization of a soluble form of B7-1. *Immunity* **2000**, *12*, 51–60.

293 Stevens, F.J. and Y. Argon. Pathogenic light chains and the B-cell repertoire. *Immunol Today* **1999**, *20*, 451–457.

294 Stevens, F.J. and Y. Argon. Protein folding in the ER. *Cell Dev Biol* **1999**, *10*, 443–454.

295 Hendershot, L., D. Bole, G. Kohler and J.F. Kearney. Assembly and secretion of heavy chains that do not associate posttranslationally with immunoglobulin heavy chain-binding protein. *J Cell Biol* **1987**, *104*, 761–767.

296 Ma, J., J.F. Kearney and L.M. Hendershot. Association of transport-defective light chains with immunoglobulin heavy chain binding protein. *Mol Immunol* **1990**, *27*, 623–630.

297 Wei, J. and L.M. Hendershot. Protein folding and assembly in the endoplasmic reticulum. *EXS* **1996**, *77*, 41–45.

298 Skowronek, M.H., L.M. Hendershot and I.G. Haas. The variable domain of nonassembled Ig light chains determines both their half-life and binding

to the chaperone BiP. *Proc Natl Acad Sci USA* **1998**, *95*, 1574–1578.

299 Hellman, R., M. Vanhove, A. Lejeune, F. J. Stevens and L. M. Hendershot. The *in vivo* association of BiP with newly synthesized proteins is dependent on the rate and stability of folding and not simply on the presence of sequences that can bind to BiP. *J Cell Biol* **1999**, *144*, 21–30.

300 Dul, J. L., O. R. Burrone and Y. Argon. A conditional secretory mutant in an Ig L chain is caused by replacement of tyrosine/phenylalanine 87 with histidine. *J Immunol* **1992**, *149*, 1927–1933.

301 Melnick, J., J. L. Dul and Y. Argon. Sequential interaction of the chaperones BiP and GRP94 with immunoglobulin chains in the endoplasmic reticulum. *Nature* **1994**, *370*, 373–375.

302 Davis, D. P., R. Khurana, S. Meredith, F. J. Stevens and Y. Argon. Mapping the major interaction between binding protein and Ig light chains to sites within the variable domain. *J Immunol* **1999**, *163*, 3842–3850.

303 Dul, J. L., D. P. Davis, E. K. Williamson, F. J. Stevens and Y. Argon. Hsp70 and antifibrillogenic peptides promote degradation and inhibit intracellular aggregation of amyloidogenic light chains. *J Cell Biol* **2001**, *152*, 705–716.

304 Davis, D. P., G. Gallo, S. M. Vogen, J. L. Dul, K. L. Sciarretta, A. Kumar, R. Raffen, F. J. Stevens, et al. Both the environment and somatic mutations govern the aggregation pathway of pathogenic immunoglobulin light chain. *J Mol Biol* **2001**, *313*, 1043–1056.

305 Gidalevitz, T., C. Biswas, H. Ding, D. Schneidman-Duhovny, H. J. Wolfson, F. Stevens, S. Radford and Y. Argon. Identification of the N-terminal peptide binding site of glucose-regulated protein 94. *J Biol Chem* **2004**, *279*, 16543–16552.

306 Argon, Y. and B. B. Simen. GRP94, an ER chaperone with protein and peptide binding properties. *Semin Cell Dev Biol* **1999**, *10*, 495–505.

307 Gardner, A. M., S. Aviel and Y. Argon. Rapid degradation of an unassembled immunoglobulin light chain is mediated by a serine protease and occurs in a pre-Golgi compartment. *J Biol Chem* **1993**, *268*, 25940–25947.

308 Melnick, J., J. L. Dul and Y. Argon. Sequential interaction of the chaperones BiP and GRP94 with immunoglobulin chains in the endoplasmic reticulum. *Nature* **1994**, *370*, 373–375.

309 Eulitz, M., D. T. Weiss and A. Solomon. Immunoglobulin heavy-chain-associated amyloidosis. *Proc Natl Acad Sci USA* **1990**, *87*, 6542–6546.

310 Solomon, A., D. T. Weiss and C. Murphy. Primary amyloidosis associated with a novel heavy-chain fragment (AH amyloidosis). *Am J Hematol* **1994**, *45*, 171–176.

311 Tan, S. Y., I. E. Murdoch, T. J. Sullivan, J. E. Wright, O. Truong, J. J. Hsuan, P. N. Hawkins and M. B. Pepys. Primary localized orbital amyloidosis composed of the immunoglobulin gamma heavy chain C_H3 domain. *Clin Sci (Lond)* **1994**, *87*, 487–491.

312 Murdoch, I. E., T. J. Sullivan, I. Moseley, P. N. Hawkins, M. B. Pepys, S. Y. Tan, A. Garner and J. E. Wright. Primary localised amyloidosis of the orbit. *Br J Ophthalmol* **1996**, *80*, 1083–1086.

313 Mai, H. L., D. Sheikh-Hamad, G. A. Herrera, X. Gu and L. D. Truong. Immunoglobulin heavy chain can be amyloidogenic: morphologic characterization including immunoelectron microscopy. *Am J Surg Pathol* **2003**, *27*, 541–545.

314 Copeland, J. N., P. A. Kouides, M. Grieff and T. Nadasdy. Metachronous development of nonamyloidogenic lambda light chain deposition disease and IgG heavy chain amyloidosis in the same patient. *Am J Surg Pathol* **2003**, *27*, 1477–1482.

315 Yazaki, M., T. Fushimi, T. Tokuda, F. Kametani, K. Yamamoto, M. Matsuda, H. Shimojo, Y. Hoshii, et al. A patient with severe renal amyloidosis associated with an immunoglobulin gamma-heavy chain fragment. *Am J Kidney Dis* **2004**, *43*, e23–e28.

21
Transthyretin

Ana Margarida Damas and Maria João Saraiva

21.1
Introduction

Transthyretin (TTR) is a 55-kDa protein which is involved in the transport of thyroid hormones and also in the binding to the retinol-binding protein (RBP) during the transport of vitamin A in the plasma. The protein, initially known as pre-albumin since it migrates faster than serum albumin on electrophoresis of the whole plasma, is now named for its function: it transports (trans) the hormone thyroxine (thy) and retinol (retin).

TTR is synthesized mainly in the liver, but also in the choroid plexus of brain, in the retina and in the pancreas. The protein is encoded by a single gene on human chromosome 18 [1], and the gene sequence [2, 3] and three-dimensional structure [4, 5] are well known. The crystallographic structure revealed a homotetrameric protein with an extensive β-sheet conformation and a channel running along the molecule.

There are several TTR-related amyloid diseases. Corino de Andrade, in 1952, reported for the first time about a hereditary systemic amyloidosis found in a Portuguese patient, which he described as a "peculiar form of peripheral neuropathy with special involvement of the peripheral nerves" [6]. The biochemical basis of this pathology was revealed later by Costa et al. [7], who discovered that the main protein component of amyloid fibrils was TTR. In 1983, M. Saraiva described the first variant form of TTR, with methionine replacing a valine at position 30 of the polypeptide chain, that was found in amyloid fibrils extracted from Portuguese patients with familial amyloidotic polyneuropathy (FAP) [8]. Soon after Andrade's description, an American kindred was reported [9] and, since then, the number of kindreds that are now known to be afflicted with TTR amyloidosis has increased. In addition to Portugal, kindreds have been identified mainly in Sweden and Japan, although some other minor foci have been reported worldwide.

At present, more than 80 single-site amyloidogenic variants have been described in the literature as promoters of FAP [10]. Moreover, the wild-type protein was described as the causative agent for senile systemic amyloidosis (SSA), a disease characterized by late deposition of amyloid in the heart [11].

Amyloid Proteins. The Beta Sheet Conformation and Disease. J. D. Sipe

ISBN: 3-527-31072-X

Understanding the amyloid diseases at the molecular level guides the discovery of new drugs for diagnostic and therapy. It is important to understand what triggers the conversion of the soluble TTR protein from a physiological to a pathogenic conformation, prone to aggregate, and how changes in the protein molecular structure cause aggregation and how the pathogenic units associate to form the fibrillar aggregates (for recent reviews, see [12–14]).

It is possible, and probable, that the mutations affect the stability of TTR, inducing conformational changes that lead to aggregation. However, since in SSA only the wild-type protein is present in the amyloid fibrils, TTR seems to have intrinsic characteristics that drive it towards amyloidogenesis. Most probably, a number of other factors, in addition to the mutations, lead the protein towards an amyloidogenic conformation, with the mutations only increasing the intrinsic propensity of the protein for an amyloidogenic behavior. Other mechanisms that may alter secondary protein structure *in vivo* are the interaction with other proteins, protein degradation and post-translational modifications such as phosphorylation or glycation.

Interestingly, there are reports suggesting that TTR may have a role in the catabolism of amyloid β protein, which is the main component of the amyloid deposits present in the brain of Alzheimer's patients. Also, there are reports about the binding of TTR to Aβ and its role on preventing Aβ amyloid formation [15]; recently, the study of several TTR variants suggested that TTRs with decreased binding to Aβ may contribute to Aβ amyloid formation *in vivo* [16].

21.2 Gene Structure and Regulation

The gene encoding human TTR is a single copy gene mapped to chromosome 18 [3] and assigned to 18q11.2–q12.1 [17]. The gene spans about 7.0 kb with four exons of 95, 131, 136 and 253 bp, respectively [18]. The first exon encodes a signal peptide of 20 amino acid residues and the first 3 amino acids of the mature protein.

The 5′-flanking region contains a TATA box-like sequence at position –30, a GC-rich region of about 20 bp and a CAAT box at position –101. Two sequences homologous to glucocorticoid responsive elements have been identified at positions –224 and –212; additionally two $(CA)_n$ dinucleotide repeats were located further upstream. The polyadenylation site in the 3′-untranslated region is located at position 123.

The three introns span 934, 2090 and 3308 bp. Two independent open reading frames (ORFs) with the same transcription direction as TTR and putative consensus regulatory sequences for transcription have been described in the first and third introns.

ORF1 has two putative initiation codons, the alternative reading frames encoding polypeptides of 60 and 37 amino acids. A TATA sequence is located 106 nucleotides upstream from the first initiation codon and a CAAT sequence 55 nucleotides further upstream. The polyadenylation signal lies 93 nucleotides downstream of the termination codon of the first alternative frame. ORF2 has the potential to code for a polypeptide of 49 amino acids or, alternatively, 69 amino acids, depending upon the reading frame of the initiation codon. Sixty nucleotides upstream of the first putative initiation codon there is a possible TATA equivalent motif (TATATAT) and a CAAT sequence 44 nucleotides upstream from that motif. Two alternative polyadenylation signals are located 45 and 243 nucleotides downstream the stop codon of the second frame. However, no evidence was found for novel transcripts containing productively spliced products of either ORF or for shorter transcripts using the promoter and polyadenylation signals associated with them. ORF1 transcription products were identified in liver, pancreas and brain, and correspond to TTR transcripts in which intron 1 had not been removed; the transcripts containing ORF2 may represent TTR hnRNA. Neither ORF is productively expressed as part of a larger transcript or as an independent polypeptide [19].

The rat and mouse genes show a gene organization similar to that in humans with about 80% sequence homology. The 5′-flanking region contains elements for hepatocyte-specific expression, including binding sites for HNF-1, C/E BP, HNF-3 and HNF-4 [20]. The binding sites for HNF-3 and HNF-4 are well conserved in the human TTR gene. Other liver-specific nuclear protein potential binding sites were identified in the human gene at positions –216 to –221 and –199 to –204. The binding motif (TGG7AA7CC7T) is common to factors Tf-LF1, Tf-LF2 and LF-A1. A tissue-specific enhancer was identified in the –3.5 to –3.6 region of the human TTR gene, which contained a binding site for HNF-4. In addition, two sequences homologous to AFP1, a hepatoma nuclear protein were described in the –6 kb region. These elements might be involved in the regulation of gene expression in liver by *cis-* and *trans-*acting factors.

21.3 Function

TTR is involved in the transport of thyroid hormones and, through its interaction with RBP, vitamin A. Plasma TTR is responsible for the transport of all of the RBP, but only approximately 20% of the thyroid hormones [21]. In fact, the thyroid hormones are also bound by thyroxine-binding globulin (approximately 70%), serum albumin (approximately 10%), and, in a small number of species, to lipoproteins and vitamin D-binding protein [22]. TTR is therefore only one of the proteins responsible for thyroid hormone transport in plasma; this seems likely to explain why TTR knock-out mice show no major abnormalities and seem to compensate with alternative mechanisms [23].

The pool of TTR which is synthesized in the choroid plexus is secreted into the cerebrospinal fluid, where it represents the main thyroid hormone-binding protein [24]. Studies in TTR-null mice clearly show that the protein is not indispensable for thyroid hormone's entry into the brain and other tissues [25]. The importance of TTR in the transport of vitamin A, which is vital for the normal development of tissues in the body, was also assessed in mice lacking TTR [26]. The animals showed no abnormalities in their retinal anatomy and function, confirming the existence of multiple mechanisms for the transport of retinol.

Recently, it was reported that TTR seems to have a chymotrypsin-like activity, being able to cleave the C-terminus of apolipoprotein AI (ApoAI) [27]. This new role of TTR as a plasma protease, which is not yet completely characterized, may have an impact not only in amyloid diseases but also in other disorders, namely those related to lipid metabolism.

21.4
Three-dimensional Structure of TTR

TTR is a tetrameric protein of four identical subunits, the monomers, each 127 amino acids in length [28]. Each monomer has a β-sandwich fold composed of two four-stranded β-sheets: DAGH and CBEF. The structure of the TTR tetramer is shown in Fig. 21.1 (together with the nomenclature for the β-strands as well as the interaction of the protein with thyroxine and RBP). Two monomers associate edge to edge through hydrogen bonds forming a dimer composed of two β-sheets, with strands DAGHH′G′A′D′ and CBEFF′E′B′C′. Hydrophobic interactions and hydrogen bonding between the AB loops from one dimer and strand H from a different dimer lead to the soluble and functional tetrameric protein. The CBEF sheets are oriented towards the external surface while the

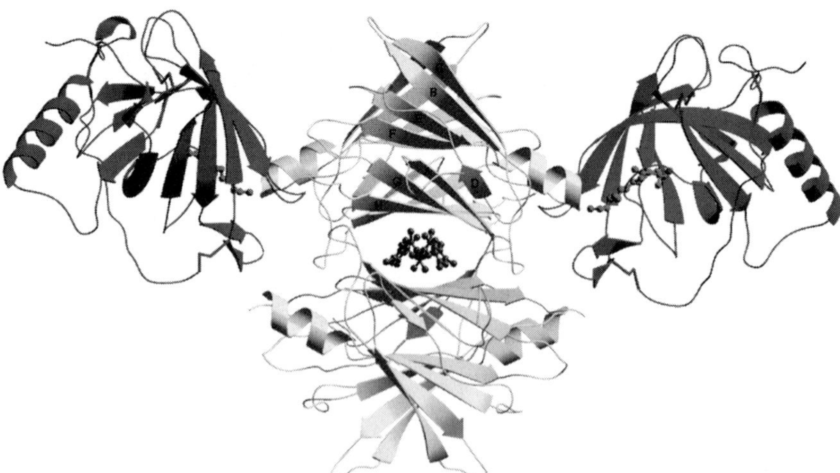

Fig. 21.1 The interaction between TTR (yellow) and two RBP molecules (blue). The T4 molecules bound in the TTR channel and the retinol bound to RBP are shown in a ball and stick representation. The labeling system for β-strands is depicted for one of the monomers.

DAGH sheets line a central channel where two identical binding sites for the thyroid hormones are located [29, 30].

Although the association between TTR and the retinol binding protein, for the transport of vitamin A, was known for a long time, the exact location of the binding sites was not known until 1995, when the structure of the complex between human TTR and chicken RBP was solved by X-ray crystallography [31]. The structure of this complex showed that the C-terminal of hRBP, which is 8 amino acids longer than chicken RBP and is characteristic of all mammalian RBPs, is involved in the protein–protein recognition interface [32]. The X-ray crystal structure revealed complexes of RBP and TTR in a stoichiometry of 2:1 (Fig. 21.1).

Over recent years, the three-dimensional structures of several variants have been studied in order to shed some light on the structure–function relationship as well as the structural details leading to amyloid deposition. The state of the art regarding the three-dimensional structure of wild-type TTR and some of its mutants has been reviewed [14, 33]. All the reported crystallographic structures show the same quaternary structure, and no large modifications between the wild-type and the variant proteins were observed. Moreover, a comparative analysis of 23 structures of amyloidogenic TTRs indicated that the reported structural differences were non-significant [34].

Over the years, a number of studies on the effect of point mutations over protein structure and function support the conclusion that, unless the mutations occur in residues that are crucial for the protein function, the effects of the substitutions are usually benign, except in a few cases that usually lead to conformational diseases. In the case of TTR variants, the largest structural change occurs in Leu55Pro TTR, where the D strand adopts a coil conformation and becomes part of a long loop that connects strands C and E (Fig. 21.2). This long loop is located at the edge of the β-sandwich and consequently leads to a region of the molecule that becomes more exposed to the solvent [35].

It is also important to point out that most of the crystallographic structures have been determined at high resolution but, so far, only the X-ray crystallographic structure of Thr119Met TTR has been determined at atomic resolution

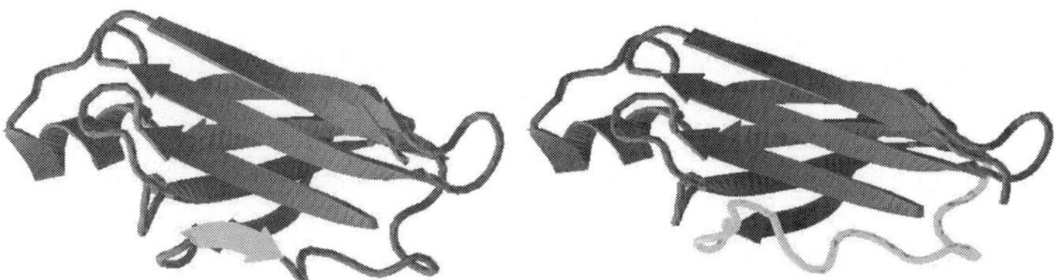

Fig. 21.2 The differences in the TTR monomer for the wild-type protein (A) and Leu55Pro TTR (B). Strand D represented in red in the wild-type TTR becomes part of a long loop in the highly amyloidogenic Leu55Pro TTR.

[36]. This structure reveals in detail not only the interactions between the observed amino acids, but also alternative conformations for a large number of the protein side-chains. Considering that a discrete barrier separates the normally folded and functional protein and the abnormal state that triggers a cascade of events leading to amyloid formation, it is conceivable that the very fine details that distinguish the two forms of the protein are not detectable in low- to medium-resolution structures and may become evident only when the comparison is performed using data with atomic resolution.

In addition to the long list of human TTR structures, most of them being single-site variant proteins characterized by their amyloidogenic or non-amyloidogenic behavior, the structures of rat [37], chicken [38] and sea bream [39, 40] TTRs have also been determined. All of the non-human proteins reveal a high degree of conservation of the tertiary and quaternary structures, although differences in discrete regions of the molecule explain distinctive binding to RBP and the thyroid hormones. As well as the β-sheets that compose a large extension of the TTR molecule, each monomer exhibits a small helical region [5, 37] which, in chicken TTR, is substituted by a loop. The high affinity for L-thyroxine and negligible affinity for RBP in piscine TTR were both explained by the X-ray crystal structure of the sea bream protein.

21.5 TTR Amyloidosis (ATTR)

Transthyretin amyloidosis results from the self-association of the protein into amyloid fibrils which deposit in tissues at the extracellular level. A large number of mutations related to FAP have been described in the literature. They promote tetramer dissociation and affect monomer stability, favoring an amyloidogenic intermediate structure [41]. After the first discovery of a variant, Val30Met TTR, a large number of inherited variants were detected in the following years. While some of the variants affect thyroxine and/or RBP binding, the majority are associated with FAP and only a small number have non-amyloidogenic characteristics. In fact, there are two cases where heterozygous patient carriers of the amyloidogenic Val30Met mutation and a non-amyloidogenic mutation, Thr119Met or Arg104His, show a more benign form of the disease [42, 43]. It seems that these two mutations introduce alterations that compensate for the effects produced by the amyloidogenic mutations by increasing the resistance of the tetramer to dissociation into monomers.

Although a "hotspot region", corresponding to amino acids 42–55 in the TTR monomer, was initially considered to be the region where most of the amyloidogenic mutations occurred [44], it is now more evident that the amyloidogenic mutations are distributed over the entire molecule.

The role that TTR mutations play in TTR amyloidogenesis is not completely understood. So far, the solved amyloidogenic TTR structures point to a clear destabilization of the tetrameric form of the protein, which dissociates more easily and originates an amyloidogenic intermediate species. This intermediate is probably very close to the Leu55Pro TTR monomer, which in the crystalline form consists of two β-sheets, composed of strands AGH and CBEF, and a long loop connecting strands C and E [35]. Interestingly, the region in the TTR molecule previously identified with part of the chain where a large number of amyloidogenic mutations occur, "the hotspot region", corresponds to strands C and D of the two β-sheets, which are the edge strands in the dimer.

We believe that the modifications introduced by the amyloidogenic mutations contribute to a destabilization of the protein quaternary structure leading to its dissociation and an increase in the number of available units, with an amyloidogenic structure, which is responsible for amyloid formation.

Several molecular models concerning TTR amyloid fibrils have been proposed. While some researchers consider that the fibrils could be built up of tetramers linked by disulfide bridges [45], others consider that dimers [46] or monomers [47] are the assembly units present in amyloid fibrils. This question regarding the composition of the units in amyloid fibrils was addressed using constructed mutants with varied quaternary structure stability [48]. The conclusion was that functional tetramers in which the monomers were linked by disulfide bridges did not form amyloid *in vitro* and the same happened for another variant in which the two dimers resulted from disulfide linked monomers. Moreover, the same proteins in reducing conditions immediately formed the fibrillar material. These observations clearly indicated that monomers rather than dimers or tetramers are the building blocks in amyloid fibrils.

The association between the amyloidogenic monomers may occur due to the exposure of a newly formed hydrophobic surface or/and polar amino acids allowing for new salt bridges, both acting as sticky patches that lead to protein aggregation. In fact, it was observed that hydrophobic interactions contribute significantly to the TTR tetramer stability and, on the contrary, the electrostatic repulsions lead to its destabilization [49]. Protein aggregation may also arise due to domain swapping if two monomers exchange part of their structure and form a tight association between the two molecules. A detailed molecular description of amyloid fibrils is still not available.

The dissociation of the TTR tetramer into non-native monomers using conditions very close to physiological has been observed [50, 51]. In fact, *in vitro* studies using protein concentrations in the serum ranges and at nearly physiological pH show that TTR dissociates in an irreversible way into non-native monomers and small aggregate species, which may be the driving units for amyloid. Later, transmission electron microscopy and mass per length measurements confirmed that fibrillogenesis results from an assembly of monomers [52], as previously proposed by Inouye et al. [47] when they interpreted the diffraction pattern of vitreous TTR amyloid samples.

The structure of the non-native monomer that assembles into amyloid fibrils remains to be described. It is probably close to the structure of the Leu55Pro TTR monomer, as revealed by X-ray crystallography, since it was reported that crystals were grown from a solution of aggregates with amyloidogenic properties [53].

Deuterium–proton exchange was used to compare the monomeric and tetrameric wild-type TTR structures at neutral pH [54, 55]. The authors concluded that the monomer retained its native-like conformation upon dissociation. However, the TTR monomer at pH 4.5, the precursor of amyloid fibrils *in vitro*, revealed a stable DAGH β-sheet and a conformational instability of the CBEF β-sheet [56].

Molecular dynamics simulations were also performed on wild-type, Val30Met and Leu55Pro TTR [57]. In all cases, comparison of the protein monomeric forms with the structure of the monomer as it is within the tetrameric structure reveals large displacements in AB, BC, FG and GH loops. In the case of wild-type and Val30Met TTR, the E and F strands, the end of B and the beginning of C strands also undergo conformational alterations. In the Leu55Pro monomer, substantial changes were observed in the loop region between the C and E strands. As pointed out from the X-ray crystallographic results of the Leu55Pro TTR variant, the mutation leads to the breakage of the links between strands D and A, resulting in a surface loop that connects strands C and E. The molecular dynamics study reveals that this loop seems to undergo a more substantial change in the monomeric form of the variant, resulting in large conformational changes of strands A, G and H.

A different technique, X-ray absorption near edge spectroscopy (XANES), was used for comparison of the TTR protein in amyloid fibrils and in the corresponding protein solutions. This led to the conclusion that Cys10 and Met13, located just preceding and within strand A, become more exposed to the solvent when the protein is in a fibrillar form [58]. This result is consistent with the molecular dynamics simulations for the Leu55Pro TTR monomer that reveal significant alterations in the CE loop and the A strand. Interestingly, it was reported that strand D of wild-type TTR provides high protection for the Val14 and Val16 residues of strand A [56]. Therefore, when strand D becomes a loop, it is possible that the two valine residues become exposed and lead to a hydrophobic patch that acts as an aggregation agent. The possibility of amyloid formation due to domain swapping is also a probable mechanism considering the high liability of the CE loop and the newly exposed hydrophobic patch.

21.6
TTR Amyloid Inhibitors

Since TTR is produced mainly by the liver, liver transplantation may be used to correct the metabolic deficiency by producing wild-type TTR. The first liver transplantation for an FAP patient was performed in 1990 in Sweden [59] and this procedure is now carried out worldwide. It is associated with considerable risks for the patients and a less invasive therapy is necessary.

It is almost consensual that a destabilization of the TTR tetramer leads to its dissociation, thereby forming the amyloidogenic intermediate species. Therefore, a possible intervention is the prevention of protein dissociation. The potential use of small molecules that bind in the TTR channel and stabilize the native fold of the protein is under study [60–64]. Thus far, it has been demonstrated that molecules such as thyroxine, flufenamic acid, diclofenac and diflunisal are capable of preventing the conformational changes that lead to amyloid fibril formation. However, the use of these compounds in animal models has not been proved yet and might be compromised due to the presence in blood of proteins, other than TTR, capable of binding these ligands. Recent studies addressed the binding of these proposed compounds in relation to their selectivity towards TTR binding and inhibition potency for amyloid formation; there was a much higher efficiency of novel diflunisal derivatives to displace T4 from TTR in plasma. This displacement was preferential over albumin and thyroxine binding globulin, and stabilization of the tetrameric structure of Val30Met TTR in *ex vivo* assays was clearly demonstrated [65].

A different therapeutic strategy may involve the discovery of drugs capable of preventing protein association. In fact, blockage of the interaction sites will impair the association of the amyloidogenic units into fibrils.

A strategy involving the disaggregation of the mature amyloid fibrils has been proposed, based on the finding of agents that promote the disruption of TTR amyloid fibrils; such is the case of iododoxyrubicin (IDOX) [66, 67] and tetracyclines [68, 69]. These agents also act on other amyloids.

Until recently, it was generally considered that the cytotoxic effects of amyloid TTR deposits were due to the mature fibrils. However, nerves from FAP patients were examined for TTR deposition and non-fibrillar TTR deposition was found in asymptomatic carriers of the Val30Met mutation. Moreover, expression of inflammatory and oxidative stress products was observed in the tissues with non-fibrillar TTR deposition, suggesting that oligomeric TTR deposition may be cytotoxic; in cell culture, insoluble oligomeric TTR species induced cytotoxicity, whereas mature TTR fibrils did not [70].

Recently it was reported that pre-fibrillar aggregates of the *Escherichia coli* protein HypF, a protein not associated with a clinical disease, produce similar cytotoxic effects in murine fibroblasts as those due to similar aggregates containing amyloidogenic proteins [71]. The authors suggest that the toxicity of the misfolded aggregates results from the alteration of specific cellular parameters, which depend upon the type of molecular aggregate rather than on the specific protein.

A deeper understanding on the structural organization that is responsible for cell toxicity is necessary for the design of drugs that (1) act over the formation of the early aggregates or (2) bind to these aggregates, inhibiting their toxicity. Studies should address cytotoxicity of disaggregated products resulting from the action of amyloid fibril disrupters; in this regard, recent studies showed that in the case of TTR, disaggregated fibrils obtained after tetracycline treatment did not induce cell toxicity [69].

While a complete characterization of the amyloid pathway at the molecular level is not available, the interpretation of results concerning possible therapeutic strategies remains difficult, since the *in vivo* processes are multifactorial.

21.7 Ligand Binding

TTR has a channel that runs through the center of the molecule where two identical thyroxine-binding sites are located. The first detailed description of the binding of a thyroid hormone metabolite to TTR was reported for 3,3′-diiodo-L-thyronine [72]. The structure of TTR complexed with thyroxine was reported later by Wojtczac et al. [30]. Both hormones bind to TTR in the same location, which has been fully described [29]. The binding site has hydrophobic and hydrophilic regions that seem to be ideal to accommodate the T4 and T3 molecules.

Although the two binding sites are identical, negative cooperativity between them was reported. This behavior is explained by a series of structural events that, upon ligand binding in the first site, lead to the collapse of the newly occupied site and subsequent opening of the second binding site [73].

Crocodile TTR binds triiodothyronine (T3) with higher affinity than thyroxine (T4) and, using a chimeric construct, it was demonstrated that it is the N-terminal region of TTR that modulates T4 and T3 binding [74]. A systematic change seems to have occurred during the evolution of the N-terminal region of TTR since the predominant region where mutations occurred was within the first 10–13 amino acids from the N-terminus, leading to changes in length and hydropathy of the N-terminal regions. This region in reptilian and avian TTRs is more hydrophobic than in the case of mammalians.

Retinol transport from the liver to the peripheral tissues occurs when the molecule is bound to RBP, which in turn binds TTR for prevention of glomerular filtration of RBP in the kidneys [75]. Each TTR tetramer binds two RBP proteins and the T4-binding sites are not affected by the RBP binding. Several surface regions of the TTR molecule, including the C-terminus, are involved in the binding to RBP [31].

Approximately 1–2% of plasma TTR circulates bound to high-density lipoproteins (HDLs) and the association of TTR with the HDL vesicle occurs via ApoAI [76]. TTR uptake experiments using multiple cell lines and experimental systems suggest a possible common pathway between lipoproteins and TTR uptake. Thus, megalin, a member of the low-density lipoprotein (LDL) family is a receptor for TTR uptake, as demonstrated *in vitro* by cell lines expressing megalin and by surface plasmon resonance (SPR) analysis. Moreover, *in vivo*, megalin-null mice excrete TTR and lack lysosomal accumulation of TTR in renal tubules, demonstrating the role of megalin in TTR uptake in kidney [77]. Furthermore, patients with renal tubular impairment excrete TTR. Members of the LDL family have in common the ability to bind the receptor-associated protein (RAP), a protein that antagonizes binding of ligands to this family of lipoprotein receptors. In liver and other tissues, uptake of TTR was found to be inhibited by RAP and affected by lipoproteins, suggesting that TTR bind to an yet unidentified cell surface receptor that is RAP sensitive [78].

TTR aggregates and TTR uncomplexed with RBP have been described as binding RAGE, the receptor for advanced glycation end products [79]. Moreover, the interaction of RAGE with TTR aggregates may contribute to the cellular stress and toxicity observed in FAP patients [80]. The relevance of this interaction will be elucidated as soon as a detailed characterization of the receptor and its interaction with TTR is available.

Exposure to environmental pollutants is a serious public health problem. Experiments in rodents [81, 82] showed a decrease of thyroxine in the plasma levels when the animals were exposed to polychlorinated biphenyls (PCBs). In fact, X-ray crystallography studies did show that PCBs bind in the T4 binding site, thereby displacing the thyroxine hormone [82].

Binding studies of chlorinated dibenzo-*p*-dioxins and dibenzofurans to TTR were performed using molecular graphics simulations, which also revealed that both compounds may bind in the TTR channel, thereby inhibiting the binding of the hormones [83]. These metabolites can lead to changes in the plasma transport and metabolism of thyroid hormones in the animals. Furthermore, it is possible that, as described for other proteins, a number of non-specific and low-affinity binding sites for small compounds exist on the protein surface, and these may also have a role in the process of protein aggregation [84].

21.8 Post-translational Modifications

Post-translational modifications of purified TTR have been studied by mass spectrometry. Human TTR is modified predominantly via the thiol moiety due to the presence of the Cys10 residue; four major modifications have been observed: *S*-sulfonated and *S*-thiolated, conjugated to cysteine, glutathione and cysteinylglycine [85, 86]. Rat TTR shows exactly the same modifications apart from the cysteinylglycine moiety, which is not observed [87].

It is known that over 70% of homocysteine in plasma is disulfide-bonded to protein cysteine residues and the functional consequences of protein homocysteinylation are only now emerging. In fact, hyperhomocysteinemia is a risk factor for cardiovascular diseases as well as for Alzheimer's disease; also, transthyretin-Cys10-*S*-*S*-homocysteine was identified in the plasma of patients with end-stage renal disease and homocystinuria [88].

Plasma samples from individuals with different ages were also analyzed in order to determine the impact of age upon thiol conjugation and whether there is a possible relationship between these modifications and amyloid formation. It was reported that thiol conjugation in TTR is dependent upon age, and also that amyloidogenic variant proteins have higher susceptibility to thiol conjugation than the wild-type protein [89].

The modifications of the SH functionality of Cys10, i.e. TTR-Cys, TTR-GSH, TTR-CysGly and *S*-sulfonated TTR, were studied in order to understand their possible influence over the stability and/or amyloidogenic potential of the protein [90]. The authors concluded that *S*-sulfonated TTR was less amyloidogenic and formed fibrils more slowly than wild-type TTR under acidic conditions and, on the contrary, the other conjugated forms had increased amyloidogenicity.

It is possible that the amyloidogenic variants are prone to translational modifications which in turn may increase the protein amyloidogenic potential.

21.9 Evolution

TTR synthesis starts very early in development; TTR mRNA is detected in hepatocytes and the epithelial cells of choroid plexus in the eighth week of gestation, and in pancreatic islets from mid-pregnancy [91]. TTR is first expressed in the tela choroidea, the forerunner of choroid plexus, followed by expression in the liver [92]. Expression of TTR in liver and choroid plexus is observed in eutherians, birds, diprotodont marsupials and some polyprotodont marsupials. In reptilians, monotremes and some polyprotodont marsupials, TTR is only expressed in the choroid plexus [22]. TTR synthesis seems to have first occurred in the brain of reptiles [93], arising subsequently and independently in the liver of birds, eutherians and diprotodont marsupials.

Comparison of the amino acid sequence of TTR from mammals and marsupials with avian, amphibian and piscine TTR revealed the presence of three additional residues at the N-terminus in the latter [22]. The 5′ region of the gene seems to have undergone distinct changes during evolution in eutherians and birds. Changes in the splice site of intron 1 led to the production of a TTR molecule with a shorter and more hydrophilic N-terminal sequence [94]. This change might relate to distinct affinities for thyroid hormones among vertebrates [95]; the longer, more hydrophobic N-termini correlate with preferential binding to T3, whereas the shorter, more hydrophilic N-termini, to T4. As the rest of the molecule had not changed significantly during vertebrate evolution, the gene coding for TTR must have evolved prior to the divergence of the vertebrates from the non-vertebrates [96].

Interestingly, while TTR has been identified only in vertebrates, a number of proteins related to TTR, the TTR-related family, are found in a wide range of species that includes bacteria, plants and animals. Sequence alignment of these proteins with human TTR indicate that most probably the characteristic TTR tertiary and quaternary structures are conserved [97].

References

1 Wallace, M. R., Naylor, S. L., Kluve-Beckerman, B., Long, G. L., McDonald, L., Shows, T. B. and Benson, M. D. Localization of the human prealbumin gene to chromosome 18. *Biochem Biophys Res Commun* **1985**, *129*, 753–758.

2 Mita, S., Maeda, S., Shimada, K. and Araki, S. Cloning and sequence analysis of cDNA for human prealbumin. *Biochem Biophys Res Commun* **1984**, *124*, 558–564.

3 Whitehead, A. S., Skinner, M., Bruns, G. A., Costello, W., Edge, M. D., Cohen, A. S. and Sipe, J. D. Cloning of human prealbumin complementary DNA. Localization of the gene to chromosome 18 and detection of a variant prealbumin allele in a family with familial amyloid polyneuropathy. *Mol Biol Med* **1984**, *2*, 411–423.

4 Blake, C. C., Swan, I. D., Rerat, C., Berthou, J., Laurent, A. and Rerat, B. An X-ray study of the subunit structure of prealbumin. *J Mol Biol* **1971**, *61*, 217–224.

5 Blake, C. C., Geisow, M. J., Oatley, S. J., Rerat, B. and Rerat, C. Structure of prealbumin: secondary, tertiary and quaternary interactions determined by Fourier refinement at 1.8 Å. *J Mol Biol* **1978**, *121*, 339–356.

6 Andrade, C. A peculiar form of peripheral neuropathy: a typical generalized amyloidosis with special involvement of the peripheral nerves. *Brain* **1952**, *75*, 408–427.

7 Costa, P. P., Figueira, A. S. and Bravo, F. R. Amyloid fibril protein related to prealbumin in familial amyloidotic polyneuropathy. *Proc Natl Acad Sci USA* **1978**, *75*, 4499–4503.

8 Saraiva, M. J., Costa, P. P. and Goodman, D. S. Studies on plasma transthyretin (prealbumin) in familial amyloidotic polyneuropathy, Portuguese type. *J Lab Clin Med* **1983**, *102*, 590–603.

9 Falls, H. F., Jackson, J., Carey, J. H., Rukavina, J. G. and Block, W. D. Ocular manifestations of hereditary primary systemic amyloidosis. *Am Arch Ophthalmol* **1955**, *54*, 660–664.

10 Saraiva, M. J. Transthyretin mutations in familial amyloidotic polyneuropathy. *Hot Thyroidol* **2004**, *1*, www.hotthyroidology.-com.

11 Westermark, P., Sletten, K., Johansson, B. and Cornwell III, G. G. Fibril in senile systemic amyloidosis is derived from normal transthyretin. *Proc Natl Acad Sci USA* **1990**, *87*, 2843–2845.

12 Brito, R. M. M., Damas, A. M. and Saraiva, M. J. Amyloid formation by transthyretin: from protein stability to protein aggregation. *Curr Med Chem- Immunol Endocr Metab Ag* **2003**, *3*, 349–360.

13 Hamilton, J. A. and Benson, M. D. Transthyretin: a review from a structural perspective. *Cell Mol Life Sci* **2001**, *58*, 1491–1521.

14 Damas, A. M. and Saraiva, M. J. Review: TTR amyloidosis – structural features leading to protein aggregation and their implications on therapeutic strategies. *J Struct Biol* **2000**, *130*, 290–299.

15 Schwarzman, A. L., Gregori, L., Vitek, M. P., Lyubski, S., Strittmatter, W. J., Enghilde, J. J., Bhasin, R., Silverman, J., et al. Transthyretin sequesters amyloid beta protein and prevents amyloid formation. *Proc Natl Acad Sci USA* **1994**, *91*, 8368–8372.

16 Schwarzman, A. L., Tsiper, M., Wente, H., Wang, A., Vitek, M. P., Vasiliev, V. and Goldgaber, D. Amyloidogenic and anti-amyloidogenic properties of recombinant transthyretin variants. *Amyloid: J Protein Folding* **2004**, *11*, 1–9.

17 Sparkes, R. S., Sasaki, H., Mohandas, T., Yoshioka, K., Klisak, I., Sakaki, Y., Heinzmann, C. and Simon, M. I. Assignment of the prealbumin (PALB) gene (familial amyloidotic polyneuropathy) to human chromosome region 18q11.2–q12.1. *Hum Genet* **1987**, *75*, 151–154.

18 Sasaki, H., Yoshioka, N., Takagi, Y. and Sakaki, Y. Structure of the chromosomal gene for human serum prealbumin. *Gene* **1985**, *37*, 191–197.

19 Soares, M., Centola, M., Chae, J., Saraiva, M. J. and Kastner, D. L. Human transthyretin intronic open reading frames are not independently expressed *"in vivo"*

or part of functional transcripts. *Biochem Biophys Acta* **2003**, *1626*, 65–74.

20 Costa, R. H., Van Dyke, T. A., Yan, C., Kuo, F. and Darnell, J. E. Jr. Similarities in transthyretin gene expression and differences in transcription factors: liver and yolk sac compared to choroid plexus. *Proc Natl Acad Sci USA* **1990**, *87*, 6589–6593.

21 Benvenga, S., Lapa, D. and Trimarchi, F. Thyroxine binding to members and nonmembers of the serine protease inhibitor family. *J Endocrinol Invest* **2002**, *25*, 32–38.

22 Schreiber, G. and Richardson, S. J. The evolution of gene expression. Structure and function of transthyretin. *Comp Biochem Physiol* **1997**, *116B*, 137–160.

23 Palha, J. A., Episkopou, V., Maeda, S., Shimada, K., Gottesman, M. E. and Saraiva, M. J. Thyroid hormone metabolism in a transthyretin-null mouse strain. *J Biol Chem* **1994**, *269*, 33135–33139.

24 Weisner, B. and Kauerz, U. The influence of the choroid-plexus on the concentration of prealbumin in CSF. *J Neurol Sci* **1983**, *61*, 27–35.

25 Palha, J. A. Transthyretin as a thyroid hormone carrier: function revisited. *Clin Chem Lab Med* **2002**, *40*, 1292–1300.

26 Bui, B. V., Armitage, J. A., Fletcher, E. L., Richardson, S. J., Schreiber, G. and Vingrys, A. J. Retinal anatomy and function of the transthyretin null mouse. *Exp Eye Res* **2001**, *73*, 651–659.

27 Liz, M. A., Faro, C. J., Saraiva, M. J. and Sousa, M. M. Transthyretin, a new cryptic protease. *J Biol Chem* **2004**, *279*, 21431–21438.

28 Kanda, Y., Goodman, W. S., Canfield, R. E. and Morgan, F. J. The amino acid sequence of human plasma prealbumin. *J Biol Chem* **1974**, *249*, 6796–6805.

29 De La Paz, P., Burridge, J. M., Oatley, S. J. and Blake, C. C. F. Multiple modes of binding of thyroid hormones and other iodothyronines to human plasma transthyretin. In *The Design of Drugs to Macromolecular Targets*, Beddell, C. R. (ed.). Wiley, New York, **1992**, pp. 120–157.

30 Wojtczak, A., Cody, V., Luft, J.R. and Pangborn W. Structures of human transthyretin complexed with thyroxine at 2.0 Å resolution and 3′,5′-dinitro-*N*-acetyl-L-thyronine at 2.2 Å resolution. *Acta Crystallogr D* **1996**, *52*, 758–765.

31 Monaco, H. L., Rizzi, M. and Coda, A. Structure of a complex of two plasma proteins: transthyretin and retinol-binding protein. *Science* **1995**, *268*, 11039–10410.

32 Naylor, H. M. and Newcomer, M. E. The structure of human retinol-binding protein (RBP) with its carrier protein transthyretin reveals an interaction with the carboxy terminus of RBP. *Biochemistry* **1999**, *38*, 2647–2653.

33 Eneqvist, T. and Sauer-Eriksson, A. E. Structural distribution of mutations associated with familial amyloidotic polyneuropathy in human transthyretin. *Amyloid: J Protein Folding* **2001**, *8*, 149–168.

34 Hornberg, A., Eneqvist, T., Olofsson, A., Lundgren, E. and Sauer-Eriksson, A. E. A comparative analysis of 23 structures of the amyloidogenic protein transthyretin. *J Mol Biol* **2000**, *302*, 649–669.

35 Sebastião, M. P., Saraiva, M. J. and Damas, A. M. The crystal structure of amyloidogenic Leu55 → Pro transthyretin variant reveals a possible pathway for transthyretin polymerization into amyloid fibrils. *J Biol Chem* **1998**, *273*, 24715–24722.

36 Sebastião, M. P., Lamzin, V., Saraiva, M. J. and Damas, A. M. Transthyretin stability as a key factor in amyloidogenesis: X-ray analysis at atomic resolution. *J Mol Biol* **2001**, *306*, 733–744.

37 Wojtczak, A. Crystal structure of rat transthyretin at 2.5 Å resolution: first report on a unique tetrameric structure. *Acta Biochim Pol* **1997**, *44*, 505–517.

38 Sunde, M., Richardson, S. J., Chang, L., Pettersson, T. M., Schreiber, G. and Blake, C. C. F. The crystal structure of transthyretin from chicken. *Eur J Biochem* **1996**, *236*, 491–499.

39 Folli, C., Pasquato, N., Ramazzina, I., Battistutta, R., Zanotti, G. and Berni, R. Distinctive binding and structural properties of piscine transthyretin. *FEBS Lett* **2003**, *555*, 279–284.

40 Eneqvist, T., Lundberg, E., Karlsson, A., Huang, S., Santos, C. R. A., Power, D. M. and Sauer-Eriksson, E. High resolution crystal structures of piscine transthyretin reveal different binding modes for triiodothyronine and thyroxine. *J Biol Chem* **2004**, *279*, 26411–26416.

41 Hammarstrom, P., Wiseman, R. L, Powers, E. T. and Kelly, J. W. Prevention of transthyretin amyloid disease by changing protein misfolding energetics. *Science* **2003**, *299*, 713–716.

42 Alves, I., Hays, M. T. and Saraiva, M. J. Comparative stability and clearance of [Met30] transthyretin and [Met119] transthyretin. *Eur J Biochem* **1997**, *249*, 662–668.

43 Almeida, M. R., Alves, I. L., Terazaki, H., Ando, Y. and Saraiva, M. J. Comparative studies of two transthyretin variants with protective effects on familial amyloidotic polyneuropathy: TTR R104H and TTR T119M. *Biochem Biophys Res Commun* **2000**, *270*, 1024–1028.

44 Serpell, L. C., Goldsteins, G., Dacklin, I., Lundgren, E. and Blake, C. F. The "edge strand" hypothesis: prediction and test of a mutational "hot-spot" on the transthyretin molecule associated with FAP amyloidogenesis. *Amyloid: Int J Exp Clin Invest* **1996**, *3*, 75–85.

45 Terry, C., Damas, A. M., Oliveira, P., Saraiva, M. J., Alves, I., Costa, P., Sakaki, Y. and Blake, C. Structure of Met30 variant of transthyretin and its amyloidogenic implications. *EMBO J* **1993**, *12*, 735–741.

46 Schormann, N., Murrel, J. R. and Benson, M. D. Tertiary structures of amyloidogenic and non-amyloidogenic transthyretin variants: new model for amyloid fibril formation. *Amyloid: Int J Exp Clin Invest* **1998**, *5*, 175–187.

47 Inouye, H., Domingues, F. S., Damas, A. M., Saraiva, M. J., Lundgren, E., Sandgren, O. and Kirschner, D. Analysis of X-ray diffraction patterns from amyloid of biopsied vitreous tumor and kidney of transthyretin (TTR) Met30 familial amyloidotic polyneuropathy (FAP) patients: axially arrayed TTR monomers constitute the protofilament. *Amyloid: Int J Clin Invest* **1998**, *5*, 163–174.

48 Redondo, C., Damas, A. M. and Saraiva, M. J. Designing transthyretin mutants affecting tetrameric structure: implications in amyloidogenicity. *Biochem J* **2000**, *348*, 167–172.

49 Hammarstrom, P., Jiang, X., Deechongkit, S. and Kelly, J. W. Anion shielding of electrostatic repulsions in transthyretin modulates stability and amyloidosis: insight into the chaotrope unfolding dichotomy. *Biochemistry* **2001**, *40*, 11453–11459.

50 Quintas, A., Saraiva, M. J. and Brito, R. M. M. The amyloidogenic potential of transthyretin variants correlates with their tendency to aggregate in solution. *FEBS Lett* **1997**, *418*, 297–300.

51 Quintas, A., Saraiva, M. J. and Brito, R. M. M. The tetrameric protein transthyretin dissociates to a non-native monomer in solution. *J Biol Chem* **1999**, *274*, 32943–32949.

52 Cardoso, I., Goldsbury, C. S., Muller, S. A., Olivieri, V., Wirtz, S., Damas, A. M., Aebi, U. and Saraiva, M. J. Transthyretin fibrillogenesis entails the assembly of monomers: a molecular model for *in vitro* assembled transthyretin amyloid-like fibrils. *J Mol Biol* **2002**, *317*, 683–695.

53 Sebastião, M. P., Merlini, G., Saraiva, M. J. and Damas, A. M. The molecular interaction of 4′-iodo-4′-deoxydoxorubicin with Leu-55Pro transthyretin "amyloid-like" oligomer leading to disaggregation. *Biochem J* **2000**, *351*, 273–279.

54 Liu, K., Cho, H. S., Hoyt, D. W., Nguyen, T. N., Olds, P., Kelly, J. W. and Wemmer, D. E. Deuterium–proton exchange on the native wild-type transthyretin tetramer identifies the stable core of the individual subunits and indicates mobility at the subunit interface. *J Mol Biol* **2000**, *303*, 555–565.

55 Liu, K., Kelly, J. W. and Wemmer, D. E. Native state hydrogen exchange study of suppressor and pathogenic variants of transthyretin. *J Mol Biol* **2002**, *320*, 821–832.

56 Liu, K., Cho, H. S., Lashuel, H. A., Kelly, J. W. and Wemmer, D. E. A glimpse of a possible amyloidogenic intermediate of transthyretin. *Nat Struct Biol* **2000**, *7*, 754–757.

57 Yang, M., Lei, M. and Huo, S. Why is Leu55 → Pro55 transthyretin variant the most amyloidogenic: insights from molecular dynamics simulations of transthyretin monomers. *Protein Sci* **2003**, *12*, 1222–1231.

58 Gales, L., Cardoso, I., Fayard, B., Quintanilha, A., Saraiva, M.J. and Damas, A.M. X-ray absorption spectroscopy reveals a substantial increase of sulfur oxidation in transthyretin (TTR) upon fibrillization. *J Biol Chem* **2003**, *278*, 11654–11660.

59 Lewis, W.D., Skinner, M., Simms, R.W., Jones, L.A., Cohen, A.S. and Jenkins, R.L. Orthotopic liver transplantation for familial amyloidotic polyneuropathy. *Clin Transplant* **1994**, *8*, 107–110.

60 Peterson, S.A., Klabunde, T., Lashuel, H.A., Purkey, H., Sacchettini, J.C. and Kelly, J.W. Inhibiting transthyretin conformational changes that lead to amyloid formation. *Proc Natl Acad Sci USA* **1998**, *95*, 12956–12960.

61 Oza, V.B., Petrassi, H.M., Purkey, H.E. and Kelly, J.W. Synthesis and evaluation of anthranilic acid-based transthyretin amyloid fibril inhibitors. *Bioorg Med Chem Lett* **1999**, *9*, 1–6.

62 Baures, P.W., Peterson, S.A. and Kelly, J.W. Discovering transthyretin amyloid fibril inhibitors by limited screening. *Bioorg Med Chem* **1998**, *6*, 1389–1401.

63 Oza, V.B., Smith, C., Raman, P., Koepf, E.K., Lashuel, H.A., Petrassi, H.M., Chiang, K.P., Powers, E.T., et al. Synthesis, structure and activity of diclofenac analogues as transthyretin amyloid fibril formation inhibitors. *J Med Chem* **2002**, *45*, 321–332.

64 Klabunde, T., Petrassi, M.P., Oza, V.B., Raman, P., Kelly, J.W. and Sacchettini, J. Rational design of potent human transthyretin amyloid diseases inhibitors. *Nat Struct Biol* **2000**, *7*, 312–321.

65 Almeida, M.R., Macedo, B., Cardoso, I., Alves, I., Valencia, G., Arsequelle, G., Planas, A. and Saraiva, M.J. Effective and selective action of an iodinated diflunisal derivative in transthyretin binding and tetramer stabilization in serum from familial amyloidotic polyneuropathy patients. *Biochem J* **2004**, *381*, 351–356.

66 Merlini, G., Ascari, E., Amboli, N., Bellotti, V., Arbustini, E., Perfetti, V., Ferrari, M., Zorzoli, I., et al. Interaction of the anthracycline 4′-iodo-4′-deoxydoxorubicin with amyloid fibrils: inhibition of amyloidogenesis. *Proc Natl Acad Sci USA* **1995**, *92*, 2959–2963.

67 Palha, J.A., Ballinari, D., Amboldi, N., Cardoso, I., Fernandes, R., Bellotti, V., Merlini, G. and Saraiva, M.J. 4′-Iodo-4′-deoxydoxorubicin (I-DOX) disrupts the fibrillar structure of transthyretin amyloid. *Am J Pathol* **2000**, *156*, 1919–1925.

68 Tagliavini, F., Forloni, G., Colombo, L., Rossi, G., Girola, L., Canciani, B., Angeretti, N., Giampaolo, L., et al. Tetracycline affects abnormal properties of synthetic PrP peptides and PrP^{Sc} *in vitro*. *J Mol Biol* **2000**, *300*, 1309–1322.

69 Cardoso, I., Merlini, G. and Saraiva, M.J. 4′-Iodo-4′-deoxydoxorubicin and tetracyclines disrupt transthyretin amyloid fibrils *in vitro* producing non-cytotoxic species. Screening for TTR fibril disrupters. *FASEB J* **2003**, *17*, 803–809.

70 Sousa, M.M., Cardoso, I., Fernandes, R., Guimaraes, A. and Saraiva, M.J. Deposition of transthyretin in early stages of familial Amyloidotic polyneuropathy: evidence for toxicity of nonfibrillar aggregates. *Am J Pathol* **2001**, *159*, 1993–2000.

71 Bucciantini, M., Calloni, G., Chiti, F., Formigli, L., Nosi, D., Dobson, C.M. and Stefani, M. Pre-fibrillar amyloid protein aggregates share common features of cytotoxicity. *J Biol Chem* **2004**, *279*, 31374–31382.

72 Wojtczak, A., Luft, J. and Cody, V. Mechanism of molecular recognition. Structural aspects of 3,3′-diiodo-L-thyronine binding to human serum transthyretin. *J Biol Chem* **1992**, *267*, 353–357.

73 Neumann, P., Cody, V. and Wojtczak, A. Structural basis of negative cooperativity in transthyretin. *Acta Biochim Pol* **2001**, *48*, 867–875.

74 Prapunpoj, P., Richardson, S.J. and Schreiber, G. Crocodile transthyretin: structure, function and evolution. *Am J Physiol Reg Integr Comp Physiol* **2002**, *283*, R885–R896.

75 Monaco, H. L. The transthyretin–retinol-binding protein complex. *Biochim Biophys Acta* **2000**, *1482*, 65–72.

76 Sousa, M. M., Berglund, L. and Saraiva, M. J. Transthyretin in high density lipoproteins – association via apolipprotein AI. *J Lipid Res* **2000**, *41*, 58–65.

77 Sousa, M. M., Norden, A. G. W., Jacobsen, C., Willnow, T. E., Christensen, E. I., Thakker, R. V., Verroust, P. J., Moestrup, et al. Evidence for the role of megalin in renal uptake of transthyretin. *J Biol Chem* **2000**, *275*, 38176–38181.

78 Sousa, M. M. and Saraiva, M. J. M. Internalization of transthyretin: evidence for a novel yet unidentified receptor-associated protein (RAP)-sensitive receptor. *J Biol Chem* **2001**, *276*, 14420–14425.

79 Sousa, M. M., Yan, S. D., Stern, D. and Saraiva, M. J. Interaction of the receptor for advanced glycation end products (RAGE) with transthyretin triggers nuclear transcription factor κB (NF-κB) activation. *Lab Invest* **2000**, *80*, 1101–1110.

80 Sousa, M. M., Du Yan, S., Fernandes, R., Guimaraes, A., Stern, D. and Saraiva, M. J. Familial amyloidotic polyneuropathy: receptor for advanced glycation end products-dependent triggering of neuronal inflammatory and apoptotic pathways. *J Neurosci* **2001**, *21*, 7576–7586.

81 Brouwer, A. Inhibition of thyroid hormone transport in plasma of rats by polychlorinated biphenyls. *Arch Toxicol Suppl* **1989**, *13*, 440–445.

82 Lans, M. C., Damas, A., Blake, C. C. F., Saraiva, M. J., Klasson-Wehler, E. and Brouwer, A. Structural basis for the binding of hydroxylated polychlorobiphenyl (PCB) metabolites to human transthyretin. In *Thyroid Hormone Binding Proteins as Novel Targets for Hydroxylated Polyhalogenated Aromatic Hydrocarbons (PHAHs): Possible Implications For Toxicity*, Lans, M. C. (ed.). Landbouw Universiteit, Wageningen, **1995**, pp. 93–113.

83 McKinney, J. D., Chae, K., Oatley, S. J. and Blake, C. C. F. Molecular interactions of toxic chlorinated dibenzo-*p*-dioxins and dibenzofurans with thyroxine binding prealbumin. *J Med Chem* **1985**, *28*, 375–381.

84 Bellotti, V., Mangione, P. and Stoppini, M. Biological activity and pathological implications of misfolded proteins. *Cell Mol Life Sci* **1999**, *55*, 977–991.

85 Terazaki, H., Ando, Y., Suhr, O., Ohlsson, P. I., Obayashi, K., Yamashita, T., Yoshimatsu, S., Suga, M., et al. Post-translational modifications of transthyretin in plasma. *Biochem Biophys Res Commun* **1998**, *249*, 26–30.

86 Lim, A., Prokaeva, T., McComb, M. E., Connors, L. H., Skinner, M. and Costello, C. E. Identification of *S*-sulfonation and *S*-thiolation of a novel transthyretin Phe33Cys variant from a patient diagnosed with familial transthyretin amyloidosis. *Protein Sci* **2003**, *12*, 1775–1785.

87 Tajiri, T., Ando, Y., Hata, K., Kamide, K., Hashimoto, M., Nakamura, M., Terazaki, H., Yamashita, T., et al. Amyloid formation in rat transthyretin: effect of oxidative stress. *Clin Chim Acta* **2002**, *323*, 129–137.

88 Lim, A., Sengupta, S., McComb, M. E., Theberge, R., Wilson, W. G., Costello, C. E. and Jacobsen, D. W. *In vitro* and *in vivo* interactions of homo cysteine with human plasma transthyretin. *J Biol Chem* **2003**, *278*, 49707–49713.

89 Suhr, O. E., Svendsen, I. H., Ohlsson, P., Lendoire, J., Trigo, P., Tashima, K., Ranløv, P. J. and Ando, Y. Impact of age and amyloidosis on thiol conjugation of transthyretin in hereditary transthyretin amyloidosis. *Amyloid: Int J Exp Clin Invest* **1999**, *6*, 187–191.

90 Zhang, Q. and Kelly, J. W. Cys10 mixed disulfides make transthyretin more amyloidogenic under mildly acidic conditions. *Biochemistry* **2003**, *42*, 8756–8761.

91 Jacobsson, B. *In situ* localization of transthyretin-mRNA in the adult human liver, choroids plexus and pancreatic islets and in endocrine tumours of the pancreas and gut. *Histochemistry* **1989**, *91*, 299–304.

92 Harms, P. J., Tu, G. F., Richardson, S. J., Aldred, A. R., Jaworowski, A. and Schreiber, G. Transthyretin (prealbumin) gene expression in choroid plexus is strongly conserved during evolution of vertebrates. *Comp Biochem Physiol B* **1991**, *99*, 239–249.

93 Achen, M. G., Duan, W., Pettersson, T. M., Harms, P. J., Richardson, S. J., Lawrence, M. C., Wettenhall, R. E., Aldred, A. R., et al. Transthyretin gene expression in choroid plexus first evolved in reptiles. *Am J Physiol* **1993**, *265*, R982–R989.

94 Aldred, A. R., Prapunpoj, P. and Schreiber, G. Evolution of shorter and more hydrophilic transthyretin N-termini by stepwise conversion of exon 2 into intron 1 sequences (shifting the 3′ splice site of intron 1). *Eur J Biochem* **1998**, *252*, 612.

95 Chang, L., Munro, S. L., Richardson, S. J. and Schreiber, G. Evolution of thyroid hormone binding by transthyretins in birds and mammals. *Eur J Biochem* **1999**, *259*, 534–542.

96 Richardson, S. J. The evolution of transthyretin synthesis in vertebrate liver, primitive eukaryotes and in bacteria. *Clin Chem Lab Med* **2002**, *40*, 1191–1199.

97 Eneqvist, T., Lundberg, E., Nilsson, L., Abagyan, R. and Sauer-Eriksson, A. E. The transthyretin-related protein family. *Eur J Biochem* **2003**, *270*, 518–532.

22
High-Density Lipoprotein Amyloid Proteins

Barbara Kluve-Beckerman

22.1
Introduction

To date, more than 20 proteins are known to form amyloid *in vivo*. While they comprise a diverse array of proteins, three of them are apolipoproteins associated with high-density lipoprotein (HDL): serum amyloid A (SAA), apolipoprotein (Apo) AI and ApoAII. This chapter reviews aspects of SAA, ApoAI and ApoAII that relate to their role as amyloid-forming proteins. While deposition of each of these proteins is associated with a clinically distinct disease, the three proteins share features that suggest that common or related pathogenic mechanisms are involved. Since each is associated with HDL, the three proteins likely share amphipathic properties of lipid binding and aqueous solvent accessibility. None of the three is predicted, or has been shown, to have a significant degree of β-sheet structure, a conformational property that they, or fragments derived from them, must assume in order to assemble as β-structured amyloid fibrils. Finally, the amyloid subunits derived from these proteins have usually undergone partial proteolysis (SAA and ApoAI), often in combination with structural alteration due to a genetic mutation (ApoAI and ApoAII). Such alterations presumably confer upon the proteins a structure favorable to, or even required for, participation in amyloid fibril formation.

22.2
SAA [Secondary, Reactive, Amyloid A (AA) Amyloidosis]

22.2.1
Background

Using the newly described water extraction method published by Pras et al. in 1968, several research groups in the early 1970s successfully pioneered the isolation of amyloid fibrils and biochemical characterization of their chief protein constituent [1]. The protein they isolated was distinct from the previously identi-

Amyloid Proteins. The Beta Sheet Conformation and Disease. J. D. Sipe

ISBN: 3-527-31072-X

fied immunoglobulin amyloid fibril protein and was termed amyloid A (AA). Soon thereafter, a protein with immunologic identity to AA was discovered in serum, specifically in the HDL fraction, of patients experiencing an acute-phase response. SAA, as this protein was designated, has been extensively studied for more than 30 years with regard to its role as a precursor to AA amyloid, also called reactive or secondary amyloid because it most often develops in association with chronic or recurrent infection or inflammation. In contrast to most other types of amyloid, reactive amyloidosis occurs naturally in species other than humans, and can also be induced experimentally in various laboratory animals and cell culture models. As such, studies of AA amyloid pathogenesis have served as the prototype for research into other, more commonly occurring types of amyloid. The discovery, and chemical and physiological characterization of the amyloid A fibril and the SAA proteins have been thoroughly reviewed [2–4].

AA amyloidosis is a systemic condition that involves deposition of amyloid fibrils in multiple organs and tissues of the body (commonly kidney, spleen, liver, intestine), but not in brain [5]. Subunit proteins comprising the amyloid fibrils are N-terminal cleavage fragments of SAA. While AA fragments commonly consist of SAA residues 1–76, fragments in many fibril preparations have ragged C-termini. N-terminal AA fragments containing 45–83 residues have been characterized (reviewed in [2]). SAA, precursor to AA fibrils, is very highly conserved in mammals, and is also present in birds and fish. In addition to laboratory animals in which AA amyloidosis can be experimentally induced (e.g. mouse, hamster, mink, rabbit, monkey), a variety of animals, both domestic and wild, including cat, dog, horse, cow, sheep, goat, chicken, duck, cheetah and gazelle, have been diagnosed with AA amyloidosis resulting from natural causes [2, 6–9]. Comparison of acute-phase SAA proteins in eight species, including duck, reveals invariant residues at positions (based on alignment with human SAA): 6, 10, 13, 16–17, 19–21, 23–24, 26–27, 33–45, 47–51, 53–55, 58–60, 63, 70, 72, 75, 79–80, 82–83, 86–87, 89, 92–94, 96–97, 99–101 and 104 [10].

22.2.2
Gene and Protein (Primary) Structure

A number of the early studies into AA and SAA structure provided clues suggesting the polymorphic nature of SAA. It soon became apparent that SAA comprises a family of closely related proteins encoded by a cluster of genes. Although this chapter is limited to human and mouse SAA proteins, most mammalian species studied to date, including horse, dog, mink, rabbit and hamster, are known to have multiple acute-phase, hepatically produced SAA isoforms.

Mapping of the human SAA genes within constructs containing large segments of chromosome 11 DNA provided definitive understanding of the composition and organization of this multi-gene family. Four distinct SAA genes have been characterized and localized to chromosome 11p15.1, spanning a 150-kb region [11]. Genes for acute-phase proteins SAA1 and SAA2 are located around

15–20 kb apart in divergent (head-to-head) transcriptional orientations and share over 90% nucleotide identity in exons and introns [11–13]. The encoded proteins have an 18-amino acid signal peptide and a mature sequence of 104 residues with a molecular weight of around 12 kDa (Table 22.1). They are synthesized by the liver and circulate in plasma associated with HDL. The gene for SAA4, which is expressed constitutively at low levels and present in normal plasma at concentrations of 40–60 μg/ml, is 9 kb downstream of and in the same orientation as *SAA2*. *SAA4* is significantly divergent from *SAA1* and *SAA2*, sharing only 67–76% nucleotide identity in exons, no identity in introns and 54% identity at the protein level. SAA4 protein is composed of 112 residues; based on alignment with SAA1 and SAA2, SAA4 has an 8-amino acid insertion between residues 69 and 70 within which there is a glycosylation site. Non-glycosylated and glycosylated forms of SAA4 have molecular masses of 14 and 19 kDa, respectively, and are present on HDL [14]. Human *SAA3* is 110 kb downstream of *SAA4* and its orientation is unknown. Human SAA3 has been regarded as a pseudogene since it contains a single base insertion that shifts the reading frame, causing premature termination, and the corresponding protein product has not been identified in plasma [15]. However, expression of SAA3 mRNA has recently been demonstrated via RT-PCR methodology in human mammary epithelial cells following stimulation with prolactin or lipopolysaccharide (LPS) [16]. With the possible exception of human *SAA3*, SAA genes in all species studied, share the same structure of four exons and three introns: exon 1 [5′-untranslated region (UTR)], intron 1, exon 2 (5′-UTR plus codons –1 through 12), intron 2, exon 3 (codons 12 through 58), intron 3 and exon 4 (codons 58 through 104 or 112 plus 3′-UTR).

The organization and structure of the mouse SAA gene family parallels that of humans [17, 18]. The family, which includes four active genes and one pseudogene, spans about 45 kb on the proximal arm of chromosome 7 in a region syntenic with human chromosome 11p. Most strains of mice have two highly homologous, cytokine-inducible genes (*Saa1* and *Saa2*) encoding hepatically produced acute-phase SAA proteins; they also have a cytokine-inducible gene (*Saa3*) expressed in liver and extrahepatically by various cell types, most predominantly macrophages [19, 20]. Mouse SAA1 and SAA2 proteins, after cleavage of a 19-residue signal peptide, are 103-amino acids long and, in the reference strain BALB/c, differ from each other at only nine positions (residues 6, 7, 27, 30, 31, 60, 63, 76 and 101). The most notable recognized difference between the two proteins is that only SAA1.1-derived AA protein (previously designated SAA2) is present in amyloid fibrils. The CE/J mouse strain is distinct from BALB/c in having a single SAA gene, the sequence of which is a hybrid of *Saa1* and *Saa2*. The SAA protein in CE/J mice (SAA2.2; Table 22.2) matches amyloidogenic SAA1.1 at positions 27, 30, 31 and 76, and non-amyloidogenic SAA2.1 at positions 6, 7, 60, 63 and 101, and differs from both at position 11. CE/J mice are distinct in being resistant to the development of experimentally induced amyloidosis [21, 22]. Mouse SAA3 protein, which shares only around 64% identity with SAA1 and SAA2, has eluded isolation, but it has been detected immunochemically in various tissues and on HDL from LPS-stimulated mice [23]. A fourth gene (*Saa4*) is also present in mice [24]. *Saa4* appears to be a homologue of the human SAA4 gene and is significantly divergent from *Saa1* and *Saa2*. Mouse SAA4 protein is present constitutively on HDL and has an eight-residue insert like human SAA4 [25].

22.2.3
Polymorphisms and Amyloidogenicity

The nomenclature of human and mouse SAA proteins, as set forth by the International Nomenclature Committee on Amyloidosis in 1998 [26], is presented in Figs. 22.1 and 22.2 and Tables 22.1 and 22.2, and is based on a multitude of studies in which SAA (and AA) proteins, cDNAs and genes from numerous individuals were isolated and sequenced. As shown in Table 22.1, five alleles of *SAA1* (1.1, 1.2, 1.3, 1.4 and 1.5) and two alleles of *SAA2* (2.1 and 2.2) have been identified thus far. In the vast majority of publications cited below, SAA1 alleles were differentiated by codons 52 and 57, and designated as α (Val52, Ala57), β (Ala52, Val57) or γ (Ala52, Ala57). The recommended revised nomenclature is used below; it has been assumed that alleles SAA1α, 1β and 1γ correspond to SAA1.1, 1.5 and 1.3, respectively. However, according to the revised nomenclature, SAA1β can actually be SAA1.2, 1.4 or 1.5 since these alleles cannot be distinguished on the basis of codons 52 and 57 alone. Nucleotides at polymorphic sites in codons 60 and 72 are also needed to differentiate among alleles 1.2 (Ala52, Val57, Asp60, Asp72), 1.4 (Ala52, Val57, Asn60, Gly72) and 1.5 (Ala52, Val57, Asp60, Gly72). Two non-coding polymorphisms in the SAA1.1 allele, one in intron 4 and one in the 3′-UTR of exon 4, also have been identified [27].

	10	20	30
SAA1.1	R S F F S F L G E A	F D G A R D M W R A	Y S D M R E A N Y I
SAA4	E S W R S F F K E A	L Q G V G D M G R A	Y M D I M I S M H Q
	40	**50**	**60**
SAA1.1	G S D K Y F H A R G	N Y D A A K R G P G	G V W A A E A I S D
SAA4	N S N R Y L Y A R G	N Y D A A Q R G P G	G V W A A K L I S R
	70	**80**	**90**
SAA1.1	A R E N I Q R F F G	H G A E D S L A D Q	A A N E W G R S G K
SAA4	S R V Y L Q G L I S	T V L E D S K S N E	K A E E W G R S G K
	(D Y Y L F G N S)		
	100		
SAA1.1	D P N H F R P A G L	P E K Y	
SAA4	D P D R F R P D G L	P K K Y	

Fig. 22.1 Amino acid sequence of human SAA1.1 and SAA4 [26].

	10	20	30
SAA1.1	G F F S F I G E A F	Q G A G D M W R A Y	T D M K E A G W K D
SAA3	R W V Q F M K E A G	Q G S R D M W R A Y	S D M K K A N W K N
SAA4	D G W Y S F F R E A V	Q G T W D L W R A Y	R D N L E A N Y Q N

	40	50	60
SAA1.1	G D K Y F H A R G N	Y D A A Q R G P G G	V W A A E K I S D A
SAA3	S D K Y F H A R G N	Y D A A R R G P G G	A W A A K V I S D A
SAA4	A D Q Y F Y A R G N	Y E A Q Q R G S G G	I W A A K I I S T S

	70	80	90
SAA1.1	R E S F Q E F F G R	G H E D T M A D Q E	A N R H G R S G K D
SAA3	R E A V Q K F T G H	G A E D S R A D Q F	A N E W G R S G K D
SAA4	R K Y F Q G L L N H	G L E T L Q A T Q K	A E E W G R S G K N
	(N R Y Y F G I R)		

	100	
SAA1.1	P N Y Y R P P G L P	A K Y
SAA3	P N H F R P A G L P	K R Y
SAA4	P N H F R P E G L P	E K F

Fig. 22.2 Amino acid sequence of mouse SAA1.1, SAA3, and SAA4. SAA1.1 (formerly SAA2) and SAA4 (formerly SAA5) are the revised, recommended nomenclature [26].

Table 22.1 Residues differentiating human acute phase SAA proteins*

	52	57	60	68	69	71	72	84	90
SAA1.1 (1α)	V	A	D	F	F	H	G	E	K
SAA1.2 (1β)	A	V	D	F	F	H	D	E	K
SAA1.3 (1γ)	A	A	D	F	F	H	G	E	K
SAA1.4 (1δ)	A	V	N	F	F	H	G	E	K
SAA1.5 (1β)	A	V	D	F	F	H	G	E	K
SAA2.1 (2α)	A	V	N	L	T	H	G	K	R
SAA2.2 (2β)	A	V	N	L	T	R	G	K	R

* Revised, recommended nomenclature is in bold; former designation is in parentheses [26].

The recognition that there were multiple genes and alleles for human SAA, together with the fact that only a small percent of patients with inflammatory conditions predisposing to reactive amyloidosis actually develop the disease, prompted investigators to explore the possibility that particular SAA polymorphisms are associated with amyloidosis, i.e. protein products of some SAA alleles are more likely than others to form amyloid. Precedence for amyloid-forming iso-

Table 22.2 Residues differentiating SAA1 and SAA2 in inbred strains of mice

	6	7	11	27	30	31	60	63	76	101
SAA1.1 (SAA2)*	I	G	Q	G	D	G	A	S	M	A
SAA1.2 (SJL/J)	I	G	Q	G	D	G	A	S	M	D
SAA2.1 (SAA1)*	V	H	Q	N	N	S	G	A	I	D
SAA2.2 (CE/J)	V	H	L	G	D	G	G	A	M	D

* SAA1.1 and SAA2.1 are in BALB/C mice. Revised, recommended nomenclature is in bold; former designation is in parentheses [26].

types is seen in other species where amyloid fibrils are derived preferentially or exclusively from a particular SAA gene product (e.g. most notably mouse SAA1.1 [28]).

Based on a preponderance of human SAA1-derived AA proteins extracted from amyloid fibrils, it is generally agreed that SAA1 is the predominant amyloidogenic isoform in humans [29]. Hence, efforts have focused on determining which, if any, of the SAA1 alleles correlate with development of amyloidosis. The two clinical populations most commonly affected by AA amyloid are patients with familial Mediterranean fever (FMF) and patients with rheumatoid arthritis (RA), and these are the predominant groups that have been studied with respect to SAA1 genotype. Analyses of SAA gene sequences in Japanese RA patients, in whom the detected incidence of AA amyloid is higher than in US RA patients, led to the identification of the SAA1.3 (γ) allele [30]. Among Japanese subjects, the frequency of the SAA1.3 allele and homozygosity for *SAA1.3*, in particular, is significantly higher in RA patients with amyloidosis than in RA patients without amyloidosis or in the general population. Concordant data from several Japanese laboratories support the conclusion that the SAA1.3 allele is a risk factor for development of AA amyloidosis in Japanese RA patients [31–34]. More recently, one of these groups found amyloid development in this population to be more strongly associated with a single nucleotide polymorphism (SNP) in the 5′-flanking region of *SAA1.* Their data suggest that the presence of C rather than T at nucleotide –13 is a susceptibility factor for amyloidosis [33]. In the Japanese patients, –13T showed linkage to SAA1.3 and SAA1.5, but not to SAA1.1. A similar correlation between –13T and reactive amyloidosis was reported in a study of American Caucasian AA amyloid patients having a variety of predisposing conditions; however, –13T in these patients was linked to SAA1.1 [34]. The mechanism by which a nucleotide in a non-coding region of the SAA1 gene could influence the likelihood for amyloid development is unclear.

In studies of other ethnic populations, basic differences between Japanese and Caucasians with respect to SAA and amyloidosis have become apparent. First, in contrast to the Japanese in whom the frequency of SAA1.1, SAA1.3 and SAA1.5 alleles is almost equal, the frequency of the SAA1.3 allele in Caucasians is extremely low, regardless of health status (around 5%). Second, the allele that best correlates with amyloidosis in non-Japanese groups is *SAA1.1* [34, 35]. A study in the UK reported a 90.2% frequency of the SAA1.1 allele in juvenile RA (JRA) patients with amyloidosis, compared to 56.3% in JRA patients without amyloidosis and 75.8% in healthy controls [35]. An increased incidence of AA amyloidosis also has been reported in Armenian FMF patients homozygous for *SAA1.1*. The association was seen regardless of genotype for *MEFV*, the gene responsible for FMF, although it was most marked in patients having the Met694Val variant of MEFV [36]. Investigators in Lebanon studying 70 FMF patients (30 with and 40 without amyloidosis) similarly found *SAA1.1* and the Met694Val allele of *MEFV* to be more prevalent in patients with amyloidosis, while *SAA1.3* and *SAA1.5* were more prevalent in patients without amyloid [37]. A statistically significant association between the occurrence of amyloidosis and *SAA1.1* homozygosity has also been noted in Israel where the presence of –13T was rare and not related to amyloidosis [38]. Likewise, a preponderance of the SAA1.1 allele, in particular *SAA1.1* homozygosity, has also been reported in several studies of Turkish patients [39, 40]. Among 74 Turkish patients with FMF (all on colchicine treatment), eight also had amyloidosis and seven of the eight were homozygous for *SAA1.1*. In a group of 16 American Caucasian patients with amyloidosis, the frequency of the SAA1.1 allele was 91% compared with 72% in the general population [34]. It remains to be determined how *SAA1.1* in Caucasians and *SAA1.3* in Japanese (both of which show linkage to –13 T in the respective population) may increase susceptibility to amyloidosis in patients with predisposing conditions.

22.2.4
Protein Structure (Three-dimensional)

While the amino acid sequence of SAA has been known for 20 years, little progress has been made regarding its three-dimensional structure. Early circular dichroism (CD) studies performed on human SAA determined an α-helix content of 32%, which increased to 40% upon incubation with phospholipids [41]. Analyses of the sequence by computer algorithms predicted an α-helical content of 40%, and 45% total β-structure, as well as a Ca^{2+}-binding site (Gly48–Pro49–Gly50–Gly51) and lipid-binding N-terminal domain [42]. Three consecutive heptad segments within the first 23 residues of SAA likely form an amphipathic helix. Subsequent studies provided experimental evidence that the N-terminal region is required for HDL binding and a determinant of amyloid fibril formation [43, 44]. CD measurements indicated that mouse SAA1.1, SAA2.1 and SAA2.2 have α-helical contents of 15–20, 32 and 14–18%, respectively [21, 45]. Binding of heparan sulfate proteoglycan (HSPG) in the presence of Ca^{2+} increased the β-sheet content of amyloidogenic SAA1.1, but had no effect on the β-structure of SAA2.1 or SAA2.2. This finding suggests that HSPG can influence SAA folding and thereby play an important role in amyloid pathogenesis, as is expected based on the presence of HSPG in all AA amyloid deposits, as well as in amyloid of all types (see also Chapter 7).

Mouse SAA2.2 in aqueous solution has recently been shown to have a hexameric quaternary structure with a putative central channel [46]. Further investigations using the combined techniques of glutaraldehyde cross-linking, polyacrylamide gel electrophoresis (PAGE) and sedimentation velocity analytical ultracentrifugation have demonstrated a temperature-sensitive equilibrium between hexameric and monomeric forms of lipid-free SAA, with a shift from hexameric to monomeric SAA with increase in temperature from 20 to 37 °C. Evidence was also provided that SAA is in monomeric form when it binds to HDL at 37 °C. At temperatures above 37 °C, some dissociation of SAA from HDL occurred, raising the possibility that at fever-like temperatures *in vivo* (41 °C), HDL-bound and free forms of SAA may co-exist [47].

The insolubility of SAA in aqueous solvents and its tendency to aggregate have precluded the formation of crystals suitable for X-ray diffraction. Various activities, however, have been localized to distinct regions in the SAA sequence. Cholesterol binding has been localized to residues 1–18 and residues 40–63, and shown to promote dimerization of SAA [48]. Laminin binding occurs through residues 24–76 and cell adhesion via residues 29–42. The heparin/heparan sulfate-binding site associated with residues 78–104 is not usually present in amyloid fibrils (reviewed in [49]).

22.2.5
Induction of Protein Synthesis

Acute-phase SAA1 and SAA2 are produced primarily in the liver in response to cytokines released by inflammatory cells [50]. The cytokines are primarily interleukin (IL)-1, IL-6 and tumor necrosis factor (TNF) which act on hepatocytes to trigger a cascade involving NFκB, C/EBP, YY1 and SEF transcription factor families to induce transcription of SAA genes (reviewed in [3, 51]). Expression of SAA genes is regulated by both positive and negative *cis*- and *trans*-acting promoter elements to effect tissue specific expression; production of SAA proteins is also regulated by message stabilization ([52, 53], reviewed in [49]). Within 4 h of an acute inflammatory stimulus increases in acute-phase SAA mRNA can be measured in liver; peak levels are reached by 15–20 h. Increases in the plasma concentration of SAA follow a similar time course, reaching levels as much as 500- to 1000-fold above normal. Highest levels are seen during acute bacterial infections when SAA can replace ApoAI as the most prominent protein on HDL. Upon cessation of the inflammatory stimulus, plasma levels gradually return to baseline over the course of 5–7 days [50].

Chronically elevated levels of SAA are associated with the development of reactive amyloidosis. While this is clearly the most defined risk factor, it is not known if the degree, duration or frequency of elevated SAA levels is more influential in predisposing to amyloid development. Diseases currently most often associated with the development of reactive amyloidosis include rheumatoid arthritis, Crohn's disease, ankylosing spondylitis, FMF, TNF receptor-associated periodic fever syndrome (TRAPS), tuberculosis and leprosy [5].

22.2.6 Association with HDL

Remodeling of HDL with SAA during an acute-phase response to injury is expected to influence HDL metabolism and cholesterol transport. Two hypotheses have emerged regarding the role of HDL-SAA. Both propose that HDL-SAA is targeted to cells/tissues damaged by inflammation. According to one, HDL-SAA goes to such sites to remove cholesterol released during tissue injury, while the other states that HDL-SAA delivers cholesterol needed for rebuilding of injured tissue (reviewed in [54]). With respect to SAA's role in lipid metabolism, studies have shown that SAA can bind cholesterol and enhance its cellular uptake [48]; HDL-SAA has higher affinity for macrophages and lower affinity for hepatocytes than normal HDL [55, 56]; and the activity of several enzymes involved in cholesterol metabolism is altered during an acute phase response. Increased plasma SAA has been shown to correlate with decreased lecithin-cholesterol acyl transferase (LCAT) activity (for which ApoAI is cofactor) and increased unesterified cholesterol levels, possibly reflecting the decreased level of ApoAI on SAA-rich HDL [57]. Increased activity of neutral cholesterol ester hydrolase in the presence of SAA, favoring free cholesterol formation, also has been demonstrated [58]. Moreover, increased hydrolysis of HDL-SAA relative to normal HDL by non-pancreatic secretory phospholipase A2 (sPLA2) has been reported [59].

22.2.7 Catabolism, Macrophages and Amyloidogenesis

While there has been much progress toward understanding intricate mechanisms of SAA synthesis at the molecular level, less has been learned about SAA catabolism. Early studies showed that SAA is degraded in liver and has a much shorter half-life in plasma than other HDL apolipoproteins [60, 61]. Proteases implicated in the degradation of SAA include serum proteases, proteases secreted by neutrophils and macrophages, cell-associated proteases of lymphocytes, neutrophils and macrophages, and tissue homogenates from spleen, liver and kidney (reviewed in [62]). In some studies, SAA was degraded to completion, while in others cleavage into discreet fragments was observed. Cleavage sites within the N-terminal half of SAA have been identified for the serine protease elastase and lysosomal aspartic proteases including cathepsin D; since the sites are within the region found in AA proteins, cleavage by them would preclude amyloid formation. Cathepsin K, which can degrade SAA completely, has been implicated in amyloid resorption based on its secretion from multinucleated giant cells. In contrast, cathepsin B, a lysosomal cysteine protease, has been shown *in vitro* to cleave human SAA between residues 76 and 77, transiently generating AA protein. Proteolytic events specifically occurring *in vivo*, however, have yet to be defined.

A growing number of studies indicate that macrophages are active participants in SAA metabolism. HDL-SAA binds to macrophages via saturable (specific) receptors [55, 63]. SAA is endocytosed, and traffics through endosomes and into lysosomes for degradation [63–65]. Whether SAA internalization by macrophages relates to a particular function or solely reflects its catabolic pathway remains to be determined. Signal transduction receptors not associated with uptake also have been shown to bind SAA; these include the receptor for advanced glycation end-products (RAGE) [66] and FPRL-1 (also known as lipoxin A4 receptor) which mediates chemotactic activity of SAA for monocytes and neutrophils [67].

Aberrant catabolism of SAA in macrophages has long been considered a contributing factor in amyloidogenesis. Thirty years ago the presence of dense fibrillar inclusions in mononuclear cells of tissues from amyloid-laden mice was described and proposed to represent intralysosomal formation of amyloid fibrils [68]. Since then SAA, as well as AA fibrils, have been identified by immunoelectron and confocal microscopy in lysosome-derived organelles, along cell membranes and in intercellular aggregates [64, 69–72]. A direct link between macrophages and amyloid formation was first provided by studies using peritoneal cells obtained from amyloidotic mice. Cultures of these cells maintained in SAA-rich medium or ascetic fluid developed tiny, transient amyloid masses [73]. More recently, amyloid formation has been achieved in a model that employs resident peritoneal macrophages cultured in the presence of mouse recombinant SAA1.1. These cultures demonstrate extensive and sustained amyloid production that is initiated intracellularly in vesicles [74]. Pulse–chase studies using the macrophage culture model have shown that while the vast bulk of endocytosed SAA is completely degraded, a small percent accumulates in lysosomal-derived vesicles, undergoes limited C-terminal cleavage, and forms nascent amyloid which is eventually extruded to the cell surface. The process requires metabolically active cells, with cell death seen only after extensive accumulation of amyloid on cell surfaces [72].

Although the exact mechanism of macrophage involvement in AA amyloid formation is not yet known, roles in concentrating SAA and inducing structural alterations in SAA seem likely based on ideas that fibrillogenesis is a nucleation-dependent process requiring a critical protein concentration and that even the most amyloidogenic SAA (mouse SAA1.1) lacks sufficient β-sheet structure to form amyloid. A nucleation type of process is consistent with the biphasic kinetics of AA amyloid induction seen in the well-established mouse model in which amyloid is induced via repeated injections of an inflammatory stimulus [75]. A slow pre-deposition or lag phase believed to involve accumulation and nucleation of precursor protein is followed by a more rapid fibril deposition phase. The duration of the first phase in AA amyloidogenesis can be tremendously shortened by administration of amyloid-enhancing factor (AEF) together with an inflammatory agent [76]. AEF, originally prepared as a glycerol extract of spleen from amyloidotic animals, has been proposed to act as a nidus about which fibrils polymerize. AA fibrils extracted from amyloid deposits, as well as short synthetic peptides corresponding various amyloidogenic proteins, also have AEF bioactivity [77]. Moreover, acceleration of AA amyloid formation has recently been achieved by oral administration of

AEF/AA leading to the proposal that AA amyloid be considered a transmissible disease akin to prion-associated disorders [78]. AEF activity is destroyed by treatment with denaturing agents, and therefore it has been thought to be conformation dependent [76]. Evidence implicating β-sheet structure as the essence of AEF bioactivity has been provided by the demonstration that intravenous injections of denatured, sonicated silk can substitute for AEF in accelerating AA amyloid formation in the mouse model [79].

Conformational changes in SAA leading to β-sheet structure have been proposed to be acid-induced based on *in vitro* studies in which fibril formation from SAA is achieved under very low pH conditions [44, 80]. Hence, lysosomes, while thought to play a predominant role in SAA catabolism, can be considered likely sites of fibrillogenesis. While the normal pathway for SAA in lysosomes ends in complete degradation, it appears that with increased uptake of SAA by macrophages due to increased catabolic demands following an inflammatory episode, there may be increased concentration of SAA in lysosomes and an increased chance that a few molecules upon acid denaturation may escape degradation, re-fold into β-sheet structure, and undergo assembly into amyloid protofibrils.

22.3 ApoAI Amyloidosis

22.3.1 Background

ApoAI comprises around 70% of the protein on HDL and is present in serum at a concentration of 1.0–1.5 mg/ml. Based on the long-recognized association between decreased plasma levels of HDL and increased risk of atherosclerotic cardiovascular disease, great efforts have been made to understand the function and structure of ApoAI (reviewed in [81, 82]). As the major protein component, ApoAI plays a key role in defining HDL structure and solubility, and also governs the all-important HDL process of reverse cholesterol transport in which cholesterol is transferred from peripheral cells to the liver for catabolism (reviewed in [83–86]). ApoAI functions in this process by activating lecithin cholesterol acyltransferase (LCAT), which esterifies HDL cholesterol, and ApoAI then helps deliver cholesteryl ester-loaded HDL to cells by interacting with the receptors involved in selective cholesterol uptake. In addition to promoting the passive efflux of cholesterol, ApoAI also activates cholesterol efflux that occurs via the ATP-binding cassette transporter A1 (ABCA1), an integral membrane protein. Thus, of all the constituents of HDL (e.g. triglycerides, cholesterol, phospholipids, ApoAI, ApoAII, ApoCII and SAA), it is ApoI that confers HDL with its anti-atherosclerotic properties.

22.3.2
Gene and Protein Structure

The ApoAI gene, located on chromosome 11q23, comprises four exons interrupted by three introns: exon 1 (5′-UTR), intron 1 (197 bp), exon 2 (5′-UTR plus codons –24 through –10), intron 2 (185 bp), exon 3 (codons –10 through codon 43), intron 3 (588 bp) and exon 4 (codons 43 through codon 243 plus 3′-UTR) [87, 88]. ApoAI is synthesized in the liver and intestine as a 267-amino-acid protein containing a 24-amino acid pre-pro-peptide. Intracellularly, an 18-residue segment is removed yielding a 249-amino acid propeptide. The six-residue pro-segment of unknown function is stable through secretion and removed by a serum protease specific for an unusual Gln–Gln–Asp–Glu site, cleaving between Gln and Asp (Fig. 22.1). This cleavage generates mature ApoAI, a 243-residue, non-glycosylated protein with a molecular mass of around 28 kDa [89]. Human ApoAI lacks cysteine and exists as a monomer, while mouse ApoAI has one cysteine which forms a disulfide bond to generate dimeric ApoAI molecules [90]. The amino acid sequence of human ApoAI was first determined by Brewer et al. in 1978 [91], and confirmed by cloning and characterization of ApoAI cDNAs by several groups in the early 1980s [92–94] (Fig. 22.3).

The secondary and tertiary structure of ApoAI has been the subject of intense investigation for decades. Repeats within the amino acid sequence, spanning residues 44–243 (exon 4), are thought to be associated with formation of 10 α-helices having amphipathic character that allows for interaction with lipids via hydrophobic faces and aqueous environments via hydrophilic faces. The repeats, most of which are demarcated by a proline residue, include eight 22mers (helices 2 and 9) and two 11mers (helices 1, 3–8 and 10) [95–97]. There is general consensus that the last three helices are essential for lipid binding. Various models have been proposed for both lipid-bound and lipid-free structures of ApoAI (reviewed in [84, 98]). Numerous studies have utilized naturally occurring ApoAI mutants and recombinant ApoAI with specific deletions or mutations to elucidate structure–function relationships within defined regions of the protein (reviewed in [84, 99–101]). The challenge currently at hand is to define ApoAI tertiary structure. Awaiting resolution are the (1) comparative structures of ApoAI in the lipid-free state, in nascent discoidal HDL and in mature spherical HDL, (2) features of ApoAI that allow for interconversion of HDL among the various forms, and (3) structures within ApoAI responsible for its various activities. Studies have indicated that ApoAI undergoes major structural changes when going from a free to lipid-bound state and this conformational plasticity is essential to the various activities of ApoAI.

The first crystal structure of ApoAI was obtained using a truncated form of the human protein consisting of residues 43–243, the lipid-binding portion of the protein corresponding to exon 4 and designated Δ1–43 [102]. Recombinant Δ1–43 was found to be similar to native ApoAI in its ability to bind lipids, be incorporated into HDL and activate LCAT. The continuously curved, amphipathic helical conformation adopted by Δ1–43 more closely mimicked the struc-

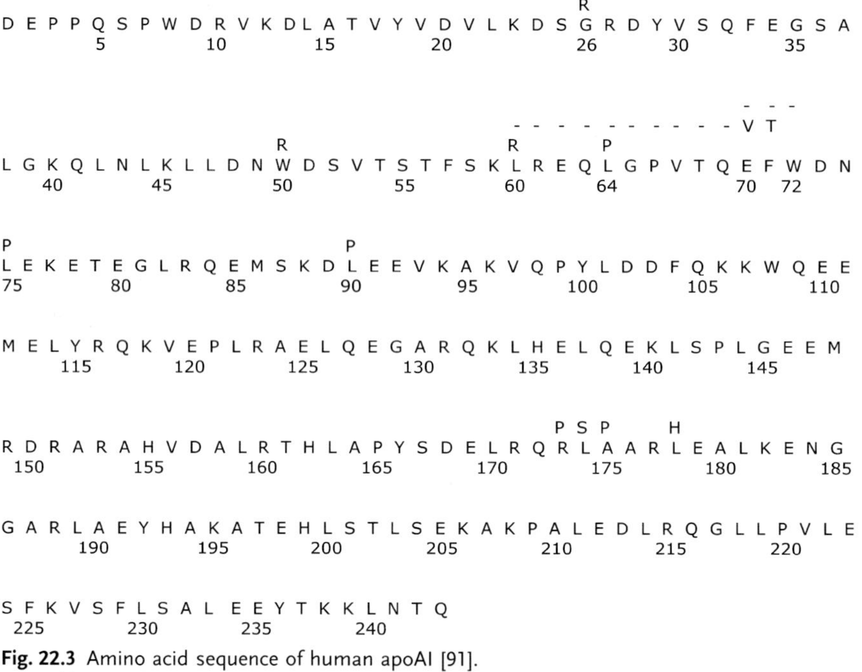

Fig. 22.3 Amino acid sequence of human apoAI [91]. See Table 22.3 for amino acid substitutions in variant ApoAI proteins.

ture of lipid-bound rather than lipid-free native ApoAI. Crystals of Δ1–43 in the original study were formed under high salt conditions in the absence of lipid. Designated form I, these crystals indicated a single pseudo-continuous, amphipathic α-helix formed from helices 1–9 and punctuated by kinks at regularly spaced proline residues causing the molecule to adopt an unusual "horseshoe-like" shape. In the crystal, ApoAIΔ1–43 existed as an elliptically shaped dimer formed from two antiparallel, tightly interacting molecules. Subsequently, form II crystals of Δ1–43 have been formed at much lower ionic strength and in the presence of mild detergent [103]. Proposed models accommodated by form II crystals include compact (four-helix bundle) and extended (two-helix coiled coil) structures. A recent determination of ApoAI structure by nuclear magnetic resonance in a lipid-mimetic solution has shown the protein to be composed of 11 α-helices (residues 8–32, 45–64, 67–77, 82–86, 90–97, 100–118, 122–140, 146–162, 167–205, 210–216 and 221–239) and to exist as a monomer in sodium dodecylsulfate (SDS) with no evidence of interhelical interactions [104].

22.3.3 Association with HDL

Newly synthesized ApoAI is found in plasma transiently in the form of lipid-poor, nascent small discoidal HDL particles, designated pre-β HDL containing only ApoAI and phospholipids (reviewed in [105]). Pre-β HDL ApoAI, in either lipid-poor or lipid-free state, is also generated continually by HDL remodeling through the action of cholesterol transfer protein (CETP) or hepatic lipase which act reciprocally of LCAT to maintain a balance between lipid-poor and lipid-bound pools of ApoAI in plasma (reviewed in [105]). Nascent lipid-poor HDL particles mature as they acquire phospholipid and cholesterol either from other lipoprotein particles or from cells via passive efflux or active processes employing ABCA1 (reviewed in [86]). The cholesterol added to HDL is esterified by LCAT (for which ApoAI serves as cofactor) and then partitions into the HDL core. This process results in spherical HDL particles with neutral lipid cores and outer shells containing two or more apolipoproteins. Designated α-HDL, these particles are the most abundant form in plasma. Cholesterol esterification by LCAT decreases the amount of unesterified cholesterol in plasma and thereby sets up a concentration gradient that favors cholesterol efflux from cells. Cholesteryl ester-rich HDL particles are targeted to cells involved in cholesterol metabolism (degradation and/or steroidogenesis) and via ApoAI bind to scavenger receptor class B-1 (SRB-1) receptors (prominent in liver and adrenal gland) ([106], reviewed in [107]). Cholesteryl ester is transferred into the cells, while lipid-depleted HDL is released back into extracellular fluid in a process referred to as selective lipid uptake. The lipid-depleted HDL particles can then re-acquire cholesterol from peripheral cells or undergo catabolism by the kidney through glomerulofiltration, reabsorption and degradation in proximal tubules.

22.3.4 Amyloidogenic Variants of ApoAI

To date, 12 mutations in the human ApoAI gene have been identified that are associated with the development of amyloidosis (Table 22.3). All but one of the mutations have been confirmed in at least two individuals. DNA analysis of affected and at-risk individuals indicates an autosomal dominant mode of inheritance. All patients to date have been heterozygous for the mutant gene. The vast majority of studies reported thus far are descriptive in nature, presenting clinical manifestations, changes in the sequence of ApoAI at DNA and protein levels, methods to test for these changes, and characterization of ApoAI fragments comprising the amyloid. With minor exceptions, the amyloid is composed solely of fragments comprising approximately the first 90 residues of variant ApoAI; as a rule, neither full-length nor normal ApoAI is present. The various ApoAI substitutions and the amyloid diseases associated with them will be discussed in order of their occurrence on the polypeptide chain in an N- to C-terminal direction.

Table 22.3 Amyloidogenic variants of apoAI

Mutation	Geographic location	Clinical features	Effect on protein	Amyloid fragment (residues)	References
G26R	USA, UK	Renal failure, Peripheral neuropathy	+1 charge	1–83	108–115
W50R	UK	Renal failure	+1 charge	1–86, 1–92, 1–93	115, 116
L60R	UK	Splenic, hepatic, renal deposition	+1 charge	1–88, 1–92, 1–93, 1–94	115, 117
L64P	USA, UK (Italian)	Renal failure	Proline: helix-breaker?	1–96	115, 118
Del 60–71 Ins V-T	Spain	Fatal liver failure	+1 charge	1–83, 1–92	119
Del 70–72	South Africa	Renal failure; massive visceral deposits w/o clinical consequences	+1 charge	No tissue available	115, 120
L75P	Italy	Slowly progressive renal disease	Proline: helix-breaker?	1–96	121-123
L90P	France	Cutaneous, cardiac deposition	Proline: helix-breaker?	1–88	124
R173P	USA, UK	Cutaneous, laryngeal, cardiac deposition	Proline: helix-breaker?	7 kDa, 9 kDa	115, 125
L174S	Italy	Massive cardiac deposition		1–93, traces of full-length	126, 127
A175P	UK		Proline: helix-breaker?		115
L178H	France	Cardiac, cutaneous, laryngeal deposition	+1 charge	N-terminal fragments, full-length apoAI; TTR	128

22.3.4.1 **Gly26Arg**

Amyloid now known to be associated with this variant was clinically characterized by Van Allen et al. in 1968 [108]. The affected family from Iowa (USA) presented with peripheral neuropathy and renal failure – the ultimate cause of death. In 1988, amyloid was isolated from the kidney of one of the family members and studied biochemically [109]. N-terminal sequence analysis revealed for the first time ApoAI as the amyloid fibril subunit protein. A Gly to Arg substitution was identified at position 26. Complete sequencing of the fibril subunit showed that it consisted of the N-terminal 83 residues of ApoAI. Studies at the gene level confirmed the Gly26Arg substitution by identifying a G to C transversion in codon 26 (GGC → CGC) in exon 3 [110].

Family members heterozygous for the mutation were found to have HDL cholesterol levels below the 10th percentile. Metabolic studies carried out in individuals of the Iowa family utilizing iodinated ApoAI revealed a much shorter plasma residence time for variant ApoAI versus normal ApoAI, suggesting that metabolic factors are operative in pathogenesis [111]. Additional data raised the possibility that the variant was being sequestered extravascularly; although the variant was more rapidly cleared from plasma than the normal protein, the cumulative urinary excretion of radioactivity in patients injected with Gly26Arg ApoAI was 44% of the injected dose at 2 weeks, compared to 78% in patients injected with normal ApoAI. Metabolic studies carried out in rabbits confirmed faster clearance/catabolism of Gly26Arg ApoAI and suggested its association with HDL differed from that of normal ApoAI [112]. The replacement of Gly with Arg in the ApoAI sequence confers the variant protein with an additional +1 charge, allowing for its detection in plasma by electrophoresis and immunoblotting. Since its original identification in the Iowa family, Gly26Arg has been detected other families in the USA as well as the UK with renal disease as its chief manifestation [113–115].

22.3.4.2 **Trp50Arg**

Amyloidosis associated with this mutation was first described in 1995 in the UK in a patient who presented with non-neuropathic symptoms at age 34 and died at age 45 after renal transplantation [116]. The father of this individual had died at age 45 and was found to be positive for amyloid in the liver and kidney. DNA analysis identified a T to C transition in codon 50 (TGG → CGG) in exon 4. Amyloid subunits were ApoAI fragments comprising residues 1–86, 1–92 and 1–93. Relative to normal ApoAI, the Leu50Arg variant has an additional +1 charge. Two more patients with this mutation have now been identified in the UK [115].

22.3.4.3 **Leu60Arg**

This variant was first identified in 1992 in an English family who presented with progressive hypertension, thrombocytopenia and easy bruising, but no neuropathy [117]. DNA sequencing of the ApoAI gene revealed a T to G transversion in codon 60 (CTG → CGG) in exon 4. Amyloid fibrils isolated from spleen contained an ApoAI fragment of around 10 kDa. Electrospray mass spectrometry of the 10-kDa material showed components with masses corresponding to N-terminal fragments of 88, 92, 93 and 94 residues, each with a Leu to Arg substitution that was confirmed by direct protein sequencing. The Leu60Arg variant has an extra +1 charge compared to normal ApoAI. Five additional cases of this variant have been detected in the UK since the original report [115].

22.3.4.4 **Leu64Pro**

This variant was identified in the USA and also was one of seven detected at the National Amyloidosis Centre in the UK over the course of a 5-year period [115, 118]. Both cases were reported in abstracts in 2004. The UK patient, a middle-aged Italian man, presented with renal failure and had asymptomatic deposits in liver and spleen, while renal deposition was reported in the US patient. This variant was shown to arise from a T to C transition in codon 64 (CTC → CCC) in exon 4. Amyloid fibrils extracted from kidney of the US patient contained 96-residue fragments of ApoAI with a Leu64Pro substitution. The variant protein accounted for less than 10% of the total ApoAI in plasma.

22.3.4.5 Del 60–71, Ins Val–Thr

Fatal liver amyloidosis with little or no symptomatic renal involvement characterizes the phenotype of a Spanish family with this ApoAI variant reported in 1996 [119]. Affected family members have died between the ages of 48 and 66 years. A 35-bp deletion in exon 4 specifies a loss of residues 60–71 (Leu-Arg-Glu-Gln-Leu-Gly-Pro-Val-Thr-Gln-Glu-Phe). An insertion of five nucleotides (GTCAC) adds Val (GTC) and Thr (AC plus the residual third position C of codon 71) in exon 4. Splenic amyloid subunits were determined by electrospray mass spectrometry to be variant ApoAI peptides comprising residues 1–83 and 1–92. The variant ApoAI carries an additional +1 charge. Levels of ApoAI and HDL in plasma of heterozygous individuals were below normal, reflecting significantly reduced levels of variant ApoAI. Among 11 family members examined, the patient with the lowest ratio of variant to normal ApoAI in plasma had massive amyloid deposits in the liver and elsewhere.

22.3.4.6 Del 70–72

Renal failure due to massive amyloid has been the typical presentation in a South African family expressing del 70–72 ApoAI [120]. Reported in 1998, this variant is associated with early-onset, aggressive deposition in the liver and spleen, as well as kidneys. Family members who have had renal transplants have remained clinically well for nearly 20 years, despite the continued presence of massive visceral deposits. Their prolonged survival following transplantation possibly reflects the lack of heart involvement and slow tempo of deposition in the transplanted kidney. Phenotypic variability is noted as several affected family members have experienced only mild symptoms. Although no tissue has been available for biochemical analysis, complete concordance has been demonstrated between the occurrence of amyloid and the presence of a 9-bp deletion spanning codons 70–73 [GA(G–TTC–TGG–GA)T] (deleted bases are within parentheses) in exon 4. The net result is a loss of residues 70–72 (Glu–Phe–Trp) and a variant ApoAI carrying an extra +1 charge. Variant ApoAI was significantly less abundant in plasma than normal ApoAI; total ApoAI levels in all heterozygous individuals were below normal. An additional person with del 70–72 ApoAI amyloid was recently reported in the UK [115].

22.3.4.7 Leu75Pro

Mild, slowly progressing symptoms characterize the amyloid associated with this variant, first reported in 2001 [121, 122]. Amyloid was discovered in 11 unrelated individuals in northern Italy via liver biopsy performed on the basis of elevated liver enzymes. Immunoelectron microscopy indicated fibrils of ApoAI. DNA sequencing of the ApoAI gene revealed a T to C transition in codon 75 (CTG → CCG) in exon 4. The heterozygous mutation was found in all affected patients, but not in controls. Nine patients eventually suffered renal failure. Hypogonadism due to testicular involvement was also observed. Variant ApoAI was about 10% of the total protein in HDL. A family in the USA with Leu75Pro ApoAI was reported in 2003 [123]. With the exception of elevated liver enzymes, no clinical symptoms were noted. The Leu75Pro substitution was confirmed at the protein level by sequence analysis of plasma ApoAI and amyloid fibrils extracted from liver biopsy specimens. The subunit protein consisted predominantly of an approximately 96-residue N-terminal fragment of variant ApoAI. Variant ApoAI constituted less than 10% of total plasma ApoAI.

22.3.4.8 **Leu90Pro**

A French family presenting with cutaneous amyloid in the form of yellow maculopapular skin lesions and subsequently developing restrictive cardiomyopathy leading to death was described in 1999 [124]. Laryngeal, esophageal and rectal biopsies also contained amyloid, but no renal deposition was seen. DNA sequencing of the ApoAI gene revealed a T to C transition in codon 90 (CTG → CCG) in exon 4. Amyloid isolated from cardiac tissue contained peptides of around 7 and 9 kDa as determined by SDS-PAGE; sequence analysis identified peptides corresponding to normal ApoAI through residue 88. The peptide comprising residues 89–94 was also present, but in very low amounts and contained Pro rather than Leu at position 90. Thus, the vast majority of amyloid consisted of ApoAI peptides that did not extend C-terminal enough to include the substitution.

22.3.4.9 **Arg173Pro**

Cutaneous, laryngeal and cardiac amyloid deposition was also described in an American family in 1999 [125]. Skin rashes having the appearance of acanthosis nigricans and hoarseness due to laryngeal deposits have developed as early as age 20. Amyloid in dermal papillae and surrounding blood vessels has been associated with petechial hemorrhages and thickened skin of the neck and hands. Cardiomyopathy is slowly progressive and has been fatal in several cases. Amyloid isolated from limited amounts of skin biopsy tissue was analyzed by SDS–PAGE and N-terminal sequencing. ApoAI peptides of around 7 and 9 kDa were identified, indicating an amyloid subunit peptide comprising about 90–100 residues. Direct DNA sequencing of the ApoAI gene revealed a G to C transversion in codon 173 (CGC → CCC) in exon 4. Three additional cases of Arg173Pro ApoAI amyloid have been identified in the UK [115].

22.3.4.10 **Leu174Ser**

Massive heart involvement was the overriding clinical phenotype of an Italian family reported in 1999 [126]. Gene analysis detected a T to C transition in codon 174 (TTG → TCG) in exon 4. Amyloid-laden cardiac tissue became available after heart transplantation. Immunoelectron microscopy revealed anti-ApoAI-reactive fibrils. Purified amyloid, analyzed by SDS-PAGE, N-terminal sequencing and mass spectrometry, consisted of ApoAI fragments spanning residues 1–93. Levels of HDL and ApoAI in plasma were lower than in non-affected family members. An unrelated Italian patient having a similar clinical presentation was reported in 2001 with Leu174Ser ApoAI [127]. Biochemical analyses of amyloid fibrils isolated from heart tissue identified residues 1–93 of ApoAI as the subunit protein. Fibril material from both patients also contained trace amounts of full-length ApoA1 with an apparent prevalence of the normal over variant protein. X-ray diffraction patterns were also obtained for *ex vivo* amyloid fibrils. This analysis showed typical cross-β reflections, as well as reflections suggesting the presence of a well-defined coiled coil helical structure oriented within the fibrils. In plasma of the second patient, Leu174Ser ApoAI accounted for around 25% of the total ApoAI.

22.3.4.11 **Ala175Pro**

This variant was identified in patients studied at the National Amyloidosis Centre in the UK over the course of a 5-year period and reported in a 2004 abstract [115].

22.3.4.12 **Leu178His**

Cardiac, laryngeal and cutaneous amyloid deposition with relatively early onset in two siblings of a French family was reported in 2000 [128]. The proband, a 41-year old woman, presented with yellow, maculopapular skin lesions, laryngeal deposits, and signs of heart involvement and peripheral neuropathy; her brother had died at age 39 of cardiac amyloid. Immunohistochemistry of biopsy tissue revealed the presence of transthyretin (TTR) in addition to massive deposition of ApoAI. Fibrils extracted from skin biopsies were shown by SDS-PAGE and Western analysis to contain full-length ApoAI and N-terminal fragments of ApoAI. TTR immunoreactivity was detected in an 11-kDa fragment and a high-molecular-weight fragment. Sequencing of the ApoAI gene identified a T to A transversion in codon 178 (CTT → CAT) in exon 4. Normal and variant ApoAI were detected in plasma by isoelectric focusing and immunoblotting. Consistent with the predicted His for Leu substitution, the variant ApoAI carries an extra +1 charge. Ratios of the two forms in plasma were not reported. Plasma cholesterol and ApoAI levels in the proband were normal. No mutations in TTR were identified by single-strand conformation polymorphism (SSCP) analysis of the gene or by mass spectrometry of immunoprecipitated TTR.

A survey of the various mutations reveals that those in the N-terminal portion of ApoAI are associated with renal and/or hepatic amyloid deposition, while those in the C-terminal portion result in cardiac, cutaneous, and/or laryngeal deposition. It also is notable that six of the 12 ApoAI variants (Gly26Arg; Trp50Arg; Leu60Arg; del 60–71, ins Val-Thr; del70–72; Leu178His) carry an extra +1 charge relative to normal ApoAI. The possibility that this charge difference in some way enhances amyloidogenicity has been proposed. Five of the other six variants contain proline in place of the normal residue (Leu64Pro; Leu75Pro; Leu90Pro; Arg173Pro; Ala175Pro). In a study unrelated to amyloidogenic ApoAI variants, ApoAI charge and conformation were shown to regulate the *in vivo* clearance of reconstituted HDL [129]. The most positively charged lipoprotein AI particles were found to have the fastest clearance rate, while exposure of an epitope at residues 98–121 which increased the negative charge was associated with slower clearance. Consistent with these findings, and as discussed above, the more positively charged Gly26Arg ApoAI variant has been shown to undergo faster clearance than normal ApoAI [111]. In light of proline's tendency to disrupt α-helical structure, it would not be surprising to find that the five TTR variants with proline substitutions have decreased lipid affinity and hence faster clearance similar to those with +1 charge. It seems reasonable to propose that ApoAI prior to undergoing amyloid formation is cleared from the plasma and initially embarks on the pathway toward catabolism, which presumably routes the protein into an environment conducive to unfolding. However, for reasons yet to be determined, rather than being completely degraded, ApoAI undergoes refolding and partial cleavage. Thus, for proteins such as ApoAI, SAA, and ApoAII that have inherent amyloidogenic potential, factors favoring clearance and catabolism could also favor amyloidogenesis.

22.4 ApoAII Amyloidosis

22.4.1 Background

ApoAII is the second most abundant protein on HDL, accounting for around 20% of the protein. It is present in plasma at a concentration of about 30–40 mg/dl. In contrast to ApoAI, whose anti-atherogenic properties are well accepted, the physiologic function of ApoAII remains unclear. While there is evidence that ApoAII plays a role in HDL remodeling and metabolism, its particular effects on reverse cholesterol transport and atherogenesis are complex and controversial. In some systems and under certain conditions, the presence of ApoAII correlates with enhanced reverse cholesterol transport, while in other studies, ApoAII demonstrates properties considered to be pro-atherogenic (reviewed in [130–132]). The only known case of ApoAII deficiency in humans is that of two sisters, both homozygous for an intron 3/exon 3 splice site mutation in the ApoAII gene. Despite their lack of ApoAII, they had HDL cholesterol levels in the normal range and did not manifest coronary heart disease [133]. *In vi-*

tro approaches to decipher ApoAII function include the use of synthetically constructed lipoprotein particles or native isolated lipoprotein particles in cell culture systems; *in vivo* studies have utilized genetically modified mice that either overexpress human or mouse ApoAII or are deficient in mouse ApoAII. In order to measure effects of ApoAII on certain activities, conditions and/or reconstituted lipoprotein particles are employed that occur rarely or not at all *in vivo* (e.g. HDL having artificially high levels of ApoAII) (see Chapters 6 and 10). As a consequence, data among the different studies are difficult to compare, and consensus is often lacking.

22.4.2
Gene and Protein Structure

The human ApoAII gene has been cloned, sequenced, and mapped to chromosome 1 in the 1q21–1q23 region ([134–137], reviewed in [138]). It has the same basic structure of four exons and three introns as genes for other apolipoproteins, including SAA and ApoAI: exon 1 (5′-UTR), intron 1 (169 bp), exon 2 (5′-UTR, plus codons –23 through –6), intron 2 (293 bp), exon 3 (codons –6 through codon 39), intron 3 (395 bp) and exon 4 (codons 39 through codon 77 plus 3′-UTR). Several polymorphisms in the ApoAII gene locus have been identified. A polymorphic microsatellite GT repeat is present in intron 2 and has been used as a marker in linkage studies [139]. Three single nucleotide polymorphisms include a G to A transition affecting intron 3 splicing (resulting in ApoAII deficiency as noted above) [133], a C to T transition in intron 3 [140] and a C to T transition in the 3′-flanking region [141]. Other than these polymorphisms, mutations in the human ApoAII (except for those discussed in this chapter) have not been reported, possibly because, if they do exist, they do not result in an altered phenotype. This makes those associated with amyloidosis – four single base mutations in the ApoAII stop codon (out of a total of eight possible single base mutations), all described since 2001 – especially intriguing (Fig. 22.4).

ApoAII is synthesized in the liver as a 100-amino-acid pre-pro-polypeptide similar to ApoAI. An 18-residue pre-segment is removed co-translationally, and cleavage of the five-residue pro-segment is accomplished by a thiol protease after secretion. Mature human ApoAII contains 77 residues and is present in plasma as a homodimer formed by disulfide linkage of cysteines at position 6 (Fig. 22.2); the dimer has a molecular mass of around 17.4 kDa (reviewed in [138], [142]). Mouse ApoAII, lacking cysteine, circulates as a monomer [143]. Mouse and human ApoAII proteins share about 55% amino acid identity. Both are predicted to have significant α-helical content, 68% in mouse and 62% in human, with helices having amphipathic properties of both polar and apolar faces appropriate for the observed ApoAII–phospholipid interaction. In human ApoAII, residues 7–30, 39–50 and 51–70 have been predicted to form class A amphipathetic helices; lipid-associating domains have been identified within residues 12–31 and residues 40–77 [144]. As measured by CD, the α-helical content of ApoAII, like ApoAI and SAA, is increased upon its association with phospholipids. Although ApoAII is

→ATG AAG CTG CTC GCA GCA
M K L L A A
-23 (M), -20 (L)

ACT GTG CTA CTC CTC ACC ATC TGC AGC CTT GAA GGA →GCT TTG GTT CGG AGA →CAG GCA
T V L L L T I C S L E G A L V R R Q A
-15 (L), -10 (C), -6 (G), -5 (A), -1 (R), +1 (Q)

AAG GAG CCA TGT GTG GAG AGC CTA GTT TCT CAG TAC TTC CAG ACC GTG ACT GAC TAT
K E P C V E S L V S Q Y F Q T V T D Y
5 (P), 10 (L), 15 (F), 21 (Y)

GGC AAG GAC CTG ATG GAG AAG GTC AAG AGC CCA GAG CTT CAG GCC GAG GCC AAG TCT
G K D L M E K V K S P E L Q A E A K S
25 (L), 30 (K), 35 (Q), 40 (S)

TAC TTT GAA AAG TCA AAG GAG CAG CTG ACA CCC CTG ATC AAG AAG GCT GGA ACG GAA
Y F E K S K E Q L T P L I K K A G T E
45 (S), 50 (T), 55 (K)

CTG GTT AAC TTC TTG AGC TAT TTC GTG GAA CTT GGA ACA CAG CCT GCC ACC CAG TGA
L V N F L S Y F V E L G T Q P A T Q STOP
60 (L), 65 (S), 70 (L), 75 (A), 77 (Q)
(GGA) G
(TCA) S
(CGA) R
(AGA) R

AGT GTC CAG ACC ATT GTC TTC CAA CCC CAG CTG GCC TCT AGA ACA CCC ACT
S V Q T I V F Q P Q L A S R T P T
80 (V), 85 (F), 90 (A), 95 (T)

GGC CAG TCC TAG AGC TCC TGT CCC TAC CCA CTC TTT GCT ACA ATA AAT
G Q S STOP
98 (S)

Fig. 22.4 Nucleotide and amino acid sequence of normal and amyloid-associated apoAII. Mutations in stop codon and resulting amino acid substitutions and extension of translated sequence are shown. Arrows indicate in order start of prepropeptide, propeptide, and mature apoAI.

present in humans, rodents and fish, it is either absent or expressed at a low level in dogs, pigs, chickens and rabbits [145].

22.4.3
Association with HDL and Potential Function

ApoAII is mainly associated with HDL in the subclass containing both ApoAI and ApoAII (HDL-LpAI/AII). Only minor amounts of ApoAII are associated with chylomicrons, very-low-density lipoprotein and HDL particles lacking ApoAI (HDL-LpAII). Formation of HDL-LpAI/AII appears to occur in plasma possibly by fusion of HDL-LpAI and HDL-LpAII or by displacement of ApoAI from HDL by ApoAII (reviewed in [130]). It is fairly well accepted that plasma HDL cholesterol and ApoAI levels are governed by the rate of ApoAI catabolism, whereas ApoAII levels are determined by synthetic rate (reviewed in [131]).

Data about the effects of ApoAII on HDL structure and metabolism are often contradictory (reviewed in [130, 131]). There is some consensus that HDL-AI is more active than HDL-AI/AII in overall reverse cholesterol transport. Addition of ApoAII to reconstituted HDL-AI particles has been shown to correlate with decreased negative surface charge and marked increase in rate of ApoAI clearance from plasma, supporting the view that ApoAII acts in a pro-atherogenic manner. However, turnover studies in humans have generated seemingly contradictory data showing slower clearance of ApoAI from HDL-AI/AII than from HDL-AI.

The possibility that ApoAII influences the ability of HDL to efflux cholesterol from peripheral cells has been explored using transgenic mice and cell culture systems (reviewed in [130, 131, 146]). In the majority of studies, HDL-AI was more effective in capturing cell-derived cholesterol than was HDL-AI/AII. In one study, progressive increases in the mass ratio of ApoAII to ApoAI on HDL were associated with significantly decreased rates of cholesterol efflux. In another study, cholesterol efflux from cultured cells was measured using serum from genetically engineered mice as the cholesterol acceptor. Serum from ApoAI-deficient mice which expressed human ApoAI, but not serum from ApoAI-deficient mice expressing human ApoAII showed improved cholesterol efflux compared to control serum from ApoAI-deficient mice. Other studies, however, have shown HDL-AI and HDL-AI/AII to be equally effective in promoting cholesterol efflux. Results are obviously influenced by the structure and composition of ApoAII-containing lipoprotein particles or serum used as cholesterol acceptor (as they can vary significantly depending on the nutritional status of their source) and by the type of cell relinquishing cholesterol.

The capacity of ApoAII to modulate the activity of enzymes involved in HDL metabolism, including LCAT, hepatic lipase (HL) and CEPT, remains unclear (reviewed in [130, 131]). In general, data suggest an inhibitory effect of ApoAII on LCAT activity. It has been proposed that ApoAII exerts it effects by decreasing LCAT binding to HDL-LpAI/AII. Consistent with this, a larger proportion of LCAT in human plasma is bound to HDL-LpAI than to HDL-LpAI/AII; a similar preference of LCAT for HDL-LpAI was observed in the plasma of mice expressing human proteins LCAT, ApoAI and ApoAII. Consensus is lacking as to whether ApoAII activates, inhibits or has no effect on HL. Similar to ApoAI, ApoAII appears to either inhibit or not influence CEPT activity.

Regarding the role of ApoAII in selective lipid uptake, greater uptake of cholesteryl esters from HDL-AI than from HDL-AI/AII has been demonstrated. ApoAII, like ApoAI, was a good ligand for SR-B1, but appeared to impair uptake via an antagonist effect. In contrast, another group found that HDL-AI/AII had less affinity than HDL-AI for SR-BI, while they observed no decrease in selective cholesterol ester uptake from HDL-AI/AII relative to that from HDL-AI. Conflicting findings may derive from the use of non-physiologic components, e.g. lipid-poor ApoAII or ApoAII-rich HDL, and differences in cell systems (reviewed in [130, 131]).

Recent studies have investigated potential biochemical and genetic linkage between ApoAII and familial combined hyperlipidemia (FCH), insulin resistance, and Type 2 diabetes (reviewed in [147]). It has been reported that transgenic mice overexpressing mouse ApoAII exhibit several traits associated with the insulin resistance syndrome, including increased atherosclerosis, hypertriglyceridemia, obesity and decreased fatty acid oxidation, while ApoAII-deficient mice display opposite phenotypes. Moreover, the location of *APOAII* on chromosome

1q21–1q23 places the gene within a region that is being intensely searched for genes associated with these complex, often overlapping diseases. Chromosomal loci associated with plasma ApoAII levels have been identified in families with FCH. The loci, however, do not overlap with those associated with FCH itself or with the ApoAII gene itself, underscoring the complexity of potential metabolic interactions contributing to FCH.

22.4.4 Amyloidogenic Variants of Human ApoAII

Aside from its enigmatic physiologic role, ApoAII (like HDL proteins SAA and ApoAI) is now known to cause amyloidosis. Amyloid-forming ApoAII proteins are variants that have a 21-amino acid extension at the C-terminus resulting from one of several mutations in the stop codon of the ApoAII gene. While the peptide extension obviously confers amyloidogenic potential to ApoAII, this potential can only be realized when ApoAII is free from HDL. Thus, ApoAII metabolism, especially as it relates to HDL, is expected to be a key determinant of amyloid formation. At the same time, metabolism of variant ApoAII is likely to be altered from normal due to changes in secondary structure. Although the peptide extension is predicted to contain an α-helix, the helix differs from the class A amphipathic helices which are present in normal ApoAII and have been shown to be important for lipid binding [148]. It seems reasonable to speculate that metabolic changes, even slight shifts favoring dissociation of ApoAII from HDL, could favor amyloid fibril formation. Currently, there are three known amyloidogenic variant ApoAII proteins encoded by four different stop codon mutations.

22.4.4.1 Stop78Gly

Amyloidosis now known to be associated with this variant was first described clinically in 1973 in the USA [149]. Two sisters presented with renal failure, and died at 47 and 52 years of age, despite kidney transplants. Autopsies revealed amyloid in vessel walls throughout various organs and giant cells in the kidney that looked to be reabsorbing amyloid. Two brothers in the next generation also suffered renal failure. Amyloid in the walls of small vessels and massive glomerular deposition was detected in biopsy specimens. All genes known at that time to be associated with hereditary amyloidosis were screened for mutations, but none were found. Tissues were made available when one of the individuals died. In sharp contrast to biopsies examined 14 years prior to death, renal autopsy specimens lacked identifiable glomerular structures and, instead, showed extensive cyst formation. The only remaining amyloid was present in vessel walls. Isolation of this material and analysis by SDS-PAGE yielded a 10-kDa protein having the N-terminal sequence of ApoAII. Complete analysis identified the amyloid subunit as ApoAII comprising the normal 77 residues plus a 21-amino-acid extension at the C-terminus. ApoAII lacking the extension was not detected, suggesting that amyloid fibril formation had occurred after reduction of the disulfide-linked ApoAII dimer to a monomeric state. Homodimers of normal and variant ApoAII, as well as heterodimers, were seen in plasma analyzed by Western analysis under non-reducing conditioning. Consistent with protein findings, ApoAII gene analysis showed a T to G transversion in the stop codon (TGA → GGA). The mutation created codon 78 which encoded Gly and was followed by 60 bases, encoding an additional 20 residues, before a new stop codon was encountered [148].

22.4.4.2 **Stop78Ser**

This variant was identified in 2001 in a patient in the USA [150]. A 42-year-old patient presented with proteinuria and elevated serum creatinine, and was found by biopsy to have amyloid in glomeruli and vessel walls. No other clinical manifestations have been noted. The ApoAII gene of this patient shows a G to C transition in the stop codon (TGA → TCA). The mutation was verified at the protein level by sequence analysis of ApoAII isolated from plasma. Both normal ApoAII having 77 residues and a variant ApoAII having the predicted Ser at position 78 and 21-residue extension were found. No other family members manifest clinical symptoms or have been tested by DNA analysis for the mutation.

22.4.4.3 **Stop78Arg**

In 2003, a third ApoAII amyloidogenic variant was identified in an Armenian-Russian family who present during middle age with slowly progressive renal insufficiency [151]. Extensive glomerular amyloid and amyloid in vessel walls was seen in kidney biopsy specimens. DNA analysis of the proband's ApoAII gene showed a T to C transition in the stop codon (TGA → CGA), predicting Arg at position 78 followed by 20 additional residues. As expected, his plasma contained a larger immunoreactive ApoAII band, as well as the normal-sized ApoAII band. Amyloid fibrils isolated from rectal biopsy tissue, however, contained only the larger variant ApoAII as revealed by Western analysis. While no other family members have been tested for this ApoAII amyloidogenic variant, two other individuals have suffered renal failure. The same ApoAII variant, Stop78Arg, was reported in 2004 in a Spanish family [152]. In this family, however, the mutation in the stop codon resulting in Arg is a T to A transition (TGA → AGA). Renal biopsies revealed substantial glomerular and vascular amyloid deposits in the Spanish family, similar to the clinical manifestation of the Russian family.

22.4.5 **Mouse ApoAII Amyloidosis**

The amyloidogenic potential of ApoAII was recognized in mice long before human ApoAII amyloidosis was identified. Spontaneous age-associated amyloid in a mouse model of accelerated senescence was reported in 1982 and isolated and identified as ApoAII in 1986 by Takeda et al. [153]. Six of 11 senescence-accelerated prone strains (designated SAMP), which they have characterized, develop severe spontaneous systemic amyloidosis (SAMP1, 2, 7, 9, 10 and 11). Early amyloid deposition occurs in liver, spleen, stomach, intestine and tongue; eventually all organs except brain in SAMP mice are affected (reviewed in [154]). The amyloid-prone SAMP strains, as well as A/J, SJL/J, SM/J and IVCS strains that also exhibit ApoAII amyloidosis, are homozygous for the *ApoAII*^{c} allele. *ApoAII*^{c} is one of seven alleles defined by fifteen polymorphisms that generate eight amino acid substitutions. Residues defining *ApoAII*^{c} include Gln5, Ser9, Gln13, Gln16, Glu20, Val26, Ala 38, Arg54 and Asn62 [155]. The particular structure of ApoAIIC has been shown to be a strong determinant of the amyloid phenotype and capable of triggering less amyloidogenic ApoAII proteins to engage in fibril formation [156]. A congenic strain (R1.P1-ApoAII^C) having the ApoAIIC allele on the genetic background of the amyloid-resistant SAMR1 strain shows a high incidence of severe amyloidosis. Moreover, injection of SAMR1 mice with ApoAIIC amyloid fibrils induces amyloid formation from the endogenous and normally non-amyloidogenic ApoAIIB isotype. Conformational rearrangement of ApoAIIB in the presence of ApoAIIC fibrils supports a nucleation-dependent mechanism of polymerization and raises the possibility of amyloid transmissibility. Studies using the ApoAII mouse model continue to provide valuable insights into the pathogenesis of amyloid diseases.

22.5 Conclusion

The proteins discussed in this chapter – SAA, ApoAI and ApoAII – normally reside in the circulation as functional, HDL-bound proteins having prominent α-helical content. Their conversion into insoluble, biologically inert β-pleated sheet amyloid fibrils is a complex pathologic process. It has become clear that very minor differences in amino acid composition, often substitution of a single residue, can confer amyloid-forming potential. Realization of that potential, however, depends on input from many genetically determined (and possibly environmental) factors. Even subtle differences in the metabolic processing of SAA isotypes or normal versus variant ApoAI and ApoAII proteins may, over time, allow their amyloidogenic structure to be manifest. Identification of the factors that "bring out the worst" in SAA, ApoAI and ApoAII begs investigation.

References

1 Pras, M., M. Schubert, D. Zucker-Franklin, A. Rimon and E. Franklin. The characterization of soluble amyloid prepared in water. *J Clin Invest* **1968**, *47*, 924–933.

2 Husby, G., G. Marhaug, B. Dowton, K. Sletten and J. D. Sipe. Serum amyloid A (SAA): biochemistry, genetics and the pathogenesis of AA amyloidosis. *Amyloid: Int J Clin Exp Invest* **1994**, *1*, 119–137.

3 Uhler, C. M. and A. S. Whitehead. Serum amyloid A, the major vertebrate acute-phase reactant. *Eur J Biochem* **1999**, *265*, 501–523.

4 Sipe, J. D. and A. S. Cohen. History of the amyloid fibril. *J Struct Biol* **2000**, *130*, 88–98.

5 Benson, M. D. Amyloidosis. In *Arthritis and Allied Conditions: A Textbook of Rheumatology*, 14th edn, W. J. Koopman (ed.). Lippincott Williams & Wilkins, Philadelphia, PA, **2001**, pp. 1866–1895.

6 Sletten, K., K. H. Johnson and P. Westermark. The amino acid sequence of protein AA from a burro (*Equus asinus*). *Amyloid: J Protein Folding Disord* **2003**, *10*, 144–146.

7 Johnson, K. H., K. Sletten, L. Munson, T. D. O'Brien, R. Papendick and P. Westermark. Amino acid sequence analysis of amyloid protein A (AA) from cats (captive cheetahs: *Acinoyx jubatus)* with a high prevalence of AA amyloidosis. *Amyloid: Int J Exp Clin Invest* **1997**, *4*, 171–177.

8 Syverson, P. V., J. Juul, G. Marhaug, G. Husby and K. Sletten. The primary structure of serum amyloid A protein in sheep: comparison with serum amyloid A in other species. *Scand J Immunol* **1994**, *39*, 88–94.

9 Kluve-Beckerman, B., F. E. Dwulet, S. P. DiBartola and M. D. Benson. Primary structures of dog and cat amyloid A proteins: comparison to human AA. *Comp Biochem Physiol* **1989**, *94B*, 175–183.

10 Woo, P., M. Edbrooke, J. Betts, G. Watson and P. Francis, Serum amyloid A gene regulation. In *Acute Phase Proteins: Molecular Biology, Biochemistry and Clinical Application*, A. Mackiewicz, I. Kushner and H. Baumann (eds). CRC Press, Boca Raton, FL, **1993**, pp. 397–408.

11 Sellar, G. C., K. Oghene, S. Boyle, W. A. Bickmore and A. S. Whitehead. Organization of the region encompassing the human serum amyloid A (SAA) gene family on chromosome 11p15.1. *Genomics* **1994**, *23*, 492–495.

12 Betts, J. C., M. R. Edbrooke, R. V. Thakker and P. Woo. The human acute-phase serum amyloid A gene family: structure, evolution, and expression in hepatoma cells. *Scand J Immunol* **1991**, *34*, 471–482.

13 Kluve-Beckerman, B. and M. Song. Genes encoding human serum amyloid A proteins SAA1 and SAA2 are located 18 kb apart in opposite transcriptional orientations. *Gene* **1995**, *159*, 289–290.

14 Whitehead, A. S., M. C. de Beer, D. M. Steel, M. Rits, J. M. Lelias, W. S. Lane and F. C. de Beer. Identification of novel members of the serum amyloid A protein superfamily as constitutive apolipoproteins of high density lipoprotein. *J Biol Chem* **1992**, *267*, 3862–3867.

15 Kluve-Beckerman B., M. L. Drumm and M. D. Benson. Non-expression of the human serum amyloid A three (*SAA3*) gene. *DNA Cell Biol* **1991**, *10*, 651–661.

16 Larson, M. A., S. H. Wei, A. Weber, A. T. Weber and T. L. McDonald. Induction of human mammary-associated serum amyloid A3 expression by prolactin or lipopolysaccharide. *Biochem Biophys Res Commun* **2003**, *301*, 1030–1037.

17 Lowell C. A., D. A. Potter, R. S. Stearman and J. F. Morrow. Structure of the murine serum amyloid A gene family. *J Biol Chem* **1986**, *261*, 8442–8452.

18 Butler, A. and A. S. Whitehead. Mapping of the mouse serum amyloid A gene cluster by long-range polymerase chain reaction. *Immunogenetics* **1996**, *44*, 468–474.

19 Stearman, R. S., C. A. Lowell, C. G. Peltzman and J. F. Morrow. The sequence and structure of a new serum amyloid A gene. *Nucleic Acids Res* **1986**, *14*, 797–809.

20 Sipe J. D., H. Rokita and F. C. de Beer. Cytokine regulation of the mouse SAA gene family. In *Acute Phase Proteins: Molecular Biology, Biochemistry and Clinical Application*, A. Mackiewicz, I. Kushner and H. Baumann (eds). CRC Press, Boca Raton, FL, **1993**, pp. 511–526.

21 de Beer, M. C., F. C. de Beer, W. D. McCubbin, C. M. Kay and M. S. Kindy. Structural prerequisites for serum amyloid A fibril formation. *J Biol Chem* **1993**, *268*, 20606–20612.

22 Sipe, J. D., I. Carreras, W. A. Gonnerman, E. S. Cathcart, M. C. de Beer and F. C. de Beer. Characterization of the inbred CE/J mouse strain as amyloid resistant. *Am J Pathol* **1993**, *143*, 1480–1485.

23 Meek, R. L., N. Ericksen and E. P. Benditt. Murine serum amyloid A3 is a high density apolipoprotein and is secreted by macrophages. *Proc Natl Acad Sci USA* **1992**; *89*, 7949–7952.

24 de Beer, M. C., F. C. de Beer, C. J. Geradot, D. R. Cecil, M. R. Webb, M. L. Goodson and M. S. Kindy. Structure of the mouse *Saa4* gene and its linkage to the serum amyloid A gene family. *Genomics* **1996**, *34L*, 139–142.

25 de Beer, M. C., M. S. Kindy, W. S. Lane and F. C. de Beer. Mouse serum amyloid A protein (SAA5) structure and expression. *J Biol Chem* **1994**, *269*, 4661–4667.

26 Sipe, J. and Committee. Editorial Part 2. Revised nomenclature for serum amyloid A (SAA). *Amyloid: Int J Exp Clin Invest* **1999**, *6*, 67–70.

27 Faulkes, D. J., J. C. Betts and P. Woo. Characterization of five human serum amyloid A_1 alleles. *Amyloid: Int J Exp Clin Invest* **1994**, *1*, 255–262.

28 Hoffman, J. S., L. H. Ericcson, E. Eriksen, K. A. Walsh and E. P. Benditt. Murine tissue amyloid protein AA, NH_2-terminal sequence identity with only one of two serum amyloid proteins (apoSAA) gene products. *J Exp Med* **1984**, *159*, 641–646.

29 Liepnieks, J. J., B. Kluve-Beckerman and M. D. Benson. Characterization of amyloid A protein in human secondary amyloidosis: the predominant deposition of serum amyloid A1. *Biochim Biophys Acta* **1995**, *1270*, 81–86.

30 Baba, A., T. Takahashi, T. Kasama, M. Fujie and H. Shirawasa. A novel polymorphism of human serum amyloid A protein, SAA1, is characterized by alanines at both residues 52 and 57. *Biochim Biophys Acta* **1992**, *282*, 615–620.

31 Baba, S., S. A. Masago, T. Takahashi, T. Kasama, H. Sugimura, S. Tsugane, Y. Tsutsui and H. Shirasawa. A novel allelic variant of serum amyloid A, SAA1 gamma: genomic evidence, evolution, frequency, and implication as a risk factor for reactive systemic AA-amyloidosis. *Hum Mol Genet* **1995**, *4*, 1083–1087.

32 Moriguchi, M., C. Terai, Y. Koseki, M. Uesato, A. Nakajima, S. Inada, M. Nishinarite, S. Uchida, A. Nakajima, S. Y. Kim, C. L. Chen and N. Kamatani. Influence of genotypes at SAA1 and SAA2 loci on the development and the length of latent period of secondary AA-amyloidosis in patients with rheumatoid arthritis. *Hum Genet* **1999**, *105*, 360–366.

33 Moriguchi, M., C. Terai, H. Kaneko, Y. Koseki, H. Kajiyama, M. Uesato, S. Inada and N. Kamatani. A novel single-nucleotide polymorphism at the 5′-lanking region of SAA1 associated with risk or type AA amyloidosis secondary to rheumatoid arthritis. *Arthritis Rheum* **2001**, *44*, 1266–1272.

34 Yamada, T., Y. Okuda, K. Takasugi, L. Wang, D. Marks, M. D. Benson and B. Kluve-Beckerman. An allele of serum amyloid A1 associated with amyloidosis in both Japanese and Caucasians. *Amyloid: J Protein Folding Disord* **2003**, *10*, 7–11.

35 Booth, D. R., S. E. Booth, J. D. Gillimore, P. N. Hawkins and M. B. Pepys. SAA1 alleles as risk factors in reactive systemic amyloidosis. *Amyloid: Int J Exp Clin Invest* **1998**, *5*, 262–265.

36 Cazeneuve, C., H. Ajrapetyan, S. Papin, F. Roudot-Thoraval, D. Genevieve, E. Mndjoyan, M. Papazian, A. Sarkisian, A. Babloyan, B. Boissier, P. Duguesnoy, J.-C. Kouyoumdjian, E. Girodon-Boulandet, G. Grateau, T. Sarkisian and S. Amselem. Identification of *MEFV*-independent modifying genetic factors for familial Mediterranean fever. *Am J Hum Genet* **2000**, *67*, 1136–1143.

37 Medlej-Hashim, M., V. Delague, E. Chouery, N. Salem, M. Rawashdeh, G. Lefranc, J. Loiselet and A. Megarbane. Amyloidosis in familial Mediterranean fever patients: correlation with *MEFV* genotype and SAA1 and *MICA* polymorphisms effects. *BMC Med Genet* **2004**, *5*, 4.

38 Gershoni-Baruch, R., R. Brik, N. Zacks, M. Shinawi, M. Lidar and A. Livneh. The contribution of genotypes at the MEFV and SAA1 loci to amyloidosis and disease severity in patients with familial Mediterranean fever. *Arthritis Rheum* **2003**, *48*, 1149–1155.

39 Bakkaloglu, A., A. Duzova, S. Ozen, B. Balci, N. Besbas, R. Topaloglu, F. Ozaltin and E. Yilmaz. Influence of serum amyloid A (SAA1) and SAA2 gene polymorphisms on renal amyloidosis, and on SAA/C-reactive protein values in patients with familial Mediterranean fever in the Turkish population. *J Rheumatol* **2004**, *31*, 1139–1142.

40 Akar, N., M. Hasipek, E. Akar, M. Ekim, F. Yalcinkaya and N. Cakar. Serum amyloid A1 and tumor necrosis factor-alpha alleles in Turkish familial Mediterranean fever patients with and without amyloidosis. *Amyloid: J Protein Folding Disord* **2003**, *10*, 12–16.

41 Bausserman, L.L., P.N. Herbert, T. Forte, R.D. Klausner, K.P. McAdam, J.C. Osborne, Jr and M. Rosseneu. Interaction of the serum amyloid A proteins with phospholipid. *J Biol Chem* **1983**, *258*, 10681–10688.

42 Turnell, W.G., R. Sarra, I.D. Glover, J.O. Baum and D. Caspi. Secondary structure of human SAA1. Presumptive identification of calcium and lipid binding sites. *Mol Biol Med* **1986**, *3*, 387–407.

43 Patel, H., J. Bramall, H. Waters, M.C. de Beer and P. Woo. Expression of recombinant human serum amyloid A in mammalian cells and demonstration of the region necessary for high-density lipoprotein binding and amyloid fibril formation by site-directed mutagenesis. *Biochem J* **1996**, *318*, 1041–1049.

44 Westermark, G.T., U. Engstrom and P. Westermark. The N-terminal segment of protein AA determines its fibrillogenic property. *Biochem Biophys Res Commun* **1992**, *182*, 27–33.

45 McCubbin, W.D., C.M. Kay, S. Narindrasorasak and R. Kisilevsky. Circular-dichroism studies on two murine serum amyloid A proteins. *Biochem J* **1988**, *256*, 775–783.

46 Wang, L., H.A. Lashuel, T. Walz and W. Colón. Murine apolipoprotein serum amyloid A in solution forms a hexamer containing a central channel. *Proc Natl Acad Sci USA* **2002**, *99*, 15947–15952.

47 Wang, L. and W. Colón. The interaction between apolipoprotein serum amyloid A and high-density lipoprotein. *Biochem Biophys Res Commun* **2004**, *317*, 157–161.

48 Liang, J.S., B.M. Schreiber, M. Salmona, G. Phillip, W.A. Gonnerman, F.C. de Beer and J.D. Sipe. Amino terminal region of acute phase, but not constitutive, serum amyloid A (apoSAA) specifically binds and transports cholesterol into aortic smooth muscle and HepG2 cells. *J Lipid Res* **1996**, *37*, 2109–2116.

49 Sipe, J.D. Serum amyloid A: from fibril to function. Current status. *Amyloid: Int J Exp Clin Invest* **2000**, *7*, 10–12.

50 Mackiewicz, A., Kushner, I. and Baumann, H. (eds). *Acute Phase Proteins: Molecular Biology, Biochemistry and Clinical Applications*. CRC Press, Boca Raton, FL, **1993**.

51 Schreiber, B.M. Editorial. Serum amyloid A; in search of function. *Amyloid: J Protein Folding Disord* **2002**, *9*, 279–280.

52 Li, L. and W.S.L. Liao. An upstream repressor element that contributes to hepatocyte-specific expression of the rat serum amyloid A1 gene. *Biochem Biophys Res Commun* **1999**, *264*, 395–403.

53 Jiang, S.-L., D. Samols, D. Rzewnicki, S.S. Macintyre, I. Greber, J. Sipe and I. Kushner. Kinetic Modeling and Mathematical analysis indicate that acute phase gene expression in Hep3B cells is regulated by both transcriptional and post-transcriptional mechanisms. *J Clin Invest* **1995**, *95*, 1253–1261.

54 Oram, J.F. Does serum amyloid A mobilize cholesterol from macrophages during inflammation? *Amyloid: Int J Exp Clin Invest* **1996**, *3*, 290–293.

55 Kisilevsky, R. and L. Subrahmanyan. Serum amyloid A changes high density lipoprotein's cellular affinity: a clue to serum amyloid A's principal function. *Lab Invest* **1992**, 66, 778–785.

56 Banka, C. L., T. Yuan, M. C. de Beer, M. Kindy, L. K. Curtiss and F. C. de Beer. Serum amyloid A (SAA): influence on HDL-mediated cellular cholesterol efflux. *J Lipid Res* **1995**, *36*, 1058–1065.

57 Steinmetz, A., G. Hocke, R. Saile, P. Puchois and J.-C. Fruchart. Influence of Serum amyloid A on cholesterol esterification in human plasma. *Biochim Biophys Acta* **1989**, *1006*, 173–178.

58 Lindhorst, E., D. Young, W. Bagshaw, M. Hyland and R. Kisilevsky. Acute inflammation, acute phase serum amyloid A and cholesterol metabolism in the mouse. *Biochim Biophys Acta* **1997**, *1339*, 143–154.

59 Pruzanski, W., E. Stefanski, F. C. de Beer, M. C. de Beer, P. Vadas, A. Ravandi and A. Kuksis. Lipoproteins are substrates for human secretory group IIA phospholipase A2: preferential hydrolysis of acute phase HDL. *J Lipid Res* **1998**, *39*, 2150–2160.

60 Bausserman, L. L., A. L. Van Saritelli, P. Zuiden, C. J. Gollaher and P. N. Herbert. Degradation of serum amyloid A by isolated perfused rat liver. *J Biol Chem* **1987**, *262*, 1583–1589.

61 Tape, C. and R. Kisilevsky. Apolipoprotein A-I and apolipoprotein SAA half-lives during acute inflammation and amyloidogenesis. *Biochim Biophys Acta* **1990**, *1043*, 295–300.

62 Rocken C. and A. Shakespeare. Pathology, diagnosis and pathogenesis of AA amyloidosis. *Virchows Arch* **2002**, *440*, 111–122.

63 Rocken, C. and R. Kisilevsky. Comparison of the binding and endocytosis of high-density lipoprotein from healthy (HDL) and inflamed (HDLSAA) donors by murine macrophages of four different mouse strains. *Virchows Arch* **1998**, *432*, 547–555.

64 Chan, S. L., S. Chronopoulos, J. Murray, D. W. Laird and Z. Ali-Khan. Selective localization of murine ApoSAA$_1$/SAA$_2$ in endosomes-lysosomes in activated macrophages and their degradation products. *Amyloid: Int J Exp Clin Invest* **1997**, *4*, 40–48.

65 Kluve-Beckerman, B., J. J. Manaloor and J. J. Liepnieks. Binding, trafficking, and accumulation of serum amyloid A in peritoneal macrophages. *Scand J Immunol* **2001**, *53*, 393–400.

66 Yan, S. D., H. Zhu, A. Zhu, A. Golabek, H. Du, A. Roher, J. Yu, C. Soto, A. M. Schmidt, D. Stern and M. Kindy. Receptor-dependent cell stress and amyloid accumulation in systemic amyloidosis. *Nat Med* **2000**, *6*, 643–651.

67 Su, S. B., W. H. Gong, J. L. Gao, W. Shen, P. M. Murphy, J. J. Oppenheim and J. M. Wang. A seven-transmembrane, G protein-coupled receptor, FPRL1, mediates the chemotactic activity of serum amyloid A for human phagocytic cells. *J Exp Med* **1999**, *189*, 395–402.

68 Shirahama, T. S. and A. Cohen. Intralysosomal formation of amyloid fibrils. *Am J Pathol* **1975**, *81*, 101–116.

69 Takahashi, M., T. Yokota, H. Kawano, T. Gondo, T. Ishihara and F. Uchino. Ultrastructural evidence for intracellular formation of amyloid fibrils in macrophages. *Virchows Arch A* **1989**, *415*, 411–419.

70 Chronopoulos, S., S. L. Chan, M. J. H. Ratcliffe and Z. Ali-Khan. Colocalization of ubiquitin and serum amyloid A and ubiquitin-bound AA in the endosomes–lysosomes: a double immunogold electron microscopic study. *Amyloid: Int J Exp Clin Invest* **1995**, *2*, 191–194.

71 Miura K. and H. Shirasawa. Amyloid A (AA) fibril formation in renal tubules occurs intracytoplasmically, possibly at the site of membrane assembling structures. *Amyloid: Int J Exp Clin Invest* **1994**, *1*, 107–113.

72 Kluve-Beckerman, B., J. Manaloor and J. J. Liepnieks. A pulse–chase study tracking the conversion of macrophage-endocytosed serum amyloid A into extracellular amyloid. *Arthritis Rheum* **2002**, *46*, 1905–1913.

73 Shirahama, T., K. Miura, S.-T. Ju, R. Kisilevsky, E. Gruys, E. and A. Cohen. Amyloid enhancing factor-loaded macrophages in amyloid fibril formation. *Lab Invest* **1990**, *62*, 61–68.

74 Kluve-Beckerman, B., J.J. Liepnieks, L. Wang and M.D. Benson. A cell culture system for the study of amyloid pathogenesis: amyloid formation by murine peritoneal cells in the presence of recombinant serum amyloid A. *Am J Pathol* **1999**, *155*, 123–133.

75 Sipe, J.D., K.P.W.J. McAdam and F. Uchino. Biochemical evidence for the biphasic development of experimental amyloidosis. *Lab Invest* **1978**, *38*, 110–114.

76 Axelrad, M.A., R. Kisilevsky, J. Willmer, S.J. Chen and M. Skinner. Further characterization of amyloid-enhancing factor. *Lab Invest* **1982**, *47*, 139–141.

77 Johan, K., G. Westermark, U. Engstrom, A. Gustavsson, P. Hultman and P. Westermark. Acceleration of amyloid protein A amyloidosis by amyloid-like synthetic fibrils. *Proc Natl Acad Sci USA* **1998**, *95*, 2558–2563.

78 Lundmark, K., G.T. Westermark, S. Nystrom, C.L. Murphy, A. Solomon and P. Westermark. Transmissibility of systemic amyloidosis by prion-like mechanism. *Proc Natl Acad Sci USA* **2002**, *99*, 6979–6984.

79 Kisilevsky, R.L., Lemieux, L. Boudreau, Y. Dun-Sheng and P. Fraser. New clothes for amyloid-enhancing factor (AEF): silk as AEF. *Amyloid: Int J Exp Clin Invest* **1999**, *6*, 98–106.

80 Yamada, T., B. Kluve Beckerman, J.J. Liepnieks and M.D. Benson. Fibril formation from recombinant human serum amyloid A. *Biochim Biophys Acta* **1994**, *1226*, 323–329.

81 Barter, P.J. and K.A. Rye. High density lipoproteins and coronary heart disease. *Atherosclerosis* **1996**, *121*, 1–12.

82 Assmann, G. and A.M. Gotto, Jr. HDL cholesterol and protective factors in atherosclerosis. *Circulation* **2004**, *109 (23 Suppl 1)*, III8–14.

83 Tall, A.R. An overview of reverse cholesterol transport. *Eur Heart J* **1998**, *19 (Suppl A)*, A31–A35.

84 Frank, P.G. and Y.L. Marcel. Apolipoprotein A-I: structure–function relationships. *J Lipid Res* **2000**, *41*, 853–872.

85 Brouillette, C.G., G.M. Anantharamaiah, J.A. Engler and D.W. Borhani. Structural models of human apolipoprotein A-I: a critical analysis and review. *Biochim Biophys Acta* **2001**, *1531*, 4–46.

86 Fielding, C.J. and P.E. Fielding. Cellular cholesterol efflux. *Biochim Biophys Acta* **2001**, *1533*, 175–189.

87 Karathanasis, S.K., V.I. Zannis and J.L. Breslow. Isolation and characterization of the human apolipoprotein A-I gene. *Proc Natl Acad Sci USA* **1983**, *80*, 6147–6151.

88 Shoulders, C.C., A.R. Kornblitt, B.S. Munro and F.E. Baralle. Gene structure of human apolipoprotein A1. *Nucleic Acids Res* **1983**, *11*, 2827–2837.

89 Zannis, V.I., S.K. Karathanasis, H.T. Keutmann, G. Goldberger and J.L. Breslow. Intracellular and extracellular processing of human apolipoprotein A-I: secreted apolipoprotein A-I isoprotein 2 is a propeptide. *Proc Natl Acad Sci USA* **1983**, *80*, 2574–2578.

90 Januzzi, J.L., N. Azrolan, A. O'Connell, K. Aalto-Setala and J.L. Breslow. Characterization of the mouse apolipoprotein Apoa-1/Apoc-3 gene locus: genomic, mRNA, and protein sequences with comparisons to other species. *Genomics* **1992**, *14*, 1081–1088.

91 Brewer, H.B. Jr, T. Fairwell, A. LaRue, R. Ronan, A. Houser and T.J. Bronzert. The amino acid sequence of human apoA-I, an apolipoprotein isolated from high density lipoproteins. *Biochem Biophys Res Commun* **1978**, *80*, 623–630.

92 Breslow, J.L., D. Ross, J. McPherson, H. Williams, D. Kurnit, A.L. Nussbaum, S.K. Karathanasis and V.I. Zannis. Isolation and characterization of cDNA clones for human apolipoprotein A-I. *Proc Natl Acad Sci USA* **1982**, *79*, 6861–6865.

93 Cheung, P. and L. Chan. Nucleotide sequence of cloned cDNA of human apolipoprotein A-I. *Nucleic Acids Res* **1983**, *11*, 3703–3715.

94 Law, S.W., G. Gray, H.B. Brewer, Jr. cDNA cloning of human apoA-I: amino acid sequence of preproapoA-I. *Biochem Biophys Res Commun* **1983**, *112*, 257–264.

95 Fitch, W.M. Phylogenies constrained by the crossover process as illustrated by human hemoglobins and a thirteen-cycle, eleven-amino-acid repeat in human apolipoprotein A-I. *Genetics* **1977**, *86*, 623–644.

96 McLachlan, A.D. Repeated helical pattern in apolipoprotein-A-I. *Nature* **1977**, *267*, 465–466.

97 Segrest, J.P., D.W. Garber, C.G. Brouillette, S.C. Harvey and G.M. Anantharamaiah. The amphipathic α-helix: a multifunctional structural motif in plasma apolipoproteins. *Adv Protein Chem* **1994**, *45*, 303–369.

98 Klon, A.E., J.P. Segrest and S.C. Harvey. Comparative models for human apolipoprotein A-I bound to lipid in discoidal high-density lipoprotein particles. *Biochemistry* **2002**, *41*, 10895–10905.

99 Narayanaswami, V. and R.O. Ryan. Molecular basis of exchangeable apolipoprotein function. *Biochim Biophys Acta* **2000**, *1483*, 5–36.

100 Sorci-Thomas, M.G. and M.J. Thomas. The effects of altered apolipoprotein A-I structure on plasma HDL concentration. *Trends Cardiovasc Med* **2002**, *12*, 121–128.

101 Marcel, Y.L. and R.S. Kiss. Structure-function relationships of apolipoprotein A-I: a flexible protein with dynamic lipid associations. *Curr Opin Lipidol* **2003**, *14*, 151–157.

102 Borhani, D.W., D.P. Rogers, J.A. Engler and C.G. Brouillette. Crystal structure of truncated human apolipoprotein A-I suggests a lipid-bound conformation. *Proc Natl Acad Sci USA* **1997**, *94*, 12291–12296.

103 Borhani, D.W., J.A. Engler and C.G. Brouillette. Crystallization of truncated human apolipoprotein A-I in a novel conformation. *Acta Crystallogr D Biol Crystallogr* **1999**, *55*, 1578–1583.

104 Okon, M., P.G. Frank, Y.L. Marcel and R.J. Cushley. Heteronuclear NMR studies of human serum apolipoprotein A-I. Part I. Secondary structure in lipid-mimetic solution. *FEBS Lett* **2002**, *517*, 139–143.

105 Rye, K.A. and P.J. Barter. Formation and metabolism of prebeta-migrating, lipid-poor apolipoprotein A-I. *Arterioscler Thromb Vasc Biol* **2004**, *24*, 421–428.

106 Rigotti, A., B.L. Trigatti, M. Penman, H. Rayburn, J. Herz and M. Krieger. A targeted mutation in the murine gene encoding the high density lipoprotein (HDL) receptor scavenger receptor class B type I reveals its key role in HDL metabolism. *Proc Natl Acad Sci USA* **1997**, *94*, 12610–12615.

107 Connelly, M.A. and D.L. Williams. Scavenger receptor BI: a scavenger receptor with a mission to transport high density lipoprotein lipids. *Curr Opin Lipidol* **2004**, *15*, 287–295.

108 Van Allen, M.W., J.A. Frohlich and J.R. Davis. Inherited predisposition to generalized amyloidosis. *Neurology* **1969**, *19*, 10–25.

109 Nichols, W.C., F.E. Dwulet, J. Liepnieks and M.D. Benson. Variant apolipoprotein AI as a major constituent of a human hereditary amyloid. *Biochem Biophys Res Commun* **1988**, *156*, 762–768.

110 Nichols, W.C., R.E. Gregg, H.B. Brewer, Jr and M.D. Benson. A mutation in apolipoprotein A-I in the Iowa type of familial amyloidotic polyneuropathy. *Genomics* **1990**, *8*, 318–323.

111 Rader, D.J., R.E. Gregg, M.S. Meng, J.R. Schaefer, M.R. Kindt, L.A. Zech, M.D. Benson and H.B. Brewer, Jr. *In vivo* metabolism of a mutant apolipoprotein, apoA-I_{Iowa}, associated with hypoalphalipoproteinemia and hereditary systemic amyloidosis. *J Lipid Res* **1992**, *33*, 755–763.

112 Genschel, J., R. Haas, M.J. Propsting and H.-J. Schmidt. Hypothesis: apolipoprotein A-I induced amyloidosis. *FEBS Lett* **1998**, *430*, 145–149.

113 Jones, L.A., J.A. Harding, A.S. Cohen and Skinner M. New USA family has apolipoprotein AI (Arg26) variant. In *Amyloid and Amyloidosis 1990: Proc VIth Int Symp on Amyloidosis*, J.B. Natvig, O. Forre, G. Husby, A. Husebeck, B. Skogen, K. Sletten and P. Westermark (eds). Kluwer, Dordrecht, **1991**, pp. 385–388.

114 Vigushin, D.M., J. Gough, D. Allan, A. Alguacil, B. Penner, N.M. Pettigrew, G. Quinonez, K. Bernstein, S.E. Booth, D.R. Booth, A.K. Soutar, P.N. Hawkins and M.B. Pepys. Familial nephropathic systemic amyloidosis caused by

apolipoprotein AI variant Arg26. *Q J Med* **1994**, *87*, 149–154.

115 Hawkins, P. N., A. Bybee, H. J. B. Goodman, H. J. Lachmann, D. Rowczenio, J. A. Gilbertson, J. O'Grady, N. Heaton, A. Stangou and M. B. Pepys. Phenotype, genotype and outcome in hereditary apoAI amyloidosis. In *Amyloid and Amyloidosis: Proc Xth Int Symp on Amyloid*, G. Grateau, R. A. Kyle and M. Skinner (eds). CRC Press, Boca Raton, FL, **2005**, p. 316.

116 Booth, D. R., S. Y. Tan, S. E. Booth, J. J. Hsuan, N. F. Totty, O. Nguyen, T. Hutton, D. M. Vigushin, G. A. Tennent, W. L. Hutchinson, N. Thomson, A. K. Soutar, P. N. Hawkins and M. B. Pepys. A new apo-lipoprotein A I variant, Trp50Arg, causes hereditary amyloidosis. *Q J Med* **1995**, *88*, 695–702.

117 Soutar, A. K., P. N. Hawkins, D. M. Vigushin, G. A. Tennent, S. E. Booth, T. Hutton, O. Nguyen, N. F. Totty, T. G. Feest, J. J. Hsuan and M. B. Pepys. Apolipoprotein AI mutation Arg-60 causes autosomal dominant amyloidosis. *Proc Natl Acad Sci USA* **1992**, *89*, 7389–7393.

118 Murphy, C. L., S. Wang, K. Weaver, T. K. Williams, M. A. Gertz, D. T. Weiss and A. Solomon. Apolipoprotein A-I amyloidosis associated with a novel mutant. In *Amyloid and Amyloidosis: Proc Xth Int Symp on Amyloid*, G. Grateau, R. A. Kyle and M. Skinner (eds). CRC Press, Boca Raton, FL, **2005**, pp. 363–365.

119 Booth, D. R., S. Y. Tan, S. E. Booth, G. A. Tennent, W. L. Hutchinson, J. J. Hsuan, N. F. Totty, O. Truong, A. K. Soutar, P. N. Hawkins, M. Bruguera, J. Caballeria, M. Sole, J. M. Campistol and M. B. Pepys. Hereditary hepatic and systemic amyloidosis caused by a new deletion/insertion mutation in the apolipoprotein AI gene. *J Clin Invest* **1996**, *97*, 2714–2721.

120 Persey, M. R., D. R. Booth, S. E. Booth, R. Van Zyl-Smit, B. K. Adams, A. B. Fattaar, G. A. Tennent, P. N. Hawkins and M. B. Pepys. Hereditary nephropathic systemic amyloidosis caused by a novel variant apolipoprotein A-I. *Kidney Int* **1998**, *53*, 276–281.

121 Obici, L., G. Palladini, V. Perfetti, S. Marciano, M. Bruno, S. Giorgetti, V. Bellotti, E. Arbustini and G. Merlini. In *Amyloid and Amyloidosis: Proc IXth Int Symp on Amyloid*, M. Bely and A. Apathy (eds). David Apathy, Hungary, **2001**, pp. 81–83.

122 Obici, L., G. Palladini, S. Giorgetti, V. Bellotti, G. Gregorini, E. Arbustini, L. Verga, S. Marciano, S. Donadei, V. Perfetti, L. Calabresi, C. Bergonzi, F. Scolari and G. Merlini. Liver biopsy discloses a new apolipoprotein A-I hereditary amyloidosis in several unrelated Italian families. *Gastroenterology* **2004**, *126*, 1416–1422.

123 Coriu, D., A. Dispenzieri, F. J. Stevens, C. L. Murphy, W. Shuching, D. T. Weiss and A. Solomon. Hepatic amyloidosis resulting from deposition of the apolipoprotein A-I variant Leu75Pro. *Amyloid: J Protein Folding Disord* **2003**, *10*, 215–233.

124 Hamidi Asl, L., J. J. Liepnieks, K. Hamidi Asl, T. Uemichi, G. Moulin, E. Desjoyaux, R. Loire, M. Delpech, G. Grateau and M. D. Benson. Hereditary amyloid cardiomyopathy caused by a variant apolipoprotein A1. *Am J Pathol* **1999**, *154*, 221–227.

125 Hamidi Asl, K., J. J. Liepnieks, M. Nakamura, F. Parker and M. D. Benson. A novel apolipoprotein A-1 variant, Arg173Pro, associated with cardiac and cutaneous amyloidosis. *Biochem Biophys Res Commun* **1999**, 257, 584–588.

126 Obici, L., V. Bellotti, P. Mangione, M. Stoppini, E. Arbustini, L. Verga, I. Zorzoli, E. Anesi, G. Zanotti, C. Campana, M. Viganò and G. Merlini. The new apolipoprotein A-I variant Leu174 → Ser causes hereditary cardiac amyloidosis, and the amyloid fibrils are constituted by the 93-residue N-terminal polypeptide. *Am J Pathol* **1999**, *155*, 695–702.

127 Mangione, P., M. Sunde, S. Giorgetti, M. Stoppini, G. Esposito, L. Gianelli, L. Obici, L. Asti, A. Andreola, P. Viglino, G. Merlini and V. Bellotti. Amyloid fibrils derived from the apolipoprotein

AI Leu174Ser variant contains elements of ordered helical structures. *Protein Sci* **2001**, *10*, 187–199.

128 Mendes de Sousa, M., C. Vital, D. Ostler, R. Fernandes, J. Pouget-Abadie, D. Carles and M.J. Saraiva. Apolipoprotein AI and transthyretin as components of amyloid fibrils in a kindred with Apo AI Leu178His amyloidosis. *Am J Pathol* **2000**, *156*, 1911–1917.

129 Braschi, S., T.A. Neville, M.C. Vohl and D.L. Sparks. Apolipoprotein A-I charge and conformation regulate the clearance of reconstituted high density lipoprotein *in vivo*. *J Lipid Res* **1999**, *40*, 522–532.

130 Blanco-Vaca, F., J.C. Escolà-Gil, J.M. Martín-Campos and J. Julve. Role of apoA-II in lipid metabolism and atherosclerosis: advances in the study of an enigmatic protein. *J Lipid Res* **2001**, *42*, 1727–173.

131 Tailleux, A., P. Duriez, J.-C. Fruchart and V. Clavey. Apolipoprotein A-II, HDL metabolism and atherosclerosis. *Atherosclerosis* **2002**, *164*, 1–13.

132 Martin-Campos, J.M., J.C. Escola-Gil, V. Ribas and F. Blanco-Vaca. Apolipoprotein A-II, genetic variation on chromosome 1q21–q24, and disease susceptibility. *Curr Opin Lipidol* **2004**, *15*, 247–253.

133 Deeb, S.S., K. Takata, R.L. Peng, G. Kajiyama and J.J. Albers. A splice-junction mutation responsible for familial apolipoprotein A-II deficiency. *Am J Hum Genet* **1990**, *46*, 822–827.

134 Moore, M.N., K.T. Kao, Y.K. Tsao and L. Chan. Human apolipoprotein A-II: nucleotide sequence of a cloned cDNA, and localization of its structural gene on human chromosome 1. *Biochem Biophys Res Commun* **1984**, *123*, 1–7.

135 Lackner, K.J., S.W. Law and H.B. Brewer, Jr. The human apolipoprotein A-II gene: complete nucleic acid sequence and genomic organization. *Nucleic Acids Res* **1985**, *13*, 4597–4608.

136 Knott, T.J., S.C. Wallis, M.E. Robertson, L.M. Priestley, M. Urdea, L.B. Rall and J. Scott. The human apolipoprotein AII gene: structural organization and sites of expression. *Nucleic Acids Res* **1985**, *13*, 6387–6398.

137 Middleton-Price, H.R., J.A. van den Berghe, J. Scott, T.J. Knott and S. Malcolm. Regional chromosomal localisation of *APOA2* to 1q21–1q23. *Hum Genet* **1988**, *79*, 283–285.

138 Li, W.-H., M. Tanimura, C.-C. Luo, S. Datta and L. Chan. The apolipoprotein multigene family: biosynthesis, structure–function relationship, and evolution. *J Lipid Res* **1988**, *29*, 245–271.

139 Weber, J.L. and P.E. May. Abundant class of human DNA polymorphisms which can be typed using the polymerase chain reaction. *Am J Hum Genet* **1989**, *44*, 388–396.

140 Dupuy-Gorce, A.M., E. Desmarais, S. Vigneron, C. Buresi, V. Nicaud, A. Evans, G. Luc, D. Arveiler, P. Marques-Vidal, F. Cambien, L. Tiret, A. Crastes de Paulet and G. Roizes. DNA polymorphisms in linkage disequilibrium at the 3′ end of the human *APOAII* gene: relationships with lipids, apolipoproteins and coronary heart disease. *Clin Genet* **1996**, *50*, 191–198.

141 Scott, J., T.J. Knott, L.M. Priestley, M.E. Robertson, D.V. Mann, G. Kostner, G.J. Miller and N.E. Miller. High-density lipoprotein composition is altered by a common DNA polymorphism adjacent to apoprotein AII gene in man. *Lancet* **1985**, *1*, 771–773.

142 Brewer, H.B., S.E. Lux, R. Ronan and K.M. John. Amino acid sequence of human apoLp-GlnII (apoA-II), an apolipoprotein isolated from the high density lipoprotein. *Proc Natl Acad Sci USA* **1972**, *69*, 1304–1308.

143 Miller, C.G., T.D. Lee, R.C. LeBoeuf and J.E. Shively. Primary structure of apolipoprotein A-II from inbred mouse strain BALB/c. *J Lipid Res* **1987**, *28*, 311–319.

144 Segrest, J.P., M.K. Jones, H. de Loof, C.G. Brouillette, Y.V. Venkatachalapathi and G.N. Anantharamaiah. The amphipathic helix in the exchangable apolipoproteins: a review of secondary structure and function. *J Lipid Res* **1992**, *33*, 141–166.

145 Chapman, M. J. Comparative analysis of mammalian plasma lipoproteins. *Methods Enzymol* **1986**, *128*, 70–143.

146 Kalopissis, A. D. and J. Chambaz. Transgenic animals with altered high-density lipoprotein composition and functions. *Curr Opin Lipidol* **2000**, *11*, 149–153.

147 Kalopissis, A. D., D. Pastier and J. Chambaz. Apolipoprotein A-II: beyond genetic associations with lipid disorders and insulin resistance. *Curr Opin Lipidol* **2003**, *14*, 165–172.

148 Benson, M. D., J. J. Liepnieks, M. Yazaki, T. Yamashita, K. Hamidi Asl, B. Guenther and B. Kluve-Beckerman. A new human hereditary amyloidosis: The result of a stop-codon mutation in the apolipoprotein AII gene. *Genomics* **2001**, *72*, 272–277.

149 Weiss, S.W. and D. L. Page. Amyloid nephropathy of Ostertag with special reference to renal glomerular giant cells. *Am J Pathol* **1973**, *72*, 447–460.

150 Yazaki, M., J. J. Liepnieks, T. Yamashita, B. Guenther, M. Skinner and M. D. Benson. Renal amyloidosis caused by a novel stop-codon mutation in the apolipoprotein A-II gene. *Kidney Int* **2001**, *60*, 1658–1665.

151 Yazaki, M., J. J. Liepnieks, M. S. Barats, A. H. Cohen and M. D. Benson. Hereditary systemic amyloidosis associated with a new apolipoprotein AII stop codon mutation Stop78Arg. *Kidney Int* **2003**, *64*, 11–16.

152 Rowczenio, D., J. A. Gilbertson, A. Bybee, D. Hernandez and P. N. Hawkins. Hereditary amyloidosis in a Spanish family associated with a novel non-stop mutation in the gene for apolipoprotein AII. In *Amyloid and Amyloidosis: Proc Xth Int Symp on Amyloid*, G. Grateau, R. A. Kyle and M. Skinner (eds). CRC Press, Boca Raton, FL, **2005**, p. 366.

153 Higuchi, K., T. Yonezu, K. Kogishi, A. Matsumura, S. Takeshita, A. Kohno, M. Matsushita, M. Hosokawa and T. Takeda. Purification and characterization of a senile amyloid-related antigenic substance ($apoSAS_{SAM}$) from mouse serum; $apoSAS_{SAM}$ is an apoA-II apolipoprotein of mouse high density lipoprotein. *J Biol Chem* **1986**, *261*, 12834–12840.

154 Higuchi, K., M. Hosokawa and T. Takeda. Senescence-accelerated mouse. *Methods Enzymol* **1999**, *309*, 674–686.

155 Kitagawa, K., J. Wang, T. Matsushita, K. Kogishi, M. Hosokawa, X. Fu, Z. Guo, M. Mori and K. Higuchi. Polymorphisms of mouse apolipoprotein A-II: seven alleles found among 41 inbred strains of mice. *Amyloid: J Protein Folding Disord* **2003**, *10*, 207–214.

156 Xing, Y, A. Nakamura, T. Korenaga, Z. Guo, J. Yao, X. Fu, T. Matsushita, K. Kogishi, M. Hosokawa, F. Kametani, M. Mori and K. Higuchi. Induction of protein conformational change in mouse senile amyloidosis. *J Biol Chem* **2002**, *277*, 33164–33169.

23
Gelsolin

Hadar Benyamini, Kannan Gunasekaran, Haim Wolfson and Ruth Nussinov

23.1
Physiology, Pathology and Genetics

23.1.1
Gelsolin Amyloidosis

Familial amyloidosis of the Finnish type (FAF) was first described in 1969 as a new type of autosomal-dominant amyloidosis by Meretoja [1]. FAF patients present neurologic, ophthalmologic and dermatological symptoms, mainly corneal lattice dystrophy, progressive cranial and peripheral neuropathy, and skin changes. Amyloid deposition is found in various tissues including skin, cornea, vascular walls and perineurium, particularly in the facial nerve. Homozygosity is associated with severe outcome including earlier onset, renal failure due to glomerular amyloid deposition and premature death [2, 3].

23.1.2
Normal and Mutant Protein Function

Gelsolin-like proteins participate in regulating the cytoskeleton structure and cell movement through actin filament severing, capping and nucleating in eukaryotes. Intracellular gelsolin regulates the architecture and motility of cells [4], while plasma gelsolin participates in the actin scavenging system [5, 6] which maintains blood viscosity by controlling the ratio of monomeric (G) to filamentous (F) actin [7].

Gene knock-out experiments have shown that while gelsolin is not required for survival, it is required for the rapid movement of dynamic cells such as fibroblasts and platelets [8], and the contraction of neurons [9]. Gelsolin is involved in cell signaling and apoptosis [10], and its levels have been found to be decreased in many cancers, including bladder, breast, prostate, colon and small cell lung carcinomas ([10] and references therein). It has also been shown to act as a metastasis [11] and a tumor suppressor [12].

Amyloid Proteins. The Beta Sheet Conformation and Disease. J. D. Sipe

ISBN: 3-527-31072-X

Gelsolin-related amyloidosis is caused by a point mutation at domain 2 of the protein followed by the formation of a proteolytic amyloidogenic fragment [13]. The function of the mutant versus the normal protein was studied. Westberg et al. [14] have transfected the wild-type and mutant gelsolin (G654A) into a neural cell line. Neural differentiation was inhibited upon overexpression of the wild-type, but not the mutant, protein. The role of gelsolin in neural differentiation is also implied in the expression pattern of gelsolin in the developing rat brain [15]. Weeds et al. [16] have shown impaired actin severing activity of mutant gelsolin in patient's plasma which can be attributed to the aberrant proteolysis rather than the point mutation. Kangas et al. [17] have expressed the wild-type and mutant gelsolin proteins in mouse embryonic gelsolin-null fibroblasts. The intracellular actin modulating activity was not influenced by the mutation.

In summary, gelsolin FAF mutations might affect neuronal differentiation, but not intracellular actin-modulating activity in fibroblasts. FAF symptoms appear to be caused mainly due to amyloid deposition and not due to loss of cytoplasmic protein function [17].

23.1.3 Gelsolin Amyloid Genetics

The gene for human gelsolin is located on the human chromosome 9q32–q34 [18]. The gene codes for both the intracellular and the plasma gelsolin that has an additional N-terminus of 23 residues. Both forms are ubiquitously expressed in adult

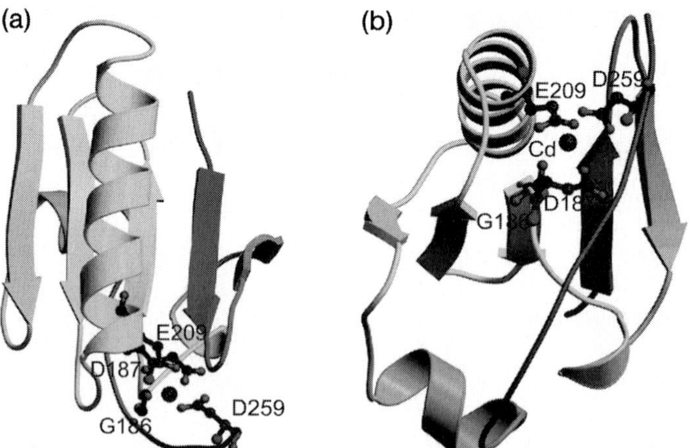

Fig. 23.1 Ribbon representation of the X-ray crystal structure of gelsolin domain 2 (residues 159–261 shown; PDB code: 1KCQ) in two different orientations [26]. The residues involved in the metal binding are shown in ball-and-stick representation. Mutations at position Asp187 to either Asn or Tyr cause disruption of the calcium-binding site and thus lead to proteolysis.

tissues [19]. FAF is a dominantly inherited disease, exclusively associated with mutations in Asp187 of gelsolin domain 2. Although originally discovered in Finland, the mutation was later found in families of American, Dutch, Danish, Czech, Japanese ([20] and references therein) and Portuguese origin [21]. Haplotype analysis has shown that the mutations occurred independently at different locations [22]. The amyloidogenic mutations are either G654A or G654T, yielding Asp187Asn or Asp187Tyr substitutions, respectively (Fig. 23.1). Huff et al. have reproduced *in vitro* the FAF cellular phenotype by mutating Glu209. However, *in vivo* all mutations occur at the same position (Asp187), a fact that indicates that C653G654 is a hotspot for mutations in gelsolin [23].

23.2 Mechanism of Amyloid Formation by Gelsolin

23.2.1 Cell Biology

The full-length gelsolin protein contains 755 amino acids and is folded into six structurally homologous domains. The amyloid deposits found in FAF patients correspond to a 71-residue proteolytic fragment containing residues 173–243. In a few cases, an amyloidogenic fragment of 53 residues was found, corresponding to residues 173–225 [13]. The enzyme that performs the N-terminal cleavage was recently identified as furin [24], a pro-protein convertase that cleaves after an RXXR sequence, matching Arg169XXArg172 in gelsolin. When isolated from the wild-type gelsolin, the proteolytic fragments found in FAF are equally amyloidogenic [25], indicating that the amyloidogenic effect of the mutation is exerted by allowing the proteolysis of the deposited fragment. To a lesser extent, in COS cells transfection of furin and gelsolin the wild-type is also processed by the enzyme [24], i.e. the wild-type is also a furin substrate. However, *in vivo*, its cleavage site is not accessible to proteolysis. The reason for this difference in furin processing was found with the determination of the crystal structure of wild-type domain 2 [26]. The structure revealed that Asp187 is part of a binding site of Ca^{2+}, a binding that significantly stabilizes the domain [23] (Fig. 23.1). Lacking this site due to Asp187 substitutions, the FAF mutants are unstable, exposing the proteolytic site. These mutants are prone to cleavage and formation of the amyloidogenic fragment, whereas the wild-type protein is stabilized by Ca^{2+} binding and thus protected from cleavage. The role of Ca^{2+} stabilization of the domain was confirmed by Huff et al. [23] who mutated Glu209, which participates in the same Ca^{2+}-binding site, but is not related to FAF. In transfected cells, the Glu209Gln protein has the same characteristics as the FAF Asp187 mutants.

Interestingly, amyloid formation by mutated gelsolin occurs via a normal physiological process. The same mechanism of furin proteolysis followed by fibril formation acts in Pmel17 fiber formation that takes place during melanosome biogenesis [27]. Furin is similarly involved in the processing of the protein BRI in the initiation of familial British dementia [28].

23.2.2
Domain Stability and Amyloid Formation

Some mechanisms were either suggested or shown through which globularly folded proteins transform into amyloid fibrils. These include the secondary structure switch [29–31], domain swapping [32–35] and proteolysis [13, 28]. Amyloid formation is also facilitated by the loss of edge strand protection [36, 37], which might be executed via proteolysis or domain swapping. Common to all the suggested mechanisms is the destabilization of the native globular monomer in order to obtain an intermediate conformation that is more prone to

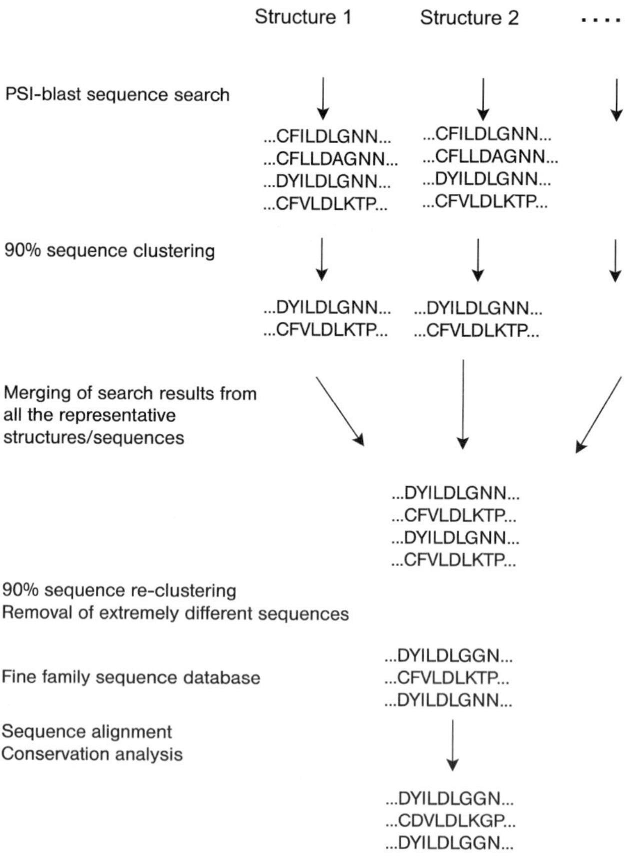

Fig. 23.2 Flow-chart of the scheme used for constructing the gelsolin sequence database and alignment. Sequences of the representative structures were used to initiate an iterative Psi-BLAST search of the non-redundant database (nrdb).

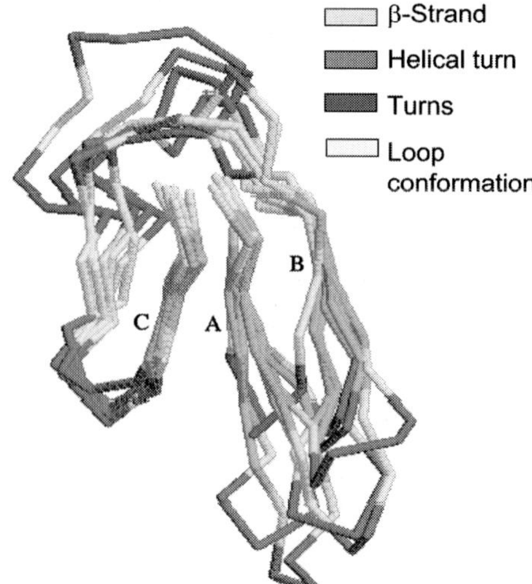

Fig. 23.3 Superposition of the gelsolin representative structures for the fragment containing strands A, B and C. The turns that connect strand B are variable in length and conformation.

form amyloids. This means that within a certain globular domain, the unstable segments are the key regions for initiating the amyloidogenic process. For the gelsolin-related FAF and the β_2-microglobulin (β_2M)-related dialysis-related amyloidosis (DRA), we looked for a relationship between the relative conservation and stability of different domain segments [38, 39]. We measured conservation

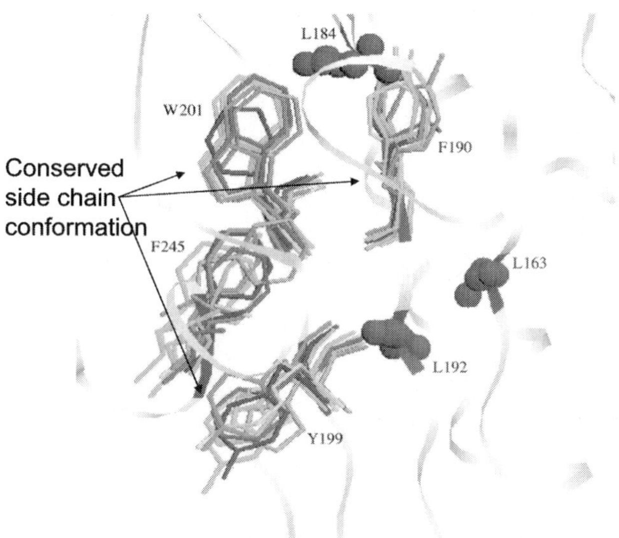

Fig. 23.4 Conserved aromatic and hydrophobic residues in the gelsolin family obtained from the alignment of the eight representative structures.

by multiple sequence as well as multiple structural alignments of protein family members (Fig. 23.2). We hypothesized that conservation of a protein segment reflects its relative stability within the protein domain. In both studied cases of the gelsolin and β_2M amyloidoses, we found that less-conserved segments are indeed involved in initiation of amyloid formation. In the case of gelsolin, it was one of the exposed edge β-strands, strand B, that was found to have different lengths and conformations in different family members, and was also the least sequentially and structurally conserved (Fig. 23.3). Aromatic residues located on the central strands C and D were found to be highly conserved in the gelsolin family (Fig. 23.4). In the β_2M case, three strands were found as unstable (strands A, D and G). Docking of β_2M monomers, without strands A and G, leads to fibril modeling (Fig. 23.5). We used the docking programs BUDDA (backbone unbound docking application) and PPD (protein–protein docking), which were developed by our research group [40, 41], to predict the interaction between two consecutive monomers in the fibril. Based on optimization of the geometric complementarity between the two input molecules, the programs predict the complexed conformation. We started by finding the best docking of two β_2M monomers and iteratively applied the obtained three-dimensional transfor-

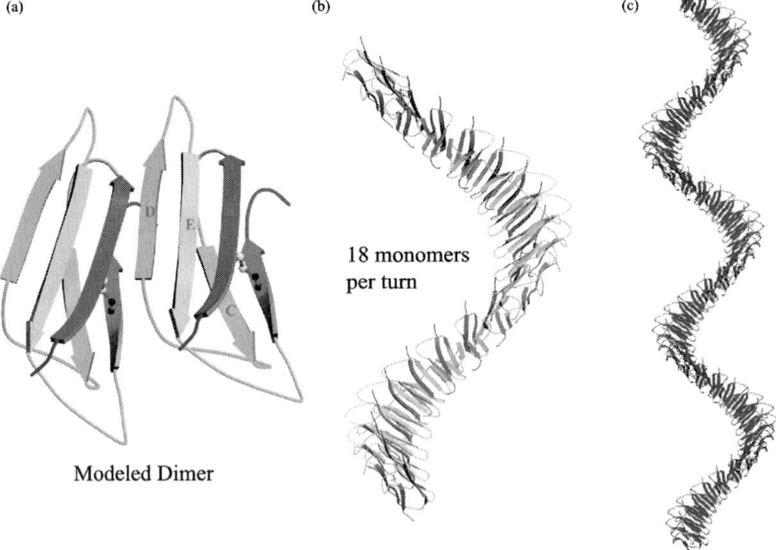

Fig. 23.5 Assembly of multiple monomers of β_2M to a cross-β-helical fibril. The assembly of the monomers was generated by an iterative application of the three-dimensional transformation that was suggested for monomer–monomer interaction shown in (a). (b) Generated with 20 monomers. (c) Generated with 50 monomers. In total, 52 β-strands constitute one complete turn along the fibril axis.

mation on additional monomers in order to examine the resulting putative fibril. The resulting fibril model is consistent with the cross-β structure. In both gelsolin and β_2M, the conservation analyses were supported by experimental methods that assessed the stability of domain segments, like limited proteolysis, urea denaturation and hydrogen–deuterium exchange. The crystal structure of the metal-bound gelsolin domain 2 [26] and molecular dynamics (MD) simulations performed on the structure further support strand B instability, rather than global destabilization of the whole domain as was previously suggested [42]. MD simulations suggested that this strand is the first to be affected from domain destabilization [26]. Comparison of the metal-bound and free gelsolin domain 2 shows the significant deviation encompassing strand B, including the cleavage site for furin which produces the amyloidogenic fragment.

23.3
Conclusions

FAF is caused by mutation of Asp187 of domain 2 of gelsolin. The mutation disrupts a Ca^{2+}-binding site and thus destabilizes the domain, exposing a proteolytic site for furin cleavage on the unstable strand B between residues 172 and 173. A second, yet unrecognized, enzyme completes the formation of the proteolytic amyloidogenic fragment deposited in FAF patients. Furin enzyme inhibition appears a promising direction for therapy.

Acknowledgments

We would like to thank Yuval Inbar for help in the docking and Max Shatsky for the multiple structure comparison program. This research was supported in part by the "Center of Excellence in Geometric Computing and its Applications" funded by the Israel Science Foundation (administered by the Israel Academy of Sciences). The research of H. J. W. is partially supported by the Hermann Minkowski-Minerva Center for Geometry at Tel Aviv University. The research of R. N. has been funded in whole or in part with Federal funds from the National Cancer Institute, National Institutes of Health, under contract NO1-CO-12400. The content of this publication does not necessarily reflect the view or policies of the Department of Health and Human Services nor does mention of trade names, commercial products or organizations imply endorsement by the US Government.

References

1 Meretoja J. Familial systemic paramyloidosis with lattice dystrophy of the cornea, progressive cranial neuropathy, skin changes and various internal symptoms. A previously unrecognized heritable syndrome. *Ann Clin Res* **1969**, *1*, 314–324.

2 Kiuru S. Familial amyloidosis of the Finnish type (FAF). A clinical study of 30 patients. *Acta Neurol Scand* **1992**, *86*, 346–353.

3 Kiuru S, Matikainen E, Kupari M, Haltia M, Palo J. Autonomic nervous system and cardiac involvement in familial amyloidosis, Finnish type (FAF). *J Neurol Sci* **1994**, *126*, 40–48.

4 Yu FX, Sun HQ, Janmey PA, Yin HL. Identification of a polyphosphoinositide-binding sequence in an actin monomer-binding domain of gelsolin. *J Biol Chem* **1992**, *267*, 14616–114621.

5 Vasconcellos CA, Lind SE. Coordinated inhibition of actin-induced platelet aggregation by plasma gelsolin and vitamin D-binding protein. *Blood* **1993**, *82*, 3648–3657.

6 Haddad JG, Harper KD, Guoth M, Pietra GG, Sanger JW. Angiopathic consequences of saturating the plasma scavenger system for actin. *Proc Natl Acad Sci* **1990**, *87*, 1381–1385.

7 Yin HL. Gelsolin: calcium-and polyphosphoinositide-regulated actin-modulating protein. *BioEssays* **1987**, *7*, 176–179.

8 Witke W, Sharpe AH, Hartwig JH, Azuma T, Stossel TP, Kwiatkowski DJ. Hemostatic, inflammatory, and fibroblast responses are blunted in mice lacking gelsolin. *Cell* **1995**, *81*, 41–51.

9 Lu M, Witke W, Kwiatkowski DJ, Kosik KS. Delayed retraction of filopodia in gelsolin null mice. *J Cell Biol* **1997**, *138*, 1279–1287.

10 Kwiatkowski DJ. Functions of gelsolin: motility, signaling, apoptosis, cancer. *Curr Opin Cell Biol* **1999**, *11*, 103–108.

11 Fujita H, Okada F, Hamada J, Hosokawa M, Moriuchi T, Koya RC, Kuzumaki N. Gelsolin functions as a metastasis suppressor in B16-BL6 mouse melanoma cells and requirement of the carboxyl-terminus for its effect. *Int J Cancer* **2001**, *93*, 773–780.

12 Sagawa N, Fujita H, Banno Y, Nozawa Y, Katoh H, Kuzumaki N. Gelsolin suppresses tumorigenicity through inhibiting PKC activation in a human lung cancer cell line, PC10. *Br J Cancer* **2003**, *88*, 606–612.

13 Maury CP. Gelsolin-related amyloidosis. Identification of the amyloid protein in Finnish hereditary amyloidosis as a fragment of variant gelsolin. *J Clin Invest* **1991**, *87*, 1195–1199.

14 Westberg JA, Zhang KZ, Andersson LC. Regulation of neural differentiation by normal and mutant (G654A, amyloidogenic) gelsolin. *FASEB J* **1999**, *13*, 1621–1626.

15 Tanaka J, Sobue K. Localization and characterization of gelsolin in nervous tissues: gelsolin is specifically enriched in myelin-forming cells. *J Neurosci* **1994**, *14*, 1038–1052.

16 Weeds AG, Gooch J, McLaughlin P, Maury CP. Variant plasma gelsolin responsible for familial amyloidosis (Finnish type) has defective actin severing activity. *FEBS Lett* **1993**, *335*, 119–123.

17 Kangas H, Ulmanen I, Paunio T, Kwiatkowski DJ, Lehtovirta M, Jalanko A, Peltonen L. Functional consequences of amyloidosis mutation for gelsolin polypeptide – analysis of gelsolin–actin interaction and gelsolin processing in gelsolin knock-out fibroblasts. *FEBS Lett* **1999**, *454*, 233–239.

18 Kwiatkowski DJ, Westbrook CA, Bruns GA, Morton CC. Localization of gelsolin proximal to ABL on chromosome 9. *Am J Hum Genet* **1988**, *42*, 565–572.

19 Paunio T, Kangas H, Kiuru S, Palo J, Peltonen L, Syvnen AC. Tissue distribution and levels of gelsolin mRNA in normal individuals and patients with gelsolin-related amyloidosis. *FEBS Lett* **1997**, *406*, 49–55.

20 Maury CP, Liljestrom M, Boysen G, Tornroth T, de la Chapelle A, Nurmiaho-Lassila EL. Danish type gelsolin related amyloidosis: 654G–T mutation is associated with a disease pathogenetically

and clinically similar to that caused by the 654G–A mutation (familial amyloidosis of the Finnish type). *J Clin Pathol* **2000**, *53*, 95–99.

21 Conceicao I, Sales-Luis ML, De Carvalho M, Evangelista T, Fernandes R, Paunio T, Kangas H, Coutinho P, Neves C, Saraiva MJ. Gelsolin-related familial amyloidosis, Finnish type, in a Portuguese family: clinical and neurophysiological studies. *Muscle Nerve* **2003**, *28*, 715–721.

22 Paunio T, Sunada Y, Kiuru S, Makishita H, Ikeda S, Weissenbach J, Palo J, Peltonen L. Haplotype analysis in gelsolin-related amyloidosis reveals independent origin of identical mutation (G654) of gelsolin in Finland and Japan. *Hum Mutat* **1995**, *6*, 60–65.

23 Huff ME, Page JL, Balch WE, Kelly JW. Gelsolin domain 2 Ca^{2+} affinity determines susceptibility to furin proteolysis and familial amyloidosis of Finnish type. *J Mol Biol* **2003**, *334*, 119–127.

24 Chen CD, Huff ME, Matteson J, Page JL, Phillips R, Kelly JW, Balch WE. Furin initiates gelsolin familial amyloidosis in the Golgi through a defect in Ca^{2+} stabilization. *EMBO J* **2001**, *20*, 6277–6287.

25 Ratnaswamy G, Koepf E, Bekele H, Yin H, Kelly J. The amyloidogenicity of gelsolin is controlled by proteolysis and pH. *Chem Biol* **1999**, *6*, 293–304.

26 Kazmirski SL, Isaacson RL, An C, Buckle A, Johnson CM, Daggett V, Fersht AR. Loss of metal-binding site in gelsolin leads to familial amyloidosis-Finnish type. *Nat Struct Biol* **2002**, *9*, 112–116.

27 Berson JF, Theos AC, Harper DC, Tenza D, Raposo G, Marks MS. Proprotein convertase cleavage liberates a fibrillogenic fragment of a resident glycoprotein to initiate melanosome biogenesis. *J Cell Biol* **2003**, *161*, 521–533.

28 Kim SH, Wang R, Gordon DJ, Bass J, Steiner DF, Lynn DG, Thinakaran G, Meredith SC, Sisodia SS. Furin mediates enhanced production of fibrillogenic ABri peptides in familial British dementia. *Nat Neurosci* **1999**, *2*, 984–988.

29 Pan KM, Baldwin M, Nguyen J, Gasset M, Serban A, Groth D, Mehlhorn I, Huang Z, Fletterick RJ, Cohen FE, Prusiner S. Conversion of alpha-helices into beta-sheets features in the formation of the scrapie prion proteins. *Proc Natl Acad Sci USA* **1993**, *90*, 10962–10966.

30 Andreola A, Bellotti V, Giorgetti S, Mangione P, Obici L, Stoppini M, Torres J, Monzani E, Merlini G, Sunde M. Conformational switching and fibrillogenesis in the amyloidogenic fragment of apolipoprotein A-I. *J Biol Chem* **2003**, *278*, 2444–2451.

31 Soto C, Castano EM, Frangione B, Inestrosa NC. The alpha-helical to beta-strand transition in the amino-terminal fragment of the amyloid beta-peptide modulates amyloid formation. *J Biol Chem* **1995**, *270*, 3063–3067.

32 Bennett MJ, Schlunegger MP, Eisenberg D. 3D domain swapping: a mechanism for oligomer assembly. *Protein Sci* **1995**, *4*, 2455–2468.

33 Sinha N, Tsai CJ, Nussinov R. A proposed structural model for amyloid fibril elongation: domain swapping forms an interdigitating beta-structure polymer. *Protein Eng* **2001**, *14*, 93–103.

34 Lee S, Eisenberg D. Seeded conversion of recombinant prion protein to a disulfide-bonded oligomer by a reduction–oxidation process. *Nat Struct Biol* **2003**, *10*, 725–730.

35 Sanders A, Jeremy Craven C, Higgins LD, Giannini S, Conroy MJ, Hounslow A, Waltho JP, Staniforth RA. Cystatin forms a tetramer through structural rearrangement of domain-swapped dimers prior to amyloidogenesis. *J Mol Biol* **2004**, *336*, 165–178.

36 Richardson JS, Richardson DC. Natural beta-sheet proteins use negative design to avoid edge-to-edge aggregation. *Proc Natl Acad Sci USA* **2002**, *99*, 2754–2759.

37 Wang W, Hecht M. Rationally designed mutations convert de novo amyloid-like fibrils into monomeric beta-sheet proteins. *Proc Natl Acad Sci USA* **2002**, *99*, 2760–2765.

38 Benyamini H, Gunasekaran K, Wolfson H, Nussinov R. Conservation and amyloid formation: a study of the gelsolin-like family. *Proteins* **2003**, *51*, 266–282.

39 Benyamini H, Gunasekaran K, Wolfson H, Nussinov R. β_2-Microglobulin amyloidosis: insights from conservation analysis and fibril modelling by protein docking techniques. *J Mol Biol* **2003**, *330*, 159–174.

40 Polak V. BUDDA: backbone unbound docking application. *Masters Thesis.* School of Computer Science, Tel-Aviv University, **2002**.

41 Norel R, Lin SL, Wolfson HJ, Nussinov R. Molecular surface complementarity at protein–protein interfaces: the critical role played by surface normals at well placed, sparse points in docking. *J Mol Biol* **1995**, *252*, 263–273.

42 Isaacson RL, Weeds AG, Fersht AR. Equilibria and kinetics of folding of gelsolin domain 2 and mutants involved in familial amyloidosis-Finnish type. *Proc Natl Acad Sci USA* **1999**, *96*, 11247–11252.

24
Lysozyme

Mireille Dumoulin, Vittorio Bellotti and Christopher M. Dobson

24.1
Introduction

Human lysozyme is a protein of 130 residues belonging to the c-type class of lysozymes. It is encoded by a gene located on chromosome 12, and is organized as four exons and three introns [1]. Human lysozyme is expressed at high levels in hematopoietic cells where it is found in granulocytes, monocytes and macrophages as well as in their bone marrow precursors [2]. The protein is present at high concentrations in various tissues and fluids, including liver, articular cartilage, plasma, saliva, tears and milk [2]. Lysozyme is a bacteriolytic enzyme that hydrolyzes preferentially the β1,4 glycosidic linkages between the *N*-acetylmuramic acid and *N*-acetylglucosamine groups that occur in the peptidoglycan cell wall component of certain microorganisms. Thus, lysozyme appears to have a role in host defense [3].

There are five known natural mutations in the human lysozyme gene, and these give rise to six variant proteins: I56T, F57I, W64R, D67H, T70N and F57I/T70N. All of them except the T70N variant have been found to be associated with a non-neuropathic amyloid disease [4–8]. Wild-type lysozyme as well as the I56T, D67H and T70N variants have been successfully expressed in large quantities in heterologous organisms [7, 9, 10], permitting extensive studies of their properties. Comparison of the properties of the variant proteins with those of wild-type lysozyme has shown how the various mutations affect structural and functional properties such as activity, stability, structure, folding, dynamics and aggregation. In this chapter, we summarize the data that have been reported concerning the normal behavior of lysozyme and the clinical manifestations of lysozyme amyloidosis, the properties of amyloid fibrils formed *in vivo* and *in vitro*, and the effects of the mutations *in vitro* on the properties of the protein. Finally, we describe a mechanism for lysozyme fibril formation that has emerged from all of these data, and discuss the significance of the mechanism for understanding the nature and origins of the group of amyloid diseases associated with protein aggregation.

Amyloid Proteins. The Beta Sheet Conformation and Disease. J. D. Sipe

ISBN: 3-527-31072-X

24.2
Lysozyme in Healthy Subjects

The normal concentration of lysozyme in plasma ranges from 4 to 13 mg/l and only traces are detectable in the urine in healthy subjects [11]. The turnover of wild-type lysozyme was investigated in the 1970s in animal models [12] and in humans [13]. The lifetime of lysozyme in plasma is very short: approximately 75% of the total lysozyme content of plasma is eliminated in 1 h, mainly through clearance via the kidneys. Absorption of lysozyme after filtration through the glomerula occurs mainly through the renal proximal tubules [14]. Megalin, which is an endocytic receptor abundantly expressed at the apical membrane of renal proximal tubules [15], is likely to be involved in the endocytic reabsorption of lysozyme and other low-molecular-weight proteins such as insulin and β_2-microglobulin [16]. Reabsorbed lysozyme is probably degraded to soluble low-molecular-weight peptides in the acidic environment of lysosomal compartments [14]. On the basis of the concentration and turnover in plasma, it has been estimated that at least 500 mg of lysozyme is produced per day by a normal subject [13]. Greatly increased concentrations of lysozyme in plasma and urine have been correlated with several pathological conditions, and have been considered for many years to be a marker of monocytic leukemia [11]. In patients with myeloproliferative disorders, but normal renal function, the production of lysozyme is increased by a factor up to 4.

24.3
Clinical Manifestations of Lysozyme Amyloidosis

The causative role of lysozyme in some familial forms of amyloid disease was first demonstrated in 1993, from studies of two English families affected by hereditary systemic amyloidosis and carrying either the mutation D67H or I56T [4]. The molecular diagnosis in both families was made on the basis of immunochemical staining, N-terminal amino acid sequencing of protein extracted from purified *ex vivo* fibrils and nucleotide sequencing of the lysozyme gene [4]. As in other forms of hereditary amyloidosis, the disease is transmitted through an autosomal dominant mode. In the family carrying the mutation D67H, the clinical presentation was characterized by hepatic hemorrhage caused by the massive amyloid deposition (up to kilograms) in the liver and spleen. In the second family, whose affected members carried the mutation I56T, the amyloid deposition was systemic, but the main clinical feature was a cutaneous amyloidosis characterized by the presence of dermal petechiae in all affected subjects.

Following this initial observation, another kindred related to the original family affected by lysozyme amyloidosis associated with the D67H variant and a new family, apparently not related to the first one, but presenting the same lysozyme mutation were discovered [17]. In the new kindred, renal failure was the most distinctive clinical feature, although all of the affected subjects also displayed massive hepatic and splenic amyloid deposits. It is worth noting, however, that despite the deposition of amyloid aggregates in the liver and spleen, these patients had no clinical symptoms, such as spontaneous hepatic haemorrhage and rupture, that are usually associated with the involvement of these two organs and that were observed in the patients of the first family carrying the D67H mutation.

Two other amino acid mutations, giving rise to the F57I and W64R lysozyme variants, have been found recently to be associated with familial amyloidosis in French and Italian/Canadian families, respectively [6, 8]. The diagnosis in these cases was based on the positive immunostaining of the tissues containing amyloid deposits with an antibody specific for lysozyme (for the W64R variant only) and the identification of the mutation in the lysozyme gene (for both variants). As in the case of the families carrying either the I56T or the D67H mutation, the affected subjects were heterozygous for the F57I or W64R mutation [6, 8]. One affected individual from the family carrying the F57I mutation is heterozygous for both the F57I variant allele and the T70N allele [8]. No affected tissues from patients carrying either F57I or the W64R mutations have, however, become available for extraction and analysis of the lysozyme amyloid protein they contain.

The F57I mutation causes a systemic amyloidosis with a prominent renal involvement in the affected kindred [8]. The first patient identified as carrying the W64R mutation displayed gastroenteric and hepatic localization associated with renal involvement [6]. The most common and prominent feature in the affected members of this family is, however, a progressive renal failure caused by glomerular amyloid deposits. Of particular interest in this case were the attempts to identify the W64R variant in urine and plasma by a combination of chromatographic and mass spectrometry [6]. Despite the sensitivity of the techniques used, only wild-type lysozyme was found in the biological fluids and no traces of the pathogenic variant protein were detected. This finding could result from selective and highly efficient incorporation of the amyloidogenic variant protein into the amyloid deposits, or from a higher level of intracellular or extracellular degradation of the variants relative to the wild-type protein.

Finally, a fifth mutation, T70N, has been discovered and found to occur in 5–6% of the human population [5, 8]. Despite an investigation designed to identify patients affected with amyloidosis associated with this mutation [7], no case has yet been reported. Even in the patient presenting the combination of T70N and the pathogenic F57I alleles, the mutation at position 70 has no detectable effect on the clinical phenotype [8]. In fact, in the family bearing the double mutation F57I/T70N, the onset of the disease in one affected member lacking the polymorphism was even earlier than that of the double heterozygous patient.

In summary, the phenotype of lysozyme amyloidosis, as well as the age at which the onset of the disease occurs, is very variable both within and between families [4, 6, 8, 17]. Such differences have been reported for other systemic amyloid disorders, such as transthyretin amyloidosis, and the underlying basis for phenotypic variation remains largely unknown [17].

24.4
Characteristics of *Ex Vivo* and *In Vitro* Amyloid Fibrils

The fibrils deposited in tissues of patients with systemic amyloidosis associated with the I56T or D67H mutations have been found to contain only the full-length variant proteins, although the patients are heterozygous [4, 9]. Thus, while both the wild-type lysozyme and the amyloidogenic variants are produced, the former appears not to be converted into amyloid fibrils under *in vivo* conditions. Moreover, after extraction from fibril preparations, the D67H variant was shown to refold to the native state under appropriate conditions and, indeed, unlike the D67H fibrils themselves, to exhibit enzymatic activity [9]. However, when the disulfide bonds within the chain were reduced with 2-mercaptoethanol during the extraction procedure, the protein did not refold to an active form [9]. These results suggest that neither cleavage of the polypeptide chain nor reduction of disulfide bounds is required for, or results from, fibril formation. The D67H variant, therefore, must fold and form its disulfide bonds correctly in the cell, and the subsequent deposition in tissue does not involve the reduction of disulfide bonds formed during *de novo* folding. This is also likely to be the case with the I56T variant [4]. As mentioned above, tissues from patients having the F57I or the W64R mutation have not become available for extraction of lysozyme amyloid fibrils, preventing characterization of the protein contained in the deposits [6, 8]. In the case of the W64R variant it has been suggested that the protein may not persist as a full-length molecule *in vivo*, but could, for example, be proteolyzed at an early stage after its synthesis [6].

When observed by electron microscopy, after negative staining with uranyl acetate, most of the *ex vivo* D67H lysozyme fibrils are wavy in nature and their diameters range from 8 to 13 nm (Fig. 24.1) [18]. X-ray fiber diffraction of such fibrils showed a meridional reflection at 4.6–4.8 Å and a broad equatorial reflection at 8–14 Å, characteristic of the amyloid cross-β structure [19]. No reflections could be attributed to helical structure, suggesting that, if helices persist after transformation of the soluble protein to the fibrillar form, they are not regularly ordered. Cryo-electron microscopy of *ex vivo* D67H fibrils suggest that the fibrils are made up of five or six protofilaments and that they have a hollow core [18, 20].

The I56T and D67H variants, and indeed wild-type lysozyme, have been shown to be able to form amyloid fibrils *in vitro* (Fig. 24.1). The three proteins most readily produce fibrils under conditions such as low pH [21], high temperature [22], moderate concentrations of denaturant [23] or after pressure treatment [24] where a significant population of partially unfolded protein is present. The two amyloidogenic variants form amyloid fibrils under similar conditions, but more extreme conditions are required for the wild-type protein (Fig. 24.1). Fibrils formed *in vitro* have been analyzed by a wide variety of techniques including electron microscopy [21–24], Congo red birefringence [9, 21], thioflavin T fluorescence [21, 23] and X-ray fiber diffraction [21, 23], all of which unequivocally demonstrate their amyloid character. The morphology of the fibrils depends on the conditions under which fibril formation takes place (Fig. 24.1). Most fibrils derived from the D67H variant

Conditions of formation	Ref	
Ex vivo (from the D67H variant)	Most of the fibrils are wavy. Their diameter ranges from 8-13nm. A small minority of fibrils are wider and twister and are thought to arise from the association of several wavy fibrils. 	18
I56T/D67H: 10 mM glycine buffer, pH 2, 37 °C 10 mg/ml, 6 days Wild-type: The same but 57°C	Fibrils form the I56T variant show clear polymorphism with a major population consisting of several protofilaments assembled in a twisted manner. Other fibrils appear to result from lateral self assembly of individual protofibrils into larger structures in the form of flat ribbons of 25-30 nm in diameter. The fibrils of the wild type protein and the D67H variant appear narrower having a diameter of 6-7 nm, and have a more uniform appearance (see picture). Fibrils from the wild-type, protein in particular appear bent or curved. Under these conditions of incubation, some protein degradation may occur prior to fibril formation [42]. 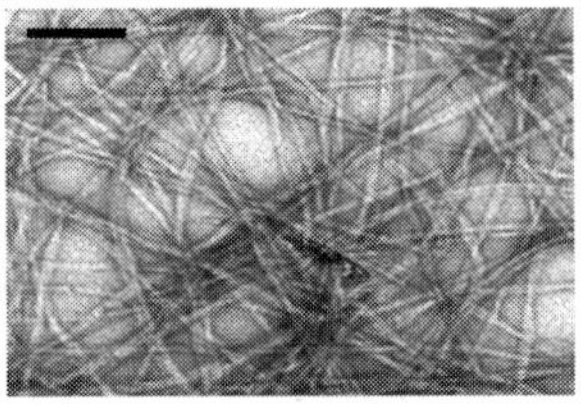	21
I56T/D67H: 0.1 M citrate buffer, pH 5.5, 3M urea, 48 °C 0.1 mg/ml, stirring, 6 h Wild-type: The same but 55°C	Most fibrils appear to be incorporated into large bundles of 100-400 nm in diameter. Some well resolved fibrils of approximately 6 nm in diameter are also observed. No significant difference between the morphology of the fibrils formed from the three protein species have been observed. 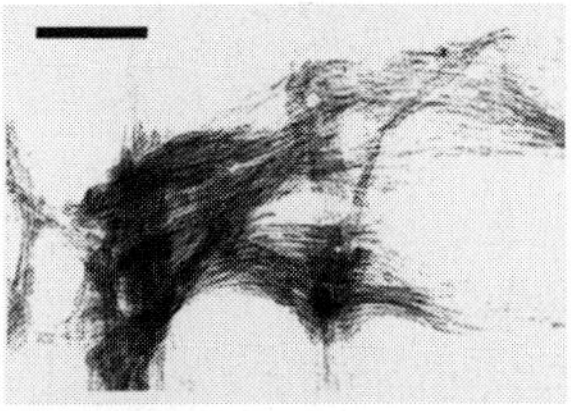	23
I56T/D67H: sodium acetate buffer, pH 5, 65 ° C, 0.1 mg/ml, stirring, 3h	The fibrils appear very similar to formed by incubation at pH 5.5 in the presence of 3M urea at 48°C. 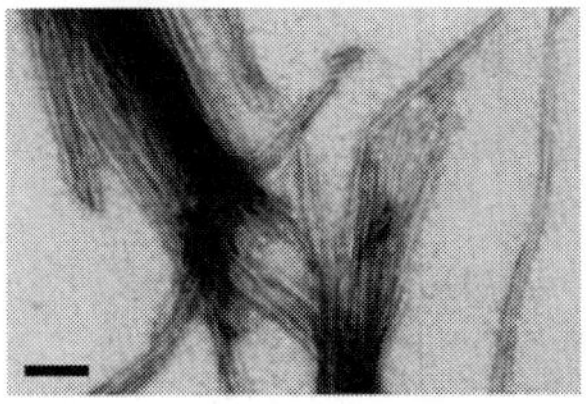	22

Fig. 24.1 Amyloid fibrils formed *in vivo* and *in vitro* from I56T, D67H and wild-type lysozymes, observed by electron microscopy after negative staining with uranyl acetate. *In vitro*, the two variant proteins form fibrils under similar conditions. Under the same buffer conditions the wild-type lysozyme require incubation at a higher temperature to form fibrils. The scale bar represents 50 nm.

at pH 2.0 and 37 °C are narrow, having a diameter of 6–7 nm in contrast to the ribbon-like structures with a diameter of 100–400 nm observed for most fibrils in samples of the same protein incubated at pH 5.5 in presence of 3 M urea and 48 °C (Fig. 24.1). Similar morphological variations have been reported for other amyloid fibril systems [25–27], and are thought to result at least in part from variations in the number and arrangements of protofilaments.

The fact that the wild-type protein has been shown to form fibrils *in vitro* demonstrates that the ability to form ordered aggregates is an intrinsic property of human lysozyme and does not require the presence of specific mutations in its primary structure. This observation is in full agreement with the idea that the ability to form amyloid fibrils is a generic property of proteins, resulting from stable interactions involving the main chain atoms that are common to all polypeptides [28, 29]. Fibril formation by D67H and wild-type lysozymes can be greatly accelerated by seeding the solutions with preformed fibrils made from the I56T variant protein as well as those from their own sequence [21]. Similarly, the formation of fibrils by the D67H and I56T variants can be greatly accelerated by seeding the solution with preformed fibrils derived from wild-type lysozyme [21]. Together, these results suggest that the fact that the wild-type protein has not been found in *ex vivo* deposits from patients suffering from lysozyme-associated amyloidosis is likely to be related to the higher stability of the wild-type native structure relative to that of the variant proteins, rather than from any direct structural effects of the mutations (see below).

24.5 *In Vitro* Studies of the Properties of Variant Lysozymes

24.5.1 Effects of Mutations on the Native Structure of Lysozyme

Human lysozyme in its native state is folded with approximately 10% of its residues in β-structure, 30% in α-helices and approximately another 20% in short 3_{10} helices. The protein exhibits two structural domains (Fig. 24.2a): the α domain made up of four α-helices and two 3_{10} helices, and the β domain made up of a triple-stranded β-sheet, a 3_{10} helix and a long loop. The protein contains four disulfide bonds, of which two are located in the α domain, one in the long loop of the β domain and one connecting the two domains. The active site is located in the cleft that is formed between the two domains.

X-ray crystallographic data show that the I56T, D67H and T70N variants in their native states have wild-type folds and all four disulfide bonds correctly formed [9, 30, 31]. The structure of the I56T variant is virtually identical to that of the wild-type protein (Fig. 24.2a). The D67H mutation, however, disrupts a series of hydrogen bonds in the β domain of the mutant protein, resulting in a very significant movement (up to 11 Å) of some of the residues in the region of the sequence that includes residues 42–55 and 66–75 (Fig. 24.2b) [9]. In the T70N variant, a conformational displacement in the same region of the structure has also been observed; in this case, however, it encompasses fewer residues and its amplitude is lower (Fig. 24.2c). Moreover, in this variant protein, the hydrogen bond networks in the β domain that are characteristic of the wild-type protein are essentially conserved [7, 31]. These findings suggest that there

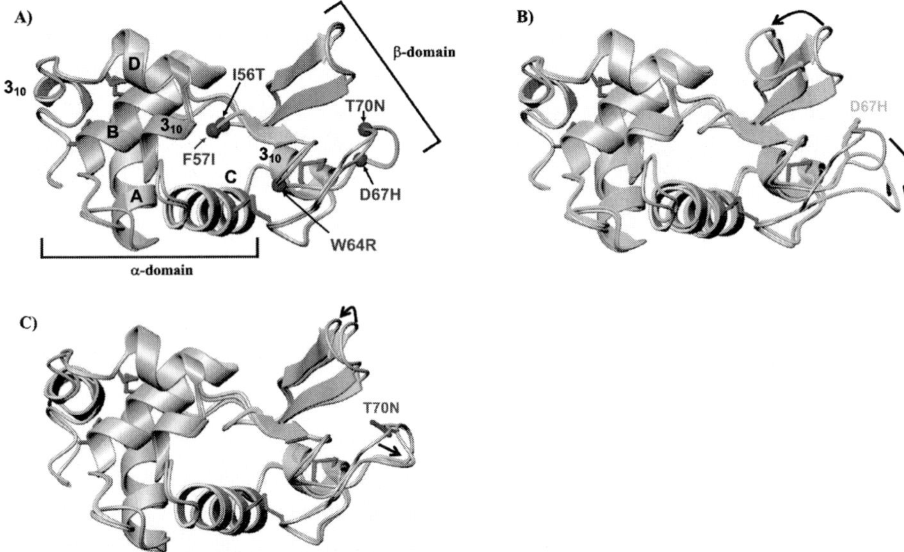

Fig. 24.2 Overlay of ribbon diagrams representing the structures of the various lysozyme species. (A) Wild-type human lysozyme (grey) and the I56T variant (pink). The structure of the I56T variant is virtually indistinguishable from that of the wild-type protein. The α-helices are labeled A–D and the three 3_{10} helices are also indicated. The four disulfide bonds are shown in red and the natural mutations in blue. (B) Wild-type human lysozyme (grey) and the D67H variant (green). The amyloidogenic mutation D67H is shown in green. The black arrows indicate the relative movements in the positions of residues 42–55 and 66–75 in the D67H structure relative to wild-type lysozyme. (C) Wild-type human lysozyme (grey) and the T70N variant (blue). The non-amyloidogenic mutation T70N is shown in blue. The black arrows indicate the relative movements in the positions of residues 45–51 and 68–75 in the T70N structure relative to those of wild-type lysozyme. The lysozyme structures were generated from coordinates determined by X-ray diffraction [9, 31] [PDB 1LYY (D67H), 1LOZ (I56T), 1LZ1 (wild-type) and 1W08 (T70N)] and produced using MOLMOL [48].

is no direct correlation between the structural changes induced by the mutations and the propensity of the variant proteins to form amyloid fibrils.

Closer inspection of the crystal structures of the variant proteins suggests that the key to the amyloidogenic behavior of the I56T and D67H mutations could, however, lie at least in part in the effect they produce at the interface between the α and β domains. In case of the I56T variant, the substitution of the isoleucine residue by threonine introduces a polar hydrophilic group into a hydrophobic pocket and, thereby, disrupts long-range hydrophobic contacts that bridge the in-

Table 24.1 Comparative summary of biochemical properties of the wild-type and the three lysozyme variants so far studied

	Wild-type	I56T	D67H	T70N
Enzymatic parameters [a)]				
K_m (μM)	16.5	18.5	38.0	30.0
K_{cat} (M/s)	14.5	15.0	9.5	10.0
ΔG (kJ/mol) [b)]	57.4	27.1	37.6	41.8
ΔΔG (kJ/mol)	–	30.3	19.8	15.6
T_m (°C) [c)]	78±1	68±1	68±1	74±1
ANS fluorescence (% relative to that of the amyloidogenic variants) [d)]	32	100	100	48

a) Enzymatic activities determined with PNP-(GlcNAc)$_5$ at pH 5.0 and 37 °C [7].
b) Determined at pH 6.5 and 20 °C [7].
c) Determined by far UV-CD measurements in 0.1 M sodium acetate buffer pH 5.0 [22, 31].
d) Determined in 0.1 M sodium acetate buffer pH 5.0 (Kumita et al., unpublished results).

terface between the α and β domains in the wild-type protein [9]. In the case of the D67H variant, the changes in the orientation of the β-strands and the long loop (Fig. 24.2 b) are transmitted to the domain interface resulting in the side-chain of I56 adopting a different rotameric state and the indole ring of W64 packing further from the C-helix. In contrast, in the T70N variant, only minor deviations are found in the α/β domain interface region [31]. Although the side-chain of I59T adopts a new rotameric state, the side-chains of I56 and W64 are both in their wild-type conformations. Taken together these observations suggest that the domain interface in the amyloidogenic variant proteins is less stable than in T70N and wild-type lysozymes.

Although all three variants (I56T, D67H and T70N) have been found to be enzymatically active, the D67H and T70N lysozymes both have a lower activity than the wild-type protein (Table 24.1) [7, 9]. This lowered activity results from a combination of a lower substrate affinity and a lower turnover number, and is perhaps the result of structural distortions in the active site induced by the movement of the β domain in these two variant proteins (see above).

24.5.2
Effects of the Mutations on the Folding of Lysozyme

24.5.2.1 Equilibrium Unfolding

Chemically induced equilibrium unfolding studies carried out at pH 4.0 and 10 °C [30, 32] and at pH 6.5 and 20 °C [7] have shown that the three variant proteins, I56T, D67H and T70N, like the wild-type protein, appear to unfold with a single cooperative transition characteristic of a two-state unfolding process. Moreover, in all cases, the denaturation is completely reversible. The three variants, however, are all destabilized relative to the wild-type protein (Table 24.1). The I56T protein is the most destabilized ($\Delta\Delta G = -30.3$ kJ/mol) with the D67H variant being destabilized just slightly more than the T70N protein ($\Delta\Delta G = -19.8$ and –15.6 kJ/mol for the D67H and T70N variants, respectively).

Thermal unfolding experiments carried out at pH 2.5–5.0 show, however, that the I56T and D67H variants are decreased in thermostability relative to the wild-type protein to a similar extent; their unfolding transitions being about 10 °C lower at all pH values [9, 21–23, 30, 32]. On the other hand, the T70N variant is less thermostable than wild-type lysozyme (its T_m is about 4 °C lower, Table 1), but is more stable than the amyloidogenic variants. Most importantly, the thermal unfolding at pH 5.0 of the I56T and D67H variant lysozymes is substantially less cooperative than that of the T70N variant and the wild-type lysozyme [7, 9, 22]. A large increase in 8-anilino-1-naphthalene-sulfonic acid (ANS) fluorescence is observed near to the denaturation temperature of the I56T and D67H proteins at pH 5.0 [9, 22]. As ANS binds to accessible hydrophobic clusters in proteins, this finding indicates that a partially unfolded intermediate is significantly populated during unfolding of these proteins. The T70N variant (Kumita et al., unpublished results) and the wild-type protein [9, 22] also bind ANS near to the temperature of the mid-point of their unfolding transition at pH 5.0. The intensity of ANS binding is, however, only about 48 and 32% of that observed for the amyloidogenic variants for the T70N variant and the wild-type protein, respectively. These results suggest that a partially unfolded intermediate species is present under these conditions, but is populated to lower levels in the case of the T70N variant and particularly the wild-type protein. The wild-type protein has, however, been found to populate more significantly an intermediate state during thermal denaturation at pH 2 [21, 33]. These results indicate that the ability to populate an intermediate species is an intrinsic property of the lysozyme fold and is not restricted to the amyloidogenic variants. These latter proteins, however, populate such intermediate species under much milder conditions and to a greater extent than do the T70N variant and wild-type lysozyme. Very interestingly, the fact that the T70N variant is not associated with amyloidosis despite its destabilization relative to the wild-type protein suggests that a fine balance exists between a benign mutation and one that induces amyloid disease [7, 31].

24.5.2.2 **Kinetics of Unfolding and Refolding**

The kinetics of unfolding of the I56T, D67H and T70N variants and wild-type lysozyme in high concentrations of guanidinium chloride at pH 4.0 and 10 °C [30, 32] and at pH 5.0 and 20 °C [7, 34] have been found to fit single exponential functions. The unfolding of the three variants is, however, much faster than that of the wild-type protein. For example, the T70N, I56T and D67H unfold, respectively, about 3, 30 and 160 times faster than wild-type lysozyme at pH 5 and 20 °C in the presence of 5.4 M guanidinium chloride [7, 34].

A range of techniques, including stopped-flow circular dichroism, fluorescence and pulsed hydrogen–deuterium exchange analyzed by mass spectrometry and nuclear magnetic resonance (NMR) spectroscopy, has been used to monitor in detail the refolding of wild-type lysozyme and the I56T and D67H variants from their guanidinium chloride-denatured states [34, 35]. These studies have shown that wild-type human lysozyme refolds in the presence of its native disulfide bonds via multiple parallel tracks and through a series of well-defined intermediates. For the majority of the molecules that follow a slow track, a series of distinct steps is involved in the development of the fully native structure (Fig. 24.3) [34, 35]. In the first step, the A and B helices and the C-terminal 3_{10} helix are stabilized in a locally cooperative manner. This step is followed by the cooperative folding of the helices C and D, and finally by the folding of the β domain. As for the homologous hen lysozyme [36], it is likely that a final step involving the docking of the two domains is required to generate the native close-packed structure with a functional active site. About 10% of the molecules fold along a fast track that arises from a population of molecules that are able to form the native state more efficiently than the remainder of the molecules. For these rapidly folding molecules, the β domain becomes structured concomitantly with the formation of the α domain [34, 35].

No significant differences have been observed in the refolding behavior of the T70N variant and the wild-type protein as monitored by fluorescence experiments at pH 5 and 20 °C [7]. The refolding of both the I56T and D67H variants *in vitro* has been studied in greater detail [30, *32*, 34], and has also been found to be closely similar to the wild-type protein, i.e. it occurs via multiple parallel pathways and through a series of well-defined intermediates. The rates of folding are, however, very different for the two variants (Fig. 24.3) [34]. The rate of the first refolding step, i.e. the development of structure in the A and B helices and the C-terminal 3_{10} helix, is not affected by either of the two mutations. This result is fully consistent with the fact that neither of the mutated residues is in or near this region of the native fold. For the D67H variant, the subsequent steps in structure formation, i.e. the coalescence first of the helices C and D, and then of the β domain, also occur on similar time scales as for the wild-type protein. For the I56T protein, however, all these subsequent refolding steps take place more slowly, reflecting the location of residue I56 at the interface region between the α and β domains, and the importance of the development of structure in this region of the protein for all these events.

To summarize, the lower stability of the D67H variant is almost entirely the result of an increase in the unfolding rate compared to the wild type protein, whereas the reduction in stability of the I56T variant results from a combination of changes in both unfolding and refolding rates.

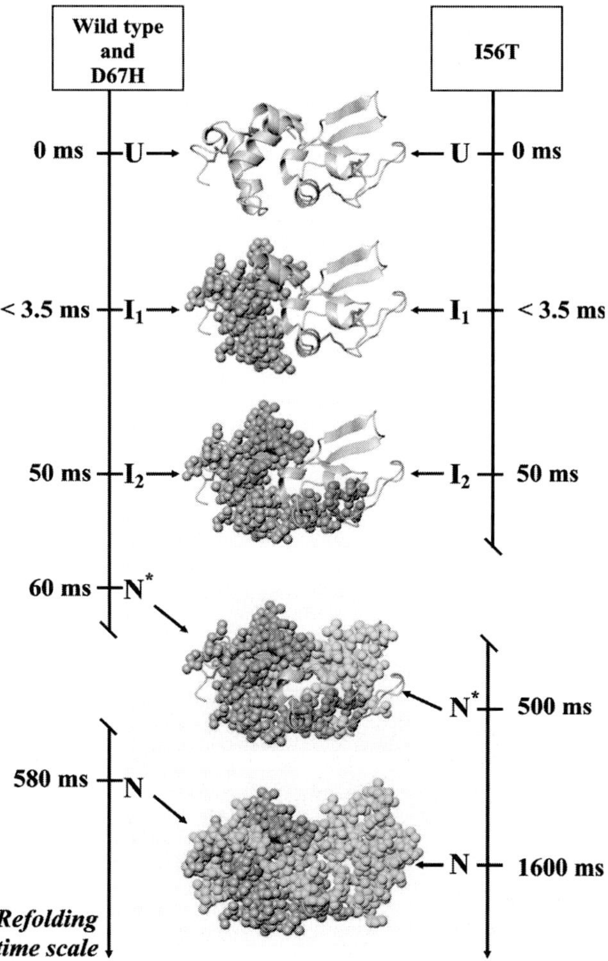

Fig. 24.3 Representation of the different steps and average time constants for the slow refolding track occurring in wild-type lysozyme and its I56T and D67H variants. The data for the D67H variant and the wild-type protein are shown together because of the close similarity of their kinetic behavior during refolding. N* represents a species which has native-like protection, but has not undergone the final rearrangement step to generate the native protein. The nature of the later step is not, however, clear and it may be that only a relatively small proportion of molecules requires such a rearrangement process. For each species identified along the folding pathway, a structural model generated using the program MOLMOL [48] is shown with the regions showing high protection against hydrogen exchange displayed in space-filled mode colored according to the time at which such protection occurs: blue, <3.5 ms; pink, 3.5–50 ms; yellow >50 ms; green, the remainder. The four disulfide bonds are shown in red. (Adapted from [34].)

24.5.3
Effect of the Mutations on the Conformational Dynamics of Lysozyme

The internal folding dynamics taking place on a fast time scale have been investigated by NMR spectroscopy for both the I56T and D67H variants, and compared to those of the wild-type protein [37]. The results show that the behavior of the I56T variant is closely similar to that of the wild-type protein. By contrast, the D67H variant has a region of substantially increased flexibility in the vicinity of the mutation (residues 65-80). Because the I56T variant shows no such dynamical behavior, the higher flexibility of a region of the β domain in the na-

A)

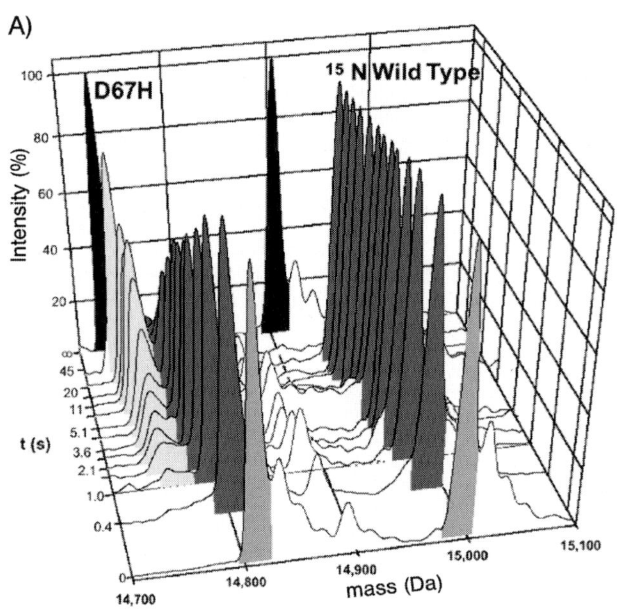

B)

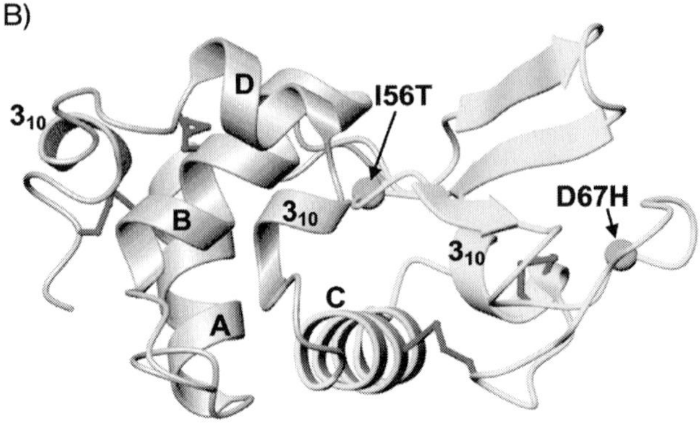

tive state of the D67H variant does not, therefore, appear to play a fundamental role in the process of conversion to amyloid fibrils [37].

The conformational dynamics on a slower time scale for wild-type lysozyme and the I56T and D67H variant proteins have been investigated by using a combination of electrospray ionization tandem mass spectrometry (ESI-MS) and NMR spectroscopy to monitor the solution hydrogen–deuterium exchange properties of the labile amide and side-chain hydrogens [22, 38]. These studies (Fig. 24.4) show that both the I56T and D67H proteins can transiently populate at a significant level a partially folded intermediate that is in dynamic equilibrium with the native protein under physiologically relevant values of pH and temperature. In the partially structured species formed transiently in this way by the I56T and D67H variants, the β domain and the adjacent C-helix have been found to be simultaneously unfolded while the remainder of the protein appears to retain a largely native-like structure (Fig. 24.4 b). The similarity in the behavior of the I56T and D67H variant proteins is remarkable given the different locations of the two mutations and their different effects on the native structure of the protein (Fig. 24.2 a and b). It can, however, be rationalized from the structural data describe above, as these suggest that in both variants the crucial interface region between the α and β domains is destabilized relative to the wild-type protein, and that this perturbation of the interface is likely to be an

Fig. 24.4 (A) ESI-MS of wild-type lysozyme and the D67H variant showing the local unfolding step that can lead to aggregation. Mass spectra were recorded for an approximately equimolar mixture of ^{15}N-labeled wild-type lysozyme and D67H variant following exposure to hydrogen exchange conditions at pH 8.2 and 37 °C for various periods of time. Both wild-type lysozyme and the D67H variant were initially exposed to D_2O so as to replace all the labile hydrogens with deuterium atoms; the exchange process therefore involved the replacement of these deuterium atoms with hydrogen atoms from the solvent H_2O. The peaks colored yellow are observed in spectra of the D67H variant, but not in that of the wild-type protein, and represent those molecules where hydrogen exchange occurs as a locally cooperative unfolding of the β domain and C-helix. The peaks colored red reveal a gradual loss of deuterium as a result of independent local fluctuations in the structure. The peaks in the spectra of control samples recorded before exchange (time zero) and after complete exchange (time infinite) are colored green and black, respectively. (From [38].) The behavior of the I56T variant is remarkably similar to that observed for the D67H variant under similar conditions [22]. (B) Ribbon diagram of lysozyme produced using MOLMOL [48] showing in yellow the region of the molecule that unfolds transiently in a locally cooperative manner in both the I56T and D67H amyloidogenic variants. The amyloidogenic mutations are shown in green.

important factor in the reduction in global cooperativity of the native proteins. Interestingly, the intermediate that is sampled occasionally in this way for the variant proteins under equilibrium conditions resembles that populated in the normal refolding of the protein from a highly denatured state (see above), emphasizing the close link between normal and aberrant folding behavior [38].

24.6 Mechanism of Fibril Formation

As a result of the findings from the various biophysical studies described above, a mechanism for the formation of lysozyme fibrils can be proposed. The ability of the amyloidogenic variants of lysozyme to form amyloid fibrils *in vivo* is thought to be primarily a result of the reduced stability of their native states relative to that of the wild-type protein and, especially, to the reduction of their global cooperativity. The relatively low level of destabilization caused by the mutations appears to permit the variant proteins to fold efficiently enough to avoid degradation in the endoplasmic reticulum, and hence to be secreted into extracellular space where the protein functions [28]. In contrast, other variants that are destabilized more substantially will be degraded and not secreted in significant quantities. It appears that the amyloidogenic mutations that have been reported so far destabilize the native state to such an extent that these proteins can easily be converted to a partially unfolded state under physiological conditions, whereas the wild-type protein cannot. Calculations based on hydrogen-exchange protection suggest that the population of the partially unfolded protein under physiological conditions is nearly 100 times greater for the amyloidogenic I56T and D67H variants than for wild-type lysozyme [38]. Interestingly, T70N lysozyme appears not to be destabilized to an extent sufficient to enable it to form significant quantities of amyloid fibrils *in vivo*, showing that there is only a narrow window of stability that results in amyloidogenic behavior.

In the partially unfolded intermediate species populated transiently by the I56T and D67H variants, the β domain and the sequentially-adjacent C-helix are substantially destabilized, whereas the remaining regions of the molecule largely retain their native like properties (Fig. 24.4 b). The formation of this transient intermediate is thought to be the critical event that initiates the aggregation event that ultimately leads to the formation of fibrils. Indeed, as part of this region of the protein forms β-sheet structure in the native protein, initial steps in the aggregation process could simply involve the formation of intermolecular hydrogen bonds in the local region of the β domain, rather than the intramolecular ones that characterize the native structure (Fig. 24.5). This initial step can perhaps be considered as a rudimentary form of domain swapping, whereby oligomers with predominantly native-like interactions form prior to further conformational reorganization associated with the formation of extensive β-sheet structure. Indeed, such species have been observed in other systems to form during the early stages of aggregation prior to reorganization to form the familiar amyloid structure [39–41].

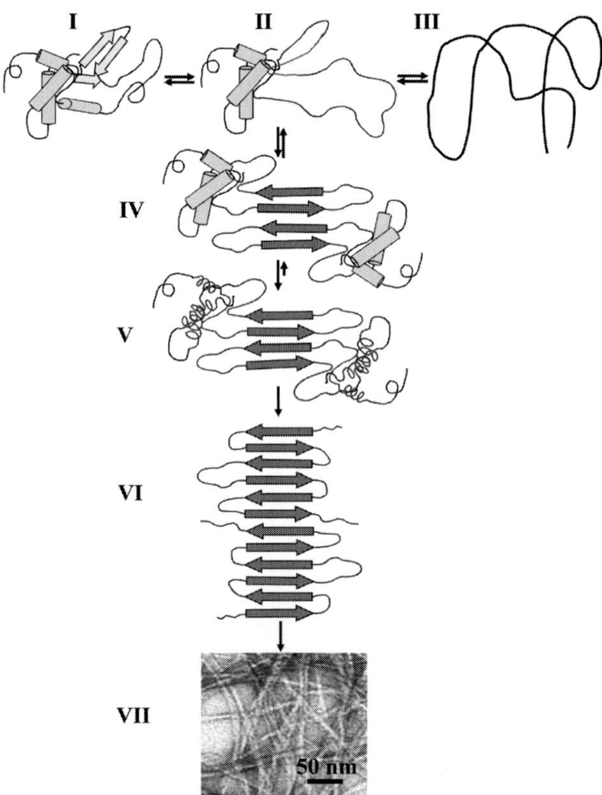

Fig. 24.5 Proposed mechanism for amyloid fibril formation by lysozyme. Under physiological relevant conditions, the variant proteins (I) populate transiently an intermediate species (II), in which the α domain and the C-helix are cooperatively unfolded. These intermediate species then self-associate (IV) through the newly exposed aggregation prone regions via the formation of intermolecular interactions to initiate fibril formation. Further rearrangement (V and VI) is likely to occur in the remainder of the structure, including the recruitment of additional regions of the polypeptide chain into the β-sheet structure prior to the formation of mature fibrils (VII). Note also that the disulfide bridges are not represented on this scheme although they are present in all the species represented here including the fibrils. (Adapted from [9].)

The critical involvement of the partially unfolded species discussed above in the initiation of the aggregation process is supported by two recent studies. First, an important observation made with hen egg white lysozyme, a protein homologous to human lysozyme with 60% sequence identity, is that fragments corresponding to parts of the β domain and all the residues of the C-helix, are readily cleaved from the protein by proteolysis at low pH and elevated temperatures where the native state is significantly destabilized [42]. These fragments correspond almost exactly to the regions of the human I56T and D67H variants that unfold to form the intermediates described above. In addition, they rapidly form amyloid fibrils in contrast to fragments corresponding to the remainder of the α domain that remain largely soluble under the same conditions. This result indicates that the region corresponding to the β domain and the C-helix not only can unfold with local cooperativity, but also has a higher intrinsic propensity to aggregate than do other regions of the protein. Thus, it is likely that, following exposure to the solvent as the result of partial unfolding, this region of the variant proteins readily initiates the aggregation event that ultimately leads

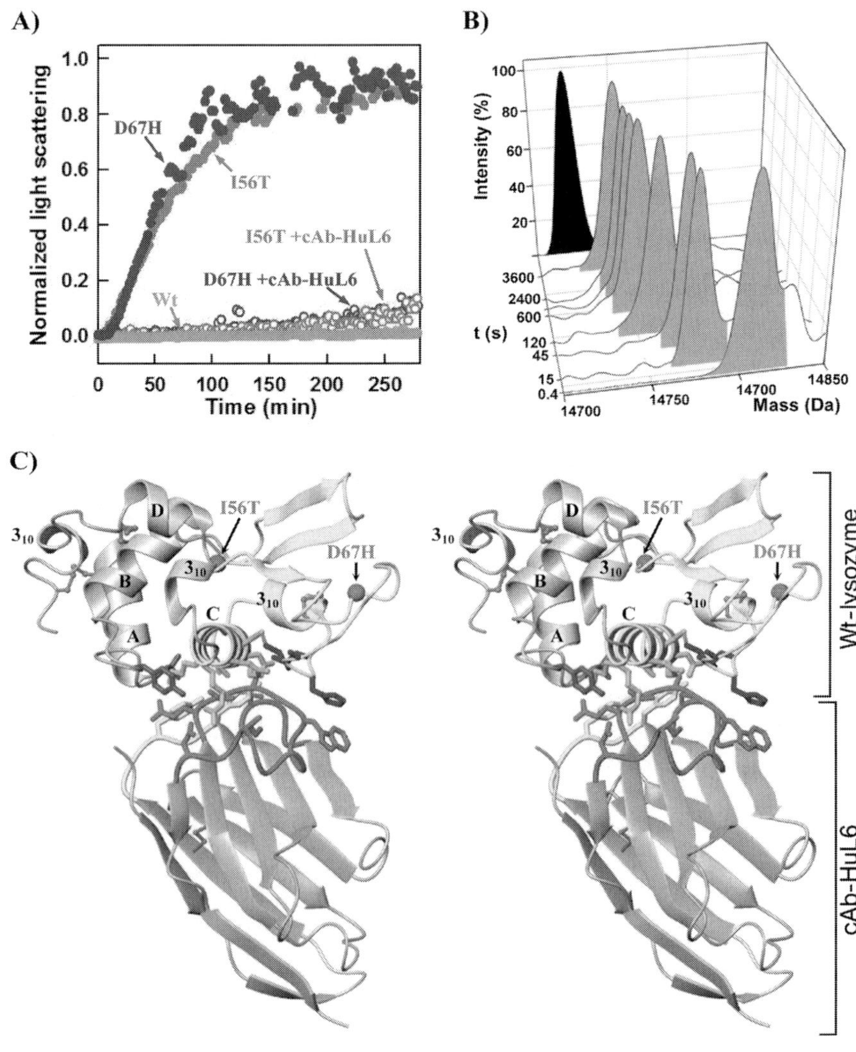

to the formation of fibrils. Indeed, the highly amyloidogenic character of the destabilized region is probably the reason why all the known amyloidogenic mutations are localized in the β domain of the protein.

Second, we have shown recently that binding a camelid antibody fragment, cAb-HuL6, to both the I56T and D67H variant proteins inhibits dramatically their ability to aggregate *in vitro* and hence to form amyloid fibrils [22, 23] (Fig. 24.6 a). The effect of the binding of the antibody fragment on the global cooperativity of the variant proteins was investigated by pulsed labeling hydrogen/deuterium exchange monitored by mass spectrometry (Fig. 24.6 b). In the presence of the antibody fragment, on the time scale of the experiment (up to

Fig. 24.6 (A) Time course of the aggregation of the I56T and D67H variant lysozymes in the absence (red and blue solid circles for the I56T and D67H variants, respectively) and presence of an equimolar quantity of cAb-HuL6 (red and blue open circles for the I56T and D67H variants, respectively) as monitored by light scattering. Data are also shown for wild-type lysozyme in the absence of cAb-HuL6 (green solid circles). The lysozyme concentration was 0.1 mg/ml, and the data were recorded at 65 °C and pH 5.0 while the solutions were stirred. (Adapted from [22].) (B) ESI-MS of D67H lysozyme in the presence of an equimolar concentration of the antibody fragment. The peak observed in the spectra of a control sample after complete hydrogen–deuterium exchange is colored black. A single peak whose mass deceases with the length of time for which the exchange was allowed to proceed is observed. The peaks of the species of lower mass observed in the spectra of the free D67H variant (peaks colored yellow in Fig. 24.4 a), and that result from a locally cooperative unfolding of the β domain and the C-helix [38], are therefore not observed in the spectra of the D67H protein in the presence of the antibody fragment (right hand panel). This result suggests that the binding of the antibody fragment to the D67H variant restores the global cooperativity that is characteristic of the wild-type lysozyme. (From [23].) A similar restoration of the global cooperativity was observed for the I56T variant upon binding the antibody fragment [22]. (C) Stereo view of a ribbon representation of the X-ray structure of wild-type lysozyme complexed with the cAb-HuL6 fragment. The β-strands in cAb-HuL6 are colored green. The lysozyme molecule is shown in light blue and the helices are labeled. The sites of mutation are shown in red and the region of the molecule found to unfold transiently in a locally cooperative manner in the amyloidogenic variants is colored yellow. The disulfide bridges are colored orange. The side-chains of residues constituting the epitope are shown in violet, light blue and dark blue for the α domain, C-helix and β domain, respectively. Those residues constituting the paratope are shown in yellow, pink and red, and are from the CDR1, CDR2 and CDR3 loops of the cAb-HuL6 structure, respectively. The structure was produced using MOLMOL [48].

1 h) virtually none of the molecules of the I56T [22] and D67H [23] variant proteins undergoes even a single locally cooperative unfolding event of the type observed for the lysozyme variants when unbound (see above and Fig. 24.4). These results indicate that the frequency of such fluctuations in both variants is drastically reduced as a result of binding to the antibody fragment. In accord with this finding, real-time MS and NMR experiments have shown that the binding of the antibody restores the high degree of global cooperativity that is characteristic of the wild-type protein [22]. These results, therefore, provide further evidence that the formation of a partially unfolded species with a high propensity to aggregate, resulting from the locally cooperative unfolding of the β domain and the C-helix, is the critical event that triggers the aggregation process that occurs in the absence of the antibody fragment.

Structural studies of the complex between the wild-type protein and the antibody fragment reveal that the epitope on the lysozyme molecule includes neither the site of the mutations nor most of the residues in the region of the protein structure that is destabilized by the mutations (Fig. 24.6c) [23]. Thus, the effects of binding are not simply to mask the entire region of the protein destabilized by the mutation and hence prevent its unfolding from the remainder of the structure. Rather, it appears that the binding of the antibody fragment restores by a more subtle mechanism the global cooperativity of the lysozyme structure that is disrupted by these two amyloidogenic mutations. Analysis of the NMR chemical shift changes of lysozyme resulting from binding of the antibody fragment shows that restoration of the global cooperativity occurs, at least in part, through the transmission of long-range conformational effects to the interface between the two structural domains [22, 23]. These results strongly suggest that there is a close link between the ability of an amyloidogenic variant to populate an intermediate that results from the perturbation of the interface between the two lysozyme domains and its propensity to convert into amyloid fibrils. This conclusion reinforces the view that the remarkable cooperativity of native protein structures is an essential evolutionary development to enable otherwise marginally stable structures to resist aggregation under conditions in which they exert their biological function [43].

Despite the very close similarity of the intermediate species transiently populated by the I56T and D67H variants, it is unlikely that the formation of this specific structure is a prerequisite of amyloid formation. The common, perhaps universal, ability of polypeptides to form amyloid fibrils with similar morphological properties indicates that the specific structure of any partially folded intermediate formed by an amyloidogenic system is unlikely to be critical in defining the characteristic cross-β structure of the mature fibrillar assembly. The properties of the intermediate are, however, likely to be extremely important in determining the regions of the polypeptide chain involved in the core of the cross-β structure and in defining the properties and rate of formation of the initial oligomers that ultimately undergo structural reorganization to form the mature fibrils [38].

24.7
Conclusion and Future Perspectives

Human lysozyme is the best characterized, in terms of stability, dynamics, folding and aggregation, of any of the proteins associated with amyloid disease. Our knowledge, at a molecular level, of the effects of natural amyloidogenic mutations on the properties of the protein has enabled the reduction of global cooperativity to be identified as a major factor underlying the amyloidogenicity of pathogenic lysozyme mutations. The formation of amyloid fibrils *in vivo*, from at least the I56T and D67H variants, results from a series of events that by themselves might not be seriously harmful but together can be catastrophic. (1) The production in significant amounts of the proteins that are able to fold correctly to an extent that permits them to escape the quality control system and be secreted in the extracellular space where they normally function. (2) The mutations decrease sufficiently the stability and the global cooperativity of the proteins to enable them to populate under some conditions partially unfolded states in which at least part of the main chain and the hydrophobic core is exposed. (3) The region exposed in this manner is highly prone to aggregation and readily initiates the aggregation events that ultimately leads to the formation of amyloid fibrils.

As a result of the increasingly serious impact of amyloid disorders, such as Alzheimer's disease and related disorders of protein folding such as Parkinson's disease, on the aging population of the developed world, a range of strategies for the prevention and treatment of these diseases is being very actively explored [44]. One of the most attractive therapeutic options is to prevent the formation and accumulation of aggregation-prone species, thus inhibiting at the earliest step the process that leads to amyloid formation. This can be achieved most efficiently by stabilizing the native state of an amyloidogenic protein by binding small synthetic ligands [45, 46] or, as discussed here, antibody fragments [22, 23]. The sucess of the latter strategy suggests that exploration of a series of antibodies raised against a given protein antigen could well allow the identification of species able to overcome the effects of a wide range of protein misfolding diseases. Indeed, recent studies report the success of both active and passive vaccination approaches to slowing and/or reversing the aggregation process and its pathological consequences in mouse models of light chain amyloidosis, Alzheimer's disease and the mammalian prion diseases [47]. The detailed characterization of the behavior of lysozyme, as discussed in this article can therefore provide a framework for developing a comprehensive understanding of the fundamental principles that underlie the onset, development and prevention of amyloid disease in general.

Acknowledgments

The authors wish to thank Janet Kumita and Russell Johnson for their critical reading of the manuscript. M. D. and C. M. D. would like to acknowledge the European Commission, the BBRSC, the Wellcome Trust, the Leverhulme Trust and the Belgium Government, under the framework of the Interuniversity Attraction Poles, for their support of those parts of our own research that are described in this chapter.

References

1 Peters C. W., Kruse U., Pollwein R., Grzeschik K. H. and Sippel A. E. The human lysozyme gene. Sequence organization and chromosomal localization. *Eur J Biochem* 1989, *182*, 507–516.

2 Reitamo S., Klockars M., Adinolfi M. and Osserman E. F. Human lysozyme (origin and distribution in health and disease). *Ric Clin Lab* 1978, *8*, 211–231.

3 Jolles P. and Jolles J. What's new in lysozyme research? *Mol Cell Biochem* 1984, *63*, 165–189.

4 Pepys M. B., et al. Human lysozyme gene mutations cause hereditary systemic amyloidosis. *Nature* 1993, *362*, 553–557.

5 Booth D. R., Pepys M. B. and Hawkins P. N. A novel variant of human lysozyme (T70N) is common in the normal population. *Hum Mutat* 2000, *16*, 180.

6 Valleix S., et al. Hereditary renal amyloidosis caused by a new variant lysozyme W64R in a French family. *Kidney Int* 2002, *61*, 907–912.

7 Esposito G., et al. Structural and folding dynamics properties of T70N variant of human lysozyme. *J Biol Chem* 2003, *278*, 25910–25918.

8 Yazaki M., Farrell S. A. and Benson M. D. A novel lysozyme mutation Phe57Ile associated with hereditary renal amyloidosis. *Kidney Int* 2003, *63*, 1652–1657.

9 Booth D. R., et al. Instability, unfolding and aggregation of human lysozyme variants underlying amyloid fibrillogenesis. *Nature* 1997, *385*, 787–793.

10 Spencer A., Morozova-Roche L. A., Noppe W., MacKenzie D. A., Jeenes D. J., Joniau M., Dobson C. M. and Archer D. B. Expression, purification, and characterization of the recombinant calcium-binding equine lysozyme secreted by the filamentous fungus *Aspergillus niger*: comparisons with the production of hen and human lysozymes. *Protein Expr Purif* 1999, *16*, 171–180.

11 Osserman E. F. and Lawlor D. P. Serum and urinary lysozyme (muramidase) in monocytic and monomyelocytic leukemia. *J Exp Med* 1966, *124*, 921–952.

12 Hansen N. E., Karle H. and Andersen V. Lysozyme turnover in the rat. *J Clin Invest* 1971, *50*, 1473–1477.

13 Hansen N. E., Karle H., Andersen V. and Olgaard K. Lysozyme turnover in man. *J Clin Invest* 1972, *51*, 1146–1155.

14 Hysing J. and Tolleshaug H. Quantitative aspects of the uptake and degradation of lysozyme in the rat kidney *in vivo*. *Biochim Biophys Acta* 1986, *887*, 42–50.

15 Christensen E. I. and Birn H. Megalin and cubilin: multifunctional endocytic receptors. *Nat Rev Mol Cell Biol* 2002, *3*, 256–266.

16 Orlando R. A., Rader K., Authier F., Yamazaki H., Posner B. I., Bergeron J. J. and Farquhar M. G. Megalin is an endocytic receptor for insulin. *J Am Soc Nephrol* 1998, *9*, 1759–1766.

17 Gillmore J. D., Booth D. R., Madhoo S., Pepys M. B. and Hawkins P. N. Hereditary renal amyloidosis associated with variant lysozyme in a large English family. *Nephrol Dial Transplant* 1999, *14*, 2639–2644.

18 Jimenez J. L., Tennent G., Pepys M. B. and Saibil H. R. Structural diversity of *ex vivo* amyloid fibrils studied by cryo-electron microscopy. *J Mol Biol* 2001, *311*, 241–247.

19 Sunde M., Serpell L.C., Bartlam M., Fraser P.E., Pepys M.B. and Blake C.C.F. Common core structure of amyloid fibrils by synchrotron X-ray diffraction. *J Mol Biol* 1997, *273*, 729–739.

20 Serpell L.C., Sunde M., Benson M.D., Tennent G.A., Pepys M.B. and Fraser P.E. The protofilament substructure of amyloid fibrils. *J Mol Biol* 2000, *300*, 1033–1039.

21 Morozova-Roche L.A., Zurdo J., Spencer A., Noppe W., Receveur V., Archer D.B., Joniau M. and Dobson C.M. Amyloid fibril formation and seeding by wild-type human lysozyme and its disease-related mutational variants. *J Struct Biol* 2000, *130*, 339–351.

22 Dumoulin M., et al. Reduced global cooperativity is a common feature underlying the amyloidogenicity of pathogenic lysozyme mutations. *J Mol Biol* 2005, in press.

23 Dumoulin M., et al. A camelid antibody fragment inhibits the formation of amyloid fibrils by human lysozyme. *Nature* 2003, *424*, 783–788.

24 De Felice F.G., Vieira M.N., Meirelles M.N., Morozova-Roche L.A., Dobson C.M. and Ferreira S.T. Formation of amyloid aggregates from human lysozyme and its disease-associated variants using hydrostatic pressure. *FASEB J* 2004, *18*, 1099–1101.

25 Bauer H.H., Aebi U., Haner M., Hermann R., Muller M. and Merkle H.P. Architecture and polymorphism of fibrillar supramolecular assemblies produced by *in vitro* aggregation of human calcitonin. *J Struct Biol* 1995, *115*, 1–15.

26 Goldsbury C.S., et al. Polymorphic fibrillar assembly of human amylin. *J Struct Biol* 1997, *119*, 17–27.

27 McParland V.J., Kad N.M., Kalverda A.P., Brown A., Kirwin-Jones P., Hunter M.G., Sunde M. and Radford S.E. Partially unfolded states of beta2-microglobulin and amyloid formation *in vitro*. *Biochemistry* 2000, *39*, 8735–8746.

28 Dobson C.M. The structural basis of protein folding and its links with human disease. *Phil Trans R Soc Lond B* **2001**, *356*, 133–145.

29 Dobson C.M. Protein folding and misfolding. *Nature* 2003, *426*, 884–890.

30 Funahashi J., Takano K., Ogasahara K., Yamagata Y. and Yutani K. The structure, stability, and folding process of amyloidogenic mutant human lysozyme. *J Biochem (Tokyo)* 1996, *120*, 1216–1223.

31 Johnson R.J.K., et al. Rationalising lysozyme amyloidosis: insight from the structure and solution dynamics of T70N lysozyme. Submitted for publication.

32 Takano K., Funahashi J. and Yutani K. The stability and folding process of amyloidogenic mutant human lysozymes. *Eur J Biochem* 2001, *268*, 155–159.

33 Haezebrouck P., Joniau M., Van Dael H., Hooke S.D., Woodruff N.D. and Dobson C.M. An equilibrium partially folded state of human lysozyme at low pH. *J Mol Biol* 1995, *246*, 382–387.

34 Canet D., Sunde M., Last A.M., Miranker A., Spencer A., Robinson C.V. and Dobson C.M. Mechanistic studies of the folding of human lysozyme and the origin of amyloidogenic behavior in its disease-related variants. *Biochemistry* 1999, *38*, 6419–6427.

35 Hooke S.D., Radford S.E. and Dobson C.M. The refolding of human lysozyme: a comparison with the structurally homologous hen lysozyme. *Biochemistry* 1994, *33*, 5867–5876.

36 Itzhaki L.S., Evans P.A., Dobson C.M. and Radford S.E. Tertiary interactions in the folding pathway of hen lysozyme: kinetic studies using fluorescent probes. *Biochemistry* 1994, *33*, 5212–5220.

37 Chamberlain A.K., Receveur V., Spencer A., Redfield C. and Dobson C.M. Characterization of the structure and dynamics of amyloidogenic variants of human lysozyme by NMR spectroscopy. *Protein Sci* 2001, *10*, 2525–2530.

38 Canet D., et al. Local cooperativity in the unfolding of an amyloidogenic variant of human lysozyme. *Nat Struct Biol* 2002, *9*, 308–315.

39 Bouchard M., Zurdo J., Nettleton E.J., Dobson C.M. and Robinson C.V. Formation of insulin amyloid fibrils followed by FTIR simultaneously with CD and electron microscopy. *Protein Sci* 2000, *9*, 1960–1967.

40 Eakin C. M., Attenello F. J., Morgan C. J. and Miranker A. D. Oligomeric assembly of native-like precursors precedes amyloid formation by beta-2 microglobulin. *Biochemistry* 2004, *43*, 7808–7815.

41 Plakoutsi G., Taddei N., Stefani M. and Chiti F. Aggregation of the acylphosphatase from *Sulfolobus solfataricus*. The folded and partially unfolded states can be both precursors for amyloid formation. *J Biol Chem* 2004, 279, 14111–14119.

42 Frare E., Polverino de Laureto P., Zurdo J., Dobson C. M. and Fontana A. A highly amyloidogenic region of hen lysozyme. *J Mol Biol* 2004, *340*, 1153–1165.

43 Dobson C. M. Protein misfolding, evolution and disease. *Trends Biochem Sci* 1999, *24*, 329–332.

44 Dobson C. M. Protein folding and disease: a view from the first Horizon Symposium. *Nat Rev Drug Disc* 2003, *2*, 154–160.

45 Hammarstrom P., Wiseman R. L., Powers E. T. and Kelly J. W. Prevention of transthyretin amyloid disease by changing protein misfolding energetics. *Science* 2003, *299*, 713–716.

46 McCammon M. G., Scott D. J., Keetch C. A., Greene L. H., Purkey H. E., Petrassi H. M., Kelly J. W. and Robinson C. V. Screening transthyretin amyloid fibril inhibitors. Characterization of novel multiprotein, multiligand complexes by mass spectrometry. *Structure* 2002, *10*, 851–863.

47 Dumoulin M. and Dobson C. M. Probing the origins, diagnosis and treatment of amyloid diseases using antibodies. *Biochimie* 2004, *86*, 589–600.

48 Koradi R., Billeter M. and Wuthrich K. MOLMOL: a program for display and analysis of macromolecular structures. *J Mol Graph* 1996, *14*, 51–55.

25
Fibrinogen

Gilles Grateau and Marc Delpech

25.1
Introduction

Fibrinogen Aα chain was recognized as the amyloid protein responsible for a unique form of hereditary amyloidosis in 1993 by Dr Merrill Benson's group [3], which should be named AFib amyloidosis, according to the nomenclature of the International Society of Amyloidosis [35]. In all of the patients with AFib amyloidosis so far reported, the main clinical manifestation of the disease is amyloid nephropathy. AFib amyloidosis belongs to the group formerly called hereditary renal amyloidosis (HRA) or familial amyloidotic nephropathy, first described by Ostertag in 1932 and, therefore, often called familial amyloidosis of the Ostertag type [23, 24]. The phenotype of the family described by Ostertag included renal disease associated with liver and splenic involvement. After Ostertag's original report, which was the earliest description of hereditary amyloidosis, the biochemical nature of the amyloid fibril protein associated with this type of amyloidosis remained unknown for almost 60 years. At present, recent advances in protein biochemistry have enabled the identification of four other proteins for which variant forms are responsible for heredofamilial amyloidosis, in addition to the fibrinogen Aα chain: apolipoprotein AI and AII, and lysozyme [2]. These three latter forms are treated in other chapters of this book (see Chapters 22 and 24).

25.2
Clinical Manifestations

Like all other hereditary amyloid disorders, the mode of inheritance of AFib amyloidosis is autosomal dominant – a pattern remarkably consistent in the first reported kindreds. Subsequently, incomplete penetrance of the E526V mutation was reported [31] and, notably, Lachmann et al. have highlighted the predominant sporadic presentation of this mutant [17]. In another family, a sporadic case of AFib in a 7-year-old child associated with a previously unrecognized 517–522 indel *de novo* mutation has been reported [5].

Amyloid Proteins. The Beta Sheet Conformation and Disease. J. D. Sipe

ISBN: 3-527-31072-X

Age at onset of AFib is variable. Most often, disease appears in middle age, from 40 to 60 years [1, 3–4, 13, 14, 17, 19, 20, 30–34]. However, age at onset is much earlier in two kindreds, with 4897delT and 517–522 indel mutations, as the disease began in two patients at 12 and 7 years of age [5, 13]. This should be emphasized, as amyloidosis is very uncommon in children, with the exception of amyloidosis complicating familial Mediterranean fever [25].

25.2.1
Amyloid Nephropathy is the Main Clinical Feature of AFib Amyloidosis

In all cases so far reported, asymptomatic proteinuria or the nephrotic syndrome are the presenting signs of the disease. Hypertension is not constant at the beginning of the disease, but is reported in some families [4, 31]. In contrast, hypertension is a constant feature at presentation in the series of sporadic cases of E526V AFib patients reported by Lachmann et al. [17]. Renal disease is progressive and leads to renal failure within a relatively short delay. In Lachmann et al.'s series, time from presentation to renal failure was 2.3 years [17].

25.2.2
Other Manifestations of AFib Amyloidosis

Autopsy studies demonstrated the presence of amyloid deposits in the kidneys, liver, spleen, adrenals [1] and lungs [19, 20]. In the experience of the London National Amyloidosis Centre, serum amyloid P component scintigraphy showed renal amyloid in all patients and splenic deposits in most of them, whereas liver deposits were very uncommon. Clinical manifestations of extrarenal organ involvement are uncommon. Hepatosplenomegaly is lacking in AFib amyloidosis. However, spontaneous spleen rupture and liver failure have been described in three patients, suggesting possible long-term occurrence of these complications [17]. Anemia is probably related to renal failure [31, 32]. Pulmonary amyloidosis was confirmed by a biopsy in one patient with chest X-ray infiltrates [20]. Progressive peripheral neuropathy has recently been described in two sisters, with the novel E540V variant, suggesting possible amyloid neuropathy [4]. Autonomic neuropathy has been reported in another patient treated with combined liver and kidney transplantation [37].

25.2.3
Diagnosis of AFib Amyloidosis

Current diagnosis of AFib amyloidosis relies on clinical arguments, histological features and genetics. As previously mentioned, the renal presentation of AFib amyloidosis is quite constant. Furthermore, the most frequent variant, E526V, has an incomplete penetrance and often presents as sporadic (13 of 18 cases in one series) [17]. Finally, in the same series the median age of the patients at presentation is 59 years, a section where monoclonal gammopathy of undetermined significance is frequent and in fact a low-grade paraproteinemia was present in four of the 18 AFib patients. Thus, the diagnosis of AFib amyloidosis should be evoked in every patient presenting with amyloid in the kidney biopsy without arguments for AA amyloidosis. In patients with isolated kidney amyloid and a paraproteinemia, AFib should be systematically searched for before concluding AL amyloid. In the series of the National Amyloid Centre in London, 18 patients among 350 referred for AL amyloidosis (5.1%) had AFib amyloidosis finally diagnosed [17].

Kidney biopsy is always performed to disclose amyloid deposits. It is now established that, whatever the mutation is, amyloid deposits are only deposited in the glomerulus, not in the vessels, interstitium and tubules [17]. This has become an important clue for the diagnosis, particularly with regard to the discussion about AL amyloidosis. Immunohistochemistry is difficult as routine anti-fibrinogen antibodies do not constantly stain amyloid deposits. Specific antibodies directed against the new peptide purified from the deposits, in the case of the 4904delG mutation, strongly stained amyloid deposits [33]. However, these antibodies are not routinely available.

All mutations so far described to be associated with AFib amyloidosis are in the C-terminal part of the fibrinogen Aα chain protein. This clustering makes these mutations easily available by sequencing this coding part of the gene after *in vitro* amplification of genomic DNA by the polymerase chain reaction. Different techniques aimed at detecting mutation after genetic amplification can also be used, such as single-strand conformation polymorphism [34]. When the mutation modifies a restriction enzyme site, it can also be detected directly with the use of the appropriate enzyme [17].

25.2.4 Treatment

The kidney is the main target of AFib amyloidosis, whereas the fibrinogen protein is synthesized virtually only by the liver. Functional clinical assessment of both organs determines the principles of AFib amyloidosis treatment. Chronic dialysis and renal transplantation are indicated to restore renal function when end-stage renal failure has been reached. Kidney transplantation seems to be indicated as a treatment of end-stage renal failure when amyloid deposition is not clinically extent to the major organs, which could jeopardize the benefit or renal transplantation. However, some observations suggest that renal transplantation of AFib amyloidosis is followed by recurrence of amyloid deposition leading to proteinuria and renal failure within a period of years [12, 13]. Therefore, combined renal and hepatic transplantation has been performed recently by some clinical investigators [13, 30, 37]. One patient first received a kidney transplant that was followed 6 years later by recurrent amyloid deposition leading to end-stage renal failure. She subsequently developed liver failure, and liver transplantation was performed both for preserving liver function and for preventing further deposition of the toxic fibrinogen Aα protein [30]. In another series, one patient with end-stage renal failure was directly treated by combined liver and kidney transplantation, and remains well 2.5 years later [37]. The English Amyloidosis Centre has recently reported on five patients who have undergone combined liver and kidney transplantation. In two, mass spectrometry analysis showed complete elimination of the amyloidogenic variant fibrinogen A from the plasma. Overall, the outcome of hepatorenal transplantation has been excellent and all five patients are alive with functioning grafts [30].

25.3 The Fibrinogen Molecule

Fibrinogen is a 340-kDa protein that plays a major role in platelet aggregation and blood clot formation. Fibrinogen is secreted in plasma from liver, which is the sole organ of synthesis [18]. The normal plasma concentration is 2–4 g/l and the half-life in healthy subjects is about 4 days [7]. Fibrinogen, also found in lymph and interstitial fluids (at a concentration 20–40% that of plasma), is composed of three chains Aα, Bβ and γ arranged in a dimer (Aα, Bβ, $\gamma)_2$. The polypeptide chains are organized in two large globular structures at each end of the protein and a central smaller globular domain connected together by two linear structures (α-helix polypeptide chains). The X-ray crystallographic structure of a proteolytically modified fibrinogen has been reported at 1.8-Å resolution [27].

The Aα and Bβ chains are, respectively, composed of 610 (molecular weight 66 kDa) and 461 amino acid residues [8]. There are two variants of the γ-chain, of 427 and 411 amino acid residues, respectively, resulting from an alternative splicing of the messenger RNA [10, 36]. The Bβ and γ chains are glycosylated [22], and the Aα chain is phosphorylated [29]. The fibrinogen chains are encoded by three related genes located on chromosome 4 (between q23 and q32) [16]. The fibrinogen Aα chain gene spans over 7 kb and is comprised of six exons [6, 11]; a complementary DNA encoding a mature protein of 625 amino acids has been cloned and sequenced [28].

During coagulation, A and B chains are cleaved by thrombin at the Arg16–Gly17 and Arg14–Gly15 peptide bonds, respectively. Upon cleavage, fibrinopeptides A (FpA) and B (FpB) are released, and polymerization sites are exposed, leading to polymerization of fibrin. Fibrin polymers are subsequently stabilized covalently by transglutamination catalyzed by coagulation factor XIII. A number of fibrinogen Aα chain variants have been identified in patients with blood coagulation disorders [15]. Most of the variants have an amino acid substitution in the first 19 residues of the N-terminus [9, 15].

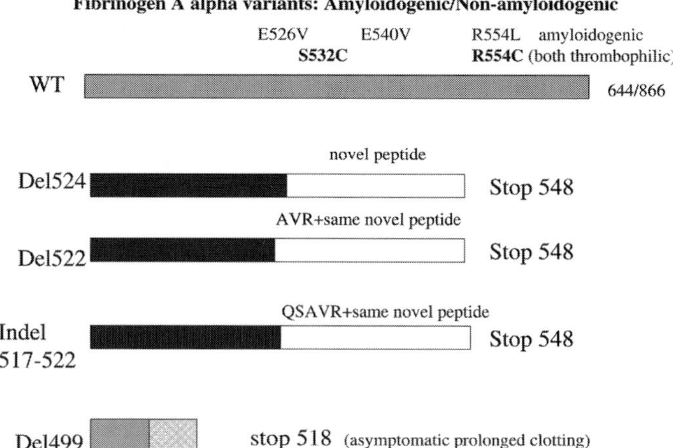

Fig. 25.1 The spectrum of fibrinogen Aα chain variants in the C-terminal part of the protein (from [5]).

25.4 The Various AFib Mutations and Related Peptides

Some data are currently available on the nature of the fibrinogen Aα chain-derived peptide which is deposited as amyloid fibrils *in vivo. In vitro* experiments aimed at evaluating the amyloidogenicity of fibrinogen Aα chain-derived peptides are lacking. Therefore, the mechanisms of amyloid fibril formation in AFib amyloidosis are far from being elucidated.

Six mutations in the fibrinogen Aα chain have been reported thus far and we can distinguish two groups of mutations. The first group includes three single amino acid substitutions which lead to a normal fibrinogen Aα chain protein, except for the substitution. The second group includes three deletions which all lead to a novel polypeptide chain in the C-terminal part of the fibrinogen Aα chain (see Fig. 25.1).

25.4.1 R554L Mutation

The first mutation in the fibrinogen Aα chain to be reported is R554L. A fragment of the fibrinogen Aα chain (amino acid residues 500–580) with an amino acid substitution of Leu for Arg at position 554 was isolated from renal amyloid deposits and a guanine to thymine transversion at the second base of codon 554 was subsequently revealed by DNA sequencing analysis [3]. This mutation was subsequently found in two other kindreds [14, 34]. In one of the families, fibrinogen was purified from plasma from two affected siblings, proteolytically digested, and fragments from position spanning amino acid residues 518–584 were isolated by high-performance liquid chromatography and submitted to amino acid sequence analysis. The ratio of recovered normal to variant fragments was 1:1 in one patient and 1.5:1 in the other [34].

25.4.2
E526V Mutation

The second fibrinogen Aα chain mutation found to be associated with amyloidosis was originnally found in two Irish-American kindreds [31]. DNA analyses revealed an adenine to thymine transversion at the second base of codon 526. This mutation is present in many populations and is certainly the most frequent fibrinogen mutation associated with hereditary amyloidosis. Haplotype analysis of the fibrinogen Aα chain gene indicated that the mutant genes in four kindreds of European origin were possibly derived from a single ancestry [1, 19, 20, 32].

DNA sequence of this variant predicts a Val for Glu substitution at the amino acid level. The amyloid peptide deposited in the kidney of patients with the E526V mutation has been shown to be similar in size to the R554L amyloid peptide (Liepniks and Benson, cited in [13]). In one patient, the plasma fibrinogen values of mutant gene carriers were within the normal range, and peptide sequencing analysis of plasma fibrinogen showed about the same amount of normal and variant protein [31].

25.4.3
E540V Mutation

A third missense mutation in the fibrinogen Aα chain gene has recently been found in a German family with a phenotype indistinguishable from that associated with the prevalent fibrinogen Aα chain variant E526V. DNA analysis disclosed an adenine to thymine transversion at the codon 540 leading to a glutamine to valine substitution [4].

25.4.4
4904delG Mutation

In a Caucasian-American kindred, DNA sequencing analysis revealed a single nucleotide deletion at the third base of codon 524 of the fibrinogen Aα chain gene (4904delG) that resulted in a frameshift and premature termination of the protein at codon 548 [33]. This deletion predicted the existence of a complete new C-terminal polypeptide chain from codon 524 to codon 548. A monoclonal antibody specific for this putative novel amyloid peptide reacted strongly with glomerular amyloid deposits. However, there was no tissue available for biochemical analysis of the amyloid fibril protein and confirmation of the novel peptide predicted from DNA sequence. Asymptomatic mutant gene carriers in this kindred had relatively low plasma fibrinogen concentrations compared to siblings without the mutation and only the variant fibrinogen was detected in their plasma.

25.4.5
4897delT Mutation

A second single nucleotide deletion in the fibrinogen Aα chain gene (4897delT) was found in a French family [13]. The frameshift resulting from the 4897delT at codon 522 results in expression of a novel peptide sequence which was proven by amino acid sequencing of amyloid fibril protein from the proband's kidney. The protein contains 49 residues, the N-terminus representing 23 residues of the normal fibrinogen Aα chain (499–521) and the remaining 26 residues representing a completely new sequence. This polypeptide corresponds to the predicted novel peptide resulting from the 4904delG mutation with three additional amino acid residues (alanine–valine–arginine) at the N-terminal part.

25.4.6
517–522 Delin Complex Mutation

A third, complex, insertion deletion mutation in the fibrinogen Aα chain gene was reported in an Asian child with systemic amyloidosis [5]. DNA sequencing analysis revealed a complex 15-base deletion and 2-base insertion at codon 517–522 of the fibrinogen Aα chain. This mutation predicts a novel polypeptide chain identical to the one associated with the 4897delT mutation except from an additional glutamine at the N-terminal site.

25.5
Mechanisms of AFib Amyloidosis

Six mutations in the fibrinogen Aα chain are currently known to be responsible for hereditary amyloidosis. The amyloid peptide deposited in tissues has not been determined in all six. Two different peptides are deposited according to the type of mutation: missense mutations that lead (or are predicted to lead) to the deposition a fragment of the C-terminal part of the fibrinogen Aα chain that differs from the wild-type chain only by a single substitution. The three known deletions lead (or are predicted to lead) to the deposition of a completely novel fragment of the C-terminal part of the fibrinogen Aα chain, with a premature stop codon at 548. This novel polypeptide lacks homology with any other mammalian-derived protein. An enigma actually lies in the ability of these two different peptides, derived from the same protein, to form amyloid aggregates. It can be hypothesized, reasonably, that the N-terminal region, which is shared by all amyloidogenic fibrinogens, plays a major role in amyloidogenicity. However, computer modeling studies predict that the C-terminal novel region has a β-strand structure that could render it particularly amyloidogenic. The respective contribution of the N-terminal and novel C-terminal region of the peptides to amyloid formation remains thus to be elucidated. Whatever the mutation, AFib amyloid peptides are similar in size, around 4500 Da, thus comparable in size to AANF and Aβ, the small amyloid proteins. The size of AFib, AANF and Aβ is small compared with the size of the amyloid fibril-forming peptides found in the other hereditary amyloidoses, which range from 8 to 14 kDa; size is important in fibril formation in the individual categories of amyloidosis. It has been shown that in blood plasma the fibrinogen Aα chain undergoes a physiological proteolytic cleavage at several sites, leading to the formation of intermediate peptides of different size [21]. This intravascular cleavage would explain the specific predominant targeting of the kidney glomerulus which is characteristic of AFib amyloidosis in relation to other types of hereditary amyloidosis. The molecular mechanisms of the pathologic cleavage of fibrinogen Aα chain leading to amyloidogenic peptides are unknown. In the family with the 4904delG mutation, gene carriers have a relatively low plasma fibrinogen concentration and the variant protein is not detected in their plasma. In two patients with the R554L mutation, the normal to variant ratio was 1:1 and 1.5:1. These data, although preliminary, suggest that amyloidogenic mutation could modify the tertiary structure of this region of the fibrinogen Aα chain molecule, resulting in abnormal proteolytic cleavage and amyloid-prone peptide generation. Isotopic studies, such as those performed with AApoAI variants, would help to confirm this metabolic hypothesis [26].

Non-amyloidogenic mutations have also been described in the C-terminal part of the fibrinogen Aα chain [15]. A deletion at codon 499 creates a novel peptide with a terminal cysteine at position 518 that results in asymptomatic prolonged clotting time. Two missense mutations, S532C and R554C, are associated with thrombophilia and not with amyloidosis. Disruption of this region of fibrinogen seems to predispose to amyloid formation, unless the mutation creates a novel unpaired cysteine. It is striking that different mutations at codon 554 can lead to two completely different disorders: thrombophilic state (R554L) or hereditary amyloidosis (R554C).

Acknowledgments

We are particularly grateful to Alison Bybee and Arie Stangou to providing us unpublished results presented at the Xth International Symposium on Amyloid and Amyloidosis, Tours, April 2004. A database for human fibrinogen variants is available at *http://www.geht.org/databaseang/fibrinogen*

References

1 Alexander, F. and E. L. Atkins. Familial renal amyloidosis, case reports, literature review and classification. *Am J Med* **1975**, *59*, 121–128.
2 Benson, M. D. Amyloidosis and other protein deposition diseases. In *Emery and Rimoin's Principles and Practice of Medical Genetics*, 4th edn. Rimoin, D. L., J. M. Connor, R. E. Pyeritz and B. R. Korf (eds). Churchill Livingstone, Edinburgh, **2002**, vol. 2, pp. 2058–2073.
3 Benson, M. D., J. Liepnieks, T. Uemichi, G. Wheeler and R. Correa. Hereditary renal amyloidosis associated with a mutant fibrinogen A-chain. *Nat Genet* **1993**, *3*, 252–255.
4 Bybee, A., M. Hollenbeck, E. R. Debusmann, D. Gopaul, J. A. Gilbertson, H. J. Lachmann, M. B. Pepys and P. N. Hawkins. Hereditary renal amyloidosis in a German family associated with fibrinogen A*α* chain (FGA) Glu540Val (E540V). In *Amyloid and Amyloidosis: Proc Xth Int Symp on Amyloidosis*, April 18–22, 2004, Tours, France. Grateau, G., R. A. Kyle and M. Skinner (eds). Life Sciences/CRC Press, Boca Raton, FL, **2005**, p. 367.
5 Bybee, A., H. G. Kang, I. S. Ha, M. S. Park, H. I. Cheong, Y. Choi and P. N. Hawkins. A novel complex indel mutation in the fibrinogen A*α* chain gene in an Asian child with systemic amyloidosis. In *Amyloid and Amyloidosis: Proc Xth Int Symp on Amyloidosis*, April 18–22, 2004, Tours, France. Grateau, G., R. A. Kyle and M. Skinner (eds). Life Sciences/CRC Press, Boca Raton, FL, **2005**, p. 315.
6 Chung, D. W., J. E. Harris and E. W. Davie. Nucleotide sequences of the three genes coding for human fibrinogen. *Adv Exp Med Biol* **1991**, *281*, 39–48.
7 Collen, D., G. N. Tytgat, H. Claeys and R. Piessens. Metabolism and distribution of fibrinogen. I. Fibrinogen turnover in physiological conditions in humans. *Br J Haematol* **1972**, *22*, 681–700.
8 Doolittle, R. F., K. W. K. Watt, B. A. Cottrell, D. D. Strong and M. Riley. The amino acid sequence of the chain of human fibrinogen. *Nature* **1979**, *280*, 464–468.
9 Ebert, R. F. *Index of Variant Human Fibrinogens*. CRC Press, Boca Raton, FL, **1994**.
10 Francis, C. W., E. Müller, A. Henschen, P. J. Simpson and V. J. Marder. Carboxyl-terminal amino acid sequences of two variant forms of the chain of human plasma fibrinogen. *Proc Natl Acad Sci USA* **1988**, *85*, 3358–3362.
11 Fu, Y. Carboxy-terminal-extended variant of the human fibrinogen a subunit: a novel exon conferring marked homology to and subunits. *Biochemistry* **1992**, *31*, 11968–11972.
12 Gillmore, J. D., D. R. Booth, M. Rela, N. D. Heaton, V. Rahman, A. J. Stangou, M. B. Pepys and P. N. Hawkins. Curative hepatorenal transplantation in systemic amyloidosis caused by the Glu526Val fibrinogen alpha-chain variant in an English family. *Q J Med* 2000, *93*, 269–275.
13 Hamidi Asl, L., J. J. Liepnieks, T. Uemichi, J.-M. Rebibou, E. Justrabo, D. Droz, C. Mousson, J.-M. Chalopin, M. D. Benson, M. Delpech and G. Grateau. Renal amyloidosis with a frame shift mutation in fibrinogen chain producing a novel amyloid protein. *Blood* **1997**, *90*, 4799–4805.
14 Hamidi Asl, L., V. Fournier, C. Billerey, E. Justrabo, D. Chevet, D. Droz, C. Pecheux, M. Delpech and G. Grateau. Fibrinogen A*α* chain mutation (Arg554 Leu) associated with hereditary renal amyloidosis in a French family. *Amyloid: Int J Exp Clin Invest* **1998**, *5*, 279–284.
15 Hanss, M. and F. Biot. A database for human fibrinogen variants. *Ann NY Acad Sci* **2000**, *936*, 89–90.
16 Kant, J. A., A. J. Fornace, D. Saxe, M. I. Simon, O. W. McBride and G. R. Crabtree. Evolution and organization of the fibrinogen locus on chromosome 4: gene duplication accompanied by transposition and inversion. *Proc Natl Acad Sci USA* **1985**, *82*, 2344–2348.
17 Lachmann, H. J., D. R. Booth, S. E. Booth, A. Bybee, J. A. Gilbertson, J. D. Gillmore, M. B. Pepys and P. N. Hawkins. Misdiagnosis of hereditary amyloidosis as AL (primary) amyloidosis *N Engl J Med* **2002**, *346*, 1786–1791.

18 Louache, F.N., E. Cramer, J. Breton-Gorius and W. Vainchenker. Fibrinogen is not synthesized by human megakaryocytes. *Blood* **1991**, *77*, 311–316.

19 Mornaghi, R., P. Rubinstein and E.C. Franklin. Studies on the pathogenesis of a familial form of renal amyloidosis. *Trans Ass Am Physicians* **1981**, *94*, 211–216.

20 Mornaghi, R., P. Rubinstein and E.C. Franklin. Familial renal amyloidosis: case reports and genetic studies. *Am J Med* **1982**, *73*, 609–614.

21 Mosesson, M.W., J.S. Finlayson, R.A. Umfleet and D. Galanakis. Human fibrinogen heterogeneities. I. Structural and related studies of plasma fibrinogens which are high solubility catabolic intermediates. *J Biol Chem* **1972**, *247*, 5210–5219.

22 Nickerson, J.M. and G.M. Fuller. Modification of fibrinogen chains during synthesis: glycosylation of B and chains. *Biochemistry* **1981**, *20*, 2818–2821.

23 Ostertag, B. Demonstration einer eigenartigen familiaren "Paramyloidose". *Zentralbl Allg Pathol* **1932**, *56*, 253–254.

24 Ostertag, B. Familiare Amyloid-Erkrankung. *Z menschl Vererb -u Konstit Lehre Bd* **1950**, *30*, 105–115.

25 Ozkaya, N. and F. Yalcinkaya. Colchicine treatment in children with familial Mediterranean fever. *Clin Rheumatol* **2003**, *22*, 314–317.

26 Rader, D.J., R.E. Gregg, M.S. Meng, J.R. Schaefer, L.A. Zech, M.D. Benson and H.B. Brewer, Jr. *In vivo* metabolism of a mutant apolipoprotein, apoA-IIowa, associated with hypoalphalipoproteinemia and hereditary systemic amyloidosis. *J Lipid Res* **1992**, *33*, 755–763.

27 Rao, S.P.S., M.D. Poojary, Jr, B.W. Elliott, L.A. Melanson, B. Oriel and C. Cohen. Fibrinogen structure in projection at 1.8 Å resolution: electron density by co-ordinated cryo-electron microscopy and X-ray crystallography. *J Mol Biol* **1991**, *222*, 89–98.

28 Rixon, M.W., W.-Y. Chan, E.W. Davie and D.W. Chung. Characterization of a complementary deoxyribonucleic acid coding for the α chain of human fibrinogen. *Biochemistry* **1983**, *22*, 3237–3244.

29 Seydewitz, H.H., C. Kaiser, H. Rothweiler and I. Witt. The location of a second *in vivo* phosphorylation site in the A-αchain of human fibrinogen. *Thromb Res* **1984**, *33*, 487–498.

30 Stangou, A.J., H.J. Lackmann, H.J.B. Goodman, A. Bybee, D. Rowzcenio, G. Tennent, S.O. Brennan, J.G. O'Gracky, N.D. Heaton, M. Rela, M.B. Pepys and P.N. Hawkins. Fibrinogen A α-chain amyloidosis: clinical features and outcome after hepatorenal or solitary kidney transplantation. In *Amyloid and Amyloidosis: Proc Xth Int Symp on Amyloidosis,* April 18–22, 2004, Tours, France. Grateau, G., R.A. Kyle and M. Skinner (eds). Life Sciences/CRC Press, Boca Raton, FL, **2005**, p. 312–314.

31 Uemichi, T., J.J. Liepnieks and M.D. Benson. Hereditary renal amyloidosis with a novel variant fibrinogen. *J Clin Invest* **1994**, *93*, 731–736.

32 Uemichi, T., J.J. Liepnieks, F. Alexa der and M.D. Benson. Molecular basis of renal amyloidosis: fibrinogen a chain Val 526 in Irish-American and Polish-Canadian kindreds. *Q J Med* **1996**, *89*, 745–750.

33 Uemichi, T., J.J. Liepnieks, T. Yamada, M.A. Gertz, N. Bang and M.D. Benson. A frame shift mutation in the fibrinogen Aα chain gene in a kindred with renal amyloidosis. *Blood* **1996**, *87*, 4197–4203.

34 Uemichi, T., J.J. Liepnieks, M.A. Gertz and M.D. Benson. Fibrinogen A alpha chain Leu 554: an African-American kindred with late onset renal amyloidosis. *Amyloid* **1998**, *5*, 188–192.

35 Westermark, P., S. Araki, M.D. Benson, A.S. Cohen, B. Frangione, C.L. Masters, M.J. Saraiva, J.D. Sipe, G. Husby, R.A. Kyle and D. Selkoe. Nomenclature of amyloid fibril proteins. Report from the meeting of the International Nomenclature Committee on Amyloidosis, August 8–9, 1998. Part 1. *Amyloid* **1999**, *6*, 63–66.

36 Wolfenstein-Todel, C. and M.W. Mosesson. Human plasma fibrinogen heterogeneity: evidence for an extended carboxyl-terminal sequence in a normal γ chain variant (γ'). *Proc Natl Acad Sci USA* **1980**, *77*, 5069–5073.

37 Zeldenrust, S., M.A. Gertz, T. Uemichi, J. Bjornsson, R. Wiesner, T. Schwab and M.D. Benson. Orthotopic liver transplantation for hereditary fibrinogen amyloidosis. *Transplantation* **2003**, *75*, 560–561.

26
β_2-Microglobulin

Thomas R. Jahn and Sheena E. Radford

26.1
Introduction: Dialysis-related Amyloidosis: A Deposition Disorder of β_2-Microglobulin (β_2M)

β_2M is one of the more than 20 currently known human proteins or peptides that aggregate into classical amyloid fibrils and thus lead to significant human suffering [1]. The disorder resulting from β_2M deposition is known as dialysis-related amyloidosis (DRA), a serious complication of renal dysfunction that becomes evident in all patients undergoing long-term hemodialysis – its prevalence increasing with the duration of dialytic therapy and the age of the patient [2, 3]. Interestingly, although its name indicates its relation to dialysis, DRA has been described in uremic patients who have not been treated by dialysis, suggesting that the dialysis procedure itself is not solely responsible for the disorder, but may exacerbate the disease process [4, 5]. The first observations pointing to this long-term complication of dialysis, in the form of carpal tunnel syndrome (CTS), were made in 1975, about 10 years after regular dialysis treatment was introduced [6]. Following these initial observations, Assenat et al. reported the presence of an unknown type of amyloid in tissues removed during carpal tunnel surgery in patients on long-term dialysis [7]. However, it was not until 1985 that β_2M was shown to be the major constituent protein of dialysis-related amyloidosis [8, 9]. Like other amyloid diseases, the deposition of β_2M into amyloid fibrils is tissue specific. In the case of DRA, β_2M amyloid deposits are found characteristically in osteoarticular tissues, leading to bone and joint destruction that give rise to the symptoms characteristic of DRA [10]. Even though the pathological processes involved in DRA are not yet understood, they clearly separate into several phases: (1) the uremic retention of the precursor molecule following renal replacement therapy, (2) deposition of β_2M-containing amyloid fibrils in joints, synovia, cartilage and bones, and (3) a later stage inflammatory reaction (Fig. 26.1) [11, 12]. In this chapter we will discuss current knowledge of the pathogenesis and clinical manifestations of DRA, followed by a description of recent insights into the structural basis of the molecular mechanism of aggregation of β_2M into amyloid fibrils based on biophysical and biochemical

Amyloid Proteins. The Beta Sheet Conformation and Disease. J. D. Sipe

ISBN: 3-527-31072-X

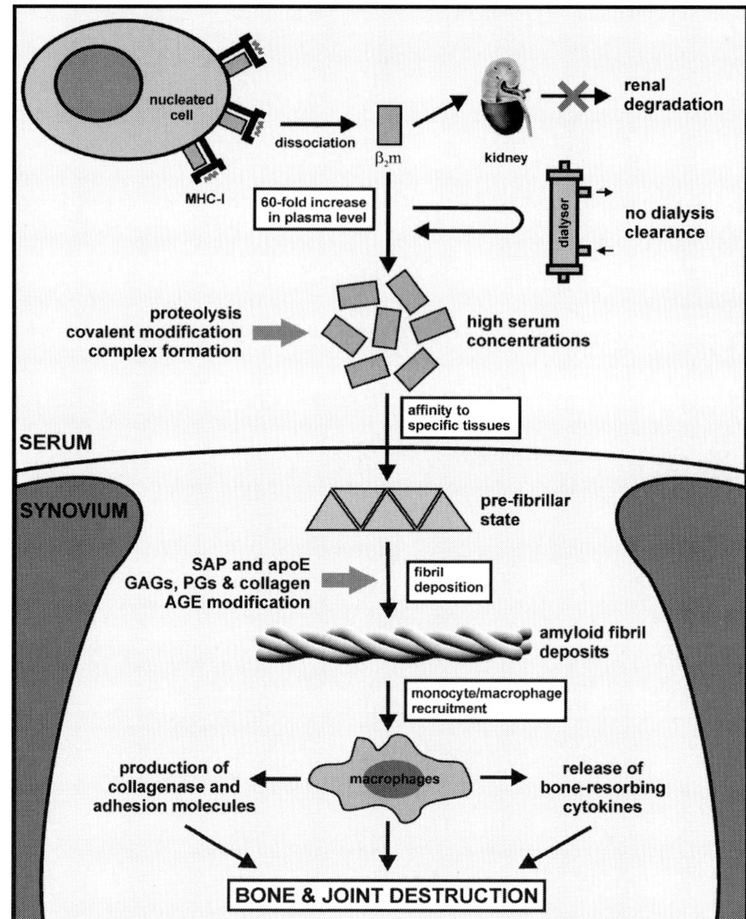

Fig. 26.1 Schematic diagram outlining events involved in the development of DRA. Several factors may exacerbate or be critically involved in the progress of the disease (Table 26.1).

studies *in vitro*. We conclude with a description of the potentials for new therapies of β_2M amyloidosis that build on structural models for the mechanism of aggregation and highlight areas that need further research if these aims are to be realized.

26.2
Current Knowledge of the Mechanism of Development of DRA *In Vivo*

26.2.1
β_2M: Normal Cellular Role

β_2M plays a central role in cellular immunology, in that the protein forms the non-covalently bound light chain of the major histocompatibility complex type I (MHC-I) that is found on the surface of all nucleated cells (Figs 26.1 and 26.2). β_2M is a small [99 amino acid (11,860 Da)] protein that has a typical immunoglobulin fold (Fig. 26.2 B) [13, 14]. A disulfide bridge between cysteine residues in positions 25 and 80 stabilizes the global β-sandwich fold [15]. Interestingly, the disulfide bond is thought to be intact in *ex vivo* β_2M fibrils [16] and is required for the elongation of seeds of β_2M fibrils *in vitro* [17, 18]. Although β_2M is not involved in any direct contact with the antigenic peptide within the fully assembled MHC-I complex, it is crucial for the folding and assembly of the whole complex, and has thus been described as a "chaperone" required for the successful folding and assembly of the MHC class I complex [19]. As part of its normal catabolic cycle, β_2M dissociates from the heavy chain of the MHC-I complex, whereupon it is carried in the blood to the kidneys where more than 95% of the protein is removed by degradation in the proximal tubule [20]. In addition to catabolism in the proximal tubule, minor amounts of the protein are removed from the serum in the peritubular tissue, while only 1% of the β_2M produced is degraded by extra-renal metabolism [21]. In healthy individuals the

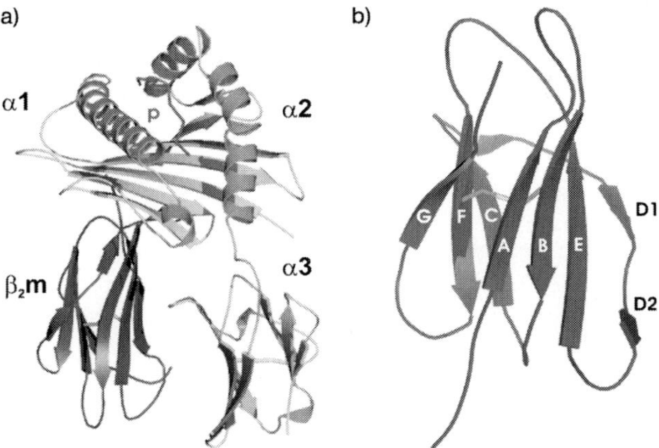

Fig. 26.2 Ribbon diagram of the crystal structure of an MHC-I complex (a) and the solution NMR structure of monomeric human β_2M (b). The structures were taken from PDB code 1UVQ [144] and 1JNJ [117], respectively. Individual subunits (α1–3), the bound peptide fragment [p (red)] as well as individual β-strands in native β_2M (A–G) are shown. Drawn using the program PyMOL [145].

Table 26.1 Factors potentially involved in the pathogenesis of β_2M amyloid fibril deposition in patients undergoing renal therapy (after [3])

Systemic factors
high precursor (β_2M) concentration in the circulation (up to 60 mg/l)
proteolytic changes of β_2M (especially truncation of the N-terminal 6 or 19 amino acids)
complex formation of circulating β_2M (e.g. α_2-Macroglobulin)
modification of β_2M by AGE or AOPP
induction of inflammatory state by the dialysis technique
Local factors
binding of GAGs, PGs and/or collagen
stabilization of fibrils by SAP and/or ApoE
crystal deposits (Al, Fe, Ca)
local generation of AGE- or AOPP-modified β_2M
inflammatory response due to normal and modified β_2M fibrils
monocyte/macrophage recruitment

plasma concentration of β_2M is kept constant at 1.5–3 mg/l by renal catabolism [12]. Using radiolabeled β_2M, Karlsson et al. [22] estimated the daily synthesis of β_2M in normal subjects to be 3.5 mg/kg/day or about 100–200 mg/day, a value that does not change significantly in uremic patients [21, 23]. Since β_2M cannot be removed from the serum by the kidney or the dialysis membrane in patients with renal dysfunction, β_2M concentrations are increased up to 60-fold in patients with end-stage renal disease (ESRD) [2, 24], which ultimately leads to the deposition of the full-length, wild-type protein into amyloid plaques. The retention of β_2M has been shown to be the main pathogenic process underlying β_2M amyloid formation [2, 3, 24]. However, elevated concentrations of β_2M are also found in other diseases that do not result in amyloid deposition [25, 26], suggesting that an increased serum concentration of β_2M alone is insufficient to cause the onset of DRA. Interestingly, no relationship exists between the plasma β_2M concentration and the apparent extent and degree of severity of DRA [27, 28], suggesting that other systemic and/or local factors could be involved in initiating the disease process. Several possibilities are discussed in Section 26.2.3 below and are summarized in Table 26.1.

26.2.2 Clinical Manifestation and Diagnosis of DRA

Clinical findings associated with β_2M amyloidosis are generally related to amyloid deposition in osteoarticular tissues, in particular involving the synovial membranes [12, 29]. Clinically relevant organ involvement is rare and largely confined to very long-term patients [30]. Two major clinical problems dominate the course of DRA: CTS and painful, chronic arthropathies that may ultimately evolve to progressive joint destruction [2, 12].

Initially reported by Charra et al. [31], CTS is a prominent and relatively early feature of β_2M amyloidosis, increasing with time on hemodialysis treatment. Clinical symptoms are rarely present before 5 years, but increase almost linearly to nearly 100% prevalence after 15 years of treatment [2, 24, 29, 32]. The deposition of amyloid is also associated with the syndrome of chronic arthralgias and arthropathy [33], although the precise pathogenic role of amyloid deposition remains speculative.

Chronic arthralgias are usually bilateral and involve the shoulders, knees, wrists and small joints of the hand (Fig. 26.3). Destructive arthropathies of large peripheral joints cause a build up of large amyloid deposits and lead to major incapacity [2, 29, 34].

Currently, there is no treatment for DRA other than organ transplantation. Whilst transplantation usually leads to symptomatic improvement within days, whether amyloid fibrils, once deposited, can subsequently undergo degradation *in vivo* has not been demonstrated conclusively to date [29, 35, 36]. Histologically, β_2M amyloid deposits may be detected only a few months after the initiation of hemodialysis [12, 24], although most amyloid deposits precede clinical mani-

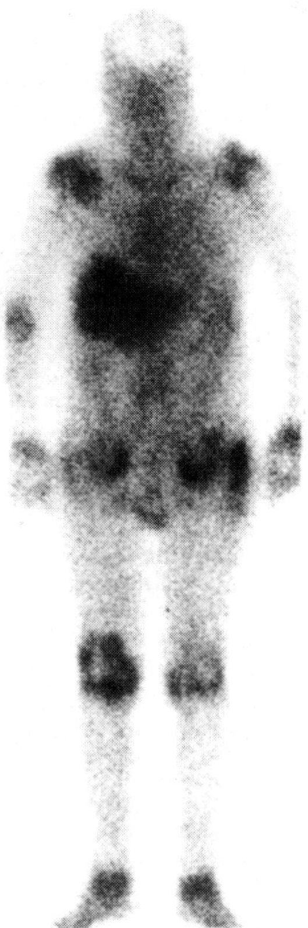

Fig. 26.3 Anterior view of a DRA patient imaged using [^{111}In]β_2M scintigraphy. Local tracer accumulations are noted in shoulders, elbows, wrists, hips, knees and feet. Tracer accumulation in the liver is a non-specific phenomenon, which is related to uptake of tracer in the reticuloendothelial system. (Reproduced from [2], with permission.)

festations by several years and some may never appear to cause clinically relevant problems.

The definitive diagnosis of β_2M amyloidosis today relies on histological findings, including Congo red staining, immunohistochemical demonstration of the precursor molecule within the deposits and the demonstration of fibrils using electron microscopy (EM) [37]. Two scintigraphic methods, employing either radiolabeled serum amyloid protein (SAP) or β_2M itself, have been introduced to detect amyloid deposits in DRA [38]. Using ^{123}I-labeled SAP, or ^{131}I- or ^{111}In-labeled β_2M, amyloid deposits have been visualized in several long-term hemodialysis patients [2, 39–41]. The scintigraphy image in Fig. 26.3 shows a particularly clear example in which a patient after 17 years of dialysis treatment accumulated significant β_2M amyloid deposits throughout the joints [2]. Why deposition of β_2M is so specific for the joints is not clear, although a weak affinity for collagen may be involved [42, 43]. Clearly, biochemical and biophysical analysis of the affinity of β_2M for different components of the joints is required both for developing a molecular understanding of DRA, as well as for the design of new therapeutic agents.

26.2.3
Composition of Dialysis-Related Amyloid (DRA)

The detailed molecular mechanism of β_2M amyloid deposition is not yet understood, and much work is required both *in vivo* and *in vitro* in order to delineate the mechanism by which clearance of the normally highly soluble protein becomes inactivated and β_2M becomes immobilized in fibrillar plaques. A schematic diagram outlining features potentially important to the development of DRA is shown in Fig. 26.1. What is known is that the length of time on dialysis is more important in determining the onset of DRA than is the concentration of β_2M in the plasma itself [27]. Additional factors (Table 26.1) may also contribute to the pathogenesis of DRA by increasing the "amyloidogenicity" of the precursor molecule and/or the stability of the resulting fibrils [3, 12]. The composition of β_2M amyloid deposits is summarized in Table 26.2 and the roles of individual factors in the development of the disorder are described in detail below.

26.2.3.1 β_2M in Amyloid Deposits and Associated Biological Factors

Studies of *ex vivo* fibril material from DRA patients has shown that the major constituent of the amyloid deposits is full-length, intact β_2M [8, 9]. This finding has been confirmed more recently by *in vitro* studies that have shown that *ex vivo* amyloid material can be refolded into native globular β_2M [16]. In addition to the full-length, wild-type protein, up to 25% of β_2M found in amyloid deposits is cleaved at the N-terminus, most commonly involving removal of 6 or 19 residues from this region [44]. It is unclear whether proteolysis occurs prior to fibril formation or after fibril deposition [45–47]. In addition, a variant form of β_2M in which Asn17 is deaminated to Asp has been isolated from *ex vivo* amyloid deposits

Table 26.2 Composition of β_2M amyloid deposits (after [83])

Component	**Reference**
β_2M	
intact	8, 9
truncated	44
deaminated	48
acidic and radical forms	48, 143
AGE and AOPP modified forms	64, 73
Extracellular matrix	
GAGs	55
PGs	58
collagen	61
SAP	83
ApoE	53
Ubiquitin	83
κ light chain	2
a_2-Macroglobulin	59
Calcium	83
Macrophages	69

[48], although studies of this protein *in vitro* show that this variant behaves very similarly in aggregation and stability assays to the wild-type protein [49]. Several authors suggest that intact β_2M can be readily induced to form fibrils under physiologically relevant conditions *in vitro*, e.g. by incubating high concentrations of β_2M in the absence of any proteolytic treatment [50] or by incubating the protein in the supernatant of peripheral mononuclear blood cells from dialysis patients (but not with cells from healthy subjects) [51]. Even in the absence of white blood cells, human urine-derived β_2M is able to form amyloid-like fibrils *in vitro* when the fibrillar form is stabilized by addition to the incubation medium of glycosaminoglycans (GAGs) as well as serum-derived amyloid P component (SAP) [52]. In addition, stabilization of fibrils by apolipoprotein E (ApoE), a cholesterol transporter protein, may play a role in the evolution of the disease [53]. Together, these studies suggest that a number of biological factors are an important feature of DRA and that interference with binding of stabilizing components may provide a therapeutic strategy to reduce or prevent DRA.

26.2.3.2 **GAGs, Proteoglycans (PGs) and Collagen**

The preferential location of β_2M amyloid fibrils in osteoarticular tissues is a fascinating feature of DRA, but is, nonetheless, only poorly understood. As found in other amyloid disorders [54], β_2M amyloid fibrils are closely associated with extracellular matrix components, including GAGs [55], PGs [56] and collagen [43]. *In vitro* assays suggest that GAGs and PGs may enhance β_2M amyloid deposition *in vivo* by binding and stabilizing β_2M fibrils, as well as by acting as a scaffold for the polymerization events themselves [57]. For example, the vascular wall and synovium of joints contain relatively high concentrations of heparan sulfate-containing PGs, whilst, in the disks and cartilage, an increase of chondroitin sulfate PGs is observed in patients undergoing dialysis [58]. Also, the role of α_2-Macroglobulin in the development of dialysis amyloid is unknown [59], although this anti-protease, which inhibits the degradation of extracellular matrix components, is often found associated both with soluble β_2M as well as with β_2M amyloid deposits [60]. A third contributing factor to β_2M amyloid deposition in joints may arise from the known affinity of β_2M for some types of collagen, as well as advanced glycation end-products (AGE)-modified collagen [42, 61]. Once in the joint and bound to collagen, β_2M may then form fibrils spontaneously, facilitated by other components such as GAGs that are abundant therein and by other features within the synovial tissue *per se* [62].

26.2.3.3 **AGE Modification**

An isoform of β_2M with a more acidic isoelectric point (acidic β_2M) has been recognized in the serum, ultrafiltrate and amyloid deposits from dialysis patients [48]. More recently it has been demonstrated that a part of acidic β_2M results from modification of the molecule with AGE [63–66]. AGE modification of β_2M has been proposed to be an important factor in the pathogenesis of β_2M amyloidosis [67]. The AGE modification of β_2M, a non-enzymatic Maillard reaction, results in a number of different products in uremic sera and in β_2M amyloid deposits, such as pentosidine, *N*-carboxymethyllysine and imidazolone [63, 65, 66]. However, based on the observation that serum AGE levels in uremia are more than 10-fold higher than those of diabetic patients [63], it is thought that AGE modification is more likely to play a role in inducing an inflammatory response, rather then affecting the formation of fibrils directly.

26.2.3.4 **Macrophages**

On the basis of the histological findings of post-mortem studies, Garbar et al. [68] proposed that the deposition of β_2M and the formation of β_2M amyloid fibrils occurs in the absence of infiltrating monocytes/macrophages or a local inflammatory response. It is in later phases of development that β_2M deposits stain positive for AGE and infiltrating macrophages can be observed [68]. Macrophages are the predominant cell type infiltrating β_2M amyloid deposits [69]. There is controversy as to whether the macrophages surrounding amyloid deposits facilitate the deposition of amyloid fibrils or serve to degrade them. Nonetheless, the observation that β_2M fibril formation can occur in peripheral blood mononuclear cell cultures [51], coupled with the finding that amyloid filaments are located in the lysosomes of macrophages surrounding amyloid deposits [70], demonstrates a role of macrophages in DRA, most probably as part of the reactive processes transforming clinically silent deposits into symptomatic osteoarticular destruction in the later stages of disease development [71, 72].

26.2.3.5 **Inflammation**

AGE-modified β_2M is thought to lead to severe tissue damage, even in the absence of large β_2M amyloid deposits, as is observed typically in CTS. This idea was confirmed by Miyata et al. [73], suggesting that AGE-modified β_2M amyloid chemotactically attracts circulating monocytes. Furthermore, their binding to AGE receptors (RAGE) induces the production and release of pro-inflammatory cytokines such as interleukin-1β and tumor necrosis factor-α [74]. More recent work has shown that human macrophages are stimulated by AGE-modified β_2M to produce all types of transforming growth factors (TGF-β1–3), whereas unmodified β_2M had no cytokine-inducing activity [75]. These observations suggest that AGE modification induces local inflammation mediated by cytokines, which then leads to matrix degradation, bone resorption and formation of bone cysts. Furthermore, the amount of cytokines secreted from macrophages is sufficient to stimulate the synthesis of collagenase in cultured synovial cells [73], and may also contribute to progressive bone loss and the formation of bone cysts. Finally, β_2M modification in the form of advanced oxidation protein products (AOPP) [76] may also be increased due to enhanced oxidative stress during hemodialysis or exposure of β_2M to hydroxyl radicals [64].

In contrast to the results outlined above, others have suggested that unmodified β_2M may also contribute to inflammatory processes at the site of amyloid deposition and may play an early role in fibrillogenesis [77, 78]. Thus, β_2M deposition results in the stimulation of cyclooxygenase-2 (COX-2) expression and, at least under *in vitro* conditions, unmodified β_2M can intensify inflammation via prostaglandin synthesis [79]. Another study by Moe et al. [80] showed that unmodified β_2M induces metalloproteinase-1 expression without concomitant release of the tissue inhibitor of metalloproteinase-1 in synovial fibroblasts, leading to the degradation of interstitial collagen with consequent destruction of articular cartilage and subchondral bone [77]. Interestingly, AGE modification of β_2M did not augment, but rather abrogated, these effects [80]. Recently, the cellular uptake of β_2M and AGE-β_2M was examined by O'Neill et al. [81]. These authors showed the pattern of β_2M and AGE-β_2M uptake to differ in both synovial fibroblasts and monocytes/macrophages. Based on these results, the authors suggest that the onset of β_2M amyloid fibril formation may be triggered by the binding of β_2M to the MHC-I complex of surrounding cells, which induces a cascade of protein induction (COX-2, metalloproteinases, adhesion molecules) [81]. This then leads to the degradation of the joint surface, inducing local amyloid deposition. Newly formed amyloid deposits are subsequently glycated, inducing the infiltration of macrophages and leading to the symptoms of DRA (see Fig. 26.1 for a summary of these processes). Clearly, much remains to be learned to fully understand this complex pathogenesis.

25.2.3.6 **Influence of Dialysis Procedure**

Because retention of β_2M is a prerequisite for amyloid formation, removal of circulating β_2M during hemodialysis is one obvious route for therapeutic intervention, although to date no clear correlation between the type of dialysis procedure used and the onset of DRA has been observed [82]. Nonetheless, improved clearance of β_2M can be achieved by the use of continuous ambulatory peritoneal dialysis or high-flux dialyzer membranes [83]. Several studies have demonstrated that a single standard high-flux hemodialysis session reduces β_2M plasma levels by 50% [84, 85], possibly by the adsorption of β_2M to the synthetic dialyzer membrane, in addition to diffusive clearance [86, 87]. The development of more biocompatible high-flux hemodialyzer membranes thus continues to be explored as one route to reduce the incidence of DRA in hemodialysis patients, reducing or even preventing the onset of β_2M amyloid-associated symptoms [88]. It remains to be seen whether any further developments in membrane technology can completely alleviate DRA.

26.3
Structure and Morphology of β_2M Amyloid Fibrils

26.3.1
Amyloid Formation from β_2M *In Vitro*

As discussed in detail in the previous section, amyloid deposits in DRA are complex macromolecular assemblies that comprise an array of different cellular components. In order to derive more detailed insights into the underlying structure of the amyloid fibrils, a number of groups worldwide have developed protocols for the successful formation of amyloid-like fibrils of β_2M *in vitro*. Perhaps most interestingly, studies from a number of laboratories have shown that incubation of a high concentration of β_2M at neutral pH is not sufficient to initiate formation of amyloid fibrils *in vitro*, at least without the addition of seeds formed from amyloid fibrils *ex vivo* [52, 89–91]. In contrast, fibril formation of wild-type β_2M *in vitro* can be induced at neutral pH by the addition of Cu^{2+} ions [92], by extending *ex vivo* seeds with β_2M in which the N-terminal six residues have been removed by mutagenesis [93] or by destabilizing the protein with 20% (v/v) trifluoroethanol [94], by concentrating and drying the protein on a dialysis membrane [50], or by specific local destabilization of the N- or C-terminal strands [95]. In parallel with these studies, incubation of full-length, wild-type β_2M under acidic conditions has been shown to result in the rapid formation of amyloid-like fibrils, both in unseeded reactions [49, 90, 96, 97], as well as in the presence of *ex vivo* seeds [89, 98].

How closely the fibrils generated *in vitro* relate structurally to those formed *in vivo* is currently unknown, since detailed structural analysis of *ex vivo* β_2M fibrils has not been carried out to date. Nonetheless, most of the fibrils generated *in vitro* conform to the definition of amyloid, in that they show red/green birefringence in the presence of Congo red, bind the dye Thioflavin (ThT), are long and unbranched polymers as seen using electron microscopy and atomic force microscopy (AFM), and they give rise to a cross-β fiber diffraction pattern [90, 96, 97]. Akin to the data obtained with other amyloidogenic proteins, the mor-

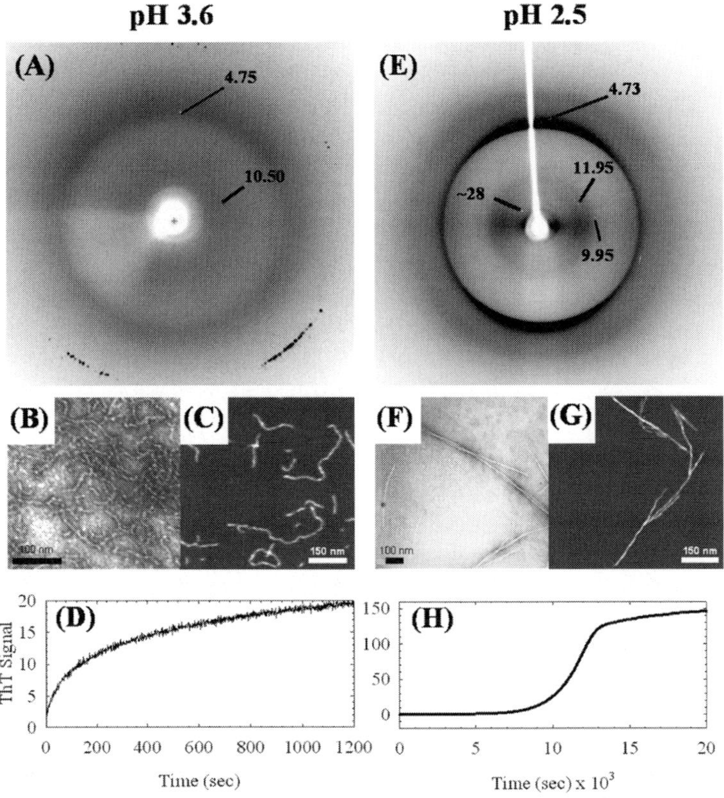

Fig. 26.4 Characterization of different types of β_2M amyloid-like fibrils formed *in vitro* at pH 3.6 at an ionic strength of 0.4 M (left-hand panels) and at pH 2.5 at low ionic strength (50 mM) (right-hand panels). (A and E) X-ray fiber diffraction images: the major reflections on the meridian and equator are labeled in Å. (B and F) Negative stain EM images and (C and G) tapping mode AFM images of fibrils dried onto mica. Growth kinetics at 37 °C monitored by the fluorescence of Thioflavin T, (D) pH 3.6, 0.4 M ionic strength (no agitation); (H) pH 2.5, 50 mM ionic strength (agitation at 1400 r.p.m.). (Adapted from [97].)

phology of β_2M fibrils formed *in vitro* depends critically on the growth conditions (Fig. 26.4). Under acidic conditions (below pH 5.0) at high ionic strength (around 200–400 mM), β_2M rapidly and spontaneously assembles into amyloid fibrils that bind Thioflavin T and Congo red, give rise to red/green birefringence, and result in an X-ray diffraction pattern consistent with a cross-β structure (Fig. 26.4 A–D) [97]. These fibrils are short (less than 600 nm) and flexible, and appear to be constructed by the juxtaposition of spherical particles about 30 nm in length and about 3.5 nm in height, as measured by AFM [49, 96]. Although these fibrils have morphological properties akin to protofibrils seen in other amyloid proteins [99], there is no evidence that these species are able to assemble further into higher order structures [96]. In contrast, at lower pH (below pH 3.0) and low ionic strength (below 50 mM), β_2M assembles in a nucleation-dependent manner to form long, straight fibrils that also give rise to a cross-β X-ray fiber diffraction pattern (Fig. 26.4 E–H). Under these conditions, various intermediate aggregates form in the lag phase of assembly, including amorphous species, rings and toroids [96], as found in the lag phase of assembly of many proteins [100]. Subsequent to these species, single protofilaments as well as lengthways and laterally extended fibrils with a range of morphologies are formed [96]. These higher-order assemblies include straight aperiodic fibrils, as well as fibrils displaying a left-handed twist with regular periodicity of about 90 nm and at least four protofilaments [96].

26.3.2
Initial Progress Towards the Structure of β_2M Amyloid Fibrils in Atomic Detail

Delineating the structure, at high resolution, of an amyloid fibril for any protein is still beyond our capabilities, despite the fact that the first definition of amyloid as the generic cross-β structure was made more than 40 years ago [101]. Major problems that need to be overcome to deduce the structure of amyloid include their insolubility, as well as their heterogeneity, even when formed within a single reaction [96]. Most importantly, the structural relationship of fibrils formed under different conditions needs to be determined, noting that these species all adhere to the generic cross-β architecture of amyloid, yet may have distinct biological activities [102]. Substantial progress has been made towards determining the structure of β_2M amyloid fibrils formed *in vitro* using a combination of limited proteolysis, hydrogen exchange and electron microscopy. Notably, both fibril types formed in acidic conditions, pH 3.6 or pH 2.5, (Fig. 26.4) depolymerize at neutral pH or after the addition of organic solvents such as dimethylsulfoxide (DMSO) [103, 104], allowing experiments to be performed in which the fibrils are digested with proteases, or exposed to deuterium oxide, and the patterns of protection later determined by analysis of the depolymerized monomeric proteins [105–108]. Using these approaches, it has been shown that the central region of the polypeptide chain, spanning residues 20–87, is protected from a range of proteases in fibrils formed at pH 4.0, suggesting that the core of the fibrils is formed from these approximately 60 amino acids, whilst the N- and C-terminal regions of the protein are exposed to solvent [106]. Perhaps most interestingly, the N- and C-terminal regions of the protein were shown to be protected from proteolysis in native monomeric β_2M, suggesting that these regions become more flexible and exposed to the solvent in the fibrillar form. Finally, and importantly, common cleavages in the fibrils formed at pH 4.0 *in vitro* were found to occur at Lys6 and Lys19, sites identical to those found to occur *in vivo*, suggesting a similarity of the fibrils formed in both environments [44, 109].

A very elegant approach to probe the conformation of β_2M in amyloid fibrils on a residue specific level was introduced by Hoshino et al. [107]. By exposing β_2M amyloid fibrils to D_2O for different lengths of time, rapidly quenching the hydrogen exchange reaction by freeze-drying and subsequent dissolution in 95% DMSO/5% D_2O, residue-specific protection patterns could be obtained using two-dimensional ^1H–^{15}N-nuclear magnetic resonance (NMR). The protection factors indicate that the β-strands involved in the core of the native protein (26.5 A), as well as their connecting loops, comprise the structured core of the amyloid fibril (Fig. 26.5 C). In accord with the structural properties of the monomeric amyloid precursor formed at pH 3.6 (Fig. 26.5 B) [110] and proteolysis data [106], the N- and C-terminal regions are relatively weakly protected from hydrogen exchange in the fibrils formed at pH 2.5 (Fig. 26.5 C) [107]. In a recent study extending this work, Goto et al. further analyzed the difference in subunit organization between long, straight fibrils formed at pH 2.5, with the nodular fibrils formed at pH 3.6 at high ionic strength shown in Fig. 26.4 (B and C) [108]. Interestingly, and in contrast with the result obtained using long, straight fibrils, residues in the vicinity of the disulfide bond (Cys25–Cys80) exchange rapidly in the nodular fibrils, suggesting that the disulfide bond is exposed to the solvent. This result was confirmed by the ability to reduce this disulfide bond with dithiothreitol, whilst this is not possible in mature fibrils or the native globular protein [17, 111]. Reduced β_2M loses its ability to form rigid fibrils, instead forming thinner nodular fibrils under all conditions studied [8, 17, 18, 112]. The power of the hydrogen-exchange method was most pronounced, however, when a detailed characterization of the rates of hydrogen exchange of different residues was considered. Instead of the usual single exponential kinetics of hydrogen exchange observed for monomeric species and homogeneous en-

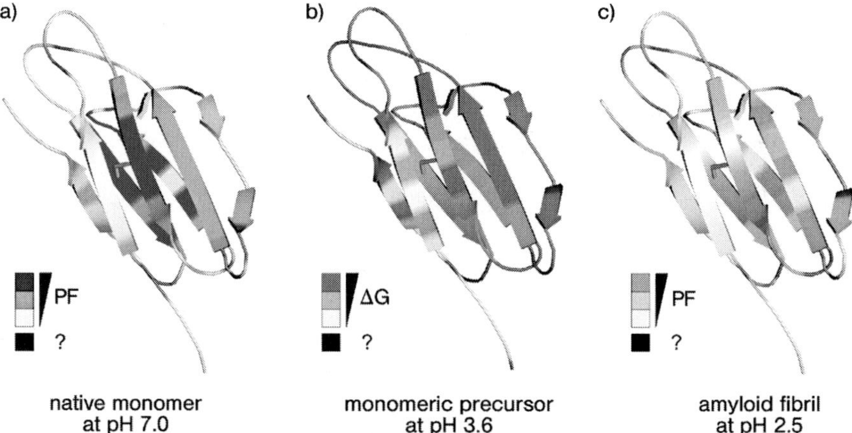

Fig. 26.5 Comparison of the stability of residual structure in different conformational states of β_2M. The protection factors (PF) were measured by native state hydrogen exchange for native β_2M at pH 7.0 (A) [128] and by hydrogen exchange of the amyloid fibrils at pH 2.5 (C) [108]. The residual stability of the monomeric precursor at pH 3.6 was determined by urea titration experiments monitored by NMR (B) [110].

sembles, hydrogen exchange of fibrils formed at pH 2.5 is bi-exponential, with a fast phase represented by relatively rapid exchange (a few days) and a slow phase with almost no observable exchange [108]. The remaining, unexchanging, core is suggested to result from the supramolecular structure of the amyloid fibrils and may be used as structural constraints for further model building. Taken together, these results suggest that the core of the long and straight fibrils formed at pH 2.5 extends to almost 80% of the entire sequence, whilst the curved, nodular fibrils are composed of a more limited region corresponding to only 30% of the amino acid sequence.

By studying the structural conservation of β_2M and by using protein-docking techniques, Nussinov et al. have recently built a model for the β_2M amyloid fibril structure based on the premise that the formation of amyloid involves the docking of native-like subunits [113]. In this model, the A and G strands of the native monomer were assumed not to participate in the fibril structure, whilst the β-bulge in strand D in the native protein (Fig. 26.2 B) was assumed to straighten, in accord with a recent X-ray structure of the monomeric human protein [114]. Inter-subunit interactions were then built between strands B and D, organized in a parallel orientation between two monomeric building blocks, forming a continuous β-sheet in a fibril with a characteristic cross-β structure. Whilst structurally appealing, there is currently no experimental validation for such a model, although it is consistent with the proteolysis and hydrogen exchange results presented above. Moreover, since β_2M at pH 2.5 is highly unfolded [90, 112, 115], then there is no reason to assume, *a priori*, that assembly should involve the stacking of native-like subunits. Clearly, more experimental data will be needed to validate or refine such models, especially in the light of data that indicate that the supramolecular organization of β_2M amyloid fibrils may involve more than one common assembly unit.

26.3.3
Mechanisms of Fibril Formation

Several groups have proposed that a nucleation-dependent polymerization model could explain the general mechanism of amyloid formation *in vitro* [99]. This model was initially tested by Naigi et al. for the assembly of β_2M amyloid at acidic pH, by analyzing the extension phase of seed elongation, monitored using Thioflavin T fluorescence [89]. The extension of β_2M seeds proceeds by a pseudo-first-order reaction with a rate constant that is maximal at pH 2.5 [89]. The authors suggest, therefore, that the extension of β_2M fibrils proceeds via the consecutive association of β_2M monomers onto the ends of existing fibrils. In more recent studies, the seed-dependent extension reaction of β_2M was also followed by monitoring Thioflavin T fluorescence with total internal reflection fluorescence microscopy [116]. These data showed that extension is predominantly unidirectional and that some seeds must be polar. The extension reaction for individual fibrils could be fitted to a single exponential curve, consistent with the extension in test tubes [89], yielding a fibril growth rate of about

47.4 ± 15.0 nm/min [116]. In contrast to these results, the lack of a lag phase in the assembly of curved, modular fibrils suggests that these assemblies may form in a nucleation-independent manner, at least over the timescales studied (Fig. 26.4 D) [90, 97].

26.4 Structural Characteristics of Monomeric Fibril Precursor States

26.4.1 Predicting Regions Key to the Formation of Amyloid by β_2M

Studies of β_2M amyloid formation *in vitro* have shown that native, monomeric protein at neutral pH is a highly soluble and stable entity that is not highly amyloidogenic [90, 91, 97, 117]. These data suggest that amyloid formation for β_2M *in vivo* involves events that destabilize the native fold to facilitate the aggregation process (Fig. 26.6). One key initiating event is the dissociation of β_2M from the MHC-I heavy chain presented at the cell surface. Many crystal structures of human β_2M bound to the heavy chain of the MHC-I complex have been solved to date (Fig. 26.2A) [118]. However, clues about the amyloidogenic transition came from studies of the free monomer both in solution and from X-ray crystallography. Whilst initial NMR studies of monomeric human β_2M suggested that the secondary structure of the protein is maintained in the isolated domain in solution [119], more recent high-resolution studies report significant structural changes compared with the bound form [117]. Thus, dissociation of β_2M from the 3α domain of the heavy chain results in increased fraying of the termini, combined with increased dynamics in the loop linking strands A and B. Increased local dynamics in these regions may then perturb the balance of hydrophilic and hydrophobic interactions in the native protein that decrease its solubility and facilitate aggregation. In the MHC-I-bound form of β_2M strand D, which lies at the edge of the β-sandwich, is divided by a two-residue β-bulge (Fig. 26.2). Such features are often found at the edge strands of β-sheet proteins and have been proposed to play a role in preventing aggregation of exposed edge strands in β-sheet containing proteins [120, 121]. Interestingly, however, in a recent crystal structure of monomeric human β_2M this feature is no longer present, but is replaced with a continuous, six-residue β-strand [114]. This conformation provides an ideal assembly surface, making the C/D edge-strand pair vulnerable to aggregation, underlined by the hydrogen-bonding pattern within the crystal lattice. Whether such conformational changes are involved in the formation of β_2M amyloid fibrils remains to be validated experimentally, but should this occur, further structural reorganization prior to assembly would be required to form the polymeric assembly of amyloid.

In order to determine which region(s) of β_2M may be important in initiating the aggregation of the protein, several groups have engaged in studies of peptide fragments equivalent to different regions of the native protein. Using this

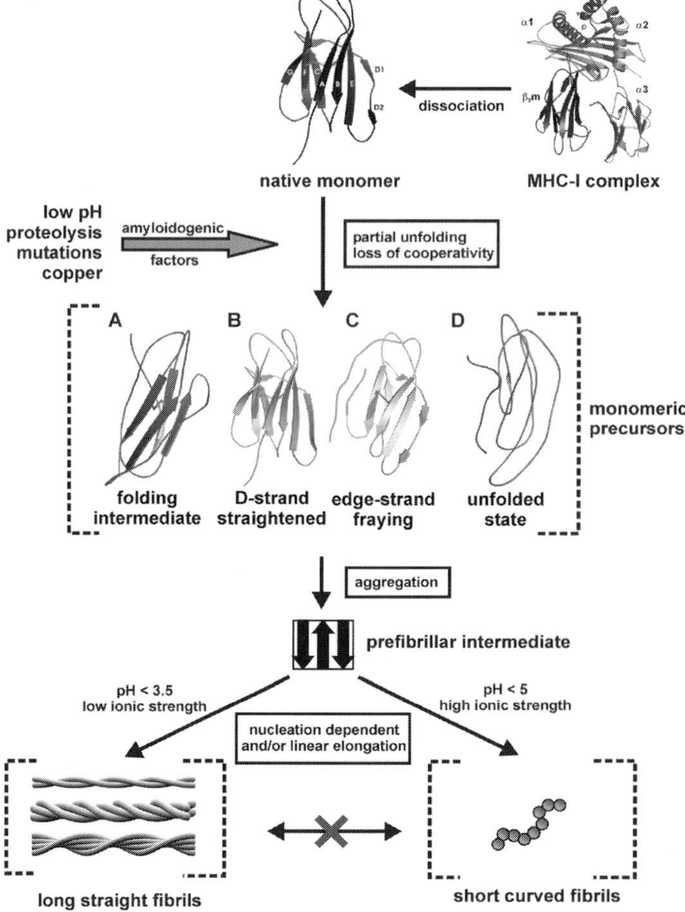

Fig. 26.6 Schematic diagram of suggested models for β_2M amyloid fibril formation *in vitro*. As indicated, under amyloidogenic conditions, native β_2M is destabilized and the population of folding intermediates (A), the loss of the β-bulge in strand D (B), the displacement of the edge-strands (C) or extensive unfolding (D) may result in the population of assembly-competent monomer species, thereby promoting amyloid formation. The exact mechanism leading to the formation of different amyloid species is not clear so far, but may be highly dependent on the growth conditions used.

approach, different peptides have been identified that are both amyloidogenic in isolation and may also facilitate aggregation of the intact wild-type protein (Fig. 26.7). Thus, peptides corresponding to the native β-strands B/C (residues Ser20–Lys31) [122, 123], E (residues Asp59–Thr71) [124] and F/G (residues Pro72–Met99) [104] have all been shown to form amyloid-like fibrils *in vitro*,

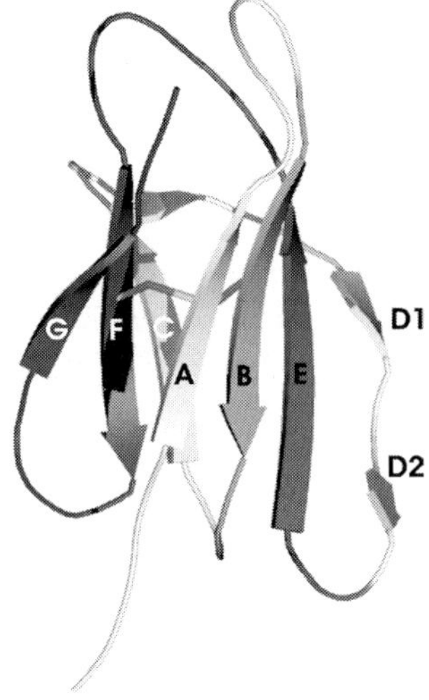

Fig. 26.7 Structure of human β_2M, indicating the location of peptide fragments shown to form amyloid-like fibrils *in vitro*: Asn21–His31 (K3) in green [122], Asp59–Thr71 in red [124] and Pro72–Met99 in blue [104]. The ribbon diagram was created using PyMOL [145] with the structure reported by Verdone et al. [117].

although, in the latter case, a high salt concentration (above 1.0 M NaCl) was needed. Interestingly, however, by creating a series of peptides corresponding to the entire sequence of β_2M, Jones et al. [124] showed that only the sequence encompassing the native strand E is amyloidogenic in isolation, whilst all other peptides, including those representing strands B or C, were not capable of forming amyloid-like fibrils in solution individually, or in pairwise mixtures, despite analyzing their properties over a wide range of pH and ionic strength conditions [124]. These data highlight an important feature of amyloid formation – the behavior of individual sequences is highly context dependent, reminiscent of the very different aggregation properties of other peptides and proteins [125]. Which features determine the unique aggregation properties of these peptides? There seems to be no simple correlation between secondary structure propensity, peptide length, p*I* or hydrophobicity and the ability of peptides to aggregate [124]. However, for peptide E at least, a high content of aromatic side-chains seems to correlate closely with the ability of this sequence to self-associate [124]. These data suggest, therefore, that the ability of a sequence to undergo favorable π–π electron stacking may be critically important in the ordered assembly of these peptides, as seen for other peptides and proteins [126]. Interestingly, the E-strand was found to be the most stable strand in the amyloid precursor state of β_2M formed at pH 3.6 [110] (Fig. 26.5 B) and is the only part of the protein that retains partial residual structure in the acid-denatured state at pH 2.5 [112, 115]. In accord with these results, aggregation predictions based on peptide studies showed that the sequence that forms the native strand E is the only region inherently prone to aggregation [125]. Whether the fibril assembly of various peptides reflects the assembly mechanism of the full-length protein and/or how these regions accomplish assembly in the context of the full-length protein is unclear. Further experiments will now be needed to determine the importance of each region in different stages of the assembly process, e.g. by combining predictions with mutagenesis studies so that the importance of different regions, or even individual residues, in the assembly process can be elucidated.

26.4.2
Partially Unfolded Species as Precursors of Amyloidosis

Several studies have now shown that partial unfolding is a prerequisite of fibril formation from β_2M *in vitro*. Circular dichroism (CD) studies have shown that titration of one or more residues with an apparent pK_a of 4.7 destabilizes native β_2M and generates partially unfolded species that are highly amyloidogenic [90]. At pH 3.6, β_2M adopts a molten globule-like species that is weakly protected from hydrogen exchange and, as is characteristic of such states, results in broad resonances in NMR spectra [110]. At a more acidic pH (pH 2.5), monomeric β_2M is largely disordered as judged by CD and NMR [90, 112]. The initial rate of fibrillogenesis is maximal around pH 3.6 at high ionic strength, occurring without a lag phase and leading to the formation of curved, nodular fibrils (Fig. 26.4 B–D). Further unfolding decreases the population of this amyloid precursor conformation and results in the development of long, straight amyloid-like fibrils with lag kinetics (Fig. 26.4 F–H) [97]. In a recent analysis of β_2M at different pH values using electrospray ionization mass spectrometry, the population of the partially unfolded and acid denatured states formed during acid denaturation have been detected and quantified, opening the door to the identification of key amyloid precursor states by combining these studies with detailed analysis of the rates of aggregation under different pH conditions [127]. Whether the morphology of fibrils formed under different pH conditions is dictated by the different kinetics of assembly or the different conformational properties of the monomeric precursor at each pH remains to be seen.

NMR analysis of the partially unfolded states of β_2M formed under acidic conditions, together with a number of other studies, have provided important clues into the structural rearrangements required for the normally soluble native protein to form amyloid-like fibrils *in vitro*. By using a urea titration approach monitored by NMR, the conformational properties of partially unfolded β_2M formed by acidification to pH 3.6 has been characterized in some detail (Fig. 26.5 B) [110]. A significant destabilization of the overall β-sandwich structure compared with the native monomer at pH 7.0 (Fig. 26.5 A) [128] was observed at low pH, resulting in a non-cooperatively stabilized ensemble in which the N-terminal region is predominantly unfolded. Although the N- and C-terminal regions of the protein are highly destabilized in this amyloid precursor, the five central β-strands form a stable substructure. Of these β-strands, the E strand forms the most stable region. This very hydrophobic strand is also structured in the acid unfolded form of β_2M at pH 2.5 [112], although the overall structure of this species is highly denatured. These data show a remarkable correspondence with the properties of β_2M in the assembled fibrils measured using hydrogen exchange as described above [108] in that the N- and C-terminal strands are the least protected regions in all of the conformations studied to date (Fig. 26.5). Whilst these studies are immensely powerful, they cannot provide the site-specific information about inter-residue contacts that are required if more detailed models of the structure of the amyloid fibrils and the conformational changes required to assemble β_2M into such a structure are to be revealed. Nonetheless, the observations presented here indicate that population of partially unfolded conformers is an important feature in β_2M amyloidosis *in vitro*, in common with many other human amyloid diseases [129].

26.4.3 Factors Facilitating Fibril Formation

The studies described above provide a wealth of data about the mechanism of association of β_2M into amyloid fibrils *in vitro* under acidic conditions. How β_2M forms amyloid *in vivo* at physiological pH commencing from the native, folded protein is much less clear, although partial unfolding of the native protein must be an important initiating event. Several investigations have recently proposed different mechanisms by which amyloidogenic precursors of β_2M are generated at pH 7.0, as well as the origin of their effect (see Fig. 26.6). Here we describe these features in turn, drawing out similarities and differences in the different models.

26.4.3.1 **Proteolysis**

Amyloid fibril deposits from patients with DRA contain both full-length β_2M, as well as proteolytically processed fragments, with the main truncated species lacking six residues at the N-terminus (ΔN6β_2M) (Table 26.2). Interestingly, such truncated species have not been detected in the soluble form of monomeric β_2M in the serum of dialysis patients, suggesting that proteolysis may be a post-deposition event [44, 109]. In contrast to these results, *in vitro* analysis of ΔN6β_2M, created by deleting the N-terminal six residues by protein engineering, showed that this modification reduces the stability of β_2M substantially, resulting in a dynamic and poorly folded species that is able to extend seeds of β_2M *ex vivo* at neutral pH [93]. Molecular dynamics simulations found that the main effect of the truncation is an increase of conformational dynamics and a subsequently increased probability of strand separation that is proposed to initiate intermolecular association [130]. In contrast to these results, proteolytic cleavage at other sites including the specific cleavage at Lys58 that is found in β_2M in the serum of patients, does not result in amyloidogenic proteins at pH 7.0 [131, 132].

26.4.3.2 **Mutational Analysis**

To address the question of whether the destabilization of native β_2M is important in the generation of amyloidogenic precursor states, a number of different mutants have been created and analyzed [95, 97, 133]. These data show that whilst destabilization of the native protein increases the pH at which fibrils can be formed *in vitro*, population of specific denatured states is still required for β_2M fibril formation, showing a clear correlation between population of different precursor states and the ultimate fibril morphology [97, 127]. These data suggest that the production of amyloid-like fibrils does not simply require destabilization of native β_2M, but the population of unfolded states with specific conformational properties. Mutations in strands A or G promote fibril formation at neutral pH, even in unseeded reactions, while similar mutations created in the remaining β-strands have little effect on the pH dependence of fibril formation [95, 97]. Together with the observation that β_2M can be induced to form fibrils at neutral pH by the removal of the N-terminal 6 amino acids [93], these data suggest that strands A and G play an important role in protecting monomeric β_2M from fibril formation. To determine the effect of individual β-strands on fibril formation, Chiba et al. introduced proline residues into each of the seven β-strands of β_2M [133]. Their data supported the conclusion that the amyloidogenicity of β_2M variants is determined by a number of factors, including the stability and unfolding rate of the native protein, as well as the stability of the amyloid fibril itself. Further analysis of the structural and dynamic properties of wild-type β_2M and its variants should give an increased understanding of conversion process towards an amyloidogenic species and the subsequent mechanism of amyloid fibril formation.

26.4.3.3 **Rare Unfolding Events**

Amyloid formation from intact wild-type β_2M has been observed in seeded reactions under precise experimental conditions. By monitoring the refolding pathway of intact monomeric β_2M at neutral pH, Chiti et al. have proposed that a partially folded state is rarely, but significantly, populated at equilibrium (Fig. 26.6) [134]. In accord with these data, capillary electrophoresis studies showed that a second species exists in equilibrium with the native state under physiological conditions and represents about 15% of the total β_2M [91]. This species is less structured than native β_2M, showing a less packed hydrophobic core and a higher affinity for both Congo red and the hydrophobic dye 1-anilino-8-naphthalene sulfonate [131]. The equilibrium population of this partially folded state and its ability to aggregate in the presence of preformed fibrils may explain the observation that β_2M has the ability to elongate pre-existing fibrils at neutral pH in the absence of any other factors [91]. Furthermore, spectroscopic data suggest a close similarity between $\Delta N6\beta_2$M [93] and the equilibrium partially unfolded state of full-length β_2M [91], suggesting that these species may represent key amyloid precursors at neutral pH.

26.4.3.4 **Copper**

The relevance of copper within the context of β_2M fibril formation in DRA as well as in the wider context of amyloid formation is the subject of an on-going debate [92, 135]. Since the majority of normal (15–25 μM) serum copper is tightly bound to plasma proteins [136], little, if any, free copper is available to facilitate β_2M amyloidosis *in vivo*. Nevertheless, the dialysis procedure itself may allow the interaction of β_2M with free Cu^{2+}, for example, in the case of in the widely used Cuprophan® cellulose membranes. Exploiting this idea, Miranker et al. have shown that Cu^{2+} binds to β_2M tightly (K_d=2.8 μM) and specifically, and results in the destabilization of the protein and the subsequent formation of amyloid fibrils [92]. Further studies have shown that Cu^{2+} binds specifically to non-native states, rationalizing the destabilization of the native protein in its presence [137]. Heteronuclear NMR experiments suggest that three out of four histidines (His13, His31 and His51) are involved in Cu^{2+} binding at pH 7.0, whereby the conformational dynamics of β-strand D are changed by Cu^{2+} binding to His51, which is then propagated to the core of the molecule, thus promoting the overall destabilization [138]. Interestingly, the mutant His31Tyr shows reduced Cu^{2+} binding [139, 140], but shows a striking alternate conformation for the N-terminal region in the mutant protein crystals [141]. Recent results suggest that Cu^{2+} addition mediates formation of a monomeric, activated state followed by a discrete oligomeric assembly of di-, tetra- and hexameric forms that ultimately lead to the formation of long amyloid assemblies [142].

26.4.3.5 **A Consensus Model?**

Whilst the models for β_2M amyloid formation under the conditions described above differ in detail, common features of the mechanism of amyloid formation *in vitro* are emerging (Fig. 26.6). Thus, dissociation of the N- and/or C-terminal strands from the core of native β_2M appears to play a key role in the initiation of amyloid fibril formation from this β-sandwich protein, whilst the remaining β-strands form the protected core in the ultimate protein fibril. Moreover, an intact disulfide bond is required both for generation of long, straight fibrils *in vitro* and in forming assembly competent precursor states at very low pH [17, 18]. In addition, the amyloid precursor states identified to date have the common features that they are destabilized states that have increased conformational dynamics and decreased cooperativity. Together, these features generate an increased population of non-native species that expose new assembly-competent surfaces that promote fibril formation. The next exciting stages of research in this area will be to elucidate the assembly mechanism at higher resolution, so that the role of individual residues in fibril assembly under different conditions can be elucidated and to determine the structure of the amyloid fibril itself. In addition, the role of biological factors in the facilitation of assembly *in vivo* must be determined and characterized in detail using biophysical and biochemical analyses *in vitro*. Finally, it should also be borne in mind that there is unlikely to be a unique pathway of fibrillogenesis and that the mechanism of assembly along different routes, the conformational properties of precursor states and the structure of the ultimate fibril itself may well be very heterogeneous and highly dependent on the experimental growth conditions employed.

26.5 Summary and Future Implications

The available data on β_2M deposition in the form of amyloid fibrils in association with end-stage renal failure suggest progression of DRA in several consecutive phases (Fig. 26.1). Initially, declining kidney function and the inability to clear β_2M by dialysis treatment leads to an increase in β_2M concentration in the serum that is a key initiating event. Second, at the preferred site of deposition, i.e. the synovial tissue of large joints, formation of amyloid fibrils comprised predominantly of full-length, unmodified β_2M takes place. Finally, fibril deposition induces a local inflammatory response by attracting macrophages and stimulating these cells to release cytokines, which then leads to severe bone destruction and the acute and painful symptoms of DRA. Whether the fibrils themselves and/or other pre-fibrillar or aggregated species are responsible for the development of the pathology of DRA is currently unknown. The experimental and clinical findings described above, however, suggest an active role of the fibril; at least since AGE-modified β_2M fibrils induce the inflammatory activities responsible for the symptoms of this disorder. Moreover, whether amyloid deposits in DRA can regress under certain conditions, e.g. after renal transplantation, remains controversial. What is clear, however, is that decreased catabolism of β_2M with a concomitant increase in serum concentration is a key initiating event in DRA. Deriving new insights into the events occurring during aggregation of β_2M *in vivo*, combined with identification of the factors involved in sequestration of β_2M in the joints over the forthcoming years, will form an important platform upon which to derive new therapeutic strategies for these patients, especially those not able to undergo renal transplant. In parallel, more research into the dialysis procedure itself, including the development of membranes with increased biocompatibility or enhanced β_2M clearance, also offers hope of amelioration or even prevention of this disorder.

As summarized in Fig. 26.6, experimental data from a number of studies have now identified several factors that promote the destabilization of native β_2M and facilitate amyloid formation *in vitro*. The regions of β_2M that could be involved in promoting fibril formation are diverse and reinforce the likely heterogeneity of the assembly mechanism, as well as the fibril morphologies produced. Intriguingly, the most vulnerable strands A, D and G are all involved in contacts with the heavy chain in the intact MHC-I complex, emphasizing the pivotal role of the heavy chain in stabilizing β_2M and preventing its self-assembly. The precise conformational changes required for amyloid formation of β_2M remain unknown and more experiments will be needed to determine the mechanisms of unfolding and assembly of the amyloid precursor state. Numerous possibilities exist for the process of monomer assembly into fibrils, including head-to-tail association, stacking of head-to-head dimers, domain swapping or assembly of initially largely unfolded polypeptide chains; only further information on the detailed structure of β_2M fibrils as well as the identification and characterization of early aggregated states may answer these questions. Despite the abundance of remaining unanswered questions, β_2M amyloid formation is now amongst the most widely and detailed studied systems *in vitro* as well as *in vivo*. Given the current momentum in this field it may not be long before such questions are answered. Delineating the mechanism of assembly of this small, all β-sheet protein into amyloid fibrils, thus, will not only help to solve the intellectual challenge of the mechanisms of protein misfolding and assembly, but may also lead to the development of therapeutic agents that prevent this acute and debilitating disease.

Acknowledgments

We thank all members of the SER amyloid group for helpful discussions and for reading this article prior to its publication. We also thank Yuji Goto for providing data about hydrogen exchange in amyloid fibrils of β_2M. T. R. J. is funded by The Wellcome Trust and S. E. R. is a BBSRC Professorial Fellow.

References

1 Westermark, P., et al. Amyloid fibril protein nomenclature – 2002. *Amyloid* **2002**, *9*, 197–200.
2 Floege, J. and G. Ehlerding. β_2-Microglobulin-associated amyloidosis. *Nephron* **1996**, *72*, 9–26.
3 Drueke, T. B. β_2-Microglobulin and amyloidosis. *Nephrol Dial Transplant* **2000**, *15 Suppl 1*, 17–24.
4 Moriniere, P., et al. Destructive spondyloarthropathy with β_2-microglobulin amyloid deposits in a uremic patient before chronic hemodialysis. *Nephron* **1991**, *59*, 654–657.
5 Zingraff, J. J., et al. β_2-Microglobulin amyloidosis in chronic renal failure. *N Engl J Med* **1990**, *323*, 1070–1071.
6 Warren, D. J. and L. S. Otieno. Carpal tunnel syndrome in patients on intermittent hemodialysis. *Postgrad Med J* **1975**, *51*, 450–452.
7 Assenat, H., et al. Hemodialysis: carpal tunnel syndrome and amyloid substance. *Nouv Presse Med* **1980**, *9*, 1715.
8 Gorevic, P. D., et al. β_2-Microglobulin is an amyloidogenic protein in man. *J Clin Invest* **1985**, *76*, 2425–2429.
9 Gejyo, F., et al. A new form of amyloid protein associated with chronic hemodialysis was identified as β_2-microglobulin. *Biochem Biophys Res Commun* **1985**, *129*, 701–706.
10 Ohashi, K. Pathogenesis of β_2-microglobulin amyloidosis. *Pathol Int* **2001**, *51*, 1–10.
11 Davison, A. M. Amyloidosis in patients with end-stage renal failure: uremia associated or dialysis related? *Contrib Nephrol* **1995**, *113*, 92–100.
12 Floege, J. and M. Ketteler. β_2-microglobulin-derived amyloidosis: an update. *Kidney Int Suppl* **2001**, *78*, S164–S171.
13 Cunningham, B. A. Structure and significance of β_2-microglobulin. *Fedn Proc* **1976**, *35*, 1171–1176.
14 Becker, J. W. and G. N. Reeke, Jr. Three-dimensional structure of β_2-microglobulin. *Proc Natl Acad Sci USA* **1985**, *82*, 4225–4229.
15 Isenman, D. E., et al. The structure and function of immunoglobulin domains: studies with β_2-microglobulin on the role of the intrachain disulfide bond. *Proc Natl Acad Sci USA* **1975**, *72*, 548–552.
16 Bellotti, V., et al. β_2-microglobulin can be refolded into a native state from *ex vivo* amyloid fibrils. *Eur J Biochem* **1998**, *258*, 61–67.

17 Smith, D. P. and S. E. Radford. Role of the single disulfide bond of β_2-microglobulin in amyloidosis *in vitro*. *Protein Sci* **2001**, *10*, 1775–1784.

18 Ohhashi, Y., et al. The intrachain disulfide bond of β_2-microglobulin is not essential for the immunoglobulin fold at neutral pH, but is essential for amyloid fibril formation at acidic pH. *J Biochem (Tokyo)* **2002**, *131*, 45–52.

19 Hill, D. M., et al. A dominant negative mutant β_2-microglobulin blocks the extracellular folding of a major histocompatibility complex class I heavy chain. *J Biol Chem* **2003**, *278*, 5630–5638.

20 Sundin, D. P., et al. Characterization of the β_2-microglobulin endocytic pathway in rat proximal tubule cells. *Am J Physiol* **1994**, *267*, 380–389.

21 Floege, J., et al. Clearance and synthesis rates of β_2-microglobulin in patients undergoing hemodialysis and in normal subjects. *J Lab Clin Med* **1991**, *118*, 153–165.

22 Karlsson, F. A., et al. Turnover in humans of β_2-microglobulin: the constant chain of HLA-antigens. *Eur J Clin Invest* **1980**, *10*, 293–300.

23 Vincent, C., et al. Kinetics of ^{125}I-$_2$-microglobulin turnover in dialysed patients. *Kidney Int* **1992**, *42*, 1434–1443.

24 Koch, K. M. Dialysis-related amyloidosis. *Kidney Int* **1992**, *41*, 1416–1429.

25 Keating, M. J. Chronic lymphocytic leukemia. *Semin Oncol* **1999**, *26*, 107–114.

26 Malaguarnera, M., et al. Serum β_2-microglobulin in chronic hepatitis C. *Dig Dis Sci* **1997**, *42*, 762–766.

27 Gejyo, F., et al. Serum levels of β_2-microglobulin as a new form of amyloid protein in patients undergoing long-term hemodialysis. *N Engl J Med* **1986**, *314*, 585–586.

28 Sethi, D. and P. E. Gower. Synovial-fluid β_2-microglobulin levels in dialysis arthropathy. *N Engl J Med* **1986**, *315*, 1419–1420.

29 Jadoul, M., et al. Does dialysis-related amyloidosis regress after transplantation? *Nephrol Dial Transplant* **1997**, *12*, 655–657.

30 Gal, R., et al. Systemic distribution of β_2-microglobulin-derived amyloidosis in patients who undergo long-term hemodialysis. Report of seven cases and review of the literature. *Arch Pathol Lab Med* **1994**, *118*, 718–721.

31 Charra, B., et al. Carpal tunnel syndrome, shoulder pain and amyloid deposits in long-term hemodialysis patients. *Proc Eur Dial Transplant Ass Eur Ren Ass* **1985**, *21*, 291–295.

32 Miyata, T., et al. β_2-microglobulin in renal disease. *J Am Soc Nephrol* **1998**, *9*, 1723–1735.

33 Bardin, T. and D. Kuntz. The arthropathy of chronic hemodialysis. *Clin Exp Rheumatol* **1987**, *5*, 379–386.

34 Ferreira, A., et al. Relationship between serum β_2-microglobulin, bone histology, and dialysis membranes in uremic patients. *Nephrol Dial Transplant* **1995**, *10*, 1701–1707.

35 Mourad, G. and A. Argiles. Renal transplantation relieves the symptoms but does not reverse β_2-microglobulin amyloidosis. *J Am Soc Nephrol* **1996**, *7*, 798–804.

36 Campistol, J. M. Dialysis-related amyloidosis after renal transplantation. *Semin Dial* **2001**, *14*, 99–102.

37 Jadoul, M., et al. Pathological aspects of β_2-microglobulin amyloidosis. *Semin Dial* **2001**, *14*, 86–89.

38 Schaeffer, J., et al. Diagnostic aspects of β_2-microglobulin amyloidosis. *Nephrol Dial Transplant* **1996**, *11 (Suppl 2)*, 144–146.

39 Nelson, S. R., et al. Imaging of haemodialysis-associated amyloidosis with ^{123}I-serum amyloid P component. *Lancet* **1991**, *338*, 335–339.

40 Linke, R. P., et al. Production of recombinant human β_2-microglobulin for scintigraphic diagnosis of amyloidosis in uraemia and haemodialysis. *Eur J Biochem* **2000**, *267*, 627–633.

41 Floege, J., et al. Imaging of dialysis-related amyloid (Aβ-amyloid) deposits with $^{131}I\beta$-$_2$-microglobulin. *Kidney Int* **1990**, *38*, 1169–1176.

42 Homma, N., et al. Collagen-binding affinity of β_2-microglobulin, a preprotein of hemodialysis-associated amyloidosis. *Nephron* **1989**, *53*, 37–40.

43 Coggi, G., et al. Coexpression of intermediate filaments in normal and neoplastic human tissues: a reappraisal. *Ultrastruct Pathol* **1989**, *13*, 501–514.

44 Linke, R. P., et al. Lysine-specific cleavage of β_2-microglobulin in amyloid deposits associated with hemodialysis. *Kidney Int* **1989**, *36*, 675–681.

45 Argiles, A., et al. Biochemical characterization of serum and urinary β_2-microglobulin in end-stage renal disease patients. *Nephrol Dial Transplant* **1992**, *7*, 1106–1110.

46 Campistol, J. M., et al. Polymerization of normal and intact β_2-microglobulin as the amyloidogenic protein in dialysis-amyloidosis. *Kidney Int* **1996**, *50*, 1262–1267.

47 Bellotti, V., et al. Dynamic of β_2-microglobulin fibril formation and reabsorption: the role of proteolysis. *Semin Dial* **2001**, *14*, 117–122.

48 Odani, H., et al. Purification and complete amino acid sequence of novel β_2-microglobulin. *Biochem Biophys Res Commun* **1990**, *168*, 1223–1229.

49 Kad, N. M., et al. β_2-Microglobulin and its deamidated variant, N17D form amyloid fibrils with a range of morphologies *in vitro*. *J Mol Biol* **2001**, *313*, 559–571.

50 Connors, L. H., et al. *In vitro* formation of amyloid fibrils from intact β_2-microglobulin. *Biochem Biophys Res Commun* **1985**, *131*, 1063–1068.

51 Campistol, J. M., et al. *In vitro* spontaneous synthesis of β_2-microglobulin amyloid fibrils in peripheral blood mononuclear cell culture. *Am J Pathol* **1992**, *141*, 241–247.

52 Ono, K. and F. Uchino. Formation of amyloid-like substance from β_2-microglobulin *in vitro*. Role of serum amyloid P component: a preliminary study. *Nephron* **1994**, 66, 404–407.

53 Gejyo, F., et al. Apolipoprotein E and alpha 1-antichymotrypsin in dialysis-related amyloidosis. *Kidney Int Suppl* **1997**, *62*, S75–S78.

54 Hirschfield, G. M. and P. N. Hawkins. Amyloidosis: new strategies for treatment. *Int J Biochem Cell Biol* **2003**, *35*, 1608–1613.

55 Athanasou, N. A., et al. Highly sulfated glycosaminoglycans in articular cartilage and other tissues containing β_2-microglobulin dialysis amyloid deposits. *Nephrol Dial Transplant* **1995**, *10*, 1672–1678.

56 Aruga, E., et al. Macromolecules that are colocalized with deposits of β_2-microglobulin in hemodialysis-associated amyloidosis. *Lab Invest* **1993**, *69*, 223–230.

57 Yamaguchi, I., et al. Glycosaminoglycan and proteoglycan inhibit the depolymerization of β_2-microglobulin amyloid fibrils *in vitro*. *Kidney Int* **2003**, *64*, 1080–1088.

58 Inoue, S., et al. Ultrastructural organization of hemodialysis-associated β_2-microglobulin amyloid fibrils. *Kidney Int* **1997**, *52*, 1543–1549.

59 Argiles, A., et al. High-molecular-mass proteins in haemodialysis-associated amyloidosis. *Clin Sci (Lond)* **1989**, *76*, 547–552.

60 Motomiya, Y., et al. Circulating level of α_2-macroglobulin-β_2-microglobulin complex in hemodialysis patients. *Kidney Int* **2003**, *64*, 2244–2252.

61 Hou, F. F., et al. Interaction between β_2-microglobulin and advanced glycation end products in the development of dialysis-related amyloidosis. *Kidney Int* **1997**, *51*, 1514–1519.

62 Moe, S. M. and N. X. Chen. The role of the synovium and cartilage in the pathogenesis of β_2-microglobulin amyloidosis. *Semin Dial* **2001**, *14*, 127–130.

63 Miyata, T., et al. Identification of pentosidine as a native structure for advanced glycation end products in β_2-microglobulin-containing amyloid fibrils in patients with dialysis-related amyloidosis. *Proc Natl Acad Sci USA* **1996**, *93*, 2353–2358.

64 Miyata, T., et al. Relevance of oxidative and carbonyl stress to long-term uremic complications. *Kidney Int Suppl* **2000**, *76*, 5120–5125.

65 Niwa, T., et al. Amyloid β_2-microglobulin is modified with imidazolone, a novel advanced glycation end product, in dialysis-related amyloidosis. *Kidney Int* **1997**, *51*, 187–194.

66 Degenhardt, T. P., et al. Technical note. The serum concentration of the advanced glycation end-product *N*-ε-(car-

boxymethyl)lysine is increased in uremia. *Kidney Int* **1997**, *52*, 1064–1067.

67 Niwa, T. Dialysis-related amyloidosis: pathogenesis focusing on AGE modification. *Semin Dial* **2001**, *14*, 123–126.

68 Garbar, C., et al. Histological characteristics of sternoclavicular β_2-microglobulin amyloidosis and clues for its histogenesis. *Kidney Int* **1999**, *55*, 1983–1990.

69 Ohashi, K., et al. Cervical discs are most susceptible to β_2-microglobulin amyloid deposition in the vertebral column. *Kidney Int* **1992**, *41*, 1646–1652.

70 Brancaccio, D., et al. Ultrastructural localization of advanced glycation end products and β_2-microglobulin in dialysis amyloidosis. *J Nephrol* **2000**, *13*, 129–136.

71 Hou, F. F., et al. β_2-Microglobulin modified with advanced glycation end products delays monocyte apoptosis. *Kidney Int* **2001**, *59*, 990–1002.

72 Hou, F. F., et al. Pathogenesis of β_2-microglobulin amyloidosis: role of monocytes/macrophages. *Semin Dial* **2001**, *14*, 135–139.

73 Miyata, T., et al. Involvement of β_2-microglobulin modified with advanced glycation end products in the pathogenesis of hemodialysis-associated amyloidosis. Induction of human monocyte chemotaxis and macrophage secretion of tumor necrosis factor-*a* and interleukin-1. *J Clin Invest* **1994**, *93*, 521–528.

74 Miyata, T., et al. The receptor for advanced glycation end products (RAGE) is a central mediator of the interaction of AGE-β_2-microglobulin with human mononuclear phagocytes *via* an oxidant-sensitive pathway. Implications for the pathogenesis of dialysis-related amyloidosis. *J Clin Invest* **1996**, *98*, 1088–1094.

75 Matsuo, K., et al. Transforming growth factor-β is involved in the pathogenesis of dialysis-related amyloidosis. *Kidney Int* **2000**, *57*, 697–708.

76 Witko-Sarsat, V., et al. Advanced oxidation protein products as a novel marker of oxidative stress in uremia. *Kidney Int* **1996**, *49*, 1304–1313.

77 Ohashi, K., et al. Increased matrix metalloproteinases as possible cause of osseoarticular tissue destruction in long-term hemodialysis and β_2-microglobulin amyloidosis. *Virchows Arch* **1996**, *428*, 37–46.

78 Chen, N. X., et al. Signal transduction of β_2-microglobulin-induced expression of VCAM-1 and COX-2 in synovial fibroblasts. *Kidney Int* **2002**, *61*, 414–424.

79 Crofford, L. J. COX-2 in synovial tissues. *Osteoarthritis Cartilage* **1999**, *7*, 406–408.

80 Moe, S. M., et al. β_2-microglobulin induces MMP-1 but not TIMP-1 expression in human synovial fibroblasts. *Kidney Int* **2000**, *57*, 2023–2034.

81 O'Neill, K. D., et al. Cellular uptake of β_2-microglobulin and AGE-β2-microglobulin in synovial fibroblasts and macrophages. *Nephrol Dial Transplant* **2003**, *18*, 46–53.

82 Jaradat, M. I. and S. M. Moe. Effect of hemodialysis membranes on β_2-microglobulin amyloidosis. *Semin Dial* **2001**, *14*, 107–112.

83 van Ypersele, C. and T. E. Drueke. *Dialysis Amyloid.* Oxford University Press, Oxford, **1996**.

84 Van Ypersele de Strihou, C. β_2-microglobulin amyloidosis: effect of ESRF treatment modality and dialysis membrane type. *Nephrol Dial Transplant* **1996**, *11 (Suppl 2)*, 147–149.

85 Lornoy, W., et al. On-line hemodiafiltration. Remarkable removal of β_2-microglobulin. Long-term clinical observations. *Nephrol Dial Transplant* **2000**, *15 (Suppl 1)*, 49–54.

86 Gejyo, F., et al. Arresting dialysis-related amyloidosis: a prospective multicenter controlled trial of direct hemoperfusion with a β_2-microglobulin adsorption column. *Artif Organs* **2004**, *28*, 371–380.

87 Abe, T., et al. Effect of β_2-microglobulin adsorption column on dialysis-related amyloidosis. *Kidney Int* **2003**, *64*, 1522–1528.

88 Winchester, J. F., et al. β_2-microglobulin in ESRD: an in-depth review. *Adv Renal Replace Ther* **2003**, *10*, 279–309.

89 Naiki, H., et al. Establishment of a kinetic model of dialysis-related amyloid fibril extension *in vitro*. *Amyloid* **1997**, *4*, 223–232.

90 McParland, V. J., et al. Partially unfolded states of β_2-microglobulin and amyloid formation *in vitro*. *Biochemistry* **2000**, *39*, 8735–8746.

91 Chiti, F., et al. A partially structured species of β_2-microglobulin is significantly populated under physiological conditions and involved in fibrillogenesis. *J Biol Chem* **2001**, *276*, 46714–46721.

92 Morgan, C.J., et al. Kidney dialysis-associated amyloidosis: a molecular role for copper in fibre formation. *J Mol Biol* **2001**, *309*, 339–345.

93 Esposito, G., et al. Removal of the N-terminal hexapeptide from human β_2-microglobulin facilitates protein aggregation and fibril formation. *Protein Sci* **2000**, *9*, 831–845.

94 Yamamoto, S., et al. Glycosaminoglycans enhance the trifluoroethanol-induced extension of β_2-microglobulin-related amyloid fibrils at a neutral pH. *J Am Soc Nephrol* **2004**, *15*, 126–133.

95 Jones, S., et al. Role of the N and C-terminal strands of β_2-microglobulin in amyloid formation at neutral pH. *J Mol Biol* **2003**, *330*, 935–941.

96 Kad, N.M., et al. Hierarchical assembly of β_2-microglobulin amyloid *in vitro* revealed by atomic force microscopy. *J Mol Biol* **2003**, *330*, 785–797.

97 Smith, D.P., et al. A systematic investigation into the effect of protein destabilization on β_2-microglobulin amyloid formation. *J Mol Biol* **2003**, *330*, 943–954.

98 Yamaguchi, I., et al. Extension of β_2-microglobulin amyloid fibrils with recombinant human β_2-microglobulin. *Amyloid* **2001**, *8*, 30–40.

99 Rochet, J.C. and P.T. Lansbury, Jr. Amyloid fibrillogenesis: themes and variations. *Curr Opin Struct Biol* **2000**, *10*, 60–68.

100 Lashuel, H.A., et al. Neurodegenerative disease: amyloid pores from pathogenic mutations. *Nature* **2002**, *418*, 291.

101 Bonar, L., et al. Characterization of the amyloid fibril as a cross-β protein. *Proc Soc Exp Biol Med* **1969**, *131*, 1373–1375.

102 Chien, P., et al. Emerging principles of conformation-based prion inheritance. *Annu Rev Biochem* **2004**, *73*, 617–656.

103 Hirota-Nakaoka, N., et al. Dissolution of β_2-microglobulin amyloid fibrils by dimethylsulfoxide. *J Biochem (Tokyo)* **2003**, *134*, 159–164.

104 Ivanova, M.I., et al. Role of the C-terminal 28 residues of β_2-microglobulin in amyloid fibril formation. *Biochemistry* **2003**, *42*, 13536–13540.

105 Myers, S.L., Ashcroft, A.E. and Radford, S.E. Unpublished results.

106 Monti, M., et al. Topological investigation of amyloid fibrils obtained from β_2-microglobulin. *Protein Sci* **2002**, *11*, 2362–2369.

107 Hoshino, M., et al. Mapping the core of the β_2-microglobulin amyloid fibril by H/D exchange. *Nat Struct Biol* **2002**, *9*, 332–336.

108 Yamaguchi, K., et al. Core and heterogeneity of β_2-microglobulin amyloid fibrils as revealed by H/D exchange. *J Mol Biol* **2004**, *338*, 559–571.

109 Stoppini, M.S., et al. Detection of fragments of β_2-microglobulin in amyloid fibrils. *Kidney Int* **2000**, *57*, 349–350.

110 McParland, V.J., et al. Structural properties of an amyloid precursor of β_2-microglobulin. *Nat Struct Biol* **2002**, *9*, 326–331.

111 Gozu, M., et al. Conformational dynamics of β_2-microglobulin analysed by reduction and reoxidation of the disulfide bond. *J Biochem (Tokyo)* **2003**, *133*, 731–736.

112 Katou, H., et al. The role of disulfide bond in the amyloidogenic state of β_2-microglobulin studied by heteronuclear NMR. *Protein Sci* **2002**, *11*, 2218–2229.

113 Benyamini, H., et al. β_2-Microglobulin amyloidosis: insights from conservation analysis and fibril modelling by protein docking techniques. *J Mol Biol* **2003**, *330*, 159–174.

114 Trinh, C.H., et al. Crystal structure of monomeric human β_2-microglobulin reveals clues to its amyloidogenic properties. *Proc Natl Acad Sci USA* **2002**, *99*, 9771–9776.

115 Platt, G.W., McParland, V.J., Kalverda, A.P., Homans, S.W. and Radford, S.E. *J Mol Biol* **2005**, *346*, 279–294.

116 Ban, T., et al. Direct observation of amyloid fibril growth monitored by thioflavin T fluorescence. *J Biol Chem* **2003**, *278*, 16462–16465.

117 Verdone, G., et al. The solution structure of human β_2-microglobulin reveals the prodromes of its amyloid transition. *Protein Sci* **2002**, *11*, 487–499.

118 Berman, H.M., et al. The Protein Data Bank and the challenge of structural genomics. *Nat Struct Biol* **2000**, *7 (Suppl)*, 957–959.

119 Okon, M., et al. ^{1}H NMR assignments and secondary structure of human β_2-microglobulin in solution. *Biochemistry* **1992**, *31*, 8906–8915.

120 Siepen, J.A., et al. β-Edge strands in protein structure prediction and aggregation. *Protein Sci* **2003**, *12*, 2348–2359.

121 Richardson, J.S. and D.C. Richardson. Natural β-sheet proteins use negative design to avoid edge-to-edge aggregation. *Proc Natl Acad Sci USA* **2002**, *99*, 2754–2759.

122 Kozhukh, G.V., et al. Investigation of a peptide responsible for amyloid fibril formation of β_2-microglobulin by *Achromobacter* protease I. *J Biol Chem* **2002**, *277*, 1310–1315.

123 Ohhashi, Y., et al. Optimum amyloid fibril formation of a peptide fragment suggests the amyloidogenic preference of β_2-microglobulin under physiological conditions. *J Biol Chem* **2004**, *279*, 10814–10821.

124 Jones, S., et al. Amyloid-forming peptides from β_2-microglobulin – insights into the mechanism of fibril formation *in vitro*. *J Mol Biol* **2003**, *325*, 249–257.

125 de la Paz, M.L. and L. Serrano. Sequence determinants of amyloid fibril formation. *Proc Natl Acad Sci USA* **2004**, *101*, 87–92.

126 Aggeli, A., et al. Responsive gels formed by the spontaneous self-assembly of peptides into polymeric β-sheet tapes. *Nature* **1997**, *386*, 259–262.

127 Borysik, A.J., et al. Co-populated conformational ensembles of β_2-microglobulin uncovered quantitatively by electrospray ionization mass spectrometry. *J Biol Chem* **2004**, *279*, 27069–27077.

128 Jahn, T.R., Homans, S.W. and Radford, S.E. Unpublished results.

129 Uversky, V.N. and A.L. Fink. Conformational constraints for amyloid fibrillation: the importance of being unfolded. *Biochim Biophys Acta* **2004**, *1698*, 131–153.

130 Ma, B. and R. Nussinov. Molecular dynamics simulations of the unfolding of β_2-microglobulin and its variants. *Protein Eng* **2003**, *16*, 561–575.

131 Heegaard, N.H., et al. Conformational intermediate of the amyloidogenic protein β_2-microglobulin at neutral pH. *J Biol Chem* **2001**, *276*, 32657–32662.

132 Heegaard, N.H., et al. Cleaved β_2-microglobulin partially attains a conformation that has amyloidogenic features. *J Biol Chem* **2002**, *277*, 11184–11189.

133 Chiba, T., et al. Amyloid fibril formation in the context of full-length protein: effects of proline mutations on the amyloid fibril formation of β_2-microglobulin. *J Biol Chem* **2003**, *278*, 47016–47024.

134 Chiti, F., et al. Detection of two partially structured species in the folding process of the amyloidogenic protein β_2-microglobulin. *J Mol Biol* **2001**, *307*, 379–391.

135 Capanni, C., et al. Investigation of the effects of copper ions on protein aggregation using a model system. *Cell Mol Life Sci* **2004**, *61*, 982–991.

136 Thomson, N.M., et al. Comparison of trace elements in peritoneal dialysis, hemodialysis, and uremia. *Kidney Int* **1983**, *23*, 9–14.

137 Eakin, C.M., et al. Formation of a copper specific binding site in non-native states of β_2-microglobulin. *Biochemistry* **2002**, *41*, 10646–10656.

138 Villanueva, J., et al. Increase in the conformational flexibility of β_2-microglobulin upon copper binding: a possible role for copper in dialysis-related amyloidosis. *Protein Sci* **2004**, *13*, 797–809.

139 De Lorenzi, E., et al. Capillary electrophoresis investigation of a partially unfolded conformation of β_2-microglobulin. *Electrophoresis* **2002**, *23*, 918–925.

140 Corazza, A., et al. Properties of some variants of human β_2-microglobulin and amyloidogenesis. *J Biol Chem* **2004**, *279*, 9176–9189.

141 Rosano, C., et al. β_2-microglobulin H31Y variant 3D structure highlights the protein natural propensity towards intermolecular aggregation. *J Mol Biol* **2004**, *335*, 1051–1064.

142 Eakin, C. M., et al. Oligomeric assembly of native-like precursors precedes amyloid formation by β_2-microglobulin. *Biochemistry* **2004**, *43*, 7808–7815.

143 Ogawa, H., et al. Detection of electrophoretic heterogeneity of serum β_2-microglobulin in chronic renal failure. *ASAIO Trans* **1988**, *34*, 196–199.

144 Siebold, C., et al. Crystal structure of HLA-DQ0602 that protects against type 1 diabetes and confers strong susceptibility to narcolepsy. *Proc Natl Acad Sci USA* **2004**, *101*, 1999–2004.

145 DeLano, W. *The PyMOL Molecular Graphics System*. DeLano Scientific, San Carlos, CA, **2002**.

27
Cystatin C

Mariusz Jaskolski and Anders Grubb

27.1
Introduction

A hereditary disorder of early brain hemorrhage, caused by a mutation in the cystatin C gene, was described by the Icelandic physician Arni Arnason in 1935 [1], 25 years before the corresponding protein, now called cystatin C, was discovered [2–4]. After the pioneering work of Arnason, it has become evident that cystatin C is involved in the pathophysiology of two different types of amyloid disorders: (1) hereditary cystatin C amyloid angiopathy (HCCAA), in which an L68Q mutant is deposited as amyloid fibrils and leads to brain hemorrhage in early adult life, and (2) amyloid disorders involving deposition of amyloid β fibrils with wild-type cystatin C as a co-precipitant. This chapter concerns the structural and biological properties of cystatin C, emphasizing particularly those structural observations that suggest possible models for the formation of amyloid fibrils in general. Possible approaches to block the pathophysiological process of HCCAA that might be also relevant for blocking the formation of amyloid fibrils in other conformational diseases are discussed.

27.2
Biochemical and Physiological Characteristics

The single polypeptide chain of mature full-length cystatin C is comprised of 120 amino acid residues, of which the proline residue at position 3 (Fig. 27.1) is partly hydroxylated [5]. The molecular mass of cystatin C, 13,343 Da (unhydroxylated) or 13,359 Da (fully hydroxylated), indicates that no other post-ribosomal modifications take place [6, 7]. Two disulfide bridges are present in the C-terminal part of the polypeptide chain [8], linking residues 73–83 and 97–117 (Fig. 27.1). The biological function of cystatin C as an inhibitor of cysteine proteases of the papain family was defined in 1984 [9], and the name "cystatin" suggested for proteins inhibiting such proteases and displaying amino acid sequences homologous with those of human cystatin C and chicken cystatin [10].

Amyloid Proteins. The Beta Sheet Conformation and Disease. J. D. Sipe

ISBN: 3-527-31072-X

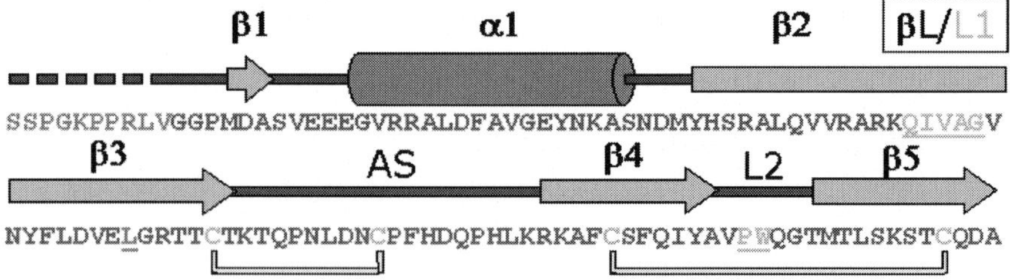

Fig. 27.1 Primary and secondary structure of human cystatin C. The secondary structure elements are shown and annotated (black) corresponding to the open conformation found in the three-dimensional domain-swapped dimer. They all correspond to the canonical cystatin fold, except for the boxed loop L1 (green), which changes conformation and becomes a segment (βL) of the long $\beta2$–βL–$\beta3$ strand. The enzyme-binding loops of the monomer are shown in underlined green lettering. The site of the L68Q mutation is underlined and red. The red N-terminal peptide is missing in the protein extracted from HCCAA amyloid. It can be also clipped off *in vitro*, by leukocyte elastase, as in the N-truncated variant of the protein discussed in Section 27.8. Even if the N-terminal peptide is intact, it cannot be fully modeled by crystallography (dashed segment) due to disorder. Two native disulfide bridges in the C-terminal part of the molecule (yellow lines) link residues 73–83 and 97–117.

The human cystatin superfamily presently comprises 13 proteins. Each cystatin has a unique inhibitory spectrum for cysteine proteases and a unique distribution in body fluids. The evolutionary advantage of this variation is possibly that a multitude of cystatins are required to control the large number of different cysteine proteases that might be produced by damaged human cells or by bacteria, fungi, viruses and parasites [11].

In addition to being an inhibitor of papain-like proteases, cystatin C has also been shown to be an efficient inhibitor of cysteine proteases of another group, called peptidase family C13, with human legumain as a representative enzyme. The cystatin C inhibitory center for legumain does not overlap with that for papain-like cysteine proteases [12].

Cystatin C has also been suggested to possess biological functions unrelated to its protease inhibiting potential. For example, it has been described as a growth factor for neural stem cells [13] and as a regulator in inflammatory processes [14, 15]. Chicken cystatin and, thus, probably human cystatin C as well upregulate nitric oxide release from peritoneal macrophages [16].

The structure of the human cystatin C gene, located on chromosome 20p11.2 [17, 18], and its promoter show that it is a "housekeeping" gene, indicating a stable production rate by most nucleated cells [19]. The presence of a hydrophobic leader sequence in pre-cystatin C strongly indicates that the protein is normally secreted [20]. Indeed, immunochemical and Northern blot studies of human tissues and cell lines have shown that cystatin C and/or its mRNA is present in virtually all investigated cell types [11]. Likewise, investigations of human cell lines in culture have demonstrated that nearly all cell lines secrete cystatin C [11].

Studies of the serum level of cystatin C in large patient cohorts have failed to correlate it with any pathophysiological states apart from those affecting the glomerular filtration rate, which is consistent with a stable secretion rate [11]. However, a recent report shows that the two less-frequent haplotypes of the promoter of the cystatin C gene cause a slightly lower production of the protein, meaning that a person homozygous for the common haplotype will have plasma concentration about 10% higher than a non-carrier [21].

Since cystatin C is produced at a stable rate by virtually all nucleated cells, the concentration of the protein in plasma will be decided by its rate of catabolism. Cystatin C, like all low-molecular-mass proteins, is catabolized by filtration through the renal glomerular filter followed by complete reabsorption into the renal tubular cells and intracellular degradation. The plasma level of cystatin C is, therefore, a good marker for the glomerular filtration rate [11]. A recent meta-analysis has shown that the plasma level of cystatin C is the best glomerular filtration rate marker described so far [22].

27.3 HCCAA

The disorder was first described by Arnason [1] as a dominantly inherited type of cerebral hemorrhage mainly affecting young adults. He noted and regretted that no diagnostic procedures were available to him when members of affected Icelandic families asked him if they carried the trait. The next step in elucidating the nature of the disease was taken when Gudmundsson et al. [23] noted that the patients suffered from heavy deposits of amyloid in the brain vasculature. Sequence studies of the deposits [24] followed by immunohistochemical investigations [25] demonstrated that cystatin C, or N-terminally truncated cystatin C, constituted the amyloid-forming material. It was also noted that patients suffering from the disease had a spinal fluid level of cystatin C reduced to one-third of that of a reference population [26]. It was subsequently observed that several clinically unsymptomatic relatives in the affected families also had an abnormally low spinal fluid level of cystatin C and that they later developed cerebral hemorrhage [27], meaning that a low spinal fluid level of cystatin C could be used to diagnose HCCAA before development of clinical signs of cerebral hemorrhage. When the complete sequence of the cystatin C polypeptide deposited as amyloid became available [28], it was found to differ in one position from the sequence previously determined for cystatin C isolated from a patient without cerebral hemorrhage [5], with a Gln residue replacing a Leu at position 68. This variation was correlated with a loss of an *Alu*I restriction site in the cystatin C gene. Consequently, the restriction fragment length polymorphism

(RFLP) method was used to show that the disease causing mutation was a single T → A substitution, by identifying it in all 22 HCCAA patients in eight families, but not in healthy relatives or in control populations [29–31]. The RFLP technique is now used to identify carriers of the HCCAA trait and as a valuable clinical test for HCCAA-based stroke.

A note on terminology is in order here, as two designation for the disease originally described by Arnason are in use, i.e. "hereditary cystatin C amyloid angiopathy" (HCCAA) and "hereditary cerebral hemorrhage with amyloidosis – Icelandic type" (HCHWA-I). The former one is preferred, as it indicates the amyloid-forming protein and agrees with the recent observations that the amyloid depositions are neither confined to cerebral vasculature [25, 32] nor to Icelandic patients [33]. It is also an appropriate designation for the condition before the first cerebral hemorrhage has occurred.

27.4 Cystatin C Oligomers *In Vivo* and *In Vitro*

Gel-filtration analysis of blood plasma and cerebrospinal fluid (CSF) from individuals without HCCAA shows that wild-type cystatin C is present in monomeric form in these body fluids. In contrast, gel filtration of these fluids from HCCAA patients demonstrates that cystatin C dimers and monomers are present in approximately equal concentrations in plasma, and that cystatin C dimers are also present in CSF, but at lower concentrations than the monomeric fraction [34]. The dimers of HCCAA patients are either homodimers of L68Q cystatin C or L68Q–wild-type cystatin C heterodimers, whereas the monomers contain little or none L68Q cystatin C. It has not been possible to isolate monomeric L68Q cystatin C from body fluids, but an *in vitro* system for production and isolation of recombinant L68Q cystatin C has been established [35]. Interestingly, L68Q cystatin C spontaneously and rapidly dimerizes in physiological buffers [35, 36], while wild-type cystatin C dimerizes only very slowly in these conditions. However, wild-type cystatin C can be induced to form dimers by the use of denaturants, like guanidinium chloride, at a low concentration [36] or by the use of elevated temperature [37].

27.5 The Phenomenon of Three-dimensional Domain Swapping

Three-dimensional domain swapping as a mechanism of protein oligomerization was described, and the term coined, by Eisenberg et al., who established it by X-ray crystallography in diphtheria toxin [38]. However, a phenomenon essentially the same in nature had been predicted over three decades earlier from ingenious, and today classic, experiments with activity recovery in dimers of ribonuclease (RNase) A with partly knocked-out active sites [39, 40]. Pre-dating the introduction of the term "three-dimensional domain swapping" were also papers by Piccoli et al. [41] and Mazzarella et al. [42], where the intertwined nature of bovine seminal RNase (BS-RNase) dimers was recognized. Three-dimensional domain swapping refers to exchange of identical structural elements or "domains" by two (or more) protein subunits. The domains can be as small as short secondary structure elements or as large as complete functional domains. In other words, in a three-dimensional domain-swapped oligomer, a structural element of one subunit takes the place of the identical structural element of another subunit and *vice versa,* leading to the recreation of the monomeric fold, but from chain segments contributed by different subunits. In a protein capable of domain swapping, there must exist a flexible hinge region, usually a loop, whose conformational change allows the molecule to partially unfold and then find another similarly open monomer (Fig. 27.2). The adhesive force of the domain-swapped oligomer resides in the "closed interface" between the swapped

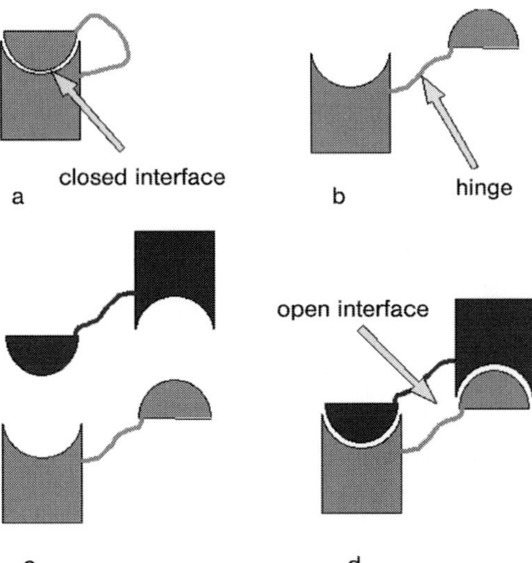

Fig. 27.2 Cartoon illustration of three-dimensional domain swapping. The compact globular structure (a) is partially unfolded (b) through a conformational change at a hinge region. The unfolding temporarily disrupts and exposes the closed interface, i.e. the contact area between the two domains. If sufficiently long-lived and if present in sufficiently high concentration (c), the unfolded chains will mutually recognize their complementary surfaces and will recreate the native contacts in a dimeric fashion (d). The dimer is not a simple sum of the two monomeric molecules. The hinge region has a new conformation and there is a new intermolecular (open) interface that was not present in the monomer. (Reprinted with permission from [58])

domains, which recreates the structure and interactions of the protomer. It is a powerful factor in the structure of the oligomer, as it has evolved to provide stability of the monomeric molecule. The oligomeric species has, however, also a new, or "open", interface between its components that is not found in the monomeric form. If the oligomer is to be more stable than the monomers, the extra stabilization energy must come from the interactions in the open interface. The new interactions must also compensate for the entropic factor (loss of translational and rotational freedom), which always favors the monomer.

Taken rigorously, three-dimensional domain swapping requires the same amino acid sequence to exist in both the closed monomeric form and as an intertwined oligomer. In practice, the usage of the term is more liberal, e.g. there is tolerance of some sequence differences, provided the folding pattern is the same, or the term could be applied even when the existence of the monomeric species is not certain.

More than 50 proteins from diverse functional and structural classes have been reported to undergo three-dimensional domain swapping. The classic example is RNase A, whose monomeric fold [43, 44] can be recreated in two types of dimers, arising either through an N-terminal [42] or a C-terminal [45] domain swap. Swapping of the latter element can also lead to cyclic RNase A trimers [46]. Even more interestingly, swapping of the N- and C-terminal segments by a central RNase A molecule with two neighbors can lead to an open-ended trimer, such as could be expected to trigger infinite linear aggregation [46].

The list of structurally characterized cases of three-dimensional domain swapping includes two amyloidogenic proteins associated with human diseases: the prion protein (PrP) [47] and human cystatin C [48]. The three-dimensional domain-swapped PrP dimer can arise only after disruption of an intramolecular S–S bridge, which is then recreated (in two copies) between the subunits. Because of this observation, the suggestion that amyloidogenic aggregation of the prion protein could involve three-dimensional domain swapping was at first regarded with skepticism [49]. However, later, in a series of ingenious redox experiments at controlled denaturing conditions, Lee and Eisenberg [50] showed that recombinant monomeric PrP can be converted not only into dimers, but also into amyloid-like fibrils, which can then be used to seed the fibril formation of fresh material.

It seems that domain swapping is a much more common phenomenon than originally believed and that, even if not always found naturally, for many proteins can be induced artificially. This reflection has a more general bearing on a fundamental canon of structural biology, one sequence–one structure, viewing protein folds as rigid invariants uniquely determined by amino acid sequences. In view of the accumulating evidence from the three-dimensional domain swapping field it may become necessary to revise those useful, but simplified, assumptions.

27.6 The Cystatin Fold

Structural details of the general fold of protein inhibitors belonging to the cystatin family were established by the X-ray crystallographic structure of the related chicken protein [51–53] with which human cystatin C shares 41% sequence identity and 62.5% similarity. The canonical features of the cystatin fold include a long $\alpha1$ helix running across a large, five-stranded antiparallel β-sheet of the following connectivity: (N)-$\beta1$–($\alpha1$)–$\beta2$–L1–$\beta3$–(AS)–$\beta4$–L2–$\beta5$-(C) (Fig. 27.1), where AS, a broad "appending structure", is positioned on the opposite ("back-side") end of the β-sheet relative to the "active" edge consisting of the N-terminus and loops L1 and L2, which together form the structural epitope responsible for inhibition of papain-like proteases. Like all other type 2 cystatins [10, 54], human cystatin C contains four characteristic disulfide-paired cysteine residues. The disulfide bridges are formed in the C-terminal half of the molecule (Fig. 27.1), stabilizing the structure of the random-coil (AS) region between strands $\beta3$ and $\beta4$ and connecting the ends of the $\beta4$–$\beta5$ hairpin.

27.7 Three-dimensional Domain Swapping in Full-length Cystatin C

Crystallization of full-length human cystatin C was reported from slightly acidic (pH 4.8) solutions of monomeric protein prepared by gel filtration in the final isolation step [55]. The crystal structure [48] revealed, however, that the molecules have aggregated to form 2-fold-symmetric dimers via three-dimensional domain swapping. This result is consistent with the view that local high concentration (as in the crystallization droplet) is necessary for the formation of three-dimensional domain-swapped oligomers. The two components are related by a perfect 180° rotation, as the dimers are constrained by exact crystallographic symmetry. In the dimers, the monomeric fold defined by the crystal structure of chicken cystatin is reconstructed with high fidelity but, as in all three-dimensional domain-swapped oligomers, from parts belonging to different polypeptide chains (Figs. 27.3 and 27.4). This confirms earlier nuclear magnetic resonance (NMR) results indicating that the secondary structure elements are preserved upon dimerization [37, 56]. Analysis of a single polypeptide chain "extracted" from the dimeric context (Fig. 27.5) reveals that the monomeric molecule underwent partial unfolding through an opening movement of loop L1, one of the inhibitory elements located at the edge of the monomeric structure. This hinge movement produces an unnaturally looking conformation, ready for swapping domains with another unfolded chain, in order to bury the exposed surfaces that are not evolved to interact with water. By changing its conformation, the L1 segment now became part of a long β-strand running from the beginning of $\beta2$ to the end of $\beta3$. In addition to the monomer-type closed interface, the dimers also contain the new open interface (Fig. 27.4). It is formed through β-sheet in-

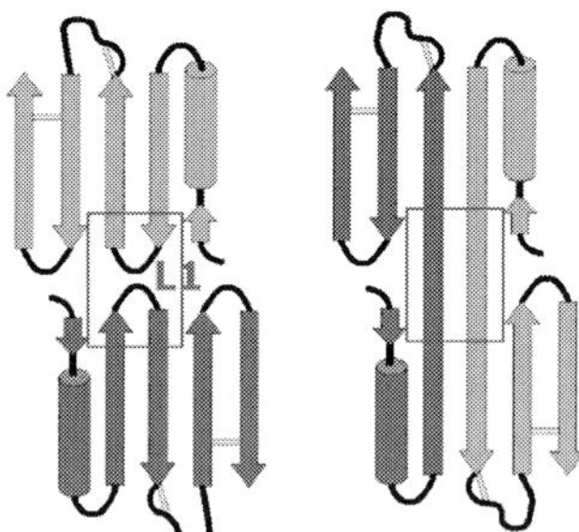

Fig. 27.3 Schematic illustration showing how a 2-fold-symmetric pair of cystatin C molecules (left) exchange domains through a conformational change of the β-hairpin loop L1 (red box, left). Note that in the domain-swapped dimer (right) a very long intermolecular antiparallel β-sheet is formed, the central part of which (red box, right) is the newly created open interface that makes the dimer energetically advantageous. The yellow lines indicate the disulfide bridges conserved in the C-terminal part of all type 2 cystatins. (Adapted from [58])

teractions in the βL region (former L1) leading to the creation of an unusually long contiguous antiparallel β-sheet formed by two copies of strands Y42–T74 (β2–βL–β3) which cross from one domain to the other with as many as 34 hydrogen bonds between the main chains and extra hydrogen bonds involving side-chains. The two domains that undergo swapping are separated by the L1 hinge. The N-terminal domain is mostly α-helical, but includes also the N-terminal β-strand (β1) and the strand that immediately follows the α1-helix, β2. The C-terminal domain has purely β conformation and consists of the remaining part of the β-sheet, strands β3–β5. This means that disruption of the monomeric fold requires not only separation of the α-helix from the β-sheet by which it is wrapped, but also tearing one of the seams of the β-sheet (β2–β3). Likewise, the closed interface rebuilt in the dimer is cemented by these same interactions. The faithfulness of the reconstruction of the fold by the swapped domains may be illustrated by the presence of a tandem of β-bulges at residues A46 and V49 in strand β2 that shift the β register by two residues. These β-bulges, as well as another one at L112 in strand β5, have their counterparts in monomeric chicken cystatin and they must be present to introduce the curvature into the β-sheet that is required for its wrapping around the α-helix. Finally, it may be observed that the two disulfide bridges introducing rigidity into the fold of this small protein are both present in the C-terminal domain and in consequence not only do not interfere with the domain-swapping process, but help to maintain the integrity of the C-terminal domain during the transition period when the protein is partially unfolded.

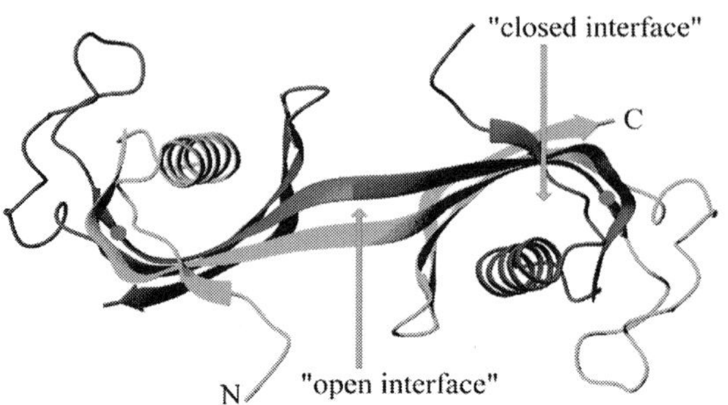

Fig. 27.4 Domain-swapped dimer of cystatin C viewed along its 2-fold axis. This view emphasizes the intermolecular β-sheet in the domain switch region (open interface) and the site of the L68Q mutation (red dot). (Reprinted with permission from [48]. © 2001 Nature Publishing Group)

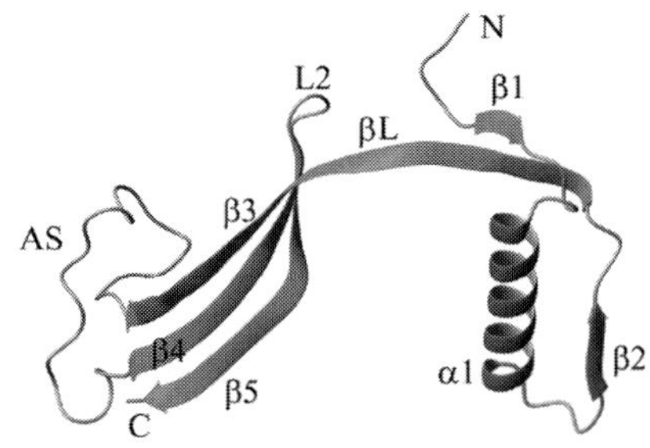

Fig. 27.5 A cystatin C subunit "extracted" from the domain-swapped context of the dimer to emphasize its unnatural, partially unfolded conformation. Standard labeling of cystatin topology is shown. The former L1 loop (as found in chicken cystatin) serves now as a linker (or hinge) and is labeled βL. The illustration is based on the coordinates deposited in the PDB with the accession code 1G96. (Reprinted with permission from [58])

It is not obvious why the highly conserved L1 loop should be predisposed to destabilization. In chicken cystatin [51] it forms a tight five-residue β-hairpin, the central element of which, Ser56, is on the border of a generously allowed Ramachandran region. However, since this serine represents a deviation from the conserved sequence, it is again not obvious that the L1 loop in monomeric cystatin C should be particularly unstable. The source of monomer instability may be, however, located elsewhere, for instance at the α–β interface, as discussed below.

The disappearance of loop L1 in the dimeric structure and, consequently, the disruption of this functional element of the protein agree with the observation that the dimers have absolutely no inhibitory effect on papain-type proteases [35, 37]. On the other hand, loop 39–41, which connects helix α1 with strand β2 and contains asparagine 39 that is crucial for inhibition of mammalian legumain, is not affected by dimerization. This is in agreement with the observation that dimeric human cystatin C is as active in inhibiting porcine legumain as the monomeric protein [12].

It has to be admitted that this is not a strict *bona fide* three-dimensional domain-swapping case as defined by Eisenberg, because the structure of monomeric human cystatin C is not precisely known. However, we know that such monomers do exist and we can be quite confident that their structure closely resembles that of the chicken homolog.

27.8 Three-dimensional Domain Swapping in N-truncated Cystatin C

When proteins are isolated from amyloid deposits of human cystatin C, a variant truncated at the N-terminus is found that lacks the first 10 amino acids. A recombinant protein with analogous N-terminal truncation has been crystallized and its crystallographic structure analyzed by Janowski et al. [57]. The crystal structure is composed of eight polypeptide chains, A–H. In contrast to full-length cystatin C, the crystals of the N-truncated variant were formed at basic pH (8.0) using entirely different conditions with respect to temperature and precipitating agent. Nevertheless, the full length and truncated molecules have aggregated into analogous dimers via the same mechanism of three-dimensional domain swapping. None

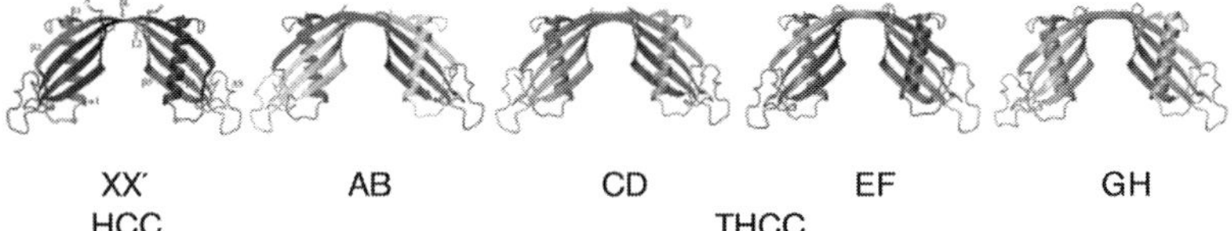

Fig. 27.6 A gallery of three-dimensional domain-swapped dimers formed by human cystatin C. XX′ is the two-fold symmetric dimer observed in the crystal structure of full-length human cystatin C (HCC). The dimers AB, CD, EF and GH, formed by the N-truncated protein (THCC), have only approximate symmetry although their overall architecture is very similar. The illustration is based on the coordinates deposited in the PDB with the accession codes 1G96 (HCC) and 1R4C (THCC).

Table 27.1 Pair-wise superpositions of the Cα traces of the different dimers formed by N-truncated cystatin C

	AB	**CD**	**EF**	**GH**
AB	0.83 (179.4)	0.50	**1.18**	**1.46**
CD	0.75	0.71 (180.0)	**1.16**	**1.51**
EF	**1.15**	**1.09**	0.57 (179.8)	0.56
GH	**1.42**	**1.42**	0.77	1.00 (179.9)

Each comparison is characterized by an r.m.s. deviation (Å) between the superimposed Cα atoms. Because the dimers are not perfectly symmetric, each superposition can be done in two ways (e.g. AB/CD and AB/DC), depicted in the upper and lower triangles of the table. The diagonal contains the self-rotation of each dimer (°, in parentheses) and the corresponding r.m.s. deviation.

of the independent dimers (AB, CD, EF and GH) has the exact symmetry of the full-length dimer (XX′), although the overall geometry of all those aggregates is very similar (Fig. 27.6). Despite the lack of exact symmetry, the dimers of the N-truncated protein are characterized by nearly ideal 180° rotations, corresponding to r.m.s. deviations between the corresponding Cα atoms (Table 27.1) within 1 Å. Table 27.1 also shows that the dimers are of two types (AB and CD versus EF and GH), such that superpositions across these types are characterized by high r.m.s. deviations (above 1 Å, bold). This is confirmed by Table 27.2, where the dimers of N-truncated cystatin C, or their subunits (individual chains), are compared with the XX′ dimer of the full-length protein.

Fig. 27.7 illustrates an overlay of dimers AB and EF calculated by superposing only one “half” (a globular domain) of the dimers to emphasize the differences in the other “half”. It reveals that the β-strands and the α-helices that are not

Table 27.2 Comparison of the three-dimensional domain-swapped dimers formed by N-truncated cystatin C (AB, CD, EF and GH) with that formed by the full-length protein (XX′), presented as r.m.s. deviations (Å) of the superimposed Cα atoms

	AB	CD	EF	GH
XX′	0.85	0.74	1.50	1.85
X [a)]	0.66/0.92	0.76/0.67	1.50/1.37	1.72/1.84
1/2 XX′ [b)]	0.52/0.42	0.47/0.52	0.59/0.84	0.57/0.78

a) Superposition of the single chain (X) of the full-length cystatin C dimer on the individual chains forming the dimers of the N-truncated protein.
b) Superposition of "half" (domain) of the XX′ dimer on the domains of each of the dimers formed by the N-truncated protein.

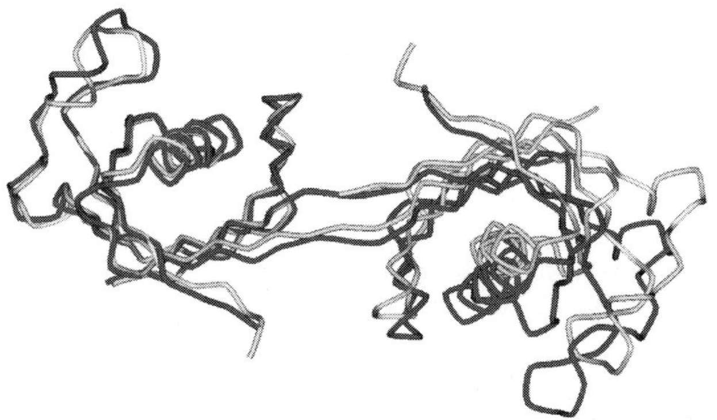

Fig. 27.7 Superposition of the AB (dark grey) and EF (grey) dimers of N-truncated cystatin C viewed along their approximate dyads. The superposition was calculated using the Cα atoms from only one half (left-hand side) of the molecules. The differences in the dimer geometry are therefore emphasized in the right-hand side domain.

constrained by superposition show a systematic rotation between the two molecules. The rotation arises from a conformational change in the hinge region, which is propagated to the periphery of the molecule, resulting in maximum Cα deviations in excess of 7 Å. This rotation within the dimers indicates that there is a degree of flexibility at the dimer-specific open interface. On a larger scale, this flexibility is of key importance for the hypothetical mechanism of cystatin C dimerization, which assumes that domain swapping is preceded by partial unfolding of the protein chain occurring at the L1 β-hairpin loop [48, 58].

The hinge movement does not affect the structure of the globular domain (i.e. "half" of the dimer) as illustrated in Table 27.2 (last row), where the results of such comparisons with the globular domain of the full-length dimer are shown. The small values of the r.m.s. deviations in all these comparisons (less than 0.6 Å on average) again attest to the high fidelity with which the monomeric cystatin fold is recreated, regardless of the intermolecular context. The exception is the AS loop, which is mobile and assumes variable conformation.

27.9
Structural Implications for L68Q Cystatin C

In both crystal structures of human cystatin C (full-length and N-truncated), Leu68 is located on the central strand $\beta 3$ of the β-sheet, on its concave face, and is covered by the $\alpha 1$-helix (Fig. 27.4). In the crystallographic three-dimensional domain-swapped dimers, the $\beta 3$ and $\alpha 1$ elements are contributed by different subunits, but since this interaction is part of the closed interface, it is logically assumed that identical contacts exist in monomeric cystatin C. In the hydrophobic core of the protein, L68 occupies a niche formed by the surrounding residues on the β-sheet and the hydrophobic face of the helix (Fig. 27.8). The closest distances

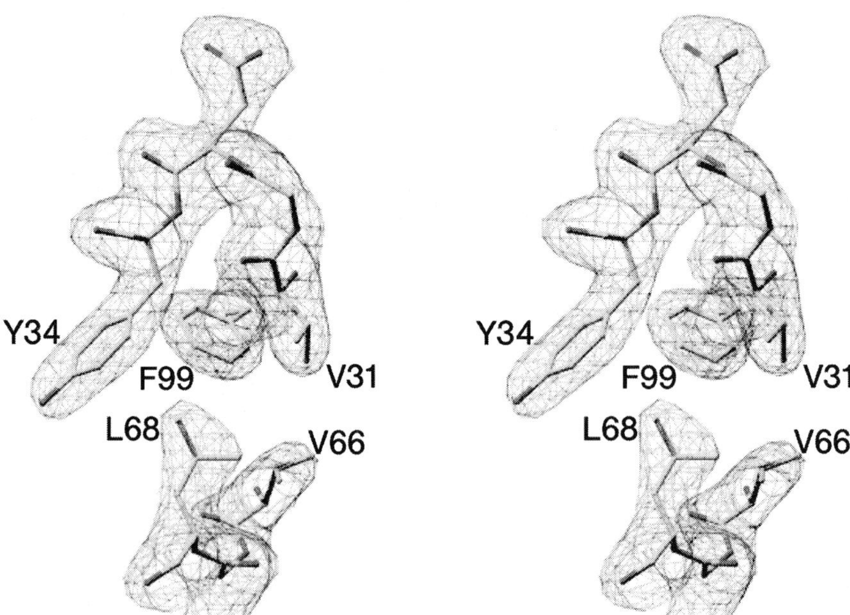

Fig. 27.8 Stereoview of Leu68 in the structure of full-length cystatin C [48]. The $2F_o{-}F_c$ map (contoured at the 1.5σ level) shows the placement of L68 in a hydrophobic pocket formed by residues of the β-sheet of one molecule and the α-helix of the other molecule of the dimer. (Reprinted with permission from [48]. © 2001 Nature Publishing Group)

in this area represent typical hydrophobic contacts. Replacement of the leucine side-chain by the longer glutamine side-chain, as in the naturally occurring pathological L68Q variant of human cystatin C, would not only make those contacts prohibitively close, but would also place the mutated hydrophilic chain in a hydrophobic environment. This would definitely destabilize the molecular α–β interface and lead to repulsive interactions expelling the α-helix, together with the intervening strand β2, from the compact molecular core and forcing the molecule to unfold into the α and β domains. This explains the increased dynamic properties of the L68Q mutant compared with wild-type cystatin C as observed by NMR spectroscopy [56, 59]. Under the assumption that the refolded dimer recreates the topology of monomeric cystatin C, those destabilizing effects would be similar in both cases. However, the dimeric structure may be more resistant to disruption because of the extra stabilization contributed by the β-interactions in the linker region or more generally in the long β2–βL–β3 region. A hydrophilic substitution at the α–β interface would be also expected to reduce the unfavorable solvent contacts of the newly exposed surface. A speculative diagram illustrating the thermodynamic relations in monomer–dimer equilibria of wild-type and L68Q cystatin C is presented in Fig. 27.9. The above discussion of the effect of the L68Q substitution on cystatin C dimerization is supported by the observation that the mutated variant forms dimers in blood plasma much more easily than wild-type cystatin C [34].

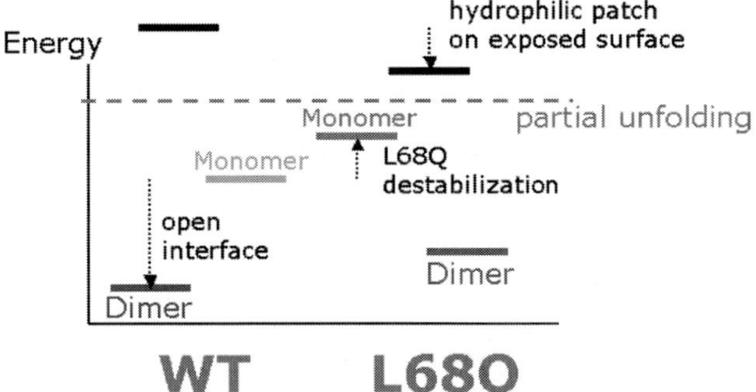

Fig. 27.9 Hypothetical thermodynamics of wild-type (left) and L68Q (right) cystatin C. In normal conditions, the wild-type monomer is probably sufficiently stable and the energy of the unfolded intermediate sufficiently high to make transitions to the three-dimensional domain-swapped form rare, even though it may be energetically favored. The monomeric form of the L68Q mutant is destabilized by the repulsive interactions at the Q68 side-chain and partial unfolding may be achieved more easily. Additionally, the hydrophilic Q68 at the solvent-exposed surface would make the unfolded intermediate less unstable. Once formed, the L68Q cystatin C dimer would be sufficiently stable (and the wild-type dimer even more so), to make spontaneous dissociation impossible. (Adapted from [58])

It is generally accepted that three-dimensional domain swapping itself and the precise orientation of the subunits are influenced by the interactions in the open interface [60], thus lending a possibility to control the overall structure through careful mutations designed to change the nature of the open interface. Those remarks are of general validity and have been illustrated with many examples. On the other hand, the example of cystatin C and its L68Q mutant suggests that mutations in the closed interface have also to be considered in the interplay of the kinetic and thermodynamic factors controlling the formation of three-dimensional domain-swapped oligomers. As pointed out by Perutz in a short, but inspirational, note [61], mutations of "internal" residues might result in a loss in free energy of stabilization which, even if small, might lead to a disruptive, "loosening" effect on the native structure. In the case of cystatin C, the L68Q substitution decreases the energy necessary for the transition from the monomeric to the dimeric form by destabilizing the monomer (higher energy) and by lowering the energy of the partially unfolded state (less unfavorable interactions with solvent in the open conformation).

27.10 Higher Oligomers Observed by Crystallography and Other Methods

The solvent content of the full-length cystatin C crystals is high (71%), while in the crystals of the N-truncated protein only moderate (51%). Yet, the crystal packing of the two forms shows remarkable resemblances, in that the three-dimensional domain-swapped dimers enter into similar interactions leading to the formation of higher aggregates. The most conspicuous species are octameric assemblies composed of four three-dimensional domain-swapped dimers each [58]. In the octamer of full-length cystatin C, two tetrad-related dimers interact through a rich system of hydrogen bonds (duplicated in two copies) involving the back-side loops ($\alpha1$–$\beta2$ and AS) of one dimer, and the solvent-accessible surface of the β-sheet of another. Additionally, the side-chain of M41 is locked in a hydrophobic pocket formed by residues on the $\beta3$ and $\beta4$ strands of the complementary dimer. The octamer appears to have a stable structure, as the total number of hydrogen bonds between the dimers is 72. The pattern of connectivity within each octamer is both dimeric and circular, and leads to full utilization of all the elements available for interactions (all back-side loops and all β-sheets). In this sense, the octamers are closed, sphere-like assemblies (Fig. 27.10 a) and, as such, are rather difficult to reconcile with infinite aggregation, characteristic of amyloid fibrils. On the other hand, it is very intriguing that practically identical octamers exist in the crystal structure of the N-truncated protein. In this case, two independent octamers are formed, each generated by crystallographic symmetry (primed molecules) from a pair of three-dimensional domain-swapped dimers: ABCDA′B′C′D′ and EFGHG′H′E′F′.

In both structures, the octamers form similar hydrophobic contacts with their crystal-packing neighbors. These interactions involve residues from the linker regions (unfolded L1 loops), which are fairly exposed in the dimer structure

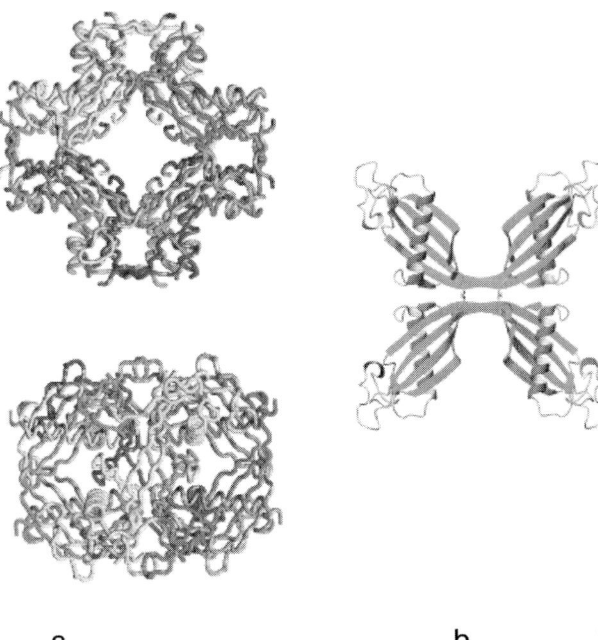

Fig. 27.10 Aggregates formed by the three-dimensional domain-swapped dimers of cystatin C. (a) Two perpendicular views of an octameric aggregate composed of four domain-swapped dimers formed via β-sheet···back-side loops interactions. (b) Dimer···dimer interactions involving hydrophobic contacts between the β-sheets of the linker regions. These interactions interconnect the octameric aggregates into infinite three-dimensional networks. (Reprinted with permission from [58])

(Fig. 27.10b). Each patch of interactions includes several short C···C contacts between residues I56, A58, and V60. In N-truncated cystatin C, the interactions are less symmetric than those depicted in Fig. 27.10b because the strands $\beta 2$, βL and $\beta 3$ from the interacting octamers are tilted by different angles and shifted, but this does not change the overall association mode. This type of association is not very attractive from the point of view of amyloid aggregation, but, again, it is intriguing why this mode of interactions is preserved in all copies of the three-dimensional domain-swapped dimers of human cystatin C.

The crystal structure of N-truncated cystatin C also reveals a unique mode of interaction, an interoctameric β-sheet (Fig. 27.11). It is of significance for amyloid fibril formation because it leads to extension of the β-sheet of the domain-swapped dimer into intermolecular context [57]. Intermolecular β-sheets of infinite propagation, called cross-β-structures and deduced from fiber diffraction, are believed to be the fundamental motif of amyloid fibril architecture [62, 63].

In analogy to human cystatin C, chicken cystatin has been shown by NMR spectroscopy to undergo similar dimerization via three-dimensional domain swapping

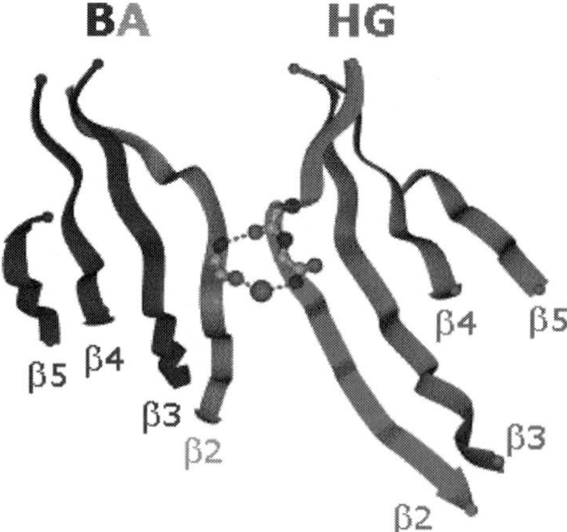

Fig. 27.11 β-Sheet-type interaction between dimers AB and GH in the crystal structure of N-truncated cystatin C. This interaction extends the antiparallel β-sheets of the dimers into intermolecular context. The intermolecular interaction is parallel and is mediated by a water molecule.

[64]. Incubation in destabilizing, but pre-denaturing, conditions (including elevated temperature) leads to higher assemblies with mass increments equivalent to one dimer. From ^{1}H-NMR and circular dichroism (CD) studies of the tetrameric species formed during the pre-exponential phase of fibril formation, Sanders et al. [65] have concluded that they are based on the same domain-swapping principle as the dimers. Interestingly, in one of the hypothetical models, the authors show a tetramer of human cystatin C formed by circular domain swapping, with a mutual organization of the "dimeric" subunits as observed in the crystal structures of full-length and N-truncated human cystatin C (Fig. 27.10b).

27.11
In Vivo Amyloid Deposits Containing Cystatin C

The amyloid deposits in HCCAA seem to be composed predominantly of L68Q cystatin C fibrils, which can be extracted in significant amounts [24]. However, it has not been excluded that wild-type cystatin C constitutes part of the HCCAA amyloid fibrils. An N-terminal truncation of 10 amino acid residues of the cystatin C polypeptide chain in the HCCAA amyloid has been observed [24], but it is unknown whether this truncation occurs before or after the deposition of the fibrils, or perhaps during the extraction procedures. Immunohistochemical evidence for the presence of small amounts of serum amyloid P component and amyloid β protein in the amyloid deposits of HCCAA has also been obtained. The amyloid deposits in HCCAA are mainly situated in the media of the vessel walls of arteries and arterioles in the brain vasculature [66], but sensitive histochemical techniques have also revealed the presence of such deposits in the vasculature outside the central nervous system, e.g. in arterioles of the skin and adrenal cortex, and also in non-vascular tissues such as lymph nodes, spleen and salivary glands [66].

Cerebral amyloid angiopathy with wild-type (or variant) amyloid β protein as the major amyloid constituent is a condition with high prevalence in the elderly and also commonly found in patients with Alzheimer's disease or Down's syndrome [67–71]. The condition is associated with cerebral hemorrhage and may account for more than 10% of the brain hemorrhage cases in the elderly [67]. Immunohistochemical investigations of the amyloid deposits have demonstrated that all, or a considerable portion of them, display cystatin C immunoreactivity in addition to their amyloid β-protein immunoreactivity [67, 68, 70–73]. Quantitative estimations have generally indicated, however, that cystatin C is a minor constituent of the deposits [67, 68, 70, 73]. Efforts to demonstrate the presence of cystatin C variants, e.g. L68Q cystatin C, in the amyloid deposits have been so far unsuccessful [73–76]. Although these observations suggest that cystatin C might be involved in the pathophysiological processes of *inter alia* Alzheimer's disease, its exact role is so far unknown. However, recent data show that cystatin C displays a specific, saturable and high affinity binding to the amyloid β fibril forming peptides Aβ1–42 and Aβ1–40 and to the amyloid β precursor protein [77]. Moreover, wild-type cystatin C seems to be able to inhibit the *in vitro* formation of amyloid fibrils from Aβ1–42 and Aβ1–40 [77].

27.12 Formation of Cystatin C Amyloid Fibrils *In Vitro*

It has been observed that both wild-type and L68Q cystatin C form amyloid fibrils *in vitro*, but due to the small amounts of L68Q cystatin C available, most published studies have concerned wild-type cystatin C [36, 78]. Although several different *in vitro* systems for the production of wild-type cystatin C amyloid fibrils have been tested, only those involving mildly denaturing conditions have been proven to be useful. For instance, the fibrils shown in Fig. 27.12 have been obtained by incubation of cystatin C at 3 mg/ml concentration in 10 mM glycine buffer, pH 2.0, at 48 °C under constant stirring. Interestingly, denaturing conditions are also known to induce dimerization of wild-type cystatin C [36].

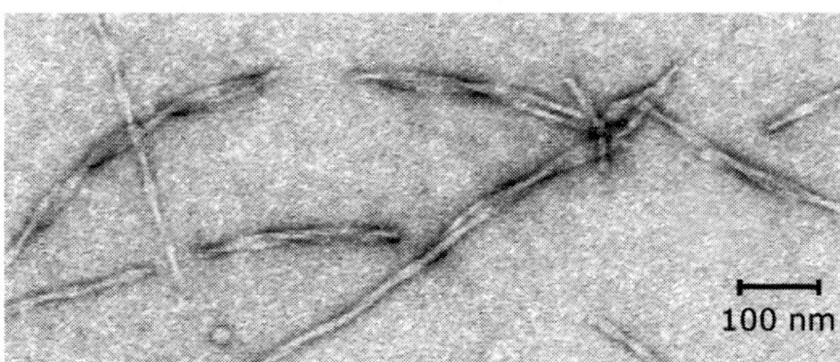

Fig. 27.12 Amyloid fibrils of wild-type cystatin C. The fibrils were formed by incubating wild-type cystatin C at 3 mg/ml for 7 days in 10 mM glycine buffer, pH 2.0, at 48 °C.

27.13
Inhibition of Dimerization and Fibril Formation by Protein Engineering

Since it has been suggested, based upon structural studies, that cystatin C amyloid fibrils might be formed by means of a process involving three-dimensional domain swapping, it was of interest to investigate if stabilization of the monomeric form of wild-type and L68Q cystatin C would stop or decrease the formation of dimers or amyloid fibrils from the monomeric proteins. Two variants of both wild-type and L68Q cystatin C were produced, with cysteine mutations strategically inserted into the amino acid sequence to inhibit domain swapping by intramolecular disulfide bridges, but without affecting the biological function of the protein as inhibitor of papain-like enzymes [36]. One of the disulfide bridges was designed to link the α1-helix and the β-sheet, and the other to connect strands β2 and β3 (Fig. 27.13), and their formation could be confirmed by mass spectrometry. The two monomeric domain-stabilized forms of both wild-type and L68Q cystatin C were then incubated at conditions known to transform virtually all monomers of cystatin C into dimers and it was observed (Fig. 27.14) that no dimers were formed by any of the stabilized proteins [36]. It must be noted that those experiments were carried in oxidative atmosphere (air) where the stabilizing S–S bonds are not easily disrupted.

The formation of amyloid fibrils from the two domain-stabilized monomers of wild-type cystatin C was also tested by incubation at conditions known to produce large amounts of amyloid fibrils. The stabilization against domain swap-

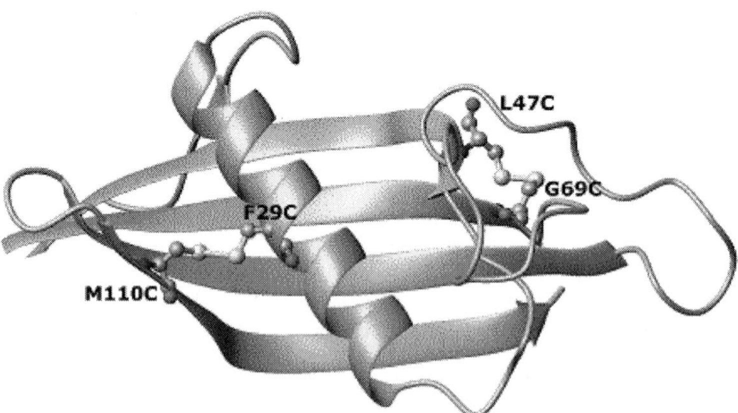

Fig. 27.13 A model illustrating the positioning of the S–S bridges introduced to stabilize the monomeric fold of cystatin C. One pair of cysteine mutations introduces a new disulfide bond linking the strands β2 and β3 of the β-sheet. The other pair of mutations create a link between the α-helix and strand β5. In the domain swapped dimer of cystatin C, strand β2 and the α-helix are contributed by one of the monomers, and strands β3–β5 by the other.

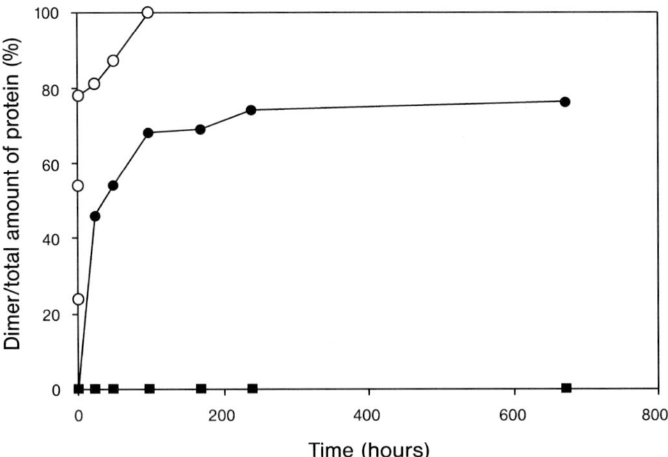

Fig. 27.14 Production of dimers from non-stabilized and stabilized monomeric forms of wild-type and L68Q cystatin C. Unmodified wild-type cystatin C (•) and L68Q cystatin C (○), as well as their variants with the monomeric fold stabilized by disulfide bridges (■), were incubated at conditions inducing dimer formation. The amounts of dimers were analyzed by agarose gel electrophoresis.

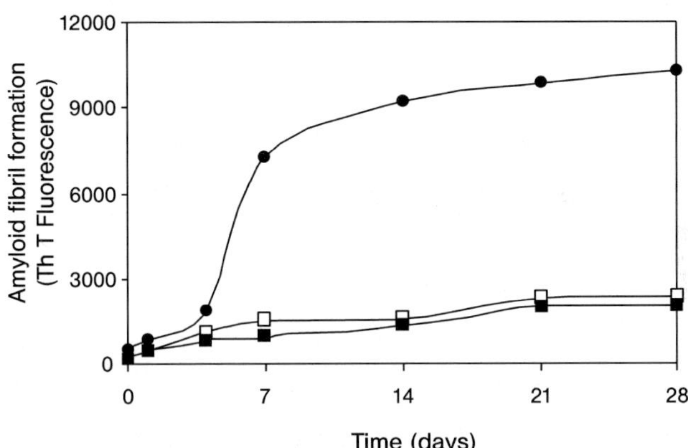

Fig. 27.15 Production of amyloid fibrils from non-stabilized and stabilized monomeric wild-type cystatin C. Fibrils were produced from solutions of wild-type cystatin C (•) or of its two S–S stabilized mutants (□ and ■) at 3 mg/ml concentration in 10 mM glycine buffer, pH 2.0. The solutions were stirred at 48 °C for various periods of time and the amyloid fibrils were quantified by the Thioflavin T (ThT) fluorescence assay [36].

ping reduced the amyloid fibril formation by about 80% for both forms of stabilized wild-type cystatin C (Fig. 27.15). Although this significant reduction of the amount of amyloid fibrils agrees with the notion that domain swapping might be of fundamental importance in amyloid fibril formation, the observation that some residual amounts of fibrils are nevertheless formed requires further consideration. One explanation would be that three-dimensional domain swapping (or unfolding of the protein, for that matter) is not absolutely necessary for aggregation into amyloid fibrils. Another, more likely, possibility is that, under the harsh conditions of the experiment, some of the stabilizing S–S bridges are disrupted, allowing the molecules to enter the fibril-formation pathway.

27.14 Inhibition of Dimerization by Monoclonal Antibodies and Carboxymethylpapain

Although the production of cystatin C variants stabilized against domain swapping is very useful in demonstrating that prevention of domain swapping has the capacity to inhibit formation of dimers and amyloid fibrils, it is obvious that, if treatment strategies based upon prevention of domain swapping are to be developed, exogenous agents stabilizing the monomeric form of cystatin C must be sought. It is therefore of interest that catalytic amounts of a monoclonal antibody raised against wild-type cystatin C have been shown [36] to inhibit the dimerization process of both wild-type and L68Q cystatin C (Fig. 27.16). This observation is related to the discovery that antibodies to surface epitopes of

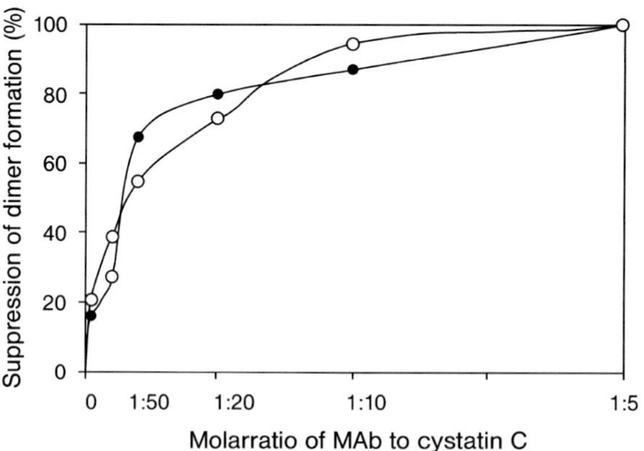

Fig. 27.16 Suppression of dimer formation of wild-type cystatin C and L68Q cystatin C by a monoclonal antibody. Wild-type cystatin C (•) or L68Q cystatin C (○) was incubated at conditions inducing dimer formation with different amounts of a monoclonal antibody against cystatin C. The degree of suppression of dimer formation was calculated from the amounts of dimers formed without additions of the monoclonal antibody.

PrP^C can inhibit the generation of the amyloidogenic species PrP^{Sc} and thereby interfere with prion biogenesis [79]. Subsequent studies of the monoclonal antibody against cystatin C have shown that also monovalent Fab fragments of the antibody efficiently inhibit dimerization of wild-type and L68Q cystatin C. The reason why the antibody is so efficient in preventing dimerization even at very low molar ratio is not clear, but one possibility is that the antibody reacts with some intermediate form of the protein on its dimerization pathway.

Cysteine proteases of the papain family can only interact with the monomeric form of cystatin C and might, therefore, also be potential agents for the stabilization of the monomeric form. Indeed, recent results show that carboxymethyl-papain, an active-site-alkylated derivative of the enzyme, inhibits the formation of dimers from both monomeric wild-type and L68Q cystatin C analogously to the monoclonal antibody [36].

27.15 Outlook

Is the three-dimensional domain swapping seen in the crystals of two variants of human cystatin C, grown at completely different conditions, a mere coincidence or does it tell us about the protein's fundamental property responsible for its pathological aggregation?

Although three-dimensional domain swapping has been proposed as a mechanism of amyloid fibril formation [38, 80, 81], there is still no direct experimental proof that, in amyloid fibrils, the protein molecules are associated via domain exchange. However, the striking parallels between the two phenomena, and the experimental findings regarding three-dimensional domain swapping in proteins involved in human amyloidosis, provide very strong circumstantial evidence. Thus, with the current knowledge about protein aggregation, the mechanism of three-dimensional domain swapping seems to be a plausible possibility. In clinical situations, detection of aberrant, non-physiological oligomers might be a diagnostic signal warning that abnormal conformational changes are taking place and that the risk of amyloid formation is high. Indeed, it has been shown that dimers of the L68Q mutant of human cystatin C are present in the blood plasma of patients with the trait for hereditary cystatin C amyloid angiopathy [34]. However, even if one accepts that three-dimensional domain swapping is involved in the formation of amyloid fibrils, two essentially different scenarios should be considered. On the one hand, the oligomers could arise by swapping of the same domains as within the dimers, but occurring in an open-ended, linear fashion. In this situation, formation of symmetrical three-dimensional domain-swapped dimers would be a dead-end on the oligomerization pathway, depleting the concentration of unfolded molecules and, at a given stage, temporarily preventing or, rather, delaying amyloid formation.

In the second scenario, the domain-swapped dimers themselves could be the building blocks of the growing fibrils. Some arguments for this scenario are provided by the experiments with chicken cystatin [65] and, structurally, by the crystallographic study of N-truncated human cystatin C [57], where the closed-ended dimeric assemblies are demonstrated to be compatible with the formation of intermolecular β-sheets, expected from the cross-β structure model of amyloid fibrils. The results of the experiments with engineered cystatin C mutants resistant to three-dimensional domain swapping are ambiguous in this respect. While dimerization of the disulfide-stabilized molecules is completely abolished, they still show some residual ability to form fibrillar aggregates. The most likely explanation is that this observation is an artifact of the harsh fibril-formation conditions, capable of disrupting the stabilized structure, but further experiments, with covalently stabilized dimers, are necessary to shed more light on the relation between the three-dimensional domain swapped dimers and the amyloid fibrils.

A separate, not yet fully understood, question concerns the ability of exogenous agents, such as monoclonal antibodies or carboxymethylpapain, to prevent dimerization of wild-type and L68Q cystatin C even at catalytic concentrations. It is obvious that structural studies of these agents in complexes with cystatin C are necessary to explain the details of this inhibition.

Acknowledgments

This work was supported by grants from the Swedish Science Research Council (project 05196), the Medical Faculty of the University of Lund, A. Påhlsson's, A. Österlund's and G. and J. Kock's Foundations, and the Polish State Committee for Scientific Research (4 T09A 039 25), and by a subsidy from the Foundation for Polish Science to M. J.

References

1 Arnason, A. Apoplexie und ihre Vererbung. *Acta Psychiatr Neurol* **1935**, *Suppl VII*, 1–180.

2 Clausen, J. Proteins in normal cerebrospinal fluid not found in serum. *Proc Soc Exp Biol Med* **1961**, *107*, 170–172.

3 MacPherson, C. F. C. and B. R. Cosgrove. Immunochemical evidence for a gamma globulin peculiar to cerebrospinal fluid. *Can J Biochem* **1961**, *39*, 1567–1574.

4 Butler, E. A. and F. V. Flynn. The occurrence of post-gamma protein in urine: a new abnormality. *J Clin Pathol* **1961**, *14*, 172–178.

5 Grubb, A. and H. Löfberg. Human γ-trace, a basic microprotein: amino acid sequence and presence in the adenohypophysis. *Proc Natl Acad Sci USA* **1982**, *79*, 3024–3027.

6 Grubb, A. Diagnostic value of analysis of cystatin C and protein HC in biological fluids. *Clin Nephrol* **1992**, *38 (Suppl 1)*, 20–27.

7 Grubb, A. Cystatin C. In *Human Protein Data*, Haeberli, A. (ed.). VCH, Weinheim, **1993**.

8 Grubb, A., H. Löfberg and A. J. Barrett. The disulphide bridges of human cystatin C (γ-trace) and chicken cystatin. *FEBS Lett* **1984**, *170*, 370–374.

9 Barrett, A., M. E. Davies and A. Grubb. The place of human γ-trace (cystatin C)

amongst the cysteine proteinase inhibitors. *Biochem Biophys Res Commun* **1984**, *120*, 631–636.

10 Barrett, A. J., et al. Nomenclature and classification of the proteins homologous with the cysteine-proteinase inhibitor chicken cystatin. *Biochem J* **1986**, *236*, 312.

11 Grubb, A. O. Cystatin C – properties and use as diagnostic marker. *Adv Clin Chem* **2000**, *35*, 63–99.

12 Alvarez-Fernandez, M., A. J. Barrett, B. Gerhartz, P. M. Dando, J. Ni and M. Abrahamson. Inhibition of mammalian legumain by some cystatins is due to a novel second reactive site. *J Biol Chem* **1999**, *274*, 19195–19203.

13 Taupin, P., R. Jasodhara, W. H. Fischer, S. T. Suhr, K. Hakansson, A. Grubb and F. H. Gage. FGF-2-responsive neural stem cell proliferation requires CCg (glycosylated cystatin C), a novel autocrine/paracrine cofactor. *Neuron* **2000**, *28*, 385–397.

14 Leung-Tack, J., C. Tavera, M. C. Gensac, J. Martinez and A. Colle. Modulation of phagocytosis-associated respiratory burst by human cystatin C: role of the N-terminal tetrapeptide Lys–Pro–Pro–Arg. *Exp Cell Res* **1990**, *188*, 16–22.

15 Leung-Tack, J., C. Tavera, J. Martinez and A. Colle. Neutrophil chemotactic activity is modulated by human cystatin C, an inhibitor of cysteine proteases. *Inflammation* **1990**, *14*, 247–258.

16 Verdot, L., et al. Cystatins up-regulate nitric oxide release from interferon-gamma-activated mouse peritoneal macrophages. *J Biol Chem* **1996**, *271*, 28077–28081.

17 Abrahamson, M., M. Q. Islam, J. Szpirer, C. Szpirer and G. Levan. The human cystatin C gene (CST3), mutated in hereditary cystatin C amyloid angiopathy, is located on chromosome 20. *Hum Genet* **1989**, *82*, 223–226.

18 Schnittger, S., V. V. Rao, M. Abrahamson and I. Hansmann. Cystatin C (CST3), the candidate gene for hereditary cystatin C amyloid angiopathy (HCCAA), and other members of the cystatin gene family are clustered on chromosome 20p11.2. *Genomics* **1993**, *16*, 50–55.

19 Abrahamson, M., et al. Structure and expression of the human cystatin C gene. *Biochem J* **1990**, *268*, 287–294.

20 Abrahamson, M., A. Grubb, I. Olafsson and A. Lundwall. Molecular cloning and sequence analysis of cDNA coding for the precursor of the human cysteine proteinase inhibitor cystatin C. *FEBS Lett* **1987**, *216*, 229–233.

21 Eriksson, P., et al. Human evidence that the cystatin C gene is implicated in focal progression of coronary artery disease. *Arterioscler Thromb Vasc Biol* **2004**, *24*, 551–557.

22 Dharnidharka, V. R., C. Kwon and G. Sevens. Serum cystatin C is superior to serum creatinine as a marker of kidney function: a meta-analysis. *Am J Kidney Dis* **2002**, *40*, 221–226.

23 Gudmundsson, G., J. Hallgrimsson, T. A. Jonasson and O. Bjarnason. Hereditary cerebral haemorrhage with amyloidosis. *Brain* **1972**, *95*, 387–404.

24 Cohen, D. H., H. Feiner, O. Jensson and B. Frangione. Amyloid fibril in hereditary cerebral hemorrhage with amyloidosis (HCHWA) is related to the gastroentero-pancreatic neuroendocrine protein, gamma trace. *J Exp Med* **1983**, *158*, 623–628.

25 Löfberg, H., A. Grubb, E. Nilsson, O. Jensson, G. Gudmundsson, H. Blöndal, A. Arnason and L. Thorsteinsson. Immunohistochemical characterization of the amyloid deposits and quantitation of pertinent cerebrospinal fluid proteins in hereditary cerebral hemorrhage with amyloidosis. *Stroke* **1987**, *18*, 431–440.

26 Grubb, A., O. Jensson, G. Gudmundsson, A. Arnason, H. Löfberg and J. Malm. Abnormal metabolism of γ-trace alkaline microprotein. The basic defect in hereditary cerebral hemorrhage with amyloidosis. *N Engl J Med* **1984**, *311*, 1547–1549.

27 Jensson, O., et al. Hereditary cystatin C (γ-trace) amyloid angiopathy of the CNS causing cerebral hemorrhage. *Acta Neurol Scand* **1987**, *76*, 102–114.

28 Ghiso, J., O. Jensson and B. Frangione. Amyloid fibrils in hereditary cerebral hemorrhage with amyloidosis of Icelandic type is a variant of γ-trace basic pro-

tein (cystatin C). *Proc Natl Acad Sci USA* **1986**, *83*, 2974–2978.

29 Palsdottir, A., M. Abrahamson, L. Thorsteinsson, A. Arnason, I. Olafsson, A. Grubb and O. Jensson. Mutation in cystatin C gene causes hereditary brain hemorrhage. *Lancet* **1988**, *ii*, 603–604.

30 Levy, E., C. Lopez-Otin, J. Ghiso, D. Geltner and B. Frangione. Stroke in Icelandic patients with hereditary amyloid angiopathy is related to a mutation in the cystatin C gene, an inhibitor of cysteine proteases. *J Exp Med* **1989**, *169*, 1771–1778.

31 Abrahamson, M., S. Jonsdottir, I. Olafsson, O. Jensson and A. Grubb. Hereditary cystatin C amyloid angiopathy: identification of the disease-causing mutation and specific diagnosis by polymerase chain reaction based analysis. *Hum Genet* **1992**, *89*, 377–380.

32 Benedikz, E., H. Blondal and G. Gudmundsson. Skin deposits in hereditary cystatin C amyloidosis. *Virchows Arch A Pathol Anat Histopathol* **1990**, *417*, 325–331.

33 Graffagnino, C., M.H. Herbstreith, D.E. Schmechel, E. Levy, A.D. Roses and M.J. Alberts. Cystatin C mutation in an elderly man with sporadic amyloid angiopathy and intracerebral hemorrhage. *Stroke* **1995**, *26*, 2190–2193.

34 Bjarnadottir, M., et al. The cerebral hemorrhage-producing cystatin C variant (L68Q) in extracellular fluids. *Amyloid* **2001**, *8*, 1–10.

35 Abrahamson, M. and A. Grubb. Increased body temperature accelerates aggregation of the Leu → Gln mutant cystatin C, the amyloid-forming protein in hereditary cystatin C amyloid angiopathy. *Proc Natl Acad Sci USA* **1994**, *91*, 1416–1420.

36 Nilsson, M., et al. Prevention of domain swapping inhibits dimerization and amyloid fibril formation of cystatin C. Use of engineered disulfide bridges, antibodies and carboxymethylpapain to stabilize the monomeric form of cystatin C. *J Biol Chem* **2004**, *279*, 24236–24245.

37 Ekiel, I. and M. Abrahamson. Folding-related dimerization of human cystatin C. *J Biol Chem* **1996**, *271*, 1314–1321.

38 Bennett, M.J., S. Choe and D.S. Eisenberg. Domain swapping: Entangling alliances between proteins. *Proc Natl Acad Sci USA* **1994**, *91*, 3127–3131.

39 Crestfield, A.M., W.H. Stein and S. Moore. On the aggregation of bovine pancreatic ribonuclease. *Arch Biochem Biophys* **1962**, *1 (Suppl)*, 217–222.

40 Crestfield, A.M., W.H. Stein and S. Moore. Properties and conformation of the histidine residues at the active site of ribonuclease. *J Biol Chem* **1963**, *238*, 2421–2428.

41 Piccoli, R., M. Tamburrini, G. Piccialli, A. Di Donato, A. Parente and G. D'Alessio. The dual-mode quaternary structure of seminal RNase. *Proc Natl Acad Sci USA* **1992**, *89*, 1870–1874.

42 Mazzarella, L., S. Capasso, D. Demasi, G. Di Lorenzo, C.A. Mattia and A. Zagari. Bovine seminal ribonuclease: structure at 1.9 Å resolution. *Acta Crystallogr D* **1993**, *49*, 389–402.

43 Wlodawer, A., R. Bott and L. Sjolin. The refined crystal structure of ribonuclease A at 2.0 Å resolution. *J Biol Chem* **1982**, *257*, 1325–1332.

44 Wlodawer, A., L.A. Svansson, L. Sjolin and G.L. Gilliland. Structure of phosphate-free ribonuclease A refined at 1.26 Å resolution. *Biochemistry* **1988**, *27*, 2705–2717.

45 Liu, Y., G. Gotte, M. Libonati and D. Eisenberg. A domain-swapped RNase A dimer with implications for amyloid formation. *Nat Struct Biol* **2001**, *8*, 211–214.

46 Liu, Y., G. Gotte, M. Libonati and D. Eisenberg. Structures of the two 3D domain-swapped RNase A trimers. *Protein Sci* **2002**, *11*, 371–380.

47 Knaus, K.J., M. Morillas, W. Swietnicki, M. Malone, W.K. Surewicz and V.C. Yee. Crystal structure of the human prion protein reveals a mechanism for oligomerization. *Nat Struct Biol* **2001**, *8*, 770–774.

48 Janowski, R., M. Kozak, E. Jankowska, Z. Grzonka, A. Grubb, M. Abrahamson and M. Jaskolski. Human cystatin C, an amyloidogenic protein, dimerizes through three-dimensional domain swapping. *Nat Struct Biol* **2001**, *8*, 316–320.

49 Nicholson, E. M., H. Mo, S. B. Prusiner, F. E. Cohen and S. Marqusse. Differences between the prion protein and its homolog Doppel: a partially structured state with implications for scrapie formation. *J Mol Biol* **2002**, *316*, 807–815.

50 Lee, S. and D. Eisenberg. Seeded conversion of recombinant prion protein to a disulfide-bonded oligomer by a reduction–oxidation process. *Nat Struct Biol* **2003**, *10*, 725–730.

51 Bode, W., et al. The 2.0 Å X-ray crystal structure of chicken egg white cystatin and its possible mode of interaction with cysteine proteinases. *EMBO J* **1988**, *7*, 2593–2599.

52 Dieckmann, T., L. Mitschang, M. Hofmann, J. Kos, V. Turk, E. A. Auerswald, R. Jaenicke and H. Oschkinat. The structure of native phosphorylated chicken cystatin and of recombinant unphosphorylated variant in solution. *J Mol Biol* **1993**, *234*, 1048–1059.

53 Engh, R. A., T. Dieckmann, W. Bode, E. A. Auerswald, V. Turk, R. Huber and H. Oschkinat. Conformational variability of chicken cystatin. Comparison of structures determined by X-ray diffraction and NMR spectroscopy. *J Mol Biol* **1993**, *234*, 1060–1069.

54 Rawlings, N. D. and A. J. Barrett. Evolution of proteins of the cystatin superfamily. *J Mol Evol* **1990**, *30*, 60–71.

55 Kozak, M., E. Jankowska, R. Janowski, Z. Grzonka, A. Grubb, M. Alvarez-Fernandez, M. Abrahamson and M. Jaskolski. Expression of a selenomethionyl derivative and preliminary crystallographic studies of human cystatin C. *Acta Crystallogr D* **1999**, *55*, 1939–1942.

56 Ekiel, I., et al. NMR structural studies of human cystatin C dimers and monomers. *J Mol Biol* **1997**, *271*, 266–277.

57 Janowski, R., M. Abrahamson, A. Grubb and M. Jaskolski. 3D domain swapping in N-truncated human cystatin C. *J Mol Biol* **2004**, *341*, 151–160.

58 Jaskolski, M. 3D Domain swapping, protein oligomerization and amyloid formation. *Acta Biochim Pol* **2001**, *48*, 807–827.

59 Gerhartz, B., I. Ekiel and M. Abrahamson. Two stable unfolding intermediates of the disease-causing L68Q variant of human cystatin C. *Biochemistry* **1998**, *37*, 17309–17317.

60 Liu, Y., P. J. Hart, M. P. Schlunegger and D. Eisenberg. The crystal structure of a 3D domain-swapped dimer of RNase A at 2.1 Å resolution. *Proc Natl Acad Sci USA* **1998**, *95*, 3437–3442.

61 Perutz, M. F. Mutations make enzymes polymerize. *Nature* **1997**, *385*, 773–774.

62 Glenner, G. G. Amyloid deposits and amyloidosis – the β-fibrilloses 1. *N Eng J Med* **1980**, *302*, 1283–1292.

63 Glenner, G. G. Amyloid deposits and amyloidosis – the β-fibrilloses 2. *N Eng J Med* **1980**, *302*, 1333–1343.

64 Staniforth, R. A., S. Giannini, L. D. Higgins, M. J. Conroy, A. M. Hounslow, R. Jerala, C. J. Craven and J. P. Waltho. Three-dimensional domain swapping in the folded and molten-globule states of cystatins, an amyloid-forming structural superfamily. *EMBO J* **2001**, *20*, 4774–4781.

65 Sanders, A., C. J. Craven, L. D. Higgins, S. Giannini, M. J. Conroy, A. M. Hounslow, J. P. Waltho and R. A. Staniforth. Cystatin forms a tetramer through structural rearrangement of domain-swapped dimers prior to amyloidogenesis. *J Mol Biol* **2004**, *336*, 165–178.

66 Olafsson, I. and A. Grubb. Hereditary cystatin C amyloid angiopathy. *Amyloid: Int J Exp Clin Invest* **2000**, *7*, 70–79.

67 Itoh, Y., M. Yamada, M. Hayakawa, E. Otomo and T. Miyatake. Cerebral amyloid angiopathy: a significant cause of cerebellar as well as lobar cerebral hemorrhage in the elderly. *J Neurol Sci* **1993**, *116*, 135–141.

68 Maruyama, K., S. Ikeda, T. Ishihara, D. Allsop and N. Yanagisawa. Immunohistochemical characterization of cerebrovascular amyloid in 46 autopsied cases using antibodies to beta protein and cystatin C. *Stroke* **1990**, *21*, 397–403.

69 Greenberg, S. M. Cerebral amyloid angiopathy: prospects for clinical diagnosis and treatment. *Neurology* **1998**, *51*, 690–694.

70 Vinters, H. V., G. S. Nishimura, D. L. Secor and W. M. Pardridge. Immunoreactive A4 and gamma-trace peptide colocalization in amyloidotic arteriolar lesions

in brains of patients with Alzheimer's disease. *Am J Pathol* **1990**, *137*, 233–240.

71 Haan, J., M.L. Maat-Schieman, S.G. van Duinen, O. Jensson, L. Thorsteinsson and R.A. Roos. Co-localization of beta/A4 and cystatin C in cortical blood vessels in Dutch, but not in Icelandic hereditary cerebral hemorrhage with amyloidosis. *Acta Neurol Scand* **1994**, *89*, 367–371.

72 Benedikz, E., H. Blondal, G. Johannesson and G. Gudmundsson G. Dementia with non-hereditary cystatin C angiopathy. *Prog Clin Biol Res* **1989**, *317*, 517–522.

73 Maruyama, K., F. Kametani, S. Ikeda, T. Ishihara and N. Yanagisawa. Characterization of amyloid fibril protein from a case of cerebral amyloid angiopathy showing immunohistochemical reactivity for both beta protein and cystatin C. *Neurosci Lett* **1992**, *144*, 38–42.

74 Anders, K.H., et al. Giant cell arthritis in association with cerebral amyloid angiopathy: immunohistochemical and molecular studies. *Hum Pathol* **1997**, *28*, 1237–1246.

75 Nagai, A., et al. No mutations in cystatin C gene in cerebral amyloid angiopathy with cystatin C deposition. *Mol Chem Neuropathol* **1998**, *33*, 63–78.

76 Itoh, Y. and M. Yamada. Cerebral amyloid angiopathy in the elderly: the clinicopathological features, pathogenesis and risk factors. *J Med Dent Sci* **1997**, *44*, 11–19.

77 Sastre, M., et al. Binding of cystatin C to Alzheimer's amyloid β inhibits *in vitro* amyloid fibril formation. *Neurobiol Aging* **2004**, *25*, 1033–1043.

78 Calero, M., et al. Distinct properties of wild-type and the amyloidogenic human cystatin C variant of hereditary cerebral hemorrhage with amyloidosis, Icelandic type. *J Neurochem* **2001**, *77*, 628–637.

79 White, A.R. and S.H. Hawke. Immunotherapy as a therapeutic treatment of neurodegenerative disorders. *J Neurochem* **2003**, *87*, 801–808.

80 Klafki, H.-W., et al. Reduction of disulfide bonds in an amyloidogenic Bence Jones protein leads to formation of "amyloid-like" fibrils *in vitro*. *Biol Chem Hoppe-Seyler* **1993**, *372*, 1117–1122.

81 Cohen, F.E. and S.B. Prusiner. Pathologic conformations of prion proteins. *Annu Rev Biochem* **1998**, *67*, 793–819.

Hormone

28
Endocrine Amyloid

Gunilla T. Westermark

28.1
Nomenclature for Endocrine Amyloid

The following hormones have been accepted by the Nomenclature Committee of the International Society of Amyloidosis as amyloid precursor proteins: atrial natriuretic factor (ANF), calcitonin (Cal), islet amyloid polypeptide (IAPP), insulin (Ins) and prolactin (Pro), and their designations as amyloid proteins are AANF, ACal, AIAPP, AIns and APro [1].

28.2
When and Why do Proteins form Amyloid?

These questions, asked for other amyloid proteins, are also applicable for endocrine amyloid proteins; furthermore, not much is known about the physiologic circumstances under which hormones misfold and form amyloid fibrils. Polypeptide hormones are good candidates for amyloid precursor proteins. They are usually small polypeptides that are synthesized as pro-hormones, which, after post-translational processing, become biological active. Prior to secretion, the hormones are stored at low pH in secretory granules, usually at very high concentrations. The turnover of endocrine granules is regulated by crinophagy, a process where hormone-containing secretory granules not released from the cell by exocytosis are fused with the lysosomes and degraded by proteases [2]. This is a process that could give rise to undesirable, more amyloidogenic fragments. The incidence of endocrine amyloid increases with aging [3–9] and identical de-

Amyloid Proteins. The Beta Sheet Conformation and Disease. J. D. Sipe

ISBN: 3-527-31072-X

posits can also be present in endocrine tumors which often allow an augmented synthesis of the precursor [10–14].

During the last 15 years, islet amyloid has attracted much experimental attention, such that, today, numerous results indicate that occurrence of amyloid derived from IAPP in the islets of Langerhans is of significance for the integrity of β cells and also for the development of Type 2 diabetes [15–17]. Recent published data support a connection between isolated atrial amyloidosis (IAA) derived from ANF and cardiac disease [18–20]. These are essential findings and upcoming research will provide more knowledge about the importance of endocrine amyloid.

Current research on endocrine amyloid involves studies of the consequences of aberrant post-translational processing of the pro-hormone, overexpression of the pro-hormone and degradation of the mature peptide on the amyloidogenic process. Furthermore, characterization of amyloid deposits of unknown origin is expected to add new members to the endocrine amyloid group.

All forms of endocrine amyloid are localized and deposition is restricted to the organ in which the precursor is synthesized. Endocrine amyloid is mainly deposited extracellularly, but intracellular localized amyloid may also be present. This is especially common in the pituitary. At high magnification, the amyloid fibrils can be seen arranged in bundles that, when in contact with the cells, occur in invaginations of the cell membrane (Fig. 28.1). In malignant endocrine tumors, e.g. medullary carcinoma of the thyroid, the amyloid can be present in metastases [21, 22]. The morphological appearance of amyloid after Congo red staining varies – this is especially true for endocrine amyloid. Deposits with an amorphous appearance are present, but more condensed deposits occur where

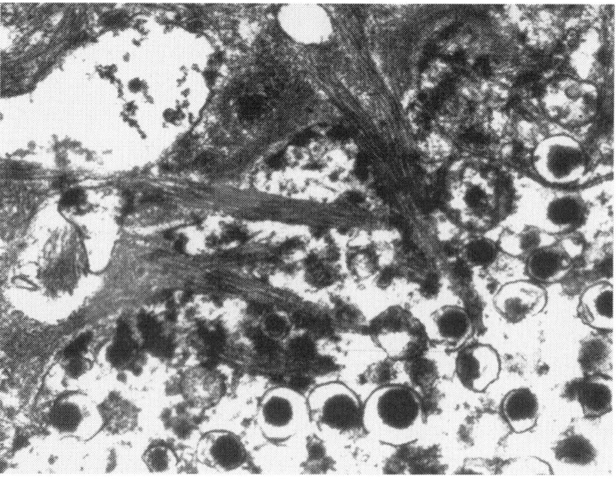

Fig. 28.1 Electron micrograph of IAPP assembled into fibril bundles and deposited extracellularly in invaginations of the β cell.

the fibrils are arranged in such a way that Maltese cross [23–25], star-like, or a defined spheroid formations also develop [24–26]. Both amorphous and condensed forms can be present at the same time. Because the affinity for Congo red and green birefringence remain, the different deposition patterns should represent variation in fibril aggregation and not variation in aggregation of monomers into amyloid fibrils.

Immunohistochemistry, performed with antibodies raised against the native hormone, has been used for characterization of endocrine amyloid. This strategy has not always proven to be successful and false-negative results may result in the case of deposition patterns in which the antigenic epitope is hidden. False-positive results are not uncommon either. Therefore, biochemical analysis of the amyloid protein in combination with immunohistochemistry is a requirement for verifying the origin and chemical identity of endocrine amyloid. New techniques that allow sequencing of formalin-fixed material will be useful for this work in the future.

28.3 Amyloid in Cardiac Atria

28.3.1 Heart as an Endocrine Organ

The heart is not only a pump, but also an endocrine organ. The existence of endocrine granules in the atrial myocytes [27–29], together with the natriuretic and diuretic effects obtained after intravenous injections of extracts from atrial myocytes, pointed towards an endocrine function for the atria [30]. The hormone was isolated, sequenced [31, 32] and named ANF (also known as ANP or A-type natriuretic peptide). ANF was the first to be characterized in a family of peptides that exert natriuretic and diuretic effects, and also smooth muscle relaxant activity. Other members of this family are brain natriuretic factor (BNF; also known as BNP or B-type natriuretic peptide) [33, 34] and cardiac natriuretic factor (CNF; also known as CNP or C-type natriuretic peptide) [35, 36]. The abbreviations most commonly used today are ANP, BNP and CNP. ANP and BNP are both expressed in the heart; expression of ANP during normal conditions is restricted to the atria, while BNP expression occurs in atrial and ventricle myocytes [37–41]. The main source of CNP is brain and endothelial cells, where it is believed to participate in the complicated regulation of blood pressure [42–44].

The three peptides are structurally related and synthesized as pre-pro-hormones. ANF is expressed as a 151-amino-acid pre-pro-peptide from which the signal peptide is enzymatically removed after entering into the endoplasmic reticulum of the myocyte; diarginine residues are removed from the C-terminus of the propeptide by endopeptidase cleavage [31, 32]. The pro-ANP (pro-ANP1–126) is stored in the secretory granules of the atrial myocyte, and is processed into the C-terminal fragment of 28-residue α-ANP and the N-terminal pro-

ANP1–98 at the time of secretion. The processing occurs between residues 98 and 99, catalyzed by the transmembrane serine peptidase Corin [45, 46]. Processing and release of the peptide occurs in response to extension of the myocyte.

BNP is expressed as a 132-residue pre-pro-peptide [34, 47, 48] and, contrary to pro-ANP, the 108-residue pro-peptide is processed in the secretory granules where the 32-residue long mature hormone is stored [49]. The enzyme responsible for BNP processing is not known. Release of BNP from the ventricle myocytes is through the constitutive pathway [50] because, at this location, secretory granules do not reside in myocytes.

CNF has not yet been reported to have any implications in the development of amyloid and is, therefore, excluded from this discussion.

28.3.2
Atrial Amyloid

The aging heart is often affected by amyloid and isolated atrial amyloid (IAA) is one of the more frequently occurring forms of cardiac amyloidosis. As evident from Table 28.1, IAA is an age-related form of amyloidosis. It has never been reported to occur before the age of 20, but, thereafter, the prevalence increases by 10–20% for each decade [3, 4, 6, 51] of life. Women are affected more frequently than men; after 80 years of age, atrial amyloid is present in all women studied and in 89% of men [4]. The amyloid deposition is restricted to the cardiac atria and is most abundant in the atrial appendages. Deposition occurs in both atria, but can, in some cases, be restricted to a single atrium; if so, the left atrium is more often involved than the right atrium [6, 20].

It is significant that two different forms of age-dependent amyloid occur in the heart [51]. One is IAA and the other is senile systemic amyloidosis. The latter is systemic, but cardiac involvement often predominates and the condition is sometimes erroneously called “senile cardiac amyloid”. This form of amyloid is comprised of wild-type transthyretin fibrils. These two types are not distinguished in older literature.

Table 28.1 Occurrence of isolated amyloid in the atrium

No. patients	Percent with amyloid	Age group	Reference
100	91	65–89	3
25	100	>80	
200	51.5	0–90	4
20	95	>80	
39	54	>70	5
167	16	25–60	6

Data are presented as percentage of amyloid in studied subjects. The data published by Kawamura et al. [3] and Steiner [4] is presented for the entire studied material and separately for the age group over 80.

The amyloid deposits in IAA are found in the subendothelial atrial wall and, after Congo red staining, the amyloid is visible as spotty aggregates between the myocytes and is also present perivascularly [52]. Immunohistochemistry with antibodies specific for ANP showed reactivity with the amyloid, suggesting an endocrine origin for this amyloid [53–55].

28.3.3
Isolation and Characterization of Atrial Amyloid

Biochemical characterization of a low-molecular-weight component extracted from amyloid-containing atria, but absent in non-amyloid-containing tissue, revealed an amino acid sequence identical with ANP at position 99–126, α-ANP. Antibodies produced against this low-molecular-weight peptide labeled atrial amyloid and the secretory granules of the myocardial cells. Immunoelectron microscopy with ANP-specific antibodies discloses intracellular amyloid in the myoendocrine cells [55, 56] (Fig. 28.2).

In an immunohistochemical study performed by Pucci et al. [54], antibodies specific for either the N- or C-terminus of pro-ANP were demonstrated to label atrial amyloid. Immunoreactivity with antibodies against the N-terminal pro-ANP fragments in amyloid indicates that amyloid is partly made up by unprocessed ANP1–126. However, the first biochemical characterization of atrial amyloid revealed only the amino acid sequence for α-ANP [56]. This could result

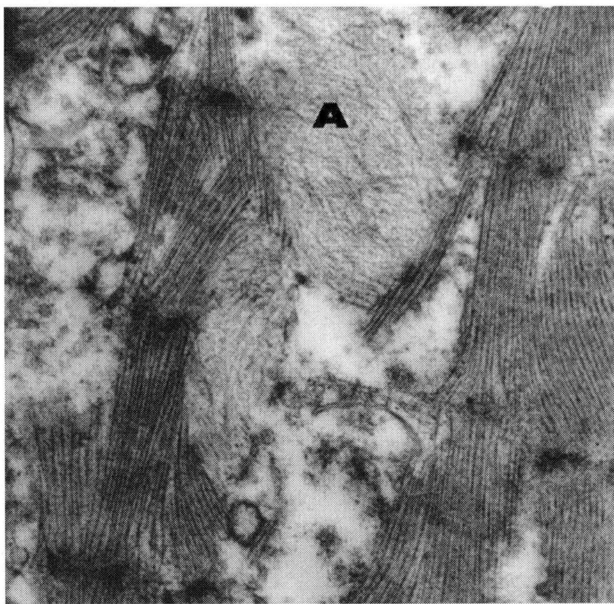

Fig. 28.2 Electron micrograph of intracellular amyloid (A) in an atrial myocyte.

from selection of the low-molecular-weight fraction for sequencing where pro-ANP was present in another fraction. Linke et al. presented data on 7-kDa ANP immunoreactive material, isolated from atrial amyloid deposits [53, 56]. The size of this protein could correspond to pro-ANP. In addition, immunoreactivity of atrial amyloid has been reported with pro-BNP-specific antibodies [54]. The possible presence and co-localization of the pro-ANP and pro-BNP within atrial amyloid will require verification by protein isolation and sequence analyses.

In *in vitro* studies on fibrillogenesis of synthetic α-ANP, it was shown that the peptide has amyloidogenic properties and can assemble into amyloid-like fibrils [52, 57].

28.3.4 Are there Clinical Implications for IAA?

The presence of amyloid deposits in the atrium may have effects on atrial conduction and hinder the secretion of ANP from cells. ANP secretion has been found to be increased during ischemia and acute myocardial infarction, but these conditions are not known to be associated with IAA. Instead, IAA has been reported to be associated with atrial fibrillation [19], a condition that increases the risk for thrombus formation, and with chronic rheumatic heart disease [6, 18, 58]. The incidence of IAA was particularly high in patients undergoing mitral valve replacement [19]. Defects in the mitral valve could result in increased back pressure into the left atrium, and thereby stimulate increased synthesis and release of ANP, which may be an important risk factor for IAA. The prevalence of IAA is also increased in conjunction with cardiac insufficiency [6], a condition also leading to increased stretching of atrial myocytes.

Hitherto, atrial amyloid has exclusively been found in humans and has not been reported to occur in other species, despite the presence of ANP. This could be due to the lack of studies in old animals, but, in a study on hearts tissue sections taken from old rats at the age 48 months, no atrial amyloid was present [59]. Pig ANP is identical to human and rat ANP differs only at one residue (Fig. 28.3). Therefore, it is not impossible that atrial amyloid may develop in other species than human.

```
Human ANF  SLRRSSCFGGRMDRIGAQSGLGCNSFRY
Rat   ANF  SLRRSSCFGGRIDRIGAQSGLGCNSFRY
Pig   ANF  SLRRSSCFGGRMDRIGAQSGLGCNSFRY
```

Fig. 28.3 The amino acid sequences for human, rat and pig ANF are aligned. The sequence for pig ANF is identical to the human sequence, while the rat ANF is substituted at one position.

28.4 Endocrine Amyloid in the Thyroid

The thyroid contains a small population of cells of neuroendocrine nature. These so-called C- or parafollicular cells express the calcitonin/CGRP gene. The gene is comprised of 6 exons that, by alternative splicing, are translated into calcitonin or calcitonin gene-related peptide (CGRP) [60]. Calcitonin is expressed as pre-pro-calcitonin and pro-calcitonin is post-translationally processed at dibasic amino acid positions to yield bioactive peptides. At the N-terminal region, proteolytic cleavage occurs at Lys–Arg and the C-terminal processing site is at Gly–Lys–Lys. The C-terminal amino acid Gly is modified by amidation in the mature peptide, a prerequisite for a biologically active molecule. The enzymes responsible for the processing of pro-calcitonin are not known. Immunohistochemistry with antibodies specific for the processing converting enzymes PC1/3 and PC2 show that both PCs are expressed in the C cells [61]. Both calcitonin and CGRP are processed by proteolytic cleavage at dibasic amino acids and it is possible that these enzymes are active in the C cells.

The nucleotide sequences of two distinct genes encoding calcitonin have been determined, as has the amino acid sequence of the amyloid protein isolated from the medullary carcinoma derived from the CALC-I gene.

28.4.1 Amyloid in Medullary Carcinomas

C cells can be transformed to the malignant tumor medullary carcinoma (MCT). Calcitonin-derived amyloid is very frequently present in this tumor [12, 13, 62] (Table 28.2). Like normal C cells, neoplastic C cells secrete calcitonin and the secretion can be augmented compared to the normal production [63]. Due to the close proximity of the C cells and the amyloid, calcitonin was an early amyloid protein candidate. The first biochemical characterization of MCT amyloid revealed a protein with a molecular mass of 5700 Da [64, 65]. This corresponds to a protein with a size larger than calcitonin, but the partial amino acid sequence was identical to residues 9–19 of human calcitonin. The amino acid composition suggested that the isolated protein was derived from a pro-calcitonin, although, at the time of characterization the existence of pro-calcitonin was

Table 28.2 Occurrence of endocrine amyloid in patients with medullary carcinoma of the thyroid

No. patients	Percent with amyloid		Reference
153	69	sporadic cases	12
50	85	familial cases	
109		all	62
4	100	all	13

not recognized. In a subsequent study, mature full-length calcitonin was identified in the amyloid deposits from another individual [66].

The application of immunohistochemistry with antibodies specific for calcitonin has given contradictory results; some studies showed positive labeling, while in other studies no reactivity of amyloid with the antibodies could be detected [13, 21, 67]. A senile form of calcitonin-derived amyloid in the thyroid gland has not yet been reported.

28.4.2
Can Amyloid be of Benefit?

There are at least two papers presenting studies of prognostic risk factors in patients with MCT. This kind of cancer is rare, but, in the studies taken together, 350 different tumors were analyzed [12, 62]. It was found that the presence of amyloid in the tumor correlated with a more favorable prognosis than non-amyloid containing tumors. In the paper by Bergholm et al. [12], it was shown that the 10-year survival rate in patients with amyloid-containing tumors was 81%, as compared to 49% for patients with tumors without amyloid. There was also a positive correlation with high calcitonin immunoreactivity. Viewed in the light of current thinking, it could be speculated that a high expression level of precursor protein increases the risk for amyloid formation. During assembly, amyloid fibrils or protofibrils may exert cytotoxicity leading to apoptotic effects on tumor cells [68–70]. Amyloid formation may thereby have a protective role by automatically reducing the tumor mass. That would be an unexpected positive side-effect for a substance always considered to be detrimental.

28.5
Amyloid Deposits in the Pituitary

Amyloid deposits in the anterior lobe of the pituitary are almost always found by the age of 80 and above [7–9]. Adenomas at this location also frequently house amyloid deposits [10, 11] (Table 28.3).

Table 28.3 Occurrence of endocrine amyloid in aging pituitary and pituitary adenomas

No. patients	Aging/adenoma	Percent with amyloid	Age	Reference
250	aging	32	0–100	7
300	aging	52	0–100	8
22	aging	96	>80	8
108	aging	80	>85	9
215	adenoma	41		10
48	adenoma	71		11

More often than seen with other types of amyloidosis, pituitary amyloid deposits are characterized by distinct different morphological appearances between cases. The amyloid can appear as an amorphous deposition [71, 72] together with perivascular localization; intracellular amyloid is also present in some cases [23]. Amyloid present in the aging pituitary usually exerts this pattern. Two additional forms of amyloid with defined boundaries have been described to occur in adenomas [24, 25]. The most frequently occurring of these deposition patterns is stellate depositions with stars of variable sizes (10–40 μm). The second, rarer form of deposition, is the dense spherical form, in which the depositions can have a diameter up to 1500 μm [11, 23, 24, 73]. In adenomas, often a combination of vascular and intracellular amyloid deposition is present together with one of these better-defined forms of aggregates. The spherule depositions have been reported to be a feature of some prolactin-secreting adenomas, but others have failed to find this correlation between the morphology of the deposition and type of adenoma [24, 74, 75].

Immunohistochemistry has been performed with many different antibodies raised against pituitary hormones, including prolactin, growth hormone (GH), adrenocorticotropic hormone and thyroid-stimulating hormone. The results thereof are inconsistent and, in most studies, the amyloid was not immunolabeled, despite definite labeling of the surrounding endocrine cells [8, 73, 74]. Stellate-shaped amyloid deposited in GH-producing adenomas has been shown in two different studies to immunolabel with anti-GH antibodies [71, 76]. Before GH can be added to the list of amyloid-forming hormones, the presence of GH has to be verified by biochemical methods such as amino acid sequencing.

28.5.1 Prolactin as an Amyloid Fibril Protein

Biochemical characterization has been performed on amyloid isolated from a prolactin-producing adenoma [75] and from amyloid recovered from pituitaries of old individuals [5]. Amyloid fibril proteins with amino acid sequences starting at position 1 of prolactin and extending to at least residue 32 were identified in these materials. In Western blot analysis of the amyloid, larger prolactin fragments were seen. Amyloid in tissue sections of pituitaries from old individuals was not stained by commercially available anti-prolactin antibodies. Antibodies were therefore raised against three synthetic peptides, corresponding to residues 7–21, 20–34 and 45–57 of prolactin. Interestingly, amyloid was only labeled with antibodies raised against the synthetic peptide corresponding to position 45–57. Taking into account the results of amino acid sequencing, Western blot analysis and the immunoreactivity observed with the antibodies raised against peptide 45–57, the prolactin amyloid likely consists of the first approximately 50 residues of prolactin [5].

The pituitary could be expected to be the site of at least two distinct forms of endocrine amyloid deposition, since amyloid is found in 71% of prolactin-secreting adenomas and in 79% of GH-secreting adenoma [11].

GH and prolactin are synthesized by the somatotropic and lactotropic cells that constitute 50% of the hormone-producing cells in the anterior pituitary. These two hormones have evolved from a single ancestral gene; the human genome contains a cluster of GH genes, in which five GH-like genes, one of which is expressed in the pituitary, have been localized to chromosome 17 [77, 78]. Alternative splicing of the pituitary GH mRNA results in two isoforms differing in molecular weight [79]. There is a single prolactin gene. GH and prolactin show about 16% amino acid sequence homology and have some structural similarities. Both hormones are synthesized as pro-hormones, and the mature size is 191 residues for GH and 199 residues for prolactin. Variants and splicing products from both hormones occur [80].

28.5.2
Prolactin Deposited as Amyloid in the Aged Pituitary

The biochemical characterization of amyloid from the aging pituitary showed that the amyloid consists of prolactin. In contrast to the other hormones secreted from anterior pituitary, prolactin is constantly secreted from the lactotropic cells; the release is dominant negatively regulated by dopamine released from hypothalamus. Dopamine, a prolactin-inhibiting factor (PIT), binds to pituitary D2 receptors on the lactotrophs with the result that synthesis and release of prolactin is inhibited [81]. Studies in young and adult rats showed, in some strains, an age-dependent decrease in dopamine concentration and also aberrant signaling through the D2 receptor [82]. If this is true in humans as well, it could result in an age-related increased synthesis of the prolactin precursor that potentiates amyloid formation.

There was an interesting discovery that treatment of adenoma with bromocriptine, a dopamine agonist, prior to resection increased the amount of amyloid [11]. This may be due to a process in which bromocriptine hinders the release, but not synthesis, of the amyloid fibril precursor [83].

28.6
Endocrine Amyloid in the Islets of Langerhans

The increasing incidence of Type 2 diabetes is a global problem. This form of diabetes has a multifactorial trait including genetic background and lifestyle components, such as physical activity and diet. A large group of these patients are overweight and the disease is spreads following the course of global "westernization". The disease usually develops after the age of 40, but today, with increasing weight and obesity of the younger population, the age of onset is expected to decrease [84, 85]. Type 2 diabetes has two major components of pathology – the initial peripheral insulin resistance with an increasing demand for insulin to maintain the blood glucose level and the β cell "stress" in response to the hyperglycemic condition that can cause what has been called "β cell exhaustion".

28.6.1 IAPP Amyloid and its Putative Role for the Development of Type 2 Diabetes

Islet amyloid was described in 1901 by Opie [86] and is a common finding in the islets of Langerhans at autopsy. Prevalence studies on islet amyloid have shown that the occurrence of amyloid increases with age, but, more importantly, the prevalence of islet amyloid found in subjects with Type 2 diabetes is increased [87–89]. The incidence of islet amyloid varies among studies, but it has been reported to occur in up to 95% of the individuals with Type 2 diabetes (Table 28.4). This is in contrast to below 15% in non-diabetic individuals (Table 28.4). The number of affected islets is higher and the amount of amyloid in each islet is also greater in the diabetic pancreas, as compared to that seen in the non-diabetic organ. The degree of islet amyloid correlates with the severity of the disease in that patients treated with insulin have more islet amyloid as compared with patients able to control their disease with diet and oral anti-diabetic drugs [88]. Islet amyloid is also present in insulin-producing islet tumors [13, 14] and has been reported to occur in more than 50% of insulinomas (Table 28.5).

Electron microscopy studies of islet amyloid revealed a close proximity between amyloid fibrils and the insulin-producing β cells [90, 91]. This was not true for the other islet cells and, therefore, the amyloid was thought to derive from the β cells; for a long time insulin was considered to be the likely amyloid fibril precursor [92]. Thus, it was a surprise when the biochemical characterization of the main protein component of amyloid, isolated from a human insulinoma, disclosed the amino acid sequence of a novel peptide [93, 94]. The obtained peptide was named islet amyloid polypeptide IAPP [93]. Later the same

Table 28.4 Prevalence of islet amyloid in patients with and without Type 2 diabetes

Patients with diabetes		Patients without diabetes		References
No. patients	Percent with islet amyloid	No. patients	Percent with islet amyloid	
27	59	142	12	97
12	100	9	60	15
235	40	533	3	88

Table 28.5 Prevalence of endocrine amyloid in insulinomas

No. patients	Percent tumors with amyloid	Reference
12	58	13
20	65	14

peptide was isolated from amyloid deposited in the pancreata from patients with Type 2 diabetes and referred to by some as amylin [95]. The peptide was also known as diabetes-associated peptide (DAP) for a short period [91].

IAPP amyloid occurs in islets throughout the pancreas, but it is found to a lesser degree in islets residing in the caput region. Studies of islet amyloid have mostly been performed using post-mortem pancreas tissue in which amyloid is deposited extracellularly between β cells and capillaries in the so-called perivascular area [96–98]. The endocrine cellular composition of pancreatic islets is distorted by the amyloid and the number of β cells can be decreased by 40–60% [97, 99]. The amyloid fibril or oligomeric IAPP aggregates formed during the early stage of fibril assembly could be a putative cause for this cell reduction. In addition, islets with amyloid display a proliferation in glucagon-producing α cells [97].

28.6.2 IAPP

In humans, IAPP is principally synthesized in the β cells in the pancreatic islets [100], although extrapancreatic expression occurs in the gastric and intestinal mucosa [101]. In β cells, IAPP is stored together with insulin in the secretory granules [100, 102] and, intragranularly, IAPP is localized to the halo region while insulin comprises the core region [103].

IAPP expression has been more thoroughly studied in rodents, and, in these species, IAPP is produced both by the β cells and the somatostatin-producing δ cells [104]. IAPP immunoreactivity is also present to a higher extent in the gut region of rodents [105] and has also been detected in the nervous system, where extra-endocrine reactivity has been shown in the dorsal root ganglia and peripheral sensory nerves [106].

IAPP is phylogenetically well preserved, and has been detected in the pancreas of all studied mammals [107–112], in chicken islets [113], and in the Brockmann body of the teleostean fishes *Cottus scorpius* and *Salmo salar* [114] (Fig. 28.4). It is notable that antibodies that detect IAPP in bony fish do not detect IAPP in the pancreas from Raya, a cartilage fish.

IAPP is a polypeptide of neuroendocrine origin [115, 116] and it is structurally related to CGRP, calcitonin and adrenomedullin [117–119]. IAPP shares approximately 50% identity with CGRP. Human IAPP is derived from a single gene located on the short arm of chromosome 12 [120]. This region is evolutionarily related to the short arm of chromosome 11 that contains the calcitonin and CGRP genes and the adrenomedullin gene. The human IAPP gene consists of three exons, of which the first is non-coding; exon 2 encodes the 22-residue signal peptide and part of the N-terminal flanking peptide, and exon 3 encodes the remaining part of the 89-amino-acid pre-pro-IAPP molecule [115, 116, 121]. Pro-IAPP (67 residues) undergoes post-translational processing to yield biological activity; this involves enzymatic processing at dibasic amino acid residues where a 12-residue N-terminal flanking peptide and a 15-residue C-terminal flanking peptide are removed, and C-terminal amidation occurs at the tyrosine residue. The mature

```
Human               KCNTATCATQRLANFLVHSSNNFGAILSSTNVGSNTY
Cat                 KCNTATCATQRLANFLIRSSNNLGAILSPTNVGSNTY
Monkey              KCNTATCATQRLANFLVRSSNNFGTILSSTNVGSDTY
Dog                 KCNTATCATQRLANFLVRTSNNLGAILSPTNVGSNTY
Chicken             KCNTATCVTQRLADFLVRSSSNIGAIYSPTNVGSNTY
Degu                KCNTATCATQRLTNFLVRSSHNLGAALPPTKVGSNTY
Mouse               KCNTATCATQRLANFLVRSSNNLGPVLPPTNVGSNTY
Chinese hamster     KCNTATCATQRLANFLVHSNNNLGPVLSPTNVGSNTY
Hedgehog            RCNTATCATQRLVNFLSRSSNNLGAILSPTDVG
Rabbit              KCNTVTCATQRLANFLIHSSNNFGAIFSPPSVG
Hare                       VTQRLANFLIHSSNNFGAIFSPPN
Red kangaroo                TQRLADFLVRSNNNMGAIFSPTNVG
Cottus Scorpius                       SNSTIGTVNAPTNVGS
Cougar              KXNTATXATQRLANFLIRSSXNLGAILSX
Tamarin               NTATCSMHRLADFLGRSSNNFGAILSPTNVGS
Sheep                 GTATCETQRLANFLAPSSNKLGAIFSPTKMGS
Cow                   GTATCETQCLANFLAPSSNKLGAIFSPTKMGS
Pig                    MATCATQHLANFLDRSRNNLGTIFSPTKVGS
```

Fig. 28.4 Comparison of the amino acid sequences for IAPP from different species.

IAPP (37 amino acid residues) holds a disulfide bridge between cysteine residues 2 and 7 [115]. This post-translational processing occurs in the secretory granules at the same location where pro-insulin is processed into insulin by the pro-hormone converting enzymes PC2 and PC1/3. Post-translational processing studies on recombinant and synthetic pro-IAPP showed that PC2 and PC1/3 process pro-IAPP into IAPP [122, 123]. Pro-IAPP processing studies in mouse made deficient for either PC2 or PC1/3 show that the N-terminus is processed by PC2 [124], while the C-terminus is processed by PC1/3 [125]. It was also shown that, in the absence of PC1/3, PC2 could process pro-IAPP into IAPP1–37 [124].

28.6.3 Expression of IAPP

A major cause of the development of islet amyloidosis may be an overexpression of IAPP; therefore, it is important to understand the regulation of IAPP synthesis and release. The co-localization of IAPP and insulin results in a simultaneous release of the peptides after stimulation; therefore glucose and other secretagogs that stimulate insulin secretion will concurrently stimulate IAPP secretion.

The plasma levels of IAPP at fasting state are 2–10 pmol/l [126–128] in humans, while the normal circulating levels of IAPP in rodents are 30–200 pmol/l [129–131]. However, expression and/or translation of IAPP and insulin mRNAs are independently regulated. A decreased secretion of IAPP versus insulin, in response to glucose load, has been determined in both diabetic and non-diabetic humans [132–134], and a relative increase in IAPP secretion versus insulin has been reported in humans after dexamethasone treatment [135–137]. In several animal models, disproportionate ratios of IAPP to insulin occur [136, 138, 139], e.g. in the genetically obese OB/OB mouse strain, insulin was increased 46-fold, while IAPP increased 4-fold over a 45-week period [140]. In NMRI mice fed a high fat diet over a 6-month period, IAPP plasma levels increased 5 times as compared to control animals. Only a moderate insulin increase was observed and this resulted in a change of the IAPP/insulin ratio from 6% in control animals to 20% in the experimental group [141]. Studies in C57BL/6J mice on the effect that high fat diets exert on IAPP and insulin mRNA expression showed that, after 48 weeks on the high fat diet, the IAPP mRNA was unchanged, although insulin mRNA decreased. The change in mRNA expression did not mirror the plasma levels and no increase in IAPP was recorded in this study [142].

The local islet concentration of IAPP is expected to be much higher than in the circulation, but, in isolated β cell granules, the IAPP concentration has been determined to only about 2% of that of insulin [103]. This is in accordance with high-performance liquid chromatography analysis of isolated islets where IAPP concentration was estimated to be 1–2% of insulin [143]. Double immunolabeling of human islets, with antibodies specific for IAPP and insulin, revealed the existence of a small subpopulation of β cells containing 2 times more IAPP than the average cell. Insulin was found be evenly distributed between the cells [144]. It is uncertain whether granules with increased IAPP content have any influence on the circulating IAPP levels, but, in an autocrine or paracrine situation, they could be sufficient for elevating the local IAPP concentration.

Circulating IAPP is eliminated through the kidneys and, in conditions with renal failure, augmented IAPP plasma levels occur [145]. Intracellular insulin degradation is performed by insulin-degrading enzyme (IDE) [146]. This cytosolic metallothiolproteinase degrades other peptides (ANP, glucagon and Aβ) all known to form amyloid or amyloid-like fibrils. It has been shown that IAPP is a substrate for IDE and that this degradation is prevented by specific inhibitors [147].

28.6.4 Biological Activity of IAPP

Over the years, IAPP has been ascribed a large variety of physiological effects, but several of these effects are acquired after administration of pharmacological concentrations of IAPP.

28.6.4.1 Autocrine or Paracrine Effect on the Islet Cells

β Cells IAPP has been shown to act upon the β cells and operate as an inhibitor of basal or glucose-stimulated insulin secretion [148, 149]. The N-terminal truncated IAPP peptide (IAPP8–37) is an inhibitor for IAPP1–37 and, in its presence, insulin secretion is enhanced. That this effect is physiological is supported by the increased glucose clearance that occurs in the IAPP-deficient mouse strain [150].

***α* Cells** IAPP has an inhibitory effect on the α cells and suppresses glucagon secretion [151, 152]. *In vivo*, an opposite correlation between IAPP and glucagon secretion has been shown in both healthy and Type 2 diabetic individuals [153]. Thus, the reduced numbers of β cells and proliferation of α cells described in the case of amyloid-containing islets can result in a situation where low levels of IAPP are insufficient for inhibition of glucagon and be a risk factor for development of the hyperglucagonemia present in Type 2 diabetes.

***δ* Cells** IAPP has a negative effect on somatostatin release and, in studies on isolated rat islets, exogenous IAPP leads to decreased somatostatin secretion in a concentration-dependent manner [154]. These results are in accordance with the effect obtained after immuno-neutralization of IAPP, where arginine-stimulated somatostatin secretion was enhanced from incubated islet [155].

28.6.4.2 Calcium Metabolism

It has been suggested that IAPP plays a role in the regulation of calcium metabolism, but it has been difficult to provide evidence for this as many contradictory results have been published. IAPP has been shown to lower plasma calcium levels by inhibition of osteoclasts [156–158] or by stimulation of osteoblast proliferation [159]. In a recent publication of a study performed with IAPP-deficient mice, the animals were found to have a lower bone mass as a result of increased bone resorption [160]. IAPP may exert its effect through activation of the calcitonin receptor on the osteoclasts and thereby regulate osteoclastogenesis [161, 162].

28.6.4.3 IAPP and Satiety

IAPP can affect feeding behavior and can act as an anorectic substance [163, 164]. The effect is achieved after either peripheral or local administration of IAPP directly into the brain. The effect is present after both acute and chronic IAPP infusion, and operates by reducing the size of meal taken [165]. Specific receptors for IAPP and CGRP are present in the area postrema/nucleus of the solitary tract (AP/NTS) [166] and damage in this area reduces the anorectic effect of IAPP. Specific blockage of the receptors with the IAPP antagonist 8–32 results in an increased food intake [167]. In many of these studies the effect has been achieved after administration of IAPP at concentrations corresponding to or close to physiological levels.

IAPP is also active on gastric mobility and lowers gastric emptying, thereby decreasing glucose influx [168]. This effect is interesting and a putative target for a drug which could be used for weight reduction [169].

28.6.5
Amyloidogenic Properties of the IAPP Molecule

Human IAPP is one of the most amyloidogenic proteins that exist. It readily forms amyloid-like fibrils even in highly chaotropic environments such as 5 M guanidine hydrochloride, a feature that causes problems in designing studies of the folding mechanisms of the polypeptide. When the amino acid sequence for the amyloidogenic human IAPP is compared to the non-amyloidogenic rat/mouse IAPP sequence, a high degree of identity between the molecules is seen, but, relative to the human IAPP1–37 sequence, Gly20 and Ser28 are substituted with prolines in the rat/mouse molecule. Secondary structure prediction of human IAPP indicated a β-sheet structure in the region 20–29 of the molecule [170, 171] and synthetic peptides corresponding to the sequence 20–29 of human IAPP assemble readily into amyloid-like fibrils [171]. The corresponding rat/mouse sequence does not form amyloid. The fibril-forming capacity of the human peptide can be changed by the introduction of proline into positions 24 and 28. Particularly, proline at position 28 is believed to be of importance as an inhibitor of amyloid fibril formation.

The peptide GAILS that corresponds to position 24–28 of human IAPP forms amyloid-like fibrils [172, 173]. Fibrils have also been formed *in vitro* from peptides corresponding to other regions of IAPP. The sequences corresponding to residues 1–20 of human and rodent IAPP both form fibrils [174]. This region contains a single amino acid difference, with His in the human sequence replaced by Arg in rodents. At the C-terminal part of the molecule (residues 30–37) the sequence is identical in human and rat; a peptide with this primary structure has been shown to form amyloid-like fibrils [174]. Still, *in vivo*, rodent IAPP does not aggregate and assemble into amyloid fibrils, indicating that the important region for fibril formation is within the 20–29 segment.

Different model systems for IAPP amyloid fibril formation have been established; in one model, human IAPP is separated into three β-strands (positions 8–20, 24–29 and 32–37), which form intramolecular sheet structures when the molecule is folded. Amyloid fibrils are formed when these monomers are assembled either in a parallel or antiparallel fashion. The introduction of a proline residue at position 28 in this model would disrupt the β-strand structures and obstruct the intermolecular strand formation [175].

Aggregation into amyloid is considered to be a pathological event and is not believed to be included in IAPP's normal function. During the early phase of Type 2 diabetes, an increased release of pro-insulin and aberrantly processed pro-insulin takes place [176, 177]. Because pro-IAPP and pro-insulin are processed by the same enzymes at the same location, pro-IAPP is expected to undergo a similar abnormal processing. This would give rise to increased concentrations of pro-IAPP and pro-IAPP intermediates. As mentioned above, IAPP aggregates and forms amyloid-like fibrils spontaneously. *In vitro* studies have shown that insulin can act as a potent inhibitor and prevent fibril formation [103, 178]. The activity is exerted in a concentration-dependent manner. If this mechanism operates in the secretory granules, any changes in IAPP and insulin ratio might trigger intracellular amyloid formation.

Several mutations have been described for the human IAPP gene. One of these mutations is seen within the coding region for the mature IAPP and results in an amino acid substitution at position 20 where Ser is substituted with Gly. This mutation has only been reported to occur in Asian populations and it is linked to an increased risk for the development of diabetes [179–181]. *In vitro* studies on IAPP S20G show that this molecule is more fibrillogenic compared to wild-type IAPP [182]. When the mutation is included into the IAPP fibril model suggested by Jaikaran et al. [174], the glycine residue at position 20 should result in a more flexible molecule due to the larger size of serine compared to glycine.

The IAPP sequences of monkey, cat and dog are similar to the human IAPP sequence, and all lack proline residues at position 24 and 28. Cat and monkey are two species that develop islet amyloid in association with a form of diabetes that resembles human Type 2 diabetes [183–185]. IAPP-derived islet amyloid also occurs in raccoon, but knowledge of the diabetes in these animals is limited. Interestingly, IAPP amyloid does not develop in dogs, in spite of the presence of an amyloidogenic IAPP sequence. IAPP amyloid does, however, occur in canine islet cell tumors [109].

28.6.6 Pathogenesis of Islet Amyloid and Cellular Effects of Aggregated IAPP

The amyloid-containing tissue material collected post-mortem most often derives from subjects with long-standing diabetes with advanced amyloid deposits and severe cell destruction. Biochemical characterization of islet amyloid has revealed only IAPP1–37, but immunolabeling with antibodies specific for the flanking peptides has shown the presence of pro-IAPP within the amyloid deposits [186, 187]. When tissue material with better morphology than usually obtained in autopsy specimens was studied, e.g. insulinomas recovered at surgery, intracellular amyloid was seen together with the extracellular deposition [14]. Similarly, normal human islets, transplanted into nude mice, rapidly develop IAPP amyloid that is initially intracellular [188].

Human pro-IAPP (recombinant and synthetic) has been shown to be amyloidogenic and to aggregate into amyloid-like fibrils [189]. Expression of pro-IAPP in Cos-1 cells, a cell line that lacks processing enzymes and, thus, does not processes pro-IAPP into IAPP, results in the accumulation of intracellular amyloid [190]. To further investigate the effect of processing, we have expressed human pro-IAPP in different cell lines: β cells that allow processing of pro-IAPP into IAPP, AtT-20 cells that lack PC1/3 and will only process pro-IAPP at the C-terminal processing site, and GH4 cells, a cell line deficient of both processing enzymes. The results of the expression show that complete processing of pro-IAPP into IAPP, which occurs in β cells, prevents amyloid formation, while intracellular amyloid develops in cells in which incomplete or no processing of pro-IAPP occurs.

All of these findings, taken together, may indicate that the initial IAPP amyloid assembly takes place intracellularly from pro-IAPP. This first amyloid may lead to cell death and thereafter act as a template for fibril elongation, now from mature IAPP.

Histological examination of pancreatic tissue with islet amyloid does not indicate any immunological response to the amyloid or any signs of severe cell death. In studies of fibril toxicity, amyloid-like fibrils produced from human IAPP1–37 applied to "normal" β cells isolated from rat or human islets induced cell death. It was shown in that system that cell death was accompanied by apoptotic signs such as DNA condensation and fragmentation [68]. Application of amyloid-like IAPP fibrils to RINm5F and HIT-T15 cells, two β cell-derived cell lines, led to apoptosis of 100% of the RIN cells within 24 h, while HIT cells remained unaffected [191]. This result shows that cells can have different reaction patterns to amyloid-like fibrils; the basis for the observed differences could be that HIT cells are resistant to free radical formation. Further studies have shown that amyloid peptides can be incorporated into lipid layers and at this location create ion-leaking pores [69]. During assembly into mature amyloid fibrils, various intermediates appear and these can be more toxic to cells than the fully grown fibrils [70, 192]. Thus, "protofibrils" (immature fibrils) consisting of oligomers possess a higher degree of toxicity than the aged, fully assembled amyloid fibrils. These results are to some degree supported by the electron microscopic appearance of human islets where amyloid sometimes is present in very deep invaginations of the plasma membrane without being associated with apoptotic signs.

28.6.7 **Transgenic Animals**

The lack of good animal models useful for islet amyloid studies was overcome by establishing transgenic mice expressing the gene for human IAPP. At the beginning of the 1990s, such mouse strains were generated by multiple groups. The strategy differed and various transgenes were used containing the human IAPP gene driven by the rat insulin I [193, 194] or II [195] promoter. Other constructs contained the rat insulin II promoter with a cDNA fragment containing the entire coding sequence for human IAPP [196]. One strain expressed the human IAPP gene driven by the human promoter [197]. The expression of the human IAPP gene varied between the different established strains, but, on the whole, islet amyloid occurred in only one of these strains. However, when isolated islets derived from transgenic animals were cultured *in vitro* for 1–2 weeks, intracellular amyloid could be detected and the amount of amyloid corresponded to the glucose concentration. The occurrence of islet amyloid in transgenic mice increased when the human IAPP expression was introduced into animals with diabetic traits [198, 199]. It was interesting that amyloid appeared in transgenic animals fed a diet high in fat [200]. In one strain, expressing the gene for human IAPP but deficient in endogenous IAPP, islet amyloid occurred subsequent to increased fat intake, and was found intra- and extracellularly [201]. The β cells affected by amyloid had pyknotic nuclei and the intracellular amyloid immunolabeled with antibodies specific for the processing regions of pro-IAPP [202]. This, again, indicates that the initial amyloid is intracellular and contains pro-IAPP.

28.7
Insulin as an Amyloid-forming Protein

The capacity of insulin to form fibrils has been recognized for very long time [203, 204]. Such fibrillar metamorphosis demands repeated heating and freezing of an acidic insulin solution. The similarity of these synthetic fibrils with naturally occurring amyloid fibrils was noted by Glenner et al. [92], and Westermark [205] and Glenner used these synthetic fibrils to resolve the amyloid fibril structure.

As a rare event, insulin-derived amyloid has been iatrogenically induced at the site of injection in diabetic patients [206]. These depositions, containing complete insulin molecules, have not been shown to have any clinical implications. By continuous subcutaneous insulin infusion in Wistar rats, localized insulin amyloid deposits were apparent by as early as 6 weeks [207]. Interestingly, this procedure did not include harsh manipulations such as heating, freezing or acidification used for the *in vitro* production of insulin-like amyloid fibrils. It is unknown whether the injected or infused insulin preparations contained preformed fibrils or protofibrils capable as acting as seed, and thereby induce and speed up the amyloid formation.

In contrast to what is seen in humans, the islet amyloid in the new world hystricomorph rodent degu (*Octodon degus*) derives from endogenously produced insulin [208]. The amino acid sequence of degu insulin differs significantly from that of human insulin and from insulin of most other rodents. The degu insulin A-chain contains two additional amino acid residues, and the sequence differs at 8 and 9 residues compared with rat and human, respectively. The B-chain lacks a residue and consists of 29 instead of 30 residues. The B-chain sequence differs at 11 and 10 positions as compared to rat and human, respectively. The non-amyloidogenic degu IAPP amino acid sequence contains proline at position 28, which has been suggested to be of great importance for the prevention of IAPP amyloid formation, and as seen in the non-amyloidogenic sequence found in rat and mouse [209]. Most importantly, insulin-derived islet amyloidosis in the degu is associated with diabetes.

28.8
Can Other Islet Hormones Aggregate and Form Amyloid?

Antiserum specific for somatostatin provided evidence of immunolabeled amyloid deposited in a somatostatin-producing tumor [210]. Therefore, somatostatin has been suggested to be an amyloid fibril protein. However, biochemical characterization remains to be performed.

Islet amyloid has also been described to occur in a glucagon-producing tumor, but no report on the nature of this amyloid protein exists. *In vitro*, glucagon can form amyloid-like fibrils [92], and amyloid fibrils made up by glucagon exert toxic effects on PC12 cells by the activation of caspase-3 and thus induce cell death [211].

28.9
Other Amyloids with Possible Endocrine Origin

28.9.1
Parathyroid Gland

Amyloid arises in the parathyroid gland hyperplasia and in adenomas with hyperparathyroidism, and also occurs in association with aging [212]. The reported incidence of amyloid varies; Andersson and Ewen [213] reported a prevalence of 16% in a group of glands from individuals with clinically diagnosed hyperparathyroidism and 46% in parathyroid glands in a semi-non-selected autopsy material; Iwata et al. [214] investigated glands from 128 post-mortem cases and found amyloid in 41% of them. In the same material, the incidence of amyloid in the age group above 70 years was 60% and amyloid was not found to be present before the age of 30. Parathyroid glands are made up of chief (or principal) cells and oxyphil cells. The chief cells secrete parathyroid hormone (PTH) that regulates the Ca^{2+} and PO_4^{3-} balance in the blood. PTH is synthesized as a 115-residue pre-pro-hormone and, after the removal of the signal peptide, the 90-residue pro-hormone is processed into the 84-residue PTH. The pro-hormone converting enzyme 7 (PC7) [215], a recently discovered processing enzyme, and furin are both expressed by the chief cells, and have been shown to be able to process pro-PTH into PTH. PC7 and furin are both active at neutral pH and process peptides in the *trans*-Golgi compartment. The biological activity of PTH resides in the N-terminus of the molecule and PTH undergoes fragmentation at position 33 or 26 in the secretory granules to yield an N- and C-terminal fragments.

There are no publications on the biochemical nature of parathyroid amyloid, but PTH is a natural candidate since it is synthesized in the gland and occurs there at high concentration. PTH can aggregate and form amyloid-like fibrils that stain with Congo red and reveal green birefringence when viewed in polarized light [216]. Immunohistochemistry with antibodies against PTH on an amyloid-producing parathyroid tumor showed reactivity with both chief and oxyphil cells, while staining for calcitonin and thyroglobulin was negative [217]. In this publication the authors did not state whether the amyloid labeled with the antibodies. Thus, this amyloid still has to be characterized.

Acknowledgments

Supported by the Swedish Research Council.

References

1 Westermark P, Benson M, Buxbaum J, Cohen A, Frangione B, Ikeda S, Masters C, Merlini G, Saraiva M, Sipe J. Amyloid fibril protein nomenclature – 2002. *Amyloid* **2002**, *9*, 197–200.

2 Schnell A, Swenne I, Borg L. Lysosomes and pancreatic islet function. A quantitative estimation of crinophagy in the mouse pancreatic B-cell. *Cell Tissue Res* **1988**, *252*, 9–15.

3 Kawamura S, Takahashi M, Ishihara T, Uchino F. Incidence and distribution of isolated atrial amyloid: histologic and immunohistochemical studies of 100 aging hearts. *Pathol Int* **1995**, *45*, 335–342.

4 Steiner I. The prevalence of isolated atrial amyloid. *J Pathol* **1987**, *153*, 395–398.

5 Westermark P, Eriksson L, Engström U, Eneström S, Sletten K. Prolactin-derived amyloid in the aging pituitary gland. *Am J Pathol* **1997**, *150*, 67–73.

6 Looi L. Isolated atrial amyloidosis: a clinicopathologic study indicating increased prevalence in chronic heart disease. *Hum Pathol* **1993**, *24*, 602–607.

7 Saeger W, Warner R, Missmal H. Amyloidosen der Hypophyse im Sektionsgut. *Pathologe* **1983**, *4*, 177–182.

8 Tashima T, Kitamoto T, Tateishi J, Ogomori K, Nakagaki H. Incidence and characterization of age related amyloid deposits in the human anterior pituitary gland. *Virchows Arch A Path Anat* **1988**, *412*, 323–327.

9 Röcken C, Eick B, Saeger W. Senile amyloidoses of the pituitary and adrenal glands. Morphological and statistical investigations. *Virchows Arch* **1996**, *429*, 293–299.

10 Saeger W, Gerigk C, Missmahl P, Ludecke D. Amyloid deposits in hypophyseal adenomas. Polarization optic, immunohistologic and electron microscopy studies. *Pathologe* **1983**, *4*, 183–189.

11 Saitoh Y, Mori H, Matsumoto K, Ushio Y, Hayakawa T, Mori S, Arita N, Mogami H. Accumulation of amyloid in pituitary adenomas. *Acta Neuropathol* **1985**, *68*, 87–92.

12 Bergholm U, Adami HO, Auer G, Bergström R, Backdahl M, Grimelius L, Hansson G, Ljungberg O, Wilander E. Histopathologic characteristics and nuclear DNA content as prognostic factors in medullary thyroid carcinoma. A nationwide study in Sweden. The Swedish MTC Study Group. *Cancer* **1989**, *64*, 135–142.

13 Westermark P, Grimelius L, Polak JM, Larsson LI, Van Noorden S, Wilander E, Pearse AG. Amyloid in polypeptide hormone-producing tumors. *Lab Invest* **1977**, *37*, 212–215.

14 O'Brien T, Butler A, Roche P, Johnson K, Butler PC. Islet amyloid polypeptide in human insulinomas. Evidence for intracellular amyloidogenesis. *Diabetes* **1994**, *42*, 329–336.

15 Westermark P, Grimelius L. The pancreatic islet cells in insular amyloidosis in human diabetic and non-diabetic adults. *Acta Pathol Microbiol Scand* **1973**, *81*, 291–300.

16 Clark A, Charge SB, Badman MK, MacArthur DA, de Koning EJ. Islet amyloid polypeptide: actions and role in the

pathogenesis of diabetes. *Biochem Soc Trans* **1996**, *24*, 594–599.

17 Höppener JWM, Nieuwenhuis MG, Vroom TM, Ahren B, Lips CJM. Role of islet amyloid in type 2 diabetes mellitus: consequence or cause? *Mol Cell Endocrinol* **2002**, *197*, 205–212.

18 Johansson B, Westermark P: Isolated atrial amyloidosis. Increased frequency in patients with congestive cardiac failure. In *Amyloid and Amyloidosis*, J Natvig, Ö Förre, G Husby, A Husebeck, B Skogen, K Sletten, P Westermark (eds). Kluwer, Dordrecht, **1990**, pp. 473–476.

19 Röcken C, Peters B, Juenemann G, Saeger W, Klein HU, Huth C, Roessner A, Goette A. Atrial amyloidosis: an arrhythmogenic substrate for persistent atrial fibrillation. *Circulation* **2002**, *106*, 2091–2097.

20 Leone O, Boriani G, Chiappini B, Pacini D, Cenacchi G, Martin Suarez S, Rapezzi C, Bacchi Reggiani ML, Marinelli G. Amyloid deposition as a cause of atrial remodelling in persistent valvular atrial fibrillation. *Eur Heart J* **2004**, *25*, 1237–1241.

21 Butler M, Khan S. Immunoreactive calcitonin in amyloid fibrils of medullary carcinoma of the thyroid gland. An immunogold staining technique. *Arch Pathol Lab Med* **1986**, *110*, 647–649.

22 Schröder S, Bocker W, Baisch H, Burk C, Arps H, Meiners I, Kastendieck H, Heitz P, Kloppel G. Prognostic factors in medullary thyroid carcinomas. Survival in relation to age, sex, stage, histology, immunocytochemistry, and DNA content. *Cancer* **1988**, *61*, 806–816.

23 Kubota T, Kuroda E, Yamashima T, Tachibana O, Kabuto M, Yamamoto S. Amyloid formation in prolactinoma. *Arch Pathol Lab Med* **1986**, *110*, 72–75.

24 Landolt AM, Kleihues P, Heitz PU. Amyloid deposits in pituitary adenomas. Differentiation of two types. *Arch Pathol Lab Med* **1987**, *111*, 453–458.

25 Landolt AM, Heitz PU. Differentiation of two types of amyloid occurring in pituitary adenomas. *Pathol Res Pract* **1988**, *183*, 552–554.

26 Frankel K. Amyloid spherules in prolactinoma. *Acta Cytol* **2000**, *44*, 276–277.

27 Cantin M, Gutkowska J, Thibault G, Milne R, Ledoux S, MinLi S, Chapeau C, Garcia R, Hamet P, Genest J. Immunocytochemical localization of atrial natriuretic factor in the heart and salivary glands. *Histochemistry* **1984**, *80*, 113–127.

28 Forssmann W, Birr C, Carlquist M, Cristmann M, Finke R, Henschen A, Hock D, Kirchheim H, Kreye V, Lottspeich Fea. The auricular myocardiocytes of the heart constitute an endocrine organ. Characterization of a porcine cardiac peptide hormone, cardiodilatin-126. *Cell Tissue Res* **1984**, *236*, 425–430.

29 Maldonado C, Saggau W, Forssmann W. Cardiodilatin-immunoreactivity in specific atrial granules of human heart revealed by the immunogold stain. *Anat Embryol* **1986**, *173*, 295–298.

30 de Bold A. Heart atria granularity effects of changes in water-electrolyte balance. *Proc Soc Exp Biol Med* **1979**, *161*, 508–511.

31 Oikawa S, Imai M, Ueno A, Tanaka S, Noguchi T, Nakazato H, Kangawa K, Fukuda A, Matsuo H. Cloning and sequence analysis of cDNA encoding a precursor for human atrial natriuretic polypeptide. *Nature* **1984**, *309*, 724–726.

32 Nakayama K, Ohkubo H, Hirose T, Inayama S, Nakanishi S. mRNA sequence for human cardiodilatin-atrial natriuretic factor precursor and regulation of precursor mRNA in rat atria. *Nature* **1984**, *310*, 699–701.

33 Seilhamer J, Arfsten A, Miller J, Lundquist P, Scarborough R, Lewicki J, Porter J. Human and canine gene homologs of porcine brain natriuretic peptide. *Biochem Biophys Res Commun* **1989**, *165*, 650–658.

34 Sudoh T, Maekawa K, Kojima M, Minamino N, K. K, Matsuo H. Cloning and sequence analysis of cDNA encoding a precursor for human brain natriuretic peptide. *Biochem Biophys Res Commun* **1989**, *159*, 1427–1434.

35 Tawaragi Y, Fuchimura K, Tanaka S, Minamino N, Kangawa K, Matsuo H. Gene and precursor structures of human C-type natriuretic peptide. *Biochem Biophys Res Commun* **1991**, *175*, 645–651.

36 Ogawa T, Vatta M, Bruneau B, de Bold A. Characterization of natriuretic peptide production by adult heart atria. *Am J*

Physiol Heart Circ Physiol **1999**, *276*, 1977–1986.

37 Gardner D, Hedges B, Wu J, LaPointe M, Deschepper C. Expression of the atrial natriuretic peptide gene in human fetal heart. *J Clin Endo Metab* **1989**, *69*, 729–737.

38 Takemura G, Fujiwara H, Yoshida H, Mukoyama M, Saito Y, Nakao K, Fujiwara T, Uegaito T, Imura H, Kawai C. Identification and distribution of atrial natriuretic polypeptide in ventricular myocardium of humans with myocardial infarction. *J Pathol* **1990**, *161*, 285–292.

39 Dagnino L, Drouin J, Nemer M. Differential expression of natriuretic peptide genes in cardiac and extracardiac tissues. *Mol Endocrinol* **1991**, *5*, 1292–1300.

40 Benvenuti L, Aiello V, Higuchi Mde L, Palomino S. Immunohistochemical expression of atrial natriuretic peptide (ANP) in the conducting system and internodal atrial myocardium of human hearts. *Acta Histochem* **1997**, *99*, 187–193.

41 Doyama K, Fukumoto M, Takemura G, Tanaka M, Oda T, Hasegawa K, Inada T, Ohtani S, Fujiwara T, Itoh H. Expression and distribution of brain natriuretic peptide in human right atria. *J Am Coll Cardiol* **1998**, *32*, 1832–1838.

42 Ueda S, Minamino N, Aburaya M, Kangawa K, Matsukura S, Matsuo H. Distribution and characterization of immunoreactive porcine C-type natriuretic peptide. *Biochem Biophys Res Commun* **1991**, *175*, 759–767.

43 Horio T, Tokudome T, Maki T, Yoshihara F, Suga S-i, Nishikimi T, Kojima M, Kawano Y, Kangawa K. Gene expression, secretion, and autocrine action of C-type natriuretic peptide in cultured adult rat cardiac fibroblasts. *Endocrinology* **2003**, *144*, 2279–2284.

44 Pelisek J, Kuhnl A, Rolland PH, Mekkaoui C, Fuchs A, Walker G, Ogris M, Wagner E, Nikol S. Functional analysis of genomic DNA, cDNA, and nucleotide sequence of the mature C-type natriuretic peptide gene in vascular cells. *Arterioscler Thromb Vasc Biol* **2004**, *24*, 1646–1651.

45 Yan W, Sheng N, Seto M, Morser J, Wu Q. Corin, a mosaic transmembrane serine protease encoded by a novel cDNA from human heart. *J Biol Chem* **1999**, *274*, 14926–14935.

46 Wu F, Yan W, Pan J, Morser J, Wu Q. Processing of pro-atrial natriuretic peptide by corin in cardiac myocytes. *J Biol Chem* **2002**, *277*, 16900–16905.

47 Aburaya M, Hino J, Minamino N, Kangawa K, Maatsuo H. Isolation and identification of rat brain natriuretic peptides in cardiac atrium. *Biochem Biophys Res Commun* **1989**, *163*, 226–232.

48 Saito Y, Nakao K, Itoh H, Yamada T, Mukoyama M, Arai H, Hosoda K, Shirakami G, Suga S, Minamino N. Brain natriuretic peptide is a novel cardiac hormone. *Biochem Biophys Res Commun* **1988**, *158*, 360–368.

49 Hino J, Tateyama H, Minamino N, Kangawa K, Matsuo H. Isolation and identification of human brain natriuretic peptides in cardiac atrium. *Biochem Biophys Res Commun* **1990**, *167*, 693–700.

50 Bloch K, Seidman J, Naftilan J, Fallon J, Seidman C. Neonatal atria and ventricles secrete atrial natriuretic factor via tissue-specific secretory pathways. *Cell* **1986**, *47*, 695–702.

51 Westermark P, Johansson B, Natvig JB. Senile cardiac amyloidosis: evidence of two different amyloid substances in the ageing heart. *Scand J Immunol* **1979**, *10*, 303–308.

52 Johansson B, Westermark P. The relation of atrial natriuretic factor to isolated atrial amyloid. *Exp Mol Pathol* **1990**, *52*, 266–278.

53 Linke R, Voigt C, Storkel F, Eulitz M. N-terminal amino acid sequence analysis indicates that isolated atrial amyloid is derived from atrial natriuretic peptide. *Virchows Arch B Cell Pathol Mol Pathol* **1988**, *55*, 125–127.

54 Pucci A, Wharton J, Arbustini E, Grasso M, Diegoli M, Needleman P, Vigano M, Polak J. Atrial amyloid deposits in the failing human heart display both atrial and brain natriuretic peptide-like immunoreactivity. *J Pathol* **1991**, *165*, 235–241.

55 Takahashi M, Hoshii Y, Kawano H, Gondo T, Yokota T, Okabayashi H, Shimada I, Ishihara T. Ultrastructural evidence for the formation of amyloid fibrils within cardiomyocytes in isolated atrial amyloid. *Amyloid* **1998**, *5*, 35–42.

56 Johansson B, Wernstedt C, Westermark P. Atrial natriuretic peptide deposited as atrial amyloid fibrils. *Biochem Biophys Res Commun* **1987**, *148*, 1087–1092.

57 Maioli E, Torricelli C, Santucci A, Pacini A. Molecular assembly of endogenous and synthetic big atrial natriuretic peptide (ANP) and its amyloidogenic implications. *Biochim Biophys Acta* **2000**, *1500*, 31–40.

58 Looi L. Raised prevalence of isolated atrial amyloidosis in chronic heart disease. In *Amyloid and Amyloidosis*, J Natvig, Ö Förre, G Husby, A Husebeck, B Skogen, K Sletten, P Westermark (eds). Kluwer, Dordrecht, **1990**, pp. 470–473.

59 Cornwell GG III, Thomas B, Snyder D. Myocardial fibrosis in aging germ-free and conventional Lobund-Wistar rats: the protective effect of diet restriction. *J Gerontology* **1991**, *46*, 167–170.

60 Steenbergh P, Hoppener J, Zandberg J, Visser A, Lips C, Jansz H. Structure and expression of the human calcitonin/ CGRP gene. *FEBS Lett* **1986**, *209*, 97–103.

61 Kurabuchi S, Tanaka S. Immunocytochemical localization of prohormone convertases PC1 and PC2 in the mouse thyroid gland and respiratory tract. *J Histochem Cytochem* **2002**, *50*, 903–910.

62 Scopsi L, Sampietro G, Boracchi P, Del Bo R, Gullo M, Placucci M, Pilotti S. Multivariate analysis of prognostic factors in sporadic medullary carcinoma of the thyroid. A retrospective study of 109 consecutive patients. *Cancer* **1996**, *78*, 2173–2183.

63 Emmertsen K. Medullary thyroid carcinoma and calcitonin. *Danish Med Bull* **1985**, *32*, 1–28.

64 Westermark P. Amyloid of medullary carcinoma of the thyroid; partial characterization. *Uppsala J Med Sci* **1975**, *80*, 88–92.

65 Sletten K, Westermark P, Natvig JB. Characterization of amyloid fibril proteins from medullary carcinoma of the thyroid. *J Exp Med* **1976**, *143*, 993–998.

66 Khurana R, Agarwal A, Bajpai VK, Verma N, Sharma AK, Gupta RP, Madhusudan KP. Unraveling the amyloid associated with human medullary thyroid carcinoma. *Endocrinology* **2004**, *145*, 5465–5470.

67 Berger G, Berger N, Guillaud M, Trouillas J, Vauzelle J. Calcitonin-like immunoreactivity of amyloid fibrils in medullary thyroid carcinomas. An immunoelectron microscope study. *Virchows Arch A Pathol Anat* **1988**, *412*, 543–551.

68 Lorenzo A, Razzaboni B, Weir G, Yankner B. Pancreatic islet cell toxicity of amylin associated with type-2 diabetes mellitus. *Nature* **1994**, *368*, 756–760.

69 Mirzabekov TA, Lin M-c, Kagan BL. Pore formation by the cytotoxic islet amyloid peptide amylin. *J Biol Chem* **1996**, *271*, 1988–1992.

70 Qahwash I, Weiland KL, Lu Y, Sarver RW, Kletzien RF, Yan R. Identification of a mutant amyloid peptide that predominantly forms neurotoxic protofibrillar aggregates. *J Biol Chem* **2003**, *278*, 23187–23195.

71 Osamura R, Watanabe K, Komatsu N, Ohya M, Kageyama N. Amorphous and stellate amyloid in functioning human pituitary adenomas: histochemical, immunohistochemical and electron microscopic studies. *Acta Pathol Jpn* **1982**, *32*, 605–611.

72 Canda T, Sengiz S, Canda MS, Acar UD, Erbayraktar RS, Yilmaz HS. Histochemical and immunohistochemical features of a case showing association of meningioma and prolactinoma containing amyloid. *Brain Tumor Pathol* **2002**, *19*, 1–3.

73 Paetau A, Partanen S, Mustajoki P, Valtonen S, Pelkonen R, Wahlstrom T. Prolactinoma of the pituitary containing amyloid. *Acta Endocrinol* **1985**, *109*, 176–180.

74 Bononi P, Martinez A, Nelson P, Amico J. Amyloid deposits in a prolactin-producing pituitary adenoma. *J Endocrinol Invest* **1993**, *16*, 339–343.

75 Hinton DR, Polk RK, Linse KD, Weiss MH, Kovacs K, Garner JA. Characterization of spherical amyloid protein from a prolactin-producing pituitary adenoma. *Acta Neuropathol* **1997**, *93*, 43–49.

76 Mori H, Mori S, Saitoh Y, Moriwaki K, Lida S, Matsumoto K. Growth hormone-producing pituitary adenoma with crystal-like amyloid immunohistochemically

positive for growth hormone. *Cancer* **1985**, *55*, 96–102.

77 Miller W, Eberhardt N. Structure and evolution of the growth hormone gene family. *Endocr Rev* **1983**, *4*, 97–130.

78 Hirt H, Kimelman J, Birnbaum M, Chen E, Seeburg P, Eberhardt N, Barta A. The human growth hormone gene locus: structure, evolution, and allelic variations. *DNA* **1987**, *6*, 59–70.

79 Baumann G. Growth hormone heterogeneity in human pituitary and plasma. *Hormone Res* **1999**, *51*, 2–6.

80 Forsyth I, Wallis M. Growth hormone and prolactin – molecular and functional evolution. *J Mammary Gland Biol Neopl* **2002**, *7*, 291–312.

81 Elsholtz H, Lew A, Albert P, Sundmark V. Inhibitory control of prolactin and Pit-1 gene promoters by dopamine. Dual signaling pathways required for D2 receptor-regulated expression of the prolactin gene. *J Biol Chem* **1991**, *266*, 22919–22925.

82 Esquifino AI, Cano P, Jimenez V, Reyes Toso CF, Cardinali DP. Changes of prolactin regulatory mechanisms in aging: 24-h rhythms of serum prolactin and median eminence and adenohypophysial concentration of dopamine, serotonin, γ-aminobutyric acid, taurine and somatostatin in young and aged rats. *Exp Gerontol* **2004**, *39*, 45–52.

83 Osamura R, Teramoto A, Watanabe K. Ultrastructural localization of prolactin in the human pituitary prolactinomas and its changes by bromocriptine treatment. *Acta Pathol Jpn* **1986**, *36*, 1123–1130.

84 Wiegand S, Maikowski U, Blankenstein O, Biebermann H, Tarnow P, Gruters A. Type 2 diabetes and impaired glucose tolerance in European children and adolescents with obesity – a problem that is no longer restricted to minority groups. *Eur J Endocrinol* **2004**, *151*, 199–206.

85 Centers for Disease Control and Prevention Primary Prevention Working Group. Primary prevention of type 2 diabetes mellitus by lifestyle intervention: implications for health policy. *Ann Intern Med* **2004**, *140*, 951–957.

86 Opie E. The relation of diabetes mellitus to lesions of the pancreas: hyaline degeneration of the islets of Langerhans. *J Exp Med* **1001**, *5*, 527–540.

87 Westermark P, Wilander E. The influence of amyloid deposits on the islet volume in maturity onset diabetes mellitus. *Diabetologia* **1978**, *15*, 417–421.

88 Maloy A, Longnecker D, Greenberg E. The relation of islet amyloid to the clinical type of diabetes. *Hum Pathol* **1981**, *12*, 917–922.

89 Zhao H-L, Lai FMM, Tong PCY, Zhong D-R, Yang D, Tomlinson B, Chan JCN. Prevalence and clinicopathological characteristics of islet amyloid in chinese patients with type 2 diabetes. *Diabetes* **2003**, *52*, 2759–2766.

90 Westermark P. Fine structure of islets of Langerhans in insular amyloidosis. *Virchows Arch A Pathol Anat* **1973**, *359*, 1–18.

91 Clark A, Cooper G, Lewis C, Morris J, Willis A, Reid K, Turner R. Islet amyloid formed from diabetes-associated peptide may be pathogenic in type-2 diabetes. *Lancet* **1987**, *2*, 231–234.

92 Glenner G, Eanes E, Bladen H, Linke R, Termine J. Beta-pleated sheet fibrils. A comparison of native amyloid with synthetic protein fibrils. *J Histochem Cytochem* **1974**, *22*, 1141–1158.

93 Westermark P, Wernstedt C, Wilander E, Sletten K. A novel peptide in the calcitonin gene related peptide family as an amyloid fibril protein in the endocrine pancreas. *Biochem Biophys Res Commun* **1986**, *140*, 827–831.

94 Westermark P, Wernstedt C, O'Brien TD, Hayden DW, Johnson KH. Islet amyloid in type 2 human diabetes mellitus and adult diabetic cats contains a novel putative polypeptide hormone. *Am J Pathol* **1987**, *127*, 414–417.

95 Cooper GJ, Willis AC, Clark A, Turner RC, Sim RB, Reid KB. Purification and characterization of a peptide from amyloid-rich pancreases of type 2 diabetic patients. *Proc Natl Acad Sci USA* **1987**, *84*, 8628–8632.

96 Westermark P. Quantitative studies on amyloid in the islets of Langerhans. *Uppsala J Med Sci* **1972**, *77*, 91–94.

97 Clark A, de Koning EJ, Hattersley AT, Hansen BC, Yajnik CS, Poulton J. Pancreatic pathology in non-insulin dependent diabetes (NIDDM). *Diabetes Res Clin Pract* **1995**, *28 (Suppl)*, S39–S47.

98 O'Brien TD, Wagner JD, Litwak KN, Carlson CS, Cefalu WT, Jordan K, Johnson KH, Butler PC. Islet amyloid and islet amyloid polypeptide in cynomolgus macaques (*Macaca fascicularis*): an animal model of human non-insulin-dependent diabetes mellitus. *Vet Pathol* **1996**, *33*, 479–485.

99 Stefan Y, Orci L, Malaisse-Lagae F, Perrelet A, Patel Y, Unger R. Quantitation of endocrine cell content in the pancreas of nondiabetic and diabetic humans. *Diabetes* **1982**, *31*, 694–700.

100 Johnson KH, O'Brien TD, Hayden DW, Jordan K, Ghobrial HK, Mahoney WC, Westermark P. Immunolocalization of islet amyloid polypeptide (IAPP) in pancreatic beta cells by means of peroxidase–antiperoxidase (PAP) and protein A–gold techniques. *Am J Pathol* **1988**, *130*, 1–8.

101 Toshimori H, Narita R, Nakazato M, Asai J, Mitsukawa T, Kangawa K, Matsuo H, Matsukura S. Islet amyloid polypeptide (IAPP) in the gastrointestinal tract and pancreas of man and rat. *Cell Tissue Res* **1990**, *262*, 401–406.

102 Lukinius A, Wilander E, Westermark GT, Engstrom U, Westermark P. Co-localization of islet amyloid polypeptide and insulin in the B cell secretory granules of the human pancreatic islets. *Diabetologia* **1989**, *32*, 240–244.

103 Westermark P, Li ZC, Westermark GT, Leckstrom A, Steiner DF. Effects of beta cell granule components on human islet amyloid polypeptide fibril formation. *FEBS Lett* **1996**, *379*, 203–206.

104 Ahren B, Sundler F. Localization of calcitonin gene-related peptide and islet amyloid polypeptide in the rat and mouse pancreas. *Cell Tissue Res* **1992**, *269*, 315–322.

105 Mulder H, Lindh AC, Ekblad E, Westermark P, Sundler F. Islet amyloid polypeptide is expressed in endocrine cells of the gastric mucosa in the rat and mouse. *Gastroenterology* **1994**, *107*, 712–719.

106 Mulder H, Leckstrom A, Uddman R, Ekblad E, Westermark P, Sundler F. Islet amyloid polypeptide (amylin) is expressed in sensory neurons. *J Neurosci* **1995**, *15*, 7625–7632.

107 Nishi M, Chan SJ, Nagamatsu S, Bell GI, Steiner DF. Conservation of the sequence of islet amyloid polypeptide in five mammals is consistent with its putative role as an islet hormone. *Proc Natl Acad Sci USA* **1989**, *86*, 5738–5742.

108 Betsholtz C, Christmanson L, Engstrom U, Rorsman F, Jordan K, O'Brien TD, Murtaugh M, Johnson KH, Westermark P. Structure of cat islet amyloid polypeptide and identification of amino acid residues of potential significance for islet amyloid formation. *Diabetes* **1990**, *39*, 118–122.

109 Jordan K, Murtaugh MP, O'Brien TD, Westermark P, Betsholtz C, Johnson KH. Canine IAPP cDNA sequence provides important clues regarding diabetogenesis and amyloidogenesis in type 2 diabetes. *Biochem Biophys Res Commun* **1990**, *169*, 502–508.

110 Nishi M, Bell GI, Steiner DF. Sequence of a cDNA encoding Syrian hamster islet amyloid polypeptide precursor. *Nucleic Acids Res* **1990**, *18*, 6726.

111 Johnson KH, Wernstedt C, O'Brien TD, Westermark P. Amyloid in the pancreatic islets of the cougar (*Felis concolor*) is derived from islet amyloid polypeptide (IAPP). *Comp Biochem Physiol B* **1991**, *98*, 115–119.

112 Miyazato M, Nakazato M, Shiomi K, Aburaya J, Kangawa K, Matsuo H, Matsukura S. Molecular forms of islet amyloid polypeptide (IAPP/amylin) in four mammals. *Diabetes Res Clin Pract* **1992**, *15*, 31–36.

113 Fan L, Westermark G, Chan SJ, Steiner DF. Altered gene structure and tissue expression of islet amyloid polypeptide in the chicken. *Mol Endocrinol* **1994**, *8*, 713–721.

114 Westermark GT, Falkmer S, Steiner DF, Chan SJ, Engstrom U, Westermark P. Islet amyloid polypeptide is expressed in the pancreatic islet parenchyma of the teleostean fish, *Myoxocephalus (cottus) scorpius*. *Comp Biochem Physiol B Biochem Mol Biol* **2002**, *133*, 119–125.

115 Sanke T, Bell GI, Sample C, Rubenstein AH, Steiner DF. An islet amyloid peptide is derived from an 89-amino acid precursor by proteolytic processing. *J Biol Chem* **1988**, *263*, 17243–17246.

116 Betsholtz C, Svensson V, Rorsman F, Engstrom U, Westermark GT, Wilander E, Johnson K, Westermark P. Islet amyloid polypeptide (IAPP): cDNA cloning and identification of an amyloidogenic region associated with the species-specific occurrence of age-related diabetes mellitus. *Exp Cell Res* **1989**, *183*, 484–493.

117 Rosenfeld M, Amara S, Roos B, Ong E, Evans R. Altered expression of the calcitonin gene associated with RNA polymorphism. *Nature* **1981**, *290*, 63–65.

118 Kitamura K, Sakata J, Kangawa K, Kojima M, Matsuo H, Eto T. Cloning and characterization of cDNA encoding a precursor for human adrenomedullin. *Biochem Biophys Res Commun* **1993**, *194*, 720–725.

119 Edbrooke M, Parker D, McVey J, Riley J, Sorenson G, Pettengill O, Craig R. Expression of the human calcitonin/CGRP gene in lung and thyroid carcinoma. *EMBO J* **1995**, *4*, 715–724.

120 Mosselman S, Hoppener J, Zandberg J, van Mansfeld A, Geurts van Kessel A, Lips C, Jansz H. Islet amyloid polypeptide: identification and chromosomal localization of the human gene. *FEBS Lett* **1988**, *239*, 227–232.

121 Mosselman S, Hoppener J, Lips C, Jansz H. The complete islet amyloid polypeptide precursor is encoded by two exons. *FEBS Lett* **1989**, *247*, 154–158.

122 Badman MK, Shennan KI, Jermany JL, Docherty K, Clark A. Processing of pro-islet amyloid polypeptide (pro-IAPP) by the prohormone convertase PC2. *FEBS Lett* **1996**, *378*, 227–231.

123 Higham CE, Hull RL, Lawrie L, Shennan KI, Morris JF, Birch NP, Docherty K, Clark A. Processing of synthetic pro-islet amyloid polypeptide (pro-IAPP) "amylin" by recombinant prohormone convertase enzymes, PC2 and PC3, *in vitro*. *Eur J Biochem* **2000**, *267*, 4998–5004.

124 Wang J, Xu J, Finnerty J, Furuta M, Steiner DF, Verchere CB. The prohormone convertase enzyme 2 (PC2) is essential for processing pro-islet amyloid polypeptide at the NH_2-terminal cleavage site. *Diabetes* **2001**, *50*, 534–539.

125 Marzban L, Trigo-Gonzalez G, Zhu X, Rhodes CJ, Halban PA, Steiner DF, Verchere CB. Role of beta-cell prohormone convertase (PC)1/3 in processing of pro-islet amyloid polypeptide. *Diabetes* **2004**, *53*, 141–148.

126 Nakazato M, Asai J, Kangawa K, Matsukura S, Matsuo H. Establishment of radioimmunoassay for human islet amyloid polypeptide and its tissue content and plasma concentration. *Biochem Biophys Res Commun* **1989**, *164*, 394–399.

127 Butler PC, Chou J, Carter WB, Wang YN, Bu BH, Chang D, Chang JK, Rizza RA. Effects of meal ingestion on plasma amylin concentration in NIDDM and nondiabetic humans. *Diabetes* **1990**, *39*, 752–756.

128 van Jaarsveld B, Hackeng W, Lips C, Erkelens D. Plasma concentrations of islet amyloid polypeptide after glucagon administration in type 2 diabetic patients and non-diabetic subjects. *Diabetic Med* **1993**, *10*, 327–330.

129 Jamal H, Bretherton-Watt D, Suda K, Ghatei M, Bloom S. Islet amyloid polypeptide-like immunoreactivity (amylin) in rats treated with dexamethasone and streptozotocin. *J Endocrin Invest* **1990**, *126*, 425–429.

130 Gill A, Yen T. Effects of ciglitazone on endogenous plasma islet amyloid polypeptide and insulin sensitivity in obese-diabetic viable yellow mice. *Life Sci* **1991**, *48*, 703–710.

131 Tokuyama Y, Kanatsuka A, Ohsawa H, Yamaguchi T, Makino H, Yoshida S, Nagase H, Inoue S. Hypersecretion of islet amyloid polypeptide from pancreatic islets of ventromedial hypothalamic-lesioned rats and obese Zucker rats. *Endocrinology* **1991**, *128*, 2739–2744.

132 Nagamatsu S, Carroll R, Grodsky G, Steiner D. Lack of islet amyloid polypeptide regulation of insulin biosynthe-

sis or secretion in normal rat islets. *Diabetes* **1990**, *39*, 817–874.

133 Novials A, Sarri Y, Casamitjana R, Rivera F, Gomis R. Regulation of islet amyloid polypeptide in human pancreatic islets. *Diabetes* **1993**, *42*, 1514–1519.

134 Mulder H, Ahren B, Sundler F. Islet amyloid polypeptide (amylin) and insulin are differentially expressed in chronic diabetes induced by streptozotocin in rats. *Diabetologia* **1996**, *39*, 649–657.

135 Bretherton-Watt D, Ghatei M, Bloom S, Jamal H, Ferrier G, Girgis S, Legon S. Altered islet amyloid polypeptide (amylin) gene expression in rat models of diabetes. *Diabetologia* **1989**, *32*, 881–883.

136 O'Brien T, Westermark P, Johnson K. Islet amyloid polypeptide and insulin secretion from isolated perfused pancreas of fed, fasted, glucose-treated, and dexamethasone-treated rats. *Diabetes* **1991**, *40*, 1701–1706.

137 Ludvik B, Clodi M, Kautzky-Willer A, Capek M, Hartter E, Pacini G, Prager R. Effect of dexamethasone on insulin sensitivity, islet amyloid polypeptide and insulin secretion in humans. *Diabetologia* **1993**, *36*, 84–87.

138 Mitsukawa T, Takemura J, Asai J, Nakazato M, Kangawa K, Matsuo H, Matsukura S. Islet amyloid polypeptide response to glucose, insulin, and somatostatin analogue administration. *Diabetes* **1990**, *39*, 639–642.

139 Rink T, Beaumont K, Koda J, Young A. Structure and biology of amylin. *Trends Pharmacol Sci* **1993**, *14*, 113–118.

140 Leckstrom A, Lundquist I, Ma Z, Westermark P. Islet amyloid polypeptide and insulin relationship in a longitudinal study of the genetically obese (*ob*/*ob*) mouse. *Pancreas* **1999**, *18*, 266–273.

141 Westermark GT, Leckstrom A, Ma Z, Westermark P. Increased release of IAPP in response to long-term high fat intake in mice. *Horm Metab Res* **1998**, *30*, 256–258.

142 Mulder H, Martensson H, Sundler F, Ahren B. Differential changes in islet amyloid polypeptide (amylin) and insulin mRNA expression after high-fat diet-induced insulin resistance in C57BL/6J mice. *Metabolism* **2000**, *49*, 1518–1522.

143 Nishi M, Sanke T, Nagamatsu S, Bell GI, Steiner DF. Islet amyloid polypeptide. A new beta cell secretory product related to islet amyloid deposits. *J Biol Chem* **1990**, *265*, 4173–4176.

144 Paulsson J, Westermark G. Differences in distribution of insulin and IAPP on the cellular level. In *IX International Symposium on Amyloidosis*, M Bely and A Apathy (eds). Budapest, Hungary, **2001**, pp. 424–426.

145 Leckström A, Björklund K, Permert J, Larsson R, Westermark P. Renal elimination of islet amyloid polypeptide. *Biochem Biophys Res Commun* **1997**, *239*, 265–268.

146 Bennett RG, Duckworth WC, Hamel FG. Degradation of amylin by insulin-degrading enzyme. *J Biol Chem* **2000**, *275*, 36621–36625.

147 Bennett RG, Hamel FG, Duckworth WC. An insulin-degrading enzyme inhibitor decreases amylin degradation, increases amylin-induced cytotoxicity, and increases amyloid formation in insulinoma cell cultures. *Diabetes* **2003**, *52*, 2315–2320.

148 Silvestre RA, Peiro E, Degano P, Miralles P, Marco J. Inhibitory effect of rat amylin on the insulin responses to glucose and arginine in the perfused rat pancreas. *Reg Pept* **1990**, *31*, 23–31.

149 Kogire M, Ishizuka J, Thompson JC, Greeley GH, Jr. Inhibitory action of islet amyloid polypeptide and calcitonin gene-related peptide on release of insulin from the isolated perfused rat pancreas. *Pancreas* **1991**, *6*, 459–463.

150 Gebre-Medhin S, Mulder H, Pekny M, Westermark G, Törnell J, Westermark P, Sundler F, Ahren B, Betsholtz C. Increased insulin secretion and glucose tolerance in mice lacking islet amyloid polypeptide (amylin). *Biochem Biophys Res Commun* **1998**, *250*, 271–277.

151 Panagiotidis G, Salehi AA, Westermark P, Lundquist I. Homologous islet amyloid polypeptide: effects on plasma levels of glucagon, insulin and glucose in the mouse. *Diabetes Res Clin Pract* **1992**, *18*, 167–171.

152 Akesson B, Panagiotidis G, Westermark P, Lundquist I. Islet amyloid polypeptide inhibits glucagon release and exerts a dual action on insulin release from isolated islets. *Reg Pept* **2003**, *111*, 55–60.

153 Ludvik B, Thomaseth K, Nolan JJ, Clodi M, Prager R, Pacini G. Inverse relation between amylin and glucagon secretion in healthy and diabetic human subjects. *Eur J Clin Invest* **2003**, *33*, 316–322.

154 Fehmann H, Weber V, Goke R, Goke B, Eissele R, Arnold R. Islet amyloid polypeptide (IAPP; amylin) influences the endocrine but not the exocrine rat pancreas. *Biochem Biophys Res Commun* **1990**, *167*, 1102–1108.

155 Wang F, Adrian TE, Westermark GT, Ding X, Gasslander T, Permert J. Islet amyloid polypeptide tonally inhibits beta-, alpha-, and delta-cell secretion in isolated rat pancreatic islets. *Am J Physiol Endocrinol Metab* **1999**, *276*, E19–E24.

156 Zaidi M, Alam AS, Shankar VS, Bax BE, Moonga BS, Bevis PJ, Pazianas M, Huang CL. A quantitative description of components of *in vitro* morphometric change in the rat osteoclast model: relationships with cellular function. *Eur Biophys J* **1992**, *21*, 349–355.

157 Datta HK, Zaidi M, Wimalawansa SJ, Ghatei MA, Beacham JL, Bloom SR, MacIntyre I. *In vivo* and *in vitro* effects of amylin and amylin-amide on calcium metabolism in the rat and rabbit. *Biochem Biophys Res Commun* **1989**, *162*, 876–881.

158 Alam AS, Moonga BS, Bevis PJ, Huang CL, Zaidi M. Amylin inhibits bone resorption by a direct effect on the motility of rat osteoclasts. *Exp Physiol* **1993**, *78*, 183–196.

159 Cornish J, Callon KE, Cooper G, Reid I. Amylin stimulates osteoblast proliferation and increases mineralized bone volume in adult mice. *Biochem Biophys Res Commun* **1995**, *207*, 133–139.

160 Dacquin R, Davey RA, Laplace C, Levasseur R, Morris HA, Goldring SR, Gebre-Medhin S, Galson DL, Zajac JD, Karsenty G. Amylin inhibits bone resorption while the calcitonin receptor controls bone formation *in vivo*. *J Cell Biol* **2004**, *164*, 509–514.

161 Christmanson L, Westermark P, Betsholtz C. Islet amyloid polypeptide stimulates cyclic AMP accumulation via the porcine calcitonin receptor. *Biochem Biophys Res Commun* **1994**, *205*, 1226–1235.

162 Sexton PM, Houssami S, Brady CL, Myers DE, Findlay DM. Amylin is an agonist of the renal porcine calcitonin receptor. *Endocrinology* **1994**, *134*, 2103–2107.

163 Lutz TA, Geary N, Szabady MM, Del Prete E, Scharrer E. Amylin decreases meal size in rats. *Physiol Behav* **1995**, *58*, 1197–1202.

164 Arnelo U, Blevins JE, Larsson J, Permert J, Westermark P, Reidelberger RD, Adrian TE. Effects of acute and chronic infusion of islet amyloid polypeptide on food intake in rats. *Scand J Gastroenterol* **1996**, *31*, 83–89.

165 Arnelo U, Permert J, Adrian TE, Larsson J, Westermark P, Reidelberger RD. Chronic infusion of islet amyloid polypeptide causes anorexia in rats. *Am J Physiol* **1996**, *271*, R1654–R1659.

166 Lutz TA, Senn M, Althaus J, Del Prete E, Ehrensperger F, Scharrer E. Lesion of the area postrema/nucleus of the solitary tract (AP/NTS) attenuates the anorectic effects of amylin and calcitonin gene-related peptide (CGRP) in rats. *Peptides* **1998**, *19*, 309–317.

167 Reidelberger RD, Haver AC, Arnelo U, Smith DD, Schaffert CS, Permert J. Amylin receptor blockade stimulates food intake in rats. *Am J Physiol Reg Integr Comp Physiol* **2004**, *287*, R568–574.

168 Young AA, Gedulin B, Vine W, Percy A, Rink TJ. Gastric emptying is accelerated in diabetic BB rats and is slowed by subcutaneous injections of amylin. *Diabetologia* **1995**, *38*, 642–648.

169 Weyer C, Maggs D, Young A, Kolterman O. Amylin replacement with pramlintide as an adjunct to insulin therapy in type 1 and type 2 diabetes mellitus: a physiological approach to-

ward improved metabolic control. *Curr Pharm Des* **2001**, *7*, 1353–1373.

170 Betsholtz C, Christmansson L, Engstrom U, Rorsman F, Svensson V, Johnson KH, Westermark P. Sequence divergence in a specific region of islet amyloid polypeptide (IAPP) explains differences in islet amyloid formation between species. *FEBS Lett* **1989**, *251*, 261–264.

171 Westermark P, Engstrom U, Johnson KH, Westermark GT, Betsholtz C. Islet amyloid polypeptide: pinpointing amino acid residues linked to amyloid fibril formation. *Proc Natl Acad Sci USA* **1990**, *87*, 5036–5040.

172 Tenidis K, Waldner M, Bernhagen J, Fischle W, Bergmann M, Weber M, Merkle M-L, Voelter W, Brunner H, Kapurniotu A. Identification of a penta- and hexapeptide of islet amyloid polypeptide (IAPP) with amyloidogenic and cytotoxic properties. *J Mol Biol* **2000**, *295*, 1055–1071.

173 Reches M, Porat Y, Gazit E. Amyloid fibril formation by pentapeptide and tetrapeptide fragments of human calcitonin. *J Biol Chem* **2002**, *277*, 35475–35480.

174 Jaikaran ET, Higham CE, Serpell LC, Zurdo J, Gross M, Clark A, Fraser PE. Identification of a novel human islet amyloid polypeptide beta-sheet domain and factors influencing fibrillogenesis. *J Mol Biol* **2001**, *308*, 515–525.

175 Jaikaran ET, Clark A. Islet amyloid and type 2 diabetes: from molecular misfolding to islet pathophysiology. *Biochim Biophys Acta* **2001**, *1537*, 179–203.

176 Rachman J, Levy J, Barrow B, Manley S, Turner R. Relative hyperproinsulinemia of NIDDM persists despite the reduction of hyperglycemia with insulin or sulfonylurea therapy. *Diabetes* **1997**, *46*, 1557–1562.

177 Porte D, Jr, Kahn SE. Beta-cell dysfunction and failure in type 2 diabetes: potential mechanisms. *Diabetes* **2001**, *50 (Suppl 1)*, S160–S163.

178 Janciauskiene S, Eriksson S, Carlemalm E, Ahren B. B cell granule peptides affect human islet amyloid polypeptide (IAPP) fibril formation *in vitro*. *Biochem Biophys Res Commun* **1997**, *236*, 580–585.

179 Sakagashira S, Sanke T, Hanabusa T, Shimomura H, Ohagi S, Kumagaye KY, Nakajima K, Nanjo K. Missense mutation of amylin gene (S20G) in Japanese NIDDM patients. *Diabetes* **1996**, *45*, 1279–1281.

180 Seino S. S20G mutation of the amylin gene is associated with Type II diabetes in Japanese. Study Group of Comprehensive Analysis of Genetic Factors in Diabetes Mellitus. *Diabetologia* **2001**, *44*, 906–909.

181 Lee SC, Hashim Y, Li JK, Ko GT, Critchley JA, Cockram CS, Chan JC. The islet amyloid polypeptide (amylin) gene S20G mutation in Chinese subjects: evidence for associations with type 2 diabetes and cholesterol levels. *Clin Endocrinol* **2001**, *54*, 541–546.

182 Ma Z, Westermark GT, Sakagashira S, Sanke T, Gustavsson A, Sakamoto H, Engstrom U, Nanjo K, Westermark P. Enhanced *in vitro* production of amyloid-like fibrils from mutant (S20G) islet amyloid polypeptide. *Amyloid* **2001**, *8*, 242–249.

183 Johnson KH, Hayden DW, O'Brien TD, Westermark P. Spontaneous diabetes mellitus-islet amyloid complex in adult cats. *Am J Pathol* **1986**, *125*, 416–419.

184 Westermark P, Wernstedt C, Wilander E, Hayden DW, O'Brien TD, Johnson KH. Amyloid fibrils in human insulinoma and islets of Langerhans of the diabetic cat are derived from a neuropeptide-like protein also present in normal islet cells. *Proc Natl Acad Sci USA* **1987**, *84*, 3881–3885.

185 de Koning EJ, Bodkin NL, Hansen BC, Clark A. Diabetes mellitus in Macaca mulatta monkeys is characterised by islet amyloidosis and reduction in beta-cell population. *Diabetologia* **1993**, *36*, 378–384.

186 Westermark P, Engstrom U, Westermark GT, Johnson KH, Permerth J, Betsholtz C. Islet amyloid polypeptide (IAPP) and pro-IAPP immunoreactivity in human islets of Langerhans. *Diabetes Res Clin Pract* **1989**, *7*, 219–226.

187 Clark A, Lloyd J, Novials A, Hutton JC, Morris JF. Localisation of islet amyloid polypeptide and its carboxy terminal flanking peptide in islets of diabetic man and monkey. *Diabetologia* **1991**, *34*, 449–451.

188 Westermark P, Eizirik DL, Pipeleers DG, Hellerstrom C, Andersson A. Rapid deposition of amyloid in human islets transplanted into nude mice. *Diabetologia* **1995**, *38*, 543–549.

189 Kapurniotu A. Amyloidogenicity and cytotoxicity of islet amyloid polypeptide. *Biopolymers* **2001**, *60*, 438–459.

190 O'Brien T, Butler P, Kreutter D, Kane L, Eberhardt N. Human islet amyloid polypeptide expression in COS-1 cells. A model of intracellular amyloidogenesis. *Am J Pathol* **1995**, *147*, 609–616.

191 Janciauskiene S, Ahren B. Fibrillar islet amyloid polypeptide differentially affects oxidative mechanisms and lipoprotein uptake in correlation with cytotoxicity in two insulin-producing cell lines. *Biochem Biophys Res Commun* **2000**, *267*, 619–625.

192 Hartley DM, Walsh DM, Ye CP, Diehl T, Vasquez S, Vassilev PM, Teplow DB, Selkoe DJ. Protofibrillar intermediates of amyloid beta-protein induce acute electrophysiological changes and progressive neurotoxicity in cortical neurons. *J Neurosci* **1999**, *19*, 8876–8884.

193 de Koning EJ, Morris ER, Hofhuis FM, Posthuma G, Hoppener JW, Morris JF, Capel PJ, Clark A, Verbeek JS. Intra- and extracellular amyloid fibrils are formed in cultured pancreatic islets of transgenic mice expressing human islet amyloid polypeptide. *Proc Natl Acad Sci USA* **1994**, *91*, 8467–8471.

194 Janson J, Soeller WC, Roche PC, Nelson RT, Torchia AJ, Kreutter DK, Butler PC. Spontaneous diabetes mellitus in transgenic mice expressing human islet amyloid polypeptide. *Proc Natl Acad Sci USA* **1996**, *93*, 7283–7288.

195 Fox N, Schrementi J, Nishi M, Ohagi S, Chan SJ, Heisserman JA, Westermark GT, Leckstrom A, Westermark P, Steiner DF. Human islet amyloid polypeptide transgenic mice as a model of non-insulin-dependent diabetes mellitus (NIDDM). *FEBS Lett* **1993**, *323*, 40–44.

196 D'Alessio DA, Verchere CB, Kahn SE, Hoagland V, Baskin DG, Palmiter RD, Ensinck JW. Pancreatic expression and secretion of human islet amyloid polypeptide in a transgenic mouse. *Diabetes* **1994**, *43*, 1457–1461.

197 Yagui K, Kanatsuka A, Makino H. Construction of transgenic mouse system expressing human islet amyloid polypeptide (IAPP)/amylin. *Nippon Rinsho* **1994**, *52*, 2746–2750.

198 Soeller WC, Janson J, Hart SE, Parker JC, Carty MD, Stevenson RW, Kreutter DK, Butler PC. Islet amyloid-associated diabetes in obese A(vy)/a mice expressing human islet amyloid polypeptide. *Diabetes* **1998**, *47*, 743–750.

199 Höppener JW, Oosterwijk C, Nieuwenhuis MG, Posthuma G, Thijssen JH, Vroom TM, Ahren B, Lips CJ. Extensive islet amyloid formation is induced by development of Type II diabetes mellitus and contributes to its progression: pathogenesis of diabetes in a mouse model. *Diabetologia* **1999**, *42*, 427–434.

200 Verchere CB, D'Alessio DA, Palmiter RD, Weir GC, Bonner-Weir S, Baskin DG, Kahn SE. Islet amyloid formation associated with hyperglycemia in transgenic mice with pancreatic beta cell expression of human islet amyloid polypeptide. *Proc Natl Acad Sci USA* **1996**, *93*, 3492–3496.

201 Westermark GT, Gebre-Medhin S, Steiner DF, Westermark P. Islet amyloid development in a mouse strain lacking endogenous islet amyloid polypeptide (IAPP) but expressing human IAPP. *Mol Med* **2000**, *6*, 998–1007.

202 Westermark GT, Steiner DF, Gebre-Medhin S, Engstrom U, Westermark P. Pro islet amyloid polypeptide (proIAPP) immunoreactivity in the islets of Langerhans. *Uppsala J Med Sci* **2000**, *105*, 97–106.

203 Waugh D. A mechanism for the formation of fibrils from protein molecules. *J Cell Physiol* **1957**, *49*, 145–164.

204 Burke M, Rougvie M. Cross-protein structures. I. Insulin fibrils. *Biochemistry* **1972**, *11*, 2435–2439.

205 Westermark P. On the nature of the amyloid in human islets of Langerhans. *Histochemistry* **1974**, *38*, 27–33.

206 Dische FE, Wernstedt C, Westermark GT, Westermark P, Pepys MB, Rennie JA, Gilbey SG, Watkins PJ. Insulin as an amyloid-fibril protein at sites of repeated insulin injections in a diabetic patient. *Diabetologia* **1988**, *31*, 158–161.

207 Storkel S, Schneider H, Muntefering H, Kashiwagi S. Iatrogenic, insulin-dependent, local amyloidosis. *Lab Invest* **1983**, *48*, 108–111.

208 Hellman U, Wernstedt C, Westermark P, O'Brien TD, Rathbun WB, Johnson KH. Amino acid sequence from degu islet amyloid-derived insulin shows unique sequence characteristics. *Biochem Biophys Res Commun* **1990**, *169*, 571–577.

209 Nishi M, Steiner DF. Cloning of complementary DNAs encoding islet amyloid polypeptide, insulin, and glucagon precursors from a New World rodent, the degu, *Octodon degus*. *Mol Endocrinol* **1990**, *4*, 1192–1198.

210 Ohsawa H, Kanatsuka A, Tokuyama Y, Yamaguchi T, Makino H, Yoshida S, Horie H, Mikata A, Kohen Y. Amyloid protein in somatostatinoma differs from human islet amyloid polypeptide. *Acta Endocrinol* **1991**, *124*, 45–53.

211 Onoue S, Ohshima K, Debari K, Koh K, Shioda S, Iwasa S, Kashimoto K, Yajima T. Mishandling of the therapeutic peptide glucagon generates cytotoxic amyloidogenic fibrils. *Pharmacol Res* **2004**, *7*, 1274–1283.

212 Lieberman A, DeLellis R. Intrafollicular amyloid in normal parathyroid glands. *Arch Pathol Med* **1973**, *95*, 422–423.

213 Anderson T, Ewen S. Amyloid in normal and pathological parathyroid gland. *J Clin Pathol* **1974**, *27*, 656–663.

214 Iwata T, Imada N, Nakamura H, Fujihara S, Yamashita Y, Yokota T, Kamei T, Uchino F. Amyloidosis – intrafollicular amyloid in normal parathyroid glands. *Acta Pathol Jpn* **1981**, *31*, 513–519.

215 Canaff L, Bennett HPJ, Hou Y, Seidah NG, Hendy GN. Proparathyroid hormone processing by the proprotein convertase-7: comparison with furin and assessment of modulation of parathyroid convertase messenger ribonucleic acid levels by calcium and 1,25-dihydroxyvitamin D_3. *Endocrinology* **1999**, *140*, 3633–3642.

216 Kedar I, Ravid M, Sohar E. *In vitro* synthesis of "amyloid" fibrils from insulin, calcitonin and parathormone. *Isr J Med Sci* **1976**, *12*, 1137–1140.

217 Ordonez N, Ibanez M, Samaan N, Hickey R. Immunoperoxidase study of uncommon parathyroid tumors. Report of two cases of nonfunctioning parathyroid carcinoma and one intrathyroid parathyroid tumor-producing amyloid. *Am J Surg Pathol* **1983**, *7*, 535–542.

Glossary of Terms

Advanced glycation end product (AGE) Products derived from the nonenzymatic reaction of glucose and proteins in vivo that exhibit a yellow-brown pigmentation and an ability to participate in protein-protein crosslinking. These substances are involved in biological processes relating to protein turnover and it is believed that their excessive accumulation contributes to the chronic complications of diabetes mellitus.

Advanced lipoxidation end product (ALE) Products derived from the reaction of lipid peroxides with cysteine, histidine and lysine residues of proteins [Chapter 5].

Alloform (Other form); Peptides or proteins of identical provenance but differing structure, e.g., proteins encoded by the same structural gene but differing in primary structure as a result of posttranslational processing in vivo or designed chemical synthesis in vitro.

Amphipathic Having both hydrophilic and hydrophobic domains.

Amylin Islet amyloid polypeptide (IAPP).

Amyloid Starch like [Reference 11, Chapter 1].
A type of extracellularly (usually) deposited substance composed of an amyloid protein and additional components including heparan sulfate proteoglycan, serum amyloid p component, and apolipoprotein E.

Amyloid enhancing factor Tissue extract of amyloid-laden organs that accelerates amyloidogenesis in animal models of AA amyloidosis.

Amyloid fibril Insoluble fibril that is deposited in tissues; 8–10 or 6–12 nm in diameter, nonbranching, congophilic and cross beta pleated sheet by X-ray diffraction or other physical methods.

Amyloidosis A disease or disorder that results from a sequence of changes in protein folding that leads to the deposition of insoluble amyloid fibrils, mainly in the extracellular spaces of organs and tissues.

Amylog Amyloid like fibrils that are created in vitro [Reference 183, Chapter 1].

Basement membrane Near ubiquitous extracellular structures that delineate regions of different histological cell type, e.g. epithelial cells and connective tissue fibroblasts. Absent in cartilage and bone, except for walls of blood vessels in bone [chapter 7].

Birefringence The property of nonisotropic media, such as crystals, whereby a single incident beam of light traverses the medium as two beams, each plane-polarized, the planes being at right angles to each other.

Chaperone Proteins that reversibly bind unfolded segments of polypeptides susceptible to aggregation or diversion.

Configuration Combination and arrangement of atoms in a molecule.

Amyloid Proteins. The Beta Sheet Conformation and Disease. J. D. Sipe

ISBN: 3-527-31072-X

Conformation The three dimensional shape of a molecule.

Conformer A conformation of a molecule; generally at an energy minimum.

Domain A three dimensional region or segment of a molecule.

Entropy A measure of the disorder of a system and that part of the heat or energy of a system which is not available to perform work. Entropy increases in all natural (spontaneous and irreversible) processes.

Enthalpy A thermodynamic property of a system defined as the sum of the internal energy of a body and the product of its volume multiplied by the pressure.

Epimer A stereoisomer that differs only in the configuration at a single atom.

Epitope The region on an antigenic molecule that interacts with the binding site of an immunoglobulin.

Extracellular matrix A meshwork-like substance, which supports cells and to which they adhere. Found within the extracellular space and in association with the basement membrane of the cell surface.

Glycosaminoglycan (GAG) Heteropolysaccharides which contain an N-acetylated hexosamine in a characteristic repeating disaccharide unit. The repeating structure of each disaccharide involves alternate 1,4- and 1,3-linkages consisting of either N-acetylglucosamine or N-acetylgalactosamine.

Hyaline Transparent and glassy, or nearly so.

Hydrotactoid Structure of water molecules forming hydrates surrounding nonpolar polymerized amino acids [Reference 15, Chapter 3].

Isoform Two or more different forms of a protein that may be produced from different genes, or from the same gene by alternative splicing. Also used to distinguish conformations of the product of a single gene, particularly PrP.

Isomer Two or more forms of a molecule that have the same number and kind of atoms and the same atomic arrangement, but differ in spatial relationships.

Kinetics The study of the rate of change of a specific factor.

Metastable A state of pseudo-equilibrium that has a free energy higher than that of the true equilibrium state but from which a system does not change spontaneously.

Micelle Electrically charged colloidal particle consisting of oriented molecules held together by non-covalent bonds.

Microscopy, Polarization Microscopy using polarized light in which phenomena due to the preferential orientation of optical properties with respect to the vibration plane of the polarized light are made visible and correlated parameters are made measurable.

Molecular dynamics (MD) Computational tool for used for definition of protein structure.

Mucopolysaccharide Glycosaminoglycan.

Native state/structure The conformation(s) in which a polypeptide spends most of its time and is functional. [Reference 20, Chapter 3].

Nidus A central point or locus of folding.

Neurofibrillary tangles Abnormal structures located in various parts of the brain and composed of dense arrays of paired helical filaments (neurofilaments and microtubules). These double helical stacks of transverse subunits are twisted into left-handed ribbon-like filaments that likely incorporate the following proteins: (1) the intermediate filaments: medium- and high-molecular-weight neurofilaments; (2) the microtubule-associated proteins map-2 and tau; (3) actin; and (4) ubiquitins. As one of the hallmarks of Alzheimer's Disease, the neurofibrillary tangles eventually occupy the whole of the cytoplasm in certain classes of cell in the neocortex, hippocampus, brain stem, and diencephalon.

Nucleation The initial stage in self assembly of polypeptides that are in or are about to assume the beta pleated sheet conformation; it is evidenced by the formation of small particles (nuclei) which are capable of growing into amyloid fibrils.

Nucleus A core around which polypeptide precursors of amyloid fibrils aggregate.

Oligomer A protein species composed of more than one folded monomer chain, usually of the same kind. Some proteins in the native state are oligomeric, e.g. transthyretin (TTR). However, the term is also used to refer to the self assembly of amyloidogenic monomers into small soluble aggregates.

Paranucleus A body resembling the cell nucleus sometimes seen in cytoplasm near the nucleus.

Pentraxin A protein that assembles into a doughnut like quaternary structure, with five identical polypeptide chains that form the ring.

Protofilament Beta pleated sheet strands.

Protofibril Soluble oligomer that forms subsequent to an amyloid nucleus and is an intermediate of fibril formation.

Racemization Conversion of one enantiomorph into another, for example, a *d*-amino acid to the corresponding *l*-amino acid.

Random coil The continuously changing and disordered conformation of a polypeptide chain, usually as a result of being dissolved in a solvent that favors exposure of all of the amino acid monomers to the solvent.

Reticulin A scleroprotein fibril consisting mostly of type III collagen. Reticulin fibrils are extremely thin, with a diameter of between 0.5 and 2 μm. They are involved in maintaining the structural integrity in a variety of organs.

Sporadic Occurring occasionally in a random or isolated manner.

Tau proteins Microtubule-associated proteins that are mainly expressed in neurons. Aggregation of specific sets of tau proteins in filamentous inclusions is the common feature of intraneuronal and glial fibrillar lesions.

Thermodynamics Mathematical analysis of energy relationships between heat, work, temperature and equilibrium.

Wild-type Occurring in nature, not associated with loss of function.

This glossary of selected terms was compiled using the chapters in this volume, Dorlands Illustrated Medical Dictionary, 18th edition. W.B. Saunders, Phildadelphia, and the Gateway database of the National Library of Medicine.

Subject Index

a

Amyloid Proteins. The Beta Sheet Conformation and Disease. J. D. Sipe

ISBN: 3-527-31072-X